Introduction

Welcome to your new course! This book is designed to act as a comprehensive course book, covering both the core material and all the options you might take while studying for the IB Diploma in Biology. It will also help you to prepare for your examinations in a thorough and methodical way.

Content

As you will see when you look at the table of contents, there is a chapter for each of the core topics, another for each of the HL topics and another for each of the options you might choose to take. Within each chapter, at the end of each numbered section, there are numbered exercises for you to practise and apply the knowledge which you have gained. They will also help you to assess your progress. Sometimes, there are worked examples which show you how to tackle a particularly tricky or awkward question. Here is an example.

Worked example

Compare two groups of barnacles living on a rocky shore. Measure the width of their shells to see if a significant size difference is found depending on how close they live to the water. One group lives between 0 and 10 metres from the water level. The second group lives between 10 and 20 metres above the water level.

Measurement was taken of the width of the shells in millimetres. 15 shells were measured from each group. The mean of the group closer to the water indicates that living closer to the water causes the barnacles to have a larger shell. If the value of t is 2.25, is that a significant difference?

Solution

The degree of freedom is 28 ($15 + 15 - 2 = 28$).

2.25 is just above 2.05.

Referring to the bottom of this column in the table, $p = 0.05$ so the probability that chance alone could produce that result is only 5%.

The confidence level is 95%. We are 95% confident that the difference between the barnacles is significant. Barnacles living nearer the water have a significantly larger shell than those living 10 metres or more away from the water.

At the end of each chapter, there are practice questions taken from past exam papers. These also include a markscheme. Towards the end of the book, just before the index, you will find pages with suggested answers to all the exercises and practice questions that have been included. The answers are grouped into Exercises and Practice questions for each chapter.

After the options chapters, you will find a Theory of Knowledge chapter, which should stimulate wider research and the consideration of moral and ethical issues in the field of biology.

Finally, there are three short chapters offering advice on internal assessment, on writing extended essays, and on developing examination strategies.

Information boxes

Throughout the book you will see a number of coloured boxes interspersed through each chapter. Each of these boxes provides different information and stimulus as follows.

Assessment statements

2.1.1 Outline the cell theory.

2.1.2 Discuss the evidence for the cell theory.

2.1.3 State that unicellular organisms carry out all the functions of life.

You will find a box like this at the start of each section in each chapter. They are the numbered objectives for the section you are about to read and they set out what content and aspects of learning are covered in that section.

In families where genetic disease has been passed down generation after generation, serious questions arise.

- If someone is a carrier, should they have children knowing that they might transmit the gene?
- How would it feel knowing – or not knowing – if you had the gene?
- How might parents feel when they find out they have passed the gene on to their offspring?
- Is it difficult for people who have genetic diseases to find someone to marry? A job? Life insurance?

In addition to the Theory of Knowledge chapter, there are TOK boxes throughout the book. These boxes are there to stimulate thought and consideration of any TOK issues as they arise and in context. Often they will just contain a question to stimulate your own thoughts and discussion.

Fish have a 2-chambered heart and amphibians have a 3-chambered heart. Reptiles, birds and mammals all have a 4-chambered heart (the ventricles of a reptile heart are only partially divided).

These boxes contain interesting information which will add to your wider knowledge but which does not fit within the main body of the text.

Once pyruvate is obtained, the next pathway is determined by the presence of oxygen. If oxygen is present, pyruvate enters the mitochondria and aerobic respiration occurs. If oxygen is not present, anaerobic respiration occurs in the cytoplasm. In this case, pyruvate is converted to lactate in animals, and ethanol and carbon dioxide in plants.

These are key facts which are drawn out of the main text and highlighted. This makes them easily identifiable for quick reference. The boxes also enable you to identify the core learning points within a section.

To what degree do you think the following are factors in malnutrition?

- poverty and wealth;
- cultural differences concerning dietary preference;
- climatic conditions (i.e. annual rainfall);
- poor distribution of food (i.e. insufficient roads, bridges, railways);
- a nomadic lifestyle;
- wars;
- corrupt politicians misusing agriculture or aid money;
- lack of healthcare leading to a cycle of disease and poverty.

A global perspective is important to the International Baccalaureate. These boxes indicate examples of internationalism within the area of study. The information given offers you the chance to think about how biology fits into the global landscape. The boxes also cover environmental and political issues raised by biology.

Examiner's hint: When showing crosses and offspring, don't forget to include such important details as:

- genotypes of the parents;
- a key of what the letters mean (although many questions state this);
- phenotypes of the offspring;
- alleles found in the gametes.

Examiner's hints can be found alongside questions, exercises and worked examples. They provide insight into how to answer a question in order to achieve the highest marks in an examination. They also identify common pitfalls when answering such questions and suggest approaches that examiners like to see.

For an idea of the vastness and diversity of life on Earth as well as the daunting nature of trying to classify it, visit heinemann.co.uk/hotlinks, enter the express code 4242P and click on Weblink 15.5.

These boxes direct you to the Heinemann website, which in turn will take you to the relevant website(s). On the web pages you will find background information to support the topic, perhaps video simulations and the like.

Now you are ready to start. Good luck with your studies!

1 Statistical analysis

Introduction

This is an Africanized honey bee (AHB). AHBs have spread to the USA from Brazil. They are now in competition with the local bee population, which are European honey bees (EHBs). EHBs were brought to America by European colonists in the 1600s. AHBs are now out-competing EHBs in areas the former invade.

This mixed oak forest is a mature stage in the development of plant communities surrounding bodies of water. Studying the growth rate of the trees such as maple, beech, oak and hickory gives us evidence of the health of these plant communities.

This is the common bean plant used by many students in their classrooms. Bean plants grow in about 30 days under banks of artificial lights. Seeds are easy to obtain. Germinated seeds can be placed in paper cups with sterilized soil. Many factors can be tested to determine whether or not they affect the growth of the bean plants.

In this chapter, you will learn how scientists analyse the evidence they collect when they perform experiments. You will be designing your own experiments, so this information will be very useful to you. You will be learning about:

- means;
- error bars;
- t-test;
- standard deviation;
- significant difference;
- causation and correlation.

Have your calculator by you to practise calculations for standard deviation and t-test so that you can use these methods of analysing data when you do your own experiments.

1.1 Statistics

Assessment statements

1.1.1 State that error bars are a graphical representation of the variability of data.
1.1.2 Calculate the mean and standard deviation of a set of values.
1.1.3 State that the term standard deviation is used to summarize the spread of values around the mean, and that 68% of values fall within one standard deviation of the mean.
1.1.4 Explain how the standard deviation is useful for comparing the means and spread of data between two or more samples.
1.1.5 Deduce the significance of the difference between two sets of data using calculated values for *t* and the appropriate tables.
1.1.6 Explain that the existence of a correlation does not establish that there is a causal relationship between two variables.

Reasons for using statistics

Statistics can be used to describe conditions which exist in countries around the world. The data can heighten our awareness of global issues. To learn about United Nations poverty facts and consumption statistics, visit heinemann.co.uk/hotlinks, insert the express code 4242P and click on Weblinks 1.1a and 1.1b.

Biology examines the world in which we live. Plants and animals, bacteria and viruses all interact with one another and the environment. In order to examine the relationships of living things to their environments and each other, biologists use the scientific method when designing experiments. The first step in the scientific method is to make observations. In science, observations result in the collection of measurable data. For example: What is the height of bean plants growing in sunlight compared to the height of bean plants growing in the shade? Do their heights differ? Do different species of bean plants have varying responses to sunlight and shade? After we have observed, we then decide which of these questions to answer. Assume we want to answer the question, 'Will the bean plant, *Phaseolus vulgaris*, grow taller in sunlight or the shade?' We must design an experiment which can try to answer this question.

How many bean plants should we use in order to answer our question? Obviously, we cannot measure every bean plant that exists. We cannot even realistically set up thousands and thousands of bean plants and take the time to measure their height. Time, money, and people available to do the science are all factors which determine how many bean plants will be in the experiment. We must use samples of bean plants which represent the population of all bean plants. If we are growing the bean plants, we must plant enough seeds to get a representative sample.

The World Health Organization (WHO) website has statistical information on infectious diseases. Over the next year, the system aims to provide a single point of access to data, reports and documents on:

- major diseases of poverty including malaria, HIV/AIDS, tuberculosis;
- diseases on their way towards eradication and elimination such as guinea worm, leprosy, lymphatic filariasis;
- epidemic-prone and emerging infections such as meningitis, cholera, yellow fever and anti-infective drug resistance.

Statistics is a branch of mathematics which allows us to sample small portions from habitats, communities, or biological populations, and draw conclusions about the larger population. Statistics mathematically measures the differences and relationships between sets of data. Using statistics, we can take a small population of bean plants grown in sunlight and compare it to a small population of bean plants grown in the shade. We can then mathematically determine the differences between the heights of these bean plants. Depending on the sample size that we choose, we can draw conclusions with a certain level of confidence. Based on a statistical test, we may be able to be 95% certain that bean plants grown in sunlight will be taller than bean plants grown in the shade. We may even be able to say that we are 99% certain, but nothing is 100% certain in science.

Mean, range, standard deviation and error bars

Statistics analyses data using the following terms:

- mean;
- range;
- standard deviation;
- error bars.

For more information and access to the WHO website, visit heinemann.co.uk/hotlinks, insert the express code 4242P and click on Weblink 1.2.

Mean

The mean is an average of data points. For example, suppose the height of bean plants grown in sunlight is measured in centimetres at 10 days after planting. The heights are 10, 11, 12, 9, 8 and 7 centimetres. The sum of the heights is 57 centimetres. Divide 57 by 6 to find the mean (average). The mean is 9.5 centimetres. The mean is the central tendency of the data.

For more statistics on the AIDS epidemic, visit heinemann.co.uk/hotlinks, insert the express code 4242P and click on Weblink 1.3.

Range

The range is the measure of the spread of data. It is the difference between the largest and the smallest observed values. In our example, the range is 12 – 7 = 5. The range for this data set is 5 centimetres. If one data point were unusually large or unusually small, this very large or small data point would have a great effect on the range. Such very large or very small data points are called outliers. In our sample there is no outlier.

Standard deviation

The standard deviation (SD) is a measure of how the individual observations of a data set are dispersed or spread out around the mean. Standard deviation is determined by a mathematical formula which is programmed into your calculator. You can calculate the standard deviation of a data set by using the SD function of a graphic display or scientific calculator.

Error bars

Error bars are a graphical representation of the variability of data. *Error bars can be used to show either the range of data or the standard deviation on a graph.* Notice the error bars representing standard deviation on the histogram in Figure 1.1 and the line graph in Figure 1.2. The value of the standard deviation above the mean is shown extending above the top of each bar of the histogram and the same standard deviation below the mean is shown extending below the top of each bar of the histogram. Since each bar represents the mean of the data, the standard deviation for each type of tree will be different, but the value extending above and below one bar will be the same. The same is true for the line graph. Since each point on the graph represents the mean data for each day, the bars extending above and below the data point are the standard deviations above and below the mean.

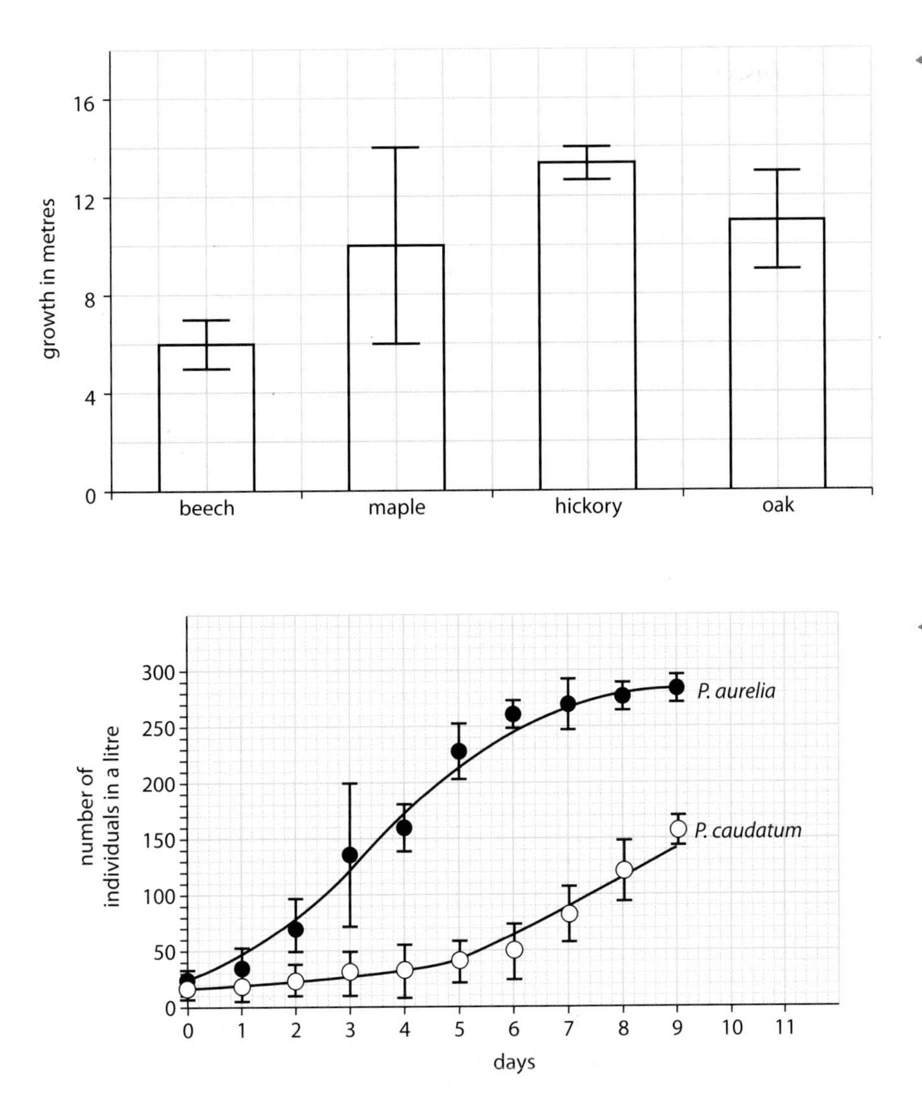

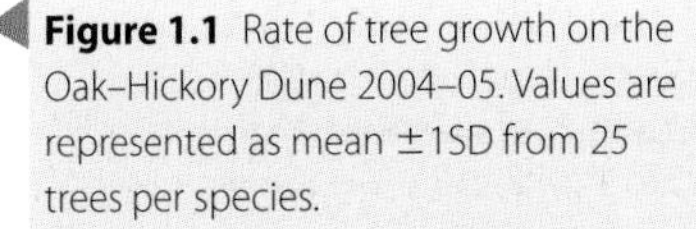

Figure 1.1 Rate of tree growth on the Oak–Hickory Dune 2004–05. Values are represented as mean ±1SD from 25 trees per species.

Figure 1.2 Mean population density (±1SD) of two species of *Paramecium* grown in solution.

Standard deviation

We use standard deviation to summarize the spread of values around the mean and to compare the means and spread of data between two or more samples.

Summarizing the spread of values around the mean

In a normal distribution, about 68% of all values lie within ±1 standard deviation of the mean. This rises to about 95% for ±2 standard deviations from the mean.

To help understand this difficult concept, let's look back to the bean plants growing in sunlight and shade. First, the bean plants in the sunlight: suppose our sample is 100 bean plants. Of that 100 plants, you might guess that a few will be very short (maybe the soil they are in is slightly sandier). A few may be much taller than the rest (possibly the soil they are in holds more water). However, all we can measure is the height of all the bean plants growing in the sunlight. If we then plot a graph of the heights, the graph is likely to be similar to a bell curve (see Figure 1.3). In this graph, the number of bean plants is plotted on the *y* axis and the heights ranging from short to medium to tall are plotted on the *x* axis.

Many data sets do not have a distribution which is this perfect. Sometimes, the bell-shape is very flat. This indicates that the data is spread out widely from the mean. In some cases, the bell-shape is very tall and narrow. This shows the data is very close to the mean and not spread out.

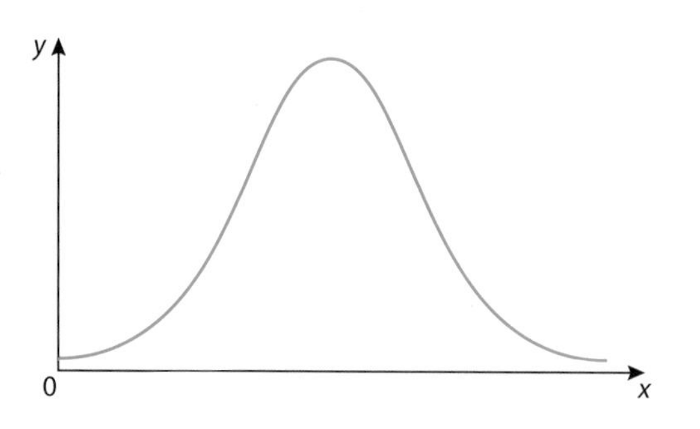

Figure 1.3 This graph shows a bell curve.

The standard deviation tells us how tightly the data points are clustered around the mean. When the data points are clustered together, the standard deviation is small; when they are spread apart, the standard deviation is large. Calculating the standard deviation of a data set is easily done on your calculator.

Look at Figure 1.4. This graph of normal distribution may help you understand what standard deviation really means. The dotted area represents one standard deviation in either direction from the mean. About 68% of the data in this graph is located in the dotted area. Thus, we say that for normally distributed data, 68% of all values lie within ±1 standard deviation from the mean. Two standard deviations from the mean (the dotted and the cross-hatched areas) contain about 95% of the data. If this bell curve were flatter, the standard deviation would have to be larger to account for the 68% or 95% of the data set. Now you can see why standard deviation tells you how widespread your data points are from the mean of the data set.

How is this useful? For one thing, it tells you how many extremes are in the data. If there are many extremes, the standard deviation will be large; with few extremes the standard deviation will be small.

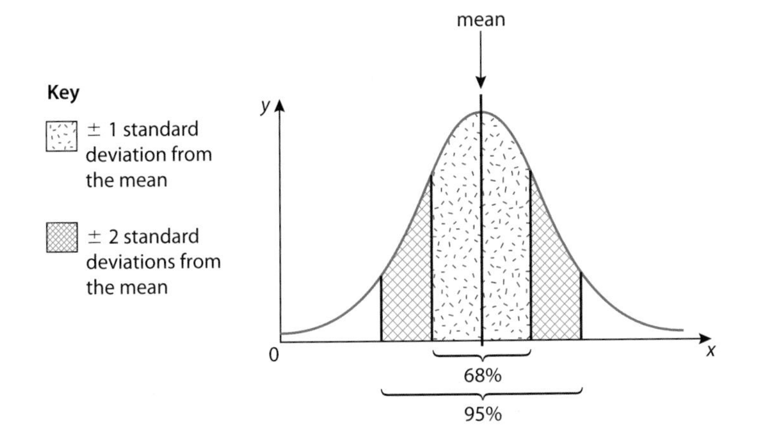

Figure 1.4 This graph shows a normal distribution.

For directions on how to calculate standard deviation with a TI-86 calculator, visit: heinemann.co.uk/hotlinks, insert the express code 4242P and click on Weblink 1.4a.

If you have a TI-83 calculator, visit: heinemann.co.uk/hotlinks, insert the express code 4242P and click on Weblink 1.4b.

Comparing the means and spread of data between two or more samples

Remember that in statistics we make inferences about a whole population based on just a sample of the population. Let's continue using our example of bean plants growing in the sunlight and shade to determine how standard deviation is useful for comparing the means and the spread of data between these two samples. Here are the raw data sets for bean plants grown in sunlight and in shade.

Height of bean plants in the sunlight in centimetres ±0.1 cm	Height of bean plants in the shade in centimetres ±0.1 cm
124	131
120	60
153	160
98	212
123	117
142	65
156	155
128	160
139	145
117	95
Total 1300	Total 1300

First, we determine the mean for each sample. Since each sample contains 10 plants, we can divide the total by 10 in each case. The resulting mean is 130.0 centimetres for each condition.

Of course, that is not the end of the analysis. Can you see there are large differences between the two sets of data? The height of the bean plants in the shade is much more variable than that of the bean plants in the sunlight. The means of each data set are the same, but the variation is not the same. This suggests that other factors may be influencing growth in addition to sunlight and shade.

How can we mathematically quantify the variation that we have observed? Fortunately, your calculator has a function that will do this for you. All you have to do is input the raw data. For practice, find the standard deviation of each raw data set above before you read on.

The standard deviation of the bean plants growing in sunlight is 17.68 centimetres while the standard deviation of the bean plants growing in the shade is 47.02 centimetres. Looking at the means alone, it appears that there is no difference between the two sets of bean plants. However, the high standard deviation of the

To use an online calculator to do the t-test go to: heinemann.co.uk/hotlinks, insert the express code 4242P and click on Weblink 1.5a or 1.5b.

bean plants grown in the shade indicates a very wide spread of data around the mean. The wide variation in this data set makes us question the experimental design. Is it possible that the plants in the shade are also growing in several different types of soil? What is causing this wide variation in data? This is why it is important to calculate the standard deviation in addition to the mean of a data set. If we looked at only the means, we would not recognize the variability of data seen in the shade-grown bean plants.

Significant difference between two data sets using the t-test

In order to determine whether or not the difference between two sets of data is a significant (real) difference, the t-test is commonly used. The t-test compares two sets of data, for example heights of bean plants grown in the sunlight and heights of bean plants grown in the shade. Look at the bottom of the table of t values (opposite), and you will see the probability (p) that chance alone could make a difference. If $p = 0.50$, we see the difference is due to chance 50% of the time. This is not a significant difference in statistics. However, if you reach $p = 0.05$, the probability that the difference is due to chance is only 5%. That means that there is a 95% chance that the difference is due (in our bean example) to one set of the bean plants being in the sunlight. A 95% chance is a significant difference in statistics. Statisticians are never completely certain but they like to be at least 95% certain of their findings before drawing conclusions.

When comparing two groups of data, we use the mean, standard deviation and sample size to calculate the value of t. When given a calculated value of t, you can use a table of t values. First, look in the left-hand column headed 'Degrees of freedom', then across to the given t value. The degrees of freedom are the sum of sample sizes of each of the two groups minus 2.

If the degree of freedom is 9, and if the given value of t is 2.60, the table indicates that the t value is just greater than 2.26. Looking down at the bottom of the table, you will see that the probability that chance alone could produce the result is only 5% (0.05). This means that there is a 95% chance that the difference is significant.

Worked example 1.1

Compare two groups of barnacles living on a rocky shore. Measure the width of their shells to see if a significant size difference is found depending on how close they live to the water. One group lives between 0 and 10 metres from the water level. The second group lives between 10 and 20 metres above the water level.

Measurement was taken of the width of the shells in millimetres. 15 shells were measured from each group. The mean of the group closer to the water indicates that living closer to the water causes the barnacles to have a larger shell. If the value of t is 2.25, is that a significant difference?

Solution

The degree of freedom is 28 ($15 + 15 - 2 = 28$). 2.25 is just above 2.05.

Referring to the bottom of this column in the table, $p = 0.05$ so the probability that chance alone could produce that result is only 5%.

The confidence level is 95%. We are 95% confident that the difference between the barnacles is significant. Barnacles living nearer the water have a significantly larger shell than those living 10 metres or more away from the water.

Table of t values

Degrees of freedom	*t* values					
1	1.00	3.08	6.31	12.71	63.66	636.62
2	0.82	1.89	2.92	4.30	9.93	31.60
3	0.77	1.64	2.35	3.18	5.84	12.92
4	0.74	1.53	2.13	2.78	4.60	8.61
5	0.73	1.48	2.02	2.57	4.03	6.87
6	0.72	1.44	1.94	2.45	3.71	5.96
7	0.71	1.42	1.90	2.37	3.50	5.41
8	0.71	1.40	1.86	2.31	3.367	5.04
9	0.70	1.38	1.83	2.26	3.25	4.78
10	0.70	1.37	1.81	2.23	3.17	4.590
11	0.70	1.36	1.80	2.20	3.11	4.44
12	0.70	1.36	1.78	2.18	3.06	4.32
13	0.69	1.35	1.77	2.16	3.01	4.22
14	0.69	1.35	1.76	2.15	2.98	4.14
15	0.69	1.34	1.75	2.13	2.95	4.07
16	0.69	1.34	1.75	2.12	2.92	4.02
17	0.69	1.33	1.74	2.11	2.90	3.97
18	0.69	1.33	1.73	2.10	2.88	3.92
19	0.69	1.33	1.73	2.09	2.86	3.88
20	0.69	1.33	1.73	2.09	2.85	3.85
21	0.69	1.32	1.72	2.08	2.83	3.82
22	0.69	1.32	1.72	2.07	2.82	3.79
24	0.69	1.32	1.71	2.06	2.80	3.75
26	0.68	1.32	1.71	2.06	2.78	3.71
28	0.68	1.31	1.70	2.05	2.76	3.67
30	0.68	1.31	1.70	2.04	2.75	3.65
35	0.68	1.31	1.69	2.03	2.72	3.59
40	0.68	1.30	1.68	2.02	2.70	3.55
45	0.68	1.30	1.68	2.01	2.70	3.52
50	0.68	1.30	1.68	2.01	2.68	3.50
60	0.68	1.30	1.67	2.00	2.66	3.46
70	0.68	1.29	1.67	1.99	2.65	3.44
80	0.68	1.29	1.66	1.99	2.64	3.42
90	0.68	1.29	1.66	1.99	2.63	3.40
100	0.68	1.29	1.66	1.99	2.63	3.39
Probability (p) that chance alone could produce the difference	0.50 (50%)	0.20 (20%)	0.10 (10%)	0.05 (5%)	0.01 (1%)	0.001 (0.1%)

Correlation does not mean causation

We make observations all the time about the living world around us. We might notice, for example, that our bean plants wilt when the soil is dry. This is a simple observation. We might do an experiment to see if watering the bean plants prevents wilting. Observing that wilting occurs when the soil is dry is a simple correlation, but the experiment gives us evidence that the lack of water is the cause of the wilting. Experiments provide a test which shows cause. Observations without an experiment can only show a correlation.

Africanized honey bees

The story of Africanized honey bees (AHBs) invading the USA includes an interesting correlation. In 1990, a honey bee swarm was found outside a small town in southern Texas. They were identified as AHBs. These bees were brought from Africa to Brazil in the 1950s, in the hope of breeding a bee adapted to the South American tropical climate. But by 1990, they had spread to the southern US. Scientists predicted that AHBs would invade all the southern states of the US, but this hasn't happened. Look at Figure 1.5: the bees have remained in the southwest states (area shaded in yellow) and have not travelled to the south-eastern states. The edge of the areas shaded in yellow coincides with the point at which there is an annual rainfall of 137.5 cm (55 inches) *spread evenly throughout the year*. This level of *year-round wetness* seems to be a barrier to the movement of the bees and they do not move into such areas.

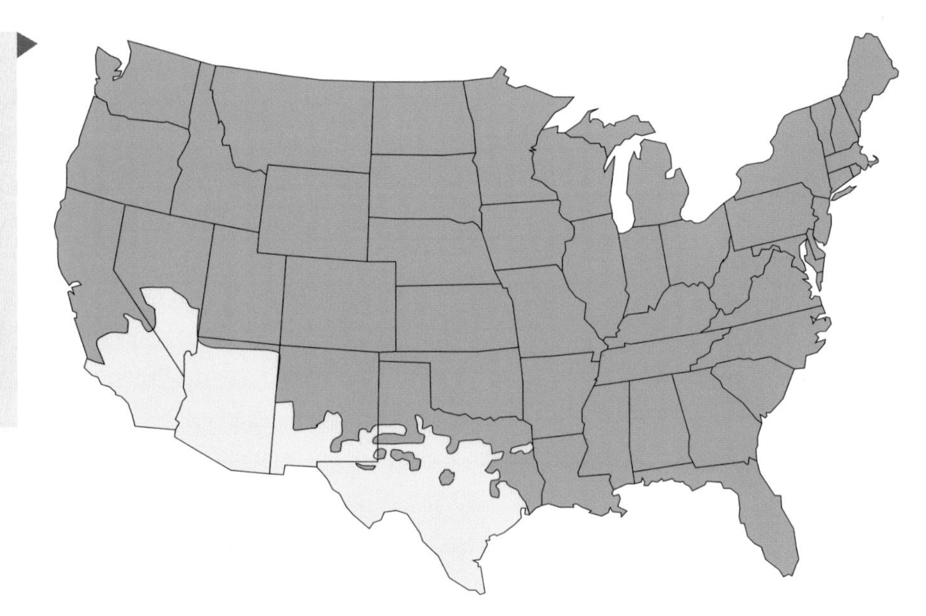

Figure 1.5 AHBs have not moved beyond the areas shaded yellow in the last 10 years. So, states in the south east (Louisiana, Florida, Alabama and Mississippi) seem unlikely to be bothered by AHBs if the 137.5 cm (55 inches) of rain correlation holds true. This is an example of a mathematical correlation and is not evidence of a cause. In order to find out if this is a cause, scientists must design experiments to explain mechanisms which may be the cause of the observed correlation.

Cormorants

When using a mathematical correlation test, the value of r signifies the correlation. The value of r can vary from +1 (completely positive correlation) to 0 (no correlation) to –1 (completely negative correlation). For an example, we can measure in millimetres the size of breeding cormorant birds to see if there is a correlation between the sizes of males and females which breed together.

Pair numbers	Size of female cormorants	Size of male cormorants
1	17.1	16.5
2	18.5	17.4
3	19.7	17.3
4	16.2	16.8
5	21.3	19.5
6	19.6	18.3
$r = 0.88$		

The r value of 0.88 shows a positive correlation between the sizes of the two sexes: large females mate with large males. However, correlation is not cause. To find the cause of this observed correlation requires experimental evidence. There may be a high correlation but only carefully designed experiments can separate causation from correlation.

Just because event X is regularly followed by event Y, it does not necessarily follow that X causes Y. Biologists are often faced with the difficult challenge of determining whether or not events that appear related are causally associated. For example, there may be an association between large numbers of telephone poles in a particular geographic area and the number of people in that area who have cancer, but that does not mean that the telephone poles cause cancer. Carefully designed experiments are needed to separate causation from correlation.

The tobacco companies used to say that there was a statistical correlation between smoking and lung cancer, but they insisted that there was no causal connection. In other words, they said X did not follow Y in this case. In fact, they said that people who thought that smoking caused cancer were committing a fallacy of *post ergo proper hoc* (one thing follows another). However, we now have lots of other evidence that smoking does cause lung cancer.

So, try assessing the following statements as to whether you think they represent a correlation or if there is a causal connection between the two things in each case.

1 Cars with low mileage per gallon/litre of fuel cause global warming.
2 Drinking red wine protects against heart disease.
3 Tanning beds can cause skin cancer.
4 UV rays increase the risk of cataracts.
5 Vitamin C cures the common cold.

Use of tobacco by adolescents is a major public health problem in all six WHO regions. Worldwide, more countries need to develop, implement, and evaluate tobacco-control programmes to address the use of all types of tobacco products, especially among girls.

You can find many interesting statistics if you visit heinemann.co.uk/hotlinks, insert the express code 4242P and click on Weblink 1.6.

It can be difficult to imagine what it was like when people did not have some of the knowledge that we take for granted today. In the mid-19th century, people had a different paradigm of what caused disease; they thought there were many causes of disease but they did not know about germs (microorganisms). The modern concept that germs can cause disease, was introduced by Louis Pasteur. Further evidence of the germ theory was demonstrated by Robert Koch.

To find information on Pasteur and Koch, and discover answers to the questions below, visit heinemann.co.uk/hotlinks, insert the express code 4242P and click on Weblink 1.7.

1 What was the paradigm of disease for people in the mid-19th century? What did they think caused disease? Were they looking at causation or correlation?
2 How did Louis Pasteur's work change this paradigm?
3 Explain how Robert Koch's work gave evidence which was required to show that a bacterium plays a causal role in a certain disease.

Exercises

1 Define an error bar.
2 Define standard deviation.
3 Explain the use of standard deviation when comparing the means of two sets of data.
4 If you are given a calculated value for t, what can be deduced from a t-table?
5 Give a specific example of a correlation.
6 Explain the relationship between cause and correlation.

Practice questions

As this chapter covers new syllabus material, there are no past examination papers available. Hence, there is no markscheme for these questions.

1 What is standard deviation used for?

A To determine that 68% of the values are accurate.
B To deduce the significant difference between two sets of data.
C To show the existence of cause.
D To summarize the spread of values around the means.

2 What is an error bar?

A A graphical representation of the variability of data.
B A graphical representation of the correlation of data.
C A calculated mean.
D A histogram.

3 What is the t-test used for?

A Comparing the spread of data between two samples.
B Deducing the significant difference between two sets of data.
C Explaining the existence of cause and correlation.
D Calculating the mean and the standard deviation of a set of values.

4 What is the relationship between cause and correlation?

A A correlation does not establish a causal relationship.
B A correlation requires a scientific experiment, while cause does not.
C A correlation requires collection of data, cause does not.
D A causal relationship does not establish correlation.

5 An experiment using 0.5 grams of fresh garlic and crushed garlic root, leaf and bulb from garlic cloves sprouted for 2 days was performed. This garlic was used in a bioassay on lettuce seedlings to see if growth of the lettuce seedlings was inhibited compared to a control. Data recorded was seedling length in millimetres.

Fresh garlic	Crushed sprouted root	Crushed sprouted leaf	Crushed sprouted bulb	Control (no treatment)
4	4	3	3	20
5	5	4	4	19
6	5	5	5	17
4	5	5	4	20
3	4	5	5	21
4	3	6	4	22
3	3	4	5	19
5	5	5	4	17
4	4	4	4	16
3	3	3	5	20

(a) Find the mean and standard deviation of each of these data sets.
(b) Discuss the variability of the garlic data.
(c) Compare the means of each group of data.
(d) **(i)** If the calculated t for fresh garlic compared to the control is $t = 13.9$, is the difference between the two groups a significant difference?
(ii) What is the probability that the difference between the groups is due to chance?

(e) **(i)** If the calculated t for the sprouted bulb compared to the fresh garlic is $t = 0.33$, is the difference significant?

(ii) What is the probability that the difference between the two groups is due to chance?

6 An experiment using fertilizer on bean plants was performed. Germinated seeds were planted in sterile soil to which different amounts of a commercial fertilizer were added. Height of the plants was measured in centimetres 25 days after planting. Data recorded was bean plant height in centimetres (±0.5 cm).

Group A No fertilizer	Group B 0.001% fertilizer	Group C 0.01% fertilizer	Group D 0.1% fertilizer
10	8	7	12
7	8	10	13
8	7	8	15
7	10	10	10
8	10	8	10
10	7	9	15
9	8	9	10
8	8	8	10
7	9	6	14
9	9	7	10

(a) Find the mean and standard deviation of each set of data.

(b) Discuss the variability of the bean plant data.

(c) Compare the means of each group of data.

(d) **(i)** If the calculated t for Group B compared to the the control is $t = 0.60$, is the difference between the two groups a significant difference?

(ii) What is the probability that the difference between the two groups is due to chance?

(e) **(i)** If the calculated t for Group D compared to the the control is $t = 2.90$, is the difference between the two groups significant?

(ii) What is the probability that the difference is due to chance?

2 Cells

Introduction

Whether organisms are extremely small or extremely large, it is imperative to understand their smallest functional units. These units are known as cells. Organisms range in size from a single cell to thousands of cells. To better understand plants and all the organisms around us, we must study their cells.

Look at the picture. Human nerve cells (neurones) are essential to our lives. Because of these cells, we are able to acknowledge and respond to our surroundings. Neurones are usually very efficient but sometimes things go wrong. Will a greater understanding and better treatment of conditions such as depression result from an improved comprehension of how these cells function?

This is an artist's impression of human nerve cells.

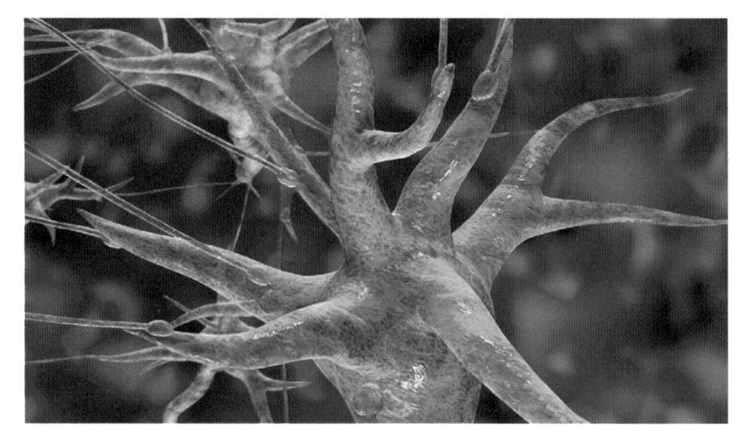

2.1 Cell theory

Assessment statements

2.1.1 Outline the cell theory.
2.1.2 Discuss the evidence for the cell theory.
2.1.3 State that unicellular organisms carry out all the functions of life.
2.1.4 Compare the relative sizes of molecules, cell membrane thickness, viruses, bacteria, organelles and cells, using the appropriate SI unit.
2.1.5 Calculate the linear magnification of drawings and the actual size of specimens in images of known magnification.
2.1.6 Explain the importance of the surface area to volume ratio as a factor limiting cell size.
2.1.7 State that multicellular organisms show emergent properties.
2.1.8 Explain that cells in multicellular organisms differentiate to carry out specialized functions by expressing some of their genes but not others.
2.1.9 State that stem cells retain the capacity to divide and have the ability to differentiate along different pathways.
2.1.10 Outline one therapeutic use of stem cells.

Cell theory

It has taken several hundred years of research to formulate modern cell theory. Many scientists have contributed to developing the three main principles of this theory. These are:

- all organisms are composed of one or more cells;
- cells are the smallest units of life;
- all cells come from pre-existing cells.

This theory has amassed tremendous credibility, largely through use of the microscope – an important tool. Robert Hooke first described cells in 1665 while observing cork with a microscope he built himself. A few years later, Antonie van Leeuwenhoek observed the first living cells and referred to them as 'animalcules', meaning little animals. In 1838, botanist Mathias Schleiden stated that plants are made of 'independent, separate beings' called cells. One year later, the zoologist Theodor Schwann made a similar statement about animals.

The second principle continues to gain support today, as we have not been able to find any living entity that is not made of at least one cell.

Some very famous scientists, such as Louis Pasteur in the 1860s, have performed experiments to support the last principle. After sterilizing chicken broth by boiling, Pasteur showed that living organisms would not 'spontaneously' reappear. Only after exposure to pre-existing cells was life able to re-establish itself in the sterilized chicken broth.

Theories are developed after the accumulation of much data. Sometimes, theories are completely abandoned because of conflicting evidence.

Viruses are not considered to be living. They can not carry out the functions of life on their own. However, they do utilize cells to perpetuate themselves.

Functions of life

All organisms exist in either a unicellular or a multicellular form. And all organisms carry out all the functions of life. These functions include:

- metabolism;
- growth;
- reproduction;
- response;
- homeostasis;
- nutrition.

All these functions are tied together to produce a functioning living unit.

- Metabolism includes all the chemical reactions that occur within an organism.
- Growth may be limited but is always evident in one way or another.
- Reproduction involves hereditary molecules that can be passed to offspring.
- Response to the environment is imperative to the survival of the organism.
- Homeostasis refers to maintaining a constant internal environment. Examples of constant internal environments may involve temperature and acid–base levels.
- Nutrition is all about providing a source of compounds with many chemical bonds which can be broken to provide the organism with the energy and the nutrients necessary to maintain life.

The functions of life are manifested in different ways in the various types of organism, but all life forms maintain the same general functions.

You may see different terms for these functions in other sources.

Cells and sizes

Cells are made up of a number of different subunits. These subunits are often of a particular size, but all are microscopically small. In most cases, microscopes with high magnification and resolution are needed to observe cells and especially their subunits. Resolution refers to the clarity of a viewed object.

Most cells are up to 100 micrometres; organelles are up to 10 micrometres.

Bacteria are up to 1 micrometre.

Viruses are up to 100 nanometres.

Membranes are 10 nanometres thick, and molecules are near 1 nanometre.

All of these objects are three-dimensional.

Light microscopes use light, which passes through the living or dead specimen, to form an image. Stains may be used to improve viewing of parts. Electron microscopes provide us with the greatest magnification (over 100 000×) and resolution. These use electrons passing through a specimen to form an image. It is hard to understand or visualize very small sizes, so it is important to appreciate relative size. Cells are relatively large, and then in decreasing size order are:

- organelles;
- bacteria;
- viruses;
- membranes;
- molecules.

Drawings or photographs are often enlarged. To calculate the magnification, you need this formula:

- magnification = size of image divided by size of specimen.

If you want to calculate the actual size of a specimen seen with a microscope, you need to know the diameter of the microscope's field of vision. This may be calculated with a special micrometer or with a simple ruler on a light microscope. The size of the specimen can then be calculated in the field.

Scale bars are often used with a micrograph or drawing so that actual size can be determined.

Limiting cell size

So, the cell is a small object. You may wonder why cells do not grow to larger sizes, especially since growth is one of the functions of life. There is a factor called the surface area to volume ratio that effectively limits the size of cells. In the cell, the rate of heat and waste production and rate of resource consumption are functions of (depend on) its volume. Most of the chemical reactions occur in the interior of the cell and its size affects the rate of these reactions. The surface of the cell, the membrane, controls what materials move in and out of the cell. Cells with more surface area per unit volume are able to move more materials in and out of the cell, for each unit volume of the cell.

As the width of an object such as a cell increases, the surface area also increases but at a much slower rate than the volume. This is shown by the following table in which you can see that the volume increases by a factor calculated by cubing the radius; at the same time, the surface area increases by a factor calculated by squaring the radius.

Sphere formulas:

- surface area = (four)(pi)(radius squared) = $4\pi r^2$
- volume = (four-thirds)(pi)(radius cubed) = $\frac{4}{3}\pi r^3$

Cell radius (*r*)	0.25 units	0.5 units	1.25 units
Surface area	0.79 units	3.14 units	7.07 units
Volume	0.06 units	0.52 units	1.77 units
Surface area : volume	13.17 : 1	6.04 : 1	3.99 : 1

This means that a large cell has relatively less surface area to bring in needed materials and to rid the cell of waste, than a small cell. Because of this, cells are limited as to the size they can attain and still be able to carry out the functions of life. Thus, large animals do not have larger cells, they have more cells.
Cells that are larger in size have modifications that allow them to function efficiently. This is accomplished by shape changes such as from spherical to long and thin. Also, some larger cells have infoldings or outfoldings to increase their surface relative to their volume.

Cell reproduction and differentiation

One of the functions that many cells retain is the ability to reproduce themselves. In multicellular organisms, this allows the possibility of growth. It also allows for the replacement of damaged or dead cells.

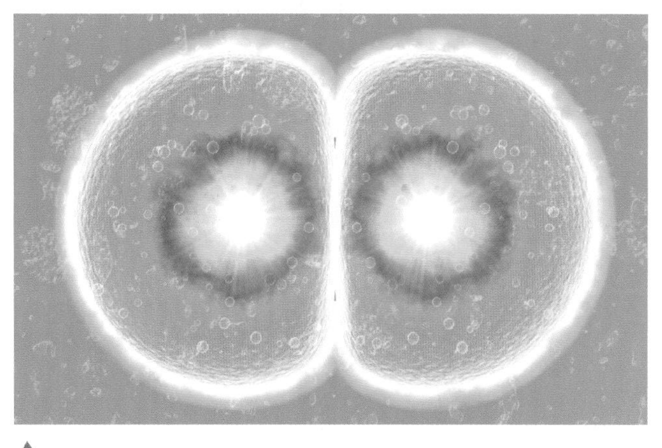

This is a computer artwork of an egg cell fertilized during in vitro fertilization and now undergoing the first cell division.

Multicellular organisms like ourselves usually start out as a single cell after some type of sexual reproduction. This single cell has the ability to reproduce at a very rapid rate, and the resulting cells then go through a differentiation process to produce all the required cell types that are necessary for the well-being of the organism. The number of different cell types from the one original cell may indeed be staggering. This differentiation process is the result of the expression of certain specific genes but not others. Genes, segments of DNA on a chromosome, allow for the production of all the different cells in the organism. Therefore, each cell contains all the genetic information for the production of the complete organism. However, each cell becomes a specific type of cell dependent on which DNA segment becomes active.

Some cells have a greatly, or even completely, diminished ability to reproduce once they become specialized. Nerve and muscle cells are prime examples of this type of cell. Other cells, such as epithelial cells like skin, retain the ability to rapidly reproduce throughout their life. The offspring of these rapidly reproducing cells then differentiate into the same cell type as the parent.

Stem cells

There are populations of cells within organisms that retain their ability to divide and differentiate into various cell types. These cells are called stem cells.

Plants contain such cells in regions of meristematic tissue. Meristematic areas occur near root and stem tips and are composed of rapidly reproducing cells that produce new cells capable of becoming various types of tissue within the root or stem. Gardeners take advantage of these cells when they take cuttings from stems or roots and use them to produce new plants.

In the early 1980s, scientists found pluripotent or embryonic stem cells in mice. These stem cells retain the ability to form any type of cell in an organism and can even form a complete organism.

When stem cells divide to form a specific type of tissue, they also produce some cells that remain as stem cells. This allows for the continual production of a particular type of tissue. Medical experts immediately noted the possibilities of such cells in treating certain human diseases. But a problem discovered early in the research was that stem cells cannot be distinguished by their appearance. They can only be isolated from other cells on the basis of their behaviour.

Stem cell research and treatments

Some of the most promising research recently has been directed towards growing large numbers of embryonic stem cells in culture so that they could be used to replace differentiated cells lost due to injury and disease. This involves therapeutic cloning. Parkinson's disease and Alzheimer's disease are caused by loss of brain cells, and it is hoped that implanted stem cells could replace many of these lost brain cells thus relieving the disease symptoms. Certain forms of diabetes deplete the pancreas of essential cells and it is hoped that a stem cell implant in this organ could have positive effects. As most of the research at present is occurring in mice, it is likely to be quite a time before this treatment approach becomes possible in humans.

However, there is a type of stem cell treatment that has been proceeding successfully in humans for many years. Besides embryonic or pluripotent stem cells, there are tissue-specific stem cells. These stem cells reside in certain tissue types and can only produce new cells of that particular tissue. For example, blood stem cells have been routinely introduced into humans to replace the damaged bone marrow of some leukaemia patients.

Internationally, there has been much sharing of data involving stem cell research. Many nations have banned or limited research in this area due to local, cultural and religious traditions.

Where do you stand in the debate about the nature of stem cell research? How do you feel about the source of pluripotent stem cells?

There are important ethical issues involved in stem cell research. Especially controversial is the use of embryonic or pluripotent stem cells. This is because these cells come from embryos often obtained from laboratories carrying out in-vitro fertilization (IVF). To gather these cells involves death of the embryo and opponents argue that this represents the taking of a human life. On the other hand, it is argued that this research could result in the significant reduction of human suffering and is, therefore, totally acceptable.

How the scientific community conveys information concerning its research to wider society is very important. This information must be accurate, complete and understandable so that society can make informed decisions as to the appropriateness of the research. For example, in 2005, stem cells successfully helped to restore the lost insulation of nerve cells in rats thus resulting in greater mobility of the animals.

But there is a need to balance the very great opportunities of this type of research with the potential risks. For example, there is recent evidence that some types of cancer may be caused by stem cells undergoing a malignant transformation. This shows possible risk in the implantation of stem cells.

Exercises

1 How would the excretion of metabolic wastes from cells be related to the concept of surface area to volume ratio?
2 Name two disadvantages of using an electron microscope.
3 How does specialization in muscle and nerve cells affect their ability to reproduce?
4 What would prevent stem cells from other species being successful in humans?

2.2 Prokaryotic cells

Assessment statements

2.2.1 Draw and label a diagram of the ultrastructure of *Escherichia coli* (*E. coli*) as an example of a prokaryote.
2.2.2 Annotate the diagram with the functions of each named structure.
2.2.3 Identify structures from 2.2.1 in electron micrographs of *E. coli*.
2.2.4 State that prokaryotic cells divide by binary fission.

What is a prokaryotic cell?

After exhaustive studies of cells, it is apparent that all cells use common molecular mechanisms. There are huge differences between the forms of life but cells are the basic unit of life and have many characteristics in common. Cells are often divided into certain groups based on major characteristics. One such grouping divides cells into two groups: prokaryotic and eukaryotic. Prokaryotic cells are much smaller and simpler than eukaryotic cells. In fact, most prokaryotic cells are less than 1 micrometre in diameter. Because of this and many other reasons to be discussed later, the prokaryotic cells are thought to have appeared on Earth first. As bacteria are prokaryotic cells, you can see that such cells play a large role in the world today.

The word 'prokaryotic' is from the Greek pro, which means 'before', and karyon, which means 'kernel', referring to the nucleus.

Features of prokaryotic cells

Study Figure 2.1 and be sure you can identify:

- the cell wall;
- the plasma membrane;
- flagella;
- ribosomes;
- the nucleoid (a region containing free DNA).

• **Examiner's hint:** Practise drawing a typical prokaryotic cell. Label clearly all the parts stated in the text. Once you have completed the drawing and labelling, annotate the diagram with the functions of each named structure.

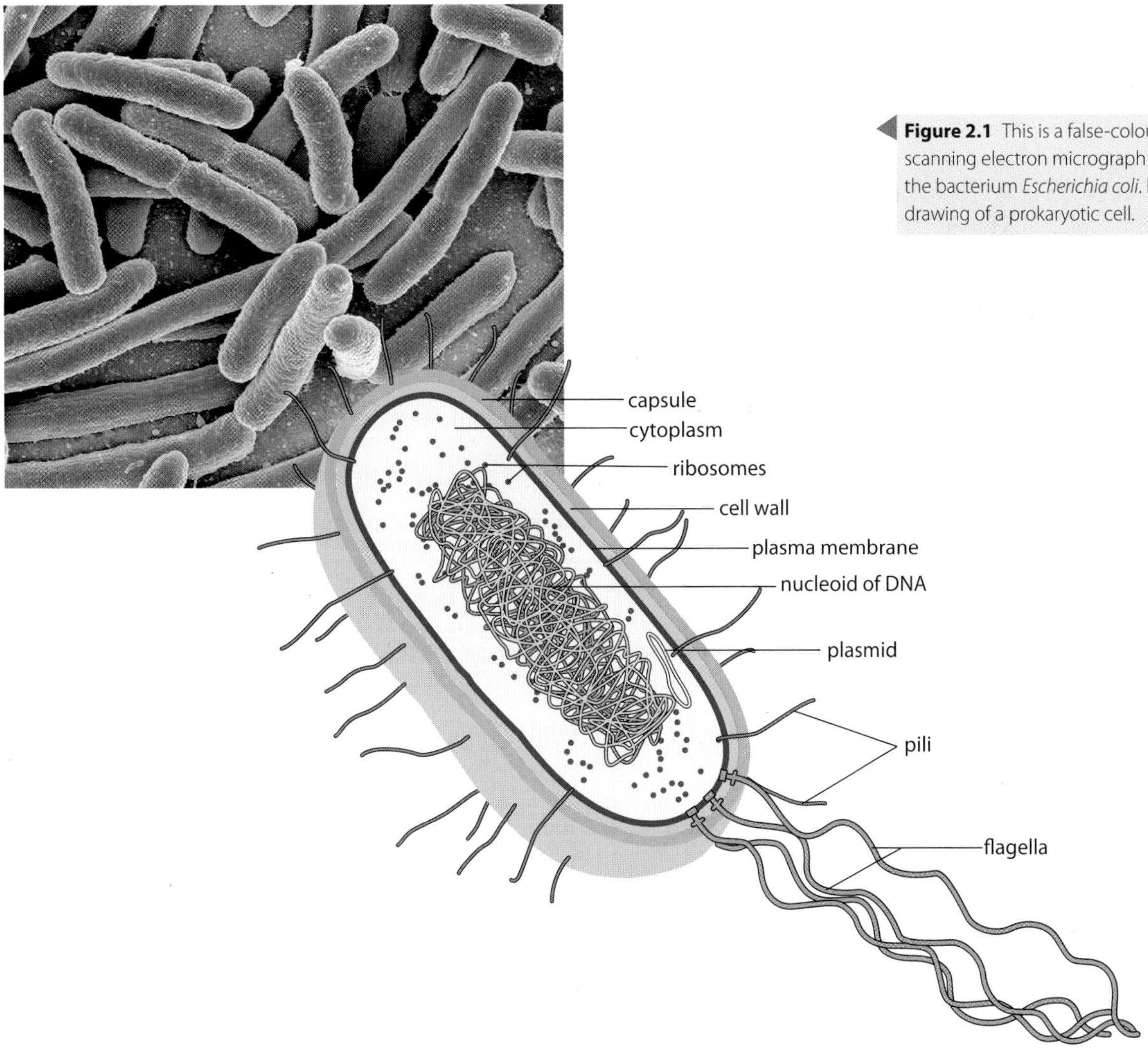

Figure 2.1 This is a false-colour scanning electron micrograph (SEM) of the bacterium *Escherichia coli*. Below is a drawing of a prokaryotic cell.

Cell wall and plasma membrane

The prokaryotic cell wall protects and maintains the shape of the cell. In most prokaryotic cells, this wall is composed of a carbohydrate–protein complex called peptidoglycan. Some bacteria have an additional layer of a type of polysaccharide outside the cell wall. This layer makes it possible for some bacteria to adhere to structures such as teeth, skin and food.

The plasma membrane is just inside the cell wall and has a composition similar to the membranes of eukaryotic cells. It controls the movement of materials in and out of the cell to a large extent, and it plays a role in binary fission of the prokaryotic cell. The cytoplasm occupies the complete interior of the cell. The most visible structure with a microscope capable of high magnification is the chromosome or molecule of DNA. There is no compartmentalization in the cytoplasm. Therefore, all cellular processes in prokaryotic cells occur in the cytoplasm.

Pili and flagella

Some bacterial cells contain hair-like growths on the outside of the cell wall. These structures are called pili and are used for attachment. However, their main function is in joining bacterial cells in preparation for the transfer of DNA from one cell to another (sexual reproduction).

If a bacterium has flagella (plural) or a flagellum (singular), they are longer than pili. The flagella allow cell motility.

Ribosomes

Ribosomes occur in all prokaryotic cells and they function as sites of protein synthesis. These small structures occur in very large numbers in cells with high protein production, and when numerous, impart a granular appearance to an electron micrograph of the prokaryotic cell.

The nucleoid region

The nucleoid region of the bacterial cell is non-compartmentalized and contains a single, long, continuous, circular thread of DNA. Therefore, this region is involved with cell control and reproduction. In addition to the bacterial chromosome, bacteria may also contain plasmids. These small, circular, DNA molecules are not connected to the main bacterial chromosome. The plasmids replicate independently of the chromosomal DNA. Plasmid DNA is not required by the cell under normal conditions but it may help the cell adapt to unusual circumstances.

Binary fission

Prokaryotic cells divide by a very simple process called binary fission. During this process, the DNA is copied, the two daughter chromosomes become attached to different regions on the plasma membrane, and the cell divides into two genetically identical daughter cells. This divisional process includes an elongation of the cell and a partitioning of the newly produced DNA by microtubule-like fibres made of protein called FtsZ.

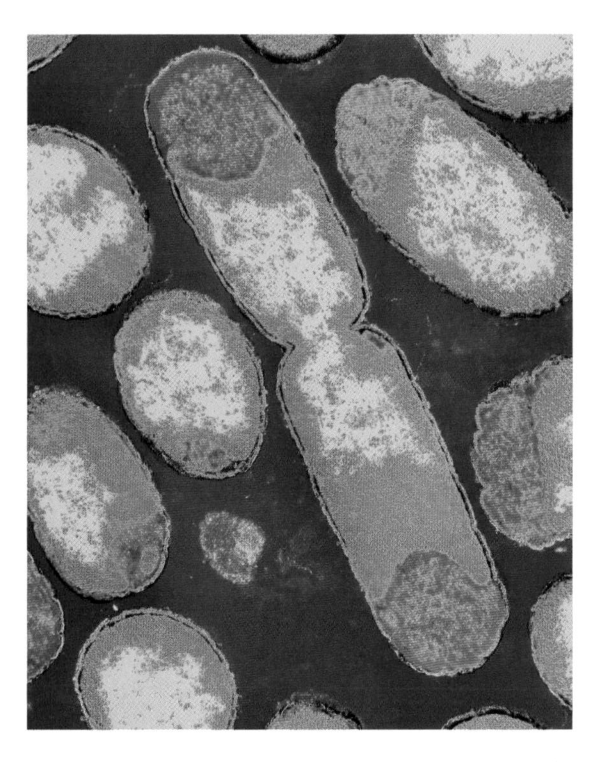

This is a false-colour transmission electron micrograph (TEM) showing *E. coli* dividing by binary fission.

Summary

Here is a list of the major distinguishing characteristics of prokaryotic cells.

1 Their DNA is not enclosed within a membrane and is one circular chromosome.
2 Their DNA is free; it is not attached to proteins.
3 They lack membrane-bound organelles. Ribosomes are complex structures within the plasma membrane, but they have no exterior membrane.
4 Their cell wall is made up of a unique compound called peptidoglycan.
5 They usually divide by binary fission, a simple form of cell division.
6 They are characteristically small in size, usually 1–10 micrometres.

For more information on bacterial cell structure, visit heinemann.co.uk/hotlinks, insert the express code 4242P and click on Weblink 2.1.

Exercises

5 What is a disadvantage of the prokaryotic cells having their DNA free in the cytoplasm without a nuclear membrane?

6 What structures are involved in sexual reproduction in prokaryotic cells?

2.3 Eukaryotic cells

Assessment statements

2.3.1 Draw and label a diagram of the ultrastructure of a liver cell as an example of an animal cell.
2.3.2 Annotate the diagram with the functions of each named structure.
2.3.3 Identify structures from 2.3.1 in electron micrographs of liver cells.
2.3.4 Compare prokaryotic and eukaryotic cells.
2.3.5 State three differences between plant and animal cells.
2.3.6 Outline two roles of extracellular components.

What is a eukaryotic cell?

Whereas prokaryotic cells occur in the bacteria, eukaryotic cells occur in organisms such as algae, protozoa, fungi, plants and animals (see Figures 2.2, 2.3 and 2.4).

Figure 2.2 Look at this drawing of a typical animal cell and compare it with Figure 2.3.

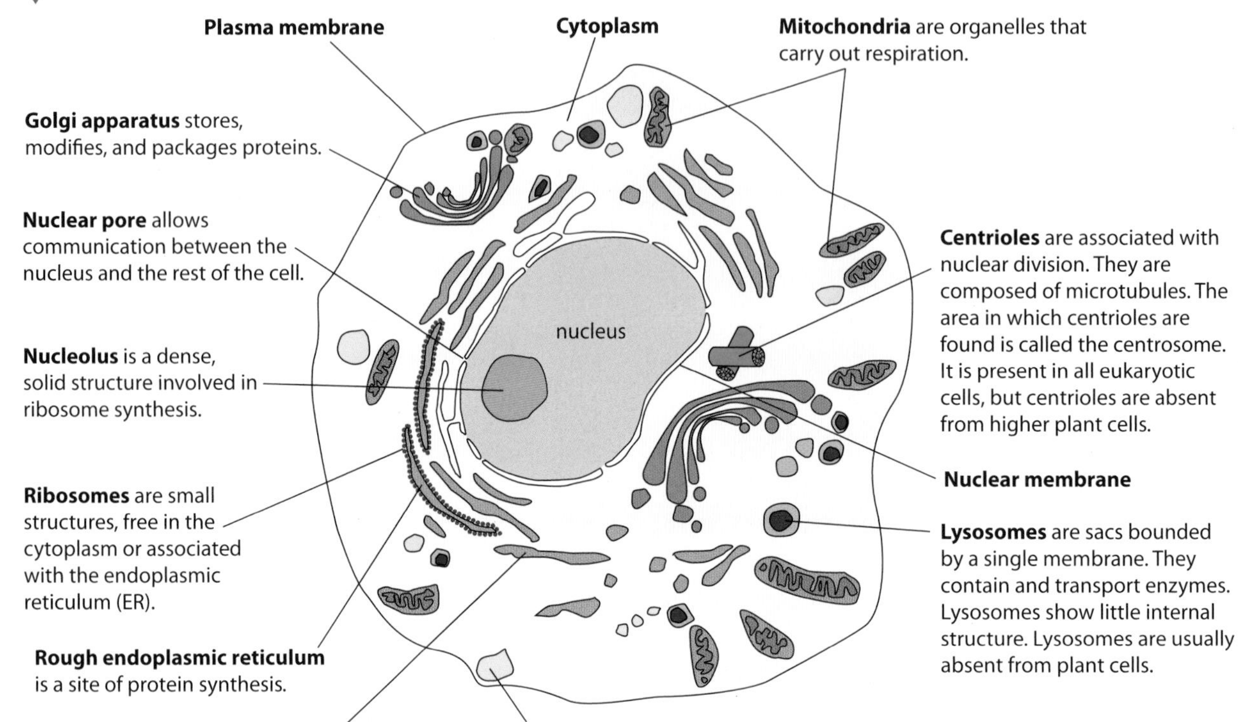
Plasma membrane
Cytoplasm
Mitochondria are organelles that carry out respiration.
Golgi apparatus stores, modifies, and packages proteins.
Nuclear pore allows communication between the nucleus and the rest of the cell.
Centrioles are associated with nuclear division. They are composed of microtubules. The area in which centrioles are found is called the centrosome. It is present in all eukaryotic cells, but centrioles are absent from higher plant cells.
nucleus
Nucleolus is a dense, solid structure involved in ribosome synthesis.
Nuclear membrane
Ribosomes are small structures, free in the cytoplasm or associated with the endoplasmic reticulum (ER).
Lysosomes are sacs bounded by a single membrane. They contain and transport enzymes. Lysosomes show little internal structure. Lysosomes are usually absent from plant cells.
Rough endoplasmic reticulum is a site of protein synthesis.
Smooth endoplasmic reticulum is ER without ribosomes.
Vacuoles are smaller than those found in plant cells.

Eukaryotic cells range in diameter from 5 to 100 micrometres. Also, usually noticeable is the 'kernel' or nucleus in the cytoplasm. Other organelles may be visible in the cell if you have a microscope with high enough magnification and resolution. Organelles are non-cellular structures that carry out specific functions (a bit like organs in multicellular organisms). Different types of cell often have different organelles. These structures bring about compartmentalization in eukaryotic cells. This is not a characteristic of prokaryotic cells. Compartmentalization allows chemical reactions to be separated. This is especially important when the adjacent chemical reactions are incompatible. It also allows chemicals for specific reactions to be isolated. This isolation results in increased efficiency.

The term 'eukaryote' comes from the Greek words meaning 'true kernel,' or nucleus.

Organelles of eukaryotic cells

Common organelles include the following (see Figures 2.2 and 2.3):

- endoplasmic reticulum;
- ribosomes;
- lysosomes (not usually in plant cells);
- Golgi apparatus;
- mitochondria;
- nucleus;
- chloroplasts (only in plant and algal cells);
- centrosomes (in all eukaryotic cells but centrioles are not found in higher plant cells);
- vacuoles.

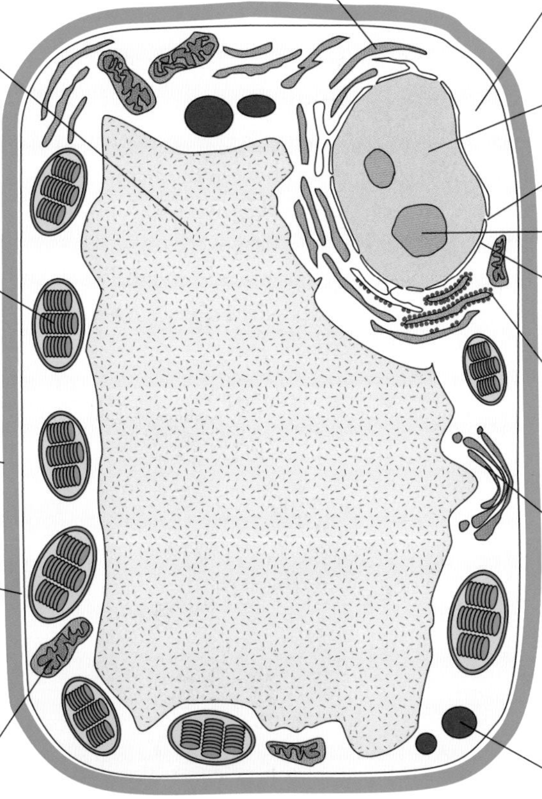

Endoplasmic reticulum (ER) is a network of tubes and flattened sacs. ER connects with the plasma membrane and the nuclear membrane and may be smooth or have attached ribosomes (rough ER).

Central vacuole has storage and hydrolytic functions

Cytoplasm contains dissolved substances, enzymes, and the cell organelles.

Nucleus contain most of the cell's DNA.

Nuclear pore

Nucleolus

Chloroplasts are specialized plastids containing the green pigment chlorophyll. They consist of grana within the colourless stroma. They are the sites for photosynthesis.

Nuclear membrane is a double-layered structure.

Cell wall is a semi-rigid structure composed mainly of cellulose.

Ribosomes are small (20 nm) structures which manufacture proteins. They may be free in the cytoplasm or associated with the surface of the endoplasmic reticulum.

Plasma membrane is inside the cell wall.

Golgi apparatus

Mitochondria are bounded by a double membrane. They are energy transformers.

Starch granules are composed of carbohydrate stored in amyloplasts.

Figure 2.3 What is different and what is similar between this typical plant cell and Figure 2.2?

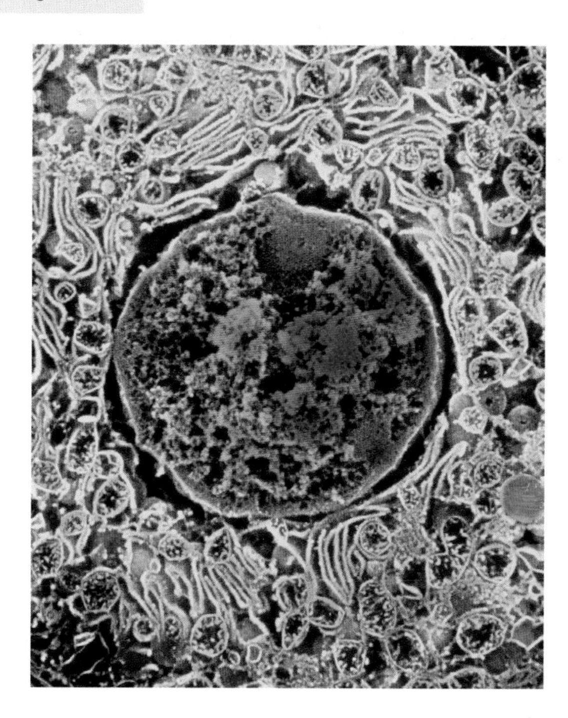

Figure 2.4 Carefully examine this false-colour TEM of part of a liver cell. Can you tell it is an animal cell? Locate as many of the structures of an animal cell as you can. How do the structures of this cell reflect the overall functions of the liver?

Cytoplasm

All eukaryotic cells have a region called the cytoplasm that occurs inside the plasma membrane or the outer boundary of the cell. It is in this region that the organelles occur. The fluid portion of the cytoplasm between the organelles is referred to as the cytosol.

Endoplasmic reticulum

The endoplasmic reticulum (ER) is an extensive network of tubules or channels that extends almost everywhere in the cell from the nucleus to the plasma membrane. Its structure enables its function which is the transportation of materials throughout the internal region of the cell. The endoplasmic reticulum is of two general types: smooth and rough (see Figure 2.5). Smooth ER does not have any of the organelles called ribosomes on its exterior surface. Rough ER has ribosomes on its exterior.

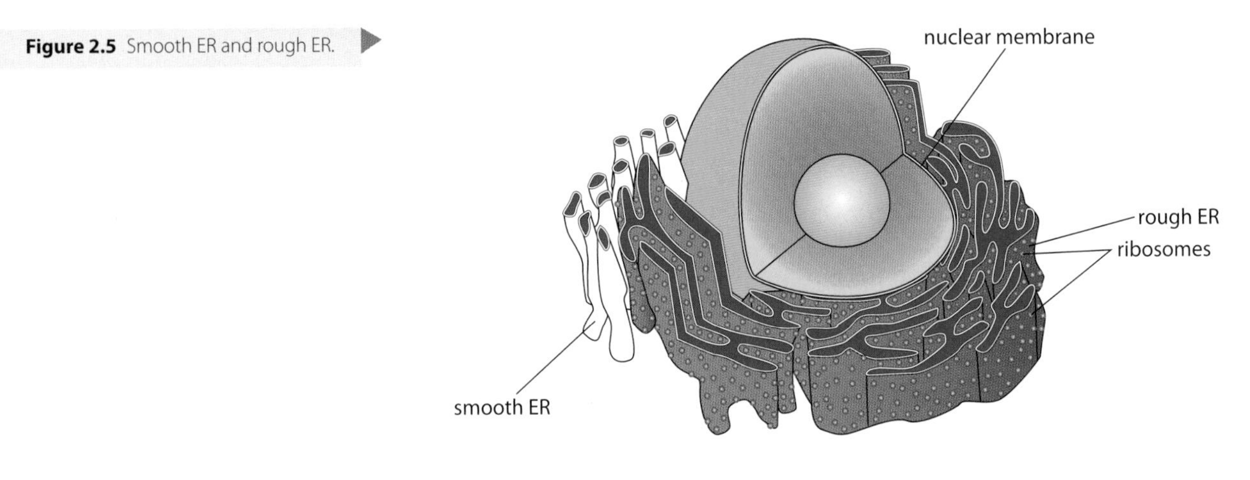

Figure 2.5 Smooth ER and rough ER.

The smooth ER has many unique enzymes embedded on its surface. Its functions are:

- production of membrane phospholipids and cellular lipids;
- production of sex hormones such as testosterone and oestrogen;
- detoxification of drugs in the liver;
- storage of calcium ions needed for contraction in muscle cells;
- transportation of lipid-based compounds;
- to aid the liver in releasing glucose into the bloodstream when needed.

Rough ER has ribosomes on the exterior of the channels. These ribosomes are involved in protein synthesis. Therefore, this type of ER is involved in protein development and transport. These proteins may become parts of membranes, enzymes, or even messengers between cells. Most cells contain both types of ER with the rough ER being closer to the nuclear membrane.

Ribosomes

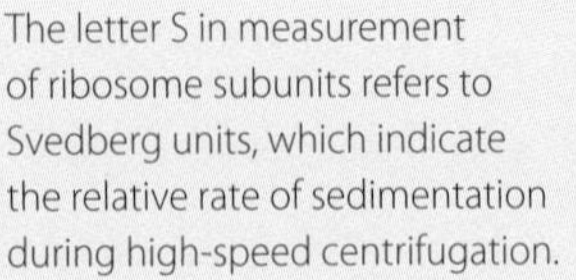

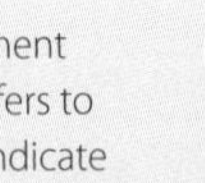

The letter S in measurement of ribosome subunits refers to Svedberg units, which indicate the relative rate of sedimentation during high-speed centrifugation.

Ribosomes are unique structures that do not have an exterior membrane. They carry out protein synthesis in the cell. These structures may be found free in the cytoplasm or they may be attached to the surface of endoplasmic reticulum. They are always composed of a type of RNA and protein. You will recall that prokaryotic cells also contain ribosomes. However, the ribosomes of eukaryotic cells are larger and denser than those of prokaryotic cells. Eukaryotic ribosomes are composed of two subunits. These subunits together equal 80S. The ribosomes in prokaryotic cells are also of two subunits, but they only equal 70S.

Lysosomes

Lysosomes are intracellular digestive centres that arise from the Golgi apparatus. The lysosome lacks any internal structures. They are sacs bounded by a single membrane that contain as many as 40 different enzymes. The enzymes are all hydrolytic and catalyse the breakdown of proteins, nucleic acids, lipids and carbohydrates. Lysosomes fuse with old or damaged organelles from within the cell to break them down so that recycling of the components may occur. Also, lysosomes are involved with the breakdown of materials that may be brought into the cell by phagocytosis (see Section 2.4 Membranes). The interior of a functioning lysosome is acidic. This acidic condition is necessary for the enzymes to hydrolyse large molecules.

Golgi apparatus

The Golgi apparatus consists of what appears to be flattened sacs called cisternae, which are stacked on top of one another (see Figure 2.6). This organelle functions in the collection, packaging, modification, and distribution of materials synthesized in the cell. One side of the apparatus is near the rough ER, called the *cis* side. It receives products from the ER. These products move into the cisternae of the Golgi apparatus. Movement then continues to the discharging or opposite side, the *trans* side. Small sacs called vesicles can be seen coming off the *trans* side. These vesicles carry modified materials to wherever they are needed inside or outside the cell. This organelle is especially prevalent in glandular cells such as those in the pancreas, which manufacture and secrete substances.

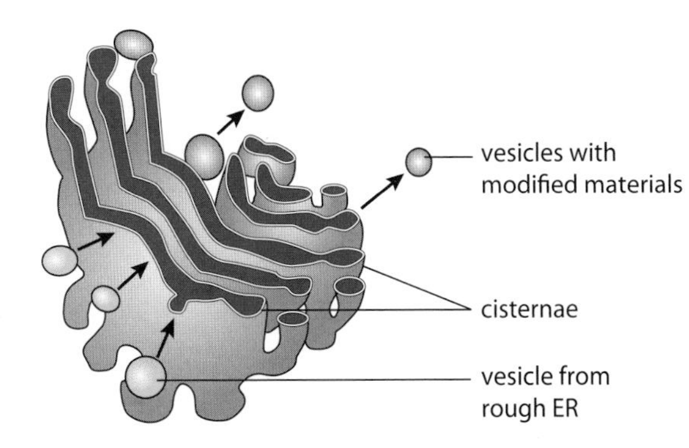

Figure 2.6 In this drawing of the Golgi apparatus, the movement of the vesicles is shown by arrows. Can you identify which side is the *cis* side and which is the *trans* side?

Mitochondria

Mitochondria (singular, mitochondrion) are rod-shaped organelles that appear throughout the cytoplasm (see Figure 2.7 overleaf). Their size is close to that of a bacterial cell. Mitochondria have their own DNA, a circular chromosome similar to that in bacterial cells, allowing them some autonomy within the cell. They have a double membrane. The outer membrane is smooth, but the inner membrane is folded into cristae (singular, crista). Inside the inner membrane is a semi-fluid substance called matrix. An area called the inner membrane space lies between the two membranes. The cristae provide a huge internal surface area for the chemical reactions characteristic of the mitochondria to occur. Most mitochondrial reactions involve the production of usable cellular energy called ATP. Because of this, the mitochondrion is often called the 'cell powerhouse'. This organelle also produces and contains its own ribosomes. These ribosomes are of the 70S type. Cells that have high energy requirements, such as muscle cells, have large numbers of mitochondria.

Figure 2.7 Compare this drawing of a mitochondrion with the false-colour TEM of a mitochondrion.

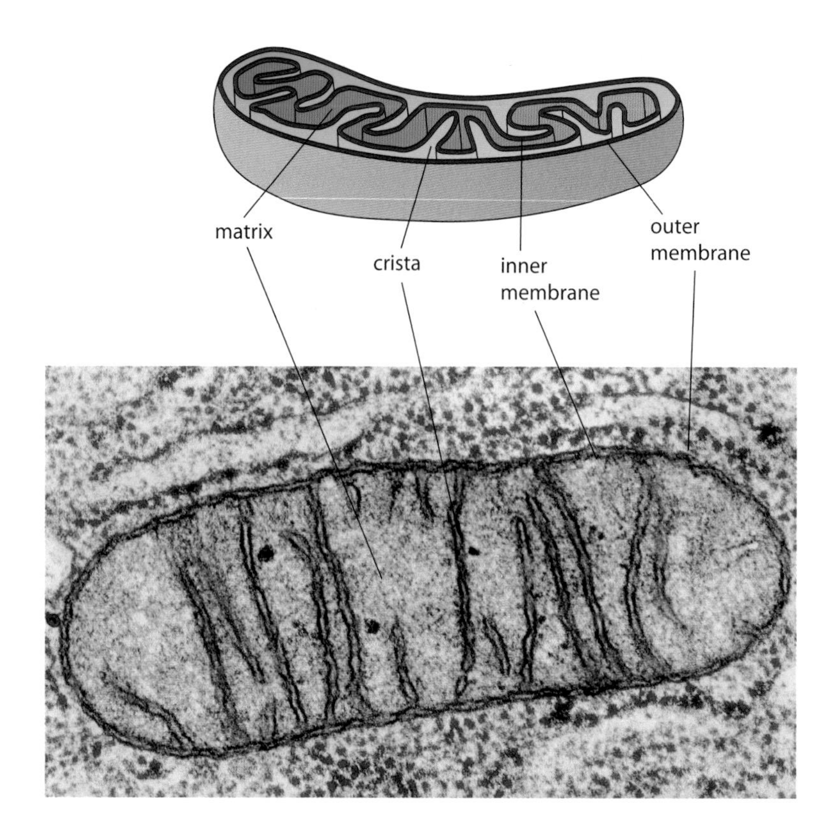

Nucleus

The nucleus in eukaryotic cells is an isolated region where the DNA resides. It is bordered by a double membrane referred to as the nuclear envelope. This membrane allows compartmentalization of the eukaryotic DNA, thus providing an area where DNA can carry out its functions and not be affected by processes occurring in other parts of the cell. The nuclear membrane does not provide complete isolation as it has numerous pores that allow communication with the cell's cytoplasm (see Figure 2.8).

Figure 2.8 The nucleus has a double membrane with pores and contains a nucleolus.

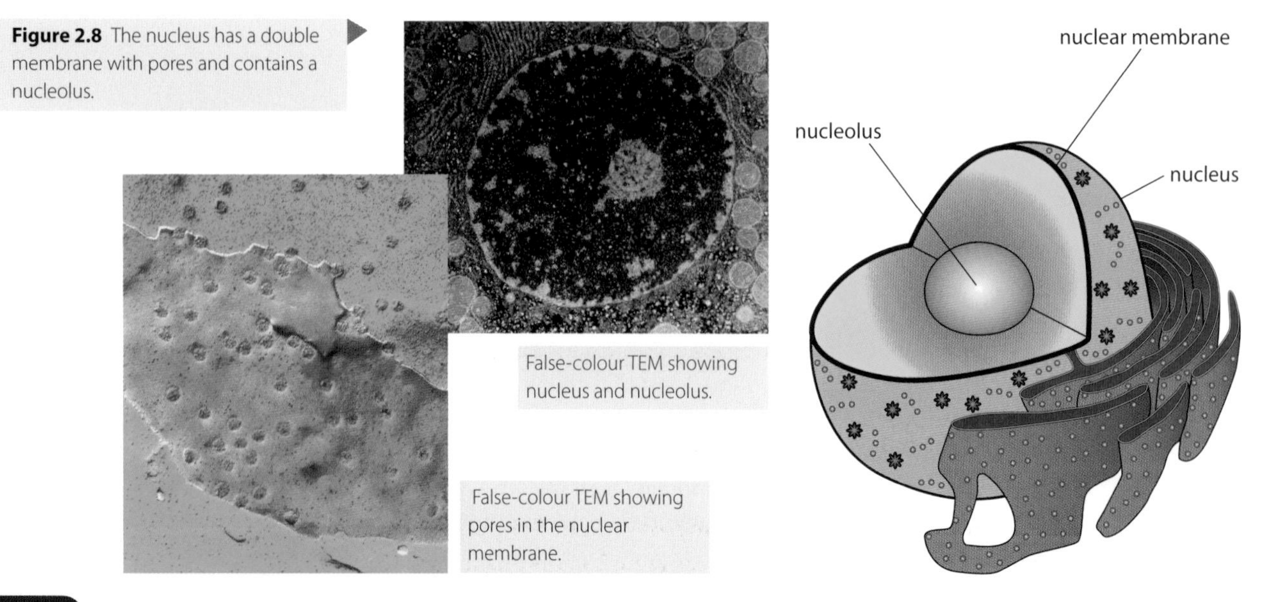

False-colour TEM showing nucleus and nucleolus.

False-colour TEM showing pores in the nuclear membrane.

The DNA of eukaryotic cells often occurs in the form of chromosomes; chromosomes vary in number depending on the species. Chromosomes carry all the information necessary for the cell to exist. This allows for the survival of the organism, whether unicellular or multicellular. DNA is the genetic material of the cell. It enables certain traits to be passed to the next generation. When the cell is not in the dividing process, the chromosomes are not present as visible structures. During this phase, the cell's DNA is in the form of chromatin. Chromatin is formed of strands of DNA and proteins called histones. This DNA and histone combination often results in structures called nucleosome. A nucleosome consists of eight histones with a strand of DNA wrapped around them and secured with a ninth histone (see also Chapter 7, page 195). This produces a structure that resembles a string of beads. A chromosome is a highly coiled structure of many nucleosomes (see Figure 2.9).

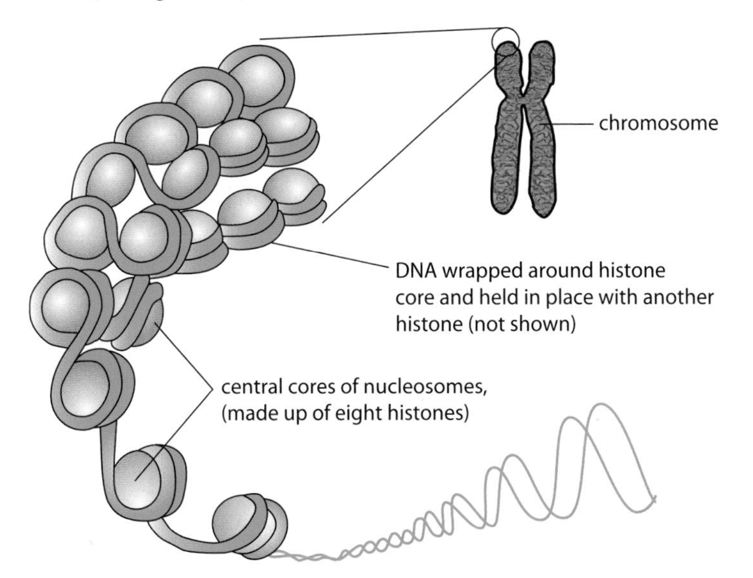

Figure 2.9 This drawing shows how DNA is packaged into chromosomes.

The nucleus is often located centrally in the cell's cytoplasm, although in some cell types it is pushed to one side. The side position is characteristic of plant cells since these cells often have a large central vacuole. Most eukaryotic cells possess a single nucleus, but some do not have a nucleus, and some have multiple nuclei. Without a nucleus, a cell cannot reproduce. Loss of reproductive ability is often paired with increased specialization to carry out a certain function. For example, human red blood cells do not have nuclei; they are specialized to transport respiratory gases. Most nuclei also include one or more dark areas called nucleoli (singular, nucleolus). Molecules of the cell ribosomes are manufactured in the nucleolus. The molecules pass through the nuclear envelope before assembly as ribosomes.

Chloroplasts

Chloroplasts occur only in algae and plant cells. The chloroplast contains a double membrane and is about the same size as a bacterial cell. Like the mitochondrion, it contains its own DNA and 70S ribosomes. The DNA of the chloroplast is in the form of a ring.

You should note all the characteristics that chloroplasts and mitochondria have in common with prokaryotic cells.

Besides DNA and ribosomes, the interior of the chloroplast includes the grana (singular, granum), the thylakoids, and the stroma (see Figure 2.10). A granum is made up of numerous thylakoids stacked like a pile of coins. The thylakoids are flattened membrane sacs with components necessary for the absorption of light. Absorption of light is the first step in the process of photosynthesis. The fluid stroma is similar to the cytosol of the cell. It occurs outside the grana but within the double membrane. Stroma contains many enzymes and chemicals necessary to complete the process of photosynthesis. Like mitochondria, chloroplasts are capable of reproducing independently of the cell.

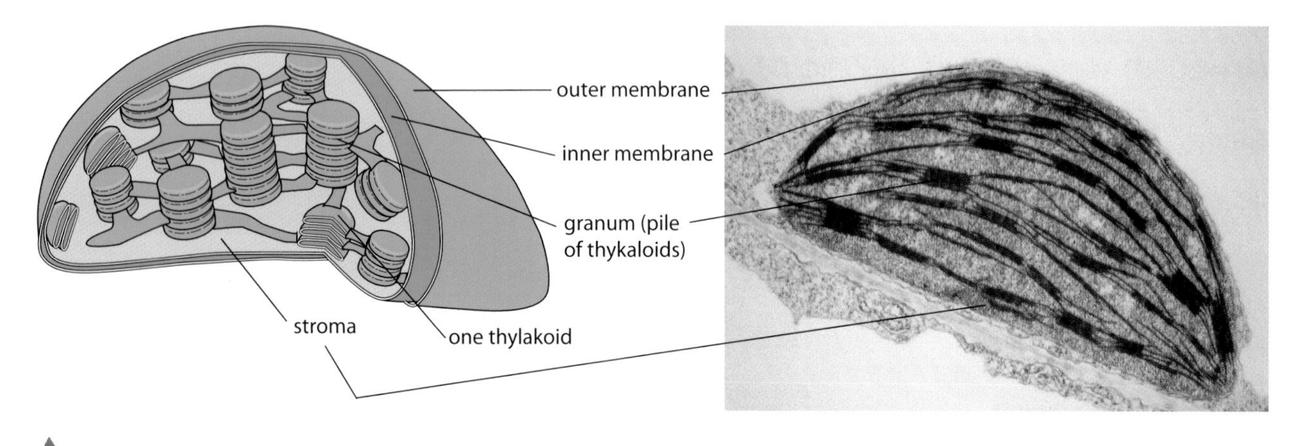

Figure 2.10 Compare this drawing of a chloroplast with the TEM of a chloroplast.

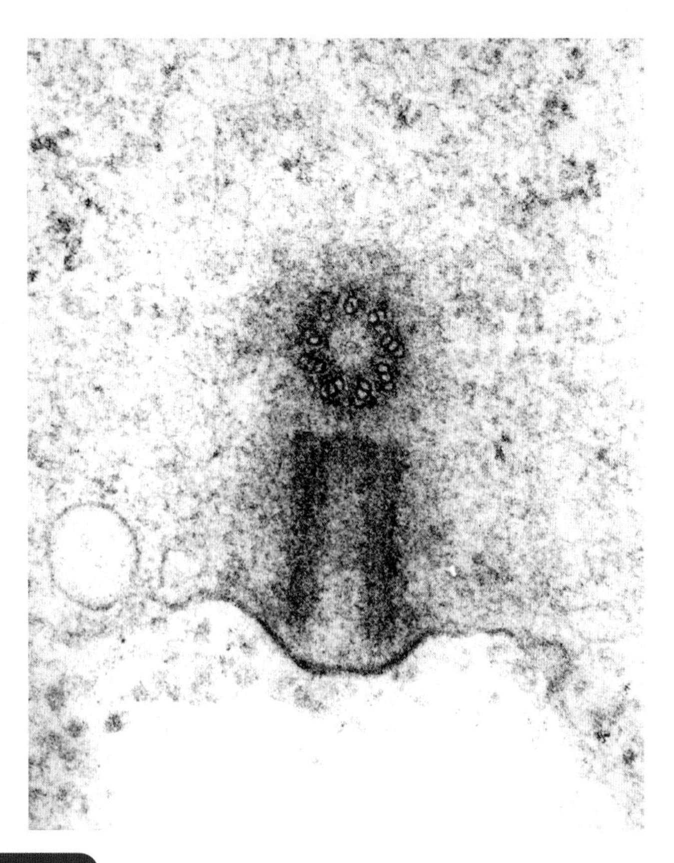

This TEM shows the two centrioles of a centrosome.

Centrosome

The centrosome occurs in all eukaryotic cells. Generally, it consists of a pair of centrioles at right angles to one another. These centrioles are involved in assembling microtubules, which are important to the cell in providing structure and allowing movement. Microtubules are also important to cell division. Higher plant cells produce microtubules even though they do not have centrioles. The centrosome is located at one end of the cell close to the nucleus.

Vacuoles

Vacuoles are storage organelles that usually form from the Golgi apparatus. They are membrane-bound and have many possible functions. They occupy a very large space inside the cells of most plants. They may store a number of different substances including potential food (to provide nutrition), metabolic wastes and toxins (to be expelled from the cell), and water. Vacuoles enable cells to have higher surface area to volume ratios even at larger sizes. In plants, they allow an uptake of water that provides rigidity for the organism.

Comparison of prokaryotic and eukaryotic cells

A table may best be used to summarize the differences between prokaryotic and eukaryotic cells.

Prokaryotic cells	Eukaryotic cells
DNA in a ring form without protein	DNA with proteins as chromosomes/ chromatin
DNA free in the cytoplasm (nucleoid region)	DNA enclosed within a nuclear envelope (nucleus)
no mitochondria	mitochondria present
70S ribosomes	80S ribosomes
no internal compartmentalization to form organelles	internal compartmentalization present to form many types of organelle
size less than 10 micrometres	size more than 10 micrometres

• **Examiner's hint:** When comparing, be certain to state the characteristic in each compared type, as shown in this table comparing prokaryotic and eukaryotic cells.

If you are asked to state the similarities between the two types of cell, you should be certain to include the following:

- both types of cell have some sort of outside boundary that always involves a plasma membrane;
- both types of cell carry out all the functions of life;
- DNA is present in both cell types.

Comparison of plant and animal cells and their extracellular components

Let's look at how to compare two general types of eukaryotic cell, plant and animal. A table like the one below can be used to emphasize the differences. However, don't forget to also recognize the similarities between these two cell types later.

Plant cells	Animal cells
Exterior of cell includes an outer cell wall with a plasma membrane just inside.	Exterior of cell includes only a plasma membrane. There is no cell wall.
Chloroplasts are present in the cytoplasm.	There are no chloroplasts.
Possess large centrally located vacuoles.	Vacuoles are usually not present or are small.
Store carbohydrates as starch.	Store carbohydrates as glycogen.
Do not contain centrioles within a centrosome area.	Contain centrioles within a centrosome area.
Because a rigid cell wall is present, this cell type has a fixed, often angular, shape.	Without a cell wall, this cell is flexible and more likely to be a rounded shape.

Models are a way scientists explain complex structures such as cellular membranes. Models are based on the knowledge available at the time. Even though early models of cell membranes were wrong (because of incomplete data), they helped in the development of the presently accepted model of cell membranes.

Membrane structure

As early as 1915, scientists were aware that the structure of cell membranes included proteins and lipids. Further research established that the lipids were phospholipids. Early theories mostly centred on the phospholipids forming a bilayer with the proteins forming thin layers on the exterior and interior of the bilayer. Then, in 1972, S. J. Singer and G. Nicolson proposed that the proteins are inserted into the phospholipid layer rather than forming a layer on its surfaces. They believed the proteins formed a mosaic floating in a fluid layer of phospholipids. Since 1972, further evidence has been gathered about membranes and only slight changes to the Singer and Nicolson model have been made.

The present agreed model of the cellular membrane is the fluid mosaic model. It is shown in Figure 2.13. You should note that all cellular membranes, whether plasma membranes or organelle membranes, have the same general structure.

Phospholipids

In Figure 2.13, note that the 'backbone' of the membrane is a bilayer produced from huge numbers of molecules called phospholipids. Each phospholipid is composed of a 3-carbon compound called glycerol. Two of the glycerol carbons have fatty acids attached. The third carbon is attached to a highly polar organic alcohol that includes a bond to a phosphate group. Fatty acids are not water soluble because they are non-polar. On the other hand, because the organic

Figure 2.13 In the fluid mosaic model of the cell membrane there is a double layer of lipids (fats) arranged with their tails facing inwards. Proteins are thought to 'float' in the lipid bilayer.

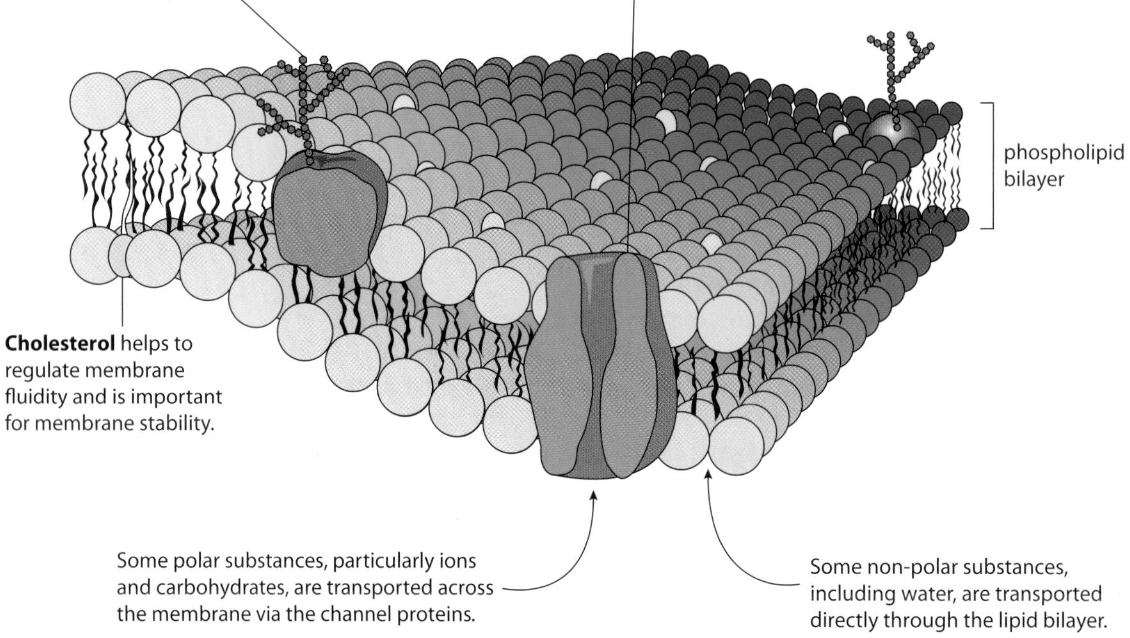
Glycoproteins are composed of carbohydrate chains attached to peripheral proteins. They play a role in recognition of like cells and are involved in immune responses.
Integral proteins completely penetrate the lipid bilayer. They control the entry and removal of specific molecules from the cell.
phospholipid bilayer
Cholesterol helps to regulate membrane fluidity and is important for membrane stability.
Some polar substances, particularly ions and carbohydrates, are transported across the membrane via the channel proteins.
Some non-polar substances, including water, are transported directly through the lipid bilayer.

Models are a way scientists explain complex structures such as cellular membranes. Models are based on the knowledge available at the time. Even though early models of cell membranes were wrong (because of incomplete data), they helped in the development of the presently accepted model of cell membranes.

Membrane structure

As early as 1915, scientists were aware that the structure of cell membranes included proteins and lipids. Further research established that the lipids were phospholipids. Early theories mostly centred on the phospholipids forming a bilayer with the proteins forming thin layers on the exterior and interior of the bilayer. Then, in 1972, S. J. Singer and G. Nicolson proposed that the proteins are inserted into the phospholipid layer rather than forming a layer on its surfaces. They believed the proteins formed a mosaic floating in a fluid layer of phospholipids. Since 1972, further evidence has been gathered about membranes and only slight changes to the Singer and Nicolson model have been made.

The present agreed model of the cellular membrane is the fluid mosaic model. It is shown in Figure 2.13. You should note that all cellular membranes, whether plasma membranes or organelle membranes, have the same general structure.

Phospholipids

In Figure 2.13, note that the 'backbone' of the membrane is a bilayer produced from huge numbers of molecules called phospholipids. Each phospholipid is composed of a 3-carbon compound called glycerol. Two of the glycerol carbons have fatty acids attached. The third carbon is attached to a highly polar organic alcohol that includes a bond to a phosphate group. Fatty acids are not water soluble because they are non-polar. On the other hand, because the organic

Figure 2.13 In the fluid mosaic model of the cell membrane there is a double layer of lipids (fats) arranged with their tails facing inwards. Proteins are thought to 'float' in the lipid bilayer.

Glycoproteins are composed of carbohydrate chains attached to peripheral proteins. They play a role in recognition of like cells and are involved in immune responses.

Integral proteins completely penetrate the lipid bilayer. They control the entry and removal of specific molecules from the cell.

Cholesterol helps to regulate membrane fluidity and is important for membrane stability.

Some polar substances, particularly ions and carbohydrates, are transported across the membrane via the channel proteins.

Some non-polar substances, including water, are transported directly through the lipid bilayer.

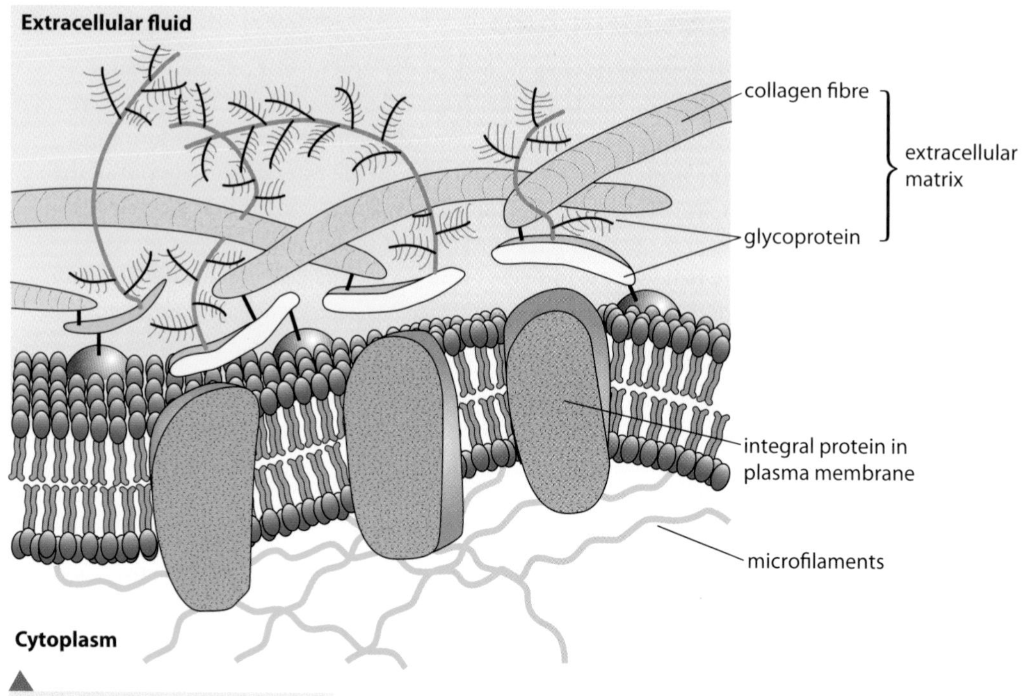

Figure 2.12 This is a drawing of the extracellular matrix of an animal cell.

Exercises

7 Why do muscle cells have a large number of mitochondria?

8 Name two organelles that are similar to prokaryotic cells.

9 If plant cells have chloroplasts for photosynthesis, why do they also need mitochondria?

2.4 Membranes

Assessment statements

2.4.1 Draw and label a diagram to show the structure of a membrane.

2.4.2 Explain how the hydrophobic and hydrophilic properties of phospholipids help to maintain the structure of cell membranes.

2.4.3 List the functions of membrane proteins.

2.4.4 Define *diffusion* and *osmosis*.

2.4.5 Explain passive transport across membranes by simple diffusion and facilitated diffusion.

2.4.6 Explain the role of protein pumps and ATP in active transport across membranes.

2.4.7 Explain how vesicles are used to transport materials within a cell between the rough endoplasmic reticulum, Golgi apparatus and plasma membrane.

2.4.8 Describe how the fluidity of the membrane allows it to change shape, break and re-form during endocytosis and exocytosis.

Most cellular organelles are present in both plant and animal cells. When an organelle is present in both types of cell, it usually has the same structure and function. For example, both cell types contain mitochondria that possess cristae, matrix, and the double membrane. Also, in both cell types, the mitochondria function in the production of ATP for use by the cell.

The outermost region of various cell types is often unique, as shown by the following table.

Cell	Outermost part
bacteria	cell wall of peptidoglycan
fungi	cell wall of chitin
yeasts	cell wall of glucan and mannan
algae	cell wall of cellulose
plants	cell wall of cellulose (see Figure 2.11)
animals	no cell wall; plasma membrane secretes a mixture of sugar and proteins called glycoproteins that forms the extracellular matrix

Figure 2.11 This drawing of a section through plant cells shows the primary walls, middle lamella and secondary walls.

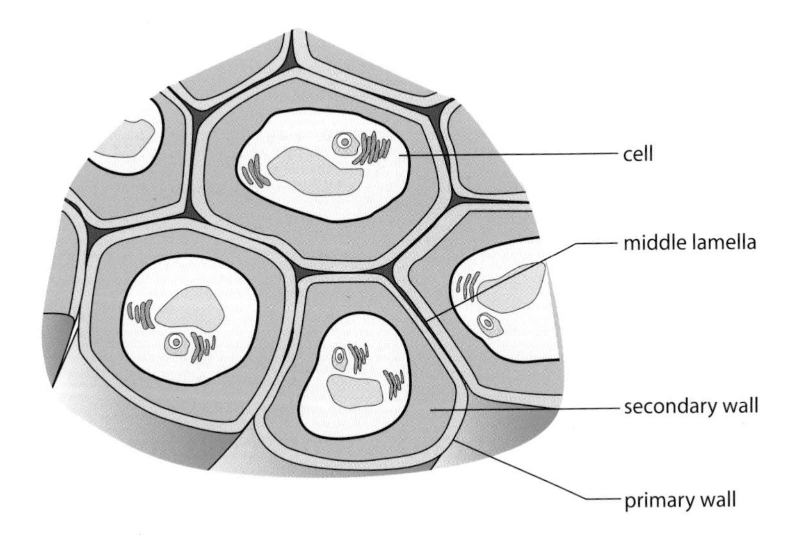

Whenever the cell wall is present, it is involved in maintaining cell shape. It also helps regulate water uptake. Because of its rigidity, it will only allow a certain amount of water to enter the cell. In plants, when an adequate amount of water is inside the cell, there is pressure against the cell wall. That pressure serves to support the plant's upright position.

For more information on cell anatomy, visit heinemann.co.uk/hotlinks, insert the express code 4242P and click on Weblinks 2.2a and 2.2b.

The extracellular matrix (ECM) of many animal cells is composed of collagen fibres plus a combination of sugars and proteins called glycoproteins (see Figure 2.12). These form fibre-like structures that anchor the matrix to the plasma membrane. This strengthens the plasma membrane and allows attachment between adjacent cells. The ECM allows for cell-to-cell interaction, possibly altering gene expression and bringing about coordination of cell action within the tissue. Many researchers think the ECM is involved in directing stem cells to differentiate. Cell migration and movement also appear to be, at least partially, the result of interactions in this area.

Vacuoles

Vacuoles are storage organelles that usually form from the Golgi apparatus. They are membrane-bound and have many possible functions. They occupy a very large space inside the cells of most plants. They may store a number of different substances including potential food (to provide nutrition), metabolic wastes and toxins (to be expelled from the cell), and water. Vacuoles enable cells to have higher surface area to volume ratios even at larger sizes. In plants, they allow an uptake of water that provides rigidity for the organism.

Comparison of prokaryotic and eukaryotic cells

A table may best be used to summarize the differences between prokaryotic and eukaryotic cells.

Prokaryotic cells	Eukaryotic cells
DNA in a ring form without protein	DNA with proteins as chromosomes/ chromatin
DNA free in the cytoplasm (nucleoid region)	DNA enclosed within a nuclear envelope (nucleus)
no mitochondria	mitochondria present
70S ribosomes	80S ribosomes
no internal compartmentalization to form organelles	internal compartmentalization present to form many types of organelle
size less than 10 micrometres	size more than 10 micrometres

• **Examiner's hint:** When comparing, be certain to state the characteristic in each compared type, as shown in this table comparing prokaryotic and eukaryotic cells.

If you are asked to state the similarities between the two types of cell, you should be certain to include the following:

- both types of cell have some sort of outside boundary that always involves a plasma membrane;
- both types of cell carry out all the functions of life;
- DNA is present in both cell types.

Comparison of plant and animal cells and their extracellular components

Let's look at how to compare two general types of eukaryotic cell, plant and animal. A table like the one below can be used to emphasize the differences. However, don't forget to also recognize the similarities between these two cell types later.

Plant cells	Animal cells
Exterior of cell includes an outer cell wall with a plasma membrane just inside.	Exterior of cell includes only a plasma membrane. There is no cell wall.
Chloroplasts are present in the cytoplasm.	There are no chloroplasts.
Possess large centrally located vacuoles.	Vacuoles are usually not present or are small.
Store carbohydrates as starch.	Store carbohydrates as glycogen.
Do not contain centrioles within a centrosome area.	Contain centrioles within a centrosome area.
Because a rigid cell wall is present, this cell type has a fixed, often angular, shape.	Without a cell wall, this cell is flexible and more likely to be a rounded shape.

alcohol with phosphate is highly polar, it is water soluble. This structure means that membranes have two distinct areas when it comes to polarity and water solubility (see Figure 2.14). One area is water soluble and polar, and is referred to as hydrophilic (water loving). This is the phosphorylated alcohol side. The other area is not water soluble and is non-polar. It is referred to as hydrophobic (water fearing).

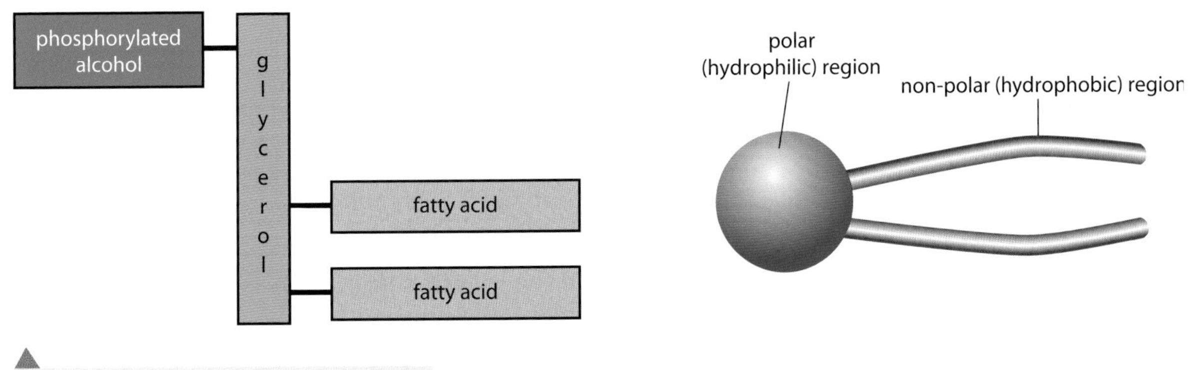

Figure 2.14 This is a model of a phospholipid.

The hydrophobic and hydrophilic regions cause phospholipids to always align as a bilayer if there is water present and there is a large number of phospholipid molecules (see Figure 2.15). Because the fatty acid 'tails' do not strongly attract one another, the membrane tends to be fluid or flexible. This allows animal cells to have a variable shape and also allows the process of endocytosis (which is discussed below). What maintains the overall structure of the membrane is the tendency water has to form hydrogen bonds.

Figure 2.15 This model of a phospholipid bilayer shows how phospholipid molecules behave in two layers. Both layers have the phosphorylated alcohol end of the molecules towards the outside and the fatty acid tails oriented toward each other in the middle.

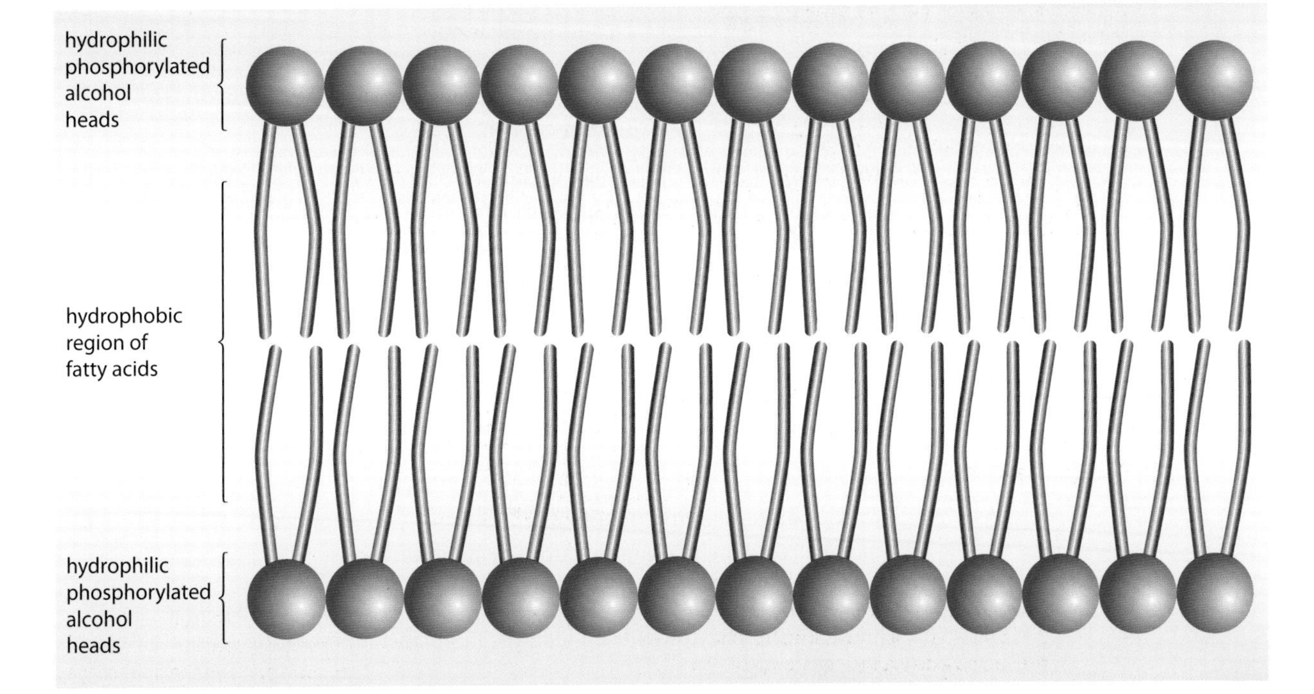

Cholesterol

Membranes must be fluid to function properly. They are a bit like olive oil in their consistency. At various locations in the hydrophobic region (fatty acid tails) in animal cells are cholesterol molecules. These molecules have a role in determining membrane fluidity, which changes with temperature. The cholesterol molecules allow effective membrane function at a wider range of temperatures than if they were not present. Plant cells do not have cholesterol molecules; they depend on saturated or unsaturated fatty acids to maintain proper membrane fluidity.

Proteins

The last major components of cellular membranes are the proteins. It is the proteins that create extreme diversity in membrane function. Proteins of various types are embedded in the fluid matrix of the phospholipid bilayer. This creates the mosaic effect referred to in the fluid mosaic model. There are usually two major types of protein. One is referred to as integral protein and the other type is referred to as peripheral protein. Integral proteins have both hydrophobic and hydrophilic regions in the same protein. The hydrophobic region, with non-polar amino acids, is in the mid-section of the phospholipid membrane, holding the protein in place. Their hydrophilic region is exposed to the water solutions on either side of the membrane. Peripheral proteins, on the other hand, do not protrude into the middle hydrophobic region, but remain bound to the surface of the membrane. Often these peripheral proteins are anchored to an integral protein. You can see the location of these proteins if you look back to Figure 2.13.

• **Examiner's hint:** Be certain you can draw and label all the parts of a membrane as described in this section on the fluid mosaic model.

Membrane protein functions

As you will recall, it is the membrane proteins that impart different functions to different membranes. There are many different proteins but they tend to have six general functions:

- hormone binding sites;
- enzymatic action;
- cell adhesion;
- cell-to-cell communication;
- channels for passive transport;
- pumps for active transport.

Proteins that serve as hormone binding sites have specific shapes exposed to the exterior that fit the shape of specific hormones. The attachment between the protein and the hormone causes a change in shape in the protein and results in a message being relayed to the interior of the cell.

Cells have enzymes attached to membranes that catalyse many chemical reactions. The enzymes may be on the interior or the exterior of the cell. Often they are grouped so that a sequence of metabolic reactions, called a metabolic pathway, may occur.

Cell adhesion is provided by proteins when they hook together in various ways to provide permanent or temporary connections. These connections, referred to as junctions, may include gap junctions or tight junctions.

Many of the cell-to-cell communication proteins include attached molecules of carbohydrate. They provide an identification label representing cells of different types or species.

Some proteins contain channels that span the membrane providing passageways for substances to pass through. When this transport is passive, a material moves through the channel from an area of high concentration to an area of lower concentration.

In active transport, proteins shuttle a substance from one side of the membrane to another by changing shape. This process requires the expenditure of energy in the form of ATP. It does not require a difference in concentration to occur.

Passive and active transport

There are two general means of cellular transport:

- passive transport;
- active transport.

Passive transport does not require energy (in the form of ATP), but active transport does. Passive transport occurs in situations where there are areas of different concentration of a particular substance. Movement of the substance occurs from an area of high concentration to an area of lower concentration. Movement is said to occur along a concentration gradient.

When active transport occurs, the substance is moved against a concentration gradient, so energy expenditure must occur.

Passive transport: diffusion and osmosis

Examine Figure 2.16; it shows chemical diffusion.

Diffusion

Diffusion is one type of passive transport. Particles of a certain type move from a region of high concentration to a region of low concentration. However, in a living system, diffusion often involves a membrane. For example, oxygen gas moves from outside a cell to inside. Oxygen is used by the cell when its mitochondria carry out respiration, thus creating a relatively lower oxygen concentration inside the cell than outside. Oxygen diffuses into the cell as a result. Carbon dioxide diffuses in the opposite direction to the oxygen because carbon dioxide is produced as a result of mitochondrial respiration.

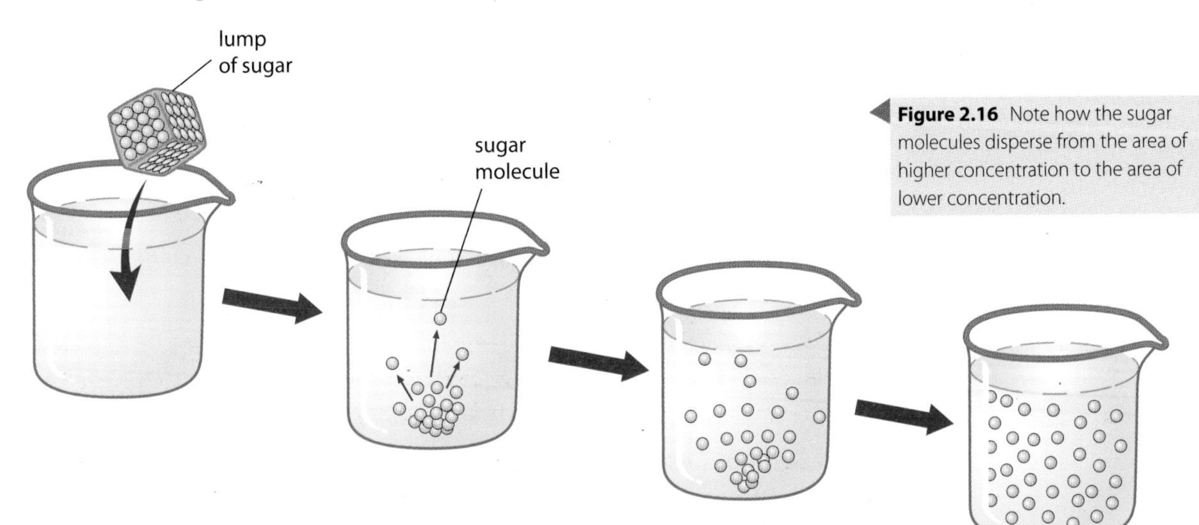

Figure 2.16 Note how the sugar molecules disperse from the area of higher concentration to the area of lower concentration.

An example of a disease involving a failure of facilitated diffusion is cystinuria. It occurs when the protein that carries the amino acid cysteine is absent from kidney cells. The result is a build-up of amino acids in the kidney resulting in very painful stones.

Facilitated diffusion

Facilitated diffusion is a particular type of diffusion involving a membrane with specific carrier proteins that are capable of combining with the substance to aid its movement. The carrier changes shape to accomplish this task but does not require energy. It is evident from this explanation that facilitated diffusion is very specific and depends on the carrier protein. You should note that the rate of facilitated diffusion will level off when total saturation of available carriers occurs.

Partially permeable membranes are also called selectively permeable membranes

Osmosis

Osmosis is another type of passive transport, so movement occurs along a concentration gradient. However, osmosis involves only the passive movement of water molecules across a partially permeable membrane. A partially permeable membrane is one which only allows certain substances to pass through (a permeable membrane would allow everything through). The concentration gradient of water that allows the movement to occur is the result of a difference between solute concentrations on either side of a partially permeable membrane. A hyperosmotic solution has a higher concentration of total solutes than a hypo-osmotic solution. Water therefore moves from a hypo-osmotic solution to a hyperosmotic solution across a partially permeable membrane (see Figure 2.17). If iso-osmotic solutions occur on either side of a partially permeable membrane, no net movement of water is evident.

Passive transport continues until there is an equal concentration of the substance in both areas involved. This is called equilibrium.

This table summarizes diffusion and osmosis across cellular membranes.

Simple diffusion	Substances other than water move between phospholipid molecules or through proteins which possess channels.
Facilitated diffusion	Non-channel protein carriers change shape to allow movement of substances other than water.
Osmosis	Only water moves through the membrane using aquaporins which are proteins with specialized channels for water movement.

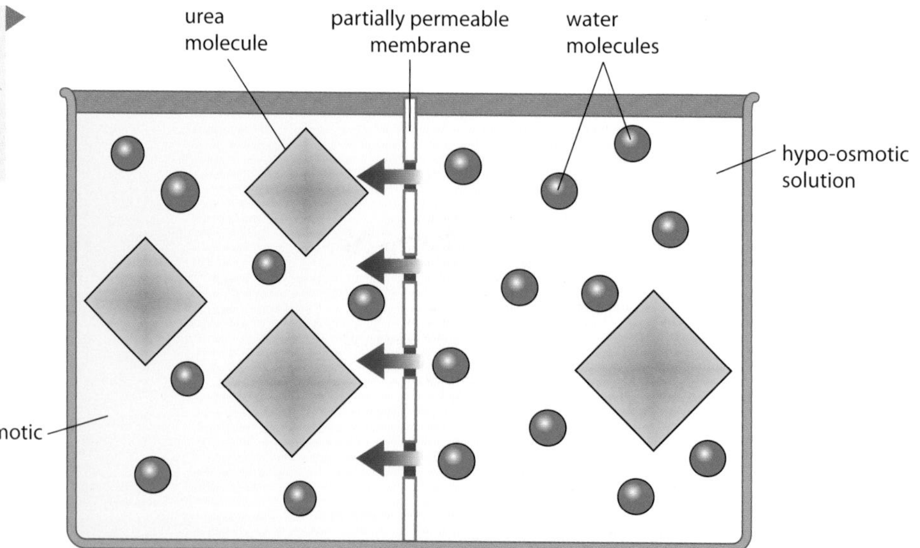

Figure 2.17 The partially permeable membrane between two solutions of different osmotic concentrations allows water molecules to pass from the hypo-osmotic solution to the hyperosmotic solution.

Size and charge

The size and polarity of molecules determine the ease with which various substances can cross membranes. These characteristics and the ability of molecules to cross membranes are arranged along a continuum like this:

small and non-polar molecules cross membranes easily ⟷ large and polar molecules cross membranes with difficulty

Substances that can move across a membrane passively are influenced by two major factors: size and charge. Substances that are small in size and non-polar move across membranes with ease. Substances that are polar, large in size, or both, do not. Examples of small, non-polar substances are gases such as oxygen, carbon dioxide, and nitrogen. Ions such as chloride ions, potassium ions, and sodium ions would have a great deal of difficulty crossing membranes passively, as would large molecules such as glucose and sucrose. Molecules such as water and glycerol are small, uncharged polar molecules that can fairly easily cross membranes.

Active transport and the cell

As you will recall, active transport requires work to be performed. This means energy must be used, so ATP is required. Active transport involves the movement of substances against a concentration gradient. This process allows the cell to maintain interior concentrations of molecules that are different from exterior concentrations. Animal cells have a much higher concentration of potassium ions than their exterior environment, whereas sodium ions are more concentrated in the extracellular environment than in the cells. The cell allows the maintenance of this condition by pumping potassium ions into the cell and by pumping sodium ions out of it. Along with energy, a membrane protein must be involved for this active transport to occur. We can now explain how the sodium–potassium pump works.

There are many other examples of active transport in cells besides the sodium–potassium pump. Liver cells use the process to accumulate glucose molecules from blood plasma even though the liver cell has a higher glucose concentration.

Cystic fibrosis is a human genetic disease in which the membrane protein that transports chloride ions is missing. This causes high concentrations of water inside the cells that line the lungs and leads to the production of very thickened mucus. This results in a serious condition.

The sodium–potassium pump

This mechanism for actively moving sodium and potassium ions has five stages.

1 A specific protein binds to three intracellular sodium ions (see Figure 2.18).

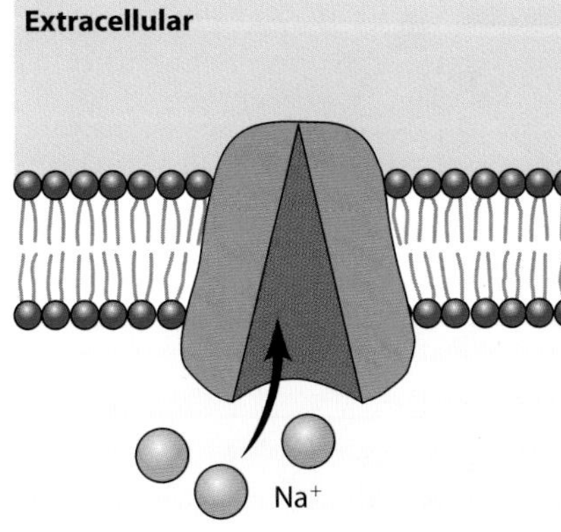

Figure 2.18 Stage 1: A protein in a phospholipid bilayer opens to the intracellular side and attaches three sodium ions.

2 Binding of sodium ions causes phosphorylation by ATP (see Figure 2.19).

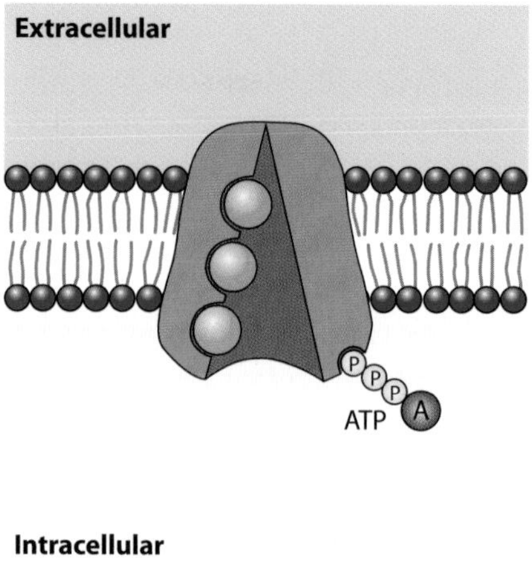

Extracellular

ATP

Intracellular

Figure 2.19 Stage 2: ATP attaches to the protein.

3 The phosphorylation causes the protein to change its shape, thus expelling sodium ions to the exterior (see Figure 2.20).

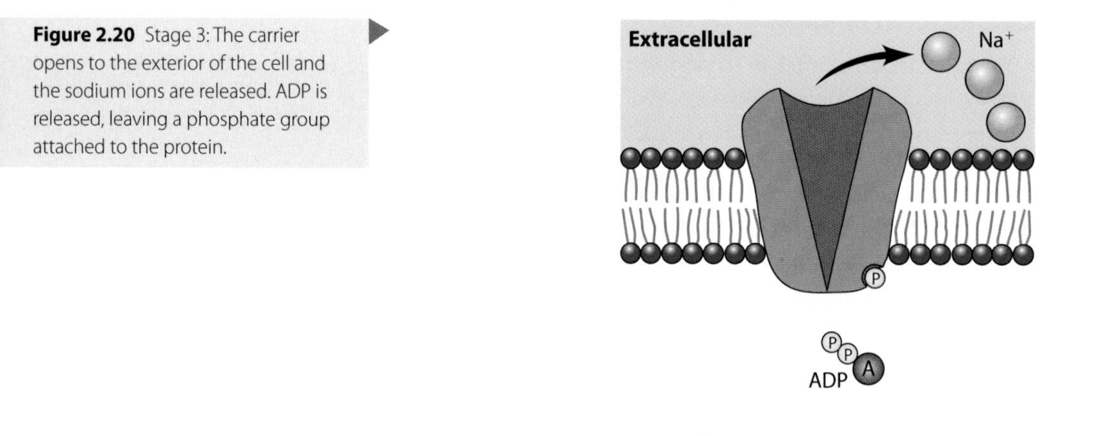

Figure 2.20 Stage 3: The carrier opens to the exterior of the cell and the sodium ions are released. ADP is released, leaving a phosphate group attached to the protein.

4 Two extracellular potassium ions bind to different regions of the protein and this causes the release of the phosphate group (see Figures 2.21 and 2.22).

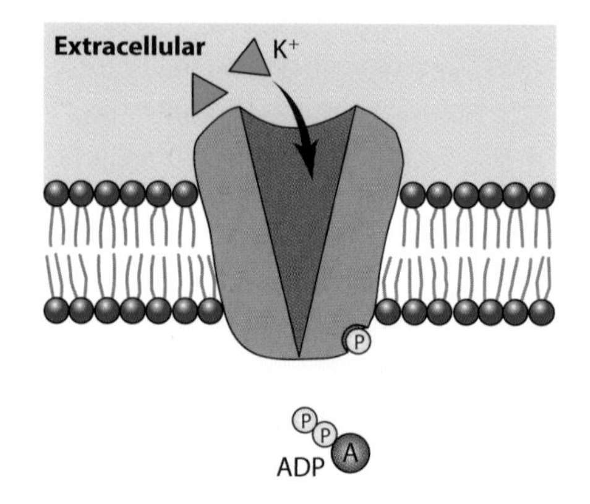

Figure 2.21 Stage 4: Extracellular potassium ions attach to the protein.

5 Loss of the phosphate group restores the protein's original shape thus causing the release of the potassium ions into the intracellular space (see Figure 2.22).

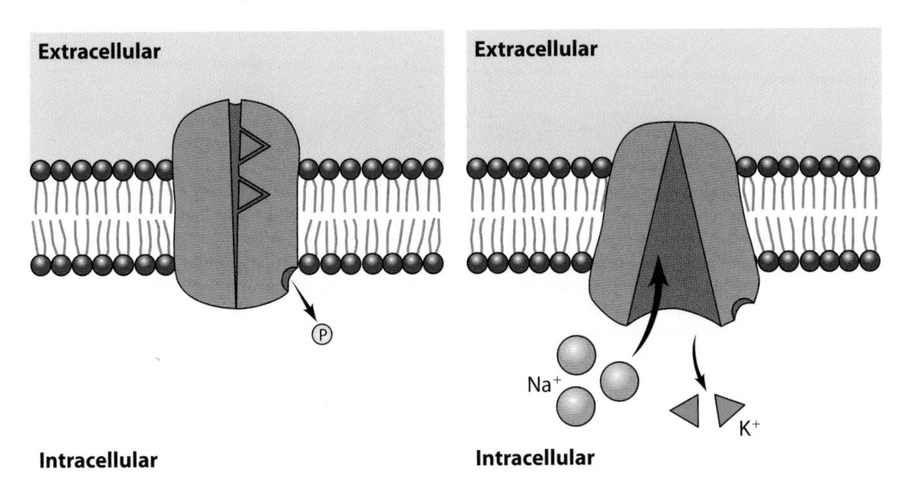

Figure 2.22 Stage 5: The protein opens towards the cell interior again and releases the potassium ions into the interior.

The sodium–potassium pump shows how important and active specific proteins are in the active transport of particular substances. It is also clear how ATP plays a crucial role in active transport.

Endocytosis and exocytosis

Endocytosis and exocytosis are processes that allow larger molecules to move across the plasma membrane. Endocytosis allows macromolecules to enter the cell, while exocytosis allows molecules to leave. Both processes depend on the fluidity of the plasma membrane. It is important to recall why the cell membranes are fluid in consistency: the phospholipid molecules are not closely packed together largely due to the rather 'loose' connections between the fatty acid tails. It is also important to remember why the membrane is rather stable: the hydrophilic and hydrophobic properties of the different regions of the phospholipid molecules cause them to form a stable bilayer in an aqueous environment.

Endocytosis occurs when a portion of the plasma membrane is pinched off to enclose macromolecules or particulates. This pinching off involves a change in the shape of the membrane. The result is the formation of a vesicle that then enters the cytoplasm of the cell. The ends of the membrane reattach because of the hydrophobic and hydrophilic properties of the phospholipids and the presence of water. This could not occur if it weren't for the fluid nature of the plasma membrane.

Examples of endocytosis include:

- phagocytosis – the intake of large particulate matter;
- pinocytosis – the intake of extracellular fluids.

Exocytosis is essentially the reverse of endocytosis, so the fluidity of the plasma membrane and the hydrophobic and hydrophilic properties of its molecules are just as important as in endocytosis. Exocytosis usually begins in the ribosomes of the rough ER and progresses through a series of four steps until the produced substance is secreted to the environment outside the cell.

1 Protein produced by the ribosomes of the rough ER enters the lumen of the ER.
2 Protein exits the ER and enters the *cis* side or face of the Golgi apparatus; a vesicle is involved (see Figure 2.23).
3 As the protein moves through the Golgi apparatus, it is modified and exits on the *trans* face inside a vesicle (see Figure 2.23).
4 The vesicle with the modified protein inside moves to and fuses with the plasma membrane – this results in the secretion of the contents from the cell.

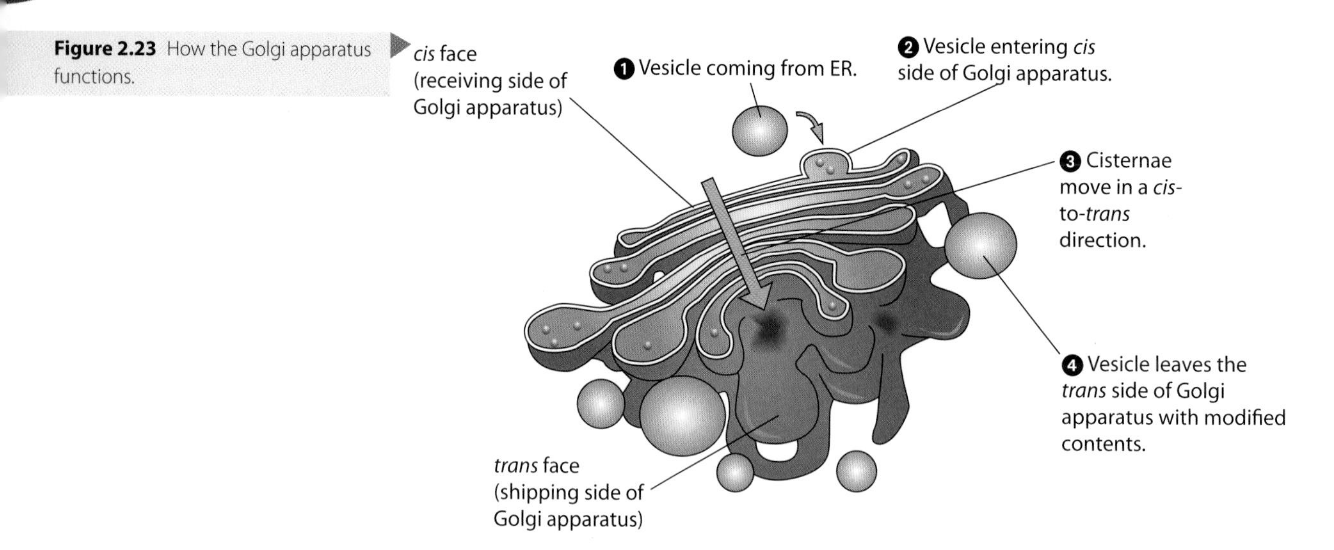
cis face
(receiving side of
Golgi apparatus)
1 Vesicle coming from ER.
2 Vesicle entering cis
side of Golgi apparatus.
3 Cisternae
move in a cis-
to-trans
direction.
4 Vesicle leaves the
trans side of Golgi
apparatus with modified
contents.
trans face
(shipping side of
Golgi apparatus)

Figure 2.23 How the Golgi apparatus functions.

Examples of exocytosis occur when:

- pancreas cells produce insulin and secrete it into the bloodstream (to help regulate blood glucose levels);
- neurotransmitters are released at the synapse in the nervous system.

The fluidity of the plasma membrane is essential to allow fusion and subsequent secretion of the vesicle contents. At this point the vesicle membrane is actually a part of the plasma membrane.

Exercises

10 Why is the term 'equilibrium' used with passive but not active transport?

11 What type of amino acids are present where integral proteins attach to cell membranes?

12 Why are exocytosis and endocytosis known as examples of active transport?

2.5 Cell division

Assessment statements

2.5.1 Outline the stages in the cell cycle, including interphase (G_1, S, G_2), mitosis and cytokinesis.

2.5.2 State that tumours (cancers) are the result of uncontrolled cell division and that these can occur in any organ or tissue.

2.5.3 State that interphase is an active period in the life of a cell when many metabolic reactions occur, including protein synthesis, DNA replication and an increase in the number of mitochondria and/or chloroplasts.

2.5.4 Describe the events that occur in the four phases of mitosis (prophase, metaphase, anaphase, and telophase).

2.5.5 Explain how mitosis produces two genetically identical nuclei.

2.5.6 State that growth, embryonic development, tissue repair and asexual reproduction involve mitosis.

The cell cycle

The cell cycle describes the behaviour of cells as they grow and divide. In most cases, the cell produces two cells that are genetically identical to the original. These are called daughter cells. The cell cycle integrates a growth phase with a divisional phase. Sometimes, cells multiply so rapidly that they form a solid mass of cells called a tumour. We refer to this disease state as cancer. It appears that any cell can lose its usual orderly pattern of division because we have found cancer in almost all tissues and organs.

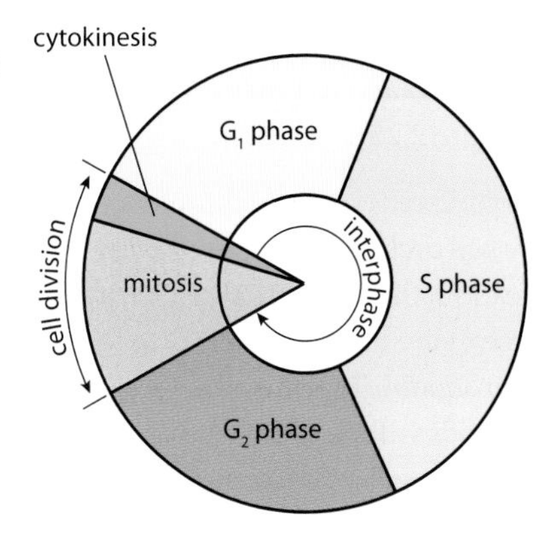

Figure 2.24 The cell cycle in eukaryotes.

You may wonder what causes a cell to go out of control. To answer this question, we must first understand the ordinary cell cycle. Usually, the life of a cell involves two major phases. In one phase, growth is a major process. The other phase involves division. The cell cycle begins and ends as one cell so it can be represented by a circle divided into various named sections (see Figure 2.24).

Interphase

The largest part of the cell cycle in most cells is interphase. It is the longest and most variable of the cell cycle phases. Interphase includes three phases: G_1, S, and G_2. During G_1 the major event is growth of the cell. At the beginning of G_1, the cell is the smallest it will ever be. After G_1 comes the S phase in which the main activity is replication of the DNA of the cell, the chromosomes. This phase is sometimes referred to as the synthesis phase. Once the chromosomes have replicated, the cell enters its second growth phase, G_2. During this phase, the cell grows and makes preparations for mitosis or the M phase. During G_2, organelles may increase in number, DNA begins to condense from chromatin to chromosomes, and microtubules may begin to form.

Interphase is a very active time in a cell's life. It involves metabolic reactions, DNA replication, and an increase in the number of organelles.

Because interphase involves growth, it is essential that protein synthesis occurs at a rapid rate during this phase.

Mitosis

Once all preparations are made and the DNA has replicated, the cell moves into mitosis or the M phase. During mitosis the replicated chromosomes separate and move to opposite poles of the cell thus providing the same genetic material to each of these locations. When the chromosomes are at the poles of the cell, the cytoplasm divides to form two distinct cells from the larger parent. These two cells have the same genetic material and are referred to as daughter cells.

Examiner's hint: To remember the correct order of phases in the cell cycle and mitosis, remember the word 'shipmate.'
If you take away the word 'she,' you get ipmat (interphase, prophase, metaphase, anaphase, and telophase).

Mitosis involves four phases. They are, in proper sequence:

- prophase;
- metaphase;
- anaphase;
- telophase.

Before considering a detailed description of these phases, it is essential to understand the chromosome. As you will recall, during the second growth phase, G_2, the chromatin (elongated DNA and histones) begins to condense. This condensation is accomplished via a process called supercoiling. First, the DNA wraps around histones to produce nucleosomes (see also Chapter 7, page 195). The nucleosomes are further wrapped into a solenoid. Solenoids group together in looped domains, and then a final coiling occurs to produce the chromosome (see Figure 2.25).

Eukaryotic cells contain chromosomes which, before replication in the S phase of the cell cycle, are composed of one molecule of DNA (see Figure 2.26). After replication, the chromosome includes two molecules of DNA. These two identical molecules are held together by the centromere, and each molecule is referred to as a chromatid. Together, they are called sister chromatids (see Figure 2.26). The chromatids will eventually separate during the process of mitosis. When they do, each is then called a chromosome and each has its own centromere.

Now that we are familiar with the structure of a chromosome, we can explain the four phases of mitosis. Remember, when we enter mitosis, replication of DNA has, so the chromosomes are each composed of two sister chromatids.

Figure 2.25 This diagram shows you how DNA is packaged by supercoiling from a single double helix to nucleosomes, to solenoids, to looped domains and finally to a chromosome.

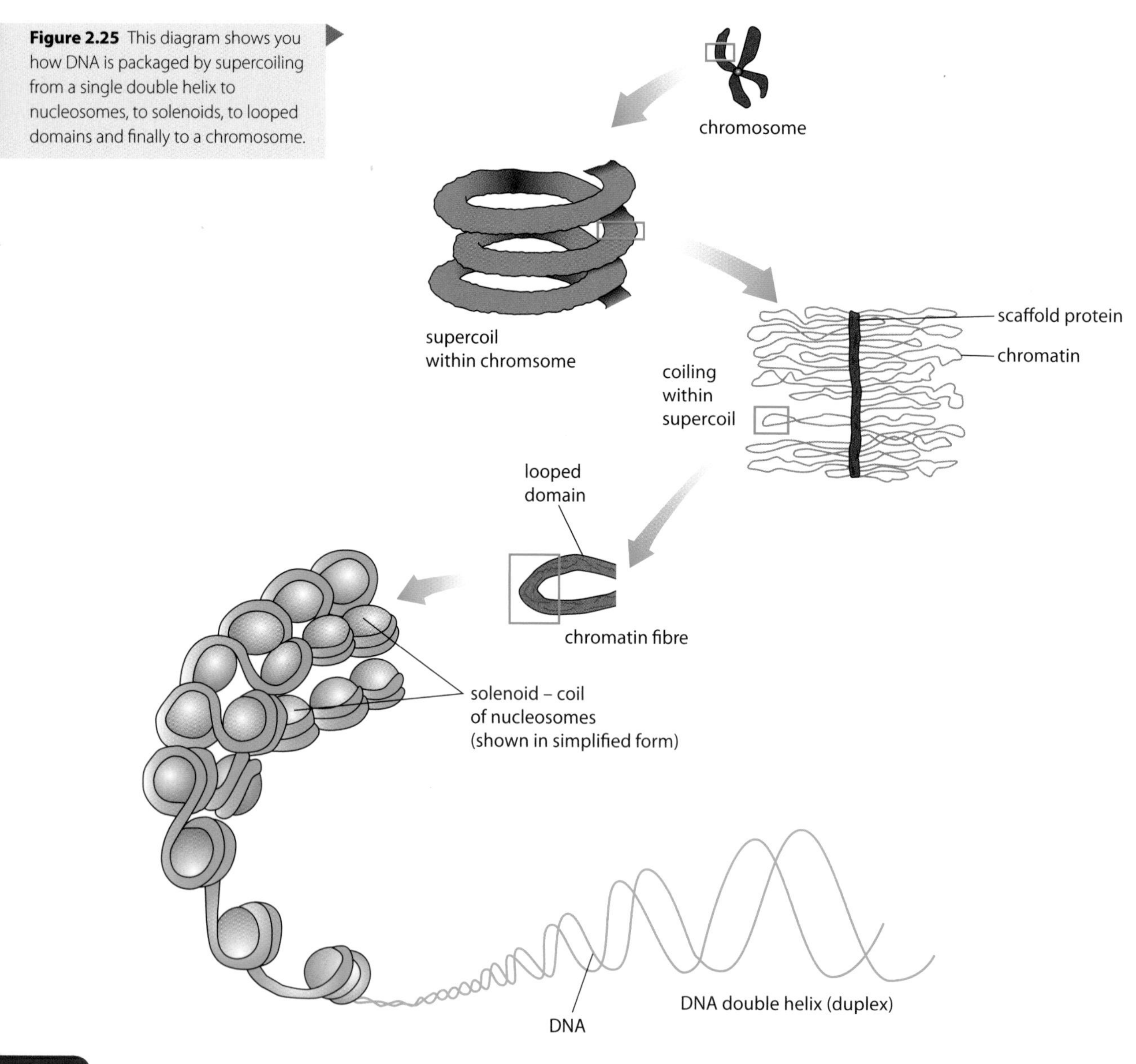

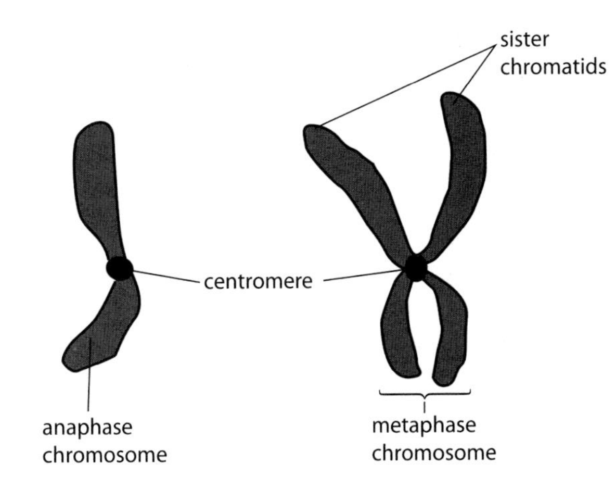

Figure 2.26 An anaphase chromosome is a single molecule of DNA and has a centromere. A metaphase chromosome has sister chromatids attached at the centromere.

Prophase

Examine Figure 2.27.

1 The chromatin fibres become more tightly coiled to form chromosomes.
2 The nuclear envelope disintegrates and nucleoli disappear.
3 The mitotic spindle begins to form and is complete at the end of prophase.
4 The centromere of each chromosome has a region called the kinetochore that attaches to the spindle.
5 The centrosomes move toward opposite poles of the cell due to lengthening microtubules.

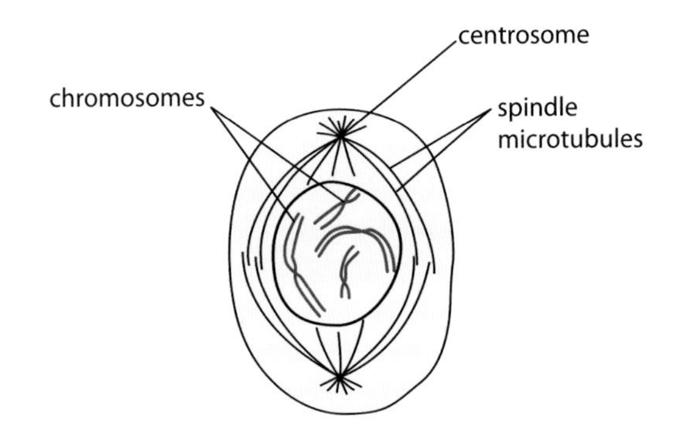

Figure 2.27 This animal cell is in prophase. For clarity, only a small number of chromosomes are shown.

Metaphase

Examine Figure 2.28.

1 The chromosomes are moved to the middle or equator of the cell. This is referred to as the metaphase plate.
2 The chromosome's centromeres lie on the plate.
3 The movement of chromosomes is due to the action of the spindle which is made of microtubules.
4 The centrosomes are now at the opposite poles.

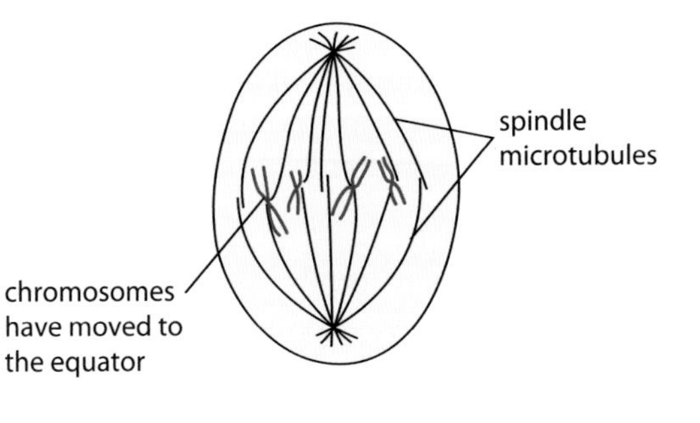

Figure 2.28 The cell is now in metaphase. Again, only a small number of chromosomes are shown.

Anaphase

Examine Figure 2.29.

1 This is usually the shortest phase of mitosis. It begins when the two sister chromatids of each chromosome are split.
2 These chromatids, now chromosomes, move toward the opposite poles of the cell.
3 The chromatid movement is due to shortening of the microtubules of the spindle.
4 Because the centromeres are attached to the microtubules, they move towards the poles first.
5 At the end of this phase, each pole of the cell has a complete, identical set of chromosomes.

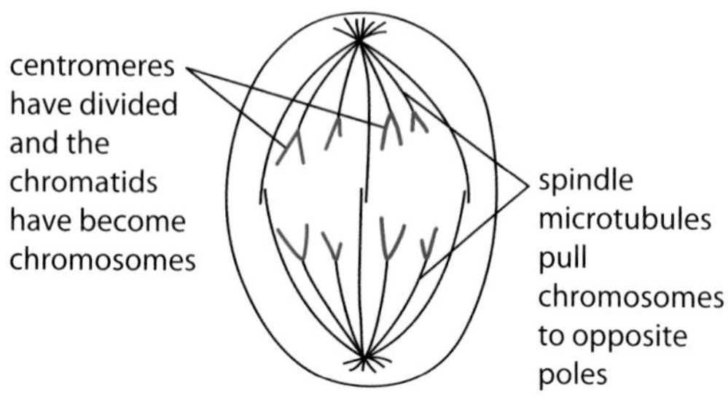
centromeres have divided and the chromatids have become chromosomes
spindle microtubules pull chromosomes to opposite poles

Figure 2.29 The cell is now in anaphase. Again, only a small number of chromosomes are shown.

Telophase

Examine Figure 2.30.

1 The chromosomes are at each pole.
2 A nuclear membrane (envelope) begins to re-form around each set of chromosomes.
3 The chromosomes start to elongate to form chromatin.
4 Nucleoli reappear.
5 The spindle apparatus disappears.
6 The cell is elongated and ready for cytokinesis.

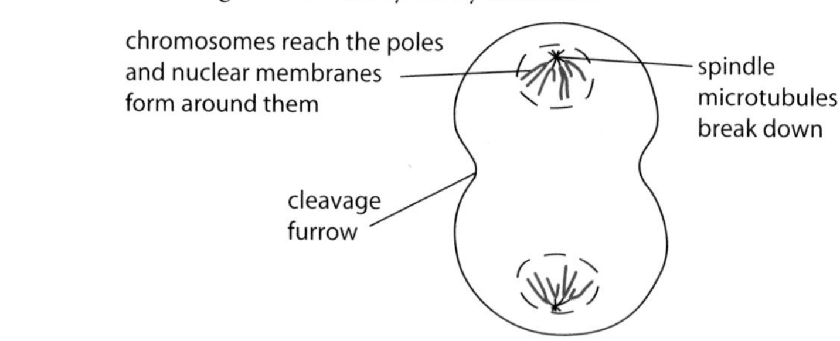
chromosomes reach the poles and nuclear membranes form around them
spindle microtubules break down
cleavage furrow

Figure 2.30 Finally, the cells enters telophase.

Cytokinesis

As you can see, the phases of mitosis involve nuclear division. It appears that the process of mitosis occurs in discrete stages. But this is not at all true – they occur in a continuum. We only use the separate stages to help understand the overall process. Once nuclear division has occurred, the cell undergoes cytokinesis. You should note that cytokinesis in animal cells involves an inward pinching of the fluid plasma membrane to form cleavage furrows. However, plant cells have a relatively firm cell wall and they form a cell plate. The cell plate occurs midway between the two poles of the cell and moves outward toward the sides of the cell from a central region. Both processes result in two separate daughter cells that have genetically identical nuclei.

Growth of organisms, development of embryos, tissue repair, and asexual reproduction all involve mitosis. Mitosis does not happen by itself. It is a part of the cell cycle.

Exercises

13 A chemical called colchicine disrupts the formation of microtubules. What effect would this drug have on a cell going through mitosis?

14 If a parent cell has 24 chromosomes, how many chromatids would be present during metaphase of mitosis?

15 Explain where cytokinesis occurs in the cell cycle.

Practice questions

1 What is/are the advantage(s) of using an electron microscope?

I Very high resolution
II Very high magnification
III The possibility of examining living material

A I only
B I and II only
C II and III only
D I, II and III *(1 mark)*

2 Which phases of mitosis are shown in diagrams I and II?

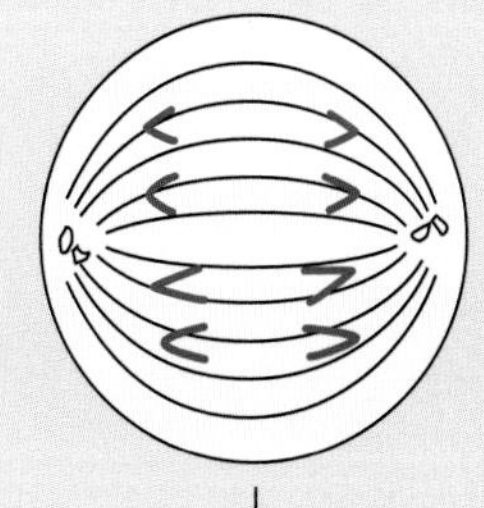

I

II

A metaphase and prophase
B metaphase and telophase
C anaphase and prophase
D anaphase and metaphase *(1 mark)*

3 The key below can be used to identify some of the structures in the cytoplasm of liver cells. Which structures are ribosomes?

1	Enclosed in a membrane	go to 2
	Not enclosed in a membrane	go to 3
2	Diameter less than 100 nm	A
	Diameter greater than 100 nm	B
3	Composed of one globular structure	C
	Composed of two subunits	D

(1 mark)

4 Which of the following is required for osmosis to occur?

A An enzyme
B A fully permeable membrane
C ATP
D A solute concentration gradient *(1 mark)*

5 In the diagram below macromolecules are being transported to the exterior of a cell.

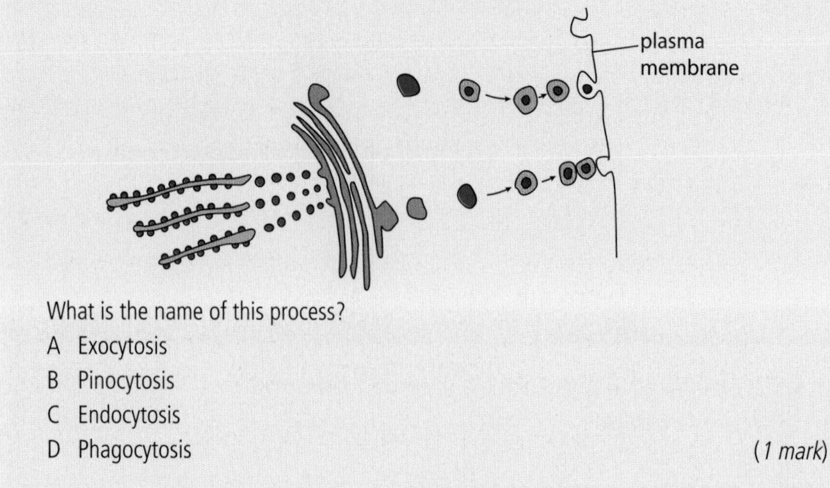

What is the name of this process?

A Exocytosis

B Pinocytosis

C Endocytosis

D Phagocytosis (*1 mark*)

6 A study was carried out to determine the relationship between the diameter of a molecule and its movement through a membrane. The graph below shows the results of the study.

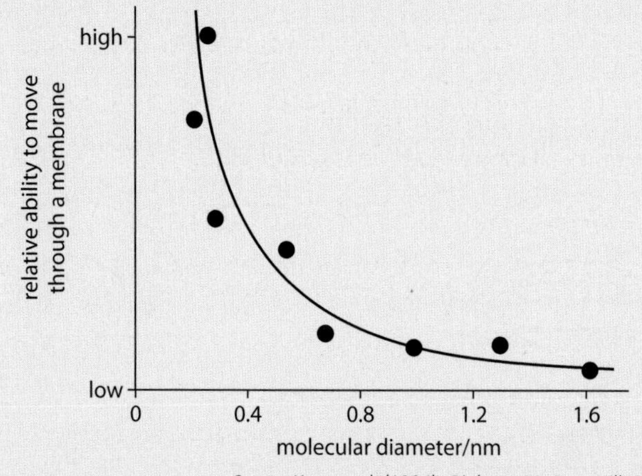

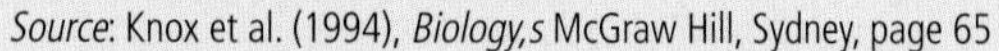
Source: Knox et al. (1994), *Biology,s* McGraw Hill, Sydney, page 65

(a) From the information in the graph alone, describe the relationship between the diameter of a molecule and its movement through a membrane. (*2*)

A second study was carried out to investigate the effect of passive protein channels on the movement of glucose into cells. The graph below shows the rate of uptake of glucose into erythrocytes by simple diffusion and facilitated diffusion.

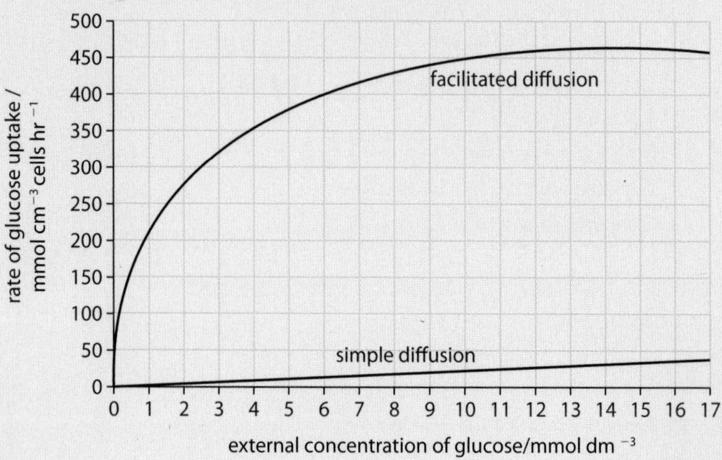

(b) Identify the rate of glucose uptake at an external glucose concentration of 4 mmol dm^{-3} by:

(i) simple diffusion **(ii)** facilitated diffusion (*2*)

(c) **(i)** Compare the effect of increasing the external glucose concentration on glucose uptake by facilitated diffusion and by simple diffusion. (*3*)

(ii) Predict, with a reason, the effect on glucose uptake by facilitated diffusion of increasing the external concentration of glucose to 30 mmol dm^{-3}. (*2*)

(*Total 9 marks*)

7 **(a)** An organelle is a discrete structure with a specific function within a cell. In the table below, identify the missing organelles and outline the missing functions.

Name of organelle	Structure of organelle	Function of organelle
Nucleus	Region of the cell containing chromosomes, surrounded by a double membrane, in which there are pores.	Storage and protection of chromosomes.
Ribosome	Small spherical structures consisting of two subunits.	
	Spherical organelles, surrounded by a single membrane and containing hydrolytic enzymes.	Digestion of structures that are not needed within cells.
	Organelles surrounded by two membranes, the inner of which is folded inwards.	

(*4*)

(b) The table above shows some of the organelles found in a particular cell. Discuss what type of cell this could be. (*2*)

(*Total 4 marks*)

8 **(a)** Distinguish between diffusion and osmosis. (*1*)

(b) Explain how the properties of phospholipids help to maintain the structure of the cell surface membrane. (*2*)

(c) State the composition and the function of the plant cell wall. (*2*)

(*Total 5 marks*)

3 The chemistry of life

Introduction

Organic chemistry is the chemistry of carbon compounds. Biochemistry is the branch of organic chemistry which attempts to explain the chemistry in living organisms. Fortunately, biochemistry is rather similar from one organism to another. The majority of organic molecules found in living organisms can be classified into one of four types:

- carbohydrates;
- lipids;
- proteins;
- nucleic acids.

In this chapter, you will be introduced to each of these types of molecule and begin to understand their importance in living things. In addition, you will begin your study of some of the more important biochemical pathways in organisms.

3.1 Chemical elements and water

Assessment statements

3.1.1 State that the most frequently occurring chemical elements in living things are carbon, hydrogen, oxygen and nitrogen.
3.1.2 State that a variety of other elements are needed by living organisms, including sulfur, calcium, phosphorus, iron and sodium.
3.1.3 State one role for each of the elements mentioned in 3.1.2.
3.1.4 Draw and label water molecules to show their polarity and hydrogen bond formation.
3.1.5 Outline the thermal, cohesive and solvent properties of water.
3.1.6 Explain the relationship between the properties of water and its uses in living organisms as a coolant, medium for metabolic reactions and transport medium.

Elements found in living organisms

A typical person could survive for about a month without food. They would survive only a week or less without water.

It could be argued from a purely scientific viewpoint that any living organism is merely a collection of elements in the form of atoms, ions and molecules. The four most common elements found in living things are carbon, hydrogen, oxygen and nitrogen. These elements are used in the molecular structures of all carbohydrates, proteins, lipids and nucleic acids.

In addition, living organisms contain a variety of other elements which are extremely important, but are less common. Here is a table showing some of the less common elements and an example role that each plays in living organisms.

Element	Example role in plants	Example role in animals	Example role in prokaryotes
sulfur	in some amino acids	in some amino acids	in some amino acids
calcium	co-factor in some enzymes	co-factor in some enzymes and component of bones	co-factor in some enzymes
phosphorous	phosphate groups in ATP	phosphate groups in ATP	phosphate groups in ATP
iron	in cytochromes	in cytochromes and in haemoglobin	in cytochromes
sodium	in membrane function	in membrane function and sending nerve impulses	in membrane function

Structure of water

Water is the solvent of life. Virtually all cells have water within (cytoplasm) and water in the surrounding environment (intercellular fluid, pond water, etc.). Water is an incredibly abundant substance on Earth and has some very interesting properties. Many of these properties depend on the structure of water molecules (see Figure 3.1).

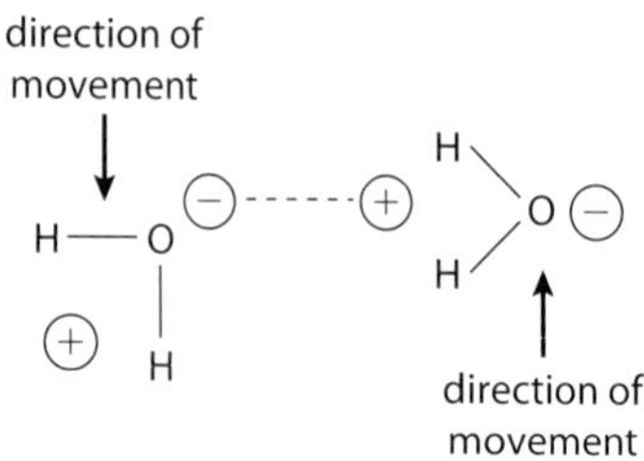

Figure 3.1 In liquid water, molecules form a 'split-second' hydrogen bond (dotted line) even though each may be moving in a different direction.

You can float a paper clip on water because of water's surface tension.

Basilisk lizards may be as long as 0.8 metres, but they can run across the top of bodies of water. The relatively large surface area of their toes does not break through the surface tension of the water as long as they keep running.

The hydrogen and oxygen atoms in a single water molecule are held together by a type of bond called a polar covalent bond. Polar covalent bonding results from an unequal sharing of electrons. In water, the single oxygen atom is bonded to two different hydrogen atoms. Each oxygen–hydrogen bond is a polar covalent bond and results in a slight negative charge at the oxygen end of the molecule and a slight positive charge at the end with the two hydrogens. Because the two ends of each water molecule have opposite charges, water molecules interact with each other in very interesting ways.

Properties of water and living organisms

Water has a number of unique properties important to living organisms.

Thermal properties

Water has thermal properties that are important to living things. One of those thermal properties is high specific heat. In simple terms, this means that water can absorb or give off a great deal of heat without changing temperature greatly. Think of a body of water on a very cold night; even though the air may be very cold, the body of water is relatively stable in temperature. All living things are composed of a great deal of water and thus you can think of your water content as a temperature stabilizer. Water also has a high heat of vaporization. This means that water absorbs a great deal of heat when it evaporates. Many organisms, including ourselves, use this as a cooling mechanism. Internal body heat results in perspiration, the perspiration then evaporates from your skin. Much of the heat that turned the water molecules from the liquid phase to the vapour phase came from your body and thus sweating not only makes you feel cooler, it really does lower your temperature.

To access a good primer on the properties of water, visit heinemann.co.uk/hotlinks, insert the express code 4242P and click on Weblink 3.1.

Water evaporates from leaves through small openings called stomata. As shown here, each stoma is surrounded by two guard cells which, when swollen with water, hold the stoma open. One benefit to the plant is the cooling effect that evaporation provides.

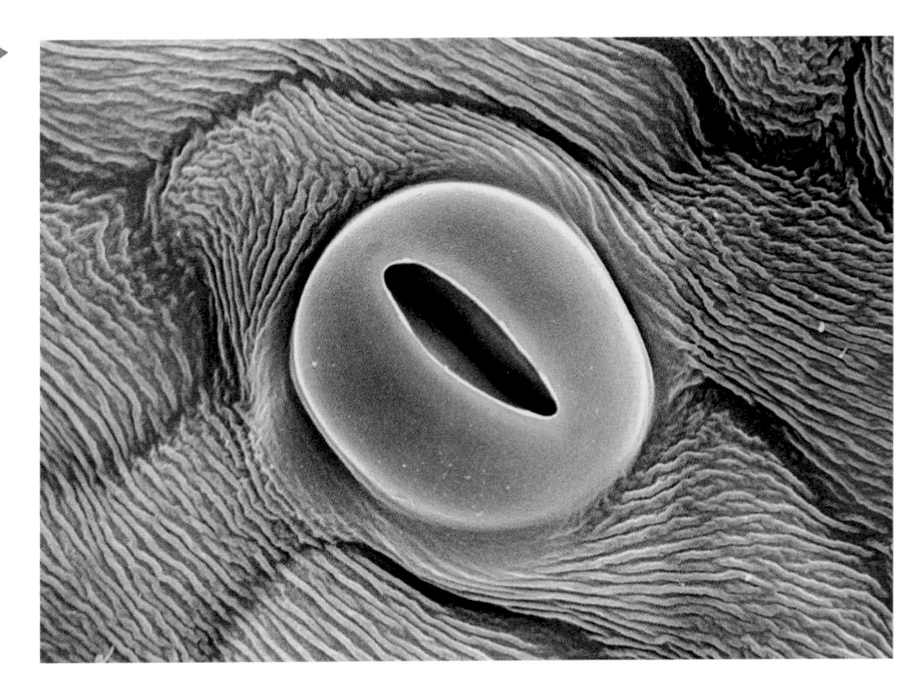

Cohesive properties

Water molecules are highly cohesive. Cohesion is when molecules of the same type are attracted to each other. This attraction is due to the polar covalent bonding mentioned earlier. Water molecules have a positive end and a negative end. Whenever two water molecules are near each other the positive end of one attracts the negative end of another. When water cools below the freezing point, molecular motion has slowed to the point where these polar attractions become locked into place and an ice crystal forms. Liquid water has molecules with a much faster molecular motion and the water molecules are able to influence each other, but not to the point where molecules stop their motion. This cohesion between liquid water molecules explains a variety of events, including:

- why water forms into droplets when spilled;
- why water has a surface tension that allows some organisms to 'walk on water' (for some this is 'run on water');
- how water is able to move as a column in the vascular tissues of plants;
- why water has a high heat capacity and high heat of vaporization as discussed earlier.

Solvent properties

Water is an excellent solvent of other polar molecules. You may remember from earlier science classes that 'like dissolves like'. The vast majority of molecules typically found inside and outside of most cells are also polar molecules. This includes carbohydrates, proteins, and nucleic acids (DNA and RNA). Most types of lipids are relatively non-polar and thus most organisms have special strategies to deal with the transport and biochemistry of lipids.

Because water is an excellent solvent for biochemically important molecules, it is also the medium in which most of the biochemistry of a cell occurs. A cell contains a wide variety of fluids, all of which are primarily water. We refer to such solutions as aqueous solutions. Here is a table of some common aqueous solutions in which specific biochemical reactions take place.

Aqueous solution	Location	Common reactions
cytoplasm	fluid inside cell but outside organelles	glycolysis / protein synthesis reactions
nucleoplasm	fluid inside nuclear membrane	DNA replication / transcription
stroma	fluid inside chloroplast membrane	light-independent reactions of photosynthesis
blood plasma	fluid in arteries, veins and capillaries	loading and unloading of respiratory gases / clotting

Examples of water as a solvent in plants and animals

The properties of water make it an excellent medium for transport. Vascular tissue in plants carries water and a variety of dissolved substances. More specifically, xylem carries water and dissolved minerals up from the root system to the leaves of a plant. Phloem then transports dissolved sugars from the leaves to the stems, roots and flowers of a plant.

Blood is the most common transport medium in animals and is largely made up of water. Blood is a transport medium for red blood cells, white blood cells, platelets and a wide variety of dissolved molecules. The liquid portion of blood is called blood plasma. Some of the more common solutes in blood plasma are:

- glucose (blood sugar);
- amino acids;
- fibrinogen (protein involved in blood clotting);
- hydrogencarbonate ions (as a means of transporting CO_2).

Exercises

1. Choose any specific aquatic or terrestrial animal and make a list of all of the ways that water is important to this animal.
2. How are the properties of water discussed above involved in any item on your list?

3.2 Carbohydrates, lipids and proteins

Assessment statements

3.2.1 Distinguish between organic and inorganic compounds.
3.2.2 Identify amino acids, glucose, ribose and fatty acids from diagrams showing their structure.
3.2.3 List three examples each of monosaccharides, disaccharides and polysaccharides.
3.2.4 State one function of glucose, lactose and glycogen in animals, and of fructose, sucrose and cellulose in plants.
3.2.5 Outline the role of condensation and hydrolysis in the relationships between monosaccharides, disaccharides and polysaccharides; between fatty acids, glycerol and triglycerides; and between amino acids and polypeptides.
3.2.6 State three functions of lipids.
3.2.7 Compare the use of carbohydrates and lipids in energy storage.

Introduction to carbohydrates, lipids, proteins and nucleic acids

Living things are composed of an amazing array of molecules. We can start to make sense of all of these molecules by classifying them into a molecule type. Molecules of the same type have certain qualities in common and become fairly easy to recognize with a little practice. The following table shows some of the more common biochemically important molecules and their subcomponents (or building blocks).

Molecule	Subcomponents
carbohydrates	monosaccharides
lipids	glycerol and fatty acids
proteins (polypeptides)	amino acids
nucleic acids	nucleotides

Many of the carbons found in foods that you eat (such as various carbohydrates) will be eliminated from your body in the molecules of carbon dioxide that you breathe out.

Common biochemicals and their structure

Molecules can be classified as being either inorganic or organic. All organic molecules contain the element carbon, although not all carbon-containing molecules are organic. Carbon dioxide is a common example of a molecule that contains carbon that is not organic. In your study of biochemistry, you will encounter many types of organic molecule that are important to living things. The table which follows summarizes some of the most important categories and some examples.

You will notice that virtually all images you see of atoms and molecules are in the form of models. Why are models used? What do the real atoms and molecules look like?

Category	Subcategory	Example molecules
carbohydrates	monosaccharides	glucose, galactose, fructose
	disaccharides	maltose, lactose, sucrose
	polysaccharides	starch, glycogen, cellulose
proteins		enzymes, antibodies
lipids		triglycerides, phospholipids
nucleic acids		DNA, RNA

To access many visual images of biochemical molecular models, visit heinemann.co.uk/hotlinks, insert the express code 4242P and click on Weblink 3.2.

You should learn to recognize the structures of some of the most common molecules in biochemistry. Study Figures 3.2–3.6 (the notes are given to help you in the recognition and appreciation of the molecule).

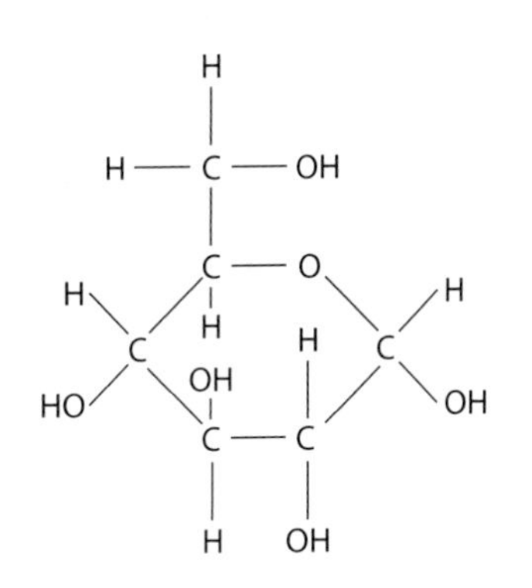

Figure 3.2 Ring structure of glucose, a 6-carbon monosaccharide.

Note: Glucose is a common 6-carbon monosaccharide. Plants store glucose as the polysaccharide starch; animals store glucose as the polysaccharide glycogen.

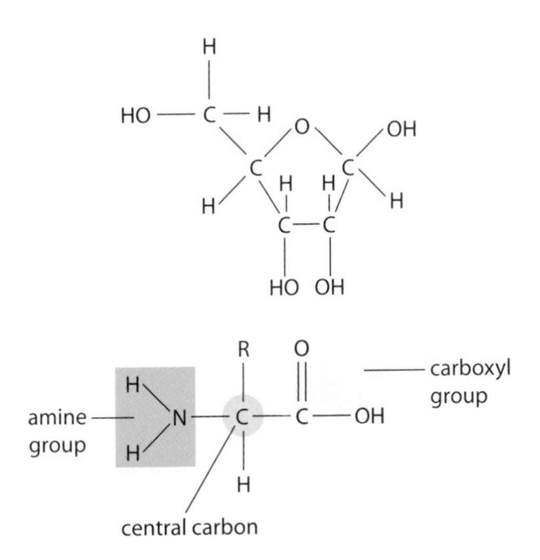

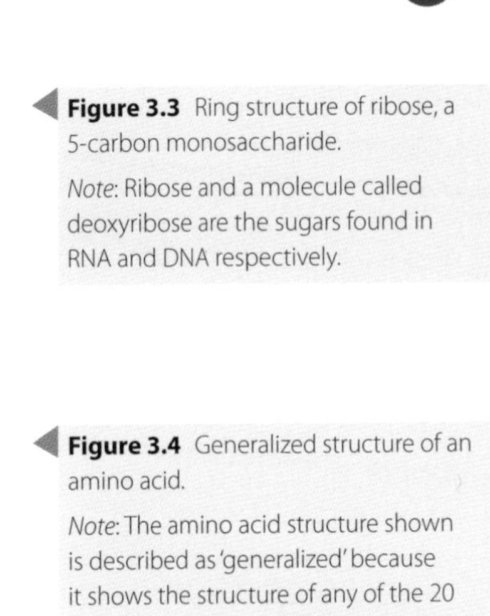

Figure 3.3 Ring structure of ribose, a 5-carbon monosaccharide.

Note: Ribose and a molecule called deoxyribose are the sugars found in RNA and DNA respectively.

Figure 3.4 Generalized structure of an amino acid.

Note: The amino acid structure shown is described as 'generalized' because it shows the structure of any of the 20 amino acids. Each of the 20 contains a different, specific molecular structure in place of the R group. For example, the amino acid alanine has a CH_3 group in place of the generalized R.

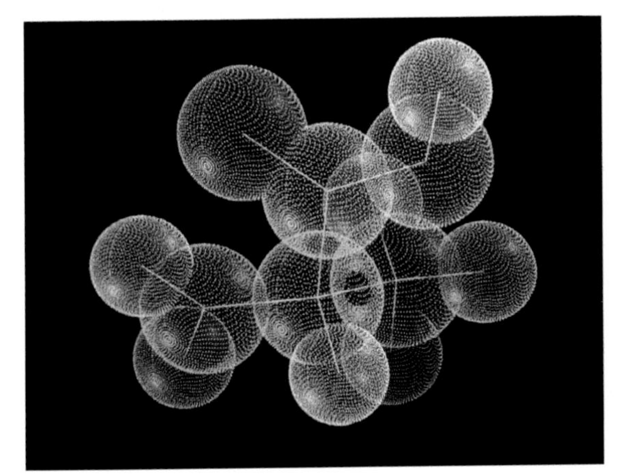

This is a colour-coded molecular model of the amino acid, alanine (green = carbon, pink = oxygen, blue = nitrogen, white = hydrogen).

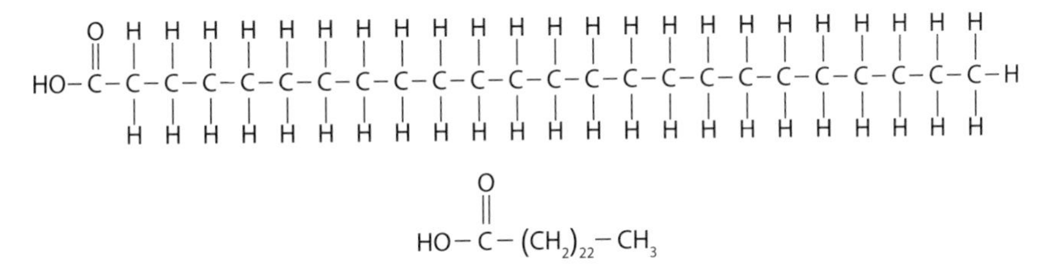

Figures 3.5 and 3.6 Structure of a typical fatty acid.

Note: Two different structures are shown representing the same fatty acid. Specific fatty acids differ from each other by the total number of carbons and by the presence and location of any double bonds between carbons. Notice that the two structures shown have the same total number of carbons and neither contains a double bond between carbons. Therefore, they are the same fatty acid.

Functions of carbohydrates in animals and plants

Carbohydrates are among the most commonly found biochemical molecules found in both animals and plants. You have seen earlier that carbohydrates exist in different 'sizes' – monosaccharides, disaccharides, and polysaccharides. All of these carbohydrates serve many functions in living organisms; example functions of a variety of carbohydrates are shown in the following tables.

This table shows the importance of carbohydrates in animals.

Name	Type	One function
glucose	monosaccharide	chemical fuel for cell respiration
lactose	disaccharide	makes up some of the solutes in milk
glycogen	polysaccharide	stores glucose in liver and muscles

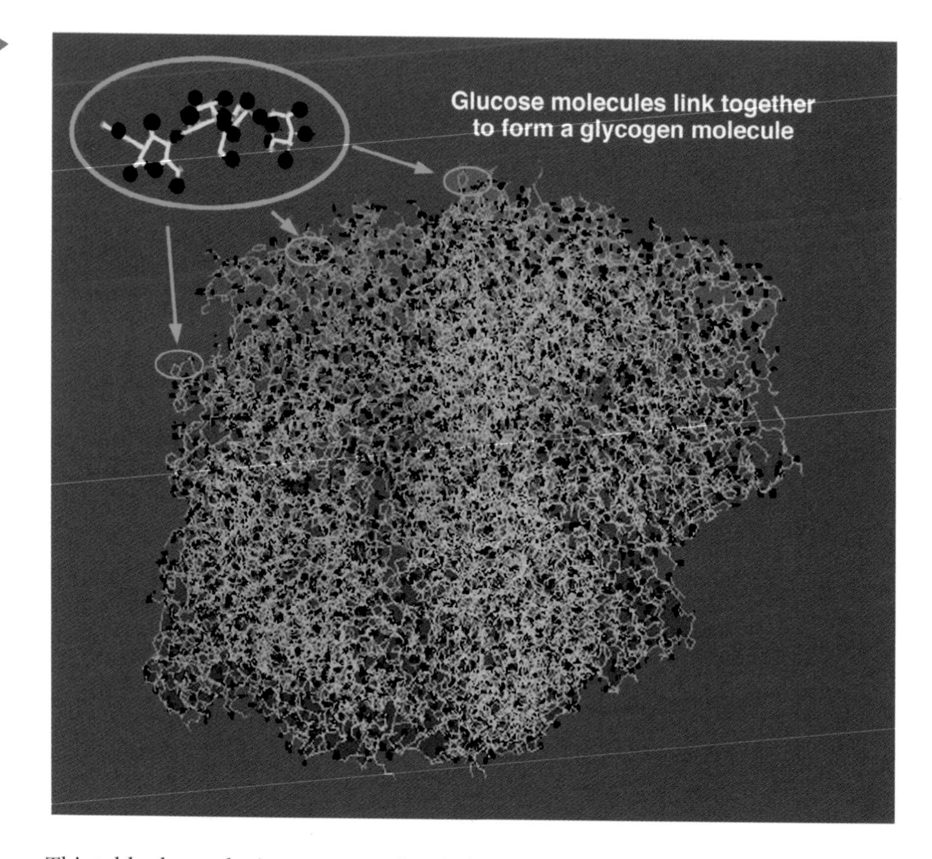
Glucose molecules link together to form a glycogen molecule

This model of glycogen shows how the monosaccharide building blocks are dwarfed by the overall molecular size of the polysaccharide.

This table shows the importance of carbohydrates in plants.

Name	Type	One function
fructose	monosaccharide	found in many fruits (makes them sweet)
sucrose	disaccharide	often transported from leaves of plants to other locations in plants by vascular tissue
cellulose	polysaccharide	one of the primary components of plant cell walls

Role of lipids

Lipids are biochemically important molecules that serve many functions. We refer to triglyceride lipids in solid form as fats. In liquid form, triglycerides are called oils. Everyone is aware of the role that fat plays in energy storage. If you eat more food than you burn, your body will store much of the excess as fat in adipose cells. Each adipose cell gets smaller or larger depending on how much lipid is being stored. People either gain or lose weight depending on how much lipid is being stored at any given time in a fairly fixed number of adipose cells. Lipids are very efficient molecules for storing energy. As seen earlier, carbohydrates are also used for storing energy in living organisms. Glycogen is a carbohydrate used by animals to store energy and starch is used by plants. Laboratory studies have shown that in an equal mass of carbohydrate and lipid, the lipid stores approximately twice as much chemical energy as the carbohydrate.

This drawing shows a fat cell (adipocyte) becoming larger as lipids are stored in it.

Lipids are also important for thermal insulation. A good reminder of this is to study the amount of blubber (fat) that cold-climate animals form in order to stay warm; 30% or more of the body mass of some seals may be due to the blubber layer beneath their skin.

A special category of lipid called phospholipid makes up the double layer of all cell membranes. These phospholipid molecules have a polar end turned towards water and a non-polar end which turns away from water.

Condensation and hydrolysis reactions

Many organisms, including all animals, rely on the foods they eat to gain the building block molecules which make up their own larger molecules. When animals eat foods, the food is digested (or hydrolysed) into the building blocks. After these building blocks are transported to body cells, they are bonded together to form larger molecules once again.

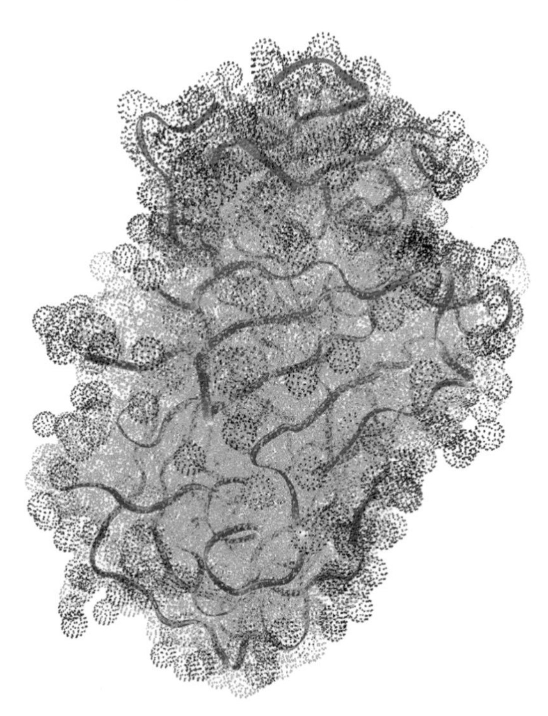

This computer graphic image shows pepsin, an enzyme which helps to digest proteins. Thus, pepsin is an example of a hydrolysing enzyme.

Let's explore what happens to ingested foods. Foods are chemically digested in your alimentary canal. The digestive enzymes that accomplish this are hydrolysing enzymes. Each reaction is called a hydrolysis and requires a molecule of water as a reactant. This is a good way to recognize hydrolysis reactions – water is always 'split' as part of the reaction. Examples of hydrolysis include:

- hydrolysis of a disaccharide to two monosaccharides (see Figure 3.7);
  lactose + water → glucose + galactose
- hydrolysis of a polysaccharide to many monosaccharides;
  starch + (many) water → (many) glucose
- hydrolysis of a triglyceride lipid to glycerol and fatty acids (see Figure 3.8);
  triglyceride + 3 water → glycerol + 3 fatty acids
- hydrolysis of a polypeptide (protein) to amino acids.
  protein + (many) water → (many) amino acids

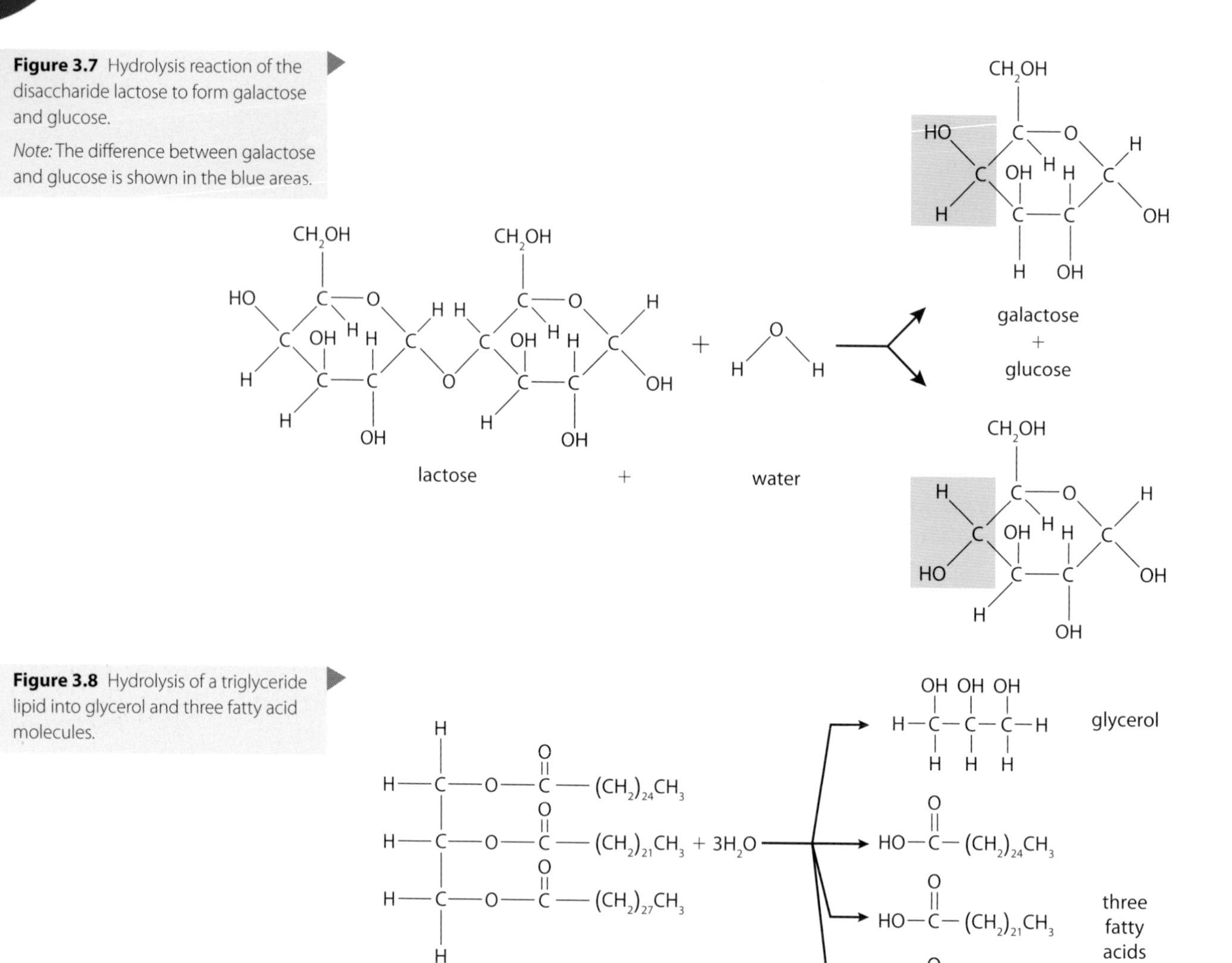

Figure 3.7 Hydrolysis reaction of the disaccharide lactose to form galactose and glucose.

Note: The difference between galactose and glucose is shown in the blue areas.

Figure 3.8 Hydrolysis of a triglyceride lipid into glycerol and three fatty acid molecules.

Condensation reactions are in many ways the reverse of hydrolysis reactions. In cells, condensation reactions occur to re-form the larger biochemically important molecules. In the four examples on page 53, simply reverse the reaction arrow and each example shows a condensation reaction. For example:

- condensation of amino acids to form a polypeptide.

  (many) amino acids → protein + (many) water

Notice in condensation reactions that water molecule(s) are products rather than reactants. Condensation reactions require a different type of enzyme, one that is capable of catalysing reactions in which covalent bonds are created rather than broken.

Exercises

3 **(a)** Write a word equation showing the formation of any specific disaccharide.
(b) Classify this reaction as a hydrolysis or condensation.

4 **(a)** Write a word equation showing the formation of a triglyceride.
(b) What type of enzyme would catalyse this reaction?
(c) Is water a reactant or a product of this reaction?
(d) How many water molecules are involved in this reaction?

3.3 DNA structure

Assessment statements

3.3.1 Outline DNA nucleotide structure in terms of sugar (deoxyribose), base and phosphate.

3.3.2 State the names of the four bases in DNA.

3.3.3 Outline how DNA nucleotides are linked together by covalent bonds into a single strand.

3.3.4 Explain how a DNA double helix is formed using complementary base pairing and hydrogen bonds.

3.3.5 Draw and label a simple diagram of the molecular structure of DNA.

Nucleotides are the building blocks of DNA

DNA (deoxyribonucleic acid) is not just a long molecule – it is an incredibly long molecule. In order to make sense of the structure of DNA, you must learn to recognize and work with the subcomponents of DNA called nucleotides (see Figure 3.9). Each nucleotide of DNA is composed of a phosphate group, a sugar called deoxyribose and a molecule that is called a nitrogenous base.

For many years most scientists everywhere believed it was protein, not DNA, that contained our genetic information. Research conducted in the first few decades of the 20th century proved that DNA contains our genetic blueprint.

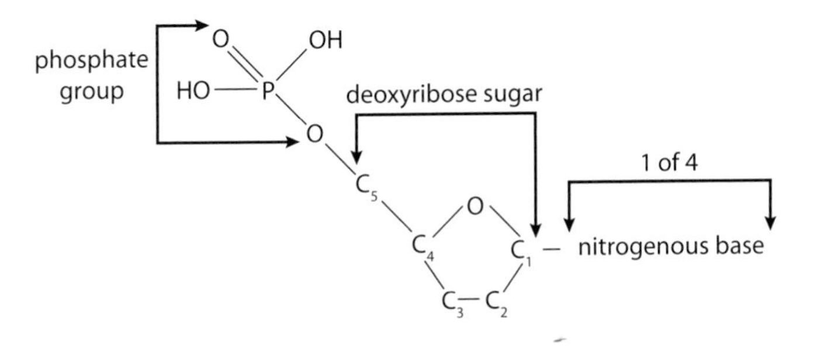

Figure 3.9 Structure of a single DNA nucleotide.

There are four possible nitrogenous bases in the nucleotides of DNA. The four bases are adenine, thymine, cytosine and guanine. It is very common to see these bases shortened to their abbreviations (A, T, C and G). Notice in Figure 3.10 that all nucleotides are exactly the same except for the nitrogenous base. Abbreviated forms of all the components have been used.

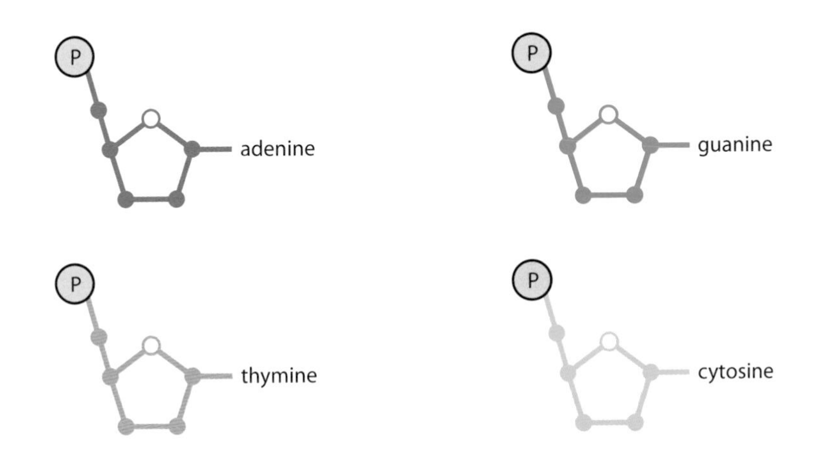

Figure 3.10 The four nucleotides found in DNA molecules.

Each strand of DNA is composed of nucleotides covalently linked

When is science not a science? Visit heinemann.co.uk/hotlinks, insert the express code 4242P and click on Weblink 3.3. Think about whether this is science, art, or something else.

DNA molecules are often described as having the shape of a double helix. This means that DNA is composed of two strands and each of the strands is shaped like a spiral staircase. Let's first explore how each of the nucleotides in a single strand are covalently bonded together. Figure 3.11 shows five nucleotides bonded together to form the beginning of a single DNA strand. Each adjoining nucleotide has been drawn in a different colour for emphasis of the nucleotide structure. No attempt has been made to draw the helical shape of the strand.

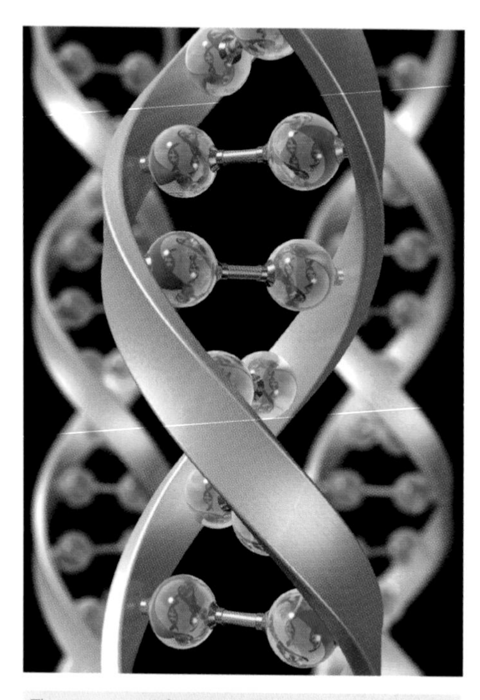

The two sugar–phosphate strands form a double helix with nitrogenous bases (shown in blue) found inside the helix shape.

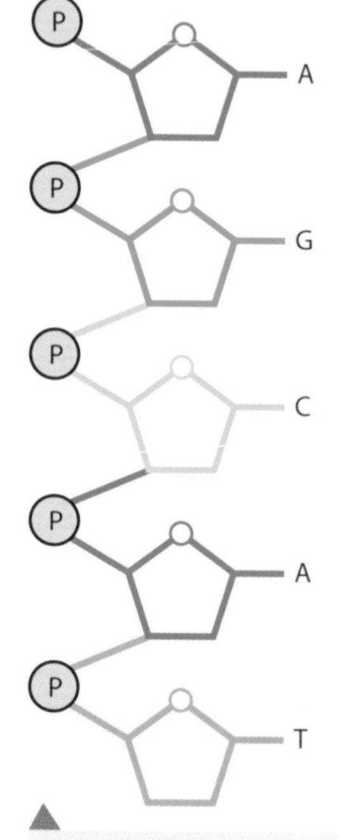

Figure 3.11 Five nucleotides bonded to form a very small section of one strand of DNA.

Complementary base pairs and hydrogen bonds help form the double helix

Now we are ready to look at how the two single strands of DNA interact to help form the double helix. Imagine a double-stranded DNA molecule as a ladder (see Figure 3.12). The two sides of the ladder are made up of the phosphates and deoxyribose sugars. The rungs of the ladder (what you step on) are made up of the nitrogenous bases. Since the ladder has two sides, there are two bases making up each rung. The two bases making up one rung are bases that are said to be complementary to each other. The complementary base pairs are adenine–thymine and cytosine–guanine.

Adenine and thymine are held together by two hydrogen bonds. Cytosine and guanine are held together by three hydrogen bonds. Because adenine and guanine are twice the size of thymine and cytosine, complementary base pairing is the only arrangement that gives a consistent distance from one strand across to the other strand and also leads to bonding between the bases. We can now use all of this information to construct a simple, yet accurate drawing of DNA.

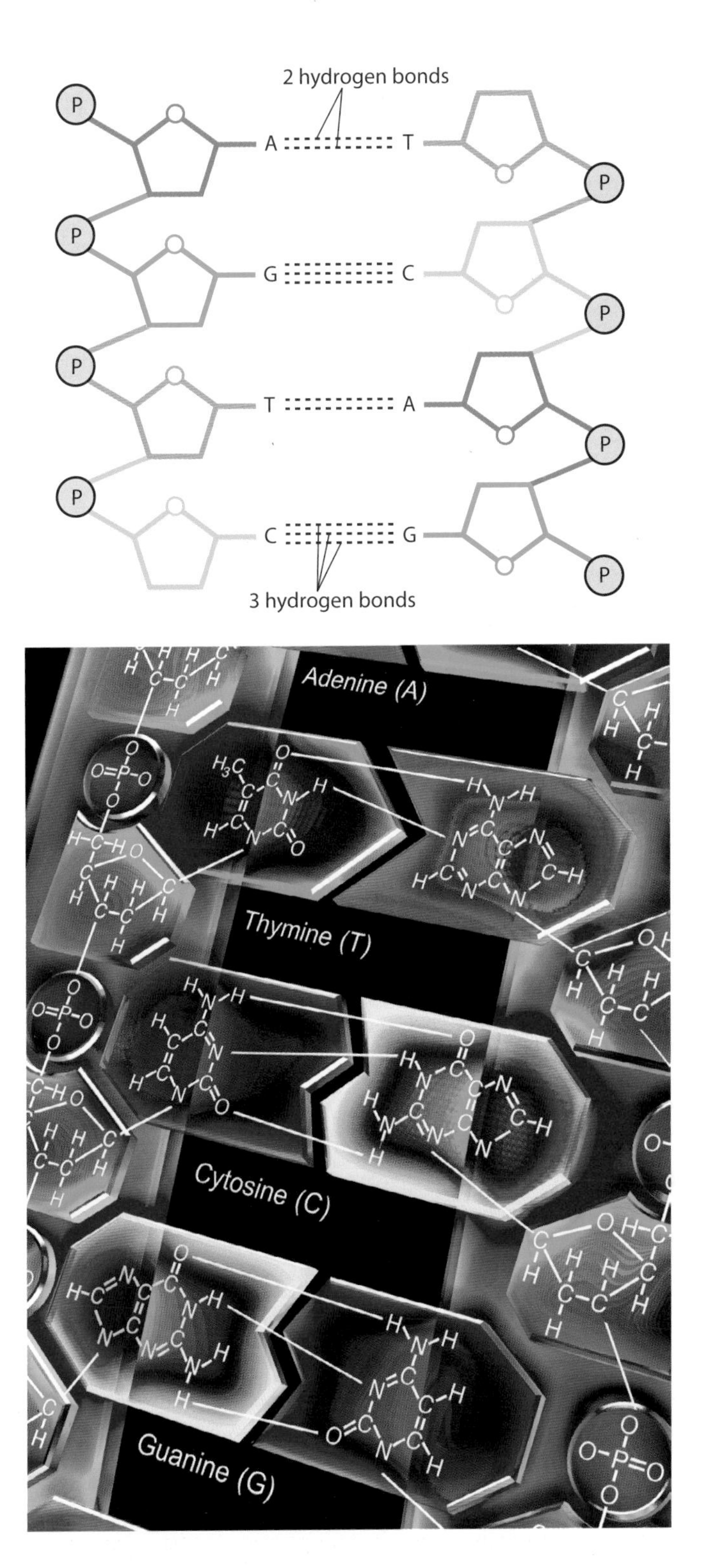

Figure 3.12 Small section of a double-stranded DNA molecule showing hydrogen bonds between complementary nitrogenous bases. The two single strands that make up the double-stranded molecule run in opposite directions to each other. The term that describes this is 'antiparallel'. Thus we say that the two strands of the double helix are antiparallel and complementary to each other.

This artwork shows complementary base pairs and hydrogen bonding in DNA. Note that thymine and cytosine are much smaller molecular structures than adenine and guanine.

W The Human Genome Project (see Chapter 4) was set up to map all the genes (the genome) of humans. If you are interested in this international project and would like to know about its results and future goals, visit heinemann.co.uk/hotlinks, insert the express code 4242P and click on Weblink 3.4.

Exercises

5 Why do researchers often give DNA information as the sequence of nitrogenous bases without indicating the presence of the phosphate group and sugar component of each nucleotide?

6 Starting with a blank piece of paper, practise drawing a ladder diagram of DNA in which the nitrogenous base sequence of one strand is C, T, G, G, A, T. Be sure to include a representation of the phosphate groups and deoxyribose sugar in each nucleotide.

3.4 DNA replication

Assessment statements

3.4.1 Explain DNA replication in terms of unwinding the double helix and separation of the strands by helicase, followed by formation of the new complementary strands by DNA polymerase.
3.4.2 Explain the significance of complementary base pairing in the conservation of the base sequence of DNA.
3.4.3 State that DNA replication is semiconservative.

DNA replication involves 'unzipping'

Helicase may catalyse the unzipping of DNA at a rate measured in hundreds of base pairs per second.

Cells must prepare for a cell division by doubling the DNA content of the cell in a process called DNA replication. This process doubles the quantity of DNA and also ensures that there is an exact copy of each DNA molecule. You should try to picture the environment in which the DNA is actually replicating. This is the environment of the nucleus during interphase of the cell cycle. During interphase, there is a nuclear membrane which separates the fluid of the nucleus (nucleoplasm) from the cytoplasm. The DNA is in the form of chromatin (not tightly coiled chromosomes). Among the variety of molecules present in the nucleoplasm are two types that are particularly important for the process of DNA replication; they are:

- enzymes needed for replication – these include helicase and a group of enzymes collectively called DNA polymerase;
- free nucleotides – these are nucleotides that are not yet bonded and are found floating freely in the nucleoplasm, some contain adenine, some thymine, some cytosine and some guanine (free nucleotides are more correctly called nucleoside triphosphates).

One of the early events of DNA replication is the separation of the double helix into two single strands. You should remember that the double helix is held together by the hydrogen bonds between complementary base pairs (A and T, C and G). The enzyme that initiates this separation into two single strands is called helicase. Helicase begins at a point in or at the end of a DNA molecule and moves one complementary base pair at a time, breaking the hydrogen bonds so the double-stranded DNA molecule becomes two separate strands.

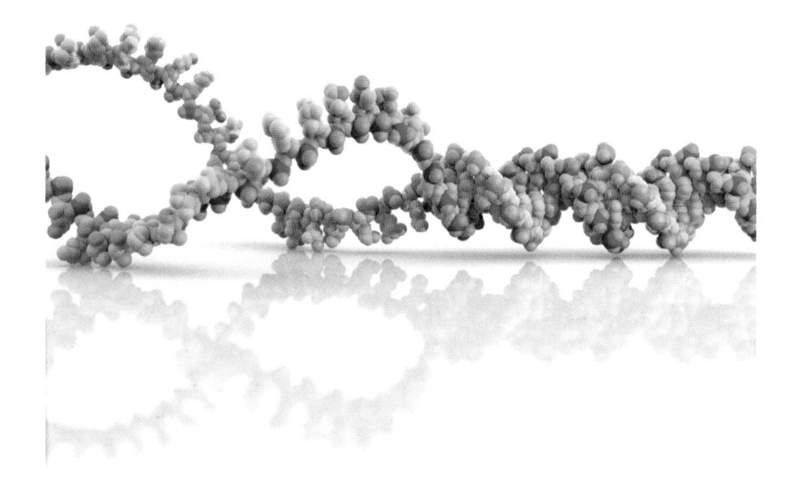

Helicase (currently at about the half-way point in this image of a DNA double helix being unzipped) would have started on the left and be moving towards the right.

The unpaired nucleotides on each of these single strands can now be used as a template to help create two double-stranded DNA molecules identical to the original. Some people use the analogy of a zipper for this process. When you pull on a zipper, the slide mechanism is like helicase. The separation of the two sides of the DNA molecule are like the two opened sides of a zipper (see Figure 3.13).

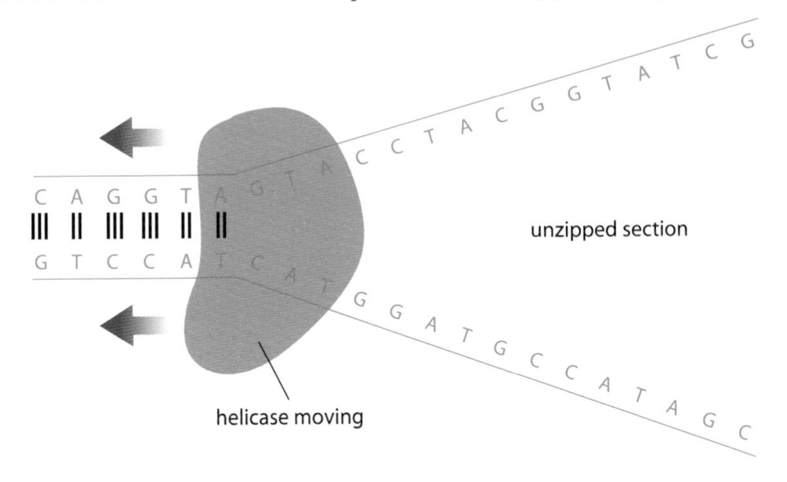

Figure 3.13 The first step of DNA replication is helicase unzipping the double-stranded DNA molecule forming a section with two single strands.

Formation of two complementary strands

As shown in Figure 3.13, once DNA has become unzipped, the nitrogenous bases on each of the single strands are unpaired. In the environment of the nucleoplasm, there are many free-floating nucleotides. These nucleotides are available to form complementary pairs with the single-stranded nucleotides of the unzipped molecule. This does not happen in a random fashion. A free nucleotide locates on one opened strand at one end and then a second nucleotide can come in to join the first. This will require that these two nucleotides become covalently bonded together as they are the beginning of a new strand. The formation of a covalent bond between two adjoining nucleotides is catalysed by one of the DNA polymerase enzymes that is important in this process.

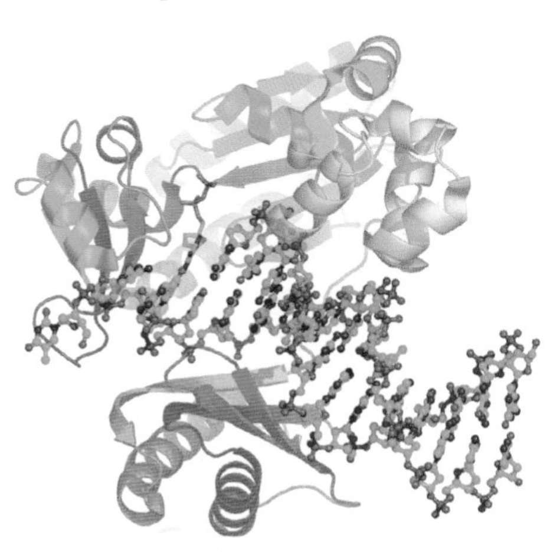

A small section of DNA (shown in the centre of this artwork) is seen in a DNA polymerase enzyme.

A third nucleotide then joins the first two and the process continues in a repetitive way for many nucleotides. The other unzipped strand also acts as a template for the formation of another new strand. This strand forms in a similar fashion, but in the opposite direction to the first strand. Notice that one strand is replicating in the same direction as helicase is moving and the other strand is replicating in the opposite direction.

Who should decide how fast and how far humans should go with our study of DNA and the technology that is rapidly emerging?

Significance of complementary base pairing

The pattern of DNA replication ensures that two identical copies of DNA are produced from one. Figure 3.14 illustrates a very small section of DNA replicating.

Figure 3.14 DNA replication

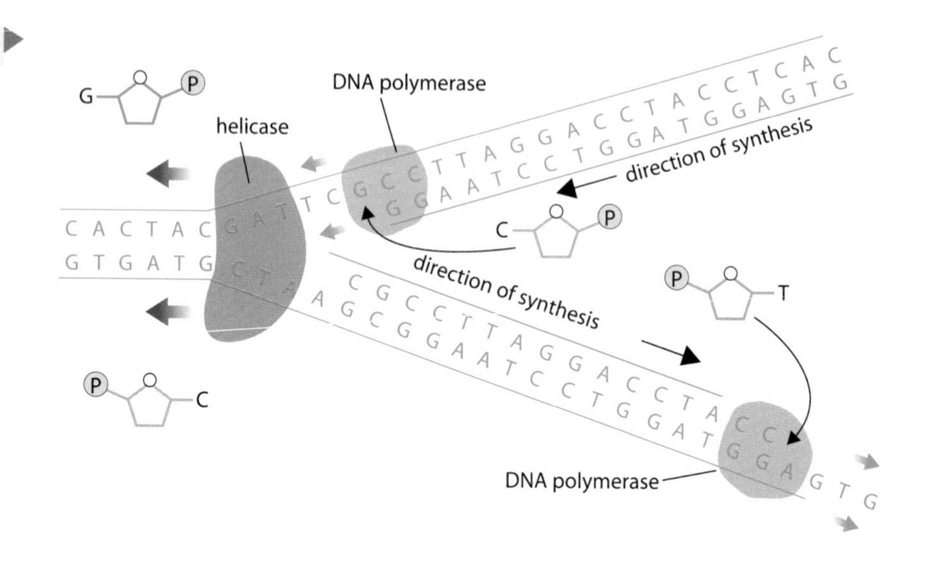

The experimental work which determined that DNA replication was semiconservative is often called 'the most beautiful experiment in biology'. To see how this classic experiment was conducted, visit heinemann.co.uk/hotlinks, insert the express code 4242P and click on Weblink 3.5.

Notice that in the area where replication has already taken place, the two strands are absolutely identical to each other. This is because the original double–stranded molecule had complementary pairs of nucleotides and it was the complementary nucleotides that used the unzipped single-stranded areas as templates.

This also means that no DNA molecule is ever completely new. Every DNA molecule after replication consists of a strand that was 'old' now paired with a strand that is 'new'. DNA replication is described as a semiconservative process because half of a pre-existing DNA molecule is always conserved (saved).

Exercises

7 The concept of semiconservative DNA replication has some interesting repercussions. For example, one can argue that there never is such a thing as a 'new' DNA molecule. How long has your DNA been in you? In your family lineage?

8 Most DNA mutations occur during DNA replication. Suggest how a mutation called a deletion could occur. Suggest how a mutation called a substitution could occur.

3.5 Transcription and translation

Assessment statements

3.5.1 Compare the structure of RNA and DNA.

3.5.2 Outline DNA transcription in terms of the formation of an RNA strand complementary to the DNA strand by RNA polymerase.

3.5.3 Describe the genetic code in terms of codons composed of triplets of bases.

3.5.4 Explain the process of translation, leading to polypeptide formation.

3.5.5 Discuss the relationship between one gene and one polypeptide.

Protein synthesis introduction

You probably took your first life science class several years ago. In that class, you probably learned that the nucleus of the cell was the 'control centre' and that the nucleus contained DNA. This information is not wrong – it is just incomplete. The control that DNA has over a cell is by a process called protein synthesis. In simplest terms, DNA controls the proteins produced in a cell. Some of the proteins produced are enzymes. The production (or lack of production) of a particular enzyme can have a dramatic effect on the overall biochemistry of the cell. Thus, DNA indirectly controls the biochemistry of carbohydrates, lipids, and nucleic acids by the production of enzymes.

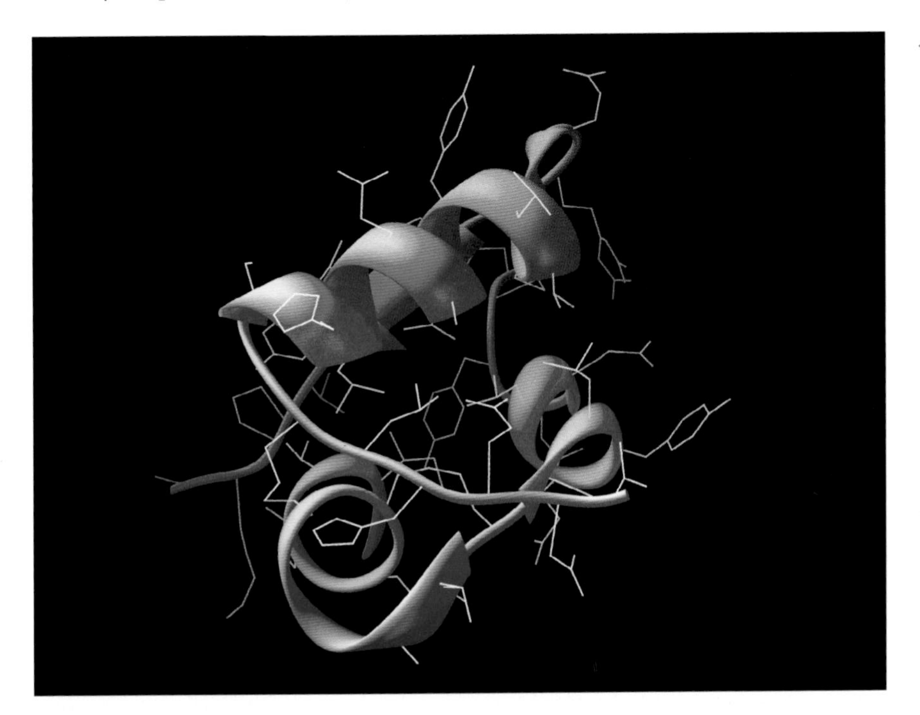

This computer graphic shows an insulin molecule. Insulin is a protein hormone and is produced by protein synthesis.

Protein synthesis has two major sets of reactions called transcription and translation. Both either produce or require a type of nucleic acid called RNA (ribonucleic acid).

This table compares DNA and RNA.

DNA	RNA
contains a 5-carbon sugar	contains a 5-carbon sugar
5-carbon sugar is deoxyribose	5-carbon sugar is ribose
each nucleotide has one of four nitrogenous bases	each nucleotide has one of four nitrogenous bases
the nitrogenous bases are cytosine, guanine, adenine, and thymine	the nitrogenous bases are cytosine, guanine, adenine, and uracil
double-stranded molecule	single-stranded molecule

• **Examiner's hint:** The command term 'compare' requires both similarities and differences. Answers may be given in the form of a table. In the table shown here, there are five comparisons (not ten).

Transcription produces RNA molecules

The sections of DNA that code for polypeptides are called genes. Any one gene is a specific sequence of nitrogenous bases found in a specific location in a DNA molecule. Molecules of DNA are found within the confines of the nucleus, yet

Is there significance to the fact that the structure of DNA is universal among all living things?
Is there further significance that all living organisms use the same genetic code?

proteins are synthesized outside the nucleus in the cytoplasm. This means there has to be an intermediary molecule which carries the message of the DNA (the code) to the cytoplasm where the enzymes, ribosomes, and amino acids are found. This intermediary molecule is called messenger RNA or mRNA.

The nucleoplasm (fluid in the nucleus) contains free nucleotides as discussed earlier. In addition to the free nucleotides used for DNA replication, the nucleoplasm also contains free RNA nucleotides. Each of these is different from its DNA counterpart as RNA nucleotides contain the sugar ribose (not deoxyribose). Another major difference is that no RNA nucleotides contain thymine, instead there is a nitrogenous base unique to RNA and called uracil.

Transcription process

The process of transcription begins when an area of DNA of one gene becomes unzipped (see Figure 3.15). This is very similar to the unzipping process involved in DNA replication, but in this case only the area of the DNA where the particular gene is found is unzipped. The two complementary strands of DNA are now single-stranded in the area of the gene. Recall that RNA (including mRNA) is a single-stranded molecule. This means that only one of the two strands of DNA will be used as a template to create the mRNA molecule. An enzyme called RNA polymerase is used as the catalyst for this process.

As RNA polymerase moves along the strand of DNA acting as the template, RNA nucleotides float into place by complementary base pairing. The complementary base pairs are the same as in double-stranded DNA, with the exception that adenine on the DNA is now paired with uracil on the newly forming mRNA molecule. Consider the following facts concerning transcription:

- only one of the two strands of DNA is 'copied', the other strand is not used;
- mRNA is always single-stranded and shorter than the DNA that it is copied from as it is a complementary copy of only one gene;
- the presence of thymine in a molecule identifies it as DNA (the presence of deoxyribose is another clue);
- the presence of uracil in a molecule identifies it as RNA (the presence of ribose is another clue).

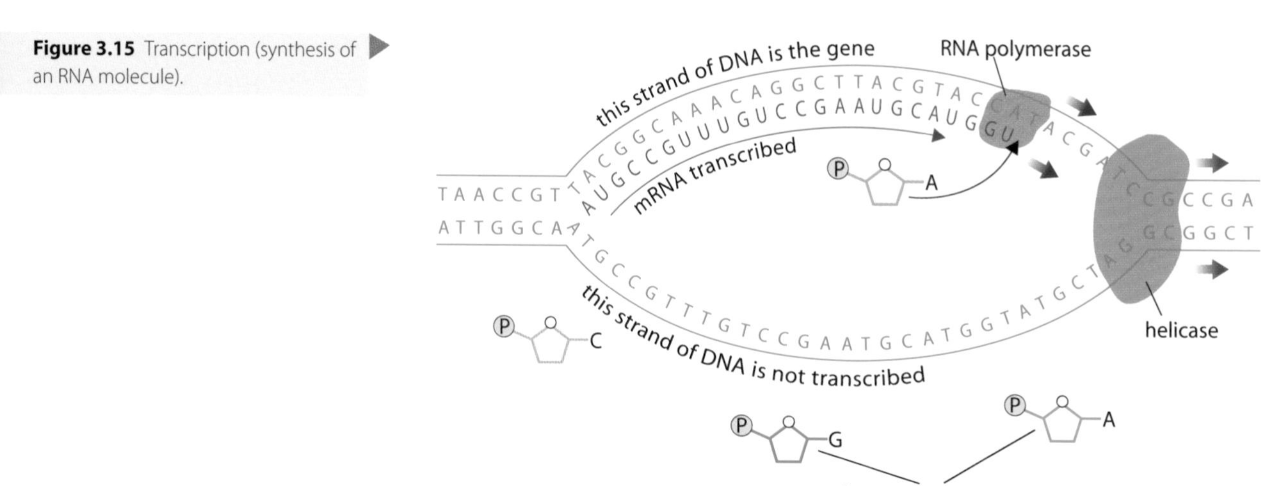

Figure 3.15 Transcription (synthesis of an RNA molecule).

The genetic code is written in triplets

The mRNA molecule produced by transcription represents a complementary copy of one gene of DNA. The sequence of mRNA nucleotides is the transcribed version of the original DNA sequence. This sequence of nucleotides making up the length of the mRNA is typically enough information to make one polypeptide. As you will recall, polypeptides are composed of amino acids covalently bonded together in a specific sequence. The message written into the mRNA molecule is the message that determines the order of the amino acids. Researchers found experimentally that the genetic code is written in a language of three bases. In other words, every three bases is enough information to code for 1 of the 20 amino acids. Any set of three bases that determines the identity of one amino acid is called a triplet. When a triplet is found in a mRNA molecule, it is called a codon or codon triplet.

Translation results in the production of a polypeptide

There are three different kinds of RNA molecules. They are all single-stranded and each is transcribed from a gene of DNA.

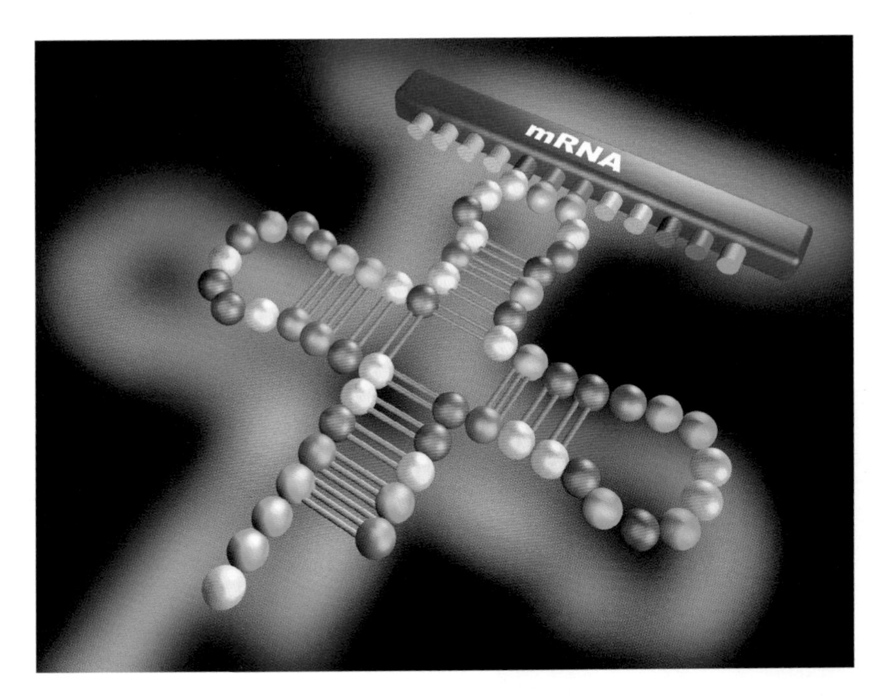

In this model, you can see mRNA (upper right) and tRNA (clover shape). The amino acid which would be bonded to the tRNA is not shown.

Here is a quick summary of each RNA type:

- mRNA – as described above, each mRNA is a complementary copy of a DNA gene and is enough genetic information to code for a single polypeptide;
- rRNA – ribosomal RNA, each ribosome is composed of rRNA and ribosomal protein;
- tRNA – transfer RNA, each type of tRNA transfers 1 of the 20 amino acids to the ribosome for polypeptide formation.

Figure 3.16 (overleaf) shows a typical tRNA molecule. Notice that the three bases in the middle loop are called the anticodon bases and they determine which of the 20 amino acids is attached to the tRNA.

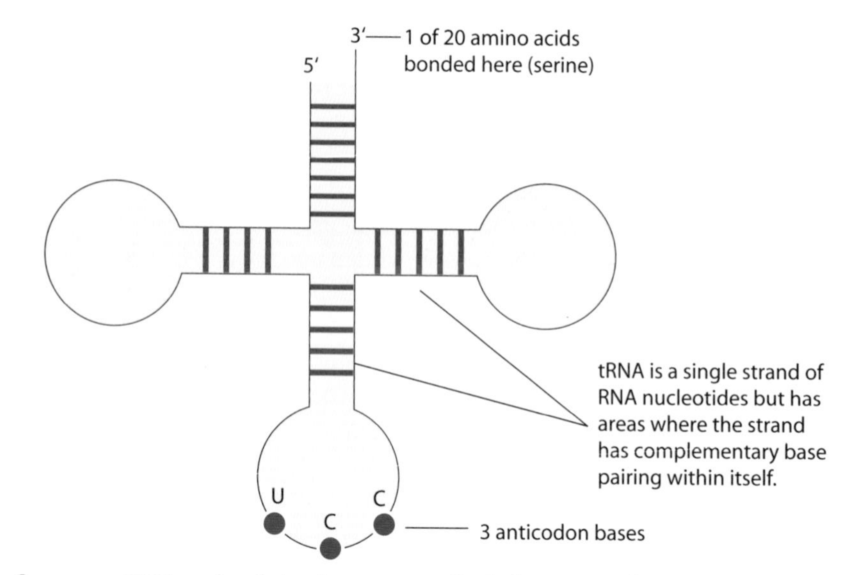

Figure 3.16 Structure of a tRNA molecule.

Once an mRNA molecule has been transcribed, the mRNA detaches from the single-strand DNA template and floats free in the nucleoplasm. At some point, the mRNA will float through one of the many holes in the nuclear membrane (nuclear pores) and will then be in the cytoplasm.

Translation process

The mRNA will locate a ribosome and align with it so that the first two codon triplets are within the boundaries of the ribosome.

A specific tRNA molecule now floats in – its tRNA anticodon must be complementary to the first codon triplet of the mRNA molecule. Thus, the first amino acid is brought into the translation process. It is not just any amino acid – its identity was originally determined by the strand of DNA that transcribed the mRNA being translated. While the first tRNA 'sits' in the ribosome holding the first amino acid, a second tRNA floats in and brings a second (again specific) amino acid. The second tRNA matches its three anticodon bases with the second codon triplet of the mRNA. As you can see in Figure 3.17, two specific amino acids are now being held side by side. An enzyme now catalyses a condensation reaction between the two amino acids and the resulting covalent bond between them is called a peptide bond.

This image shows a DNA molecule overlaid on a chart showing the triplet genetic code.

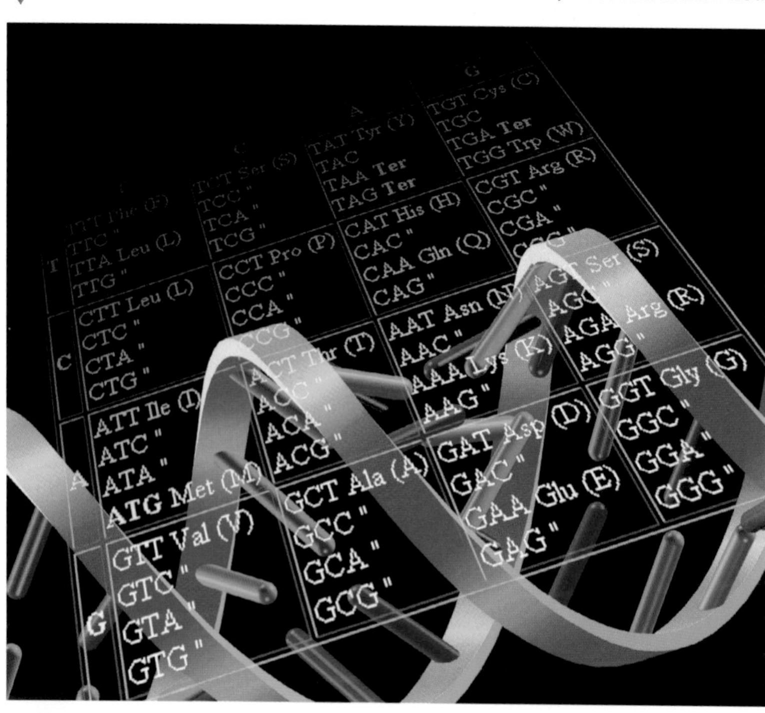

The next step in the translation process involves the breaking of the bond between the first tRNA molecule and the amino acid that it transferred in. This bond is no longer needed as the second tRNA is currently bonded to its own amino acid and that amino acid is covalently bonded to the first amino acid. The first tRNA floats away into the cytoplasm and invariably reloads with another amino acid of the same type. The ribosome that has only one tRNA in it now moves one codon triplet down the mRNA molecule. This, in effect, puts the second tRNA in the ribosome position that the first originally occupied and creates room for a third tRNA to float in bringing with it a third specific amino acid. The process now becomes somewhat repetitive as another peptide bond forms, the ribosome moves on by another triplet and so on. The process continues until the ribosome gets to the last codon triplet. The final codon triplet will be a triplet that does not act as a code for an amino acid, it signals 'stop' to the process of translation. The entire polypeptide breaks away from the final tRNA molecule, and becomes a free floating polypeptide in the cytoplasm of the cell.

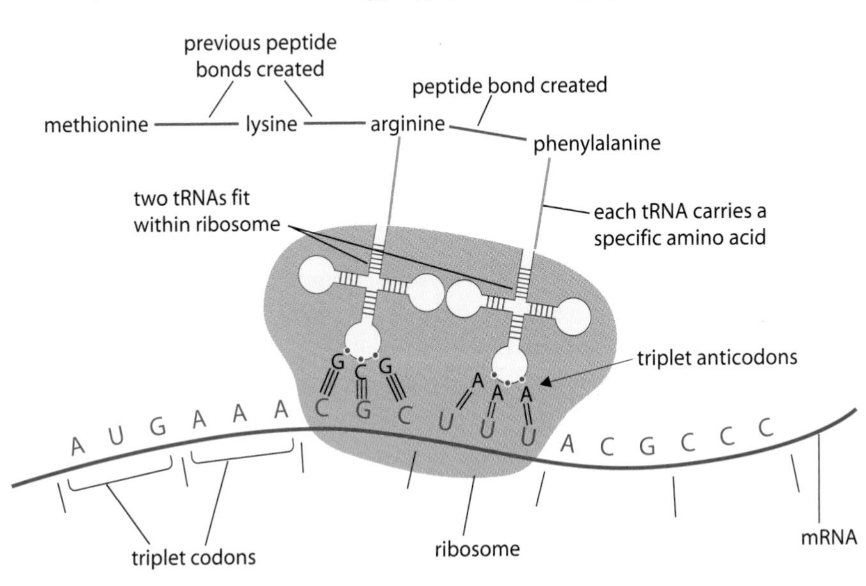

Figure 3.17 Events of translation (synthesis of a polypeptide).

The one gene/one polypeptide hypothesis

In the early 1940s, experimental work was performed which led to the hypothesis that every one gene of DNA produced one enzyme. This was soon amended to include all proteins and not just enzymes. It was later discovered that many proteins are actually composed of more than one polypeptide and it was proposed that each individual polypeptide required a separate gene. Thus for many years students learned the 'one gene, one polypeptide' hypothesis.

In the last few years, researchers have discovered that at least some genes are not quite that straightforward. For example, one gene may lead to a single mRNA molecule, but the mRNA molecule may then be modified in many different ways. Each modification may result in the production of a different polypeptide during the translation portion of protein synthesis.

As the Human Genome Project progressed, the number of genes thought to make up a human being dropped dramatically. Early estimates were about 100 000 genes; current estimates are below 30 000 genes. One of the reasons for this change in thinking is the presence of genes that produce mRNA which is modified in different ways to produce different polypeptides.

Exercises

Imagine that an mRNA leaves the nucleus of a eukaryotic cell with the following base sequence:
AUGCCCCGCACGUUUCCAAGCCCCGGG
Locate an mRNA codon chart. If you do not have a text that includes a codon chart, you can access one if you visit heinemann.co.uk/hotlinks, insert the express code 4242P and click on Weblink 3.6.

9 Determine the amino acids in sequence that are coded for by the above mRNA molecule.

10 Determine the DNA code sequence which gave rise to the above mRNA codons.

11 What would the amino acid sequence be if the first cytosine of the mRNA molecule was replaced with a uracil? (This would be due to a substitution mutation occurring to the DNA molecule which transcribed this mRNA.)

3.6 Enzymes

Assessment statements

3.6.1 Define *enzyme* and *active site*.
3.6.2 Explain enzyme–substrate specificity.
3.6.3 Explain the effects of temperature, pH and substrate concentration on enzyme activity.
3.6.4 Define *denaturation*.
3.6.5 Explain the use of lactase in the production of lactose-free milk.

Enzymes are organic molecules which act as catalysts

Enzymes are proteins. Thus, enzymes are long chains of amino acids that have taken on a very specific three-dimensional shape. Think of a flexible metal wire that someone bends many times into what is called a globular shape. This shape is complex and at first glance appears to be random, but in enzymes (and other proteins) the complex shape is not random: it is very specific. Somewhere in the three-dimensional shape of the enzyme is an area that is designed to match a specific molecule known as that enzyme's substrate. This area of the enzyme is called the active site. The active site of an enzyme matches the substrate in a similar way to the way a glove fits over a hand. In this analogy, the glove represents the active site and the hand represents the substrate.

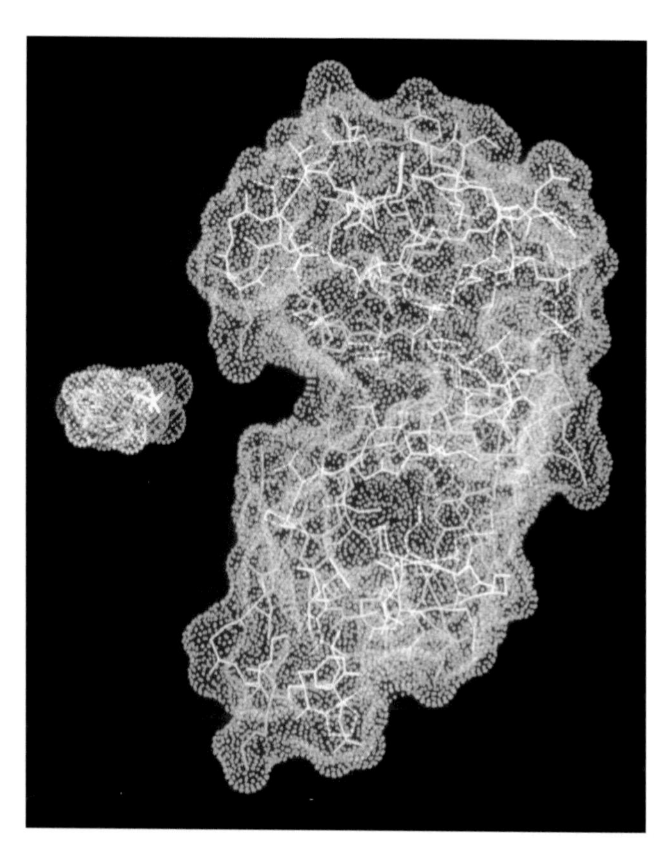

This computer graphic image shows an enzyme (larger molecule on right) and its substrate. Notice the active site on the left side of the enzyme.

Another analogy that is very commonly used for enzyme–substrate activity is a lock and key. In this analogy, the lock represents the enzyme's active site and the key represents the substrate. Because the three-dimensional shape of the internal portion of the lock is complex and specific, only one key will fit. The same principle is generally true for enzymes and their substrates – they are specific for each other.

As catalysts, enzymes influence the rate of reactions. As a general rule, a set of reactants in the presence of an enzyme will form product(s) at a faster rate than without the enzyme. Enzymes cannot force reactions to occur that would not otherwise occur. The real role of an enzyme in a reaction is to lower the energy level needed to start the reaction. This energy is referred to as the activation energy of the reaction. Thus, enzymes lower the activation energy of reactions. Enzymes are not considered reactants and are not used up in the reaction. An enzyme can function as a catalyst many, many times.

Factors affecting enzyme-catalysed reactions

These reaction rates are affected by temperature, pH and substrate concentration.

Effect of temperature

Imagine an enzyme and its substrate floating freely in a fluid environment. Both the enzyme and substrate are in motion and the rate of that motion is dependent on the temperature of the fluid. Fluids with higher temperatures will have faster-moving molecules (more kinetic energy). Reactions are dependent on molecular collisions and, as a general rule, the faster molecules are moving the more often they collide and with greater energy. Reactions with or without enzymes will increase their reaction rate as temperature (and thus molecular motion) increases. Reactions which use enzymes do have an upper limit however (see Figure 3.18). That limit is based on the temperature at which the enzyme (as a protein) begins to lose its three-dimensional shape due to intramolecular bonds being stressed and broken. When an enzyme loses its shape, including the shape of the active site, it is said to be denatured. Denaturation is sometimes permanent and sometimes only temporary until the molecule re-forms its normal shape.

Whether or not an enzyme is permanently destroyed by denaturation is largely dependent on whether the covalent bonds, such as peptide bonds, in the molecule have broken. DNA determined the order of these amino acids and, without the control of DNA, the amino acids have no way of reattaching in the original order.

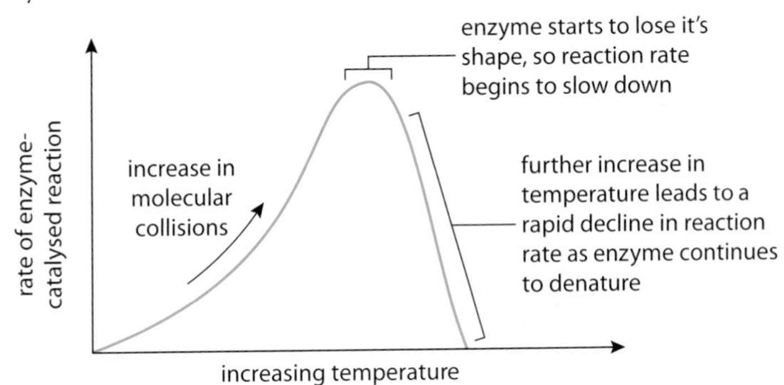

Figure 3.18 The effect of increasing temperature on the rate of an enzyme-catalysed reaction.

Effect of pH

The pH of a solution is dependent on the relative number of hydrogen ions (H^+) compared to hydroxide ions (OH^-) in the same solution. Any substance that gives off hydrogen ions, for example HCl, is an acid and results in a solution with a pH below 7. Any substance that gives off hydroxide ions, for example NaOH, is a base and results in a solution with a pH higher than 7. Pure water has a neutral pH of 7 because when water dissociates (splits), it results in an equal number of hydrogen ions and hydroxide ions.

The pH scale is a logarithmic scale. This means that each whole number on the pH scale represents an increase or decrease in a power of 10. Thus, a solution with a pH of 4 has ten times more hydrogen ions than one with a pH of 5. A solution with a pH of 4 has 100 times more hydrogen ions than one with a pH of 6.

The active site of an enzyme typically includes many amino acids of that protein. Some amino acids have areas that are charged either positively or negatively. The negative and positive areas of a substrate must match the opposite charge when the substrate is in the active site of an enzyme in order for the enzyme to have catalytic action. When a solution has become too acidic, the relatively large number of hydrogen ions (H^+) can bond with the negative charges of the enzyme or substrate and not allow proper charge matching between the two. A similar scenario occurs when a solution has become too basic; the relatively large number of hydroxide ions (OH^-) can bond with the positive charges of the substrate or enzyme and once again not allow proper charge matching between the two. Either of these scenarios will result in an enzyme becoming less efficient and sometimes becoming completely inactive in extreme situations. One further possibility is that the numerous extra positive and negative charges of acidic and basic solutions can result in the enzyme losing its shape and thus becoming denatured.

There is no one pH that is best for all enzymes (see Figure 3.19). Many of the enzymes active in the human body are most active when in an environment that is near neutral. There are exceptions to this; for example, pepsin is an enzyme that is active in the stomach. The environment of the stomach is highly acidic and pepsin is most active in an acidic pH.

Figure 3.19 The effect of pH on the rate of an enzyme-catalysed reaction.

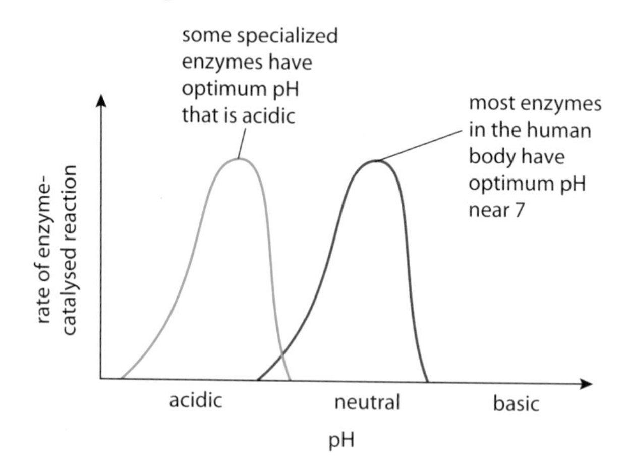

Effect of substrate concentration

If there is a constant amount of enzyme, as the concentration of a substrate increases, the rate of reaction will increase as well (see Figure 3.20). This is explained by the idea of increased molecular collisions. If you have more reactant molecules, there are more to collide. There is a limit to this however. The limit

Figure 3.20 The effect of increasing substrate concentration on the rate of an enzyme-catalysed reaction.

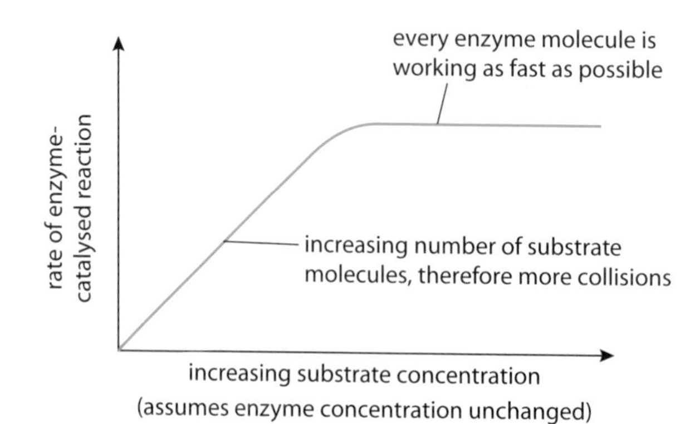

is due to the fact that enzymes have a maximum rate at which they can work. If every enzyme molecule is working as fast as possible, adding more substrate to the solution will not further increase the reaction rate (see Figure 3.20).

Use of lactase to help solve the problem of lactose intolerance

Almost all humans on Earth are born with the ability to digest lactose, one of the most common sugars found in milk. The reason for this is that we are born with the ability to produce the enzyme lactase in our digestive tract. Lactase is the enzyme that digests the disaccharide lactose into two monosaccharides. The monosaccharides are much more readily absorbed into the bloodstream. Most people on Earth lose the ability to produce lactase as they get older, and by adulthood no longer produce any significant amount of lactase. Normal milk and milk products enter their digestive tract and are not digested; instead bacterial colonies in their intestines feed directly on the lactose. This leads to such symptoms as cramping, excessive gas, and diarrhoea.

There are more people on Earth with lactose intolerance than there are without it. Thus, it can be argued that lactose intolerance has become the norm.

Milk and milk products can be treated with lactase before consumption. When this treatment occurs, the nutrients in the milk are not affected, but the person is able to absorb the sugars as they have been predigested. The technology to treat milk and milk products in this way must be improved in order for it to be useful on a large scale. Lactose intolerance has been shown to have an extremely high incidence in some ethnic groups and be relatively low in others. This is a good example of natural variation in a population.

Exercises

12 From memory, draw each of the following with proper axes:

a graph showing effect of increasing temperature on the rate of reaction of an enzyme-catalysed reaction;

b graph showing effect of substrate concentration on the rate of reaction of an enzyme-catalysed reaction.

13 In your own words, explain why each of the above graphs has the shape it has. Check the validity of your graphs before offering an explanation.

3.7 Cell respiration

Assessment statements

3.7.1 Define *cell respiration.*

3.7.2 State that, in cell respiration, glucose in the cytoplasm is broken down by glycolysis into pyruvate, with a small yield of ATP.

3.7.3 Explain that, during anaerobic cell respiration, pyruvate can be converted in the cytoplasm into lactate, or ethanol and carbon dioxide, with no further yield of ATP.

3.7.4 Explain that, during aerobic cell respiration, pyruvate can be broken down in the mitochondrion into carbon dioxide and water with a large yield of ATP.

Cell respiration is used by all cells to produce ATP

Organic molecules contain energy in their molecular structures. Each covalent bond in a glucose, amino acid or fatty acid represents stored chemical energy. When we burn wood in a fire, we are releasing that stored chemical energy in the form of heat and light. Burning is the release of chemical energy called rapid oxidation. Rapid oxidation is not controlled by enzymes and results in the breaking of many, many covalent bonds in a very short period of time and thus a nearly uncontrolled energy release.

Cells break down (or metabolize) their organic nutrients by way of slow oxidation. A molecule, such as glucose, is acted on by a series of enzymes. The function of these enzymes is to catalyse a sequential series of reactions in which the covalent bonds are broken (oxidized) one at a time. Each time a covalent bond is broken, a small amount of energy is released. The ultimate goal of releasing energy in a controlled way is to trap the released energy in the form of ATP molecules. If a cell does not have glucose available, other organic molecules may be substituted, such as fatty acids or amino acids.

This is a computer graphic of glucose. The backbone of the molecule is shown in stick form. The spheres represent the relative sizes of the individual atoms ($C_6H_{12}O_6$).

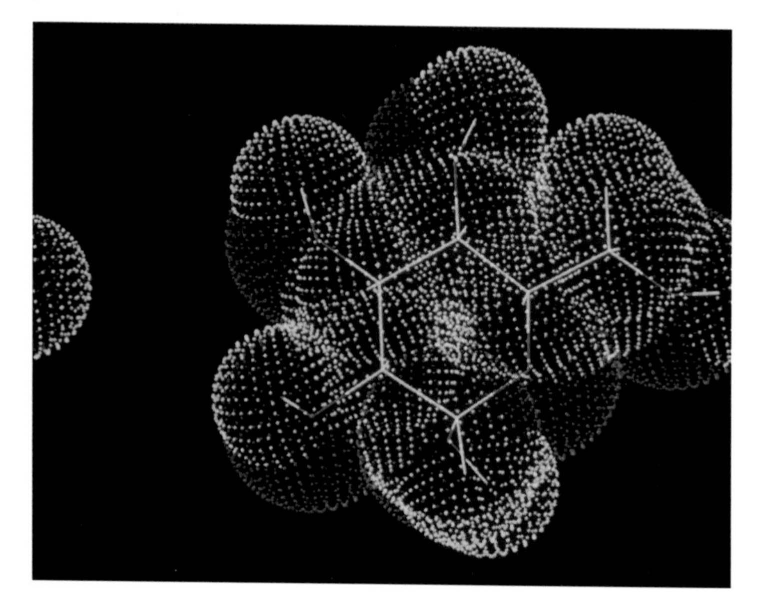

Glycolysis is the first step in the cell respiration process

Assuming that glucose is the organic nutrient being metabolized, all cells begin the process of cell respiration in the same way. Glucose enters a cell through the plasma membrane and floats in the cytoplasm. An enzyme modifies the glucose slightly, then a second enzyme modifies this molecule even more. This is followed by an entire series of reactions which ultimately cleave the 6-carbon glucose into two 3-carbon molecules. Each of these 3-carbon molecules is called pyruvate. Certainly not all, but some, of the covalent bonds in the glucose were broken during this series of reactions. Some of the energy that was released from the breaking of these bonds was used to form a small number of ATP molecules. Notice in Figure 3.21 that two ATP molecules are needed to begin the process of glycolysis and a total of four ATP molecules are formed. This is referred to as a net gain of two ATP (gain of four ATP minus the two ATP needed to start).

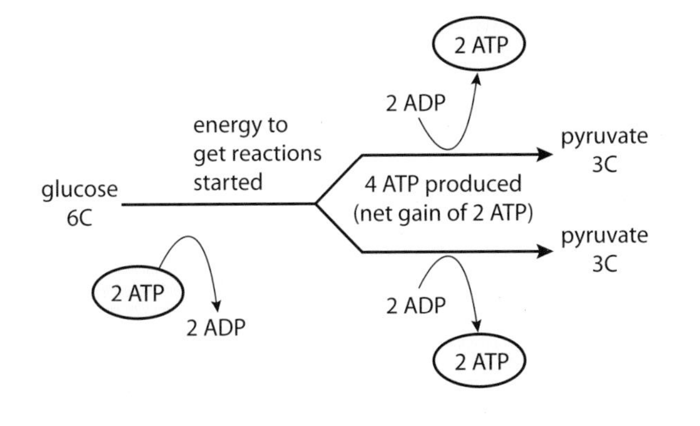

Figure 3.21 Simplified version of the events of glycolysis.

Some cells use anaerobic respiration for ATP production

The term 'cell respiration' refers to a variety of biochemical pathways that can be used to metabolize glucose. All of the pathways start with glycolysis. In other words, glycolysis is the metabolic pathway that is common to all organisms on Earth. Some organisms derive their ATP completely without the use of oxygen and are referred to as anaerobic. The breakdown of organic molecules for ATP production in an anaerobic way is also called fermentation. There are two main anaerobic pathways which will be discussed separately: alcoholic fermentation and lactic acid fermentation.

All alcohol for drinking is ethanol. Beer, wine, and spirits contain different proportions of ethanol and other ingredients for flavouring.

Alcoholic fermentation

Yeast is a common single-celled fungus that uses alcoholic fermentation for ATP generation (see Figure 3.22). You will recall that all organisms use glycolysis to begin the cell respiration sequence. Thus, yeast cells take in glucose from their environment and generate a net gain of two ATP by way of glycolysis. The organic products of glycolysis are always two pyruvate molecules. Yeast then converts both of the 3-carbon pyruvate molecules to molecules of ethanol. Ethanol is a 2-carbon molecule, so a carbon atom is 'lost' in this conversion. The 'lost' carbon atom is given off in a carbon dioxide molecule. Both the ethanol and carbon dioxide that are produced are waste products to the yeast and are simply given off into the environment. Bakers' yeast is added to bread products for baking as the generation of carbon dioxide helps the dough to rise. It is also common to use yeast in the production of ethanol as drinking alcohol.

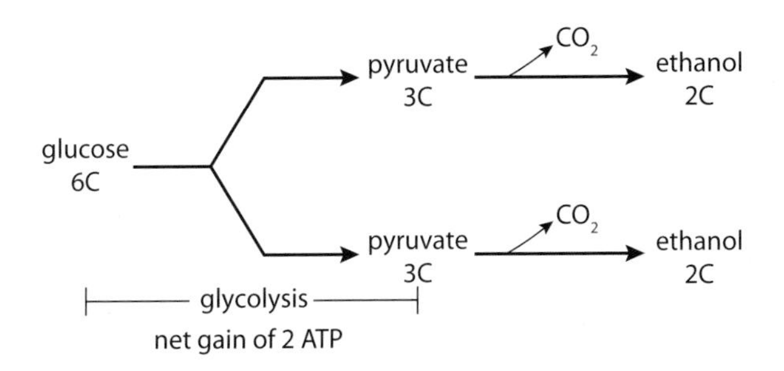

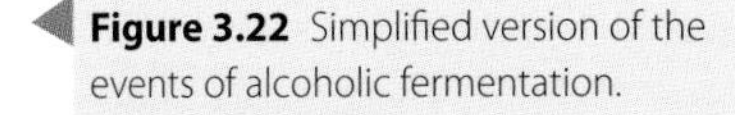

Figure 3.22 Simplified version of the events of alcoholic fermentation.

Lactic acid fermentation

Organisms that use an aerobic cell respiration pathway sometimes find themselves in a metabolic situation where they cannot supply enough oxygen to their cells. A good example of this is a person pushing beyond their normal exercise pattern

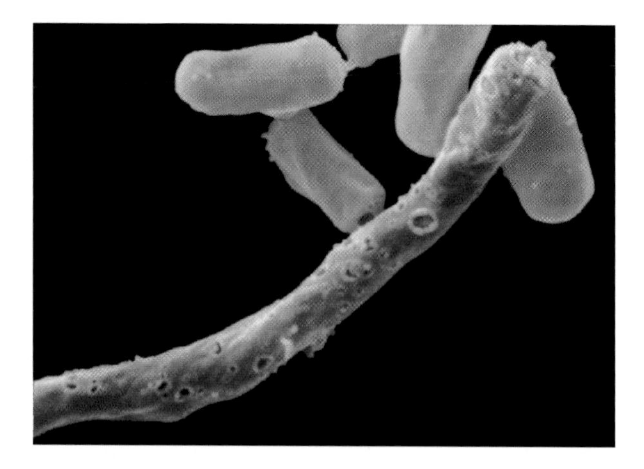

This false-colour SEM shows an anaerobic species of bacterium. It is *Clostridium perfringens* (×14 065).

or routine. In this situation, the person's pulmonary and cardiovascular systems (lungs and heart) supply as much oxygen to their cells as is physically possible. If the person's exercise rate exceeds their capability of supplying oxygen, then at least some of the glucose entering into cell respiration will follow the anaerobic pathway called lactic acid fermentation (see Figure 3.23).

Once again, recall that glycolysis is used by all cells to begin the cell respiration sequence. Also remember that glycolysis:

- takes place in the cytoplasm;
- results in the net gain of two ATP per glucose;
- results in the production of two pyruvate molecules.

Cells that are aerobic normally take the two pyruvate molecules and further metabolize them in an aerobic series of reactions. But this is not a normal situation, this is a cell that is not getting a sufficient amount of oxygen for the aerobic pathway. In a low-oxygen situation, excess pyruvate molecules are converted into lactic acid molecules. Like pyruvate, lactic acid molecules are 3-carbon molecules so there is no production of carbon dioxide. What benefit does this serve? It allows glycolysis to continue with the small gain of ATP generated in addition to the ATP which is already being generated through the aerobic pathway.

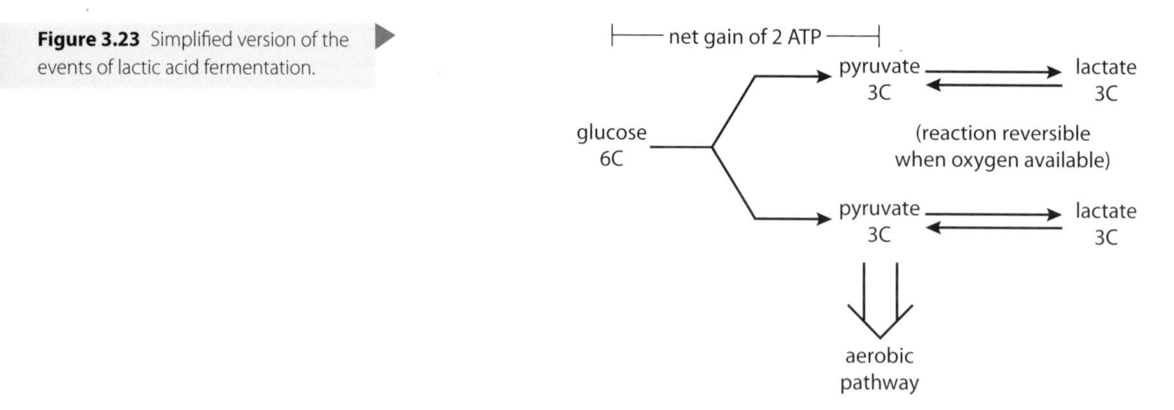

Figure 3.23 Simplified version of the events of lactic acid fermentation.

Aerobic cell respiration is the most efficient pathway

Cells that have mitochondria use an aerobic pathway for cell respiration. This pathway also begins with glycolysis and thus a net gain of two ATP is generated as well as two pyruvate molecules. The two pyruvate molecules now enter a mitochondrion and are further metabolized.

Mitochondria have their own DNA and ribosomes.

Each pyruvate first loses a carbon dioxide molecule and becomes a molecule known as acetyl-CoA. Each acetyl-CoA enters into a series of reactions called the Krebs cycle. During this series of reactions, two more carbon dioxide molecules are produced from each original pyruvate that entered. The Krebs cycle series of reactions is a cycle because each time it returns to the molecule that once again reacts with another incoming acetyl-CoA molecule (see Figure 3.24).

This high-resolution, false-colour SEM shows a single mitochondrion. Any cell containing mitochondria uses aerobic cell respiration as its primary cell respiration pathway.

Some ATP is directly generated during the Krebs cycle and some is indirectly generated through a later series of reactions directly involving oxygen. Aerobic cell respiration breaks down (or completely oxidizes) a glucose molecule and the end-products are carbon dioxide and water. The reason aerobic cell respiration is so much more efficient than anaerobic cell respiration is that anaerobic pathways do not completely oxidize the glucose molecule. This explains why ethanol and lactic acid are generated – both ethanol and lactic acid represent portions of the original glucose that were not oxidized. Aerobic cell respiration leaves no such by-products and results in a yield of ATP per glucose that is much higher than either of the anaerobic pathways.

The rate of cell respiration in an aerobic organism is measured by a respirometer. This device works by measuring the oxygen intake of an animal while the carbon dioxide output is eliminated by the use of a chemical. This causes a drop in the partial pressure in the device and, when put under water, the rate at which water enters the device is a reflection of the oxygen consumed. To view a respirometer (scroll down to see the pictures): visit heinemann.co.uk/hotlinks, insert the express code 4242P and click on Weblink 3.7.

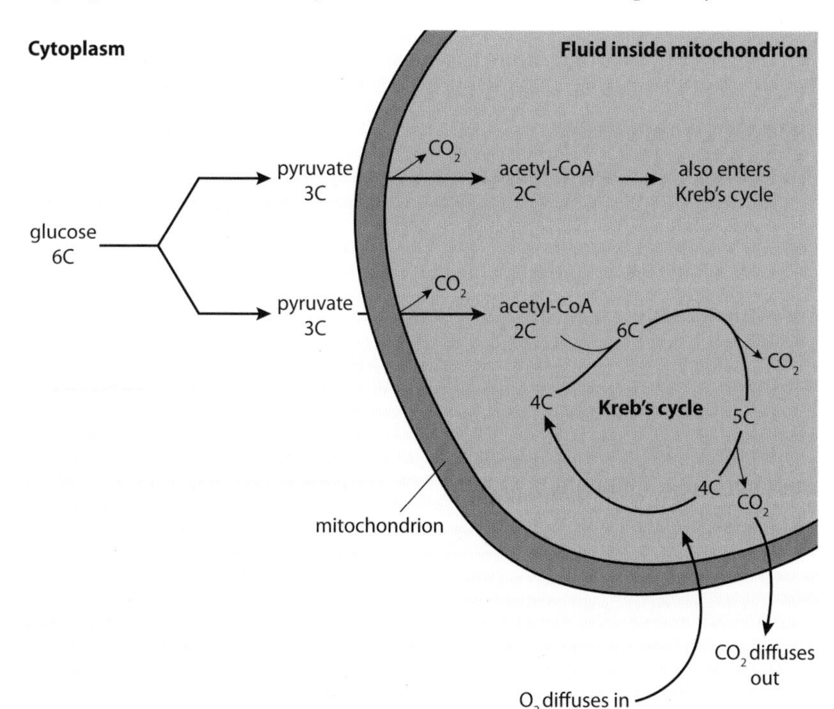

Figure 3.24 Aerobic cell respiration.

Exercises

14 Which stage of cell respiration is common to all types of cell respiration?

15 Where does this stage of cell respiration occur in a cell?

16 Why does that make sense?

17 Why do we inhale oxygen and exhale carbon dioxide?

3.8 Photosynthesis

Assessment statements

3.8.1 State that photosynthesis involves the conversion of light energy into chemical energy.

3.8.2 State that light from the Sun is composed of a range of wavelengths (colours).

3.8.3 State that chlorophyll is the main photosynthetic pigment.

3.8.4 Outline the differences in absorption of red, blue and green light by chlorophyll.

3.8.5 State that light energy is used to produce ATP, and to split water molecules (photolysis) to form oxygen and hydrogen.

3.8.6 State that ATP and hydrogen (derived from the photolysis of water) are used to fix carbon dioxide to make organic molecules.

3.8.7 Explain that the rate of photosynthesis can be measured directly by the production of oxygen or the uptake of carbon dioxide, or indirectly by an increase in biomass.

3.8.8 Outline the effects of temperature, light intensity and carbon dioxide concentration on the rate of photosynthesis.

Photosynthesis converts light energy into chemical energy

Plants and other photosynthetic organisms produce foods that begin food chains. We count on the Sun to be a constant energy source for both warmth and food production for all of planet Earth. The sunlight that strikes our planet must be converted into a form of chemical energy in order to be useful to all non-photosynthetic organisms. The most common chemical energy produced from photosynthesis is the molecule glucose. If you recall, glucose is also the most common molecule that organisms use for fuel in the process of cell respiration.

Plants use the pigment chlorophyll to absorb light energy

Leaves contain a mixture of different photosynthetic pigments (not just chlorophyll). This mixture of pigments can be separated using a technique called paper chromatography. To see how, visit heinemann.co.uk/hotlinks, insert the express code 4242P and click on Weblink 3.8.

The vast majority of plant leaves appear green to our eyes. If you were able to zoom into leaf cells and look around, you would see that the only structures in a leaf that are actually green are the chloroplasts. Plants contain a variety of pigments in chloroplasts. The photosynthetic pigment that dominates in most plant species is the molecule chlorophyll.

Plants make use of the same part of the electromagnetic spectrum that our eyes are able to see. We call this the visible portion of the spectrum. Sunlight is actually a mixture of different colours of light. You can see these colours when you let sunlight pass through a prism. The prism separates the colours because each of the colours is a different wavelength and is refracted in the prism to a slightly different angle. Water droplets are natural prisms, which is why the different colours of sunlight can be seen in a rainbow.

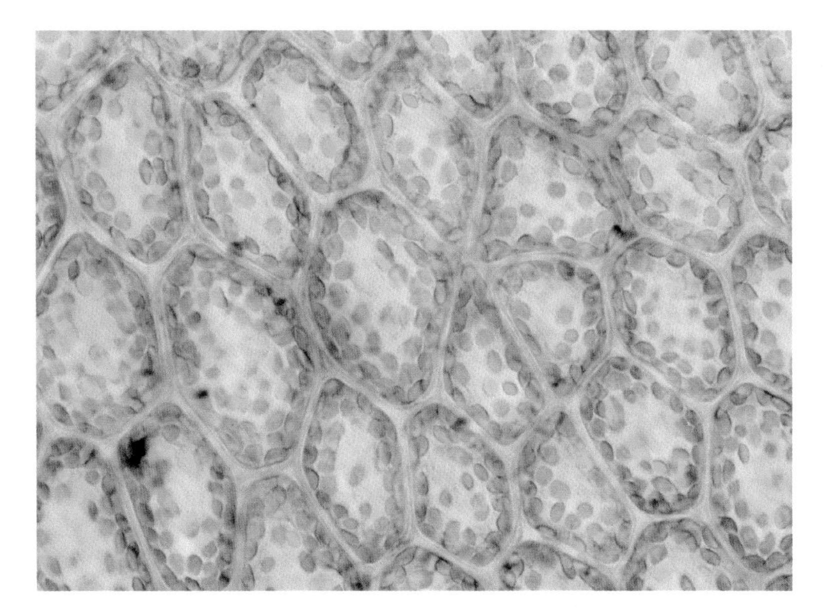

Inside each of these plant leaf cells are many green chloroplasts. Each chloroplast is loaded with chlorophyll.

The visible light spectrum includes many colours, but, for the purpose of considering how chlorophyll absorbs light energy, we are going to consider three regions of the spectrum:

- red end of spectrum;
- green middle of spectrum;
- blue end of spectrum.

Substances can do one of only two things when they are struck by a particular wavelength (colour) of light; they can:

- absorb that wavelength (if so, energy is being absorbed and may be used);
- reflect that wavelength (if so, the energy is not being absorbed and you will see that colour).

The advice to wear light-coloured clothing in warm months is good advice based on the principle that lighter colours reflect more energy and therefore keep you cooler.

Worked example

You are walking outside with a friend who is wearing a red and white shirt. Explain why the shirt appears red and white.

Solution

Sunlight is a mixture of all of the wavelengths (colours) of visible light. When sunlight strikes the red pigments in the shirt, the blue and the green wavelengths of light are absorbed, but the red wavelengths are reflected. Thus, our eyes see red. When sunlight strikes the white areas of the shirt, all the wavelengths of light are reflected and our eyes and brain interpret this mixture as white.

Now, let's apply this information to how chlorophyll absorbs light for photosynthesis. Chlorophyll is a green pigment. This means that chlorophyll reflects green light and therefore must absorb the other wavelengths of the visible light spectrum. When a plant leaf is hit by sunlight, the red and blue wavelengths of light are absorbed by chlorophyll and used for photosynthesis. Almost all the energy of the green wavelengths is reflected, not absorbed. So, do not try to grow plants in only green light.

Photosynthesis occurs in two stages

Photosynthesis produces sugar molecules as a food source for the plant. Sugars, such as glucose, are held together by covalent bonds. It requires energy to create those covalent bonds and the source of that energy can ultimately be traced back to the Sun.

The first stage of photosynthesis is a set of reactions that 'trap' light energy and convert it to the chemical energy of ATP. The second stage of photosynthesis is a set of reactions in which ATP is used to help bond carbon dioxide and water molecules together to create a sugar, such as glucose.

The first stage of photosynthesis

The first stage of photosynthesis is a set of reactions typically referred to as the light-dependent reactions (see Figure 3.25). In this set of reactions, chlorophyll (and other photosynthetic pigments) absorb light energy and convert that energy to a form of chemical energy, specifically ATP. In addition, light energy is also used to accomplish a reaction that is called photolysis of water. In this reaction, a water molecule is split into its component elements: hydrogen and oxygen.

Figure 3.25 Functions of light during photosynthesis.

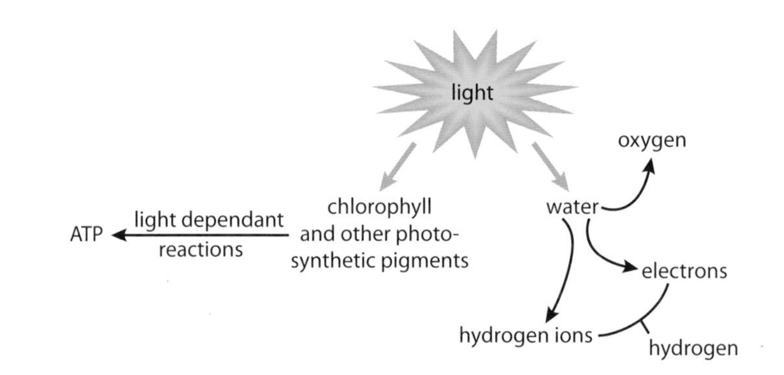

This is an SEM (with false colour added) of an upper leaf section. These cells are very active in photosynthesis as is shown by the large number of chloroplasts.

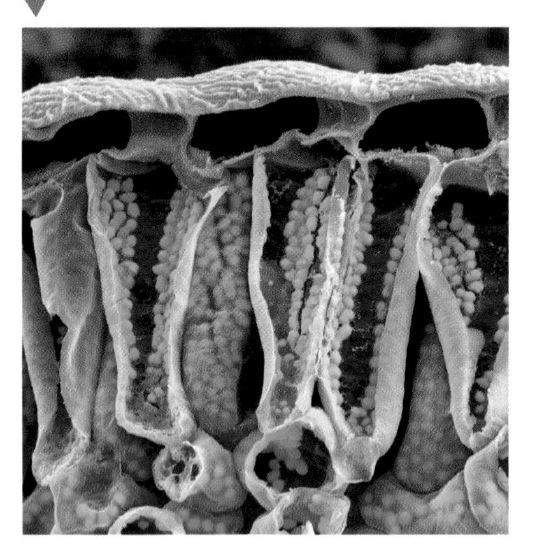

The oxygen that is split away due to the photolysis of water is typically released from the plant leaf as a waste product. It is very useful for us and the millions of different organisms that need oxygen for aerobic cell respiration. However, from the plant's perspective, the useful products formed during this stage of photosynthesis are ATP and hydrogen.

The second stage of photosynthesis

The second stage of photosynthesis is a series of reactions collectively referred to as the light-independent reactions. ATP and hydrogen are used as forms of chemical energy to convert carbon dioxide and water into useful organic molecules for the plant. Carbon dioxide is one of the few molecular forms of carbon that is considered to be inorganic. Glucose, a typical product of photosynthesis, is an organic molecule. It requires six inorganic carbon dioxide molecules to form one glucose molecule.

$$6CO_2 + 6H_2O \rightarrow C_6H_{12}O_6 + 6O_2$$

This conversion of an inorganic form of an element to an organic form is known as a fixation. Therefore, photosynthesis can be described as a series of reactions in which carbon dioxide and water are fixed into glucose, and oxygen is produced as a by-product.

The fixation reaction described above requires energy. The energy to create the glucose comes directly from the ATP and hydrogen created in the first stage of photosynthesis. Ultimately, this energy can be traced back to sunlight. It is also important to note that glucose is only one of the many possible organic molecules that can form from photosynthesis.

Measuring the rate of photosynthesis

Let's take a moment to revisit the summary reaction for photosynthesis.

$$6CO_2 + 6H_2O \rightarrow C_6H_{12}O_6 + 6O_2$$

This balanced equation shows us that carbon dioxide molecules are reactants and oxygen molecules are products of photosynthesis. If you now recall some information you learned earlier about cell respiration, you will see that the reverse is true for that process. In other words, for cell respiration, oxygen is a reactant and carbon dioxide is a product. Knowing that plants do both photosynthesis and cell respiration might lead you to the conclusion that the two processes 'cancel each other out' when looking at oxygen or carbon dioxide levels. But this thinking is a bit simplistic.

At any given time of year, any one plant has a fairly consistent rate of cell respiration. Not only is this rate consistent throughout the day and night, it is also at a relatively low level. Plants do not have muscle and other ATP-demanding tissues as do animals. They need ATP for various biochemical processes, but the level is typically far below that required by an animal.

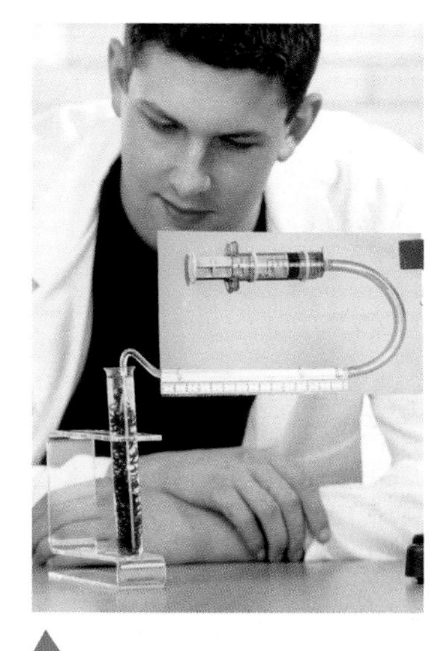

This student is measuring oxygen produced by an aquatic plant. The rate of O_2 produced is a direct reflection of the rate of photosynthesis.

The same consistency is not true concerning the rate of photosynthesis. Photosynthetic rate is highly dependent on many environmental factors including intensity of light and air temperature. During the daytime, especially on a warm sunny day, the rate of photosynthesis may be very high for a particular plant. If so, the rate of carbon dioxide taken in by the plant and the rate of oxygen released will both also be very high. As the plant is also doing cell respiration, a correction should be made for that in both carbon dioxide and oxygen levels. At night, the rate of photosynthesis may drop to zero. At that time, a given plant may be giving off carbon dioxide and taking in oxygen to maintain its relatively low and consistent rate of cell respiration (see Figure 3.26).

Measuring the rate of oxygen production or carbon dioxide intake is considered to be a direct measurement of photosynthetic rate as long as a correction is made for cell respiration. Another common approach is to measure photosynthesis by keeping track of the change in biomass of experimental plants. Massing of plants is considered to be an indirect reflection of photosynthetic rate as the increase or decrease in biomass may be traced to a whole variety of factors besides photosynthetic rate.

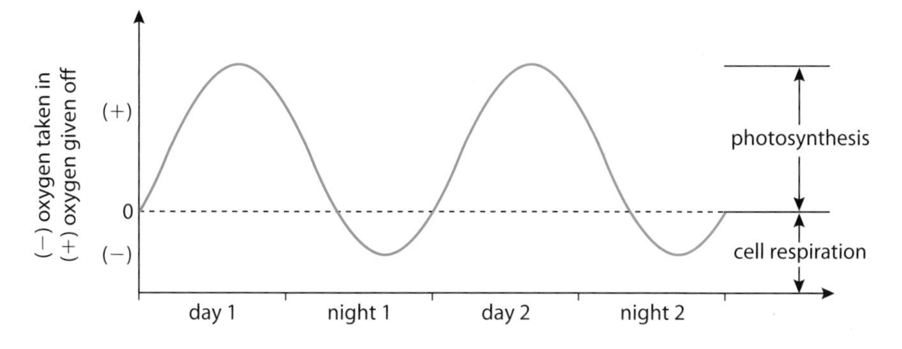

Figure 3.26 Graph showing the oxygen given off and taken in by a hypothetical plant over a 48-hour period. When the line intersect is at 0, the oxygen generated by photosynthesis is equal to the oxygen needed for cell respiration.

The effects of changing environmental factors on rate of photosynthesis

Let's look at the patterns that emerge when three common environmental factors are varied and how that is predicted to change the rate of photosynthesis in a generalized plant (see Figures 3.27–3.29).

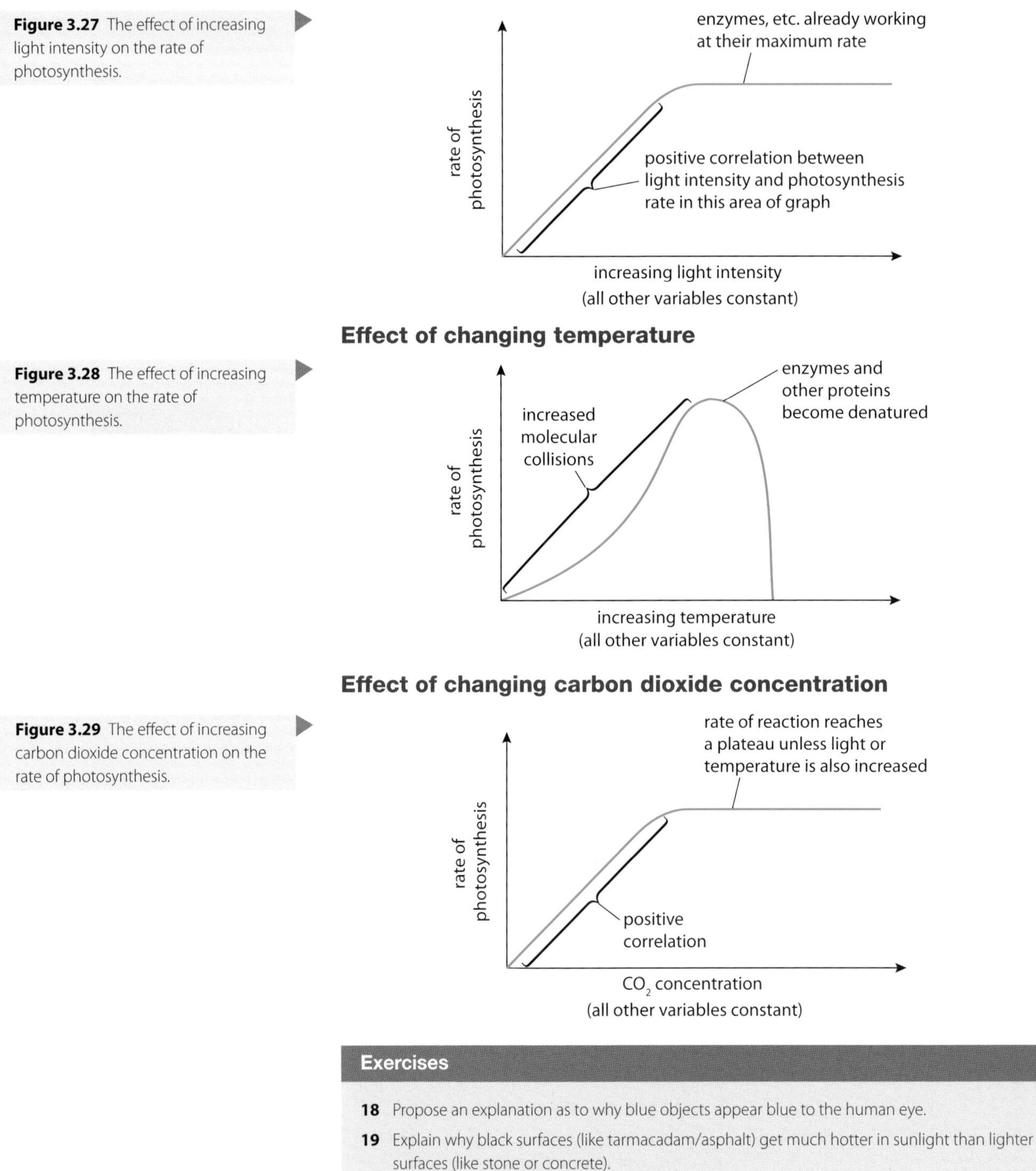

Figure 3.27 The effect of increasing light intensity on the rate of photosynthesis.

Figure 3.28 The effect of increasing temperature on the rate of photosynthesis.

Figure 3.29 The effect of increasing carbon dioxide concentration on the rate of photosynthesis.

Exercises

18 Propose an explanation as to why blue objects appear blue to the human eye.

19 Explain why black surfaces (like tarmacadam/asphalt) get much hotter in sunlight than lighter surfaces (like stone or concrete).

20 Plants produce sugars by photosynthesis. What do plants do with the sugars after that?

21 Why do most plants produce an excess of sugars in some months of the year?

Practice questions

1 Biosphere 2, an enormous greenhouse built in the Arizona desert in the USA, has been used to study five different ecosystems. It is a closed system so measurements can be made under controlled conditions. The effects of different factors, including changes in carbon dioxide concentration in the greenhouse, were studied. The data shown below were collected over the course of one day in January 1996.

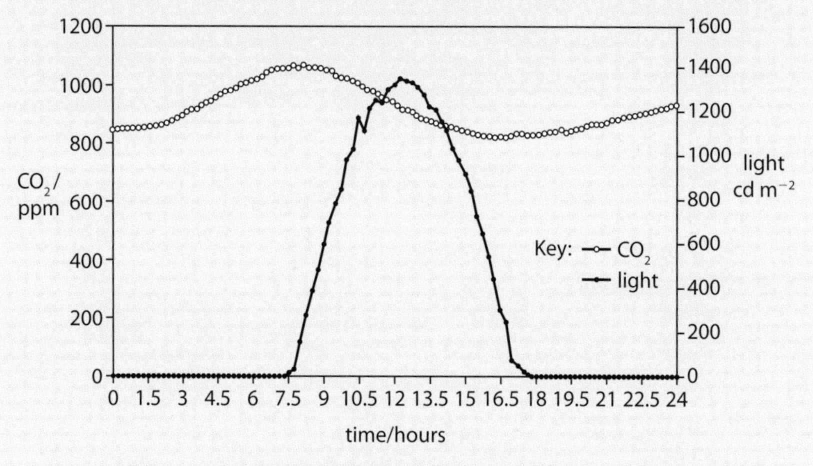

Source: www.ldeo.columbia.edu/martins/climate_water/labs/lab6/labinstr6./html

(a) **(i)** Identify the time of day when the sun rose. *(1)*

(ii) Identify the time of minimal CO_2 concentration. *(1)*

(b) Determine the maximum difference in the concentration of CO_2 over the 24-hour period. *(1)*

(c) Suggest reasons for changes in CO_2 concentration during the 24-hour period. *(2)*

(Total 5 marks)

2 Describe the significance of water to living organisms. *(6 marks)*

3 Draw the structure of a fatty acid. *(1 mark)*

4 Outline how monosaccharides are converted into polysaccharides. *(2 marks)*

5 The graph below shows the effect of changing the substrate concentration on an enzyme controlled reaction.

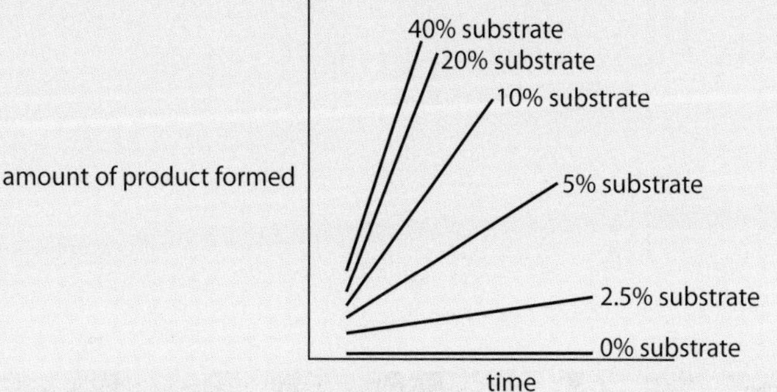

What is the correct interpretation of these data?

A The rate of reaction increases continuously with increase in substrate concentration.

B The rate of reaction decreases continuously with increase in substrate concentration.

C The rate of reaction increases up to a point and then remains constant.

D The rate of reaction is not affected by any change in the substrate concentration.

(1 mark)

• **Examiner's hint:** When / (a backward slash) is used in a unit expression, it is not read as 'per'. You read the slash as 'expressed as'. For example, CO_2 / ppm is read as 'carbon dioxide expressed as parts per million'.

• **Examiner's hint:** When you are being asked to read a value from a graph, use a straight edge to determine the value. Do not just 'eyeball' the value. You should bring a metric ruler as a straight edge to the exam site.

• **Examiner's hint:** Take a couple of minutes to plan an essay such as that in question 2. You know that you must try to earn 6 marks. Think about what these marks may be for, before beginning your writing and consider writing them down as a plan. If you do not do this, your writing is more likely to wander off track.

• **Examiner's hint:** 'Outline' does not mean create a true outline. 'Outline' is a command term that means 'give a brief account or summary'. Use the number of marks available as a guide to how brief you can afford to be.

• **Examiner's hint:** Multiple-choice questions on IB examinations always have four choices. There is no penalty for incorrect answers, so you should answer all of them even if you guess on a few.

6 Which of the following are connected by a hydrogen bond?

A The hydrogen and oxygen atoms of a water molecule.

B A base pair of a DNA molecule.

C Two amino acid molecules of a dipeptide.

D Two glucose molecules in a disaccharide. *(1 mark)*

7 Where do transcription and translation occur in eukaryotic cells?

	Transcription	Translation
A	cytoplasm	cytoplasm
B	cytoplasm	mitochondria
C	nucleus	cytoplasm
D	nucleus	nucleus

(1 mark)

8 Of the following products, which is produced by both anaerobic respiration and aerobic respiration in humans?

I pyruvate

II ATP

III lactate

A I only

B I and II only

C I, II and III

D II and III only *(1 mark)*

9 Pigments are extracted from the leaves of a green plant. White light is then passed through the solution of pigments. What effect do the leaf pigments have on the white light?

A Green wavelengths are absorbed and red and blue wavelengths are transmitted.

B Red and blue wavelengths are absorbed and green wavelengths are transmitted.

C Blue wavelengths are absorbed and green and red wavelengths are transmitted.

D Green and red wavelengths are absorbed and blue wavelengths are transmitted.

(1 mark)

4 Genetics 1

Introduction

Imagine the power of being able to predict the future. One of the fundamental reasons for studying science in general and genetics in particular is to be able to do just that. Here are some of the kinds of question that geneticists try to answer:

- What will my baby brother look like?
- Will my children be able to see the difference between red and green even though I can't?
- If there is a history of genetic disease in my family, what are the chances that my future baby might be affected?

Other questions include:

- How is it possible that I have blue eyes whereas everyone else in my family has brown eyes?
- Can we find out who was at the crime scene by analysing the DNA left behind?
- How can crops be genetically changed to improve their quality and quantity?
- Is it possible to clone humans?

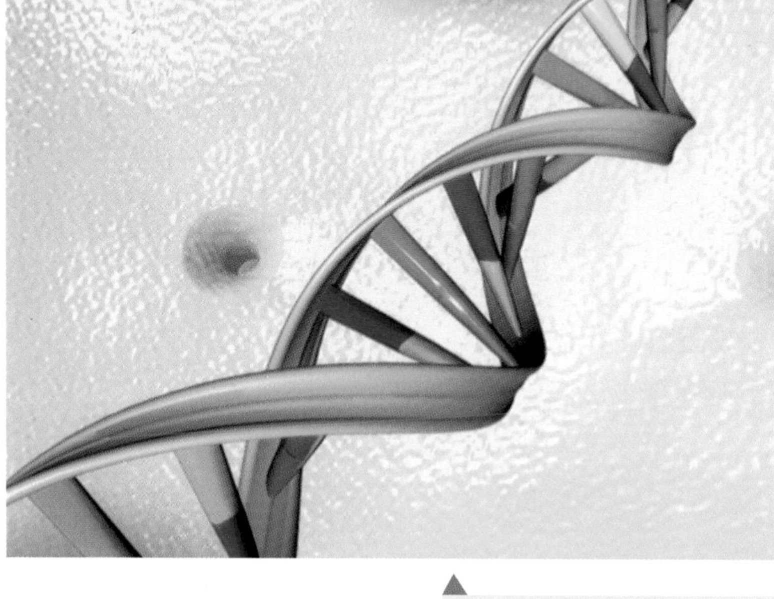

A computer model of a short strand of DNA against the nuclear membrane showing a nuclear pore.

To answer these questions, we must understand the mechanisms of genetics. Genetics is the science of how inherited information is passed on from one generation to the next using genetic material – genes made of DNA.

4.1 Chromosomes, genes, alleles and mutations

Assessment statements

4.1.1 State that eukaryote chromosomes are made of DNA and proteins.
4.1.2 Define *gene*, *allele* and *genome*.
4.1.3 Define *gene mutation*.
4.1.4 Explain the consequence of a base substitution mutation in relation to the processes of transcription and translation, using the example of sickle cell anaemia.

How DNA is organized

Chromosomes are bundles of long strands of DNA. If you could unwind a chromosome, it would be like unravelling a ball of string. A typical human cell contains enough DNA to stretch for nearly two metres.

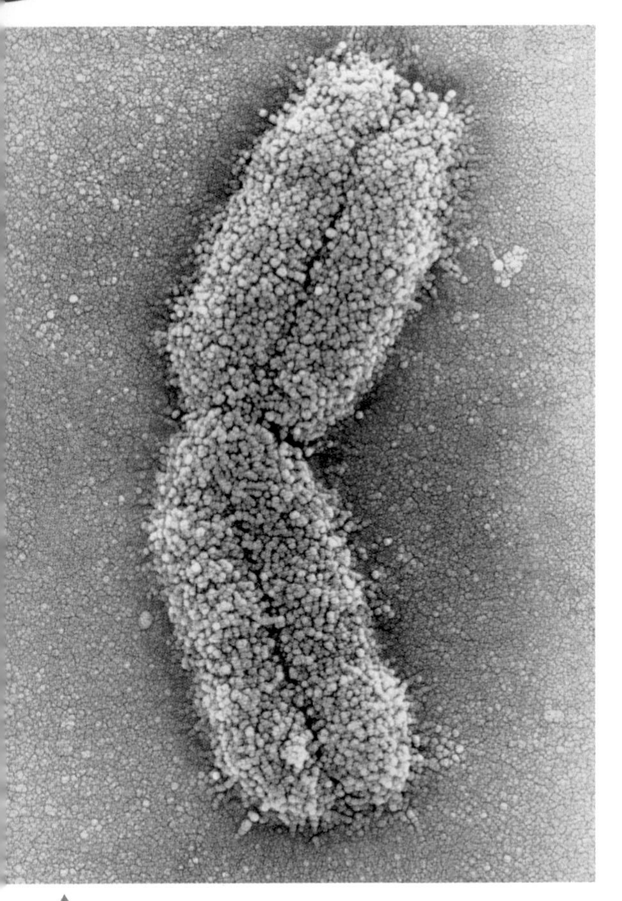

Each vertical rod of a chromosome is called a chromatid. The area where the chromatids are joined together is called the centromere. This coloured electron micrograph shows two chromatids joined at a centrally positioned centromere.

In eukaryotes that reproduce sexually, chromosomes always come in pairs (except in sex cells or gametes). Humans have 46 chromosomes in 23 pairs. The DNA in eukaryotes is associated with proteins which help to keep the DNA organized.

Prokaryotes have only one chromosome, and the DNA is not associated with proteins.

Genes

A gene is a heritable factor that controls a specific characteristic. 'Heritable' means passed on from parent to offspring, and 'characteristic' refers to genetic traits such as your hair colour or your blood type. The estimated 30 000 genes which you possess are organized into chromosomes.

The genes which determine eye colour have more than one form. Some people have genes which give them brown eyes, others have genes for blue or green eyes. Such variations of a gene are called alleles. An allele is one specific form of a gene, differing from other alleles by one or a few bases. Alleles of the same gene occupy a corresponding place (locus) on each chromosome of a pair (see page 91).

In some people, earlobes are attached and in others, they are not. The gene for this trait comes in two possible forms: one allele for attached earlobes and one allele for non-attached earlobes.

In order to find out which gene does what, a list must be made showing the order of all the letters in the DNA code. Researchers use highly specialized laboratory equipment to locate and identify sequences of bases. The complete set of an organism's base sequences is called its genome.

The complete genome of a few organisms have been fully written out. Among these are the fruit fly and the bacterium *E. coli* because these two organisms have been used extensively in genetics experiments for decades.

If each DNA nucleotide base pair (A–T, G–C) were typed out, a human cell would contain more letters than 10 sets of encyclopaedias.

Mutations

A mutation is a random, rare change in genetic material. One type involves a change of the sequence of bases in DNA. If DNA replication works correctly, this should not happen (see Chapter 3, page 58). But nature sometimes makes mistakes. For example, the base thymine (T) might be put in the place of adenine (A) along the sequence. When this happens, the corresponding bases along mRNA are altered during transcription.

Base substitution mutation

The consequence of changing one base could mean that a different amino acid is placed in the growing polypeptide chain. This may have little or no effect on the organism or it may have a major influence on the organism's physical characteristics. Look at the photograph of the fruit flies – it shows the consequence of a mutation. The fruit fly on the right has an extra pair of wings. Mutations in fruit flies can also change their eye colour, the number of legs, the shape of the wings as well as scores of other traits.

A normal fruit fly and one with additional wings.

In humans, a mutation is sometimes found in the gene which creates haemoglobin for red blood cells. This mutation gives a different shape to the haemoglobin molecule. The difference leads to red blood cells which look very different from the usual flattened disk pinched in the middle.

The mutated red blood cell with the characteristic curved shape made its discoverers think of a sickle (a curved knife used to cut tall plants). The condition which results from this mutation is therefore called sickle cell anaemia.

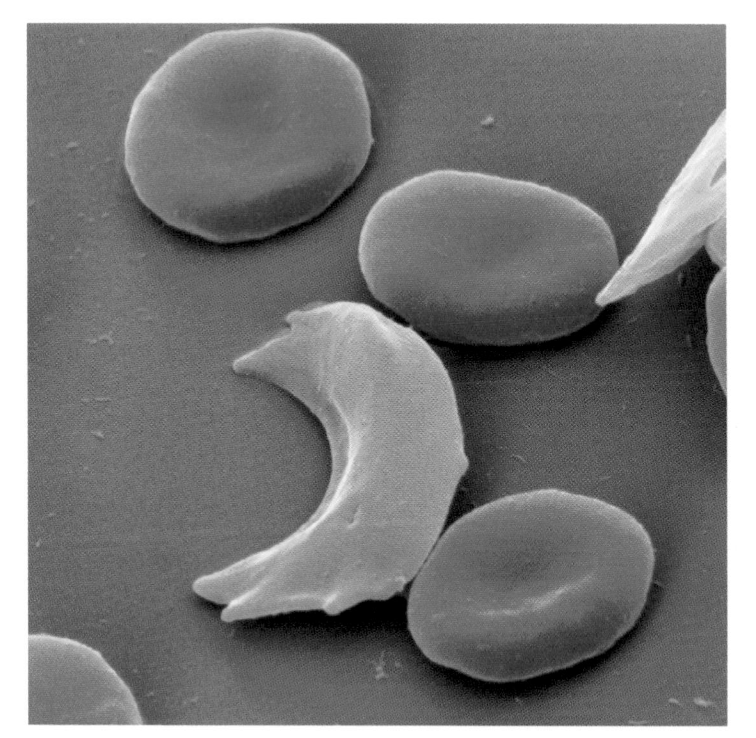

Three typical red blood cells and one sickle-shaped one.

The kind of mutation which causes sickle cell anaemia is called a base substitution mutation. In this case, one base is substituted for another so that the codon GAG becomes GTG. So, during translation instead of adding glutamic acid, which is the intended amino acid, valine is added instead. Since valine has a different shape and different properties from glutamic acid, the shape of the resulting polypeptide chain is modified. As a result, the haemoglobin molecule has a different shape and so does the red blood cell.

- How do you think the symptoms of sickle cell anaemia might affect a young person's life?
- What do you think would be the psychological effects of being told you have such a gene mutation?
- If you had the gene for sickle cell anaemia, would you want to know the chances of your future children getting it? Would it affect your decision of whether or not to marry someone who may also carry the gene?
- If you had the opportunity to genetically screen (test) the embryos of your future children, would you do it? If you say yes, what would you do to the embryos which were discovered to be affected by sickle cell anaemia?
- Do you think that parents should have the opportunity to abort if they knew their unborn baby was affected?
- Should carriers of the condition be obliged to get advice from a genetic counsellor before having children?

The symptoms of sickle cell anaemia are weakness, fatigue and shortness of breath. Oxygen cannot be carried as efficiently by the irregularly-shaped red blood cells. In addition, the haemoglobin tends to crystallize within the red blood cells, causing them to be less flexible. The affected red blood cells can get stuck in capillaries so blood flow can be slowed or blocked.

People affected by sickle cell anaemia have a risk of passing the mutated gene to their offspring. From a demographics point of view, the mutated gene is mostly found in populations originating from West Africa or from the Mediterranean.

Malaria is one of the most deadly diseases in the world. However, the parasite which causes malaria has difficulty infecting a person who has sickle cell anaemia. As a result, people affected by the condition have a natural resistance to malaria.

Statistical analysis of the number of cases of malaria and the prevalence of sickle cell anaemia shows that there is a strong correlation and this is no coincidence – there is a causal link. The reason why the mutated gene has been passed on successfully is that it has helped people to avoid dying of malaria. Natural selection has ensured that this gene's frequency in a population is balanced with the normal gene's frequency so that not everyone has sickle cell anaemia and not everyone dies of malaria.

For an explanation of natural selection, see page 140.

Exercises

1. Draw and label a chromosome. Include the following labels: chromatid, centromere. Indicate an example of a locus.
2. What is the difference between an allele and a gene?
3. Compare and contrast prokaryotic DNA and eukaryotic DNA.
4. Explain why eukaryotic chromosomes always come in pairs.

4.2 Meiosis

Assessment statements

4.2.1 State that meiosis is a reduction division of a diploid nucleus to form haploid nuclei.

4.2.2 Define *homologous chromosomes*.

4.2.3 Outline the process of meiosis, including pairing of homologous chromosomes and crossing over, followed by two divisions, which results in four haploid cells.

4.2.4 Explain that non-disjunction can lead to changes in chromosome number, illustrated by reference to Down's syndrome (trisomy 21).

4.2.5 State that, in karyotyping, chromosomes are arranged in pairs according to their size and structure.

4.2.6 State that karyotyping is performed using cells collected by chorionic villus sampling or amniocentesis, for pre-natal diagnosis of chromosome abnormalities.

4.2.7 Analyse a human karyotype to determine gender and whether non-disjunction has occurred.

Meiosis

Meiosis is a form of cell division which results in gametes (sex cells). Although meiosis has some similarities to mitosis (see Chapter 2, page 39), it is important to understand that there are some fundamental differences.

One characteristic which makes meiosis unique is that each new cell which results from it has only half the number of chromosomes that a typical cell in that organism has. For example, humans have 46 chromosomes in their cells, but in the sperm and egg cells, there are only 23 chromosomes in each cell. Cells which contain half the chromosome number are called haploid cells. Cells with the full chromosome number are called diploid cells.

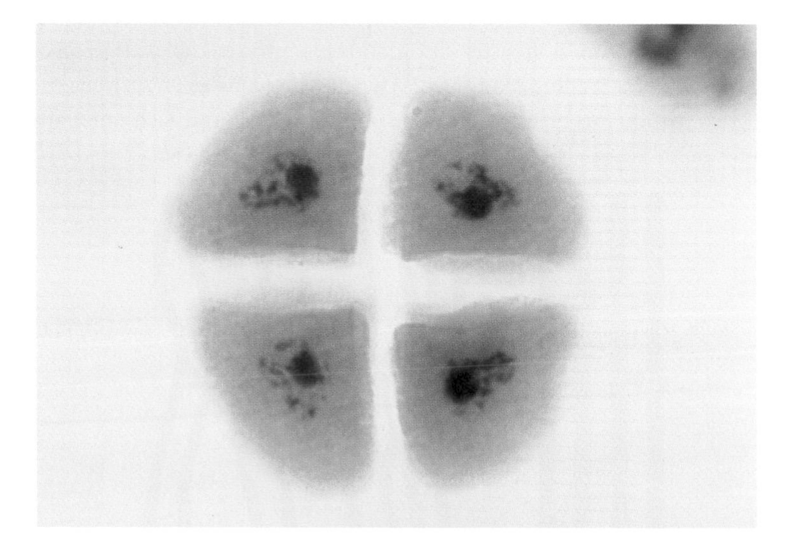

Four haploid pollen cells formed by meiosis.

This type of cell division is called a reduction division because the number of chromosomes has been reduced. This reduction is necessary in gamete production because during sexual reproduction, each parent contributes 50% of the genetic information.

The cells formed from cell division are referred to as daughter cells. Meiosis generates four haploid daughter cells and each cell has a unique mix of half of the genetic information of the parent cell.

Homologous chromosomes

In a diploid human cell, the 46 chromosomes can be grouped into 23 pairs of chromosomes called homologous chromosomes. Homologous means similar in shape and size and it means that the two chromosomes carry the same genes. The reason there are two of each is that one came from the father and the other from the mother.

Although a pair of homologous chromosomes carry the same genes, they are not identical because the alleles for the genes from each parent could be different. We use the letter n to denote the number of unique chromosomes in an organism. In eukaryotes, there are n pairs of chromosomes. With two of each, that makes a total of $2n$ per cell. This is a shorthand way of writing the chromosome number for haploid (n) and diploid ($2n$) cells. An egg and a sperm cell of a human each contain n or 23 chromosomes, so when they unite, there are $2n$ or 46.

The phases of meiosis

Meiosis is a step-by-step process by which a diploid parent cell produces four haploid daughter cells. Before the steps begin, DNA replication allows the cell to make a complete copy of its genetic information during interphase. This results in each chromatid having an identical copy, or sister chromatid, attached to it at the centromere.

In order to produce a total of four cells, the parent cell must divide two times: the first meiotic division makes two cells and then each of these divides during the second meiotic division to make a total of four cells.

One of the characteristics which distinguishes meiosis from mitosis (see Chapter 2, page 39) is that during the first step, called prophase I, there is an exchange of genetic material between non-sister chromatids in a process called crossing over (see Figure 4.1). This trading of segments of genes happens when sections of two homologous chromatids break at the same point, twist around each other and each connects to the other's initial position.

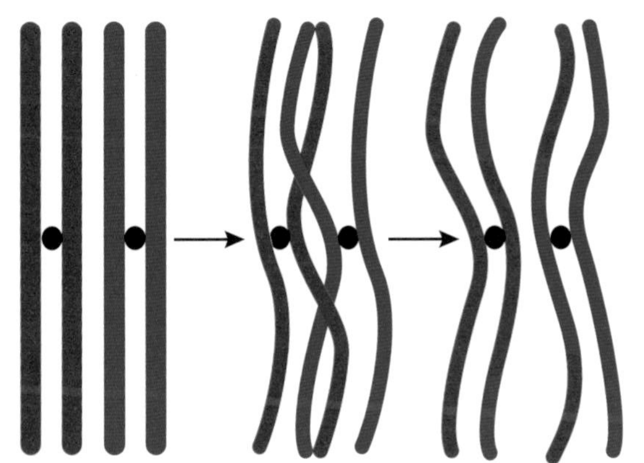

Figure 4.1 Crossing over in a pair of homologous chromosomes.

Crossing over allows DNA from a person's maternal chromosomes to mix with DNA from the paternal chromosomes. In this way, the recombinant chromatids which end up in the sperm or the egg cells are a mosaic of the parent cell's original chromatids.

Meiosis I takes place in order to produce two cells with a single set of chromosomes each (see Figure 4.2).

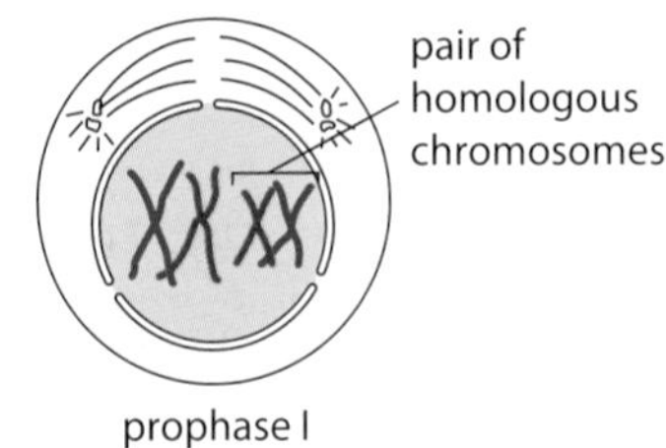

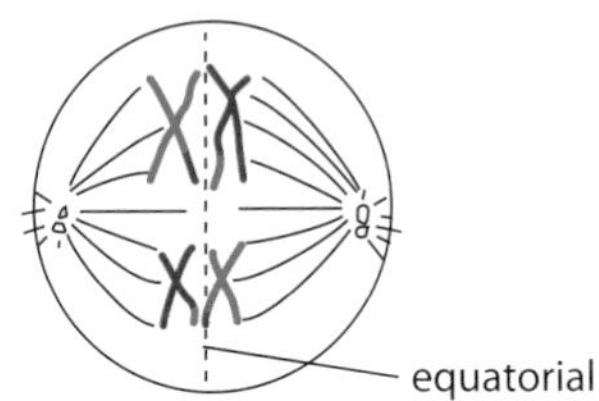

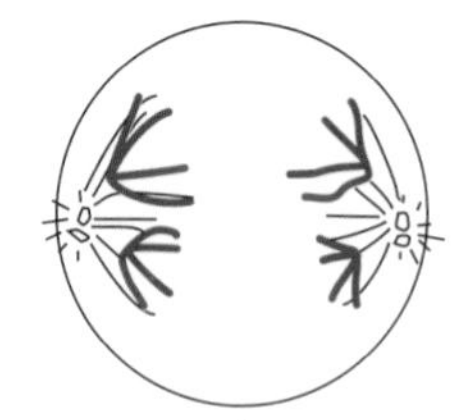

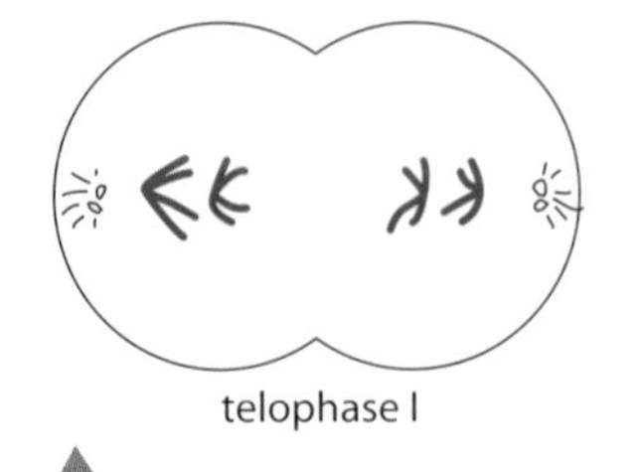

Figure 4.2 The stages of meiosis I.

Prophase I

1 Chromosomes become visible as the DNA becomes more compact.
2 Homologous chromosomes, also called homologues, are attracted to each other and pair up – one is from the individual's father, the other from the mother.
3 Crossing over occurs.
4 Spindle fibres made from microtubules form.

Metaphase I

1 The bivalents (another name for the pairs of homologous chromosomes) line up across the cell's equator.
2 The nuclear membrane disintegrates.

Anaphase I

1 Spindle fibres from the poles attach to chromosomes and pull them to opposite poles of the cell.

Telophase I

1 Spindles and spindle fibres disintegrate.
2 Usually, the chromosomes uncoil and new nuclear membranes form.
3 Many plants do not have a telophase I stage.

At the end of meiosis I, cytokinesis happens: the cell splits into two separate cells. The cells at this point are haploid because they contain only one chromosome of each pair. However, each chromatid still has its sister chromatid attached to it, so no S phase is necessary.

Now meiosis II takes place in order to separate the sister chromatids (see Figure 4.3).

Prophase II

1 DNA condenses into visible chromosomes again.
2 New meiotic spindle fibres are produced.

Metaphase II

1 Nuclear membranes disintegrate.
2 The individual chromosomes line up along the equator of each cell in no special order; this is called random orientation.
3 Spindle fibres from opposite poles attach to each of the sister chromatids at the centromeres.

Anaphase II

1 Centromeres of each chromosome split, releasing each sister chromatid as an individual chromosome.
2 The spindle fibres pull individual chromatids to opposite ends of the cell.
3 Because of random orientation, the chromatids could be pulled towards either of the newly forming daughter cells.
4 In animal cells, cell membranes pinch off in the middle, whereas in plant cells, new cell plates form to demarcate the four cells.

Telophase II

1 Chromosomes unwind their strands of DNA.
2 Nuclear envelopes form around each of the four haploid cells, preparing them for cytokinesis.

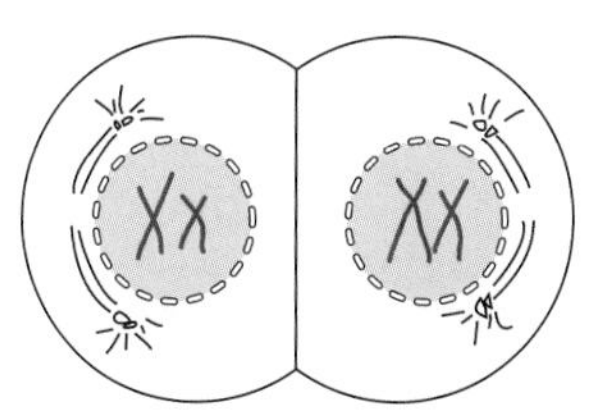
prophase II

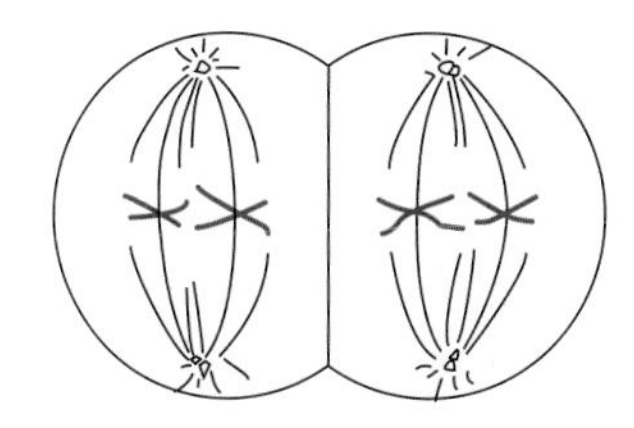
metaphase II

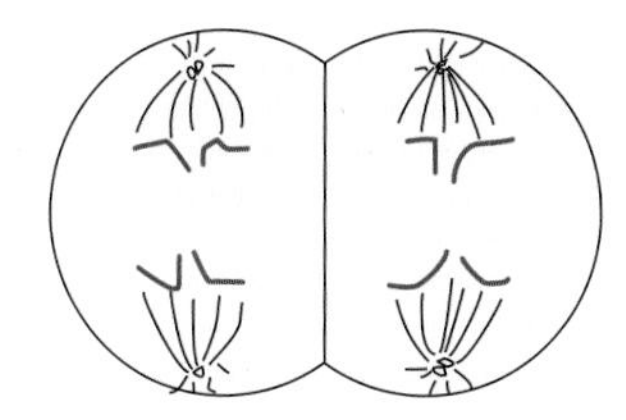
anaphase II

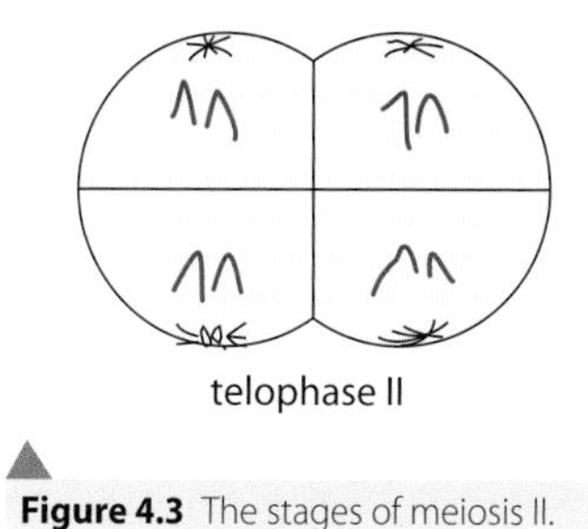
telophase II

Figure 4.3 The stages of meiosis II.

Down's syndrome

Sometimes chromosomes do not separate the way they are expected to during the first or second meiotic division. This results in an unequal distribution of chromosomes. In humans, this means that an egg cell or a sperm cell might have 24 instead of 23 chromosomes. This unexpected distribution of chromosomes is due to a non-disjunction, a process by which two or more homologous chromosomes stick together instead of separating.

In the case of Down's syndrome, non-disjunction happens in the 21st pair of chromosomes: the child receives 3 instead of 2. Such an anomaly is called a trisomy and Down's syndrome is also referred to as trisomy 21. Having an additional chromosome brings about malformations of the digestive system and causes differing degrees of learning difficulties. Children with Down's syndrome often follow specialized education programmes adapted for their needs.

The child on the right has Down's syndrome.

Down's syndrome is the most common chromosomal anomaly and affects approximately 1 birth in 800. The risk of Down's syndrome increases as the age of the mother increases, particularly over the age of 35. Non-disjunctions can happen with other chromosomes, and all of them can have a major impact on a child's development. Some developmental consequences are so severe that the fetus may not survive beyond a few weeks or months.

Karyotypes

A karyotype is a photograph of the chromosomes found in a cell arranged according to a standard format, as in the photograph opposite. The chromosomes are placed in order according to their size and shape. The shape depends mainly on the position of the centromere. A karyotype is made by the following steps.

1 The cells are stained and prepared on a glass slide to see their chromosomes under a light microscope.
2 Photomicrograph images are obtained of the chromosomes during mitotic metaphase.
3 The images are cut out and separated, a process which can be done using scissors or using a computer.
4 The images of each pair of chromosomes are placed in order by size and the position of their centromeres.

Obtaining cells for karyotyping

An unborn baby's cells can be extracted in one of two ways: either by a process called amniocentesis or by removing them from the chorionic villus. Amniocentesis involves using a hypodermic needle to extract some of the amniotic fluid around the developing baby. Inside the liquid, some of the baby's cells can be found and used for the preparation of a karyotype. In the second method, cells are obtained by chorionic villus sampling, which involves obtaining a tissue sample from the placenta's finger-like projections into the uterus wall.

In either case, among the cells collected are white blood cells which are then grown in the laboratory. The preparation of a karyotype is an expensive and invasive procedure. It is usually used for seeing if an unborn baby has any chromosomal anomalies: 45 or 47 chromosomes instead of 46. If the parents or doctors are concerned about the chromosomal integrity of an unborn child (for example, if an expectant mother is over the age of 35), a karyotype is recommended.

Analysis of a human karyotype

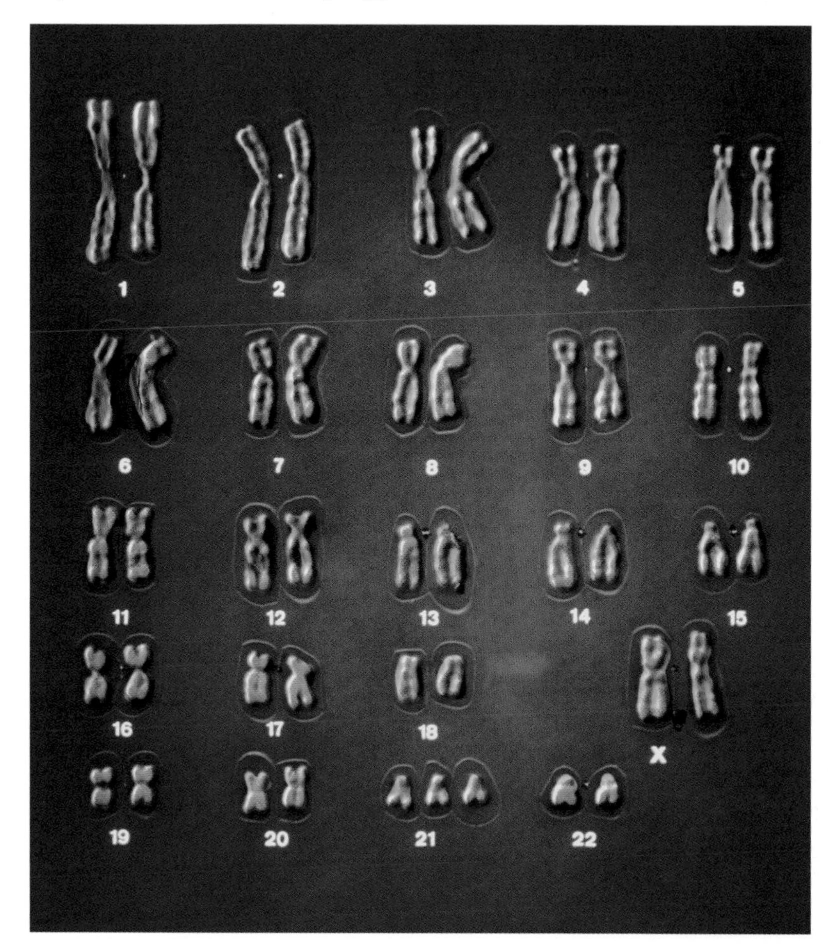

This karyotype was prepared using false colour imagery.

Look at the karyotype. We can tell whether this person is a male or a female and if any non-disjunction has happened.

To determine if the individual is a male or a female, the last pair of chromosomes must be examined. In this case, two big X chromosomes indicate that this is a female. To see if a non-disjunction has occurred, the pairs of chromosomes must be checked to see if there are more or fewer than two for any pair. In this case, the 21st pair has three chromosomes indicating Down's syndrome. What cannot be determined from this information is how severely the girl will be affected by the physical or mental effects of the condition. Some children are only mildly affected, whereas others are very severely affected.

There are many countries where it would be possible to ask that a pregnancy be terminated if parents discover early in the pregnancy that their child has an atypical number of chromosomes (more or fewer than 46). This is not only a legal question, it is a medical and ethical one, often bringing in philosophical and religious points of view.

- Should there be laws governing this?
- Should expectant mothers over a certain age be obliged to get a karyotype?
- Should it be a question of personal freedom?
- If one government were to make karyotyping illegal, should it also be illegal to travel to another country to have a karyotype done?

Since the extraction of cells from an unborn baby is an invasive procedure, there is a risk that it may harm the child or even cause a miscarriage. Doctors and parents-to-be must consider which is more important, finding out if the baby has a chromosomal anomaly or providing safe conditions for its development without any extraction of cells.

- Who should decide when a karyotype is necessary? Who should decide whether to keep or to abort the babies which present chromosomal anomalies? The doctors? The future parents? Both? What do you think?
- If you found out that your future child had a chromosomal anomaly that would make him or her very different from other children, what would you do?

Karyotyping can also be used to see if the child has the gender that the parents want their child to have. If parents are disappointed, they may choose not to keep the baby and try again later for another child with the desired gender. Do you think this is acceptable?

Exercises

5 Why is meiosis referred to as a reduction division?

6 Explain why meiosis rather than mitosis is necessary for gamete production.

7 State the name of a type of cell in your body which is haploid.

8 Draw and label the stages of meiosis II.

9 Explain the reason why more women aged 35 to 40 have karyotypes of their unborn babies prepared than do pregnant women who are aged 25 to 30.

4.3 Theoretical genetics

Assessment statements

4.3.1 Define *genotype, phenotype, dominant allele, recessive allele, codominant alleles, locus, homozygous, heterozygous, carrier* and *test cross*.

4.3.2 Determine the genotypes and phenotypes of the offspring of a monohybrid cross using a Punnett grid.

4.3.3 State that some genes have more than two alleles (multiple alleles).

4.3.4 Describe ABO blood groups as an example of codominance and multiple alleles.

4.3.5 Explain how the sex chromosomes control gender by referring to the inheritance of X and Y chromosomes in humans.

4.3.6 State that some genes are present on the X chromosome and absent from the shorter Y chromosome in humans.

4.3.7 Define *sex linkage*.

4.3.8 Describe the inheritance of colour blindness and haemophilia as examples of sex linkage.

4.3.9 State that a human female can be homozygous or heterozygous with respect to sex-linked genes.

4.3.10 Explain that female carriers are heterozygous for X-linked recessive alleles.

4.3.11 Predict the genotypic and phenotypic ratios of offspring of monohybrid crosses involving any of the above patterns of inheritance.

4.3.12 Deduce the genotypes and phenotypes of individuals in pedigree charts.

Who was Gregor Mendel?

In 1865, an Austrian monk named Gregor Mendel published results of his experiments on how garden pea plants passed on their characteristics. At the time, the term 'gene' did not exist (he used the term 'factors' instead) and the role that DNA played would not be discovered for nearly another century. Among the questions Mendel asked were:

- How can I be sure that I will get only smooth peas and no wrinkled ones?
- How can I be sure that the resulting plants will be short or tall?
- How can I be sure to obtain only flowers of a certain colour?

Mendel used artificial pollination in a series of experiments in which he carefully chose the pollen of various plants to fertilize other plants. He used a small brush to place the pollen on the reproductive parts of the flowers, thus replacing the insects which do it naturally. This technique takes away the role of chance because the experimenter knows exactly which plants are fertilized by which pollen.
In one cross, he wanted to see what would happen if he bred tall plants with short plants. The result was that he got all tall plants. But then when he crossed the resulting tall plants with each other, some of the offspring in the new generation were short. We are now going to look at Mendel's laws of genetics in the light of modern understanding of DNA. But to understand how it all works, you must master the vocabulary below.

Gregor Mendel (1822–84) studied the genetics of garden pea plants.

Visit heinemann.co.uk/hotlinks, insert the express code 4242P and click on Weblink 4.1. The Genetics Web Lab Directory has online simulations where you can cross plants the same way Mendel did. Look for the one called Mendel's Peas. This is also a good resource for other genetics concepts and evolution.

Sometimes it is possible to figure out a person's genotype by looking at their physical appearance (phenotype). Other times, it is not possible to tell because the person could be hiding one of their alleles (one allele could be recessive).

Key terminology

Genotype – The symbolic representation of pair of alleles possessed by an organism, typically represented by two letters.

Examples: **Bb**, **GG**, **tt**.

Phenotype – The characteristics or traits of an organism.

Examples: five fingers on each hand, colour blindness, type O blood.

Dominant allele – An allele that has the same effect on the phenotype whether it is paired with the same allele or a different one. Dominant alleles are always expressed in the phenotype.

Example: the genotype **Aa** gives the dominant **A** trait because the **a** allele is masked. The **a** allele is not transcribed and translated during protein synthesis.

Recessive allele – An allele that has an effect on the phenotype only when present in the homozygous state.

Example: **aa** gives rise to the recessive trait because no dominant allele is there to mask it.

Codominant alleles – Pairs of alleles that both affect the phenotype when present in a heterozygote.

Example: a parent with curly hair and a parent with straight hair can have children with different degrees of hair curliness as both alleles influence hair condition when both are present in the genotype.

Locus – The particular position on homologous chromosomes of a gene (see Figure 4.4). Each gene is found at a specific place on a specific pair of chromosomes.

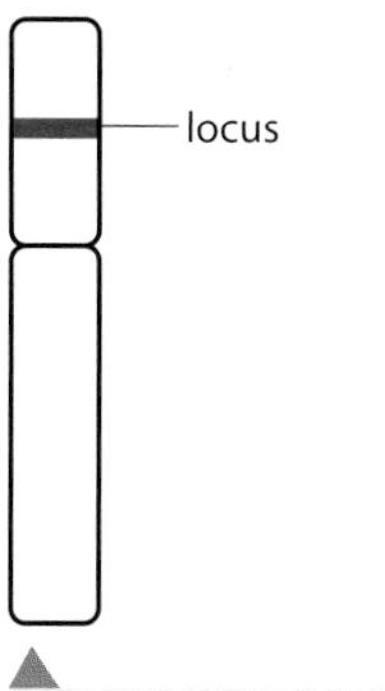

Figure 4.4 This drawing shows you the locus of a gene on a chromosome.

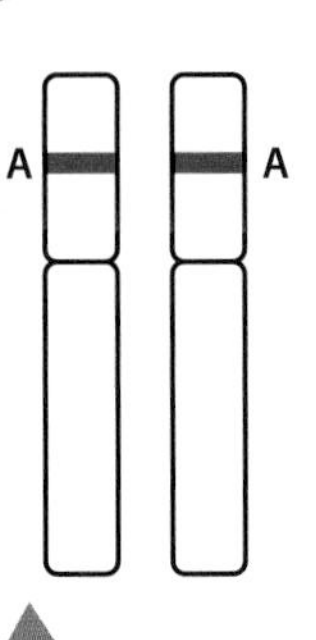

Figure 4.5 This drawing shows you a pair of chromosomes showing a homozygous state, **AA**.

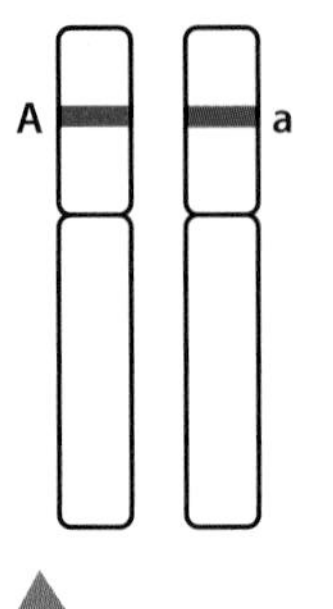

Figure 4.6 This drawing shows you a pair of chromosomes showing a heterozygous state, **Aa**.

Homozygous – Having two identical alleles of a gene (see Figure 4.5).

Examples: **AA** is a genotype which is homozygous dominant whereas **aa** is the genotype of someone who is homozygous recessive for that trait.

Heterozygous – Having two different alleles of a gene (see Figure 4.6). This results from the fact that the paternal allele is different from the maternal one.

Example: **Aa** is a heterozygous genotype.

Carrier – An individual who has a recessive allele of a gene that does not have an effect on their phenotype.

Example: **Aa** carries the gene for albinism (see below) but has pigmented skin – an ancestor must have been albino and some offspring might be, too. If both parents are unaffected by a recessive condition yet both are carriers, some of their progeny could be affected (**aa**).

Test cross – Testing a suspected heterozygote plant or animal by crossing it with a known homozygous recessive (**aa**). Since a recessive allele can be masked, it is often impossible to tell if an organism is **AA** or **Aa** until they produce offspring which have the recessive trait.

Constructing a Punnett grid

This is a Punnett grid.

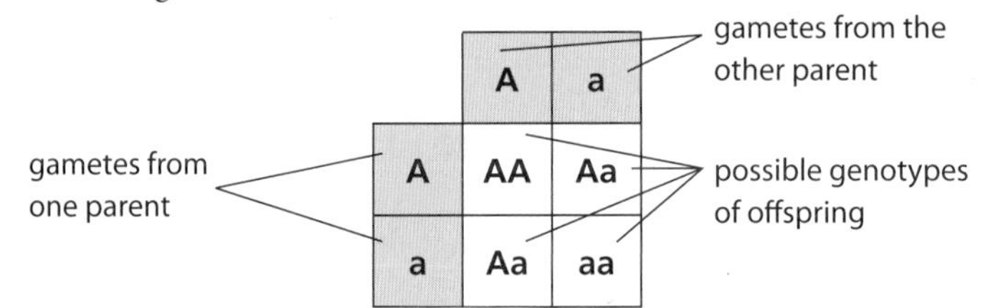

It is a diagram whose purpose it is to show all possible combinations of genetic information for a particular trait. All the Punnett grids in this chapter are for monohybrid crosses, which means they show the results for one trait only.

Albino animals lack pigmentation so this penguin does not have the black markings characteristic of most penguins.

Let's consider a condition called albinism. Most people and animals are unaffected by albinism and have pigmented skin, hair, eyes, fur or feathers. But some people and animals lack pigmentation. An individual with little or no pigmentation is called an albino. We can trace the inheritance of albinism with a Punnett grid.

In order to set up a Punnett grid, the following steps must be followed.

Step 1 – Choose and indicate a letter to show the alleles

Use the big and small versions of letters to represent different alleles. Usually, a capital letter represents the dominant allele and the lower case version represents the recessive allele. For example:

- **A** = dominant allele, allows pigments to form;
- **a** = recessive allele, albinism – fewer or no pigments.

Get used to saying 'big A' and 'little a' when reading alleles and genotypes. Also, do not mix letters: you cannot use **P** for pigmented and **a** for albino, for example. Once you have chosen a letter, write down what it means so that it is clear which allele is which.

• **Examiner's hint:** Watch out when choosing letters. Nearly half the letters of the alphabet should be avoided because they are too similar in their capital and lower case forms. Don't use: Cc, Ff, Kk, Oo, Pp, Ss, Uu, Vv, Ww, Xx, Yy, Zz.

Step 2 – Determine the parents' genotypes

To be sure that no possibilities are forgotten, write out all three possibilities and decide by elimination which genotype or genotypes fit each parent.

The three possibilities here are:

- homozygous dominant (**AA**) – in this case, the phenotype shows pigmentation;
- heterozygous (**Aa**) – in this case, the phenotype shows pigmentation but the heterozygote is a carrier of the albino allele;
- homozygous recessive (**aa**) – in this case, the phenotype shows albinism.

The easiest genotype to determine by simply looking at the person or animal is **aa**. The other two are more of a challenge. To determine for sure if the individual is **AA** or **Aa**, we have to look for evidence that the recessive gene was received from an albino parent or was passed on to the individual's offspring. In effect, the only way to produce an albino is for each parent to donate one **a**.

Step 3 – Determine the gametes which the parents could produce

An individual with a genotype **AA** can only make gametes with the allele **A** in them. Carriers can make **A**-containing gametes or gametes with **a**. Obviously, individuals whose genotype is **aa** can only make sperms or eggs which contain the **a** allele. So you can record and label with **A** or **a** all the possible gametes.

Step 4 – Draw a Punnett grid

Once all the previous steps have been completed, drawing the actual grid is simple. The parents' gametes are placed on the top and side. As an example, consider a monohybrid cross involving a female carrier **Aa** crossed with a male albino **aa**.

You might guess that, since there are three **a** alleles and only one **A**, there should be a 75% chance of seeing offspring with the recessive trait. But this is not the case. Here is a grid with the parents gametes.

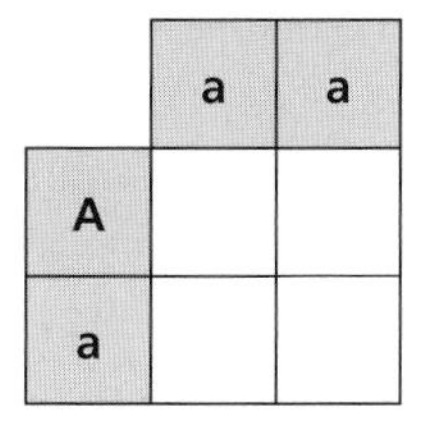

	a	a
A		
a		

Now you can fill in the empty squares with each parent's possible alleles by copying the letters from the top down and from left to right. When letters of different sizes end up in the same box, the big one goes first.

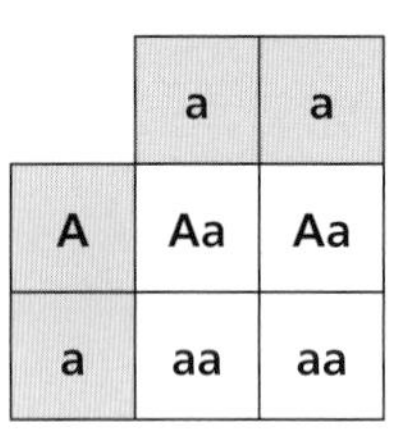

	a	a
A	Aa	Aa
a	aa	aa

Step 5 – Deduce the chances for each genotype and phenotype

In a grid with four squares, each square can represent one of two possible statistics:

- the chance that these parents will have offspring with that genotype – here, each square represents a 25% chance;
- the likely proportion of offspring that will have the resulting genotypes – this only works for large numbers of offspring.

The results from the above example show the following: there is a 50% chance of producing offspring with genotype **Aa** and a 50% chance of producing offspring with genotype **aa**. Since humans tend to produce a small number of offspring, this interpretation should be used. If the example were about plants which produce hundreds of seeds, the results could be interpreted in the following way: 50% of the offspring should be **Aa** and the other half should be **aa**.

Finally, the phenotypes can be deduced by looking at the genotypes. For example, **Aa** offspring will have a phenotype showing pigmentation so they will not be affected by albinism whereas all the **aa** offspring will be albinos.

Counting kernels of corn can give percentages of offspring with a certain phenotype since each kernel is one offspring.

Multiple alleles

So far, only two possibilities have been considered for a gene: dominant, **A** or recessive, **a**. With two alleles, three different genotypes are possible which produce two different phenotypes. Genetics is not always this simple; sometimes there are three or more alleles for the same gene. This is the case for the alleles which determine the ABO blood type in humans.

Blood type

The ABO blood type system in humans has four possible phenotypes: A, B, AB and O. To create these four blood types there are three alleles of the gene. These three alleles can produce six different genotypes.

The gene for the ABO blood type is represented by the letter **I**. To represent more than just two alleles (**I** and **i**) superscripts are introduced. As a result, the three alleles for blood type are written as follows: $\mathbf{I^A}$, $\mathbf{I^B}$ and **i**. The two capital letters with superscripts represent alleles which are codominant:

- $\mathbf{I^A}$ = the allele for type A blood;
- $\mathbf{I^B}$ = the allele for type B blood;
- **i** = the recessive allele for type O.

Crossing these together in all possible combinations creates six genotypes which give rise to the four phenotypes listed earlier:

- $\mathbf{I^AI^A}$ or $\mathbf{I^Ai}$ gives a phenotype of type A blood;
- $\mathbf{I^BI^B}$ or $\mathbf{I^Bi}$ gives type B blood;
- $\mathbf{I^AI^B}$ gives type AB blood (due to codominance);
- **ii** gives type O blood.

Notice how the genotype $\mathbf{I^AI^B}$ clearly shows codominance. Neither allele is masked; both show expression in the phenotype of type AB blood.

The sex chromosomes: X and Y

The 23rd pair of chromosomes are called the sex chromosomes because they determine if a person is a male or a female. The X chromosome is longer than the Y chromosome and contains many more genes. Unlike the other 22 pairs of chromosomes, this is the only pair in which it is possible to find two chromosomes that are very different in size and shape.

For Nile crocodiles, it is not chromosomes which dictate the sex of the offspring. Rather, the average temperature of the sand around the eggs during the 3-month incubation period determines if the babies will be males or females.

In human females there are two X chromosomes. When women produce gametes, each egg will contain one X chromosome. Human males have one X chromosome and one Y chromosome. When males produce sperm cells, half of them contain one X chromosome and half contain one Y chromosome. As a result, when an egg cell meets a sperm cell during fertilization, there is always a 50% chance that the child will be a boy and a 50% chance that the child will be a girl (see Figure 4.7):

- XX = female;
- XY = male.

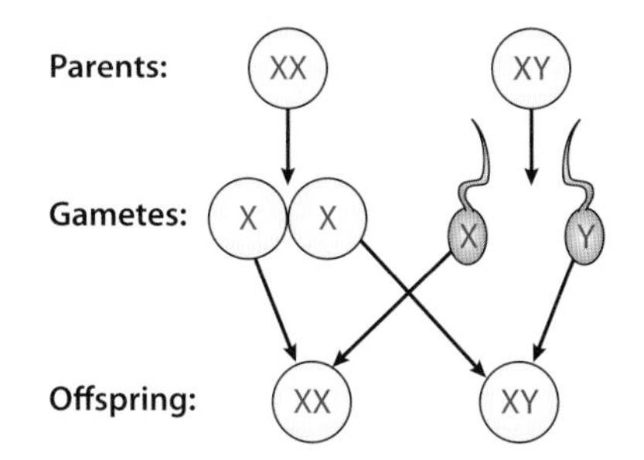

Figure 4.7 Will the child be a boy or a girl?

The chances remain the same no matter how many boys or girls the family already has.

Genes carried on the sex chromosomes

Because the Y chromosome is significantly smaller than the X chromosome, it has fewer loci and therefore fewer genes than the X chromosome. This means that sometimes alleles present on the X chromosome have nothing to pair up with. For example, a gene whose locus is at an extremity of the X chromosome would have no counterpart on the Y chromosome because the Y chromosome does not extend that far from its centromere.

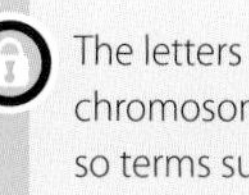

The letters X and Y refer to chromosomes and not to alleles, so terms such as dominant and recessive do not apply.

Sex linkage

Any genetic trait whose allele has its locus on the X or the Y chromosome is said to be sex linked. Often genetic traits which show sex linkage affect one gender more than the other. Two examples of genetic traits which have this particularity are colour blindness and haemophilia.

- Colour blindness is the inability to distinguish between certain colours, often green and red. To people who are colour blind, the two colours look the same; they would not see the difference between a green apple and a red apple, for example.
- Haemophilia is a disorder in which blood does not clot properly. For most people, a small cut or scrape on their skin stops bleeding after a few minutes and eventually a scab forms. This process is called clotting. People with haemophilia risk bleeding to death from what most people would consider a minor injury such as a bruise, which would rupture many tiny blood vessels. Such bleeding can also occur in internal organs. Medical treatments such as special injections help to give people affected by haemophilia a better quality of life.

Because the scientist John Dalton had red–green colour blindness, the condition is sometimes referred to as daltonism and people who have it are said to be daltonian. He asked that his eyes be dissected after his death to verify his hypothesis that the liquid inside them was blue. It was not. However, the eyes were kept in 1844 for study and, a century and a half later, scientists used the tissue samples to identify the gene for colour blindness.

Alleles and genotypes of sex-linked traits

Since the alleles for both colour blindness and haemophilia are found only on the X chromosome, the letter X is used in representing them:

- $\mathbf{X^b}$ = recessive allele for colour blindness;
- $\mathbf{X^B}$ = allele for the ability to distinguish colours;
- $\mathbf{X^h}$ = allele for haemophilia;
- $\mathbf{X^H}$ = allele for the ability to clot blood.

In both cases, there is no allele on the Y chromosome, so Y is written alone without any superscript. Here are all the possible genotypes for colour blindness:

- $\mathbf{X^BX^B}$ gives the phenotype of a non-affected female;
- $\mathbf{X^BX^b}$ gives the phenotype of a non-affected female who is a carrier;
- $\mathbf{X^bX^b}$ gives the phenotype of an affected female;
- $\mathbf{X^BY}$ gives the phenotype of a non-affected male;
- $\mathbf{X^bY}$ gives the phenotype of an affected male.

In the above list, **B** and **b** could be replaced by **H** and **h** to show the genotypes for haemophilia.

Carriers of sex-linked traits

Sex-linked recessive alleles such as $\mathbf{X^b}$ are rare in most populations of humans worldwide. For this reason, it is unlikely to get one and much less probable to get two. This is why so few women are colour blind: their second copy of the gene is likely to be the dominant allele for full-colour vision and mask the recessive allele. The same is true for haemophilia.

As you have seen, there are three possible genotypes for females but only two possibilities for males. Only women can be heterozygous, $\mathbf{X^BX^b}$, and as a result, they are the only ones who can be carriers. If you are in any doubt here, look again at the list of genotypes for colour blindness.

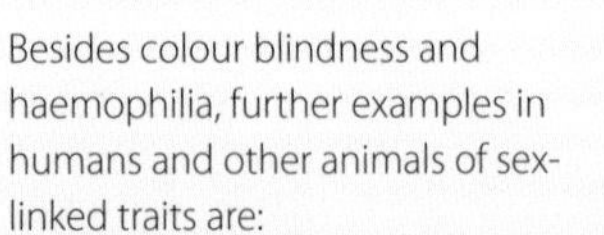

Besides colour blindness and haemophilia, further examples in humans and other animals of sex-linked traits are:

- Duchenne muscular dystrophy;
- white eye colour in fruit flies;
- calico–tortoiseshell fur colour in cats.

Since men do not have a second X chromosome, there are only two possible genotypes $\mathbf{X^BY}$ or $\mathbf{X^bY}$ in relation to colour blindness. With just the one recessive allele **b**, a man will be colour blind. This is contrary to what you have seen up to now concerning recessive alleles: usually people need two to have the trait and with one, they are carriers. In this case, the single recessive allele in males determines the phenotype. Men cannot be carriers for X-linked alleles.

Applications of the Punnett grid

Using the five steps of the Punnett grid method (see page 93), we are going to examine the theoretical chances of how genetic traits can be passed on from one generation to the next.

Short or tall pea plants?

Let's first consider a cross that Gregor Mendel did with his garden pea plants. He took purebred tall plants and crossed them with purebred short plants. Purebred means that the tall plants' parents were known to be all tall and the short plants' parents were known to be all short. In other words, he knew that none of the plants were heterozygous. He wanted to find out if he would get all tall plants, some tall and some short or all short.

The answer took months for Mendel to confirm but a Punnett grid can get the answer in seconds (see Key facts box): the result was 100% tall plants. Why? Because in garden pea plants, the allele for tall is dominant over the allele for short plants, thus masking the short trait.

The five steps of the Punnett grid method.

- Step 1 – Choose a letter:
  T = allele for a tall plant,
  t = allele for a short plant.
- Step 2 – Parents' genotypes:
  TT for the purebred tall and **tt** for the short.
- Step 3 – Determine gametes: the purebred tall parent can only give **T**, short parent can only give **t**.
- Step 4 – Draw Punnett grid.

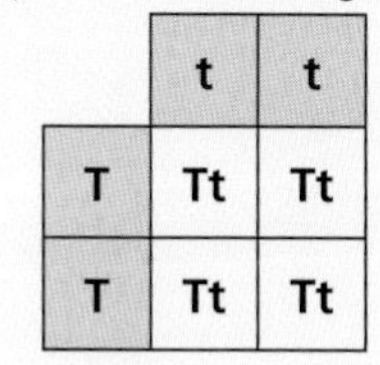

	t	t
T	Tt	Tt
T	Tt	Tt

- Step 5 – Interpret grid: 100% **Tt** and will be tall, so 0% will be short.

The name given to the generation produced by a cross such as this is the first filial generation, usually referred to as the F_1 generation. What would happen if tall plants from the F_1 generation were crossed to make a second filial generation (F_2)? The Punnett grid gives us the result.

	T	t
T	TT	Tt
t	Tt	tt

This grid can be interpreted in two ways:

- there is a 75% chance of producing tall offspring and a 25% chance of producing short offspring;
- 75% of the offspring will be tall and 25% of the offspring will be short.

Although 75% of the plants are tall, they have differing genotypes. Some tall plants are homozygous dominant and others are heterozygous.

Also, in a real experiment, it is unlikely that exactly 25% of the offspring would be short plants. The reason is essentially due to chance. For example, if 90 F_2 peas were produced and all of them were planted and grew into new plants, there is no mathematical way that exactly 25% of them would be short. At the very best, 23 out the 90 plants would be short and that gives 25.56%; that is as close as it is possible to get to 25% in this case.

Even if a convenient number of plants were produced, such as 100 plants, farmers and breeders would not be surprised if they got 22, 26 or even 31 short plants instead of the theoretical 25. If the results of hundreds of similar crosses were calculated, the number would probably be very close to 25%. The same phenomenon can be seen in the gender of human children. Although the theoretical percentage is calculated to be 50% girls and 50% boys, in reality, few families have exactly half and half. This is due to chance.

When examined closely by experts in statistics, some of Gregor Mendel's results seem too good to be true. His numbers do not show the expected variations which are typically found by farmers and researchers when breeding plants. What happened? Did he think that the unexpected results were due to mistakes and so he omitted them from his findings? Or did he purposefully change the numbers so they would fit with what he wanted to show? Such a practice is called fudging the data and it is considered unethical. No one knows why Mendel's numbers are so perfect and the mystery may never be elucidated. How can we be sure that modern scientific studies are free from fudged data?

Inheriting your blood group

Punnett grids can also be set up to predict blood type inheritance or the chance of a couple's next child having haemophilia or albinism.

Worked example

Is it possible for a couple to have four children, each child showing a different blood type?

Solution

There is only one way for this to happen: one parent must have type A blood but be a carrier of the allele for type O blood and the other parent must have type B blood and also be a carrier of the of the allele for type O blood (if necessary, refresh your memory of the blood group alleles on page 95).

The cross would be $\mathbf{I^Ai} \times \mathbf{I^Bi}$ and the grid is shown below. See if you can determine the phenotype of each child before reading on.

	I^A	i
I^B	I^AI^B	I^Bi
i	I^Ai	ii

So, would it be possible for this couple to have four children and all of them have a different blood group? Yes. Would it be possible for the same couple to have four children and all of them have type AB blood? Yes, but it would not be likely. This question is similar to 'Could a couple have 10 children, all of them girls?' It is possible but statistically unlikely.

Gregor Mendel is considered to be the founding father of genetics. And yet, like many other scientists with radically new ideas, he was not recognized or commended for his work during his lifetime. Why did other scientists in the second half of the 19th century not take notice of Mendel when he tried to point out the importance of the mathematical ratios in offspring? He showed his results from years of cross breeding plants as well as all his painstaking calculations and statistics and yet very few people took his ideas seriously.

Why is it that the scientific community is so reluctant to take on new ideas and paradigm shifts? Does this reluctance have a positive or negative impact on the advancement of science?

Is it acceptable that a scientist's new ideas are not adopted just because he or she is not a well-known person with an established reputation? Inversely, should the new ideas of a well-known scientist be embraced solely because he or she has a solid reputation? How is it possible to determine which new ideas are valid and which ones should be rejected?

Pedigree charts

These are the symbols used in pedigree charts.
○ circle = female
□ square = male
● filled circle = a female who possesses the trait being studied
■ filled square = a male who possesses the trait being studied
| vertical line = the relationship parents–offspring
— horizontal line between a man and a woman means they are the parents who had the offspring

The term 'pedigree' refers to the record of an organism's ancestry. A pedigree animal such as a racehorse or a show dog is an animal whose owner can prove the true descent of the horse or dog based on official documents about its parents and grandparents.

Pedigree charts are diagrams which are constructed to show biological relationships. In genetics, they are used to show how a trait can pass from one generation to the next. Used in this way for humans, a pedigree chart is similar to a family tree complete with parents, grandparents, aunts, uncles and cousins.

To build such a chart, symbols are used to represent people. The Key fact box (left) shows you these symbols and what each represents. Preparing a pedigree chart helps in preparing Punnett grids for predicting the likely outcome of the next generation.

Example 1 – Huntington's disease

Huntington's disease (Huntington's chorea) is caused by a dominant allele which we will refer to by the letter **H**. This genetic condition causes severely debilitating nerve damage but the symptoms do not show until the person is about 40. As a result, someone who has the gene for Huntington's disease does not know it for certain until they have started a career and possibly started a family.

The symptoms include difficulty walking, speaking, and holding objects. Within a few years, the person loses complete control of his or her muscles and dies an early death. Since it is dominant, all it takes is one **H** in the person's genetic makeup to cause the condition.

In the pedigree chart in Figure 4.8, the symbols indicate that the unaffected members of the family are the mother, the first child (a girl) and the fourth child (a boy). Those who are affected are the father, the second child (a boy) and the third child (a girl). To work out if the father is **HH** or **Hh**, consider the fact that some of his children do not have the trait. This proves that he must have given one **h** to each of them. Hence, he can only be **Hh** and not **HH**.

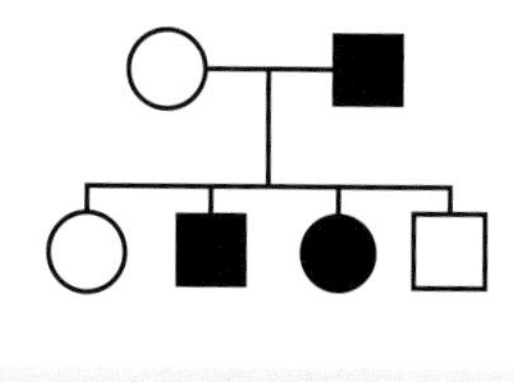

Figure 4.8 This is a pedigree chart showing members of a family affected by Huntington's disease.

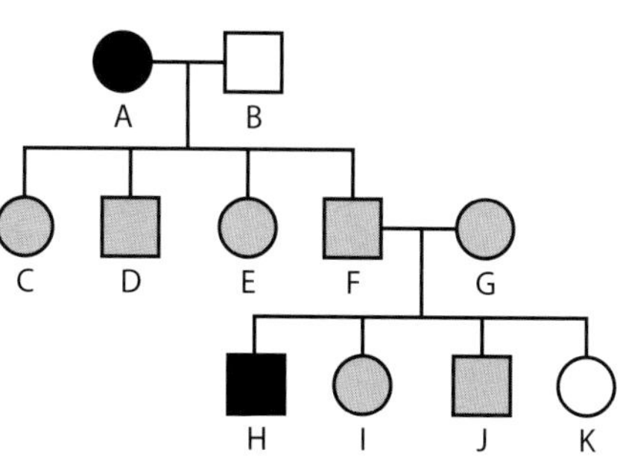

This pedigree chart shows how pink flowers can arise in pure-bred snapdragon plants. Black shapes represent snapdragon plants with red flowers, white shapes represent white-flowered plants and grey shapes represent plants with pink flowers. Can you determine the genotypes of individuals A to K?

Example 2 – Codominance in flower colour

Codominance in certain flowers can create more than two colours, so a pedigree can help keep track of how the offspring got their phenotypes. For example, in purebred snapdragon flowers, sometimes white × red = pink.

The system of letters for showing flower colour in snapdragon flowers uses a prefix **C**, which refers to the gene that codes for flower colour, plus a superscript which refers to the specific colour, **R** (red) or **W** (white).

So the alleles for codominant flower colour are:

- $\mathbf{C^R}$ for red flowers;
- $\mathbf{C^W}$ for white flowers.

The genotypes and their phenotypes are:

- $\mathbf{C^RC^R}$ makes red flowers;
- $\mathbf{C^WC^W}$ makes white flowers;
- $\mathbf{C^RC^W}$ makes pink flowers.

For codominant traits, grey is used in pedigree charts rather than black or white.

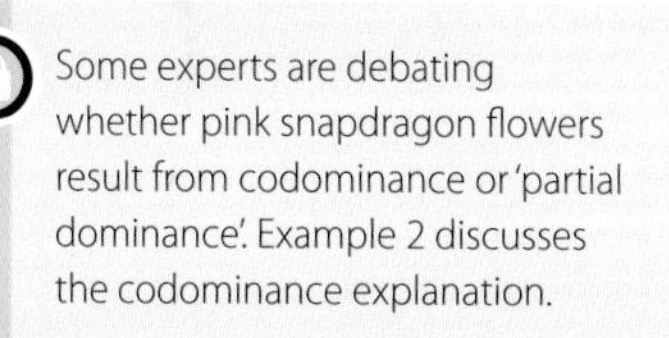

Some experts are debating whether pink snapdragon flowers result from codominance or 'partial dominance'. Example 2 discusses the codominance explanation.

Pure-bred red-flowered and white-flowered snapdragon plants can sometimes produce pink-flowered offspring.

Example 3 – Codominance in the shape of red blood cells

The prefix **Hb** refers to the gene that codes for haemoglobin (the molecule in red blood cells). The superspript letter is for the typical shape of haemoglobin, **A** for normal and **S** for the shape that causes sickle cells (see page 83). The genotypes and phenotypes are as follows:

- $\mathbf{Hb^AHb^A}$ genotype generates haemoglobin that results in the phenotype with disk-shaped red blood cells (the person has normal blood);
- $\mathbf{Hb^SHb^S}$ genotype generates haemoglobin that results in the phenotype with curved red blood cells (the person has severe anaemia and sickle-shaped cells);
- $\mathbf{Hb^AHb^S}$ genotype generates some of each type of haemoglobin because the alleles show codominance (the person, who is said to have sickle cell trait, has fewer sickle-shaped cells so the anaemia is much less severe this person also has some resistance to malaria).

As in the system for showing blood groups, we use different superscript letters for different codominant alleles.

In families where genetic disease has been passed down generation after generation, serious questions arise.

- If someone is a carrier, should they have children knowing that they might transmit the gene?
- How would it feel knowing – or not knowing – if you had the gene?
- How might parents feel when they find out they have passed the gene on to their offspring?
- Is it difficult for people who have genetic diseases to find someone to marry? A job? Life insurance?
- Should employers or insurance companies have the right to know if someone has a genetic disease or is that a medical secret?
- When, if ever, is screening appropriate (getting a genetic test done on an unborn child)?

Exercises

10 Explain why more men are affected by colour blindness than women.

11 Using the $\mathbf{C^R}$ and $\mathbf{C^W}$ alleles for codominance in snapdragon flower colour, show how two plants could have some white-flowered offspring, some pink-flowered offspring and some red-flowered offspring within one generation.

12 Draw a pedigree chart of the two generations described in exercise 11.

13 Look at the grid below showing the chances that a couple's children might have haemophilia.

a State the genotype of the mother and father.

b State the possible genotypes of the girls and boys.

c State the phenotypes of the girls and boys.

d Who are the carriers in this family?

e What are the chances that the parents' next child will be a haemophiliac?

	X^H	Y
X^H	X^HX^H	X^HY
X^h	X^HX^h	X^hY

4.4 Genetic engineering and biotechnology

Assessment statements

4.4.1 Outline the use of polymerase chain reaction (PCR) to copy and amplify minute quantities of DNA.

4.4.2 State that, in gel electrophoresis, fragments of DNA move in an electric field and are separated according to their size.

4.4.3 State that gel electrophoresis of DNA is used in DNA profiling.

4.4.4 Describe the application of DNA profiling to determine paternity and also in forensic investigations.

4.4.5 Analyse DNA profiles to draw conclusions about paternity or forensic investigations.

4.4.6 Outline three outcomes of the sequencing of the complete human genome.

4.4.7 State that, when genes are transferred between species, the amino acid sequence of polypeptides translated from them is unchanged because the genetic code is universal.

4.4.8 Outline a basic technique used for gene transfer involving plasmids, a host cell (bacterium, yeast or other cell), restriction enzymes (endonucleases) and DNA ligase.

4.4.9 State two examples of the current uses of genetically modified crops or animals.

4.4.10 Discuss the potential benefits and possible harmful effects of one example of genetic modification.

4.4.11 Define *clone*.

4.4.12 Outline a technique for cloning using differentiated animal cells.

4.4.13 Discuss the ethical issues of therapeutic cloning in humans.

Exploring DNA

DNA is at the very core of what gives animals and plants their uniqueness. We are now going to look at the astounding genetic techniques, developed in the past few decades, which enable scientists to explore and manipulate DNA. These include:

- copying DNA in a laboratory – the polymerasechain reaction (PCR);
- using DNA to reveal its owner's identity – DNA profiling;
- mapping DNA by finding where every A, T, C and G is – the Human Genome Project;
- cutting and pasting genes to make new organisms – gene transfer;
- cloning cells and animals.

These techniques offer new hope for obtaining treatments and vaccines for diseases; for creating new plants for farmers; for freeing wrongly convicted people from prison (thanks to DNA tests proving their innocence).

Techniques such as gene transfer and cloning have sparked heated debate. Is it morally and ethically acceptable to manipulate nature in this way? Are the big biotech companies investing huge sums of money into this research to help their fellow citizens or are they just in it for the economic profit? Concerning cloning and stem cell research, is it morally and ethically acceptable to create human embryos solely for scientific research?

Part of being a responsible citizen is making informed decisions relating to these difficult questions. It is not just technical complexity that makes these questions difficult, it is also because humans have never had to face them before.

Look at this image. What do you think about the idea expressed in the caption?

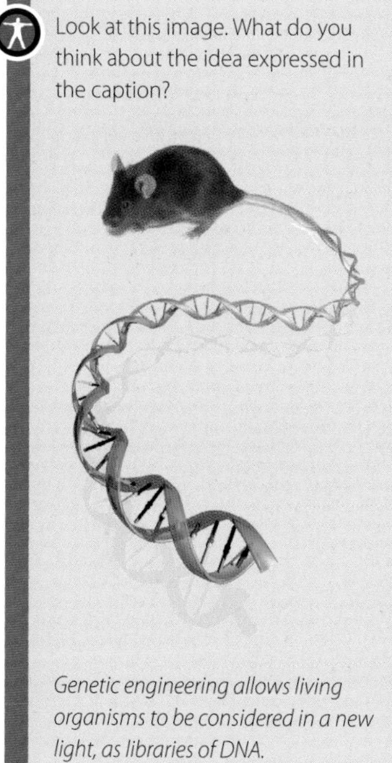

Genetic engineering allows living organisms to be considered in a new light, as libraries of DNA.

Polymerase chain reaction (PCR)

PCR is a laboratory technique which takes a very small quantity of DNA and copies all the nucleic acids in it to make millions of copies of the DNA (see Figure 4.9). PCR is used to solve a very simple problem: how to get enough DNA to be able to analyse it.

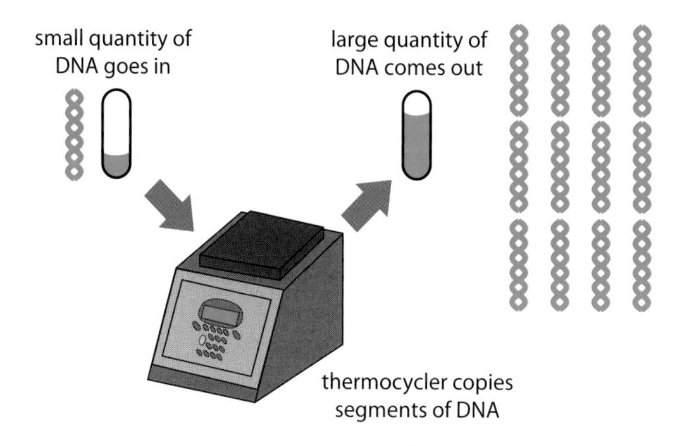

Figure 4.9 Analysis is impossible with the DNA from just one or a few cells. PCR is a way of ensuring that enough DNA for analysis can be generated.

When collecting DNA from the scene of a crime or from a cheek smear, often only a very limited number of cells are available. By using PCR, forensics experts or research technicians can obtain millions of copies of the DNA in just a few hours. Such quantities are large enough to get results from, notably using gel electrophoresis (see overleaf).

Gel electrophoresis

This laboratory technique is used to separate fragments of DNA in an effort to identify its origin. Enzymes are used to chop up the long filaments of DNA into varying sized fragments. The DNA fragments are placed into small wells (holes) in the gel which are aligned along one end. The gel is exposed to an electric current – positive on one side and negative on the other.

The effect is that the biggest, heaviest and least charged particles do not move easily through the gel so they get stuck very close to the wells they were in at the origin. The smallest, least massive and most charged particles pass through the gel to the other side with little difficulty. Intermediate particles are distributed in between. In the end, the fragments leave a banded pattern of DNA like the one shown in the photograph.

These banded lines were formed from nine different DNA samples during gel electrophoresis.

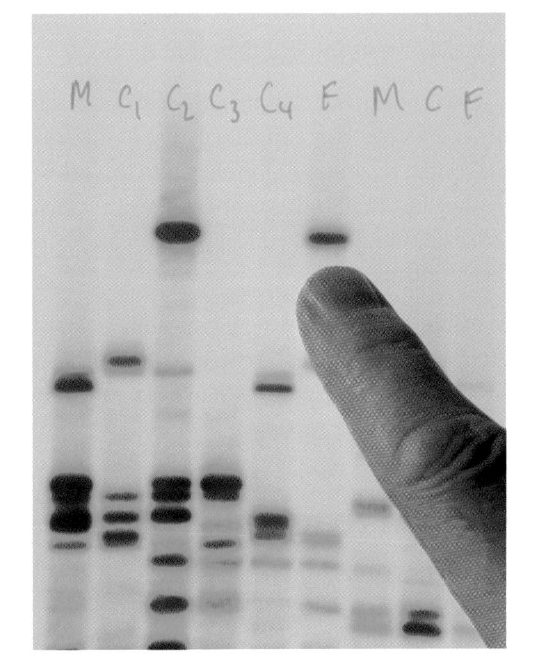

DNA profiling

The process of matching an unknown sample of DNA with a known sample to see if they correspond is called DNA profiling. This is also sometimes referred to as DNA fingerprinting because of some of the similarities with identifying fingerprints but the techniques are very different.
If, after separation by gel electrophoresis, the pattern of bands formed by two samples of DNA fragments are identical, it means that both most certainly came from the same individual. If the patterns are similar, it means that the two individuals are most probably related.

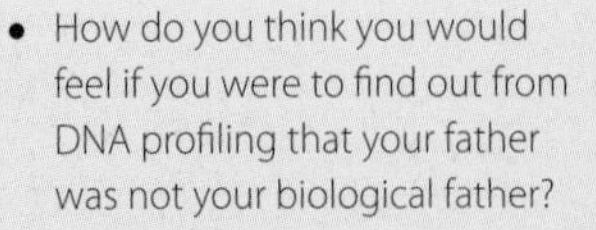

- How do you think you would feel if you were to find out from DNA profiling that your father was not your biological father?
- How would a man feel if he found out he was not the child's father?
- What effect would such a result have on the relationships between siblings or between spouses?
- What kind of emotions might someone feel after spending 18 years in prison, and then being freed thanks to a DNA test?

Applications of DNA profiling

DNA profiling can be used in paternity suits when the identity of someone's biological father must be known for legal reasons.

At a crime scene, forensics specialists can collect samples such as blood or semen which contain DNA. Gel electrophoresis is used to compare the collected DNA with that of suspects. If they match, the suspect has a lot of explaining to do. If not, the suspect is probably not the person wanted for the crime. Criminal cases are sometimes reopened many years after a judgement in order to consider new DNA profiling results. In the United States, this has lead to the liberation of many individuals who had been wrongly sent to jail for crimes they did not commit.

DNA profiling is used in other circumstances too. For example, in studies of ecosystems, when scientists use DNA samples taken from birds, whales and other organisms to clarify which individuals are related. This has helped establish a better understanding of social relationships, migrating patterns and nesting habits. In addition, the study of DNA in the biosphere has given new credibility to the ideas of evolution: DNA evidence can often reinforce previous evidence of common ancestry based on anatomical similarities between species.

For many years, tests for blood groups were done to determine paternity. In what ways do you think DNA profiling is more reliable than a blood test?

Is it possible for two people to have the same DNA profile? The technique does not look at individual base sequences of A, T, C, G – only at clumps of sequences. Do you think it is possible to calculate the chances that two people have the same profile?

Should 100% confidence be placed in DNA testing? Do you think it would be right to convict a person solely on DNA evidence?

How DNA profiles are analysed

In the photo showing gel electrophoresis of nine samples of DNA (page 102) the line marked C_2 (child number 2) and the one being pointed to, F (father), show similarities in their banding patterns. However, the children marked C_1, C_3 and C_4 do not show many similarities.

From this DNA evidence, it should be clear that person F is much more likely to be the father of child number 2, than of any of the other children. Similar techniques are used to analyse the similarities and differences between DNA collected at a crime scene and DNA samples taken from suspects.

The techniques have been perfected to a point where it is possible to determine the identity of someone by examining cells found in the traces of saliva left on the back of a postage stamp on a letter.

The Human Genome Project

In 1990, an international cooperative venture called the Human Genome Project set out to sequence the complete human genome. Because the genome of an organism is a catalogue of all the bases it possesses, the Human Genome Project hoped to determine the order of all the bases A, T, C and G in human DNA. In 2003, the Project announced that it had succeeded in achieving its goal. Now, scientists are working on deciphering which sequences represent genes and which genes do what. The human genome can be thought of as a map which can be used to show the locus of any gene on any one of the 23 pairs of chromosomes.

Dr Francis Collins, one of the leaders of the Human Genome Project team.

As you have seen, some diseases are sex linked, so it is relatively easy to determine which chromosome the gene responsible for the disease is found on; often the locus is on the X chromosome. In traits which show no sex linkage, it is difficult to know which of the 22 other chromosomes carries the gene. With genome libraries of genetic diseases, doctors can find out exactly where to look if they think one of their patients might possess a disease-carrying allele.

Another advantageous use of the human genome is the production of new medications. This idea involves several steps:

- find beneficial molecules which are produced naturally in healthy people;
- find out which gene controls the synthesis of a desirable molecule;
- copy that gene and use it as instructions to synthesize the molecule in a laboratory;
- distribute the beneficial molecule as a new medical treatment.

This is not science fiction; genetic engineering firms are finding such genes regularly.

Dr Craig Venter, one of the leaders of the Human Genome Project team.

The Human Genome Project gives a new facet to the philosophical question, 'What does it mean to be human?' Can humanity be reduced to a sequence of nitrogenous bases represented as A, T, G and C? The figure of 98.5% is often given to describe how much DNA humans have in common with chimpanzees but research in 2002 revealed that it is probably more like 95%. In any case, both species have strikingly similar genomes but that does not make chimps human.

There is also the question of intellectual property. Theoretically, the information in the human genome should belong to everyone. Yet some private biotech research companies have patented genes they have found along the sequence of the human genome, claiming that since they found what the gene codes for, they should have the legal intellectual rights to the use of that gene in biotechnology. This means that if other laboratories wanted to use the genes, they would have to pay a royalty or licensing fee. Do you think this is right?

In addition, by comparing the genetic makeup of populations around the world, countless details could be revealed about ancestries and how humans have migrated and mixed their genes with other populations over time. Without knowing it, you are carrying around in each one of your cells a library of information about your past.

Delving into human genetics confirms two major themes:

- we are all the same;
- we are all different.

On the one hand, the Human Genome Project shows that there is a very small number of DNA bases which make one person different from any other person in the world. This creates a feeling of unity, of oneness with all people. From peanut farmers in West Africa to computer technicians in California to fishermen in Norway to businesswomen in Hong Kong, all humans carry inside them a common genetic heritage.

On the other hand, the human genome shows that the small differences which do exist are important ones which give each person his or her uniqueness in terms of skin colour, facial features or resistance to disease. These differences should be appreciated and celebrated as strengths. Unfortunately, they are often the basis of discrimination and misunderstanding.

Can one genetic group be considered genetically superior to another? History has shown that many people think so, yet genetics shows that this is not the case. All human populations, whatever slight differences their genomes may have, deserve equal esteem as human beings.

Gene transfer

The technique of taking a gene out of one organism (the donor organism, e.g. a fish) and placing it in another organism (the host organism, e.g. a tomato) is a genetic engineering procedure called gene transfer. Just such a transfer was done to make tomatoes more resistant to cold and frost.

It is possible to put one species' genes into another's genetic makeup because DNA is universal: as you will recall (Chapter 3, page 63), all known living organisms use the bases A, T, C and G to code for proteins. The codons they form always code for the same amino acids, so transferred DNA codes for the same polypeptide chain in the host organism as it did in the donor organism. In the example above, proteins used by fish to resist the icy temperatures of arctic waters are now produced by the modified tomatoes to make them more resistant to cold.

Another example of gene transfer is found in Bt-corn, which has been genetically engineered to produce toxins that kill the bugs which attack it. The gene, as well as the name, come from a soil bacterium, *Bacillus thuringiensis*, which has the ability to produce a protein that is fatal to the larvae of certain crop-eating pests.

Pests such as this corn earworm are responsible for reduced yields in traditional corn crops.

The manipulation of genes raises some challenging questions. For many of these questions, there is not enough conclusive scientific data to reach a satisfactory answer.

- Is it ethically acceptable to alter an organism's genetic integrity?
- If the organism did not have that gene in the first place, could there be a good reason for its absence?
- Why are people so worried about this new technology? In selective breeding, thousands of genes are mixed and matched. With GMOs, only one gene is changed. Is that not less risky and dangerous than artificial selection?
- Would strict vegetarians be able to eat a tomato which has a fish gene in it?
- Does research involving GM animals add a whole new level to animal cruelty and suffering in laboratories?
- If Bt-crops kill insects, what happens to the local ecosystem which relies on the insects for food or pollination?

Cutting, copying and pasting genes

Although the laboratory techniques are complex, the concepts are not difficult.

Cutting and pasting DNA

The 'scissors' used for cutting base sequences are enzymes. Restriction enzymes called endonucleases find and recognize a specific sequence of base pairs along the DNA molecule. Some can locate target sequences which are sets of four base pairs, others locate sets of six pairs. The endonucleases cut the DNA at specified points. If both the beginning and the end of a gene are cut, the gene is released and can be removed from the donor organism. For pasting genes, the enzyme used is called DNA ligase. It recognizes the parts of the base sequences that are supposed to be clicked together, called the sticky ends, and attaches them.

Copying DNA (DNA cloning)

This is more complex because a host cell is needed in addition to the cutting and pasting enzymes described above. Although yeast cells can be used as host cells, the most popular candidate in genetic engineering is the bacterium *Escherichia coli*.

Like other prokaryotes, most of the genetic information for *E. coli* is in the bacterium's single chromosome. However, some DNA is found in structures called plasmids. Plasmids are small circles of extra copies of DNA floating around inside the cell's cytoplasm. To copy a gene, it must be glued into a plasmid.

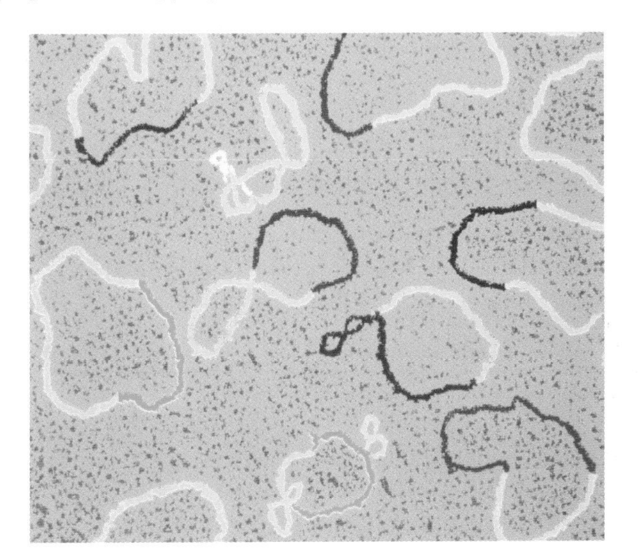

This is a false-colour electronmicrograph of plasmids.

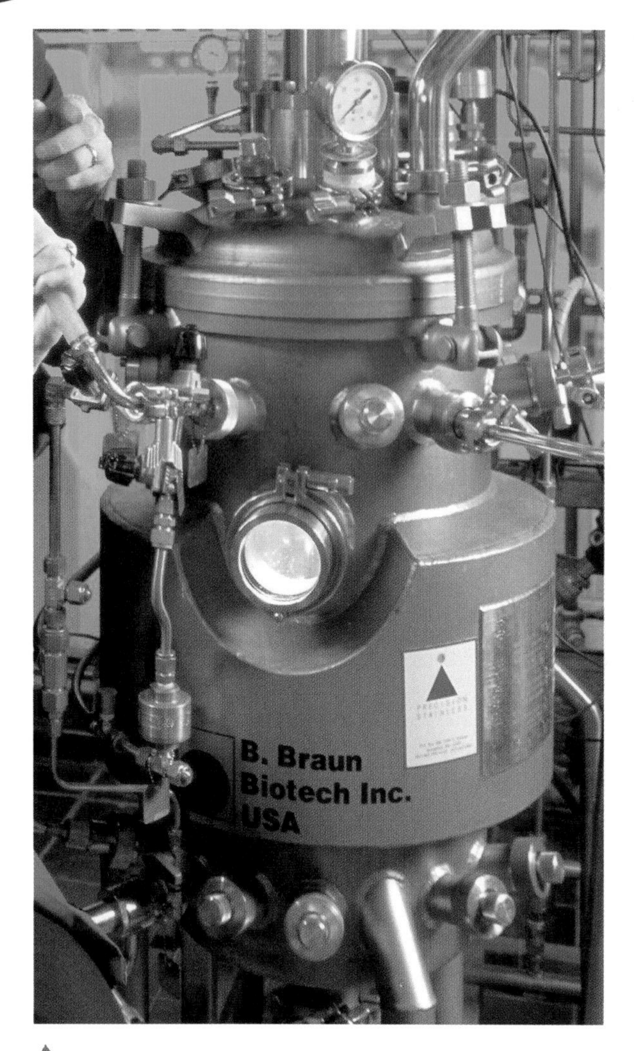

This is a bioreactor and is used to grow bacteria which have recombinant DNA.

To do this, a plasmid is removed from the host cell and cut open using a restriction endonuclease. The gene to be copied is placed inside the open plasmid. This process is sometimes called gene splicing. The gene is pasted into the plasmid using DNA ligase. The plasmid is now called a recombinant plasmid and it can be used as a vector, a tool for introducing a new gene into an organism's genetic makeup.

In the final step needed for copying (or cloning) the gene, the vector is placed inside the host bacterium and the bacterium is given its ideal conditions to grow and proliferate. This is done by putting the bacterium into a bioreactor, a vat of nutritious liquid kept at a warm temperature.

Not only does the host cell make copies of the gene as it reproduces, but since the gene is now in its genetic makeup, the modified *E. coli* cell expresses the gene and synthesizes whatever protein the gene codes for. This process has been used successfully in getting *E. coli* to make human insulin, a protein needed to treat diabetes (Chapter 6, page 184). The older technique for obtaining insulin involves extracting it from cow and pig carcasses generated by the meat industry, but this has caused allergy problems. Using recombinant human DNA avoids that problem.

Genetically modified organisms

A genetically modified organism (GMO) is one that has had an artificial genetic change using the techniques of genetic engineering such as gene transfer or recombinant DNA described above. One of the main reasons for producing a genetically modified organism is for it to be more competitive in food production.

One company controls nearly all of the worldwide GM market. To grow their GM crops, farmers must pay a yearly licensing fee to the company. The same company sells the herbicides which are compatible with the GM plants.

This raises concerns that too much economic power will be in the hands of one corporation. What effects will this have on farmers and on local economies?

Transgenic plants

The simplest kind of GM food is one in which an undesirable gene is removed. In some cases, another more desirable gene is put in its place but in other instances, only the introduction of a new gene is needed, no DNA has to be removed.

Whichever technique is applied, the end result is either that the organism no longer shows the undesired trait or that it shows one which genetic engineers want. The first commercial example of a GM food was the 'Flavr Savr' tomato. It was first sold in the US in 1994 and had been genetically modified to delay the ripening and rotting process so that it would stay fresher longer. Although it was an ingenious idea, the company lost so much money from the project that it was abandoned a few years later.

Another species of tomato was modified by a bioengineering company to make it more tolerant to higher levels of salt in the soil. This makes it easier to grow in certain regions of high salinity. One of the claims of the biotech industry is that GM foods will help solve the problem of world hunger by allowing farmers to grow foods in various otherwise unsuitable conditions. Critics point out that the problem of hunger in the world is one of food distribution, not food production.

Another plant of potential interest to the developing world is a genetically modified rice plant which has been engineered to produce beta carotene in the rice grains. The aim is that the people who eat this rice will not have deficiencies in vitamin A (the body uses beta carotene to form vitamin A).

Transgenic animals

One way of genetically engineering an animal is to get it to produce a substance which can be used in medical treatment. Consider the problem faced by some people with haemophilia – a blood condition in which their blood does not clot because they lack a protein called factor IX. If such people can be supplied with factor IX, their problem will be solved. The least expensive way of producing large amounts of factor IX is to use transgenic sheep. If a gene which codes for the production of factor IX is associated with the genetic information for milk production in a female sheep, she will produce that protein in her milk.

In the future, a wide variety of genetic modifications may be possible. Perhaps inserting genes to make animals more resistant to parasites, to make sheep produce pre-dyed wool of any chosen colour, to produce prize-winning show dogs, faster racehorses ... the possibilities seem almost boundless and it is difficult to imagine what the future might be like.

Is genetic engineering a good or a bad thing?

Genetic engineering raises many profound social and ethical questions. As you read through the ideas below, note which ones you agree with. Can you justify your opinions?

Benefits, promises, and hopes for the future

- GM crops will help farmers by improving food production.
- GM crops which produce their own pest-control substances will be beneficial to the environment because fewer chemical pesticides will be needed.
- Using GMOs to produce rare proteins for medications or vaccines could be, in the long run, less costly and produce less pollution than synthesizing such proteins in laboratories.
- Farmers can be more in control of what crops or livestock they produce. There is always some randomness in breeding; genetic modification makes the process less of a gamble. It is also much quicker than selective breeding.
- The multinational companies who make GM plants claim that they will enable farmers in developing nations to help reduce hunger by using pest-resistant crops or GM plants which require less water.

Is altering a plant or animal's DNA justified by the benefit that it may bring to humans? Just how far is it acceptable to go in manipulating a species' genes? Should experiments be stopped if they start to modify the ecosystems they are in? Who should decide?

Harmful effects, dangers, and fears

- No one knows the long-term effects of GMOs in the wild. Efforts to keep GM plants under control in well-defined areas have failed and pollen from GM crops has escaped to neighbouring fields. Genes from GM plants could be integrated into wild species giving them an unnatural advantage over other species and an ability to take over the habitat.
- There is a danger that the genes could cross species. It has been proven possible in laboratories, so there is a possibility in nature too. Again, no one knows the consequences of genes crossing species.

If there are risks associated with GM plants and animals, should all research and experiments be banned? How is it possible to know how severe the risks are? In order to assess the risks, experiments need to be done. The experiments themselves, however, present risks.

The technology is so complex that it is often difficult for scientists to explain what the risks are to farmers, consumers and lawmakers. As a result, it is a challenge for all citizens to make an educated decision about whether or not GM foods are safe to eat and safe for the environment. Without complete information, it is easy to fall into the trap of irrational fear or complete apathy.

Some groups who have made up their minds that GM crops are dangerous have decided to destroy the plants in the fields where they are being grown and tested. Does such action make the world a safer place?

- Bt-crops which produce toxins to kill insects could be harmful to humans because, unlike chemical pesticides which are only applied to the outer surface, the toxins are found throughout the plant.
- There are risks for allergies: if someone is not allergic to natural tomatoes but is allergic to GM tomatoes, they will need to know which one they are eating. But there is no difference in the outward appearance of the fruit and food labelling is not always clear.
- Critics are worried that large portions of the human food supply will be the property of a small number of corporations.
- High-tech solutions are not necessarily better than simpler solutions. Crop production could be increased by teaching farmers how to use water and natural pest-control systems more efficiently.
- A proliferation of genetically modified organisms may lead to a decrease in biodiversity.

Clones and cloning

The definition of a clone is a group of genetically identical organisms or a group of cells artificially derived from a single parent. In either case, the resulting cells or organisms were made using laboratory techniques. In farming, clones have been made for decades by regenerating plant material or by allowing an in-vitro fertilized egg to divide to make copies of itself.

Until recently, cloning was only possible using genetic information from an egg cell. Fertilized eggs are not differentiated (specialized) yet. After dividing many times, some of the cells will specialize into muscle cells, others into nerves, others into skin and so on until a fetus forms. For a long time, it was thought that once a cell has gone through differentiation, it cannot be used to make a clone. But then there was Dolly.

Cloning using a differentiated animal cell

In 1996, a sheep by the name of Dolly was born. She was the first clone whose genetic material did not originate from an egg cell. Here is how researchers at the Roslin Institute in Scotland produced Dolly (see also Figure 4.10).

This is Dolly with Ian Wilmut, a member of her cloning team.

1 From the original donor sheep to be cloned, a somatic cell (non-gamete cell) from the udder was collected and cultured. The nucleus was removed from a cultured cell.
2 An unfertilized egg was collected from another sheep and its nucleus was removed.
3 Using a zap of electrical current, the egg cell and the nucleus from the cultured somatic cell were fused together.
4 The new cell developed in vitro in a similar way to a zygote and started to form an embryo.
5 The embryo was placed in the womb of a surrogate mother sheep.
6 The embryo developed normally.
7 Dolly was born, and was presented to the world as a clone of the original donor sheep.

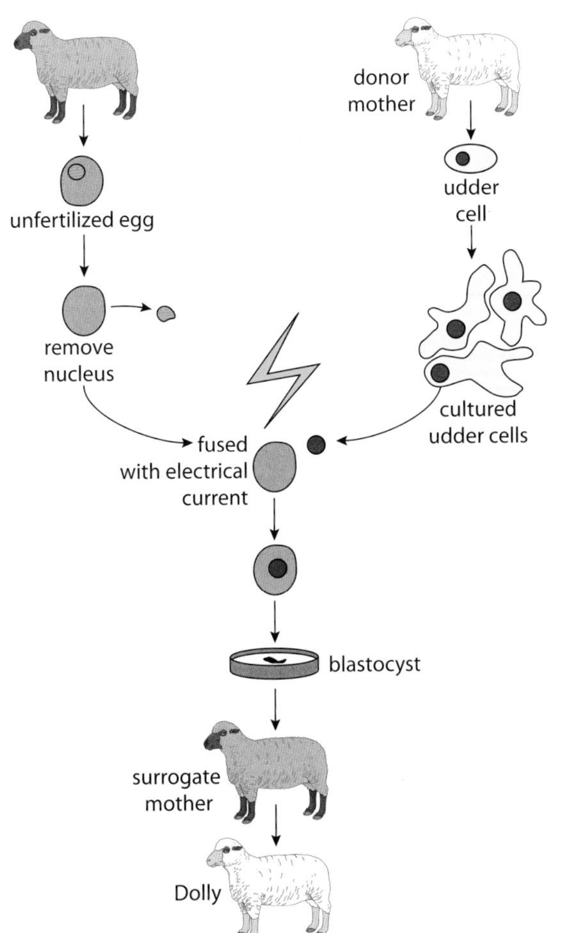

Figure 4.10 The step-by-step process of how the clone Dolly was made.

This kind of cloning is called reproductive cloning because it makes an entire individual.

Cloning using undifferentiated cells

In some cases, scientists are not interested in making an organism but simply in making copies of cells. This second type of cloning is called therapeutic cloning and its aim is to develop cells which have not yet gone through the process of differentiation. Since the first technique in this area involved using embryos, the cells are referred to as embryonic stem cells, and the branch of lab work which investigates therapeutic cloning is called stem cell research.

The idea of cloning often provokes strong negative reactions from people, especially when the only information they have comes from science fiction or horror films.

When making ethical decisions about what is good and bad, or right and wrong, it is important to be as well informed as possible.

In dealing with the ethical issues of cloning, it should be stressed that there are two distinct forms of cloning:

- reproductive cloning – making copies of entire organisms;
- therapeutic cloning – making copies of embryonic stem cells.

Some people think that both are unacceptable, others think both are fine and some are in favour of one but not the other. Where do you stand?

Ethical issues surrounding therapeutic cloning

Since therapeutic cloning starts with the production of human embryos, it raises fundamental issues of right and wrong. Is it ethically acceptable to generate a new human embryo for the sole purpose of medical research? In nature, embryos are created only for reproduction and many people believe that using them for experiments is unnatural and wrong.

However, the use of embryonic stem cells has lead to major breakthroughs in the understanding of human biology. What was once pure fiction is coming closer and closer to becoming an everyday reality thanks to stem cell research:

- growing skin to repair a serious burn;
- growing new heart muscle to repair an ailing heart;
- growing new kidney tissue to rebuild a failing kidney.

With very rare exceptions, the vast majority of researchers and medical professionals are against the idea of reproductive cloning in humans. However, there is a growing popularity for the pursuit of therapeutic cloning since the promises of stem cell research are so enticing.

Exercises

14 Explain why PCR is necessary.

15 Explain the central ethical issue concerning stem cell research.

16 Justify whether the benefits outweigh the risks in genetically modifying plants and animals.

17 Look at the foods in your house. Are food labels today effective at indicating when and how much of the food is genetically modified? Justify your answer.

Practice questions

1 Describe the consequence of a base substitution mutation with regards to sickle cell anaemia. *(7 marks)*

2 The diagram below shows the pedigree of a family with red–green colour blindness, a sex-linked condition.

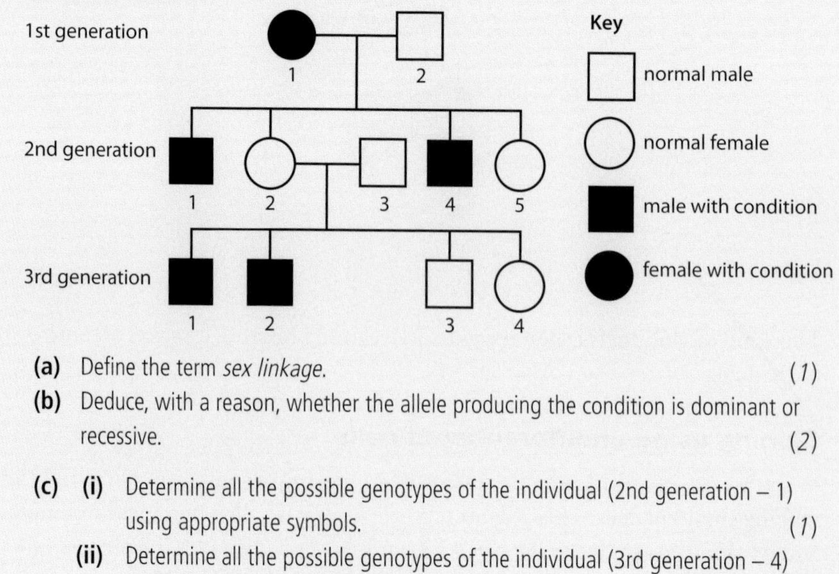

(a) Define the term *sex linkage.* *(1)*

(b) Deduce, with a reason, whether the allele producing the condition is dominant or recessive. *(2)*

(c) **(i)** Determine all the possible genotypes of the individual (2nd generation – 1) using appropriate symbols. *(1)*

(ii) Determine all the possible genotypes of the individual (3rd generation – 4) using appropriate symbols. *(1)*

(Total 5 marks)

3 Outline the differences between the behaviour of the chromosomes in mitosis and meiosis. (*5 marks*)

4 **(a)** Define the term *codominance*. (*1*)

(b) A man of blood type AB and a woman of blood type B are expecting a baby. The woman's mother had blood type O. Deduce the possible phenotypes of the offspring from the cross. Include the parents' genotypes, the gametes, the F1 genotypes and the F_1 phenotypes. (*4*)

(*Total 5 marks*)

5 Discuss the potential benefits and possible harmful effects of genetic modification. (*7*)

6 Outline DNA profiling (genetic fingerprinting), including one way in which it has been used. (*5*)

7 Karyotyping involves arranging the chromosomes of an individual into pairs. Describe one application of this process, including the way in which the chromosomes are obtained. (*5*)

5 Ecology and evolution

Introduction

What do you think these are? Can you answer the questions in the introduction for these organisms?

This chapter is all about making sense of the millions of organisms found on Earth. Approximately 1.5 million different species have been identified by biologists, over half of which are insects. No one knows how many species actually exist on the planet, but estimates are that there could easily be 10 million or more.

Simply finding a species and giving it a name is not enough to understand it. Biologists are interested in the following kinds of question when they study a living organism.

- What does it eat?
- How does it reproduce?
- Where does it prefer to live?
- How does it interact with other species?
- How does it interact with its environment?
- How are human activities affecting this organism's way of life?
- How is this organism related to other organisms which look similar to it?
- How has this species evolved over millions of years?

We are going to investigate how scientists go about answering these questions and some of the discoveries they have made.

5.1 Communities and ecosystems

Assessment statements

5.1.1 Define *species*, *habitat*, *population*, *community*, *ecosystem* and *ecology*.
5.1.2 Distinguish between *autotroph* and *heterotroph*.
5.1.3 Distinguish between *consumers*, *detritivores* and *saprotrophs*.
5.1.4 Describe what is meant by a food chain, giving three examples, each with at least three linkages (four organisms).
5.1.5 Describe what is meant by a food web.
5.1.6 Define *trophic level*.
5.1.7 Deduce the trophic level of organisms in a food chain and a food web.
5.1.8 Construct a food web containing up to 10 organisms, using appropriate information.
5.1.9 State that light is the initial energy source for almost all communities.
5.1.10 Explain the energy flow in a food chain.
5.1.11 State that energy transformations are never 100% efficient.
5.1.12 Explain reasons for the shape of pyramids of energy.
5.1.13 Explain that energy enters and leaves ecosystems, but nutrients must be recycled.
5.1.14 State that saprotrophic bacteria and fungi (decomposers) recycle nutrients.

Key terms

In order to understand this chapter, it is essential to learn the following terms and their definitions.

- *Species* – A group of organisms that can interbreed and produce fertile offspring. Members of the same species have a common gene pool (i.e. a common genetic background).
- *Habitat* – The environment in which a species normally lives or the location of a living organism. Examples: tree branch, cliff face, seashore or even your large intestine, which is the habitat of certain bacteria and parasites.
- *Population* – A group of organisms of the same species which live in the same area at the same time.
- *Community* – A group of populations living and interacting with each other in an area. Examples: the soil community in a forest, the fish community in a river.
- *Ecosystem* – A community and its abiotic environment. This idea is similar to a habitat except that it refers to where a group of interacting populations live instead of where a single species lives. Examples: forest, pond or ocean.
- *Ecology* – The study of relationships between living organisms and between organisms and their environment.

Shorelines and oceans make up habitats and ecosystems which contain diverse populations and communities of many species.

To understand the idea of 'fertile offspring', consider what happens when two different but similar species mate and produce offspring. For example, a female horse and a male donkey can mate and produce a mule. However, mules cannot usually mate to make more mules. Since the offspring (mules) are not fertile, no new species has been created. Instead, a mule is called an interspecific hybrid. When a male lion and a female tiger are crossed, a liger is the hybrid formed.

The word 'environment', which is in three of the definitions above, refers to everything which surrounds an organism. The environment can be broken up into four main components:

- the hydrosphere (water);
- the atmosphere (gases);
- the lithosphere (rocks);
- the biosphere (all living organisms).

The first three are abiotic (non-living) components of the environment and the last one is the biotic component.

Examples of biotic components	Examples of abiotic components
plants, fungi, animals, protoctists, bacteria ...	light, heat, minerals, oxygen, humidity ...

▲ Can you identify the biotic and abiotic components in this photo?

Autotrophs and heterotrophs

A sheep eating grass is an example of a heterotroph feeding on an autotroph.

Autotrophs

Some organisms are capable of making their own organic molecules as a food source. These are called autotrophs and they synthesize their organic molecules from simple inorganic substances. This involves using photosynthesis. In other words, they can take light energy from the Sun, combine it with inorganic substances and obtain chemical energy in the form of organic compounds. Since autotrophs make food which is often used by other organisms, they are producers.

Examples of autotrophs:

- cyanobacteria;
- algae;
- grass;
- trees.

Heterotrophs

Heterotrophs cannot make their own food from inorganic matter and must obtain organic molecules from other organisms. They get their chemical energy from the autotrophs or from other heterotrophs. Since they rely on others for food, they are referred to as consumers. They ingest organic matter that is living or has been recently killed.

Examples of heterotrophs:

- zooplankton;
- fish;
- sheep;
- insects.

Detritivores and saprotrophs

If detritivores and saprotrophs did not exist, waste would pile up (see Decomposers, page 121).

Detritivores

Some organisms eat non-living organic matter. Detritivores, for example, eat dead leaves, faeces or carcasses. Earthworms, woodlice and dung beetles are detritivores found in the soil community. Many, but not all, bottom feeders in rivers, lakes and oceans are detritivores.

A dung beetle is a detritivore.

Saprotrophs

Organisms called saprotrophs live on or in non-living organic matter, secreting digestive enzymes into it and absorbing the products of digestion. They play an important role in the decay of dead organic materials. The fungi and bacteria which make up saprotrophs are also called decomposers since their role is to break down waste material.

Fungi are saprotrophs.

Food chains

When studying feeding habits, it is convenient to write down which organism eats which by using an arrow. Thus, herring → seal indicates that the seal eats the herring. When the seal's eating habits are investigated and the herring's diet is considered, new organisms can be added: copepods (a common form of zooplankton) are eaten by the herring and great white sharks eat seals. Lining up organisms with arrows between them is how food chains are represented. Here are three examples of food chains from three different ecosystems.

Grassland ecosystem:

grass → grasshoppers → toad → hognose snake → hawk

River ecosystem:

algae → mayfly larva → juvenile trout → kingfisher

Marine ecosystem:

diatoms → copepods → herring → seal → great white shark

The definition of a food chain is a sequence showing the feeding relationships and energy flow between species. In other words, it answers the question, 'Who eats whom?' The direction of each arrow shows which way the energy flows.

Food webs

A food web is an interconnecting series of food chains. Since one organism often eats more than just one type of food, a simple food chain does not tell the whole story.

For example, mayfly larvae are food for many organisms, not just juvenile trout. Sculpin (a type of fish eaten by trout), adult trout and stonefly larvae all eat mayfly larvae. As a result, more than one arrow can emanate from – or arrive at – an organism in a food web.

Similarly, the kingfisher does not eat exclusively juvenile trout. It eats other small fish and sometimes small mammals, birds and insects, too. Food webs can become very complex, very quickly (see Figure 5.1).

Trophic levels

An organism's trophic level refers to its position in a food chain. Trophic levels offer a way of classifying organisms by their feeding relationships with the other organisms in the same ecosystem. The first trophic level is occupied by the producers. The second trophic level is occupied by the primary consumer and so on as shown in the table below – to read it, start at the bottom and work your way up.

● **Examiner's hint:** Although primary consumers are plant eaters and secondary consumers eat flesh, the terms herbivore or carnivore should be used only to describe their diet, not to describe their trophic levels.

Trophic level	
T5	quaternary consumer
T4	tertiary consumer
T3	secondary consumer
T2	primary consumer
T1	producer

Three trophic levels can be seen in this photograph: a producer, a primary consumer and a secondary consumer.

In order for food chains to function, there must be a large number of producers and fewer and fewer members of each subsequent trophic level. In a day of feeding, for example, a herring might eat thousands of copepods, the seal might eat tens of herring but a great white shark might eat only one seal.

Determining an organism's trophic level

As shown in the table above, a food chain starts with a producer. Now, look at this food chain:

algae → mosquito larva → dragonfly larva → fish → raccoon

It is another example from a river ecosystem. In order to determine the trophic level of each organism, start with the producer – it is the only organism in the chain which photosynthesizes: the algae.

The second trophic level, which is occupied by the primary consumer, is found by determining which organism eats the producers: here, the mosquito larva. The next levels are occupied by the secondary then tertiary consumers and so on. The highest trophic level in a food chain is occupied by a top predator, which in this case is a raccoon. Top predators are animals such as birds of prey, bears, sharks, or humans.

Some food chains have six trophic levels, but most have four. The number of levels can be limited by how much energy enters the ecosystem. Since so much is lost at each level, low energy at the start will quickly dissipate whereas abundant energy for producers can sustain several trophic levels.

Trophic levels exist in food webs as well. Since food webs are more complex, determining an organism's trophic level is more challenging. The best way to go about it is to isolate a single food chain inside the web and analyse it in the same manner as above.

Look at the food web in Figure 5.1. It is a river food web which includes the food chain:

algae → mayfly larva → juvenile trout → kingfisher

You should be able to determine why the juvenile trout, for example, is a secondary consumer (that is, it is on the third trophic level).

Constructing a food web

To construct a food web, you start with the producer at the bottom of the diagram. In a river, it is often algae or plants. It can also be debris from leaves which have settled at the bottom of the river. Above the producers, include all the primary consumers: the mayfly larva, the caddis fly larva, and the blackfly larva. Continue until you get to the top predator, in this case, the kingfisher or the adult trout. Without looking again at Figure 5.1, try to construct this food web:

- algae is eaten by the mayfly larva;
- leaf debris is eaten by the caddis fly larva and the blackfly larva;
- the mayfly larva, caddis fly larva and the blackfly larva are eaten by juvenile trout;
- the juvenile trout is eaten by the kingfisher;
- the mayfly larva is also eaten by the sculpin and the stonefly larva, all of which are eaten by adult trout.

This is a relatively simple food web. Specialists in ecology spend years studying food webs which can involve dozens of organisms.

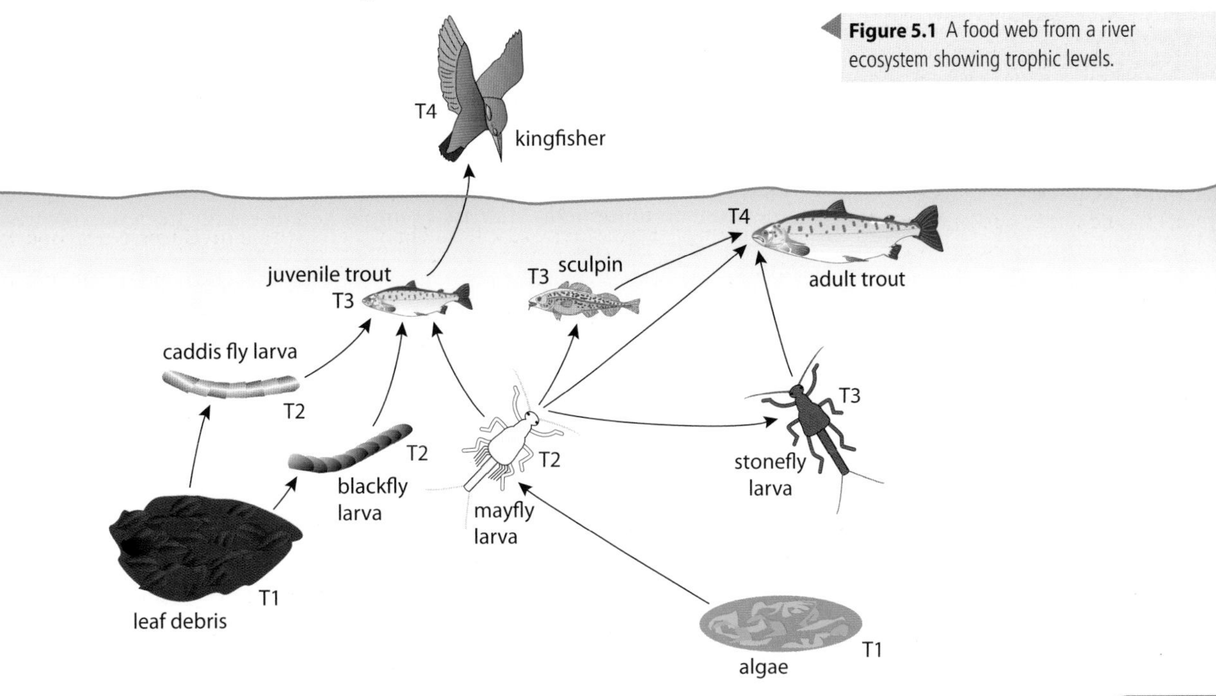

Figure 5.1 A food web from a river ecosystem showing trophic levels.

Some difficulties arise when putting organisms into trophic levels in a food web. First of all, some organisms occupy more than one trophic level or take their food from multiple trophic levels. Secondly, many ecosystems are not fully understood and the feeding preferences of all the trophic levels are simply not known. Thirdly, a food web does not show changes over time as populations change, notably because of seasonal differences.

Energy – the importance of light

The most important organisms in any food chain are the producers because without them, the next trophic level would have nothing to eat. Since photosynthetic organisms occupy the first trophic level, sunlight is the initial energy source for almost all communities.

Sunlight is the initial source of energy for all vegetation.

Some food chains can start without sunlight. For example, deep sea hydrothermal vents provide chemical energy instead of light energy to start a food chain.

If you think about the vast majority of the foods you eat, you can trace the energy they contain to sunlight. Milk, for example, gets its energy from the cow which produced it. The cow got its energy from grass, which got its energy from sunlight. In marine ecosystems, most food chains start with phytoplankton which get their energy from sunlight.

- **Examiner's hint:** Be sure to be precise with your vocabulary in an exam: 'Sun' is not the same as 'sunlight'.

Leaves are a food source when they are green and photosynthesizing, and continue to be a source of food when they have fallen to the ground and turned brown. Decomposers and detritus feeders such as soil organisms or some of the bottom feeders in a river depend on this dead organic matter for food.

1816 is sometimes called the year with no summer. It was cruelly cold for farmers. A combination of natural phenomena, including exceptionally high volcanic activity, reduced some of the Sun's light energy. The consequence was crop failure which in some cases lead to famine.

Energy flow

Once light energy has been absorbed by producers, the chemical energy obtained by photosynthesis is available to the next trophic level. Energy is transferred from one organism to the next when carbohydrates, lipids or proteins are digested.

When grass is eaten by a cow, chemical energy is transferred to the cow. However, if a clump of grass dies without being grazed on, decomposers such as fungi will use the energy it has to offer.

Inside the cow, the chemical energy is used for cellular respiration. Any heat generated by cellular respiration is lost to the environment. If the cow is eaten, some of the chemical energy in its body can be passed on to the next trophic level. If it dies and is not eaten, detritivores and decomposers will use its available energy.

The decomposers, too, perform cellular respiration and as a result, any heat thus produced will also be lost to the environment. This is not the only source of energy loss from one trophic level to the next, as we will see.

Energy loss

Only chemical energy can be used by the next trophic level (see Figure 5.2). And only a small amount of the energy which an organism absorbs is converted into chemical energy. In addition, no organism can utilize 100% of the energy present in the organic molecules of the food it eats. Typically, only 10–20% of the energy is used from the previous step in the food chain.

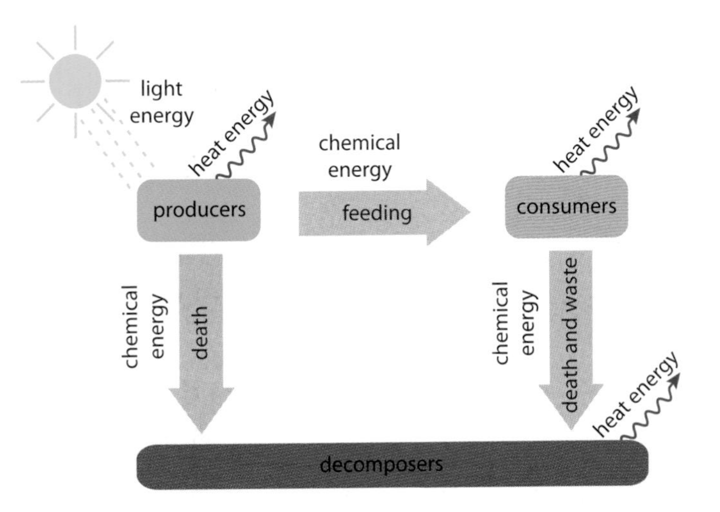

Figure 5.2 Energy flow and energy loss.

Here are the main reasons why not all of the energy present in an organism can be used by the organism in the next trophic level.

- Not all of the organism is swallowed as a food source – some parts are abandoned and will decay.
- Not all the food swallowed can be absorbed and used in the body – owls, for example, spit up the hair and bones of the animals they eat, and undigested seeds can be found in the faeces of fruit-eating animals.
- Some organisms die before being eaten by an organism from the next trophic level.
- There is considerable heat loss due to cellular respiration at all trophic levels (shown by the wavy arrows in Figure 5.2).

The loss of heat varies from one type of organism to the next. Most animals have to move, for example, which requires much more energy than a fixed plant uses. Warm-blooded animals must use a considerable amount of energy to maintain their body temperature.

Pyramid of energy

A pyramid of energy is used to show how much and how fast energy flows from one trophic level to the next in a community (see Figure 5.3). The units used are energy per unit area per unit time: kilojoules per square metre per year, kJ m^{-2} yr^{-1}. Since time is part of the unit, energy pyramids take into account the rate of energy production, not just the quantity.

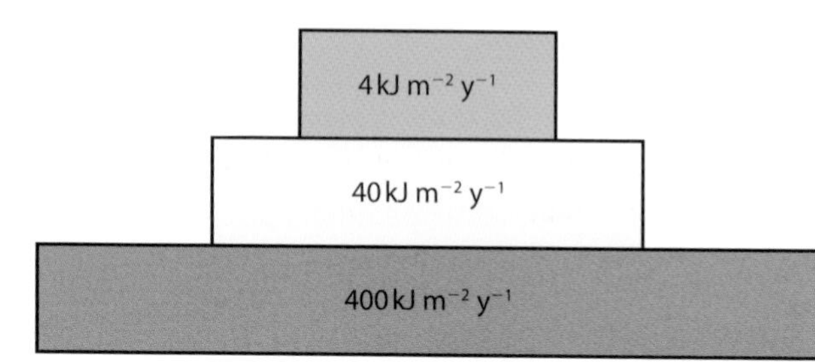

Figure 5.3 Pyramid of energy (not to scale).

• **Examiner's hint:** When drawing a pyramid of energy for an exam, do not forget to label it with numbers and be sure you know the proper units to write.

Because energy is lost, as explained in the previous section, each level is always smaller than the one before. It would be impossible to have a 3rd trophic level wider than a 2nd trophic level, for example, because organisms cannot create energy, they can only transfer it inefficiently.

Be careful not to confuse pyramids of energy with pyramids of numbers which show the population sizes of each trophic level, not energy.

How many trophic levels are shown in this picture from Peru?

Energy and nutrients

In an ecosystem, energy enters in the form of light, is converted into chemical energy by producers and transferred to consumers in the various trophic levels. Most of it is lost as heat. The term 'lost' is used to refer to the fact that organisms cannot recycle heat energy. Once it has been radiated into the environment, it cannot be collected back and used as an energy source by the ecosystem. In short, energy enters as light and exits as heat. As long as the Sun continues to shine and bring new light energy every day, the ecosystem can continue to use up and lose heat energy this way.

Even though tons of space dust fall on the Earth, there is not enough to meet the mineral needs of the biosphere. As a result, organisms must recycle the carbon, nitrogen and other elements and compounds necessary for life to exist. For this, organisms must find what they need within the materials available in their own habitat. The problem is that organisms absorb valuable minerals and organic compounds and use them to build their cells. These resources are then locked up and unavailable to others – except, of course, by eating or decomposition.

• **Examiner's hint:** In exams, students often make mistakes such as claiming that plants make energy or that energy is recycled. Be sure that you fully understand the differences between how energy and nutrients are used.

Decomposers

An effective way to unlock the precious nutrients stored in the cells of plants and animals is through decay. Decomposers (saprotrophs and detritivores) break down the body parts of dead organisms. The digestive enzymes of decomposers convert the organic matter into a more usable form for themselves and for other organisms. For example, proteins from a dead organism are broken down into ammonia and, in turn, ammonia can have its nitrogen converted into useful nitrates by bacteria.

In this way, decomposers recycle nutrients so that they are available and not locked inside the bodies or wastes of the other members of the ecosystem. Decomposers play a major role in the formation of soil, for example, without which plant growth would be greatly impaired if not impossible.

Saprotrophs on a dead log.

Exercises

1 Distinguish between habitat and ecosystem.

2 Look at these food chains again. Determine the trophic levels for each organism listed.

(a) Grassland ecosystem:
grass → grasshoppers → toad → hognose snake → hawk

(b) River ecosystem:
algae → mayfly larva → juvenile trout → kingfisher

(c) Marine ecosystem:
diatoms → copepods → herring → seal → great white shark

3 From the following information, construct a food web:

- grass is eaten by rabbits, grasshoppers, and mice;
- rabbits are eaten by hawks;
- grasshoppers are eaten by toads and mice and garter snakes;
- mice are eaten by hawks;
- toads are eaten by hognose snakes;
- hognose snakes are eaten by hawks;
- garter snakes are eaten by hawks.

4 From the food web in exercise 3, determine the trophic level of the toad.

5.2 The greenhouse effect

Assessment statements

- 5.2.1 Draw and label a diagram of the carbon cycle to show the processes involved.
- 5.2.2 Analyse the changes in concentration of atmospheric carbon dioxide using historical records.
- 5.2.3 Explain the relationship between rises in concentrations of atmospheric carbon dioxide, methane and oxides of nitrogen and the enhanced greenhouse effect.
- 5.2.4 Outline the precautionary principle.
- 5.2.5 Evaluate the precautionary principle as a justification for strong action in response to the threats posed by the enhanced greenhouse effect.
- 5.2.6 Outline the consequences of a global temperature rise on arctic ecosystems.

The carbon cycle

To understand the greenhouse effect, we need to appreciate the importance of carbon and how it is recycled. Carbon is such a crucial element to living organisms that it is part of the basis of the definition of a living thing. You will recall from Chapter 3, page 50, that the term 'organic' implies that carbon must be present. Hence, life on Earth is referred to as carbon-based life.

The environment can be broken up into four main components:

- the hydrosphere (water);
- the atmosphere (gases);
- the lithosphere (rocks);
- the biosphere (all living organisms).

Not only is carbon found in the biosphere in carbohydrates, proteins, lipids and vitamins, it is also found in the atmosphere as carbon dioxide and in the lithosphere as carbonates and fossil fuels in rocks. Petroleum, from which such products as gasoline, kerosene and plastics are made, is rich in carbon because it originated from partially decomposed organisms which died millions of years ago.

Just like other elements needed for life, carbon is constantly recycled in the environment. For example, carbon in the form of atmospheric carbon dioxide (CO_2) is absorbed by photosynthetic organisms such as photosynthetic bacteria, phytoplankton, plants and trees. As you will recall, these producers are eaten by consumers which use the carbon in their bodies. Cellular respiration from all trophic levels and from decomposers produces carbon dioxide, which is released back into the atmosphere.

Figure 5.4 shows the carbon cycle on Earth plus additional sources of carbon dioxide in Earth's atmosphere, notably humans burning wood or non-renewable resources such as petroleum products. There is another natural source of carbon dioxide which is not shown: volcanic activity.

Marine organisms take carbon out of the water and use some of it to make their carbonate shells. An accumulation of these shells as sediments at the bottom of the ocean can trap carbon for millions of years in limestone. When cement is made by humans for construction, limestone is used as one of the main ingredients. In the process, some of the carbon is released as carbon dioxide.

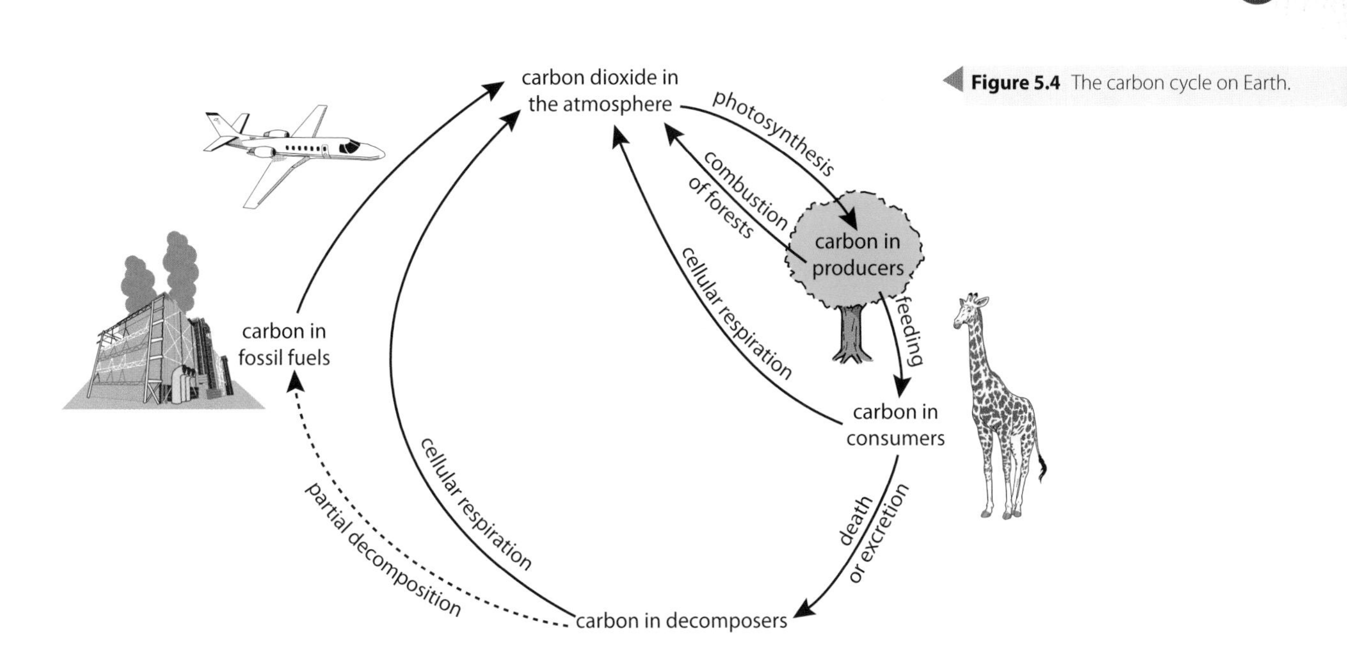

Figure 5.4 The carbon cycle on Earth.

First proposed by James Lovelock, the Gaia hypothesis views all of the biosphere on Earth as a single living organism. This idea comes from the fact that Earth has characteristics of self-regulation which maintain the conditions necessary to support life, including temperature regulation, atmospheric oxygen and carbon dioxide regulation, among others.

Other scientists prefer to look at these interactions as the cogs on the wheels of a finely tuned machine.

In what ways do you think these metaphors are useful in understanding the complex systems of this planet? In what ways do you think they are valid models and in what ways are they not valid models?

Can planet Earth be considered one giant living organism?

Changes in atmospheric carbon dioxide level

Ever since machines started replacing hand tools in the 1800s, humans have produced increasing quantities of carbon dioxide from factories, transportation and other uses of fossil fuels such as coal and oil. In addition, burning forests to make way for farmland or burning wood for cooking or heating has contributed to this increase.

Over the decades, human pollution has produced enough carbon dioxide to considerably raise its percentage in the planet's atmosphere. Estimates suggest that the level of carbon dioxide in the atmosphere has increased by over 25% compared to its pre-industrial revolution levels.

NASA data on carbon dioxide levels in the atmosphere 1958–2004. The up-and-down pattern is caused by seasonal fluctuations in activities such as photosynthesis.

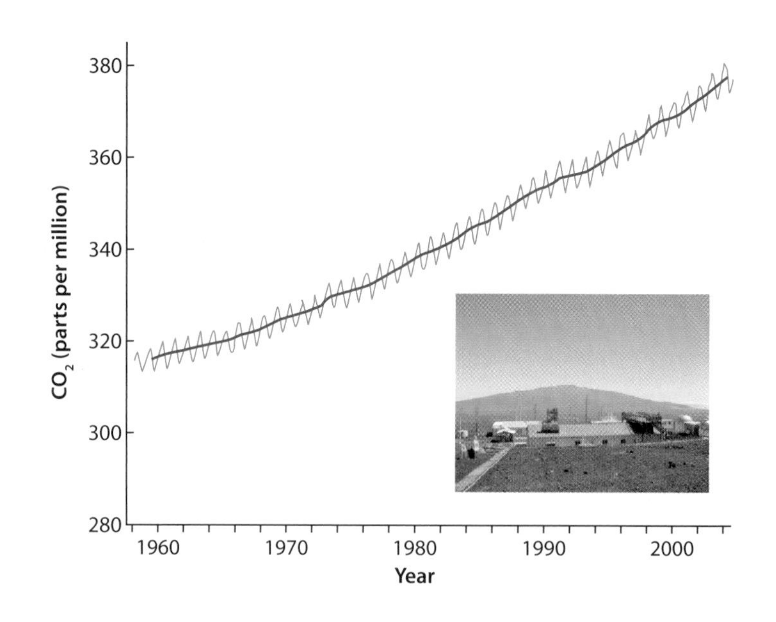

● **Examiner's hint:** When asked to label a specific point on a graph which is presented in an exam, be sure to make the mark in a clear, unambiguous way. A circle or an arrow is best.

Each vehicle produces its own mass in carbon dioxide every year.

Since plants, phytoplankton and photosynthetic bacteria are generally more active in the summer months, more carbon dioxide is extracted from the atmosphere. The recordings in the NASA graph show that in this 47-year time period alone, there has been a 19.4% increase in the mean annual concentration of carbon dioxide.

The enhanced greenhouse effect

There was a natural greenhouse effect on Earth long before humans started to generate excess carbon dioxide levels. What is worrisome today is that as a result of human pollution, Earth's natural greenhouse effect is intensifying, causing the overall temperature to increase in a phenomenon known as global warming.

To understand the greenhouse effect, you need to know how a greenhouse works. A greenhouse's walls and roof are made of glass. Sunlight penetrates through the glass and warms up the plants inside. Sunlight, which is made up of relatively short wavelengths, in itself is not warm. The temperature of outer space between the Sun and Earth is hundreds of degrees below freezing.

It is only when sunlight hits an object that some of the energy is transformed into heat, which has longer wavelengths. The glass of the greenhouse is not as transparent to heat energy as it is to light energy so some of the heat is trapped inside the greenhouse. The result is that the temperature inside the greenhouse is warmer than outside. This helps plants to grow better when it is cold outside.

The greenhouse effect on a planet is caused by its atmosphere's ability to retain heat in the same way that a greenhouse does. Greenhouse gases such as water vapour and carbon dioxide play the role of the glass. Such gases are less transparent to heat than to light. As a result, some of the heat radiated from the surface is trapped. This keeps the atmosphere warm. Here's how it works.

1 Sunlight enters Earth's atmosphere because the gases of the atmosphere are transparent to light.
2 Most of the sunlight reflects off the surface of Earth and travels back out of the atmosphere – that's why astronauts can see Earth from space.
3 Some of the light energy is transformed into heat energy and warms up the mountains, oceans, and forests on the surface. These, in turn, radiate much of the heat back to the atmosphere.
4 Greenhouse gases retain some of the heat and trap it in the atmosphere.
5 The end result of Earth's natural greenhouse effect is that the atmosphere is warmer than outer space.

Seen from space along the edge of Earth's curve, the atmosphere is a surprisingly thin, almost insignificant looking layer of gases.

Like a warm blanket to protect life on Earth, the atmosphere ensures that there are only small differences in temperature from day to night.

Climate experts have confirmed that Earth is undergoing global warming because of an enhanced greenhouse effect, also known as a runaway greenhouse effect. Increasing levels of the main greenhouse gases (caused by human-generated pollution) are causing the atmosphere to retain more and more heat.

The greenhouse gases

The gases produced by human activity which retain heat the most are carbon dioxide, methane, and oxides of nitrogen. Such gases are referred to as greenhouse gases. The concentrations of these gases in the atmosphere are naturally low, which normally prevents too much heat retention.

● **Examiner's hint:** Often in exams, students confuse the greenhouse effect with the destruction of ozone. Although both are caused by human activity and both influence the atmosphere, they are not interchangeable ideas.

The problem is that human production of greenhouse gases shows little sign of slowing. As consumer demands for fuel and food increase, so the excess production of waste gases increases. For example, oxides of nitrogen are produced by human activities such as:

- burning fossil fuels (e.g. gasoline in cars) and using catalytic converters for their exhaust systems;
- using organic and commercial fertilizers to help crops grow better;
- industrial processes (e.g. the production of nitric acid).

Methane (CH_4) is another major greenhouse gas. Its flammable properties make it useful for cooking and heating. Human activities which produce it include:

- cattle ranching – cows and bulls produce methane in their digestive tracts and release it into the atmosphere;
- waste disposal in landfills – organic wastes such as uneaten food materials decompose and release methane;
- production and distribution of natural gas – when it is made, bottled or piped to homes, leaks in containers and conduits release methane into the air.

▲ Consumer demands push industries to produce more, which means burning more energy and releasing increasing amounts of greenhouse gases.

As the human population increases, so do its demands for meat and dairy products, its use of fossil fuels and its production of decaying wastes. It is estimated that the vast majority of the methane which is present in Earth's atmosphere today is the result of human activity.

Summary

In conclusion, a rise in human population and human activity has lead to an increase in the concentrations of carbon dioxide, nitrous oxide and methane. Since all three atmospheric gases have a high potential for absorbing heat, climate experts are concerned that these gases are intensifying Earth's natural greenhouse effect to

Consumer demands for wood products such as housing, firewood, furniture and paper lead to massive deforestation. ▶

In high concentrations, nitrous oxide can be used in dental care or surgery as an anaesthetic. It is also known as laughing gas because it can make a patient feel giddy but its role in the enhanced greenhouse effect is no laughing matter.

a point where it is being thrown off balance. Climate, however, is a highly complex phenomenon and climate change is not fully understood. From what is known at present, a rise in global temperatures will most likely have the following impacts:

- an increase in photosynthetic rates;
- changes in climate with varying effects on ecosystems;
- extinction of certain species;
- melting glaciers (land ice);
- a rise in sea level which would result in flooding of coastal areas.

The precautionary principle

The Intergovernmental Panel on Climate Change (IPCC), was set up in 1988 to find out if human activities have an impact on climate. In February of 2007, their fourth report came to several conclusions, among which are the following.

- Global temperatures are increasing; global warming is not a hypothesis but a confirmed reality.
- There is over a 90% chance that the cause of this increase in temperature is due to the production of greenhouse gases by human activity. The chance that these fluctuations could be caused by natural phenomena alone is less than 5%.
- Within the coming century, sea level is expected to rise between 18 cm and 59 cm.
- It is likely that severe weather events such as heat waves, drought or heavy rains will increase.

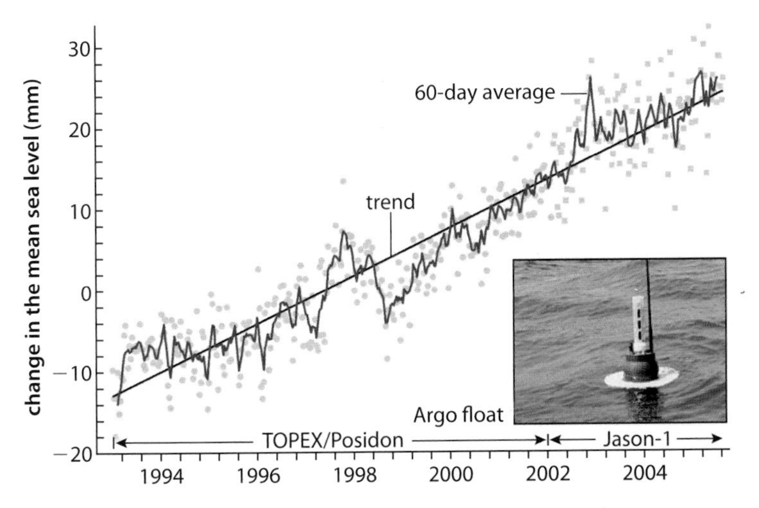

NASA data on sea level and the Argo marker which is used to make the measurements.

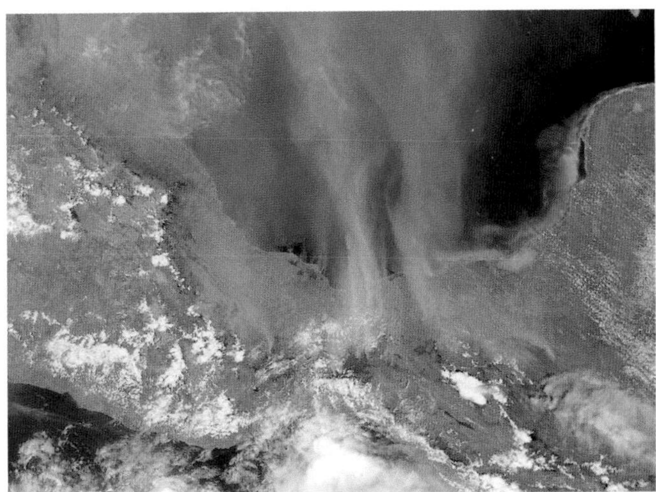

This is a satellite photo of forest fires in the Yucatan peninsula in 1998. The forests were drier than usual that year. Forest fires release large quantities of carbon dioxide into the atmosphere.

What the IPCC's report cannot say is what specific impact these consequences will have on the world's ecosystems or on specific economic activities such as the fishing industry. Although observers are predicting mostly negative impacts, no one knows precisely where, what or how severe they may be.

How sure do scientists have to be in order to justify the banning of a product or a procedure? How severe do the consequences have to be in order to take action? If the activity did not have any immediate, visible consequences, or if the consequences were only affecting remote and distant areas of the globe, how can scientists be sure that there is truly a causal relationship? The real challenge with applying the precautionary principle is when to apply it, where and to what extent.

However, there are enough warning signs to lead experts to invoke the precautionary principle. This is an ethical theory which says that action should be taken to prevent harm even if there is not sufficient data to prove that the activity will have severe negative consequences. It also stipulates that if people wish to engage in an activity that may cause changes in the environment, they must first prove that it will not do harm.

Without the precautionary principle, industries and consumers tend to proceed with their activities until it becomes clear that harm is being done to the environment. When irrefutable proof is furnished, usually action is taken to reduce the activity in question. For example, the use of a pesticide called DDT was prohibited in North America when it was proven to accumulate in ecosystems and reduce populations of birds of prey such as the bald eagle.

Evaluating the precautionary principle

With regards to global warming, tenets of the precautionary principle say that preventative action should be taken now to reduce carbon emissions and greenhouse gas production before it is too late. In addition, the principle holds that those who wish to continue producing excess greenhouse gases should prove that there are no harmful effects before continuing.

One far-reaching consequence of global warming is the melting of glaciers, often hundreds or thousands of kilometres from the cause.

In response, farmers, manufacturers, and transportation providers, among others, wonder why they should invest money in new techniques which reduce greenhouse gases if scientists are not 100% sure how an enhanced greenhouse effect is going to be harmful to the environment. Industries which make the effort to invest in such measures may find themselves less economically viable than their polluting competitors.

Consequently, unless preventative measures are taken across the board by countries worldwide, there will always be polluting competitors who can offer products at a lower price. The risk is that they will drive the ecologically conscious companies out of business because they do not use any of their capital on ecological measures.

Ideally, well-informed consumers could choose products or services which are provided by ecologically minded companies. If this were done on a massive scale, companies would provide eco-friendly products and services to attract customers and those companies which did not, would be shunned by consumers and go out of business.

Prevention is better than cure

Concerned scientists say that money spent now on preventative measures is not money wasted. Rather, it is an investment for a sustainable future. It is often much less expensive to prevent a problem than to fix it.

The precautionary principle is used in the field of medicine where it is preferable to prevent a disease rather than to wait for a patient to become ill. Why not apply this principle to the health of the planet Earth, too? It is up to each consumer, each manufacturer and each government to make a decision.

There is an ethical issue here: one of the most unjust results of human pollution is that many of those who suffer its consequences are not the ones creating and contributing to the problem. This is not only true for people who live lifestyles which do not generate excessive pollution but also for the countless organisms in the biosphere which are adversely affected by human activity.

Let's now consider one example of such impacts.

Human impact on Arctic ecosystems

The Arctic consists of those parts of North America, Greenland, Iceland, Norway and Russia which are north of the Arctic Circle. It also includes the Arctic Ocean, over which floats a huge mass of ice surrounding the North Pole.

The ecosystems which are there show signs of transformation and the people who live there, as well as the scientists who study the region, have noticed the following changes over the years:

- more and more ice is melting every year;
- there is less snow and more frozen rain in the winter;
- some regions which never had them before are now populated with mosquitoes;
- certain woody shrubs are proliferating on warmer soils where once there were only mosses and lichens on tundra;
- bird species such as robins have moved into areas where they are so foreign to the local people that a name for them does not exist in their language.

The consequence of global warming on the Arctic is that the ecosystems are changing. For example, intact ice has the capacity to harbour algae on its underside. The surface area gives the algae a place to attach and the transparency of the ice allows sunlight to pass through, making photosynthesis possible. Since algae are one of the most important producers in the Arctic ecosystem, any drop in their population will have repercussions in the rest of the food web which depends on them. As Arctic ice melts, there is less and less surface for the algae to stick to, hence there is a reduction in food produced for the next trophic level, even top predators such as the polar bear are eventually affected.

Polar bears are affected in other ways too. They rely on seals as their main food supply. One of their preferred hunting techniques is to stand on the ice near a hole and, when a seal comes up for air, to swoop it out of the water.

With less ice, this technique is less productive because the seals have plenty of open water in which to come up for air. Fewer kills means less food for the polar bears. In some regions, the number of polar bear cubs has dropped by 10% from the figures of two decades ago.

As temperatures rise, the habitats of organisms from more temperate climates extend northwards. This is how mosquitoes and robins have arrived in parts of the Arctic where they were once unknown. It is feared that with new species arriving, new pathogens will be introduced. Lastly, detritus which is frozen in the tundra will thaw and begin to decompose, releasing even more carbon dioxide and methane into the air.

To see the NASA website about Arctic ice melting, visit heinemann.co.uk/hotlinks, insert the express code 4242P and click on Weblink 5.1. The site includes an animation of the receding ice.

Exercises

5 Distinguish between how a garden greenhouse works and how the greenhouse effect on Earth works.

6 In what ways could you reduce your consumption of fossil fuels on a day-to-day basis?

7 With the help of outside sources, make a list of the 10 countries with the highest annual carbon dioxide emissions.

5.3 Populations

Assessment statements

5.3.1 Outline how population size is affected by natality, immigration, mortality and emigration.

5.3.2 Draw and label a graph showing a sigmoid (S-shaped) population growth curve.

5.3.3 Explain the reasons for the exponential growth phase, the plateau phase and the transitional phase between these two phases.

5.3.4 List three factors that set limits to population increase.

Population dynamics

Let's consider the major volcanic catastrophe at Mount Saint Helens on the west coast of the United States in 1980. After the massive eruption, little was left of the forest and rivers which once abounded on and around the mountain. The blast from the eruption knocked over massive adult trees as if they were toothpicks.

Fires and hot gases burned everything in sight. Volcanic ash rained down smothering the destroyed forest and covering the carcasses of the animals which died there. Countless species which could escape, fled the area. Although thousands of people had been evacuated, a handful of people perished that day, some of whom were photographers trying to get the photo of a lifetime.

These trees were knocked down by the Mount Saint Helens blast in 1980.

And yet, within months of the total destruction and eradication of populations, life was back. Seeds dropped from birds or blown by the wind germinated in the fertile volcanic ash. Little by little, insects then birds then small mammals moved in. Within a couple of decades, a grassland and shrub ecosystem had reappeared.

From this example, it can be deduced that there are four main factors which affect population size:

- *natality* – the number of new members of the species due to reproduction;
- *mortality* – the number of deaths;
- *immigration* – members arriving from other places;
- *emigration* – members leaving the population.

In the example of Mount Saint Helens, the massive mortality rate due to the eruption caused the populations of birds, trees, elk and just about everything else at close range to be reset to zero. Emigration before and immediately following the eruption greatly decreased populations in the wider vicinity surrounding the volcano. But immigration and natality are improving the numbers dramatically today.

Shortly after total destruction, life started to repopulate the volcanic region.

Population growth curve

The case of Mount Saint Helens shows that even from a non-existent or very small population of individuals, there can soon be a dramatic increase in numbers. Over the years, the number of trees and birds near Mount Saint Helens will rise at ever-growing rates as the organisms reproduce and occupy the available space.

Eventually, when a complete forest has grown again and all habitats are occupied, the numbers of organisms will stabilize and not get any bigger (see Figure 5.5).

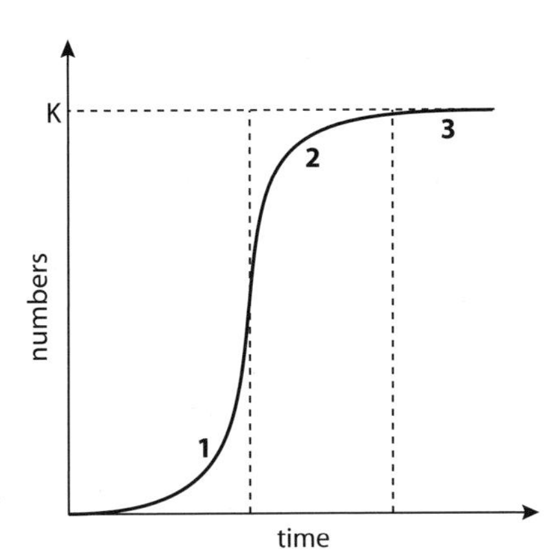

Figure 5.5 Population growth curve.

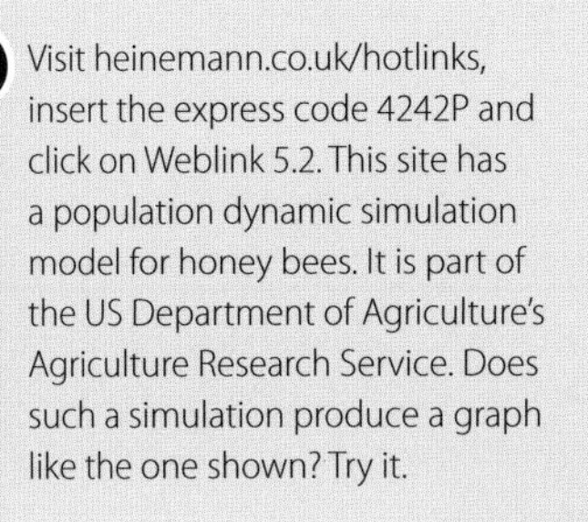

Visit heinemann.co.uk/hotlinks, insert the express code 4242P and click on Weblink 5.2. This site has a population dynamic simulation model for honey bees. It is part of the US Department of Agriculture's Agriculture Research Service. Does such a simulation produce a graph like the one shown? Try it.

● **Examiner's hint:** When asked to draw an S-shaped or sigmoid curve, remember to label it.

The sigmoid (S-shaped) curve of the graph in Figure 5.5 shows three stages in population growth.

1 The *exponential growth phase*, also called the logarithmic phase, in which the number of individuals increases at a faster and faster rate.

2 The *transitional phase*, in which the growth rate slows down considerably – the population is still increasing but at a slower and slower rate.

3 The *plateau phase* or stationary phase, in which the number of individuals has stabilized – there is no more growth.

Note: The letter K at the top left of Figure 5.5 is the carrying capacity and is explained on the next page.

So what causes the phases of the population curve?

Causes of the exponential phase

In ideal conditions, a population can double on a regular basis. Not counting mortality, for example, a population of bacteria can theoretically double its population every few hours: 1, 2, 4, 8, 16, 32, 64, 128, 256, 512, 1024, and so on. Without predators, introduced species, such as cane toads in Australia, have taken over habitats with uncontrolled population growth. The reasons for this first phase of exponential growth are:

- plentiful resources such as food, space or light;
- little or no competition from other inhabitants;
- favourable abiotic factors such as temperature or dissolved oxygen levels;
- little or no predation or disease.

Causes of the transitional phase

Eventually, after the exponential increase in the number of individuals of a population, some of the factors listed above are no longer true. This leads to the transitional phase. The causes of the transitional phase are:

- with so many individuals in the population, there is increasing competition for resources;
- predators, attracted by a growing food supply, start to move in to the area;
- because of large numbers of individuals living together in a limited space, opportunities for diseases to spread within the population increase.

Causes of the plateau phase

Two centuries ago, there were only about one billion humans on Earth. Today there are over six billion. In which phase of the S-curve is the human population?

Consider the land around Mount Saint Helens slowly being taken over by vegetation. Once all the fertile ground is covered with plants, the space available will be occupied to its maximum. Thus, there is gradually less and less available space for seeds, which the plants produce, to germinate and the number of plants stabilizes.

With increasing numbers of herbivores, there is a limited supply of food. In response to limited food supplies, animals tend to have smaller numbers of offspring.

Predators and disease increase mortality and the growth curve tends to level off.

In this phase, the number of births plus the number of immigrations is balanced with the number of deaths plus the number of emigrations.

Carrying capacity (K)

No habitat can accommodate an unlimited number of organisms – populations cannot continue to grow and grow forever. As you have just seen, there comes a time in the growth of a population when its numbers stabilize. This number, the maximum number of individuals that a particular habitat can support, is called the carrying capacity and it is represented by the letter K (see Figure 5.5, page 131).

Consider, for example, a given area of soil in a forest. There is a maximum number of trees which can grow there. This number is attained when enough trees are present to catch all the sunlight, leaving every square metre of the forest floor in shade. New tree seedlings trying to grow under the adult trees, will have difficulty getting sunlight.

But many young trees store up energy for years with very little vertical growth until a big tree dies, leaving a hole in the canopy. The young trees then race up towards the opening to take the old tree's place. Those that lose the race usually die. In any case, for a young tree to join the population, an old one must die and give up a space.

Limiting factors, which define the carrying capacity of a habitat include:

- availability of resources such as water, food, sunlight, shelter, space, or oxygen (notably in aquatic habitats);
- build-up of waste such as excrement or excess carbon dioxide;
- predation;
- disease.

Many biologists, environmental groups, economists and governments wonder what the carrying capacity of planet Earth is for the human population. Will the number of people continue its exponential growth phase or will diseases, climate change or competition for resources lead to a transitional phase or a plateau? Only time will tell.

Exercises

8 Define carrying capacity, K.

9 Look again at the photograph of flowers growing in the ash on page 131. Suggest one way the seeds could have arrived there.

The penguins in Antarctica may be facing many of the limiting factors for their habitat.

5.4 Evolution

Assessment statements

5.4.1 Define *evolution*.
5.4.2 Outline the evidence for evolution provided by the fossil record, selective breeding of domesticated animals and homologous structures.
5.4.3 State that populations tend to produce more offspring than the environment can support.
5.4.4 Explain that the consequence of the potential overproduction of offspring is a struggle for survival.
5.4.5 State that the members of a species show variation.
5.4.6 Explain how sexual reproduction promotes variation in a species.
5.4.7 Explain how natural selection leads to evolution.
5.4.8 Explain two examples of evolution in response to environmental change; one must be antibiotic resistance in bacteria.

Charles Darwin (1809–82). Darwin was not the only person to develop a theory of evolution. He was surprised to discover in 1858 that Alfred Russel Wallace had independently developed a nearly identical theory. The two men jointly presented their ideas to the Linnaean Society in 1858.

At the age of 22, Charles Darwin had the opportunity to travel on board the *HMS Beagle* for a scientific exploration mission starting in 1831 and lasting for 5 years. Little did he know, it would allow him to see nature in a new way and come up with what would become one of the most important, controversial and misinterpreted ideas in biology: the theory of evolution by natural selection.

Evolution is defined as *the process of cumulative change in the heritable characteristics of a population.* The word 'heritable' means that the changes must be passed on genetically from one generation to the next, which implies that evolution does not happen overnight. The word 'cumulative' is in the definition to stress the fact that one change is usually not enough to have a major impact on the species. Finally the word 'population' is in the definition because the changes do not affect just one individual.

Over time, if enough changes occur in a population, a new species can arise. The members of the new population will be different enough from the pre-existing one they came from that they will no longer be able to interbreed. Such a process is rarely observable during a human lifetime. Once you begin to understand evolution, it should become clear that all of life on Earth is unified by its common origins.

In addition, it has been argued that once evolution by natural selection is understood, many of the mysteries of nature are elucidated. We can answer questions such as, 'Why do certain animals have spots or stripes whereas others do not?' 'Why are some flowers red and others blue?'

Evidence for evolution

Although there are others, we will examine three phenomena that provide evidence for the theory of evolution by natural selection.

Fossil record

It is impossible to travel back in time, but the best clues scientists have about what life was like millions of years ago are from fossils. Palaeontologists have been collecting and classifying fossils in an organized fashion for almost two centuries. If you have ever been to a museum full of fossils classified by their age, you may have noticed a few things which palaeontologists have discovered and which provide convincing evidence for Earth's evolutionary past.

- Overall, life which existed more than 500 million years ago was very different from life today.
- Although the planet Earth has had extensive oceans for most of its existence, fish fossils have only been found in rocks 500 million years old or younger (less than 15% of the history of life).
- Although most of the top predators today are mammals such as bears, orca whales, big cats, wolves and the like, none of them existed at the time of the dinosaurs or before.
- Apart from organisms such as certain types of sharks, cockroaches or ferns, many living organisms today have no identical form in the fossil record.

One conclusion that can be drawn from observing fossils is that life on Earth is constantly changing. Most of the changes have been over huge timescales (hundreds of thousands or millions of years) that humans find difficult to appreciate.

Artificial selection

The fossil record is far from complete but the art and science of breeding domesticated animals (e.g. cattle, horses, dogs, sheep, pigeons) provides a good record of recent changes in heritable characteristics.

By watching which males mated with which females, breeders could see which characteristics the offspring would have. Of the offspring produced, not all would be equally valuable in the eyes of the breeder. Certain cows produced better milk or meat. Over the years, breeders learned to choose the males and females with the most desirable genetic characteristics and breed them together.

After practising selective breeding for dozens and sometimes hundreds of generations, farmers and breeders realized that certain varieties of animals had unique combinations of characteristics which did not exist before. Today, the

Cows which exist today are the result of dozens of years of selective breeding. This farmer knows which cows produce the highest quality milk and meat.

meat or milk available to us is very different from that which was produced a few generations ago, thanks to the accumulation of small changes in the genetic characteristics chosen by breeders.

Although this is evidence that evolution is happening due to an accumulation of small changes over time, the driving force is, of course, human choice. The farmers and breeders choose which animals will reproduce and which will not. This is called artificial selection and it should be obvious that it is certainly not the driving force of evolution in natural ecosystems.

Homologous anatomical structures

Other evidence for evolution comes in the form of homologous anatomical structures which are similar in form and function but which are found in seemingly dissimilar species. One of the most striking examples of this is the five-fingered limb found in animals as diverse as humans, whales, and bats (see Figure 5.6). Such limbs are called pentadactyl limbs because 'penta' means five and 'dactyl' refers to fingers. Although the shape and number of the bones may vary, the general format is the same, despite the fact that the functions of the limbs may be very different. Darwin explained that homologous structures were not just a coincidence but they are evidence that the organisms in question have a common ancestor.

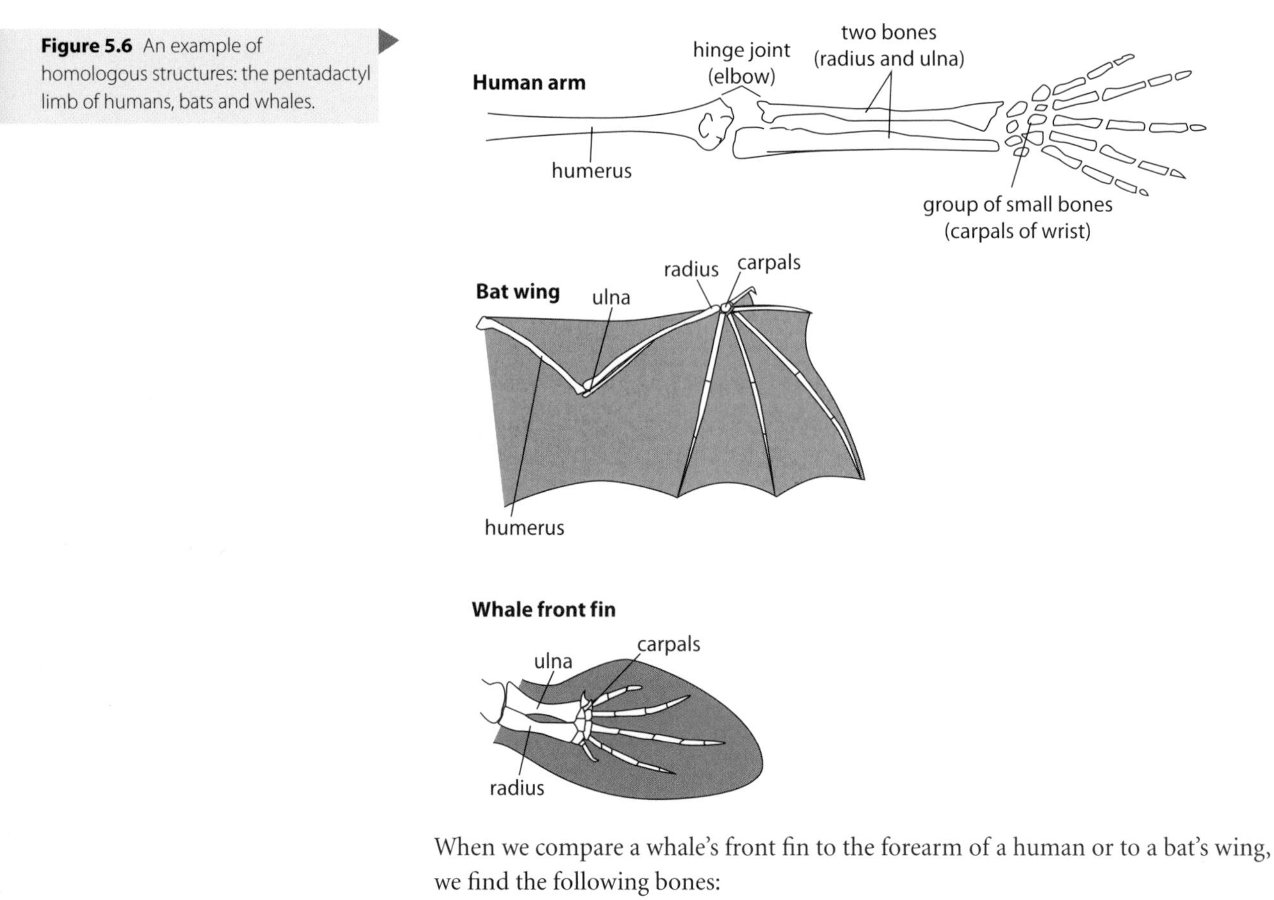

Figure 5.6 An example of homologous structures: the pentadactyl limb of humans, bats and whales.

When we compare a whale's front fin to the forearm of a human or to a bat's wing, we find the following bones:

- humerus;
- radius;
- ulna;
- carpals.

It is true that they are of differing sizes and there are variations in morphology between the three species we are considering, but the basic shape and position of the limb bones are the same. This would suggest that all five-fingered organisms have a common ancestor.

Whales, for example, could most probably swim just as well with a different number of fingers in their front fins, so the fact that there are five suggests that there is another reason besides swimming efficiency: that of common ancestry with other five-fingered organisms.

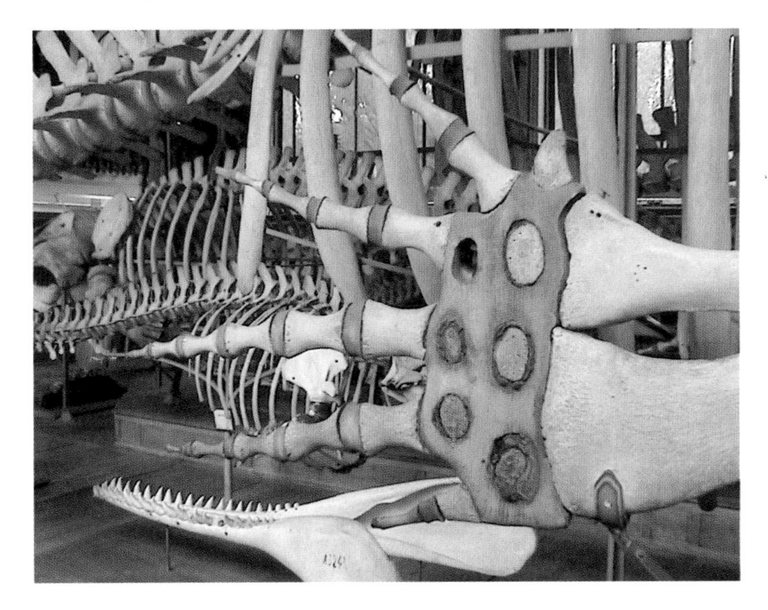

This is the front right fin of a southern right whale showing five articulated fingers.

Mechanism for evolution

Besides providing evidence for evolution, Darwin and Wallace suggested a mechanism: natural selection. How does this work? Largely by over production of offspring and the presence of natural variation.

Too many offspring

Charles Darwin noticed that plants and animals produce far more offspring than could ever survive. Plants often produce hundreds or thousands more seeds than necessary to propagate the species. Mushrooms produce millions more spores than ever grow into new mushrooms. A female fish lays hundreds or thousands of eggs but only a handful survive to adulthood.

This seems paradoxical because production of seeds, eggs and spores involves energy and nutrients which are vital to the parents' survival. Why are such valuable resources squandered on so many excess cells that are never going to give rise to viable offspring? This is what we are going to consider next.

The result of having too many offspring and not enough resources is a problem of supply and demand. There is high demand for water, space, nutrients and sunlight but there is a limited supply. The consequence is competition for these resources in order to stay alive. This is called the struggle for survival.

Many species of animals are territorial and possessive of food supplies, they spend a great deal of time and energy defending their resources. Trees, too, defend their resources by having active compounds such as tannins and alkaloids in their trunks to ward off attackers such as insects.

Competition for resources such as food can lead to adaptive behaviours which might be considered unkind or underhand by human standards. For example, birds such as cuckoos, which are brood parasites, lay their eggs in other species' nests and it is not uncommon for a newly hatched cuckoo chick to push its foster parent's own eggs over the side of the nest so that it will have more to eat.

A tree competing for sunlight would profit from a quick growth spurt early in its life in an attempt to win the race out of the shadows under other trees' leaves. But sometimes struggling for survival means cooperating with other species. Flowers provide insects with food and in return the insects help transport the flowers' reproductive pollen.

One of the greatest struggles for survival today is when animals and plants try to compete with humans for habitat space and water. Deforestation to make way for human farms or to harvest wood, for example, is driving thousands of organisms to extinction each year.

Variation within populations

Organisms such as bacteria reproduce simply by making a copy of their genetic information and then splitting in two. The result is that the second generation is identical to the first. In fact, future generations will be identical or show very little change. There is little chance for the DNA to be modified.

The story is very different for organisms which reproduce sexually. When a cat has kittens, for example, each one is slightly different. From a distance, all pigeons may look alike but close inspection of small details in their feathers, beaks and feet reveal that each one is unique.

Variation and success

Variation is closely related to how successful an organism is. A baby bird which has pigments that give it a colour matching its surroundings will have a better chance of not being seen by a predator. A fish with a slightly different shaped mouth might be able to feed from parts of a coral reef that other fish are not able to access. A plant which produces a different shaped flower might have a better chance of attracting insects for pollination.

It should be clear that a young bird with a colour that makes it very conspicuous to predators has little chance of surviving to adulthood. On the other hand, it might be more attractive to mates. A fish with an oddly shaped mouth may, in fact, be incapable of feeding adequately and die of starvation. A plant which produces flowers that are not attractive to insects will not have the flowers pollinated and will not produce any offspring.

Variation can be seen in this population of guinea pigs.

Often there are sufficient differences within a species to distinguish different races, as is the case for dogs, horses and cattle. More and more, the idea of applying the concept of race to the human population is being regarded as unscientific and controversial. It is true that variations within the human species exist due to evolutionary pressures from various climates. Darker skin protects populations living near the equator from intense sunlight whereas populations further from the equator do not need such protection. But does that constitute enough of a difference to distinguish races?

Few people agree on a clear definition of race as applied to humans. For some, the idea revolves around morphology or visible genetic characteristics. For others, it depends on geographical distribution or cultural heritage. Still others refer to behavioural characteristics.

The only thing which is consistent in all of these definitions is the fact that they are all socially constructed so that one group can separate themselves from another group. As a result, it can be argued that the only practical application for the idea of race in humans is in either promoting or fighting racism. More appropriate alternatives to the idea of race are concepts of ethnic background, cultural history and geographic or genetic origins.

Causes of variety

Besides choice of mate (the reason for extravagant courtship displays), there are two main reasons why organisms show variation:

- mutations in DNA;
- sexual reproduction greatly promotes variation in a species.

Mutation

Mutations may produce genes which lead to diseases such as phenylketonuria (PKU) in humans (Chapter 12, page 323). However, sometimes a mutation can produce a characteristic which is advantageous, perhaps a slightly faster growing tree or better frost resistance for a plant. A beneficial mutation for a bird or mammal might result in a different camouflage which better matches a changing habitat. In each generation, only a few genes mutate and most mutations produce effects that are neither useful nor harmful. As a result, sexual reproduction is a much more powerful source of variation in a population because thousands of genes are mixed and combined.

Sexual reproduction

In sexual reproduction there are two ways in which genes are mixed:

- first, meiosis (Chapter 4, page 85);
- secondly, fertilization (Chapter 11, page 312).

When an egg cell is made during meiosis, it receives 50% of the mother's genetic information. The same is true about the father's genes found in his sperm cells: 50% are present. Because of the random distribution of the chromosomes when the cells split during meiosis, each egg has a different combination of chromosomes from other eggs and each sperm has a different combination of half of the man's genes.

Part of what determines whether or not a female animal gets pregnant is that all the conditions must be right inside her body and that sperm cells must be present at the opportune moment when an egg is ready. Of the many sperm cells that may be present, only one will penetrate the egg. In determining exactly when the egg and sperm cell will meet and fuse together, a certain amount of chance and luck are involved.

The two ways that genes are mixed during sexual reproduction can be thought of as a double lottery:

- lottery 1: meiosis;
- lottery 2: fertilization.

Natural selection

Evolution is not just based on chance. In a situation where there are too many organisms for limited resources, it is obvious that some individuals will succeed in accessing those resources and the rest will fail. In other words, there is a selection. Exactly which individuals survive and which ones do not is determined by their surroundings and the compatibility of their characteristics with those surroundings. The steps of evolution by natural selection are as listed here.

- Overproduction of offspring and, in those offspring, natural variation due to genetic differences (e.g. body size, morphology, pigmentation, visual acuity, resistance to disease). In the offspring:
  - useful variations allow an individual to have a better chance of survival (e.g. hiding from predators, fleeing danger or finding food);
  - harmful variations make it difficult to survive (e.g. inappropriate colour for camouflage, heavy bones for birds, having such a big body size that there is not enough food to survive).
- Individuals with genetic characteristics that are poorly adapted for their environment tend to be less successful at accessing resources and have less of a chance of surviving to maturity.
- Individuals with genetic characteristics that are well adapted for their environment tend to be more successful at accessing resources and have a better chance of surviving to maturity.
- Because they survive to adulthood, the successful organisms have a better chance to reproduce and to pass on their successful genetic characteristics to the next generation.
- Over many generations, the accumulation of changes in the heritable characteristics of a population results in evolution – the gene pool has changed.

The spotted camouflage of this cheetah helps it to sneak up on prey unseen. A cheetah born without spots would have more difficulty hunting.

As you can see, it is impossible to sum up all these concepts in one catchy phrase such as 'the law of the jungle'. Although Darwin himself eventually adopted the phrase 'survival of the fittest', the idea of evolution by natural selection is more complex than that. In addition, many people have the misconception that what Darwin said was, 'Only the strongest survive.' This is simply not true.

The theory of evolution by natural selection is full of nuances. This could be one of the reasons it is so widely misunderstood by the general public. For example, an organism that is well adapted to its environment is not guaranteed success, it simply has a better chance. Dinosaurs such as the sauropods were the biggest, strongest animals ever to walk the planet. But they did not survive environmental changes that drove them to extinction. In fact, the fossil record indicates that more than 99.99% of all life which has ever existed on Earth is now extinct.

W

Visit heinemann.co.uk/hotlinks, insert the express code 4242P and click on Weblink 5.3, to see how the physical appearance of male fish (in this case guppies) can influence survival.

Examples of natural selection

Multiple antibiotic resistance in bacteria and pesticide resistance in rats are both carefully studied examples of natural selection. What is striking about these examples is their rapidity. Although evolution is generally considered to be a long-term process, the mechanism of natural selection can sometimes be quick, taking place over months, years or decades rather than millennia.

Antibiotic resistance in bacteria

Antibiotics are medications such as penicillin which kill or inhibit the growth of bacteria. They are given to patients suffering from bacterial infections. They are also sometimes given to people who are suffering from something else and, because their immune system is weak, are at greater risk of a bacterial infection. However, overuse of antibiotics can lead to the production of resistant strains of bacteria.

Antibiotic resistance in bacteria develops in several steps. Consider the following scenario.

1 A person gets sick from a bacterial infection such a tuberculosis.
2 Her doctor gives her an antibiotic to kill the bacteria.
3 She gets better because the bacteria are largely destroyed.
4 By a modification of its genetic makeup, however, one bacterium is resistant to the antibiotic.
5 That bacterium is not killed by the antibiotic and it later multiplies in the patient's body to make her sick again.
6 She goes back to the doctor and gets the same antibiotic.
7 This time, no results – she is still sick and asks her doctor what is wrong.
8 The doctor prescribes a different antibiotic which (hopefully) works.

Examiner's hint: In an exam, be careful not to say that bacteria become immune to the antibiotics. Although the term 'immunity' is sometimes interchangeable with 'resistance', it is not the case here. Immunity develops within a lifetime; resistance is a change which evolves from generation to generation.

But if the bacterium continues to change its genetic makeup, it could become resistant to all the antibiotics available.

SEM of the bacteria which can cause tuberculosis.

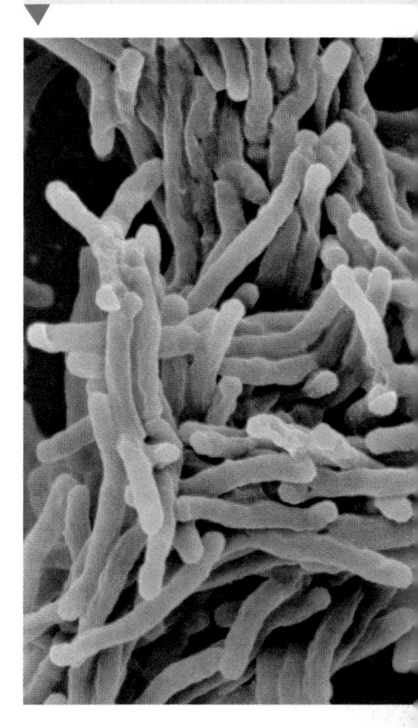

Since bacteria reproduce asexually, they generally do not change very often. However there are two sources of possible change in their genetic makeup:

- mutations (as seen on page 82);
- plasmid transfer.

Plasmid transfer involves one bacterium donating genetic information to another in a ring of nucleotides called a plasmid. Both the donating and receiving cells open their cell walls to pass the genetic material from the donor to the receiver.

The development of antibiotic-resistant bacteria has happened in several cases. New strains of syphilis, for example, have adapted to antibiotics and show multiple resistance. Some strains of tuberculosis are resistant to as many as nine different antibiotics. There is no cure for people who get sick from such super-resistant germs and they must rely on their immune system to save them.

Finding new antibiotics would only be a temporary solution and pharmaceutical companies cannot find new medications fast enough to treat these super-resistant germs. As a result, the best way to curb their expansion is to make sure that doctors minimize the use of antibiotics and that patients realize that antibiotics are not always the best solution to a health problem.

Pesticide resistance in rats

Pesticides are chemicals which kill animals that are regarded as pests. Farmers use them to eradicate pests such as rats which eat their crops. Consider the following scenario.

1 Once applied in the fields, pesticides kill all the rats ... or so the farmer thinks.
2 Due to natural variations, a few rats are slightly different and are not affected by the poison.
3 The resistant rats survive and reproduce, making a new population in which some or all of the members possess the genetic resistance.
4 Seeing rats again, the farmer puts out more poison; this time fewer rats die.
5 To kill the resistant rats, a new pesticide must be used.

The verb 'to adapt' is often misinterpreted in this context – the organism has no desire or plan to change. Wishful thinking and careful planning are not the driving forces at work in evolution by natural selection. Instead, pressures from the environment generate modifications in populations.

This example clearly illustrates how a population can adapt to its environment and how humans can be responsible for creating super-resistant creatures. Although humans are involved in both the examples we've considered here, this is not a case of artificial selection. Here, humans are the source of the environmental changes which create evolutionary pressure on the organisms to adapt or perish.

In these examples, the bacteria and the rats do not choose to adapt; adaptation happens as a result of environmental conditions. So, selection happens on an individual basis but the resulting adaptions are significant at the level of the population.

Exercises

10 In what ways are polar bears well adapted for their habitat?

11 Distinguish between artificial selection and natural selection.

12 Explain how a population of insects could develop resistance to the insecticides sprayed on them.

13 Ground-nesting birds such as wirebirds lay their eggs in a nest made on the ground. The eggs of this species are generally speckled dark brown. If a mutation occurred, causing the eggs to be brightly coloured, how would the change in colour affect their chances of survival?

5.5 Classification

Assessment statements

5.5.1 Outline the binomial system of nomenclature.

5.5.2 List seven levels in the hierarchy of taxa – kingdom, phylum, class, order, family, genus and species – using an example from two different kingdoms for each level.

5.5.3 Distinguish between the following phyla of plants, using simple external recognition features: Bryophyta, Filicinophyta, Coniferophyta and Angiospermophyta.

5.5.4 Distinguish between the following phyla of animals, using simple external recognition features: Porifera, Cnidaria, Platyhelminthes, Annelida, Mollusca and Arthropoda.

5.5.5 Apply and design a key for a group of up to eight organisms.

Binomial nomenclature

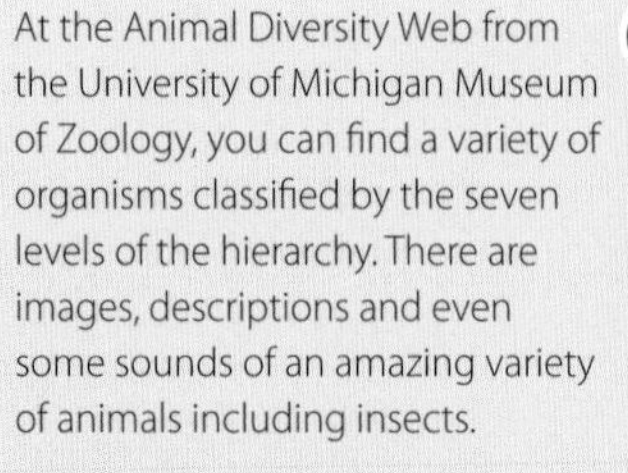

At the Animal Diversity Web from the University of Michigan Museum of Zoology, you can find a variety of organisms classified by the seven levels of the hierarchy. There are images, descriptions and even some sounds of an amazing variety of animals including insects.

Visit heinemann.co.uk/hotlinks, insert the express code 4242P and click on Weblink 5.4.

You have a name that you were given when you were born but you also have a scientific name based on your species, *Homo sapiens*. This system of naming organisms using two names is called binomial nomenclature. 'Bi-' means two, 'nomial' means name and 'nomenclature' refers to a system used to name things.

Myrmecophaga tridactyla is a name which literally means 'eater of ants' + 'with three fingers'. In case you have not guessed, it refers to an anteater and this one happens to be the giant anteater of Central and South America.

The first name in the binomial nomenclature system is always capitalized and it refers to the genus; the second name always begins with a small letter and refers to the species. Both are always written in italics when typed or underlined when written by hand. Most words used in binomial nomenclature are Latin or Greek in origin. *Lepus arcticus* is the scientific name for the Arctic hare – both terms come from Latin. This is why the term 'Latin name' is often used, although this is an oversimplification because other languages are also involved.

This system of naming organisms was consolidated and popularized by the Swedish naturalist Carolus (Carl) Linnaeus. In his book *Systema Naturae* (*The Natural World*, 1735) he listed and explained the binomial system of nomenclature for species which had been brought to him from all over the world. Although he was not the first to use the idea of genus (pl. genera), he was the first to use it along with the species name in a consistent way.

Carl Linnaeus (1707–78) perfected and systemized binomial nomenclature.

The reasons for putting living organisms into groups are numerous and include:

- trying to make sense of the biosphere;
- showing evolutionary links;
- predicting characteristics shared by members of a group.

Without a universal system, each language, culture or region may have a name for an organism. For example, the pill bug or woodlouse sound like two different organisms but they are, in fact, the same one: *Armadillidium vulgare*. The common names do not reveal anything about its evolutionary links but its scientific name does.

You have probably come across *Armadillidium vulgare* under rotting logs.

One advantage of the binomial nomenclature system is that scientists from all over the world working in any language can share data about any species and be sure that they are speaking about the same organism.

In his classification of organisms, Linnaeus used physical characteristics and social behaviour to establish four groups of humans. Reading such descriptions today is shocking because by modern standards, they have a racist nature. To what extent is it necessary to consider the social context of scientific work when evaluating ethical questions about research?

The hierarchy of classification

Humans like to see similarities and differences in the objects which surround them: hot or cold, delicious or foul-tasting, dangerous or safe, and so on. In the early days of classification, all known organisms were classified into only two kingdoms: plants and animals.

As the centuries went by and as the study of biology became more systematic, tens of thousands of new species were discovered in forests, deserts and oceans, some of which showed characteristics of both plants and animals and some were not like either plants or animals. For example, mushrooms grow on the forest floor the way plants do and yet they do not have leaves or roots and they do not photosynthesize – they get their energy from digesting dead organic matter. So mushrooms cannot be classified as plants because they are not autotrophs but they are certainly not animal-like either, one reason being that they have cell walls made of chitin.

With the invention of the microscope in the mid-1600s, many new creatures were discovered which were nothing like plants or animals. In effect, the microscope revealed that there is an entire parallel world of invisible organisms living throughout the biosphere.

Five kingdoms

Many systems of classifying all these new organisms have been tried over the years, but one of the most widely accepted systems today is that of the five kingdoms:

- Kingdom Plantae (the plants);
- Kingdom Animalia (the animals);
- Kingdom Fungi (the fungi and moulds);
- Kingdom Protoctista (the protozoa and algae);
- Kingdom Prokaryotae (the bacteria).

Kingdom Protoctista is one with which you may be least familiar. It contains protoctists such as *Paramecium* and *Amoeba* (which used to be called protozoa) as well as the algae.

Seven taxa

Because there are hundreds of thousands of different types of animals and plants as well as organisms in other kingdoms, there are several subdivisions, called taxa (singular, taxon) of the kingdoms. A seven-level hierarchy of taxa is commonly used (the last two you have encountered already: genus and species):

- kingdom;
- phylum;
- class;
- order;
- family;
- genus;
- species.

- **Examiner's hint:** To help you remember the seven levels of taxa in the correct order, you can use this mnemonic (memory trick): King Philip Came Over For Good Soup.

Here are two examples of the full identification of two species according to the seven taxa we have just named.

Taxa	Human	Garden pea
Kingdom	Animalia	Plantae
Phylum	Chordata	Angiospermae
Class	Mammalia	Dicotyledoneae
Order	Primate	Rosales
Family	Hominidae	Papilionaceae
Genus	*Homo*	*Pisum*
Species	*sapiens*	*sativum*

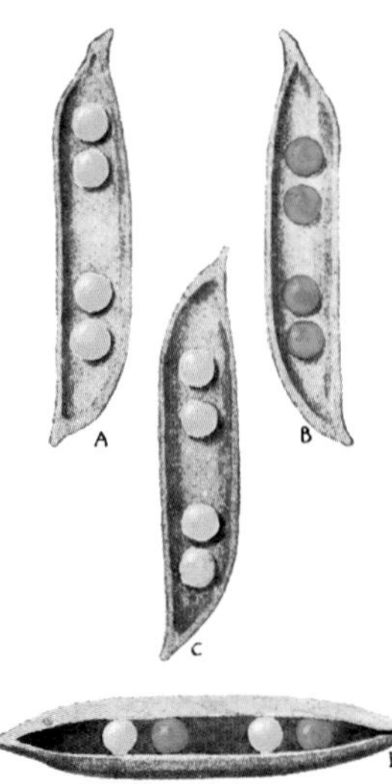

The garden pea is the plant Gregor Mendel studied.

Other classifications

The system of kingdoms and taxa is used for identifying and naming organisms but there are countless other ways to classify organisms. Here are some examples:

- by feeding habits: carnivore/herbivore
- by habitat: land dwelling/aquatic
- by daily activity: nocturnal/diurnal
- by risk: harmless/venomous
- by anatomy: vertebrates/invertebrates.

No single classification system is the 'right' way. Think of all the ways that the students of a class can be put into different groups: by eye colour, by shoe size, by birth date, by academic results, by favourite musical group, by alphabetical order, by length of fingernails, by what they had for breakfast. What is important for a system of classification is that it is clear, consistent, easily implemented and that there is a general consensus to apply it.

Some plant phyla

Of the several phyla of plants, four represent many of the types of plants you are probably most familiar with.

1. Bryophyta – the bryophyte phylum includes plants of very short stature such as mosses.
2. Filicinophyta – this phylum includes ferns and horsetails, among others.
3. Coniferophyta – the conifer phylum includes cedar, juniper, fir and pine trees, among others.
4. Angiospermophyta – the angiosperm phylum includes all plants which make flowers and which have their seeds surrounded by a fruit.

A liverwort is an example of a bryophyte.

Trees which produce seed cones and have needle-like leaves are conifers.

This moss growing on the bark of a tree branch is also a bryophyte.

To distinguish between these four phyla, two categories of characteristics can be examined:

- vegetative characteristics such as type of leaves and stems;
- reproductive characteristics.

Vegetative characteristics

Bryophytes are referred to as non-vascular plants because they do not have true vascular transport tissue inside them such as xylem tissue (which transports water and nutrients up from the roots) or phloem tissue (which transports water and nutrients from the leaves towards the stem and roots).

Members of the Filicinophyta, on the other hand, are vascular plants, as are the other two phyla described in this section.

Conifers can be recognized by the fact that all of them produce woody stems and their leaves are in the form of needles or scales.

The most obvious vegetative characteristic which allows angiosperms (i.e. members of the Angiospermophyta) to be identified quickly are their flowers and fruit. If the fruit has any seeds inside, the plant it came from is an angiosperm.

Reproductive characteristics

The mosses, liverworts and hornworts which make up the bryophytes do not produce flowers or seeds. Instead, they produces spores, which are microscopic reproductive structures. Bryophyte spores are transported by rainwater and ground humidity which is one of the reasons they are found most abundantly in damp habitats such as a forest floor. The same is true for the plants that are filicinophytes.

In contrast, all species of conifer use wind to help them reproduce by pollination. Most species of conifer produce seed cones with seed scales.

Chances are you have eaten an angiosperm today: wheat, corn, apple, oranges are examples of angiosperm seeds and their coverings.

Although angiosperms also produce seeds, they do not produce cones and they are not always pollinated by wind. Many flowering plants rely on birds, insects and sometimes mammals to transport their pollen from one flower to the next.

The sexual reproductive organs of angiosperms are their flowers. The fruit, which is the enlarged ovary of the plant, holds the seeds.

Some animal phyla

Of all the phyla of animals, we will consider six here. Some of these you may be familiar with but others you probably do not know much about. None of the animals in these six phyla in this section has a backbone; they are all invertebrates.

1 Porifera – this phylum consists of the sponges.
2 Cnidaria – this phylum includes sea jellies (jellyfish) and coral polyps, among others.
3 Platyhelminthes – this phylum is made up of flatworms.
4 Annelida – this phylum is made up of segmented worms.
5 Mollusca – this phylum contains snails, clams and octopuses, among others.
6 Arthropoda – this phylum includes insects, spiders and crustaceans, among others.

Porifera

Sponges are very simple marine animals which are sessile (i.e. they are stuck in place). They do not have mouths or digestive tracts. Rather, they feed by pumping water through their tissues to filter out food. They have no muscle or nerve tissue and no distinct internal organs.

A yellow tube sponge, one of the members of phylum Porifera.

Cnidaria

Cnidarians are a diverse group: corals, sea anemones, jellyfish, sea jellies) hydra or floating colonies such as the Portuguese man-of-war. This diversity makes it difficult to give an overall description of common characteristics. One feature which unites cnidarians is that they all have stinging cells called nematocysts.

The largest jellyfish, *Cyanea artica*, has tentacles which can reach 40 metres long.

Some of these organisms are sessile, others are free-swimming and some can be both depending on the period of their life cycle. To digest the food they catch in their tentacles, they have a gastric pouch with only one opening. Some of the free-floating species are carried by the current but others are agile swimmers.

Platyhelminthes

Flatworms have only one body cavity: a gut with one opening for food to enter and waste to exit. They have no heart and no lungs. One of the most famous – or infamous – members of this phylum is the parasitic tapeworm which can infest the intestines of mammals including humans. The reason for flatworms' flat shape is that all the cells need to be close to the surface to be able to exchange gases by diffusion.

The common earthworm is an annelid.

Annelida

Annelids are the segmented worms such as earthworms, leeches and worms called polychaetes. Here, the word 'segmented' refers to the fact that their bodies are divided up into sections separated by rings. Annelids have bristles on their bodies, although these are not always easily visible. Like the next two phyla, annelids have a gastric tract with a mouth at one end and the intestines have an opening at the other end where wastes are released.

You are sure to find molluscs in a seafood restaurant because all shellfish such as mussels and clams are molluscs.

Mollusca

Most molluscs are aquatic and include snails, clams, and octopuses. Many produce a shell using calcium. Unlike annelids, their bodies are not segmented.

Arthropoda

Arthropods have a hard exoskeleton made with chitin, segmented bodies, and limbs that can bend because they are jointed. Although the limbs are often for walking, some are adapted for swimming, others can form mouthparts.

Arthropods include insects, spiders, scorpions as well as crustaceans such as crabs and shrimp. They are true champions of diversity and adaptation because they have conquered most habitats worldwide and there are over a million species of arthropod. They vary in size from the most minute mites, just over 100 micrometres long, to the Japanese giant spider crab, which is 4 metres in length.

Spiders are arthropods.

Using a dichotomous key

To identify this moth, a dichotomous key can be used.

When a biologist encounters an organism such as a moth, the first question that comes to mind is, 'What kind of moth is it?' To find which order, genus or species it belongs to, the organism must be observed very closely for characteristics which will help to reveal its identity.

If you have ever played the guessing game in which the rule is that you can only ask 'yes' or 'no' questions, then you already know how a dichotomous key works. Here is the basic principle.

1. You look at the first section of the key which has a pair of sentences describing a characteristic.
2. Next, you look at the organism to see if the particular characteristic described in the first line is present in the organism.
3. If the answer is yes, then go to the end of its line and find the number of the next pair of statements to look at, follow the number given and continue until the end.
   If the answer is no, then go to the second statement just below it and that one should be true, so go to the end of its line and find the number of the next pair of statements to look at, follow the number given and continue until the end.
4. Keep going until the end of the line has a name instead of a number – if you have answered each question correctly, that will be the name of your organism.

Here is a simple dichotomous key for eight members of Kingdom Plantae.

1 Vascular tissue	
Does not have vascular tissue	2
Contains vascular tissue for conducting fluids	3
2 Presence of lobes on leaves	
Does not possess lobes on leaves	**moss**
Possesses lobes on leaves	**liverwort**
3 Seeds or spores	
Produces seeds	4
Produces spores	7
4 Seed covering	
Seeds encased in a sweet fruit	5
Seeds encased in a cone	6
5 Sweet fruit	
Fruit contains many small seeds	**apple**
Fruit contains one large pit	**cherry**
6 Seeds in a cone	
Long needles in a brush-like formation	**pine tree**
Leaves are flat scales	**cedar**
7 Spore-producing plants	
Has many small flat leaves	**fern**
Has no flat leaves	**horsetail**

Working your way through a key is like going through a maze. By deciding which way to go at each intersection, you can find your way. But if you take a wrong turn early on, you will not reach your goal. If you wanted to identify a moss, for example, you would have to know what 'lobes on leaves' means. This is one of the biggest challenges with using dichotomous keys: the vocabulary can sometimes be quite technical and if you do not know what the words are referring to, there is no way to answer the questions.

You also need to be sure you are using the right key, since no key can identify all species. For example, holly does not appear in the above example key. You would need to use another key.

W For some informative descriptions of many phyla of plants and animals of the past and the present, visit heinemann.co.uk/hotlinks, insert the express code 4242P and click on Weblink 5.5.

Making a dichotomous key

To make a key of your own, start out by putting things in groups and ask yourself, 'Why did I put that in that group?' In other words, identify the characteristics which make up each group. Then invent statements which divide things up into the groups you created.

Using one or two examples of animals given in each of the six phyla above, try making your own dichotomous key. You may need help with identifying the characteristics which make them different from each other. Your school library or online resources may be useful.

Exercises

14 List the five kingdoms. Determine which kingdom each of the following organisms belongs to: algae, hydra, spider, mushroom, yeast, bacterium.

15 Suggest one reason why viruses do not fit in the five-kingdom system.

16 Make a table with four columns headed Bryophyta, Filicinophyta, Coniferophyta and Angiospermophyta. Make two rows labelled Vegetative characteristics and Reproductive characteristics. Add a third row for named examples. Complete the 12 boxes of the table.

17 In the seven-taxa system, state the order which you belong to.

18 Using ten different objects found in your school bag, design a dichotomous key.

Practice questions

1 State the importance of decomposers in an ecosystem. (*1 mark*)

2 Explain how energy and nutrients enter, move through, and exit a food chain in an ecosystem. (*8 marks*)

3 In communities, groups of populations live together and interact with each other. Outline the importance of plants to populations of other organisms in a community. (*6 marks*)

4 Look at this graph.

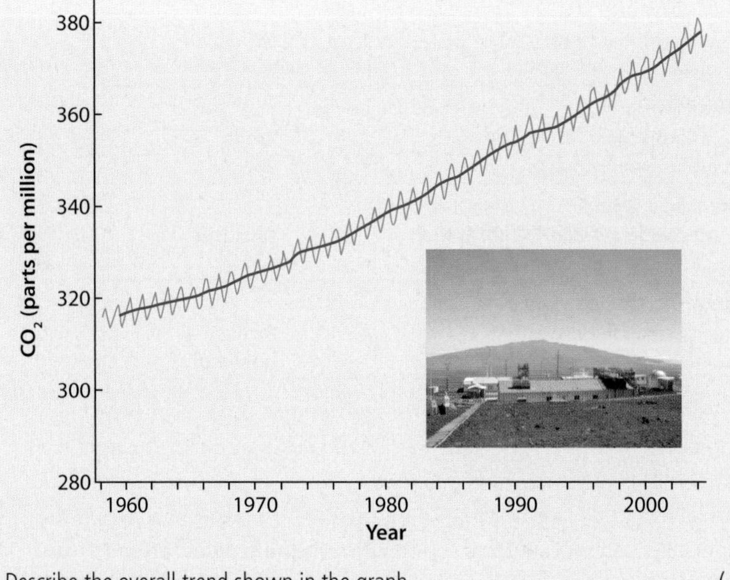

(a) Describe the overall trend shown in the graph. (*1*)
(b) Suggest a cause of the overall trend throughout the period 1970–99. (*1*)
(c) Using a clear label, identify any one point on the graph which shows the carbon dioxide level in mid-summer. (*1*)
(d) Explain why the concentration of carbon dioxide varies with the seasons. (*2*)
(*Total 5 marks*)

5 Describe with the aid of a diagram the phases of a sigmoid population growth curve. (*4 marks*)

6 Explain the reasons for the sizes of animal populations within communities changing and the reasons for them remaining constant. (*8 marks*)

7 Describe how sexual reproduction promotes genetic variation within a species. (*4 marks*)

8 With reference to one example, discuss the theory of evolution by natural selection. (*8 marks*)

9 Which statement most accurately describes the plants *Clarkia cylindrica*, *Clarkia deflexa* and *Clarkia similis*?

A All three belong to the same species.
B Each belongs to a different group.
C All three belong to the same genus.
D Each belongs to a different genus. (*1 mark*)

6 Human health and physiology 1

Introduction

The human body is composed of cells organized into tissues, tissues organized into organs, and organs organized into organ systems. The anatomy and physiology of the human body approach a complexity that is likely to keep humans investigating new things for many decades to come. In this chapter, you will learn about some of the major organ systems of the body and how those organ systems interact with each other. Overall, this is known as physiology.

6.1 Digestion

Assessment statements

6.1.1 Explain why digestion of large food molecules is essential.
6.1.2 Explain the need for enzymes in digestion.
6.1.3 State the source, substrate, products and optimum pH conditions for one amylase, one protease and one lipase.
6.1.4 Draw and label a diagram of the digestive system.
6.1.5 Outline the function of the stomach, small intestine and large intestine.
6.1.6 Distinguish between *absorption* and *assimilation*.
6.1.7 Explain how the structure of the villus is related to its role in absorption and transport of the products of digestion.

Why do we digest food?

When you eat a snack or a meal, you begin a set of events that leads to your body cells being provided with needed nutrients. In a very basic format, here is the series of events in order:

- ingestion – you eat the food;
- digestion – a series of chemical reactions, whereby you convert the ingested food to smaller and smaller molecular forms;
- absorption – small molecular forms are absorbed through cells of your digestive system and pass into nearby blood or lymphatic vessels;
- transport – your circulatory system delivers the small molecular nutrients to your body cells.

Digestion solves a problem of molecular size

Many of the foods that we ingest have very large molecules – too large to pass across any cell membrane. Yet to get into your bloodstream, molecules must pass through the cell membranes of your intestines and then through the cell

W Want to make sense of food labels that allow you to know the exact components of various food products? If you live in the US or buy US food products, visit heinemann.co.uk/hotlinks, insert the express code 4242P and click on Weblink 6.1.

membrane of a capillary. Any food that we eat must therefore be chemically digested to a suitable size. The table below shows the molecular type of some food categories and the molecules they consist of before and after digestion.

Molecule type	Molecular form ingested	Molecular form after digestion
protein	protein	amino acids
lipids	triglycerides	glycerol and fatty acids
carbohydrates	polysaccharides, disaccharides, monosaccharides	monosaccharides
nucleic acids	DNA, RNA	nucleotides

Digestion allows you to turn molecules into 'your own'

Think about the last meal you ate. Perhaps you had a vegetable or two, and fish or chicken or steak; you may have had fruit. All of these foods are composed of either plant cells or animal cells, and thus contain molecules characteristic of a living organism that is not a human being. Plant cells characteristically store excess carbohydrates in the form of starch, whereas animals store excess carbohydrates as glycogen. Each type of living organism also has its own unique set of proteins. Cow muscle (steak) is similar to human muscle, but there are differences in the amino acid sequences. You also eat the nucleic acids (DNA and RNA) of other organisms, yet your nucleic acids are a unique sequence.

When we digest food molecules, we break them down (hydrolyse them) into their smallest components (as in the right-hand column of the table above). The components then can be reassembled into larger molecules (macromolecules) that are useful to you.

Summary

Let's look more closely at one molecule type as an example. Perhaps you had an egg for breakfast this morning. In the egg white (albumin) is an amino acid called serine. Inside your stomach and small intestine, the egg albumin would be chemically digested and the serine left floating freely in the fluid environment of the small intestine. Serine is small enough as a molecule to diffuse through the cells of the small intestine and then into a small capillary blood vessel. The blood vessel joins your overall circulation and the serine is taken to your pancreas. There, the serine leaves your bloodstream and enters a pancreas cell. The cell uses the genetic code of your DNA to build the serine into, for example, the protein hormone, insulin. Thus the serine becomes a 'building block' molecule used for protein synthesis. Digestion of the protein albumin provides two benefits:

- serine would not be able to leave the intestine without digestion of the albumin to release the serine molecule which is able to diffuse into the bloodstream;
- serine would not be useful to you if it stayed in the egg albumin protein; you need the individual amino acid to help synthesize your own protein under the control of your DNA.

The role of enzymes during digestion

The very first enzyme added to ingested food is amylase. Amylase is found in saliva which is produced by three pairs of glands in the area of the mouth.

As foods move through your alimentary canal, many digestive enzymes are added along the way. Each digestive enzyme is specific for a specific food type. For example, lipase is an enzyme specific for lipid molecules and amylase is specific for amylose (otherwise known as starch).

As you may recall, enzymes are protein molecules which act as catalysts for reactions. As catalysts, the real function of enzymes is to lower the activation energy of the reactions that they catalyse. This means that reactions occurring with an enzyme can occur with a lower input of energy than the same reaction without the aid of an enzyme. The input of energy is typically in the form of heat. Enzyme-catalysed reactions proceed at higher reaction rates at a lower temperature than the same reaction without an enzyme. This is a tremendous advantage for living organisms. Digestive reaction rates suitable for sustaining life require the use of enzymes. Many of the reactions which represent the digestive process would need far higher temperatures than we are able to maintain safely if enzymes were not involved. Humans maintain a stable body temperature of 37 °C. This temperature is warm enough to maintain good molecular movement and, with the aid of enzymes, it provides enough activation energy for metabolic reactions including digestion.

Digestive enzymes all help to catalyse hydrolysis reactions. You will recall that the whole idea of the digestive process is to convert large macromolecules to smaller molecules that can be absorbed and then used. Let's look at the role of one specific enzyme as an example of a typical hydrolytic catalyst. Figure 6.1 illustrates what happens when amylase hydrolyses a starch molecule.

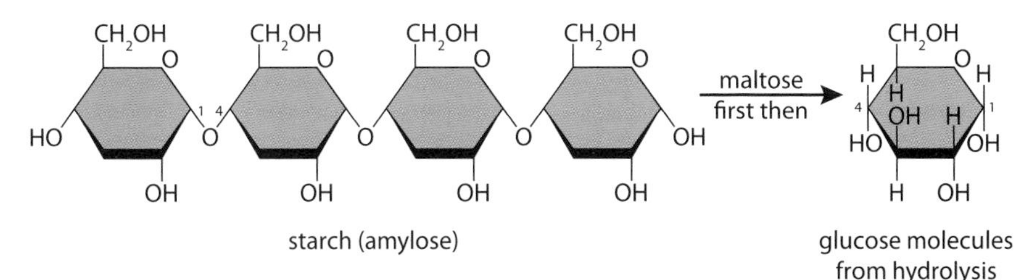

Figure 6.1 Hydrolysis of starch to maltose and glucose. Only a small portion of a starch molecule is shown.

The role of amylase is to temporarily hold the starch in its active site and put stress on the covalent bonds that bind the glucose molecules together within the polysaccharide. When these bonds are stressed, it is more likely that the surrounding thermal energy (normal body temperature) will provide enough molecular motion to break the bonds. Notice that the enzyme (amylase) does not cause the reaction, it just makes the reaction more likely to occur at physiologically normal temperature.

Examples of digestive enzymes

There are many enzymes that help us digest foods. Some of these are the individual enzymes that are specific to many of the types of carbohydrate we ingest. We also produce many different protease enzymes that collectively help us to digest proteins. Some of these protease enzymes work within a protein by recognizing specific amino acid pairs, and some digest proteins from the outer ends and work 'inwards'.

The table below shows three common digestive enzymes and some important information concerning each.

	Salivary amylase	Pepsin (a protease)	Pancreatic lipase
Source	salivary glands	stomach cells	pancreas cells
Substrate	amylose (starch)	proteins (polypeptides)	lipids
Products	maltose and glucose	amino acids	glycerol and fatty acids
Optimum pH	neutral (pH 7)	acidic (pH 3)	neutral (pH 7)

Human digestive system

Much of the human digestive system is a tube called the alimentary canal (see Figure 6.2). In order, the alimentary canal consists of:

- mouth;
- oesophagus;
- stomach;
- small intestine;
- large intestine (colon);
- rectum.

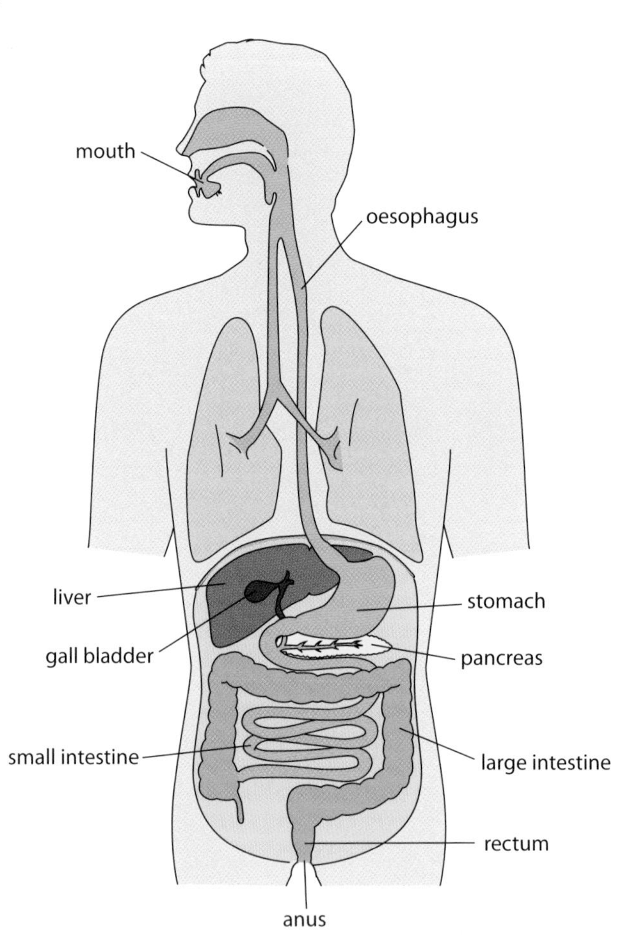

Figure 6.2 The human digestive system.

Any foods that you ingest must *either* be digested and absorbed for use by the body *or* remain undigested and be eliminated as solid waste (faeces).

Role of stomach, small intestine and large intestine

These three areas of the alimentary canal perform different functions.

A person hanging upside down could swallow food and it will go up to their stomach. This is because the food is moved by peristalsis, not by gravity.

Stomach

Food is brought to your stomach by a muscular tube called the oesophagus. When you swallow, the food is forced down to your stomach by a sequential series of smooth muscle contractions called peristalsis. Once in the stomach, the food is held for a period of time in order to mix it with a variety of secretions collectively known as gastric juice. Gastric juice is a mixture of three secretions from the cells of the stomach inner lining:

- pepsin – a protease enzyme most active in acidic pH;
- hydrochloric acid – helps degrade and break down foods and creates the acidic pH necessary for pepsin to be active;
- mucus – lines the inside of the stomach wall to prevent stomach damage from the hydrochloric acid.

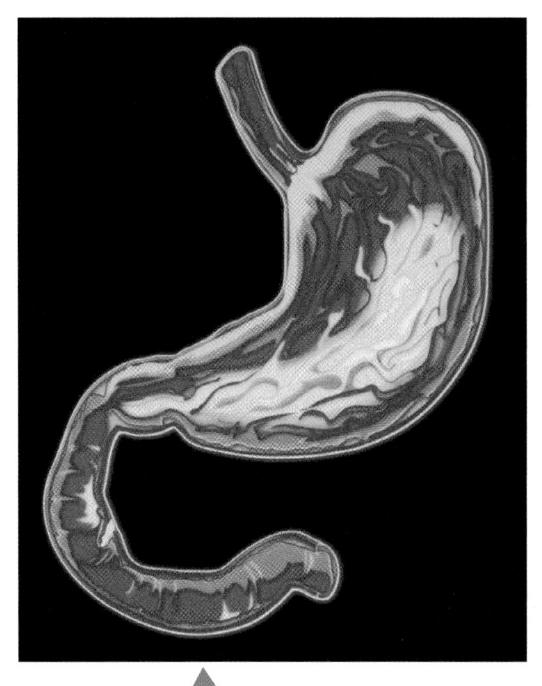

This computer graphic shows the oesophagus, stomach and duodenum.

The muscular wall of the stomach creates a churning motion in order to mix the food with the gastric juice. After a period of time, a valve at the lower end of the stomach opens and the food enters the small intestine.

Small intestine

The first portion of the small intestine is called the duodenum. Here, three different accessory organs secrete juices into the small intestine in order to continue the digestive process. These secretions include:

- bile from the liver and gall bladder;
- trypsin (a protease), lipase, amylase and bicarbonate from the pancreas.

As the digestive process continues in the small intestine, molecules are produced that are small enough to be absorbed. The inner wall of the small intestine is made up of thousands of finger-like extensions called villi (see Figure 6.3). Each villus contains a capillary bed and a lacteal. A lacteal is a small vessel of your lymphatic system just like a capillary is a small vessel of your circulatory system. If the inner lining of your small intestine were smooth, you would have a fairly limited membrane surface area for absorption. The function of the villi is to greatly increase the surface area for absorption of molecules such as glucose, amino acids, and fatty acids.

It is estimated that over 800 million people do not get enough food on a day-to-day basis.

Who can and/or should take responsibility?

Most of the molecules absorbed are taken into the capillary bed within each villus (except fatty acids which are more efficiently absorbed into the lacteal). All the absorbed molecules are taken to a wide variety of body cells by the circulatory system. Once a nutrient molecule has reached a body cell, it leaves the bloodstream by way of another capillary bed and enters a body cell. Within the body cell, the nutrient molecule may be used for energy (e.g. glucose) or it may be used as a component to help build a larger molecule inside the cell (e.g. amino acids). If the nutrient molecule is used for building larger molecules, the process of bringing the nutrient to a body cell and then using it is called assimilation.

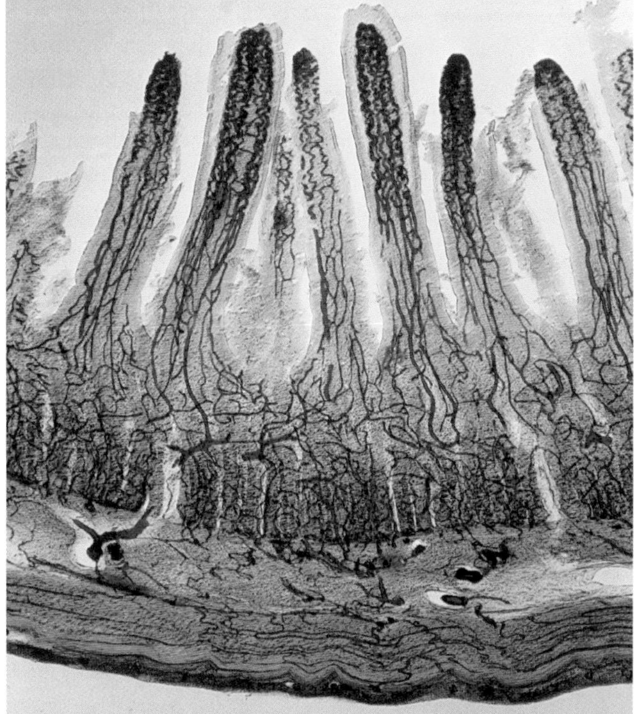

In this transverse section light micrograph, capillary beds within villi are clearly seen. Smooth muscle layers in the wall of the small intestine are seen at the bottom.

Figure 6.3 Structure of an intestinal villus.

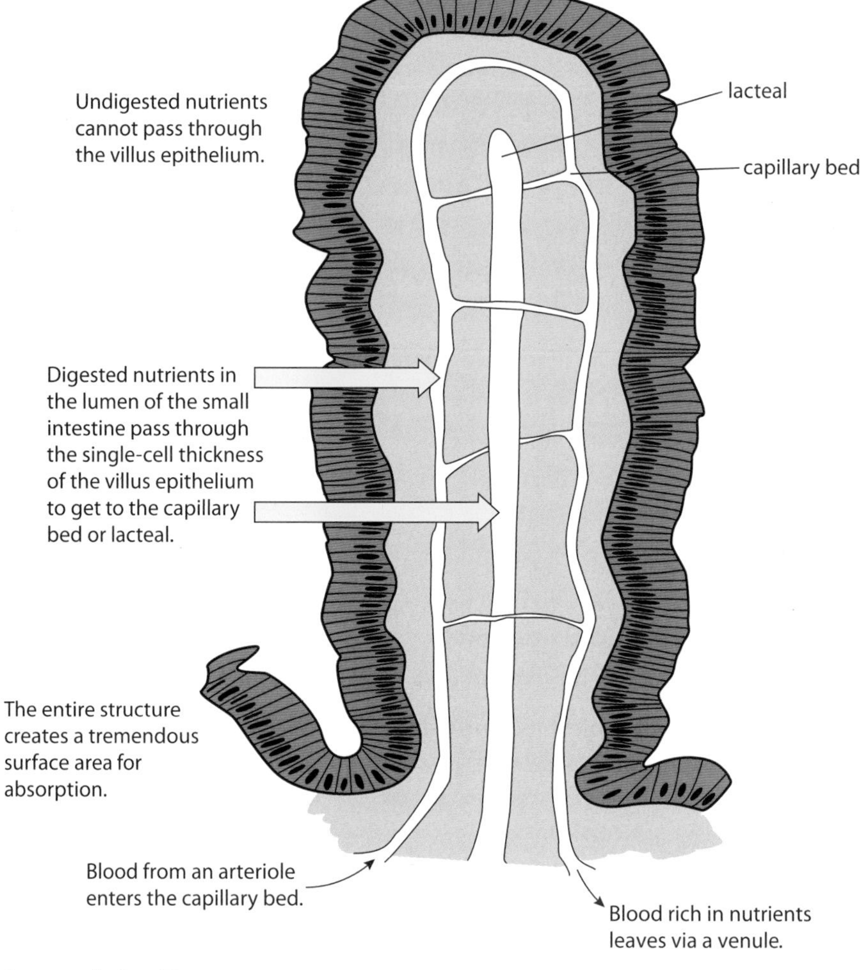
Undigested nutrients cannot pass through the villus epithelium.
lacteal
capillary bed
Digested nutrients in the lumen of the small intestine pass through the single-cell thickness of the villus epithelium to get to the capillary bed or lacteal.
The entire structure creates a tremendous surface area for absorption.
Blood from an arteriole enters the capillary bed.
Blood rich in nutrients leaves via a venule.

Large intestine

The vast majority of useful nutrients are absorbed while food is still inside the small intestine. What remains of the original food at the end of the small intestine is undigested (and therefore unabsorbed). Much of the water that we drink or that is naturally contained in many foods is also still present. This material then passes into the large intestine. The primary function of the large intestine is water absorption. Leaving the water in the alimentary canal as long as possible is beneficial because it keeps the moving food in a fluid environment. The large intestine is also home to a very large number of naturally occurring bacteria including *Escherichia coli*. These bacteria are examples of mutualistic organisms within us. We provide nutrients, water, and a warm environment for them while they synthesize vitamin K and maintain a healthy overall environment for us in our large intestines. Any food undigested by us or the bacteria is eliminated from the body as solid waste or faeces.

Exercises

1 A single sandwich is likely to contain carbohydrates, lipids and proteins. From a biochemical viewpoint, what will happen to each type of molecule on digestion?

2 You ingest a glucose molecule in the starch of a breakfast cereal. State as many specific locations as you can for this single glucose molecule from the time it is in your mouth to the time it enters a muscle cell of your right forearm.

6.2 The transport system

Assessment statements

6.2.1 Draw and label a diagram of the heart showing the four chambers, associated blood vessels, valves, and the route of blood through the heart.

6.2.2 State that the coronary arteries supply heart muscle with oxygen and nutrients.

6.2.3 Explain the action of the heart in terms of collecting blood, pumping blood, and opening and closing of valves.

6.2.4 Outline the control of the heartbeat in terms of myogenic muscle contraction, the role of the pacemaker, nerves, the medulla of the brain and adrenaline.

6.2.5 Explain the relationship between the structure and function of arteries, capillaries and veins.

6.2.6 State that blood is composed of plasma, erythrocytes, leucocytes (phagocytes and lymphocytes) and platelets.

6.2.7 State that the following are transported by the blood: nutrients, oxygen, carbon dioxide, hormones, antibodies, urea and heat.

The human heart

The human heart is designed as a pair of side-by-side pumps. Each side of the heart has a collection chamber for blood that is moving slowly in from the veins. These thin-walled, muscular chambers are called atria. Each side also has a thick-walled muscular pump (called a ventricle) which builds up enough pressure to send the blood out from the heart with a force we refer to as blood pressure. This double-sided pump works every minute of every day of your life. The blood that is pumped out from the heart typically makes a circuit through the following range of blood vessels:

- a large artery;
- smaller artery branches;
- an arteriole (smallest type of artery);
- a capillary bed;
- a venule (smallest type of vein);
- larger veins;
- a large vein which takes blood back to the heart to be pumped out once again.

The two sides of the heart allow for there being two routes for blood to flow along (see Figure 6.4). The right side of the heart sends blood along a route that is called your pulmonary circulation. On this route, the capillary bed is in one of your lungs, and blood picks up oxygen and releases carbon dioxide.

The left side of the heart sends blood along a route that is called your systemic circulation. The artery that emerges from your heart for this route is your aorta. Branches of the aorta carry blood to almost every organ and cell type in your body. On this route, the capillary bed is in one of your organs or tissues, and blood picks up carbon dioxide and releases oxygen.

Fish have a 2-chambered heart and amphibians have 3-chambered heart. Reptiles, birds and mammals all have a 4-chambered heart (the ventricles of a reptile heart are only partially divided).

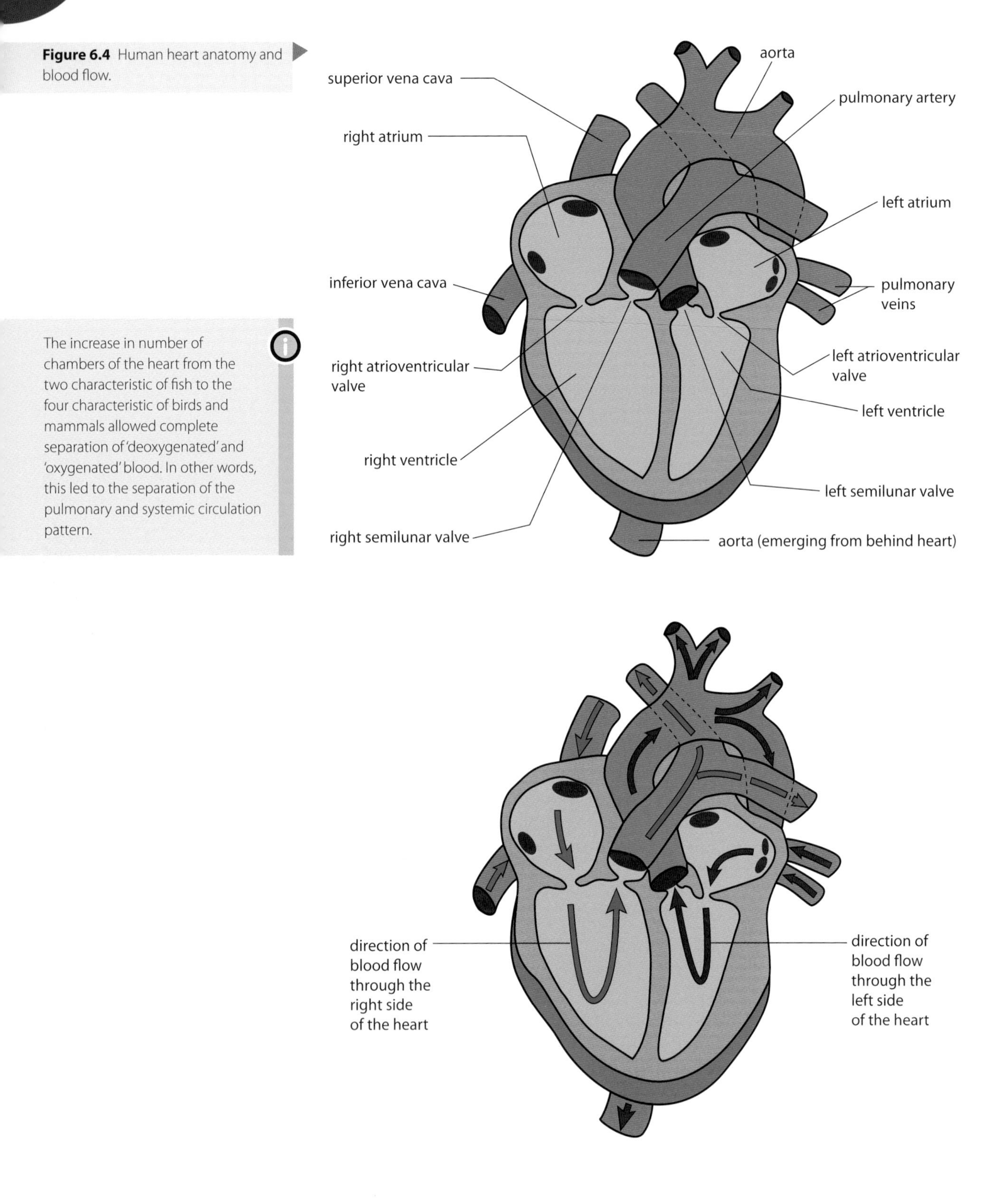

Figure 6.4 Human heart anatomy and blood flow.

The increase in number of chambers of the heart from the two characteristic of fish to the four characteristic of birds and mammals allowed complete separation of 'deoxygenated' and 'oxygenated' blood. In other words, this led to the separation of the pulmonary and systemic circulation pattern.

Pulmonary circulation

The easiest way to follow the circulation pattern within your heart and body is to think of the journey that could be taken by one red blood cell (RBC). We will start following our imaginary RBC in the systemic circulation as it is coming back from body cells.

The blood cell is first found in a large vein that is bringing blood to the right atrium. Because this RBC has already been out to the body tissues, it is in need of oxygen before beginning another trip round the body. A volume of blood collects within the right atrium (including our RBC) and begins moving down into the right ventricle through an open valve. This is the right atrioventricular valve. The right atrium contracts in order to force any remaining blood into the right ventricle. Once a volume of blood has accumulated in the right ventricle, it begins to contract. This contraction initiates several events including:

- closure of the atrioventricular valve to prevent backflow to the right atrium (it is the closing of valves that produces the characteristic 'lub dub' sounds heard through a stethoscope);
- dramatic increase in blood pressure inside the right ventricle which opens the right semilunar valve and allows blood to enter the pulmonary artery;
- due to the increase in pressure, blood leaves the heart through the pulmonary artery.

Our chosen RBC is now in an artery leading to one of the two lungs. As it approaches and then enters a lung, our RBC will be moving along smaller and smaller arteries. The smallest of these arteries is called an arteriole. Any one arteriole leads to a capillary bed. Capillaries are blood vessels that have a very small diameter and are typically only a single cell thick. This is why 'exchanges' (molecular movements in and out of the bloodstream) only occur while blood is in capillaries. Our RBC will probably pick up oxygen molecules while in the lungs and perhaps give up carbon dioxide molecules. Then the RBC will be on its way back to the heart. It will pass into larger and larger veins until the largest of those veins takes our RBC directly into the left atrium.

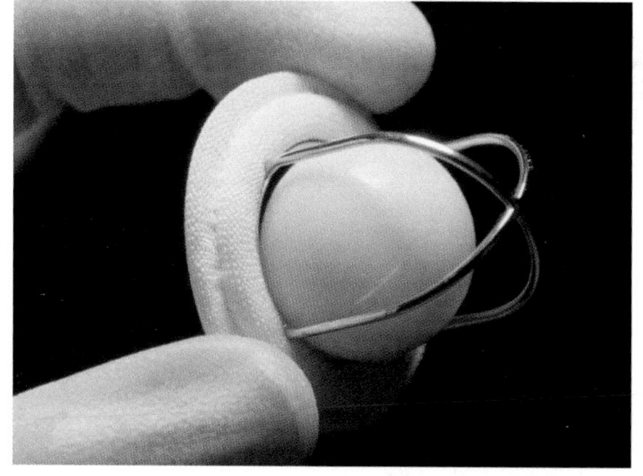

A malfunctioning heart valve can sometimes be replaced by surgery using an artificial valve. One example is shown above. The function of heart valves is to prevent backflow.

Systemic circulation

When our RBC enters the left atrium, a set of events occurs that is similar to when it entered the right atrium. In fact, the right and left sides of the heart are acting in unison – both atria contract at the same time and both ventricles contract at the same time. Blood, now including our RBC, accumulates in the left atrium and then enters the open, left atrioventricular valve. Our RBC passes into the left ventricle as this chamber fills with blood. When the left ventricle contracts, this initiates the following events:

- closure of the atrioventricular valve to prevent backflow into the left atrium;
- dramatic increase in blood pressure inside the left ventricle which opens the left semilunar valve and allows blood to enter the aorta;
- due to the increase in pressure, blood leaves the heart through the aorta.

Our chosen RBC now finds itself in the largest artery in the human body. The aorta has many branches which lead to all tissues in the body. One of the first branches from the aorta allows blood to enter the coronary arteries. The coronary arteries branch out into the heart muscle itself and supply the heart with oxygen and nutrients.

Obviously, our RBC can only take one branch at each possible branch point and ultimately will find itself in a capillary bed somewhere in the body. Let's imagine that on this circuit, our RBC has passed into a muscle in your hand. This is where

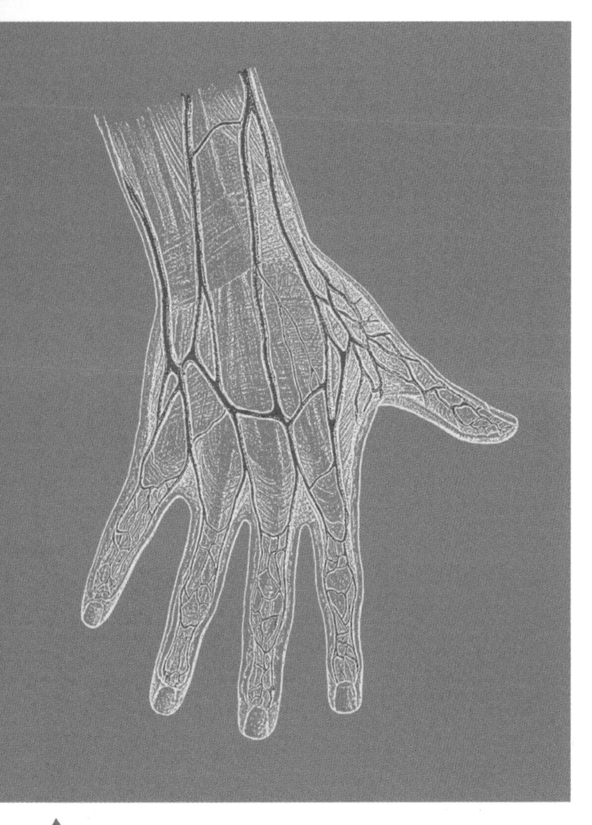

Drawing showing some of the larger blood vessels in the hand. Smaller vessels like capillaries cannot be seen without magnification.

oxygen is given off and carbon dioxide may be taken in by the blood. Our RBC will then begin its journey through larger and larger veins back to our original starting point.

Each complete circuit round the body includes both the systemic route or circuit *and* the pulmonary route or circuit. Each complete circuit typically takes no longer than a minute or two.

Control of heart rate

The majority of the tissue making up the heart is muscle. More specifically, it is cardiac muscle. Cardiac muscle spontaneously contracts and relaxes without nervous system control. This is known as myogenic muscle contraction. The myogenic activity of the heart needs to be controlled in order to keep the timing of the contractions unified and useful.

The right atrium contains a mass of tissue within its walls known as the sinoatrial node (SA node). This mass of tissue acts as the pacemaker for the heart; it sends out an 'electrical' signal to initiate the contraction of both atria. For a person with a resting heart rate of 72 beats a minute, the signal from the SA node is sent out every 0.8 seconds. Also within the right atrium, is another mass of tissue known as the atrioventricular node (AV node). The AV node receives the signal from the SA node, waits approximately 0.1 seconds and then sends out another 'electrical' signal. This second signal goes to the much more muscular ventricles and results in their contraction. This explains why both atria and then, later, both ventricles contract together (see Figure 6.5).

As a general rule, smaller animals have faster myogenic (resting) heart rates than larger animals. The heart of a mouse beats at a rate of about 700 beats a minute. An elephant has a heart rate of about 30 beats a minute.

During times of increased body activity such as exercise, the heart rate needs to increase above the resting heart rate. This is because there is an increased demand for oxygen for cell respiration during periods of heavy exercise or body activity. There is also a need to get rid of the increased levels of carbon dioxide that accumulate in the bloodstream. As exercise begins and carbon dioxide levels begin to rise, an area of your brainstem called the medulla chemically senses the increase in carbon dioxide. The medulla then sends a signal through a cranial nerve (called the cardiac nerve) to increase heart rate to an appropriate level. This signal is sent to the SA node; it does not change the mechanism of how the heart beats, just the timing. After exercise, the level of carbon dioxide in the bloodstream begins to decrease and another signal is sent from the medulla. This time the signal is

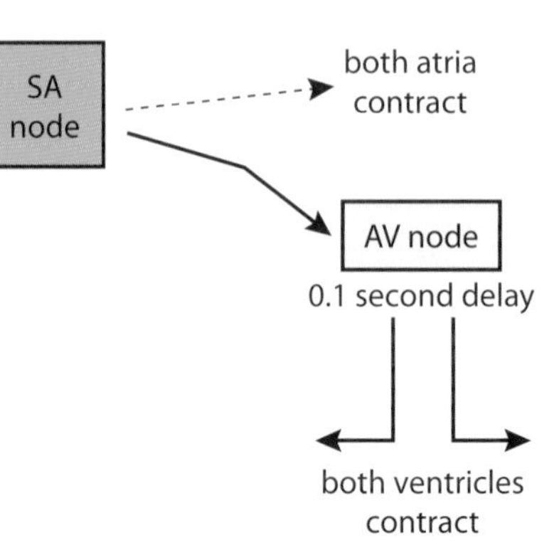

Figure 6.5 Myogenic control of heart rate.

carried by a different cranial nerve called the vagus nerve. Ultimately, electrical signals from the vagus nerve result in the SA node once again taking over the timing of the heart rate.

The heart rate can also be influenced by chemicals. One of the most common is adrenaline. During periods of high stress or excitement, your adrenal glands secrete adrenaline into the bloodstream. Among other effects, adrenaline causes the SA node to 'fire' more frequently than it does at the resting heart rate and thus heart rate increases.

Statistics for coronary heart disease (CHD) vary widely in various parts of the world. Currently, studies are trying to determine how much of the variance is due to genetics, diet, stress, and other factors.

Arteries, capillaries and veins

Arteries are blood vessels taking blood away from the heart that has not yet reached a capillary. Veins are blood vessels that collect blood from capillaries and return it to the heart. Identifying a blood vessel as being an artery or a vein has nothing to do with whether the blood is oxygenated or deoxygenated. For example, blood leaving the right ventricle is flowing through pulmonary arteries, even though it needs to be reoxygenated in the capillaries of the lung tissue. These blood vessels are pulmonary arteries because they are between the heart and the capillary bed. The newly oxygenated blood will be brought back to the heart by way of pulmonary veins.

Arteries have a relatively thick smooth muscle layer that is used by your autonomic nervous system to change the inside diameter of the blood vessels. This helps in regulating blood pressure. Blood in arteries is at high pressure as arteries are the vessels that are directly connected to the ventricles of the heart. When blood leaves an arteriole (the smallest of the arteries), it enters a capillary bed rather than a single capillary. A capillary bed is a network of capillaries that typically all drain into a single venule.

When blood enters a capillary bed much of the pressure is lost. Blood cells make their way through capillaries one cell at a time. Chemical exchanges always occur through the single-cell thickness of capillaries because the walls of arteries and veins are too thick to efficiently allow molecules in or out. Veins receive blood at relatively low pressure from the capillary beds. Because this blood has lost a great deal of blood pressure, the blood flow through veins is slower than through arteries. To account for this, veins have thin walls and a larger internal diameter. Veins also have many internal passive valves that help keep the slow-moving blood consistently moving towards the heart.

The following tables will help you organize what you have learned about blood and the circulatory system.

Quick comparison of arteries, capillaries and veins

Artery	Capillary	Vein
thick walled	wall is 1 cell thick	thin walled
no exchanges	all exchanges occur	no exchanges
no internal valves	no internal valves	have internal valves
internal pressure high	internal pressure low	internal pressure low

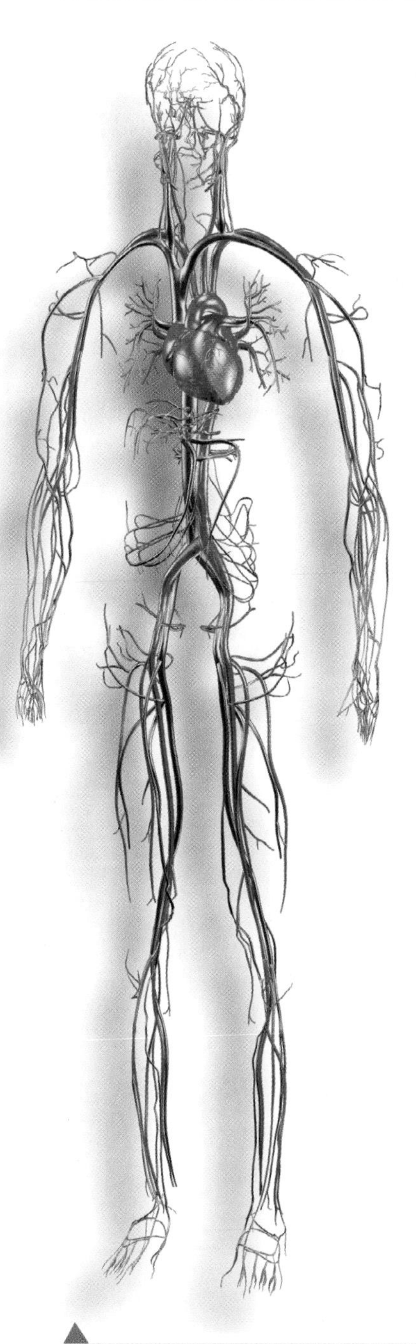

If all tissue except the blood vessels and heart were removed from the body, the body shape would still be clearly visible.

Components of blood

Component	Description
plasma	liquid portion of blood
erythrocytes	red blood cells (carry oxygen and carbon dioxide)
leucocytes	white blood cells (phagocytes and lymphocytes)
platelets	cell fragments (assist in blood clotting)

Transport by blood

What is transported	What it is or does
nutrients	glucose, amino acids, etc.
oxygen	reactant needed for aerobic cell respiration
carbon dioxide	waste product of aerobic cell respiration
hormones	transported from gland to target cells
antibodies	protein molecules involved in immunity
urea	nitrogenous waste (filtered out of the blood by kidneys)
heat	skin arterioles (can change diameter in order to gain or lose heat)

Exercises

3 Identify all of the heart chambers, valves, and blood vessels for one complete circuit of blood. Name all of these in the order blood passes through them, starting with the right atrium.

4 Before birth, a human fetus has a hole between the right atrium and left atrium. Deduce how that changes the blood flow within the fetal circulation and why fetal circulation has evolved such a pattern.

6.3 Defence against infectious disease

Assessment statements

6.3.1 Define *pathogen*.

6.3.2 Explain why antibiotics are effective against bacteria but not against viruses.

6.3.3 Outline the role of skin and mucous membranes in defence against pathogens.

6.3.4 Outline how phagocytic leucocytes ingest pathogens in the blood and in body tissues.

6.3.5 Distinguish between *antigens* and *antibodies*.

6.3.6 Explain antibody production.

6.3.7 Outline the effects of HIV on the immune system.

6.3.8 Discuss the cause, transmission and social implications of AIDS.

Pathogens cause disease

Our bodies are exposed to many disease-causing agents. Any living organism or virus that is capable of causing a disease is called a pathogen. Pathogens include: viruses, bacteria, protozoa, fungi, and worms of various types. Yet exposure to the vast majority of the pathogens does not result in a disease. Primarily, this is because we are too well defended for most pathogens to enter our bodies and, in the case of those that do enter, we have often previously developed an immunity to that pathogen. For some, such as bacteria, there are chemicals, called antibiotics, that can work against the living bacterial cells but do not affect our body cells.

The availability of antibiotics varies widely in various parts of the world. In some places, infections from bacteria are devastating to the population, whereas people in other areas barely give these same or similar infections a second thought.

How antibiotics work against bacteria

In order to understand how antibiotics work against bacteria, you need to recall that bacteria are prokaryotic cells and our body cells are eukaryotic cells. Among the many differences between the two cell types are differences in biochemical reactions and pathways. For example, protein synthesis is similar in both types of cell, but not exactly the same. Also, bacteria have a cell wall, a structure not characteristic of our body cells. Antibiotics are chemicals that take advantage of the differences between prokaryotic and eukaryotic cells. There are many categories of antibiotic. One type of antibiotic may selectively block protein synthesis in bacteria, but have no effect on our cells' ability to manufacture proteins. Another type may inhibit the production of a new cell wall by bacteria, thus blocking their ability to grow and divide.

The overuse of antibiotics by many developed nations has led to bacterial strains that are resistant to many antibiotics. Bacteria do not honour country borders and thus these resistant bacterial strains become a problem everywhere in the world.

This also explains why antibiotics have no effect on viruses. Viruses make use of our own body cells' metabolism to create new viruses. Any chemical that could inhibit this would also be damaging to our own body cells. Thus, antibiotics are chemicals with the ability to damage or kill prokaryotic cells, but not damage eukaryotic cells or their metabolism.

Preventing pathogens from entering our bodies

The best way to stay healthy is to prevent pathogens from having the chance to cause disease. One way you can do this is to try to stay away from sources of infection. This is why it is still common to isolate (or quarantine) people who have certain transmittable diseases. Obviously, it is not possible to isolate yourself from every possible source of infection. The human body has some ingenious ways to make it difficult for pathogens to enter and start an infection.

Skin

Skin is a barrier to infection. Think of your skin as having two primary layers. The underneath layer is called the dermis and is very much alive. It contains sweat glands, capillaries, sensory receptors and dermal cells that give structure and strength to the skin. The layer on top of this is called the epidermis. This epidermal layer is constantly being replaced as underlying dermal cells die and are moved upwards. This layer of mainly dead cells is a good barrier against most pathogens because it is not truly alive. As long as our skin remains intact, we are protected from most pathogens entering living tissues. This explains why it is important to cleanse and cover cuts and abrasions in the skin when they do occur.

Stomach acid

However, there are many pathogens that enter the body in other ways. Some pathogens enter the body in food and water. The very acidic environment of the stomach helps to kill most of these ingested pathogens.

Mucus

Other pathogens enter in the air we breathe (through either the nasal passageways or mouth). This route of entry is lined with a type of tissue known as mucous membrane. The table below shows the location of some mucous membranes.

Area with mucous membrane	What it is and does
trachea	tube which carries air to and from the lungs
nasal passages	tubes which allow air to enter the nose and then the trachea
urethra	tube which carries urine from bladder to the outside
vagina	reproductive tract leading from uterus to the outside

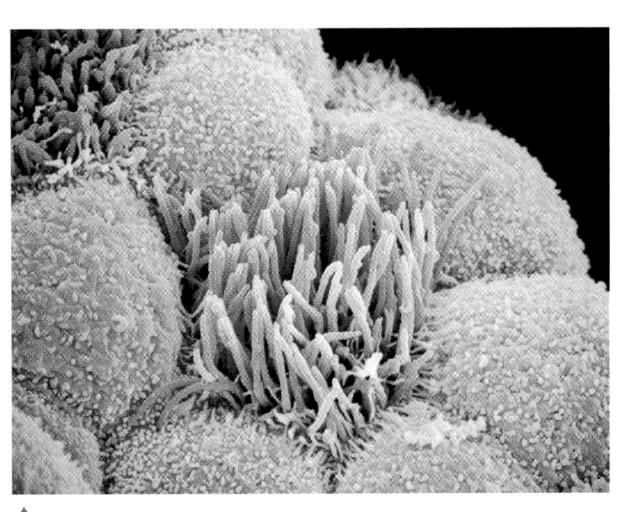

False-colour SEM of the inner lining of the trachea. The large white cells are goblet cells which secrete mucus. The hair-like cilia (in pink) are also visible.

Cells of mucous membranes produce and secrete a lining of sticky mucus. This can trap incoming pathogens and so prevent them from reaching cells they could infect. Some mucous membrane tissue is lined with cilia. Cilia are hair-like extensions capable of a wave-like movement. This movement moves trapped pathogens up and out of mucous-lined tissues like your trachea. In addition, the cells that secrete the mucus also secrete an enzyme called lysozyme. This enzyme is able to chemically damage many pathogens.

What happens when pathogens do get in?

Our bodies have yet another layer of defence for when pathogens do successfully enter.

Role of phagocytic leucocytes

Leucocytes, also known as white blood cells, are the cells in our bloodstream that help us fight off pathogens that enter our bodies and also provide us with an immunity for many pathogens we encounter a second time. There are many different types of white blood cell and they have many different roles in keeping us healthy.

One type of leucocyte that gets involved very early in the process of fighting off a pathogen is called a macrophage. Macrophages are large white blood cells that are able to change their cellular shape to surround an invader and take it in through the process of phagocytosis. This movement of the cell membrane is very similar to the movement an amoeba makes and is quite often referred to as amoeboid motion. Even though we think of blood cells as being within the blood vessels, that is not always true for these macrophages. Because macrophages can easily change their shape by amoeboid movement, they are able to squeeze their way in and out of small blood vessels. Therefore, it is not unusual for a macrophage to first encounter an invader completely outside the bloodstream.

When a macrophage does meet a cell, it recognises whether the cell is a natural part of the body and therefore 'self', or not part of the body and therefore 'not-self'. This recognition is based on the protein molecules that make up part of the surface of all cells and viruses. If the collection of proteins the macrophage encounters is determined to be 'self', then the cell is left alone. If the determination is 'not-self', the macrophage engulfs the invader by phagocytosis. All phagocytes typically contain many lysosome organelles in order to help chemically digest whatever has been engulfed. This type of response by the body is called non-specific because the identity of the pathogen has not been determined at this point, just the fact that it is something that is 'not-self' and therefore should be removed.

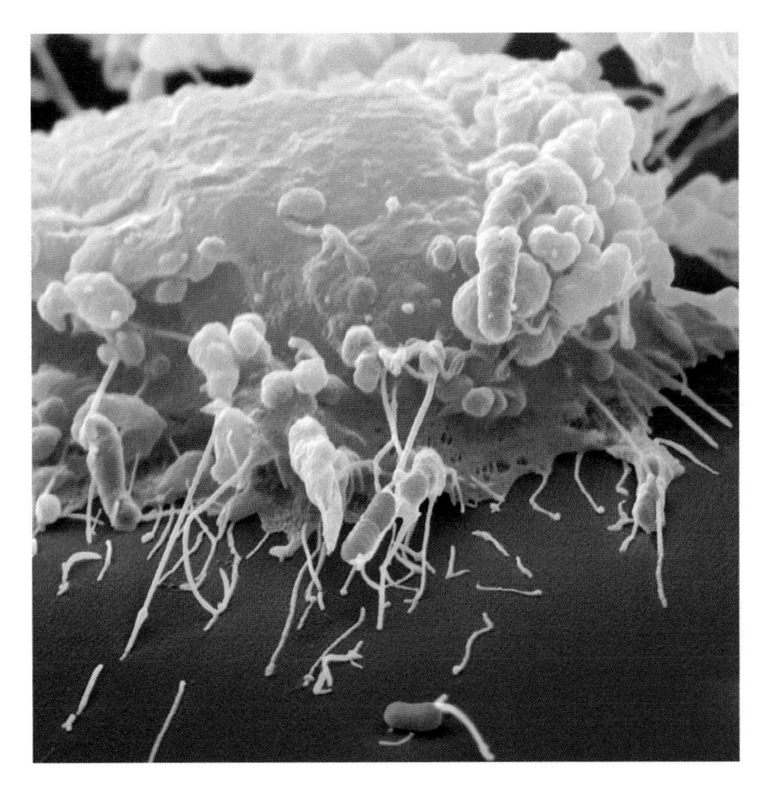

False-colour SEM showing a macrophage (yellow) engulfing *E. coli* bacteria (pink rods). This process is called phagocytosis.

Antibodies are produced in response to a specific pathogen

Antibodies are protein molecules that we produce in response to a specific type of pathogen. In other words, if you had a measles infection, you would produce one type of antibody and another type if you contract a virus that gives you flu symptoms. Each type of antibody is different because each type has been produced in response to a different pathogen. Each pathogen is made up of either cells with cell membranes or, in the case of a virus, a protein coat. The cellular invaders, like bacteria, have proteins that are embedded into their outer surface. In the language of the immune system, these foreign proteins are called antigens. In the previous section, you learned that 'not-self' proteins triggered an immune response. All of these 'not-self' proteins are antigens. Most pathogens have several different antigens on their surface and therefore may trigger the production of many different types of antibody.

Antigens are molecules that our immune system considers to be 'not-self'.

Even though each type of antibody is different and is specific for just one type of antigen, antibodies as a group of molecules are amazingly similar to each other. Each antibody is a protein that is Y shaped. At the end of each of the forks of the Y is a binding site. The binding site is where the antibody attaches itself to an antigen. Because the antigen is a protein on the surface of a pathogen (like a bacterium), the antibody thus becomes attached to the pathogen (see Figure 6.6).

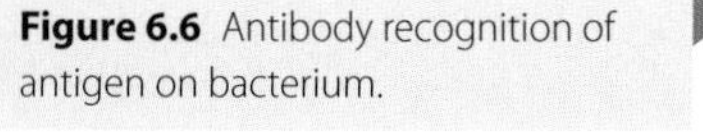
Figure 6.6 Antibody recognition of antigen on bacterium.

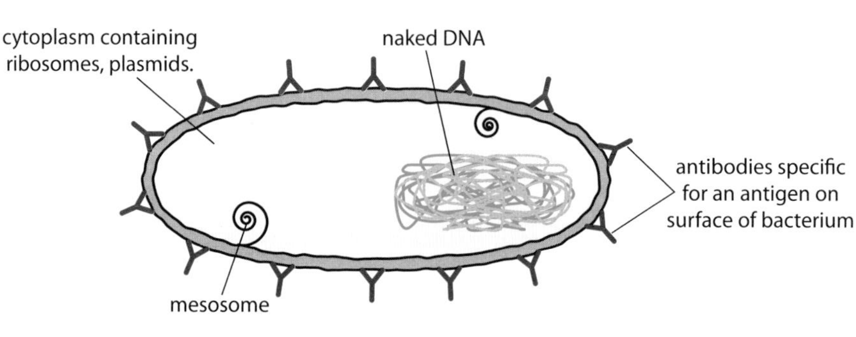

The leucocytes that produce antibodies are called B lymphocytes. Each of us has many different types of B lymphocyte and as a general rule, each type of B lymphocyte can produce one type of antibody. The problem is, each cell produces only a relatively small number of antibodies in comparison to the massive infection that may be already going on. However, our continually evolving immune response has a way of producing many antibodies. Here are the steps of a typical immune response, which tell you how this is accomplished.

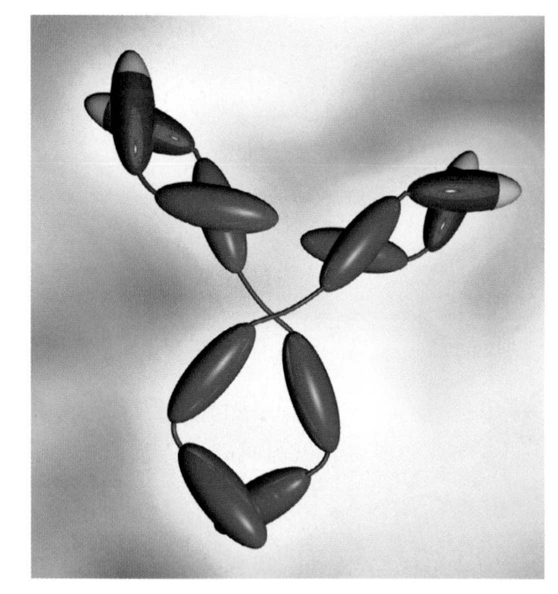
Computer artwork showing an antibody molecule. The molecule forms a Y shape with two antigen binding sites (shown in red and yellow).

1. A specific antigen type is identified (e.g. a particular cold virus).
2. A specific B lymphocyte is identified that can produce an antibody which will bind to the antigen (proteins on the cold virus).
3. The B lymphocyte and several identical B lymphocytes clone themselves (divide repeatedly by mitosis) to rapidly increase the number of the same type of B lymphocytes.
4. The newly formed 'army' begins antibody production.
5. Newly released antibodies circulate in the bloodstream and eventually find their antigen match (the proteins of the pathogen).
6. Using various mechanisms, the antibodies help eliminate the pathogen.
7. Some of the cloned antibody-producing lymphocytes remain in the bloodstream and give immunity from a second infection by the same pathogen. They are called memory cells.

How HIV damages the immune system

Human immunodeficiency virus (HIV) is the virus that eventually results in the set of symptoms collectively called acquired immune deficiency syndrome (AIDS). All viruses must find a type of cell in the body that matches their own proteins in a complementary way. This is why only certain body cells are damaged by certain viruses as is typically reflected in the symptoms associated with the particular infection. For example, a cold virus locates the proteins on mucous membrane cells in your nasal region and ultimately damages those cells. This results in symptoms including swelling of the area and excessive mucus production. Other cells of the body do not have the same protein and therefore the cold virus does not affect them.

The same specificity concept holds true for HIV. Only certain cells in the body have the protein in their membranes that HIV recognizes. One of those is a cell type that functions as a communicator cell in the bloodstream. This cell is known as a helper-T cell and this is the cell that HIV infects. Because HIV is a type of virus that has a latency period (infection occurs, but cells remain alive), it is usually many years after HIV infection that the symptoms called AIDS develop. Helper-T cells are the cells that communicate which cells need to undergo the cloning process and begin antibody production. When the helper-T cells begin to die, the communication between cells no longer occurs and antibodies do not get produced. At this stage, the individual no longer fights off pathogens as they did before and the symptoms of AIDS start to appear. It is one or more of the secondary infections that ultimately takes the life of someone with AIDS.

Issues related to AIDS

AIDS has been and will probably continue to be a difficult disease for humans to deal with. In the previous section we looked at the cause of AIDS, specifically HIV. It continues to be very difficult to find a vaccine or cure for the infection caused by this virus. HIV 'hides away' inside its host cells for years. During this time, the body's immune responses continue to work against other pathogens, but not to combat the HIV because it is already inside body cells waiting for some chemical signal(s) to become active. The virus also mutates relatively quickly for a virus. The body's immune responses or vaccines may not even recognize HIV after it has mutated several times.

Enzyme-linked immunosorbent assays (ELISA) are often used to detect the presence or absence of a particular protein. For example, people recently infected with HIV will initially produce antibodies against HIV. These antibodies can be detected by ELISA and a patient can then be told whether they are HIV positive or HIV negative.

Adding to the difficulty of developing medication is the association of HIV with sexual activity and drug abuse. This initially led to some reluctance in allocating money for HIV research. Today, huge sums of money are allocated for HIV/AIDS research, but this is a relatively new development.

The transmission of HIV has another historical significance which has affected how society responded to the disease. HIV is transmitted from person to person by body fluids. This includes body fluid exchanges during sex and the ill-advised practice of reusing unsterile syringe needles for legal or illegal drug injections. At one time, blood for transfusions was not tested for blood-borne diseases like HIV. Unfortunately, more than a few people became HIV positive from blood transfusions. Today, at least in countries with reasonable medical care, blood is routinely tested for the presence of blood-borne diseases and immediately destroyed if pathogens are found.

On the night of 28 August 1987, a small home in the town of Arcadia, Florida (USA) was burned in a fire that was determined to have been started on purpose.

To read how this event relates to the story of three small boys infected with HIV by blood transfusions, visit heinemann.co.uk/hotlinks, enter the express code 4242P and click on Weblink 6.2.

AIDS was originally labelled as a disease affecting homosexuals and drug abusers. We now know that AIDS is rapidly spreading by way of heterosexual encounters and everyone is at risk. Individuals who have been diagnosed as being HIV positive may be discriminated against in terms of employment, insurance, education access, social acceptance and many other forms of discrimination.

We should also remember that not every country has the education and medical facilities to deal with this disease. In some countries, inadequate medical care sometimes leads to an increase in infection rates as patients with a variety of ailments are often grouped together in large 'wards' and this leads to an exchange of diseases between them.

Until a cure for AIDS is found, perhaps the best that can be accomplished is to continue to lengthen the life-span of those infected, and to educate people on how to decrease their risk of exposure to HIV. This disease is truly a global problem and effective treatment and education must not be limited to certain countries.

Exercises

5 When a person is bitten by a poisonous snake, they are typically administered an antivenom. This antivenom is composed of antibodies which recognize the venom protein molecules and bind to them. The injection of this antivenom is referred to as an example of passive immunity and does not confer long-term immunity to the individual. Why not?

6 Why are some viruses potentially lethal (e.g. HIV, ebola) while others cause fairly mild symptoms?

6.4 Gas exchange

Assessment statements

6.4.1 Distinguish between *ventilation*, *gas exchange* and *cell respiration*.
6.4.2 Explain the need for a ventilation system.
6.4.3 Describe the features of alveoli that adapt them to gas exchange.
6.4.4 Draw and label a diagram of the ventilation system, including trachea, lungs, bronchi, bronchioles and alveoli.
6.4.5 Explain the mechanism of ventilation of the lungs in terms of volume and pressure changes caused by the internal and external intercostal muscles, the diaphragm and abdominal muscles.

Overview of our respiratory system

Our lungs act in concert with our heart and blood vessels to ensure that body cells are well supplied with oxygen and are able to give up carbon dioxide. Most people never seriously consider why we need oxygen, but everyone knows that we do. The process that requires oxygen (and gives off carbon dioxide) is aerobic cell respiration. In brief, this is a biochemical pathway in which the chemical bonds within a glucose molecule are sequentially broken to release energy. Much of this energy is then stored as molecules of ATP. In aerobic organisms, the process requires oxygen molecules and each of the six carbons of glucose are given off as a carbon dioxide molecule.

Throughout our lives we continuously repeat filling our lungs with air and then breathing that air out. This is called ventilation. Even though the air we breathe is inside our lungs for only a short period of time, it is enough for diffusion of gases to occur. Oxygen in the lung tissues typically diffuses into the bloodstream and carbon dioxide from the bloodstream typically diffuses into the lung tissues. Each breath in and out replenishes the gases within the lung tissues so that diffusion continues.

The movement (diffusion) of the gases is also called gas exchange. There are two locations where gas exchange occurs (see Figure 6.7):

- first, in the lungs where oxygen moves from the air of the lungs into the bloodstream (and carbon dioxide moves in the opposite direction);
- second, in a capillary bed elsewhere in the body where the opposite gas exchange occurs – oxygen diffuses out of the bloodstream and into a body cell (and carbon dioxide diffuses out of the body cell into the capillary bed).

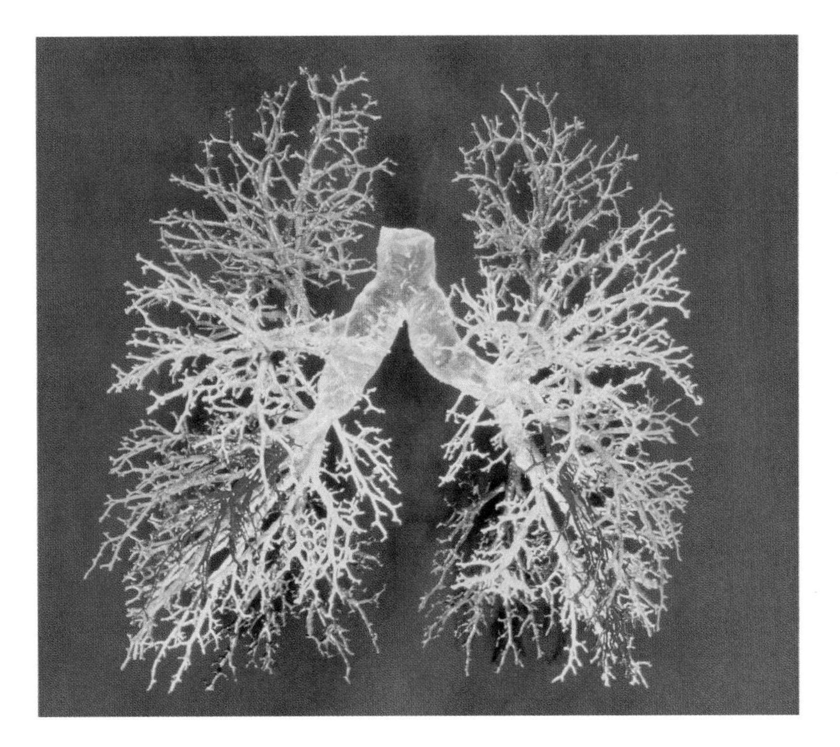

This is a resin cast image of airways in lungs. The trachea divides into the right and left primary bronchi. Many further branchings lead to smaller and smaller bronchioles.

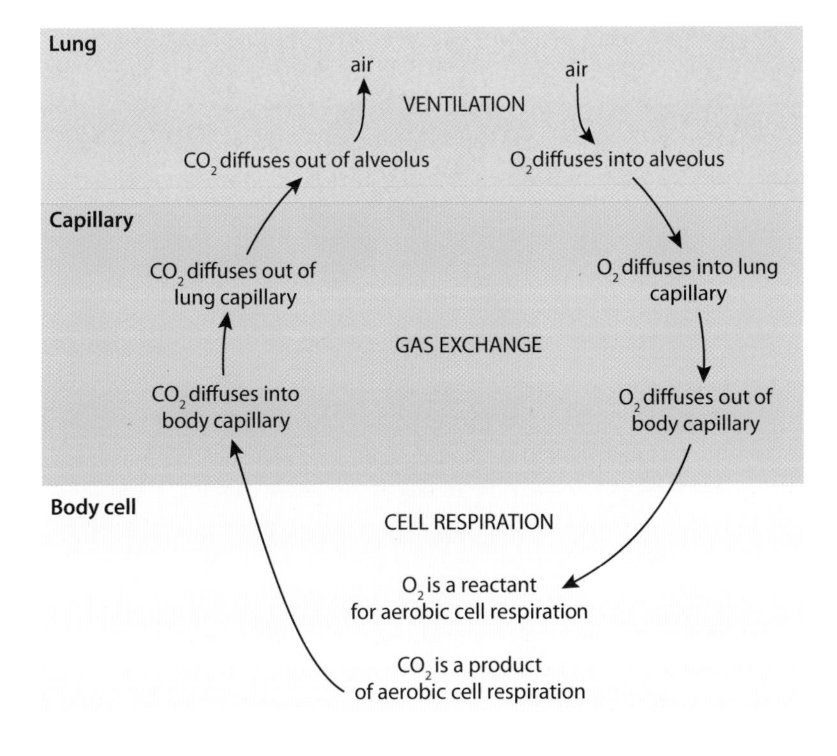

Figure 6.7 Relationship between ventilation, gas exchange and cell respiration.

Why do we need a ventilation system?

An aerobic single-celled organism such as an amoeba, survives perfectly well without a ventilation system. This is not because the amoeba does not need gas exchange, rather that it does not need a system to help accomplish gas exchange. As long as the amoeba is in an environment in which oxygen is in higher concentration outside the cell than inside the cell, oxygen will continue to diffuse into the cell.

So why do we need an entire system to aid in this process? The answer to this is because our bodies are so thick. Only our outside cells are directly exposed to air. If you were to count the number of cells inside one of your thighs, it would easily be in the millions. All of these millions of cells are too far away from the outside air to directly make use of diffusion to supply oxygen (or remove carbon dioxide). This also explains why we have a circulatory system. Your ventilation (respiratory) system and circulatory system function together to pick up oxygen molecules in the inner portions of the lungs and transport that oxygen to body cells deep in your body tissues.

Another closely related reason for having a ventilation system is to ensure that the concentration of the respiratory gases within the lungs encourages the diffusion of each gas in a direction that is beneficial to the body. For example, if a person were not breathing, it would not take long for the concentration of oxygen in the air in their lungs to become equal to or lower than the concentration of oxygen in the blood in the lung capillary bed. Thus, oxygen would not diffuse at all or would diffuse in a direction that is directly contrary to sustaining life. This does not occur because our continuous action of breathing replenishes the oxygen concentration in the lungs and keeps the oxygen in the air in the lungs at a high level in comparison to the oxygen level in the blood capillaries in the lungs.

Anatomy of the ventilation system

Figure 6.8 shows you the ventilation system of a human being.

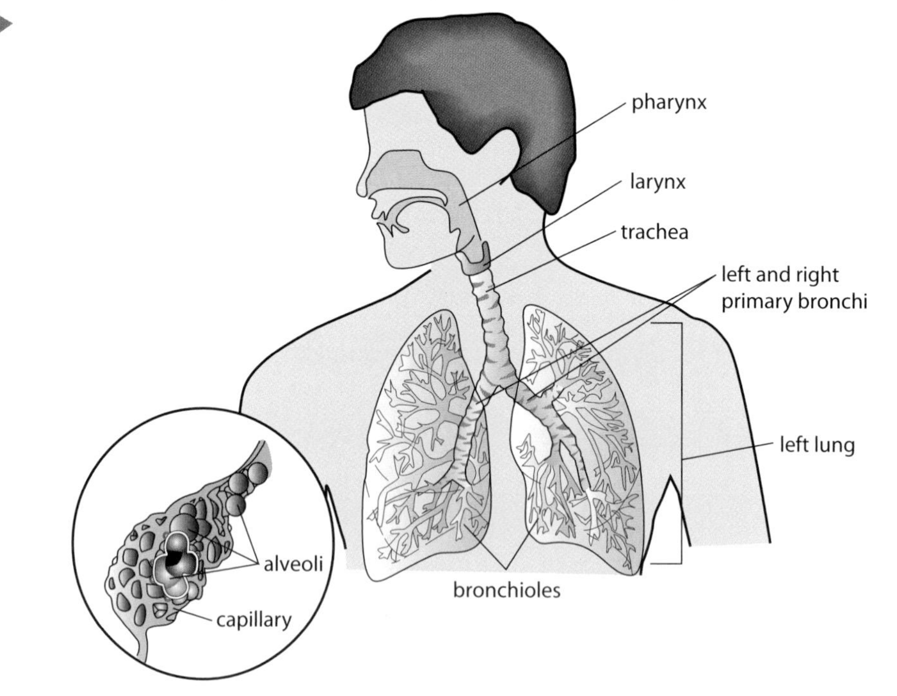

Figure 6.8 Anatomy of the human respiratory system.

Gas exchange occurs within the alveoli

When you take in air through your mouth or your nasal passages:

- the air first enters your trachea;
- then your right and left primary bronchi;
- then smaller and smaller branches of the bronchi;
- then very small branches called bronchioles;
- finally, the air enters the small air sacs in the lungs called alveoli.

Alveoli in the lungs are found as clusters at the ends of the smallest bronchioles. They have an appearance that is very similar to a bunch of grapes. There are approximately 300 million alveoli in each of your lungs. Each cluster of alveoli has a surrounding capillary bed.

The blood entering these capillary beds comes from the right ventricle via the pulmonary arteries. As you will recall, blood within the pulmonary arteries is relatively low in oxygen and high in carbon dioxide. While this blood is in the capillary bed surrounding a cluster of alveoli, oxygen diffuses from the air in each alveolus across two cell membranes. The first of these is the single cell membrane making up the wall of the alveolus and the second is the single cell membrane making up the wall of the capillary. Carbon dioxide diffuses in the opposite direction through the same two cell membranes. As long as you keep breathing and refreshing the gases within your alveoli, the concentration gradients of these two gases will ensure diffusion of each gas in the direction described.

The table below shows adaptations of alveoli that allow efficient gas exchange.

Adaptation	Advantage
spherical shape of alveoli	provides a large surface area for respiratory gases to diffuse through
flattened, single cell thickness of each alveolus	prevents respiratory gases from having to diffuse through more cell layers
moist inner lining of alveolus	allows for efficient diffusion
associated capillary bed nearby	respiratory gases do not have to diffuse far to reach single cell thick capillaries

Mechanism of ventilation

We breathe in and out continuously all our lives. Each time we take a breath, a fairly complex series of events occurs that we do not even think about or realize is happening. The tissue that makes up our lungs is passive and not muscular, therefore the lungs themselves are incapable of purposeful movement. However, there are muscles surrounding the lungs, and these include the diaphragm, muscles of the abdomen, and the intercostal muscles (surrounding your ribs).

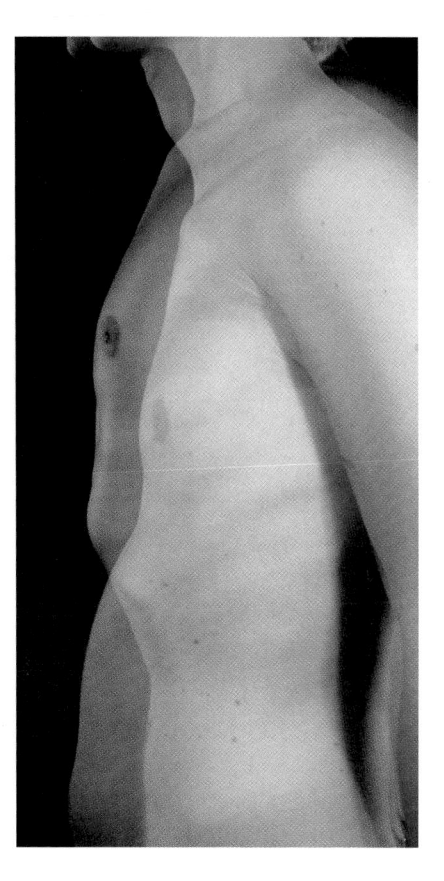

This double exposure photograph shows the positions of the chest during inspiration and expiration. Inspiration is when the rib cage is raised.

The mechanism of breathing is based on the inverse relationship between pressure and volume (see Figure 6.9). Simply put, an increase in volume will lead to a decrease in pressure. Whatever pressure does, volume will do the opposite and whatever volume does, pressure will do the opposite. Your lungs are located within your thoracic cavity (or thorax). The thoracic cavity is closed to the outside air, therefore you can think of this as a closed environment. The lungs themselves have only one opening to the outside air and that is through your trachea (via your mouth or nasal passages). Thus, we need to consider two environments that affect each other: one is the thorax and the other is the internal environment of the lungs.

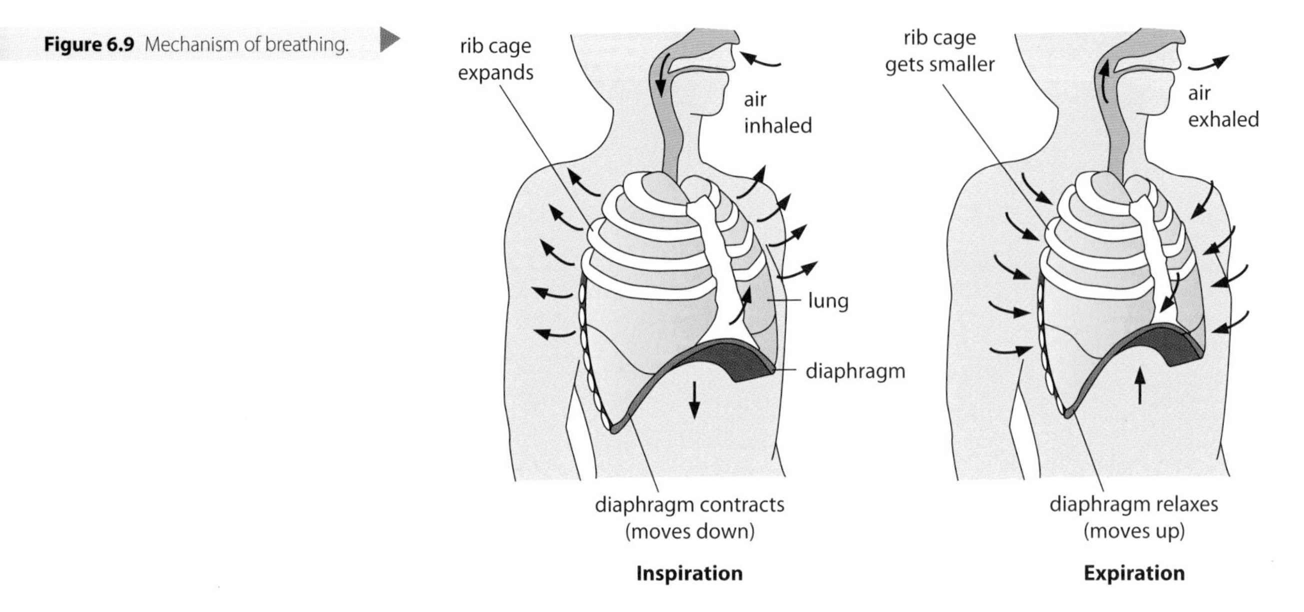

Figure 6.9 Mechanism of breathing.

Mechanism of an inspiration (breathing in)

1 The diaphragm contracts and at the same time the abdominal muscles and intercostal muscles help to raise the rib cage. Collectively, all of these actions act to increase the volume of the thoracic cavity.
2 Because the thoracic cavity has increased its volume, the pressure inside the cavity decreases. This leads to less pressure 'pushing on' the passive lung tissue.
3 The lung tissue increases its volume because there is less pressure exerted on it.
4 This leads to a decrease in pressure inside of the lungs, also known as a partial vacuum.
5 Air comes in through your open mouth or nasal passages to counter the partial vacuum within the lungs (and fills alveoli with air).

You burn about 10–14 calories every hour simply by breathing; this is because you use muscles like the diaphragm, abdominal and intercostal muscles in order to breathe.

Note: These steps are reversed for an expiration (breathing out).

All the steps become more exaggerated when you are exercising and thus breathing deeply. For example, the abdominal muscles and intercostal muscles achieve a greater initial thorax volume. This leads to deeper breathing and thus more air moving into the lungs.

Exercises

7 Create a list of steps which represent what your body would do to accomplish a single expiration (breathing out).

8 Pneumonia (excess mucus) and smoking (tar) create an extra lining inside of each of the alveoli. Describe how and why this could become life-threatening.

6.5 Nerves, hormones and homeostasis

Assessment statements

6.5.1 State that the nervous system consists of the central nervous system (CNS) and peripheral nerves, and is composed of cells called neurones that can carry rapid electrical impulses.

6.5.2 Draw and label the structure of a motor neurone.

6.5.3 State that nerve impulses are conducted from receptors to the CNS by sensory neurones, within the CNS by relay neurones, and from the CNS to effectors by motor neurones.

6.5.4 Define *resting potential* and *action potential* (depolarization and repolarization).

6.5.5 Explain how a nerve impulse passes along a non-myelinated neurone.

6.5.6 Explain the principles of synaptic transmission.

6.5.7 State that the endocrine system consists of glands that release hormones that are transported in the blood.

6.5.8 State that homeostasis involves maintaining the internal environment between limits, including blood pH, carbon dioxide concentration, blood glucose concentration, body temperature and water balance.

6.5.9 Explain that homeostasis involves monitoring levels of variables and correcting changes in levels by negative feedback mechanisms.

6.5.10 Explain the control of body temperature, including the transfer of heat in blood, and the roles of the hypothalamus, sweat glands, skin arterioles and shivering.

6.5.11 Explain the control of blood glucose concentration, including the roles of glucagon, insulin and α and β cells in the pancreatic islets.

6.5.12 Distinguish between *type I* and *type II diabetes*.

Organization of the human nervous system

Our brain and spinal cord are our central nervous system (CNS). These two structures receive sensory information from various receptors and then interpret and process that sensory information. If a response is needed, some portion of the brain or spinal cord initiates a response which is called a motor response. The cells that carry this information are called neurones. Sensory neurones bring information in to the CNS and motor neurones carry response information to muscles (see Figure 6.10, overleaf).

Together, sensory neurones and motor neurones make up the peripheral nerves. A neurone is an individual cell which carries electrical impulses from one point in the body to another and does so very quickly. When many individual neurones group together into a single structure, that structure is called a nerve. Think of a nerve as being like a telephone cable, a protective sheath surrounding many individual wires. Each wire within that cable is like a neurone.

There are two categories of peripheral nerve.

- Spinal nerves – 31 pairs (left and right) of these emerge directly from the spinal cord. They are mixed nerves, as some of the neurones within them are sensory and some are motor.
- Cranial nerves – 12 pairs of these emerge from an area of the brain known as the brainstem. One well known example is the optic nerve pair which carry visual information from the retina of the eyes to the brain.

The central nervous system (CNS) is composed of brain and spinal cord. The peripheral nervous system (PNS) consists of all the nerves and their branches which enter and leave the brain and spinal cord.

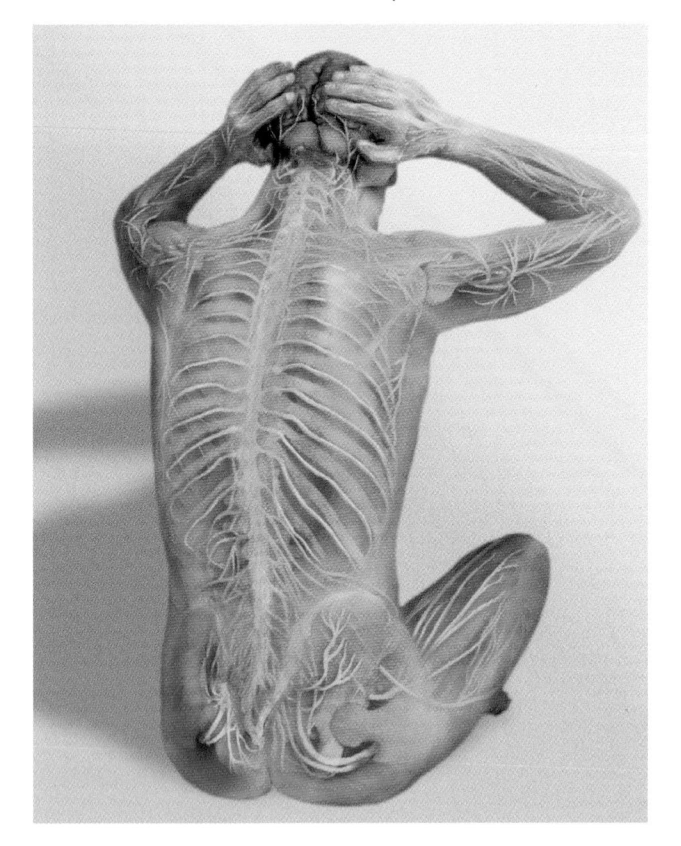

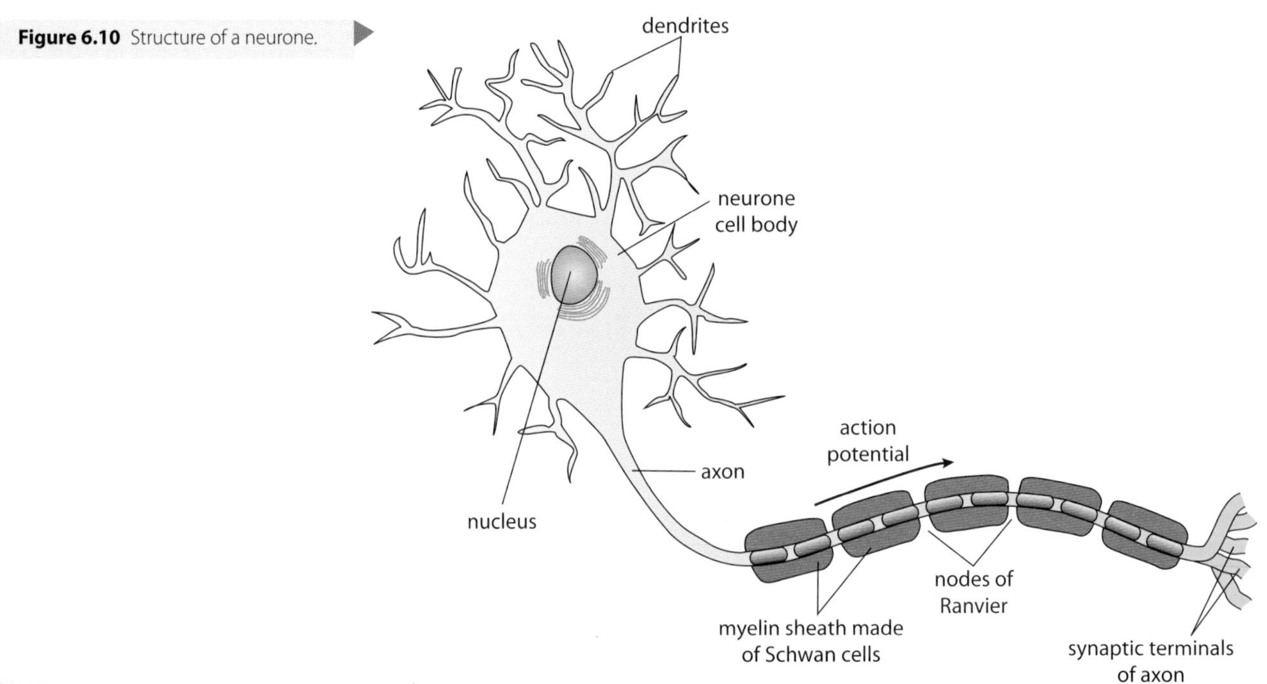

Figure 6.10 Structure of a neurone.

Example of a typical nervous system pathway

When you trace a nervous system pathway, you are following a form of an electrical impulse known as an action potential. Let's trace an example of a pathway from its initial generation to a response in the brain and the nature of that response. Imagine that you have just touched the arm of the person sitting next to you. This was an accidental touch and you immediately removed your hand.

Stimulation and interpretation

When you touched the other person's arm, a pressure receptor began an action potential or 'nerve impulse'. Each receptor in your body is designed to transform a particular kind of stimulus, in this case touch or pressure, into an action potential. The interpretation of the touch occurs in the brain (part of the CNS), so there must be a chain of neurones which take the impulse towards the CNS. This bit of sensory information reaches the spinal cord by one of the 31 pairs of spinal nerves. Sensory neurones stretch from receptors to the spinal cord. Once the action potential reaches the spinal cord, it is routed in the CNS to the appropriate area for interpretation. During the time the action potential is within the spinal cord and brain, it is carried by neurones called relay neurones. You became aware of the touch by using the interpretation made by the appropriate area of the brain.

Response

Your brain also decided to remove your hand and thus relay neurones began an action potential that ultimately resulted in withdrawal of your hand. The pathway for this action began in the brain's relay neurones and was passed down the spinal cord and eventually out along one of the spinal nerve pairs. This action potential is now on a pathway of motor neurones, neurones that are taking an impulse to a muscle. Notice in this sequence that a spinal nerve contains both sensory and motor neurones. Any one neurone can carry an action potential in only one direction, but since most nerves contain many neurones, some may be sensory and some motor. When the action potential reaches the muscle, the motor neurone sends a chemical signal to the muscle which results in a contraction and thus moves your hand away. A junction at which a neurone sends a chemical to muscle tissue is known as a motor end plate. Another name for the muscle in this example is the effector (see Figure 6.11).

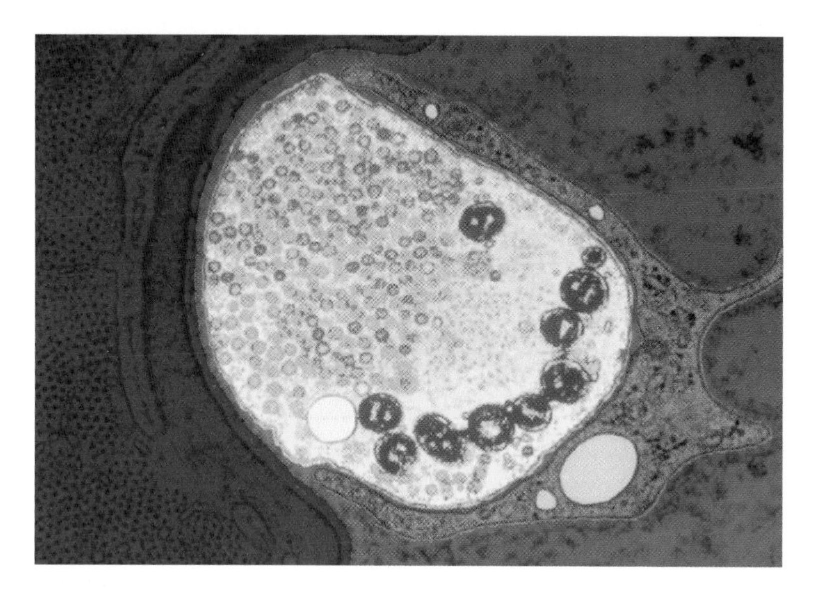

False-colour TEM showing a motor end plate. The area of the synaptic button is yellow. Note the small vesicles containing neurotransmitter. Muscle tissue is shown in red on the left.

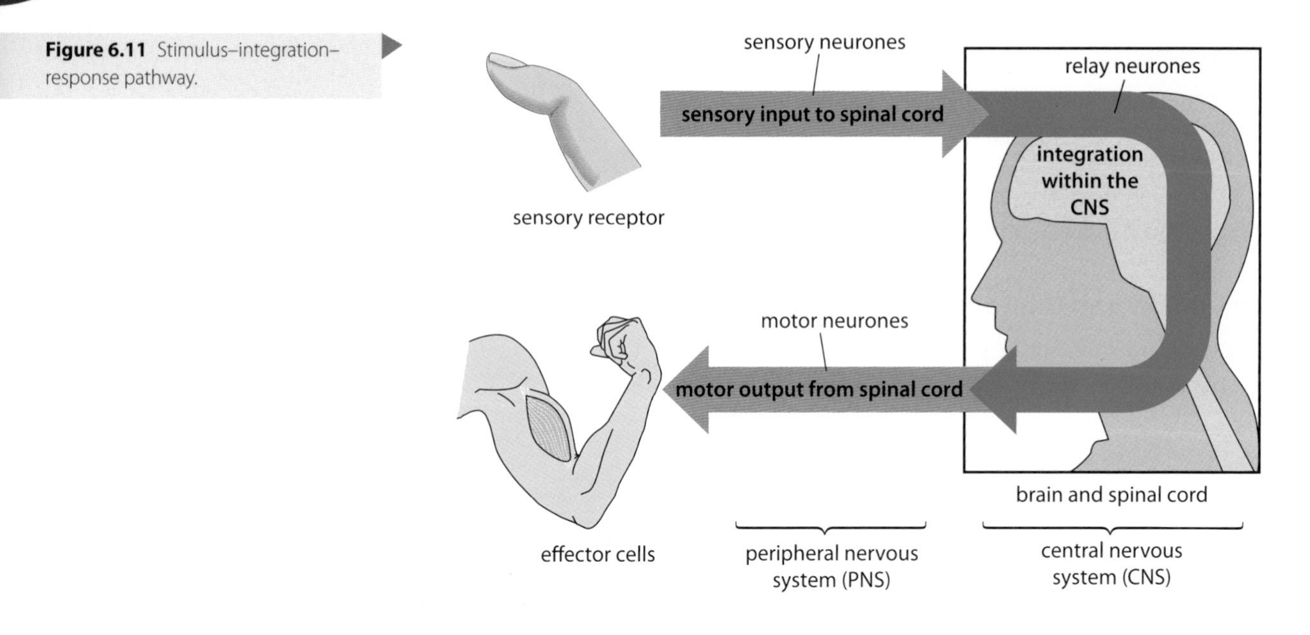

Figure 6.11 Stimulus–integration–response pathway.

What is a 'nerve impulse'?

People often equate a nerve impulse to electricity. In some ways, this is accurate as a nerve impulse can be measured in the same way as electricity is measured. For example, an action potential has a voltage, although the typical unit for this voltage is millivolts. In other ways, electricity and action potentials are very different. True electricity is typically a flow of electrons down a conductor; however, this is not the nature of an action potential. Let's look at what a nerve impulse actually is.

The term 'nerve impulse' is very misleading since a nerve doesn't carry a single impulse, the individual neurones within the nerve are each capable of carrying an action potential. It is convenient to think of the conductor of a neurone impulse as the axon. Axons of some neurones (especially sensory and motor neurones) can be extremely long. The axons of neurones in some organisms that have a very highly developed nervous system (including humans) have a surrounding membranous structure called the myelin sheath. The myelin sheath greatly increases the rate at which an action potential passes down an axon. In order to study the nature of an action potential, it is best first to study an axon which does not have a myelin sheath, otherwise known as a non-myelinated neurone.

Resting potential

Let's first look at what an axon of a neurone would be like when it is not sending an impulse. The state of being where an area of a neurone is ready to send an action potential (but is not currently sending) is called the resting potential and this area of the neurone is said to be polarized. The resting potential is characterized by the active transport of sodium ions (Na^+) and potassium ions (K^+) in two different directions. The vast majority of the sodium ions are actively transported out of the axon cell into the intercellular fluid and the majority of the potassium ions are transported into the cytoplasm. In addition, there are negatively charged organic ions permanently located in the cytoplasm of the axon. This collection of charged ions leads to a net positive charge outside the axon membrane (positive in relation to the inside) and a net negative charge inside the axon membrane (see Figure 6.12).

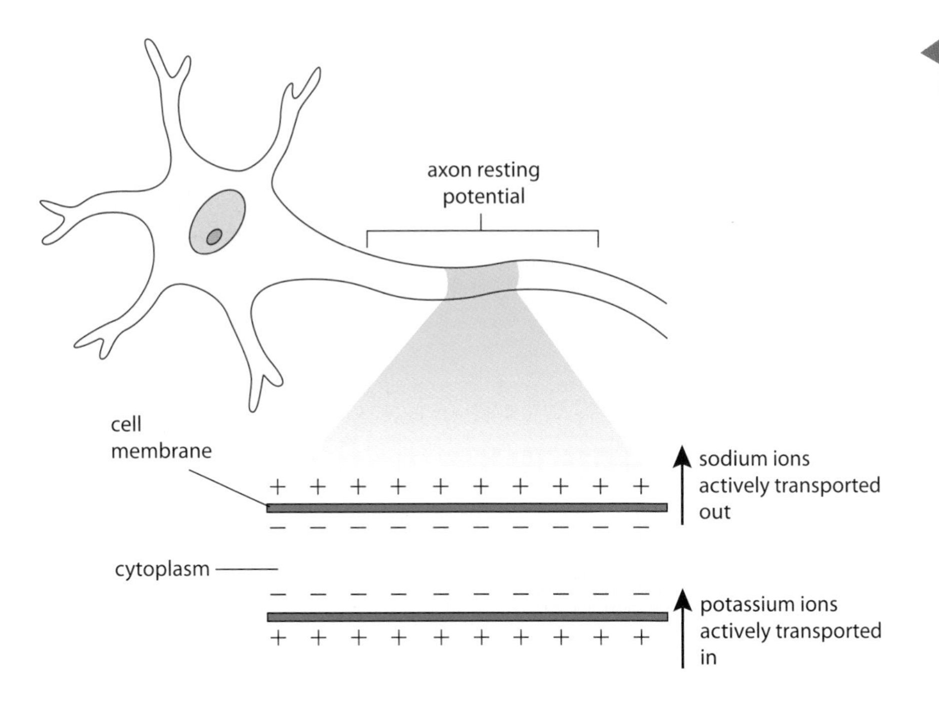

Figure 6.12 Neurone axon at resting potential.

Action potential

An action potential is often described as a self-propagating wave of ion movements in and out of the neurone membrane (see Figure 6.13). The movement of the ions is not along the length of the axon, but instead consists of ions diffusing from outside the axon to the inside and from inside the axon to the outside. The resting potential requires active transport (protein channels and ATP) to set up a concentration gradient of both sodium and potassium ions. Since sodium ions are actively transported to the outside of the membrane, they diffuse in when a channel opens for this purpose. Soon after, a channel opens for potassium ions and they diffuse out of the axon. This diffusion of sodium ions in and potassium ions out is the 'impulse' or action potential. It is a nearly instantaneous event that occurs in one area of an axon and is called depolarization. This area of the axon then initiates the next area of the axon to open up the channels for sodium, then potassium and thus the action potential continues down the axon. This is the self-propagating part of an action potential; once you start an impulse at the dendrite end of a neurone, that action potential will self-propagate itself to the far axon-end of the cell.

A wide variety of action potential animations is available online. For example, visit heinemann.co.uk/hotlinks, insert the express code 4242P and click on Weblink 6.3.

Some of the longest cells in nature are neurones. The axons of some neurones extend for a metre or more.

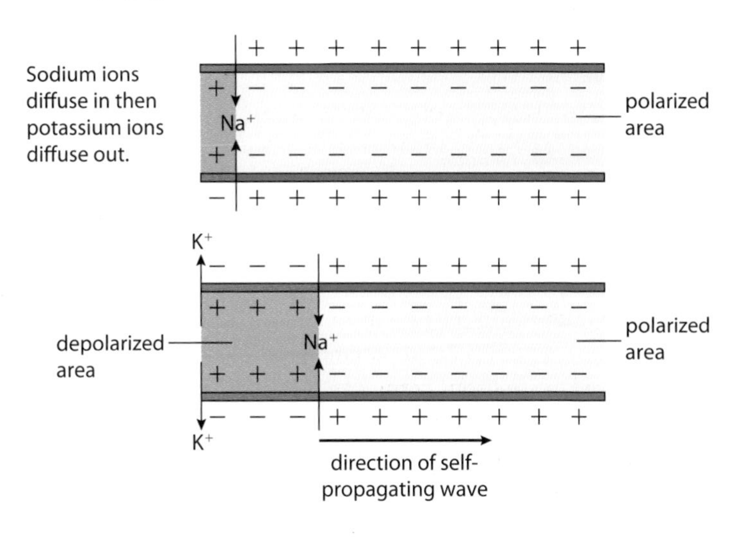

Figure 6.13 Neurone axon during action potential.

Return to the resting potential

Neurones do not send just one action potential; one neurone may send dozens of action potentials in a very short period of time. When one area of an axon has opened a channel to allow sodium ions to diffuse in and potassium ions to diffuse out, that area cannot send another action potential until the sodium and potassium ions have been restored to positions characteristic of the resting potential. Diffusion cannot do this, thus active transport is required to pump these two ions to their resting potential positions (see Figure 6.14). This is called repolarization. The time that it takes for any one neurone to send an action potential and then repolarize so it can send another is called the refractory period of that neurone.

Figure 6.14 Return to the resting potential.

Synaptic transmission: how neurones communicate with each other

If a neurone's axon is touched with an electric probe an action potential is begun which travels in both directions along the axon. However, the action potential is continued to the next neurone only in the direction in which synaptic transmission can occur.

A sensory pathway is unidirectional mainly because the sensory neurones of the pathway are lined up so that the terminal end of the axon of the first neurone adjoins the dendrites of the next neurone. From now on, we will be referring to the first neurone as the presynaptic neurone and the second as the postsynaptic neurone. This is because a chemical communication called a synapse occurs between these two neurones. All kinds of different patterns exist. Sometimes there is a one-to-one communication between the presynaptic and postsynaptic neurone. Sometimes one presynaptic neurone can form a synapse with many postsynaptic neurones and conversely many presynaptic neurones can form a synapse with one postsynaptic neurone (see Figure 6.15).

Figure 6.15 Three patterns of synaptic transmission.

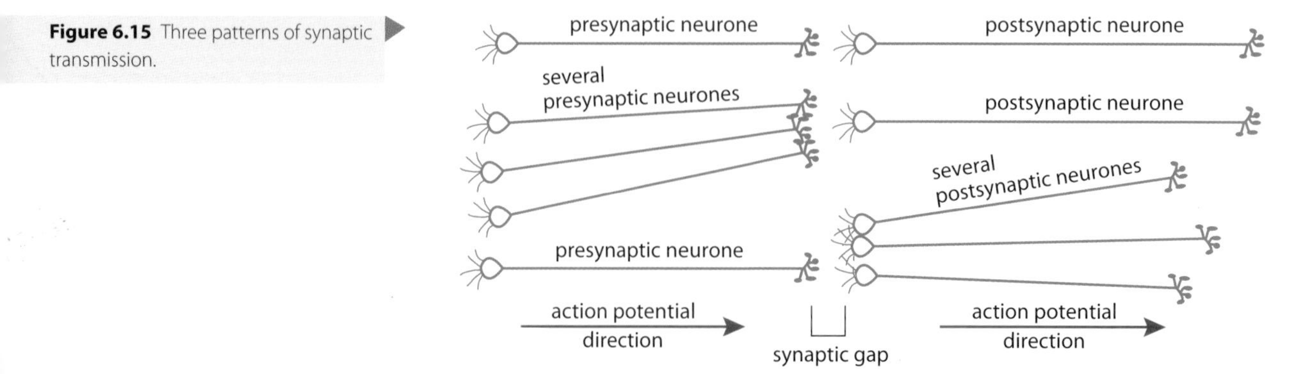

The mechanism of synaptic transmission

At the far end of axons are swollen membranous areas called terminal buttons. Within these terminal buttons are many small vesicles filled with a chemical called a neurotransmitter. The term 'neurotransmitter' is used for any chemical that is used for synaptic transmission. There are many such chemicals; a very common example in humans is acetylcholine.

When an action potential reaches the area of the terminal buttons, it initiates the following sequence of events (see Figure 6.16).

1 Calcium ions (Ca^{2+}) diffuse into the terminal buttons.

2 Vesicles containing neurotransmitter fuse with the plasma membrane and release neurotransmitter.

3 Neurotransmitter diffuses across the synaptic gap from the presynaptic neurone to the postsynaptic neurone.

4 Neurotransmitter binds with a receptor protein on the postsynaptic neurone membrane.

5 This binding results in an ion channel opening and sodium ions diffusing in through this channel (as for the action potential described on page 177).

6 This initiates the action potential to begin moving down the postsynaptic neurone because it is depolarized.

7 Neurotransmitter is degraded (broken into two or more fragments) by specific enzymes and is released from the receptor protein.

8 The ion channel closes to sodium ions.

9 Neurotransmitter fragments diffuse back across the synaptic gap to be reassembled in the terminal buttons of the presynaptic neurone.

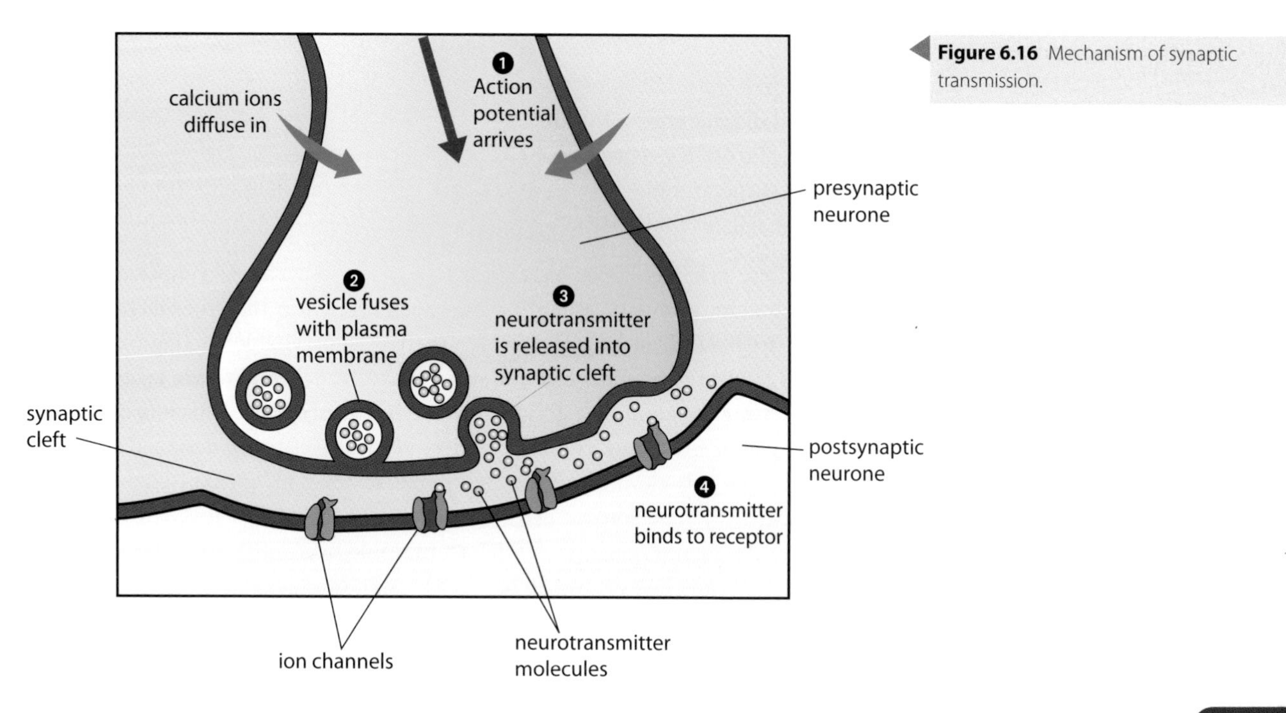

Figure 6.16 Mechanism of synaptic transmission.

In this illustration, the presynaptic neurone (upper centre) is releasing neurotransmitter to postsynaptic neurone.

Homeostasis

The human body typically stays within certain limits (often called normal limits) for many physiological variables. This is referred to as homeostasis. The variables include :

- blood pH;
- carbon dioxide concentration;
- blood glucose concentration;
- body temperature;
- water balance within tissues.

Each of these variables has an expected value or set point that is considered to be normal for homeostasis. For example, you often hear that internal body temperature is 37 °C or 98.6 °F. However, there is an inevitable fluctuation around this exact temperature. If you were to have your temperature taken soon after exercising, it would be highly unlikely to be exactly 37 °C. This is because the body would still be undergoing a variety of mechanisms, all designed to bring a temporarily elevated internal temperature down to 'normal'. The same would be true for a person out in a very cold environment. The mechanisms helping to elevate internal body temperature would be at work and internal body temperature would be expected to be close to but not exactly 37 °C.

The physiological changes that bring a value back closer to a set point, are called negative feedback mechanisms. Think of negative feedback control as working like a thermostat. The thermostat signals a set of actions required when a value rises above its set point and another set of actions required when a value falls below its set point. Thus, negative feedback functions to keep a value within the narrow range that is considered normal for homeostasis.

Your nervous system and endocrine system work cooperatively in order to ensure homeostasis. Many of the homeostatic mechanisms initiated by your nervous system are under the control of your autonomic nervous system. The endocrine system consists of numerous glands which produce a wide variety of hormones. Each hormone is transported by the bloodstream from the gland where it is produced to the specific cell types in the body that are influenced by that particular hormone.

Homeostatic control of body temperature

The biological thermostat for temperature control is an area of your brain called the hypothalamus. First, let's consider the scenario when the hypothalamus senses that your temperature is beginning to rise. This could be because you are exercising or because your environmental temperature is very warm. The hypothalamus receives information from thermoreceptors in your skin and begins to activate cooling mechanisms. These cooling mechanisms include increased activity of sweat glands and thus the subsequent evaporative cooling effect of the perspiration. In addition, arterioles in your skin dilate (get bigger) and this fills skin capillaries with blood. Heat leaves the skin capillaries by radiation so you cool down.

Each year many people die of heat stroke. Heat stroke occurs when the body's mechanisms for dissipating heat cannot keep up with the body's rising internal temperature.

When you are in a cold air environment, the hypothalamus receives appropriate information from thermoreceptors in your skin and begins to activate warming mechanisms. These include constricting skin arterioles so blood is diverted to deeper organs and tissues and less heat is loss by radiation. The hypothalamus also stimulates skeletal muscle to begin shivering. This nearly constant use of many muscles results in the generation of body heat and thus has a warming effect.

Blood glucose level must remain close to a set point

Blood glucose level is the concentration of glucose dissolved in blood plasma. Cells rely on glucose for the process of cell respiration. Cells never stop doing cell respiration and thus are constantly acting to lower the concentration of glucose in the blood. Many people eat three or more times a day, usually including foods containing glucose or carbohydrates that are chemically digested to glucose. This glucose is absorbed into the bloodstream in the capillary beds of the villi of the small intestine and thus increases blood glucose level. This 'see-saw' involving the increase and decrease of blood glucose goes on 24 hours a day, every day of your life. Blood glucose must be maintained close to the body's set point for blood glucose level, and negative feedback mechanisms ensure this.

In the intestinal villi, the glucose is routed through a multitude of capillaries, small venules and veins into the hepatic portal vein which takes the blood to the liver. Glucose concentration in the hepatic portal vein varies depending on the time of your last meal and the glucose content of the food you ate. The hepatic portal vein is the only major blood vessel in the body in which blood glucose concentration fluctuates to a large degree. All other blood vessels receive blood after it has been acted on by liver cells called hepatocytes. Hepatocytes are directed to action by two hormones (insulin and glucagon) produced in the pancreas (see Figure 6.17 for location of pancreas). These two pancreatic hormones are antagonistic; in other words, they have opposite effects on blood glucose concentration.

Figure 6.17 Pathways (ducts) of some digestive exocrine glands.

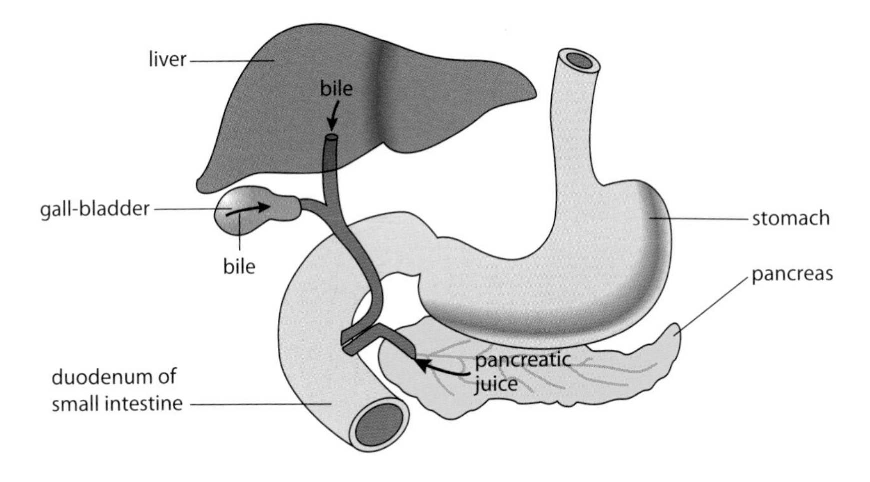

What happens when blood glucose level goes above the set point?

In the pancreas, there are cells known as β (beta) cells which produce the hormone insulin. After production, insulin is secreted by these cells and absorbed into the blood. Since all body cells chemically communicate with blood, all cells are exposed to insulin. Insulin's effect on body cells is to open protein channels in their plasma membranes. These channels allow glucose to diffuse into the cell by a process known as facilitated diffusion.

There is another important effect attributed to insulin. When blood that is relatively high in glucose enters the liver by the hepatic portal vein, insulin stimulates the hepatocytes to take in the glucose (a monosaccharide) and convert it to glycogen (a polysaccharide). The glycogen is then stored as granules in the cytoplasm of the hepatocytes. The same effect occurs in muscles.

These two effects of insulin both have the same ultimate result, which is to lower the glucose concentration in the blood or, put more briefly, to reduce blood glucose (see Figure 6.18).

What happens when blood glucose level goes below the set point?

Blood glucose level typically begins to drop below the set point when someone has not eaten for many hours or exercises vigorously for a long time. In either situation, the body now needs to use the glycogen made and stored by liver and muscle cells when blood glucose level was above the set point. Under this circumstance, α (alpha) cells of the pancreas begin to produce and secrete the hormone glucagon. The glucagon circulates in the bloodstream and stimulates hydrolysis of the granules of glycogen stored in hepatocytes and muscle cells; the hydrolysis produces the monosaccharide glucose. This glucose then enters the bloodstream. The ultimate effect is to increase the glucose concentration in the blood or, put briefly, to increase blood glucose (see Figure 6.18).

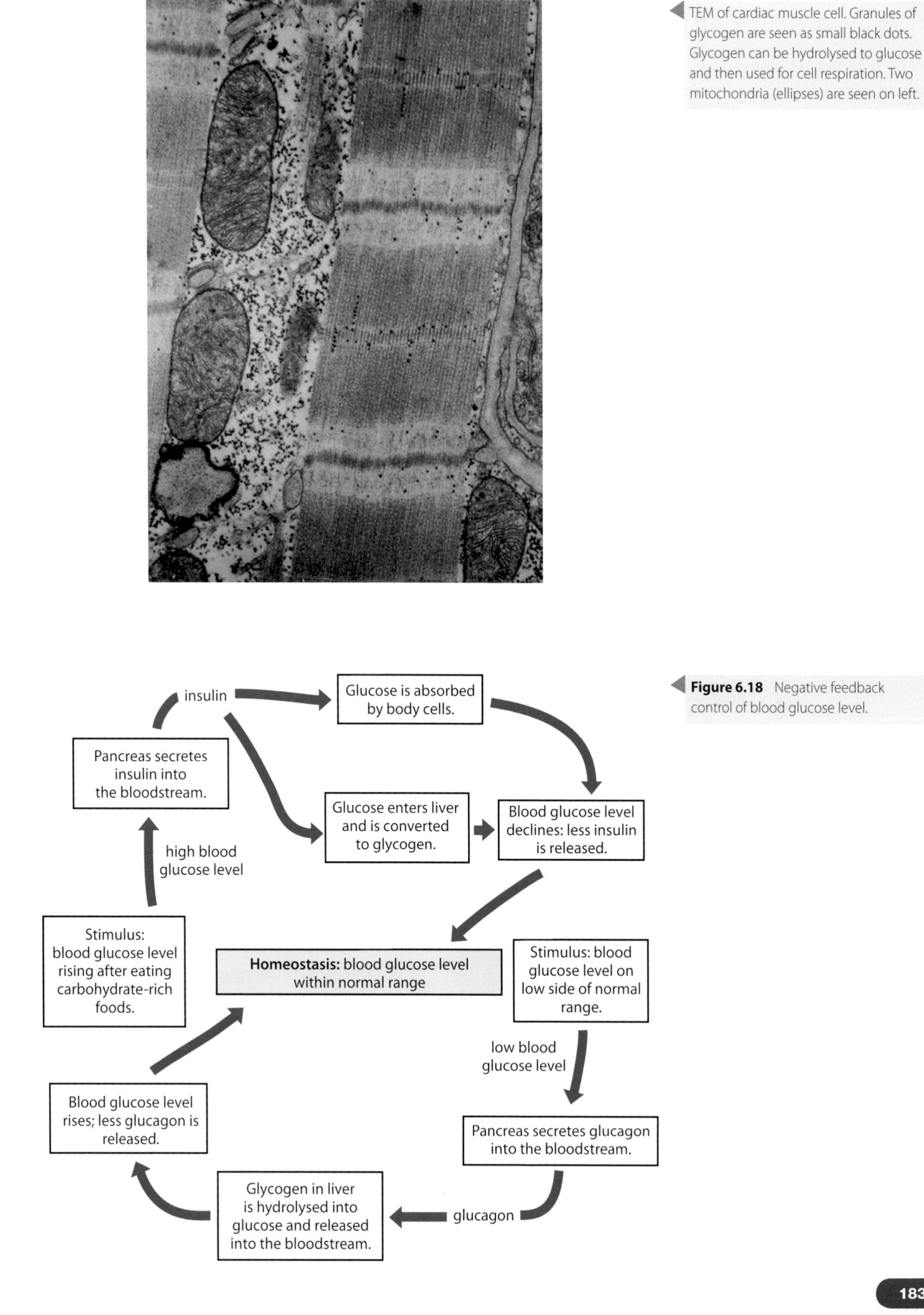
insulin
Glucose is absorbed by body cells.
Pancreas secretes insulin into the bloodstream.
Glucose enters liver and is converted to glycogen.
Blood glucose level declines: less insulin is released.
high blood glucose level
Stimulus: blood glucose level rising after eating carbohydrate-rich foods.
Homeostasis: blood glucose level within normal range
Stimulus: blood glucose level on low side of normal range.
low blood glucose level
Blood glucose level rises; less glucagon is released.
Pancreas secretes glucagon into the bloodstream.
Glycogen in liver is hydrolysed into glucose and released into the bloodstream.
glucagon

TEM of cardiac muscle cell. Granules of glycogen are seen as small black dots. Glycogen can be hydrolysed to glucose and then used for cell respiration. Two mitochondria (ellipses) are seen on left.

Figure 6.18 Negative feedback control of blood glucose level.

According to calculations by the World Health Organization (WHO), almost 3 million deaths per year worldwide are attributable to diabetes.

Diabetes

Diabetes is a disease characterized by hyperglycaemia (high blood sugar). Type I is typically caused when the β cells of the pancreas do not produce enough insulin; type II diabetes is caused by body cell receptors that do not respond properly to insulin. You will recall that the hormone insulin should cause increased facilitated diffusion of glucose (through channels) into almost all body cells. This diffusion into body cells lowers the amount of glucose in the bloodstream. People who have untreated diabetes have plenty of glucose in their blood, but not in their body cells where it is needed.

For an interesting and somewhat uplifting perspective on the life of one person with diabetes, visit heinemann.co.uk/hotlinks, insert the express code 4242P and click on Weblink 6.4.

Type I diabetes can be controlled by the injection of insulin at appropriate times. Type II diabetes is controlled by diet. Uncontrolled diabetes can lead to many serious effects including:

- damage to the retina leading to blindness;
- kidney failure;
- nerve damage;
- increased risk of cardiovascular disease;
- poor wound healing (and possibly gangrene thus making amputation necessary).

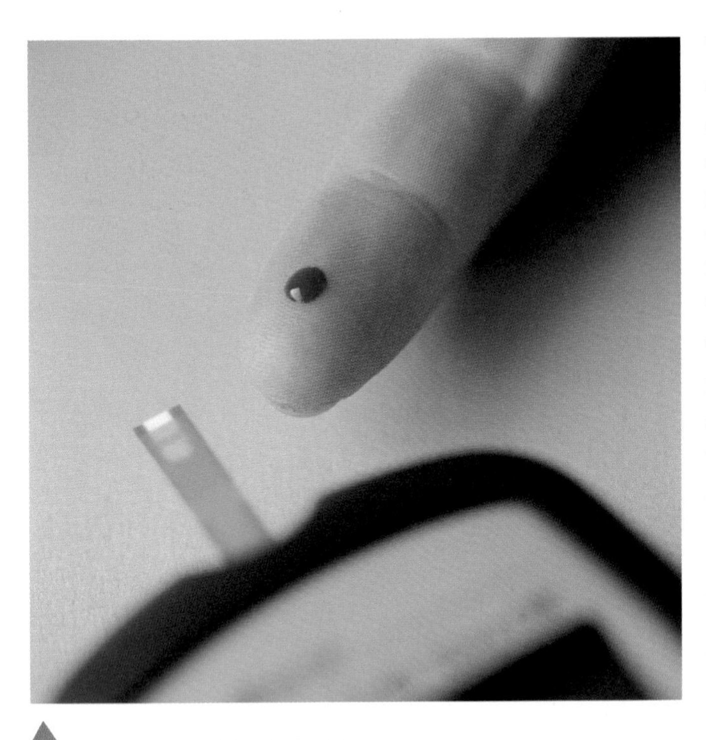

Diabetics must regularly test their blood for glucose concentration.

Type I and type II diabetes

Type I diabetes is an autoimmune disease. The body's own immune system attacks and destroys the β cells of the pancreas so little to no insulin is produced by individuals with type I diabetes. Less than 10% of diabetics have this type of the disease. Type I diabetes most often develops in children or young adults, but can develop in people of any age.

Type II diabetes is a result of the body cells no longer responding to insulin as they once did. This is known as insulin resistance. Initially, the pancreas continues to produce a normal amount of insulin, but this level may decrease after a period of time. Type II diabetes is the most common form of diabetes; approximately 90% of diabetics have this type. Type II diabetes is often associated with genetic history, obesity, lack of exercise, advanced age and certain ethnic groups. Type II diabetes is discussed in Chapter 12 (page 345). groups.

Exercises

9 The majority of homeostatic mechanisms are controlled by negative feedback. Using a hand-drawn picture of a see-saw and annotations, explain how body temperature is regulated in humans.

10 Create a flowchart showing in as much detail as possible, any one stimulus-and-response neural pathway. For example, you might want to use seeing a step on a stairway as the stimulus and raising your leg as the response. Be sure to represent how this visual sensory information would get to the CNS and how the muscles of the leg receive the motor information to contract.

6.6 Reproduction

Assessment statements

6.6.1 Draw and label diagrams of the adult male and female reproductive systems.

6.6.2 Outline the role of hormones in the menstrual cycle, including FSH (follicle stimulating hormone), LH (luteinizing hormone), oestrogen and progesterone.

6.6.3 Annotate a graph showing hormone levels in the menstrual cycle, illustrating the relationship between changes in hormone levels and ovulation, menstruation and thickening of the endometrium.

6.6.4 List three roles of testosterone in males.

6.6.5 Outline the process of in-vitro fertilization (IVF).

6.6.6 Discuss the ethical issues associated with IVF.

Human reproduction

Despite all of the 'trappings' that our societies incorporate into human reproduction, the process comes down to a male gamete (sperm) fertilizing a female gamete (egg or ovum). This cellular union ensures that half of the genetic makeup of the resulting zygote is derived from each parent. Thus, like all forms of sexual reproduction, reproduction in humans serves the bigger purpose of ensuring genetic variation in the species. In both sexes, hormones play a key role in both development of sexual dimorphism (different body forms of males and females) and regulation of sexual physiology. For instance, in males the hormone testosterone:

- determines the development of male genitalia during embryonic development;
- ensures development of secondary sex characteristics during puberty;
- maintains the sex drive of males throughout their lifetime.

Although males typically experience a slight lowering of sperm count as they age, fertility in males has been documented in individuals as old as 94 years.

The structures of the male and female reproductive system are adapted for the production and release of the gametes (see Figures 6.19 and 6.20). In addition, the female reproductive system ensures a suitable location for fertilization and provides an environment for the growth of the embryo then fetus until birth.

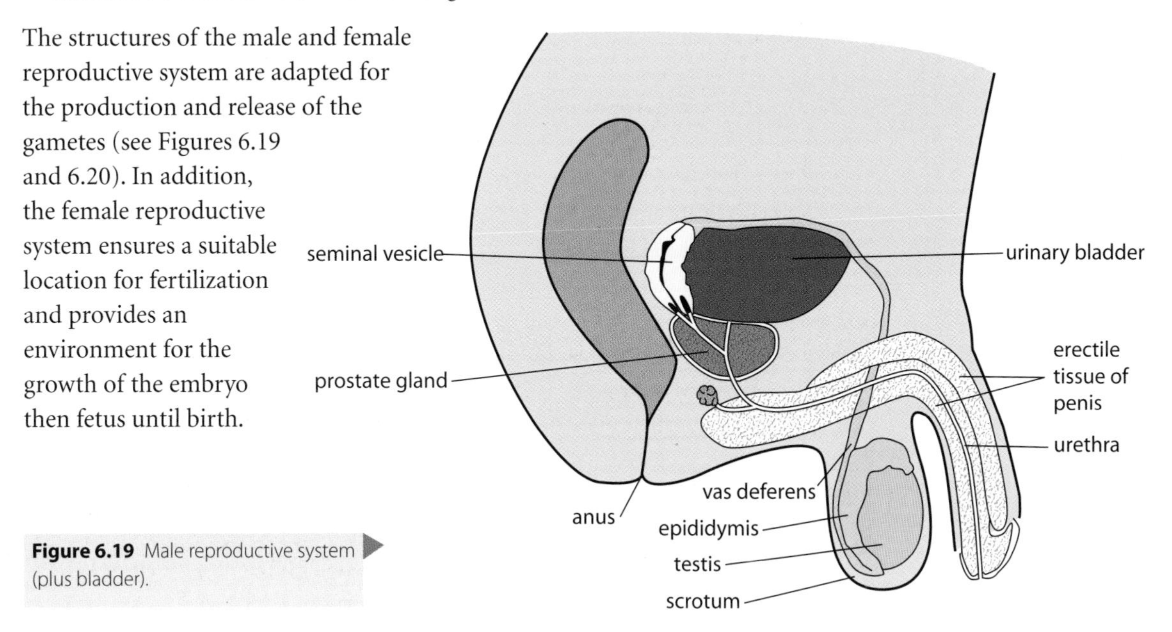

Figure 6.19 Male reproductive system (plus bladder).

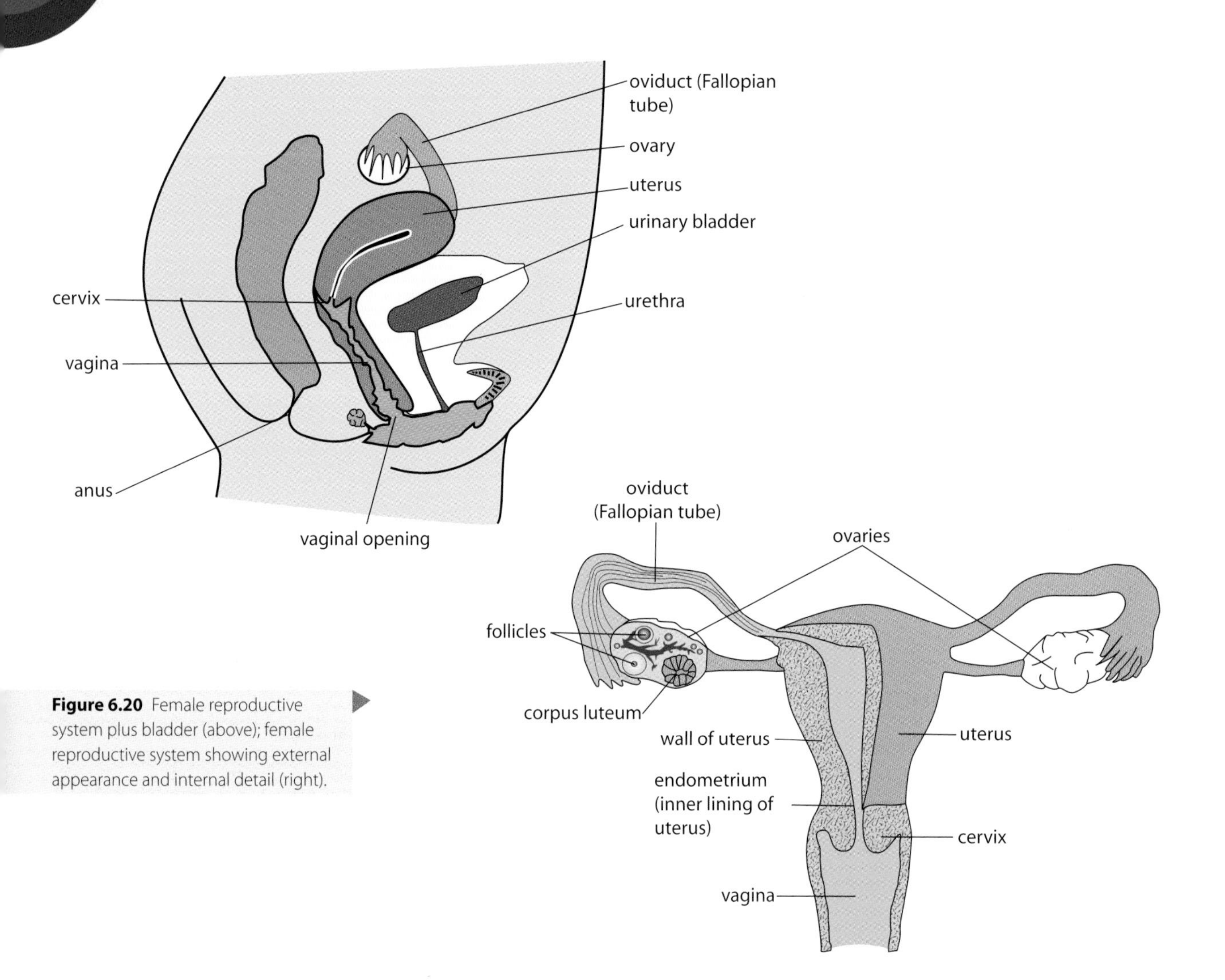

Figure 6.20 Female reproductive system plus bladder (above); female reproductive system showing external appearance and internal detail (right).

The menstrual cycle prepares the ovaries for ovulation and the uterus for implantation

Starting at the age of puberty, human females begin a hormonal cycle known as the menstrual cycle. Each cycle lasts, on average, 28 days. The purpose of the menstrual cycle is to time the release of an egg or ovum (ovulation) for possible fertilization and later implantation into the inner lining of the uterus. This implantation must occur when the uterine inner lining (the endometrium) is rich with blood vessels (it is described as highly vascular). The highly vascular endometrium is not maintained if there is no implantation. The breakdown of the blood vessels leads to the menstrual bleeding (menstruation) of a typical cycle. This menstruation is a sign that no pregnancy occurred.

Hormones from the brain

A part of a female's brainstem known as the hypothalamus is the regulatory centre of the menstrual cycle. The hypothalamus produces a hormone known as gonadotrophin releasing hormone (GnRH). The target tissue of GnRH is the nearby pituitary gland and it results in the pituitary producing and secreting two hormones into the bloodstream. These two hormones are follicle stimulating hormone (FSH) and luteinizing hormone (LH). The target tissues for these two hormones are the ovaries (see Figure 6.21).

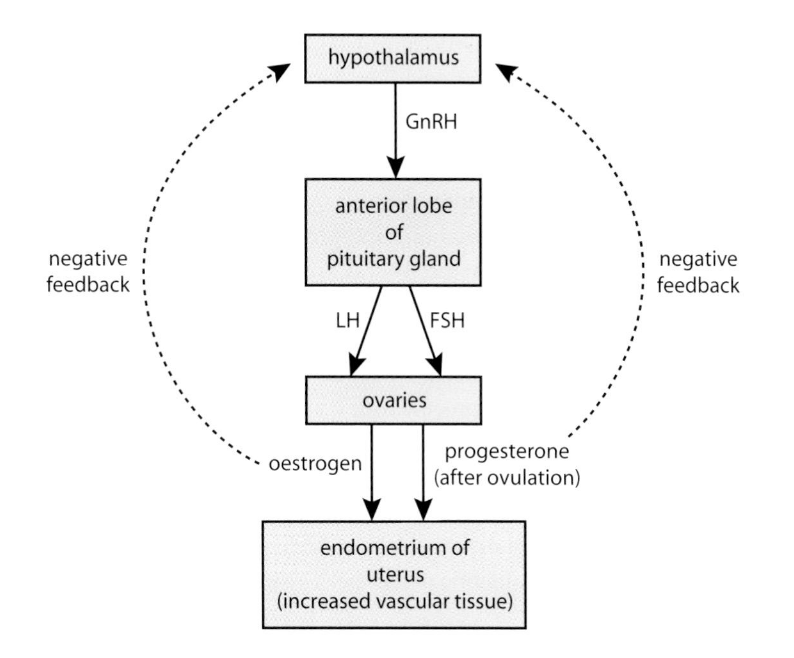

Figure 6.21 Hormonal summary of the menstrual cycle.

Effects of FSH and LH on the ovaries

The hormones FSH and LH have several effects on the ovaries. One of these effects is to increase the production and secretion of another reproductive hormone by the follicle cells of the ovary. This hormone is oestrogen. Like all hormones, oestrogen enters the bloodstream. Its target tissue is the endometrium of the uterus. The result is an increase in the blood vessels of the endometrium – or as stated earlier, the endometrium becomes highly vascular.

This light micrograph shows a human ovary section. Two Graafian follicles are visible with an oocyte at the centre of each.

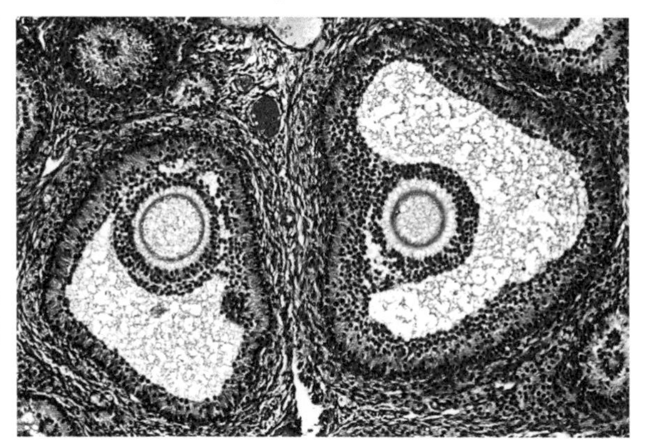

Another effect of FSH and LH is the production of structures within the ovaries known as Graafian follicles. In the ovaries are cells known as follicle cells and the true reproductive cells which are at a stage of development called oocytes. Under the chemical stimulation of FSH and LH, the somewhat randomly arranged follicle cells and oocytes take on an arrangement known as a Graafian follicle (see Figure 6.22).

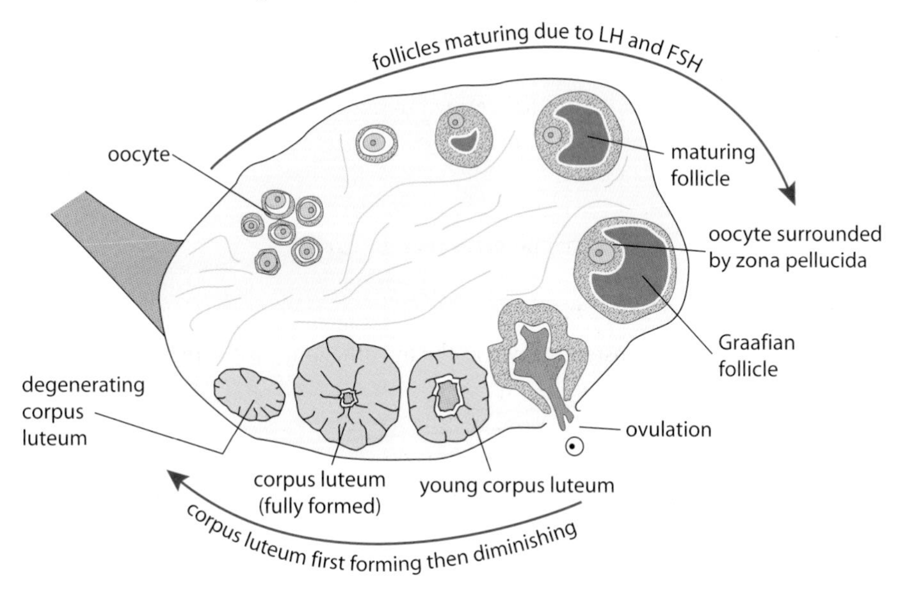

Figure 6.22 Ovary events during a single menstrual cycle (the events shown occur at differing times within a single cycle).

A spike in the level of FSH and LH leads to ovulation (release of the oocyte from the Graafian follicle). The oocyte is accompanied by the inner ring of follicle cells and a glycoprotein membrane coat known as the zona pellucida. This entire structure is known as a follicle and typically enters the Fallopian tube soon after ovulation.

The outer ring of follicle cells remains within the ovary. These follicle cells begin to produce and secrete another hormone, progesterone. The cells of this outer ring begin to divide and fill in the 'wound' area left by ovulation and this forms a glandular structure known as the corpus luteum. The corpus luteum will be hormonally active (producing progesterone) for only 10–12 days after ovulation. Progesterone is a hormone that maintains the thickened, highly vascular endometrium. As long as progesterone continues to be produced, the endometrium will not break down and an embryo will still be able to implant. In addition, the high levels of both oestrogen and progesterone are a negative feedback signal to the hypothalamus. The hypothalamus does not produce GnRH when these oestrogen and progesterone levels are high, so FSH and LH remain at levels not conducive to the production of another Graafian follicle during this time (see Figure 6.21 on page 187).

Birth control pills contain both oestrogen and progesterone. Because the pills keep the levels of these two hormones high in a woman's bloodstream, the hypothalamus does not produce GnRH. Thus, the pituitary does not produce FSH and LH and no new Graafian follicles are produced within the ovaries. The end result is ovulation does not occur.

Assuming there is no pregnancy, the corpus luteum eventually begins to break down and this leads to a decline in both progesterone and oestrogen levels. As both of these hormone levels fall, the highly vascular endometrium can no longer be maintained. The capillaries and small blood vessels begin to rupture and menstruation begins. The drop in progesterone and oestrogen also signals the hypothalamus to begin secreting GnRH and thus another menstrual cycle begins. Because the menstrual cycle is a cycle there is no true beginning or ending point. We have designated the first day of menstruation as the first day of the menstrual cycle simply because this is an event that can be readily recognized (see Figure 6.23).

Many countries have laws intended to regulate reproduction. These laws can range from control of birth control to setting a maximum 'family size'.

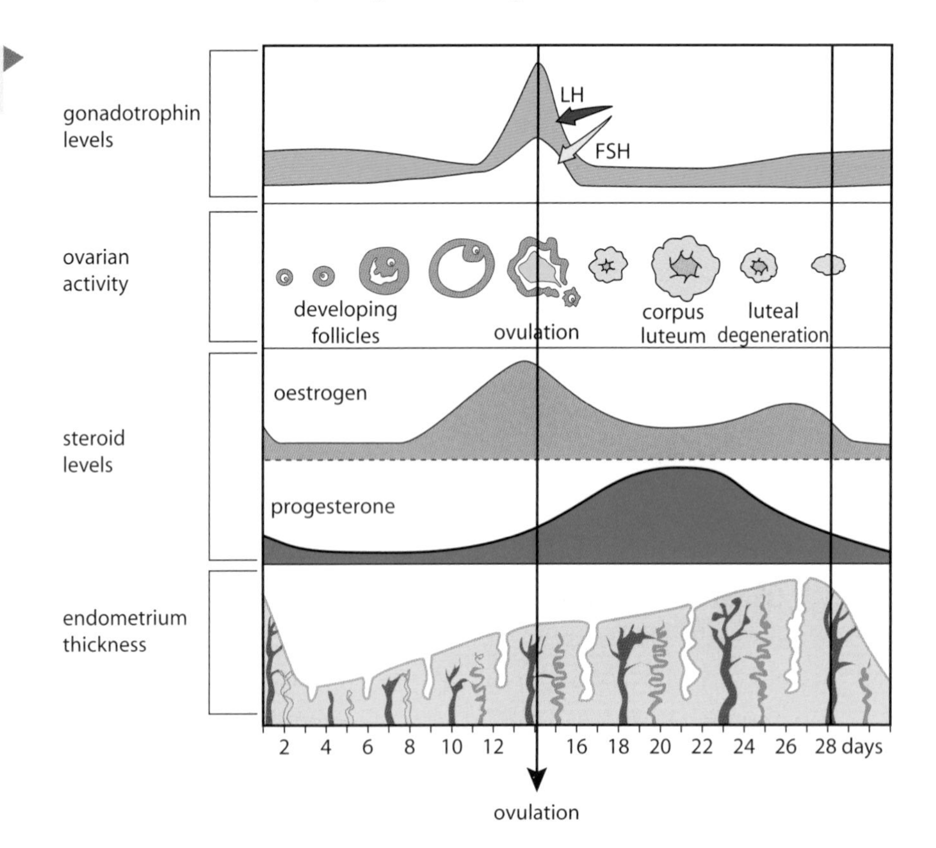

Figure 6.23 Events occurring in a 28-day menstrual cycle.

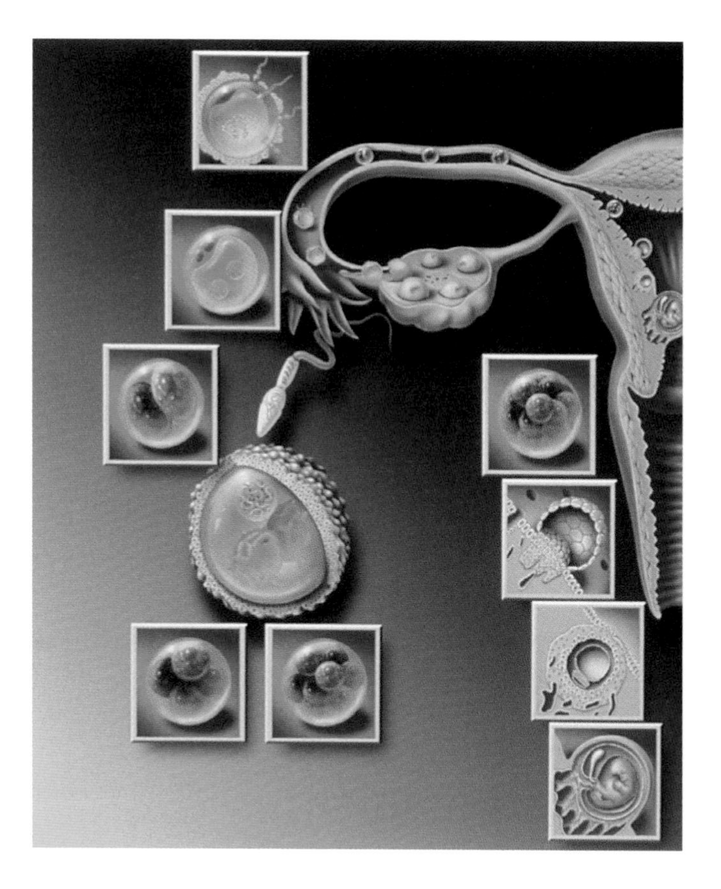

Here is an artwork summary of human reproduction.

In-vitro fertilization (IVF)

Natural fertilization typically occurs in one of a female's Fallopian tubes 24–48 hours after ovulation. The resulting zygote begins to divide by mitosis and takes several more days to travel down the Fallopian tube to the endometrium of the uterus. When the embryo reaches the endometrium, it has already mitotically divided many times. The embryo will then implant in the highly vascular tissue of the endometrium.

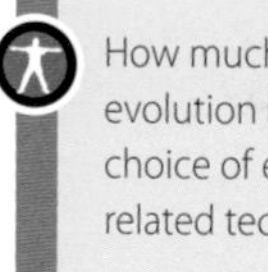

How much will future human evolution be affected by parental choice of embryos through IVF and related techniques?

If you do a websearch for IVF, many of the sites you will encounter will be from private clinics that offer IVF as a service. This doesn't mean the information on those sites is incorrect – it does mean you need to consider the possible bias behind the information.

Some couples are unable to bear children for a wide variety of possible reasons, including:

- males with low sperm counts;
- males with impotence (failure to achieve or maintain an erection);
- females who cannot ovulate normally;
- females with blocked Fallopian tubes.

Reproductive technologies have been developed to help overcome these situations. One of the most common of these new technologies is in-vitro fertilization (IVF).

Steps of an IVF procedure

To prepare for an IVF procedure, a woman is usually injected with FSH for about 10 days. This will ensure the development of many Graafian follicles within her ovaries. Several eggs (oocytes) are then harvested surgically. The man ejaculates into a container to obtain the sperm cells that are needed for fertilization. The harvested eggs are mixed with sperm cells in separate culture dishes. Microscopic observation reveals which ova are fertilized and if the early development appears

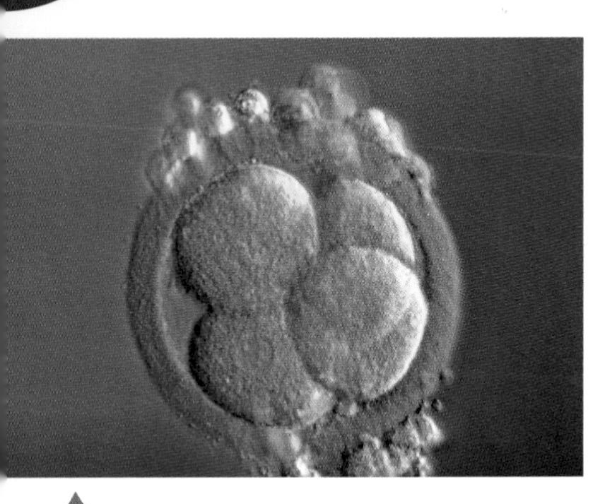

This is a light micrograph of a human 4-cell embryo two days after IVF and before implantation.

normal and healthy. Usually two or three healthy embryos are introduced into the woman's uterus for implantation. The procedure is very expensive and using only one embryo could mean a high risk of failure and having to repeat the procedure. Introducing more than one embryo increases the likelihood that at least one will implant successfully. Any healthy embryos from culturing that are not implanted can be frozen and used later if another implantation procedure is needed.

Ethical issues concerning IVF

There are many points of view on IVF.

Arguments for IVF

- It enables couples who would otherwise be unable to have children to have a family.
- Embryos that are visibly not healthy in early the stages of development can be eliminated from consideration for implantation.
- Genetic screening is possible on embryos before implantation to eliminate the chance of passing on some genetic diseases.
- IVF technology will advance and lead to further benefits in reproductive biology.

Arguments against IVF

- Embryos that are produced during culturing, but are not implanted are either frozen or destroyed.
- There are complex legal issues concerning the use of those frozen embryos when couples split up.
- Genetic screening of embryos could lead to society choosing desirable characteristics.
- Some reproductive problems of an individual are genetically passed on and IVF bypasses natures way of decreasing the genetic frequency of that reproductive problem.
- Multiple births and the problems associated with a multiple births are more likely with IVF than with natural conception.

Exercises

11 Using this book or another reference, draw on one graph the pattern shown during one menstrual cycle of each of the following hormones: FSH, LH, oestrogen, and progesterone. Close the reference and practise annotating this graph with information about what each of these hormones does during the progress of the cycle. Also include information about the origin of each of these hormones.

12 Imagine a panel put together to debate the pros and cons of IVF. Give three pro or con arguments that would be likely to be given by each of the following panel members:

- health insurance executive;
- man or woman who needs IVF to have a child;
- religious representative;
- administrator from an IVF clinic.

Practice questions

1 Poor nutrition of a woman during pregnancy has been associated with a variety of metabolic disorders later in the life of her offspring.

During the Second World War (WWII) the normally well-fed population of Holland suffered famine over a relatively short and precisely defined period. The data available from this period provided examples of fetuses that were affected by famine at specific periods during pregnancy.

Glucose tolerance was analysed in human adults 50–55 years of age who had suffered fetal famine during WWII. High glucose levels in blood plasma indicate poor glucose tolerance.

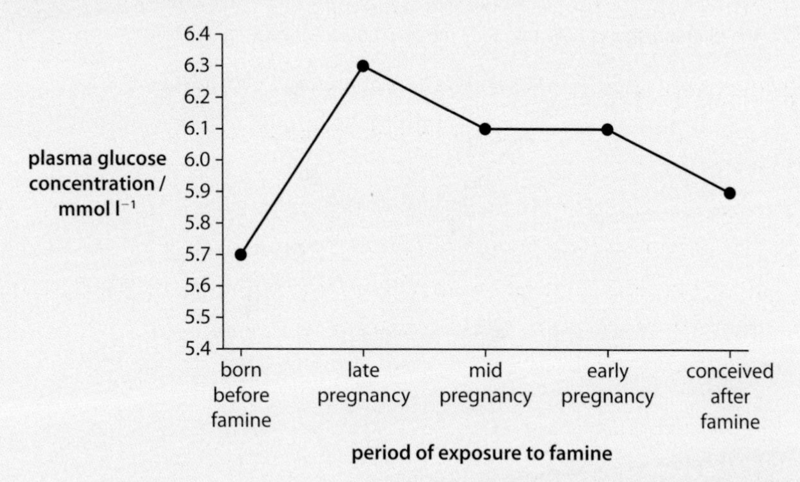

Source: N Metcalfe and P Monagham (2001), *Trends in Ecology and Evolution*, **16**, pages 254–60

(a) **(i)** Identify the period of exposure to famine that produces the greatest decrease in glucose tolerance. *(1)*

(ii) Calculate the percentage change in plasma glucose concentration after exposure to famine from early to late pregnancy. *(1)*

(iii) Suggest a reason why glucose tolerance did not return to normal in people conceived after the famine. *(1)*

(b) Outline a possible cause of poor glucose tolerance. *(1)*

(c) Suggest how poor glucose tolerance could be related to the occurrence of coronary heart disease. *(2)*

(Total 6 marks)

● **Examiner's hint:** The *y*-axis of the graph shown is read as: Plasma glucose concentration expressed as millimoles per litre.

● **Examiner's hint:** When the command term of a question is 'calculate', remember to show your work.

2 **(a)** Explain how the skin and mucous membranes prevent entry of pathogens into the body. *(3)*

(b) Explain why antibiotics are used to treat bacterial but not viral diseases. *(2)*

(Total 5 marks)

3 Discuss the ethical issues of in-vitro fertilization (IVF) in humans. *(8 marks)*

● **Examiner's hint:** The command term 'discuss' requires you to give different viewpoints. First think of different groups that would have an opinion about the topic and then relate what those opinions would be and why. Do not just give your opinion.

4 Which process decreases when the human body temperature decreases?

A Blood flow to the internal organs
B Secretion of sweat
C Secretion of insulin
D Shivering

(1 mark)

5 Which vessel carries deoxygenated blood?

A The pulmonary artery
B The coronary artery
C The aorta
D The pulmonary vein *(1 mark)*

6 Which organ secretes enzymes that are active at a low pH?

A Mouth
B Pancreas
C Stomach
D Liver *(1 mark)*

7 The diagram shows how the body regulates glucose levels in the blood.

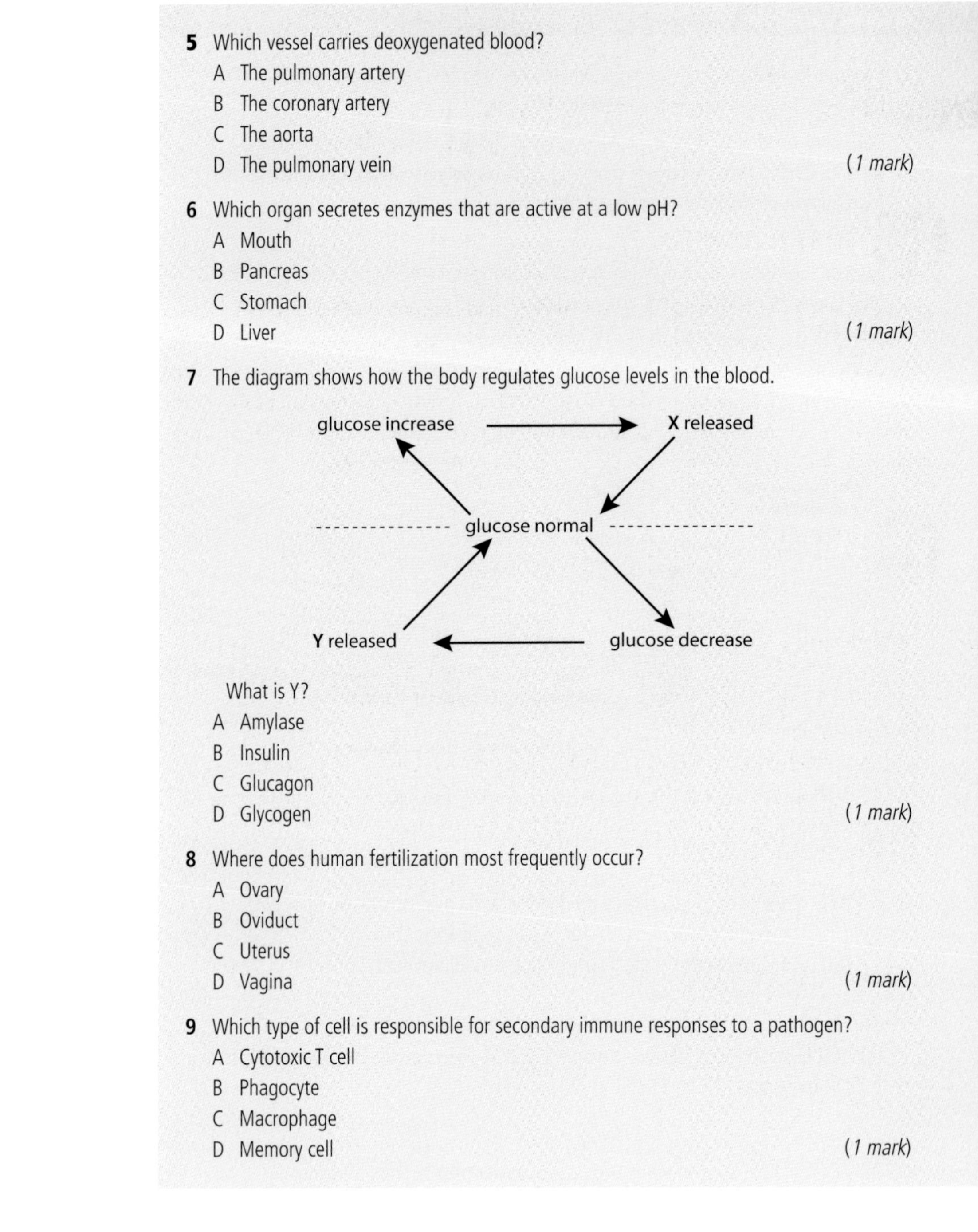

What is Y?

A Amylase
B Insulin
C Glucagon
D Glycogen *(1 mark)*

8 Where does human fertilization most frequently occur?

A Ovary
B Oviduct
C Uterus
D Vagina *(1 mark)*

9 Which type of cell is responsible for secondary immune responses to a pathogen?

A Cytotoxic T cell
B Phagocyte
C Macrophage
D Memory cell *(1 mark)*

7 Nucleic acids and proteins

Introduction

As discussed in Chapter 3, DNA is a nucleic acid composed of nucleotides. DNA molecules are some of the largest biomolecules known. To get to be an extremely large macromolecule, DNA has to include huge numbers of chemical bonds. There are two general types of chemical bond involved in DNA: covalent bonds and hydrogen bonds. These bonds allow the DNA molecule to be assembled in a very specific way.

7.1 DNA structure

Assessment statements

7.1.1 Describe the structure of DNA, including the antiparallel strands, 3′–5′ linkages and hydrogen bonding between purines and pyrimidines.
7.1.2 Outline the structure of nucleosomes.
7.1.3 State that nucleosomes help to supercoil chromosomes and help to regulate transcription.
7.1.4 Distinguish between unique or single-copy genes and highly repetitive sequences in nuclear DNA.
7.1.5 State that eukaryotic genes can contain exons and introns.

DNA structure

To understand the detailed structure of DNA, you must be familiar with the numbering of the carbon atoms in the pentose sugar of DNA, which is deoxyribose (see Figure 7.1).

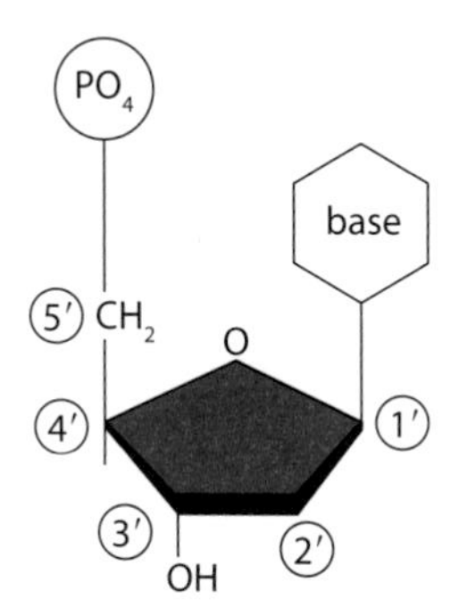

Figure 7.1 A DNA nucleotide is composed of a molecule of deoxyribose with a phosphate group attached to the 5′ (five-prime) carbon and a nitrogenous base attached to the 1′ (one-prime) carbon.

You already know that DNA is a double-stranded molecule formed in the shape of a double helix. This would be a good time for you to review what was learned in Chapter 3 by drawing a generalized single chain portion of DNA. Show the position of the covalent bonds between adjacent nucleotides. If you have difficulty doing this, you should review Chapter 3, section 3.3 pages 55–57.

Nucleotides are composed of a pentose sugar, a phosphate group, and a nitrogenous base. The nitrogenous base could be one of the purines (adenine and guanine), or one of the pyrimidines (thymine and cytosine).

How is a single chain of DNA made up?

Each strand is composed of a backbone of alternating phosphate and deoxyribose molecules. These two molecules are held together by a covalent bond called a phosphodiester bond or linkage. A phosphodiester bond in a DNA chain is arranged like this: phosphate—oxygen—carbon.

So, each nucleotide is attached to the previous one by this type of bond. This produces a chain of DNA. The reaction between the phosphate group on the 5′ carbon and the hydroxyl group on the 3′ carbon is a condensation reaction with a molecule of water released. When two nucleotides unite in this way the 2-unit polymer still has a 5′ carbon free at one end and a 3′ carbon free at the other. Each time a nucleotide is added, it is attached to the 3′ carbon end. Even when thousands of nucleotides are involved, there is still a free 5′ carbon end with a phosphate group attached and a free 3′carbon end with a hydroxyl group attached. This creates the alternating sugar–phosphate backbone of each chain.

The chain of nucleotides always grows in a 5′ to 3′ direction.

As nucleotides are linked together with covalent phosphodiester bonds, a definite sequence of nitrogenous bases develops. This sequence carries the genetic code that is essential for the life of the organism.

How are the two strands of DNA held together?

The two sugar–phosphate backbones are attached to each another by their nitrogenous bases. The two backbones or chains run in opposite directions and are described as antiparallel. One strand has the 5′ carbon on the top and the 3′ carbon on the bottom; the other strand is the opposite (see Figure 7.2).

Figure 7.2 The antiparallel strands in DNA run in opposite directions.

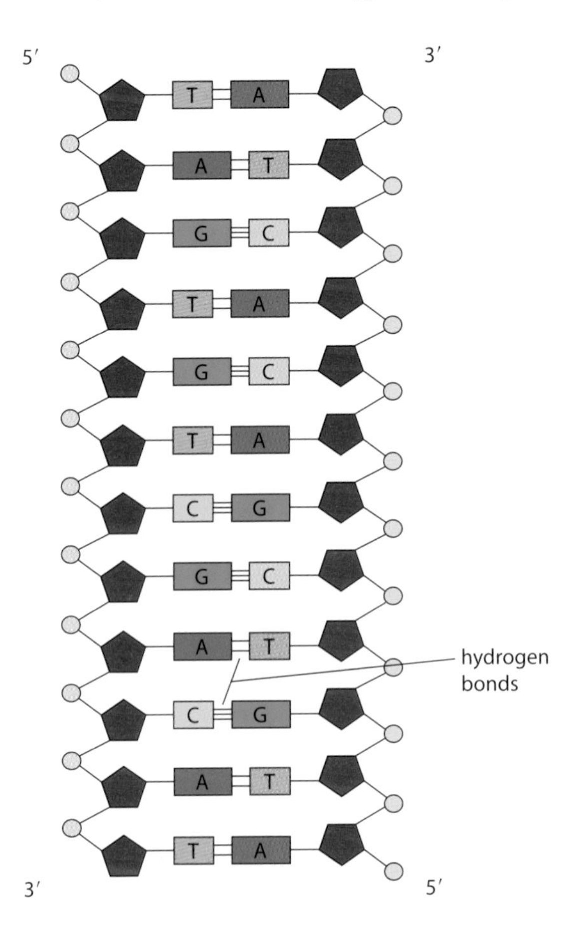

The nitrogenous bases form links by means of hydrogen bonds. There are four nitrogenous bases: adenine (A), thymine (T), cytosine (C), and guanine (G). Adenine and guanine are double-ring structures known as purines. Cytosine and thymine are single-ring structures known as pyrimidines. A single-ring nitrogenous base always pairs with a double-ring nitrogenous base. This is complementary base pairing and occurs because of the specific distance that exists between the two sugar–phosphate chains. Adenine always pairs with thymine (from the opposite chain) and cytosine always pairs with guanine (from the opposite chain). Hydrogen bonds link the nitrogenous bases together: two hydrogen bonds link adenine and thymine; three hydrogen bonds link cytosine and guanine.

To discover more about DNA structure, visit heinemann.co.uk/hotlinks, insert the express code 4242P and click on Weblink 7.1.

DNA packaging

The DNA molecules of eukaryotic cells are paired with a type of protein called histone (see Figure 7.3). Actually, there are several histones and each helps in DNA packaging. Packaging is essential because the nucleus is microscopic but a single human molecule of DNA in a chromosome may be 4 cm long.

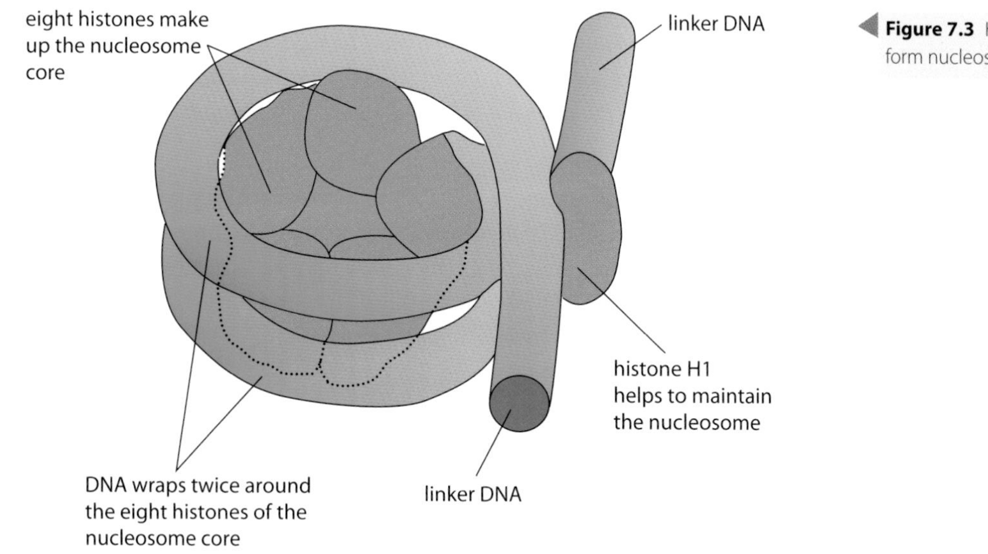
eight histones make up the nucleosome core
linker DNA
histone H1 helps to maintain the nucleosome
DNA wraps twice around the eight histones of the nucleosome core
linker DNA

Figure 7.3 Histones and DNA together form nucleosomes.

When looking at unfolded DNA with an electron microscope, you see what look like beads on a string. Each of the beads is a nucleosome. A nucleosome consists of two molecules of each of four different histones. The DNA wraps twice around these eight protein molecules. The DNA is attracted to the histones because DNA is negatively charged and histones are positively charged. Between the nucleosomes is a single string of DNA. There is often a fifth type of histone attached to the linking string of DNA near each nucleosome. This fifth histone leads to further wrapping (packaging) of the DNA molecule and eventually to the highly condensed or supercoiled chromosomes.

To discover more about chromosome structure, visit heinemann.co.uk/hotlinks, insert the express code 4242P and click on Weblink 7.2.

When DNA is wrapped around the histones and then further wrapped in even more elaborate structures, it is inaccessible to transcription enzymes. Therefore, the wrapping or packaging of DNA brings about a regulation of the transcription process. This allows only certain areas of the DNA molecule to be involved in protein synthesis.

Types of DNA sequence

Some DNA consists of highly repetitive sequences, some codes for genes and some is structural.

Highly repetitive sequences

The genomes of eukaryotes include large amounts of highly repetitive sequences. These sequences account for between 5% and 45% of the total genome. They usually are composed of 5–300 base pairs per repetitive sequence. There may be as many as 100 000 replicates of a certain type per genome. If this repetitive DNA is clustered in discrete areas, it is referred to as satellite DNA. However, this repetitive DNA is mostly dispersed throughout the genome. At the moment, we think these dispersed regions of repetitive DNA do not have any coding function. They are transposable elements, which means they can move from one genome location to another. These elements were first found by Barbara McClintock in 1950; in 1983, she received the Nobel Prize for the discovery.

The centromere of chromosomes is largely made of highly repetitive sequences called satellite DNA.

Highly repetitive sequences were originally called junk DNA. This derogatory name implied there was no reason to research these sections of DNA. Because of this, research of these sequences has been sparse.

Protein-coding genes

Within the DNA molecule of a chromosome, are the single copy genes that have coding functions. They provide the base sequences essential to produce proteins at the cell ribosomes. Any base sequence is carried from the nucleus to the ribosomes by messenger RNA. Work to determine the complete base sequence of human chromosomes (the Human Genome Project) began in the mid-1970s. In 2001, when the draft sequence of the human genome was published, it became apparent that less than 2% of the chromosomes were occupied by genes that code for protein.

A gene is not a fixed sequence of bases like the letters of a word. Genes are made of numerous fragments of protein-encoding information interspersed with non-coding fragments. The coding fragments make up what are known as exons, while the non-coding fragments make up introns. Exons and introns are discussed again in the sections involving transcription and translation.

Structural DNA

Structural DNA is highly coiled DNA that does not have a coding function. It occurs around the centromere and near the ends of chromosomes at the telomeres. Pseudogenes do not seem to have a function, probably due to mutation.

The frequency of the various types of DNA sequence are shown in the table below.

DNA sequences in the human genome	%
protein-encoding genes (exons)	1–2
introns	24
highly repetitive sequences	45
structural DNA	20
inactive genes (pseudogenes)	2
other	7–8

Telomeres are composed of many repeats of short DNA sequences. They prevent chromosome erosion and may be involved in cell aging and cancer.

Research into genomes (whole gene sets) is called genomics. It is a tremendously active area of study and is making great strides.

Exercises

1 Draw the two strands of a DNA molecule representing their antiparallel relationship.

2 Explain how nucleosomes would contribute to transcription control.

3 Would exons or would introns be more likely to contain highly repetitive sequences? Why?

7.2 DNA replication

Assessment statements

7.2.1 State that DNA replication occurs in a 5′ to 3′ direction.

7.2.2 Explain the process of DNA replication in prokaryotes, including the role of enzymes (helicase, DNA polymerase, RNA primase and DNA ligase), Okazaki fragments and deoxynucleoside triphosphates.

7.2.3 State that DNA replication is initiated at many points in eukaryotic chromosomes.

Models of DNA replication

When Watson and Crick proposed their model for the structure of DNA, they realized that the A–T and C–G base-pairing provided a way for DNA to be copied: a single strand of DNA could serve as a template to form a copy. This would account for the great accuracy necessary to pass DNA's information from one generation to the next. This idea of copying via a template was called the semiconservative model of DNA replication. There were also two other models for DNA replication; they were called the conservative model and the dispersive model.

In 1958, Matthew Meselsohn and Franklin Stahl carried out experiments involving the bacterium *Escherichia coli.* They used two isotopes of nitrogen and determined what proportions of the isotopes were present in strands of DNA after one and two replications (see Figure 7.4). After one replication of DNA, each daughter DNA molecule possessed one strand with the heavy isotope of nitrogen (^{15}N) and

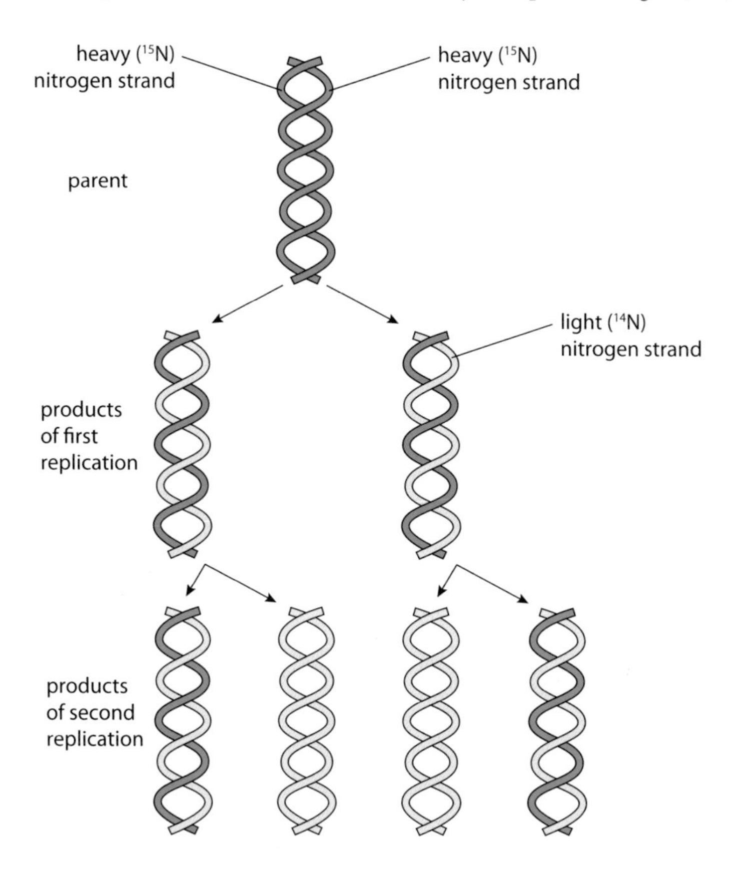

Figure 7.4 Bacteria grown in a medium with heavy nitrogen (^{15}N) have DNA that contains only heavy nitrogen. The bacteria are then placed in a medium with only light nitrogen (^{14}N). After one replication of DNA, the resulting DNA contains both light and heavy nitrogen. After another replication, the DNA is either all light or hybrid.

one strand with the light isotope of nitrogen (^{14}N) – a hybrid had formed. After a second replication, the DNA molecules were either hybrid or without the heavy isotope nitrogen. This evidence showed that the replication process of DNA is semiconservative. The other two models would not have shown these results.

Semiconservative replication

From Meselsohn and Stahl's experiment, the accepted process of DNA replication involving *E. coli* has been developed. The replication of DNA begins at special sites called origins of replication. Bacterial DNA is circular and has a single origin. Eukaryotic DNA is linear and has thousands of origins. The presence of multiple replication origins greatly accelerates the copying of large eukaryotic chromosomes.

Because the bacterial DNA molecule is much smaller than eukaryotic DNA molecules, only one origin is necessary for replication of the bacterial chromosome.

Here is a brief summary of the replication process (see Figure 7.5).

1 Replication begins at the origin which appears as a bubble because of the separation of the two strands. The separation or 'unzipping' occurs due to the action of the enzyme helicase on the hydrogen bonds between nucleotides.
2 At each end of a bubble there is a replication fork. This is where the double-stranded DNA opens to provide two parental DNA strands which are the templates necessary to produce the daughter DNA molecules by semiconservative replication.
3 The bubbles enlarge in both directions, showing that the replication process is bidirectional. The bubbles eventually fuse with one another to produce two identical daughter DNA molecules.

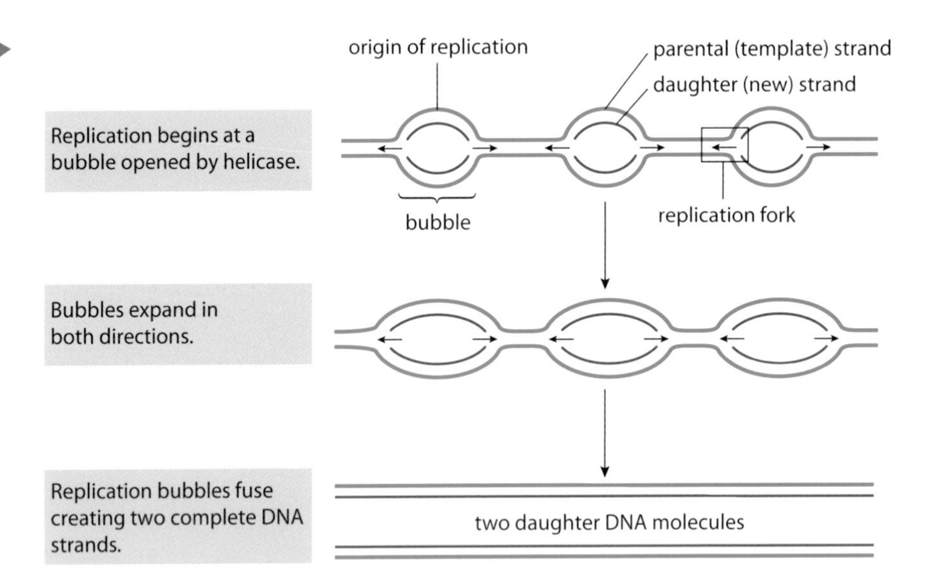

Figure 7.5 Semiconservative replication means that each parental DNA strand acts as a template to form a complementary strand so eventually two identical daughter DNA molecules are formed. Eukaryotic DNA molecules are quite long and replication begins at a large number of sites along the molecule. This allows replication to occur at a much faster rate.

Elongation of a new DNA strand

The production of a new strand of DNA using the templates exposed at the replication forks occurs in an orderly manner.

1 A primer is produced under the direction of primase at the replication fork. This primer is a short sequence of RNA, usually only 5–10 nucleotides. Primase allows joining of RNA nucleotides that match the exposed DNA bases at the point of replication.

2 The enzyme DNA polymerase III then allows the addition of DNA nucleotides in a 5′ to 3′ direction to produce the growing DNA strand.

3 DNA polymerase I also participates in the process. It removes the primer from the 5′ end and replaces it with DNA nucleotides.

Each nucleotide that is added to the elongating DNA chain is actually a deoxynucleoside triphosphate (dNTP). This molecule contains deoxyribose, a nitrogenous base (A, T, C, or G), and three phosphate groups (see Figure 7.6). As these molecules are added, two phosphates are lost. This provides the energy necessary for the chemical bonding of the nucleotides.

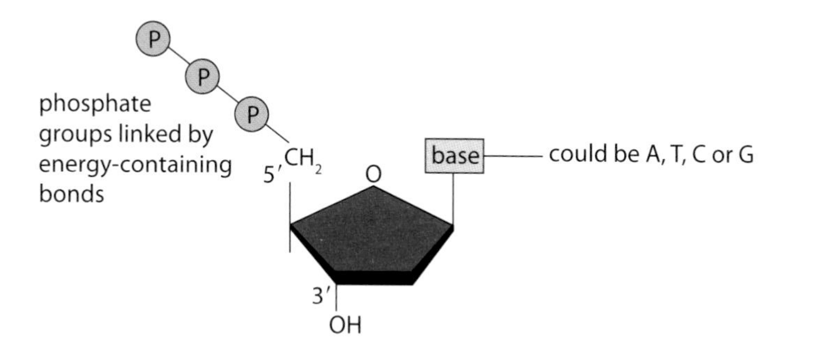

Figure 7.6 This is a generalized deoxynucleoside triphosphate molecule.

The antiparallel strands

A DNA molecule is composed of two anti-parallel strands. One strand is 5′ to 3′ and the other is 3′ to 5′. DNA strands can only be assembled in the 5′ to 3′ direction due to the action of polymerase III. Because of this, there is a difference in the process of assembling the two new strands of DNA from the templates. For the 3′ to 5′ template strand, the new DNA strand is formed as described above. The process is continuous and relatively fast, and the strand being produced is called the leading strand. The other new strand forms more slowly and is called the lagging strand.

The leading and lagging strands are assembled concurrently. However, there is a slight delay in the synthesis of the lagging strand.

Formation of the lagging strand involves fragments and an additional enzyme called DNA ligase (see Figure 7.7).

- The leading strand is being assembled continuously toward the progressing replication fork in the 5′ to 3′ direction.
- The lagging strand is assembled by way of fragments away from the progressing replication fork in the 5′ to 3′ direction.
- Primer, primase, and DNA polymerase III are required for the lagging strand just as with the leading strand.
- Because the lagging strand is assembled in fragments, these three components are necessary for the formation of each fragment. The leading strand is formed continuously, so it requires primase and primer only once.
- The fragments of the lagging strand are called Okazaki fragments, after the Japanese scientist who discovered them.
- Once the Okazaki fragments are assembled, an enzyme called DNA ligase attaches the sugar–phosphate backbones of the fragments to form a single DNA strand.

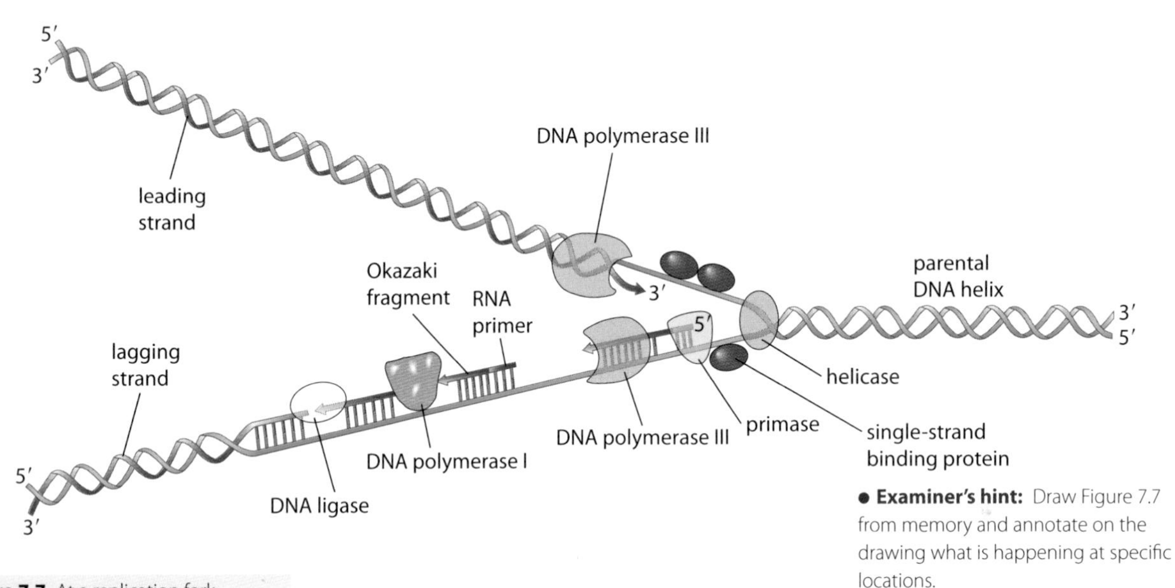

Figure 7.7 At a replication fork, helicase separates the strands of the double helix and binding proteins stabilize the single strands. There are two mechanisms for replication: continuous synthesis and discontinuous synthesis. Continuous synthesis occurs on the leading strand: primase adds an RNA primer and DNA polymerase III adds nucleotides to the 3′ end of the leading strand. DNA polymerase I then replaces the primer with nucleotides. Discontinuous synthesis occurs on the lagging strand: primase adds an RNA primer in front of the 5′ end of the lagging strand and DNA polymerase III adds nucleotides, DNA polymerase I replaces the primer and finally DNA ligase attaches the Okazaki fragment to the lagging strand.

• **Examiner's hint:** Draw Figure 7.7 from memory and annotate on the drawing what is happening at specific locations.

Replication proteins

The basic processes of DNA replication were worked out in research done with *E. coli*. This table summarises the roles of the replication proteins in *E. coli*.

Protein	Role
helicase	unwinds the double helix at replication forks
primase	synthesizes RNA primer
DNA polymerase III	synthesizes new strand by adding nucleotides onto the primer (5′ to 3′ direction)
DNA polymerase I	removes the primer and replaces it with DNA
DNA ligase	joins the ends of DNA segments and Okazaki fragments

Research involving eukaryotes has shown that DNA replication in prokaryotic and eukaryotic cells is almost identical. In eukaryotic cells, the enzyme topoisomerase stabilizes the DNA double helix when helicase unzips the molecule at multiple sites.

Speed and accuracy of replication

To discover more about DNA replication, visit heinemann.co.uk/hotlinks, insert the express code 4242P and click on Weblink 7.3.

Even though replication seems quite complicated, it occurs at a phenomenal rate: up to 4000 nucleotides are replicated per second. This speed is essential in cells like bacteria which may divide every 20 minutes. Eukaryotic cells contain huge numbers of nucleotides compared to prokaryotic cells. To accomplish rapid DNA replication in these cells, multiple replication origins are needed.

Replication of DNA is remarkably accurate. Few errors (mutations) occur, which is stunning given the huge numbers of the nucleotides that are replicated. Cells have a battery of repair enzymes that detect and correct errors when they do occur. These repair enzymes also act when chemicals or high-energy waves cause damage to existing cells.

Exercises

4 What is the energy source for the production of the complementary strand of DNA?

5 What effect would only one origin of replication on a eukaryotic chromosome have on the cell cycle?

6 Compare the number of primers needed on the leading and the lagging strand of DNA in replication.

7.3 Transcription

Assessment statements

7.3.1 State that transcription is carried out in a 5′ to 3′ direction.
7.3.2 Distinguish between the sense and antisense strands of DNA.
7.3.3 Explain the process of transcription in prokaryotes, including the role of the promoter region, RNA polymerase, nucleoside triphosphates and the terminator.
7.3.4 State that eukaryotic RNA needs the removal of introns to form mature mRNA.

The central dogma

DNA is sequestered (locked away) in the nucleus. Ribosomes are in the cytoplasm. But the two need to get together for the vital process of protein synthesis to occur. So how does the DNA code get to the ribosomes? The code is carried from the nucleus by the second type of nucleic acid called RNA (ribonucleic acid).

The set of ideas called the central dogma states that information passes from genes on the DNA to an RNA copy. The RNA copy then directs the production of proteins at the ribosome by controlling the sequence of amino acids. This mechanism is one-way and fundamental to all forms of life.

Thus, the central dogma can be summarized: DNA → RNA → protein.

The process that occurs at the first arrow in the central dogma summary is transcription. The process at the second arrow is translation. First, let's look at transcription. Translation is covered in section 7.4.

Transcription: DNA → RNA

Transcription has some similarities to replication. First, the double helix must be opened to expose the base sequence of the nucleotides. In replication, helicase unzips the DNA and both strands become templates for the formation of two daughter strands of DNA. However, in transcription, helicase is not involved. Instead, an enzyme called RNA polymerase separates the two DNA strands. The RNA polymerase also allows polymerization of RNA nucleotides as base-pairing occurs along the DNA template. To provide these functions, the RNA polymerase must first combine with a region of the DNA strand called a promoter. You will recall that, in DNA replication, DNA polymerase allows assembly only in a 5′ to 3′ direction. The same is true with RNA polymerase. The 5′ ends of free RNA nucleotides are added to the 3′ end of the RNA molecule being synthesized.

But which strand of DNA is copied?

One strand is complementary to the other, so there would be a difference in the code of the strands. You will recall that the genetic code is made up of codons (see Chapter 3, page 63). The codons are specific for certain amino acids or punctuation signals. Therefore, complementary strands mean different codons, different amino acids, and finally different proteins.

RNA polymerase binds to the promoter region of DNA. This causes the DNA to unwind. The polymerase then initiates the synthesis of an RNA molecule in a 5′ to 3′ direction.

The DNA strand that carries the genetic code is called the sense strand (or the coding strand). The other strand is called the antisense strand (or the template strand). The sense strand has the same sequence as the newly transcribed RNA except with thymine in place of uracil. The antisense strand is the strand that is copied during transcription.

The promoter region for a particular gene determines which DNA strand is the antisense strand. For any particular gene, the promoter is always on the same DNA strand. However, for other genes, the promoter may very well be on the other strand. The promoter region is a short sequence of bases that is not transcribed.

Once RNA polymerase has attached to the promoter region for a particular gene, the process of transcription begins. The DNA opens and a transcription bubble occurs. This bubble contains the antisense DNA strand, the RNA polymerase, and the growing RNA transcript (see Figure 7.8).

Figure 7.8 DNA is opened into two strands by RNA polymerase. The sense strand has the same base sequence as the new mRNA. The antisense strand is the template for transcription so the new mRNA has a base sequence that is complementary to it.

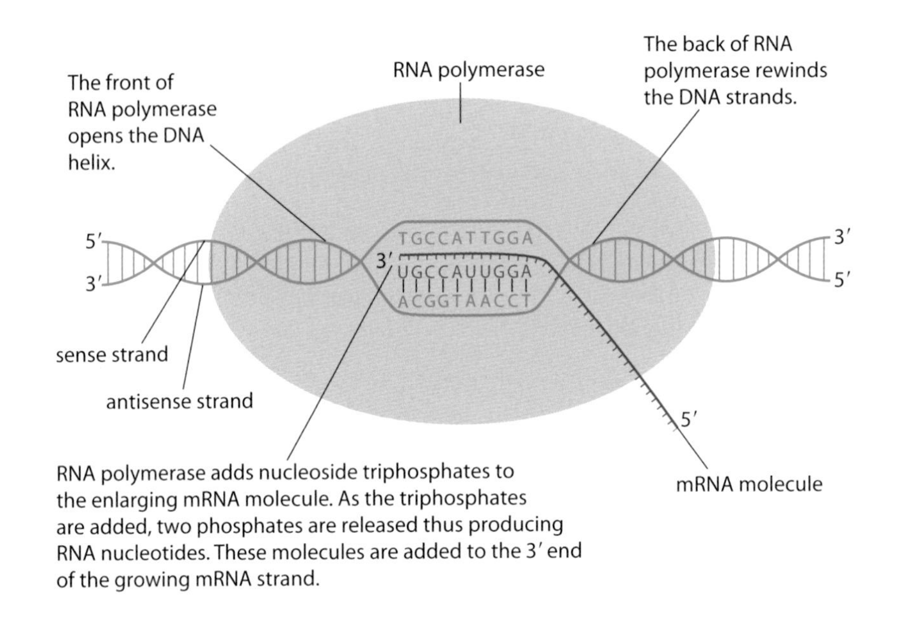

The terminator

The sections of DNA involved in transcription are: promoter → transcription unit → terminator. The transcription bubble moves from the DNA promoter region towards the terminator.

The terminator is a sequence of nucleotides that, when transcribed, causes the RNA polymerase to detach from the DNA. When this happens, transcription stops and the RNA transcript is detached from the DNA. The transcript carries the code of the DNA and is referred to as messenger RNA (mRNA).

In eukaryotes, transcription continues beyond the terminator for a significant number of nucleotides. Eventually, the transcript is released from the DNA strand.

Nucleoside triphosphates (NTPs) containing three phosphates and the 5-carbon sugar, ribose, are paired with the appropriate exposed bases of the antisense strand. Polymerization of the mRNA strand occurs with the catalytic help of RNA polymerase and the energy provided by the release of two phosphates from NTP. This portion of the transcription process is often referred to as elongation.

Post-transcription processing

Eukaryotic DNA is different from prokaryotic DNA in that within the protein-coding regions there are stretches of non-coding DNA. The stretches of non-coding DNA are called introns. To make a functional mRNA strand in eukaryotes, the introns are removed.

Prokaryotic mRNA does not require processing because no introns are present.

For more information on DNA transcription, visit heinemann.co.uk/hotlinks, insert the express code 4242P and click on Weblink 7.4.

Exercises

7 If the mRNA transcript is forming in the 5′ to 3′ direction, in what direction is the transcription bubble moving on the DNA antisense strand?

8 What type of mRNA requires processing? Explain why.

7.4 Translation

Assessment statements

7.4.1 Explain that each tRNA molecule is recognized by a tRNA-activating enzyme that binds a specific amino acid to the tRNA, using ATP for energy.

7.4.2 Outline the structure of ribosomes, including protein and RNA composition, large and small subunits, three tRNA binding sites and mRNA binding sites.

7.4.3 State that translation consists of initiation, elongation, translocation and termination.

7.4.4 State that translation occurs in a 5′ to 3′ direction.

7.4.5 Draw and label the structure of a peptide bond between two amino acids.

7.4.6 Explain the process of translation, including ribosomes, polysomes, start codons and stop codons.

7.4.7 State that free ribosomes synthesize proteins for use primarily within the cell, and that bound ribosomes synthesize proteins primarily for secretion or for lysosomes.

Ribosomes

Once mRNA has been produced from the DNA template, the process of producing the protein can begin. This process is referred to as translation because it changes the language of DNA to the language of protein. The centre of this process is the ribosome. Therefore, you need to understand the structure of this organelle.

Ribosomes can be seen with an electron microscope. Each consists of a large subunit and a small subunit. The subunits are composed of ribosomal RNA (rRNA) molecules and many distinct proteins. rRNA proteins are generally small and are associated with the core of the RNA subunits. Roughly two-thirds of ribosome mass is rRNA. The molecules of the ribosomes are constructed in the nucleolus of eukaryotic cells and exit the nucleus through the membrane pores.

Prokaryotic ribosomes are smaller than eukaryotic ribosomes. There is also a difference in molecular makeup.

Decoding of a strand of mRNA to produce a polypeptide occurs in the space between the two subunits. In this area, there are binding sites for mRNA and three sites for binding tRNA, as shown in this table (see also Figure 7.9).

Site	Function
A site	holds the tRNA carrying the next amino acid to be added to the polypeptide chain
P site	holds the tRNA carrying the growing polypeptide chain
E site	site from which tRNA that has lost its amino acid is discharged

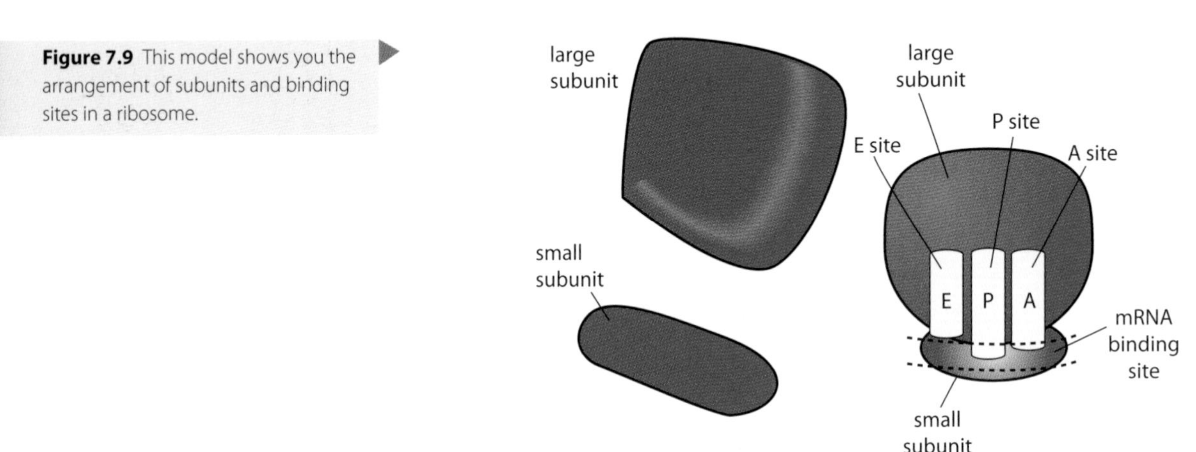

Figure 7.9 This model shows you the arrangement of subunits and binding sites in a ribosome.

Polypeptide chains are assembled in the cavity between the two subunits. This area is generally free of proteins, so binding of mRNA and tRNA is carried out by the rRNA. tRNA moves sequentially through the three binding sites: from the A site, to the P site, and finally to the E site. The growing polypeptide chain exits the ribosome through a tunnel in the large subunit core.

Translation: RNA → protein

The translation process involves several phases:

- initiation;
- elongation;
- translocation;
- termination.

Before we discuss these phases, it is important to address codons. You will recall that codons carry the genetic code from DNA to the ribosomes via mRNA. There are 64 possible codons. Three codons have no complementary tRNA anticodon – these are the stop codons. There is a start codon (AUG) that signals the beginning of a polypeptide chain. This codon also encodes the amino acid methionine. This information is key to understanding the translation process.

The triplet bases of the mRNA codon form complementary base pairs with the triplet anticodon of the tRNA.

The initiation phase

The start codon (AUG) is on the 5′ end of all mRNAs. Each codon, other than the three stop codons, attaches to a particular tRNA. The tRNA has a 5′ end and a 3′ end like all other nucleic acid strands. The

3′ end of tRNA is free and has the base sequence CCA. This is the site of amino acid attachment. Because there are complementary bases in the single-stranded tRNA, hydrogen bonds form in four areas. This causes the tRNA to fold and take on a three-dimensional structure. If the molecule is flattened, it has the two-dimensional appearance of a clover leaf (see Figure 7.10). One of the loops of the clover leaf contains an exposed anticodon. This anticodon is unique to each type of tRNA. It is this anticodon that pairs with a specific codon of mRNA.

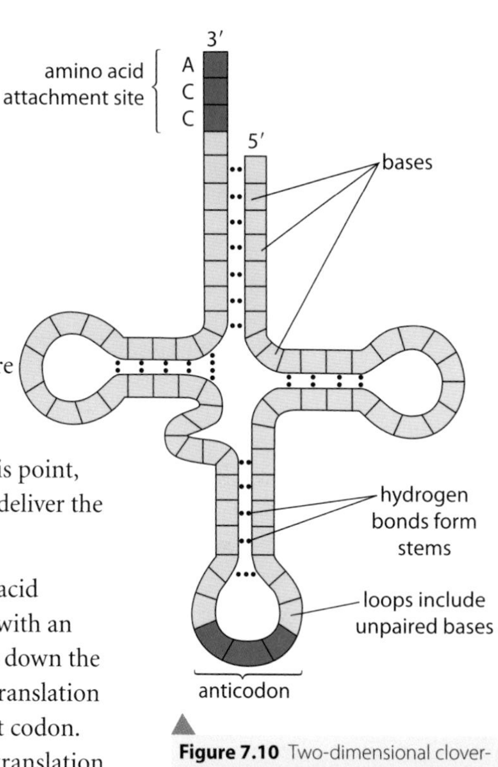

Figure 7.10 Two-dimensional clover-leaf structure of tRNA with three loops. The anticodon triplet is unique to each tRNA.

Each of the 20 amino acids binds to the appropriate tRNA due to the action of a particular enzyme. Because there are 20 amino acids there are 20 enzymes. The active site of each enzyme allows a fit only between a specific amino acid and the specific tRNA. The actual attachment of the amino acid and tRNA requires energy that is supplied by ATP. At this point, the structure is called an activated amino acid, and the tRNA may now deliver the amino acid to a ribosome to produce the polypeptide chain.

So, the first step in initiation of translation is when an activated amino acid – methionine attached to a tRNA with the anticodon UAC – combines with an mRNA strand and a small ribosomal subunit. The small subunit moves down the mRNA until it contacts the start codon (AUG). This contact starts the translation process. Hydrogen bonds form between the initiator tRNA and the start codon. Next, a large ribosomal subunit combines with these parts to form the translation initiation complex. Joining the initiation complex are proteins called initiation factors that require energy from guanosine triphosphate (GTP) for attachment. GTP is an energy-rich compound very similar to ATP.

The 20 enzymes that bind amino acids to tRNAs are grouped together and collectively called tRNA activating enzymes.

The elongation phase

Once initiation is complete elongation occurs. This phase involves tRNAs bringing amino acids to the mRNA–ribosomal complex in the order specified by the codons of the mRNA. Proteins called elongation factors assist in binding the tRNAs to the exposed mRNA codons at the A site. The initiator tRNA then moves to the P site. The ribosomes catalyse the formation of peptide bonds between adjacent amino acids brought to the polypeptide assembling area (see Figure 7.11).

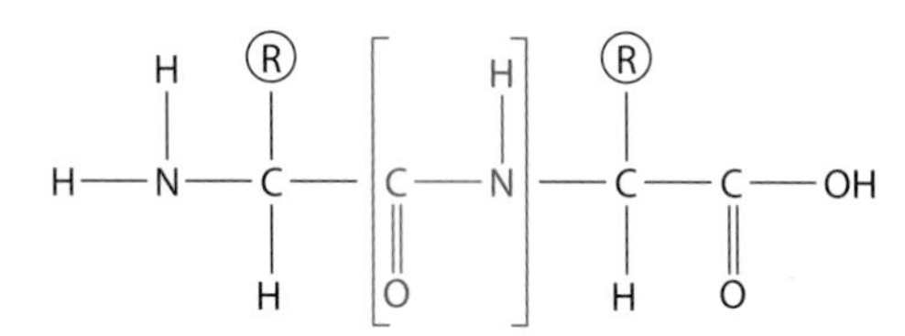

Figure 7.11 A peptide bond forms when water is given off. The process is called condensation.

The translocation phase

The translocation phase actually happens during the elongation phase. Translocation involves the movement of the tRNAs from one site of the mRNA to another. First, a tRNA binds with the A site. Its amino acid is then added to the growing polypeptide chain by a peptide bond. This causes the polypeptide chain to be attached to the tRNA at the A site. The tRNA then moves to the P site. It transfers its polypeptide chain to the new tRNA that moves into the now exposed A site. The now empty tRNA is transferred to the E site where it is released. This process occurs in the 5′ to 3′ direction. Therefore, the ribosomal complex is moving along the mRNA toward the 3′ end (see Figure 7.12). Remember, the start codon was near the 5′ end of the mRNA.

The anticodon that moves into the A site is specific for the codon of mRNA at that position. This allows the formation of a specific protein.

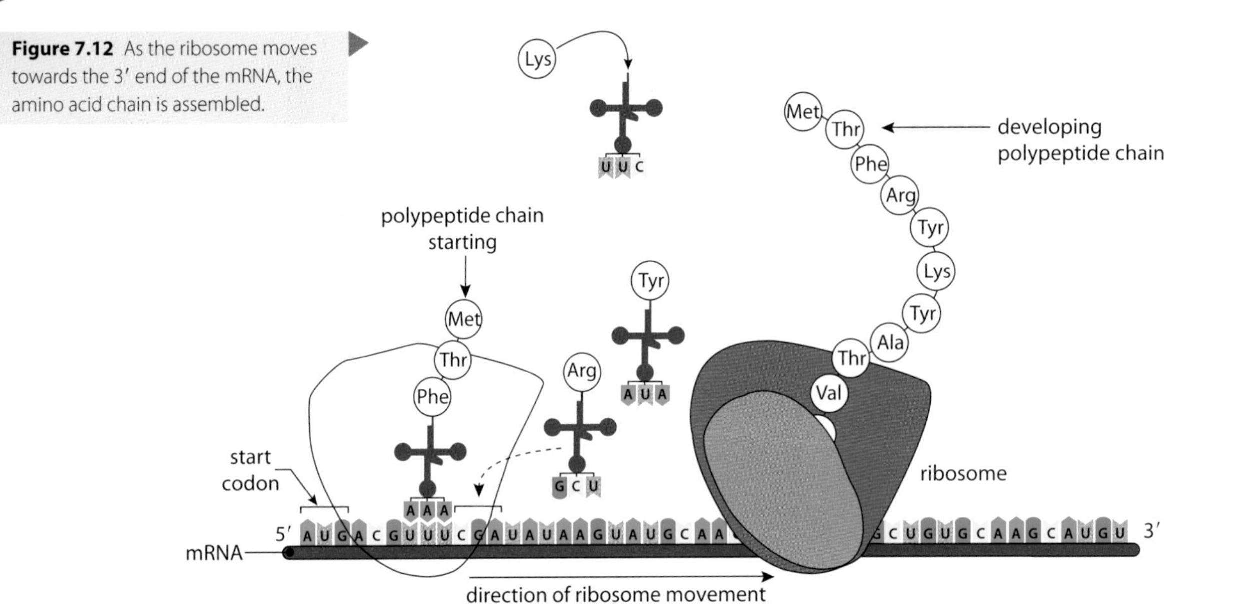

Figure 7.12 As the ribosome moves towards the 3′ end of the mRNA, the amino acid chain is assembled.

The termination phase

The termination phase begins when one of the three stop codons appears at the open A site. A protein called a release factor then fills the A site. The release factor does not carry an amino acid. It catalyses hydrolysis of the bond linking the tRNA in the P site with the polypeptide chain. This frees the polypeptide, releasing it from the ribosome. The ribosome then separates from the mRNA and splits into its two subunits.

The termination phase completes the process of translation. Proteins synthesized in this manner have several different destinations. If they are produced by free ribosomes, the proteins are primarily used within the cell. However, if the proteins are produced by ribosomes bound to the ER, they are primarily secreted from the cell or used in lysosomes.

It is common to see multiple ribosomes going through the process of translation on a single mRNA strand. This string of ribosomes is called a polysome or polyribosome.

For more information on DNA translation, visit heinemann.co.uk/hotlinks, insert the express code 4242P and click on Weblink 7.5.

Exercises

9 Describe the functions of the three types of RNA involved in the translation process.

10 Explain the value of polysomes to the cell and organism.

11 Draw a two-dimensional representation of a tRNA labelling the 3′ and 5′ ends, the anticodon, and the point of attachment to an amino acid.

7.5 Proteins

The Higher Level material in this section is also found in the first section of Option C (Chapter 14), an option available only to Standard Level students.

Assessment statements

7.5.1 Explain the four levels of protein structure, indicating the significance of each level.

7.5.2 Outline the difference between fibrous and globular proteins, with reference to two examples of each protein type.

7.5.3 Explain the significance of polar and non-polar amino acids.

7.5.4 State four functions of proteins, giving a named example of each.

Protein functions and structures

We have spent much time discussing protein production in the cell, because proteins are of very great importance. They serve many functions in the cell and organism. This table shows you just a few examples of proteins and their functions.

Protein	Function
haemoglobin	protein containing iron that transports oxygen from the lungs to all parts of the body in vertebrates
actin and myosin	proteins that interact to bring about muscle movement (contraction) in animals
insulin	a hormone secreted by the pancreas that aids in maintaining blood glucose level in vertebrates
immunoglobulins	group of proteins that act as antibodies to fight bacteria and viruses
amylase	digestive enzyme that catalyses the hydrolysis of starch

There are many internet sites that show the various structural levels of proteins. Several even show by animation how proteins go through conformational changes from primary to quaternary structure. Visit heinemann.co.uk/hotlinks, insert the express code 4242P and click on Weblink 7.6.

There are proteins that perform structural tasks, proteins that store amino acids, and some that have receptor functions so cells can respond to chemical signals. With all these functions, proteins have to be capable of assuming many forms or structures. The function of any particular protein is closely tied to its structure. There are four levels of organization to protein structure. They are called primary, secondary, tertiary, and quaternary organization.

Primary organization

The primary level of protein structure is the unique sequence of amino acids held together by peptide bonds in each protein (see Figure 7.13). There are 20 amino acids and these may be arranged in any order. The order or sequence in which the amino acids are arranged is determined by the nucleotide base sequence in the DNA of an organism. Because every organism has its own DNA, so every organism has its own unique proteins. The primary structure is simply a chain of amino acids attached by peptide bonds. Polypeptide chains may include hundreds of amino acids.

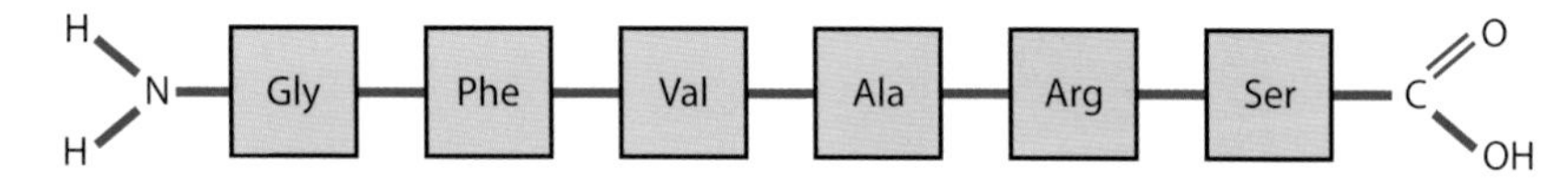

Figure 7.13 Protein primary structure.

The primary structure determines the next three levels of protein organization. Changing one amino acid in a chain may completely alter the structure and function of a protein. This is what happens in sickle cell disease. In this condition, just one amino acid is changed in the normal protein (haemoglobin) of red blood cells. The result is that the red blood cells are unable to carry oxygen, their normal function.

Secondary organization

The next level in the organization of proteins is the secondary structure. This is created by the formation of hydrogen bonds between the oxygen from the carboxyl group of one amino acid and the hydrogen from the amino group of another. Secondary structure does not involve the side chains, R groups. The two most common configurations of secondary structure are the α-helix and the β-pleated sheet (see Figure 7.14). Both have regular repeating patterns.

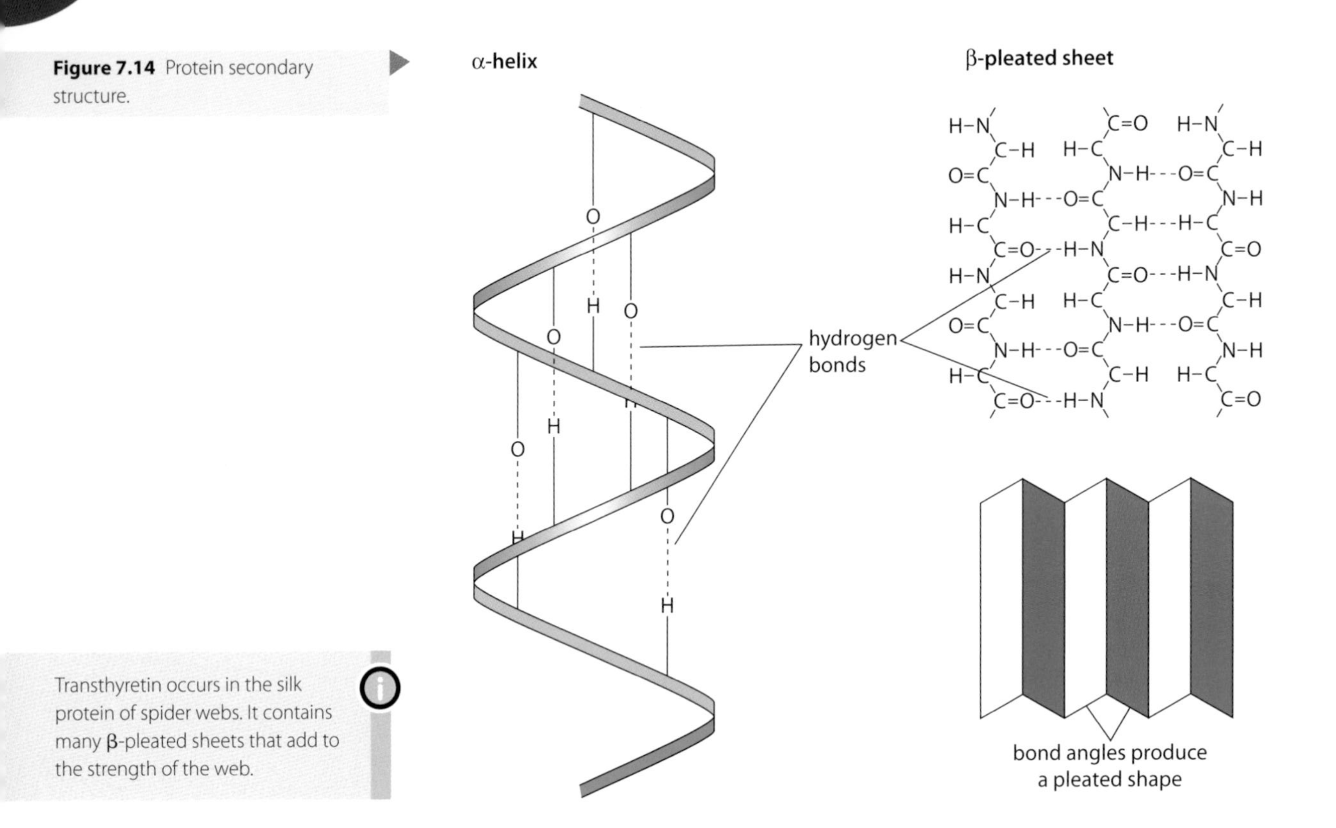

Figure 7.14 Protein secondary structure.

Transthyretin occurs in the silk protein of spider webs. It contains many β-pleated sheets that add to the strength of the web.

Tertiary organization

The third level in protein organization is the tertiary structure. The polypeptide chain bends and folds over itself because of interactions among R-groups and the peptide backbone. This results in a definite three-dimensional conformation (see Figure 7.15).

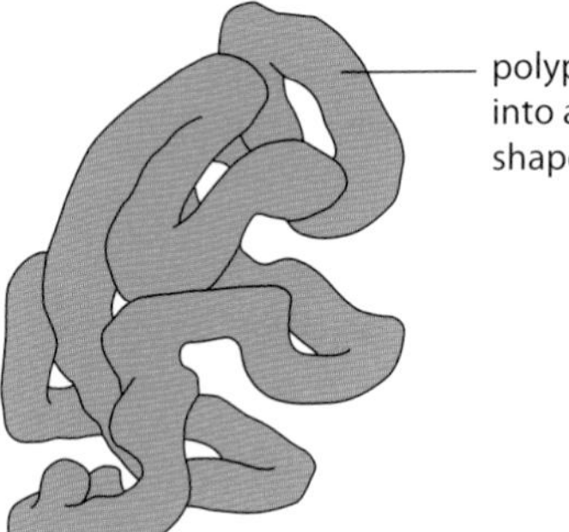

Figure 7.15 This is called a sausage model. It shows the three-dimensional conformation of lysozyme, an enzyme present in sweat, saliva and tears. Lysozyme destroys many bacteria.

Interactions that cause tertiary organization include:

- covalent bonds between sulfur atoms to create disulfide bonds – these are often called bridges because they are strong;
- hydrogen bonds between polar side chains;
- Van der Waals interactions among hydrophobic side chains of the amino acids – these interactions are rather strong because many hydrophobic side chains are forced inwards when the hydrophilic side chains interact with water towards the outside of the molecule;
- ionic bonds between positively and negatively charged side chains.

Tertiary structure is particularly important in determining the specificity of the proteins known as enzymes.

Quaternary organization

The last level is the quaternary structure. It is unique in that it involves multiple polypeptide chains which combine to form a single structure. Not all proteins consist of multiple chains, so not all proteins have quaternary structure. All the bonds mentioned in the first three levels of organization are involved in this level. Some proteins with quaternary structure include prosthetic or non-polypeptide groups. These proteins are called conjugated proteins. Haemoglobin is a conjugated protein (see Figure 7.16). It contains four polypeptide chains, each of which contains a non-polypeptide group called haem. Haem contains an iron atom that binds to oxygen.

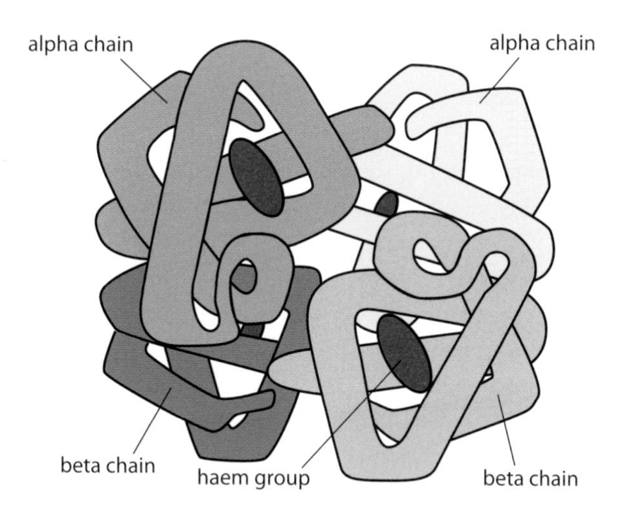

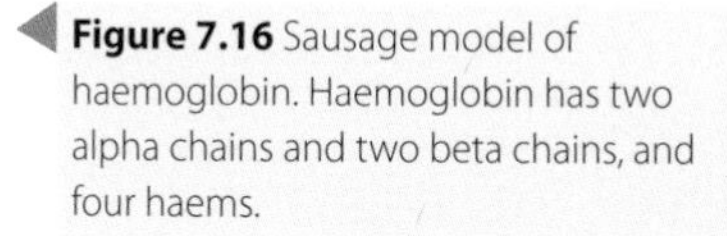

Figure 7.16 Sausage model of haemoglobin. Haemoglobin has two alpha chains and two beta chains, and four haems.

Dimers are proteins with two polypeptide subunits. Tetramers have four such units. In some cases these units are the same, but they may be different. Haemoglobin is a tetramer with two alpha chains and two beta chains.

Fibrous and globular proteins

Fibrous proteins are composed of many polypeptide chains in a long, narrow shape. They are usually insoluble in water. One example is collagen, which plays a structural role in the connective tissue of humans. Actin is another example. It is mentioned in the table above showing protein function. It is a major component of human muscle and is involved in contraction.

Globular proteins are more three-dimensional in their shape and are mostly water soluble. Haemoglobin, which delivers oxygen to body tissues, is one type of globular protein. The hormone insulin is another globular protein; it is involved in regulating blood glucose level in humans.

Polar and non-polar amino acids

Amino acids are often grouped according to the properties of their side chains (R groups). Amino acids with non-polar side chains are hydrophobic. Non-polar amino acids are found in the regions of proteins that are linked to the hydrophobic area of the cell membrane.

Polar amino acids have hydrophilic properties, and they are found in regions of proteins that are exposed to water. Membrane proteins include polar amino acids towards the interior and exterior of the membrane. These amino acids create hydrophilic channels in proteins through which polar substances can move.

Polar amino acids include serine, threonine, tyrosine, and glutamine. All of these have a group with an electrical charge in their side chain.

Non-polar amino acids have no electrical charges in their side groups. Examples are: tryptophan, leucine, alanine, and glycine.

Polar and non-polar amino acids are important in determining the specificity of an enzyme. Each enzyme has a region called the active site. Only specific substrates can combine with particular active sites. Combination is possible when 'fitting' occurs. The 'fitting' involves the general shapes and polar properties of the substrate and of the amino acids exposed at the active site.

Exercises

12 Explain why the primary level of protein organization determines the other levels.

13 What is the haem group containing iron called in the conjugated protein haemoglobin?

14 Describe how a single change in a protein's primary structure may change the protein's function.

7.6 Enzymes

The Higher Level material in this section is also found in the second section of Option C (Chapter 14), an option only available to Standard Level students.

Assessment statements

7.6.1 State that metabolic pathways consist of chains and cycles of enzyme-catalysed reactions.

7.6.2 Describe the induced-fit model.

7.6.3 Explain that enzymes lower the activation energy of the chemical reactions that they catalyse.

7.6.4 Explain the difference between competitive and non-competitive inhibition, with reference to one example of each.

7.6.5 Explain the control of metabolic pathways by end-product inhibition, including the role of allosteric sites.

Metabolism

Your metabolism is the sum of all the chemical reactions that occur in you as a living organism. The type of reaction that uses energy to build complex organic molecules from simpler ones is called anabolism. The type of reaction that breaks down complex organic molecules with the release of energy is called catabolism. This table summarises anabolic and catabolic reactions.

Anabolic reactions	Catabolic reactions
• build complex molecules	• break down complex molecules
• are endergonic	• are exergonic
• are biosynthetic	• are degradative
example: photosynthesis	example: cellular respiration

Metabolic pathways

Almost all metabolic reactions in organisms are catalysed by enzymes. Many of these reactions occur in specific sequences and are called metabolic or biochemical pathways. A very simple generalized metabolic pathway might look like this: substrate A → substrate B → final product. Each arrow represents a specific enzyme that causes one substrate to be changed to another until the final product of the pathway is formed.

Some metabolic pathways consist of cycles of reactions instead of chains of reactions. Others involve both cycles and chains of reactions. Cell respiration and photosynthesis are mentioned in Chapter 3 (pages 69 and 74) and are complex pathways with chains and cycles of reactions. Metabolic pathways are usually carried out in designated compartments of the cell where the necessary enzymes are clustered and isolated. The enzymes required to catalyse every reaction in these pathways are determined by the cell's genetic makeup.

Induced-fit model of enzyme action

Enzyme–substrate specificity was discussed in Chapter 3 (page 66). Enzyme specificity is made possible by the enzyme structure. Enzymes are very complex protein molecules with high molecular weights. The higher levels of protein structure allow enzymes to form unique areas such as the active site. The active site is the region on the enzyme that binds to a particular substrate or substrates. This binding results in the reaction occurring must faster than would be expected without the enzyme.

Enzymes are globular proteins with at least the tertiary level of organization.

In the 1890s, Emil Fischer proposed the lock-and-key model of enzyme action. He suggested that substrate molecules fit like a key into a rigid section of the enzyme 'lock'. This model provided a good explanation for the specificity of enzyme action at the time. However, as knowledge about enzyme action has increased, Fischer's model has been modified.

It is now obvious that many enzymes undergo significant changes in their conformation when substrates combine with their active sites. The accepted new model for enzyme action is called the induced-fit model. A good way to envision this model of enzyme action is to think of a hand and glove, the hand being the substrate and the glove being the enzyme. The glove looks somewhat like the hand. However, when the hand actually is placed in the glove there is an interaction that results in a conformational change of the glove, thus providing an induced fit.

The conformational changes and induced fit are due to changes in the R-groups of the amino acids at the active site of the enzyme as they interact with the substrate or substrates.

Scientific truths are often pragmatic. They are accepted because they predict why some process works. Fischer's lock-and-key model of enzyme action represents this pragmatism. The model was first presented in the 1890s. It was not until 1958 that Daniel Koshland used a larger body of knowledge to present a new model of enzyme action, now known as the induced-fit model. The new model represents a more accurate explanation of enzyme action.

Mechanism of enzyme action

1 The surface of the substrate contacts the active site of the enzyme.
2 The enzyme changes shape to accommodate the substrate.
3 A temporary complex called the enzyme–substrate complex forms.
4 Activation energy is lowered and the substrate is altered by the rearrangement of existing atoms.
5 The transformed substrate – the product – is released from the active site.
6 The unchanged enzyme is then free to combine with other substrate molecules.

Discover more about enzymes. Visit heinemann.co.uk/hotlinks, insert the express code 4242P and click on Weblinks 7.7a and 7.7b.

Enzyme action can be summarized by the following equation:

$$E + S \leftrightarrow ES \leftrightarrow E + P$$

where E is the enzyme, S is the substrate, ES is the enzyme–substrate complex, and P is the product.

Activation energy

When talking about enzyme action, we always refer to activation energy (AE). Activation energy is best understood as the energy necessary to destabilize the existing chemical bonds in the substrate of an enzyme–substrate catalysed reaction. Enzymes work by lowering the activation energy required (see Figure 7.17). That means they cause chemical reactions to occur faster because they reduce the amount of energy needed to bring about a chemical reaction.

It is important to note that even though enzymes lower activation energy of a particular reaction, they do not alter the proportion of reactants to products.

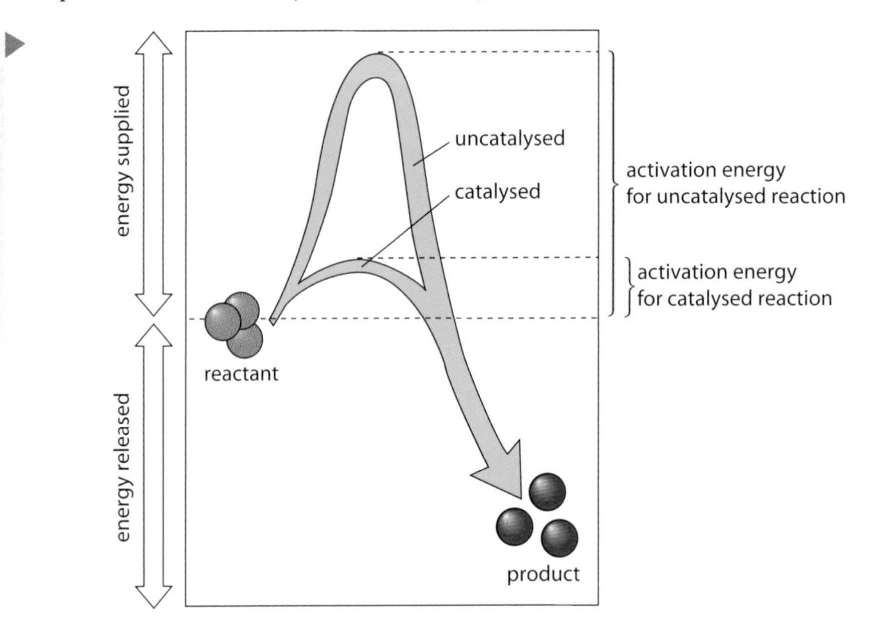

Figure 7.17 Enzymes accelerate exothermic reactions by lowering the activation energy required. The activation energy is needed to destabilize the chemical bonds in the reactant. The upper curve shows the activation energy when no enzyme is involved. The lower curve shows the activation energy required when an enzyme is present to catalyse the reaction.

Inhibition

The effects of pH, temperature, and substrate concentration on the action of enzymes are discussed in Chapter 3 (pages 67 and 68). Here, we discuss the effect of certain types of molecule on enzyme active sites. If a molecule affects the active site in some way, the activity of the enzyme may be altered.

Competitive inhibition

In competitive inhibition, a molecule called a competitive inhibitor, competes directly for the active site of an enzyme (see Figure 7.18). The result is that the substrate then has fewer encounters with the active site and the chemical reaction rate is decreased. The competitive inhibitor must have a structure similar to the substrate to be able to function in this way. An example is the use of sulfanilamide (a sulfa drug) to kill the bacteria during an infection. Folic acid is essential as a coenzyme to bacteria. It is produced in bacterial cells by enzyme action on para-aminobenzoic acid (PABA). The sulfanilamide competes with the PABA and blocks the enzyme. Because human cells do not use PABA to produce folic acid, they are unaffected by the drug.

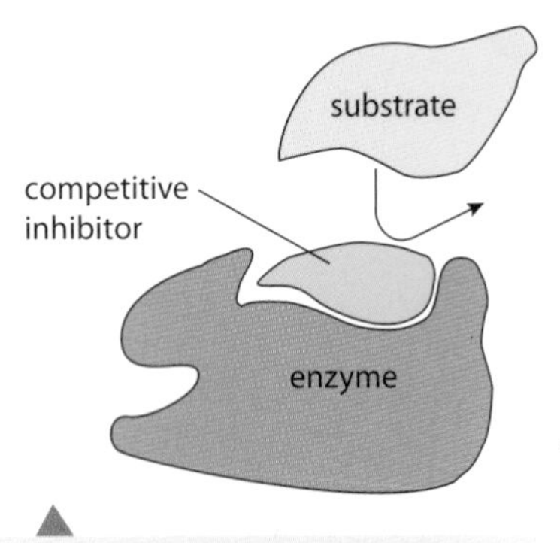

Figure 7.18 A competitive inhibitor blocks the active site of an enzyme so the substrate cannot bind to it.

Competitive inhibition may be reversible or irreversible. Reversible competitive inhibition may be overcome by increasing the substrate concentration. By doing this, there are more substrate molecules to bind with the active sites as they become available, and the chemical reaction may proceed more rapidly.

Non-competitive inhibition

Non-competitive inhibition involves an inhibitor that does not compete for the enzyme's active site. In this case, the inhibitor interacts with another site on the enzyme (see Figure 7.19). Non-competitive inhibition is also called allosteric inhibition, and the site the inhibitor binds to is called the allosteric site. Binding at the allosteric site causes a change in the shape of the enzyme's active site, making it non-functional. Examples of non-competitive inhibition include metallic ions, such as mercury, binding to the sulfur groups of component amino acids of many enzymes. This results in shape changes of the protein which causes inhibition of the enzyme.

Again, this type of inhibition may be reversible or irreversible. There are also examples of allosteric interaction activating an enzyme rather than inhibiting it.

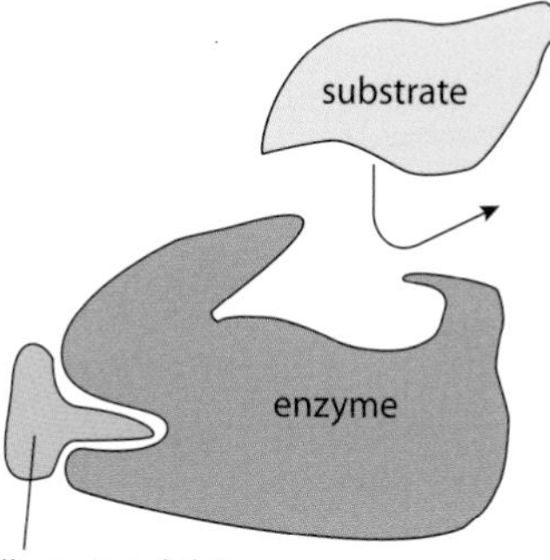

Figure 7.19 An allosteric (non-competitive) inhibitor combines with the allosteric site of an enzyme causing the active site to change shape so the substrate cannot bind to it.

End-product inhibition

End-product inhibition prevents the cell from wasting chemical resources and energy by making more of a substance than it needs. Many metabolic reactions occur in an assembly-line type of process so that a specific end-product can be achieved. Each step of the assembly line is catalysed by a specific enzyme (see Figure 7.20A). When the end-product is present in a sufficient quantity, the assembly line is shut down. This is usually done by inhibiting the action of the enzyme in the first step of the pathway (see Figure 7.20B). As the existing end-product is used up by the cell, the first enzyme is reactivated. The enzyme that is inhibited and reactivated is an allosteric enzyme. When in higher concentrations, the end-product binds with the allosteric site of the first enzyme, thus bringing about inhibition. Lower concentrations of the end-product result in fewer bindings with the allosteric site of the first enzyme and, therefore, activation of the enzyme.

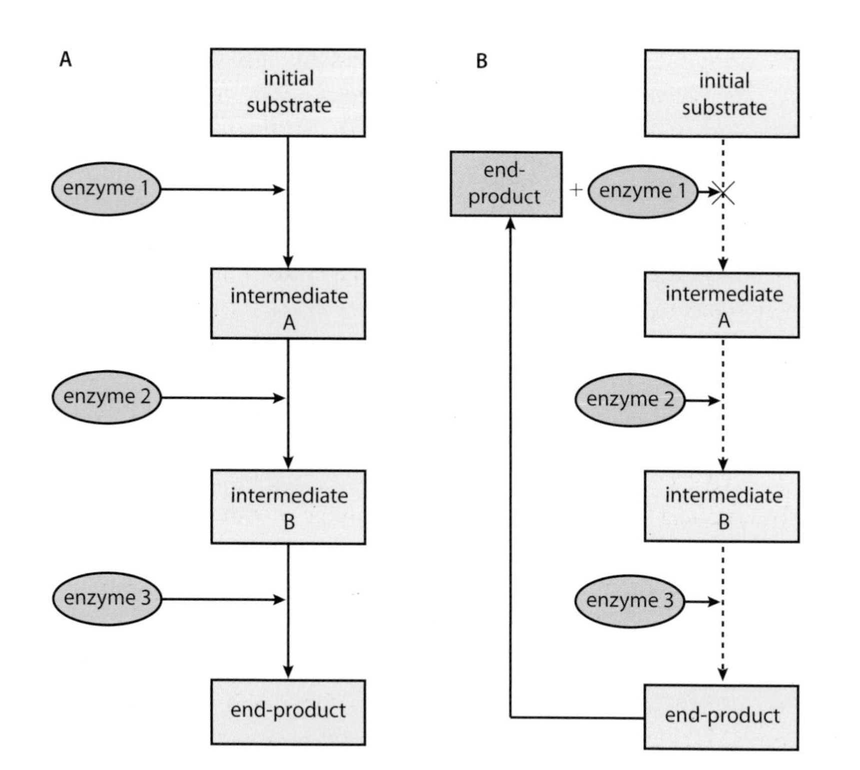

Figure 7.20 A short pathway of metabolic reactions with a specific end-product which, when in sufficient quantity, causes end-product inhibition. This is also a form of negative feedback. An example is found in control of glycolysis (see Chapter 8, page 220).

The bacterium *E. coli* uses a metabolic pathway to produce the amino acid isoleucine from threonine. It is a 5-step process. If isoleucine is added to the growth medium of *E. coli*, it inhibits the first enzyme in the pathway and isoleucine is not synthesized. This situation continues until the isoleucine is used up.

The inhibition of the first enzyme in the pathway prevents the build-up of intermediates in the cell. This is a form of negative feedback.

Exercises

15 Explain why enzymes only work with specific substrates.

16 What determines whether an enzyme is competitively or non-competitively inhibited?

Practice questions

1 In the enzyme-controlled pathway shown below, which compound is most likely to inhibit enzyme (w)?

$$\text{Precursor} \xrightarrow[w]{\text{enzyme}} \text{I} \xrightarrow[x]{\text{enzyme}} \text{II} \xrightarrow[y]{\text{enzyme}} \text{III} \xrightarrow[z]{\text{enzyme}} \text{IV}$$

A I
B II
C III
D IV

(1 mark)

2 To which parts of the deoxyribose molecule do phosphates bind in DNA?

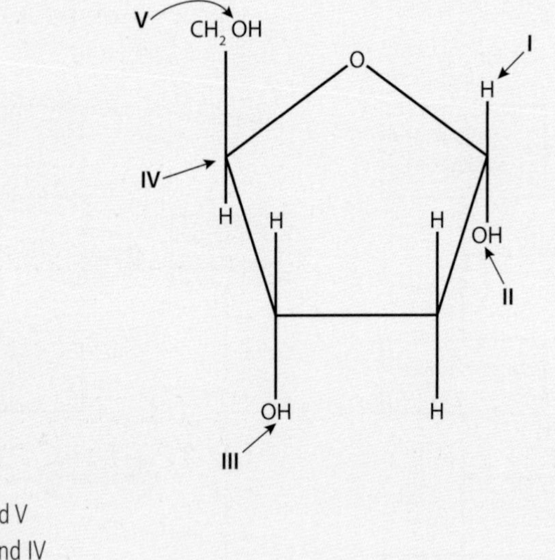

A I and V
B III and IV
C II and III
D III and V

(1 mark)

3 A biochemist isolated and purified molecules needed for DNA replication. When some DNA was added, replication occurred but the DNA molecules formed were defective. Each consisted of a normal DNA strand paired with segments of DNA a few hundred nucleotides long. Which of the following had been left out of the mixture?

A DNA ligase
B Helicase
C Nucleotides
D DNA polymerase *(1 mark)*

4 What effect do enzymes have on the activation energy of exergonic and endergonic reactions?

	Activation energy of exergonic reactions	Activation energy of endergonic reactions
A	increases	increases
B	decreases	decreases
C	increases	decreases
D	decreases	increases

(1 mark)

5 The diagram below shows a short section of DNA molecule before and after replication. If the nucleotides used to replicate the DNA were radioactive, which strands in the replicated molecules would be radioactive?

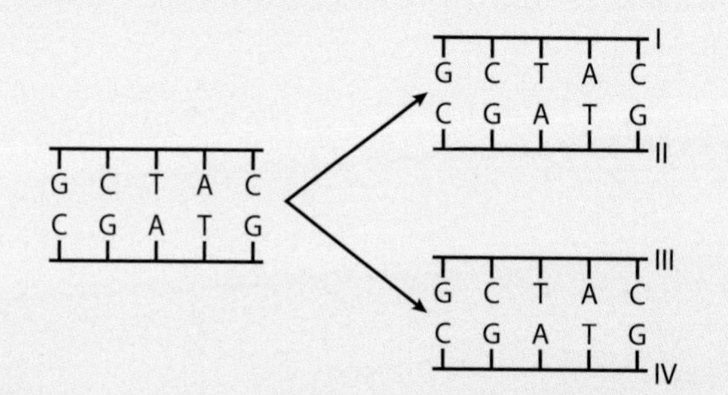

A II and III only
B I and III only
C I and II only
D I, II, III and IV *(1 mark)*

6 Which is not a primary function of protein molecules?
A Hormones
B Energy storage
C Transport
D Structure *(1 mark)*

7 Alcohol dehydrogenase is an enzyme that catalyses the reversible reaction of ethanol and ethanal according to the equation below.

$$NAD + \underset{\text{ethanol}}{CH_3CH_2OH} \rightleftharpoons \underset{\text{ethanal}}{CH_3CHO} + NADH + H^+$$

The initial rate of reaction can be measured according to the time taken for NADH to be produced.
In an experiment, the initial rate at different concentrations of ethanol was recorded (no inhibition). The experiment was then repeated with the addition of I mmol dm^{-3} 2,2,2-trifluoroethanol, a competitive inhibitor of the enzyme. A third experiment using a greater concentration of the same inhibitor (3 mmol dm^{-3}) was performed. The results for each experiment are shown in the graph overleaf.

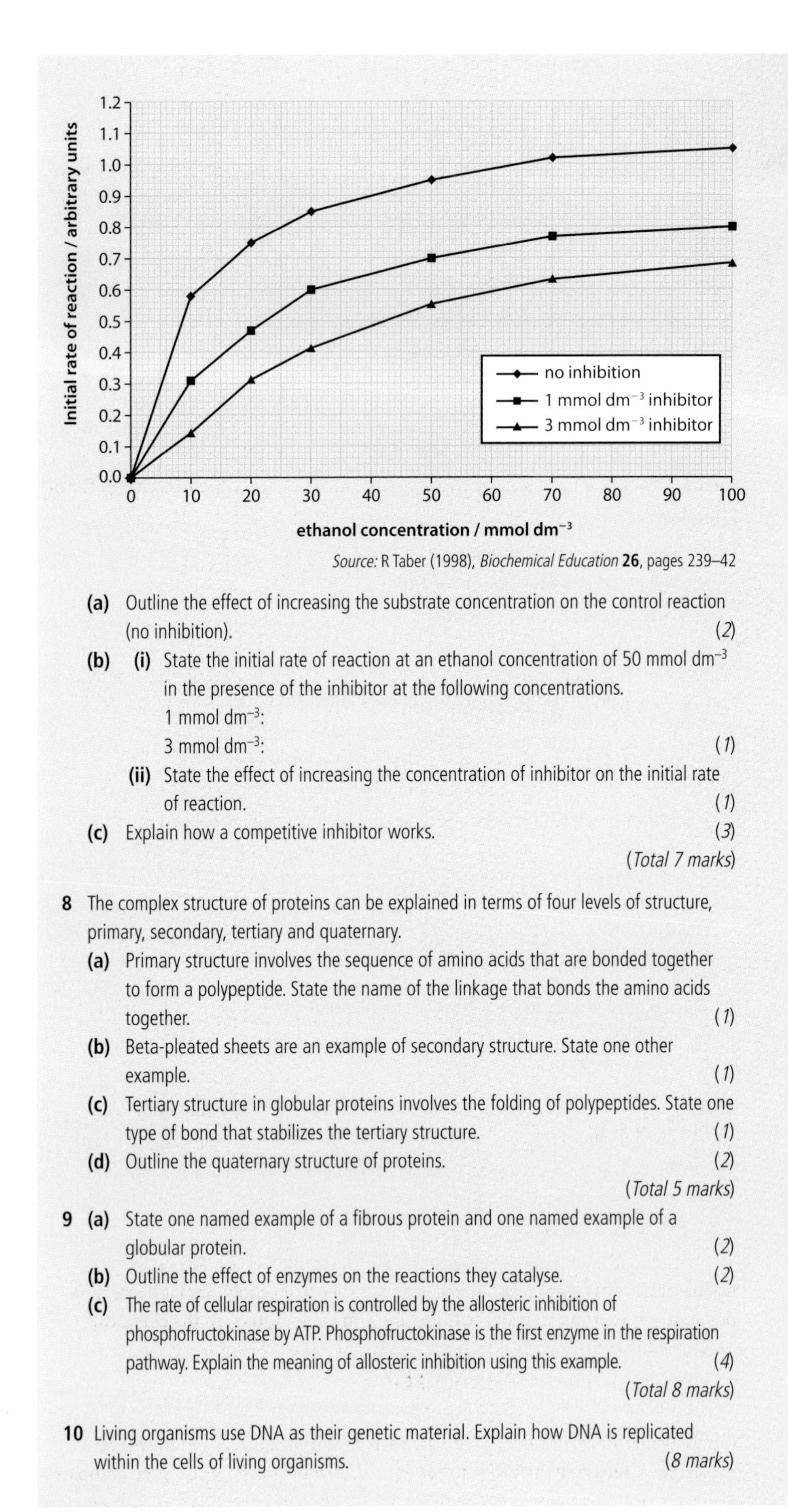

Source: R Taber (1998), *Biochemical Education* **26**, pages 239–42

(a) Outline the effect of increasing the substrate concentration on the control reaction (no inhibition). *(2)*

(b) **(i)** State the initial rate of reaction at an ethanol concentration of 50 mmol dm^{-3} in the presence of the inhibitor at the following concentrations.

1 mmol dm^{-3}:

3 mmol dm^{-3}: *(1)*

(ii) State the effect of increasing the concentration of inhibitor on the initial rate of reaction. *(1)*

(c) Explain how a competitive inhibitor works. *(3)*

(Total 7 marks)

8 The complex structure of proteins can be explained in terms of four levels of structure, primary, secondary, tertiary and quaternary.

(a) Primary structure involves the sequence of amino acids that are bonded together to form a polypeptide. State the name of the linkage that bonds the amino acids together. *(1)*

(b) Beta-pleated sheets are an example of secondary structure. State one other example. *(1)*

(c) Tertiary structure in globular proteins involves the folding of polypeptides. State one type of bond that stabilizes the tertiary structure. *(1)*

(d) Outline the quaternary structure of proteins. *(2)*

(Total 5 marks)

9 **(a)** State one named example of a fibrous protein and one named example of a globular protein. *(2)*

(b) Outline the effect of enzymes on the reactions they catalyse. *(2)*

(c) The rate of cellular respiration is controlled by the allosteric inhibition of phosphofructokinase by ATP. Phosphofructokinase is the first enzyme in the respiration pathway. Explain the meaning of allosteric inhibition using this example. *(4)*

(Total 8 marks)

10 Living organisms use DNA as their genetic material. Explain how DNA is replicated within the cells of living organisms. *(8 marks)*

8 Cell respiration and photosynthesis

Introduction

Energy is a topic of discussion every day in our modern world. We talk about the energy needed for transport. We talk about being so tired after a long day at school that we need a short nap. The need for food becomes essential at times to regain the energy level necessary for us to function. This chapter looks at energy in living systems. We discuss how plants harvest light energy and transform it into potential chemical energy, and how both plants and animals release the stored energy for life processes.

Different cultures use energy in different ways. Discuss the effect of energy shortages on both societies and individuals.

8.1 Cell respiration

The higher Level material found in this section is also found in the third section of Option C (Chapter 14), an option available only to Standard Level students.

Assessment statements

8.1.1 State that oxidation involves the loss of electrons from an element, whereas reduction involves a gain of electrons; and that oxidation frequently involves gaining oxygen or losing hydrogen, whereas reduction frequently involves losing oxygen or gaining hydrogen.

8.1.2 Outline the process of glycolysis, including phosphorylation, lysis, oxidation, and ATP formation.

8.1.3 Draw and label a diagram showing the structure of a mitochondrion as seen in electron micrographs.

8.1.4 Explain aerobic respiration: the link reaction, the Krebs cycle, the role of NADH + H^+, the electron transport chain and the role of oxygen.

8.1.5 Explain oxidative phosphorylation in terms of chemiosmosis.

8.1.6 Explain the relationship between the structure of the mitochondrion and its function.

Oxidation and reduction

In Chapter 3, the general processes of respiration and photosynthesis are discussed. In this chapter, we consider these aspects of cellular metabolism in detail. It is important to recall that metabolism is the sum of all the chemical reactions carried out by an organism. These reactions involve:

- catabolic pathways;
- anabolic pathways.

Catabolic pathways result in the breakdown of complex molecules to smaller molecules. Conversely, anabolic pathways result in the synthesis of more complex molecules from simpler ones. Cellular respiration is an example of a catabolic pathway. Photosynthesis is an example of an anabolic pathway. To understand these complex pathways, it is essential to understand two general types of chemical reaction: oxidation and reduction.

Oxidation and reduction can compared using a table like this.

Examiner's hint: If you are asked in an exam to compare oxidation and reduction, a table such as this one is an excellent way to structure the answer.

Oxidation	Reduction
loss of electrons	gain of electrons
gain of oxygen	loss of oxygen
loss of hydrogen	gain of hydrogen
results in many C—O bonds	results in many C—H bonds
results in a compound with lower potential energy	results in a compound with higher potential energy

A useful way to remember the general meaning of oxidation and reduction is to think of the words OIL RIG:

- OIL = Oxidation Is Loss (of electrons);
- RIG = Reduction Is Gain (of electrons).

These two reactions occur together during chemical reactions. Think of it in this way: one compound's or element's loss is another compound's or element's gain. This is shown by the following equation:

$$C_6H_{12}O_6 + 6O_2 \rightarrow 6CO_2 + 6H_2O + \text{energy}$$

In this equation, glucose is oxidized because electrons are transferred from it to oxygen. The protons follow the electrons to produce water. The oxygen atoms that occur in the oxygen molecules on the reactant side of the equation are reduced. Because of this reaction, there is a large drop in the potential energy of the compounds on the product side of the equation.

Because oxidation and reduction always occur together, these chemical reactions are referred to as redox reactions. When redox reactions take place, the reduced form of a molecule always has more potential energy that the oxidized form of the molecule. Redox reactions play a key role in the flow of energy through living systems. This is because the electrons that are flowing from one molecule to the next are carrying energy with them. In a similar sort of way, the catabolic and anabolic pathways mentioned earlier are also closely associated with one another. You will see this association as you work through this chapter.

An overview of respiration

Chapter 3 provided an introduction to the process of cellular respiration. Three aspects of cellular respiration were discussed:

- glycolysis;
- anaerobic respiration;
- aerobic respiration.

As you will recall, glycolysis occurs in the cytoplasm of the cell, produces small amounts of ATP and ends with the product known as pyruvate. If no oxygen is available, the pyruvate enters into anaerobic respiration. This occurs in the cytoplasm and it does not result in any further production of ATP. The products of anaerobic respiration are lactate or ethanol and carbon dioxide. If oxygen is available, the pyruvate enters aerobic respiration in the mitochondria of the cell. This process results in the production of a large number of ATPs, carbon dioxide and water.

In this chapter, we discuss cellular respiration which involves glycolysis and the three stages of aerobic respiration:

- the link reaction;
- the Krebs cycle;
- oxidative phosphorylation.

Glycolysis

The word glycolysis means 'sugar splitting' and is thought to have been one of the first biochemical pathways to evolve. It uses no oxygen and occurs in the cytosol of the cell. There are no required organelles. The sugar splitting proceeds efficiently in aerobic and anaerobic environments. Glycolysis occurs in both prokaryotic and eukaryotic cells. A hexose, usually glucose, is split in the process. This splitting actually involves many steps but we can explain it effectively in three stages.

1 Two molecules of ATP are used to begin glycolysis. In the first reaction, the phosphates from the ATPs phosphorylate glucose to form fructose-1, 6-bisphosphate (see Figure 8.1). This process involves phosphorylation.

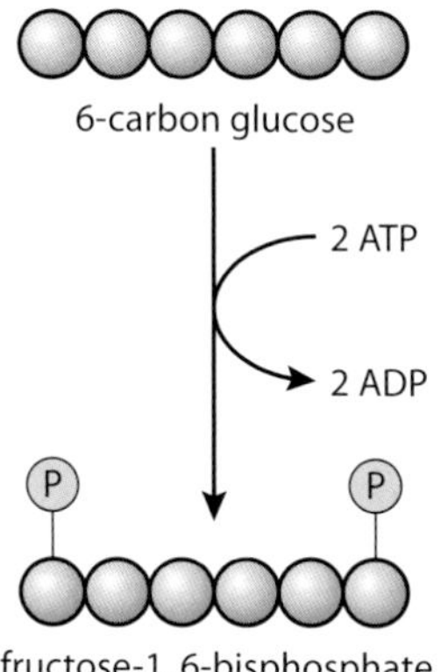

Figure 8.1 First stage of glycolysis; the circles represent carbon atoms.

2 The 6-carbon phosphorylated fructose is split into two 3-carbon sugars called glyceraldehyde-3-phosphate (G3P) (see Figure 8.2). This process involves lysis.

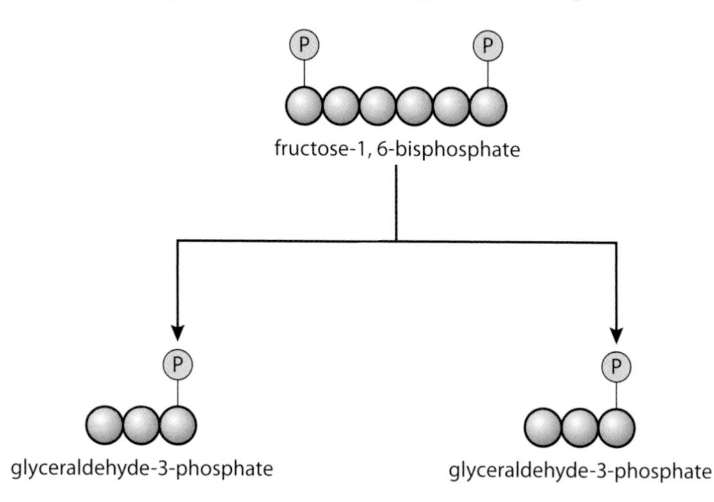

Figure 8.2 Second stage of glycolysis.

3 Once the two G3P molecules are formed, they enter an oxidation phase involving ATP formation and production of the reduced coenzyme NAD (see Figure 8.3). Each G3P or triose phosphate molecule undergoes oxidation to form a reduced molecule of NAD^+, which is NADH. As NADH is being formed, released energy is used to add an inorganic phosphate to the remaining 3-carbon compound. This results in a compound with two phosphate groups. Enzymes then remove the phosphate groups so they can be added to ADP to produce ATP. The end result is the formation of four molecules of ATP, two molecules of NADH and two molecules of pyruvate. Pyruvate is the ionized form of pyruvic acid.

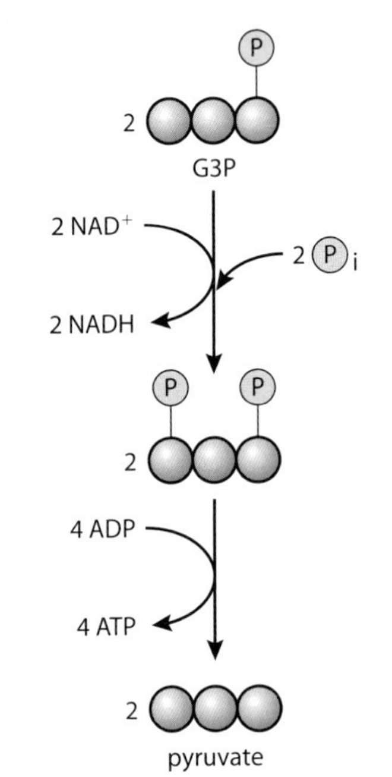

Figure 8.3 Third stage of glycolysis.

This way of producing ATP is called substrate-level phosphorylation since the phosphate group is transferred directly to ADP from the original phosphate-bearing molecule.

Once pyruvate is obtained, the next pathway is determined by the presence of oxygen. If oxygen is present, pyruvate enters the mitochondria and aerobic respiration occurs. If oxygen is not present, anaerobic respiration occurs in the cytoplasm. In this case, pyruvate is converted to lactate in animals, and ethanol and carbon dioxide in plants.

Summary of glycolysis

- Two ATPs are used to start the process.
- A total of four ATPs are produced – a net gain of two ATPs.
- Two molecules of NADH are produced.
- Involves substrate-level phosphorylation, lysis, oxidation and ATP formation.
- Occurs in the cytoplasm of the cell.
- This metabolic pathway is controlled by enzymes. Whenever ATP levels in the cell are high, feedback inhibition will block the first enzyme of the pathway (see Figure 7.20, page 213). This will slow or stop the process.
- Two pyruvate molecules are present at the end of the pathway.

Mitochondria

It is inside the mitochondria and in the presence of oxygen that the remainder of cellular respiration occurs.

We discussed the structure of the mitochondrion in Chapter 2 (page 23). You might like to refresh your memory of this because as we discuss aerobic respiration, which occurs in the mitochondrion, we will refer to parts of this organelle.

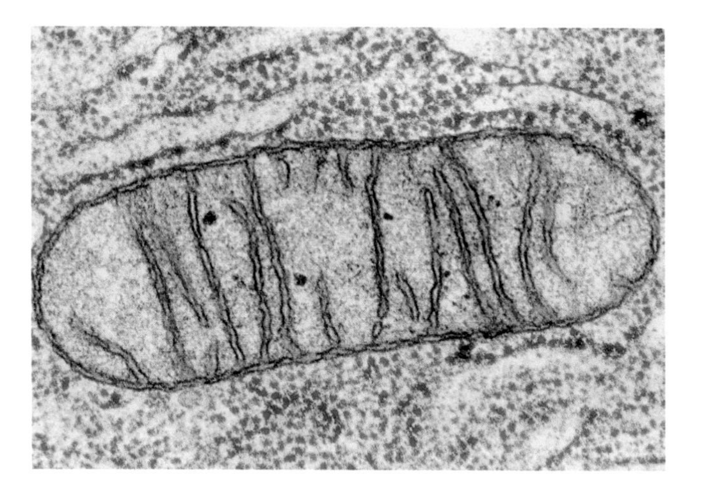

This false-colour TEM of a mitochondrion shows the internal structure. The matrix (blue) is permeated by the membranous cristae (pink).

The link reaction and the Krebs cycle

Once glycolysis has occurred and there is oxygen present, pyruvate enters the matrix of the mitochondrion via active transport. Inside, pyruvate is decarboxylated to form the 2-carbon acetyl group. This is the link reaction (see Figure 8.4). The removed carbon is released as carbon dioxide, a waste gas. The acetyl group is then oxidized with the formation of reduced NAD^+. Finally, the acetyl group combines with coenzyme A (CoA) to form acetyl CoA.

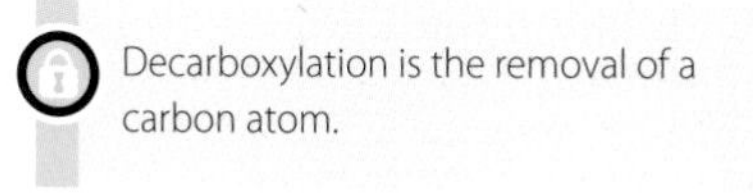

Decarboxylation is the removal of a carbon atom.

A coenzyme is a molecule that aids an enzyme in its action. Coenzymes usually act as electron donors or acceptors.

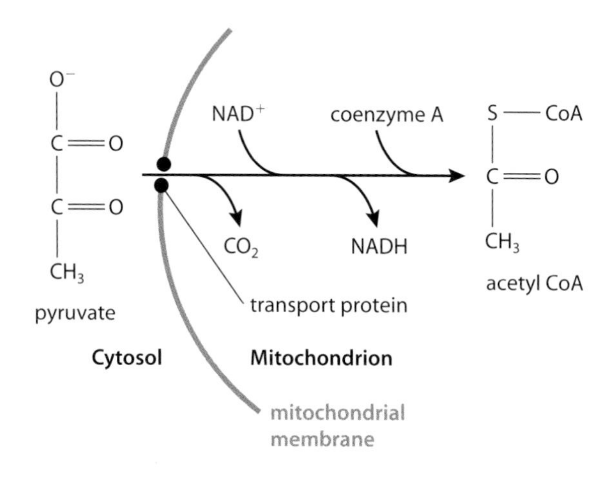

Figure 8.4 The link reaction.

The link reaction is controlled by a system of enzymes. The greatest significance of this reaction is that it produces acetyl CoA. Acetyl CoA may enter the Krebs cycle to continue the aerobic respiration process.

So far in this discussion, the respiratory substrate has been a hexose. However, in reality, acetyl CoA can be produced from most carbohydrates and fats. Acetyl CoA can be synthesized into a lipid for storage purposes. This occurs when ATP levels in the cell are high.

If cellular ATP levels are low, the acetyl CoA enters the Krebs cycle. This cycle is also called the tricarboxylic acid cycle. It occurs in the matrix of the mitochondrion and is referred to as a cycle because it begins and ends with the same substance. This is a characteristic of all cyclic pathways in metabolism. You do not need to remember the names of all the compounds formed in the Krebs cycle. However, it is important that you understand the overall process.

Let's consider the cycle as a series of steps.

1 Acetyl CoA from the link reaction combines with a 4-carbon compound called oxaloacetate. The result is a 6-carbon compound called citrate (see Figure 8.5).

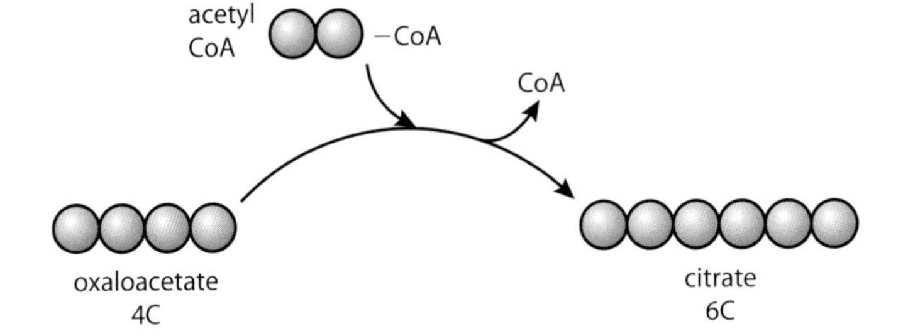

Figure 8.5 Acetyl CoA combines with oxaloacetate to form citrate.

2 Citrate (6-carbon compound) is oxidized to form a 5-carbon compound (see Figure 8.6). In this process, the carbon is released from the cell (after combining with oxygen) as carbon dioxide. While the 6-carbon compound is oxidized, NAD^+ is reduced to form NADH.

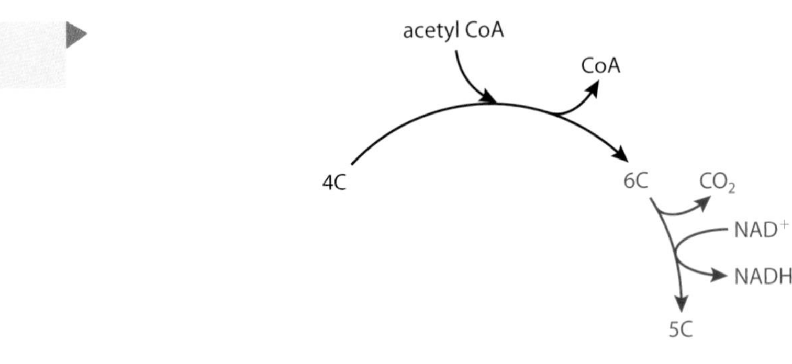

Figure 8.6 Then a 5-carbon compound is formed.

3 The 5-carbon compound is oxidized and decarboxylated to form a 4-carbon compound (see Figure 8.7). Again, the removed carbon combines with oxygen and is released as carbon dioxide. Another NAD^+ is reduced to form NADH.

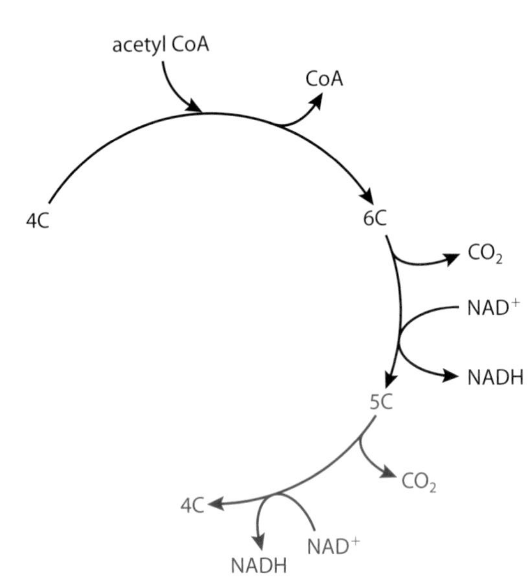

Figure 8.7 Next, a 4-carbon compound is produced.

4 The 4-carbon compound undergoes various changes resulting in several products (see Figure 8.8). One product is another NADH. The coenzyme FAD is reduced to form $FADH_2$. There is also a reduction of an ADP to form ATP.

The 4-carbon compound is changed during these steps to re-form the starting compound of the cycle, oxaloacetate. The oxaloacetate may then begin the cycle again.

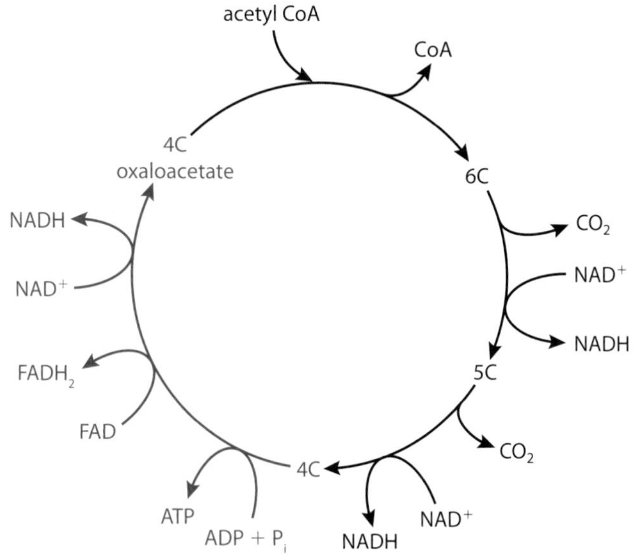

Figure 8.8 Finally, the 4-carbon compound is converted to oxaloacetate.

It is important to remember that the Krebs cycle will run twice for each glucose molecule entering cellular respiration. This is because a glucose molecule forms two pyruvate molecules. Each pyruvate produces one acetyl CoA which enters the cycle. Look again at the complete Krebs cycle (Figure 8.8) and note the following products which result from the breakdown of one glucose molecule:

- two ATP molecules;
- six molecules of NADH (allow energy storage and transfer);
- two molecules of $FADH_2$;
- four molecules of carbon dioxide (released).

So far, only four ATPs have been gained: six are generated (four from glycolysis and two from the Krebs cycle) but two are used to start the process of glycolysis. Each of these ATPs has been produced by substrate-level phosphorylation.

Ultimately, the breakdown of each glucose molecule results in a net gain of 36 ATPs. Let's now consider the phase of cellular respiration where most of the ATPs are produced. In this phase, oxidative phosphorylation is the means by which the ATPs are produced.

Two carbon dioxides are released for each glucose molecule during the link reaction. Four are released during the Krebs cycle. This accounts for all six carbon atoms that were present in the initial glucose molecule. Glucose is completely catabolized and its original energy is now carried by NADH and $FADH_2$ or is in ATP.

For an effective review of glycolysis and the Krebs cycle, visit heinemann.co.uk/hotlinks, insert the express code 4242P and click on Weblink 8.1.

Electron transport chain and chemiosmosis

The electron transport chain is where most of the ATPs from glucose catabolism are produced. It is the first stage of cellular respiration where oxygen is actually needed, and it occurs within the mitochondrion. However, unlike the Krebs cycle, which occurred in the matrix, the electron transport chain occurs on the inner mitochondrial membrane and on the membranes of the cristae.

Embedded in the involved membranes are molecules that are easily reduced and oxidized. These carriers of electrons (energy) are close together and pass the electrons from one to another due to an energy gradient. Each carrier molecule has a slightly different electronegativity and, therefore, a different attraction for electrons. Most of these carriers are proteins with haem groups and are referred to as cytochromes. One carrier is not a protein and is called coenzyme Q.

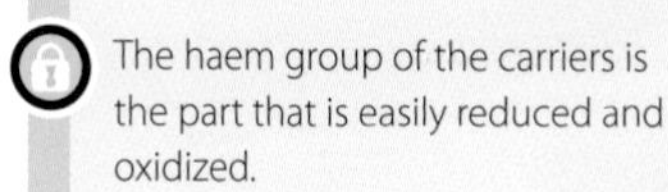
The haem group of the carriers is the part that is easily reduced and oxidized.

In this chain, electrons pass from one carrier to another because the receiving molecule has a higher electronegativity and, therefore, a stronger attraction for electrons (see Figure 8.9). In the process of electron transport, small amounts of energy are released. The sources of the electrons that move down the electron transport chain are the coenzymes NADH and $FADH_2$ from the previous stages of cellular respiration.

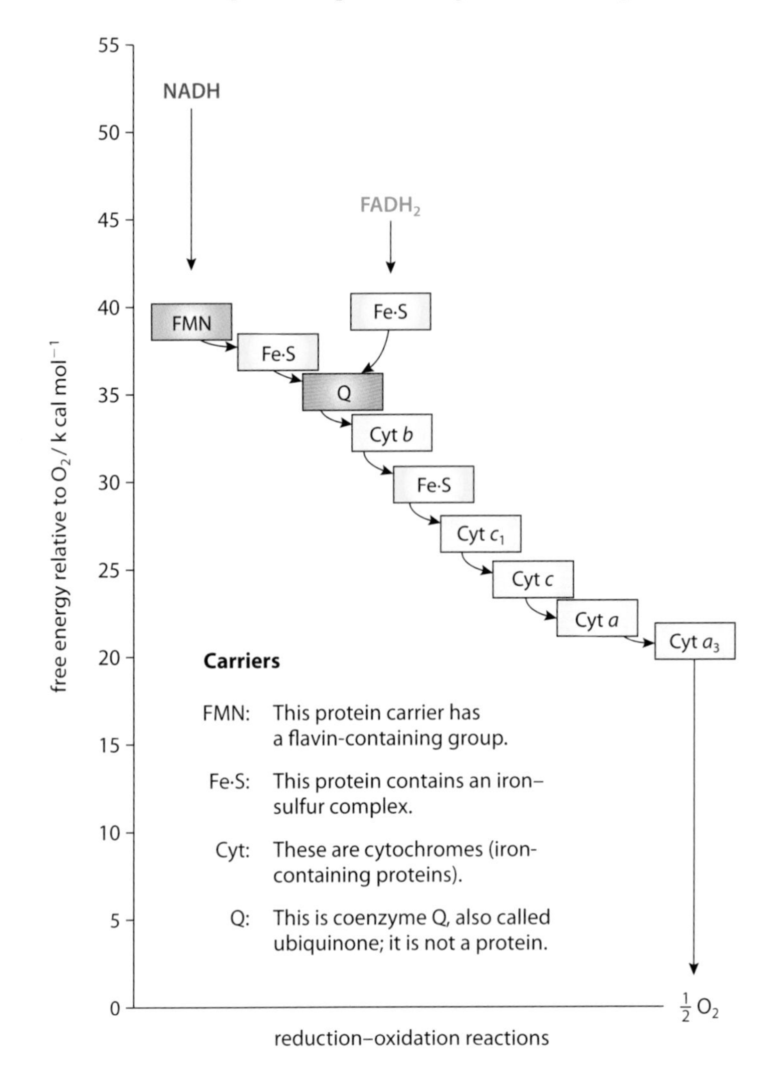

Figure 8.9 The oxidation–reduction reactions of the electron transport chain. It is not necessary for you to remember all the names of the carriers.

In Figure 8.9 it is clear that the electrons are stepping down in potential energy as they pass from one carrier to another. It is important to note that :

- $FADH_2$ enters the electron transport chain at a lower free energy level than NADH – thus, $FADH_2$ allows the production of 2ATPs while NADH allows the production of 3ATPs;
- at the very end of the chain, the de-energized electrons combine with available oxygen.

Oxygen is the final electron acceptor because it has a very high electronegativity and, therefore, a strong attraction for electrons. When the electrons combine with the oxygen, so do two hydrogen ions from the aqueous surroundings. The result is water. Because of the way this water is formed, it is referred to as water of metabolism.

The kangaroo rat of a desert region of the United States obtains 90% of its daily water intake from water of metabolism. In contrast to this, a typical human obtains only 12% of its daily water intake by way of metabolism.

It is also clear from Figure 8.9 that there are a fairly large number of electron carriers. Because of the larger number, the electronegativity difference between adjacent carriers is not so great. This means that lower amounts of energy are lost at each exchange. These lower amounts of energy are effectively harnessed by the cell to carry out phosphorylation. If the amount of energy lost at each exchange was high, much of it could not be used and damage might be done to the cell.

So energy is now available as a result of the electron transport chain. This is the energy that allows the addition of phosphate and energy to ADP to form ATP. The process by which this occurs is called chemiosmosis. Chemiosmosis involves the movement of protons (hydrogen ions) to provide energy so that phosphorylation can occur. Because this type of phosphorylation uses an electron transport chain, it is called oxidative phosphorylation. Substrate-level phosphorylation mentioned in the earlier phases of cellular respiration did not involve an electron transport chain.

No ATPs are produced directly by the electron transport chain. However, this chain is essential to chemiosmosis, which does produce the ATP.

Before we continue, it is essential to review the interior structure of the mitochondrion. In the process of cellular respiration, the structure of the mitochondrion is very closely linked to its function. The matrix is the area where the Krebs cycle occurs. The cristae provide a large surface area for the electron transport chain to function. The membranes also provide a barrier allowing for proton accumulation on one side. Embedded in the membranes are the enzymes and other necessary compounds for the processes of the electron transport chain and chemiosmosis to occur.

• **Examiner's hint:** Using any diagram or photomicrograph of a mitochondrion, annotate where the processes of respiration occur.

The inner membranes of the mitochondrion have numerous copies of an enzyme called ATP synthase. This enzyme uses the energy of an ion gradient to allow the phosphorylation of ADP. The ion gradient is created by a hydrogen ion concentration difference that occurs across the cristae membranes. Figure 8.10 shows oxidative phosphorylation.

Figure 8.10 Oxidative phosphorylation occurs at the inner membranes of the mitochondria of a cell. The pumping actions of the carriers result in a high concentration of hydrogen ions in the intermembrane space. This accumulation allows movement of the hydrogen ions through the enzyme ATP synthase. The enzyme uses the energy from the hydrogen flow to couple phosphate with ADP to produce ATP.

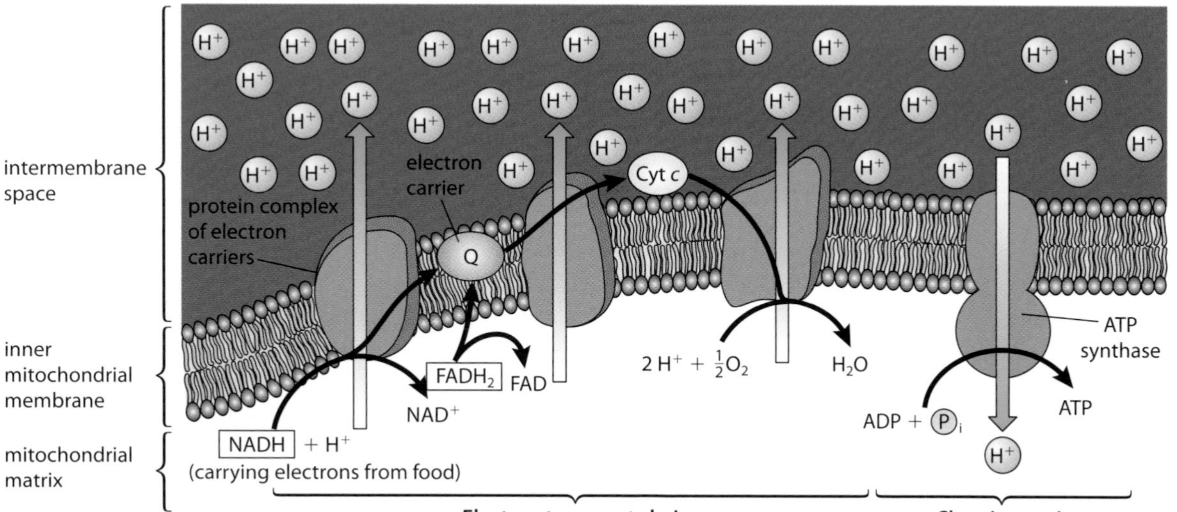

In Figure 8.10, note the three labelled areas on the left: intermembrane space, inner mitochondrial membrane, and mitochondrial matrix. Also, note that hydrogen ions are being pumped out of the matrix into the intermembrane space. The energy for this pumping action is provided by the electrons as they are de-energized moving through the electron transport chain. This creates the different hydrogen ion concentration on the two sides of the cristae membranes, mentioned above. With the higher hydrogen ion concentration in the intermembrane space, these ions begin to passively move through a channel in ATP synthase back into the mitochondrial matrix. As the hydrogen ions move through the ATP synthase channel, the enzyme harnesses the available energy thus allowing the phosphorylation of ADP.

Because of the hydrophobic region of the membrane, the hydrogen ions can only pass through the ATP synthase channel. Some poisons that affect metabolism act by establishing alternative pathways through the membrane thus preventing ATP production.

For an animation of the electron transport chain, visit heinemann.co.uk/hotlinks, insert the express code 4242P and click on Weblink 8.2.

Summary of ATP production in cellular respiration

We have now considered the complete catabolism of one molecule of glucose. The raw materials are glucose and oxygen. Many enzymes, carriers and other molecules are involved in the process. The products are carbon dioxide, water and ATP. The ATPs are essential as they provide the energy by which life is maintained. We can describe the energy flow in the general process as:

glucose → NADH or $FADH_2$ → electron transport chain → chemiosmosis → ATP

To account for the production of ATP in cellular respiration, let's look at the three main processes: glycolysis, the Krebs cycle, and the electron transport chain in a table.

Process	ATP used	ATP produced	Net ATP gain
glycolysis	2	4	2
Krebs cycle	0	2	2
electron transport chain and chemiosmosis	0	32	32
total	2	38	36

Theoretically 36 ATPs are produced by cellular respiration but in reality the number is closer to 30. This is thought to be due to some hydrogen ions moving back to the matrix without going through the ATP synthase channel. Also, some of the energy from hydrogen ion movement is used to transport pyruvate into the mitochondria. The 30 ATPs generated by cellular respiration account for approximately 30% of the energy present in the chemical bonds of glucose. The remainder of the energy is lost from the cell as heat.

Final look at respiration and the mitochondrion

Cellular respiration is the process by which ATP is provided to the organism so that life may continue. It is a very complex series of chemical reactions, most of which occur in the mitochondrion. Let's end our discussion of this essential-to-life process by looking at a table showing the parts of the mitochondrion and how those parts allow cellular respiration.

Outer mitochondrial membrane	separates the contents of the mitochondrion from the rest of the cell
Matrix	internal cytosol-like area that contains the enzymes for the link reaction and the Krebs cycle
Cristae	tubular regions surrounded by membranes increasing surface area for oxidative phosphorylation
Inner mitochondrial membrane	contains the carriers for the electron transport chain and ATP synthase for chemiosmosis
Space between inner and outer membranes	reservoir for hydrogen ions (protons), the high concentration of hydrogen ions is necessary for chemiosmosis

The overall equation for cellular respiration is:

$$C_6H_{12}O_6 + 6O_2 \rightarrow 6CO_2 + 6H_2O + \text{energy (heat or ATP)}$$

All organisms need to produce ATP for energy, so all organisms carry out respiration.

Exercises

1 Using ideal ATP production numbers, how many ATPs would an individual generate if they consumed only pyruvate and carried one pyruvate molecule through cellular respiration?

2 Striated muscles usually have more mitochondria than other cell types. Why is this important?

3 If both NAD and FAD are reduced, which would allow the greater production of ATPs via the electron transport chain and chemiosmosis?

4 If an individual took a chemical that increased the ability of hydrogen ions to move through the phospholipid bilayer of the mitochondrial membranes, what would the effect be on ATP production?

5 If ATP synthase was not present in the cristae of a mitochondrion, what would be the effect?

Photosynthesis

The Higher Level material in this section is also found in section four of Option C (Chapter 14), an option available only to Standard Level students.

Assessment statements

8.2.1 Draw and label a diagram showing the structure of a chloroplast as seen in electron micrographs.
8.2.2 State that photosynthesis consists of light-dependent and light-independent reactions.
8.2.3 Explain the light-dependent reactions.
8.2.4 Explain photophosphorylation in terms of chemiosmosis.
8.2.5 Explain the light-independent reactions.
8.2.6 Explain the relationship between the structure of the chloroplast and its functions.
8.2.7 Explain the relationship between the action spectrum and the absorption spectrum of photosynthetic pigments in green plants.
8.2.8 Explain the concept of limiting factors in photosynthesis, with reference to light intensity, temperature and concentration of carbon dioxide.

The chloroplast

Some people refer to the chloroplast as a photosynthetic machine. They are not wrong. Unlike respiration, where some of the steps occur outside the mitochondrion, all of the photosynthetic process occurs within the chloroplast. Chloroplasts, along with mitochondria, represent possible evidence for the theory of endosymbiosis. Both organelles have an extra outer membrane (indicating a need for protection in a potentially hostile environment), their own DNA, and they are very near in size to a typical prokaryotic cell (see Figure 8.11).

There are three types of plastid that occur in plant cells:
- chloroplasts are green and involved in photosynthesis;
- leucoplasts are white or 'clear' and function as energy store-houses;
- chromoplasts are brightly coloured and synthesize and store large amounts of orange, red, or yellow pigments.

All the plastids develop from a common proplastid.

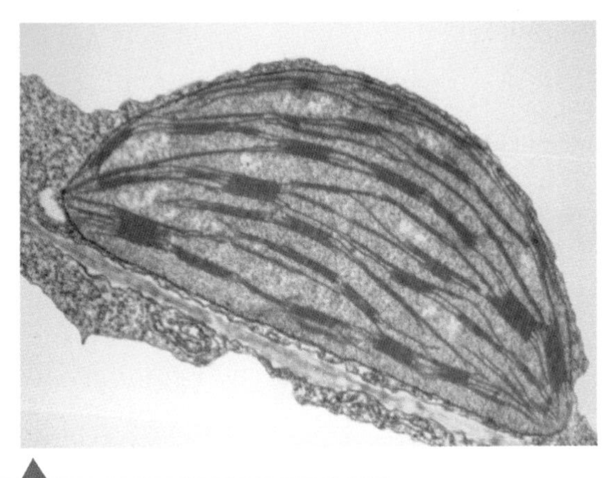

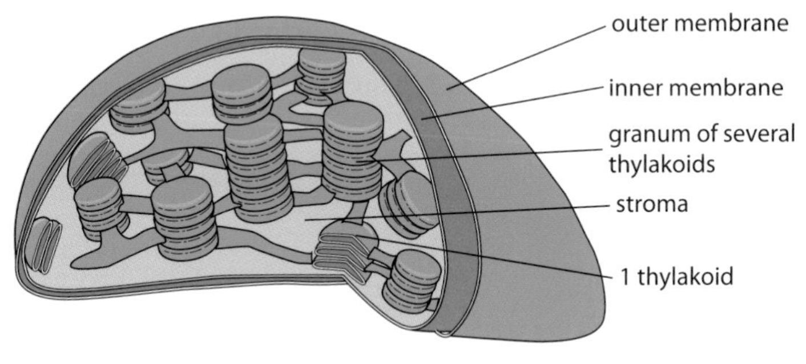

Figure 8.11 This false-colour TEM and drawing show the structure of a chloroplast. Can you find as many parts in the EM as are labelled in the drawing?

The structure of the chloroplast was discussed in Chapter 2 (page 25). You may want to return to that chapter for a brief refresher. Chloroplasts occur mostly within the cells of the photosynthetic factory of the plant, the leaves. However, some plants have chloroplasts in the cells of other organs.

The overall process of photosynthesis

During the discussion on respiration, we considered the means by which the cell breaks down chemical bonds in glucose to produce ATP. In this section, the discussion centres on the establishment of chemical bonds to produce organic compounds. Using light energy, the raw materials of photosynthesis are carbon dioxide and water. Many enzymes are involved to allow the formation of products that include glucose, more water, and oxygen. The overall equation is:

$$6CO_2 + 12H_2O \xrightarrow{\text{light}} C_6H_{12}O_6 + 6H_2O + 6O_2$$

Water occurs on both sides because 12 molecules are consumed and 6 molecules are produced. Clearly, photosynthesis is essentially the reverse of respiration. Whereas respiration is, in general, a catabolic process, photosynthesis is, in general, an anabolic process. Photosynthesis occurs in organisms referred to as autotrophs. These organisms make their own food. Non-photosynthetic and non-chemosynthetic organisms are referred to as heterotrophs. They must obtain their food (which is necessary for energy) from other organisms.

Photosynthesis involves two major stages:

- the light-dependent reaction;
- the light-independent reaction.

The light-dependent reaction

Light energy behaves as if it exists in discrete packets called photons. Shorter wavelengths of light have greater energy within their photons than longer wavelengths.

This reaction occurs in the thylakoids or grana of the chloroplast. A stack of thylakoids is called a granum (plural, grana). Light supplies the energy for this reaction to occur. The ultimate source of light is the Sun. Even though plants may survive quite well when they receive light from sources other than the Sun, most plants on our planet rely on the Sun for the energy necessary to drive photosynthesis.

To absorb light, plants have special molecules called pigments. There are several different pigments in plants and each effectively absorbs photons of light at different wavelengths. The two major groups are the chlorophylls and the carotenoids. These pigments are organized on the membranes of the thylakoids. The regions of organization are called photosystems and include:

- chlorophyll *a* molecules;
- accessory pigments;
- a protein matrix.

The reaction centre is the portion of the photosystem that contains:

- a pair of chlorophyll molecules;
- a matrix of protein;
- a primary electron acceptor.

Bacteria that carry out photosynthesis have only one type of photosystem. However, modern-day plants have two types of photosystem. Each absorbs light most efficiently at a different wavelength. Photosystem I is most efficient at 700 nanometres (nm). Photosystem II is most efficient at 680 nm. These two photosystems work together to bring about a non-cyclic electron transfer. Figure 8.12 shows the overall light-dependent reaction of photosynthesis involving non-cyclic photophosphorylation or non-cyclic electron flow.

These numbered descriptions refer to the numbered steps in Figure 8.12.

1 A photon of light is absorbed by a pigment in photosystem II and is transferred to other pigment molecules until it reaches one of the chlorophyll *a* (P680) molecules in the reaction centre. The photon energy excites one of the chlorophyll *a* electrons to a higher energy state.
2 This electron is captured by the primary acceptor of the reaction centre.
3 Water is split by an enzyme to produce electrons, hydrogen ions, and an oxygen atom. This process is driven by the energy from light and is called photolysis. The electrons are supplied one by one to the chlorophyll *a* molecules of the reaction centre.

Besides the non-cyclic electron pathway used to produce ATP by photophosphorylation, there is an alternative pathway involving a cyclic pathway. This cyclic pathway is discussed in the final section of this chapter.

Figure 8.12 This is the light-dependent reaction.

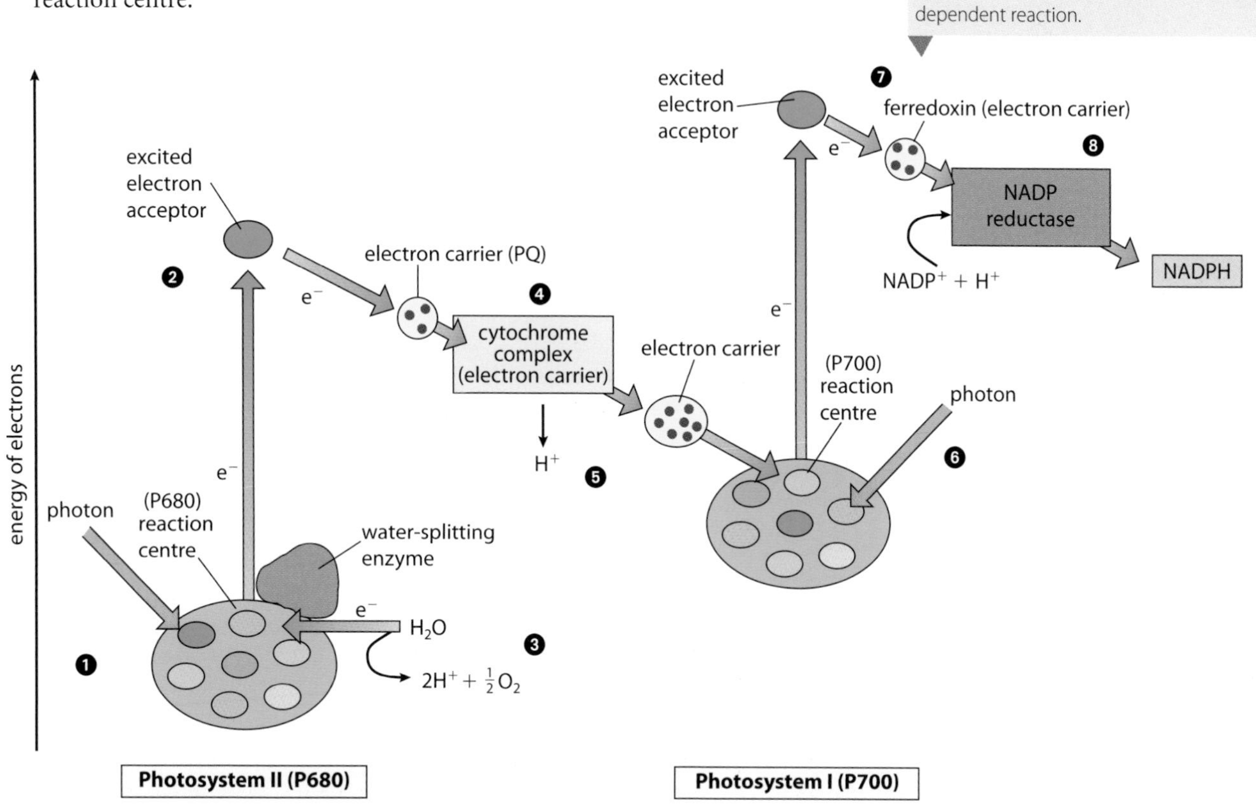

4 The excited electrons pass from the primary acceptor down an electron transport chain losing energy at each exchange. The first of the three carriers is plastoquinone (PQ). The middle carrier is a cytochrome complex.

5 The energy lost from the electrons moving down the electron transport chain drives chemiosmosis (similar to that in respiration) to bring about phosphorylation of ADP to produce ATP.

6 A photon of light is absorbed by a pigment in photosystem I. This energy is transferred through several accessory pigments until received by a chlorophyll *a* (P700) molecule. This results in an electron with a higher energy state being transferred to the primary electron acceptor. The de-energized electron from photosystem II fills the void left by the newly energized electron.

7 The electron with the higher energy state is passed down a second electron transport chain that involves the carrier ferredoxin.

8 The enzyme NADP reductase catalyses the transfer of the electron from ferredoxin to the energy carrier $NADP^{+}$. Two electrons are required to fully reduce $NADP^{+}$ to NADPH.

To view a useful website for understanding the light-dependent phase, visit heinemann.co.uk/hotlinks, enter the express code 4242P and click on Weblink 8.3.

NADPH and ATP are the final products of the light-dependent reaction. They supply chemical energy for the light-independent reaction to occur. The explanation above also shows the origin of the oxygen released by photosynthesizing plants (step 3). However, you need to know more detail about the production of ATP.

ATP production in photosynthesis is very similar to ATP production in respiration. Chemiosmosis allows the process of phosphorylation of ADP. In this case, the energy to drive chemiosmosis comes from light. As a result, we refer to the production of ATP in photosynthesis as photophosphorylation.

A comparison of chemiosmosis in respiration and photosynthesis is shown in this table.

Respiration chemiosmosis	Photosynthesis chemiosmosis
1 Involves an electron transport chain embedded in the membranes of the cristae.	1 Involves an electron transport chain embedded in the membranes of the thylakoids.
2 Energy is released when electrons are exchanged from one carrier to another.	2 Energy is released when electrons are exchanged from one carrier to another.
3 Released energy is used to actively pump hydrogen ions into the intermembrane space.	3 Released energy is used to actively pump hydrogen ions into the thylakoid space.
4 Hydrogen ions come from the matrix.	4 Hydrogen ions come from the stroma.
5 Hydrogen ions diffuse back into the matrix through the channels of ATP synthase.	5 Hydrogen ions diffuse back into the stroma through the channels of ATP synthase.
6 ATP synthase catalyses the oxidative phosphorylation of ADP to form ATP.	6 ATP synthase catalyses the photophosphorylation of ADP to form ATP.

In both cases, ATP synthase is embedded along with the carriers of the electron transport chain in the involved membranes.

In photosynthesis, the production of ATP occurs between photosystem II and photosystem I. Study Figure 8.13. Notice that the b_6-f complex, which is a cytochrome complex, pumps the hydrogen ions into the thylakoid space. This increases the concentration of these ions which then passively move through the ATP synthase channel providing the energy to phosphorylate ADP.

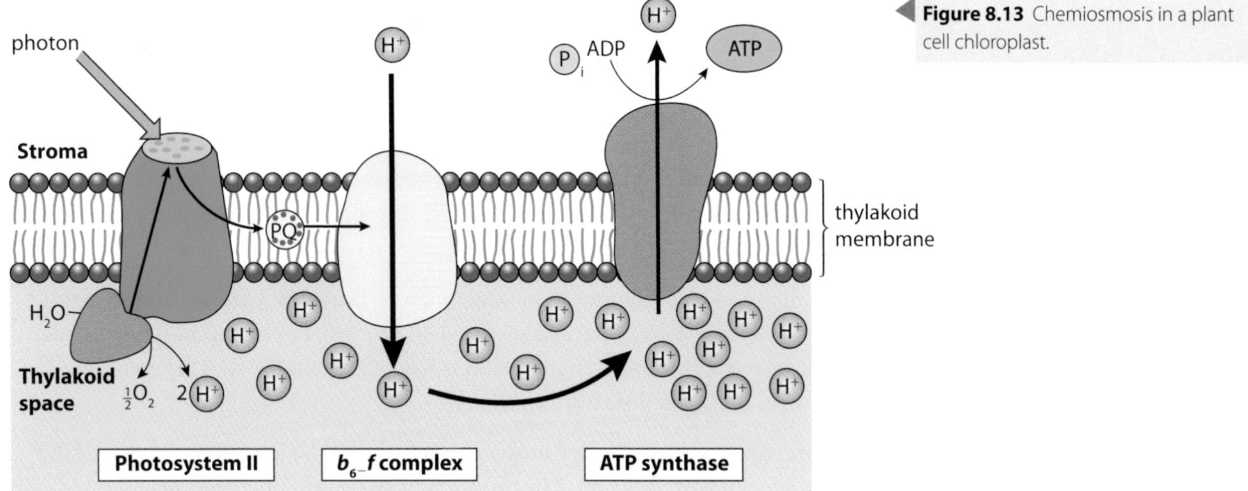

Figure 8.13 Chemiosmosis in a plant cell chloroplast.

The light-independent reaction

The light-independent reaction occurs within the stroma or cytosol-like region of the chloroplast.

The ATP and NADPH produced by the light-dependent reaction provide the energy and reducing power for the light-independent reaction to occur. Up to this point there has been no mention of carbohydrate production. Therefore, as we know glucose is a product of photosynthesis, the result of the light-independent reaction must be the production of glucose.

The light-independent reaction involves the Calvin cycle (see Figure 8.14), which occurs in the stroma of the chloroplast. Because it is a cycle, it begins and ends with the same substance. You should recall that a similar cyclic metabolic pathway occurred in respiration, the Krebs cycle.

1 Ribulose bisphosphate (RuBP), a 5-carbon compound, binds to an incoming carbon dioxide molecule in a process called carbon fixation. This fixation is catalysed by an enzyme called RuBP carboxylase (rubisco). The result is an unstable 6-carbon compound.

2 The unstable 6-carbon compound breaks down into two 3-carbon compounds called glycerate-3-phosphate.

3 The molecules of glycerate-3-phosphate are acted on by ATP and NADPH from the light-dependent reaction to form two more compounds called triose phosphate (TP). This is a reduction reaction.

Figure 8.14 The Calvin cycle. The numbered steps are described in the text.

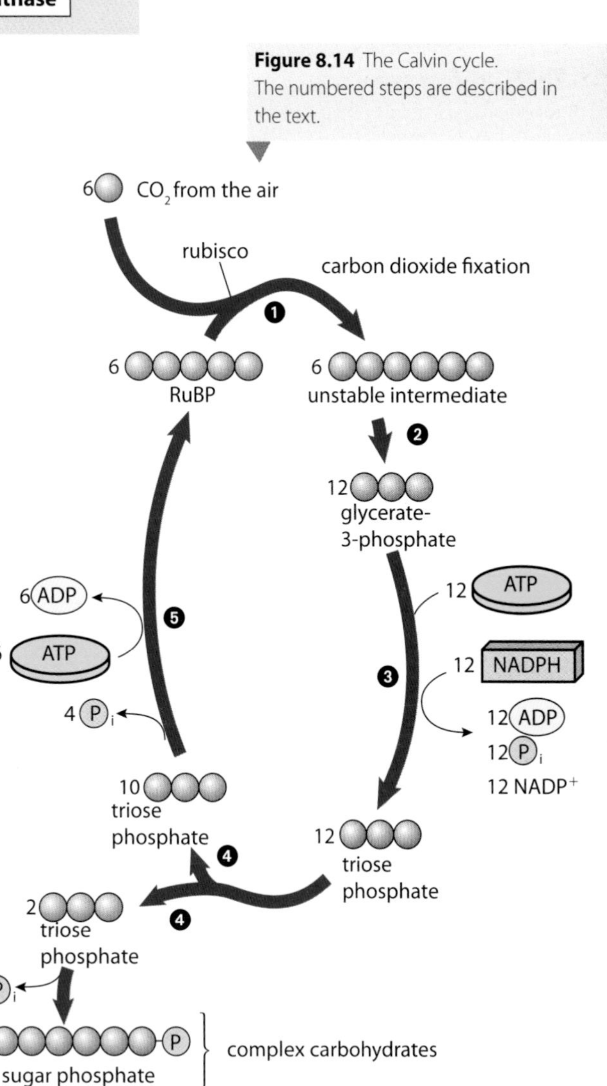

From 1945 to 1955, a team led by Melvin Calvin worked out the details of carbon fixation in a type of green algae. They used an elaborate protocol based on the 'lollipop apparatus.' How is the creation of a complex protocol such as this similar to the production of a work of art?

4 The molecules of TP may then go one of two directions. Some leave the cycle to become sugar phosphates that may become more complex carbohydrates. Most however, continue in the cycle to reproduce the originating compound of the cycle, RuBP.

5 In order to regain RuBP molecules from TP, the cycle uses ATP.

In Figure 8.14, spheres are used to represent the carbon atoms so they can be tracked through the cycle. The coefficients (numbers) in front of each compound involved show what it takes to produce one molecule of a 6-carbon sugar. It is clear that for every 12 TP molecules, the cycle produces one 6-carbon sugar and six molecules of the 5-carbon compound, RuBP. All the carbons are accounted for, and the law of conservation of mass is demonstrated. Also, it is important to note that 18 ATPs and 12 NADPH are necessary to produce 6 RuBP molecules and 1 molecule of a 6-carbon sugar.

To view a good explanation of the light-independent reaction, visit heinemann.co.uk/hotlinks, insert the express code 4242P and click on Weblink 8.4.

TP is the pivotal compound in the Calvin cycle. It may be used to produce simple sugars such as glucose, disaccharides such as sucrose, or polysaccharides such as cellulose or starch. However, most of it is used to regain the starting compound of the Calvin cycle, ribulose bisphosphate.

In summary, the process of photosynthesis includes the light-dependent and the light-independent reactions. The products of the light-dependent reaction are ATP and NADPH, which are required for the light-independent reaction to proceed. Thus, it is clear that light is needed for the light-independent reaction to occur, but not directly. Figure 8.15 summarizes the two reactions.

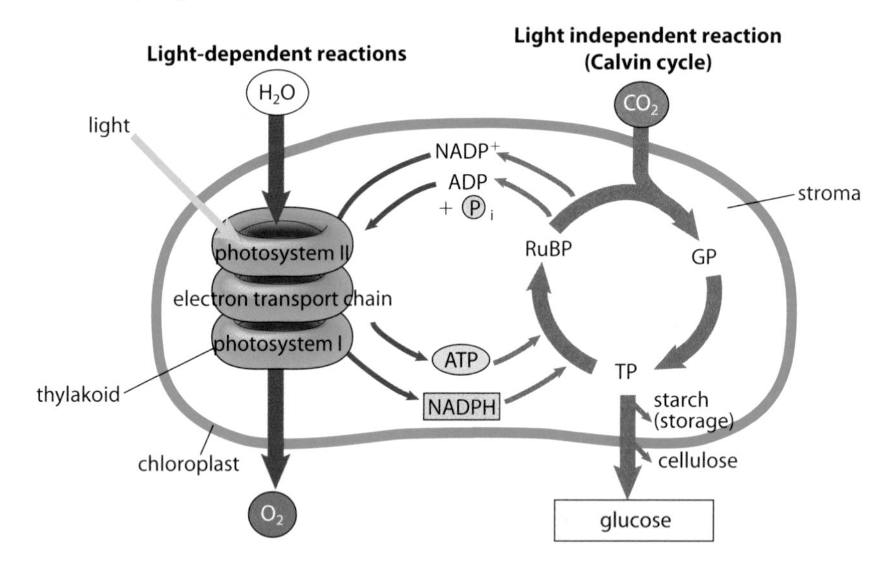

Figure 8.15 A summary of the complete process of photosynthesis.

Note that $NADP^+$ and ATP move back and forth in the chloroplast from the thylakoids to the stroma in their reduced and oxidized forms. A final summary of the two reactions is shown in this table.

Light-dependent reaction	Light-independent reaction
occurs in the thylakoids	occurs in the stroma
uses light energy to form ATP and NADPH	uses ATP and NADPH to form triose phosphate
splits water in photolysis to provide replacement electrons and H^+, and to release oxygen to the atmosphere	returns ADP, inorganic phosphate and NADP to the light-dependent reaction
includes two electron transport chains and photosystems I and II	involves the Calvin cycle

The chloroplast and photosynthesis

From the explanation of photosynthesis, it is clear how important the chloroplast is to the overall process. The structure of the chloroplast allows the light-dependent and light-independent reactions to proceed efficiently. In biology, the relationship of structure to function is a universal theme. The chloroplast and photosynthesis are no exception to this, as is shown in this table.

Chloroplast structure	Function allowed
extensive membrane surface area of the thylakoids	allows greater absorption of light by photosystems
small space (lumen) within the thylakoids	allows faster accumulation of protons to create a concentration gradient
stroma region similar to the cytosol of the cell	allows an area for the enzymes necessary for the Calvin cycle to work
double membrane on the outside	isolates the working parts and enzymes of the chloroplast from the surrounding cytosol

Action and absorption spectra of photosynthesis

The energy necessary for photosynthesis comes from light. Light is electromagnetic energy. It travels in rhythmic waves that have characteristic wavelengths. The entire range of radiation is referred to as the electromagnetic spectrum (see Figure 8.16). The specific part of this spectrum that is involved in photosynthesis is the visible light spectrum. This spectrum is visible to the human eye, and its wavelengths range from near 400 nanometres to near 740 nanometres. The shorter wavelengths of visible light have more energy than the longer wavelengths.

The wavelengths of the electromagnetic spectrum with high energy are absorbed by the ozone layer. The wavelengths with low energy are absorbed by water vapour and carbon dioxide in air.

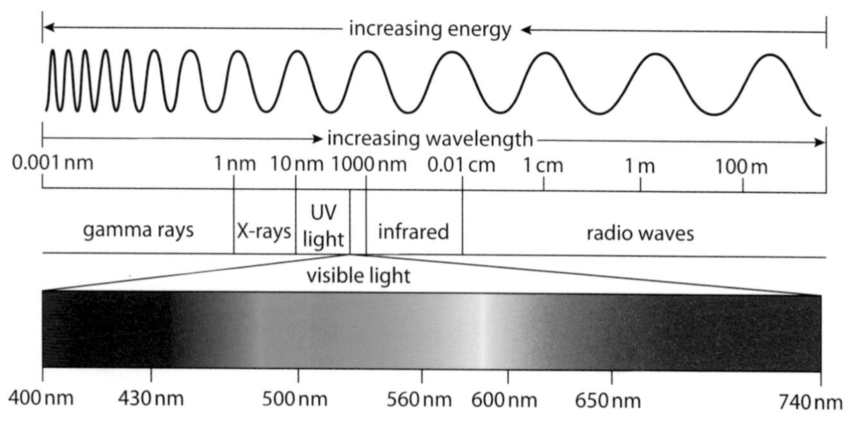

Figure 8.16 The electromagnetic spectrum.

Only about 42% of the sunlight directed towards Earth actually reaches the surface. Of this amount, only about 2% is used by plants. Of this amount, only 0.1–1.6% is incorporated into plant material.

The various pigments of photosynthesis absorb photons of light from specific wavelengths of the visible spectrum. If white light, which contains all the wavelengths of the visible light spectrum, is passed through the chloroplast of a plant cell, not all wavelengths are absorbed equally. This is because the specific pigments are present in the chloroplasts of that particular type of plant. A device called a spectrophotometer can be used to measure absorption at various light wavelengths. This results in a characteristic absorption spectrum for the plant. The absorption spectrum of a plant is the combination of all the absorption spectra of all the pigments in its chloroplasts. Figure 8.17 shows the absorption spectra of some typical photosynthetic pigments.

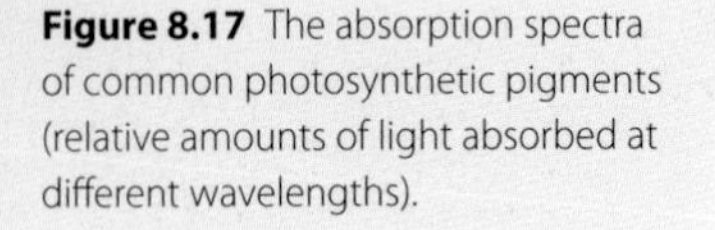

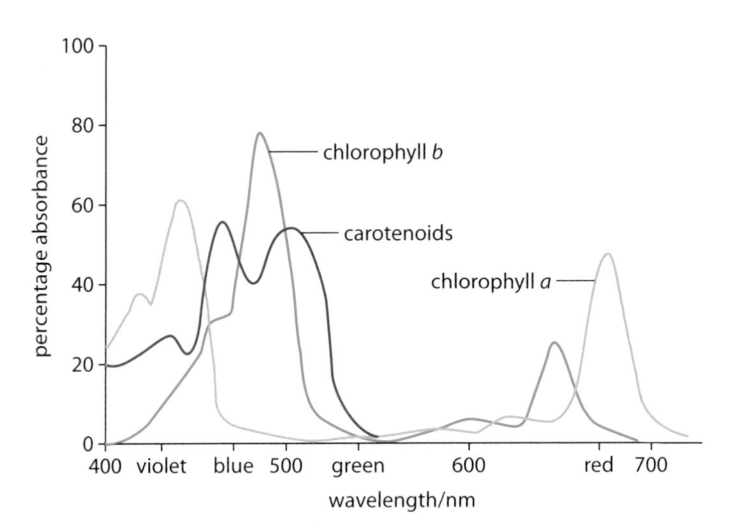

Figure 8.17 The absorption spectra of common photosynthetic pigments (relative amounts of light absorbed at different wavelengths).

Since light provides the energy to drive photosynthesis, the wavelength of the light absorbed by the chloroplasts partly determines the rate of photosynthesis. The rate of photosynthesis at particular wavelengths of visible light is referred to as the action spectrum. A common way of determining the rate of photosynthesis in order to produce an action spectrum is to measure oxygen production. High oxygen production indicates a high rate of photosynthesis. Figure 8.18 shows the action spectrum of a plant with the absorption spectra shown in Figure 8.17.

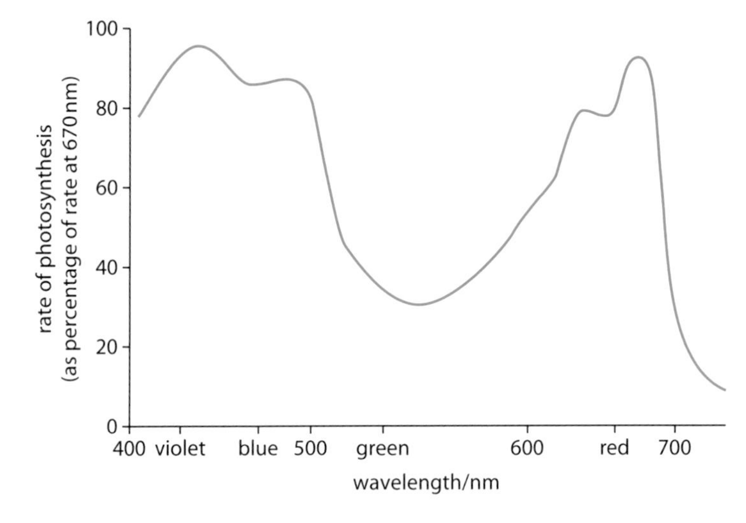

Figure 8.18 The action spectrum of photosynthesis (effectiveness of different wavelengths in fuelling photosynthesis). Note the positive correlation between this graph and Figure 8.17.

If you look at these two graphs, you can see two correlations:

- blue light and red light show the greatest absorption and they also represent the peaks in the rate of photosynthesis.
- the low absorption of green light corresponds to the lower rate of photosynthesis.

As we leave this topic, it is important to note that absorption and action spectra vary for different plants. This is due to the presence of different pigments and the different relative amounts of these pigments.

Factors affecting photosynthesis

The rate of photosynthesis can be affected by many factors. The term 'limiting factor' is used to describe the factor that controls any particular process (such as photosynthesis) at the minimum rate. Even though many factors have an effect, it is the limiting factor that actually controls the rate of the process. In photosynthesis, there are several factors that may be called limiting. Three of these are temperature, light intensity, and carbon dioxide concentration. At any one time in the life of a plant, one of these factors may be the limiting factor for photosynthesis. Whichever is the limiting factor, that factor will be altering the photosynthetic rate. Even if the other two factors change, their effect will not be evident.

Limiting factors are very evident in plant growth. Even though many different minerals are necessary for optimal plant growth, the one mineral in lowest percentage of the needed amount will limit plant growth and be referred to as the limiting factor.

The limiting factor has its effect and the others do not because of the complexity of the process. Photosynthesis has many steps that lead to the end-products. When one step is slowed by the limiting factor, the whole process is slowed. This particular step would be called the rate-limiting step. Figures 8.19, 8.20 and 8.21 show light, carbon dioxide and temperature in their roles as limiting factors. In each graph, photosynthetic rate is shown on the *y* axis. These graphs are obtained by altering only one variable at a time. This means we can draw a definite conclusion as to the cause of the change in photosynthetic rate. Try writing an explanation as to why the rate would be altered as shown for each graph.

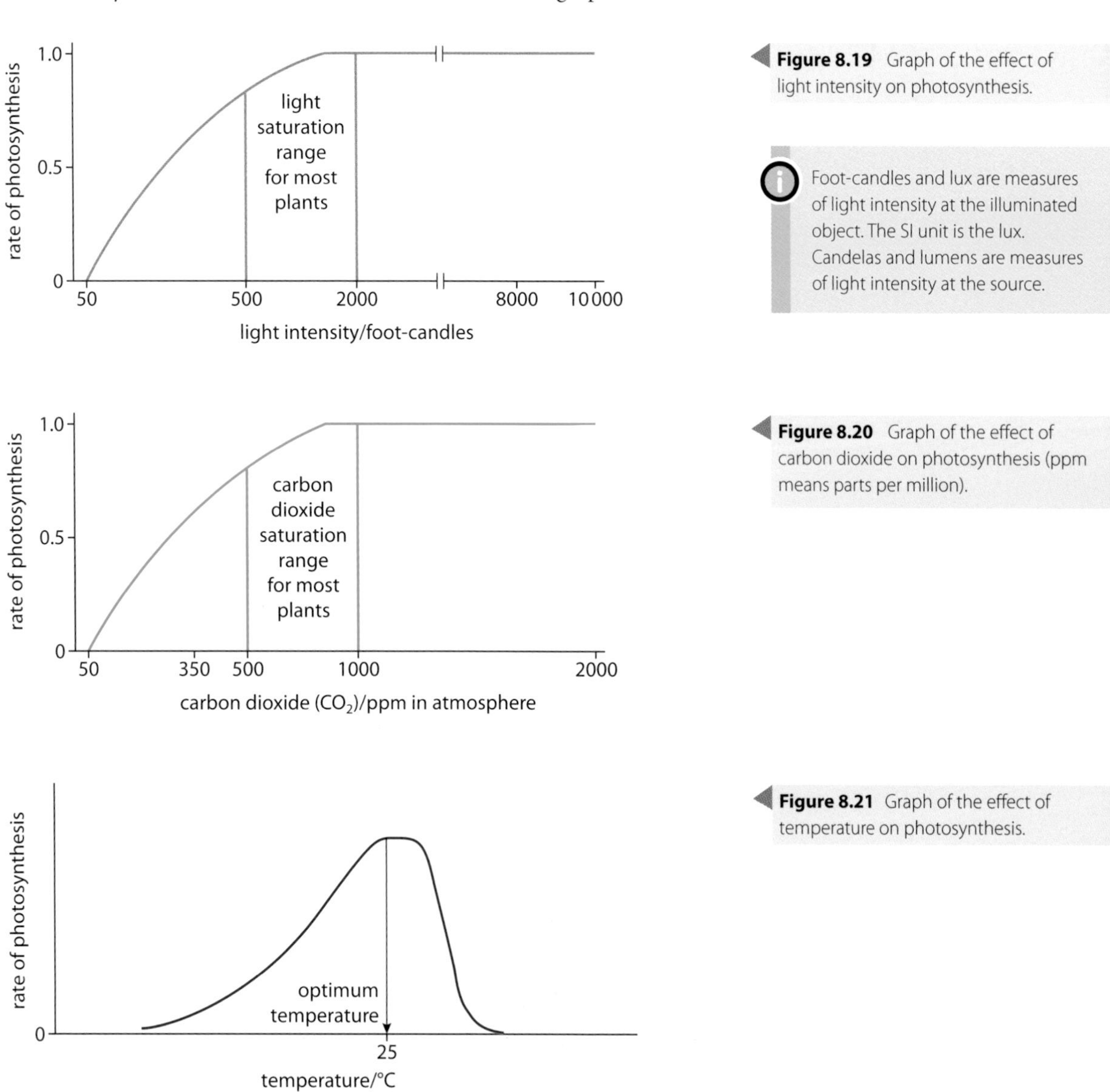

Figure 8.19 Graph of the effect of light intensity on photosynthesis.

Foot-candles and lux are measures of light intensity at the illuminated object. The SI unit is the lux. Candelas and lumens are measures of light intensity at the source.

Figure 8.20 Graph of the effect of carbon dioxide on photosynthesis (ppm means parts per million).

Figure 8.21 Graph of the effect of temperature on photosynthesis.

Controls are extremely important in experimentation. Changing one variable at a time makes it much easier to draw proper conclusions. In many research ventures, it is not always possible to control all variables.

For example, a study of middle-aged women may find that a group showed increased incidence of hypertension correlated with high sodium intake. However, the effect could be due to some other aspect of diet such as high cholesterol intake. Many medical and nutritional study results are published regularly. We must analyse the procedure and conclusions carefully.

Cyclic photophosphorylation

This is another way in which the light-dependent reaction of photosynthesis may produce ATP. It proceeds only when light is not a limiting factor and when there is an accumulation of NADPH in the chloroplast. In this process, light-energized electrons from photosystem I flow back to the cytochrome complex of the electron transport chain between photosystem II and photosystem I (see Figure 8.22). From the cytochrome complex, the electrons move down the remaining electron transport chain allowing ATP production via chemiosmosis. Thus, the electrons do not flow to the second electron transport chain that would produce NADPH. Obviously, this process is valuable since there is already an overabundance of NADPH that causes this different form of photophosphorylation to occur. The additional ATPs produced are shuttled to the Calvin cycle so that it can proceed more rapidly.

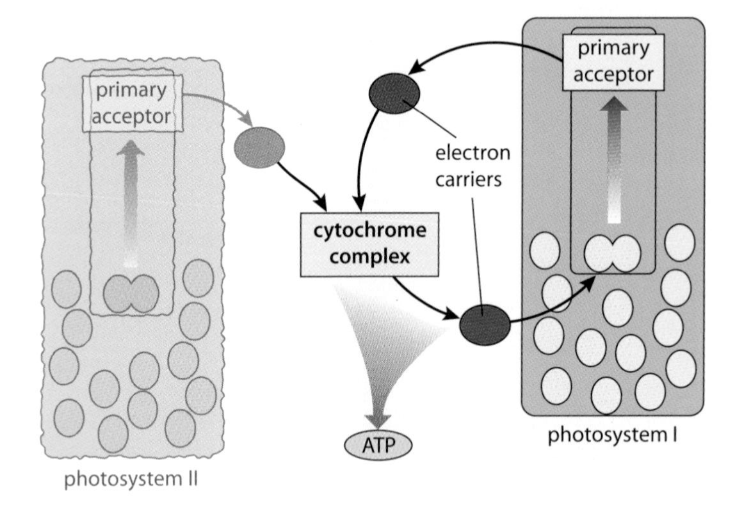

Figure 8.22 Cyclic photophosphorylation involving cyclic electron flow. Notice the electrons are cycled between photosystem I and the cytochrome complex. The result of this process is the production of ATP but no NADPH is produced.

Exercises

6 Why do plants need both mitochondria and chloroplasts?

7 You have a leaf from each of two very different plants. One leaf has more pigments than the other. Which leaf would have the greater photosynthetic rate, assuming all affecting factors are equal? Why?

8 Explain the final products of the two photosystems involved in the light-dependent reaction of photosynthesis.

9 Many scientists state that the enzyme RuBP carboxylase (rubisco) is the most ubiquitous protein on Earth. Why is there a very good chance that this is true?

10 How are the products of the light-dependent reaction important to the light-independent reaction?

Practice questions

1 Explain the process of aerobic respiration including oxidative phosphorylation. (*8 marks*)

2 Which two colours of light does chlorophyll absorb most?

A Red and yellow

B Green and blue

C Red and green

D Red and blue (*1 mark*)

3 Which is not a product of the Krebs cycle?

A CO_2

B $NADH + H^+$

C Pyruvate

D ATP (*1 mark*)

4 How many ATP molecules (net yield) are produced per molecule of glucose as a direct result of glycolysis?

A 2

B 4

C 10

D 38 (*1 mark*)

5 Which way do the protons flow when ATP is synthesized in mitochondria?

A From the inner matrix to the intermembrane space

B From the intermembrane space to the inner matrix

C From the intermembrane space to the cytoplasm

D From the cytoplasm to the intermembrane space (*1 mark*)

6 What accumulates in the inter-membrane space of the mitochondrion during electron transport?

A ATP

B Electrons

C Protons (hydrogen ions)

D Oxygen (*1 mark*)

7 Explain how the light-independent reactions of photosynthesis rely on light-dependent reactions. (*8 marks*)

8 Outline the effect of temperature, light intensity and carbon dioxide concentration on the rate of photosynthesis. (*6 marks*)

9 What is the sequence of stages during the conversion of glucose into pyruvate in glycolysis?

A lysis → phosphorylation of sugar → oxidation

B lysis → oxidation → phosphorylation of sugar

C phosphorylation of sugar → lysis → oxidation

D phosphorylation of sugar → oxidation → lysis (*1 mark*)

10 How is the proton gradient generated in chloroplasts during photosynthesis?

A Flow of electrons from carrier to carrier in the thylakoid membrane causes pumping of protons across the thylakoid membrane.

B Light causes protons to flow through protein channels in the thylakoid membrane.

C Light splits water molecules in the stroma, causing the release of protons.

D Protons are pumped across the thylakoid membrane using energy from ATP.

(*1 mark*)

9 Plant science

Introduction

It is obvious that plants are an extremely important part of life on Earth. They produce oxygen and carbohydrates while absorbing carbon dioxide from our atmosphere. The importance of oxygen and carbohydrates are clear. The removal of carbon dioxide is especially critical in the light of global warming. There are many, many other contributions of plants as well. They have been a key element in our past and will be equally important in our advancement into the future.

9.1 Plant structure and growth

Assessment statements

- 9.1.1 Draw and label plan diagrams to show the distribution of tissues in the stem and leaf of a dicotyledonous plant.
- 9.1.2 Outline three differences between the structures of dicotyledonous and monocotyledonous plants.
- 9.1.3 Explain the relationship between the distribution of tissues in the leaf and the functions of these tissues.
- 9.1.4 Identify modifications of roots, stems and leaves for different functions: bulbs, stem tubers, storage roots and tendrils.
- 9.1.5 State that dicotyledonous plants have apical and lateral meristems.
- 9.1.6 Compare growth due to apical and lateral meristems in dicotyledonous plants.
- 9.1.7 Explain the role of auxin in phototropism as an example of the control of plant growth.

This is a dogwood tree. It has many of the features of a flowering plant.

Tissue distribution in plant structures

There are many types of plant and in this chapter we are going to consider land plants. Land plants possess some common characteristics:

- they provide protection for their embryos which has increased over time;
- they have multicellular haploid and diploid phases;
- they can be compared by the presence or absence of conductive systems.

Land plants are divided into three major groups as shown in this table.

Plant group	Major features
non-vascular land plants	• no conducting tissue • often grouped together as bryophytes • usually small and grow close to the ground • include mosses, liverworts, and hornworts
seedless vascular plants	• well-developed vascular tissue • do not produce seeds • include horsetails, ferns, club mosses, and whisk ferns (were once large specimens, but most of today's representatives are relatively small)
seeded vascular plants	• most living plant species are in this group • seeds contain an embryo, a supply of nutrients, and a protective outer coat • have extensive vascular tissue and include some of the world's largest organisms

The seeded vascular plants are divided into:

- gymnosperms – have seeds that do not develop within an enclosed structure;
- angiosperms – have seeds that develop within a protective structure.

We are going to look at the angiosperms.

In general, angiosperms have three basic types of tissue:

- dermal tissue – this outer protective covering protects against physical agents and pathogenic organisms; it prevents water loss and may have specialized structures for various purposes;
- ground tissue – consists mostly of thin-walled cells that function in storage, photosynthesis, support and secretion;
- vascular tissue – xylem and phloem carry out long-distance conduction of water, minerals and nutrients within the plant and provide support.

W For great explanations and wonderful pictures of plant parts, visit heinemann.co.uk/hotlinks, insert the express code 4242P and click on Weblink 9.1.

These three tissue types all derive from meristematic tissue. Meristematic tissue is composed of aggregates of small cells that have the same function as stem cells in animals. When these cells divide, one cell remains meristematic while the other is free to differentiate and become a part of the plant body. By doing this, the population of meristematic cells is continually renewed. The cells that remain meristematic are referred to as initials, while the cells that begin differentiation are called derivatives.

Plants include organs just as complex animals do. The three major organs are roots, stems, and leaves. We are now ready to discuss how the various tissues are distributed in these organs.

Root tissues

The root's function is to absorb mineral ions and water from the soil. Roots anchor the plant and even provide food storage in some cases. They have an epidermis that forms a protective outer layer. The cortex is involved in conducting water from the soil to the interior vascular tissue. It may also be modified to carry out storage functions. There is an endodermis that surrounds the vascular tissue. The vascular tissue includes the xylem and the phloem that conduct water, nutrients, and mineral ions around the plant (see Figure 9.1).

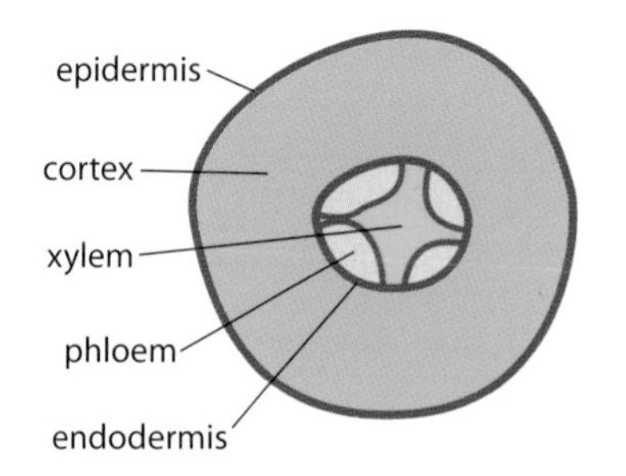

Figure 9.1 A typical plant root and its parts. Diagrams like this one of a root are plan diagrams, sometimes called low-power diagrams. The main purpose of such diagrams is to show the position of different tissues. Individual cells are not shown.

Stem tissues

The stem is the plant region where leaves are attached. The area where a leaf joins the stem is called the node, and the area between two nodes is called an internode. Leaves are arranged in a number of different ways on the stem. The arrangement of tissues in a stem is shown in Figure 9.2.

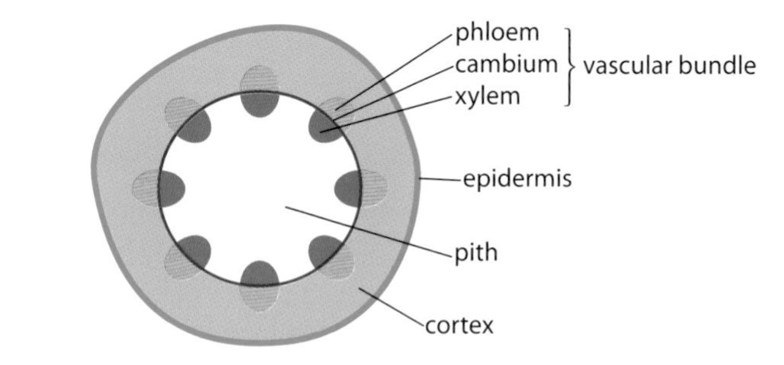

Figure 9.2 This is the distribution of tissues in a dicotyledonous stem.

The stem epidermis is mostly involved in protection. However, it may have pores (lenticels) that allow gas exchange. The transporting tissues of the stem include the xylem and phloem. They are arranged together in a circle. The xylem mainly carries water and dissolved minerals from the roots to the leaves. In woody plants, the xylem also provides support for the plant. The phloem transports organic nutrients throughout the plant. Notice in Figure 9.2 that the xylem and phloem occur together as groups and are separated by the cambium. The cambium is an area of rapidly dividing cells that differentiate into xylem and phloem. The cortex of the stem resembles that of the root. It supports and may have storage functions. The central pith region is a storage and support area. Turgid, fluid-filled cells in the cortex and pith offer support to the plant.

Leaf tissues

Leaves are involved in photosynthesis. They vary greatly in form but they generally consist of a flattened portion called the blade and a stalk called the petiole that attaches the blade to the stem. Tissue distribution in leaves is shown in Figure 9.3.

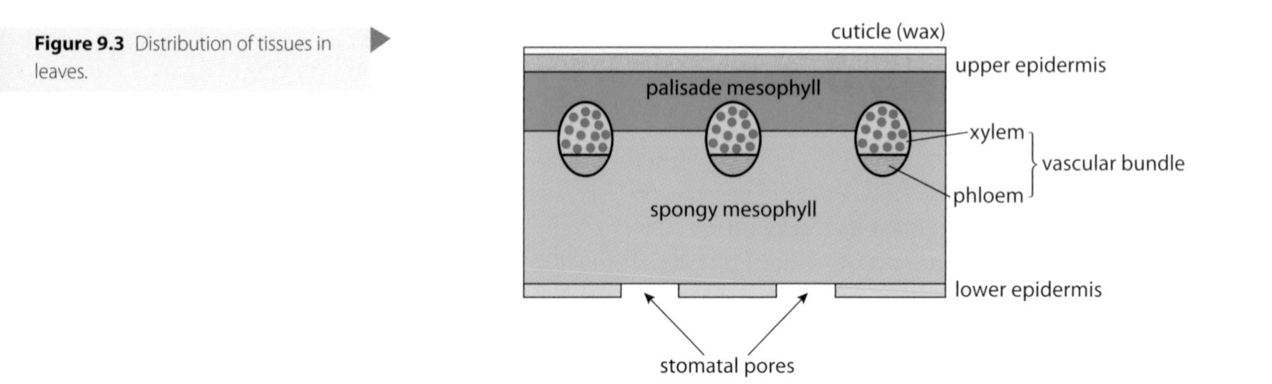

Figure 9.3 Distribution of tissues in leaves.

Many leaves have a layer of wax called the cuticle as their outermost layer. This layer protects against water loss and insect invasion. If a cuticle is not present, the outermost layer is the epidermis which protects. Like stems and roots, the leaves have vascular tissue which includes xylem and phloem. The xylem brings water to the leaves, while the phloem carries the products of photosynthesis to the rest of the plant. The xylem and phloem occur together in veins or vascular bundles. A densely packed region of cylindrical cells occurs in the upper portion of the leaf. This region is called the palisade mesophyll. The cells here contain large numbers of chloroplasts to carry out photosynthesis. The bottom portion of the leaf is composed of the spongy mesophyll. It consists of loosely packed cells with few chloroplasts. There are many air spaces in this area providing gas exchange surfaces. Stomata or stomatal pores occur on the bottom surface of leaves and they allow oxygen and carbon dioxide exchange. Specialized cells called guard cells control the opening and closing of the stomata.

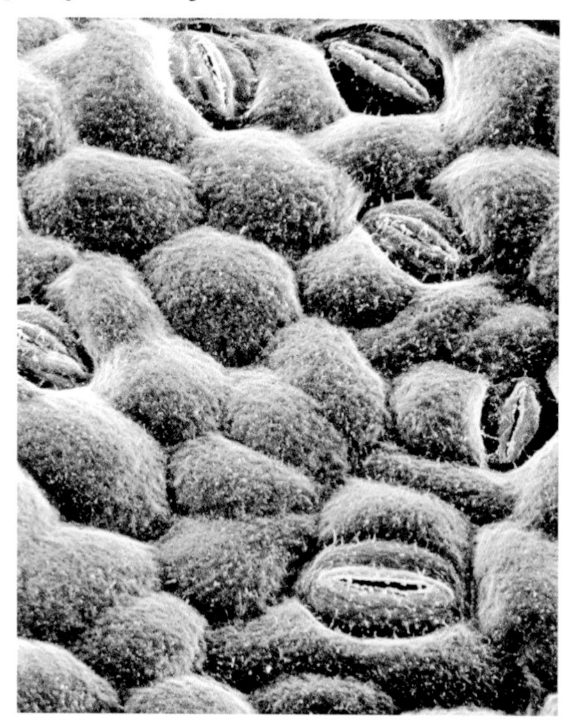

This is a lower leaf surface showing epidermal cells, stomata and guard cells.

It is important to note the functions of tissues in relation to their position in the leaf.

1 The palisade mesophyll is located in the upper portion of the leaf where light is most available. The cells of this region are chloroplast rich, thus allowing for maximal photosynthesis.

2 Veins are distributed throughout the leaf so as to transport raw materials and products of photosynthesis. The veins occur roughly in the middle of the leaf so as to be near all cells.

3 The spongy mesophyll is located just superior to the stomata allowing continuous channels for gas exchange.

4 The stomatal pores are on the bottom surface of the leaf. This area receives less light with a resulting lower temperature. The lower temperature minimizes water loss from the pores and the plant, thus the lower epidermis usually has a thinner cuticle than the upper epidermis. The positioning of the epidermis is such that the remaining structures of the leaf are protected and supported.

Monocotyledonous and dicotyledonous plants

You will recall that the land vascular plants we are considering are known as angiosperms. All angiosperms possess specialized structures called flowers. Most flowering plants coevolved with pollinators such as insects, bats, and birds. These animals allow transfer of the male pollen to the female reproductive portions of flowers so that fertilization and seed development within ovaries can occur.

For a long time, biologists have grouped angiosperms into two classes: the monocots (monocotyledonous plants) and the dicots (dicotyledonous plants). This division is based on morphological characteristics. Recently, new groupings have been emphasized due to a better understanding of the evolutionary development of the angiosperms. These new groupings involve analysis of DNA.

Based on phylogeny (evolutionary history), many biologists are now using three groups of angiosperms:

- the magnoliid complex (magnolias and laurels);
- the monocots;
- the eudicots (true dicots).

This system is likely to be dynamic as more information is analysed.

Here is a table of angiosperm differences based on morphological traits (see also Figure 9.4).

Monocots	Dicots
parallel venation in leaves	netlike venation pattern in leaves
3 flower parts or multiples of 3	4 or 5 flower parts or multiples of 4 or 5
seeds contain only one cotyledon (seed leaf)	seeds contain two cotyledons (seed leaves)
vascular bundles arranged throughout the stem	vascular bundles arranged as a ring in the stem
root system mainly fibrous	root system involves a taproot (main root)
pollen grain with one opening	pollen grain with three openings

Examiner's hint: You should be able to list and briefly describe at least three differences between monocots and dicots. Tables work well for this type of question.

Figure 9.4 Compare these drawings to the table of features differentiating between monocots and dicots.

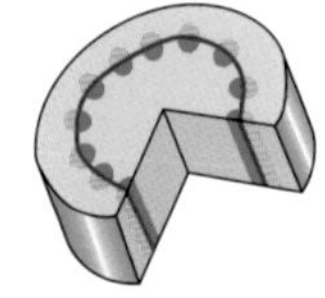

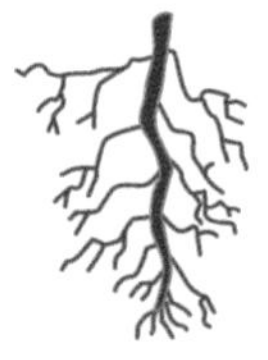

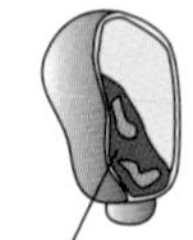

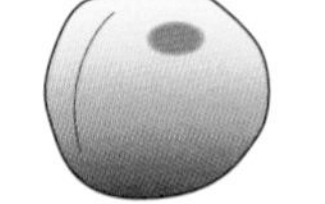

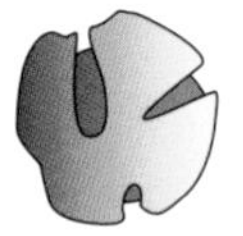

Plant organ modification

There are extensive modifications of roots, stems and leaves in different types of plant. These modifications allow plants growing in varied environments to survive.

Roots

There are two major types of root system: taproot systems and fibrous root systems. Plants with taproot systems have one main vertical root (the taproot) that develops from the embryonic root. When there is a fibrous root system, there is generally a group of thin roots spread out in the soil without a main central root.

Some possible modifications of roots are noted in this table.

Root modification	Description and example
prop roots	thick adventitious roots that grow from the lower part of the stem and brace the plant e.g. corn
storage roots	specialized cells within the root store large quantities of carbohydrates and water e.g. carrots and beets
pneumatophores (air roots)	produced by plants that live in wet places, these roots extend above the soil or water surface and facilitate oxygen uptake e.g. mangroves and cypress trees
buttress roots	large roots that develop near the bottom of trees to provide stability e.g. fig tree

Stems

Stems are capable of great modification. As they carry out the basic functions of conduction and support of photosynthetic organs, they often take on very different forms as shown in this table.

Stem modification	Description and example
bulbs	vertical, underground stems consisting of enlarged bases of leaves that store food e.g. onions
tubers	horizontally growing stems below ground that are modified as carbohydrate-storage structures e.g. potatoes
rhizomes	horizontal stems that grow just below the surface to allow plant spreading e.g. ginger plant
stolons	horizontal stems growing above ground that allow a plant to reproduce asexually e.g. strawberry plants

Leaves

Leaves also show great structural modification while maintaining their basic function of photosynthesis, as shown in this table.

Leaf modification	Description and example
tendrils	structures that coil around objects to aid in support and climbing (may also be formed from modified stems) e.g. pea plants produce tendrils from leaves
reproductive leaves	produce tiny plants along the leaf margins that fall to the ground and take root in the soil e.g. kalanchoe plants
bracts or floral leaves	coloured modified leaves that surround flowers and attract insects for pollination e.g. poinsettia
spines	reduce water loss, may be associated with modified stems that carry out photosynthesis e.g. cacti

With all these possible modifications, it is no wonder that plants occur in nearly all environments on our planet.

Meristems

Plants are different from most animals in that they show growth throughout their life. This continual pattern of growth is referred to as indeterminate. Animals (and even some plant organs such as leaves) exhibit determinate growth. This means they cease growing after reaching a certain size. Although plants continue to grow throughout their lives, they do die. Death occurs based on the plant's life cycle. Some plants are annuals and complete their life cycle in one year. Other plants are biennials and take two years to complete their life cycle before dying. Perennials live many years and when they die it is usually due to infection or some environmental factor.

Plants may show indeterminate growth because of their meristematic tissue. There are two types of this tissue in dicotyledonous plants: apical and lateral.

Apical meristems

Apical meristematic tissue, sometimes referred to as primary meristems, occurs at the tips of roots and stems. It produces primary tissues and causes primary growth (see Figure 9.5). Primary growth allows the root to extend throughout the soil. It also allows the stem to grow longer and so increases exposure to light and carbon dioxide. This type of growth results in herbaceous, non-woody stems and roots.

Lateral meristems

Lateral meristems allow growth in thickness of plants. This is referred to as secondary growth (see Figure 9.5). Most trees and shrubs (woody plants) have active lateral meristems. These plants have two types of lateral meristem.

- Vascular cambium produces secondary vascular tissue. It lies between the xylem and the phloem in the vascular bundles – on the inside it produces secondary xylem which is a major component of wood, on the outside it produces secondary phloem.
- Cork cambium occurs within the bark of a plant and produces the cork cells of the outer bark.

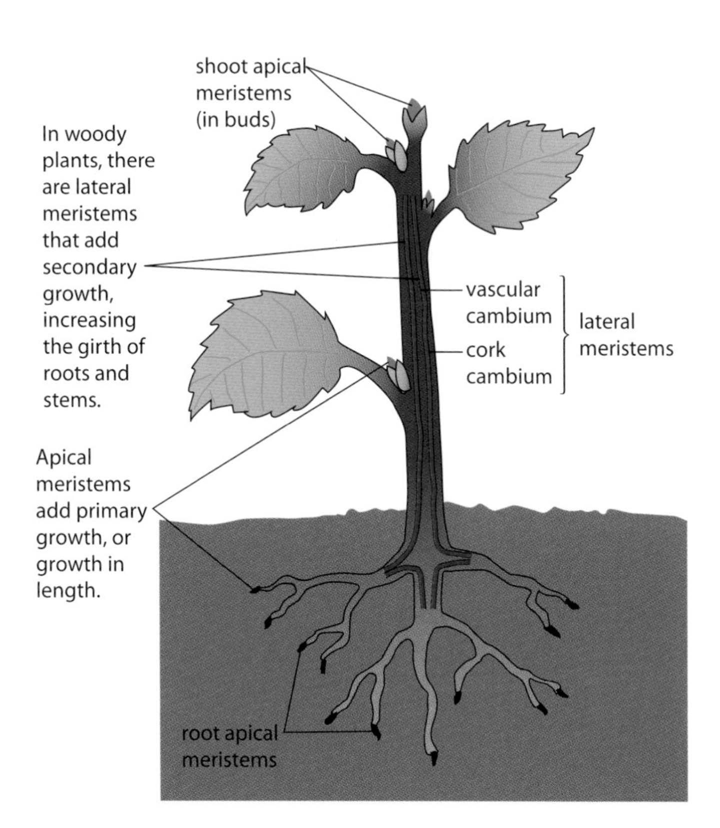
shoot apical meristems (in buds)
In woody plants, there are lateral meristems that add secondary growth, increasing the girth of roots and stems.
vascular cambium
cork cambium
lateral meristems
Apical meristems add primary growth, or growth in length.
root apical meristems

Figure 9.5 An overview of primary and secondary growth.

In seasonally growing plants, annual rings are often formed due to secondary growth. These rings may be used to determine the age of the plant, as well as to determine the relative climate of the growth year.

Auxins and phototropism

Tropisms are generally defined as growth or movement responses to directional external stimuli. Tropisms may be positive (towards the stimulus) or negative (away from the stimulus). Common stimuli for plant tropisms include chemicals, gravity, touch and light. Let's consider light as a stimulus. Phototropism means plant growth in response to light. Generally, plant stems exhibit positive phototropism and roots demonstrate negative phototropism (see Figure 9.6). Demonstrations of tropisms in plants are often simple to carry out in the laboratory.

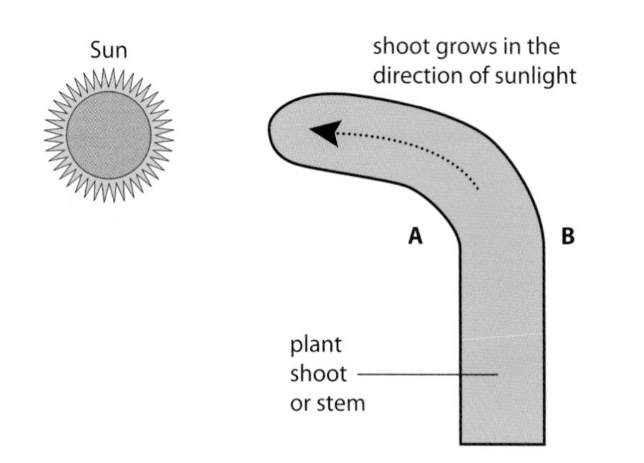
Sun
shoot grows in the direction of sunlight
A
B
plant shoot or stem

Figure 9.6 The effect of sunlight on a stem shoot. There is a higher concentration of auxin on side **B** creating increased elongation of these cells and thus growth toward the light.

The importance of phototropism to a plant is clear. If an area is crowded with plants, it is essential for seedlings to grow toward the sunlight so that photosynthesis may occur most efficiently. Auxins are plant hormones that cause the positive phototropism of plant shoots and seedlings.

Auxins are found in the embryos of seeds, the meristems of apical buds and young leaves. These hormones work only on plant cells that have auxin receptors. Auxins appear to increase the flexibility of plant cell walls in young developing shoots. This enables cell elongation on the side of the shoot necessary to cause growth towards the light.

This growth response does not appear to be due to an increased production of auxin on one side of the shoot. Rather, it seems to be due to a redistribution of available auxin especially to the stem side *away from* the light source. The result of this uneven distribution of auxin is a greater elongation of cells on the stem side away from the light and therefore curvature toward the light source. The specific plant auxin that causes this particular action is indoleacetic acid (IAA).

Exercises

1 Why would girdling, the removal of bark and vascular cambium in a narrow ring all the way around a tree, result in the death of a tree?

2 If you remove the apical meristem from a dicotyledonous plant, what would be the effect on further plant growth?

3 Of what value are tubers, a type of stem modification, to the survival of a plant species?

9.2 Transport in Angiospermophytes

Assessment statements

9.2.1 Outline how the root system provides a large surface area for mineral ion and water intake by means of branching and root hairs.

9.2.2 List ways in which mineral ions in the soil move to the root.

9.2.3 Explain the process of mineral ion absorption from the soil into roots by active transport.

9.2.4 State that terrestrial plants support themselves by means of thickened cellulose, cell turgor and lignified xylem.

9.2.5 Define *transpiration*.

9.2.6 Explain how water is carried by the transpiration stream, including the structure of xylem vessels, transpirational pull, cohesion, adhesion and evaporation.

9.2.7 State that guard cells can regulate transpiration by opening and closing stomata.

9.2.8 State that the plant hormone abscisic acid causes the closing of stomata.

9.2.9 Explain how the abiotic factors light, temperature, wind and humidity, affect the rate of transpiration in a typical terrestrial plant.

9.2.10 Outline four adaptations of xerophytes that help to reduce transpiration.

9.2.11 Outline the role of phloem in active translocation of sugars (sucrose) and amino acids from source (photosynthetic tissue and storage organs) to sink (fruit, seeds, roots).

Roots and angiosperm transport

As you will recall, roots show some very interesting and varied modifications. However, they always have the main function of providing mineral ion and water uptake for the plant. They are quite efficient in this function because of an extensive branching pattern and because of some specialized epidermal structures called root hairs (see Figure 9.7).

An adult rye plant was found to have a root system with 14 million branches totalling 630 kilometres (380 miles) in length

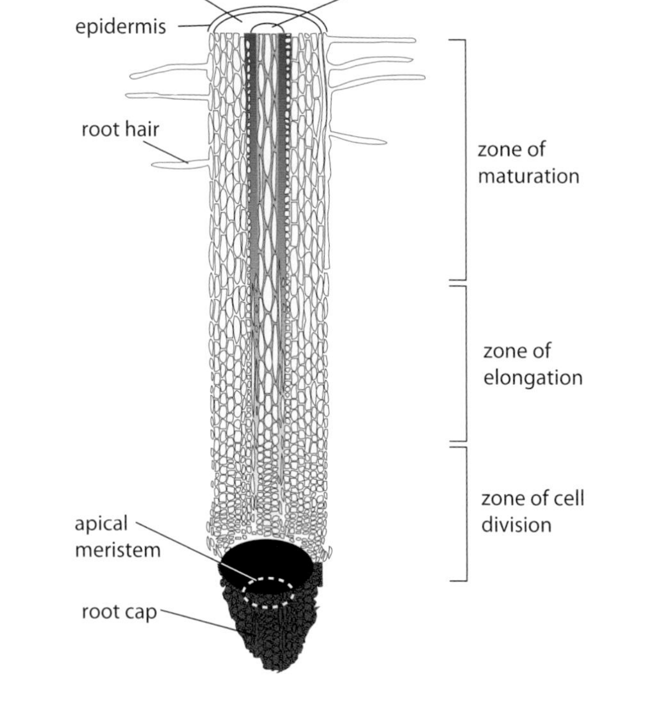

Figure 9.7 This is a root tip. Note the root hairs developing in the zone of maturation.

Root hairs increase the surface area over which water and mineral ions may be absorbed by a factor of nearly three. The root cap is very important in protecting the apical meristem during primary growth of the root through the soil. The three root zones indicate regions of cell development:

- zone of cell division – new undifferentiated cells are forming, M phase of the cell cycle;
- zone of elongation – cells are enlarging in size, corresponds to G_1 of the cell cycle;
- zone of maturation – cells are becoming functional to the plant.

Water

From Figure 9.8 (overleaf) it is evident that water must pass through several regions of the root before it enters the vascular cylinder.

epidermis → cortex → vascular cylinder

The vascular cylinder is immediately surrounded by the endodermis and the pericycle. The endodermis is a cylindrical layer of cells that separates the cortex from the vascular tissue. Just inside it is the pericycle. The pericycle is a layer of cells that can become meristematic. This is the layer that produces the lateral or branching roots of a plant. Lateral or branching roots must originate from this region to allow their connection to the vascular cylinder. The vascular cylinder includes the xylem and phloem that are involved in plant transport.

Many lands of the world have been made less productive because of agricultural over-use. Discuss ways these lands might be brought back into effective production. Also, discuss how presently productive lands could be maintained as such.

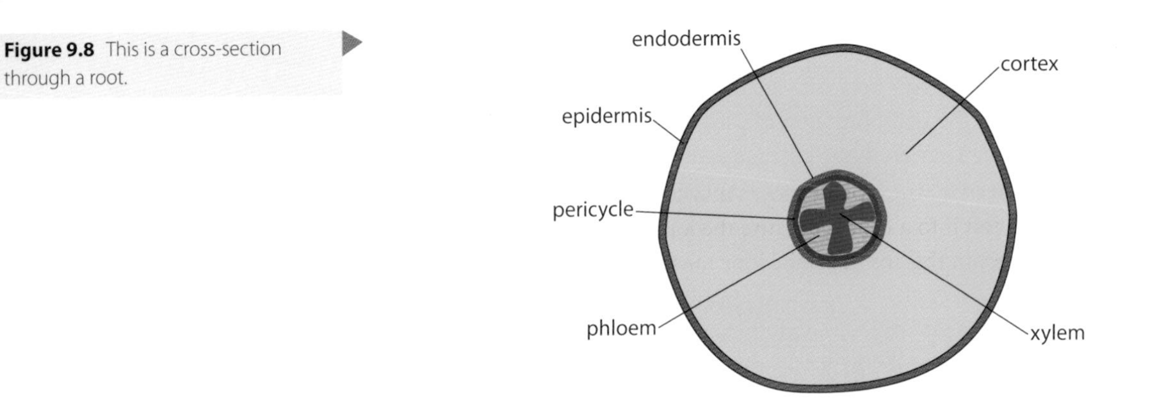

Figure 9.8 This is a cross-section through a root.

To help you gain an understanding of transport in plants, visit heinemann.co.uk/hotlinks, insert the express code 4242P and click on Weblink 9.2.

Water moves into the root hairs because they have a higher solute concentration and a lower water concentration than the surrounding soil. Therefore, the water moves through the plasma membranes into the root hair cells.

Most of the water entering a plant comes in through the root hairs by osmosis. Once in the root, water moves to the vascular cylinder by one of two possible routes:

- if the water moves from cell to cell, it is said to move by the symplastic route;
- if the water moves through the cell walls and the extracellular spaces, it is said to move by the apoplastic route.

Ions

It is essential that mineral ions move into the root as well as water. There are three major processes that allow mineral ions to pass from the soil to the root:

- diffusion of mineral ions and mass flow of water in the soil carrying these ions;
- aid provided by fungal hyphae;
- active transport.

When there is a higher concentration of a mineral outside the root than inside, the mineral moves into the root. These minerals are dissolved in and move via water. This also explains how the mass movement of water in the soil is able to carry these ions to the root.

The aid provided by fungal hyphae is unique. There is a symbiotic relationship between some roots and fungi. Large numbers of fungal filaments called hyphae form a cover over the surface of young roots. This creates an even larger surface area available for water and mineral ion absorption. This mutualistic relationship is referred to as a mycorrhiza.

Mycorrhizal fungi growing in association with roots.

Often there is a higher concentration of various mineral ions inside the plant than outside. In this situation, the passive means of transport mentioned so far are not useful. If the plant is to absorb these minerals, active transport is needed. This requires energy.

Another very common reason why a plant's roots may have to expend energy to get a particular mineral ion into a root is because the ion cannot cross the lipid bilayer of the membranes. In this instance, the ions must pass through a transport protein in the membrane. These transport proteins are specific for certain ions. They bind to the ion on one side of the membrane and then release it on the other side. This requires energy. Potassium ions move through specialized transport proteins called potassium channels.

The proton pump is the most important active transport protein in the plasma membranes of plant cells. This is how it works.

1 The proton pump uses energy from ATP to pump hydrogen ions out of the cell.
2 This results in a higher hydrogen ion concentration outside the cell than inside. This creates a negative charge inside the cell.
3 This gradient results in the diffusion of hydrogen ions back into the cell.
4 The voltage difference is called a membrane potential.
5 The hydrogen ion gradient and the membrane potential represent forms of potential energy that can be used to absorb mineral ions.

The proton pump may be used to transport mineral ions and solutes such as potassium ions, nitrogen-based ions, and even simple sugars.

This is a form of chemiosmosis, as talked about in respiration and photosynthesis (Chapter 8). In this case, however, the process is in reverse. Instead of ATP being formed by the movement of hydrogen ions, ATP is broken down to allow the hydrogen ions to move.

Support in terrestrial plants

Some terrestrial plants attain enormous sizes. There are three species of tree that are among the Earth's largest living things:

- the redwood sequoia of the west coast of the United States;
- the giant eucalyptus of the southern Australian and Tasmanian forests;
- the Douglas fir of the American Northwest.

Sequoiadendron giganteum, the redwood sequoia, is the tallest tree in the world. Individuals can grow to over 118 m (394 feet) and have a trunk diameter of over 10.5 m (36 feet). These trees require nearly 55 inches of rainfall a year to reach this size.

All three of these types of tree reach heights of over 90 m (300 feet). To grow to such a size, there has to be tremendous support within the organism. This support is provided mainly by thickened cellulose, cell turgor pressure, and lignified xylem.

The thickened cellulose occurs in the cell walls of the supporting regions of the plant. This alone is a very effective supportive adaptation. In addition to cellulose, the water-conducting cells, xylem, of terrestrial plants contain rings of a highly branched polymer called lignin. Lignified cells have much increased supportive capabilities. Turgor pressure is also important in keeping terrestrial plants upright. Turgor pressure is the pressure inside the cell that is exerted on the cell wall by the plasma membrane. This pressure is due to water that has entered the cell by osmosis. If the soil around a plant dries or gets too salty, water may no longer move into the plant. When this happens, the water content within the cells decreases and the plant wilts.

The cells that contain thickened cellulose and lignin occur in cylindrical arrangements within the plant. These act like strong rods to support the plant during high winds.

Transpiration, stomata and guard cells

Transpiration is the loss of water vapour from leaves and other aerial parts of the plant. The water is lost through openings called stomata (singular, stoma). The opening and closing of the stomata are due to specialized cells called guard cells. Transpired water has to be replaced by the intake of water at the roots. There is a continuous stream of water from the roots to the upper parts of the plant.

Transpiration

Water lost by transpiration is replaced by water absorption. More than 90% of the water taken in by the roots is lost by transpiration. This column of water throughout the plant provides minerals to the plant as well as the water necessary to carry out photosynthesis. Water lost by transpiration is important in cooling sun-drenched leaves and stems.

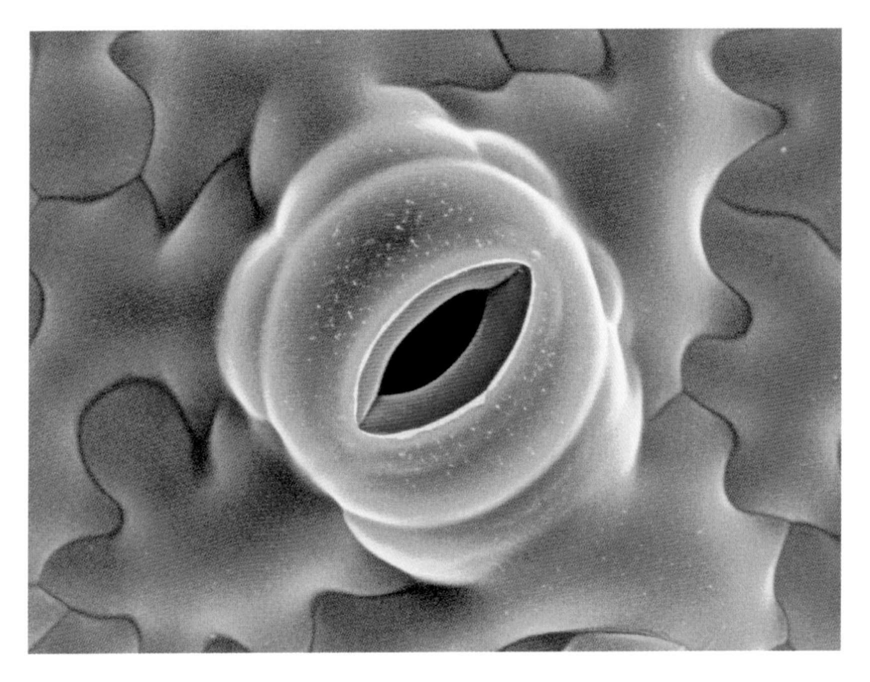

Guard cells and open stoma of a tobacco leaf.

The transpiration process is affected by a number of environmental factors, as summarized in this table.

Environmental factor	Effect
light	speeds up transpiration by warming the leaf and opening stomata
humidity	decreasing humidity increases transpiration because of the greater difference in water concentration
wind	increases the rate of transpiration because humid air near the stomata is carried away
temperature	increasing temperature causes greater transpiration because more water evaporates
soil water	if the intake of water at the roots does not keep up with transpiration, turgor loss occurs and the stomata close – this decreases transpiration
carbon dioxide	high carbon dioxide levels in the air around the plant usually cause the guard cells to lose turgor and the stomata to close

The amount of water loss to the plant because of transpiration can be enormous (see Figure 9.9). This loss could be a serious problem in plants growing in very dry conditions.

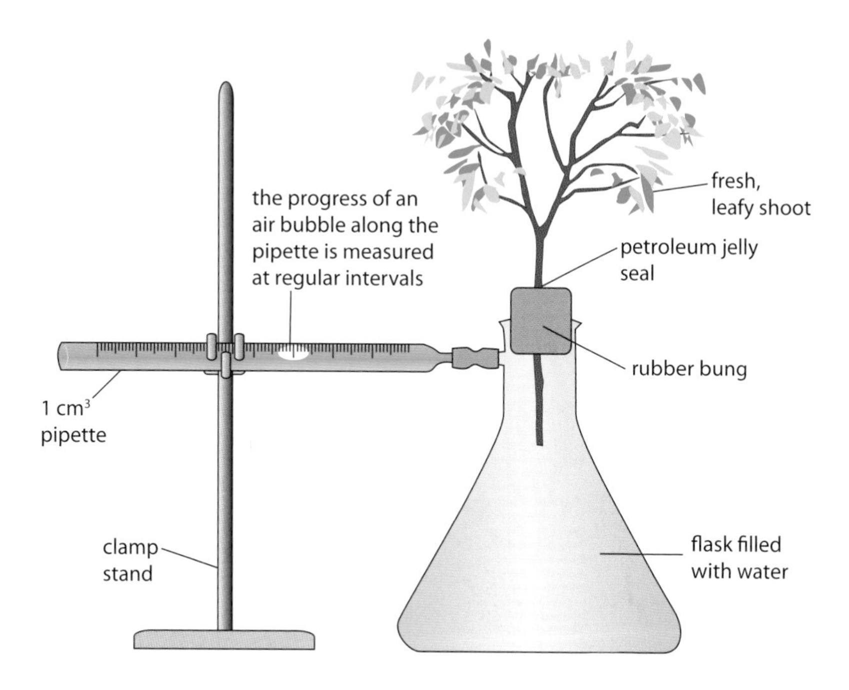

Figure 9.9 A potometer is a device for investigating transpiration rate (water loss per unit time). Transpiration is an interesting area to explore. Tests involve data logging using specialized sensors (to detect humidity, light, temperature, etc.) in the environment adjacent to the leaves of a plant.

Xerophytes are plants adapted to arid climates. They have an impressive list of modifications or adaptations to enable a decrease in transpirational water loss.

- Small, thick leaves reduce water loss by decreasing surface area.
- A reduced number of stomata decreases the openings through which water loss may occur.
- Stomata are located in crypts or pits on the leaf surface. This causes higher humidity near the stomata.
- These plants have a thickened, waxy cuticle.
- Hair-like cells on the leaf surface trap a layer of water vapour, thus maintaining a higher humidity near the stomata.
- Many desert plants shed their leaves in the driest months, and become dormant.
- Cacti exist on water the plant stores in fleshy stems. This water is restored in the rainy season.
- Alternative photosynthetic processes. There are two alternative processes called CAM photosynthesis and C_4 photosynthesis. CAM plants close stomata during the day and incorporate carbon dioxide during the night. C_4 plants have stomata open during the day but take in carbon dioxide more rapidly than non-specialised plants.

If higher temperatures result from global warming, what effect might global warming have on plant transpiration rates? How might this affect food availability in the world?

There is a final response on the part of the plant that can further decrease water loss. This is simply to close the stomata due to the action of the guard cells.

Stomata and guard cells

The closing of the stomata is only possible on a short-term basis. This is because carbon dioxide must enter the mesophyll region of the leaf so that photosynthesis can occur. The stomata open and close because of changes in the turgor pressure of the guard cells that surround them. These guard cells are cylindrical and are uneven in their cell wall thickness. As you can see in Figure 9.10, the thickened area of the guard cell wall is oriented toward the stoma. Thus, when the cells take in water and swell they bulge more to the outside. This opens the stoma. When the guard cells lose water, they sag toward each other and close the stoma.

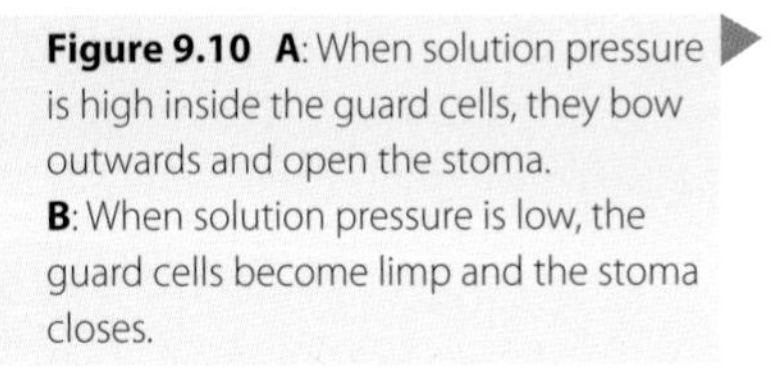

Figure 9.10 **A**: When solution pressure is high inside the guard cells, they bow outwards and open the stoma.
B: When solution pressure is low, the guard cells become limp and the stoma closes.

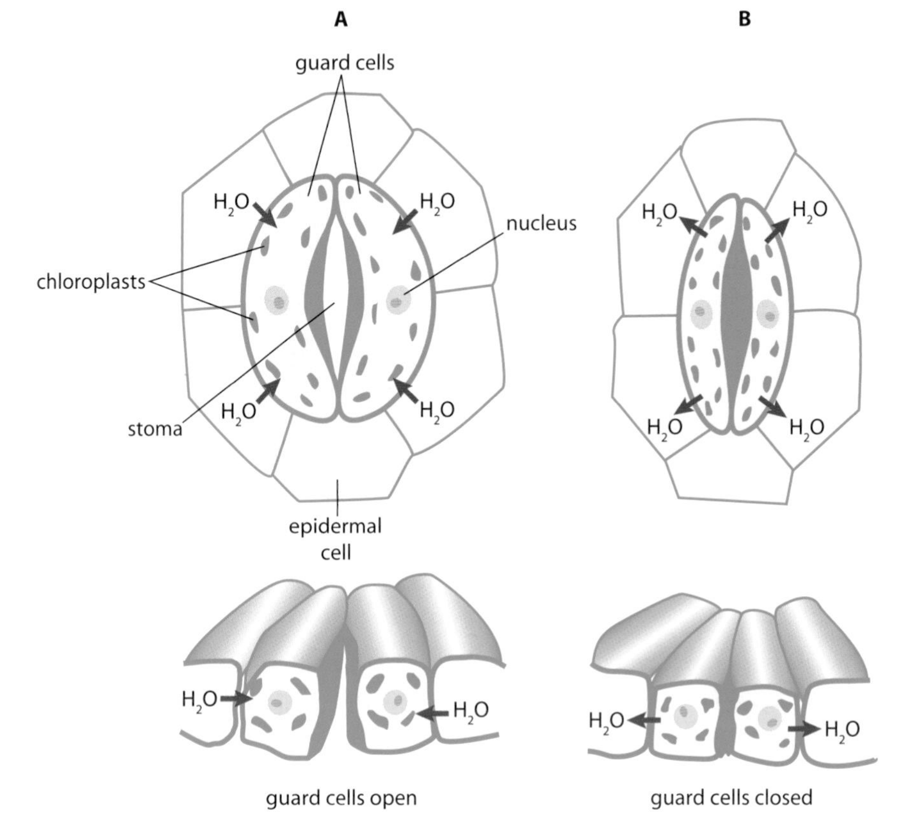

British research has shown that since 1927 there has been a decrease in stomatal density of many land plants. This is correlates with the increased atmospheric carbon dioxide levels in the 20th century due to the burning of fossil fuels.

The gain and loss of water in the guard cells is largely due to the transport of potassium ions. Light from the blue part of the spectrum triggers the activity of ATP-powered proton pumps in the plasma membrane of guard cells. This triggers the active transport of potassium into the cell. The higher solute concentration within the guard cells causes inward water movement due to osmosis.

When potassium ions passively leave the cells, water also leaves. The plant hormone abscisic acid causes potassium ions to rapidly diffuse out of the guard cells. The result is stomatal closure. This hormone is produced in the roots during times of water deficiency, for example in droughts.

Other factors such as carbon dioxide levels and even circadian rhythms (the basic 24-hour biological clock) within plants affect stomata opening and closing.

Plant water and mineral movement

There are many factors involved in the transport of water and minerals in plants. With the height of some plants, like the sequoias of western America, the transport of water from the roots to the tree top is a mammoth task. As you will recall, xylem is involved in the support as well as being the specialized water-conducting tissue of terrestrial plants.

Xylem is actually a complex tissue composed of many cell types. The two cell types largely involved in water and mineral transport are tracheids and vessel elements (see Figure 9.11). Tracheids are dead cells that taper at the ends and connect to one another to form a continuous column. Vessel elements (also called vessels) are the most important xylem cells involved in water transport. They are also dead cells and have thick, lignified secondary walls. These secondary walls are often interrupted by areas of primary wall. These areas also include pits or pores that allow water to move laterally. The vessel elements are attached end to end to form continuous columns like the tracheids. The ends of the vessel elements have perforations allowing water to move freely up the plant.

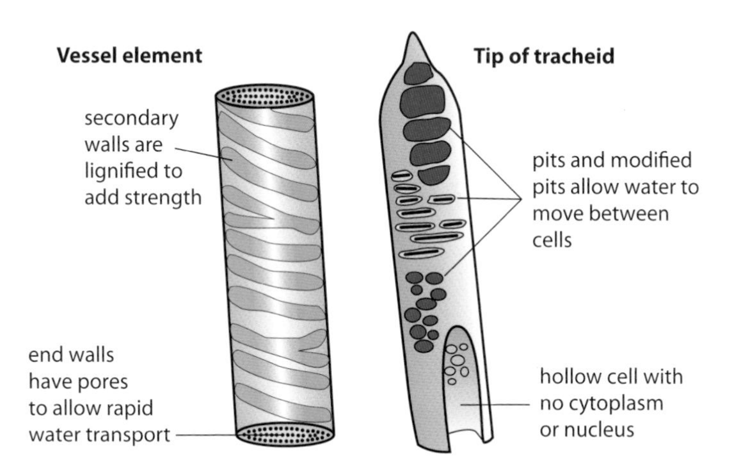

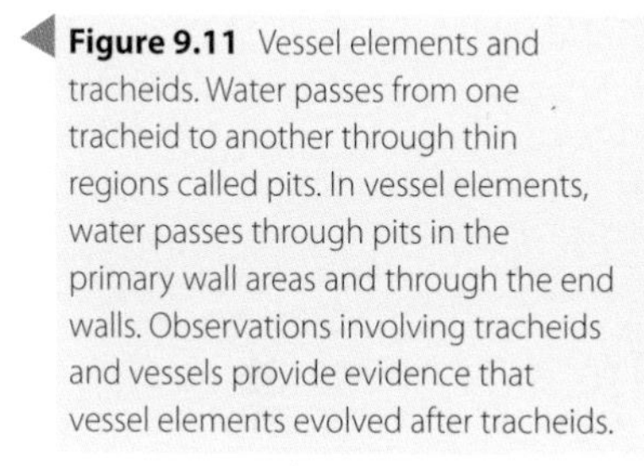
Figure 9.11 Vessel elements and tracheids. Water passes from one tracheid to another through thin regions called pits. In vessel elements, water passes through pits in the primary wall areas and through the end walls. Observations involving tracheids and vessels provide evidence that vessel elements evolved after tracheids.

Ancient flowering plants had only tracheids, while most modern flowering plants show vessel elements exclusively. Vessel elements appear to be more efficient in the transport of water.

The most widely accepted explanation for the movement of water and minerals upward in plants is referred to by several names. Some biologists call it the 'transpiration–cohesion–tension mechanism' while others call it simply the 'cohesion–tension theory'. This table and Figure 9.12 show you how it works.

Process	Explanation
1 Water moves down concentration gradients.	The spaces within the leaf have a high concentration of water vapour. Water moves from this location to the atmosphere which has a lower water concentration.
2 Water lost by transpiration is replaced by water from the vessels.	Replacing water from the vessels maintains a high water vapour concentration in the air spaces of the leaf.
3 Vessel water column is maintained due to cohesion and adhesion.	Water molecules form hydrogen bonds between one another in cohesion. Adhesion involves the hydrogen bonds that form between water molecules and the sides of the vessels; it counteracts gravity.
4 Tension occurs in the columns of water in the xylem.	This is due to the loss of water in the leaves and the replacement of it by xylem water. The columns remain continuous because of cohesion and adhesion.
5 Water is pulled from the root cortex into xylem cells.	Cohesion and adhesion maintain the columns under the tension created by transpiration.
6 Water is pulled from the soil into the roots.	This is due to tension created by transpiration and the maintenance of a continuous column of water.

Figure 9.12 Mechanism for upward movement of water in land plants.

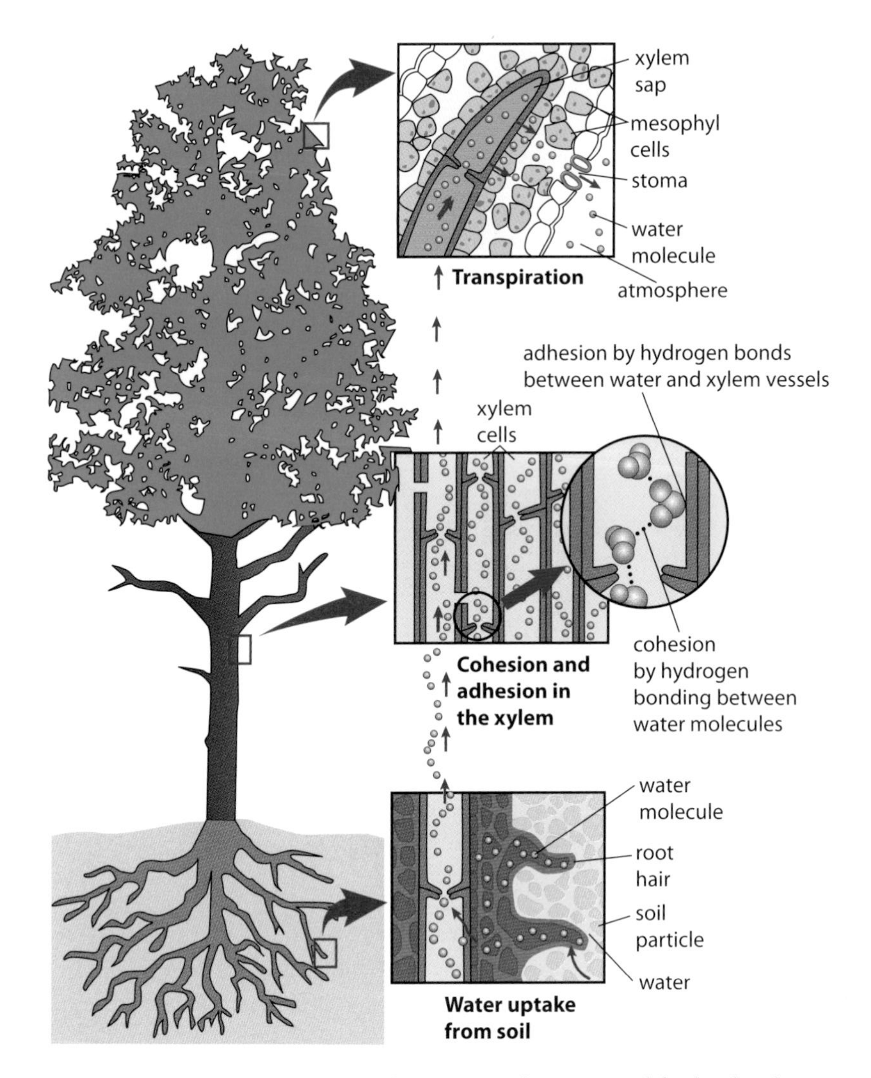

The movement of water upward in plants is mostly accounted for by the above explanation. However, besides this 'pull' there also appears to be a 'push' involved. This aiding push is root pressure. It occurs at night when ions are concentrated within the roots, thus causing water to move into the roots and into the vascular tissue. Research has shown that root pressure alone is insufficient to move water all the way up a tall tree.

The movement of organic molecules in plants

Organic molecules move via the phloem in plants. Unlike the xylem, phloem is made up of living cells – mostly sieve tube members and their companion cells (see Figure 9.13). Sieve tube members are connected to one another by sieve plates to form sieve tubes. The sieve plates have pores that allow the movement of water and dissolved organic molecules throughout the plant. Companion cells are actually connected to their sieve tube members by plasmodesmata.

The phloem lies near the outer surface of the plant stem. Earlier, we mentioned girdling (page 246). Girdling removes a section of the phloem completely around the tree. This prevents the flow of food from the leaves to the roots, and causes death of the tree.

Whereas xylem cells conduct water and minerals only upward from the roots, phloem cells transport their contents in various directions. However, the direction of movement is based on a single principle: the movement is from a source to a sink. A source is a plant organ that is a net producer of sugar either by photosynthesis or by the hydrolysis of starch. Leaves are the primary sugar sources. A sink is a plant organ that uses or stores sugar. Roots, buds, stems, seeds and fruits are all sugar sinks. It is possible for some structures to be both a source and a sink. For example, a tuber or bulb may be storing sugar or breaking down starch to provide sugar depending on the season; tubers and bulbs act as sinks in the summer and as sources in the early spring.

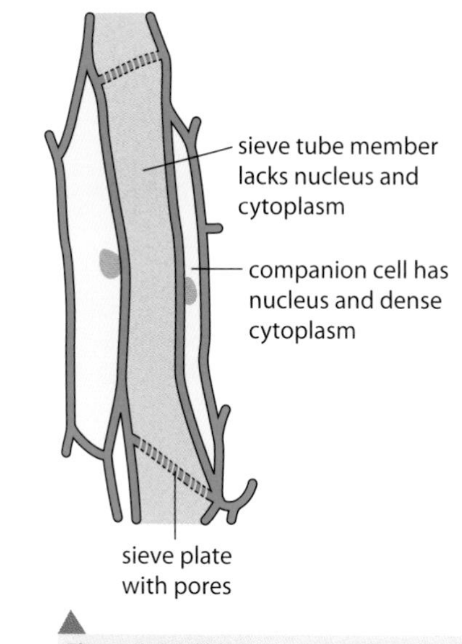

Figure 9.13 Phloem including the sieve tube member and accompanying companion cell.

The movement of organic molecules in plants is called translocation. The organic molecules are dissolved in water and the solution is referred to as phloem sap. The organic molecules of the phloem sap include:

- mostly sugars (sucrose is the most common);
- amino acids;
- plant hormones;
- mRNA (this is a recent finding and provides an explanation for communication between cells that are far apart in the plant).

The phloem sap can move as fast as 1 metre per hour. Radioactive tracers used to study this movement show that more than just diffusion and osmosis is involved. The best explanation presently for the movement of phloem sap is the pressure–flow hypothesis (see Figure 9.14). It includes the following processes.

1. Loading of sugar into the sieve tube at the source. This reduces the relative water concentration in the sieve tube members causing osmosis from the surrounding cells.
2. The uptake of water causes a positive pressure in the sieve tube that results in a flow (bulk flow) of the phloem sap.
3. This pressure is diminished by the removal of the sugar from the sieve tube at the sink. The sugars are changed at the sink to starch. Starch is insoluble and exerts no osmotic effect.
4. Xylem recycles the relatively pure water by carrying it from the sink back to the source.

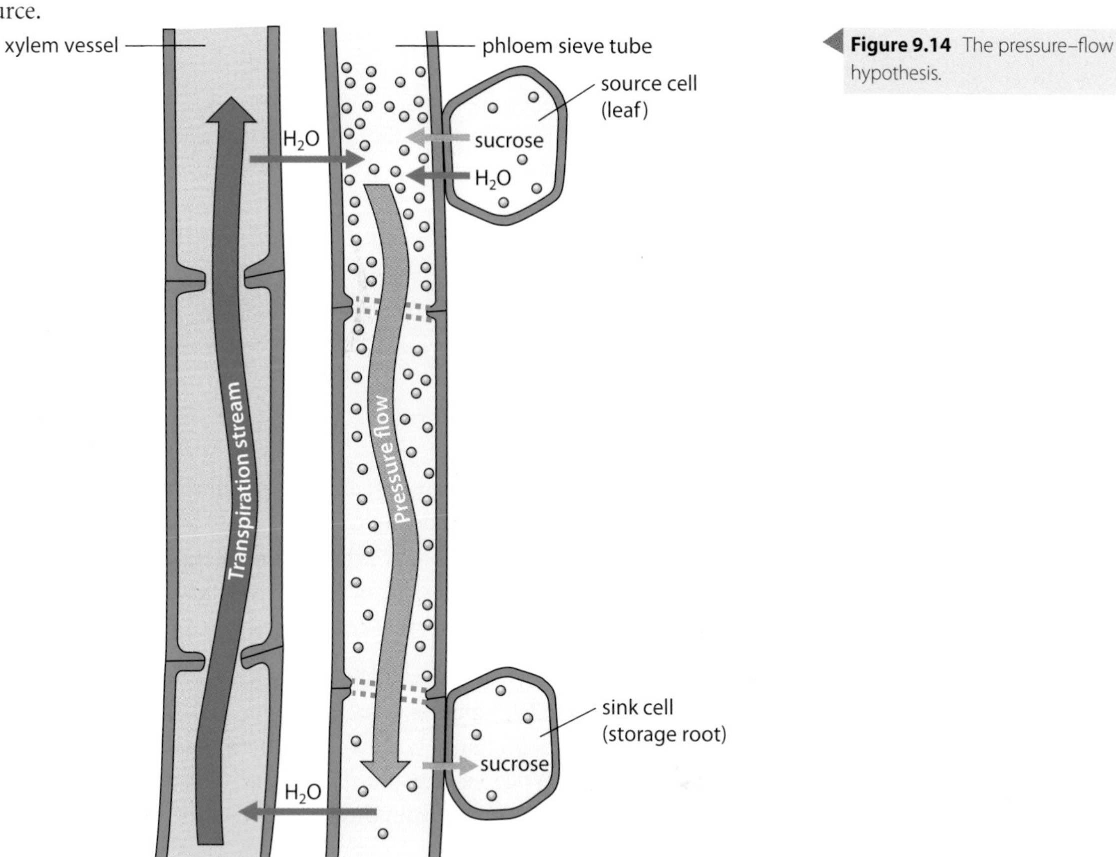

Figure 9.14 The pressure–flow hypothesis.

The loading of sugar into the sieve tube at the source and the removal of sugar at the sink is accomplished by active transport. This active transport is a chemiosmotic process involving proton pumps and cotransport proteins. The companion cells of the phloem are involved with the active transport process. *Only the loading and removal of sugar from the sieve tube members requires energy*, the actual transport in the tube is a passive process. It is essential to remember that phloem tissue occurs in all parts of the plant, not just the stem. The leaves are source regions.

The pressure that occurs within the phloem can be demonstrated with small insects called aphids. Aphids are able to penetrate a sieve tube member with their stylet (piercing mouthpart). If the stylet is then severed from the aphid, fluid from the sieve tube vessel can be seen coming out of the mouthpart remaining in the plant.

Exercises

4 When moving a plant from one place to another, why is it important to leave original soil around the roots?

5 What is the usual cause when a plant wilts?

6 Explain when a seed would be a sink and when it would be a source.

7 Why is it necessary that veins are relatively close together in leaves?

9.3 Reproduction in Angiospermophytes

Assessment statements

9.3.1 Draw and label a diagram showing the structure of a dicotyledonous animal-pollinated flower.
9.3.2 Distinguish between pollination, fertilization, and seed dispersal.
9.3.3 Draw and label a diagram showing the external and internal structure of a named dicotyledonous seed.
9.3.4 Explain the conditions needed for the germination of a typical seed.
9.3.5 Outline the metabolic processes during germination of a starchy seed.
9.3.6 Explain how flowering is controlled in long-day and short-day plants, including the role of phytochrome.

Variety in flowers

You only have to enter a flower shop or walk through a field to appreciate the tremendous variety and beauty of the reproductive structures of plants. Of course, we are talking about flowers. The flower is the hardworking and very successful reproductive structure of the angiosperms. Flowers vary greatly in size. One of the smallest is the size of a sesame seed and occurs in the aquatic plant *Wolffia*.

At the other end of the scale is the jungle flower *Rafflesia arnoldii*. This plant grows in Southeast Asia. When fully mature, it measures up to 0.9 m (3 feet) across and can weigh as much as 7 kilograms (14.5 lbs). Like many flowers, it depends on insects for pollination. When mature, this flower smells like rotting meat thus attracting flies that transfer pollen from the male reproductive structures to the female structures.

Flower structure and function

When we considered the characteristics of dicotyledonous plants, we noted that they had flower parts in 4s or 5s or multiples of these numbers. The flower parts are shown in Figure 9.15 and their functions summarized in this table.

Flower part	Function
sepals	protect the developing flower while in the bud
petals	often are colourful to attract pollinators
anther	part of stamen which produces the male sex cells, pollen
filament	stalk of stamen that holds up the anther
stigma	sticky top of carpel on which pollen lands
style	structure of the carpel that supports the stigma
ovary	base of carpel in which the female sex cells develop

Field of wild flowers in Texas, USA.

The entire female part of the flower is called the carpel. In some cases, the term 'pistil' is used to refer to a single carpel or a group of fused carpels. The entire male part of the flower is called the stamen.
Flowers occur in a myriad of colours, shapes and types:

- complete flowers contain all four basic flower parts – sepals, petals, stamen, and carpel;
- incomplete flowers lack at least one of these parts;
- staminate flowers have only stamens, not carpels;
- carpellate flowers only have carpels.

Meiosis occurs in the stamen and the carpel to produce the sex cells.

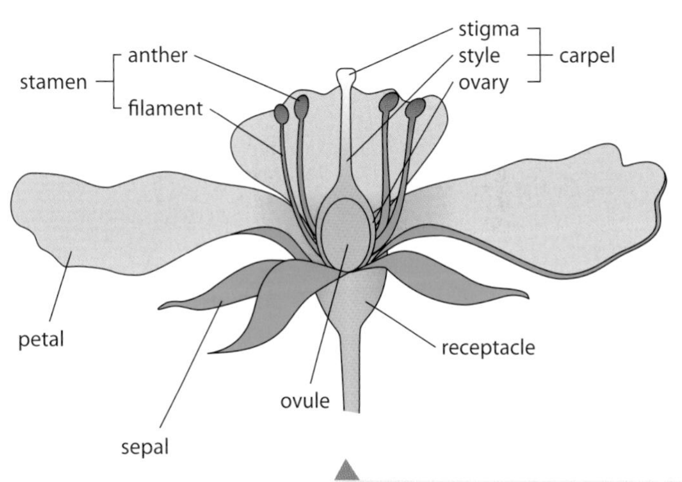

Figure 9.15 An animal-pollinated dicotyledonous flower typically shows these structures.

Pollination and fertilization

All plants show two different generations in their life cycle:

- the gametophyte generation, which is haploid;
- the sporophyte generation, which is diploid.

In plants, these two generations alternate with one another. Not surprisingly, this is called alternation of generations. The generations are named according to the reproductive cells they produce. The gametophyte generation produces the plant gametes by mitosis, whereas the sporophyte generation produces spores by meiosis. When we look at a flowering plant such as a cherry tree, we are looking at the sporophyte generation. It grew from a zygote and produces new cells by mitosis. When the cherry tree produces flowers, haploid spores are formed and develop into the haploid bodies referred to as gametophytes. Sperm form within the male gametophytes, and eggs form within the female gametophytes.
Pollination and fertilization are two very different processes in plants. Let's consider pollination first.

You can find a visual representation of a plant's life cycle if you visit heinemann.co.uk/hotlinks, insert the express code 4242P and click on Weblink 9.3.

Pollination

Pollination is the process by which pollen (containing male sex cells) is placed on the female stigma. It is the first step in the progression toward fertilization and the production of seeds. Pollen can be carried from anther to stigma by a variety of means. The early seed plants relied on wind as their pollen vector. Later, insects became a major factor in the process. It appears the first angiosperms were actually pollinated by insects. There is very convincing fossil evidence showing that the angiosperms and insects coevolved. They appear to be instrumental in each other's development.

There are, however, other vectors of pollination. These include birds, water and animals other than insects. Flowers of plants that involve insect or other animal pollinators employ various means of attraction:

- red flowers are conspicuous to birds;
- yellow and orange flowers are noticed by bees;
- heavily scented flowers are easily found by nocturnal animals;
- plants that rely on wind as their pollen vector have inconspicuous, odourless flowers.

Allergic rhinitis, commonly referred to as hay fever, is a distressing condition that is both common and widespread. It is often due to pollen in the air. People with this condition have an immune response to the proteins that project from the outer surface of the pollen.

The pollen that causes hay fever is usually from wind-pollinated plants and is, therefore, very light. Pollen from insect-attracting flowers is relatively heavy and is unlikely to be in the air and be a cause of allergic rhinitis.

There are two general types of pollination:

- self-pollination;
- cross-pollination.

In self-pollination, pollen from the anther of the same plant falls onto its own stigma. Self-pollination is a form of inbreeding and results in less genetic variation within a species.

When cross-pollination occurs, pollen is carried from the anther of one plant to the stigma of a different plant. Cross-pollination increases variation and may result in offspring with better fitness. The problem with cross-pollination is that the female stigma may not receive the male pollen because of the longer distance to travel.

Botanists may select plant genetic characteristics by controlling the process of pollination. Gregor Mendel controlled the process of pollination in garden pea plants in the development of his genetic principles.

Once pollination has occurred, the next step is fertilization.

Fertilization

Fertilization happens when the male and female sex cells unite to form a diploid zygote. The female sex cells that are fertilized by the pollen are present within the ovules of the flower. The ovules are present within the ovary of the carpel. When the pollen grain adheres to the stigma, which is covered by a sticky, sugary substance, it begins to grow a pollen tube. Pollen tube growth and fertilization occur in the following steps.

1. Pollen germinates to produce a pollen tube.
2. The pollen tube grows down the style of the carpel.
3. Within the growing pollen tube is the nucleus that will produce the sperm.
4. The pollen tube completes its growth by entering an opening at the bottom of the ovary.
5. The sperm moves from the tube to combine with the egg of the ovule to form a zygote.

Once the zygote is formed, it develops with the surrounding tissue into the seed. As the seed is developing, the ovary around the ovule matures into a fruit. The fruit encloses and helps to protect the seed.

Fertilization in the flowering plants is actually a double fertilization. One of the two sperms produced combines with the egg. The other combines with two polar nuclei within the ovary to produce the triploid ($3n$) endosperm. Endosperm has the function of storing nutrients for the early plant.

The seed

The seed is the means by which an embryo can be dispersed to distant locations. It is a protective structure for the embryo. Seeds of dicotyledonous plants usually contain the parts shown in Figure 9.16 and summarized in this table.

Seed part	Function
testa	tough, protective outer coat
cotyledons	seed leaves that function as nutrient storage structures
micropyle	scar of the opening where the pollen tube entered the ovule
embryo root and embryo shoot	become the new plant when germination occurs

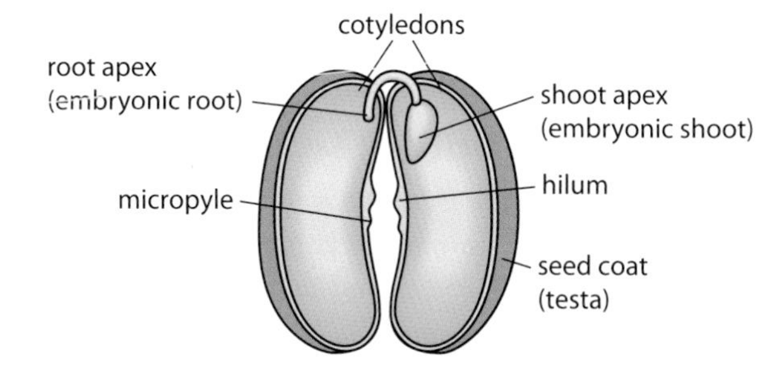

Figure 9.16 A dicotyledonous seed typically shows these structures. Note the two large cotyledons or seed leaves.

Once seeds are formed, a maturation process follows. This process involves dehydration until the water content of the seed is about 10% to 15% of its weight. At this point, the seed usually goes into a dormancy period. This is a time of very low metabolism and no growth or development. The dormancy period is quite variable for different types of seed. This is an adaptation to overcome harsh environmental conditions.

Seed dispersal is important for a number of reasons. By moving away from the parent plant, the potential new plant faces a reduction in competition for limited resources. Fruits are mature ovaries (carpels) that contain seeds. Fruits have a variety of adaptations that allow successful dispersal of seed by wind, water and animals.

If conditions become favourable, the seed will germinate (see Figure 9.17). Germination is the development of the seed into a functional plant. There are several general conditions that must be fulfilled for a seed to germinate:

- water is needed to rehydrate the dried seed tissues;
- oxygen is needed to allow aerobic respiration to produce ATP;
- an appropriate temperature for the seed is necessary (temperature is important for enzyme action).

Besides these conditions, many plants have specific conditions that must be met in order to germinate. For example, in some seeds, the testa must be disrupted or scarified before water uptake can occur. Other seeds must be exposed to fire or smoke before they germinate. The food product known as 'liquid smoke' will often cause seeds of this type to germinate. Light is generally not mentioned in discussions on seed germination because of its variable effects on the process.

In 1995, a team of biologists in China found some seeds in a dried-up lakebed. The seeds were from a type of lotus plant. After germinating some of the seeds, the biologists found them to be nearly 1300 years old. They used radiometric dating to determine this age.

Figure 9.17 Stages in the germination of a bean seed, *Phaseolus vulgaris*. The plumule includes the epicotyl and its two developing leaves. The epicotyl is the region above the attachment point of the cotyledons to the stem. The hypocotyl is the region immediately below the cotyledon attachment point.

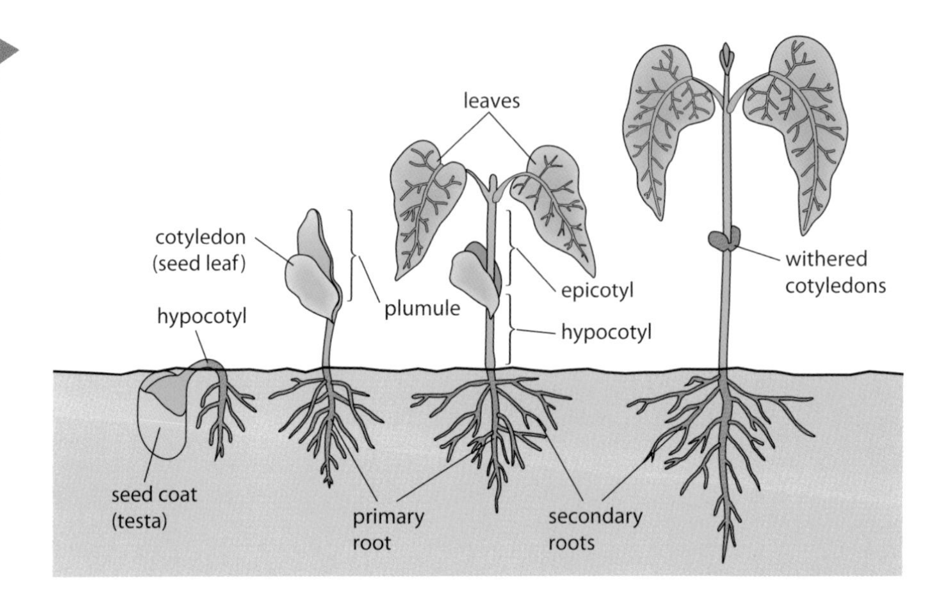

Seed germination is an uncertain time in a plant's life. The emerging seedling is fragile and will be exposed to harsh weather, parasites, predators and many other hazards. Many of the seeds will not produce a functional plant because of these threats. To compensate, plants produce large numbers of seeds and the species survives.

Seed metabolism during germination

The germination process of the seed begins with the absorption of water. The uptake of water is possible because of the relatively low water concentration of the seed. When water enters the seed, many metabolic changes occur that enable growth.

1 Gibberellin, also known as gibberellic acid, is released after the uptake of water.
2 Gibberellin is a growth substance (plant growth hormone) and it triggers the production of the enzyme amylase.
3 Amylase causes the hydrolysis of starch into maltose. The starch is present in the seed's endosperm or food reserve.
4 Maltose is then further hydrolysed into glucose that can be used for cellular respiration or may be converted into cellulose by condensation reactions.
5 The cellulose is necessary to produce the cell walls of new cells being produced.

Maltose is a pivotal compound in the process of germination as it may be used to produce a number of different compounds depending on the needs of the germinating seed. These compounds are especially needed in the meristematic regions of the early seedling. As the seedling develops and functional leaves appear, photosynthesis becomes the supplier of the nutrients needed by the plant.

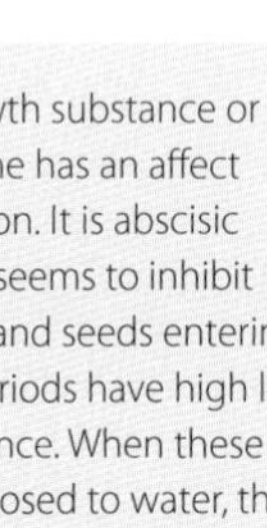

Another growth substance or plant hormone has an affect on germination. It is abscisic acid (ABA). It seems to inhibit germination and seeds entering dormancy periods have high levels of this substance. When these seeds are exposed to water, the ABA concentration is reduced and germination proceeds. In many plants, there is not a clear association between ABA and seed dormancy.

Control of flowering in angiosperms

Light is a very important factor in the life of a plant. It is required for photosynthesis, and it controls many aspects of plant growth and development. Plants are able to detect the presence of light, its direction, wavelength, and even intensity. Photoperiodism is the plant's response to light involving the relative lengths of day and night – a very important factor in the control of flowering. To ensure continued existence in an area, a plant must flower when pollinators are available and when necessary resources are plentiful. This table summarizes three categories of plant in relation to light and flowering.

Plant type	Flowering and light	Examples
long-day plants	bloom when days are longest and nights shortest (midsummer)	radishes, spinach, and lettuce
short-day plants	bloom in spring, late summer, and autumn when days are shorter	poinsettias, chrysanthemums, and asters
day-neutral plants	flower without regard to day length	roses, dandelions, and tomatoes

Even though the names refer to day length, it is actually the length of night that controls the flowering process in plants of the long-day and short-day types. The control by light is brought about by a special blue-green pigment in these plants called phytochrome. There are two forms of phytochrome. One form is inactive and is represented by P_r. The other is active and is represented by P_{fr}.

During the 1940s, experiments were done interrupting periods of darkness with brief exposures of light. These experiments indicated that periods of darkness control flowering, not periods of light. Therefore, short-day plants are actually long-night plants and long-day plants are actually short-night plants.

When red light (wavelength of 660 nm) is present in available light, the inactive form of phytochrome, P_r, is converted to the active form, P_{fr}. This conversion occurs rapidly. The active phytochrome form has the ability to absorb far-red light (wavelength of 730 nm). This P_{fr} is rapidly converted back to the inactive form in daylight. However, in darkness the conversion back to P_r is very slow (see Figure 9.18). It is thought that it is this slow conversion of P_{fr} back to P_r that allows the plant to time the dark period. This seems to be the controlling factor for flowering in the short-day and long-day plants.

In long-day plants, the remaining P_{fr} at the end of a short night stimulates the plant to flower. In other words, it acts as a promoter in these plants. However, in short-day plants the P_{fr} appears to acts as an inhibitor of flowering. For these short-day plants, enough P_{fr} has been converted to P_r to allow flowering to occur.

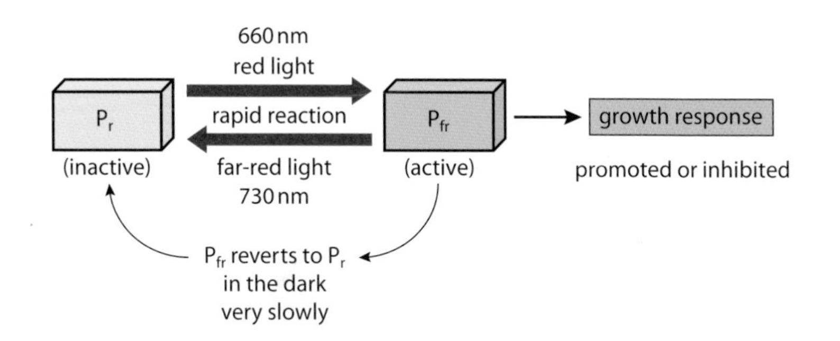

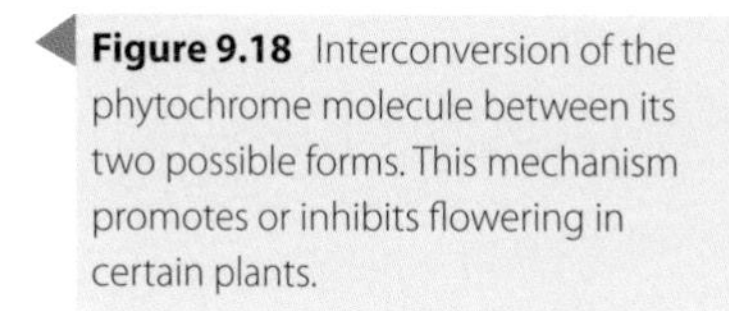
Figure 9.18 Interconversion of the phytochrome molecule between its two possible forms. This mechanism promotes or inhibits flowering in certain plants.

At the present time, it is not known if this flowering response to light is due to changes involving gene expression or to ion flows within the cells of the plant. It is thought that the active form of phytochrome causes some sort of flowering signal. Some researchers have hypothesized the presence of a specific flowering hormone called florigen. However, such a hormone has not yet been found. The mechanism represents a biological clock that is present in most plants. Such biological clocks are important in controlling a number of stages in the life of a plant.

Exercises

8 What features of dandelion seeds allow their successful dispersal?

9 Why is oxygen important for the germination of seeds?

10 The micropyle allows water to enter the testa of a seed during germination. What was the significance of the micropyle to the ovule?

11 What must gibberellin do in the cell to cause the production of amylase?

Practice questions

1 Draw the structure of a dicotyledonous animal-pollinated flower. *(6 marks)*

2 What causes movement of water through the xylem?

A Active transport in the root tissue
B Evaporation of water from leaves
C Active translocation
D Gravity *(1 mark)*

3 Explain how manipulation of day length is used in the production of flowers. *(6 marks)*

4 Fertilization, pollination and seed dispersal all occur during the reproduction of a flowering plant. In what sequence do these processes occur?

A seed dispersal → pollination → fertilization
B fertilization → pollination → seed dispersal
C pollination → fertilization → seed dispersal
D seed dispersal → fertilization → pollination *(1 mark)*

5 Explain how the abiotic factors of light, wind and humidity affect the rate of transpiration. *(8 marks)*

6 Describe the metabolic events of germination in a typical starchy seed. *(5 marks)*

7 Which of the following help(s) in supporting a terrestrial woody plant?

I Xylem tissue
II Turgor pressure
III Phloem tissue

A I only
B I and II only
C II and III only
D I, II and III *(1 mark)*

8 Seed dispersal is important in the migration of plants from one area to another area. Plants have evolved many methods, both physical and biological, by which to disperse their seeds.

50 maple seeds, which are wind dispersed, were dropped one at a time from two different heights, 0.54 m and 10.8 m respectively. The histograms below show the distribution of the distance the maple seeds travelled.

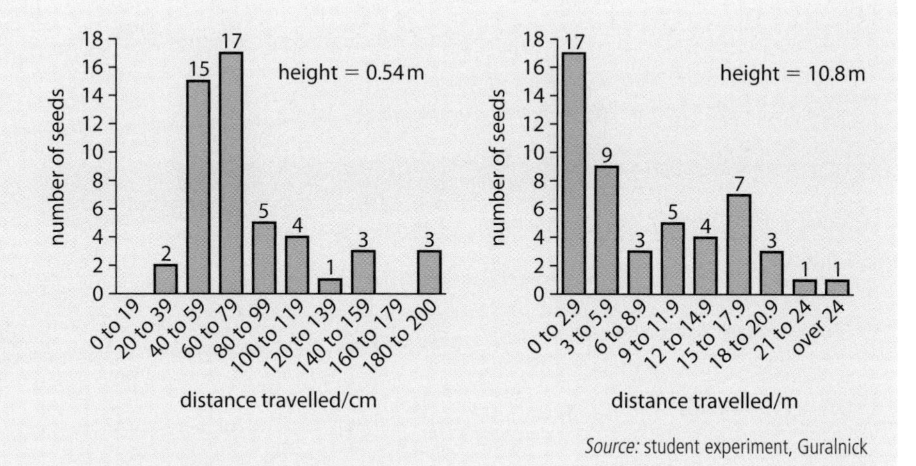

Source: student experiment, Guralnick

(a) For each height, identify the distance travelled by the greatest number of seeds.

(i) Height = 0.54 m

(ii) Height = 10.8 m *(1)*

(b) State the effect of height on seed dispersal. *(1)*

(c) Suggest two reasons for the effect of the drop height on the distance travelled by the seeds. *(2)*

The following graphs show the rate and timing of seed release from different species of grass in the same area during the summer.

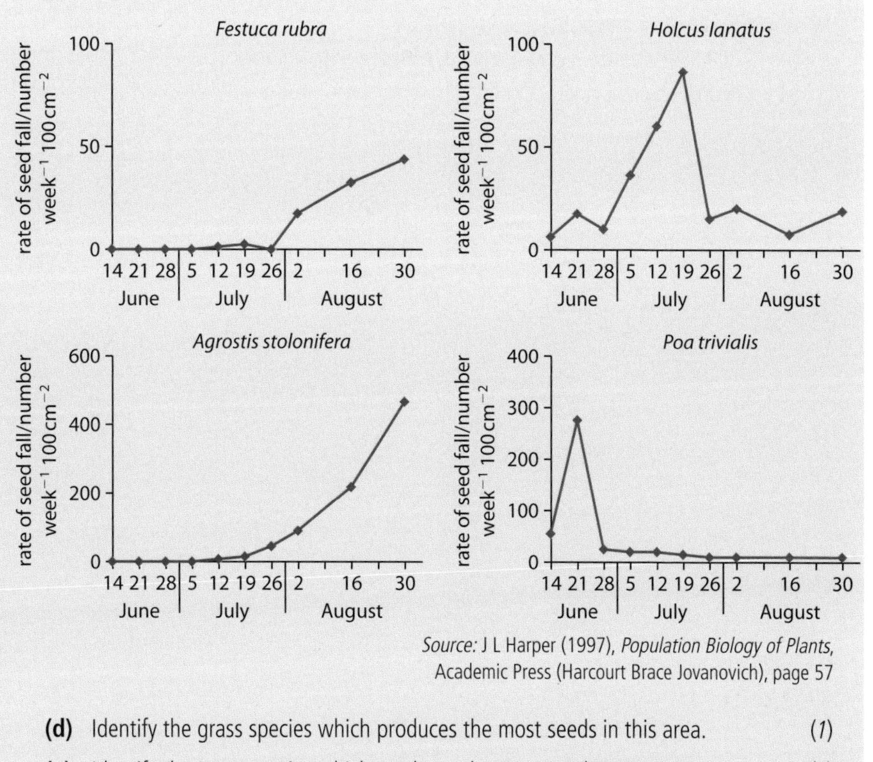

Source: J L Harper (1997), *Population Biology of Plants*, Academic Press (Harcourt Brace Jovanovich), page 57

(d) Identify the grass species which produces the most seeds in this area. *(1)*

(e) Identify the grass species which produces the most seeds in June. *(1)*

(f) Compare seed production for all species relative to the timing of their release. *(3)*

(g) Suggest two benefits for these plants in the timing of seed release. *(2)*

Biological seed dispersal is usually dependent on the nutritional content of the seed or fruit. The following table gives the nutritional content for fruits of different species in temperate and tropical climates.

	Percentage by dry weight			
Common name (genus)	Protein	Lipid	Carbohydrate	Dispersal agent
Temperate				
cranberry (*Vaccinium*)	3	6	89	birds
hawthorn (*Crataegus*)	2	2	73	birds
pin cherry (*Prunus*)	8	3	84	birds
pokeberry (*Phytolacca*)	14	2	68	birds
strawberry (*Fragaria*)	6	4	88	birds
Tropical				
bird palm (*Chamaedorea*)	14	16	55	birds
fig (*Ficus*)	7	4	79	bats
mistletoe (*Viscum*)	6	53	38	birds
monkey fruit (*Tetragastris*)	1	4	94	monkeys
wild nutmeg (*Virola*)	2	63	9	birds

Source: H Howe and L Westley (1988), *Ecological Relationship of Plants and Animals*, Oxford University Press, page 121

(h) Compare tropical fruits to temperate fruits in relation to the mean values for lipid, carbohydrate and protein content. (*2*)

(i) Explain which fruit would have the highest energy content. (*2*)

(j) Suggest one advantage and one disadvantage of dispersal of seeds by animals. (*2*)

(*Total 17 marks*)

10 Genetics 2

Introduction

What is the secret? How is it that each organism produced by the fusion of two gametes has its own genetic makeup? How is it possible that each child, each goldfish, each tree, each bird is unique, among thousands, millions or even billions?

What happens to chromosomes during the formation of egg cells and sperm cells so that offspring are always different? If your parents had dozens of other children, none would be identical to you.

Chapter 4 looked at examples of genetics with single genes which have two alleles (dominant or recessive), or single genes with multiple alleles such as blood type (which can be used to demonstrate codominance). The idea of sex-linked genes for colour blindness and haemophilia was also explored.

But what if one trait were controlled by two or five or ten genes? Could the many alleles all work together to contribute to the trait the way each instrument in an orchestra contributes to a symphony?

Meiosis

Assessment statements

10.1.1 Describe the behaviour of the chromosomes in the phases of meiosis.
10.1.2 Outline the formation of chiasmata in the process of crossing over.
10.1.3 Explain how meiosis results in an effectively infinite genetic variety in gametes through crossing over in prophase I and random orientation in metaphase I.
10.1.4 State Mendel's law of independent assortment.
10.1.5 Explain the relationship between Mendel's law of independent assortment and meiosis.

Although these apples may look the same, inside they are carrying seeds which each have a unique combination of chromosomes. This ensures that some will have a winning combination and could be successful new trees.

Chromosome behaviour during meiosis

During the first division of meiosis, there are several events which characterize it as very different from mitosis. However, meiosis II shares a certain number of similarities with mitosis (see Figure 10.1).

During prophase I

The chromosomes become more visible because they become shorter and coil up. Each chromosome already contains two chromatids.

Homologous chromosomes pair up so that maternal and paternal chromosomes are side by side. Each pair is called a bivalent and the process of their formation is called synapsis (see Figure 10.1).

When crossing over occurs, DNA is exchanged between non-sister chromatids (see 'Crossing over', below).

During metaphase I

The centromeres of each chromosome have spindle microtubules attached.

The bivalents line up randomly along the equator of the cell. This is called random orientation.

Crossing over is terminated and the exchange of DNA is complete, resulting in chromatids from the same chromosome which are usually no longer identical.

During anaphase I

Homologous chromosomes separate and are pulled to opposite poles by the spindle microtubules. The result is that there is independent assortment of genes which are not linked (see 'Linkage group', page 274).

During telophase I (depending on cell type)

The chromosomes are surrounded by two new nuclear membranes: however, some cells skip this step and start meiosis II. This phase is not shown in Figure 10.1.

In cells which do have telophase I, such as most animal cells, the chromosomes go through partial uncoiling. Each of the two daughter cells is haploid, but each chromosome is still made up of two chromatids.

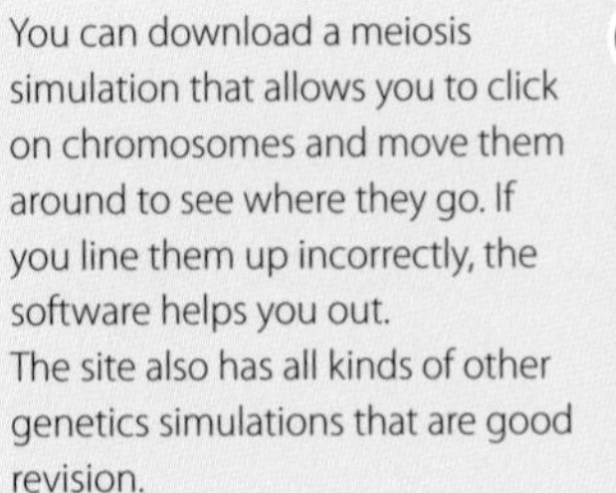
You can download a meiosis simulation that allows you to click on chromosomes and move them around to see where they go. If you line them up incorrectly, the software helps you out.
The site also has all kinds of other genetics simulations that are good revision.

Visit heinemann.co.uk/hotlinks, insert the express code 4242P and click on Weblink 10.1.

Prophase II

DNA condenses into visible chromosomes again. Note that the chromosomes are not in bivalents. In each phase of meiosis II (Figure 10.1) two daughter cells are shown.

Metaphase II

The individual chromosomes line up along the equator of each cell without respecting any special order; again, this is called random orientation.

Spindle fibres from opposite poles attach to each of the sister chromatids at the centromeres.

Anaphase II

The centromeres of each chromosome split, releasing each sister chromatid to become an individual chromosome.

The spindle microtubules pull individual chromatids to opposite ends of the cell.

Because of random orientation, the chromatids could be pulled towards either of the newly forming daughter cells.

Telophase II

The chromosomes become invisible again as they unwind their strands of DNA.

Figure 10.1 Some of the crucial steps in meiosis. Telophase I and II are not shown.

Crossing over

So you can see that the cells created by meiosis can contain a wide variety of combinations of genes from the father and from the mother (see Figure 10.2). This explains why brothers and sisters from the same parents may have family resemblances but are never identical (apart from identical twins). The reason why siblings who are not identical twins are always different is that the chances of receiving the same 23 chromosomes from each parent are so incredibly small that it is essentially impossible.

You look a lot like your sister
True, but thanks to random orientation during metaphase I and crossing over during prophase I, we are not identical

Figure 10.2 Meiotic humour.

In addition to the mixing and matching of chromatids from each parent, the process of crossing over adds more chances for variety in the offspring and results in even fewer chances of similar offspring. The result of crossing over is the exchange of bits and pieces of homologous chromosomes so that a paternal chromosome can swap a section of its DNA with its maternal homologue. The resulting chromatid contains sections of genetic material which originated in two different people.

How does crossing over work?

During prophase I of meiosis, the process of synapsis brings together two homologous chromosomes. They are the same length, they have their centromeres in the same position and generally they contain the same genes stored at the same loci. The major difference between them is that one came from the person's mother and the other chromosome in the bivalent came from the person's father. Since each parent can have different alleles for each of the genes along the chromatids, the two homologous chromosomes are by no means identical. On the other hand, the two sister chromatids from the same chromosome are strictly identical because they are the result of DNA replication during interphase.

Mixing genetic material between non-sister chromatids, in other words between paternal and maternal chromosomes, occurs when the chromatids intertwine and break. In order for crossing over to function correctly, identical breaks must occur at exactly the same position in adjacent non-sister chromatids.

Figure 10.3 Crossing over between adjacent non-sister chromatids.

Now each chromatid has a separated tip. The two segments each connect to the corresponding position on the other chromatid. The two tips are thus switched and each resulting chromatid has a segment of the other's genetic material. The place where the two connect to each other is called a chiasma (plural, chiasmata). For simplicity, Figure 10.3 shows only one chiasma, but in reality many can form along all four chromatids.

Once they are attached at their chiasmata, the chromatids repel each other and twist around to make interesting shapes depending on where and how many times they are attached (see below).

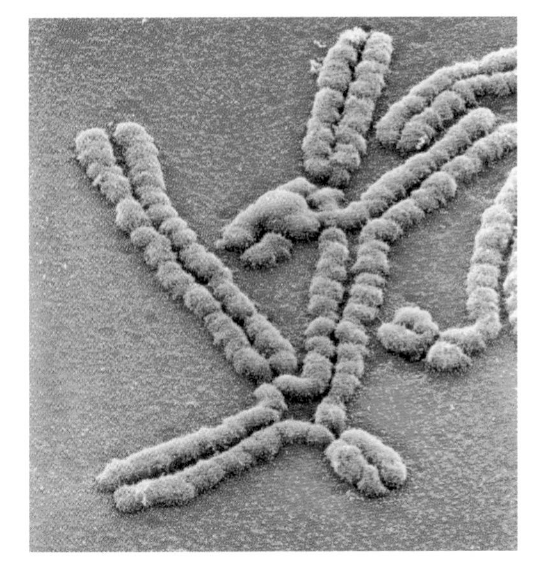

This false-colour SEM shows the cross-shaped union of chromosomes exchanging material, which is where the name chiasma (cross) comes from.

Genetic variety in gametes

The result of meiosis is the production of sex cells which produce offspring that show variety. The main ways in which gamete production is able to generate a wide variety are:

- crossing over during prophase I;
- random orientation during metaphase I.

The genetic variety in a population which reproduces sexually can partly be explained by meiosis.

One consequence of crossing over is that the two sister chromatids of a given chromosome are no longer identical to each other because one has switched a segment with its adjacent homologue. Thus, two chromatids which were once identical now have different combinations of maternal and paternal alleles at their various loci.

This means that when the chromatids get pulled to opposite poles during anaphase II, the resulting gametes will not have the same alleles even though they both received chromatids from the same chromosome. Random orientation occurs when the bivalents arrange themselves around the equator of the cell (see 'Independent assortment and meisosis', page 270).

It is difficult to calculate precisely the number of possibilities when generating sperms and eggs using meiosis. If only the number of chromosomes (n) in each haploid cell is considered, the calculation is 2^n because there are two possible chromosomes in each pair (maternal and paternal) and there are n chromosomes in all. For humans, the number is 2^{23} since there are 23 chromosomes in each gamete. Hence, the probability that a woman could produce the same egg twice is 1 in 2^{23} or 1 in 8 388 608 (see Figure 10.4). This calculation is an oversimplification, however, because it does not take into consideration the additional variety which results from crossing over.

In addition, the calculation only considers one gamete. To produce offspring, two gametes are needed and the chances that both parents produce two identical offspring (apart from identical twins) is infinitesimal.

A one-in-eight-million chance is in the same order of magnitude as the chances of you winning a lottery to become a millionaire. Although it would be theoretically possible, the chances are highly unlikely. Again, the chances are even less than this because the steps to produce gametes using meiosis also include crossing over.

Figure 10.4 In some ways, meiosis is like a lottery.

Mendel's law of independent assortment

Gregor Mendel's law of independent assortment states that when gametes are formed, the separation of one pair of alleles between the daughter cells is independent of the separation of another pair of alleles. As a general rule, one allele does not follow another when it is passed on to a gamete. The law of independent assortment implies that alleles which determine different characteristics will be transmitted independently to the next generation. Examples of Mendel's experiments are shown in section 10.2.

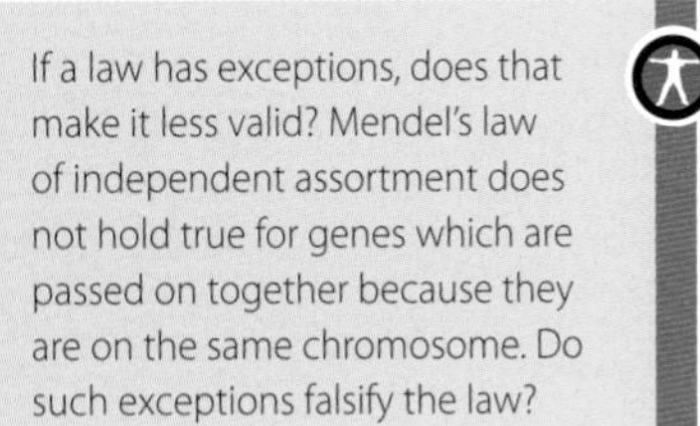

If a law has exceptions, does that make it less valid? Mendel's law of independent assortment does not hold true for genes which are passed on together because they are on the same chromosome. Do such exceptions falsify the law?

In practice, this means that just because one trait (such as a certain flower colour) is inherited from a parent it does not follow that any other specific trait of that parent (such as a specific seed colour) must be passed on as well. On the contrary, each allele in a pair can mix with either allele of another pair.

As with many rules, there are exceptions. Some genes do, in fact, go hand-in-hand so that when one is placed in a gamete during meiosis, the other follows (see 'Linkage group', below).

Independent assortment and meiosis

When Gregor Mendel was performing his experiments in the mid 1800s, he did not know anything about meiosis because it had not been discovered. Today, it is possible to answer the question which most likely went through his mind: Why do traits get passed on independently from each other?

The answer lies in an understanding of the process of meiosis. You will recall that the orientation of bivalents during metaphase I is a random process. To illustrate this, consider a cell with 4 pairs of chromosomes (see Figure 10.5). In each of the four bivalents, there is a maternal chromosome (red) and a paternal one (blue). For simplicity, no crossing over is shown.

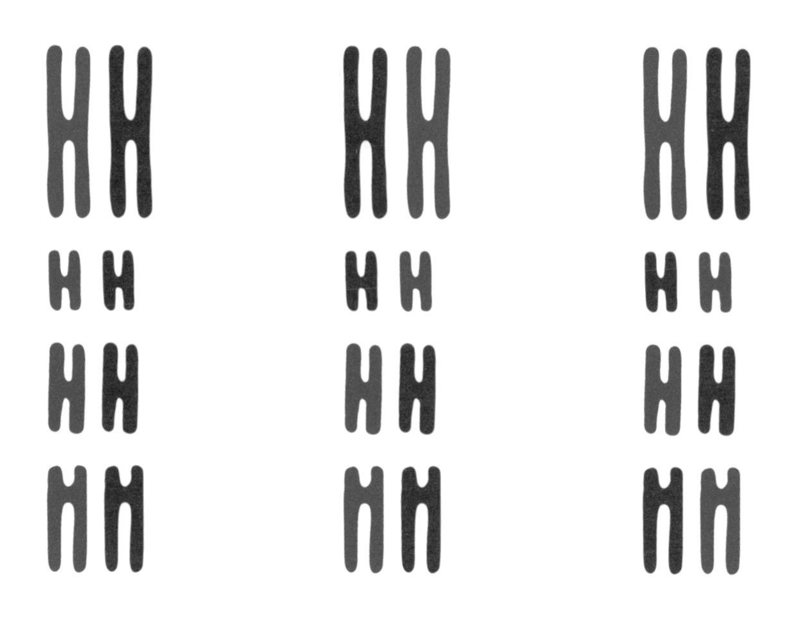

Figure 10.5 Three of the sixteen possible orientations for four bivalents. In humans there are 23 bivalents with over 8 million possible orientations.

Depending on how they line up along the equator during metaphase I, there may be more or fewer maternal or paternal chromosomes in the resulting daughter cells. In the first possibility in Figure 10.5, all four bivalents are oriented in the same direction so that the resulting daughter cells contain exclusively maternal or exclusively paternal chromosomes.

When the microtubules pull the pairs apart, all the red maternal chromosomes will be pulled to the cell that will be forming on the right and all the blue paternal chromosomes will be pulled to the left. When each of the new cells goes through meiosis II, two gametes will contain maternal chromosomes exclusively and two will contain only paternal chromosomes. Although this seems feasible with four pairs chromosomes, it is much less likely with the human number of 23.

During assortment (the distribution and separation of the chromosomes at the end of metaphase I and the beginning of anaphase I) the random positions of the chromosomes ensure a random distribution of alleles between the gametes. The first example in Figure 10.5 is unlikely – usually each gamete receives an assortment of maternal and paternal chromosomes.

In the arrangement of bivalents shown in the centre of Figure 10.5, half of the maternal chromosome are in one hemisphere and the second half in the other hemisphere. Altogether, there are 2^4 (16) possible arrangements in this example. But remember, this is a simple example cell with only 4 pairs of chromosomes instead of the 23 pairs in humans.

Exercises

1 Outline the importance of producing unique offspring in each generation.

2 Look again at Figure 10.1.
  a State the phase of meiosis during which crossing over takes place.
  b Draw telophase II using the chromatids shown.

3 Look at again at Figure 10.5.
  a State the phase of meiosis during which the bivalents line up in this way.
  b For each of the three possibilities, draw what gametes would be produced. Use the same colour coding.

10.2 Dihybrid crosses and gene linkage

Assessment statements

10.2.1 Calculate and predict the genotypic and phenotypic ratio of offspring of dihybrid crosses involving unlinked autosomal genes.
10.2.2 Distinguish between *autosomes* and *sex chromosomes*.
10.2.3 Explain how crossing over between non-sister chromatids of a homologous pair in prophase I can result in an exchange of alleles.
10.2.4 Define *linkage group*.
10.2.5 Explain an example of a cross between two linked genes.
10.2.6 Identify which of the offspring are recombinants in a dihybrid cross involving linked genes.

Dihybrid crosses

Let's consider Gregor Mendel's experiments with pea plants. In one cross, he examined the following two traits:

- seed shape: some seeds are round while others are wrinkled – the allele for round is dominant (see Figure 10.6);
- seed colour: some seeds are green inside and others are yellow – the allele for yellow is dominant (see Figure 10.6).

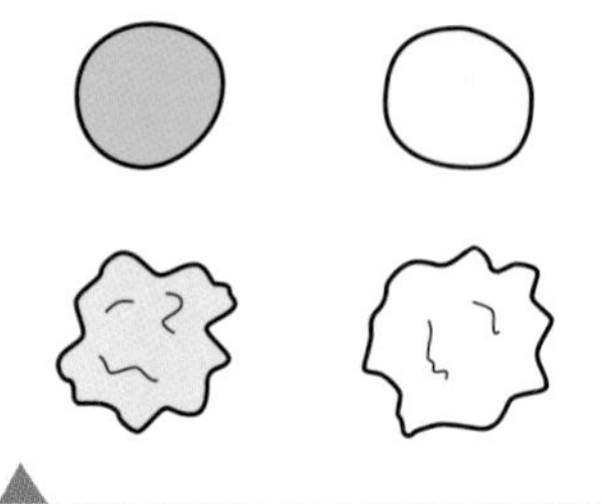

Figure 10.6 Seed colour and seed shape in peas.

Mendel crossed two true-breeding plants with each other. True-breeding means homozygous for the traits being studied – so no surprises are produced by masked recessive alleles. One parent plant was homozygous for both dominant traits (round and yellow seeds) whereas the other parent was homozygous recessive for both traits (wrinkled and green).

Monohybrid crosses:
2 × 2 = 4 possible offspring

Dihybrid crosses:
4 × 4 = 16 possible offspring

To represent the alleles, Mendel used a system of letters which is incompatible with the system we use today so, for this example, Mendel's letters will be replaced with modern conventions:

- **R** = allele for round peas;
- **r** = allele for wrinkled peas;
- **Y** = allele for yellow peas;
- **y** = allele for green peas.

Parent phenotypes:	**round yellow**	**green wrinkled**
Parent genotypes:	RRYY	rryy
Parent gametes:	RY	ry
F_1 genotypes:	RrYy	
F_1 phenotypes:	round yellow	

The F_1 generation is made up exclusively of plants which give round yellow peas. When these peas were planted, grown into adult plants and allowed to self-pollinate, Mendel expected some of the recessive traits to show up again. That did happen, and what is interesting is the ratio in which they appeared. From 15 plants, Mendel obtained 556 pea seeds in the following proportions:

- 315 round and yellow (56.6%);
- 101 wrinkled and yellow (18.2%);
- 108 round and green (19.4%);
- 32 wrinkled and green (5.8%).

If these percentages are converted to ratios, the numbers are close to the expected ratio for alleles that are passed on independently and are not found on the sex chromosomes. This ratio is 9 : 3 : 3 : 1 and is calculated using a 4 × 4 Punnett grid as shown in Figure 10.7.

	RY	Ry	rY	ry
RY	RRYY	RRYy	RrYY	RrYy
Ry	RRYy	RRyy	RrYy	Rryy
rY	RrYY	RrYy	rrYY	rrYy
ry	RrYy	Rryy	rrYy	rryy

Phenotypes

= round yellow peas × 9

= round green peas × 3

= wrinkled yellow peas × 3

= wrinkled green peas × 1

Figure 10.7 A Punnett grid for inheritance of roundness and yellowness in pea seeds.

The ratio says that for every wrinkled green pea in that generation, there should be 3 round green peas. Mendel found 3.34 times more round green peas than wrinkled green peas in his experiment. There is often a difference between the theoretical values and values obtained in experiments. If thousands of seeds were examined instead of a few hundred, the number would probably be closer to 3. A similar difference between theory and reality is seen in families where there are more or fewer boys than girls – few families have exactly 50% of each, despite the fact that the laws of genetics predict half and half.

Examiner's hint: When showing crosses and offspring, don't forget to include such important details as:

- genotypes of the parents;
- a key of what the letters mean (although many questions state this);
- phenotypes of the offspring;
- alleles found in the gametes.

Autosomes and sex chromosomes

As you saw in Chapter 4, the sex chromosomes are the X and Y chromosomes, and they are the ones which determine what sex you are. Any chromosome which is not a sex chromosome is called an autosome, or autosomal chromosome. Humans have 22 pairs of autosomes and one pair of sex chromosomes (see Figure 10.8).

Visit heinemann.co.uk/hotlinks, insert the express code 4242P and click on Weblink 10.2. Look for the online simulation (gizmo) called 'Mouse Genetics (Two Traits)'. You can breed virtual mice to see how traits are passed on.

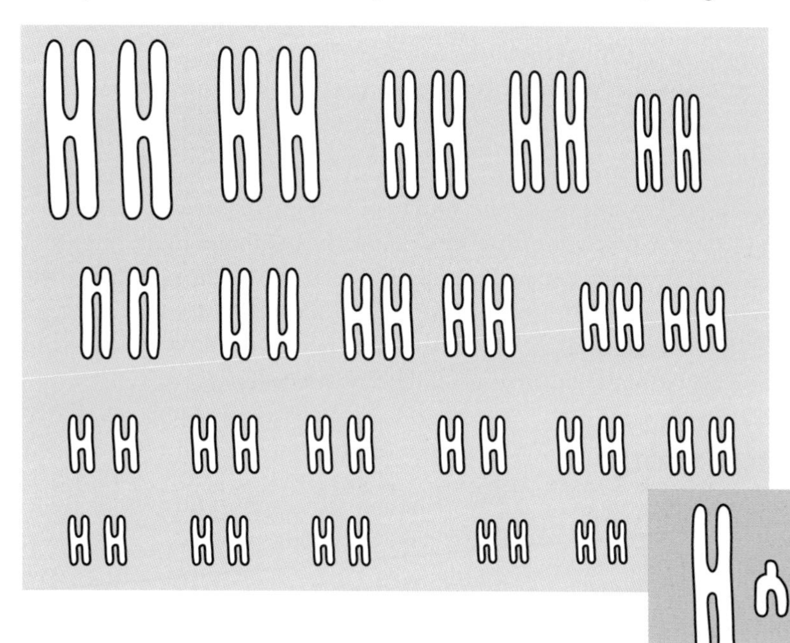

Figure 10.8 Human chromosomes: grey = autosomes, purple = sex chromosomes.

If a trait or gene is described as autosomal, its locus is on one of the 22 pairs of autosomes, not the sex chromosomes. A trait or gene which is said to be sex-linked must have its locus on a sex chromosome. Where a gene is located determines whether or not the trait it controls is more common in males or females. When a trait is more common in one sex than the other, the chances are good that the trait is sex-linked and that the locus of the gene is on either the X chromosome, the Y chromosome or both (see 'Sex linkage' in Chapter 4, page 96). If there is no pattern to the frequency of a trait between females and males, it is most likely an autosomal trait.

Exchange of alleles by crossing over

When we considered crossing over, we saw that two non-sister chromatids can swap segments of their DNA. This means that a maternal chromosome can end up with a segment of genetic material from a paternal chromosome and vice versa. Thus a chromosome originally carrying a recessive allele could end up with a dominant allele that it traded during crossing over.

For example, consider a bivalent in which the maternal chromosome has allele **B** for an autosomal trait and the paternal chromosome has the recessive allele **b**. Figure 10.9 shows a chiasma between such a gene's locus and the centromere.

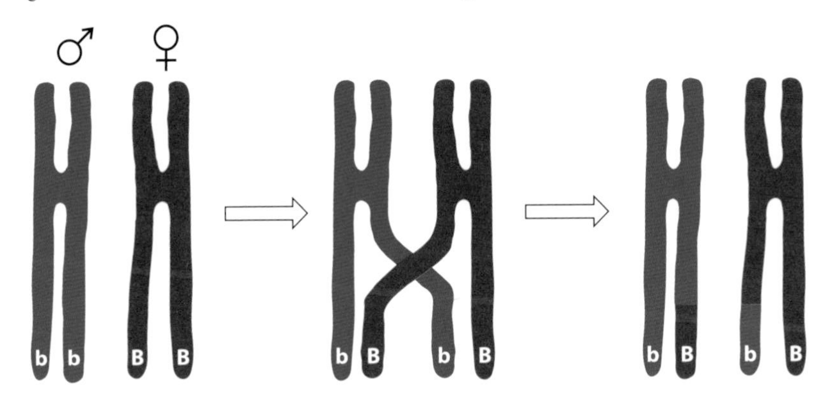

Figure 10.9 How an allele **B** from a maternal chromosome can be switched with an allele **b** on a paternal chromosome.

When crossing over is complete, the segments containing the locus of the gene have been swapped and the alleles have switched places. Now it is the 'paternal' chromosome (no longer 100% paternal) that has **B** and the 'maternal' chromosome has **b**. Note that the two paternal sister chromatids no longer carry identical alleles.

During any single crossing over event, hundreds or thousands of genes can be traded in this way between non-sister chromatids. In addition, a single bivalent can have several chiasmata producing crossing over in more than one chromatid. This is yet another source of variation in the preparation of sperm cells and egg cells. This also partially explains why, unless they are identical twins, brothers and sisters never get the same combination of their parents' alleles.

Linkage group

Any two genes which are found on the same chromosome are said to be linked to each other (see Figure 10.10). Linked genes are usually passed on to the next generation together.

A group of genes inherited together because they are found on the same chromosome are considered to be members of a linkage group. This applies to

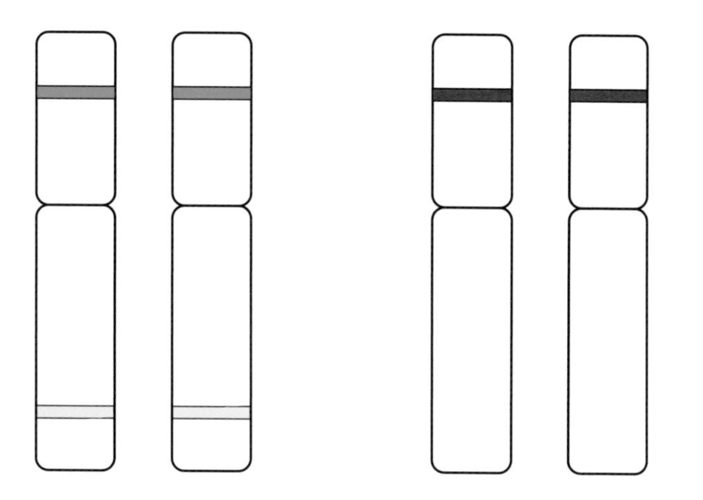

Figure 10.10 In these two pairs of chromosomes, can you see which genes are linked, and which are not?

genes found on autosomes as well as those on the sex chromosomes. In Figure 10.10, the green and yellow genes are linked. Neither is linked to the red genes.

Linked genes are the exception to Gregor Mendel's law of independent assortment. When studied together using dihybrid crosses, linked genes do not give the ratio 9 : 3 : 3 : 1. On the contrary, linked genes can give a wide variety of ratios which can be puzzling unless you understand how linkage groups work.

In his dihybrid crosses with pea plants to study seed colour and seed shape, Mendel was very fortunate that he happened to choose two traits which were not linked. The locus of the gene for seed colour is on a different chromosome from the locus of the gene for seed shape.

Linked genes

In the fruit fly, *Drosophila*, the gene for body colour (grey or black) is in the same linkage group as the gene for wing length (long or short) (see Figure 10.11). The alleles are:

- **G** = allele for grey body;
- **g** = allele for black body;
- **L** = allele for long wings;
- **l** = allele for short wings.

The genotypes of the true-breeding (homozygous) parents are:

- **GGLL** = genotype of grey-bodied, long-winged parent;
- **ggll** = genotype of black-bodied, short-winged parent.

There is nothing in the genotype's notation **GGLL** which shows that **G** must be inherited with **L**. In order to show linkage, the following notation is used:

$$\frac{G \quad L}{\overline{G \quad L}}$$

The two horizontal bars symbolize homologous chromosomes and show that the locus of **G** is on the same chromosome as **L**. One **G** is on the maternal homologue and the other **G** is on the paternal homologue. Likewise, **ggll** is shown:

$$\frac{g \quad l}{\overline{g \quad l}}$$

To read the genotype of the individual for these two linked traits, the pairs of alleles are read vertically: the above symbol's genotype is **ggll**.

• **Examiner's hint:** Be sure you don't confuse linked genes with sex linkage or polygenic inheritance or multiple alleles. You need a good command of all these ideas including:

- a precise definition of each (not one based on common knowledge or misunderstandings);
- a precise example of each which you can explain in detail.

The expected ratio for the phenotypes of offspring in a dihybrid cross is 9 : 3 : 3 : 1 but only with unlinked genes.

• **Examiner's hint:** Some books and web resources may use slightly different formats, but this notation with two horizontal bars is used in IB exams to show linkage groups.

Offspring of a dihybrid cross

A cross between a homozygous dominant true-breeding fruit fly (**GGLL**) and a homozygous recessive true-breeding fruit fly (**ggll**) would result in flies which were all heterozygous for both of the traits (**GgLl**). The flies would all be grey with long wings but they would all be carriers for the recessive alleles. Suppose such flies were accidentally put in the wrong jar in the laboratory and found themselves with another population of flies which look the same but which are all homozygous for both traits. Researchers in the lab would not be able to determine the genotype of any particular fly in that jar just by looking at it. A test cross with a known homozygous recessive would be necessary to determine whether the mystery fly's phenotype is the result of a homozygous or heterozygous genotype (see Figure 10.11).

Fruit flies are bred in laboratories to study genetics.

Figure 10.11 Test cross between a grey long-winged fly of unknown genotype with a black short-winged fly, a homozygous recessive.

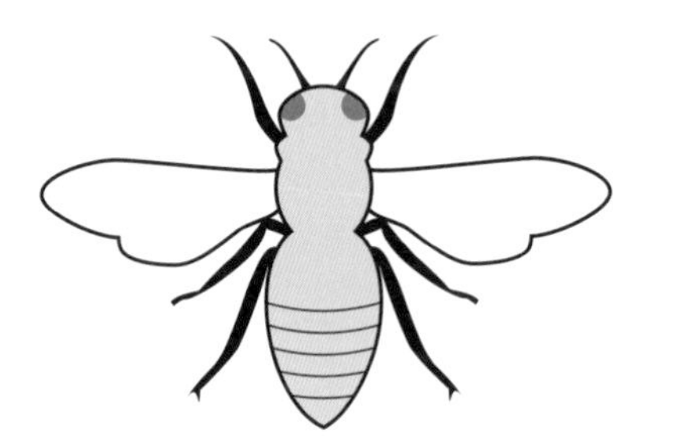

×

If the offspring of the test cross are all grey, long-winged flies, the mystery fly must be homozygous (**GGLL**) for both traits. In the case of a mystery fly with the heterozygous genotype **GgLl**, the resulting offspring would show some traits from each parent.

Here are the linked genes in the heterozygote:

G L
═══
g l

The test cross is done by mating the mystery fly (here heterozygous for both traits) with another which is known to be homozygous recessive for both traits:

g l
═══
g l

One way of showing this cross is by drawing a Punnett grid (see Figure 10.12). A full grid is not necessary since there are only four possible combinations:

	GL	Gl	gL	gl
gl	GgLl G L / g l	Gg ll G l / g l	gg Ll g L / g l	gg ll g l / g l
		(R)	(R)	

Figure 10.12 Punnett grid showing the test cross. The two offspring labelled R are the recombinants (see text for explanation). Each box shows both ways of representing the genotypes.

Another way of showing the same idea is shown in Figure 10.13.

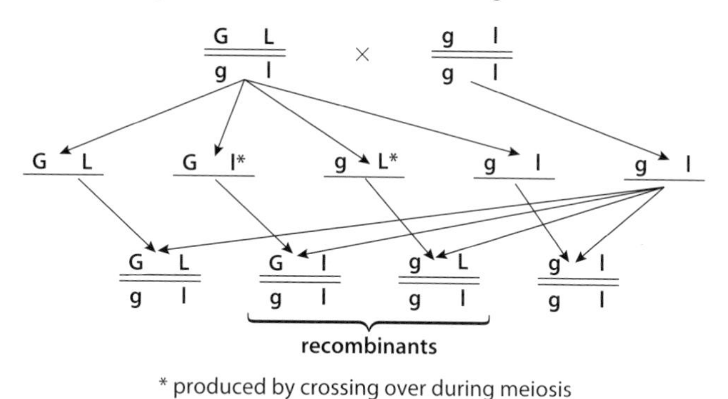

Figure 10.13 Test cross **GgLl** × **ggll**.

Look at the two possibilities for offspring in the middle of Figures 10.12 and 10.13 (the second and third possibility in each case). By examining the alleles closely, it is possible to see that those offspring are different from either parent. A new shuffling of the alleles has created a new combination which does not match either of the parents' genotypes. The term recombinant is used to describe both the new chromosome and the resulting organism.

The way these recombinants form is through the process of crossing over. Without crossing over, the allele **G** would always be inherited with **L** for the simple reason that they are linked. Thanks to crossing over, **G** sometimes gets inherited with **l**. In addition, **g** sometimes gets inherited with **L**, as seen in the recombinants in Figure 10.14

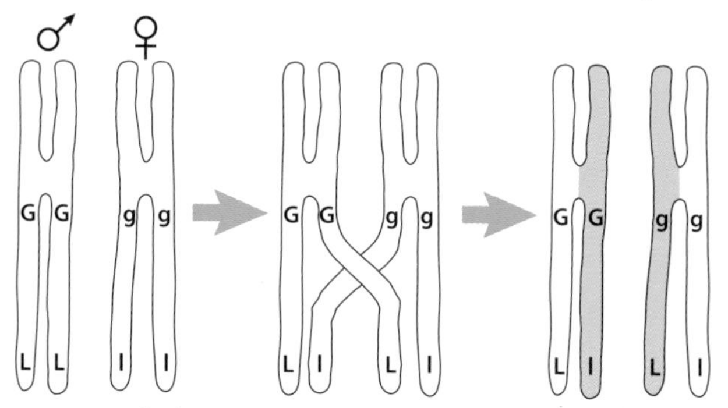

Figure 10.14 The highlighted chromatids show new combinations of alleles which were not observed in the original parents' chromosomes.

When gametes are made from the resulting bivalent shown on the right, two will contain combinations found in the parents (either **GL** or **gl**) whereas two will contain recombinants (**Gl** and **gL**). Thus, even in linked genes, nature has found a way to increase variety through crossing over.

Exercises

4 A genetic disease can be described as being an autosomal dominant disease. From this terminology, what can be deduced about the locus of the gene which causes such a disease?

5 The parents in a cross are **AABB** and **aabb** respectively.

a Draw diagrams for each showing that **A** is linked with **B** and **a** is linked with **b**.

b Show how the offspring in the cross are produced and clearly label the recombinants.

10.3 Polygenic inheritance

Assessment statements

10.3.1 Define *polygenic inheritance*.

10.3.2 Explain that polygenic inheritance can contribute to continuous variation using two examples, one of which must be human skin colour.

Polygenic inheritance defined

Polygenic inheritance involves two or more genes influencing the expression of one trait. With two or more allelic pairs found at different loci, the number of possible genotypes is greatly increased. It is believed that most human traits are too complex and show too many combinations to be determined by one gene.

This could partly explain the difficulty in finding out which genes are responsible for traits whose genetic components are poorly understood; for example, mathematical aptitude, musical talent, or susceptibility to certain illnesses or cancer.

Continuous and discontinuous variation

With dominant and recessive alleles of a single gene, the number of possible phenotypes is limited. For example, either a person has an attached earlobe or not. When multiple alleles are introduced, the number of possibilities for a single trait increases accordingly. For example, the ABO blood type has 4 alleles and 4 possible phenotypes.

Height in humans is an example of a trait which shows continuous variation.

When a second gene is introduced, the number of possible genotypes increases dramatically. With three, four or five genes determining the phenotype, the number of possibilities is so big that it is impossible to see in the phenotype the difference between certain genotypes. When an array of possible phenotypes can be produced, it is called continuous variation.

The colour of skin in humans is an example of continuous variation and we believe that the intensity of pigment in skin is due to the interaction of multiple genes, although it is unclear how many are involved.

In humans, continuous variation can also be seen in the genetic components of traits such as height, body shape, and intelligence. Each of these is also influenced by environmental components. Height, for example, is determined by whether a person inherits genes for tallness but it also depends on the person's nutrition as he or she is growing.

To help you decide whether or not a trait shows continuous variation, imagine a questionnaire to record phenotypes. In general, if it is possible to tick 'yes' or 'no' for a trait, that trait does *not* show continuous variation (e.g. attached earlobes). The same is true for a trait whose possibilities could be represented by a few choices such as blood type: A, B, AB or O.

When variation is not continuous, it is referred to as discontinuous variation. The data for discontinuous variation can be displayed as bar charts or histograms (see Figure 10.15). An unbroken transitional pattern from one group to another is not present.

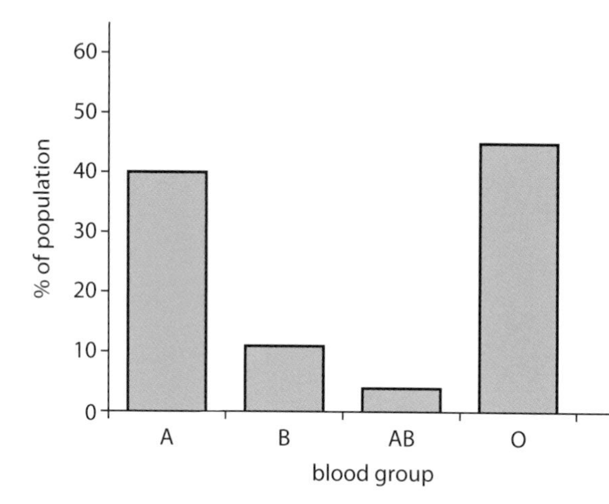

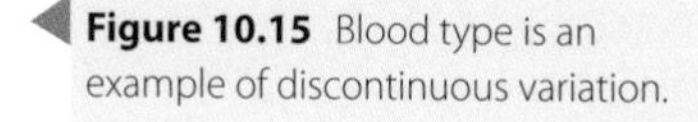
Figure 10.15 Blood type is an example of discontinuous variation.

When there are many possibilities, then the trait shows continuous variation. The graphed results produce a bell-shaped distribution curve. There is a smooth transition between the groups of frequencies (see Figure 10.16).

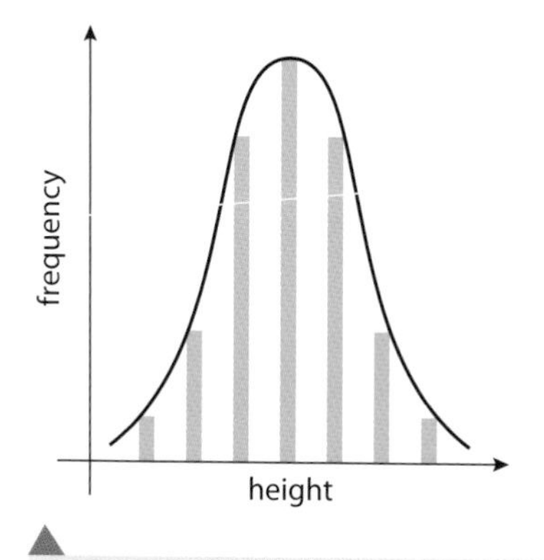

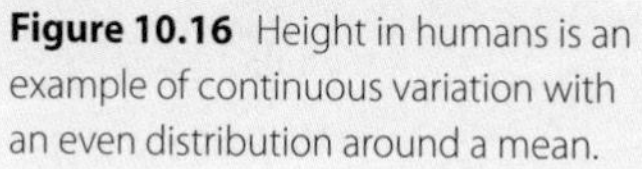
Figure 10.16 Height in humans is an example of continuous variation with an even distribution around a mean.

Does eye colour show continuous variation?

Traits such as eye colour are difficult to assess. Most people consider that there are only a few possible colours. But close examination of the iris shows that it is made up of zones, rings, streaks or speckles of different coloured pigments with varying intensities.

Although it is possible to classify eye colour into groups such as blue, brown, hazel and green, some people say that their eyes are grey or even amber. Close examination of all the people who say their eyes are brown would reveal a wide variety of colours. Even though nature shows a considerable amount of variety in eye colour, societies and governments have imposed a small number of categories. The result is that some people have trouble deciding which category they belong to. What colour do you think the eye in the photograph is?

One way to explain the complexity of eye colour is to attribute the trait to more than one pair of alleles.

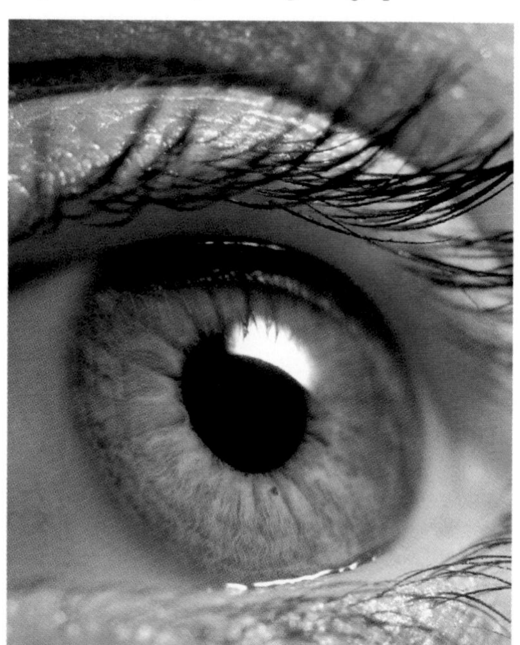

What about skin colour?

Even though there is clearly an almost boundless variety of shades of human skin colour, many societies are still fixed on labelling people with categories such as 'white' or 'black' or 'mixed'. This oversimplification serves more of an administrative purpose than a biological one.

The amount of melanin in human skin is an example continuous variation.

Melanin protects

There is a valid biological reason why humans have varying skin colour: protection from the Sun's harmful ultraviolet (UV) radiation. When sunlight is very intense, as in the tropics, people need protection from serious sunburn which can lead to melanoma, a type of skin cancer. Also, excess exposure to sunlight breaks up important nutrients such as folate, a form of B vitamin.

The best protection from UV radiation is increased melanin in the skin. Apart from albinos, all people have cutaneous melanin pigmentation. People with very dark skin have a high concentration of melanin, whereas people with very light skin have very little.

Although it is mostly determined by genetics, it is possible to increase the melanin level in your skin by exposing your body to sunlight. This process, which we call tanning, is a natural defence against the negative effects of excess sunlight. To a certain degree, the skin adapts to the amount of sunlight in order to block out dangerous UV light – the more tanned the skin is, the less UV can get through, but only up to a point.

UV light and calciferol (vitamin D)

In regions which are far from the equator, the sunlight reaching the surface is much less intense (see Figure 10.17). There are some seasons during which it shines for many hours a day and others when the Sun is up for only a few hours each day. One of the beneficial aspects of UV light is that it helps the skin to make calciferol, which is essential for proper growth and bone formation. The type of UV radiation which does this is UVB radiation (in the middle range of UV wavelengths, 280–320 nm).

Consequently, it is essential for good health to allow a moderate amount of UV radiation onto the skin. In regions of low Sun exposure, people need light-coloured skin. If they had intense concentrations of melanin, not enough UV rays would get through to allow calciferol production.

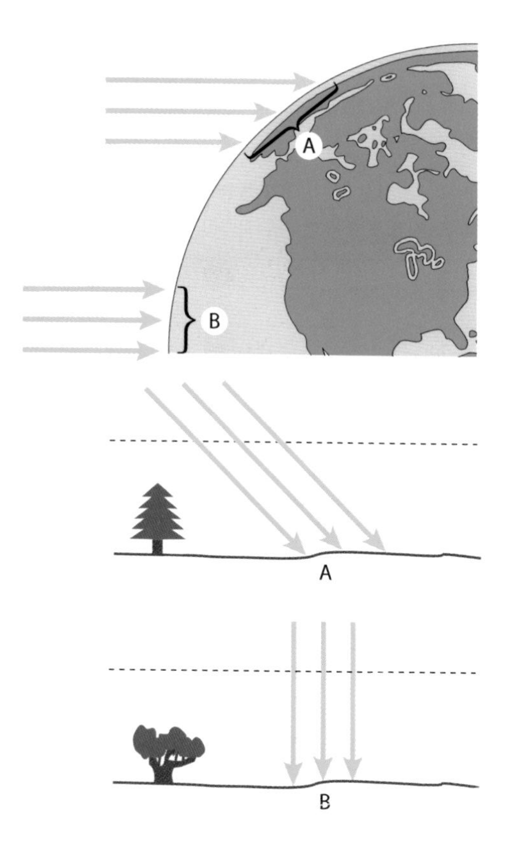

Figure 10.17 How rays from the Sun arrive on Earth in different intensities according to latitude. Far from the equator (A), rays arrive at an angle and are spread out whereas at the equator (B), rays arrive from directly overhead and are more intense. In addition, notice how the rays at (A) have to go through more of the atmosphere and so are more filtered.

The concept of skin colour in humans is a good example of biology meeting history, politics, economics, psychology and even law. Throughout history, when peoples of different skin colour meet, there have been conflicts, enslavement, violence, prejudice and racism, as well as laws passed to discriminate for or against people based on the colour of their skin. Such behaviour raises serious questions about human psychology and about social responsibility.

Human rights have progressed and considerable progress has been made to eliminate laws which promote discrimination, but there is still a long way to go to change people's attitudes and misconceptions. Perhaps one day people will look at each other and see fellow humans, no matter what degree of melanin they have in their skin.

Should there be equal esteem for all humans? Biologists now believe that variety in a population is a strength, not a weakness. So why is human diversity so often used to divide and discriminate, rather than be appreciated, respected and celebrated?

In today's society, the increase in worldwide travel and emigration has lead to a situation where more and more people are living in places where the amount of sunlight is not necessarily compatible with the amount of melanin they have in their skin. Fortunately, this is easily remedied by sun-block lotions for people with light skin living in tropical regions and vitamin D supplements for people with dark skin living in latitudes which do not receive enough sunlight.

Exercises

6 Distinguish between continuous and discontinuous variation.

7 Which of the following are examples of continuous variation?

A body mass
B shoe size
C ability to roll the tongue
D left-handedness
E finger span (i.e. distance between thumb and smallest finger when fingers are spread out)
F presence of hair between knuckles of fingers

8 In general, women tend to have lighter-coloured skin than men. Deduce with a reason the advantage this would have in relation to absorbing UVB radiation.

Practice questions

1 Describe, with the aid of a diagram, the behaviour of chromosomes in the different phases of meiosis. *(5 marks)*

2 The diagram below shows chromosomes during prophase I of meiosis. How many chromosomes are visible? How many chiasmata are visible?

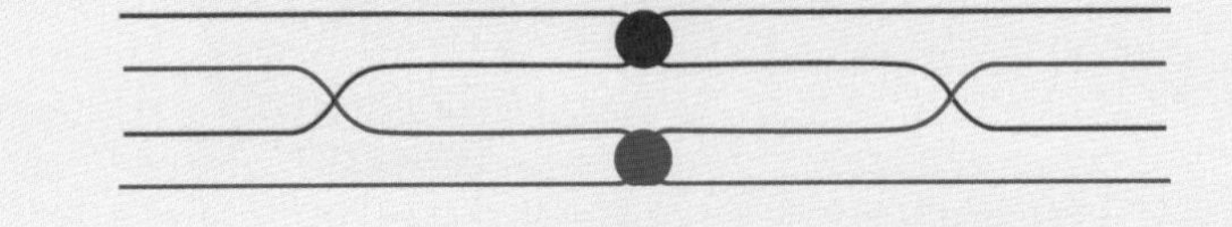

(1 mark)

3 Explain how meiosis results in great genetic variety among gametes. *(5 marks)*

4 In peas, the allele for round seed (**R**) is dominant over the allele for wrinkled seed (**r**). The allele for yellow seed (**Y**) is dominant over the allele for green seed (**y**). If two pea plants with the genotypes **YyRr** and **Yyrr** are crossed together, what ratio of phenotypes is expected in the offspring?

A 9 round yellow : 3 round green : 3 wrinkled yellow : 1 wrinkled green
B 3 round yellow : 3 round green : 1 wrinkled yellow : 1 wrinkled green
C 3 round yellow : 1 round green : 3 wrinkled yellow : 1 wrinkled green
D 1 round yellow : 1 round green : 1 wrinkled yellow : 1 wrinkled green *(1 mark)*

5 In *Zea mays*, the allele for coloured seed (**C**) is dominant over the allele for colourless seed (**c**). The allele for starchy endosperm (**W**) is dominant over the allele for waxy endosperm (**w**). Pure breeding plants with coloured seeds and starchy endosperm were crossed with pure breeding plants with colourless seeds and waxy endosperm.

(a) State the genotype and the phenotype of the F_1 individuals produced as a result of this cross. *(2)*

(b) The F_1 plants were crossed with plants that had the genotype ccww. Calculate the expected ratio of phenotypes in the F_2 generation, assuming that there is independent assortment. *(2)*

The observed percentages of phenotypes in the F_2 generation are shown below.

coloured starchy	37%	colourless starchy	14%
coloured waxy	16%	colourless waxy	33%

The observed results differ significantly from the result expected on the basis of independent assortment.

(c) Explain the reasons for the observed results of the cross differing significantly from the expected results. *(2)*

(total 6 marks)

6 A polygenic character is controlled by two genes each with two alleles. How many different possible genotypes are there fro this character: 2, 4, 9 or 16? *(1 mark)*

11 Human health and physiology 2

Introduction

This unit of study is an extension of Chapter 6 'Human health and physiology 1'. In addition to gaining depth of understanding about human reproduction and immunity, you will learn how the body accomplishes skeletal movement and filters unwanted solutes from the blood by the action of the kidneys. To gain the most from this chapter, we suggest you study it at the same time as Chapter 6, or review the relevant topics in Chapter 6 as you study this unit.

11.1 Defence against infectious disease

Assessment statements

11.1.1 Describe the process of blood clotting.
11.1.2 Outline the principle of challenge and response, clonal selection and memory cells as the basis of immunity.
11.1.3 Define *active* and *passive immunity*.
11.1.4 Explain antibody production.
11.1.5 Describe the production of monoclonal antibodies and their use in diagnosis and in treatment.
11.1.6 Explain the principle of vaccination.
11.1.7 Discuss the benefits and dangers of vaccination.

Why does blood clot?

When small blood vessels like capillaries, arterioles and venules get broken, blood escapes from the closed circulatory system. Often the damaged blood vessels are in the skin and thus pathogens have a way to gain entry into the body. Our bodies have evolved a set of responses to create a clot which 'seals' the damaged blood vessels so preventing excessive blood loss and helping to prevent pathogens from entering the body. This occurs relatively soon after the damage.

Circulating in the blood plasma are a variety of molecules called plasma proteins. These proteins serve many purposes including some that are involved in clotting. Two of the clotting proteins are prothrombin and fibrinogen. These two molecules are always present in blood plasma, but remain inactive until 'called to action' by events associated with bleeding. Also circulating in the bloodstream are cell

This false-colour SEM shows small platelets (pale green) have triggered the formation of insoluble fibrin protein fibres. Trapped in fibrin are several erythrocytes, platelets, and one leucocyte (yellow).

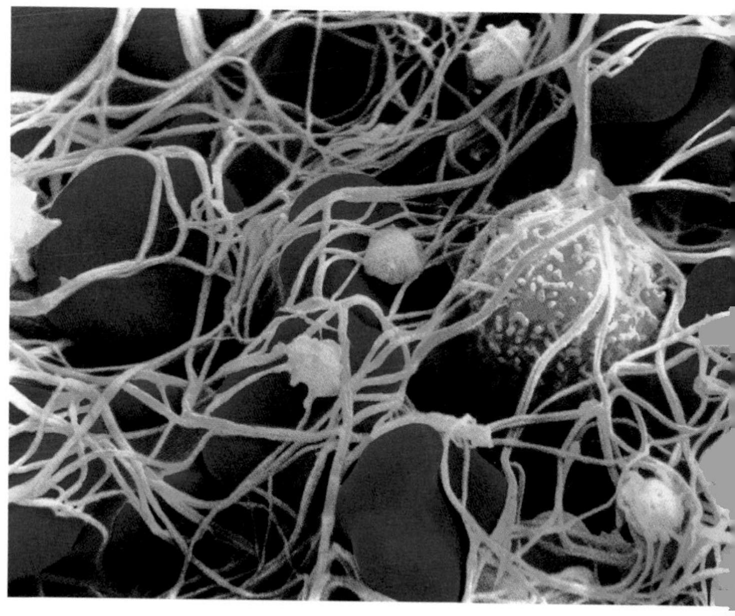

fragments known as platelets. Platelets form in bone marrow, along with red cells (erythrocytes) and white cells (leucocytes), but do not remain as entire cells. Instead, one very large cell breaks into many fragments and each of the fragments becomes a platelet. Platelets do not have a nucleus and they have a relatively short cellular lifespan of about 8–10 days.

Let's consider what happens when a small blood vessel is damaged (see Figure 11.1). The damaged cells of the blood vessel release chemicals which stimulate platelets to adhere to the damaged area. Then other platelets begin adhering to those platelets. This begins to form a plug for the damaged area. To strengthen the plug, the damaged tissue and platelets release chemicals called clotting factors which convert prothrombin into thrombin. Thrombin is an active enzyme which catalyses the conversion of soluble fibrinogen into the relatively insoluble fibrin. The appropriately named fibrin is a fibrous protein which forms a mesh-like network that helps to stabilize the platelet plug. More and more cellular debris becomes trapped in the fibrin mesh and soon a stable clot has formed preventing both further blood loss and entry of pathogens.

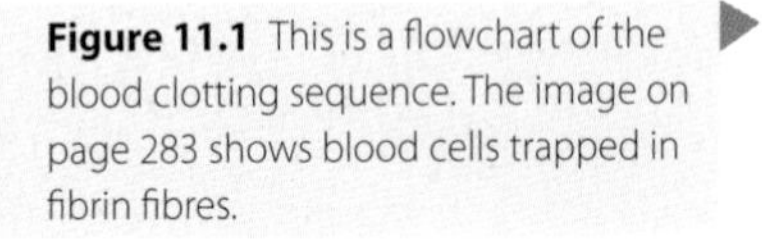

Figure 11.1 This is a flowchart of the blood clotting sequence. The image on page 283 shows blood cells trapped in fibrin fibres.

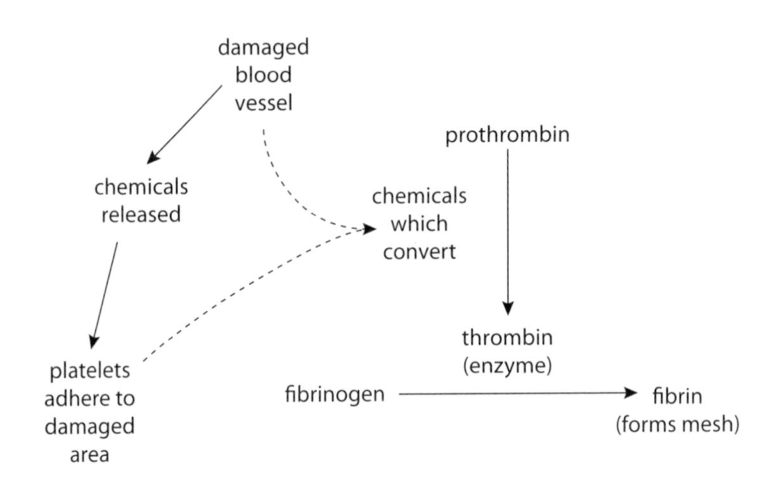

Haemophilia is an inherited blood clotting disorder. It is a sex-linked trait, and most haemophiliacs are males.

The immune response

There are more than 80 diseases affecting humans that are classified as autoimmune diseases. An autoimmune disease is the result of the immune system failing to completely recognize 'self' from 'not-self'. Lupus and multiple sclerosis are autoimmune diseases.

When a pathogen enters your body, your immune system responds and attempts to get rid of the potentially disease-causing agent or organism. Early in life, the immune system cannot recognize which invaders cause disease (e.g. polio virus) and which do not (e.g. a transplanted kidney). The recognition is simply one of 'self' or 'not-self'. Each of your body cells contains the same genetic information and all have a common set of cell membrane proteins. Some of your leucocytes are capable of recognizing that set of proteins and consider any cell with those proteins to be 'self'. A virus, bacterium, fungus, or even a transplanted organ has different proteins and thus is recognized as 'not-self'. As you may recall from Chapter 6, the collective term for all of the molecules which are 'not-self', is 'antigens'.

Antibody production

Among the leucocytes in your bloodstream are many different types of B lymphocyte. Each type of B lymphocyte or B cell is capable of synthesizing and secreting a specific antibody which binds to a specific antigen. The problem is that you cannot have enough of each type of B cell for the amount of antibody secretion that may be needed at various times. Leucocytes represent roughly 1% of all the cells in your bloodstream, so no one type of B lymphocyte is found in high numbers. Your body has cellular communication methods which lead to the cloning of the appropriate B cell type to synthesize and secrete the required antibody type to combat a specific antigen when needed.

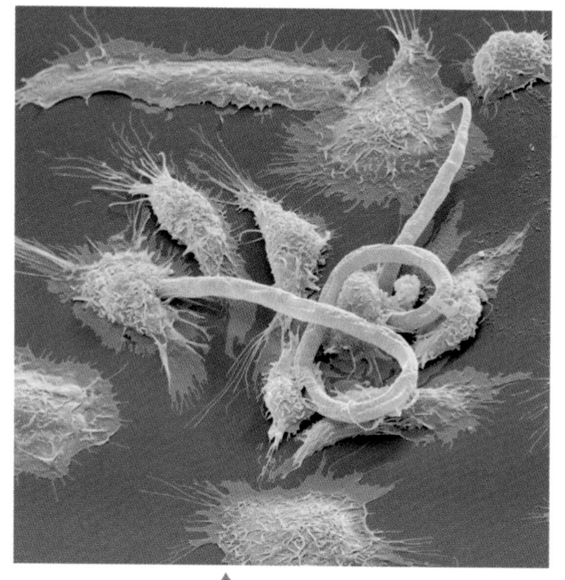

False-colour SEM showing many macrophages which have identified this parasitic nematode as 'not-self'. The macrophages are attempting phagocytosis of the worm.

The first type of leucocyte to encounter an antigen is usually the large, phagocytic cell known as a macrophage. As you may recall from Chapter 6, macrophages are found both inside and outside the bloodstream. When a macrophage encounters a 'not-self' antigen, it engulfs the possible pathogen by phagocytosis and only partially digests it. Molecular pieces of the invader are displayed on the cell membrane of the macrophage – this is known as antigen presentation. In the bloodstream, leucocytes known as helper-T cells chemically recognize the antigen being presented and become activated. Helper-T cells turn the immune response from non-specific ('not-self') to antigen-specific as the identity of the antigen is now determined. Helper-T cells chemically communicate with (activate) the specific B cell type that is able to produce the antibody needed.

Cell cloning

When a helper-T cell activates a specific B cell, the activated B cell type begins a series of cell divisions. This is known as cell cloning because all the daughter cells of these mitotic divisions are capable of producing the same antibody. There are two types of cloned B cell:

- antibody-secreting plasma cells – these cells secrete antibodies immediately and help to fight off the first (or primary) infection;
- memory cells – these cells do not secrete antibodies during the first infection, but are long-lived cells which remain circulating in the bloodstream waiting for a subsequent (or secondary) infection.

Fundamental principles of true immunity

The scenario above describes a series of events occurring in a primary infection. Your immune system helped you get rid of the pathogen, but it could not protect you from the pathogen entirely as all of the steps of the primary immune response take time. During that time, the pathogen may be causing damage and producing symptoms associated with the particular disease.

However, there is a difference in the case of a subsequent infection: the memory cells that were produced from the primary infection are circulating in your bloodstream. These very long-lived cells, now in large numbers, are capable of responding to the same pathogen very quickly.

The following principles of immunity apply for all types of infection.

- *Challenge and response*: The immune system must be challenged by an antigen during the first infection in order to develop an immunity. All the cellular events (involving macrophages, helper-T cells and B cells) are part of the response which leads to immunity to this pathogen.

- *Clonal selection*: This term best describes the identification of the leucocytes (e.g. particular plasma B cells) that can help with a specific pathogen *and* the multiple cell divisions which occur to build up the numbers of that same cell type. Simply put, your immune system selects the type of cell that will be useful and initiates cloning of that cell.
- *Memory cells*: These are the cells that provide long-term immunity. You must experience a pathogen (antigen) once in order to produce these cells and have true immunity to that specific pathogen.

Active and passive immunity

The events just described represent active immunity. Active immunity always leads to the production of memory cells and thus provides for a long-term immunity to a pathogen. Whenever the immune system is presented with an antigen and there is a full immune response, memory cells are produced.

Passive immunity is when one organism acquires antibodies which were produced in another organism. Only the organism which produced the antibodies has the memory cells and thus gains full long-term immunity. Acquiring antibodies confers only short-term benefit (as the antibodies bind to the antigen). Examples of passive immunity include the following.

- Transfer of antibodies from mother to fetus through the placenta. Memory cells are not transferred and thus only short-term protection is gained.
- Acquisition of antibodies from the mother's colostrum. Colostrum is the breast milk produced in late pregnancy and the first few days after birth. It is low in fat, but has a high antibody concentration.
- Injection of antibodies in antisera. Typical examples of such antisera are the antivenoms produced for treatment of poisonous snake and spider bites. Waiting for a primary immune response would mean massive tissue damage or possible death. Venomous animals are often 'milked' of their venom in order to inject small quantities of the venom into an animal which acts as the antibody factory.

Polyclonal and monoclonal antibodies

A primary immune response by an organism is called a polyclonal response. This is because the pathogen is typically being recognized as many antigens and not just one. For example, the capsid (protein coat) of a virus is typically made up of several different kinds of protein. Each of the protein types can cause an immune response and thus several different kinds of plasma B cell undergo clonal selection, so several different kinds of antibody are produced and several different kinds of memory cell remain after the infection. Once a polyclonal immune response has occurred, it is very difficult to separate the different kinds of antibody that have been produced.

Researchers have developed a clever and unique procedure for forming many antibodies, all of the same type. The term that applies to these 'pure' (all of the same type) antibodies is 'monoclonal'.

Production of monoclonal antibodies

The procedure for producing monoclonal antibodies begins with the injection of an antigen into a laboratory animal such as a mouse (see Figure 11.2). The choice of the antigen is very important as the antibodies which will be produced will recognize this specific antigen. The animal is given time to go through a primary immune response. As you have just seen, this response is polyclonal. After an appropriate period of time, the spleen of the lab animal is 'harvested' in order to gain access to many blood cells. At least some of the leucocytes cloned for the antigen which was recently injected will be a part of the cellular population within the spleen. There are two problems which need to be addressed at this point in the procedure:

- keeping the B cell types alive for an extended period of time;
- identification of the B cell type that produces the antibody which recognizes the desired antigen.

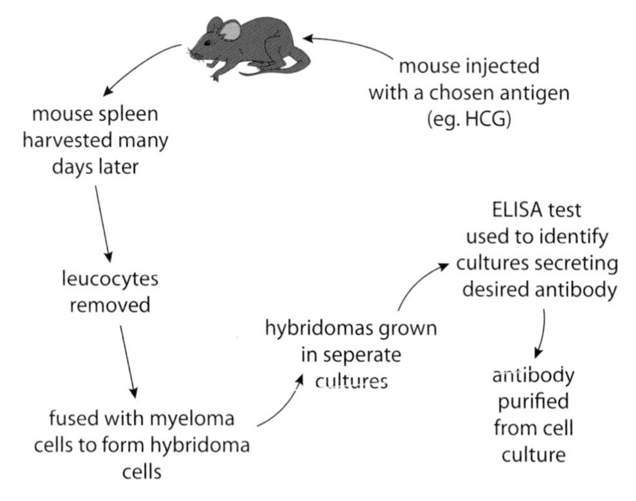
mouse injected
with a chosen antigen
(eg. HCG)
mouse spleen
harvested many
days later
leucocytes
removed
fused with myeloma
cells to form hybridoma
cells
hybridomas grown
in seperate
cultures
ELISA test
used to identify
cultures secreting
desired antibody
antibody
purified
from cell
culture

Figure 11.2 This is a flowchart for the production of monoclonal antibodies.

The B cells are kept alive by fusing them with cancerous (myeloma) cells. When B cells and myeloma cells are grown together in the proper environmental conditions, a few of the cells fuse together and become a cell called a hybridoma. These hybrid cells have characteristics of both cells; they produce antibodies of a particular type and they are very long lived (as are all cancer cells). The entire mix of cells is now transferred to an environment in which only the hybridoma cells can survive and all of the B cells and myeloma cells that did not fuse die.

Individual surviving hybridoma cells are now cultured in separate containers. Each container is tested for the presence of a particular antibody. The typical protocol to test for a specific protein (such as an antibody) is called an ELISA (enzyme-linked immunosorbent assay). An ELISA test identifies which containers hold a pure colony of B cells which are producing the type of antibody desired. These cells can be cultured for a very long period of time because they also have some of the characteristics of a tumour cell. In other words, the hybridoma cells are virtually immortal as long as they are kept in a suitable environment.

The first vaccine to be used was cowpox to protect people from smallpox. Cowpox is molecularly similar to smallpox, but has no dangerous symptoms to people. The word 'vaccine' is derived from the Latin word *vacca* meaning cow.

Uses of monoclonal antibodies

Diagnosis

Monoclonal antibodies can be used for a wide variety of diagnostic purposes. One common use is pregnancy testing. Early in pregnancy, the embryo begins to produce a hormone called human chorionic gonadotrophin (HCG). Since it is a hormone produced by the embryo, only a pregnant woman would have this hormone and the hormone shows up in small amounts in her bloodstream and urine. Hybridomas can be produced by injecting a laboratory animal with HCG.

Diseases do not recognize international borders. The health programmes and laws regarding the health of any one nation have an impact on all citizens of Earth.

The B cells that are later produced secrete antibodies that recognize HCG as an antigen. These anti-HCG antibodies are chemically bonded to an enzyme which catalyses a colour change when the antibody encounters HCG molecules. This is why pregnancy test results involve a colour indicator.

Treatment

There is increasing use of monoclonal antibodies for medical treatment as well as for diagnosis. When body cells become cancerous, they begin to produce cancer cell-specific antigens on their cell membranes. One possible treatment for cancer is to produce monoclonal antibodies that target the cancer-cell antigens. The monoclonal antibody could be chemically modified to carry with it a toxin specific for this type of cancer cell or perhaps a radioisotope for pin-point radiation therapy. The big advantage to this type of treatment is the ability to target the cancer cell directly. Far less toxin and radioisotope are needed because they are taken directly to the cancer cells.

How does a vaccine result in immunity?

One of the fundamental principles of immunity is that you cannot be immune to a pathogen before being exposed to it at least once. For some diseases, like the common cold, we simply wait for the exposure, experience the symptoms of the disease and then develop an immunity. You will probably not develop symptoms to the same cold virus ever again, but you probably will get another cold as a result of a different cold virus to which you have not yet been exposed.

For many diseases, we have developed vaccines that act as the first exposure to the pathogen. A vaccine is developed by weakening a pathogen and then injecting the pathogen into the body. There are several methods for producing a weakened pathogen: selecting a particular 'weak' strain of the pathogen, heating the pathogen, and chemical treatment of the pathogen. The leucocytes responsible for the primary immune response still recognize the weakened pathogen as 'not-self' and the primary immune response takes place. This includes the formation of memory B cells capable of producing antibodies very quickly if there is a later infection with the real pathogen.

A vaccination does not prevent an infection but, on subsequent exposure to the real pathogen, the secondary immune response is quicker and more intense than the primary immune response (see Figure 11.3). After vaccination, most people respond so quickly to the real pathogen that only very mild symptoms or perhaps no obvious symptoms at all result.

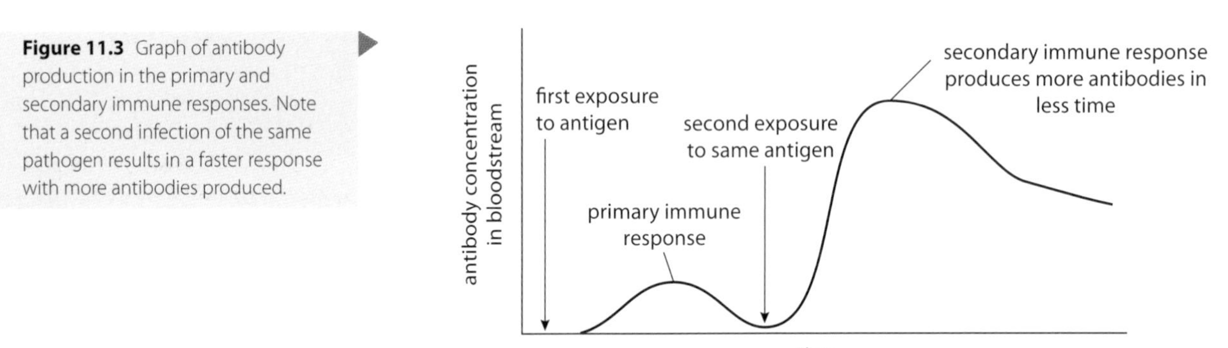

Figure 11.3 Graph of antibody production in the primary and secondary immune responses. Note that a second infection of the same pathogen results in a faster response with more antibodies produced.

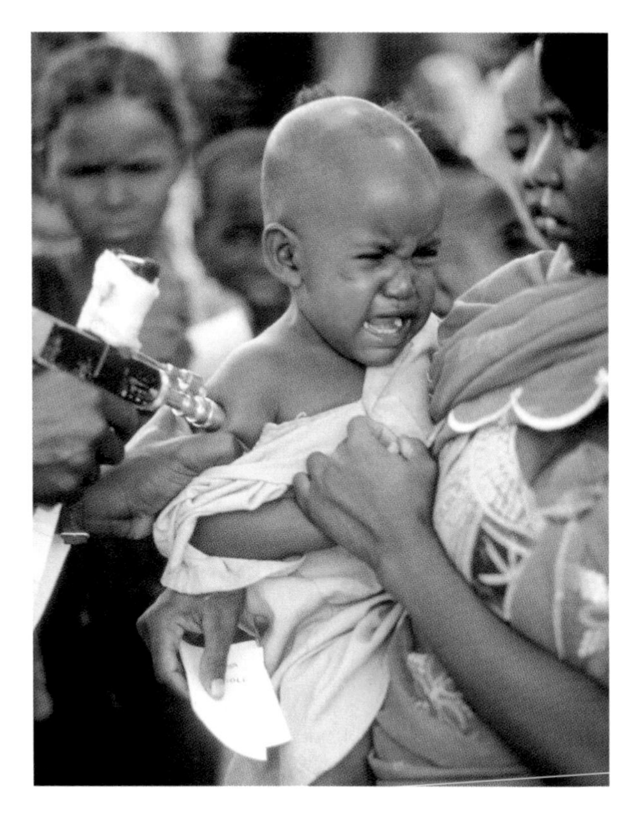

Child receiving measles vaccine.

In many nations, vaccinations against certain diseases are mandatory. Many of these vaccinations apply to schoolchildren and some apply for entry into the country. This leads to the basic question of the right of an individual to make decisions about their own healthcare.

Benefits and dangers of vaccination

Most people agree that generally speaking, vaccination has benefited humankind greatly in the battle against infectious disease. Some diseases, such as smallpox, have been eliminated from the Earth's population – there has not been a reported case of smallpox anywhere in the world since 1977. This is a direct result of a smallpox vaccination programme headed by the World Health Organization (WHO).

However, not every individual is always happy about getting a vaccination, especially if it is a requirement of a government agency. This table compares some of the benefits and dangers associated with vaccination.

Benefit	Danger
Possible total elimination of the disease. This has occurred with smallpox and many people believe it is possible to eradicate both polio and measles.	Prior to 1999, many vaccines contained thimerosal, a mercury-based preservative. Mercury has been shown to be a neurotoxin to which infants and young children are particularly susceptible.
Decrease in spread of epidemics (localized infections) and pandemics (worldwide infections). Increased international travel has made this more important than ever. An infection begun on one side of the world could be on the other side of the world in less than a day.	The perception exists that multiple vaccines given to children in a relatively short period of time may 'overload' their immune system.
Preventative medicine is typically the most cost-effective approach to healthcare. Costs associated with vaccination programmes are small compared to the costs of treating many preventable diseases.	Anecdotal evidence suggested that MMR (measles, mumps, rubella) vaccine may have a link to the onset of autism. Clinical studies have not supported this.
Each vaccinated individual benefits because the full symptoms of the disease do not have to be experienced in order to gain immunity.	Cases have been reported of vaccines leading to allergic reactions and autoimmune responses.

Exercises

1 You cut yourself slightly while shaving. List the sequenced events necessary to form a clot within one of the small damaged blood vessels.

2 Approximately 99% of your blood cells are erythrocytes and approximately 1% are leucocytes. Outline how the body produces many more of any one type of leucocyte needed when necessary.

3 Current treatments for many types of cancers include both traditional radiation and chemotherapy. What advantages do monoclonal antibody therapies offer over these treatments?

11.2 Muscles and movement

The Higher Level material in this section is also found in the first section of Option B (Chapter 13), an option available only to Standard Level students.

Assessment statements

11.2.1 State the roles of bones, ligaments, muscles, tendons and nerves in human movement.

11.2.2 Label a diagram of the human elbow joint, including cartilage, synovial fluid, joint capsule, named bones and antagonistic muscles (biceps and triceps).

11.2.3 Outline the functions of the structures in the human elbow joint named in 11.2.2.

11.2.4 Compare the movements of the hip joint and the knee joint.

11.2.5 Describe the structure of striated muscle fibres, including the myofibrils with light and dark bands, mitochondria, the sarcoplasmic reticulum, nuclei and the sarcolemma.

11.2.6 Draw and label a diagram to show the structure of a sarcomere, including Z lines, actin filaments, myosin filaments with heads, and the resultant light and dark bands.

11.2.7 Explain how skeletal muscle contracts, including the release of calcium ions from the sarcoplasmic reticulum, the formation of cross-bridges, the sliding of actin and myosin filaments, and the use of ATP to break cross-bridges and re-set myosin heads.

11.2.8 Analyse electron micrographs to find the state of contraction of muscle fibres.

Joints

A joint, also called an articulation or arthrosis, is the point where two or more bones contact one another. Arthrology is the scientific study of joints and rheumatology is the branch of medicine devoted to joint diseases and conditions. The science of kinesiology examines the movement of the human body.

Our joints provide mobility. They also hold the body together. Most joints include:

- bones;
- ligaments;
- muscles;
- tendons;
- nerves.

Bones

It is important to note that bones contain several different tissues and, therefore, they are organs. Bones have many functions:

- providing a hard framework to support the body;
- allowing protection of vulnerable softer tissue and organs;
- acting as levers so that body movement can occur;
- forming blood cells in the bone marrow;
- allowing storage of minerals, especially calcium and phosphorus.

Here we are going to concentrate on bones acting as levers for movement (see Figure 11.4).

We are able to run because our skeleton and musculature work together to produce movement.

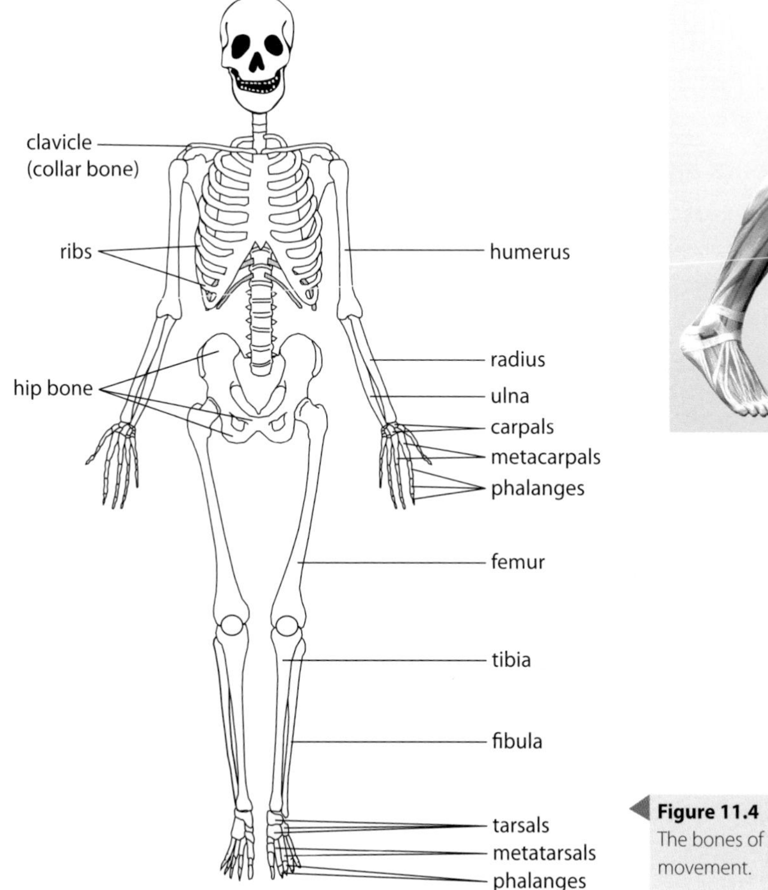

Figure 11.4 Human adults have 206 bones. The bones of the limbs are involved in movement.

Muscles and tendons

For movement to occur, it is essential that skeletal muscles are attached to bones. This attachment is provided by the tendons. Tendons are cords of dense connective tissue. It is the arrangement of the bones and the design of the joints that determine the type or range of motion possible in any particular area of the body. By acting as levers, the bones function to magnify the force provided by muscle contraction. The muscles provide the force necessary for movement by shortening the length of their fibres or cells. Since the muscles bring about movement only by shortening, it is essential they occur as antagonistic pairs (see Figure 11.5). This allows a body part to be returned to its original position after a movement.

Figure 11.5 The biceps and triceps are opposing muscle groups in the upper arm. When the biceps contracts, the triceps relaxes, and the forearm is moved upwards (flexes). When the triceps contracts, the biceps relaxes and the forearm is moved downwards (extends).

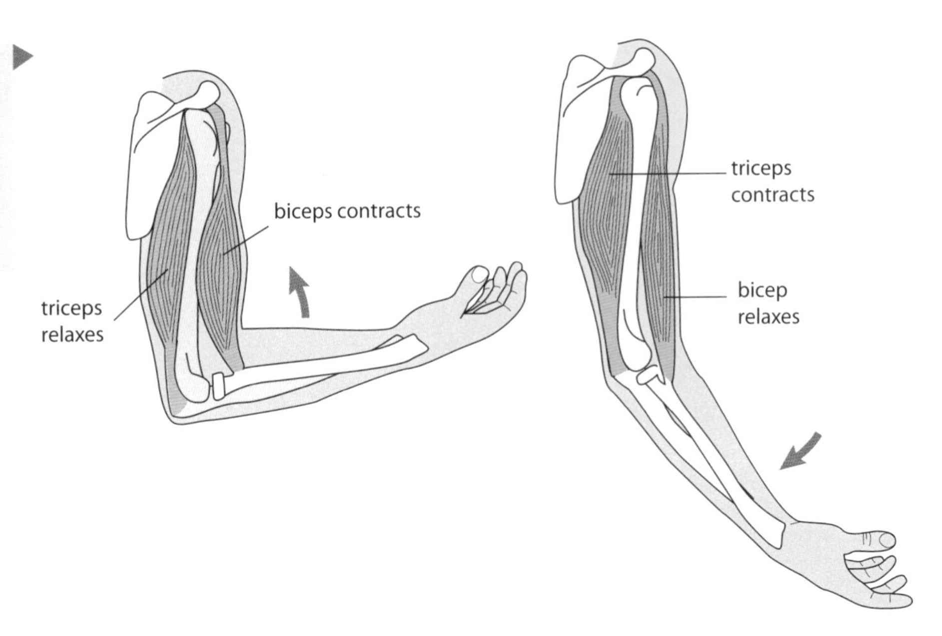

Ligaments and nerves

Ligaments are tough, band-like structures that serve to strengthen the joint. They also provide stability. The ligaments have many different types of sensory nerve endings. Proprioceptors in ligaments and muscles allow constant monitoring of the positions of the joint parts. The nerves help to prevent over-extension of the joint and its parts.

Blood supply

There is a rich supply of blood to joints. If blood vessels supplying the joint get damaged and there is local bleeding (haemorrhage), it may result in swelling of the area.

This false-colour X-ray shows a normal human elbow joint.

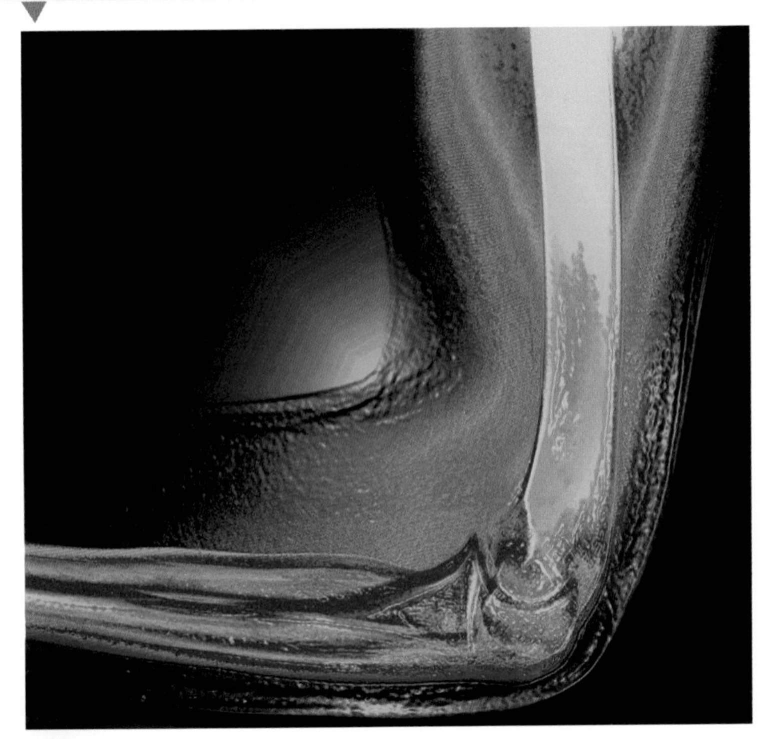

Hinge joints

Let's look at the elbow joint in detail. It is a hinge joint. This means it provides an opening-and-closing type of movement like the action of a door.

To discuss complex structures such as joints, we need a large specialist vocabulary, so you may find some of the terms used in this chapter unfamiliar. It is well worthwhile taking the time to become thoroughly familiar with any words that you are not completely sure of.

The elbow joint involves the humerus, radius and ulna bones (see Figure 11.6). The synovial fluid is present within the synovial cavity. This cavity is located within the joint capsule. The joint capsule is composed of dense connective tissue that is continuous with the membrane of the involved bones.

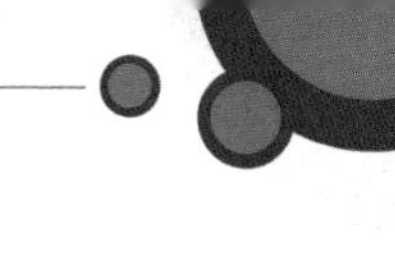

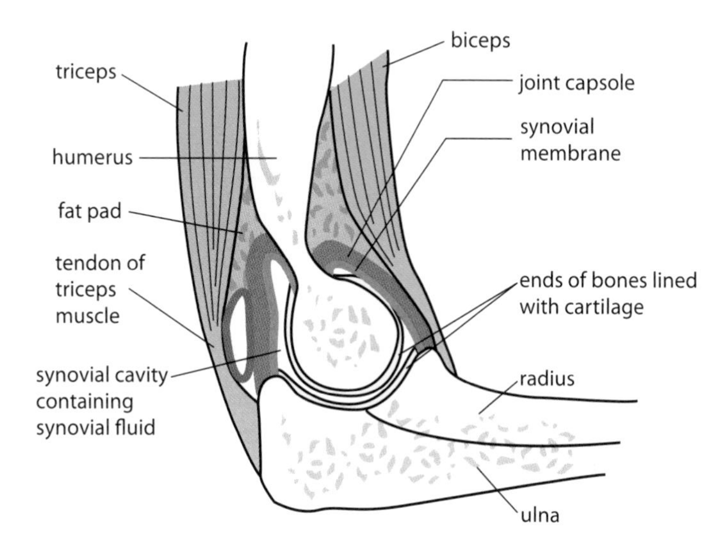

Figure 11.6 The elbow joint and its parts.

Elbow parts and their functions are summarized in this table.

Joint part	Function
cartilage	reduces friction and absorbs compression
synovial fluid	lubricates to reduce friction and provides nutrients to the cells of the cartilage
joint capsule	surrounds the joint, encloses the synovial cavity, and unites the connecting bones
tendons	attach muscle to bone
ligaments	connect bone to bone
biceps muscle	contracts to bring about flexion (bending) of the arm
triceps muscle	contracts to cause extension (straightening) of the arm
humerus	acts as a lever that allows anchorage of the muscles of the elbow
radius	acts as a lever for the biceps muscle
ulna	acts as a lever for the triceps muscle

For a further explanation and demonstration of the elbow, visit heinemann.co.uk/hotlinks, insert the express code 4242P and click on Weblink 11.1.

The elbow is called a synovial joint because of the presence of the synovial cavity. The knee is a similar joint (see Figure 11.7). These joints are freely movable. Freely movable joints are also called diarthrotic joints.

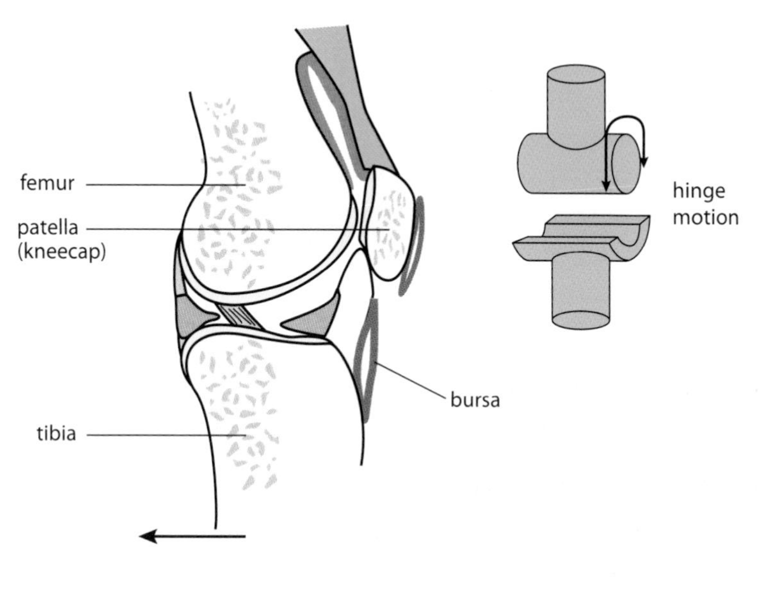

Figure 11.7 The knee joint is much like the elbow. It is a hinge joint and permits an opening-and-closing type of movement. This is movement in one direction and produces angular motion.

Ball-and-socket joints

The hip joint is also a diarthrotic joint (see Figure 11.8), but it is not a hinge joint.

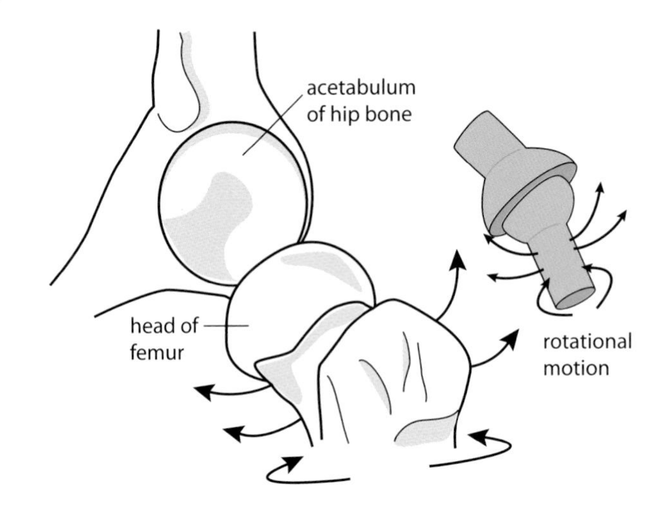

Figure 11.8 The hip joint is a ball-and-socket joint. Because of the structure of the joint, movement is possible in a number of different directions. The shoulder is also a ball-and-socket joint.

There is a great difference in the motion possible at the elbow or knee and that possible at the hip. The hip is a ball-and-socket joint and it permits movement in several directions, including rotational movement. This is because the head of the femur (a ball shape) fits into a cup-like depression of the hip bone called the acetabulum.

This table compares the hip and knee joints.

Hip joint	Knee joint
freely movable	freely movable
angular motions in many directions and rotational movements	angular motion in one direction
motions possible are flexion, extension, abduction, adduction, circumduction and rotation*	motions possible are flexion and extension
ball-like structure fits into a cup-like depression	convex surface fits into a concave surface

* Definitions:

- flexion = decrease in angle between connecting bones;
- extension = increase in angle between connecting bones;
- abduction = movement of bone away from body midline;
- adduction = movement of bone toward midline;
- circumduction = distal or far end of a limb moves in a circle;
- rotation = a bone revolves around its own longitudinal axis.

Muscle

The human body possesses three types of muscle tissue:

- skeletal or striated;
- cardiac;
- smooth or non-striated.

Here we are going to look at only striated muscle.

Striated muscle cells

Striated muscle is also called skeletal muscle because it is responsible for skeletal movement.

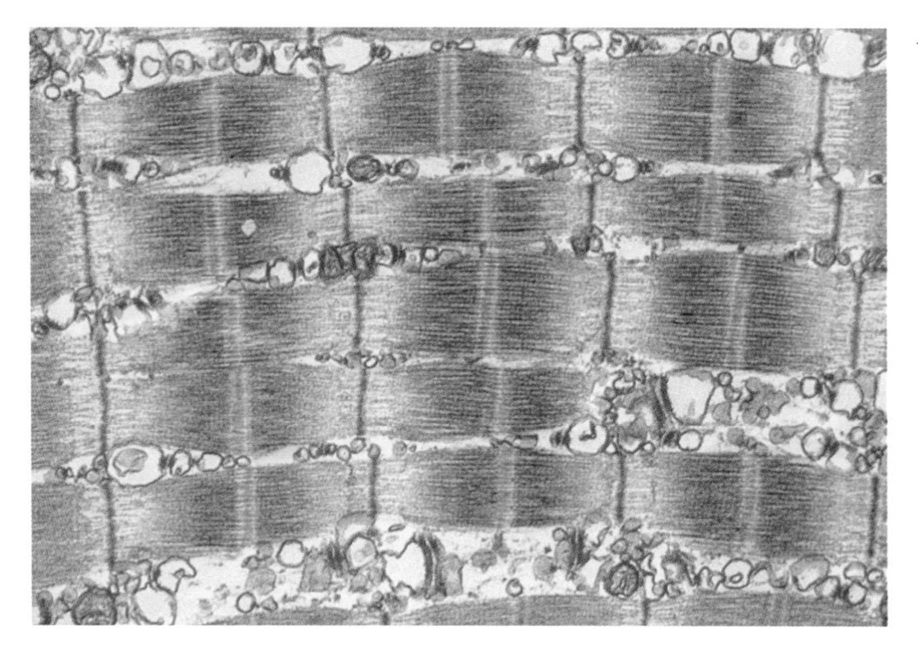

In this false-colour electron micrograph, you can clearly see why this muscle is said to be striated. The banding is quite apparent. You can see five myofibrils arranged parallel to one another horizontally. Each myofibril has a banded pattern that is repeated in its neighbours (see Figures 11.9 and 11.10). Can you identify three sarcomeres in four of the five myofibrils and two sarcomeres in the remaining myofibril?

Like any other tissue in the body, muscle is made up of cells. It is the cellular arrangement that produces the banded appearance of striated muscle. Muscles are composed of thousands of cells, which are called muscle fibres because of their elongated shape. Besides the muscle fibres, muscles include surrounding connective tissues, and blood vessels and nerves. The blood vessels and nerves penetrate the muscle body.

Muscle fibres (cells) contain multiple nuclei that lie just inside the plasma membrane, which is called the sarcolemma. The sarcolemma has multiple tunnel-like extensions that penetrate the interior of the cell. These penetrating invaginations are called transverse or T tubules.

The cytoplasm of muscle fibres is called the sarcoplasm. The sarcoplasm contains large numbers of glycosomes that store glycogen. Besides glycosomes, the sarcoplasm has large amounts of a red-coloured protein called myoglobin. Glycogen and myoglobin are discussed later when muscle movement is explained.

The sarcoplasmic reticulum is a fluid-filled system of membranous sacs surrounding the muscle myofibrils. Sarcoplasmic reticulum is much like smooth endoplasmic reticulum.

Myofibrils are rod-shaped bodies that run the length of the cell. There are many of them and they are parallel to one another. Myofibrils are closely packed and numerous mitochondria are squeezed between them. The myofibrils are the contractile elements of the muscle cells and they are the reason that striated muscle has a banded pattern.

Myofibril structure

Myofibrils are made up of sarcomeres, and sarcomeres are the units that allow movement (see Figure 11.9).

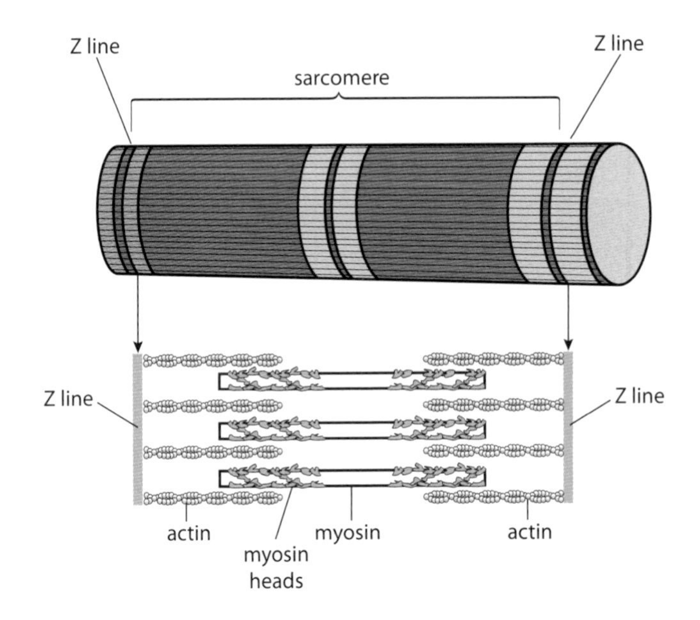

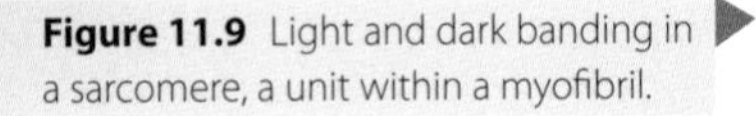
Figure 11.9 Light and dark banding in a sarcomere, a unit within a myofibril.

The sarcomere is often described as banded (see Figure 11.10).

- The Z lines mark the ends of the sarcomere.
- The A bands are dark in colour and extend the entire length of the myosin filaments. A narrow H band occurs in the middle of the A band – it contains only myosin, no actin. A supporting protein occurs in the middle of the myosin producing the M line. This protein holds the myosin filaments together.
- The I bands are light in colour and contain only actin – no myosin.

As you can see in the Figure 11.10, there are two types of filaments or myofilaments that cause the banded appearance of the muscle fibre. These myofilaments are composed of two contractile proteins, actin and myosin, and are described in the table on page 297.

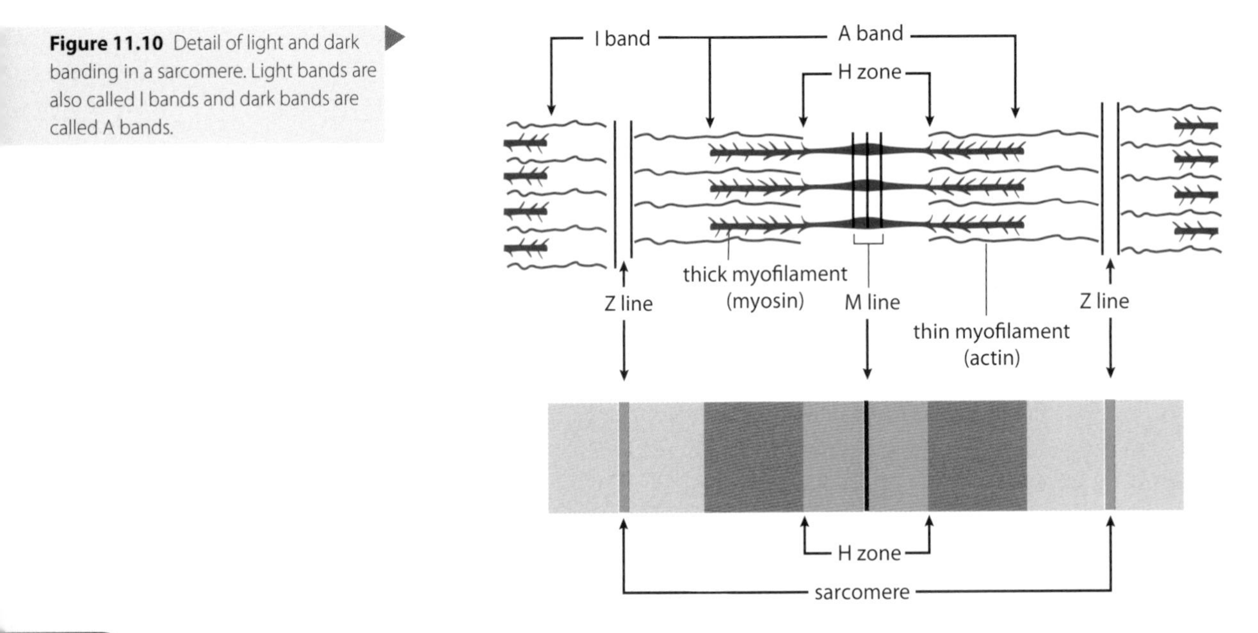

Figure 11.10 Detail of light and dark banding in a sarcomere. Light bands are also called I bands and dark bands are called A bands.

Actin	Myosin
thin filaments (8 nm in diameter)	thick filaments (16 nm in diameter)
contains myosin-binding sites	contains myosin heads that have actin-binding sites
individual molecules form helical structures	individual molecules form a common shaft-like region with outward protruding heads
includes two regulatory proteins, tropomyosin and troponin	heads are referred to as cross-bridges and contain ATP-binding sites and ATPase enzymes

In 1954, Hugh Huxley first proposed the sliding filament theory of muscle contraction.

Muscle contraction

The mechanism of muscle contraction is presently explained by the sliding filament theory. Essentially, this theory states that muscles contract when actin myofilaments slide over myosin myofilaments. The myofilaments do not actually shorten. When the actin component slides over the myosin, the sarcomere is shortened. With multiple fibres and sarcomeres in a muscle working together, this causes the movements necessary for the organism.

Botulinum toxin is produced by the bacterium *Clostridium botulinum*. The toxin blocks the release of acetylcholine, and thus prevents muscle contraction. This can affect the diaphragm, so breathing stops and death may occur.

Botulinum toxin is the active ingredient of Botox. Botox injected into the affected muscles may correct strabismus (crossed eyes) or blepharospasm (uncontrolled blinking). Botox is also used as a cosmetic to relax the muscles that cause facial wrinkles.

Key events of muscle contraction (sliding filament theory)

1. A motor neurone carries an action potential until it reaches a neuromuscular junction.
2. A neurotransmitter called acetylcholine is released into the gap between the axon terminal and the sarcolemma of the muscle fibre.
3. The acetylcholine binds to receptors on the sarcolemma.
4. Sarcolemma ion channels open and sodium ions move through the membrane.
5. This generates a muscle action potential.
6. The muscle action potential moves along the membrane and through the T tubules.
7. After generation of the muscle action potential, the acetylcholine is broken down by an enzyme called acetylcholinesterase. This ensures that one nerve action potential causes only one muscle action potential.
8. The muscle action potential moving through the T tubules causes release of calcium ions from the sarcoplasmic reticulum. The calcium ions flood into the sarcoplasm.
9. The calcium ions bind to troponin on the actin myofilaments. This exposes the myosin-binding sites.
10. The myosin heads include ATPase which splits ATP and releases energy (step 1 of Figure 11.11).
11. The myosin heads then bind to the myosin-binding sites on the actin with the help of the protein called tropomyosin (step 2 of Figure 11.11).
12. The myosin–actin cross-bridges rotate toward the centre of the sarcomere. This produces the power or working stroke (step 3 of Figure 11.11).
13. ATP once again binds to the myosin head resulting in the detachment of myosin from the actin (step 4 of Figure 11.11).
14. If there are no further action potentials, the level of calcium ions in the sacroplasm falls. The troponin–tropomyosin complex then moves to its original position, thus blocking the myosin-binding sites. The muscle then relaxes.

For some excellent representations of muscle contraction, visit heinemann.co.uk/hotlinks, enter the express code 4242P and click on Weblinks 11.2a, 11.2b and 11.2c.

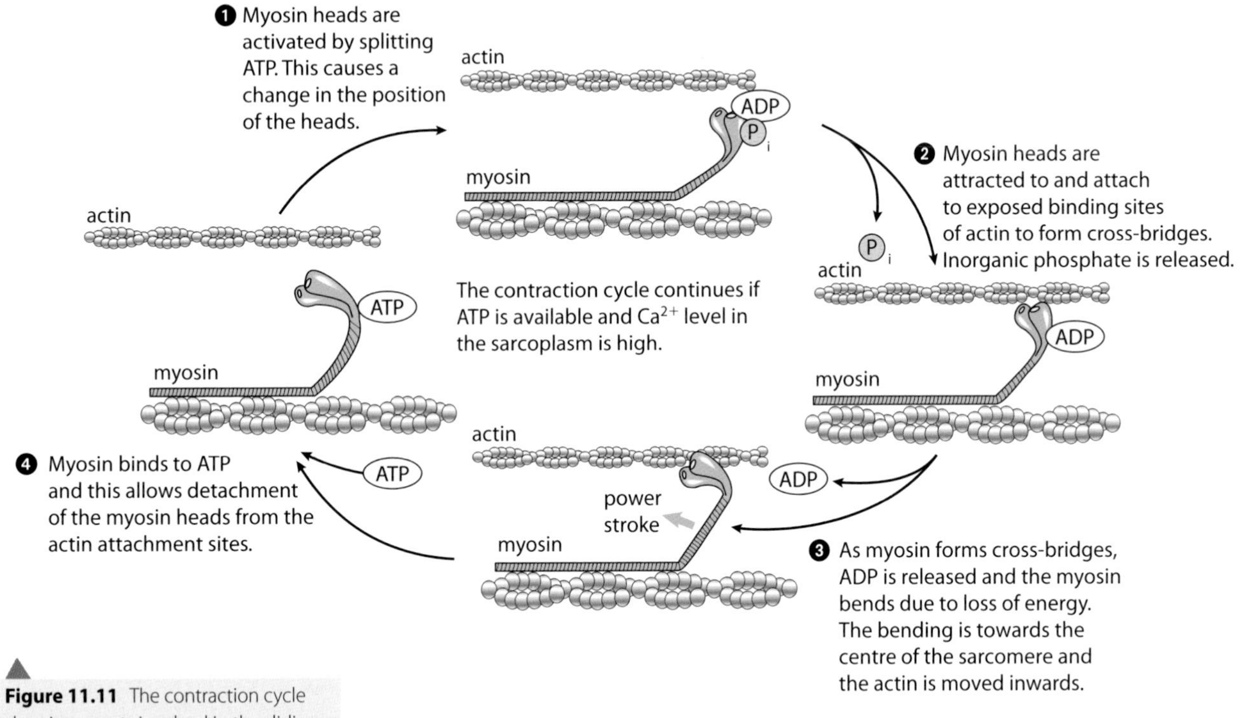
1 Myosin heads are activated by splitting ATP. This causes a change in the position of the heads.
actin
ADP
P
i
myosin
2 Myosin heads are attracted to and attach to exposed binding sites of actin to form cross-bridges. Inorganic phosphate is released.
actin
ATP
myosin
The contraction cycle continues if ATP is available and Ca2+ level in the sarcoplasm is high.
ADP
4 Myosin binds to ATP and this allows detachment of the myosin heads from the actin attachment sites.
power stroke
3 As myosin forms cross-bridges, ADP is released and the myosin bends due to loss of energy. The bending is towards the centre of the sarcomere and the actin is moved inwards.

Figure 11.11 The contraction cycle showing events involved in the sliding filament theory.

After a person dies, calcium ions leak out of the sarcoplasmic reticulum and bind to troponin. This allows the actin to slide. However, since ATP production stops at death, the myosin heads cannot detach from the actin. The result is *rigor mortis* (the rigidity of death). It lasts for about 24 hours until further muscle deterioration occurs.

Curare is a plant extract first used on darts and arrows to cause muscle paralysis and death. Similar drugs are used in modern-day surgery to relax skeletal muscles.

Actin and myosin myofilaments do not change in length during muscle contraction. Contraction is due to the rather complex sequence of events leading to the actin sliding over the myosin myofilaments. This causes the shortening of the sarcomere necessary for muscle movement. You will recall that a sarcomere extends from Z line to Z line; it is the smallest section of a muscle fibre to change in length.

Electron micrographs of sarcomeres in different degrees of contraction are clearly distinguishable. Figures 11.12–11.14 show the distinctions diagrammatically. Figure 11.12 shows two fully relaxed sarcomeres.

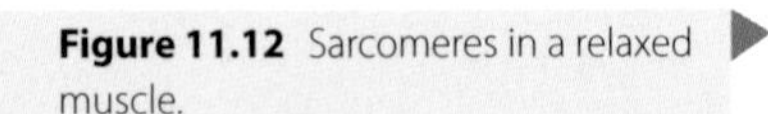

Figure 11.12 Sarcomeres in a relaxed muscle.

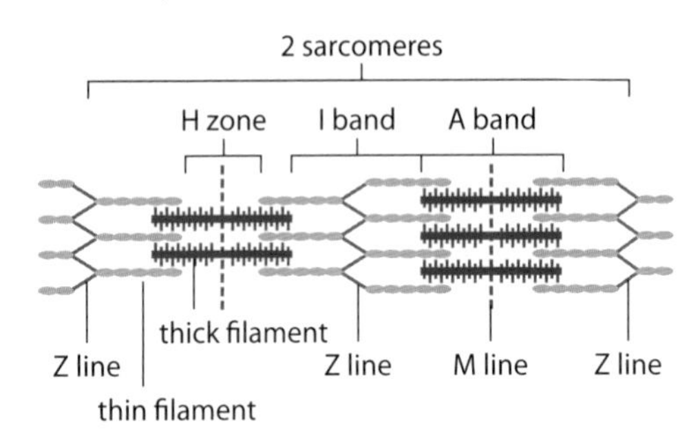
2 sarcomeres
H zone
I band
A band
thick filament
Z line
Z line
M line
Z line
thin filament

When the muscle is maximally contracted, the H zone disappears, the Z lines move closer together, the I bands are no longer present, and the A bands appear to run the complete length of the sarcomeres (see Figure 11.13).

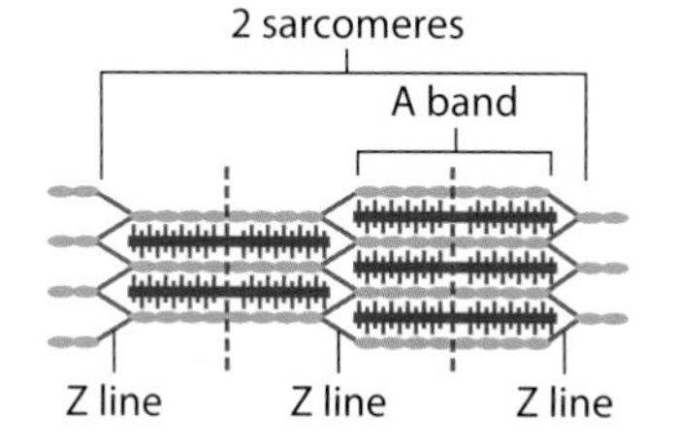

Figure 11.13 Fully contracted sarcomeres.

The muscle may also be in various states of partial contraction (see Figure 11.14). This causes a difference in the position of the sarcomere parts.

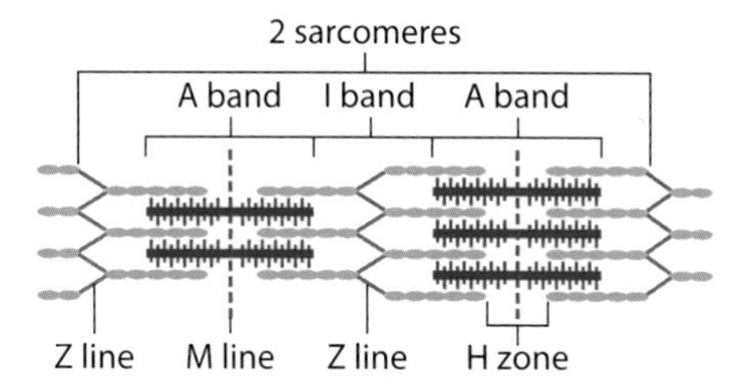

Figure 11.14 Partially contracted sarcomeres.

You can clearly see that muscles can be in many states of contraction. The number of muscle fibres in a muscle going through contraction determines the overall strength of a muscle contraction. Motor impulses from the central nervous system determine the number of muscle fibres that contract.

• **Examiner's hint:** You should be able to label parts of a muscle fibre at any stage of contraction in a photomicrograph. Study the electron micrograph on page 295 to identify the parts of the sarcomere.

Exercises

4 What would be the result of a decrease in synovial fluid at a freely movable joint?

5 Why does the H zone disappear in fully contracted sarcomeres?

6 What causes the release of calcium ions from the sarcoplasmic reticulum of a muscle fibre?

7 Of what value would a large number of mitochondria be to a muscle fibre?

8 Why must muscles occur in antagonistic pairs?

11.3 The kidney

Assessment statements

11.3.1 Define *excretion*.

11.3.2 Draw and label the structure of the kidney.

11.3.3 Annotate a diagram of a glomerulus and associated nephron to show the function of each part.

11.3.4 Explain the process of ultrafiltration, including blood pressure, fenestrated blood capillaries and basement membrane.

11.3.5 Define *osmoregulation*.

11.3.6 Explain the reabsorption of glucose, water and salts in the proximal convoluted tubule, including the roles of microvilli, osmosis and active transport.

11.3.7 Explain the roles of the loop of Henle, medulla, collecting duct and ADH in maintaining the water balance of the blood.

11.3.8 Explain the differences in the concentration of proteins, glucose and urea between blood plasma, glomerular filtrate and urine.

11.3.9 Explain the presence of glucose in the urine of untreated diabetic patients.

What is excretion?

Your blood plasma is a constantly changing solution. All of the collective reactions within your body cells are referred to as your metabolism. The bloodstream acts to supply needed substances for your metabolism and it also removes molecular waste products from the tissues. Given this constant addition of wastes such as urea, your bloodstream needs to be filtered and cleansed. Urea is a waste product from the metabolism of amino acids. Each amino acid must be deaminated (lose an amine group), and in mammals and some other animals this results in the formation of urea. It is the job of the kidneys to filter and cleanse the bloodstream of molecules like urea and other molecular wastes. Excretion is the removal from the body of the waste products of metabolic pathways.

Anatomy of a kidney

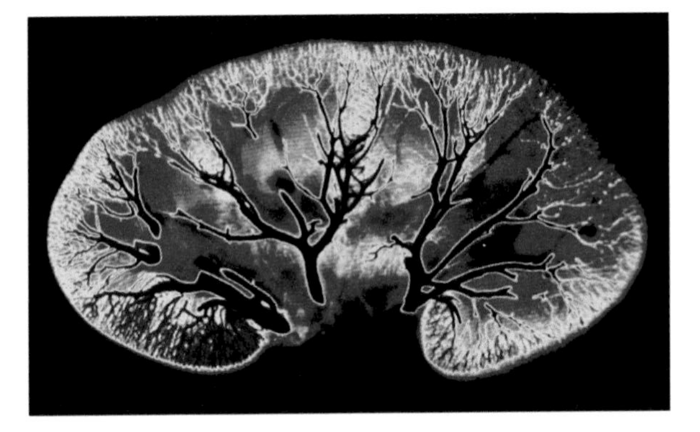

Angiogram showing the network of blood vessels within a kidney. The purpose of a kidney is to filter blood, so kidney tissue is highly vascular.

Since the function of kidneys is to filter waste products from the blood, a major blood vessel called the renal artery takes blood into each of the kidneys. Blood drains away from each kidney by a blood vessel known as the renal vein. Urine is the fluid produced by the kidneys; it consists of water and dissolved waste products which have been removed from the bloodstream. Urine collects within each kidney in an area called the renal pelvis. The renal pelvis drains urine into a tube called the ureter, which then takes the urine to the urinary bladder. When the kidney is cut in section, as shown in Figure 11.15, you can see the layer of tissue surrounding the renal pelvis, which is called the renal medulla; the layer to the outside of that is the renal cortex.

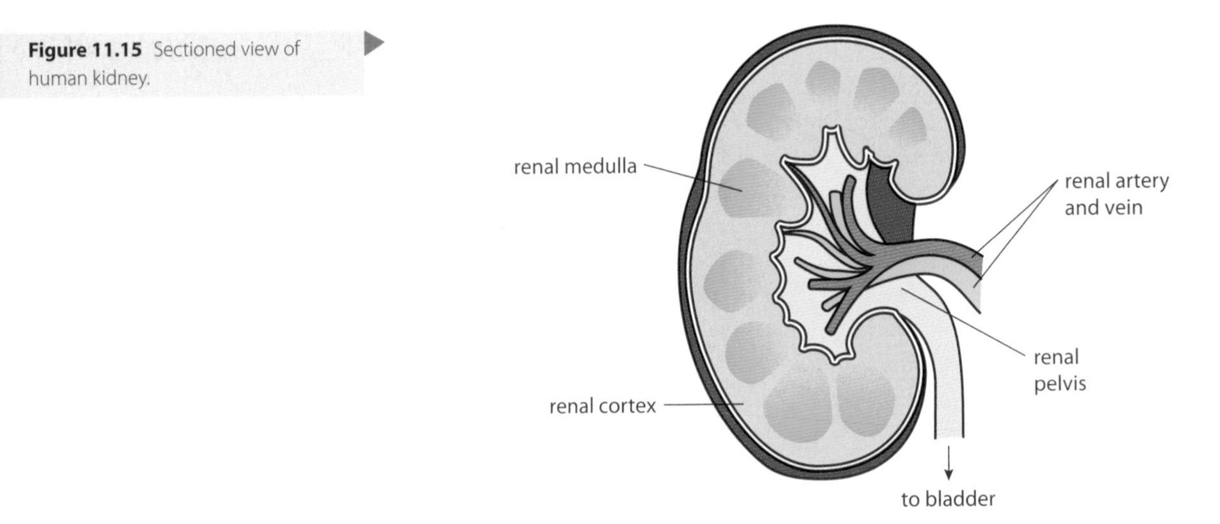

Figure 11.15 Sectioned view of human kidney.

Nephrons are the filtering units of kidneys

Each kidney is made up of about 1.25 million filtering units known as nephrons (see Figure 11.16). Each nephron consists of:

- a capillary bed, called a glomerulus, which filters various substances from the blood;
- a capsule surrounding the glomerulus called Bowman's capsule;
- a small tube (tubule) that extends from Bowman's capsule and has parts named (in this order)
  - proximal convoluted tubule
  - loop of Henle
  - distal convoluted tubule;
- a second capillary bed called the peritubular capillary bed which surrounds the three-part tubule mentioned above.

For an introduction to kidney function, visit heinemann.co.uk/hotlinks, enter the express code 4242P and click on Weblink 11.3.

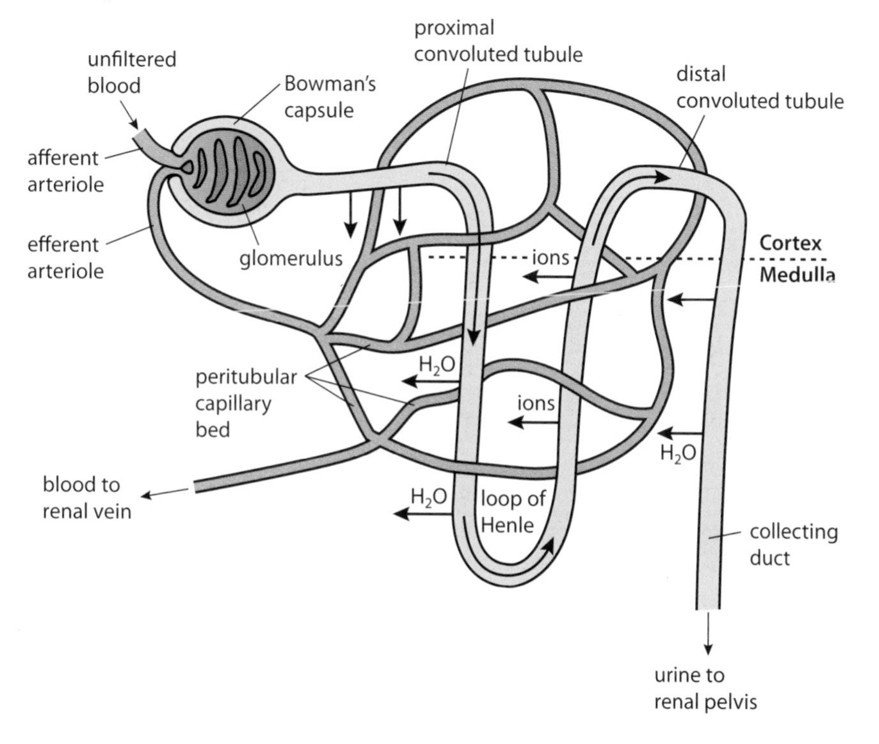

Figure 11.16 A single nephron of the human kidney.

Blood is ultrafiltered within Bowman's capsule

Each nephron contains a very small branch of the renal artery known as an afferent arteriole. This brings unfiltered blood to the nephron. Inside Bowman's capsule, the afferent arteriole branches into a capillary bed called the glomerulus. The glomerulus is similar to most other capillary beds except that the walls of the capillaries have fenestrations (very small slits) that open when blood pressure is increased. The increase in blood pressure is provided by the fact that the *efferent* arteriole, which drains blood from the glomerulus, has a smaller diameter than the *afferent* arteriole, which brings the blood to the glomerulus. This means the pressure in the glomerulus must increase (see Figure 11.17).

To view a good diagram of the glomerulus and Bowman's capsule, visit heinemann.co.uk/hotlinks, enter the express code 4242P and click on Weblink 11.4.

'Ultrafiltration' is the term used to describe the process by which various substances are filtered through the glomerulus (and its fenestrations) under the unusually high blood pressure in this capillary bed. The fluid which is ultrafiltered from the glomerulus then passes through the basement membrane which helps prevent large molecules like proteins from becoming a part of the filtrate. The filtrate next enters the proximal convoluted tubule. The blood cells, proteins, and other molecules which did not become a part of the filtrate exit Bowman's capsule by way of the efferent arteriole (see Figure 11.17).

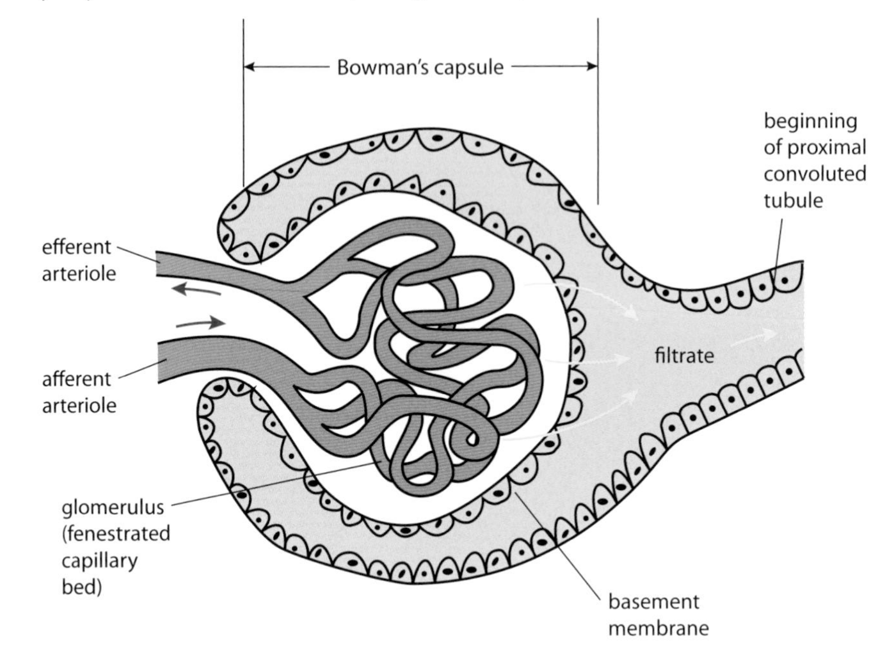

Figure 11.17 Bowman's capsule is the site of ultrafiltration.

Reabsorption recovers needed substances to the bloodstream

The filtrate that leaves Bowman's capsule contains many substances that the body cannot afford to lose in the urine, for example a great deal of water, salt ions and glucose. These substances need to be reabsorbed into the bloodstream. Much of the reabsorption process occurs from the proximal convoluted tubule. Substances leave the tubule filtrate and are taken back into the bloodstream by way of the peritubular capillary bed. This capillary bed is so named because it surrounds (peri-) the tubule (see Figure 11.16).

The entire total volume of your blood is filtered by your kidneys about 25 times each day. This shows why reabsorption is so important.

The wall of the proximal convoluted tubule is a single cell thick (see Figure 11.18). In any one area, the tubule is composed of a ring of cells. The interior of this tube is called the lumen and the filtrate flows within the lumen. The inner portion of each of the tubule cells has microvilli in order to increase the surface area for reabsorption.

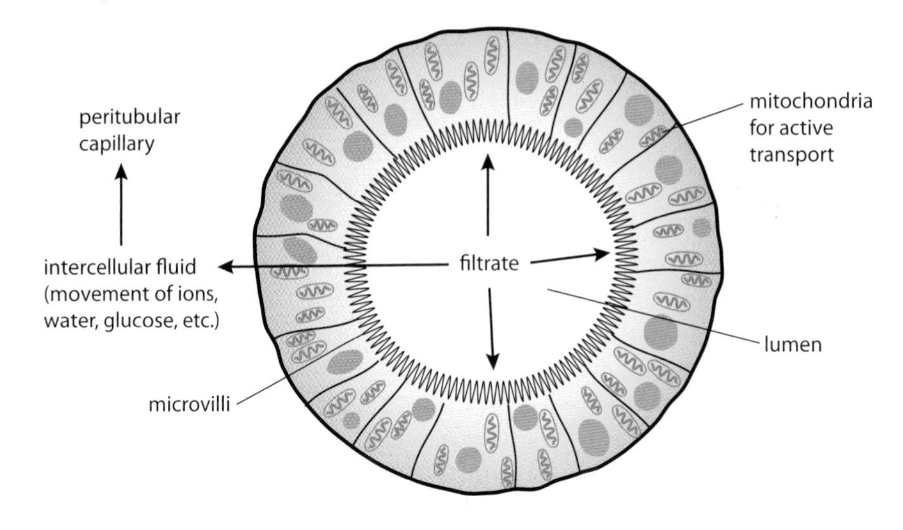

Figure 11.18 Sectioned view of a small area of the proximal convoluted tubule. Note that the tubule wall is one cell thick.

Several transport mechanisms are used in order to accomplish reabsorption. Even though any one type of molecule may be influenced by more than one transport mechanism, there are still some general patterns.

- *Salt ions*: Not all, but the majority of salt ions (e.g. Na^+, Cl^-, K^+) must leave the filtrate and be returned to the bloodstream by reabsorption. The salt ions are first actively transported into the tubule cells and then into the intercellular fluid outside the tubule. Finally, salt ions are taken into the peritubular capillary bed.
- *Water*: The movement of salt ions out of the filtrate and into the tubule cells, intercellular fluid and peritubular capillary bed induces water to follow the same route by osmosis. Recall that water moves from a hypotonic region to a hypertonic region basically following the pathway of the solutes.
- *Glucose*: In a nephron that is functioning properly, all the glucose that is in the glomerular filtrate is reabsorbed into the bloodstream. The only transport mechanism that can explain the totality of this movement is active transport. If glucose were being moved by dialysis (diffusion) or facilitated diffusion, the highest percentage that could be reabsorbed would be 50%.

Kidney nephrons and osmoregulation

Water is the solvent of life. It is the solvent in almost all body fluids including cytoplasm, blood plasma, lymph and intercellular fluid. Some water needs to be eliminated in the urine each day, but the total volume of water eliminated depends on many physiological factors. These include:

- total volume of water ingested recently as liquid and in solid foods;
- perspiration rate – exercise level and environmental temperature both influence this rate;
- ventilation rate – breathing rate is largely dependent on exercise level (a significant amount of water is exhaled when we breathe out).

The body's response mechanisms which attempt to maintain homeostatic levels of water are known as osmoregulation.

Loop of Henle creates a hypertonic environment in the medulla

Much of the water in the original filtrate remains even after the filtrate has left the proximal convoluted tubule. This water and the remaining dissolved solutes enter the descending portion of the loop of Henle. This segment of the loop of Henle is relatively permeable to water but relatively impermeable to salt ions, so some water leaves the tubule. The filtrate then enters the ascending portion of the loop of Henle, where the tubule is relatively impermeable to water, but permeable to salt ions. As the filtrate moves up the ascending segment, sodium ions are pumped out and enter the intercellular fluid (see Figure 11.19).

Photomicrograph showing numerous loops of Henle (white tubes) in the medulla region of a kidney.

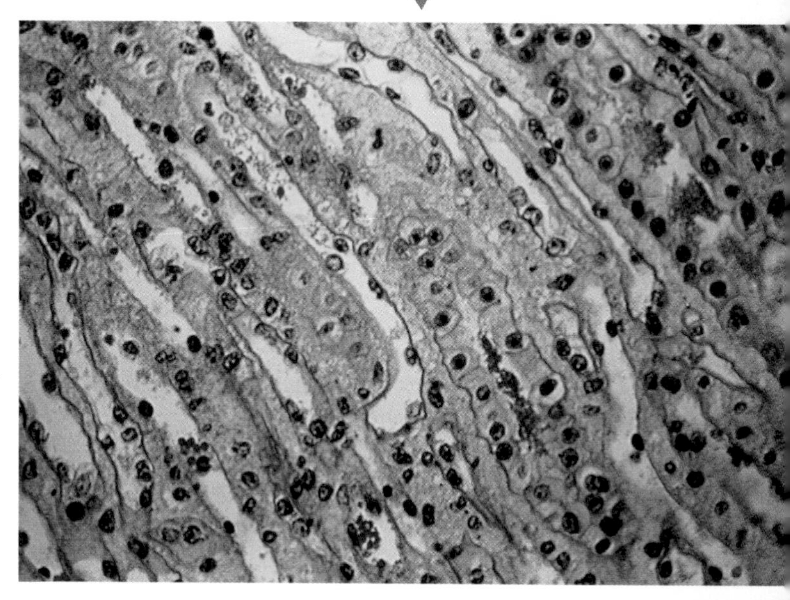

Figure 11.19 Different permeabilities in descending and ascending arms of the Loop of Henle.

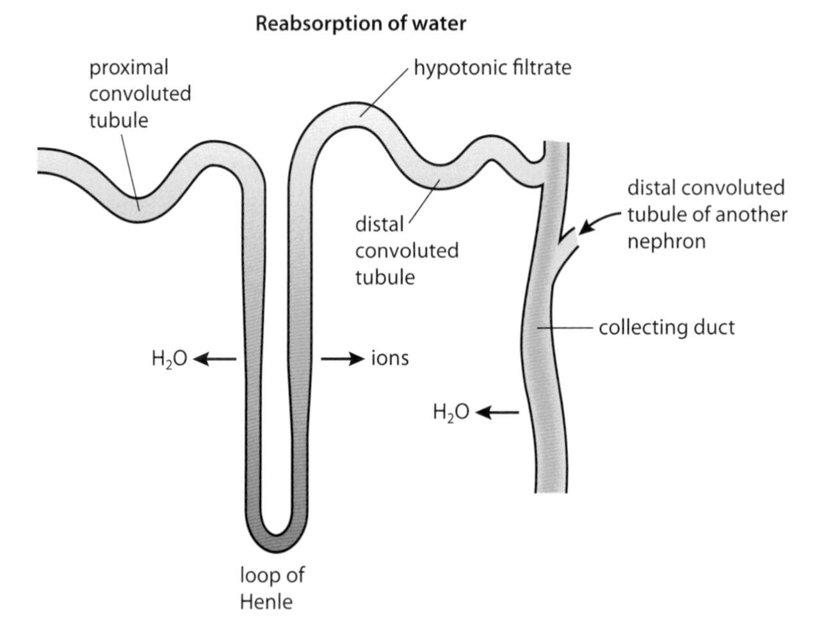

The loop of Henle of each nephron extends down into the medulla region of the kidney. Thus, the medulla is an area with many ions (a hypertonic region) in comparison to fluids within the tubules or the collecting ducts. Despite the fact that some water moves out of the descending portion of the loop by osmosis, the filtrate that moves up the ascending loop and into the distal convoluted tubule is still relatively hypotonic (has a relatively high water content).

ADH and the collecting duct in osmoregulation

The filtrate that enters the distal convoluted tubule is fine-tuned in relation to the reabsorption of solutes and then enters a nearby collecting duct. The filtrate at this point is hypotonic (has a relatively high water content). If this volume of water were to consistently leave the body as urine, an individual would need a very high water intake and not be losing water by any other means. Thus, under most circumstances, at least some of this water is reabsorbed even though all of the solutes needed have already been reabsorbed.

Figure 11.20 Mechanism of final water reabsorption in the kidney.

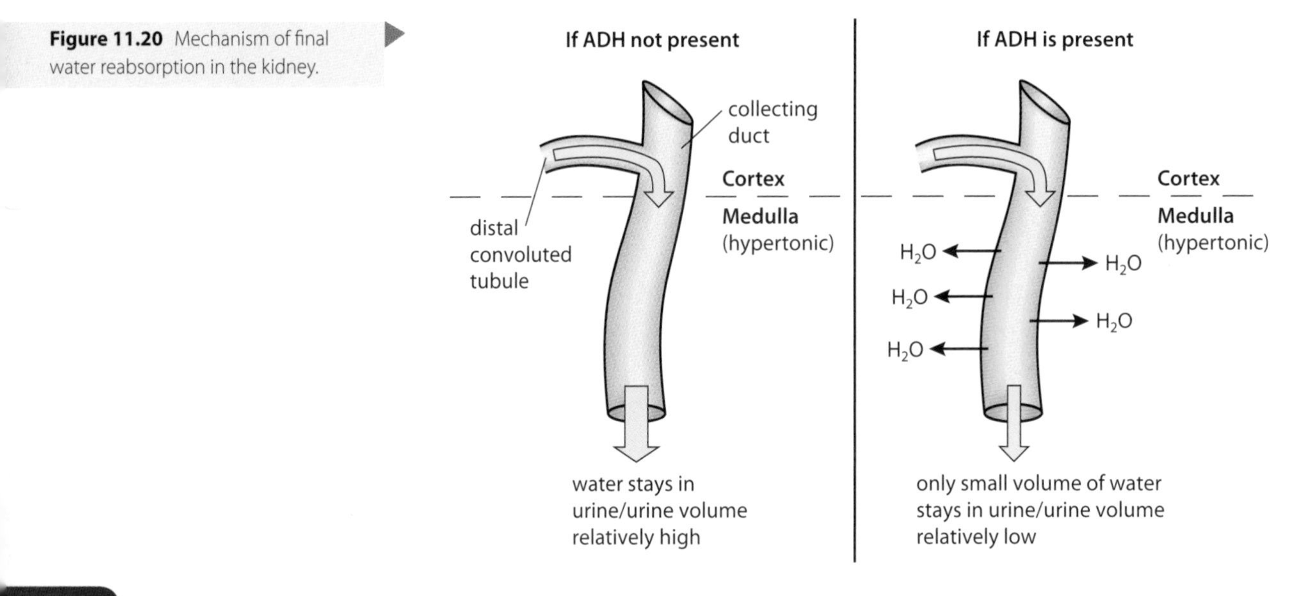

The collecting duct is differentially permeable to water. Its permeability depends on the presence or absence of antidiuretic hormone (ADH). ADH is secreted from the posterior lobe of the pituitary gland and, like all hormones, circulates in the bloodstream. The target tissue of ADH is the kidney collecting ducts (see Figure 11.20). The collecting ducts extend into the highly hypertonic medulla. If ADH is present, the collecting duct becomes permeable to water and water moves by osmosis out of the collecting duct and into the medulla intercellular fluid. From there, water enters the peritubular capillary bed and is thus returned to the bloodstream. If ADH is not present, the collecting duct becomes impermeable to water. Water then stays in the collecting duct, along with the various waste solutes and the urine is more dilute.

How do the kidneys change blood?

To appreciate what happens when blood is filtered by the kidneys, we can compare the concentrations of various molecules before, during and after the filtering process. Each type of molecule in blood plasma is different, primarily in molecular size, so the filtering process differs from one molecule to the next.

The table below compares the concentration of proteins, glucose and urea in the various fluids before, during and after the filtration process. The fluids mentioned in the table are associated with the following locations in the nephron.

- *Blood plasma*: Think of this as the blood that enters the glomerulus, it is originally from the renal artery and no filtering or reabsorption has yet occurred.
- *Glomerular filtrate*: This is the fluid (now called the filtrate) which enters into the proximal convoluted tubule after the ultrafiltration process within Bowman's capsule. No reabsorption has yet occurred.
- *Urine*: Fluid which was the filtrate that has now undergone reabsorption and osmoregulation mechanisms and is taken to the bladder for elimination.

Molecule	Amount in blood plasma in mg 100 ml^{-1}	Amount in glomerular filtrate in mg 100 ml^{-1}	Amount in urine in mg 100 ml^{-1}
proteins	> 700	0	0
glucose	> 90	> 90	0
urea	30	30	> 1800

You do not need to memorize these numbers but you should be able to explain the logic behind each change.

- Proteins are too large to fit through the basement membrane within the glomerulus. Thus, proteins do not become a part of the filtrate or urine.
- Glucose does become a part of the filtrate, but active transport (mainly in the proximal convoluted tubule) takes 100% of the glucose back into the peritubular capillary bed. In a healthy person, no glucose appears in the urine.
- Urea is not toxic unless its concentration is too high in the blood plasma. The very high concentration of urea in urine (compared to plasma and filtrate) is primarily due to the reabsorption of water – the urea content is magnified by concentration.

W For some good kidney diagrams and an interesting analogy – how blood filtration in the kidney is like making a cup of espresso – visit heinemann.co.uk/hotlinks, enter the express code 4242P and click on Weblink 11.5.

Why can diabetes lead to the presence of glucose in urine?

When a person has diabetes, their blood sugar level is not being regulated properly by the antagonistic action of the hormones insulin and glucagon. People with untreated diabetes characteristically have abnormally high levels of glucose dissolved in the blood plasma (commonly called high blood sugar or hyperglycaemia). This is especially common after eating foods high in carbohydrates.

As you will recall, glucose becomes part of the glomerular filtrate and in healthy people 100% of it is reabsorbed into the peritubular capillary bed. The transport mechanism which reabsorbs the glucose is active transport. Active transport mechanisms have a maximum rate at which they can move substances, and when the maximum threshold concentration of a molecule (like glucose) is exceeded, the active transport cannot 'keep up'. People with untreated diabetes sometimes exceed the maximum level of plasma glucose and the active transport mechanism is swamped. The result is that active transport is unable to move all the glucose back into the bloodstream and some remains in the urine.

Exercises

9 Identify all of the cellular and molecular layers that a molecule would have to pass through in order to be ultrafiltered and then reabsorbed into the bloodstream within a single nephron.

10 Some, but not all, substances can be tested for in the urine of a patient. Why do some substances present in the bloodstream show up in urine samples and others do not?

11 **a** Predict the relative amount of ADH produced in a person who has been drinking lots of water and has not been exercising recently. Justify your prediction.

b Predict the relative amount of ADH produced in a person who has been exercising vigorously and has not had a chance to completely hydrate themselves. Justify your prediction.

12 The filtering action of the kidneys does not eliminate urea from the bloodstream. Why is complete elimination of urea not necessary?

11.4 Reproduction

Assessment statements

11.4.1 Annotate a light micrograph of testis tissue to show the location and function of interstitial cells (Leydig cells), germinal epithelium cells, developing spermatozoa and Sertoli cells.

11.4.2 Outline the processes involved in spermatogenesis within the testis, including mitosis, cell growth, the two divisions of meiosis and cell differentiation.

11.4.3 State the role of LH, testosterone and FSH in spermatogenesis.

11.4.4 Annotate a diagram of the ovary to show the location and function of germinal epithelium, primary follicles, mature follicle and secondary oocyte.

11.4.5 Outline the processes involved in oogenesis within the ovary, including mitosis, cell growth, the two divisions of meiosis, the unequal division of cytoplasm and the degeneration of the polar body.

11.4.6 Draw and label the structure of a mature sperm and egg.
11.4.7 Outline the role of the epididymis, seminal vesicle and prostate gland in the production of semen.
11.4.8 Compare the processes of spermatogenesis and oogenesis, including the number of gametes and the timing of the formation and release of gametes.
11.4.9 Describe the process of fertilization, including the acrosome reaction, penetration of the egg membrane by a sperm and the cortical reaction.
11.4.10 Outline the role of HCG in early pregnancy.
11.4.11 Outline early embryo development up to the implantation of the blastocyst.
11.4.12 Explain how the structure and functions of the placenta, including its hormonal role in secretion of oestrogen and progesterone, maintain pregnancy.
11.4.13 State that the fetus is supported and protected by the amniotic sac and amniotic fluid.
11.4.14 State that materials are exchanged between the maternal and fetal blood in the placenta.
11.4.15 Outline the process of birth and its hormonal control, including the changes in progesterone and oxytocin levels and positive feedback.

Spermatogenesis produces male gametes by meiosis

The production of sperm cells occurs within the testes. The testes of human males are located outside the body in order to provide the cooler temperature necessary for production of sperm cells or spermatozoa. Inside each testis, spermatogenesis occurs within very small tubes known as seminiferous tubules. Near the outer wall of the seminiferous tubules lie germinal epithelial cells called spermatogonia. Each spermatogonium may be undergoing either mitosis or meiosis at any given time.

- *Mitosis*: Spermatogonia undergo mitosis in order to replenish their numbers. Sperm cell production starts at puberty and continues throughout life. Millions of sperm cells may be produced in a single day and mitosis replaces cells which become spermatozoa.
- *Meiosis*: Spermatogonia undergo meiosis to produce spermatozoa (singular, spermatozoon). Meiosis is also called 'reduction division' because it reduces the diploid number of chromosomes in spermatogonia to the haploid number in spermatozoa. In humans, 23 homologous pairs of chromosomes becomes 23 individual chromosomes.

W To see a section of seminiferous tubule and Sertoli cells, visit heinemann.co.uk/hotlinks, insert the express code 4242P and click Weblink 11.6.

Spermatogonia that begin either type of cell division first replicate the DNA within their still diploid nucleus. At the same time, the spermatogonia are undergoing cell growth in preparation for cell division. If any one spermatogonium undergoes mitosis, two half-size cells result, each capable of growing again for a later cell division. If a spermatogonium begins meiosis, four spermatozoa are the result. Each spermatozoon is a very small cell and contains a haploid number of chromosomes.

Spermatogenesis

Let's follow what happens to a spermatogonium during meiosis. Human spermatogonia are diploid and contain 23 homologous pairs of chromosomes (46 in total). DNA replication occurs and each of the 46 chromosomes now exists as a pair of chromatids. Meiosis I occurs (meiosis cell division one) and two half-sized cells result, each with the haploid number of chromosomes (23) because homologous pairs have been separated. Each chromosome still exists as a pair of chromatids, so there is another cell division called meiosis II. During meiosis II, the chromatids are separated. Thus four haploid cells, each containing 23 chromosomes, are created from one that originally contained 23 homologous pairs.

Meiosis is completed for these cells, but each must now differentiate into a fully functioning, motile spermatozoon. Thus, the cells stay within the interior of the seminiferous tubule for a period of time as they form the cellular structures characteristic of a spermatozoon. These structures include a flagellum for motility and an acrosome to contain enzymes necessary for fertilization. The developing sperm cells need nutrients during this period of differentiation and thus each remains attached to cells in the seminiferous tubules known as Sertoli cells. Sertoli cells also help nourish the other cell stages during meiosis.

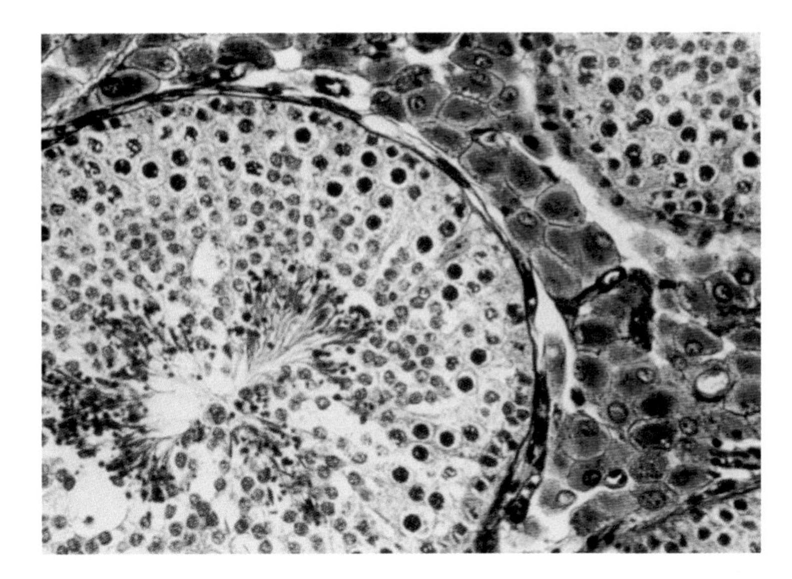

Light micrograph showing a nearly complete section of a seminiferous tubule (left). Cells near the outer edge are spermatogonia undergoing mitosis or beginning meiosis. Various meiotic stages take the cells toward the central lumen. Maturing spermatozoa are seen still attached to Sertoli cells with their flagella in the lumen. Leydig cells (brown), which produce testosterone, are shown between seminiferous tubule sections.

Each of the cell stages of meiosis has moved the resulting cell closer to the interior of the seminiferous tubule. Since the tubule is truly a small tube, there is a cavity or lumen at the centre. Once spermatozoa have completed formation of their flagella, they detach from their Sertoli cell and are carried through the lumen with the movement of fluid (see Figure 11.21). Each sperm cell is swept to the epididymis of the testis where it is stored.

Hormonal control of sperm production

Hormones influence gamete formation in males just as they do in females. Although hormonal changes are not as obviously cyclical, studies have shown that during a 24-hour period, there may be surges of production of the hormones involved in spermatogenesis. The generalized functions of these hormones are:

- leutinizing hormone (LH) stimulates Leydig cells to produce testosterone;
- follicle stimulating hormone (FSH) and testosterone stimulate the meiotic divisions of spermatogonia into spermatozoa.

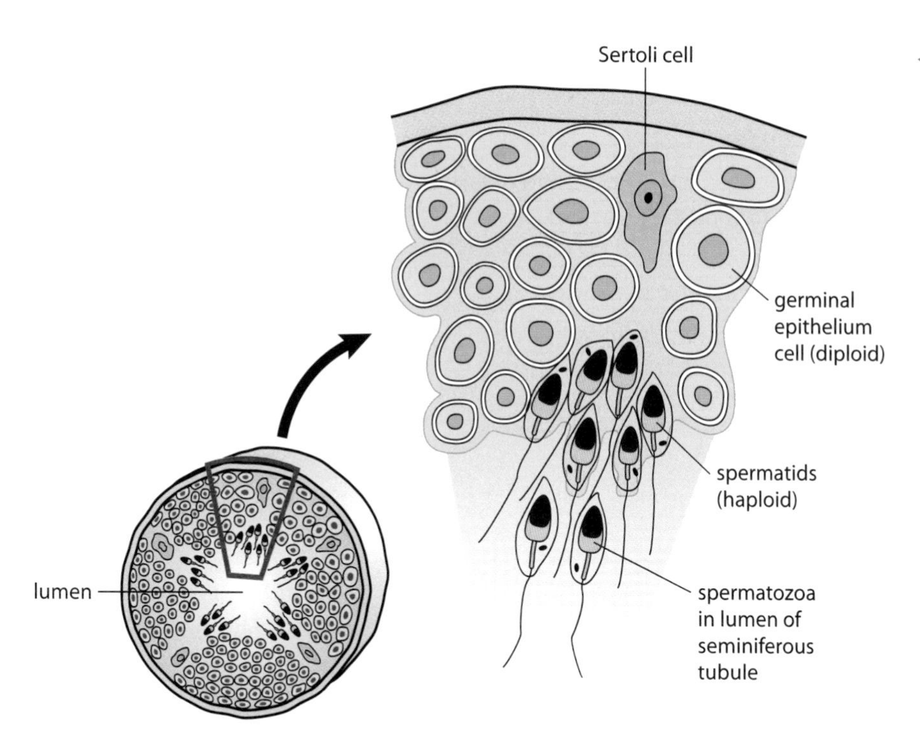

Figure 11.21 Section view of human seminiferous tubule.

Role of the epididymis, seminal vesicles and prostate

As you will recall, the sperm cells formed in the seminiferous tubules are carried through the lumen until they reach the epididymis. Here, sperm cells are stored and gain motility. On sexual arousal, a large number (millions) of sperm cells move from the epididymis into the vas deferens (see Figure 11.22). As the sperm cells move along the vas deferens near the area of the bladder, a pair of glands called the seminal vesicles add a large volume of fluid. This fluid has a high concentration of the sugar fructose, a high-energy carbohydrate needed to provide the energy for the sperm cells to swim to the ovum. Approximately 70% of the fluid in semen is added by the seminal vesicles. Near this same area, the prostate gland adds more fluid to the semen. The fluid from the prostate is alkaline and helps the spermatozoa survive the environment within the female's vagina. Approximately 30% of the semen fluid is from the prostate.

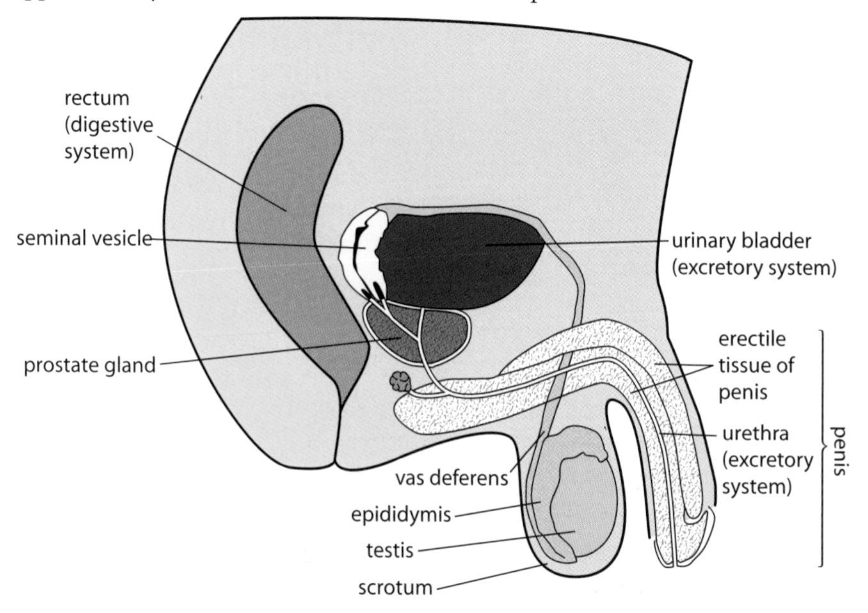

Figure 11.22 Male reproductive structures.

Oogenesis produces female gametes by meiosis

Oogenesis and spermatogenesis are the female and male processes of meiosis respectively. Thus, there are many similarities between these two processes, especially when the focus is on the behaviour of the chromosomes. Oogenesis produces four cells as the 'end-products' of meiosis, as does spermatogenesis. However, three of the four end-product cells of oogenesis are not used as gametes as they are much too small to produce a viable zygote if fertilized. These three cells are called polar bodies and their function is to be a cellular 'container' for the divided chromosomes during both meiosis I and meiosis II. The fourth haploid cell produced is very large and is the ovum. Let's look at the process of oogenesis from beginning to end (note the many similarities to spermatogenesis).

Events occurring before birth

Within the ovaries of a female fetus, cells called oogonia undergo mitosis repeatedly in order to build up the numbers of oogonia within the ovaries. These oogonia grow into larger cells called primary oocytes. Both oogonia and primary oocytes are diploid cells. The large primary oocytes begin the early steps of meiosis, but the process stops (is arrested) during prophase I.

Also within the ovaries, cells called follicle cells repeatedly undergo mitosis. A single layer of these follicle cells surrounds each primary oocyte and the entire structure is then called a primary follicle. When a female child is born, her ovaries contain nearly a half million primary follicles. These primary follicles remain relatively unchanged until the female reaches puberty and begins experiencing menstrual cycles.

Events occurring with the menstrual cycle

Each menstrual cycle, a few primary follicles finish meiosis I. The two resulting haploid cells are not even close to being equal in size. One is very large and the other is very small. The small cell is called the first polar body and simply acts as a reservoir for half of the chromosomes. The polar bodies produced during oogenesis later just degenerate. The other, very large, cell is a secondary oocyte.

You will recall that meiosis I produces haploid cells, but each cell has chromosomes existing as paired chromatids. The single ring of follicle cells begin dividing and forming a fluid. Two rings of follicle cells are formed with a fluid-filled cavity separating them. The first (inner) ring of follicle cells surrounds the oocyte, then there is the fluid-filled space, and finally the outer ring of follicle cells. The secondary oocyte begins meiosis II, but is again arrested during prophase. This entire structure is now called a Graafian follicle (see Figure 11.23). The increase in fluid between the two follicle cell layers creates a bulge on the surface of the ovary and eventually leads to ovulation.

It is a secondary oocyte with the inner ring of follicle cells that is released from the ovary at ovulation, although people often refer to this event as the release of the 'ovum' or 'egg'. The second meiotic division (meiosis II) is not completed until fertilization. The hormones FSH and LH are primarily responsible for the events leading up to and including ovulation.

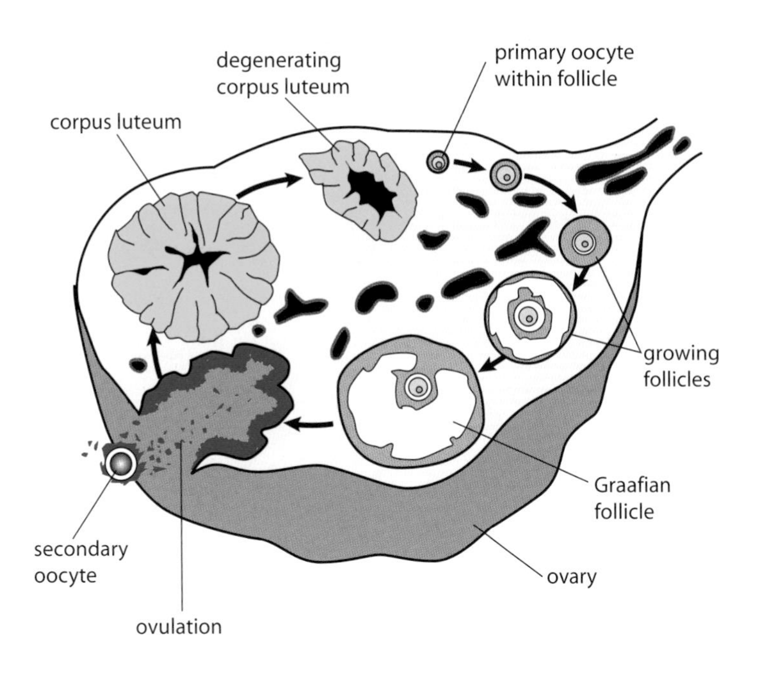

Figure 11.23 Ovary showing stages of production of a Graafian follicle leading to ovulation. This diagram is like a time-lapse photograph as not all of these stages would be occurring at the same time

Most females have an increase in body temperature of about 1°C soon after ovulation. Some couples wanting to have children chart the woman's body temperature and time sexual intercourse accordingly.

Mature sperm and 'ova'

There is no doubt that sperm cells are the end-product of spermatogenesis. It is debatable what is the end-product cell of oogenesis because the structure that is ovulated is a secondary oocyte (with a surrounding layer of follicle cells). One could argue that the process of oogenesis is not over until the fertilization process is well underway as this is when meiosis is completed. Nevertheless, it is quite common for people to say that ovulation results in an 'ovum' being released from the ovary.

The resulting male and female gametes are very well suited for their purpose. Both gametes are haploid. The spermatozoon is a very small cell with a flagellum for motility and mitochondria to provide ATP for swimming. At its anterior end, each sperm cell contains an organelle called an acrosome. The acrosome contains hydrolytic enzymes which help with the fertilization process. Sperm cells do not contain any unnecessary organelles or structures; their small size allows them to swim great distances and unnecessary structures would be a burden (see Figure 11.24).

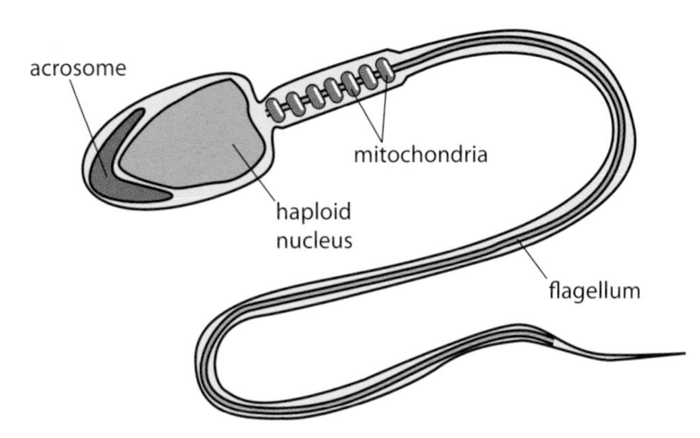

Figure 11.24 Human spermatozoon.

The egg (secondary oocyte) is the largest cell in the body, by volume (see Figure 11.25). The unequal division of the cytoplasm during meiosis ensured that one cell only would receive virtually all of the cytoplasm, nutrients and organelles necessary to start a new life. The nutrients within the ovum are collectively referred to as yolk. In addition, the cytoplasm also contains small vesicles called cortical granules which function immediately after fertilization. Just outside the plasma membrane is a layer of glycoproteins called the zona pellucida. This also has a function during fertilization.

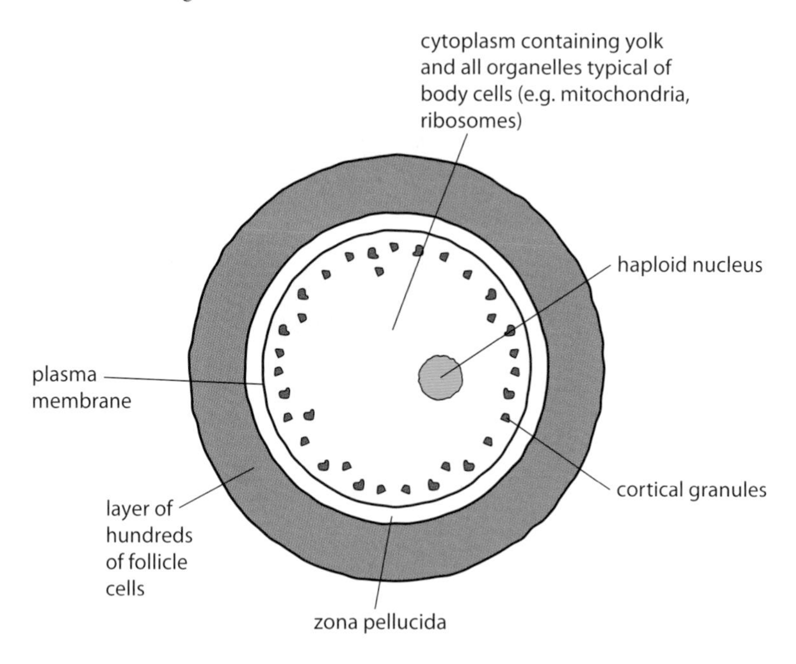

Figure 11.25 Human secondary oocyte ('ovum').

Comparison of spermatogenesis and oogenesis

The processes by which male and female gametes are produced are easily compared in a table like this one.

• **Examiner's hint:** The command term 'compare' requires you to list both similarities and differences.

Spermatogenesis	Oogenesis
Millions of sperm cells are produced every day.	Typically, one secondary oocyte is ovulated per menstrual cycle.
Four gametes are produced for each germinal cell which begins meiosis.	One gamete is produced for each germinal cell which begins meiosis (plus polar bodies).
The resulting gametes are very small.	The resulting gametes are very large.
Occurs within testis (gonad tissue).	Occurs within ovaries (gonad tissue).
Spermatozoa are released during ejaculation.	Secondary oocyte is released during ovulation.
Haploid nucleus results from meiosis.	Haploid nucleus results from meiosis.
Spermatogenesis continues all through life (starting at puberty)	Ovulation starts at puberty, occurs with each menstrual cycle, then stops during menopause.

Identical (monozygotic) twins are formed when one egg is fertilized by one sperm and the early embryo splits into two embryos. Identical twins share an identical genetic makeup.

Fraternal (dizygotic) twins are two different eggs fertilized by two different sperm cells and are thus genetically different just as non-twin siblings are different.

Fertilization

As a result of sexual intercourse, millions of sperm cells are ejaculated into a female's vagina. The motile spermatozoa absorb some of the fructose within the semen in order to 'fuel up' for what could be a very long journey. At least some of the sperm cells find their way through the cervical opening (the cervix separates the vagina and the uterus) and gain access to the uterus. They begin swimming

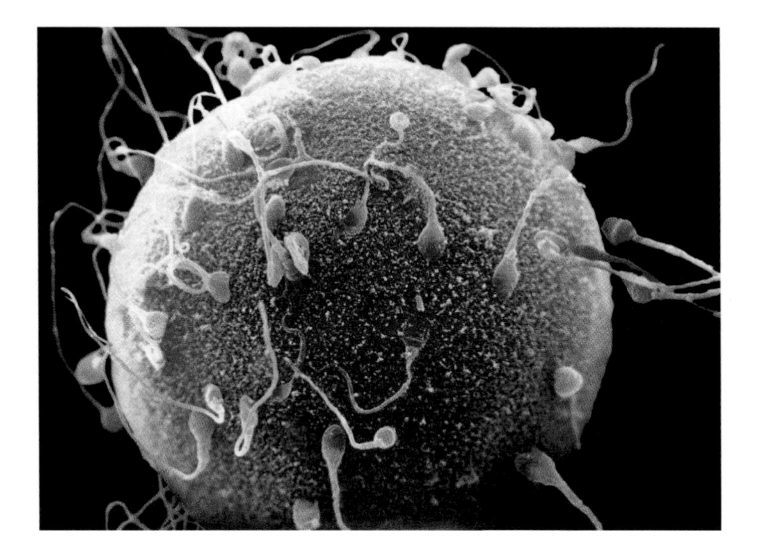

This false-colour SEM shows a human ovum (secondary oocyte) surrounded by spermatozoa. Notice the size difference between the two types of gamete.

up the endometrial lining and some of these sperm cells enter the openings of the Fallopian tubes. If the female is near the middle of her menstrual cycle, there may be a secondary oocyte within one of the two Fallopian tubes. The reason for millions of spermatozoa in each ejaculate becomes clear when you consider that only a very small percentage of the motile sperm cells will ever reach the location of the secondary oocyte.

The typical location for fertilization is within one of the Fallopian tubes. No single sperm cell can accomplish the entire act of fertilization as it takes many sperm cells to penetrate the follicle cell layer surrounding the secondary oocyte (see Figure 11.26). Several sperm cells gain access to the zona pellucida (glycoprotein gel layer) surrounding the secondary oocyte and release the hydrolytic enzymes contained in their acrosomes. One sperm cell reaches the plasma membrane of the secondary oocyte first. The plasma membranes of the two gametes fuse together. This initiates a series of events called the cortical reaction.

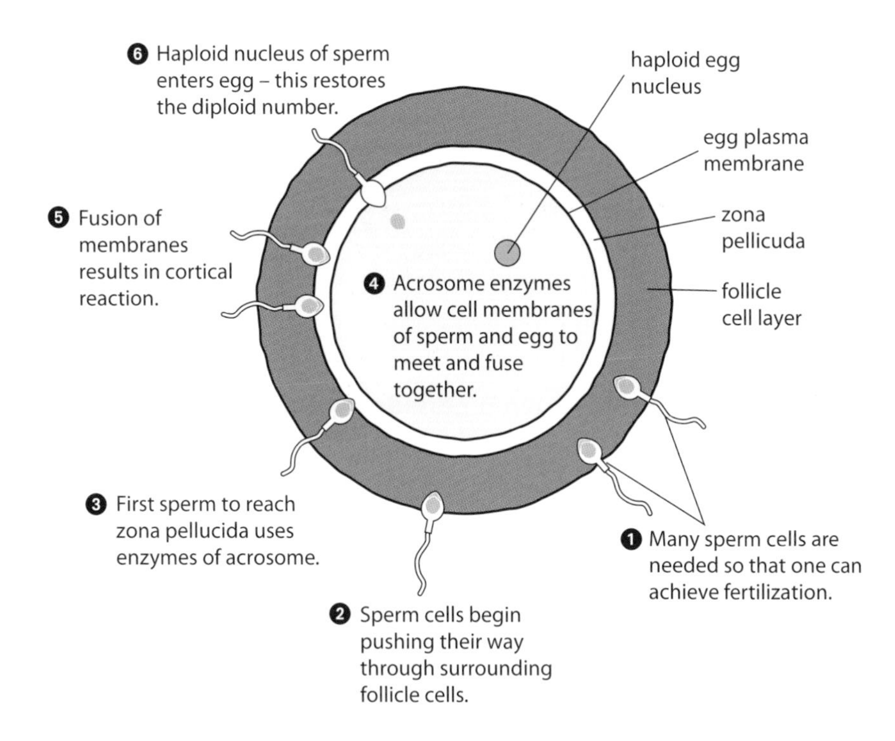

6 Haploid nucleus of sperm enters egg – this restores the diploid number.
haploid egg nucleus
egg plasma membrane
zona pellicuda
follicle cell layer
5 Fusion of membranes results in cortical reaction.
4 Acrosome enzymes allow cell membranes of sperm and egg to meet and fuse together.
3 First sperm to reach zona pellucida uses enzymes of acrosome.
2 Sperm cells begin pushing their way through surrounding follicle cells.
1 Many sperm cells are needed so that one can achieve fertilization.

Figure 11.26 Sequence of events in fertilization.

Within the cytoplasm of the secondary oocyte are many small vesicles called cortical granules; they are located all around the interior of the plasma membrane. When the two gametes fuse their plasma membranes, the cortical granules fuse with the oocyte's cell membrane and release their enzymes to the outside. These enzymes result in a chemical change in the zona pellucida making it impermeable to any more sperm cells. The cortical reaction takes place within a few seconds of the first sperm gaining access and ensures that only one sperm cell actually fertilizes the oocyte. The secondary oocyte now completes meiosis II and produces another polar body. The resulting fertilized ovum is now referred to as a zygote. The diploid condition has been restored and a new life formed.

Pregnancy

As you will recall, fertilization occurs within one of the Fallopian tubes.

Early human embryonic development

Fertilization triggers the zygote to begin a mitotic division and the first division typically occurs approximately 24 hours after fertilization. During the first 5 days or so, the early embryo is dividing by mitosis; it is also moving within the Fallopian tube and getting closer to the uterus. The rate of mitotic divisions increases and by the time the embryo reaches the uterine cavity, it is approximately 100 cells and is ready to implant itself into the endometrium of the uterus. The embryo at this stage is a ball of cells and is called a blastocyst (see Figure 11.27). A blastocyst is characterized by:

- a surrounding layer of cells called the trophoblast – this layer of cells will help form the fetal portion of the placenta;
- a group of cells on the interior known as the inner cell mass and located toward one end of the 'ball' – the inner cell mass will become the body of the embryo;
- a fluid-filled cavity.

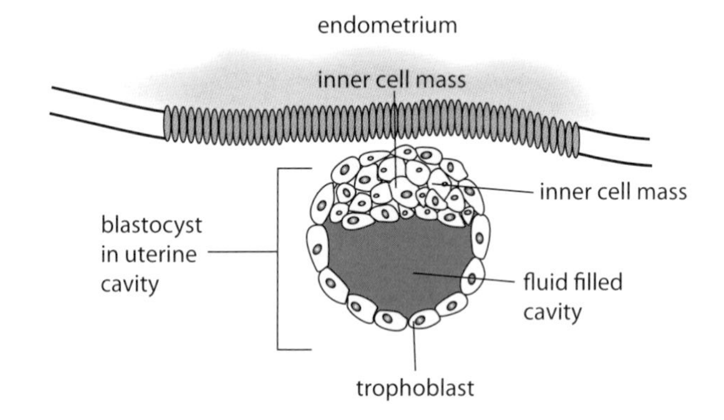

Figure 11.27 Human blastocyst in section.

Secretion of human chorionic gonadotrophin

HCG is the hormone that pregnancy diagnostic kits test for. This hormone can only be produced by an embryo and thus is a positive indicator for pregnancy.

Ovulation of the secondary oocyte left an outer ring of follicle cells within the ovary. This remaining layer of follicle cells begins mitotically dividing and also begins secreting the hormones oestrogen and progesterone. This hormone-secreting tissue has become a temporary endocrine gland known as the corpus luteum. Each menstrual cycle in which there has not been a fertilization, the corpus luteum continues to secrete progesterone for about 14 days. High levels of oestrogen and progesterone help maintain the thickened, highly vascular endometrium within the uterus. This highly vascular condition is necessary in case there has been a fertilization and an embryo will be implanting itself into the endometrium. If no fertilization has occurred, the corpus luteum ceases hormone production after 14 days and the endometrium begins breaking down.

If fertilization has occurred, the embryo enters the uterus and begins implantation about a week after fertilization. This is well within the time period that the corpus luteum is still active. Soon after implantation, the embryo begins secreting a hormone of its own. This hormone is human chorionic gonadotrophin (HCG). HCG enters the bloodstream of the mother; its target tissue is the corpus luteum; HCG acts to maintain the secretory functions of this gland beyond the length of time typical of a normal menstrual cycle. The corpus luteum continues to secrete both oestrogen and progesterone and the endometrium, in which the early embryo is embedded, is maintained. Later in pregnancy, the role of oestrogen and progesterone production is taken over by the placenta.

Role of the placenta

The primary reason that a human ovum is so large is because it contains the nutrients needed for early embryonic development. During the first two weeks after fertilization, there is no true growth of the embryo. The cell divisions that occur create an embryo of 100 or more cells, but the overall size of the embryo is no larger than that of the original egg. The nutrients stored within the egg have been used for metabolism, not for growth. When the human embryo begins implantation into the wall of the endometrium, it is rapidly running out of stored nutrients (yolk). Fortunately, as a result of implantation, the embryo and the maternal endometrium soon begin to create a structure known as the placenta.

The placenta forms from the trophoblast layer of the blastocyst. When fully formed, two fetal blood vessels within the umbilical cord carry fetal blood to the placenta. The blood within these two vessels is deoxygenated and carries waste products. This fetal blood exchanges materials with the maternal bloodstream and another fetal blood vessel (also within the umbilical cord) returns the blood to the fetus. The blood that returns to the fetus has been oxygenated and nutrients have also been added while in the placenta, as shown in this table.

Materials passed from fetus to mother within the placenta	Materials passed from mother to fetus within the placenta
carbon dioxide	oxygen
urea	nutrients (glucose, amino acids, etc.)
water	water
hormones (e.g. HCG)	hormones
	vitamins, minerals
	alcohol, many drugs, nicotine (if taken by mother during pregnancy)
	some viruses such as German measles, HIV (if mother is infected)

At no time does the blood of the fetus and the blood of the mother actually mix – there is an exchange of materials, but no exchange of blood (see Figure 11.28). In addition to the molecular exchanges shown in the table, the placenta also acts as an endocrine organ which is especially important in the second half of pregnancy. The corpus luteum does not stay active during the entire pregnancy. When the corpus luteum stops production of oestrogen and progesterone, the placenta has already begun producing and secreting these hormones. Remember that maintaining high levels of oestrogen and progesterone is necessary throughout pregnancy as these steroid hormones are needed to maintain the rich blood supply associated with the placenta.

Figure 11.28 Schematic showing blood flow pattern of the placenta.

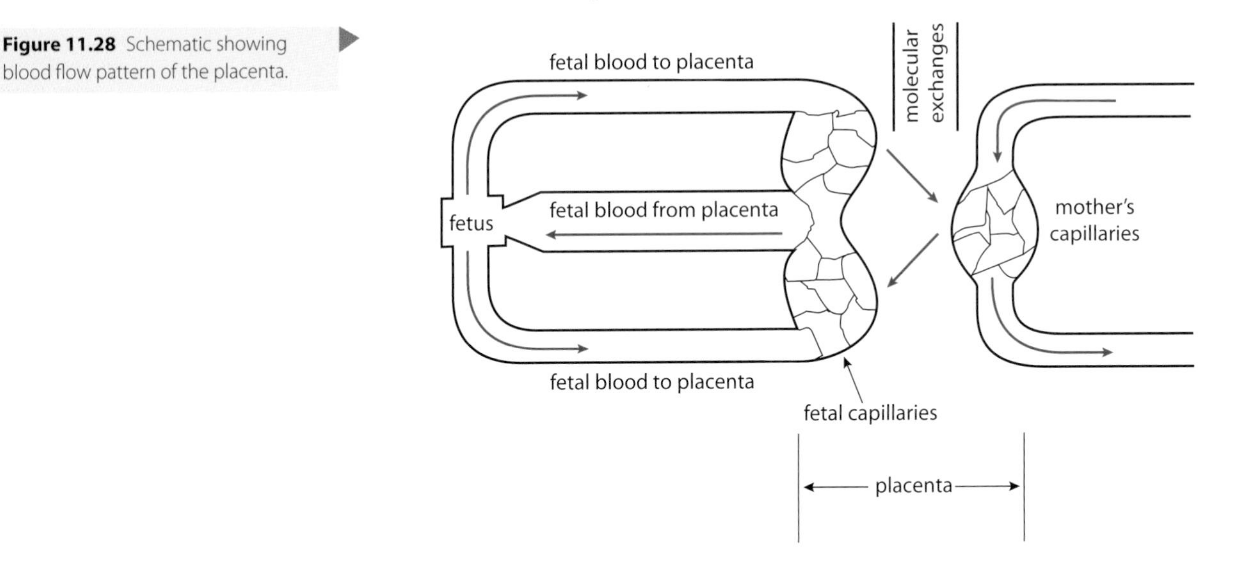

Role of amniotic fluid

Some of the tissue of a developing embryo is used to create membranous structures known as extraembryonic membranes. One of these membranous structures is the amniotic sac, which extends all of the way around the fetus. Within the sac is a fluid called amniotic fluid. The fetus floats in this fluid as it continues to grow and develop.

There are several functions of the amniotic fluid, including:

- providing a cushioning effect should a blunt force be applied to the mother's abdomen;
- providing an environment in which the fetus has free movement and therefore well balanced exercise for all developing muscles and skeleton;
- providing excellent thermal stability (amniotic fluid is mainly water and therefore has excellent temperature stability).

A 5-week-old human embryo floating in its amniotic fluid. The umbilical cord is also visible. Note that fingers and toes have already developed. The embryo at this stage is about 1 cm long.

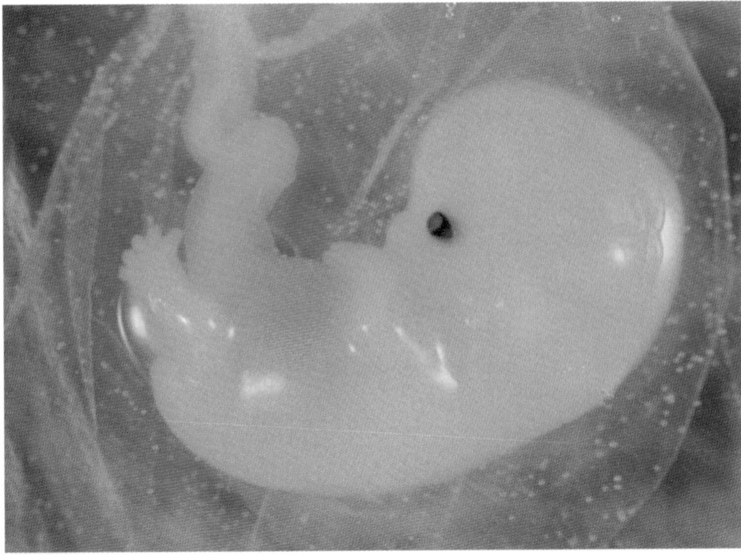

Amniotic fluid is the fluid which is sampled during the procedure known as amniocentesis. The fluid samples contain some living embryo cells which can be cultured and tested for chromosome abnormalities.

Hormonal events associated with birth

The physiological events associated with a woman's body preparing for a birth are collectively called parturition. Studies have shown that one such event is a drop in progesterone levels. At about the same time, a hormone called oxytocin is secreted

from the posterior lobe of the pituitary gland. Oxytocin is a peptide hormone that binds to protein receptors on its target cells. Some studies suggest that the decline in progesterone levels near the end of pregnancy may be associated with production of oxytocin receptors on cells of the uterus.

Production of low levels of oxytocin is associated with the beginning of labour – the first contractions of the uterus. At this stage of the birthing process, the contractions are not very intense and are not frequent. Each uterine contraction results in uterine mechanoreceptors sending signals back to the posterior lobe of the pituitary to produce more oxytocin so labour contractions become more intense and more frequent. This type of feedback mechanism is known as positive feedback and stops only when birth occurs and the uterus no longer has something to contract on.

Here is a list of the major events of vaginal childbirth:

- major hormone changes (some discussed above);
- opening of the cervix to 10 cm;
- the most typical position for the baby is head-first, face down;
- the shoulders of the baby are typically the widest part to pass through the birth canal;
- the afterbirth is the name for the expelled placenta, which occurs after the baby is born;
- lactation (breast milk production) begins soon after birth.

When a child is born in a position other than head first, it is referred to as a breech birth. A breech birth can be leg(s) or buttocks first.

Exercises

13 Why do pregnant women need to be very careful about their diet and other things they put into their body?

14 Trace the route of sperm cells and semen fluid from production to ejaculation.

15 Mitochondrial DNA (mtDNA) is of maternal origin only. Explain why this makes sense from the perspective of fertilization.

16 Why does fertilization need to occur very soon after ovulation?

Practice questions

1 A blood clot contains a network of protein. What is the protein?

A Fibrin
B Fibrinogen
C Haemoglobin
D Thrombin *(1 mark)*

2 Why do antibiotics kill bacteria but not viruses?

A Antibiotics stimulate the immune system against bacteria but not viruses.
B Viruses have a way of blocking antibiotics.
C Viruses are too small to be affected by antibiotics.
D Viruses do not have a metabolism. *(1 mark)*

3 How do the levels of oxytocin and progesterone change immediately prior to birth?

	Oxytocin	Progesterone
A	decreases	decreases
B	decreases	increases
C	increases	decreases
D	increases	increases

(*1 mark*)

4 What is the outcome for each of the following processes?

	Spermatogenesis	Oogenesis
A	4 gametes	4 gametes
B	4 gametes	1 gamete and 3 polar bodies
C	2 gametes and 2 polar bodies	2 gametes and 2 polar bodies
D	1 gamete and 3 polar bodies	4 gametes

(*1 mark*)

5 The proximal convoluted tubule is a part of the nephron (kidney tubule). Its function is selective reabsorption of substances useful to the body.

(a) Outline how the liquid that flows through the proximal convoluted tubule is produced. (*2*)

(b) **(i)** Water and salts are selectively reabsorbed by the proximal convoluted tubule. State the name of one other substance that is selectively reabsorbed. (*1*)

(ii) State the names of the processes used to reabsorb water and salts.

Water ....

Salts .... (*2*)

The drawing below shows the structure of a cell from the wall of the proximal convoluted tubule.

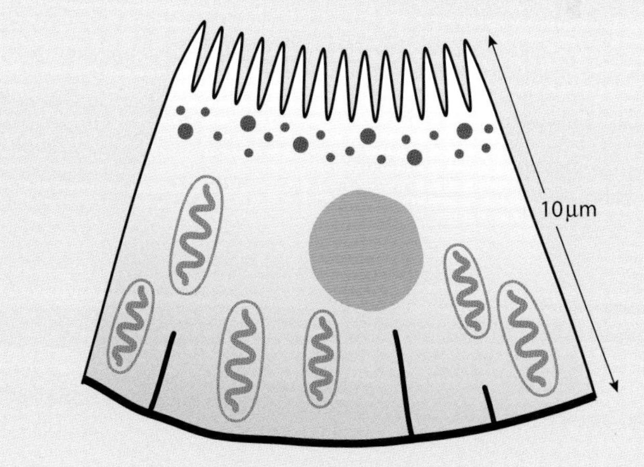

(c) The actual size of the cell is shown on the diagram. Calculate the linear magnification of the drawing. Show your working. (*2*)

(d) Explain how the structure of the proximal convoluted tubule cell, as shown in the diagram, is adapted to carry out selective reabsorption. (*2*)

(*Total 9 marks*)

● **Examiner's hint:** Remember that the command term 'discuss' requires you to give different points of view. Do not just give your opinion.

6 Discuss the benefits and dangers of immunization against bacterial and viral infections. (*8 marks*)

7 Explain how a muscle fibre contracts, following depolarization of its plasma membrane. *(6 marks)*

8 Describe the roles of nerves, muscles and bones in producing movement. *(6 marks)*

9 Compare the composition of blood arriving at the kidney with the composition of blood carried away from it. *(4 marks)*

10 Outline the process of fertilization in humans. *(6 marks)*

● **Examiner's hint:** Remember that the command term 'outline' normally allows you to give a fairly brief answer. However, note that the number of marks in the markscheme is 6, so a brief answer would not score well for question 10.

11 The plasma solute concentration, plasma antidiuretic hormone (ADH) concentration and feelings of thirst were tested in a group of volunteers. These graphs show the relationship between intensity of thirst, plasma ADH concentration and plasma solute concentration.

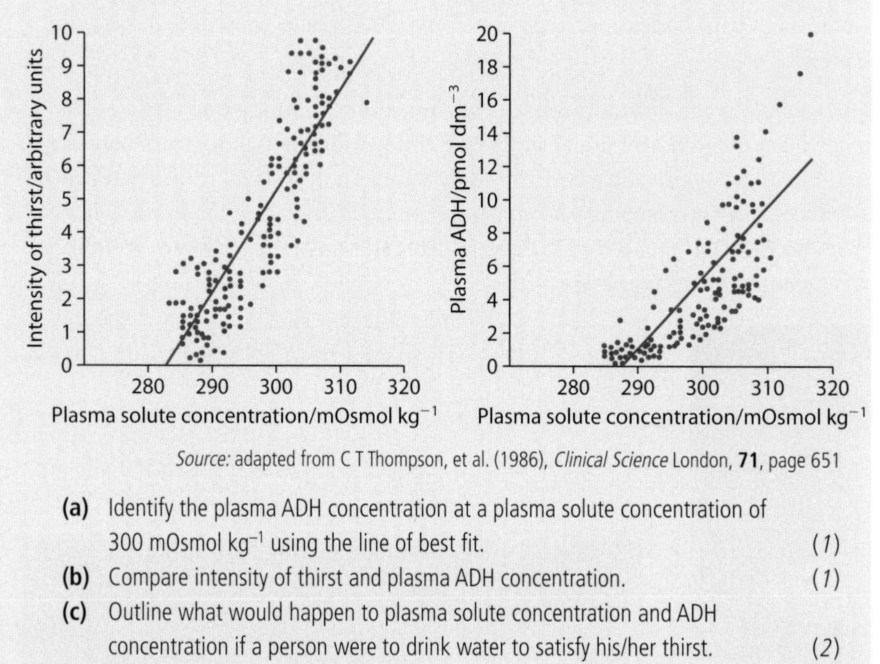

Source: adapted from C T Thompson, et al. (1986), *Clinical Science* London, **71**, page 651

(a) Identify the plasma ADH concentration at a plasma solute concentration of 300 mOsmol kg^{-1} using the line of best fit. *(1)*

(b) Compare intensity of thirst and plasma ADH concentration. *(1)*

(c) Outline what would happen to plasma solute concentration and ADH concentration if a person were to drink water to satisfy his/her thirst. *(2)*

(d) State two reasons why a person's plasma solute concentration may increase. *(2)*

(Total 6 marks)

12 Human nutrition and health (Option A)

Introduction

You have probably been told, 'You are what you eat.' And the expression really is scientifically justifiable: every bit of carbon, iron, calcium and the dozens of other elements in your body came from something you ingested sometime during your lifetime.

For example, a certain number of the calcium atoms in your bones and teeth have come from the milk you drank as a young child. And although you probably like to think of it as *your* calcium, if you really reflect on it, you will realize that you are just borrowing such elements from nature for your lifetime and then you will give them back. So take good care of these valuable substances and choose carefully what you put into your body.

All the calcium in your bones comes from foods you have consumed.

This is an X-ray of healthy bones in a human hand.

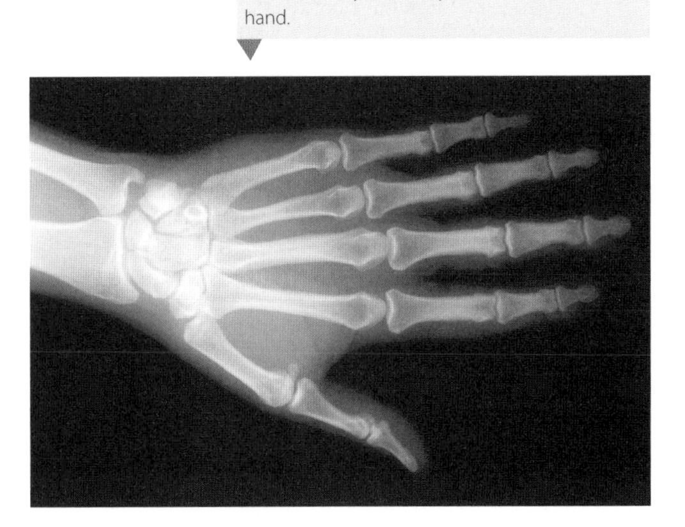

We are going to explore what is in food and how our bodies use such material. The idea of a balanced diet is considered: how do you know what foods are right for you? What happens if you get too much or not enough of certain vitamins, minerals, fats or proteins? Can foods be 'good for you' or 'bad for you'? How do you know if you are at your ideal body mass?

There are also some social issues. Why do some people not eat meat? Why do some women prefer to bottle-feed their babies rather than breastfeed? Is powdered milk for babies as good as human milk? What about food miles – are you contributing to pollution and global warming when you choose foods to buy?

A.1 Components of the human diet

Assessment statements

A.1.1 Define *nutrient* (a chemical substance found in foods that is used in the human body).

A.1.2 List the types of nutrients that are essential in the human diet, including amino acids, fatty acids, minerals, vitamins and water.

A.1.3 State that non-essential amino acids can be synthesized in the body from other nutrients.

A.1.4 Outline the consequences of protein deficiency malnutrition.

A.1.5 Explain the causes and consequences of phenylketonuria (PKU) and how early diagnosis and a special diet can reduce the consequences.

A.1.6 Outline the variation in the molecular structure of fatty acids, including saturated fatty acids, *cis* and *trans* unsaturated fatty acids, monounsaturated and polyunsaturated fatty acids.

A.1.7 Evaluate the health consequences of diets rich in the different types of fatty acid.

A.1.8 Distinguish between *minerals* and *vitamins* in terms of their chemical nature.

A.1.9 Outline two of the methods that have been used to determine the recommended daily intake of vitamin C.

A.1.10 Discuss the amount of vitamin C that an adult should consume per day, including the level needed to prevent scurvy; claims that higher intakes give protection against upper respiratory tract infections; and the danger of rebound malnutrition.

A.1.11 List the sources of vitamin D in human diets.

A.1.12 Discuss how the risk of vitamin D deficiency from insufficient exposure to sunlight can be balanced against the risk of contracting malignant melanoma.

A.1.13 Explain the benefits of artificial dietary supplementation as a means of preventing malnutrition, using iodine as an example.

A.1.14 Outline the importance of fibre as a component of a balanced diet.

Nutrients

A nutrient is a chemical substance found in foods and used in the human body. Nutrients can be absorbed to give you energy, help strengthen your bones, or even prevent you from getting a disease – as in the case of vitamins.

You will recall from Chapter 3, that a handful of organic molecules make up all living organisms. Although some of these molecules, such as certain amino acids or lipids, can be synthesized by the human body, many cannot. You must obtain from food the molecules which your body cannot make. These molecules are called essential nutrients; they are:

- essential amino acids;
- essential fatty acids;
- minerals;
- most vitamins;
- water.

• **Examiner's hint:** You might be surprised to realise that many students cannot give a precise definition for 'nutrient'. This may stem from the fact that because they eat them all the time, students assume they know what nutrients are. It is very important to memorize scientific definitions.

Carbohydrates are not essential nutrients because it is possible (though not advisable) to live on diets with very small quantities of carbohydrate, or even none at all.

Amino acids

We need 20 different amino acids to synthesize the most common human proteins. Nine of these amino acids cannot be synthesized in the body because the DNA sequences necessary to make them are not present in the human genome. They are referred to as essential amino acids. The other 11 amino acids can be synthesized in the body and are called non-essential amino acids.

Try writing out instructions about how to get to your house without using the letter e in any of the words.

That's what it is like trying to make proteins without tyrosine.

If you think about the proteins in your body as being like words in a sentence, then the amino acids are analogous to letters of the alphabet. Suppose you are typing on a keyboard which has certain letters missing, some words are impossible to write. The same is true for proteins: your body cannot make a specific protein if one of the required amino acids is not available. Such a situation could be severely detrimental to your health.

Protein deficiency

To what degree do you think the following are factors in malnutrition?

- poverty and wealth;
- cultural differences concerning dietary preference;
- climatic conditions (i.e. annual rainfall);
- poor distribution of food (i.e. insufficient roads, bridges, railways);
- a nomadic lifestyle;
- wars;
- corrupt politicians misusing agriculture or aid money;
- lack of healthcare leading to a cycle of disease and poverty.

'Deficiency' is the term used to describe a situation in which (a) a person is not getting enough of a certain nutrient and (b) this causes a health problem. Protein deficiency can lead to insufficient production of blood plasma proteins. The result of this would be the retention of fluids in certain tissues, notably the walls of the intestine. Such a condition is a good example of malnutrition.

'Malnutrition' is defined as an imbalance in the diet which leads to one or more diseases. Here, the imbalance is a deficiency but, in other circumstances, imbalance can be caused by an excess of one or more essential nutrients.

Kwashiorkor is a disease resulting from malnutrition, common among children in non-industrialized nations. The name comes from Ghana (a West African country) and refers to what happens to the first-born child in a family when a second child is born: the mother breastfeeds the younger child first and may not have enough milk for the elder child. The consequence is that the younger child has enough protein from the mother's milk, but the older child develops kwashiorkor, a type of protein deficiency disease.

There is a simple cure for protein deficiency: eat more protein. However, sources of protein can be very expensive, so this is not an easy option for many families in the world. Humanitarian aid missions to regions suffering from famine often take special high-protein biscuits for the local population so as to avoid protein deficiency. Volunteers in developing countries sometimes teach people how to raise chickens and introduce eggs into their diet – chicken eggs are one of the most complete and easily digestible sources of proteins for humans.

This child has kwashiorkor, a protein deficiency disease.

Examiner's hint: When asked for an example of a food, be sure to give the name of a food and not just a category such as starch or protein.

Phenylketonuria (PKU)

PKU is a genetic disease caused by a mutated gene on chromosome 12. The gene codes for a specific enzyme. The enzyme's function is to convert the essential amino acid phenylalanine into a different amino acid called tyrosine. When this conversion happens correctly during growth and development, a person's brain develops without difficulty.

People with PKU cannot convert the phenylalanine into tyrosine so they have concentrations of phenylalanine in their blood which are dozens of times higher than the normal level. When the body tries to eliminate the excess, toxins are produced which adversely affect brain development. The consequence of this is that babies and young children with untreated PKU can have severe mental problems and learning difficulties.

The mutated recessive gene which produces phenylketonuria has been passed on from generation to generation, so that about 1 in 18 000 babies is born a homozygous recessive with PKU. Fortunately, PKU can be diagnosed early with a blood test at birth, so parents can be informed what treatment is necessary. It is recommended for children with PKU to follow a special diet low in protein to avoid phenylalanine. This means eliminating from the diet things such as milk, meat, peanuts and cheese.

In some countries, it is required by law that food labels clearly indicate the presence of artificial additives such as aspartame, which contain phenylalanine. The presence of this artificial sweetener turns a naturally non-protein food such as a soft drink into a dangerous food for PKU sufferers.

Fatty acids

Although they have similarities in their molecular structure, not all fats are equal. What is the same between all fatty acids is that they have a carboxyl group (–COOH)at one end and a methyl group (CH_3–) at the other end. In between is a chain of hydrocarbons (hydrogen atoms and carbon atoms) which is usually between 12 and 20 carbons long.

• **Examiner's hint:** In exam questions about molecules such as fatty acids, be sure to read carefully to see whether the question is about structure or significance or the impact on a person's health.

Saturated fatty acids

In Figure 12.1, the yellow zone at the left is the carboxyl group, the white zone in the middle is the hydrocarbon chain (shown much shorter than any fatty acid in the human body), and the green zone on the right is the methyl group. The right-hand side of the molecule is also called the omega end (omega, ω, is the last letter of the Greek alphabet).

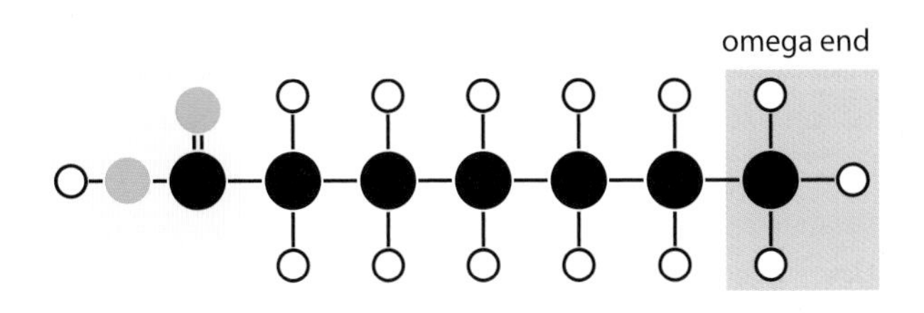

Figure 12.1 The three main sections of a fatty acid: the carboxyl group, the hydrocarbon chain and the methyl group.

Saturated fatty acids are so named because the carbons are carrying as many hydrogens as they can: they are saturated with hydrogen atoms. These molecules are typically from animal products such as butter, bacon, or the fat in red meat. They are generally solid at room temperature.

Because the carbons are carrying as many hydrogen atoms as possible, saturated fatty acids have no double bonds between the carbon atoms. The shape of the molecule is straight – there are no kinks or bends along the chain.

Monounsaturated fatty acids

If one double bond exists in the chain of hydrocarbons, the fatty acid is not saturated any more – it has two empty spaces where hydrogens could be. This type of unsaturated fatty acid is referred to as monounsaturated.

In Figure 12.2, the double bond between two carbons in the hydrocarbon chain is highlighted. Notice how the absence of two consecutive hydrogen atoms on the same side of the carbon atom chain causes the molecule to bend at the zone where the double bond is.

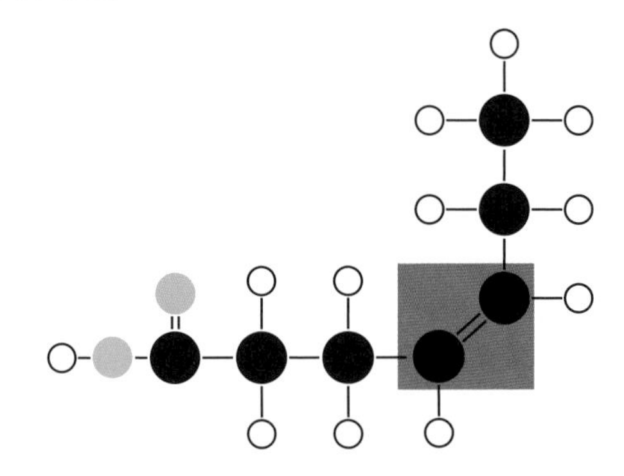

Figure 12.2 The highlighted zone in the middle of this fatty acid shows that it has a double bond. This can create a bend or kink in the shape of the molecule.

A monounsaturated hydrocarbon can exist in another form in which the missing hydrogens are on either side of consecutive carbon atoms. In this case, the molecule straightens again (see Figure 12.3).

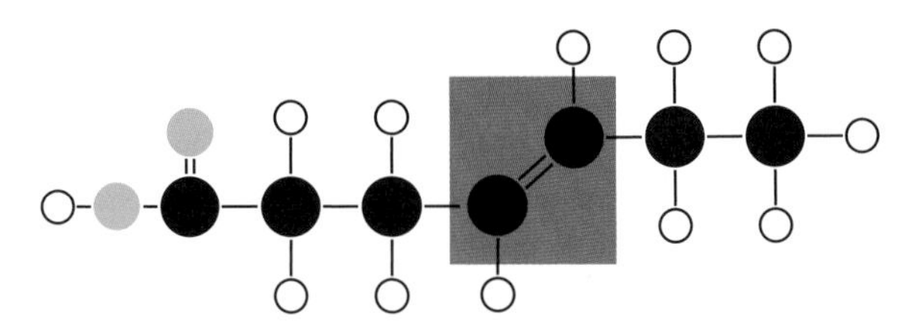

Figure 12.3 If the hydrogens are on opposite sides of the double bond, the molecule is bent back and straightened.

You will recall that one of the reasons why carbon is such a versatile atom in molecules, is that it has four places where it can bond to other atoms. The black lines in the diagrams represent bonds; you should count four coming from each carbon atom. Remember to count a double bond as two bonds.

Polyunsaturated fatty acids

Polyunsaturated fatty acids have at least two double bonds in the carbon chain. They typically come from plants such as olives. These fatty acids are called polyunsaturated because two or more carbons are not carrying the maximum number of hydrogen atoms. They tend to be liquid at room temperature.

Imagine a hydrocarbon chain several times longer than any shown in the figures so far, with several more double bonds: the molecule may have so many bends that it starts to curve over onto itself or twist around itself. This happens with polyunsaturated fatty acids, especially when the missing hydrogens are all from the same side of the chain.

Hydrogenation: cis *and* trans *fatty acids*

In processed foods such as snacks and cakes, polyunsaturated fats are often hydrogenated or partially hydrogenated. This means the double bonds (and hence the kinks) are eliminated (or partly eliminated) by adding hydrogen atoms. Hydrogenation straightens out the natural bent shape of unsaturated fatty acids. Naturally curved fatty acids are called *cis* fatty acids and the hydrogenated straightened ones are called *trans* fatty acids. The vast majority of trans fatty acids are the result of chemical transformations in food-processing factories. They are usually partially hydrogenated and still contain one or more double bonds.

One category of *cis* fatty acids is called omega–3. Most people say this without pronouncing the minus sign in the middle, although technically it should be mentioned. The name comes from the fact that the first carbon double bond to be found in this molecule is at the third carbon atom counting backwards from the omega end (see Figure 12.4). Fish are a good source of omega–3 fats.

Another category of *cis* fatty acids is called omega–6. The first carbon double bond in this molecule is found at the sixth carbon atom counting from the omega end.

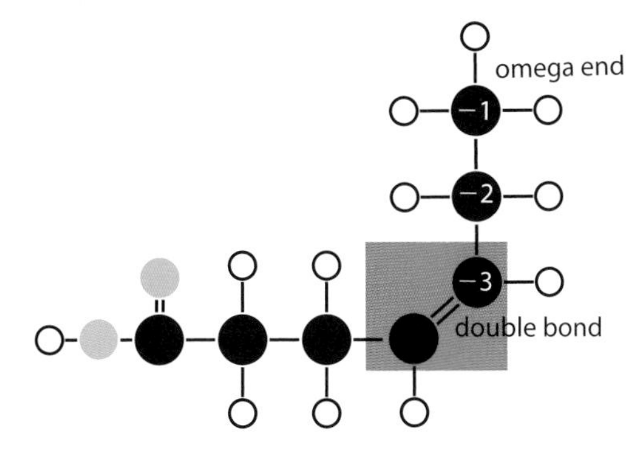

Figure 12.4 This imaginary amino acid demonstrates how we get the name omega–3. Starting at the omega end, count the carbons until you reach the first double-bond.

Figure 12.5 is a summary of the fatty acids outlined in this section. You need to understand the differences between these molecules in order to understand the next section which explores which of these are more healthful and which present risks.

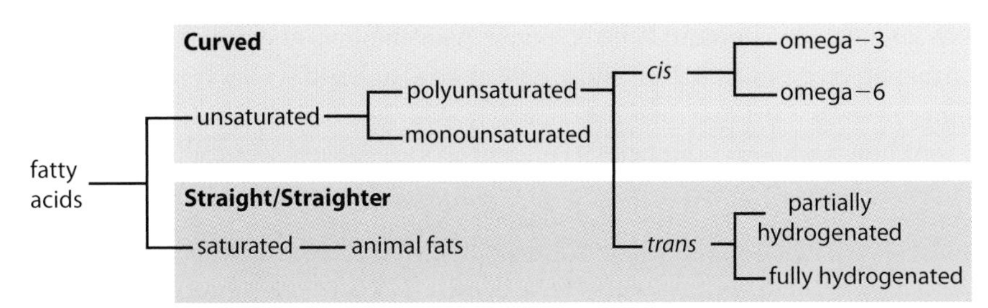

Figure 12.5 Unsaturated fatty acids, shown highlighted green, tend to have curved or curled structures. Naturally saturated fats and hydrogenated polyunsaturated *trans* fatty acids, highlighted pink, have flat, straighter molecular shapes. Remember, most *trans* fatty acids are partially hydrogenated and still contain some double bonds.

Diets rich in fats

The characteristics of different types of fatty acid are summarized in the following table.

Characteristic	Saturated fatty acids and polyunsaturated *trans* (hydrogenated) fatty acids	Monounsaturated fatty acids and polyunsaturated *cis* fatty acids (omega–3 and omega–6)
shape	straight	bent and twisted
origin	animals or artificial processing	plants
state at room temperature	solid	liquid

The shape is important because, inside your body, fatty acids which are curved are more easily picked up by the current of the blood flowing through arteries. Saturated fatty acids, which are straight, can lie flat against the walls of your arteries and are more difficult to pick up by passing blood.

You can simulate this with paper. Put a sheet of paper flat on your desk and try to move it by blowing on it. Now fold the paper in half in two different directions and unfold it onto your desk so that the corner sticks up. Try blowing again. Which is easier to move? Which state of the paper represents unsaturated and *cis* fatty acids and which represents saturated and *trans* fatty acids?

Over many years, the deposits of saturated or *trans* fatty acids combine with cholesterol to form a substance called plaque along the inner lining of your blood vessels (see Figure 12.6). Such deposits can reduce the volume of your arteries and cause your blood pressure to increase. If a chunk of plaque breaks off and is stuck in a zone which has had its diameter reduced by other deposits of plaque, no blood can get through.

Figure 12.6 Diagrammatic cutaway views of arteries with and without deposits of plaque: a substance made of saturated fatty acids and cholesterol.

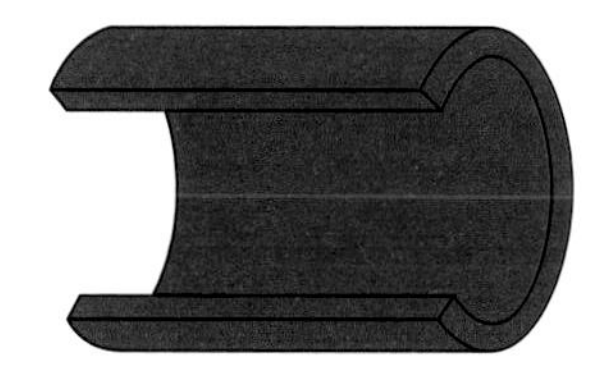

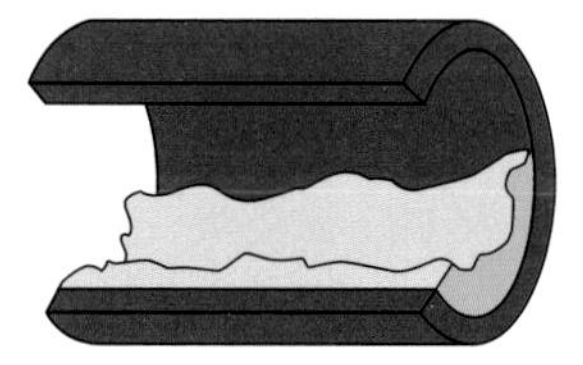

How do we know that one type of fatty acid is 'good' and another is 'bad' for our health? Is there a direct cause and effect between diet and health?

Some people live to a very old age while eating saturated fats all their lives whereas others are very careful with their diet and may still die young of a heart attack.

What are the other factors besides diet? What is it in their lifestyle or in their family history that could also have an impact? The weblink below can help.

If this happens in the region of the heart, such an event is called a heart attack. If this blockage happens in the brain, it is called a stroke. Since these are major health problems in the world, you probably know someone who has had this happen. Someone who has plaque deposits in the arteries bringing blood to the heart is suffering from a disease called coronary heart disease (CHD).

From this information, it should be clear that a diet rich in saturated fats has a much higher chance of leading to serious cardiovascular problems later in life. Diets rich in polyunsaturated fats, however, lead to lower quantities of plaque and therefore a healthier cardiovascular system. Polyunsaturated fats tend to carry away cholesterol during their journey in the bloodstream and, as a result, people who eat polyunsaturated fats reduce their cholesterol level.

Both saturated and unsaturated fats contain high amounts of energy, so consuming excess quantities of either type of fat is unhealthy. However, from the point of view of cardiovascular health, polyunsaturated fats are clearly beneficial to your health whereas saturated fats present serious dangers, especially when consumed in excess.

The American Heart Association has a healthy lifestyle webpage as well as extensive prevention measures including ways to save someone's life by recognizing early warning signs of heart attack. Visit heinemann.co.uk/hotlinks, enter the express code 4242P and click on Weblink 12.1.

There is one important point which must not be overlooked: polyunsaturated fats which have been hydrogenated and transformed into *trans* fats present dangers which are comparable to saturated fats. The green zone in Figure 12.5 shows the more healthful types of fatty acid whereas the pink zone shows the fatty acids which present higher risks of cardiovascular problems. Next time you eat a processed snack food, have a look at the label – are there any hydrogenated oils or other *trans* fats?

There are marked differences in traditional diet in human societies around the world and also differences in the rates of CHD and other diseases that have been linked to diet.

Some populations are known for their high incidence of CHD, notably those in industrialized nations. Part of the reason for this is an over-consumption of saturated fats from animal products and processed foods. People with CHD are told by their doctors to eat less fatty meat and eat more vegetables.

In contrast, the Masai people of East Africa have one of the lowest incidences of CHD ever measured. Yet much of their food comes from cows and goats. In addition, they are not farmers and do not eat many vegetables. So what is their secret? Hypotheses include a highly active outdoor lifestyle, low overall energy intake, medicinal properties of certain plants they eat, and possibly genetic factors.

Vitamins and minerals

Although vitamins and minerals are often grouped together in biology books and food labels, they are structurally very different substances. First of all, minerals are inorganic; they do not contain carbon and they are not synthesized by living organisms. Generally, minerals come from rocks, soil or sea water. Vitamins are organic. They are synthesized in living plants and animals and they always contain carbon.

The Masai of East Africa have one of the lowest incidences of heart disease in the world.

Secondly, vitamins and minerals differ in their structure: minerals are found in the form of ions, whereas vitamins are in the form of compounds (see table below).

Examples of minerals	Examples of vitamins
sodium: Na^+ calcium: Ca^{2+} iron: Fe^{2+}	vitamin A: $C_{20}H_{30}O$ vitamin C: $C_6H_8O_6$ vitamin D: $C_{27}H_{44}O$

Notice from the chemical formulae that minerals are always single elements in ionic form, whereas vitamins are always compounds of several different elements.

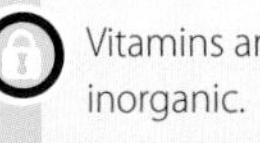

Vitamins are organic; minerals are inorganic.

So why are minerals and vitamins always grouped together? One of the main reasons is simply a question of quantity: they are both needed in very small amounts, usually only milligrams per day. In addition, both play a similar role in the body: preventing deficiency diseases (see below).

Vitamin C is important in protection against infection, helping in wound healing and in maintaining healthy gums, teeth, bones and blood vessels.

How much vitamin C do you need?

How do you know if you are getting enough vitamin C in your diet? The best way is to look up the quantity of vitamin C recommended by your governmental food and health services. The recommended level is usually between 30 and 60 mg per day. Look at labels on the foods you eat to see if you have enough.

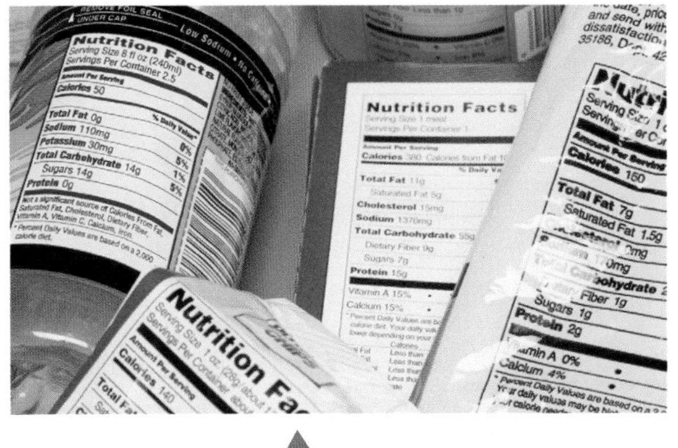

The food industry indicates on food labels the percentage of the recommended daily amount of nutrients such as vitamin C.

To determine how much vitamin C a person needs, there are two main techniques of experimentation: animal tests and tests on human subjects.

Animal testing is often carried out on laboratory mice. The aim of the experiment is to feed varying amounts of vitamin C (or other nutrient under test) to several groups of mice. Other than the amount of vitamin C, everything else in the diet and environment of the mice is the same. After a certain time, the health of the different groups of mice is compared.

Mice with insufficient amounts of the vitamin would be expected to present signs of deficiency such as increased rates of infection and illness. Those with sufficient amounts would not present signs of deficiency and would be healthier mice.

Excess vitamin C can lead to kidney stones.

Scurvy is the deficiency disease associated with lack of vitamin C. Symptoms include retention of fluid, loss of teeth, bleeding into joints, anaemia and lethargy. If untreated with vitamin C, the disease is fatal.

It is possible that some groups receive doses of vitamin C that are high enough to cause undesirable side-effects. By examining the different groups, a decision can be made as to the dose that produces the highest benefit for mice. By interpolating the results for humans, further experiments can be done to determine which dose is most beneficial.

Variables associated with the experiment:

- independent variable = the amount of vitamin C given each day;
- dependent variable = the state of health of the subjects as measured by frequency of infections, presence of scurvy or presence of undesirable side-effects;
- controlled variables = the subjects in each group and their diets should be as similar as possible (this is much harder to control with humans than with laboratory mice).

Experimentation on humans is a complex subject which raises ethical questions. Usually, experimentation is done with consenting volunteers who sign a contract acknowledging that they are aware of the potential risks before they agree to participate in the study.

During the Second World War, vitamin C experiments were done in the US using conscientious objectors (i.e. people who were drafted into military service but who refused to go into combat because of their beliefs and values). During the war, major advances were made in the science of nutrition because the military needed to know the optimal diet to keep soldiers healthy and strong.

Experimentation can expose people to grave dangers even when they are consenting volunteers. For example, in vitamin C experiments, groups which did not receive any vitamin C ran the risk of getting scurvy, which can be fatal. This raises an ethical problem for doctors: is it against the Hippocratic oath to conduct such experiments? The Hippocratic oath is a series of promises which guide doctors in deciding what is right and wrong in their profession. Part of the oath states that doctors promise to never intentionally harm a patient.

Why is it that recommended vitamin C daily doses vary between 30 and 60 mg a day. Does that mean some scientific tests are better than others? Could it be that people are physiologically different in various parts of the world and need different amounts of vitamin C? Are some scientists overly concerned about excess vitamin C intake or are others worried that the population will not get enough?

Because the results of vitamin C studies vary somewhat from country to country, it is difficult to decide how many milligrams of vitamin C you need per day. Also, the amount varies depending on age and other physical characteristics.

In general, the numbers suggested by government studies all over the world are calculated by determining:

- the minimum level to prevent scurvy;
- the minimum level to protect against upper respiratory tract infections;
- the minimum level at which undesirable side-effects, such as diarrhoea, occur.

Linus Pauling was an American chemist and biochemist whose book *How to Live Longer and Feel Better* (1986) recommended that mega doses of vitamin C (such as 1000 mg) would protect you against catching colds. This was a radical idea because vitamin C is normally regarded simply as something to prevent deficiency diseases, notably scurvy.

His ideas were not always backed up by conclusive results from clinical trials so he was criticized by other scientists. Is a man's reputation as a good scientist enough to make his ideas valid? Inversely, is it possible that an unknown scientist could come up with perfectly valid new suggestions for your health?

Can you have too much vitamin C?

If you take excessive amounts of vitamin C in your diet, your body cannot store the excess, so it is lost in the urine. However, the body gets used to excreting high levels of vitamin C and continues to do so even if the high intake stops. This level of excretion can result in low levels of ciculating vitamin C in the body and possibly even the development of scurvy (page 328). This is called rebound malnutrition.

If your diet contains a healthy balance of fresh fruits and vegetables, there should be no need to take vitamin supplements. Taking more than the recommended amount is controversial, but the fact that it is contested does not mean it cannot have beneficial effects. It is possible that some day clinical studies will shed new light on the matter and governments may re-evaluate their recommendations.

Sources of vitamin D

Vitamin D, also known as calciferol, is an important nutrient for the proper formation of bones. Without a sufficient supply of vitamin D, it is possible to develop rickets – a disease that leads to deformities in the bones. Children with rickets do not reach their optimal height during growth and their legs are often bowed inward or outward at the knees. This can also happen to lactating mothers with vitamin D deficiencies.

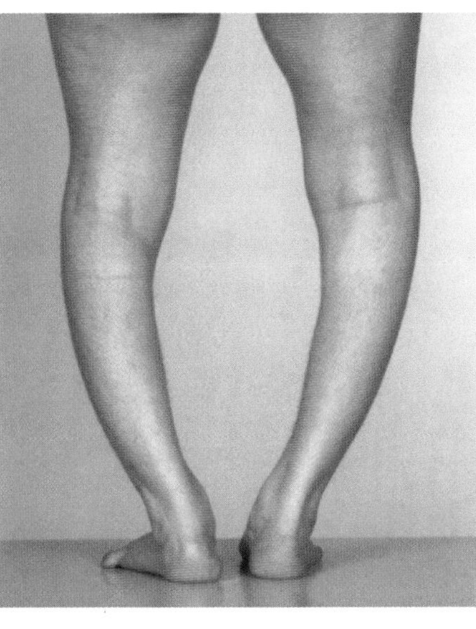

A person suffering from vitamin D deficiency can develop rickets. The consequences include severely bowed legs.

There are three ways to get vitamin D:

- exposure to sunlight;
- from foods rich in vitamin D;
- from vitamin supplements.

The epidermis of human skin contains precursors which are able to synthesize vitamin D when stimulated by the ultraviolet rays of the Sun. As long as you are not too far from the equator, about 15 minutes of exposure during the hottest time of day twice a week is enough for making all the vitamin D you need.
People who do not get enough exposure to sunlight may risk a deficiency in calciferol unless they get it from food. Foods containing vitamin D are:

- certain types of fish such as salmon, tuna and sardines;
- eggs (vitamin D is in the yolk);
- many ready-to-eat breakfast cereals are fortified with vitamin D (i.e. vitamin D has been added);
- liver;
- milk is often fortified with vitamin D (especially in the US).

Balancing two risks

The problem with relying on making vitamin D in response to exposure to sunlight is that too much exposure to ultraviolet rays from the Sun can cause sunburn and, over many years, lead to the skin cancer called malignant melanoma.

We live in a world full of risk. If you don't get enough sunlight you could develop vitamin D deficiency but if you get too much, you could develop skin cancer. Is there any such thing as an endeavour without risk? How do you decide what is the right balance of risk and benefit?

In children, the need for vitamin D is crucial, yet children's skin is extremely sensitive to sunlight and sunburn. On the one hand, we want a child to have all the vitamins necessary for optimal growth and development but, on the other hand, we also want to protect that child from the dangers of the Sun.

As a result, we must carefully balance the risks and advantages. The happy medium would be to expose the body to enough sunlight for the synthesis of vitamin D but not so much that the skin burns and increases the chances of developing skin cancer later.

In fact, in conditions of intense sunlight, exposure is not needed every day and does not need to be for very long. The rest of the time, if you want to stay outdoors, you should use sunblock to protect your skin.

After 15 minutes of absorbing UV rays for vitamin D production, people with light skin should protect themselves with a sunblock cream or lotion.

Cancer is becoming more and more of a problem. Many people now live into their 80s and 90s instead of dying in the first few decades of life due to infectious disease, as our ancestors often did. This gives cancer cells a wider window of opportunity to develop. The increasing incidence and prevalence of cancer raises many issues of economic, legal and medical dimensions.

Artificial dietary supplements

Sometimes the foods you eat are not varied enough to provide every one of the dozens of different molecules you need in order to stay healthy. For example, it can be challenging to find foods rich in iodine. Iodine is naturally present is sea water so people living near the ocean who eat seafood or kelp on a regular basis usually do not have a problem. However, people who do not have access to regular seafood are at risk of not getting enough iodine.

Iodine is a component of thyroxin, a hormone made by the thyroid gland, which is located in the neck. Hormones made by the thyroid gland regulate growth and the rate at which energy is released in the body. With insufficient iodine in the diet, an inflammation of the thyroid gland develops. Such a swelling in the neck is called a goitre. In addition, babies born to mothers with iodine deficiencies can suffer from cretinism, a condition resulting in stunted growth and varying degrees of mental problems.

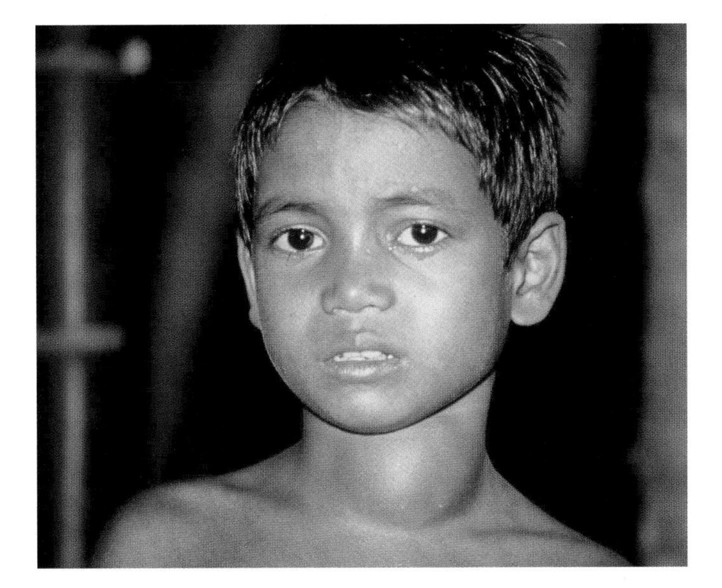

Depending on where you live, it is quite possible you have never seen a person with a goitre (the swelling on this young boy's neck). The condition is now rare in industrialized countries.

One of the main reasons for the rarity of goitres in industrialized countries is that iodine, as sodium iodide, is added to both table salt and cooking salt. This solution to the problem of a formerly widespread condition was proposed by a French chemist in the first half of the 19th century, but it was nearly a century later before the idea was finally applied in Switzerland in the early 1900s.

The US followed suit in 1924 and was able to essentially eradicate the problem of goitres in just a few decades. This was a welcome change for populations such as the inhabitants of the Great Lakes region where nearly 40% of children had visible goitres in 1924. Within just 4 years of starting to iodize salt, the number dropped by nearly 75% and by the 1950s the problem was essentially nonexistent.

Although there are risks of toxicity from excess iodine in the diet, the benefits of iodizing salt greatly outweigh the dangers of excess iodine intake. In addition, this solution is relatively simple and inexpensive as long as the society in question has a reliable system for producing, packaging and distributing salt. This is the reason why many developing nations are still affected by iodine deficiency: they lack the infrastructure necessary to implement the solution.

The importance of fibre in the diet

How many times have you been told, 'Eat up your vegetables'? Even though they might not be the most popular item on the menu, vegetables are an important part of the meals in school cafeterias and this should be true at home as well. Besides being a good source of vitamins and minerals, vegetables are an important source of fibre, although they are not the only fibre-rich foods. Fresh fruits and salads are also good sources of fibre.

Peristalsis is the name for the involuntary contractions of the gut wall that push the contents through the alimentary canal.

Fibre, also referred to as dietary fibre (or, more informally, roughage) is mostly the cellulose in plant material. It helps the human digestive system function better by providing bulk. In order for peristalsis to function optimally, the muscles which push food materials along the intestines need to have a sufficient volume of material to apply pressure to.

A diet high in fibre reduces the likelihood of constipation and can lower the chances of certain intestinal problems such as appendicitis or cancer. There is also

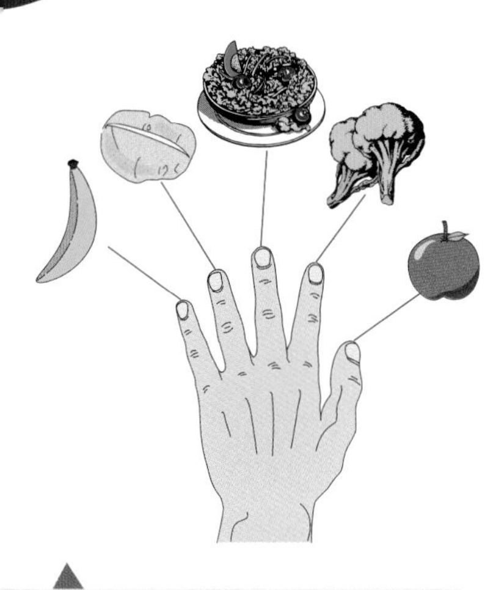

Figure 12.7 To help you remember to have at least five servings of fruits and vegetables every day, count them on your fingers.

strong evidence that such diets also reduce the risk of diabetes and cardiovascular disease. One of the possible reasons for this is that toxins bind to the fibre and are carried out of the body.

High-fibre diets also help people manage their body mass better. It is easier to lose excess weight on a diet which includes fruits and vegetables, in part because the fibre fills up the stomach to give a feeling of satiety without introducing excess energy.

A common criticism of modern diets, especially in industrialized countries, is that they do not contain enough fibre. One recommendation is to eat five to eight servings of fruit or vegetables each day (see Figure 12.7). It is estimated that only 1 in 4 people in the US actually does this.

Exercises

1. Define *essential* when used to describe fatty acids or amino acids.
2. Explain why kwashiorkor protein deficiency may affect the first child when the second child is born.
3. Distinguish between the structures of *cis* and *trans* fatty acids.
4. Explain why, despite the fact they are found in living organisms, minerals are not considered organic.
5. The children's story *Snow White and the Seven Dwarfs* depicts seven men who worked in a mine and who had significantly small stature. It has been suggested that this fictitious story has some historical basis: children who worked in mines often never grew properly. Suggest a reason why the nature of their work would cause a nutritional imbalance that affected their growth.
6. Explain why a large number of inhabitants of the Great Lakes region in the US had goitres in the early 1900s.

A.2 Energy in human diets

Assessment statements

A.2.1 Compare the energy content per 100 g of carbohydrate, fat and protein.

A.2.2 Compare the main dietary sources of energy in different ethnic groups.

A.2.3 Explain the possible health consequences of diets rich in carbohydrates, fats and proteins.

A.2.4 Outline the function of the appetite control centre in the brain.

A.2.5 Calculate body mass index (BMI) from the body mass and height of a person. $\text{BMI} = \frac{\text{(mass in kg)}}{\text{(height in m)}^2}$

A.2.6 Distinguish, using the body mass index, between being underweight, normal weight, overweight and obese.

A.2.7 Outline the reasons for increasing rates of clinical obesity in some countries, including availability of cheap high-energy foods, large portion sizes, increasing use of vehicles for transport, and a change from active to sedentary occupations.

A.2.8 Outline the consequences of anorexia nervosa.

How much energy is in nutrients?

As you can see in the table below, not all nutrients contain the same amount of energy.

Nutrient	Typical energy per mass / kJ per 100 g.
carbohydrates	1760
protein	1720
lipids	4000

Lipids contain more than twice the energy per 100 grams than the other nutrients, whereas proteins and carbohydrates have very similar energy levels. This is because the main energy-storing bonds are the ones between carbon and hydrogen. The bonds between carbon and oxygen can also store energy but less efficiently. Since lipids have a higher ratio of C—H bonds to C—O bonds than carbohydrates or lipids, they can store more energy per 100 g.

This is why it is possible to eat large quantities of fresh fruit and vegetables without too much worry of getting an excess energy intake. Such foods are high in fibre, which has virtually no usable energy, and rich in carbohydrates which contain moderate levels of energy. In addition, fruits and vegetables generally contain very small amounts of fat.

● **Examiner's hint:** It is more scientifically correct to say that a food contains lots of energy rather than to say the food contains lots of calories or joules. Saying the latter is like saying 'that bathtub has lots of litres in it' instead of 'that bathtub has lots of water in it'.

But a low-energy food such as a salad can easily be transformed into a high-fat, high-energy meal by adding things such as salad dressing, bacon, avocado, cheese or olives, all of which are rich in lipids. When it comes to a well-balanced diet, it all depends on quantities. Because lipids have so much more energy, foods which are rich in lipids should be consumed in proportionally smaller quantities.

Stay away from fatty foods if you are trying to limit your energy intake. Even a modest amount of chocolate cake, fried potatoes or buttery pastry has considerable energy because of the lipids such foods contain.

Stalks of wheat yield kernels or grains which can be ground into flour to make foods such as bread or pasta.

Sources of energy in different ethnic groups

Food energy sources around the world are markedly different in different places.

Rice

Rice is one of the most widely grown crops in the world and it is the main source of energy for 1 in 5 people on Earth. Rice is a type of grass which is native to southeast Asia and to Africa where it has been grown for several thousand years.

Among the many countries which are traditionally known for rice production and consumption are China, Japan, India, Indonesia, Bangladesh, Vietnam and Thailand. Rice is now grown on every continent except Antarctica but Asian farmers produce 90% of the rice on Earth. Most of the energy in rice comes from starch, a complex carbohydrate.

Wheat

Another widespread crop in the world is wheat. It is thought to be among the first cereals to be domesticated about 10 000 years ago. It is believed to have originated in the 'fertile crescent', a region running from the Nile river delta to Mesopotamia. Today, many cultures use wheat as their main source of energy, most notably in the form of wheat flour to make bread, cakes or pasta, semolina or couscous. Like rice, the main organic compound in wheat is carbohydrate. In Russia, for example, more than 30% of the population's energy intake comes from wheat products. When looking at the consumption per capita, other major wheat consumers include Australia, Turkey, Canada, and Iran.

It is not only from a molecular point of view that 'You are what you eat'. The same can be said from a cultural point of view.

One of the most interesting and sometimes most difficult things to do when travelling is to appreciate the foods that the local populations eat. What one group may find delectable, another may find repulsive. This is an aspect of human diversity, but also of oneness. What we eat may be different but we all eat for the same reasons and we all need the same biochemical substances, albeit in slightly different proportions.

Cassava

Cassava, also known as manioc, is a shrub which originated in South America but is now grown in many tropical regions of the world. The edible part is the root (tuber) which is an excellent source of starch. It is the staple energy source for many populations in the Caribbean. It can be cooked and eaten like a potato or transformed into flour to be used to make various dishes including tapioca. The root is exceptionally rich in carbohydrates but nearly devoid of protein so infants who eat cassava exclusively can suffer from protein deficiencies (see page 322).

This is a tuber called cassava or manioc.

Corn, also called maize, is a source of carbohydrates for certain human populations.

Maize

Like all the foods described so far in this section, corn or maize stores its energy as starch. Corn originated in Central America. Today, Mexico is one of the leading producers and consumers of corn. Nearly half the daily energy intake of Mexicans is from corn, notably in the form of tortillas.

Fish and meat

Not all populations in the world obtain their energy principally from carbohydrates. Many populations in coastal areas or on islands rely on fish as their main energy supply. This is true for the populations of the Seychelles, off the east coast of Africa. The inhabitants of these islands reportedly have the highest per capita consumption of fish in the world. The Inuit peoples of Northern America and Greenland also have a diet with a high fish intake. The energy in fish is in the form of protein, followed by fat. There are no carbohydrates in fish.

Beyond a shadow of a doubt, the world's largest per capita consumer of meat is the US. While the global average of meat consumption per person per year is approximately 37 kg, Americans eat nearly three times that amount at 90 kg per person per year.

Most of the energy from meat comes from protein and fat. Nutrition experts warn that most Americans need to reduce their meat consumption and eat more high-fibre, low-fat foods.

Health consequences of diets rich in carbohydrates, fats and proteins

Any diet other than a balanced one containing proper amounts of all food categories and requirements presents risks.

Excess carbohydrate in the diet

Eating excessive amounts of carbohydrates results in too much energy entering the body. Complex carbohydrates are broken down into simple sugars which can be used to make ATP for muscles and other tissues. If the energy from the sugars is not burned off by physical activity, the excess is stored. For this, the sugar is converted into either glycogen or fat.

When there are moderate amounts of sugar obtained from simple sugars and complex carbohydrates in the diet, glycogen is made. The body has a limited storage of glycogen, however, so if there is more glucose than can be converted to glycogen, the excess is converted into body fat for storage.

The Food and Agriculture Organization (FAO) of the UN has a webpage with a wealth of information about which foods originate from which parts of the world. Despite there being over 50 000 edible plants in the world, rice, corn and wheat provide 60% of the food energy that the world's human population needs. Visit heinemann.co.uk/hotlinks, enter the express code 4242P and click on Weblink 12.2.

Excess fat in the diet

Eating excess fat in the diet has some similar consequences to eating excess carbohydrates. But there are differences and one of the main ones is that fats contain over twice as much energy per unit mass than carbohydrates, so the problem of storing excess energy is intensified. The other difference is that depending on the nature of the fat, there could be serious health consequences. In effect, saturated fats present risks of obesity and increased risk of cardiovascular disease (see Figure 12.6, page 326). It is important to be aware of the risks of fried foods and saturated fats so that they are consumed in moderation.

From an evolutionary point of view, fat storage was necessary when our ancestors had to spend some seasons of the year with very little food. Fat was accumulated in the months when food was plentiful and burned in the winter or dry season. Today, this is still true in some places of the world but in industrialized countries, the availability of plentiful food year round has lead to problems of obesity.

Excess protein in the diet

An adult typically needs to consume no more than 50 g of protein a day. Your needs would be met by eating two or three servings of protein-rich foods in a day – perhaps meat, nuts, eggs or cooked dried beans and pulses. The risks of getting too much protein occur when people opt for a slice of steak which is so massive that it counts as two or three servings all by itself or when they eat protein-rich foods at every meal: eggs for breakfast, meat at lunch and poultry or fish at dinner.

Excess protein in the body can be harmful. We have no way to store extra protein, so the molecules which cannot be used must be destroyed and eliminated from the system. This is done by the liver and the kidneys. Overeating protein means overworking not only the digestive system but also the liver and kidneys, both of which become hypertrophied (oversized) by overcompensation.

In the chemical process of eliminating the excess protein, our kidneys use calcium. If they cannot find enough of it, they leach it from our bones. Decalcified bones break more easily. In countries where weight loss programmes are big business, some people have chosen a high-protein, low-carbohydrate diet to shed those unwanted kilos without being aware of the long-term consequences.

In many countries in the world, multinational companies make substantial profits by selling and massively advertising high-energy fast foods and soft drinks.

This raises social and ethical issues concerning how acceptable it is for a company to make economic profits while contributing to widespread health problems which have a negative effect on society.

The media is full of advice on what people should eat or should avoid. Magazines and television shows try to sell us ideas about carbohydrates being 'good' or 'bad', or protein being a miracle solution for losing weight.

How many of these claims are based on scientific findings and extensive clinical studies? Are consumers easily lead astray because the technical aspects of biochemistry are too complex? Are we given enough information in the media to make educated decisions and evaluate the claims?

Keep in mind that the main goal of magazines is to sell magazines and the main goal of television programmes is to get the most viewers possible. Do you think they would achieve their goals by informing the public in an honest, scientific and unbiased way?

And what about fibre?

Another danger of all of the above excesses (carbohydrates, fats or protein) is that if you satisfy your appetite with foods rich in such nutrients, you do not leave enough room for the high-fibre foods such as fruits and vegetables which would be a much healthier choice. These foods carry with them healthful vitamins and minerals and have been linked to reducing the risk of certain types of cancer.

Appetite control system

Hunger is the body's way of expressing its need for food and tends to be a negative sensation. Appetite is the desire to eat, a positive sensation. It is quite possible to feel hunger and yet not feel the desire to eat (i.e. to be hungry but have no appetite); for example, when you are sick. On the other hand, it is very common to not be hungry but see something that looks too good to resist.

At the end of a meal, when you have eaten a sufficient quantity of food, you have reached a state of satiety and that is when most people stop eating. Although the mechanisms of appetite and satiety are quite complex and not fully understood, they seem to be a combination of feedback loops from the nervous system, the digestive system, and the endocrine system. For example, after a meal, the pancreas releases hormones which reduce appetite. The question is, where do the feeling of hunger and the sensation of appetite originate in the body? To understand this, consider what happens when there is a problem with the system.

People who have medical complications which damage their hypothalamus (a part of the brain found at its base) can have severe appetite problems: some get very thin from a loss of appetite and others become very obese due to an insatiable appetite. From this evidence, it is clear that the hypothalamus plays an important role in regulating appetite. Although it has other functions as well, it can be said that the hypothalamus acts as your appetite control centre.

The hypothalamus is found at the base of the human brain.

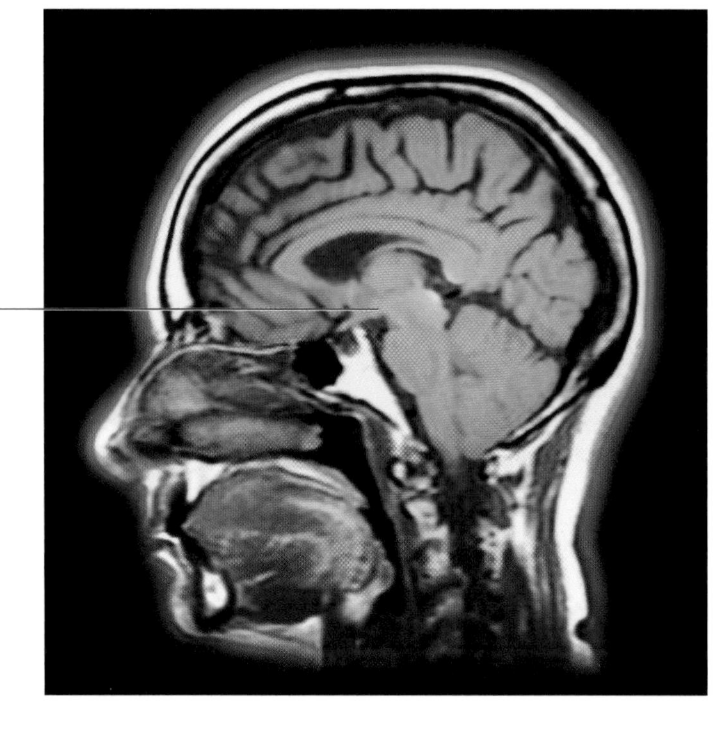

The hypothalamus controls several basic body functions which you do not have to consciously think about: hunger, thirst, body temperature and telling the body when to release certain hormones.

During a meal, your stomach fills with food, expands and stimulates nerve cells of the vagus nerve. A signal is sent to the hypothalamus to stop eating. The intestines produce various hormones to send signals about hunger and satiety to the brain.

The fat cells of adipose tissue produce a hormone called leptin which sends a message to the hypothalamus to suppress appetite. A person with more body fat produces more of this hormone so that the brain knows there are adequate energy stores. If you were to fast, your level of leptin would significantly decrease. But leptin is not the only hormone in the process, it would be a great oversimplification to think that appetite was regulated solely by leptin. In addition, other forces such as compulsive eating or persuasive advertising, seem to be able to override leptin's effects.

The feedback loops involved in appetite control can be thought of as algorithms – imagine you could listen in to the various hormonal and nervous messages being sent in your body:

- 'this sweet food tastes good, continue eating';
- 'this fatty food tastes really good, continue eating';
- 'if food is plentiful now, keep eating' (because there may not be any later);
- 'if stomach is full, stop eating';
- 'if temperature in environment increases, eat more salt' (to compensate for losses of salt in sweat);
- 'if daily sunlight hours decrease, eat more fat' (winter is coming and less food will be available);
- 'if body fat is plentiful, do not increase hunger';
- 'if under stress, eat more'.

Body mass index (BMI)

When you look in the mirror, do you like what you see? Do you consider yourself to be thin, overweight or just right? The answer given by most people is usually not very scientific because it depends on cultural and personal perceptions, feelings and expectations. The scientific answer to whether or not an individual has an appropriate mass is to calculate the body mass index (BMI).

$$\text{BMI} = \frac{(\text{mass in kg})}{(\text{height in m})^2}$$

BMI should be in the high teens to mid 20s. Here is an example of a young woman who is 17 years old, is 1.65 m tall and has a mass of 60 kg.

$$\text{BMI} = \frac{(\text{mass in kg})}{(\text{height in m})^2}$$

$$= \frac{60}{(1.65)^2}$$

$$= 22$$

Underweight, normal or overweight?

Although the ideal values for the BMI change with age, this table shows typical values for a full-grown adult.

BMI	Status
below 18.5	underweight
18.5–24.9	normal weight
25.0–29.9	overweight
30.0 and above	obese

Part of being healthy and feeling good about yourself is maintaining a healthy body mass. To have an idea of what is right for you, calculate your BMI.

At 60 kg, the young woman considered above is at a healthy mass for her age and her height because her BMI is 22. She would still be considered of healthy mass if she put on or lost up to 8 kg. If her body mass plunged to 45 kg, however, she would be dangerously underweight – her BMI would be 16.5. If her body mass increased to 75 kg but she did not grow taller, her BMI would go up to 27.5 and if she reached 82 kg, her BMI would be over 30, indicating obesity.

When talking about people's body mass, we often use the terms 'weight' and 'mass' interchangeably. In everyday life, this is acceptable but in science, we need to be aware of the difference between weight and mass:

- weight is a measure of the force pulling you down towards the ground;
- mass is the amount of matter ('stuff' made of atoms and molecules) you have in your body.

In science, the proper unit for weight is newtons and for mass it is kilograms.

- **Examiner's hint:** If you live in a country where your weight is expressed in pounds and your height in feet and inches, be careful to use metric units in exams: in this case, kilograms and metres.

Are you calling me obese?

Want to find out your BMI? Visit heinemann.co.uk/hotlinks, enter the express code 4242P and click on Weblink 12.3.

Now consider a young man of the same age but who is 1.80 m tall and who has a mass of 100 kg. According to the BMI, he would be obese. What the BMI does not tell us is how the mass is distributed. Perhaps this young man is on the rugby team and besides having a naturally big build with heavy bones, he spends many hours every week in the gym and on the field doing intense athletic training to build his muscles. Looking at him, no one would call him obese, not just because they would not dare to, but because medically it is not true.

When calculating your own BMI, keep in mind the above example. Although numbers do not 'lie', they do not always tell the whole truth. The BMI can be used as a guide but, like mass alone, it should not be used as the indicator to decide if you are at a healthy body mass. How you feel about yourself and how the people who are important in your life consider your appearance has some importance, too. In addition, your doctor or a nutrition expert is probably the most qualified to help determine what is best for your health.

Why are so many people obese?

Back in 2002, the World Health Organization's Obesity Task Force estimated that 300 million people worldwide were obese and 750 million were overweight. By 2005, the estimates rose to 400 million and 1.6 billion respectively and the numbers continue to increase. The causes for these ever-growing numbers are complex but the most obvious culprits are:

- a change in the types and quantities of food people eat;
- changes in the amount of physical activity people do on a daily basis.

Just a few generations ago, most people in the world lived on farms. The family's daily routine involved a significant amount of physical activity to care for the crops and animals. Today, a migration towards urban centres has greatly reduced the amount of daily physical activity.

How do changes in lifestyle change the way people burn food energy?

Consider a family cooking their food and heating their home in winter. What today requires only the turn of a button on a stove or a thermostat used to involve felling trees, splitting logs and transporting the wood to the home.

On modern farms, the work is greatly facilitated by machines which have replaced the agricultural workers who once filled the fields by the dozen. In urban centres,

Farming in the past …

… and farming now.

people tend to be much more sedentary (they do less physical activity). Office workers who sit at a desk or work on a computer most of their day do not use their arms and legs nearly as much as their rural ancestors did. As a result, they need less energy in their daily diet.

Paradoxically, despite the fact that people are less physically active and do not need as much energy, the availability of affordable food has greatly increased. It is now possible to have a high-energy diet all through the year.

In the past, people living on farms ate what they produced there: fresh vegetables, homemade bread, pies and jams from their fruit trees. For special occasions, an animal was slaughtered and its meat was used for the feast but most people could not afford to eat meat every day. Food had to be carefully stored for times of the year when production was less plentiful and people could not always eat enough.

Back then, fizzy drinks, fast-food hamburgers and pizzas were not available. Today, cities abound with fast-food restaurants, pizza parlours, doughnut shops and distributor machines for sugary drinks, potato chips and chocolate bars. In fast-food restaurants and ice-cream parlours, the sizes of drinks and ice-creams has increased over the decades to super-sized portions. A soft drink of 20 cl would contain a few dozen calories of energy; now there are hundreds of calories in an extra-large cup.

Thanks to drive-up windows in many urban centres, people do not even have to leave their cars to withdraw money from a bank machine or to order a fast-food meal. People burn far fewer calories sitting in a car than they do walking. This is another example of how an urban lifestyle is usually more sedentary than a rural one.

To avoid problems of obesity, the formula is simple: more physical activity, more fresh fruit and vegetables, less processed high-fat food, smaller quantities of food. More physical activity for some people means participating in a sport or going to the gym but there are much simpler solutions such as walking rather than driving, and taking the stairs instead of the elevator (lift).

During a break or a snack, people are often tempted to eat a doughnut, a piece of cake, pretzels, potato chips or a chocolate bar. Why not have an apple or a banana instead? What influences your decision? Life is full of choices and making the right ones on a day-to-day basis makes a big difference for your health now, and for your future.

The term 'anorexia' is from a Greek word meaning loss of appetite; 'nervosa' is from a Latin word meaning of nervous origin. Paradoxically, most people with anorexia nervosa are constantly hungry – a lack of appetite only comes in the later stages of the condition.

To leave out the 'nervosa' and call the condition 'anorexia' is an incorrect abbreviation.

Being thin is not the same thing as being anorexic. Anorexia nervosa is a complex medical condition involving many symptoms and can only be properly diagnosed by a doctor.

Anorexia nervosa

Anorexia nervosa is an eating disorder in which people have the firm conviction that they are overweight, even when their BMI indicates that they are, in fact, normal or underweight. They have an intense fear of gaining body mass and as a result often refuse to eat food, force themselves to vomit or do excessive exercise to burn off the energy from it. Nine out of ten people with the condition are girls or women; very few boys or men have anorexia nervosa.

One of the symptoms of anorexia nervosa is a constant fear of gaining body mass. In some respects it can be considered an obsession. Typically, anorexics are unable to admit that they have a problem so diagnosis can be difficult to obtain and accept.

Based on the fact that approximately 1% of adolescent girls have anorexia nervosa, the chances are reasonably good that someone you know has the condition. Before you start looking around and pointing your finger, be aware of other people's feelings. This is a sensitive subject and one which should be handled delicately.

There are lots of helpful websites from government organizations or women's health organizations. It is worth checking them if you or a friend has an eating disorder. Be careful in your searching, however. Some sites are not reliable and some even promote and celebrate unhealthy lifestyles. The world wide web is a real gold mine – sometimes you have to do a lot of digging to get to the good stuff.

Consequences of anorexia nervosa

The consequences of anorexia nervosa on a person's health are many. First, the endocrine system malfunctions so that the hormones responsible for controlling such things as the menstrual cycle are not produced and distributed in the body. Women who are anorexic often stop menstruating.

Other consequences involve loss of head hair, dehydration, fainting, anaemia, low blood pressure, kidney stones or kidney failure. The skin can turn yellow and fine hair can grow all over the body. Because of malnutrition, the immune system is weakened, leading to increased infections and illnesses. Due to a lack of calcium in the body, the bones are weakened and this could lead to osteoporosis.

In addition to the physical problems come a wide range of potential psychological and emotional problems affecting family relationships, friendships and romantic relationships. This is why treatment for anorexia nervosa involves psychiatric help and emotional support from family and friends as well as medical help.

A woman with anorexia nervosa who wants to have a baby would find becoming pregnant very difficult because her menstrual cycle will have stopped. If she does get pregnant, the chances of having a miscarriage are greatly increased unless she nourishes herself and her baby very carefully. However, a woman who has recovered from anorexia nervosa and is maintaining a healthy body mass will have recovered sufficiently to become a mother.

The most serious consequence of anorexia nervosa is death. Anorexics who go untreated can literally starve themselves to death. This happens in 6% to 20% of cases. This is why early diagnosis and treatment are important. Part of the treatment involves changing the person's self image. An unhealthy self image can be caused by a range of factors including:

- cultural pressure from mass media's image of what is attractive and beautiful;
- traumatic experiences in the person's life;
- any number of physiological causes from the person's genetic makeup or brain chemistry.

Exercises

Figure 12.8 Food pyramid.

Look at the food pyramid in Figure 12.8 showing recommendations by the US government for a well balanced diet and good health. Use it to answer exercises 7 to 12.

7 State the name of the nutrient in the first group on the left which contains the most energy.

8 State the name of the food group from which you would be the least likely to get any fibre.

9 There are five colour-coded named food groups but there is a sixth colour without a label. Suggest a name for this food group.

10 On the left, there is a representation of a person. Deduce what message is being conveyed by this part of the image.

11 Suggest a reason for the absence of candy or cakes on this diagram.

12 Explain why no absolute quantities (in grams) are given on the pyramid diagram.

13 Outline the causes of obesity.

14 Some books on weight loss propose a high-protein, low-carbohydrate diet. Based on the long-term health consequences such a diet generates, evaluate the effectiveness of such a diet by comparing the benefits to the health risks.

15 List three possible causes and three possible symptoms of anorexia nervosa.

A.3 Special issues in human nutrition

Assessment statements

A.3.1 Distinguish between the composition of *human milk* and *artificial milk* used for bottle-feeding babies.

A.3.2 Discuss the benefits of breastfeeding.

A.3.3 Outline the causes and symptoms of type II diabetes.

A.3.4 Explain the dietary advice that should be given to a patient who has developed type II diabetes.

A.3.5 Discuss the ethical issues concerning the eating of animal products, including honey, eggs, milk and meat.

A.3.6 Evaluate the benefits of reducing dietary cholesterol in lowering the risk of coronary heart disease.

A.3.7 Discuss the concept of food miles and the reasons for consumers choosing foods to minimize food miles.

Human milk or powdered milk for babies?

Each species of mammal produces its own milk to suit the needs of its young. This is why no mammal's milk, whether it be from a cow, a goat or any other animal, could be identical to human milk

There are some fundamental differences in the kinds of nutrient and other ingredients found in human breast milk and those found in powdered infant formula, which is usually made from cow's milk.

Human breast milk contains the following things which are not found in infant formula:

- the enzymes amylase and lipase;
- white blood cells such as macrophages;
- antibodies;
- hormones.

Breastfeeding or bottle-feeding babies with artificial infant formula is a highly debated topic.

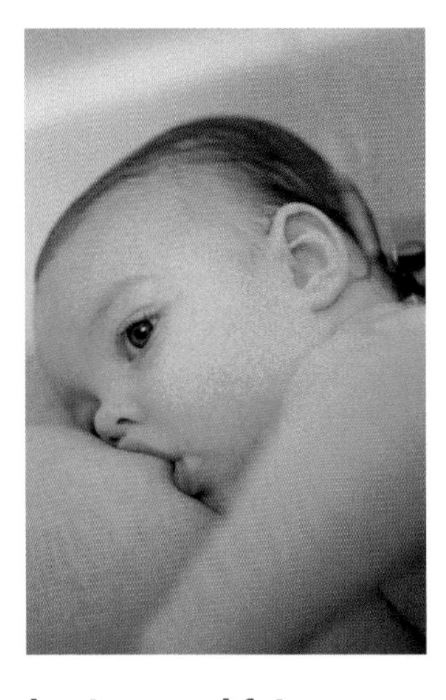

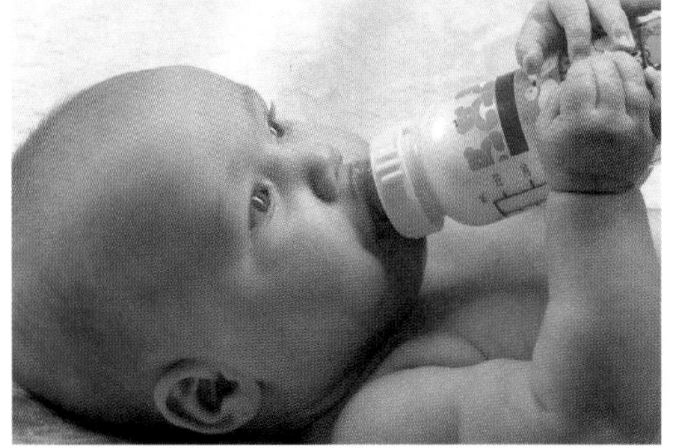

Lactose and fats

Human milk has about 50% more lactose than cow's milk. Lactose is broken down into galactose and glucose to give the baby energy. Human milk also has more cholesterol, which is essential for building new cell membranes.

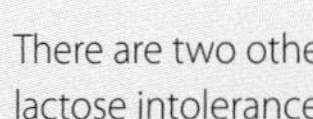

Some babies are naturally lactose intolerant – they have a negative reaction to this component of milk. This condition is extremely rare. Symptoms of lactose intolerance are vomiting, diarrhoea, abdominal pain and rashes on the body. For lactose intolerant babies a different type of baby formula is used, usually one based on soy milk.

Although adults are often told by their doctors to try to reduce fat in their diets, newborns need substantial quantities of lipids to build phospholipids for new cells and to help build nerve cells. To complete the fatty acid content in infant formula, oil is added from one or more of the following plant sources: palm, coconut, corn, soy or safflower seeds. In human milk, the amount of fats can fluctuate over the duration of the lactation to better meet the baby's needs, whereas infant formula is always the same.

There are two other types of lactose intolerance.

1 Aquired lactose intolerance can develop in childhood. It is rare before the age of 5 and unusual before the age of 10.
2 Secondary lactose intolerance can result from some common intestinal disorders of babies. This condition occurs less frequently in breastfed babies.

Protein

Human milk has naturally lower levels of protein than cow's milk for the simple reason that humans do not need to grow as fast as calves do in their first year. The amino acid taurine, however, is found in higher levels in human milk. This amino acid has been linked to brain development, something humans do better than cows.

Milk protein comes in two different forms: an easily digested liquid form called whey and a more difficult to digest solid white form called curd. The curd is made of a protein called casein. You may have noticed these two components of milk protein in cartons of plain yogurt or ricotta cheese: the solid white part is the curd and the liquid is the whey. Human milk has 65% whey and 35% casein but cow's milk has 18% whey and 82% casein, as shown in Figure 12.9.

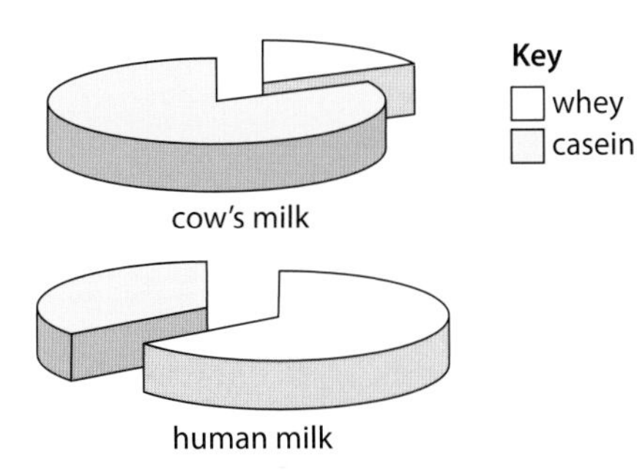

Figure 12.9 Compare the relative amounts of whey and casein in human and cow's milk.

Animal milk is heat-treated before being marketed and during the heating process some of the proteins are denatured. This increases the risk of allergic reactions in infants.

Iron

There is usually more iron found in infant formula than in human milk. This is because it is easier to absorb iron from human milk than from formula, so extra iron is added to the infant formula to compensate.

Baby formula milk is big business. To convince mothers that a particular formula is right for their babies, some companies advertise aggressively and give free samples to women shortly after they have given birth. This raises social and ethical issues about whether it is right for a company to make health claims about a food.

Criticism of big companies for their practices in both industrialized nations and developing countries has sparked boycotts of their infant formula and of other products.

Is breastfeeding right for my baby?

Nothing is more nutritious and healthy for a baby than breast milk from a healthy mother who can produce it in adequate quantities. Despite this, many women choose to feed their babies with infant formula.

There are only a few reasons why a doctor would suggest that a mother not feed her baby breast milk. Such reasons might be that the mother is not capable of making enough milk or that she could contaminate her baby with a pathogen such as HIV or tuberculosis-causing bacteria. More often, a woman makes the decision not to breastfeed for personal reasons such as convenience. Below are some of the main arguments for and against breastfeeding.

Arguments for breastfeeding

- Human milk is a liquid which is species-specific: it is tailor-made to meet a human baby's nutritional needs at each stage of development without upsetting its fragile digestive system. For example, colostrum, a thick liquid rich in carbohydrates but low in fat, is produced for the first meals of the newborn child and is easier to digest.
- The high percentage of whey and lower levels of casein proteins make human milk more digestible than cow's milk and an overall lower protein level is better suited to human babies.
- Some of the proteins in breast milk help to induce sleep in infants. These are not present in cow's or soy milk.
- The nutrients in human milk are more easily absorbed into the child's bloodstream, notably iron and vitamins.
- The risk of allergies is not present, whereas cow's milk in particular carries multiple risks of inducing allergies.
- Breastfeeding offers the potential for feeling warmth, security and emotional bonding for both the mother and the child. Although it would be difficult to prove scientifically, it can be argued that the feeling of warm skin is more reassuring than the feeling of a bottle and an artificial nipple.
- Breastfeeding her child helps the mother to return to the body mass she had before becoming pregnant because making milk requires considerable energy.
- There is no need to prepare, wash and sterilize glass or plastic bottles and synthetic nipples, as is the case with infant formula. In the preparation of bottled milk, there is a risk of the water being polluted or the bottles being contaminated with bacteria.
- Babies fed on breast milk have fewer illnesses such as ear infections or bronchitis than infants who drink powdered formula because breast milk contains white blood cells and antibodies that the mother makes and gives to her baby.
- If the cost of the bottles and formula of non-breastfed babies is compared to the cost of the additional food eaten by a lactating mother, it is less expensive to breastfeed than to buy infant formula.

Breastfeeding causes the uterus to 'return to normal' after delivery of the baby. It also inhibits the onset of menstrual cycles, so acting as a natural contraceptive and encouraging spacing between babies.

Arguments against breastfeeding

- Early on, breastfeeding can be painful for the mother. The pain can be while the baby feeds or it can be from engorgement, meaning the breasts are so full of milk that they hurt. There is also a chance of bacterial infection of the nipple when breastfeeding. Help from an experienced midwife can often overcome these problems.
- Women who carry certain pathogens are advised not to breastfeed because of the risk of passing them to their child. This is true for women who are HIV positive who can pass on the virus through their breast milk. In such a case, the benefits from breast milk would be outweighed by the dangers of the virus.

- It is not accepted in some cultures to breastfeed in public.
- It is very challenging to maintain a successful career in an industrialized nation and breastfeed at the same time, although with proper planning, many new mothers succeed in juggling their professional lives with a breastfeeding schedule. Some countries have laws which allow a mother to take breaks during her business day to breastfeed.
- In exclusive breastfeeding, the father of the child can feel left out whereas with bottle-feeding, he can share a feeling of bonding in nourishing the baby.

If asked the question, 'Which is better: your milk or powdered infant formula?', many new mothers would probably not be able to give a clear confident answer. Researchers who study the composition and benefits of milk have information to help answer the question and they publish their findings in medical journals. But not many new mothers read scientific journals. Fortunately, there are self-help books, medical pamphlets, and online resources which publish summaries of the findings of these studies. Unfortunately, not all new parents have the time to access such information and in some countries they simply do not have the resources to access them. How could the lines of communication between researchers and the public be improved?

Hint: What about the role of midwives?

Type II diabetes

There are two types of diabetes, type I and type II. Both result in the body being unable to break down sugar in the blood (the body either does not produce or does not respond to the hormone insulin). The consequence is that a person with diabetes has unusually high blood sugar levels – a condition called hyperglycaemia.

We are going to consider type II diabetes. It is by far the most common form of diabetes and it has other names:

- adult-onset diabetes (a term used less frequently today);
- diabetes mellitus type II;
- non-insulin-dependent diabetes mellitus (NIDDM).

The symptoms of type II diabetes include:

- being more thirsty than usual;
- urinating more than usual;
- feeling tired all the time;
- more infections than usual, such as skin infections.

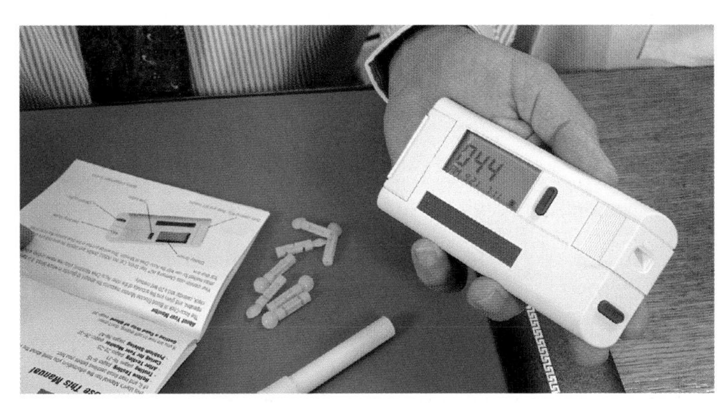

People with diabetes need to monitor their blood glucose levels – this can be done with an electronic meter.

Type II diabetes is the result of the body developing an insensitivity to insulin over many years. It is often found in people who are obese and usually after the age of 40, although it has been observed in overweight children. It is treated with a special diet (see page 346) and exercise but, unlike type I diabetes, it is usually not treated with insulin injections.

What causes type II diabetes? One cause is being overweight or obese. In addition, there are genetic factors. Evidence for this can be seen when examining the percentages of cases in different ethic groups. Some groups have a higher incidence of type II diabetes than others; these include:

- Native Australians (Aboriginal Australians);
- Native Americans;
- Maoris.

How much of type II diabetes is due to genetics and how much is due to how people live? This is another example of the famous 'nature versus nurture' debate.

Can some people who have type II diabetes just say 'I can't help it – I was born with a gene which will make me sick later in life?' If so, is there nothing they can do to prevent it from developing?

Can it be argued that such things are 100% genetic (nature) *and* 100% environmental (nurture) – that we need both interacting like two dancers in a ballet performing moves that neither one could do alone?

Similarly, there are certain families with a history of type II diabetes and other families with few or no cases. Although the disease itself is not life-threatening in the short term, over time it can lead to other problems including high blood pressure, eye damage, kidney malfunction, nerve disease and an increased chance of stroke or heart attack. Consequently, without treatment, it can be fatal.

Dietary recommendations for type II diabetes

One of the secrets to dealing with diabetes is to gain control of the glucose levels in the blood. Part of this involves having a wholesome, well-balanced diet. Since many sufferers of type II diabetes developed the disease from years of obesity, this means decreasing body mass and maintaining a healthy BMI.

People with type II diabetes should eat foods rich in fibre and complex carbohydrates and should cut down on saturated fats. Complex carbohydrates release their energy slowly as they are broken down in the digestive tube. In this way, the body can regulate blood sugar levels more steadily.

Fibre also mixes with other gut contents and allows the nutrients to be absorbed slowly and more steadily over time. If you think you might be at risk and you do not wish to develop this debilitating disease, it is important to adopt healthy eating habits now, maintain a healthy BMI and get used to eating lots of fresh fruits and vegetables.

Ethical issues and meat consumption

Although the killing and eating of animals is necessary for the survival of some species on Earth, this is not the case for most humans. Meat is not an absolute necessity in the human diet unless you live in a region such as the Arctic where there are not many other choices.

There are many reasons why some people choose not to eat meat. Such reasons might be religious, climatic, cultural, economic, or ethical. We are going to examine the ethical reasons for not eating meat.

First you need some background:

- Vegetarians do not eat the flesh of any animals: no red meat, no poultry and no fish.
- Some vegetarians do eat animal products such as eggs, milk, dairy products and honey.
- Strict vegetarians called vegans do not eat animal products such as eggs, milk, dairy products or honey.
- For some vegetarians and vegans, vegetarianism goes beyond diet and they also refrain from wearing clothing made from animal products such as leather, fur, wool, or silk.

The ethical arguments for adopting a vegetarian lifestyle are many – any of the following can be considered a motivation for not eating meat or meat products.

- Killing sentient beings is wrong, especially if it is not necessary for survival. (Sentient means having a nervous system which implies the organisms can perceive sensations such as pleasure or pain.)
- Raising animals for the sole purpose of slaughtering them for human consumption is wrong.

Visit heinemann.co.uk/hotlinks, enter the express code 4242P and click on Weblink 12.4. *The Meatrix* is a humorous animation about a serious subject: where meat comes from in the US. Use your critical thinking when watching.

- What is the bias of the authors?
- Are they showing a fair and honest representation of the meat industry?
- Do you think it will be influential in changing consumer buying habits?

- Intensive livestock production is a wasteful industry which misuses valuable resources such as water and land.
- Mass-production industrial farming uses practices which are unnecessarily cruel including penning up animals, restricting their movement, cutting off beaks, exposing animals to artificial light 24 hours a day, transporting them in uncomfortable conditions over long distances, and more.
- The fishing industry pollutes the oceans and disrupts ecosystems.

Arguments against vegetarianism include the following.

- Farm animals such as cows, sheep, pigs and chickens would not exist if not raised for food.
- Without the meat production and fishing industries, thousands of jobs would be lost.
- Meat and fish can be grown in decent conditions as in free-range, free-from-cruelty and organic practices.
- Certain nutrients are not available in a vegetarian diet, particularly a strict vegan diet.

In some places in the world, vegetarianism is widely accepted and it is easy to find restaurants and markets with vegetarian specialties. In other countries, vegetarianism is not well understood and vegetarians find themselves constantly justifying their choice. In such regions, it is difficult to find restaurants with vegetarian dishes and uncomfortable situations can arise when dinner guests are informed that there are only meat dishes on the menu.

This raises ethical issues of who is right and who is wrong because both groups are convinced that they are right and that they are giving their bodies what is needed as well as following important cultural traditions in certain cases. Such a debate is a reminder of how important it is to be mindful of the opinions of others, especially when it comes to food – what we eat is a fundamental part of our identity.

The benefits of reducing cholesterol

Cholesterol is synthesized in the liver so it can only be found in foods derived from animals and it is completely absent from foods of plant origin. Cholesterol is a type of lipid called a steroid. In moderate quantities, it is of great value for maintaining the stability of cell membranes and because it is an integral part of certain hormones, notably the male and female sex hormones.

However, increased blood cholesterol levels increase the risk of coronary heart disease and high blood pressure. This is because the arteries tend to get clogged by cholesterol and other lipids (see Figure 12.6, page 326). With less space for the blood to pass through, blood pressure goes up and so does the risk that a blood clot could get stuck and stop blood flow to the heart – thus causing a heart attack.

Although diet is important, genetics can play a major role in determining how much cholesterol people have in their blood. Whatever the case, many people have good reason to reduce their cholesterol levels. In maintaining low cholesterol levels by choosing their foods carefully, people greatly reduce their risk of developing high blood pressure and having heart attacks.

Notice the word 'risk' in statements about the correlation between health problems and high cholesterol levels in the blood. A doctor can never be 100% sure why someone had a heart attack because there are so many possible influences including genetic factors, diet, amount of exercise or level of emotional stress.

The best way forward is to maintain a healthy lifestyle. Sometimes life choices can be summarized in two questions: 'What do you want?' and 'What do you really want?' The answer to the first question might be, 'Double cheeseburger with fries, and an extra large sundae with extra caramel' but the answer to the second question might be 'To live a healthy and productive life, and live long enough to dance at my daughter's wedding and see my grandchildren grow up.' It all depends on what you *really* want for your future.

When you buy your food, do you consider how much carbon was released to bring it to your shopping cart?

The term 'food miles' was coined to express the distance from where the food was produced to where it is consumed. Why is this important? Foods which have travelled a long distance have been transported by boat, airplane, train or truck. The longer the distance, the more fuel burned and therefore the more pollutants such as carbon dioxide produced.

The idea of food miles raises the issue of personal versus global choices. On the one hand, a consumer might say, 'This is my favourite food, I don't care how far it had to travel, I am going to enjoy it.' On the other hand, a different consumer might say 'To get to this market in less than a week, this fruit has travelled more than I do in a whole year – I'm not going to financially support a system which pollutes that much. I prefer this local fruit.'

Who is right and who is wrong? Should people be free to choose the foods that they want?

Judging foods and making consumer choices based on food miles and not just on appearance or eating habits constitutes a paradigm shift. It also empowers people to feel that they can make a difference in their everyday lives to help the environment. To make informed decisions, however, people need access to clear information about how far a food has actually travelled and by what means – cargo ships, for example, consume much less fuel per kilo of food being transported than airplanes or trucks.

Food miles

Suppose you are living in a city and you want to buy some fruit. Imagine you are hesitating between peaches and apples. Like most people, you will probably make your choice based on the appearance of the fruit or on personal preference of flavour. Would you make your choice based on how far the food has travelled to get to the market?

Foods which have travelled far are not always from distant lands. In industrialized nations, the most common method of distribution today is through large supermarket chains. These organizations commonly transport goods considerable distances to one or more centralized distribution centres from which the goods are then dispatched to the supermarkets – sometimes only a few miles from where the food was first produced. Farmers who used to sell to markets near their homes are now selling to supermarkets which ship their produce all over the country, sometimes back to the region where it was produced.

All of this transportation requires fossil fuels. Burning fossil fuels releases greenhouse gases such as carbon dioxide and, as seen in Chapter 5, contributes to problems such as global warming. Consumers who buy products at large supermarkets contribute to this problem. On the other hand, by purchasing local products at small farmers' markets, consumers contribute to the economic success of their local farmers.

However, there are some positive aspects to the transport of food over long distances. For one thing, it allows consumers to have access to a wide range of food all year round. Local climate dictates that certain foods are available only at certain times of year, and some foods cannot be grown. By transporting foods from regions where the food is produced to regions where the food is needed, the population can be guaranteed a steady supply during every season of the year.

Another positive consequence of food transport is that consumers have access to a much wider variety of food, an important aspect of maintaining a healthy, well-balanced diet. In addition, it allows people to discover exotic fruits and culinary specialties from all over the world, increasing their awareness and appreciation of other cultures.

Another argument in favour of long-distance transport of food is that it allows farmers in developing countries to grow crops such as coffee or peanuts which can be sold to consumers all over the world. It is up to each consumer to make well-informed choices.

Some countries have a system of direct contracts with local farmers. The idea is that consumers help the farmer financially to invest in this year's crops and in return they are provided with a basket of fresh produce on a regular basis.

Exercises

16 Explain why babies fed with infant formula tend to have more infections and illness in their first year of life than babies fed with breast milk.

17 From the information about the types of protein in different types of milk (Figure 12.9, page 343), deduce one reason why human milk is more easily digested by babies than infant formula made from cow's milk.

18 State two reasons why a doctor might recommend infant formula to nourish a baby rather than breast milk.

19 Explain why type II diabetes is not called adult-onset diabetes anymore.

Practice questions

1 Researchers were investigating the factors that might affect life expectancy of humans or increase longevity. The graph below shows the effect of limited energy on longevity of rats. Rats were chosen based on their similar dietary requirements to humans. Control rats were fed a normal balanced diet, with sufficient energy, while experimental animals were fed a similar diet but with a total energy content 50% less than the control.

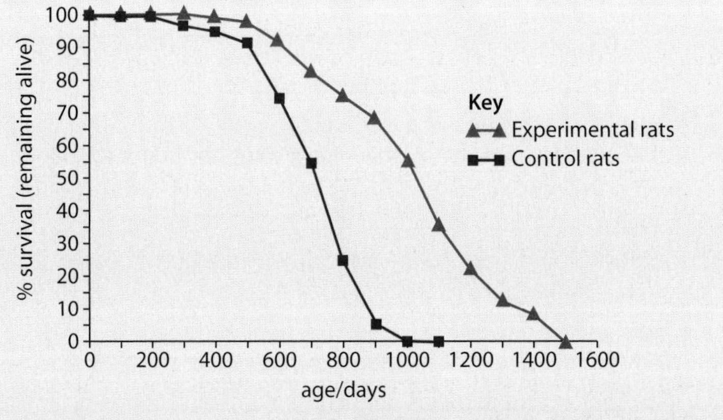

Source: Brian J Merry (1999), *Biologist*, **46** (3), pages 114–17

(a) Identify the percentage of rats that remain alive in each group after 800 days.
Control: Experimental: *(2)*

(b) Outline how the graph shows the effect of energy content on longevity. *(2)*

(c) State three constituents that must be present in both these diets. *(1)*

(d) Suggest two reasons for the difference in survival between the two groups of rats. *(2)*

(Total 7 marks)

2 (a) Define the term *nutrient.* *(1)*

(b) Discuss the relationship between nutrition and rickets. *(3)*

(Total 4 marks)

3 (a) Define the term *essential amino acids.* *(1)*

(b) Describe how a balanced diet meets the needs of the body. *(2)*

(c) Distinguish the differences between a vegan diet and a vegetarian diet. *(1)*

(d) Discuss the importance of fibre in the diet. *(3)*

(Total 7 marks)

13 Physiology of exercise (Option B)

Introduction

Talking about exercise isn't just about what our favourite professional athlete accomplished. We may wonder why certain body parts are sore after a period of activity, or what we need to do in order to be physically fit. In this chapter, we are going to look at the mechanisms that enable us to carry out everyday activities. There are many parts involved, and the means by which they act are often quite complex.

B.1 Muscles and movement

The material in this section is also found in Chapter 11, section 11.2.

Assessment statements

- B.1.1 State the roles of bones, ligaments, muscles, tendons and nerves in human movement.
- B.1.2 Label a diagram of the human elbow joint, including cartilage, synovial fluid, joint capsule, named bones and antagonistic muscles (biceps and triceps).
- B.1.3 Outline the functions of the structures in the human elbow joint named in B.1.2.
- B.1.4 Compare the movements of the hip joint and the knee joint.
- B.1.5 Describe the structure of striated muscle fibres, including the myofibrils with light and dark bands, mitochondria, the sarcoplasmic reticulum, nuclei and the sarcolemma.
- B.1.6 Draw and label a diagram to show the structure of a sarcomere, including Z lines, actin filaments, myosin filaments with heads, and the resultant light and dark bands.
- B.1.7 Explain how skeletal muscle contracts, including the release of calcium ions from the sarcoplasmic reticulum, the formation of cross-bridges, the sliding of actin and myosin filaments, and the use of ATP to break cross-bridges and re-set myosin heads.
- B.1.8 Analyse electron micrographs to find the state of contraction of muscle fibres.

Joints

A joint, also called an articulation or arthrosis, is the point where two or more bones contact one another. Arthrology is the scientific study of joints and rheumatology is the branch of medicine devoted to joint diseases and conditions. The science of kinesiology examines the movement of the human body.

Our joints provide mobility. They also hold the body together. Most joints include:

- bones;
- muscles;
- nerves.
- ligaments;
- tendons;

Bones

It is important to note that bones contain several different tissues and, therefore, they are organs. Bones have many functions:

- providing a hard framework to support the body;
- allowing protection of vulnerable softer tissue and organs;
- acting as levers so that body movement can occur;
- forming blood cells in the bone marrow;
- allowing storage of minerals, especially calcium and phosphorus.

Here we are going to concentrate on bones acting as levers for movement (see Figure 13.1).

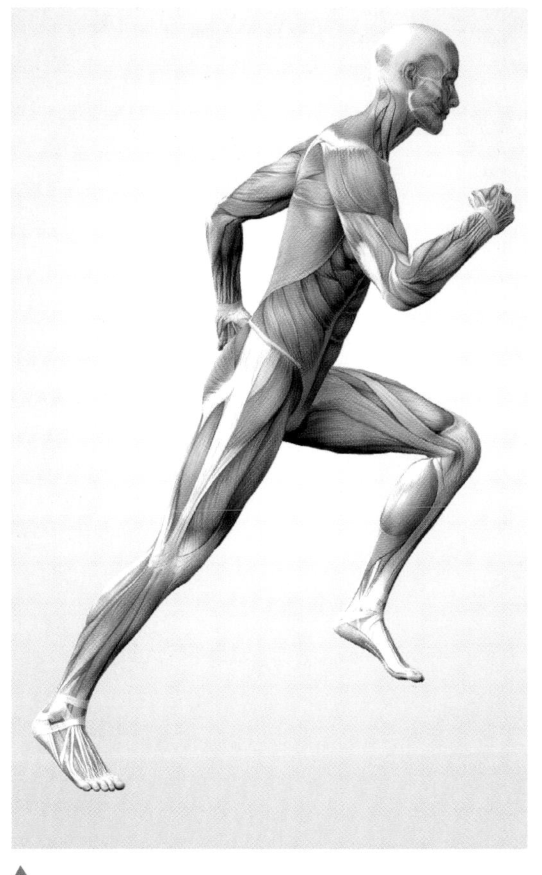

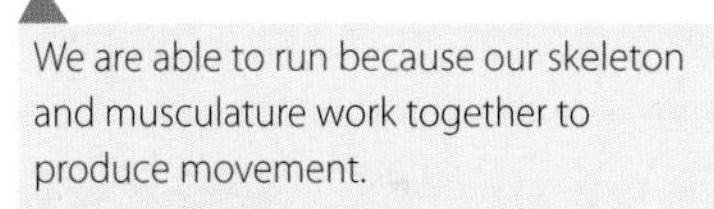

We are able to run because our skeleton and musculature work together to produce movement.

Figure 13.1 Human adults have 206 bones. The bones of the limbs are involved in movement.

Muscles and tendons

For movement to occur, it is essential that skeletal muscles are attached to bones. This attachment is provided by the tendons. Tendons are cords of dense connective tissue. It is the arrangement of the bones and the design of the joints that determine the type or range of motion possible in any particular area of the body. By acting as levers, the bones function to magnify the force provided by muscle contraction. The muscles provide the force necessary for movement by shortening the length of their fibres or cells. Since the muscles bring about movement only by shortening, it is essential they occur as antagonistic pairs (see Figure 13.2). This allows a body part to be returned to its original position after a movement.

Figure 13.2 The biceps and triceps are opposing muscle groups in the upper arm. When the biceps contracts, the triceps relaxes and the forearm moves upward (flexes). When the triceps contracts, the biceps relaxes and the forearm moves downwards (extends).

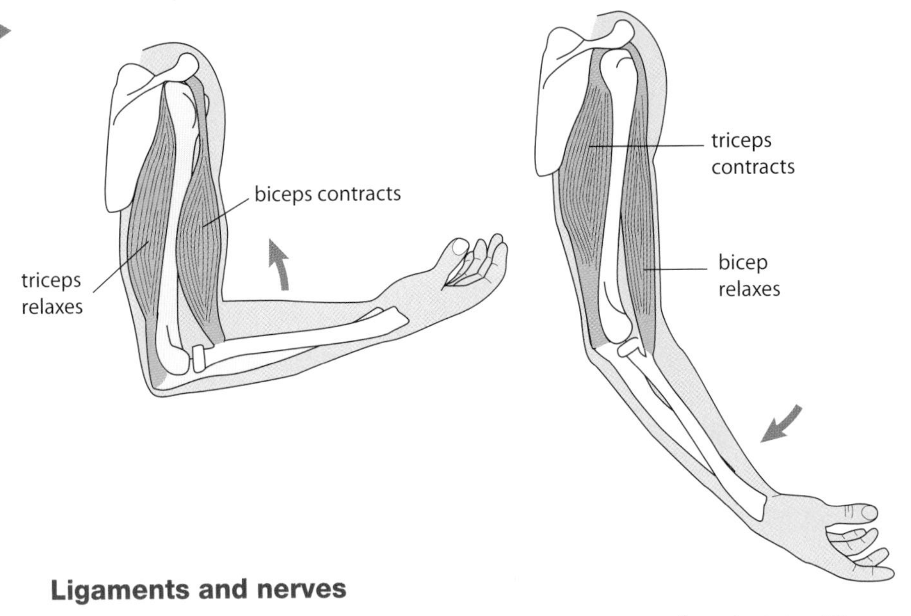

Ligaments and nerves

Ligaments are tough, band-like structures that serve to strengthen the joint. They also provide stability. The ligaments have many different types of sensory nerve endings. Proprioceptors in ligaments and muscles allow constant monitoring of the positions of the joint parts. The nerves help to prevent over-extension of the joint and its parts.

Blood supply

There is a rich supply of blood to joints. If blood vessels supplying the joint get damaged and there is local bleeding (haemorrhage), it may result in swelling of the area.

This false-colour X-ray shows a normal human elbow joint.

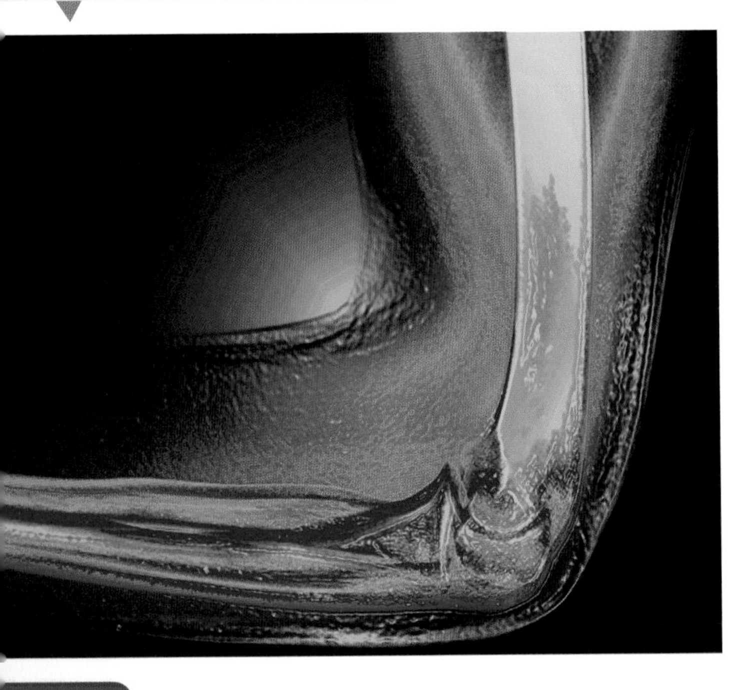

Hinge joints

Let's look at the elbow joint in detail. It is a hinge joint. This means it provides an opening-and-closing type of movement like the action of a door.

To discuss complex structures such as joints, we need a large specialist vocabulary, so you may find some of the terms used in this chapter unfamiliar. It is well worthwhile taking the time to become thoroughly familiar with any words that you are not completely sure of.

The elbow joint involves the humerus, radius and ulna bones (see Figure 13.3). The synovial fluid is present within the synovial cavity. This cavity is located within the joint capsule. The joint capsule is composed of dense connective tissue that is continuous with the membrane of the involved bones.

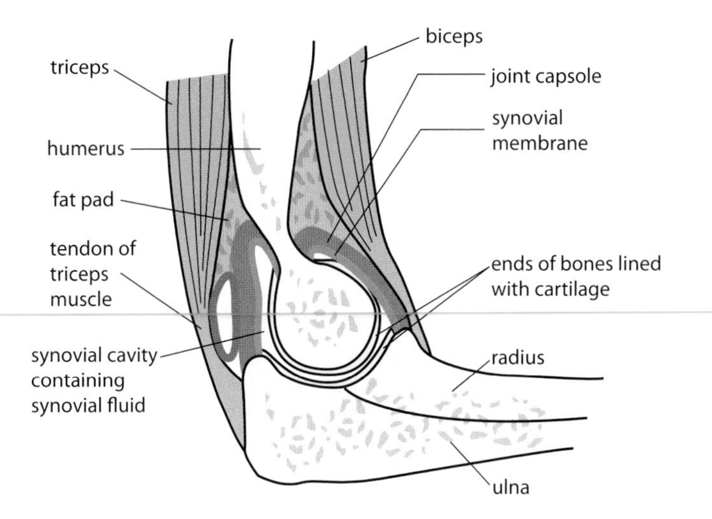

Figure 13.3 The elbow joint and its parts.

Elbow parts and their functions are summarized in this table.

Joint part	Function
cartilage	reduces friction and absorbs compression
synovial fluid	lubricates to reduce friction and provides nutrients to the cells of the cartilage
joint capsule	surrounds the joint, encloses the synovial cavity, and unites the connecting bones
tendons	attach muscle to bone
ligaments	connect bone to bone
biceps muscle	contracts to bring about flexion (bending) of the arm
triceps muscle	contracts to cause extension (straightening) of the arm
humerus	acts as a lever that allows anchorage of the muscles of the elbow
radius	acts as a lever for the biceps muscle
ulna	acts as a lever for the triceps muscle

For a further explanation of the elbow, visit heinemann.co.uk/hotlinks, insert the express code 4242P and click on Weblink 13.1.

The elbow is called a synovial joint because of the presence of the synovial cavity. The knee is a similar joint (see Figure 13.4). These joints are freely movable. Freely movable joints are also called diarthrotic joints.

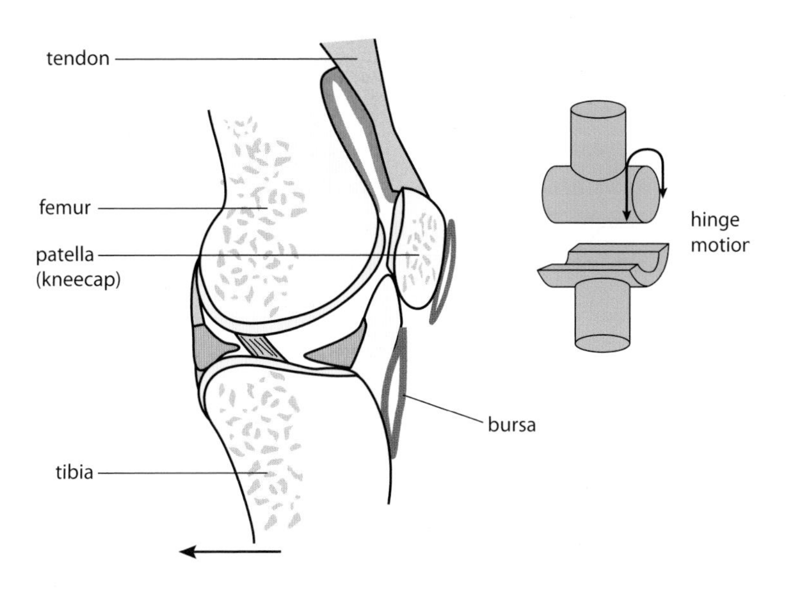

Figure 13.4 The knee joint is much like the elbow. It is a hinge joint and permits an opening-and-closing type of movement. This is movement in one direction and produces angular motion.

Ball-and-socket joints

The hip joint is also a diarthrotic joint (see Figure 13.5), but it is not a hinge joint.

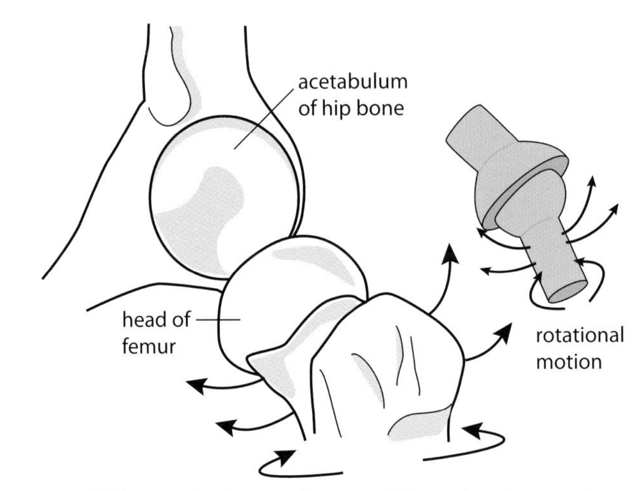

Figure 13.5 The hip joint is a ball-and-socket joint. Because of the structure of the joint, movement is possible in a number of different directions. The shoulder is also a ball-and-socket joint.

There is a great difference in the motion possible at the elbow or knee and that possible at the hip. The hip is a ball-and-socket joint and it permits movement in several directions, including rotational movement. This is because the head of the femur (a ball shape) fits into a cup-like depression of the hip bone called the acetabulum.

Comparison of hinge and ball-and-socket joints

This table compares the hip and knee joints.

Hip joint	Knee joint
freely movable	freely movable
angular motions in many directions and rotational movements	angular motion in one direction
motions possible are flexion, extension, abduction, adduction, circumduction and rotation*	motions possible are flexion and extension
ball-like structure fits into a cup-like depression	convex surface fits into a concave surface

* Definitions:

- flexion = decrease in angle between connecting bones;
- extension = increase in angle between connecting bones;
- abduction = movement of bone away from body midline;
- adduction = movement of bone toward midline;
- circumduction = distal or far end of a limb moves in a circle;
- rotation = a bone revolves around its own longitudinal axis.

Muscle

The human body possesses three types of muscle tissue:

- skeletal or striated;
- cardiac;
- smooth or non-striated.

Here we are going to look at only striated muscle.

Striated muscle cells

Striated muscle is also called skeletal muscle because it is responsible for skeletal movement.

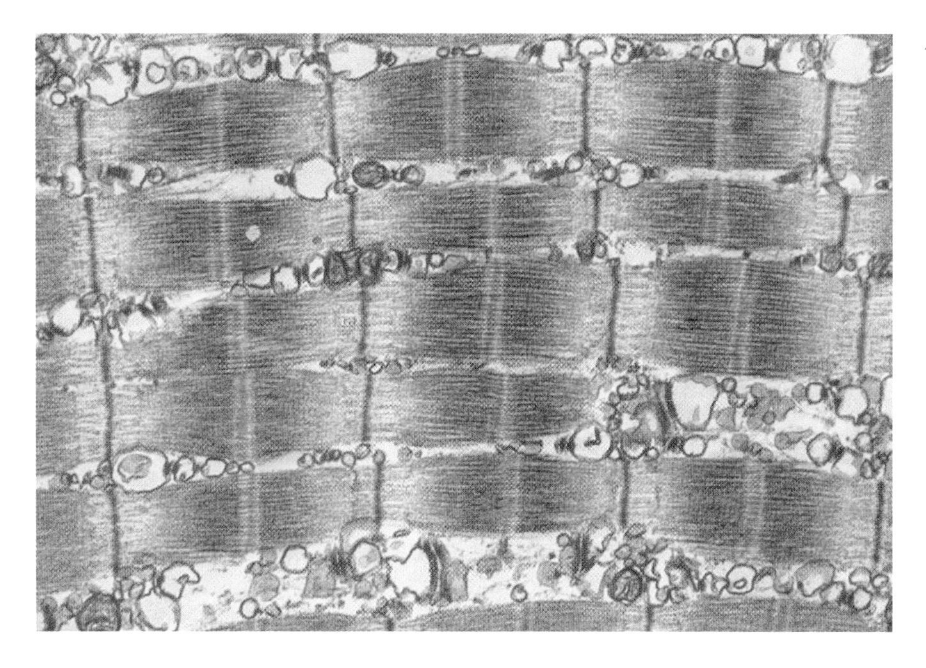

In this false-colour electron micrograph, you can clearly see why this muscle is said to be striated. The banding is quite apparent. You can see five myofibrils arranged parallel to one another horizontally. Each myofibril has a banded pattern that is repeated in its neighbours (see Figures 13.6 and 13.7). Can you identify three sarcomeres in four of the five myofibrils and two sarcomeres in the remaining myofibril?

Like any other tissue in the body, muscle is made up of cells. It is the cellular arrangement that produces the banded appearance of striated muscle. Muscles are composed of thousands of cells, which are called muscle fibres because of their elongated shape. Besides the muscle fibres, muscles include surrounding connective tissues, and blood vessels and nerves. The blood vessels and nerves penetrate the muscle body.

Muscle fibres (cells) contain multiple nuclei that lie just inside the plasma membrane, which is called the sarcolemma. The sarcolemma has multiple tunnel-like extensions that penetrate the interior of the cell. These penetrating invaginations are called transverse or T tubules.

The cytoplasm of muscle fibres is called the sarcoplasm. The sarcoplasm contains large numbers of glycosomes that store glycogen. Besides glycosomes, the sarcoplasm has large amounts of a red-coloured protein called myoglobin. Glycogen and myoglobin are discussed later when muscle movement is explained.

The sarcoplasmic reticulum is a fluid-filled system of membranous sacs surrounding the muscle myofibrils. Sarcoplasmic reticulum is much like smooth endoplasmic reticulum.

Myofibrils are rod-shaped bodies that run the length of the cell. There are many of them and they are parallel to one another. Myofibrils are closely packed and numerous mitochondria are squeezed between them. The myofibrils are the contractile elements of the muscle cells and they are the reason that striated muscle has a banded pattern.

Myofibril structure

Myofibrils are made up of sarcomeres, and sarcomeres are the units that allow movement (see Figure 13.6).

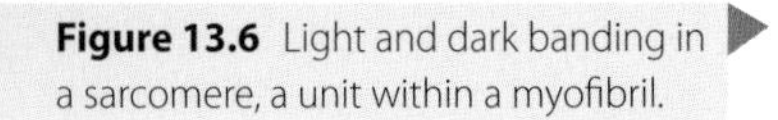
Figure 13.6 Light and dark banding in a sarcomere, a unit within a myofibril.

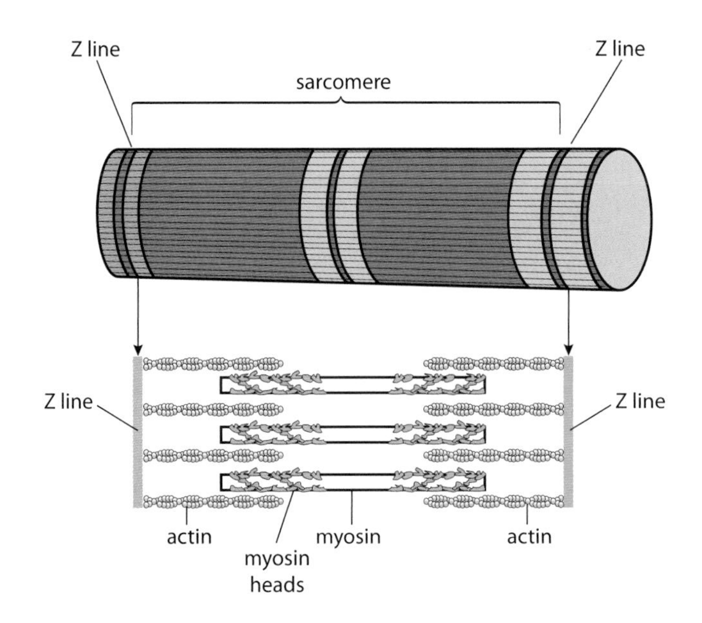

The sarcomere is often described as banded (see Figure 13.7).

- The Z lines mark the ends of the sarcomere.
- The A bands are dark in colour and extend the entire length of the myosin filaments. A narrow H band occurs in the middle of the A band – it contains only myosin, no actin. A supporting protein occurs in the middle of the myosin producing the M line. This protein holds the myosin filaments together.
- The I bands are light in colour and contain only actin – no myosin.

As you can see in Figure 13.7 there are two types of filaments or myofilaments that cause the banded appearance of the muscle fibre. These myofilaments are composed of two contractile proteins, actin and myosin, and are described in the table on page 357.

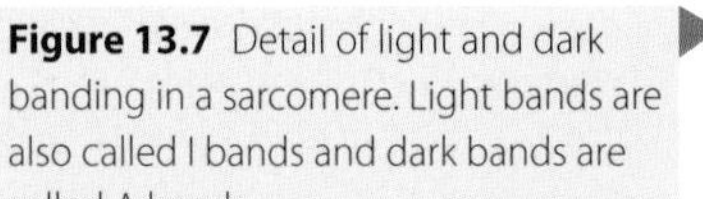
Figure 13.7 Detail of light and dark banding in a sarcomere. Light bands are also called I bands and dark bands are called A bands.

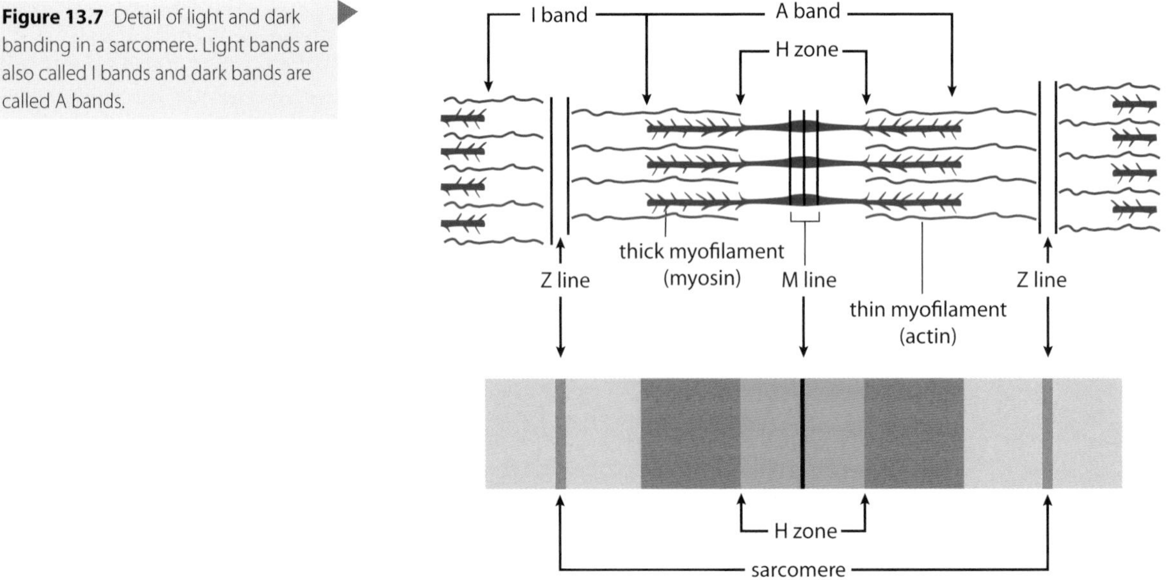

Actin	Myosin
thin filaments (8 nm in diameter)	thick filaments (16 nm in diameter)
contains myosin-binding sites	contains myosin heads that have actin-binding sites
individual molecules form helical structures	individual molecules form a common shaft-like region with outward protruding heads
includes two regulatory proteins, tropomyosin and troponin	heads are referred to as cross-bridges and contain ATP-binding sites and ATPase enzymes

In 1954, Hugh Huxley first proposed the sliding filament theory of muscle contraction.

Muscle contraction

The mechanism of muscle contraction is presently explained by the sliding filament theory. Essentially, this theory states that muscles contract when actin myofilaments slide over myosin myofilaments. The myofilaments do not actually shorten. When the actin component slides over the myosin, the sarcomere is shortened. With multiple fibres and sarcomeres in a muscle working together, this causes the movements necessary for the organism.

Botulinum toxin is produced by the bacterium *Clostridium botulinum*. The toxin blocks the release of acetylcholine, and thus prevents muscle contraction. This can affect the diaphragm, so breathing stops and death may occur.

Botulinum toxin is the active ingredient of Botox. Botox injected into the affected muscles may correct strabismus (crossed eyes) or blepharospasm (uncontrolled blinking). Botox is also used as a cosmetic to relax the muscles that cause facial wrinkles.

Key events of muscle contraction (sliding filament theory)

1. A motor neurone carries an action potential until it reaches a neuromuscular junction.
2. A neurotransmitter called acetylcholine is released into the gap between the axon terminal and the sarcolemma of the muscle fibre.
3. The acetylcholine binds to receptors on the sarcolemma.
4. Sarcolemma ion channels open and sodium ions move through the membrane.
5. This generates a muscle action potential.
6. The muscle action potential moves along the membrane and through the T tubules.
7. After generation of the muscle action potential, the acetylcholine is broken down by an enzyme called acetylcholinesterase. This ensures that one nerve action potential causes only one muscle action potential.
8. The muscle action potential moving through the T tubules causes release of calcium ions from the sarcoplasmic reticulum. The calcium ions flood into the sarcoplasm.
9. The calcium ions bind to troponin on the actin myofilaments. This exposes the myosin-binding sites.
10. The myosin heads include ATPase which splits ATP and releases energy (step 1 of Figure 13.8).
11. The myosin heads then bind to the myosin-binding sites on the actin with the help of the protein called tropomyosin (step 2 of Figure 13.8).
12. The myosin–actin cross-bridges rotate toward the centre of the sarcomere. This produces the power or working stroke (step 3 of Figure 13.8).
13. ATP once again binds to the myosin head resulting in the detachment of myosin from the actin (step 4 of Figure 13.8).
14. If there are no further action potentials, the level of calcium ions in the sarcoplasm falls. The troponin–tropomyosin complex then moves to its original position, thus blocking the myosin-binding sites. The muscle then relaxes.

For some excellent representations of muscle contraction, visit heinemann.co.uk/hotlinks, enter the express code 4242P and click on Weblinks 13.2a, 13.2b and 13.2c.

Figure 13.8 The contraction cycle showing events involved in the sliding filament theory.

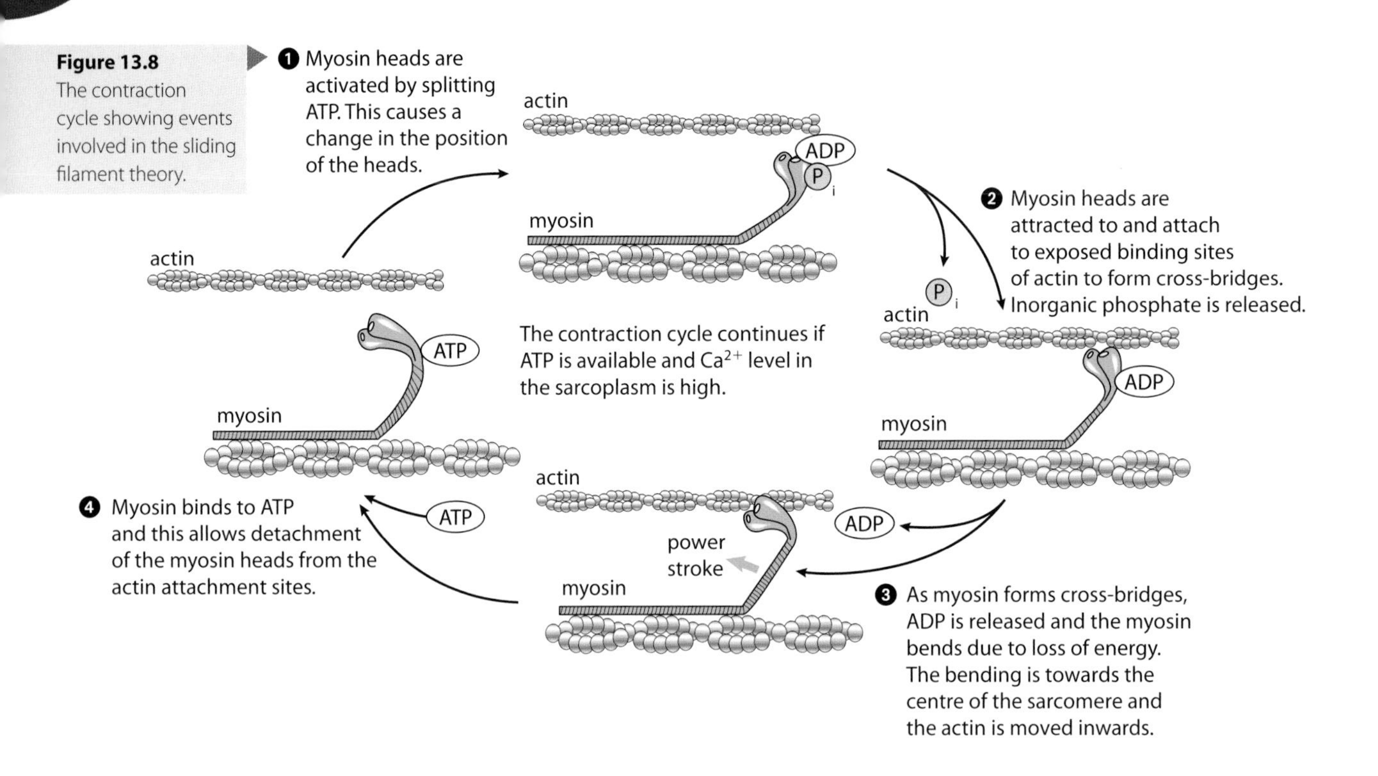

1 Myosin heads are activated by splitting ATP. This causes a change in the position of the heads.
actin
ADP
P
i
myosin
2 Myosin heads are attracted to and attach to exposed binding sites of actin to form cross-bridges. Inorganic phosphate is released.
actin
P
i
actin
ATP
The contraction cycle continues if ATP is available and Ca2+ level in the sarcoplasm is high.
ADP
myosin
myosin
actin
4 Myosin binds to ATP and this allows detachment of the myosin heads from the actin attachment sites.
ATP
ADP
power stroke
myosin
3 As myosin forms cross-bridges, ADP is released and the myosin bends due to loss of energy. The bending is towards the centre of the sarcomere and the actin is moved inwards.

Curare is a plant extract first used on darts and arrows to cause muscle paralysis and death. Similar drugs are used in modern-day surgery to relax skeletal muscles.

After a person dies, calcium ions leak out of the sarcoplasmic reticulum and bind to troponin. This allows the actin to slide. However, since ATP production stops at death, the myosin heads cannot detach from the actin. The result is *rigor mortis* (the rigidity of death). It lasts for about 24 hours until further muscle deterioration occurs.

Actin and myosin myofilaments do not change in length during muscle contraction. Contraction is due to the rather complex sequence of events leading to the actin sliding over the myosin myofilaments. This causes the shortening of the sarcomere necessary for muscle movement. You will recall that a sarcomere extends from Z line to Z line; it is the smallest section of a muscle fibre to change in length.

Electron micrographs of sarcomeres in different degrees of contraction are clearly distinguishable. Figures 13.9–13.11 show the distinctions diagrammatically. Figure 13.9 shows two fully relaxed sarcomeres.

Figure 13.9 Sarcomeres in a relaxed muscle.

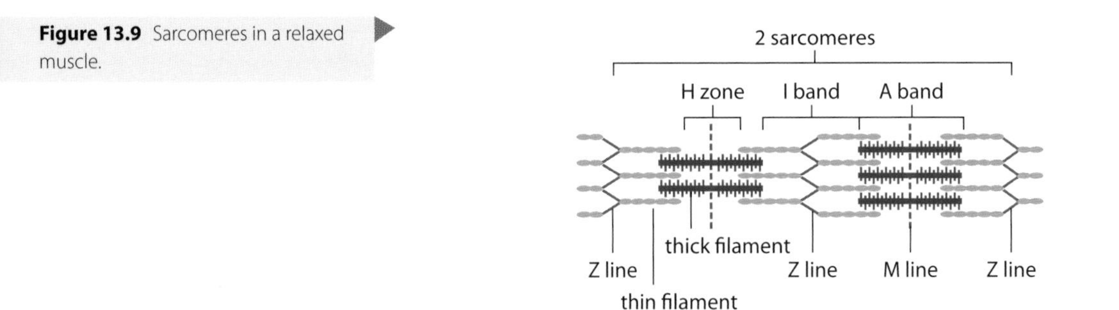

2 sarcomeres
H zone
I band
A band
thick filament
Z line
Z line
M line
Z line
thin filament

When the muscle is maximally contracted, the H zone disappears, the Z lines move closer together, the I bands are no longer present, and the A bands appear to run the complete length of the sarcomeres (see Figure 13.10).

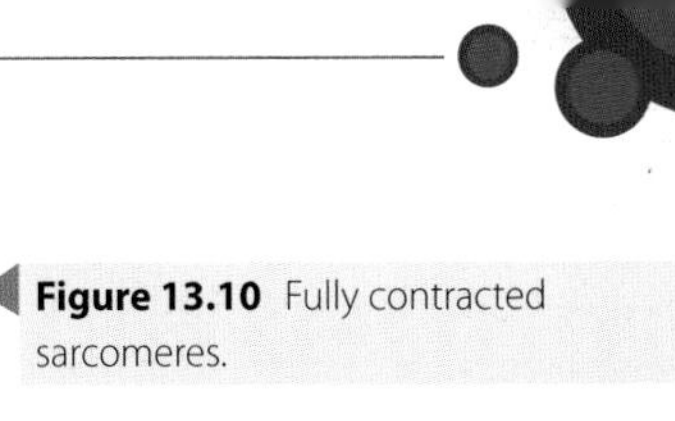

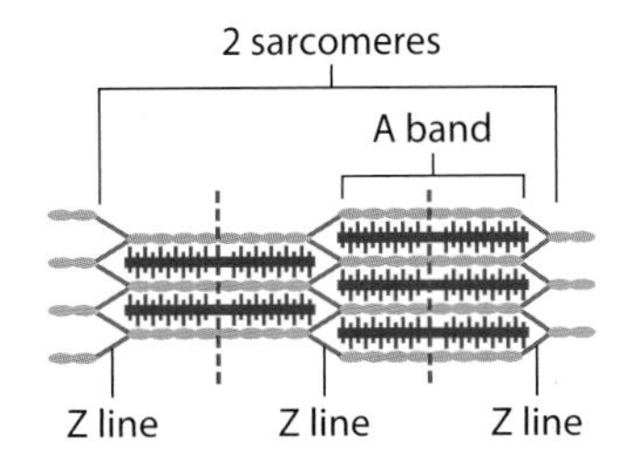

Figure 13.10 Fully contracted sarcomeres.

The muscle may also be in various states of partial contraction (see Figure 13.11). This causes a difference in the position of the sarcomere parts.

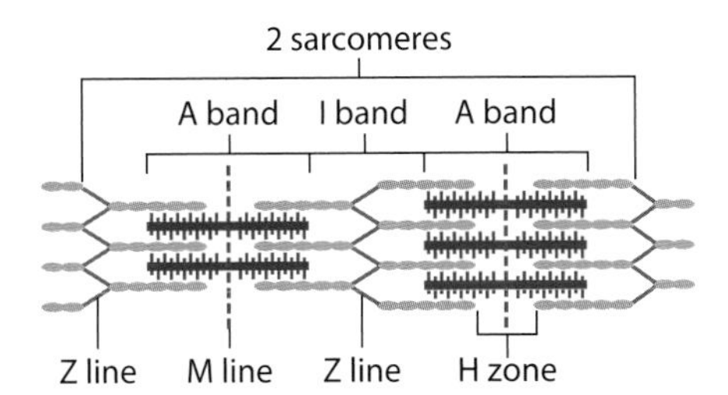

Figure 13.11 Partially contracted sarcomeres.

You can clearly see that muscles can be in many states of contraction. The number of muscle fibres in a muscle going through contraction determines the overall strength of a muscle contraction. Motor impulses from the central nervous system determine the number of muscle fibres that contract.

● **Examiner's hint:** You should be able to label parts of a muscle fibre in any stage of contraction in a photomicrograph. Study the electron micrograph on page 355 to identify the parts of the sarcomere.

Exercises

1. What would be the result of a decrease in synovial fluid at a freely movable joint?
2. Why does the H zone disappear in fully contracted sarcomeres?
3. What causes the release of calcium ions from the sarcoplasmic reticulum of a muscle fibre?
4. Of what value would a large number of mitochondria be to a muscle fibre?
5. Why must muscles occur in antagonistic pairs?

B.2 Training and the pulmonary system

Assessment statements

B.2.1 Define *total lung capacity, vital capacity, tidal volume* and *ventilation rate*.

B.2.2 Explain the need for increases in tidal volume and ventilation rate during exercise.

B.2.3 Outline the effects of training on the pulmonary system, including changes in ventilation rate at rest, maximum ventilation rate and vital capacity.

The pulmonary system

The pulmonary (respiratory) system is essential to provide an adequate flow of oxygen to the cells of the body. This same system removes the waste gas carbon dioxide from the body cells. In order to perform these functions, the cardiovascular system must also be involved. Four distinct processes are necessary for the pulmonary system to carry out its functions.

1 *Pulmonary ventilation*: This is the movement of air into and out of the lungs. It is referred to as ventilation or breathing.
2 *External respiration*: This is gas exchange between the blood and air-filled sacs of the lungs.
3 *Transport of respiratory gases*: The cardiovascular system is largely involved to move gases between the lungs and cells of the body.
4 *Internal respiration*: This refers to gas exchange between the blood and body cells.

Remember there is a third type of respiration, cellular respiration. It involves the production of ATP in the cell.

The first two processes rely solely on the pulmonary system (see Figure 13.12). The last two involve the cardiovascular system and the pulmonary system.

The cardiovascular system was discussed in Chapter 6 section 2. The lung parts and their functions were discussed in Chapter 6 section 4. You may want to review your learning of these sections of Chapter 6. In this section, you are going to look at the pulmonary system and exercise. The cardiovascular system and exercise are discussed in the next section of this chapter.

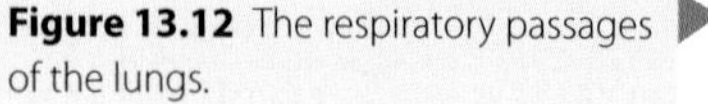

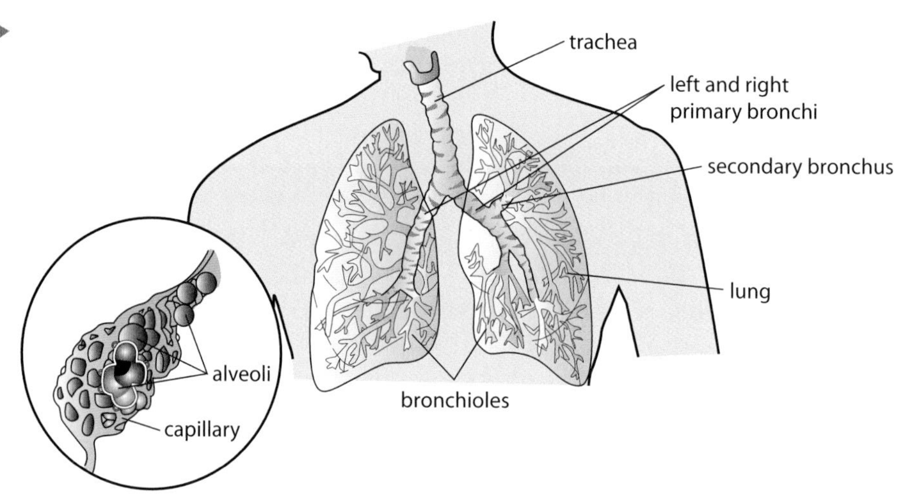

Figure 13.12 The respiratory passages of the lungs.

Lung volumes

While resting, a healthy adult averages nearly 12 breaths a minute. Each breath moves about 500 ml of air into and out of the lungs. The number of breaths a person takes each minute is referred to as their ventilation rate. The volume of air that moves in or out with each inhalation or exhalation at rest is called the tidal volume. It can be measured using a device called a spirometer or respirometer. Using spirometers on different people shows variable tidal volumes. Even in the same person, the tidal volume may show variation at different times due to changing circumstances.

Anatomical dead space includes those parts of the lungs that must be filled before air gets to the functional air sacs. If you snorkel, you will know that the snorkel tube has a limit in length. This is because it adds to the dead space. Too much dead space results in lack of fresh air getting to the lungs.

Another lung volume measurement is vital capacity. This is the maximum volume of air that can be exhaled after maximum inhalation. In young males, the vital capacity averages 4800 ml; in young females, it averages 3100 ml.

Total lung capacity is the total volume of air in the lungs after a maximum inhalation. The total lung capacity includes the vital capacity and the residual volume. Residual volume can not be measured by a spirometer since it can not be exhaled. It is the air necessary to keep the structures of the lung at least partially inflated during exhalation.

Oxygen consumption

During exercise, tremendous amounts of oxygen are consumed by the muscles so the ventilation rate must increase 10–20-fold. This rate increase is also necessary to rid the body of the large amounts of carbon dioxide produced by intensified cellular respiration. Not only does the ventilation rate increase, inhalations are deeper and more vigorous – so the tidal volume is increased. These changes do not result in significant variations in oxygen and carbon dioxide levels in the blood because they are adjustments to metabolic demands. Homeostasis of blood gases and pH is maintained by the ventilation changes. The actual control of ventilation rate and intensity during exercise is poorly understood. At present, it appears to have a strong neural control component.

When muscles receive an inadequate supply of oxygen due to intense exercise, they enter into anaerobic exercise. The result of anaerobic exercise is that the muscles produce lactic acid. Due to the presence of the lactic acid, the action of the muscles is limited. The cause of the inadequate oxygen supply to the muscles rarely involves the respiratory system. Usually, it is due to either a delivery problem of the cardiovascular system or the inability of the muscles to further increase their consumption of oxygen.

Effect of training

For the pulmonary system, the immediate results of exercise are increased ventilation rate and increased tidal volume. There are also long-term effects of regular physical activity or training.

1 *Larger vital capacity*: This is due to the development of a stronger diaphragm and stronger intercostal (between the ribs) muscles. These muscles are the primary muscles allowing ventilation to occur. The volume of the thoracic cavity is actually increased due to the greater development of these muscles.
2 *Decrease in ventilation rate at rest*: Nearly a 10% reduction is possible because of increased vital capacity and capillary development around the lung air sacs. An example of this would be a drop in ventilation rate from 14 breaths per minute (bpm) to only 12 bpm.
3 *Increase in maximum ventilation rate during exercise*: This increase may involve an increase in maximum rate of 10–15%. It is common to see a change in maximum ventilation rate with training from 40 bpm to 45 bpm.

These effects are largely due to exercise making the muscles of respiratory system more efficient (see Figure 13.13). The training required to make these changes must be aerobic. This involves exercise occurring for at least 20 minutes. The greatest benefit results when aerobic exercise occurs at least four times a week.

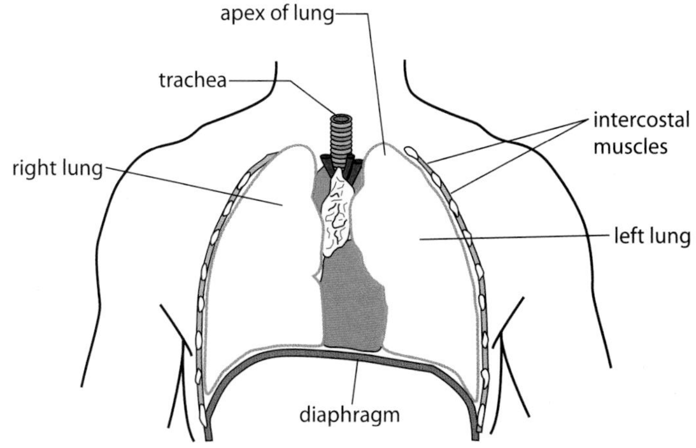

Figure 13.13 The thoracic cavity showing the presence and location of the diaphragm and the intercostal muscles.

Exercises

6 If you determined the tidal volume by measuring gas exit from a snorkel tube and gas exit directly from the mouth or nose, would there be a difference?

7 How can a snorkel tube of excess length potentially cause death due to lack of oxygen?

8 How do you think the effects of exercise on the respiratory system compare to effects caused by playing a wind instrument such as a trumpet?

9 Name some ways in which vital capacity can be increased.

B.3 Training and the cardiovascular system

Assessment statements

B.3.1 Define *heart rate, stroke volume, cardiac output* and *venous return.*

B.3.2 Explain the changes in cardiac output and venous return during exercise.

B.3.3 Compare the distribution of blood flow at rest and during exercise.

B.3.4 Explain the effects of training on heart rate and stroke volume, both at rest and during exercise.

B.3.5 Evaluate the risks and benefits of using erythropoietin (EPO) and blood transfusions to improve performance in sports.

At one time there was a rather popular theory that said when we are born we have a pre-established number of heart beats available. The theory continued with the reasoning that physical activity was bad for the individual because it utilized those available beats at a faster pace. Obviously, this idea is not widely accepted today.

The cardiovascular system

You will recall that both the pulmonary system and the cardiovascular system are involved in providing muscles with what they need during physical activity. The cardiovascular system can be improved at any age by regular exercise. The heart especially benefits from aerobic types of activity. These activities may include brisk walking, jogging, bicycling, swimming, or cross-country skiing. In the previous section, we suggested at least 20 minutes of vigorous activity at least four times a week for maximum benefit. However, it is important to mention here that not everyone is a candidate for vigorous activity. If a person has not exercised regularly, they should start off with moderate activity, and they may be wise to see their physician before starting any activity programme.

We also said in the previous section that the pulmonary system rarely is the cause of a muscle not getting what it needs to carry out or continue aerobic activities. It is usually the cardiovascular system. It is essential to understand the role of the cardiovascular system in training and how it may be improved.

Heart rate or pulse is usually between 60 and 80 beats per minute when a person is at rest.

The heart

The major component of the heart is muscle, specifically cardiac muscle (see Figure 13.14). Cardiac muscle responds to exercise just as skeletal (striated) muscle does: it increases in size, efficiency, and strength.

Some important terminology

There are some important terms you need to understand to discuss the cardiovascular system and exercise. The first is heart rate. This is the number of heart contractions per minute. It is easily measured by taking the pulse. A pulse is the alternating expansion and relaxation of elastic arteries during the cardiac cycle.

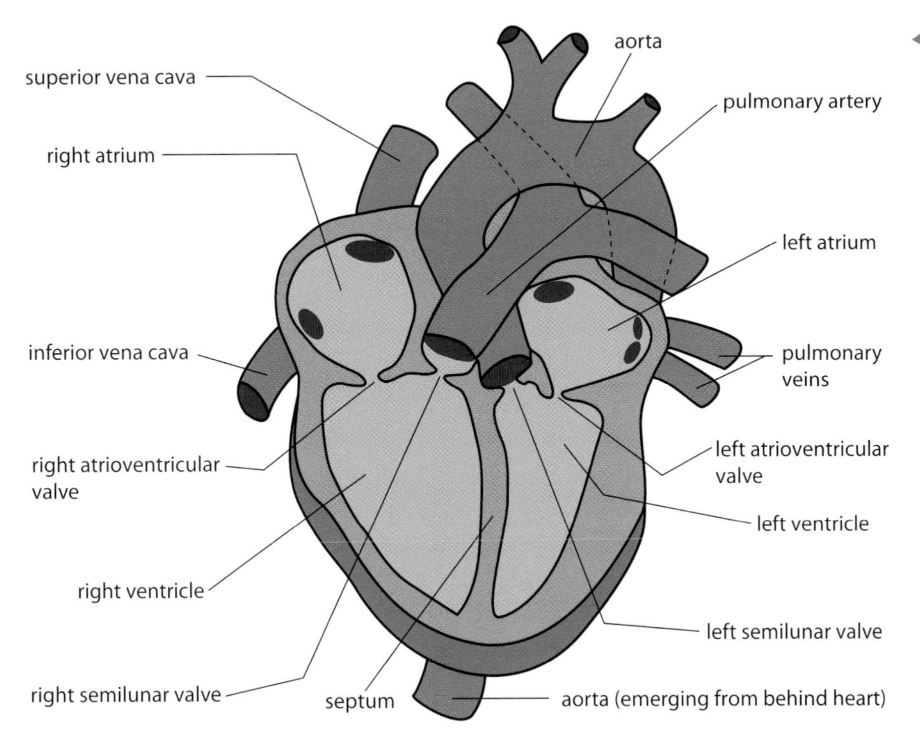

Figure 13.14 Internal anatomy of the heart.

Any artery that lies close to the surface of the body is a candidate for taking pulse. A few of the major places for determining heart rate are shown in this table.

Artery name	Artery location
temporal artery	temple region in front of the ear
facial artery	jaw on a line with the corners of the mouth
common carotid artery	next to the larynx (voice box)
radial artery	wrist area on the thumb side
dorsal artery of the foot	on top of the instep of the foot

Many areas commonly used to determine heart rate are also known as pressure points. This is because they may be compressed to stop flow to distal regions during haemorrhage.

Many factors affect heart rate or pulse, including:

- sex of the individual;
- activity of the individual;
- emotions;
- postural changes.

Heart rate can be used to assess at least one part of an individual's cardiovascular fitness.

Another important term is stroke volume. It means the volume of blood pumped out by the heart with each contraction. Usually, stroke volume is directly related to the force of ventricular contraction. Since heart contraction is known as systole, stroke volume is sometimes referred to as systolic discharge. The stroke volume for a person at rest is near 70 ml. If you multiply the stroke volume by the heart rate, the answer is known as cardiac output. Cardiac output is defined as the volume of blood pumped out of the heart per minute.

Cardiac output = CO
Stroke volume = SV
Heart rate = HR
CO (ml/min) = HR (beats/minute) × SV (ml/beat)

Arteries are blood vessels that carry blood away from the heart. Veins carry blood to the heart. The smallest vessels, the capillaries, are the sites where exchanges between the blood and body tissues occur.

If we substitute in normal resting values for HR and SV, then cardiac output is 5.25 $dm^3 min^{-1}$. This volume is approximately the amount of blood you have in your body. Therefore, each minute, all your blood circulates through the heart.

The final important term you need to understand heart action is venous return. It refers to the volume of blood that returns to the heart via the veins each minute. Venous return is affected by many factors including blood volume, heart contractions, and muscle pumps throughout the body pushing the blood in the veins back to the heart.

Heart action

Heart muscle is myogenic: it does not need nerves to cause contraction. The sinoatrial (SA) node initiates heart contraction and it is known as the pacemaker of the heart. If left to itself, this rapidly depolarizing area of the right atrium would establish and maintain a heart rate of nearly 100 beats per minute. However, neural or endocrine factors also have an effect on the heart rate. Neural control involves the cardiovascular centre of the medulla oblongata. Endocrine control involves hormones such as epinephrine and norepinephrine from the adrenal medullae.

When exercise occurs, skeletal muscles work at a higher level. The result is increased cellular respiration with higher production of carbon dioxide which is released into the blood. The added carbon dioxide lowers blood pH. The lowered blood pH is detected by chemoreceptors in the carotid arteries and the aorta. The appropriate area of the medulla oblongata is stimulated and impulses are sent via special neurones to the SA node to increase cardiac output (see Figure 13.15).

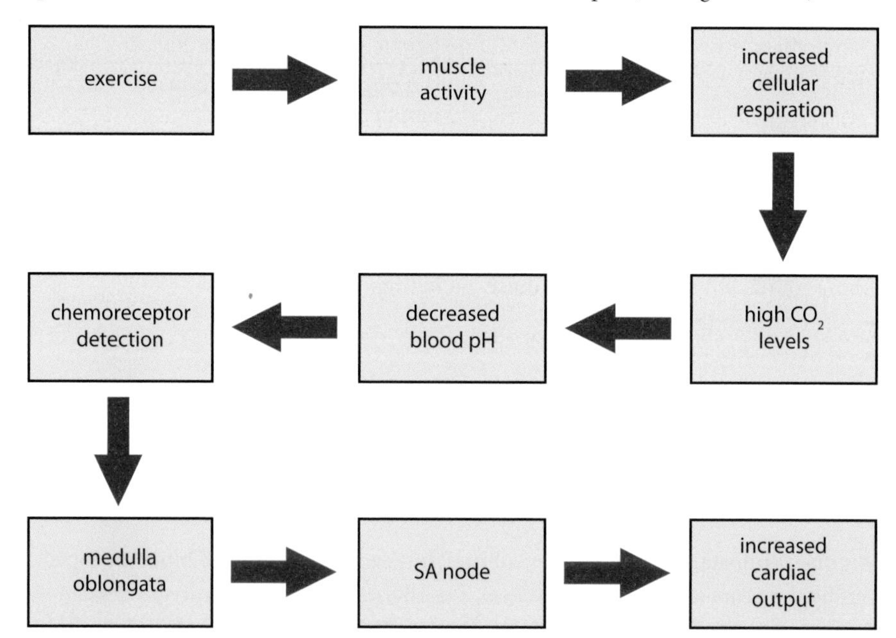

Figure 13.15 Exercise increases cardiac output.

Increased cardiac output means there must be an increased blood return to the heart by the venous system. This is accomplished by the increased squeezing action of the moving skeletal muscles on the veins that pass through them. The result is increased blood volume in the ventricles. Thus, venous return has been increased.

Blood distribution

Blood distribution is extremely important during exercise. The increased demands of the skeletal muscles must be met, but the requirements of the vital organs cannot be ignored. The blood flow to the brain is relatively constant during rest and during exercise. However, the blood flow to the structures directly involved in the activity is greatly increased. Parts of the body that experience this increase are: cardiac muscles, skeletal muscles, and the skin. Conversely, a decrease in blood flow during exercise is experienced by: kidneys, stomach, intestines, and other abdominal organs. Figure 13.16 shows this distribution pattern.

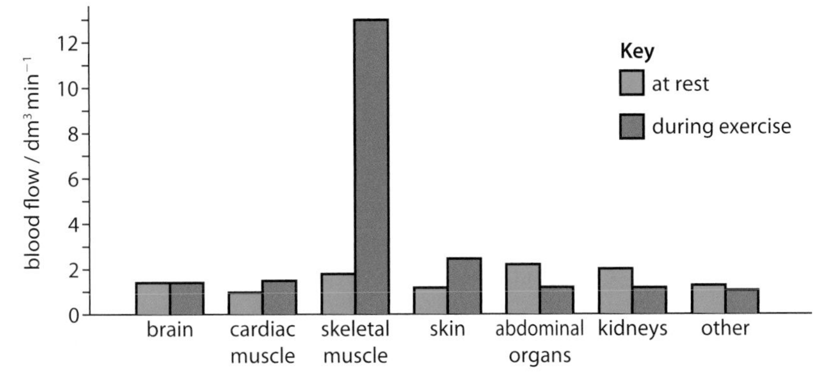

Figure 13.16 Graph showing the difference in blood distribution at rest and during exercise.

It is clear from Figure 13.16 that blood flow to the abdominal organs is significantly less during exercise. This decreases the digestive process including the absorption of nutrients from the intestines. Also, the kidneys are functioning with less blood. The result is a decreased production of urine during exercise, and conserved fluids. The skin, on the other hand, has a large increase in blood flow. This is important as it allows for increased heat dissipation. When the muscles of the body are carrying out additional activity, heat is generated and the blood carries the heat away from the source. In a lighter-skinned individual, a reddish colour of the skin will be evident due to this increased blood flow.

The additional cardiac output during exercise results in increased blood flow to the lungs as well. This serves to increase oxygen delivery to the muscles. There are also larger amounts of carbon dioxide exhaled.

With the muscles carrying out more activity, it is essential they receive a greater supply of blood. The added blood provides more oxygen so that cellular respiration and ATP production can be increased. During exercise, the heart has increased cardiac output with greater contractility. This means its muscles also need an increased supply of blood and oxygen. The increased blood supply to all these areas is also important in allowing more metabolic waste to be transported away from the areas of greater activity.

The brain is very consistent in its demand for blood supply – its neurones are negatively affected by any period of low oxygen.

Changes in blood flow during exercise allow all parts of the body to operate relatively efficiently. A homeostatic situation is maintained during exercise so that necessary body functions continue.

Effect of training on the heart

Aerobic training has a number of effects on the heart, both at rest and during exercise. Because of this training, the heart muscle undergoes hypertrophy (enlargement and strengthening). The left ventricle especially shows this. Due to the added strength, and the greater volume that accompanies it, the heart also shows an increased stroke volume at rest. The stroke volume may actually show an increase up to a maximum of 40–60% after aerobic training. Because the heart is pumping more blood with each contraction, it is common to see the heart rate decrease in individuals who undertake aerobic training. The decrease in the

To view a website that explains how physical training affects the heart, visit heinemann.co.uk/hotlinks, insert the express code 4242P and click on Weblink 13.3.

Swimmers see a smaller increase in stroke volume compared to runners or cyclists. This is thought to be due to the horizontal position of swimmers. The position while swimming leads to less pooling of blood in the lower extremities. This results in a lower stroke volume requirement.

heart rate of a resting individual may be as much as 10–20 beats per minute. Even though the heart rate has decreased, the cardiac output remains the same due to the increased stroke volume.

When a trained individual is exercising, their heart rate is lower than that of an untrained individual. Again, this is the result of more developed cardiac muscle. As at rest, the stroke volume during exercise of the trained individual is increased, so there is an increase in cardiac output. During exercise, it is possible for a well-trained athlete to have double the cardiac output of a sedentary individual.

Other effects of aerobic training include:
- more capillary networks develop in skeletal muscles;
- decrease in blood pressure;
- increase in blood volume;
- more effective blood distribution to body parts;
- increase in diameter of existing capillaries so increasing blood flow;
- decrease in recovery time for breathing and heart rate after exercise.

It is clear that exercise has some tremendous positive effects on the overall health of an individual.

Erythropoietin

No one would argue the benefits of blood transfusion when used in a medical emergency. Transfusions save countless lives in situations where blood loss is a problem. They also are used in severe cases of anaemia in which red blood cell counts are dangerously low.

However, blood transfusions have also been used for a very different purpose. Athletes might 'donate' a unit of their blood into storage. After the period of time necessary for their systems to replace the lost blood, they have the unit transfused back into their body. The transfusion adds to the oxygen delivery capabilities of their blood. This results in better performance, especially in activities that require endurance. The process artificially increases the number of red blood cells in an athlete's body to enhance performance; it is called blood doping.

Erythropoietin (EPO) has changed the procedure of blood doping. Transfusion is no longer necessary. A genetically engineered form of EPO is injected into the athlete. This stimulates the production of red blood cells in the bone marrow. With increased red blood cells comes greater oxygen carrying capability and thus more efficient muscle activity. The use of EPO has been noted among professional and even recreational athletes.

EPO is currently tested for in most major competitions. If found, the athlete risks being banned from future competitions.
- What is and what is not ethical in an athlete's preparation for an event?
- What is an acceptable level of risk in the use of possible training procedures and enhancers?

Erythropoietin is an important hormone in the human body. It is produced in the kidneys and it helps to maintain a healthy percentage of elements such as red blood cells, white blood cells and platelets in the blood. The percentage value is referred to as a haematocrit. Males normally have a haematocrit of 47% (±5%)and females have a 42% (±5%)haematocrit. Someone using EPO can raise their heamatocrit by up to 10 percentage points. Because EPO is produced naturally, it has been difficult to develop a reliable test for its abuse. However, there is now an effective test that can distinguish between natural and artificial EPO.

The benefits of EPO and blood transfusion include:
- effective treatment for patients of anaemia;
- replacement of blood lost due to injury or surgery.

The risks of EPO and blood transfusions for athletic enhancement include:

- increased blood viscosity due to high numbers of red blood cells (the blood could become so thick that it has problems passing through body capillaries) – this could lead to a stroke or heart attack;
- increased viscosity could also lead to increased blood pressure and added strain on the heart;
- increased chance of blood clotting – this could lead to heart attack or stroke;
- lower blood plasma level could lead to greater susceptibility to dehydration;
- after a period of time, the body may begin to produce antibodies against EPO – this could lead to chronic anaemia.

Acceptable athletic enhancement

There is much controversy as to what constitutes fair practice when it comes to preparing for an athletic event. Many professional athletes follow accepted practices that can increase their red blood cell count.

1 Training at high altitude to stimulate the natural mechanisms of the body to increase red blood cell numbers. Higher production of red blood cells is due to lower oxygen concentration at high altitude. The US has an Olympic training facility in the Rocky Mountains.

2 Sleeping in a low oxygen enclosure to stimulate natural erythropoietin production.

Exercises

10 Why would dehydration be a serious concern to an athlete using EPO?

11 Blood doping or EPO use may bring about higher blood pressure. Why?

12 If an Olympic event is held at a location of higher altitude, what would most likely be the effect on the times of endurance events?

13 Name some athletic events for which performance probably would not be improved by blood doping or EPO use.

B.4 Exercise and respiration

Assessment statements

B.4.1 Define VO_2 and VO_2 *max*.

B.4.2 Outline the roles of glycogen and myoglobin in muscle fibres.

B.4.3 Outline the method of ATP production used by muscle fibres during exercise of varying intensity and duration.

B.4.4 Evaluate the effectiveness of dietary supplements containing creatine phosphate in enhancing performance.

B.4.5 Outline the relationship between the intensity of exercise, VO_2 and the proportions of carbohydrate and fat used in respiration.

B.4.6 State that lactate produced by anaerobic cell respiration is passed to the liver and creates an oxygen debt.

B.4.7 Outline how oxygen debt is repaid.

VO_2 and VO_2 max

VO_2 and VO_2 max are terms used by physiologists to explain oxygen uptake or consumption.

V = volume per time
O_2 = oxygen
max = maximum

VO_2 provides valuable information demonstrating the effectiveness of the relationship between oxygen delivery and tissue metabolic demands. It also indicates the ability of muscles to use oxygen. VO_2 max is defined as a person's maximal rate of oxygen consumption. It demonstrates the ability of an individual to generate the energy required for endurance activities that last for longer than 4–5 minutes (see Figure 13.17).

Figure 13.17 Graph showing the location of VO_2 max.

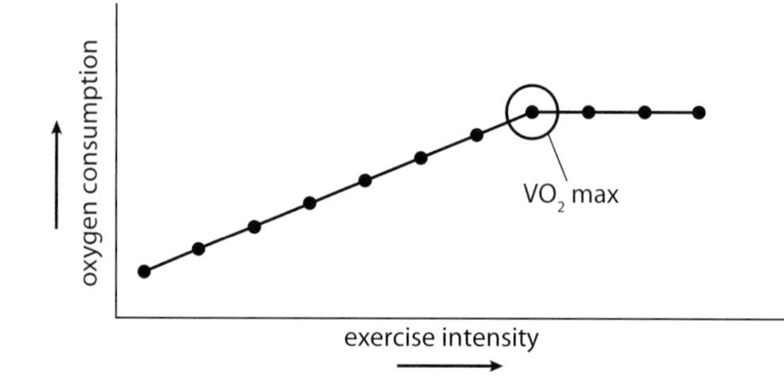

The point in Figure 13.17 where oxygen consumption first levels out is the individual's maximal aerobic capacity or VO_2 max. The units for VO_2 max are usually expressed as ml kg^{-1} min^{-1}. Oxygen and energy needs differ for relative size, which is why VO_2 is expressed relative to body weight.

VO_2 max varies a great deal between individuals, especially athletes and non-athletes, as shown in this table.

Individual	Age/years	VO_2 max for males/ ml kg^{-1} min^{-1}	VO_2 max for females/ ml kg^{-1} min^{-1}
non-athlete	10–19	47–56	38–46
non-athlete	20–29	43–52	33–42
bicyclist	18–26	62–74	47–57
swimmer	10–25	50–70	40–60
runner	18–39	60–85	50–75

Taken from Wilmore and Costill (2005) *Physiology of Sport and Exercise*, 3rd Edn. Champaign, IL. Human Kinetics

VO_2 can certainly show a significant increase due to training involving aerobic exercise. Training at 75% of aerobic power for 30 minutes, 3 times a week over 6 months increases VO_2 by an average of 15–20%. It is also true that genetics plays a major role in a person's VO_2 max.

To view a website with more information about VO_2 max, visit heinemann.co.uk/hotlinks, insert the express code 4242P and click on Weblink 13.4.

The VO_2 is important since it allows adequate oxygen for cellular respiration to occur. The best food sources for mitochondrial energy production are carbohydrates and fats. Of these two, carbohydrates are the most efficient. When exercise first begins, fats are often the source of energy in cellular respiration. However, as the intensity of exercise increases and VO_2 max is achieved, carbohydrates account for 100% of the energy source.

Energy and muscles

Myoglobin is a red pigment that occurs in many muscle cells. It is similar to haemoglobin in its oxygen-binding capabilities. It binds to oxygen when levels are high and releases it when levels are low. This provides oxygen to the mitochondria so that cellular respiration can occur most efficiently. Besides oxygen, the mitochondria need a source of organic compounds to break down to produce ATP for muscle contraction. This source is usually glucose. However, if blood glucose levels fall below normal, glycogen becomes involved. Glycogen occurs in muscle cells and the liver. It is a complex organic compound composed of a long chain of glucose molecules. Glycogen is the storage form of carbohydrates in animals.

Muscle fibres are quite different from most cells of the body. They are often switching between a low level of activity requiring small amounts of ATP and a high level of activity requiring large amounts of ATP. There is only enough ATP within a muscle cell to power it for a few seconds of activity. Therefore, muscle fibres must be able to produce ATP. They have three ways:

- from creatine phosphate;
- by anaerobic respiration;
- by aerobic respiration.

Two hormones, glucagon and adrenalin, are released when blood glucose levels are low. These hormones stimulate the breakdown of glycogen into its glucose subunits. The glucose is then available for cellular respiration.

Creatine phosphate

Anaerobic and aerobic respiration are both carried out by all the body's cells. However, only muscle cells use creatine phosphate (CP). When muscles are relaxed they have a surplus of ATP. The excess ATP is used to produce CP. Creatine kinase is the enzyme in this reaction.

$$\text{creatine} + \text{ATP} \xrightarrow{\text{creatine kinase}} \text{creatine phosphate} + \text{ADP}$$

Creatine is a small, amino acid-like molecule produced in the pancreas, liver, and kidneys. It is transported to muscle fibres. There is much more CP stored in muscle cells than ATP. When intense exercise occurs, CP is used to produce the needed ATP for muscle contraction. It provides adequate ATP for up to 10 seconds.

This table compares the three types of ATP-production by muscle cells.

Creatine phosphate	Anaerobic respiration	Aerobic respiration
ATP produced by direct phosphorylation of ADP	glycolysis and lactic acid formation produces ATP	oxidative phosphorylation produces ATP
energy source: CP	energy source: glucose	energy source: glucose, pyruvic acid, fatty acids, amino acids
no oxygen	no oxygen	oxygen
1 ATP per CP molecule	2 ATP per glucose molecule	36 ATP per glucose molecule
duration of energy: up to 10 seconds	duration of energy: up to 60 seconds	duration of energy: hours, if intensity of exercise decreases

Anaerobic respiration can only produce ATP for a limited time due to the toxicity of one of its products, lactic acid (lactate). Aerobic respiration can proceed for long periods of time as long as adequate oxygen is being delivered to the involved muscles. This oxygen delivery is provided by haemoglobin in the blood and myoglobin in the muscles. As the intensity of an activity decreases and the duration increases, the percentage of ATP provided by aerobic respiration increases. Conversely, the percentage of ATP provided by anaerobic respiration decreases as the duration of intense activity increases. At the end of an endurance event such as a marathon run, the amount of ATP being used by the muscles is nearly 100% from aerobic respiration.

Creatine phosphate as a dietary supplement

Creatine phosphate is classed as a legal performance-enhancing supplement. It may be purchased over the counter. It is thought to potentially increase athletic performance in sports involving short bursts of intense activity, such as power lifting, wrestling and sprinting. At present, studies to determine the effectiveness of this supplement are inconclusive. There does appear to be an increase in muscle size when the supplement is taken. However, that may be due to increased water retention by the muscle fibre due to the presence of creatine. Some individuals did show improved performance when taking the supplement. This improvement was only apparent in athletic events requiring repeated, intense activities. Less or no improvement was noted with sustained endurance activities. Some individuals even reported a decrease in their ability to maintain prolonged exercise when taking the supplement.

As you will recall, creatine phosphate is used by muscle cells in the first 8–10 seconds of intense exercise. The function of this natural CP is to directly phosphorylate ADP. This provides ATP for muscle activity. It is this function of CP in the muscles that provides the reason for some individuals to take supplements of artificially produced CP.

Creatine can cause side-effects when taken in doses of 20 000 mg per day or more. The side-effects include:

- high blood pressure because of water retention;
- dehydration because the water is often pulled from the rest of the body into the muscles;
- weight gain;
- diarrhoea;
- dizziness;
- muscle cramps;
- nausea;
- stomach pain.

Presently, creatine is not recommended for individuals under 18 years old. This is because too little is known about its long-term side-effects. It is also potentially very dangerous to individuals with any type of kidney disorder.

Creatine is produced in the pancreas, kidneys and liver at a rate of nearly 2 grams per day. A normal diet could also provide as much as 2 grams a day, primarily from meat and fish. If an individual takes in more creatine than is needed, most of the excess is excreted as creatinine in the urine. Recently, studies have shown taking in large amounts of creatine decreases the body's own synthesis of this substance, and it is not known if the natural synthesis process can be re-established after long-term use of the supplement.

Currently, most major athletics-governing organizations allow athletes to use creatine while competing. However, the potential dangers and the lack of conclusive studies concerning benefits of the supplement should be very seriously considered.

Anaerobic respiration

Anyone who undertakes intense physical activity knows what it means to 'go anaerobic'. Muscles begin to 'burn', breathing rate greatly increases, and intensity of exercise decreases or stops totally. These are signs that oxygen supply to the muscles is inadequate to maintain aerobic respiration. The result is anaerobic respiration and the production of limited ATP and significant amounts of lactic acid, lactate.

glucose → glycolysis → pyruvic acid → lactic acid

The lactic acid, lactate, is a problem when it is produced in a muscle. It causes changes in the chemistry of the muscle which can result in muscle fatigue. When muscle fatigue occurs, the muscle does not contract even though it is receiving stimuli. This is a physiological state in which the muscle is incapable of contraction. It may occur after intense workouts. If the situation progresses to the point where there is no ATP available, the muscle exhibits contractures. Contractures are states of continuous contraction in muscles due to lack of ATP to break the cross-bridges that form between actin and myosin myofilaments. These contractures are temporary and will end when ATP production restarts.

Muscle fatigue can easily be demonstrated by repetitive motion of the forearm until it can no longer continue the action. If the motion occurs long enough, contractures may occur. Writer's cramp is an example of contractures due to muscle fatigue.

Lactate and the oxygen debt

After muscle contraction stops, heavy breathing occurs for a period of time. This allows oxygen consumption to occur at a higher rate than usual. The period of time of increased breathing is referred to as the individual's recovery period. The length of the recovery period varies greatly depending on the intensity of the exercise and the athletic training level of the individual. Other factors affecting recovery period include the age, size, and health of the individual. The term 'oxygen debt' refers to the amount of oxygen, over and above resting oxygen consumption, which is needed to return the body to homeostatic conditions after exercise. These restorative processes include:

- converting lactate to glucose and ultimately the carbohydrate storage compound, glycogen;
- restoring proper ATP and CP levels within the muscle;
- replacing the oxygen that was removed from myoglobin.

All non-aerobic sources of ATP used during muscle activity contribute to the oxygen debt.

Much of the lactic acid produced by intense physical activity is transported to the liver via the bloodstream.

lactate in muscle → bloodstream → liver

Examiner's hint: It is essential to know the role of the liver in alleviating the oxygen debt. Be certain to note the possible pathways of lactate conversion.

In the liver, oxygen is used to convert the lactic acid back into pyruvic acid, pyruvate (see Figure 13.18). The pyruvate can then enter the Krebs cycle and follow aerobic pathways in the mitochondria of cells. The liver also possesses enzymes that allow the conversion of the lactate to glucose (see Figure 13.18). The glucose may then be released into the bloodstream to bring blood glucose to proper levels. If the glucose is not needed, the liver cells store the glucose as glycogen (see Figure 13.18).

Figure 13.18 Lactate is changed into pyruvate, glucose, and glycogen in the liver.

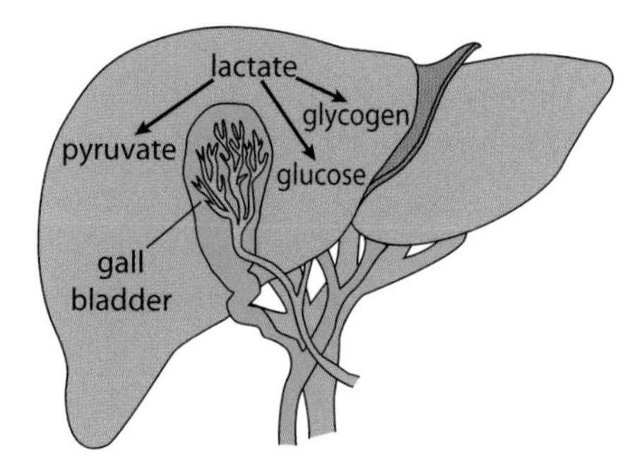

Aerobic respiration allows the net gain of 36 ATPs from one glucose molecule, while anaerobic respiration allows a net gain of only 2 ATPs.

For these conversions of lactate to occur in the liver, oxygen must be present. The availability of oxygen due to the increased breathing rate also allows the second and third restorative processes to occur. The presence of oxygen enables aerobic respiration to occur in the mitochondria. The ATP within the muscle cells is increased. Some of the ATP is used to convert creatine to CP. The CP is then available to provide energy necessary for the initial seconds of (subsequent) intense exercise. Because of the higher availability of oxygen, myoglobin is 'recharged' to allow quicker delivery of oxygen to muscle cells in times of activity.

Once all three restorative processes are completed, the oxygen debt is said to be 'paid'. At that point, ventilation returns to normal.

Exercises

14 Why do marathon runners go through a 'carbohydrate loading' process 2 or 3 days prior to the event?

15 Why can a well-trained endurance athlete carry out an activity for a long period of time without incurring an oxygen debt?

16 What food source is the most effective in providing ATP for muscle usage?

17 What is the result in the muscles when the intensity of exercise exceeds VO_2?

18 How is muscle fatigue overcome?

B.5 Fitness and training

Assessment statements

B.5.1 Define *fitness*.
B.5.2 Discuss speed and stamina as measures of fitness.
B.5.3 Distinguish between fast and slow muscle fibres.
B.5.4 Distinguish between the effects of moderate-intensity and high-intensity exercise on fast and slow muscle fibres.
B.5.5 Discuss the ethics of using performance-enhancing substances, including anabolic steroids.

Fitness

Fitness is a very interesting but poorly understood topic. We hear all the time about the importance of being fit. Many people claim to be fit, but others might politely disagree. The most straightforward definition of fitness is that it is the physical condition of your body that allows you to perform a particular exercise.

From this definition, it is clear there are many different levels of fitness. Certainly, the individual involved in an activity such as golf or walking would have a different perception of fitness from someone involved in basketball or football. To achieve fitness there needs to be, in some cases, a significant amount of training. There are three aspects to any training regimen:

- frequency (how often you exercise);
- duration (how long each exercise session lasts);
- intensity (how vigorous the exercise is).

There are exercise programmes available on the internet or through personal trainers at exercise facilities for practically every individual's tastes or needs. It is an individual's choice to select the degree of fitness they desire.

When deciding on a training programme, most exercise physiologists would say you should consider four elements:

- aerobic fitness;
- muscular fitness;
- flexibility;
- stability and balance.

Aerobic fitness

The importance of aerobic fitness exercise is that it increases the body's ability to use oxygen. As you will recall, the VO_2 max of an individual is pushed to a higher level. When the aerobic capacity of an individual is high, their heart, lungs and blood vessels are working efficiently. Not only does this fitness help you do better in the exercise of your choice, it also allows you to cope more successfully with everyday challenges. To achieve aerobic fitness, a regular exercise programme involving fairly rigorous activity is required. This aspect of training must involve elevated heart and breathing rates. However, you must be cautious beginning aerobic training if you have previously led a sedentary life. A consultation with a medical doctor would be well advised before starting.

Muscular fitness

Muscular fitness refers to the strength and endurance of your muscles. Again, this type of fitness will help in everyday activities. Achieving muscular fitness may include the use of free weights, resistance bands, weight machines, or even your own body weight (push-ups). Regularly working the muscles of the body will increase the amount of lean muscle mass and will help with weight loss.

Flexibility

Flexibility is an important part of any fitness programme. It refers to the ability to move your body's joints through their full range of motion. Stretching is the most important component in achieving this type of fitness. Activities include yoga and tai chai. Information on proper stretching techniques is available at many sources including the internet and training facilities.

Stability and balance

The muscles of the lower back, pelvis, hips and abdomen are involved in stability and balance. These areas make up what are called the core muscles. Developing these muscles will combat poor posture and also prevent falls, especially in older individuals.

● **Examiner's hint:** Be able to specifically discuss how speed and stamina can be improved by carrying out the four elements of a good fitness programme. It is a good idea to mention specific activities in your discussion.

Speed and stamina

Speed and stamina are often spoken of when discussing fitness. Certainly, what we have talked about will develop both. Speed describes the rate at which a movement is performed. Stamina or endurance refers to the ability to carry out an activity for a longer period of time. Again, depending on what activity is most important to an individual, speed or stamina may be focal points in a fitness programme.

A sprinter will concentrate on speed, while a rower or marathon runner will work largely on stamina. However, it is beneficial for both athletes to develop speed and stamina in their training regimen.

Developing and regularly practising an exercise programme that includes all of these aspects is a good aid to living a longer, healthier life. We should all be aware of the importance of developing our bodies as well as our minds. Physical exercise allows for a better functioning brain and that may equate to a higher grade on the IB biology exam!

Fast and slow muscle fibres

● **Examiner's hint:** A question in the exam about the differences between types of muscle fibre is most effectively answered with table formats. The number of characteristics presented depends on the marks awarded for the response.

Not all skeletal muscle fibres are alike. The IB syllabus requires that you understand two major types of fibre, fast and slow. There is also a third type recognized by exercise physiologists. The three types of skeletal muscle fibre are:

- slow oxidative (SO) fibres;
- fast oxidative–glycolytic (FOG) fibres;
- fast glycolytic (FG) fibres.

This table shows you the metabolic characteristics of the three types of muscle fibre.

Metabolic characteristic	SO fibres	FOG fibres	FG fibres
speed of contraction	slow	fast	fast
glycogen stores	low	intermediate	high
myoglobin content	large amount	large amount	small amount
mitochondria	many	many	few
myosin ATPase activity	slow	fast	fast
main pathway for ATP production	aerobic	aerobic	anaerobic glycolysis
rate of fatigue	slow	intermediate	fast

This table shows you the structural characteristics of these three types of muscle fibre.

Structural characteristic	SO fibres	FOG fibres	FG fibres
colour	red	red to pink	white (pale)
fibre diameter	smallest	intermediate	large
capillaries	many	many	few

This table shows you the functional characteristics of these three types of muscle fibre.

Functional characteristic	SO fibres	FOG fibres	FG fibres
capacity of generating ATP	high capacity	intermediate capacity	low capacity
rate of ATP hydrolysis	slow	fast	fast
suited activities	endurance-type activities such as distance running or rowing	sprinting and walking	short-term intense or powerful movements

Understanding and being able to reproduce the characteristics presented in these tables is best approached by remembering what type of activity each muscle fibre type is mostly involved in and then logically deducing what the characteristics are.

● **Examiner's hint:** The IB exam will be most concerned with the knowledge of the difference between the SO and the FG fibres. The SO fibres are also known as slow twitch fibres and FG fibres as fast twitch fibres.

Athletic activity and the different types of fibre

Most skeletal muscles are a mixture of the three types of fibre. However, the proportions of one type to another show significant variation from individual to individual. Genetics and a person's training regimen cause these variations (see Figure 13.19).

Clearly, our genetics determine, to a large degree, the athletic activities we have the greatest ability in. However, with training it is possible to bring about fitness in a range of activities. To bring about overall fitness, two general types of exercise are required. One is high-intensity exercise and the other is moderate-intensity exercise.

Moderate-intensity exercise is usually referred to as aerobic and involves activities such as running and swimming. This type of training will result in a transformation of some of the fast twitch or FG fibres into FOG fibres. There is no increase in the number of slow or SO fibres. These fibres will show slight increases in diameter, number of mitochondria, blood supply, and strength. The result is greater endurance or stamina of the muscles without obvious increase in muscle size. This type of activity also increases the efficiency of the cardiovascular and respiratory systems in supplying oxygen and nutrients to the muscle.

High-intensity exercise is anaerobic in nature and the result is a definite increase in the size of the FG or fast muscle fibres. The increase in size, not number, is due to an increase in the amount of actin and myosin in the contracting units, sarcomeres, of the muscles. This adds to muscle strength and results in a noticeable hypertrophy (increase in size) of the involved muscles.

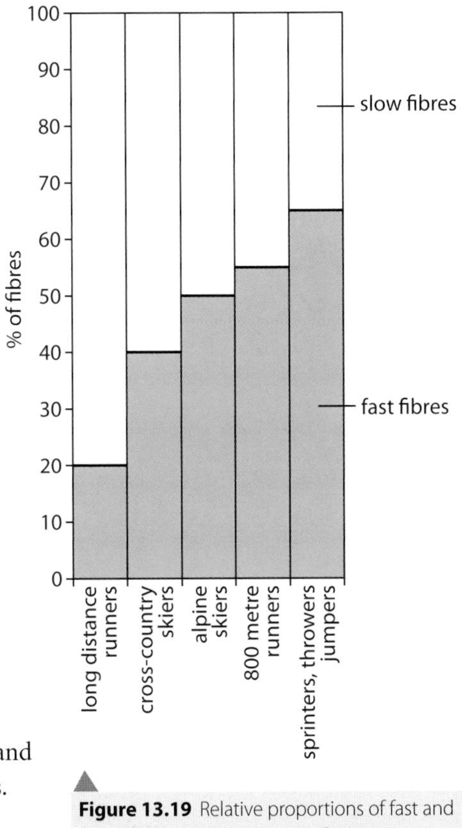

Figure 13.19 Relative proportions of fast and slow fibres in different types of athlete.

Ethics and performance-enhancing substances

You will recall that we have already discussed erythropoietin and creatine phosphate used as supplements. Anabolic steroids are another supplement taken by individuals to enhance athletic performance. Just as erythropoietin and creatine phosphate use confers some benefits for athletic ability, so do anabolic steroids. These steroids are variants of the male sex hormone, testosterone. When occurring naturally, they are responsible for the increase in muscle and bone mass plus other physical changes that accompany puberty in boys. Anabolic steroids were first manufactured by pharmaceutical companies in the early 1950s to treat certain cases of anaemia and muscle-wasting diseases. Body builders and other athletes started to use them in the 1960s and their use has continued ever since.

Athletes report several desired effects achieved by using these steroids – mainly to do with increased muscle size and greater strength. Users state that they experience improved performance during athletic events. Because these chemicals have been used for a fairly long time, definite side-effects have been noted. Some of these are:

- bloated face;
- liver cancer;
- kidney damage;
- increased risk of heart disease;
- increased aggression and irritability;
- shrivelled testes and infertility;
- wide mood swings resulting in violent actions and severe depression;
- decreased bone growth resulting in shorter stature.

Women athletes also take these steroids. Side-effects specific to women include:

- atrophy of breasts and uterus;
- menstrual irregularities;
- sterility;
- facial hair growth;
- deepening of the voice.

The list of side-effects is alarming, yet a substantial number of young people and athletes are known to use these chemicals. Because they are presently banned in most sports, it is difficult to produce exact percentages of individuals using them.

So what are the ethics of using any or all of these performance enhancers?

- Sometimes, all an athlete considers is performance on the field, track or court: positive results now are thought to be most important, not what may happen tomorrow.
- Sometimes, these enhancers are used to improve self-esteem, but the dangers of severe mood swings may be more of a problem than low self-esteem.
- Many see the supplements as an easy, fast way to improve performance; they feel more competitive and think they have a greater chance to excel. Is this fair to others competing in the sport who do not use enhancers?

Is this what we want as a society? Do we want athletes chemically groomed to produce super-human results? What about the health and well-being of athletes? Are the side-effects worth risking?

Ethics is defined by Webster's Dictionary as 'the discipline dealing with what is good and bad and with moral duty and obligation'. Despite the guidelines and legal status of performance enhancers, each individual must still decide for themselves the rights and wrongs of using performance enhancers in their preparation for competitive sports events.

Exercises

19 Compare the training that would be most effective for a weight-lifter to that which would be most effective for an endurance runner.

20 What features allow slow muscle fibres to carry out long-lasting exercise activities?

21 Even though fitness refers to a specific activity, what should be included in every person's training programme to attain fitness?

22 Explain which muscle fibres are mostly involved in the fitness measures of speed and stamina.

B.6 Injuries

Assessment statements

B.6.1 Discuss the need for warm-up routines.

B.6.2 Describe injuries to muscles and joints, including sprains, torn muscles, torn ligaments, dislocation of joints and intervertebral disc damage.

Warm-up

It is widely held that warm-up routines before physical workouts are beneficial. Most people believe they diminish the chances of injury during the intense phase of the training routine. Some studies have shown warm-ups are not as important as originally thought. However, other studies have shown rather significant benefits. At present, most exercise physiologists recommend some sort of preparation of the body's muscles before entering into an athletic event or physical workout.

It is very difficult to determine accurately the effect of no warm-ups on muscle activity and possible muscle injury. Do athletes believe what they are told without questioning it?

A warm-up may simply be a lower intensity or lower speed version of the activity that is about to occur. For example, walking before jogging; exercising with lighter weights before lifting heavy weights.

Proposed benefits of warm-ups include:

- increased movement of blood to the muscles involved in the forthcoming activity;
- delivery of more oxygen and nutrients to the involved muscles;
- preparation of the muscles for stretching;
- preparation of the heart for an increase in activity;
- improvement of coordination and reaction times;
- priming the nerve-to-muscle pathways so that the muscles are ready for exercise;
- mental preparation for the upcoming activity or event.

Logically, each of these benefits could aid the contractions of muscles involved in a workout.

Most physical trainers suggest a warm-up period of 5–10 minutes that brings about a higher rate of breathing than normal. However, breathing should not be brought to the rate it will be during the actual exercise period.

Warm-down or cool-down activities are also considered very important after the intense phase of a workout. These activities appear to aid dispersal of lactic acid and other wastes from the muscle. This decreases soreness and stiffness, and may allow exercise to be carried out again successfully after a shorter rest period.

With all of this said, many animals seem able to carry out intense exercise without a warm-up period. The value of warm-up activities for humans remains very open to debate.

Muscle and joint injuries

Injuries all too often accompany exercise regimens. Usually, these injuries involve muscles and joints and the structures associated with them.

To view a website that explains specific injuries related to physical activity, visit heinemann.co.uk/hotlinks, insert the express code 4242P and click on Weblink 13.5.

Sprains

When a sprain occurs, there is a stretching injury to the ligaments that connect bone to bone. It results in a partial tear of the ligament. Pain is often mild to moderate. Because the ligaments are stretched, there may be a feeling of looseness in the joint involved. Swelling is common.

The worst type of sprain results in completely torn ligaments. The individual usually reports a pop or snap at the time of the injury. Severe swelling and significant pain are common. The joint is quite unstable and shows extreme looseness. Torn ligaments require long healing times and may even need surgery.

Torn muscles

Muscles may be torn. This often occurs at the junction between the muscle and its tendon. Pain may be significant, especially with movement. Unless ice is applied, there may be significant haemorrhage as a result of broken blood vessels. Swelling accompanies any blood leakage.

Dislocations

Dislocations occur when a bone is pulled or pushed out of place at a joint. This type of injury is commonly seen at the knee, hip, fingers, elbow, or shoulder. When the bone moves from its normal position, it may injure soft tissues in or around the joint (ligaments, tendons, muscles, cartilage and joint capsule). Medical care should be sought.

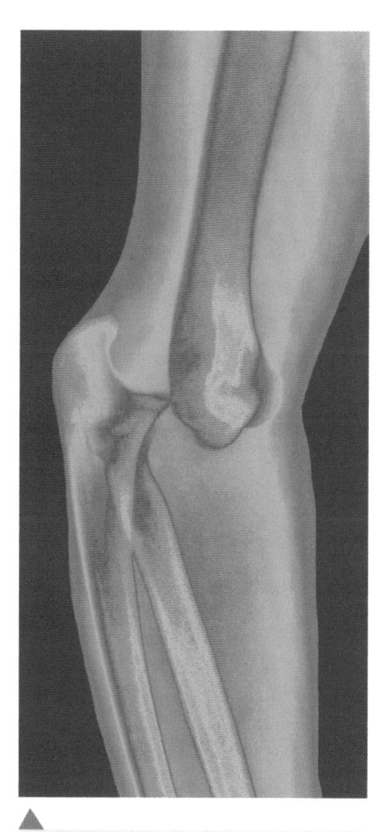

Figure 13.20 False-colour X-ray of a dislocated elbow.

Intervertebral disc damage

Intervertebral discs occur between the vertebrae. They allow flexibility of the spinal column and a degree of shock absorption in the torso. The outer part of the disc is made of tough fibrocartilagenous fibre. This surrounds an inner region of a gel-like substance. Disc injury occurs when the outer fibre is torn and the inner gel core bulges outward causing symptoms such as numbness, pain, tingling, or weakness of one or both limbs. Such injuries are often due to poor lifting techniques, abnormal movements, or lifting heavier objects. Medical attention is recommended.

First aid

Most of these injuries should be treated immediately using the RICE approach.

- Rest or immobilize the affected area.
- Ice should be applied.
- Compression to minimize blood loss and swelling.
- Elevate the affected area to decrease blood flow and prevent blood pooling.

This is only first aid and is not to take the place of any needed medical care.

Exercises

23 Using the physiology of muscle action, explain why warm-ups may decrease soft tissue injury during exercise.

24 If an individual did not conduct a warm-up period, how might their muscles respond to the early phase of their work-out?

25 Using the elbow as an example, explain what could happen with a dislocation.

Practice questions

1 **(a)** Identify the structures labelled I and II on the diagram of the elbow joint below.

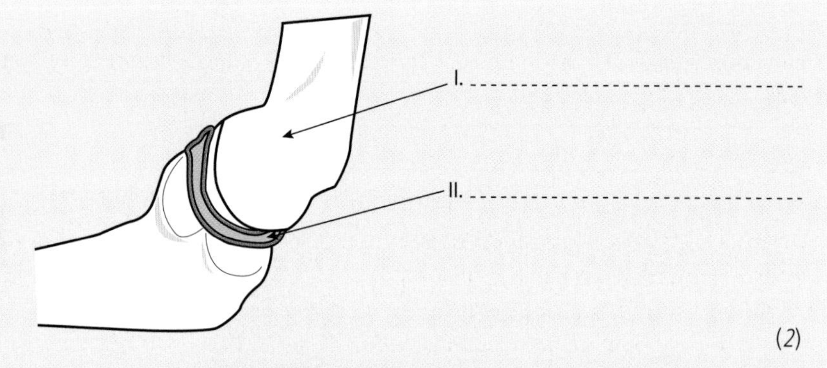

(2)

Source: R Allen and T Greenwood (2001), *Advanced Biology 2, Student Resource and Activity Manual*, 3rd edition, Biozone International Limited, page 98

(b) Explain how the action of the muscles is coordinated at this joint by the nervous system. (3)

(c) State and describe one injury that could occur in this joint. (2)

(*Total 7 marks*)

2 Explain the changes in ventilation with exercise.

(*Total 3 marks*)

3 **(a)** Outline how stamina could be used as a measure of fitness. (1)

(b) Outline two ways that training can affect the muscles. (2)

(c) Discuss the ethics of using drugs to improve sporting performances. (3)

(*Total 6 marks*)

4 **(a)** **(i)** Define the term *oxygen debt.* (1)

(ii) State the name of an organ where oxygen debt is repaid. (1)

(b) Outline two different injuries that may occur to joints. (2)

(c) Explain the need for warm-up and cool-down routines during exercise. (2)

(*Total 6 marks*)

5 A sprinter is exhausted after running a 100 m sprint, while a marathon runner is exhausted after running a marathon.

(a) State the process by which each runner obtains energy (ATP). (2)

Sprinter: Marathon runner:

(b) Outline the primary cause of muscle fatigue in a sprinter and a long distance runner. (2)

(c) Explain how lactic acid build-up during exercise relates to oxygen debt. (2)

(*Total 6 marks*)

6 A major requirement of the body is to eliminate carbon dioxide (CO_2). In the body, carbon dioxide exists in three forms: dissolved CO_2, bound as the bicarbonate ion, and bound to proteins (e.g. haemoglobin in red blood cells or plasma proteins). The relative contribution of each of these forms to overall CO_2 transport varies considerably depending on activity, as shown in the table below.

CO_2 transport in blood plasma at rest and during exercise			
		Rest	Exercise
Form of transport	Arterial blood /mmol l^{-1}	Venous blood /mmol l^{-1}	Venous blood /mmol l^{-1}
dissolved CO_2	0.68	0.78	1.32
bicarbonate ion	13.52	14.51	14.66
CO_2 bound to protein	0.3	0.3	0.24
Total CO_2 in plasma	14.50	15.59	16.22
pH of blood	7.4	7.37	7.14

Source: Geers and Gros (2000), *Physiological Reviews* **80**, pages 681–715

(a) Calculate the percentage of CO_2 found as bicarbonate ions in the plasma of venous blood at rest. *(1)*

(b) **(i)** Compare the changes in total CO_2 content in the venous plasma due to exercise. *(1)*

(ii) Identify which form of CO_2 transport shows the greatest increase due to exercise. *(1)*

(c) Explain the pH differences shown in the data. *(3)*

(Total 6 marks)

7 Physiologists measured the amount of glycogen stored in the muscle tissues of six athletes during a training period. The athletes ran 16 km each day for three consecutive days. The glycogen levels were measured before (pre-) and after (post-) each run. The athletes were allowed to rest for five days and the muscle glycogen levels were measured again. The data are shown in the scattergram below. Each athlete is shown by a different symbol.

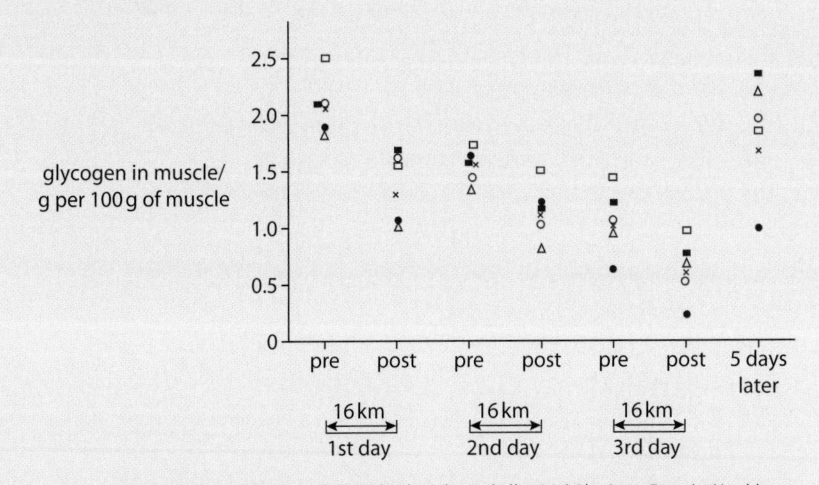

Source: W D McArdle, et al. (1987), *Physiologie de l'activité Physique. Energie, Nutrition et Performance*, Vigot, pages 51–52

(a) State the effect of running on the muscle glycogen levels. (*1*)

(b) Describe the effect of the training programme on the levels of muscle glycogen. (*2*)

Glycogen is a stored form of carbohydrate. For each athlete approximately 50% of their diet was carbohydrate.

(c) Outline the importance of carbohydrate supplies in muscle tissue during running. (*2*)

(d) Discuss the implications of this information for long distance runners training for a sports event. (*3*)

(*Total 8 marks*)

8 Explain how a muscle fibre contracts, following depolarization of its plasma membrane.

(*Total 6 marks*)

9 Describe the roles of structures at the elbow joint, including nerves, muscles and bones, in movements of the human forearm.

(*Total 8 marks*)

14 Cells and energy (Option C)

Introduction

In order to carry out necessary life functions, cells need usable energy. Usable energy comes in the form of ATP. To produce this energy currency of all organisms, enzymes are required. These enzymes, composed of proteins, catalyse the individual reactions of complex processes called photosynthesis and respiration. It is photosynthesis and respiration that allow the production of ATP so that life, in any form, is maintained. This chapter looks at bioenergetics – the study of how energy is used by living systems to maintain life.

This is a computer graphic of the enzyme known as rubisco. It is a key enzyme in photosynthesis.

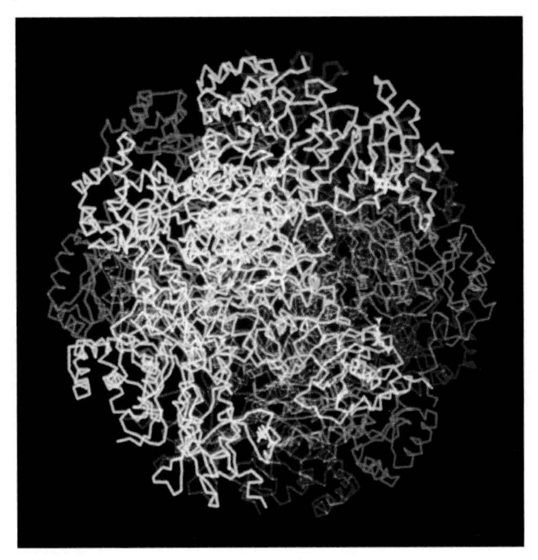

C.1 Proteins

The material in this section is also found in Chapter 7, section 7.5.

Assessment statements

C.1.1 Explain the four levels of protein structure, indicating the significance of each level.
C.1.2 Outline the difference between fibrous and globular proteins, with reference to two examples of each protein type.
C.1.3 Explain the significance of polar and non-polar amino acids.
C.1.4 State four functions of proteins, giving a named example of each.

Protein functions and structures

Proteins are an extremely important group of organic compounds found in all organisms on our planet. Many of us immediately think of muscle when the term 'protein' is used. However, proteins serve many functions in cells and organisms.

This table shows you just a few examples of proteins and their functions.

Protein	Function
haemoglobin	protein containing iron that transports oxygen from the lungs to all parts of the body in vertebrates
actin and myosin	proteins that interact to bring about muscle movement (contraction) in animals
insulin	hormone secreted by the pancreas that aids in maintaining blood glucose level in vertebrates
immunoglobulins	group of proteins that act as antibodies to fight bacteria and viruses
amylase	digestive enzyme that catalyses the hydrolysis of starch

There are many internet sites that show the various structural levels of proteins. Several even show by animation how proteins go through conformational changes from primary to quaternary structure. Visit heinemann.co.uk/hotlinks, insert the express code 4242P and click on Weblink 14.1.

There are proteins that perform structural tasks, proteins that store amino acids, and some that have receptor functions so cells can respond to chemical signals. With all these functions, proteins have to be capable of assuming many forms or structures. The function of any particular protein is closely tied to its structure.

There are four levels of organization to protein structure. They are called primary, secondary, tertiary, and quaternary organization.

Primary organization

The primary level of protein structure is the unique sequence of amino acids held together by peptide bonds in each protein (see Figure 14.1). There are 20 amino acids and these may be arranged in any order. The order or sequence in which the amino acids are arranged is determined by the nucleotide base sequence in the DNA of an organism. Because every organism has its own DNA, so every organism has its own unique proteins. The primary structure is simply a chain of amino acids attached by peptide bonds. Polypeptide chains may include hundreds of amino acids.

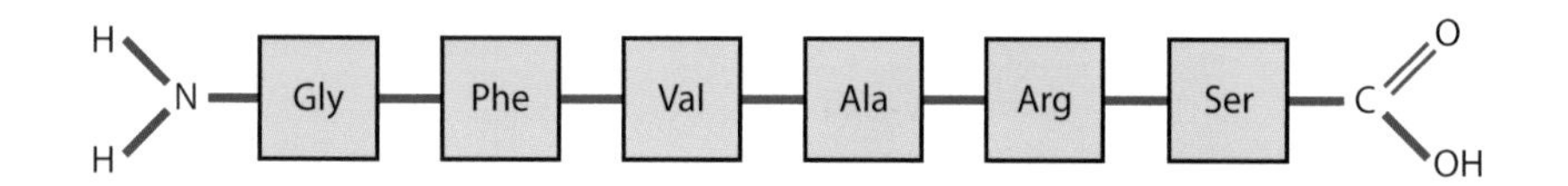

Figure 14.1 Protein primary structure: specific amino acids linked by peptide bonds.

The primary structure determines the next three levels of protein organization. Changing one amino acid in a chain may completely alter the structure and function of a protein. This is what happens in sickle cell disease. In this condition, just one amino acid is changed in the normal protein (haemoglobin) of red blood cells. The result is that the red blood cells are unable to carry oxygen, their normal function.

Secondary organization

The next level in the organization of proteins is the secondary structure. This is created by the formation of hydrogen bonds between the oxygen from the carboxyl group of one amino acid and the hydrogen from the amino group of another. Secondary structure does not involve the side chains, R-groups. The two most common configurations of secondary structure are the α-helix and the β-pleated sheet (see Figure 14.2). Both have regular repeating patterns.

Figure 14.2 Protein secondary structure.

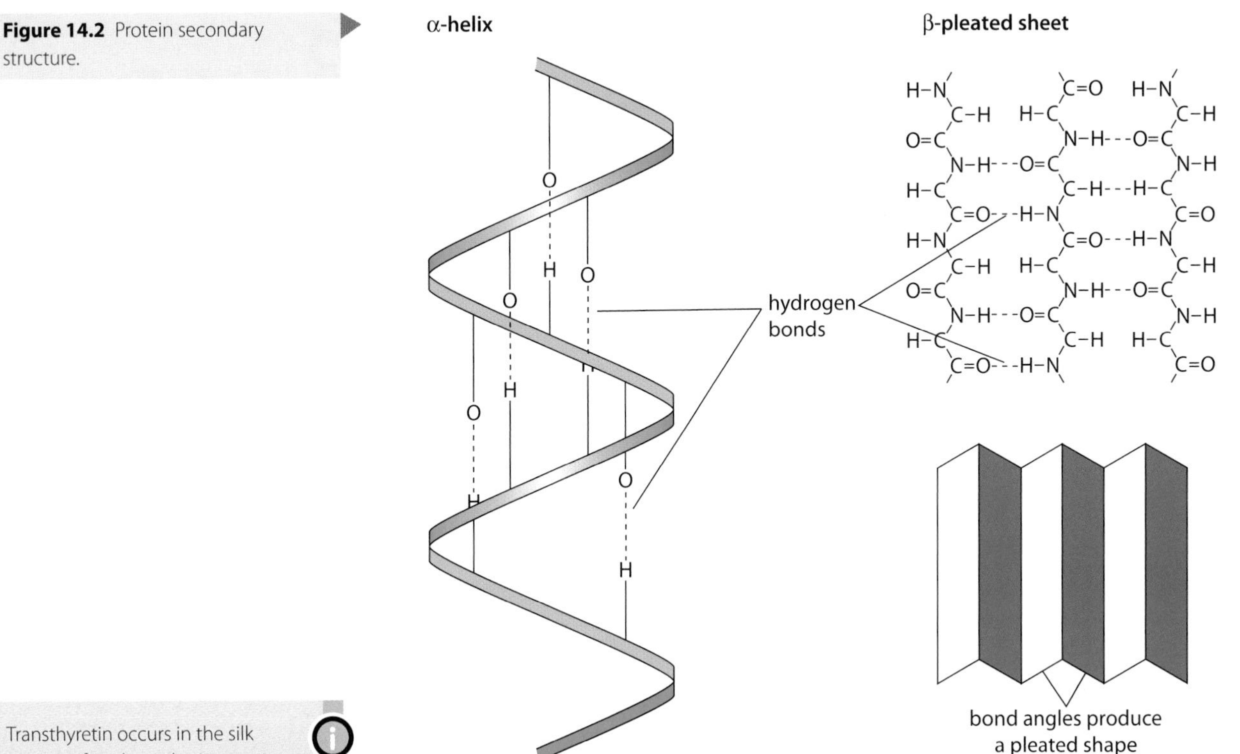

Transthyretin occurs in the silk protein of spider webs. It contains many β-pleated sheets that add to the strength of the web.

Tertiary organization

The third level in protein organization is the tertiary structure. The polypeptide chain bends and folds over itself because of interactions among R-groups and the peptide backbone. This results in a definite three-dimensional conformation (see Figure 14.3).

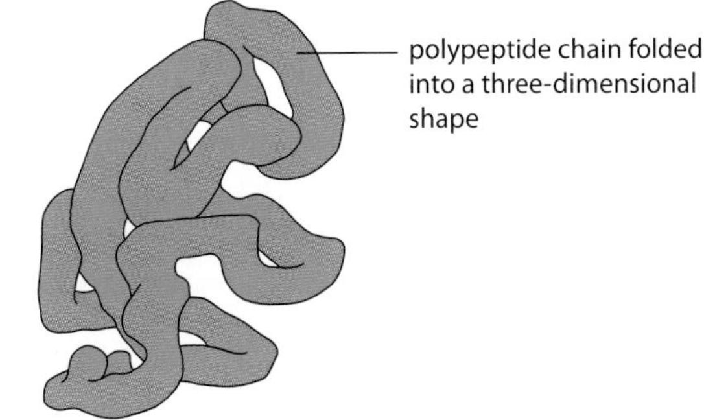

Figure 14.3 This is called a sausage model. It shows the three-dimensional conformation of lysozyme, an enzyme present in sweat, saliva and tears. Lysozyme destroys many bacteria.

Interactions that cause tertiary organization include:

- covalent bonds between sulfur atoms to create disulfide bonds – these are often called bridges because they are strong;
- hydrogen bonds between polar side chains;
- Van der Waals interactions among hydrophobic side chains of the amino acids – these interactions are rather strong because many hydrophobic side chains are forced inwards when the hydrophilic side chains interact with water towards the outside of the molecule;
- ionic bonds between positively and negatively charged side chains.

Tertiary structure is particularly important in determining the specificity of the proteins known as enzymes.

Quaternary organization

The last level is the quaternary structure. It is unique in that it involves multiple polypeptide chains which combine to form a single structure. Not all proteins consist of multiple chains, so not all proteins have quaternary structure. All the bond types mentioned in the first three levels of organization are involved in this level. Some proteins with quaternary structure include prosthetic or non-polypeptide groups. These proteins are called conjugated proteins. Haemoglobin is a conjugated protein (see Figure 14.4). It contains four polypeptide chains, each of which contains a non-polypeptide group called haem. Haem contains an iron atom that binds to oxygen.

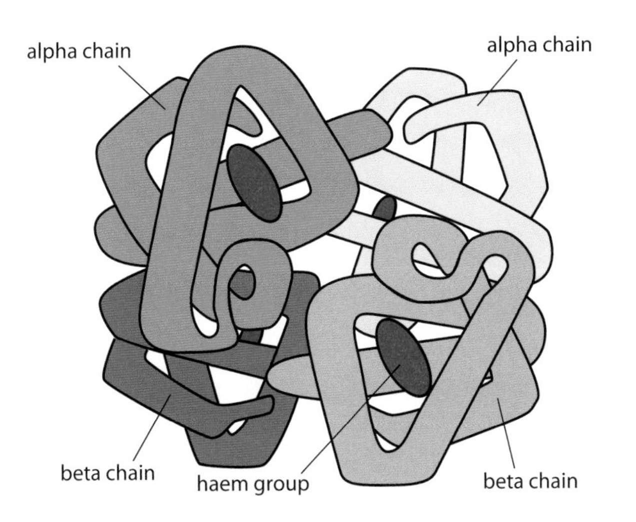

Figure 14.4 Sausage model of haemoglobin. Haemoglobin has two alpha chains and two beta chains, and four haems.

Dimers are proteins with two polypeptide subunits. Tetramers have four such units. In some cases these units are the same, but they may be different. Haemoglobin is a tetramer with two alpha chains and two beta chains.

Fibrous and globular proteins

Fibrous proteins are composed of many polypeptide chains in a long, narrow shape. They are usually insoluble in water. One example is collagen, which plays a structural role in the connective tissue of humans. Actin is another example. It is mentioned in the table above showing protein function. It is a major component of human muscle and is involved in contraction.

Globular proteins are more three-dimensional in their shape and are mostly water soluble. Haemoglobin, which delivers oxygen to body tissues, is one type of globular protein. The hormone insulin is another globular protein; it is involved in regulating blood glucose level in humans.

Polar and non-polar amino acids

Amino acids are often grouped according to the properties of their side chains (R-groups). Amino acids with non-polar side chains are hydrophobic. Non-polar amino acids are found in the regions of proteins that are linked to the hydrophobic area of the cell membrane.

Polar amino acids have hydrophilic properties, and they are found in regions of proteins that are exposed to water. Membrane proteins include polar amino acids towards the interior and exterior of the membrane. These amino acids create hydrophilic channels in proteins through which polar substances can move.

Polar amino acids include serine, threonine, tyrosine, and glutamine. All of these have a group with an electrical charge in their side chain. Non-polar amino acids have no electrical charges in their side groups. Examples are: tryptophan, leucine, alanine, and glycine.

Polar and non-polar amino acids are important in determining the specificity of an enzyme. Each enzyme has a region called the active site. Only specific substrates can combine with particular active sites. Combination is possible when 'fitting' occurs. The 'fitting' involves the general shapes and polar properties of the substrate and of the amino acids exposed at the active site.

Exercises

1 Explain why the primary level of protein organization determines the other levels.

2 What is the haem group containing iron called in the conjugated protein haemoglobin?

3 Describe how a single change in a protein's primary structure may change the protein's function.

C.2 Enzymes

The material in this section is also found in Chapter 7, section 7.6.

Assessment statements

C.2.1 State that metabolic pathways consist of chains and cycles of enzyme-catalysed reactions.

C.2.2 Describe the induced-fit model.

C.2.3 Explain that enzymes lower the activation energy of the chemical reactions that they catalyse.

C.2.4 Explain the difference between competitive and non-competitive inhibition, with reference to one example of each.

C.2.5 Explain the control of metabolic pathways by end-product inhibition, including the role of allosteric sites.

Metabolism

Your metabolism is the sum of all the chemical reactions that occur in you as a living organism. The type of reaction that uses energy to build complex organic molecules from simpler ones is called anabolism. The type of reaction that breaks down complex organic molecules with the release of energy is called catabolism. This table summarises anabolic and catabolic reactions.

Anabolic reactions	Catabolic reactions
• build complex molecules	• break down complex molecules
• are endergonic	• are exergonic
• are biosynthetic	• are degradative
example: photosynthesis	example: cellular respiration

Metabolic pathways

Almost all metabolic reactions in organisms are catalysed by enzymes. Many of these reactions occur in specific sequences and are called metabolic or biochemical pathways. A very simple generalized metabolic pathway might look like this: substrate A → substrate B → final product. Each arrow represents a specific enzyme that causes one substrate to be changed to another until the final product of the pathway is formed.

Some metabolic pathways consist of cycles of reactions instead of chains of reactions. Others involve both cycles and chains of reactions. Cell respiration and photosynthesis are mentioned in Chapter 3 (pages 69 and 74) and are complex pathways with chains and cycles of reactions. Metabolic pathways are usually carried out in designated compartments of the cell where the necessary enzymes are clustered and isolated. The enzymes required to catalyse every reaction in these pathways are determined by the cell's genetic makeup.

Induced-fit model of enzyme action

Enzyme–substrate specificity was discussed in Chapter 3 (page 66). Enzyme specificity is made possible by the enzyme structure. Enzymes are very complex protein molecules with high molecular weights. The higher levels of protein structure allow enzymes to form unique areas such as the active site. The active site is the region on the enzyme that binds to a particular substrate or substrates. This binding results in the reaction occurring must faster than would be expected without the enzyme.

Enzymes are globular proteins with at least the tertiary level of organization.

In the 1890s, Emil Fischer proposed the lock-and-key model of enzyme action. He suggested that substrate molecules fit like a key into a rigid section of the enzyme 'lock'. This model provided a good explanation for the specificity of enzyme action at the time. However, as knowledge about enzyme action has increased, Fischer's model has been modified.

Scientific truths are often pragmatic. They are accepted because they predict why some process works. Fischer's lock-and-key model of enzyme action represents this pragmatism. The model was first presented in the 1890s. It was not until 1958 that Daniel Koshland used a larger body of knowledge to present a new model of enzyme action, now known as the induced-fit model. The new model represents a more accurate explanation of enzyme action.

It is now obvious that many enzymes undergo significant changes in their conformation when substrates combine with their active sites. The accepted new model for enzyme action is called the induced-fit model. A good way to envision this model of enzyme action is to think of a hand and glove, the hand being the substrate and the glove being the enzyme. The glove looks somewhat like the hand. However, when the hand actually is placed in the glove there is an interaction that results in a conformational change of the glove, thus providing an induced fit.

The conformational changes and induced fit are due to changes in the R-groups of the amino acids at the active site of the enzyme as they interact with the substrate or substrates.

Mechanism of enzyme action

1 The surface of the substrate contacts the active site of the enzyme.
2 The enzyme changes shape to accommodate the substrate.
3 A temporary complex called the enzyme–substrate complex forms.
4 Activation energy is lowered and the substrate is altered by the rearrangement of existing atoms.
5 The transformed substrate – the product – is released from the active site.
6 The unchanged enzyme is then free to combine with other substrate molecules.

Discover more about enzymes. Visit heinemann.co.uk/hotlinks, enter the express code 4242P and click on Weblinks 14.2a and 14.2b.

Enzyme action can be summarized by the following equation:

$$E + S \leftrightarrow ES \leftrightarrow E + P$$

where E is the enzyme, S is the substrate, ES is the enzyme–substrate complex, and P is the product.

Activation energy

When talking about enzyme action, we always refer to activation energy (AE). Activation energy is best understood as the energy necessary to destabilize the existing chemical bonds in the substrate of an enzyme–substrate catalysed reaction. Enzymes work by lowering the activation energy required (see Figure 14.5). That means they cause chemical reactions to occur faster because they reduce the amount of energy needed to bring about a chemical reaction.
It is important to note that even though enzymes lower activation energy of a particular reaction, they do not alter the proportion of reactants to products.

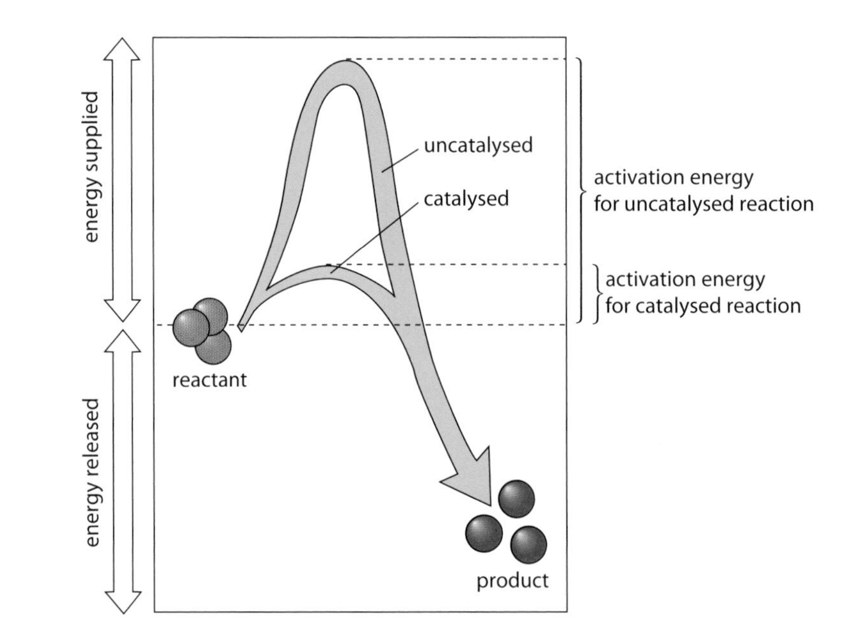

Figure 14.5 Enzymes accelerate exothermic reactions by lowering the activation energy required. The activation energy is needed to destabilize the chemical bonds in the reactant. The upper curve shows the activation energy when no enzyme is involved. The lower curve shows the activation energy required when an enzyme is present to catalyse the reaction.

Inhibition

The effects of pH, temperature, and substrate concentration on the action of enzymes are discussed in Chapter 3 (pages 67 and 68). Here, we discuss the effect of certain types of molecule on enzyme active sites. If a molecule affects the active site in some way, the activity of the enzyme may be altered.

Competitive inhibition

In competitive inhibition, a molecule called a competitive inhibitor, competes directly for the active site of an enzyme (see Figure 14.6). The result is that the substrate then has fewer encounters with the active site and the chemical reaction rate is decreased. The competitive inhibitor must have a structure similar to the substrate to be able to function in this way. An example is the use of sulfanilamide (a sulfa drug) to kill the bacteria during an infection. Folic acid is essential as a coenzyme to bacteria. It is produced in bacterial cells by enzyme action on para-aminobenzoic acid (PABA). The sulfanilamide competes with the PABA and blocks the enzyme. Because human cells do not use PABA to produce folic acid, they are unaffected by the drug.

Competitive inhibition may be reversible or irreversible. Reversible competitive inhibition may be overcome by increasing the substrate concentration. By doing this, there are more substrate molecules to bind with the active sites as they become available, and the chemical reaction may proceed more rapidly.

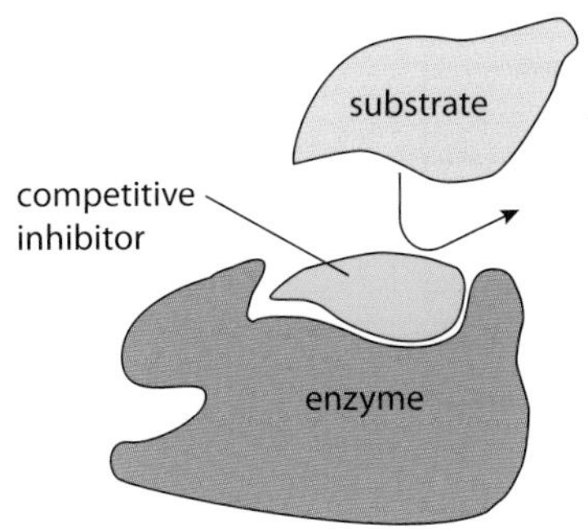

Figure 14.6 A competitive inhibitor blocks the active site of an enzyme so the substrate cannot bind to it.

Non-competitive inhibition

Non-competitive inhibition involves an inhibitor that does not compete for the enzyme's active site. In this case, the inhibitor interacts with another site on the enzyme (see Figure 14.7). Non-competitive inhibition is also called allosteric inhibition, and the site the inhibitor binds to is called the allosteric site. Binding at the allosteric site causes a change in the shape of the enzyme's active site, making it non-functional. Examples of non-competitive inhibition include metallic ions, such as mercury, binding to the sulfur groups of component amino acids of many enzymes. This results in shape changes of the protein which causes inhibition of the enzyme.

Again, this type of inhibition may be reversible or irreversible. There are also examples of allosteric interaction activating an enzyme rather than inhibiting it.

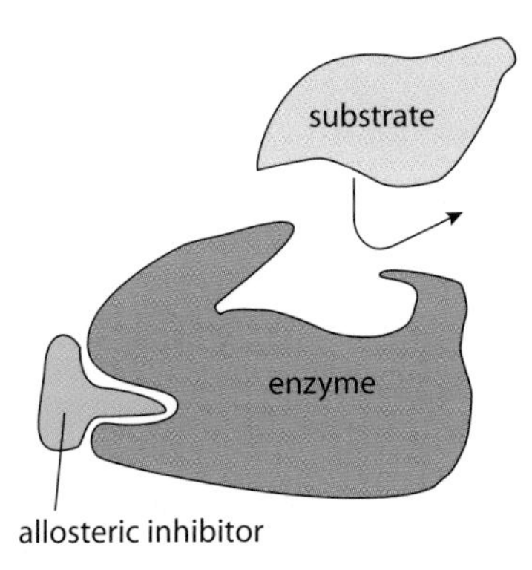

Figure 14.7 An allosteric (non-competitive) inhibitor combines with the allosteric site of an enzyme causing the active site to change shape so the substrate cannot bind to it.

End-product inhibition

End-product inhibition prevents the cell from wasting chemical resources and energy by making more of a substance than it needs. Many metabolic reactions occur in an assembly-line type of process so that a specific end-product can be achieved. Each step of the assembly line is catalysed by a specific enzyme (see Figure 14.8A). When the end-product is present in a sufficient quantity, the assembly line is shut down. This is usually done by inhibiting the action of the enzyme in the first step of the pathway (see Figure 14.8B). As the existing end-product is used up by the cell, the first enzyme is reactivated. The enzyme that is inhibited and reactivated is an allosteric enzyme. When in higher concentrations, the end-product binds with the allosteric site of the first enzyme, thus bringing about inhibition. Lower concentrations of the end-product result in fewer bindings with the allosteric site of the first enzyme and, therefore, activation of the enzyme.

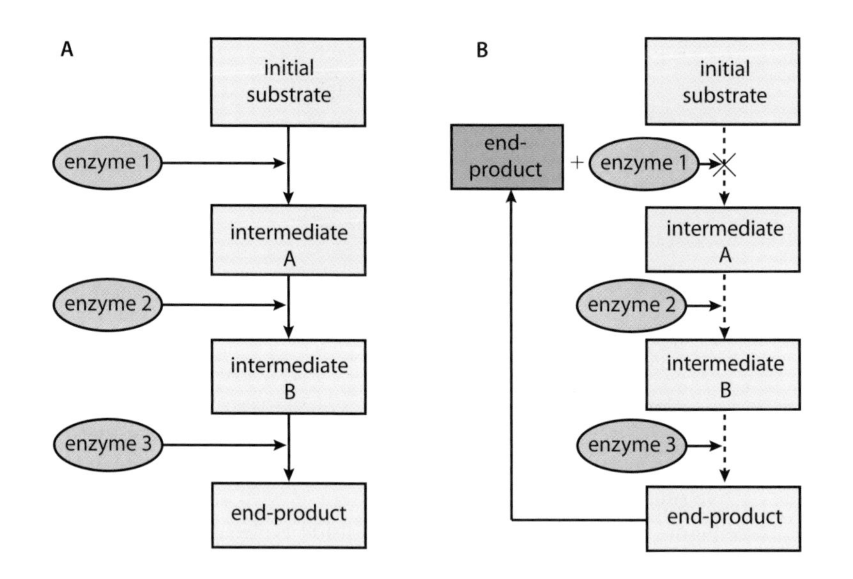

Figure 14.8 A short pathway of metabolic reactions with a specific end-product which, when in sufficient quantity, causes end-product inhibition. This is also a form of negative feedback. An example is found in control of glycolysis (see page 393).

The bacterium *E. coli* uses a metabolic pathway to produce the amino acid isoleucine from threonine. It is a 5-step process. If isoleucine is added to the growth medium of *E. coli*, it inhibits the first enzyme in the pathway and isoleucine is not synthesized. This situation continues until the isoleucine is used up.

The inhibition of the first enzyme in the pathway prevents the build-up of intermediates in the cell. This is a form of negative feedback.

Exercises

4 Explain why enzymes only work with specific substrates, and how they increase reaction rates.

5 What determines whether an enzyme is competitively or non-competitively inhibited?

6 How is the induced-fit model of enzyme action different from the lock-and-key model?

C.3 Cell respiration

The material in this section is also found in Chapter 8, section 8.1

Assessment statements

C.3.1 State that oxidation involves the loss of electrons from an element, whereas reduction involves a gain of electrons; and that oxidation frequently involves gaining oxygen or losing hydrogen, whereas reduction frequently involves losing oxygen or gaining hydrogen.

C.3.2 Outline the process of glycolysis, including phosphorylation, lysis, oxidation, and ATP formation.

C.3.3 Draw and label a diagram showing the structure of a mitochondrion as seen in electron micrographs.

C.3.4 Explain aerobic respiration, including the link reaction, the Krebs cycle, the role of NADH + H^+, the electron transport chain and the role of oxygen.

C.3.5 Explain oxidative phosphorylation in terms of chemiosmosis.

C.3.6 Explain the relationship between the structure of the mitochondrion and its function.

C.3.7 Analyse data relating to respiration.

Oxidation and reduction

In Chapter 3, the general processes of respiration and photosynthesis are discussed. In this chapter, we consider these aspects of cellular metabolism in detail. It is important to recall that metabolism is the sum of all the chemical reactions carried out by an organism. These reactions involve;

- catabolic pathways;
- anabolic pathways.

Catabolic pathways result in the breakdown of complex molecules to smaller molecules. Conversely, anabolic pathways result in the synthesis of more complex molecules from simpler ones. Cellular respiration is an example of a catabolic pathway. Photosynthesis is an example of an anabolic pathway. To understand these complex pathways, it is essential to understand two general types of chemical reaction: oxidation and reduction.

• **Examiner's hint:** If you are asked in an exam to compare oxidation and reduction, a table such as this one is an excellent way to structure the answer.

Oxidation and reduction can compared using a table like this.

Oxidation	Reduction
loss of electrons	gain of electrons
gain of oxygen	loss of oxygen
loss of hydrogen	gain of hydrogen
results in many C—O bonds	results in many C—H bonds
results in a compound with lower potential energy	results in a compound with higher potential energy

A useful way to remember the general meaning of oxidation and reduction is to think of the words OIL RIG:

- OIL = Oxidation Is Loss (of electrons);
- RIG = Reduction Is Gain (of electrons).

These two reactions occur together during chemical reactions. Think of it in this way: one compound's or element's loss is another compound's or element's gain. This is shown by the following equation:

$$C_6H_{12}O_6 + 6O_2 \rightarrow 6CO_2 + 6H_2O + \text{energy}$$

In this equation, glucose is oxidized because electrons are transferred from it to oxygen. The protons follow the electrons to produce water. The oxygen atoms that occur in the oxygen molecules on the reactant side of the equation are reduced. Because of this reaction, there is a large drop in the potential energy of the compounds on the product side of the equation.

Because oxidation and reduction always occur together, these chemical reactions are referred to as redox reactions. When redox reactions take place, the reduced form of a molecule always has more potential energy that the oxidized form of the molecule. Redox reactions play a key role in the flow of energy through living systems. This is because the electrons that are flowing from one molecule to the next are carrying energy with them. In a similar sort of way, the catabolic and anabolic pathways mentioned earlier are also closely associated with one another. You will see this association as you work through this chapter.

An overview of respiration

Chapter 3 provided an introduction to the process of cellular respiration. Three aspects of cellular respiration were discussed:

- glycolysis;
- anaerobic respiration;
- aerobic respiration.

As you will recall, glycolysis occurs in the cytoplasm of the cell, produces small amounts of ATP and ends with the product known as pyruvate. If no oxygen is available, the pyruvate enters into anaerobic respiration. This occurs in the cytoplasm and it does not result in any further production of ATP. The products of anaerobic respiration are lactate or ethanol and carbon dioxide. If oxygen is available, the pyruvate enters aerobic respiration in the mitochondrion of the cell. This process results in the production of a large number of ATPs, carbon dioxide and water.

Lactate is the material that causes muscle soreness after strenuous exercise.

In this section, we discuss cellular respiration which involves glycolysis and the three stages of aerobic respiration:

- the link reaction;
- the Krebs cycle;
- oxidative phosphorylation.

Glycolysis

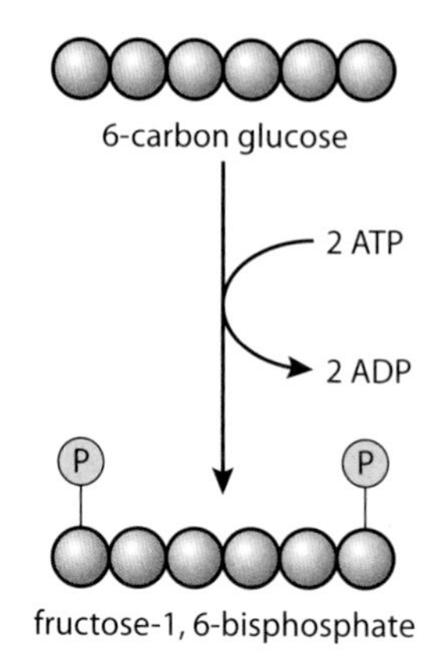

Figure 14.9 First stage of glycolysis; the circles represent carbon atoms.

The word glycolysis means 'sugar splitting' and is thought to have been one of the first biochemical pathways to evolve. It uses no oxygen and occurs in the cytosol of the cell. There are no required organelles. The sugar splitting proceeds efficiently in aerobic and anaerobic environments. Glycolysis occurs in both prokaryotic and eukaryotic cells. A hexose, usually glucose, is split in the process. This splitting actually involves many steps but we can explain it effectively in three stages.

1. Two molecules of ATP are used to begin glycolysis. In the first reaction, the phosphates from the ATPs phosphorylate glucose to form fructose-1, 6-bisphosphate (see Figure 14.9). This process involves phosphorylation.
2. The 6-carbon phosphorylated fructose is split into two 3-carbon sugars called glyceraldehyde-3-phosphate (G3P) (see Figure 14.10). This process involves lysis.

Figure 14.10 Second stage of glycolysis.

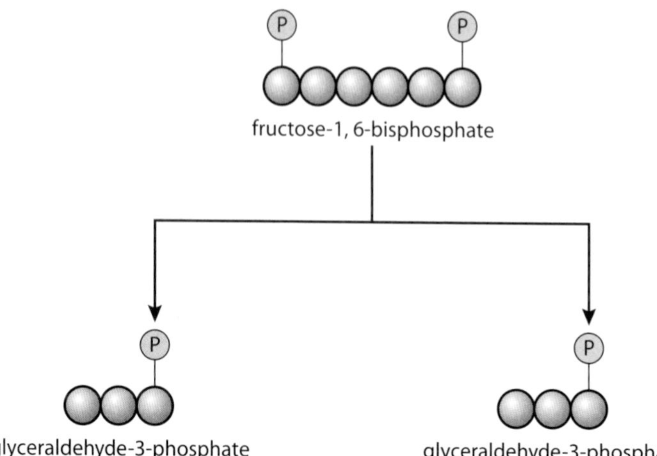

3 Once the two G3P molecules are formed, they enter an oxidation phase involving ATP formation and production of the reduced coenzyme NAD (see Figure 14.11). Each G3P or triose phosphate molecule undergoes oxidation to form a reduced molecule of NAD^+, which is NADH. As NADH is being formed, released energy is used to add an inorganic phosphate to the remaining 3-carbon compound. This results in a compound with two phosphate groups. Enzymes then remove the phosphate groups so they can be added to ADP to produce ATP. The end result is the formation of four molecules of ATP, two molecules of NADH, and two molecules of pyruvate. Pyruvate is the ionized form of pyruvic acid.

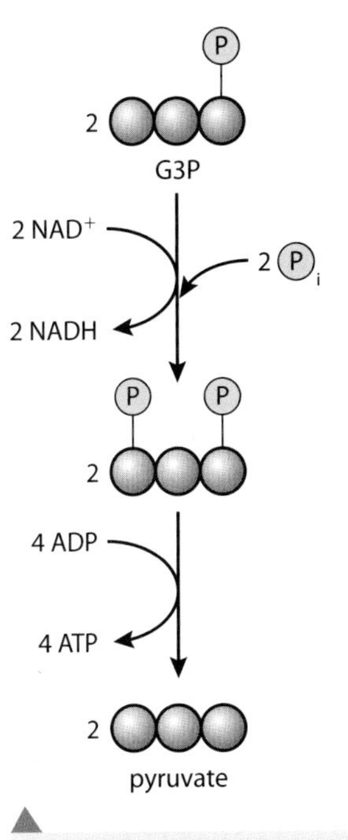

Figure 14.11 Third stage of glycolysis.

Summary of glycolysis

- Two ATPs are used to start the process.
- A total of four ATPs are produced – a net gain of two ATPs.
- Two molecules of NADH are produced.
- Involves substrate-level phosphorylation, lysis, oxidation and ATP formation.
- Occurs in the cytoplasm of the cell.
- This metabolic pathway is controlled by enzymes. Whenever ATP levels in the cell are high, feedback inhibition will block the first enzyme of the pathway (see Figure 14.8, page 390). This will slow or stop the process.
- Two pyruvate molecules are present at the end of the pathway.

This way of producing ATP is called substrate-level phosphorylation since the phosphate group is transferred directly to ADP from the original phosphate-bearing molecule.

Mitochondria

It is inside the mitochondria and in the presence of oxygen that the remainder of cellular respiration occurs.

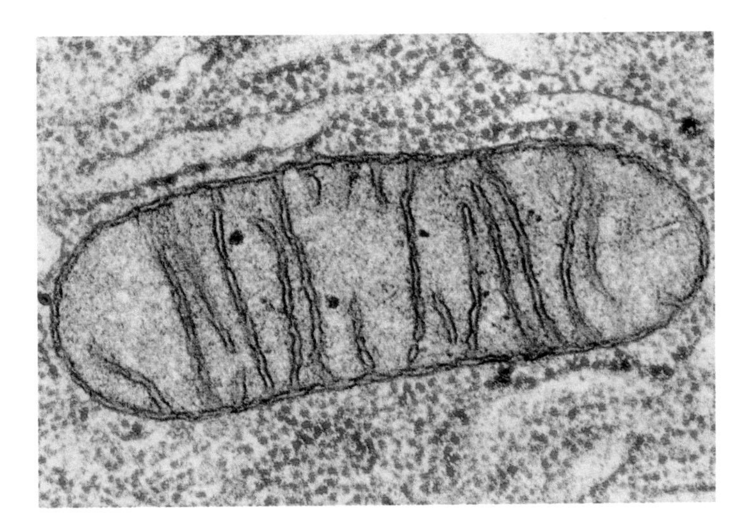

This photomicrograph of a mitochondrion shows the internal structure. The matrix (blue) is permeated by the membranes cristae (pink).

We discussed the structure of the mitochondrion in Chapter 2 (page 23). You might like to refresh your memory of this because as we discuss aerobic respiration, which occurs in the mitochondrion, we will refer to parts of this organelle.

Once pyruvate is obtained, the next pathway is determined by the presence of oxygen. If oxygen is present, pyruvate enters the mitochondria and aerobic respiration occurs. If oxygen is not present, anaerobic respiration occurs in the cytoplasm. In this case, pyruvate is converted to lactate in animals, and ethanol and carbon dioxide in plants.

The link reaction and the Krebs cycle

Once glycolysis has occurred and there is oxygen present, pyruvate enters the matrix of the mitochondrion via active transport. Inside, pyruvate is decarboxylated to form the 2-carbon acetyl group. This is the link reaction (see

Decarboxylation is the removal of a carbon atom.

Figure 14.12). The removed carbon is released as carbon dioxide, a waste gas. The acetyl group is then oxidized with the formation of reduced NAD^+. Finally, the acetyl group combines with coenzyme A (CoA) to form acetyl CoA.

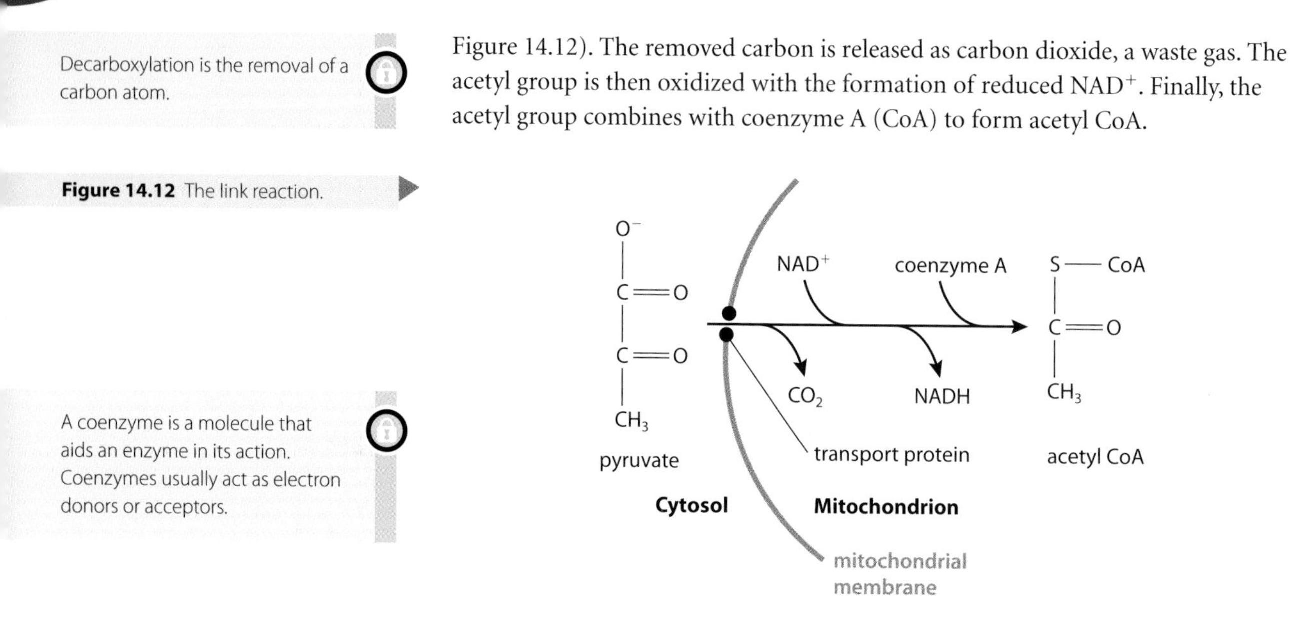

Figure 14.12 The link reaction.

A coenzyme is a molecule that aids an enzyme in its action. Coenzymes usually act as electron donors or acceptors.

The link reaction is controlled by a system of enzymes. The greatest significance of this reaction is that it produces acetyl CoA. Acetyl CoA may enter the Krebs cycle to continue the aerobic respiration process.

So far in this discussion, the respiratory substrate has been a hexose. However, in reality, acetyl CoA can be produced from most carbohydrates and fats. Acetyl CoA can be synthesized into a lipid for storage purposes. This occurs when ATP levels in the cell are high.

If cellular ATP levels are low, the acetyl CoA enters the Krebs cycle. This cycle is also called the tricarboxylic acid cycle. It occurs in the matrix of the mitochondrion and is referred to as a cycle because it begins and ends with the same substance. This is a characteristic of all cyclic pathways in metabolism. You do not need to remember the names of all the compounds formed in the Krebs cycle. However, it is important that you understand the overall process.

Let's consider the cycle as a series of steps.

1 Acetyl CoA from the link reaction combines with a 4-carbon compound called oxaloacetate. The result is a 6-carbon compound called citrate (see Figure 14.13).

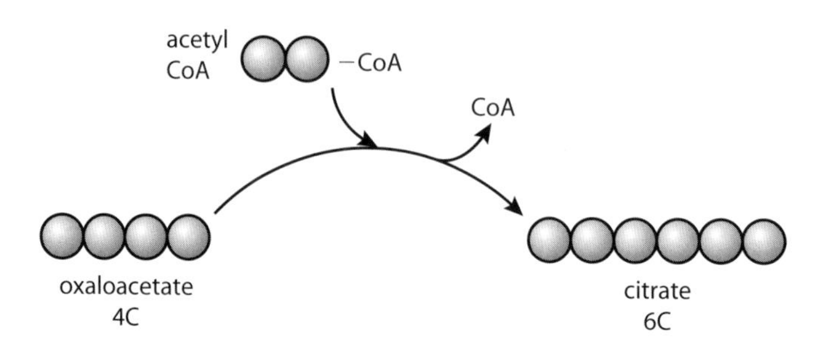

Figure 14.13 Acetyl CoA combines with oxaloacetate to form citrate.

2 Citrate (6-carbon compound) is oxidized to form a 5-carbon compound (see Figure 14.14). In this process, the carbon is released from the cell (after combining with oxygen) as carbon dioxide. While the 6-carbon compound is oxidized, NAD^+ is reduced to form NADH.

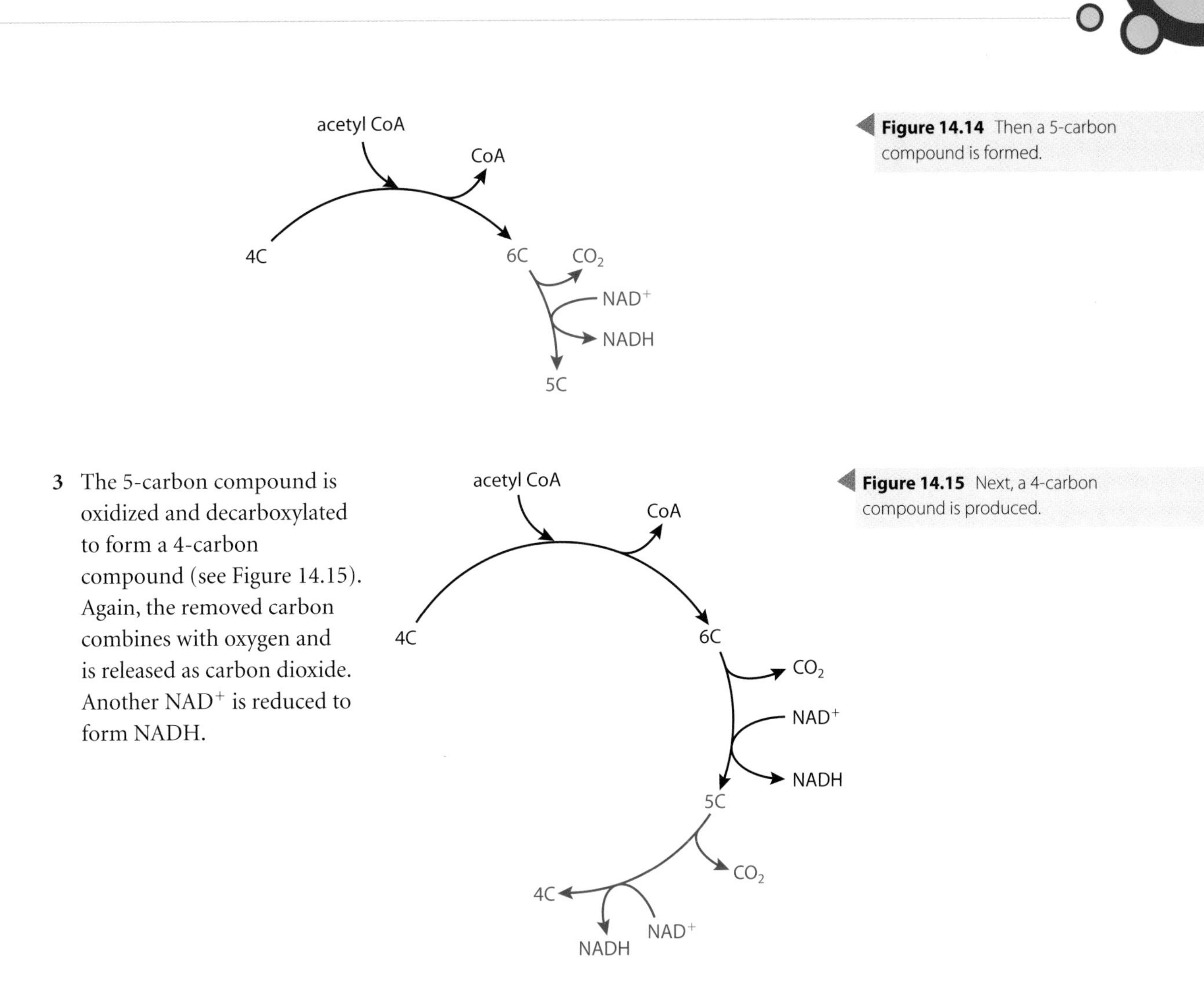

Figure 14.14 Then a 5-carbon compound is formed.

Figure 14.15 Next, a 4-carbon compound is produced.

3 The 5-carbon compound is oxidized and decarboxylated to form a 4-carbon compound (see Figure 14.15). Again, the removed carbon combines with oxygen and is released as carbon dioxide. Another NAD^+ is reduced to form NADH.

4 The 4-carbon compound undergoes various changes resulting in several products (see Figure 14.16). One product is another NADH. The coenzyme FAD is reduced to form $FADH_2$. There is also a reduction of an ADP to form ATP. The 4-carbon compound is changed during these steps to re-form the starting compound of the cycle, oxaloacetate. The oxaloacetate may then begin the cycle again.

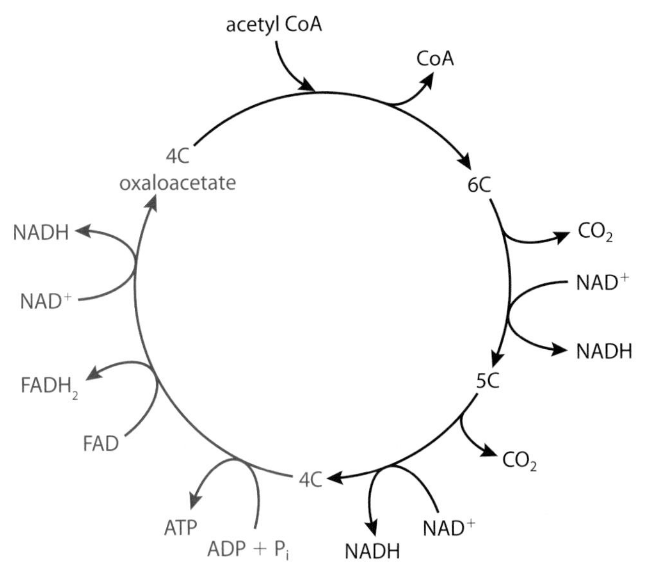

Figure 14.16 Finally, the 4-carbon compound is converted to oxaloacetate.

Two carbon dioxides are released for each glucose molecule during the link reaction. Four are released during the Krebs cycle. This accounts for all six carbon atoms that were present in the initial glucose molecule. Glucose is completely catabolized and its original energy is now carried by NADH and $FADH_2$ or is in ATP.

For an effective review of glycolysis and the Krebs cycle, visit heinemann.co.uk/hotlinks, enter the express code 4242P and click on Weblink 14.3.

It is important to remember that the Krebs cycle will run twice for each glucose molecule entering cellular respiration. This is because a glucose molecule forms two pyruvate molecules. Each pyruvate produces one acetyl CoA which enters the cycle. Look again at the complete Krebs cycle (Figure 8.8) and note the following products which result from the breakdown of one glucose molecule:

- two ATP molecules;
- six molecules of NADH (allow energy storage and transfer);
- two molecules of $FADH_2$;
- four molecules of carbon dioxide (released).

So far, only four ATPs have been gained: six are generated (four from glycolysis and two from the Krebs cycle) but two are used to start the process of glycolysis. Each of these ATPs has been produced by substrate-level phosphorylation.

Ultimately, the breakdown of each glucose molecule results in a net gain of 36 ATPs. Let's now consider the phase of cellular respiration where most of the ATPs are produced. In this phase, oxidative phosphorylation is the means by which the ATPs are produced.

Electron transport chain and chemiosmosis

The electron transport chain is where most of the ATPs from glucose catabolism are produced. It is the first stage of cellular respiration where oxygen is actually needed, and it occurs within the mitochondrion. However, unlike the Krebs cycle, which occurred in the matrix, the electron transport chain occurs on the inner mitochondrial membrane and on the membranes of the cristae.

Embedded in the involved membranes are molecules that are easily reduced and oxidized. These carriers of electrons (energy) are close together and pass the electrons from one to another due to an energy gradient. Each carrier molecule has a slightly different electronegativity and, therefore, a different attraction for electrons. Most of these carriers are proteins with haem groups and are referred to as cytochromes. One carrier is not a protein and is called coenzyme Q.

The haem group of the carriers is the part that is easily reduced and oxidized.

In this chain, electrons pass from one carrier to another because the receiving molecule has a higher electronegativity and, therefore, a stronger attraction for electrons (see Figure 14.17). In the process of electron transport, small amounts of energy are released. The sources of the electrons that move down the electron transport chain are the coenzymes NADH and $FADH_2$ from the previous stages of cellular respiration.

In Figure 14.17 it is clear that the electrons are stepping down in potential energy as they pass from one carrier to another. It is important to note that :

- $FADH_2$ enters the electron transport chain at a lower free energy level than NADH – thus, $FADH_2$ allows the production of 2ATPs while NADH allows the production of 3ATPs;
- at the very end of the chain, the de-energized electrons combine with available oxygen.

The kangaroo rat of a desert region of the United States obtains 90% of its daily water intake from water of metabolism. In contrast to this, a typical human obtains only 12% of its daily water intake by way of metabolism.

Oxygen is the final electron acceptor because it has a very high electronegativity and, therefore, a strong attraction for electrons. When the electrons combine with the oxygen, so do two hydrogen ions from the aqueous surroundings. The result is water. Because of the way this water is formed, it is referred to as water of metabolism.

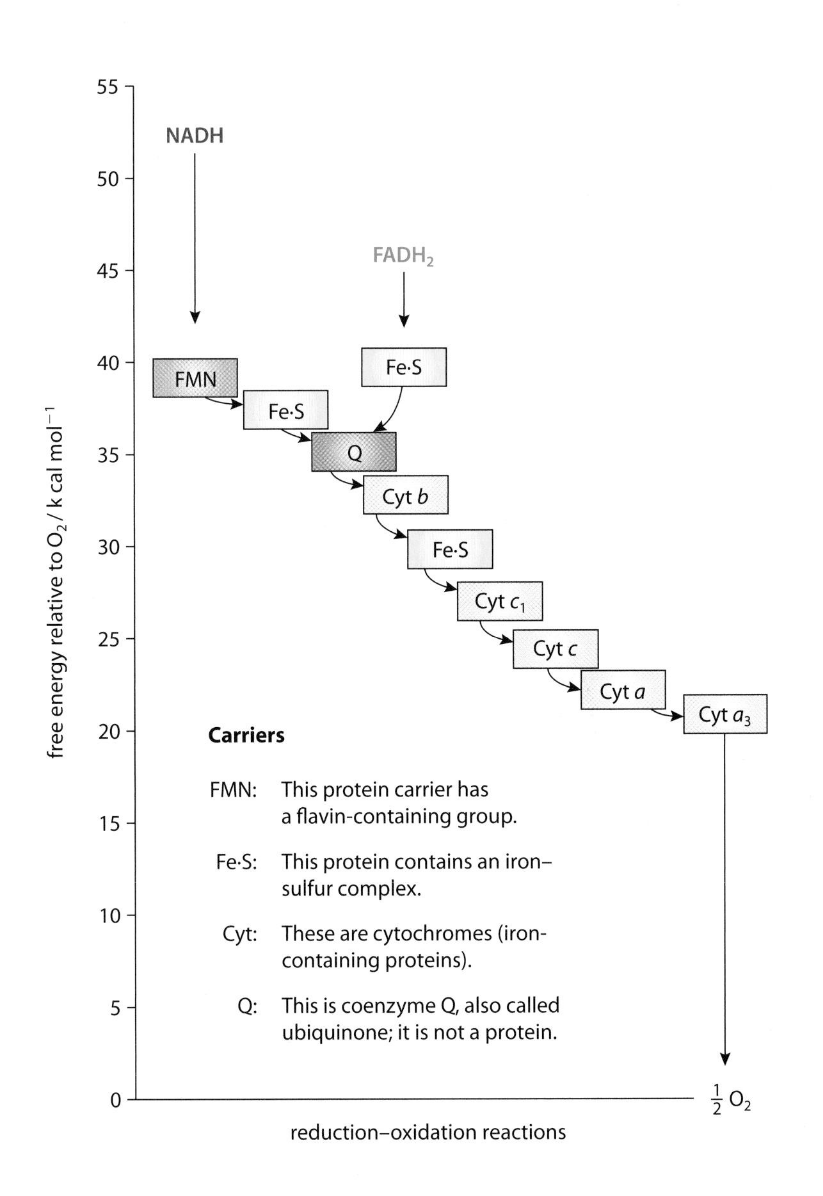

Figure 14.17 The oxidation–reduction reactions of the electron transport chain. It is not necessary for you to remember all of the names of the carriers.

It is also clear from Figure 14.17 that there are a fairly large number of electron carriers. Because of the larger number, the electronegativity difference between adjacent carriers is not so great. This means that lower amounts of energy are lost at each exchange. These lower amounts of energy are effectively harnessed by the cell to carry out phosphorylation. If the amount of energy lost at each exchange was high, much of it could not be used and damage might be done to the cell.

So energy is now available as a result of the electron transport chain. This is the energy that allows the addition of phosphate and energy to ADP to form ATP. The process by which this occurs is called chemiosmosis. Chemiosmosis involves the movement of protons (hydrogen ions) to provide energy so that phosphorylation can occur. Because this type of phosphorylation uses an electron transport chain, it is called oxidative phosphorylation. Substrate-level phosphorylation mentioned in the earlier phases of cellular respiration did not involve an electron transport chain.

No ATPs are produced directly by the electron transport chain. However, this chain is essential to chemiosmosis, which does produce the ATP.

Before we continue, it is essential to review the interior structure of the mitochondrion. In the process of cellular respiration, the structure of the mitochondrion is very closely linked to its function. The matrix is the area where

● **Examiner's hint:** Using any diagram or photomicrograph of a mitochondrion, annotate where the processes of respiration occur.

the Krebs cycle occurs. The cristae provide a large surface area for the electron transport chain to function. The membranes also provide a barrier allowing for proton accumulation on one side. Embedded in the membranes are the enzymes and other necessary compounds for the processes of the electron transport chain and chemiosmosis to occur.

The inner membranes of the mitochondrion have numerous copies of an enzyme called ATP synthase. This enzyme uses the energy of an ion gradient to allow the phosphorylation of ADP. The ion gradient is created by a hydrogen ion concentration difference that occurs across the cristae membranes. Figure 14.18 shows oxidative phosphorylation.

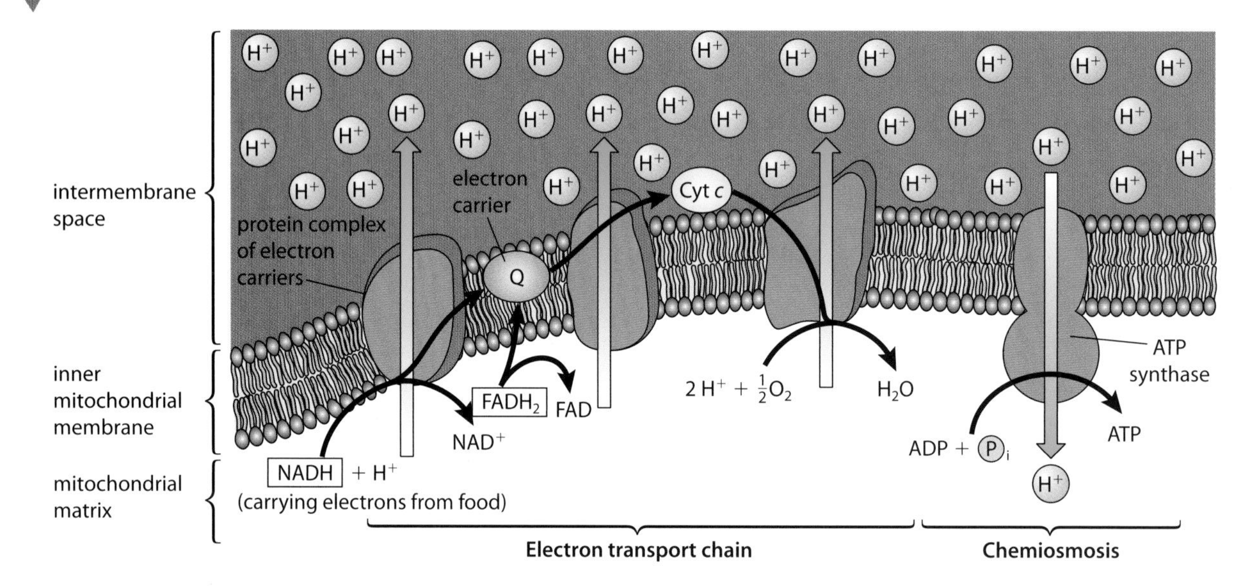

Figure 14.18 Oxidative phosphorylation occurs at the inner membranes of the mitochondria of a cell. The pumping actions of the carriers result in a high concentration of hydrogen ions in the intermembrane space. This accumulation allows movement of the hydrogen ions through the enzyme ATP synthase. The enzyme uses the energy from the hydrogen flow to couple phosphate with ADP to produce ATP.

In Figure 14.18, note the three labelled areas on the left: intermembrane space, inner mitochondrial membrane, and mitochondrial matrix. Also, note that hydrogen ions are being pumped out of the matrix into the intermembrane space. The energy for this pumping action is provided by the electrons as they are de-energized moving through the electron transport chain. This creates the different hydrogen ion concentration on the two sides of the cristae membranes, mentioned above. With the higher hydrogen ion concentration in the intermembrane space, these ions begin to passively move through a channel in ATP synthase back into the mitochondrial matrix. As the hydrogen ions move through the ATP synthase channel, the enzyme harnesses the available energy thus allowing the phosphorylation of ADP.

Because of the hydrophobic region of the membrane, the hydrogen ions can only pass through the ATP synthase channel. Some poisons that affect metabolism act by establishing alternative pathways through the membrane thus preventing ATP production.

Summary of ATP production in cellular respiration

We have now considered the complete catabolism of one molecule of glucose. The raw materials are glucose and oxygen. Many enzymes, carriers and other molecules are involved in the process. The products are carbon dioxide, water and ATP. The ATPs are essential as they provide the energy by which life is maintained. We can describe the energy flow in the general process as:

glucose → NADH or $FADH_2$ → electron transport chain → chemiosmosis → ATP

For an animation of the electron transport chain, visit heinemann.co.uk/hotlinks, enter the express code 4242P and click on Weblink 14.4.

To account for the production of ATP in cellular respiration, let's look at the three main processes: glycolysis, the Krebs cycle, and the electron transport chain in a table.

Process	ATP used	ATP produced	Net ATP gain
glycolysis	2	4	2
Krebs cycle	0	2	2
electron transport chain and chemiosmosis	0	32	32
total	2	38	36

Theoretically 36 ATPs are produced by cellular respiration but in reality the number is closer to 30. This is thought to be due to some hydrogen ions moving back to the matrix without going through the ATP synthase channel. Also, some of the energy from hydrogen ion movement is used to transport pyruvate into the mitochondria. The 30 ATPs generated by cellular respiration account for approximately 30% of the energy present in the chemical bonds of glucose. The remainder of the energy is lost from the cell as heat.

Final look at respiration and the mitochondrion

Cellular respiration is the process by which ATP is provided to the organism so that life may continue. It is a very complex series of chemical reactions, most of which occur in the mitochondrion. Let's end our discussion of this essential-to-life process by looking at a table showing the parts of the mitochondrion and how those parts allow cellular respiration.

Outer mitochondrial membrane	separates the contents of the mitochondrion from the rest of the cell
Matrix	internal cytosol-like area that contains the enzymes for the link reaction and the Krebs cycle
Cristae	tubular regions surrounded by membranes increasing surface area for oxidative phosphorylation
Inner mitochondrial membrane	contains the carriers for the electron transport chain and ATP synthase for chemiosmosis
Space between inner and outer membranes	reservoir for hydrogen ions (protons), the high concentration of hydrogen ions is necessary for chemiosmosis

The overall equation for cellular respiration is:

$$C_6H_{12}O_6 + 6O_2 \rightarrow 6CO_2 + 6H_2O + \text{energy (heat or ATP)}$$

All organisms must have the ability to produce ATP for energy and, therefore, all organisms carry out respiration.

Exercises

7 Using ideal ATP production numbers, how many ATPs would an individual generate if they consumed only pyruvate and carried one pyruvate molecule through cellular respiration?

8 Striated muscles usually have more mitochondria than other cell types. Why is this important?

9 If both NAD and FAD are reduced, which would allow the greater production of ATPs via the electron transport chain and chemiosmosis?

10 If an individual took a chemical that increased the ability of hydrogen ions to move through the phospholipid bilayer of the mitochondrial membranes, what would the effect be on ATP production?

11 If ATP synthase was not present in the cristae of a mitochondrion, what would be the effect?

C.4 Photosynthesis

The material in this section is also found in Chapter 8, section 8.2.

Assessment statements

- C.4.1 Draw and label a diagram showing the structure of a chloroplast as seen in electron micrographs.
- C.4.2 State that photosynthesis consists of light-dependent and light-independent reactions.
- C.4.3 Explain the light-dependent reactions.
- C.4.4 Explain photophosphorylation in terms of chemiosmosis.
- C.4.5 Explain the light-independent reactions.
- C.4.6 Explain the relationship between the structure of the chloroplast and its functions.
- C.4.7 Explain the relationship between the action spectrum and the absorption spectrum of photosynthetic pigments in green plants.
- C.4.8 Explain the concept of limiting factors in photosynthesis, refering to light intensity, temperature and concentration of carbon dioxide.
- C.4.9 Analyse data relating to photosynthesis.

There are three types of plastid that occur in plant cells:
- chloroplasts are green and involved in photosynthesis;
- leucoplasts are white or 'clear' and function as energy store-houses;
- chromoplasts are brightly coloured and synthesize and store large amounts of orange, red, or yellow pigments.

All the plastids develop from a common proplastid.

The chloroplast

Some people refer to the chloroplast as a photosynthetic machine. They are not wrong. Unlike respiration, where some of the steps occur outside the mitochondrion, all of the photosynthetic process occurs within the chloroplast. Chloroplasts, along with mitochondria, represent possible evidence for the theory of endosymbiosis. Both organelles have an extra outer membrane (indicating a need for protection in a potentially hostile environment), their own DNA, and they are very near in size to a typical prokaryotic cell (see Figure 14.19).

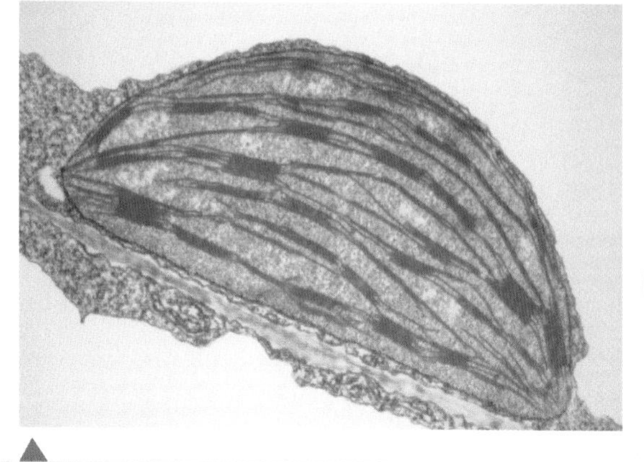

Figure 14.19 This false-colour TEM and drawing show the structure of a chloroplast. Can you find as many parts in the EM as are labelled in the drawing?

The structure of the chloroplast was discussed in Chapter 2 (page 25). You may want to return to that chapter for a brief refresher. Chloroplasts occur mostly within the cells of the photosynthetic factory of the plant, the leaves. However, some plants have chloroplasts in the cells of other organs.

The overall process of photosynthesis

During the discussion on respiration, we considered the means by which the cell breaks down chemical bonds in glucose to produce ATP. In this section, the discussion centres on the establishment of chemical bonds to produce organic compounds. Using light energy, the raw materials of photosynthesis are carbon dioxide and water. Many enzymes are involved to allow the formation of products that include glucose, more water, and oxygen. The overall equation is:

$$6CO_2 + 12H_2O \xrightarrow{\text{light}} C_6H_{12}O_6 + 6H_2O + 6O_2$$

Water occurs on both sides because 12 molecules are consumed and 6 molecules are produced. Clearly, photosynthesis is essentially the reverse of respiration. Whereas respiration is, in general, a catabolic process, photosynthesis is, in general, an anabolic process. Photosynthesis occurs in organisms referred to as autotrophs. These organisms make their own food. Non-photosynthetic and non-chemosynthetic organisms are referred to as heterotrophs. They must obtain their food (which is necessary for energy) from other organisms.

Photosynthesis involves two major stages:

- the light-dependent reaction;
- the light-independent reaction.

The light-dependent reaction

This reaction occurs in the thylakoids or grana of the chloroplast. A stack of thylakoids is called a granum (plural, grana). Light supplies the energy for this reaction to occur. The ultimate source of light is the Sun. Even though plants may survive quite well when they receive light from sources other than the Sun, most plants on our planet rely on the Sun for the energy necessary to drive photosynthesis.

Light energy behaves as if it exists in discrete packets called photons. Shorter wavelengths of light have greater energy within their photons than longer wavelengths.

To absorb light, plants have special molecules called pigments. There are several different pigments in plants and each effectively absorbs photons of light at different wavelengths. The two major groups are the chlorophylls and the carotenoids. These pigments are organized on the membranes of the thylakoids. The regions of organization are called photosystems and include:

- chlorophyll *a* molecules;
- accessory pigments;
- a protein matrix.

The reaction centre is the portion of the photosystem that contains:

- a pair of chlorophyll molecules;
- a matrix of protein;
- a primary electron acceptor.

Bacteria that carry out photosynthesis have only one type of photosystem. However, modern-day plants have two types of photosystem. Each absorbs light most efficiently at a different wavelength. Photosystem I is most efficient at 700 nanometres (nm). Photosystem II is most efficient at 680 nm. These two photosystems work together to bring about a non-cyclic electron transfer. Figure 14.20 shows the overall light-dependent reaction of photosynthesis involving non-cyclic photophosphorylation or non-cyclic electron flow.

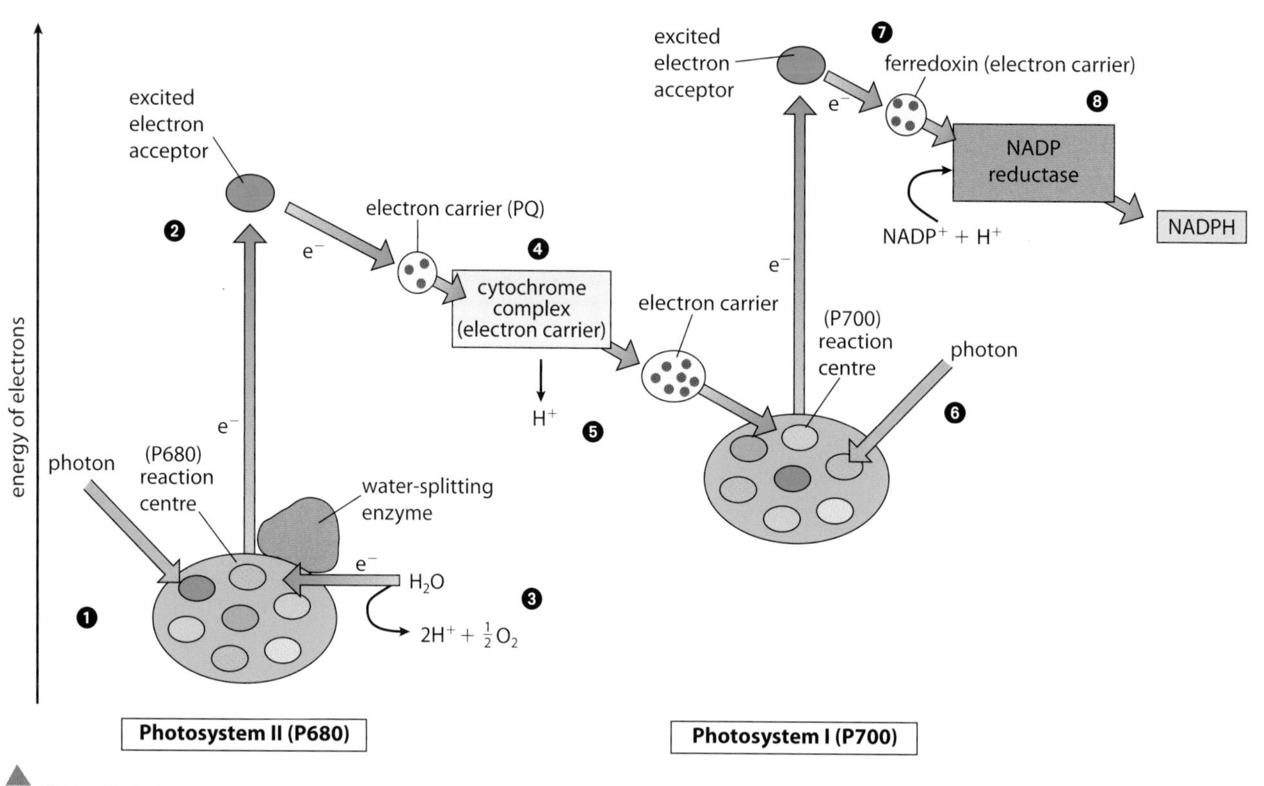

Figure 14.20 This is the light-dependent reaction.

Besides the non-cyclic electron pathway used to produce ATP by photophosphorylation, there is an alternative pathway involving a cyclic pathway. This cyclic pathway is discussed in the final section of this chapter.

These numbered descriptions refer to the numbered steps in Figure 14.20.

1 A photon of light is absorbed by a pigment in photosystem II and is transferred to other pigment molecules until it reaches one of the chlorophyll *a* (P680) molecules in the reaction centre. The photon energy excites one of the chlorophyll *a* electrons to a higher energy state.

2 This electron is captured by the primary acceptor of the reaction centre.

3 Water is split by an enzyme to produce electrons, hydrogen ions, and an oxygen atom. This process is driven by the energy from light and is called photolysis. The electrons are supplied one by one to the chlorophyll *a* molecules of the reaction centre.

4 The excited electrons pass from the primary acceptor down an electron transport chain losing energy at each exchange. The first of the three carriers shown is plastoquinone (PQ). The middle carrier is a cytochrome complex.

5 The energy lost from the electrons moving down the electron transport chain drives chemiosmosis (similar to that in respiration) to bring about phosphorylation of ADP to produce ATP.

6 A photon of light is absorbed by a pigment in photosystem I. This energy is transferred through several accessory pigments until received by a chlorophyll *a* (P700) molecule. This results in an electron with a higher energy state being transferred to the primary electron acceptor. The de-energized electron from photosystem II fills the void left by the newly energized electron.

7 The electron with the higher energy state is passed down a second electron transport chain that involves the carrier ferredoxin.

8 The enzyme NADP reductase catalyses the transfer of the electron from ferredoxin to the energy carrier $NADP^+$. Two electrons are required to fully reduce $NADP^+$ to NADPH.

NADPH and ATP are the final products of the light-dependent reaction. They supply chemical energy for the light-independent reaction to occur. The explanation above also shows the origin of the oxygen released by photosynthesizing plants (step 3). However, you need to know more detail about the production of ATP.

To view an effective website for understanding the light-dependent phase, visit heinemann.co.uk/hotlinks, enter the express code 4242P and click on Weblink 14.5.

ATP production in photosynthesis is very similar to ATP production in respiration. Chemiosmosis allows the process of phosphorylation of ADP. In this case, the energy to drive chemiosmosis comes from light. As a result, we refer to the production of ATP in photosynthesis as photophosphorylation.

A comparison of chemiosmosis in respiration and photosynthesis is shown in this table.

Respiration chemiosmosis	Photosynthesis chemiosmosis
1 Involves an electron transport chain embedded in the membranes of the cristae	1 Involves an electron transport chain embedded in the membranes of the thylakoids.
2 Energy is released when electrons are exchanged from one carrier to another.	2 Energy is released when electrons are exchanged from one carrier to another.
3 Released energy is used to actively pump hydrogen ions into the intermembrane space.	3 Released energy is used to actively pump hydrogen ions into the thylakoid space.
4 Hydrogen ions come from the matrix.	4 Hydrogen ions come from the stroma.
5 Hydrogen ions diffuse back into the matrix through the channels of ATP synthase.	5 Hydrogen ions diffuse back into the stroma through the channels of ATP synthase.
6 ATP synthase catalyses the oxidative phosphorylation of ADP to form ATP.	6 ATP synthase catalyses the photophosphorylation of ADP to form ATP.

In both cases, ATP synthase is embedded along with the carriers of the electron transport chain in the involved membranes.

In photosynthesis, the production of ATP occurs between photosystem II and photosystem I. Study Figure 14.21. Notice that the b_6-f complex, which is a cytochrome complex, pumps the hydrogen ions into the thylakoid space. This increases the concentration of these ions which then passively move through the ATP synthase channel providing the energy to phosphorylate ADP.

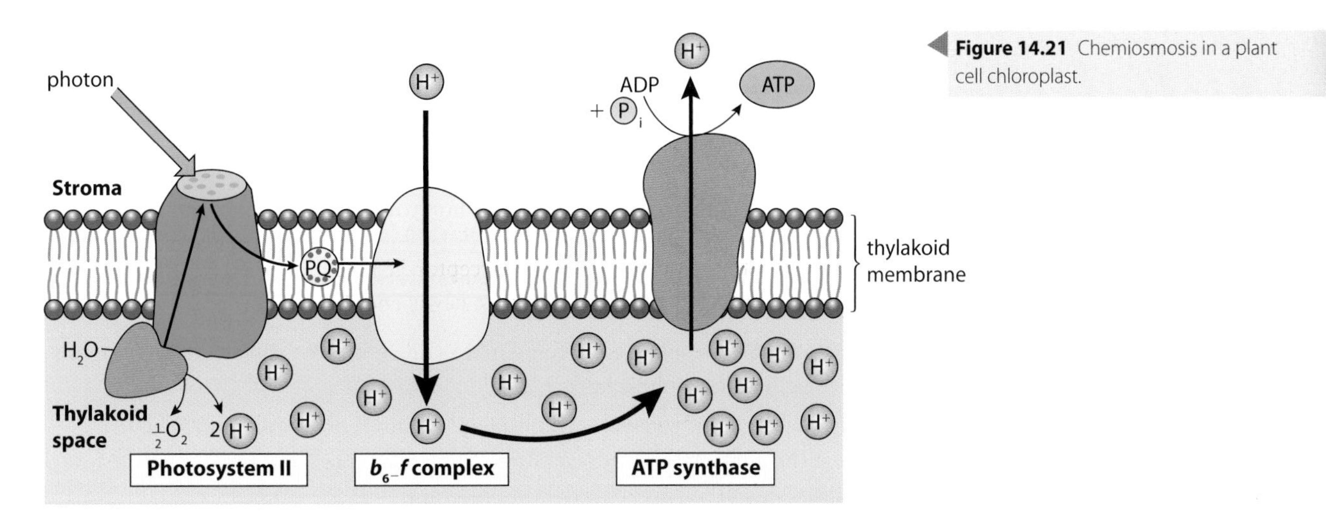

Figure 14.21 Chemiosmosis in a plant cell chloroplast.

The light-independent reaction

The light-independent reaction occurs within the stroma or cytosol-like region of the chloroplast.

The ATP and NADPH produced by the light-dependent reaction provide the energy and reducing power for the light-independent reaction to occur. Up to this point there has been no mention of carbohydrate production. Therefore, as we know glucose is a product of photosynthesis, the result of the light-independent reaction must be the production of glucose.

The light-independent reaction involves the Calvin cycle (see Figure 14.22), which occurs in the stroma of the chloroplast. Because it is a cycle, it begins and ends with the same substance. You should recall that a similar cyclic metabolic pathway occurred in respiration, the Krebs cycle.

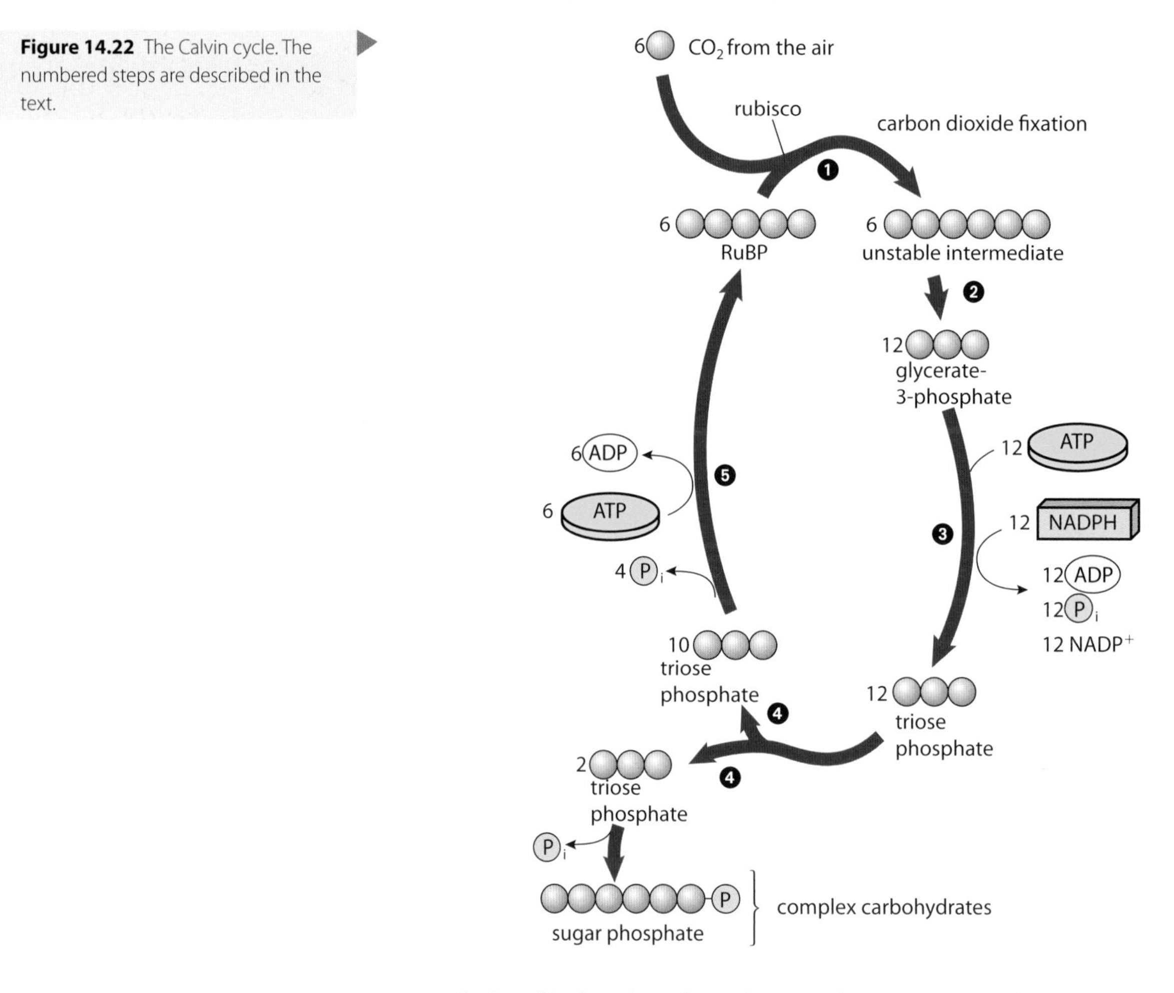

Figure 14.22 The Calvin cycle. The numbered steps are described in the text.

1 Ribulose bisphosphate (RuBP), a 5-carbon compound, binds to an incoming carbon dioxide molecule in a process called carbon fixation. This fixation is catalysed by an enzyme called RuBP carboxylase (rubisco). The result is an unstable 6-carbon compound.

2 The unstable 6-carbon compound breaks down into two 3-carbon compounds called glycerate-3-phosphate.

3 The molecules of glycerate-3-phosphate are acted on by ATP and NADPH from the light-dependent reaction to form two more compounds called triose phosphate (TP). This is a reduction reaction.

From 1945 to 1955, a team led by Melvin Calvin worked out the details of carbon fixation in a type of green algae. They used an elaborate protocol based on the 'lollipop apparatus.' How is the creation of a complex protocol such as this similar to the production of a work of art?

4 The molecules of TP may then go one of two directions. Some leave the cycle to become sugar phosphates that may become more complex carbohydrates. Most however, continue in the cycle to reproduce the originating compound of the cycle, RuBP.

5 In order to regain RuBP molecules from TP, the cycle uses ATP.

In Figure 14.22, spheres are used to represent the carbon atoms so they can be tracked through the cycle. The coefficients (numbers) in front of each compound involved show what it takes to produce one molecule of a 6-carbon sugar. It is clear that for every 12 TP molecules, the cycle produces one 6-carbon sugar and six molecules of the 5-carbon compound, RuBP. All the carbons are accounted for, and the law of conservation of mass is demonstrated. Also, it is important to note that 18 ATPs and 12 NADPH are necessary to produce 6 RuBP molecules and 1 molecule of a 6-carbon sugar.

TP is the pivotal compound in the Calvin cycle. It may be used to produce simple sugars such as glucose, disaccharides such as sucrose, or polysaccharides such as cellulose or starch. However, most of it is used to regain the starting compound of the Calvin cycle, ribulose bisphosphate.

In summary, the process of photosynthesis includes the light-dependent and the light-independent reactions. The products of the light-dependent reaction are ATP and NADPH, which are required for the light-independent reaction to proceed. Thus, it is clear that light is needed for the light-independent reaction to occur, but not directly. Figure 14.23 summarizes the two reactions.

W To view a good explanation of the light-independent reaction, visit heinemann.co.uk/hotlinks, enter the express code 4242P and click on Weblink 14.6.

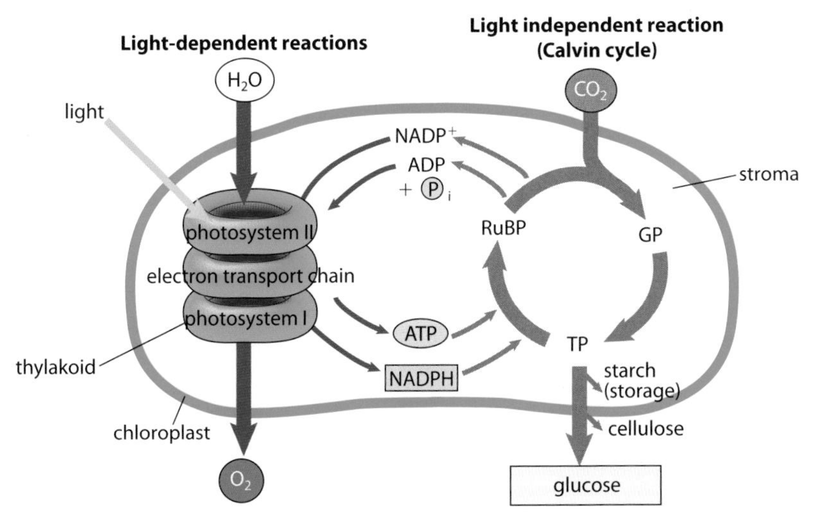

Figure 14.23 A summary of the complete process of photosynthesis.

Note that $NADP^+$ and ATP move back and forth in the chloroplast from the thylakoids to the stroma in their reduced and oxidized forms. A final summary of the two reactions is shown in this table.

Light-dependent reaction	Light-independent reaction
occurs in the thylakoids	occurs in the stroma
uses light energy to form ATP and NADPH	uses ATP and NADPH to form triose phosphate
splits water in photolysis to provide replacement electrons and H^+, and to release oxygen to the atmosphere	returns ADP, inorganic phosphate and NADP to the light-dependent reaction
includes two electron transport chains and photosystems I and II	involves the Calvin cycle

The chloroplast and photosynthesis

From the explanation of photosynthesis, it is clear how important the chloroplast is to the overall process. The structure of the chloroplast allows the light-dependent and light-independent reactions to proceed efficiently. In biology, the relationship of structure to function is a universal theme. The chloroplast and photosynthesis are no exception to this, as is shown in this table.

Chloroplast structure	Function allowed
extensive membrane surface area of the thylakoids	allows greater absorption of light by photosystems
small space (lumen) within the thylakoids	allows faster accumulation of protons to create a concentration gradient
stroma region similar to the cytosol of the cell	allows an area for the enzymes necessary for the Calvin cycle to work
double membrane on the outside	isolates the working parts and enzymes of the chloroplast from the surrounding cytosol

Action and absorption spectra of photosynthesis

The wavelengths of the electromagnetic spectrum with high energy are absorbed by the ozone layer. The wavelengths with low energy are absorbed by water vapour and carbon dioxide in air.

The energy necessary for photosynthesis comes from light. Light is electromagnetic energy. It travels in rhythmic waves that have characteristic wavelengths. The entire range of radiation is referred to as the electromagnetic spectrum (see Figure 14.24). The specific part of this spectrum that is involved in photosynthesis is the visible light spectrum. This spectrum is visible to the human eye, and its wavelengths range from near 400 nanometres to near 740 nanometres. The shorter wavelengths of visible light have more energy than the longer wavelengths.

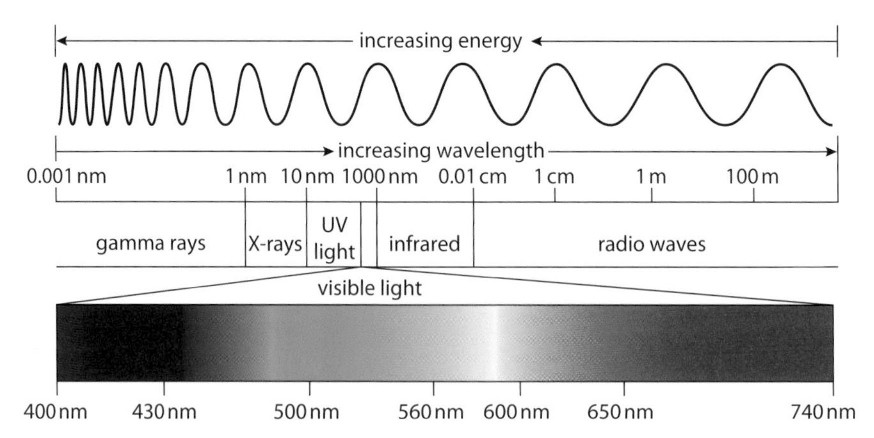

Figure 14.24 The electromagnetic spectrum.

Only about 42% of the sunlight directed towards Earth actually reaches the surface. Of this amount, only about 2% is used by plants. Of this amount, only 0.1–1.6% is incorporated into plant material.

The various pigments of photosynthesis absorb photons of light from specific wavelengths of the visible spectrum. If white light, which contains all the wavelengths of the visible light spectrum, is passed through the chloroplast of a plant cell, not all wavelengths are absorbed equally. This is because of the specific pigments present in the chloroplasts of that particular type of plant. A device called a spectrophotometer can be used to measure absorption at various light wavelengths. This results in a characteristic absorption spectrum for the plant. The absorption spectrum of a plant is the combination of all the absorption spectra of all the pigments in its chloroplasts. Figure 14.25 shows the absorption spectra of some typical photosynthetic pigments.

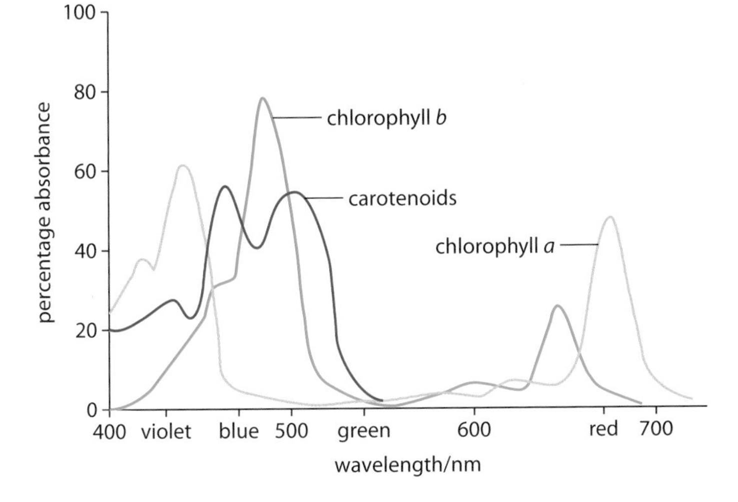

Figure 14.25 The absorption spectra of common photosynthetic pigments (relative amounts of light absorbed at different wavelengths).

Since light provides the energy to drive photosynthesis, the wavelength of the light absorbed by the chloroplasts partly determines the rate of photosynthesis. The rate of photosynthesis at particular wavelengths of visible light is referred to as the action spectrum. A common way of determining the rate of photosynthesis in order to produce an action spectrum is to measure oxygen production. High oxygen production indicates a high rate of photosynthesis. Figure 14.26 shows the action spectrum of a plant with the absorption spectra shown in Figure 14.25.

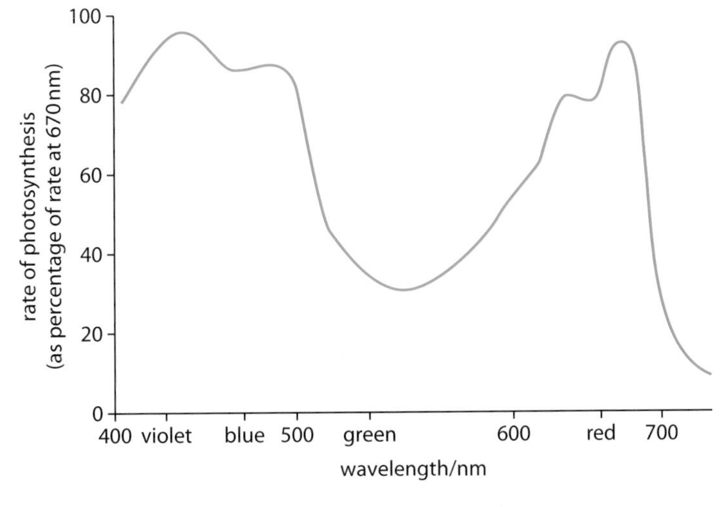

Figure 14.26 The action spectrum of photosynthesis (effectiveness of different wavelengths in fuelling photosynthesis). Note the positive correlation between this graph and Figure 14.25.

If you look at these two graphs, you can see two correlations:

- blue light and red light show the greatest absorption and they also represent the peaks in the rate of photosynthesis.
- the low absorption of green light corresponds to the lower rate of photosynthesis.

As we leave this topic, it is important to note that absorption and action spectra vary for different plants. This is due to the presence of different pigments and the different relative amounts of these pigments.

Factors affecting photosynthesis

The rate of photosynthesis can be affected by many factors. The term 'limiting factor' is used to describe the factor that controls any particular process (such

Limiting factors are very evident in plant growth. Even though many different minerals are necessary for optimal plant growth, the one mineral in lowest percentage of the needed amount will limit plant growth and be referred to as the limiting factor.

as photosynthesis) at the minimum rate. Even though many factors have an effect, it is the limiting factor that actually controls the rate of the process. In photosynthesis, there are several factors that may be called limiting. Three of these are temperature, light intensity, and carbon dioxide concentration. At any one time in the life of a plant, one of these factors may be the limiting factor for photosynthesis. Whichever is the limiting factor, that factor will be altering the photosynthetic rate. Even if the other two factors change, their effect will not be evident.

The limiting factor has its effect and the others do not because of the complexity of the process. Photosynthesis has many steps that lead to the end-products. When one step is slowed by the limiting factor, the whole process is slowed. This particular step would be called the rate-limiting step. Figures 14.27, 14.28 and 14.29 show light, carbon dioxide and temperature in their roles as limiting factors. In each graph, photosynthetic rate is shown on the *y* axis. These graphs are obtained by altering only one variable at a time. This means we can draw a definite conclusion as to the cause of the change in photosynthetic rate. Try writing an explanation as to why the rate would be altered as shown for each graph.

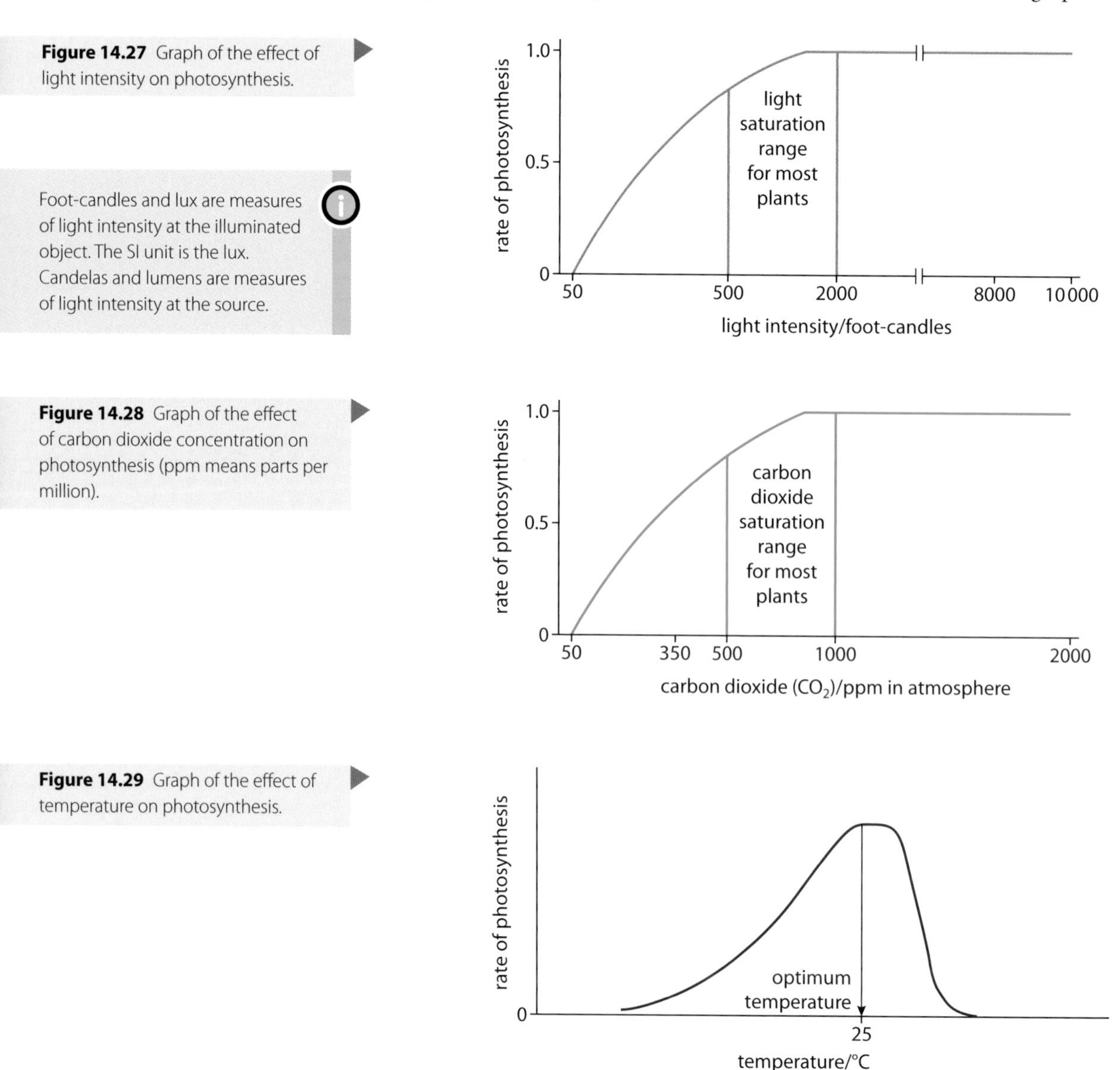

Figure 14.27 Graph of the effect of light intensity on photosynthesis.

Foot-candles and lux are measures of light intensity at the illuminated object. The SI unit is the lux. Candelas and lumens are measures of light intensity at the source.

Figure 14.28 Graph of the effect of carbon dioxide concentration on photosynthesis (ppm means parts per million).

Figure 14.29 Graph of the effect of temperature on photosynthesis.

Controls are extremely important in experimentation. Changing one variable at a time makes it much easier to draw proper conclusions. In many research ventures, it is not always possible to control all variables.

For example, a study of middle-aged women may find that a group showed increased incidence of hypertension correlated with high sodium intake. However, the effect could be due to some other aspect of diet such as high cholesterol intake. Many medical and nutritional study results are published regularly. We must analyse the procedure and conclusions carefully.

Cyclic photophosphorylation

This is another way in which the light-dependent reaction of photosynthesis may produce ATP. It proceeds only when light is not a limiting factor and when there is an accumulation of NADPH in the chloroplast. In this process, light-energized electrons from photosystem I flow back to the cytochrome complex of the electron transport chain between photosystem II and photosystem I (see Figure 14.30). From the cytochrome complex, the electrons move down the remaining electron transport chain allowing ATP production via chemiosmosis. Thus, the electrons do not flow to the second electron transport chain that would produce NADPH. Obviously, this process is valuable since there is already an overabundance of NADPH that causes this different form of photophosphorylation to occur. The additional ATPs produced are shuttled to the Calvin cycle so that it can proceed more rapidly.

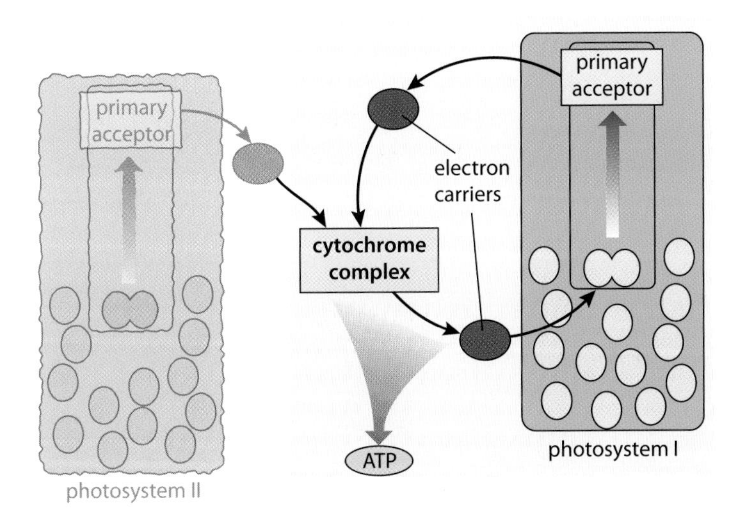

Figure 14.30 Cyclic photophosphorylation involving cyclic electron flow. Notice the electrons are cycled between photosystem I and the cytochrome complex. The result of this process is the production of ATP but no NADPH is produced.

Exercises

12 Why do plants need both mitochondria and chloroplasts?

13 You have a leaf from each of two very different plants. One leaf has more pigments than the other. Which leaf would have the greater photosynthetic rate, assuming all affecting factors are equal? Why?

14 Explain the final products of the two photosystems involved in the light-dependent reaction of photosynthesis.

15 Many scientists state that the enzyme RuBP carboxylase (rubisco) is the most ubiquitous protein on Earth. Why is there a very good chance that this is true?

16 How are the products of the light-dependent reaction important to the light-independent reaction?

Practice questions

1 There are many abiotic factors that affect the rate of photosynthesis in terrestrial plants. Wheat is an important cereal crop in many parts of the world. Wheat seedlings were grown at three different concentrations of carbon dioxide (in parts per million) and the rate of photosynthesis was measured at various light intensities.

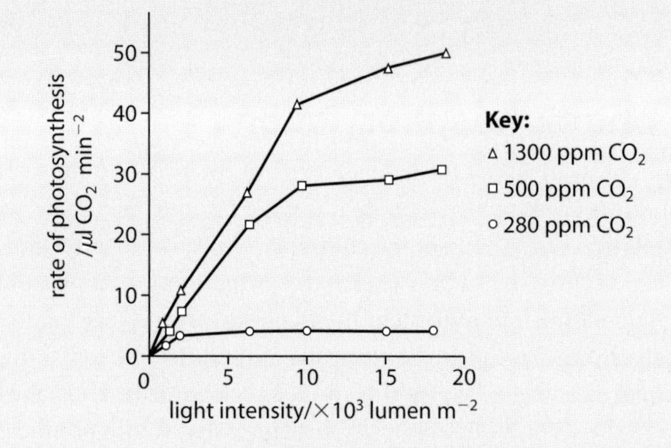

Source: Adapted from J P Kimmins, *Forest Ecology* (2nd edition) page 161

(a) Describe the relationship between the rate of photosynthesis and light intensity for wheat seedlings grown at a CO_2 concentration of 500 ppm. *(2)*

(b) Outline the effect of CO_2 concentration on the rate of photosynthesis of the wheat seedlings. *(3)*

(c) The normal atmospheric concentration of CO_2 is 370 ppm. Deduce the effect of doubling the CO_2 concentration to 740 ppm on the growth of wheat plants. *(2)*

Leaf area and chlorophyll levels were measured in sun leaves and shade leaves of *Hedera helix* (English Ivy) and *Prunus laurocerasus* (Cherry Laurel). Sun leaves developed under maximal sunlight conditions while shade leaves developed at reduced sunlight levels in the shadow of other leaves.

Species	Leaf type	Chlorophyll/μg ml^{-1}	Leaf area/cm^2
ivy	shade	4.3	72.6
	sun	3.8	62.9
laurel	shade	4.7	38.7
	sun	4.2	25.7

Source: D Curtis (1990), *Plant Ecology* independent project

(d) Calculate the percentage increase in the amount of chlorophyll in shade leaves of ivy compared to sun leaves of ivy. *(1)*

(e) Suggest a reason for the differences in chlorophyll concentration and leaf area in sun and shade leaves in these two species. *(2)*

(Total 10 marks)

2 The complex structure of proteins can be explained in terms of four levels of structure: primary, secondary, tertiary and quaternary.

(a) Primary structure involves the sequence of amino acids that are bonded together to form a polypeptide. State the name of the linkage that bonds the amino acids together. *(1)*

(b) Beta pleated sheets are an example of secondary structure. State *one* other example. (*1*)

(c) Tertiary structure in globular proteins involves the folding of polypeptides. State *one* type of bond that stabilizes the tertiary structure. (*1*)

(d) Outline the quaternary structure of proteins. (*2*)

(*Total 5 marks*)

3 Of the following products, which is produced by both anaerobic respiration and aerobic respiration in humans?

I Pyruvate
II ATP
III Lactate

A I only
B I and II only
C I, II and III
D II and III only (*1 mark*)

4 Consider the metabolic pathway shown below.

$$A \xrightarrow{1} B \xrightarrow{2} C \xrightarrow{3} D \xrightarrow{4} E$$

If there is end-product inhibition, which product (B to E) would inhibit which enzyme (1 to 4)?

	Product	Enzyme
A	C	4
B	B	3
C	B	4
D	E	1

(*1 mark*)

5 Which two colours of light does chlorophyll absorb most?

A Red and yellow
B Green and blue
C Red and green
D Red and blue (*1 mark*)

6 Anaerobic respiration occurs in the absence of oxygen while aerobic respiration requires oxygen.

(a) State *one* final product of anaerobic respiration. (*1*)

(b) Complete the table showing the differences between oxidation and reduction. (*2*)

	Oxidation	Reduction
Electrons gained or lost		
Oxygen or hydrogen gained or lost		

(c) The structure of a mitochondrion is shown in the electron micrograph on the following page.
Name the parts labelled A, B and C and state the function of each. (*3*)

(*Total 6 marks*)

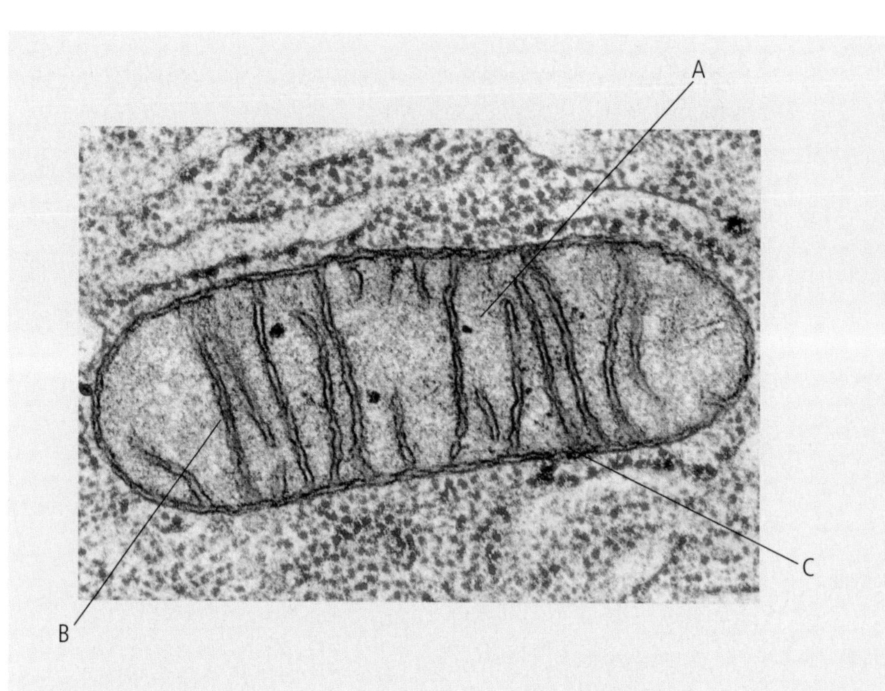

7 (a) State the site of the light-independent reactions in photosynthesis. *(1)*

The absorption spectrum of chlorophyll *a* and chlorophyll *b* are shown in the graph below.

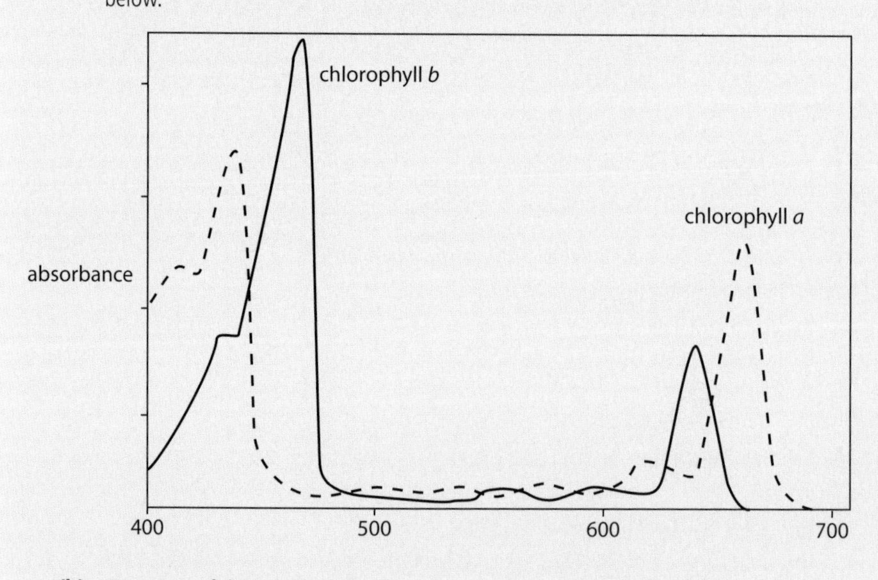

(b) On a copy of the graph above, draw the action spectrum of photosynthesis for a green plant. *(1)*

Explain photophosphorylation in terms of chemiosmosis. *(3)*

(Total 5 marks)

8 Which way do the protons flow when ATP is synthesized in mitochondria?

A From the inner matrix to the intermembrane space

B From the intermembrane space to the inner matrix

C From the intermembrane space to the cytoplasm

D From the cytoplasm to the intermembrane space *(1 mark)*

9 The diagram below shows a channel protein in a membrane. Which parts of the surface of the protein would be composed of polar amino acids.

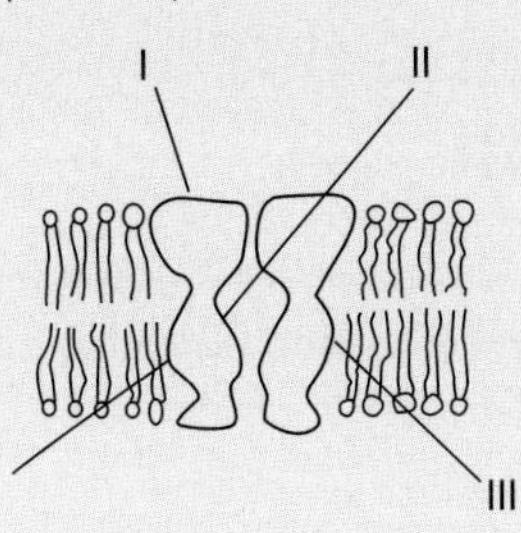

A I and II only

B II and III only

C III and IV only

D I and IV only *(1 mark)*

10 What is the sequence of stages during the conversion of glucose into pyruvate in glycolysis?

A lysis → phosphorylation of sugar → oxidation

B lysis → oxidation → phosphorylation of sugar

C phosphorylation of sugar → lysis → oxidation

D phosphorylation of sugar → oxidation → lysis *(1 mark)*

15 Evolution (Option D)

Introduction

Some of the most fundamental questions in the history of humankind are addressed in this option.

- Where do we come from?
- How did life start on Earth?
- What were our ancestors like millions of years ago?

The truly honest response to these questions is that no-one can prove beyond a shadow of a doubt that they know the answer. Each culture, each religion, each community, each school of thought has its own answers, many of which are incompatible with each other. This gives rise to intense and long-standing debates between people with different viewpoints. This chapter focuses on the answers of just one group: evolutionary biologists.

How did life start on Earth?

Is the biochemical view of how life began 'true'? What is important in science is not how 'true' an idea is, but rather how the idea was obtained and whether or not it can be verified. If it was obtained using the scientific method and its validity is supported by evidence from various sources which can be verified, then it has a good chance of being accepted by the scientific community. If an idea is not supported by verifiable evidence, it does not mean it is untrue or impossible, it just means that the idea is unscientific.

As in all scientific questions, hypotheses have been formulated based on research and observations. Some hypotheses have matured into widely accepted theories concerning how life began on Earth and developed into the forms we know today. However, since they try to answer questions about events which occurred millions or billions of years ago, some aspects of the answers are far from irrefutable.
One of the most widely accepted hypotheses about the origin of life on Earth is that it formed from non-living molecules which combined together in the appropriate conditions to form replicating molecules and, over time, the first cells.

D.1 Origin of life on Earth

Assessment statements

D.1.1 Describe four processes needed for the spontaneous origin of life on Earth.

D.1.2 Outline the experiments of Miller and Urey into the origin of organic compounds.

D.1.3 State that comets may have delivered organic compounds to Earth.

D.1.4 Discuss possible locations where conditions would have allowed the synthesis of organic compounds.

D.1.5 Outline two properties of RNA that would have allowed it to play a role in the origin of life.

D.1.6 State that living cells may have been preceded by protobionts, with an internal chemical environment different from their surroundings.

D.1.7 Outline the contribution of prokaryotes to the creation of an oxygen-rich atmosphere.

D.1.8 Discuss the endosymbiotic theory for the origin of eukaryotes.

Problems for starting life on Earth

How is it possible that a lifeless ball of rock, which the planet Earth was 3.5 billion years ago, became home to such lush vegetation and such a wide variety of bacteria, fungi, protists and animals that we see today? In order for life to be possible, some major problems had to be overcome.

The first problem is that life as we know it is based on *organic* molecules such as amino acids. Early Earth had only inorganic matter: rocks, minerals, gases and water. So where did the first organic molecules come from?

The second problem is that organisms are, by definition, organized. Monomers (single molecules) need to be connected together into polymers (long chains of molecules) such as polypeptides in order to build more complex organic compounds such as proteins.

Remember the important building up and breaking down of molecules which are essential to making life possible (see Figure 15.1).

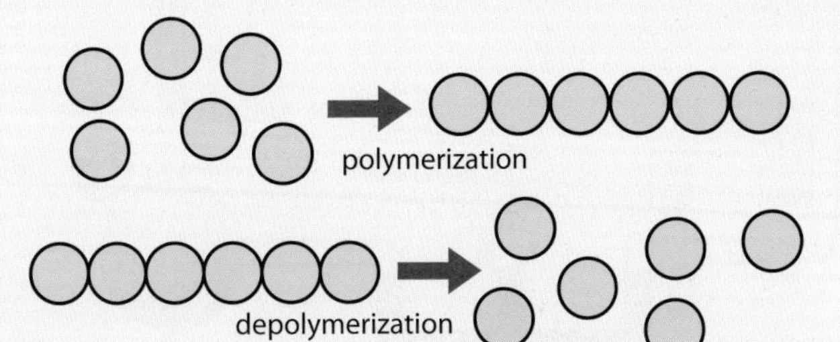

Figure 15.1 Polymerization and depolymerization.

You should be able to determine which molecules are monomers and which are polymers.

The third problem to overcome in order for life to be possible is that in order for something to be considered 'alive', it must reproduce – so self-replicating molecules were needed. Today, DNA is the molecule most used for the replication of living organisms. But DNA is a complex molecule which requires enzymes in its formation, so it is unlikely that DNA developed very early. In addition, to get polypeptides to form from DNA, RNA is needed.

The final major problem to be overcome is the fact that water, although it is credited with making life possible, tends to depolymerize molecules. Many organic compounds simply dissolve in water. This makes it difficult for molecules to become organized into polymers.

Miller and Urey

One of the criteria for deciding whether or not an idea is scientific is to decide if it is open for falsification; can it be disproved? In questions concerning events which happened millions or billions of years ago, is it possible to ever find enough evidence to know what happened? Could theories about life's origins be refuted and falsified?

Stanley Miller and Harold Urey performed a ground-breaking experiment in 1953 concerning the origin of life on Earth. In a glass sphere in their laboratory, they reproduced the environment as it was thought to have existed on Earth during the Hadean. This was a time in our planet's past in which things were very different from the way they are today. Here is how they created the 'primordial soup' environment they believed existed (see Figure 15.2):

- they introduced gases believed to be present at that time (e.g. methane, hydrogen and ammonia);
- they introduced liquid water which was heated to evaporate and cooled to condense, thus recreating the water cycle;
- they kept everything at a warm temperature;
- the apparatus was exposed to UV radiation because without an ozone layer, Earth's surface had no protection from this type of energy from the Sun;
- electric sparks were generated inside the glass sphere to represent lightning.

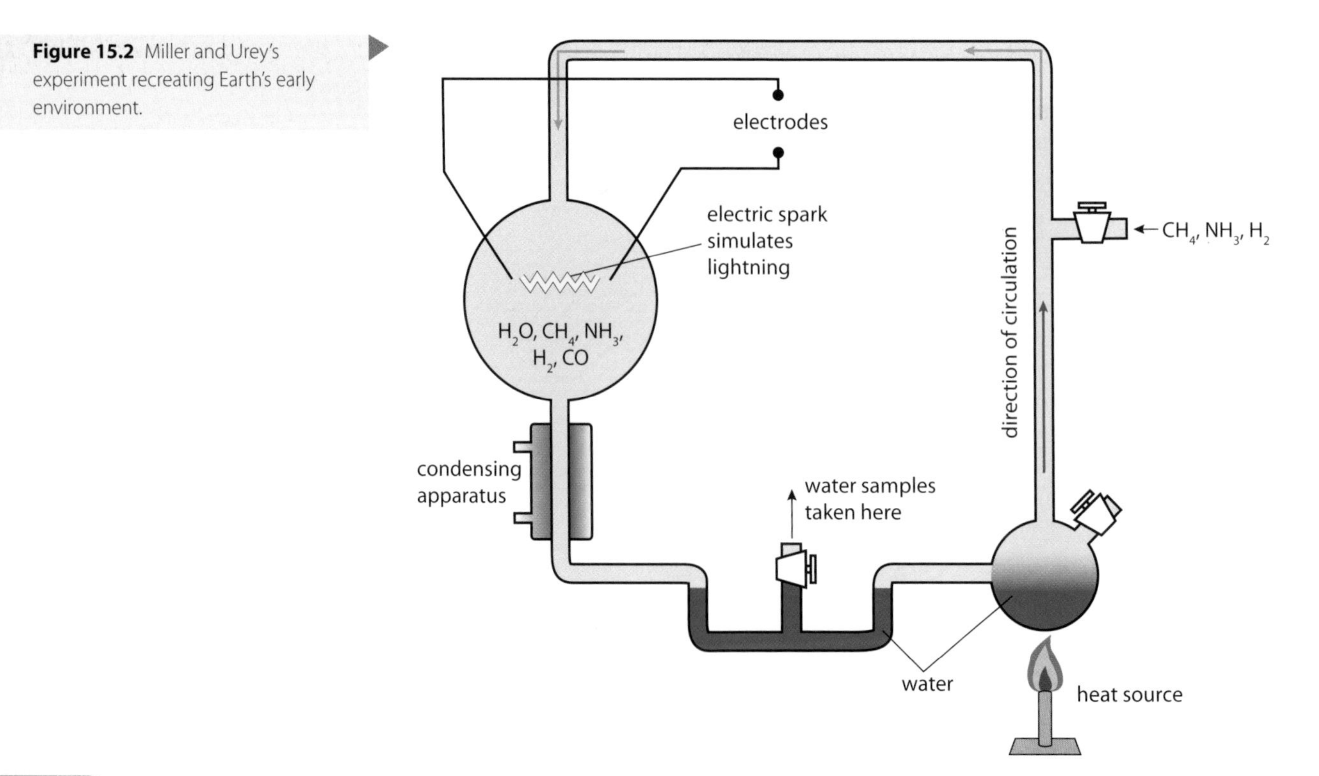

Figure 15.2 Miller and Urey's experiment recreating Earth's early environment.

The results were impressive. After a week, organic compounds such as amino acids had formed inside the 'primordial soup' model. Other molecules which had formed included certain sugars and nucleic acids.

Does this mean that they created life? No. Synthesizing organic molecules from inorganic molecules gives a possible explanation to what may have happened on early Earth to solve the first of the problems outlined earlier, but it certainly does not solve the others. Molecules such as amino acids are the building blocks of life as we know it but they cannot be described as 'alive'.

Miller and Urey's experiment is a good example of a model. Physical and conceptual models are often used in science to help support a working hypothesis. This model resulted in the successful production of organic compounds; does that prove that this is how organic compounds first formed on Earth? Could another model disprove this one and falsify the hypothesis?

Could comets have brought organic compounds to Earth?

In studying the nature and composition of comets, researchers have found that these chunks of ice wandering through space can carry organic compounds.

Geological records show that our planet was bombarded by a shower of comets and asteroids about 4000 million (4 billion) years ago in an event called the Late Heavy Bombardment. The National Aeronautics and Space Administration (NASA) in the US did experiments in 2001 showing that organic molecules hitchhiking on comets could have survived an impact on Earth's surface.

In addition, the experiments revealed that such an impact could help to polymerize certain amino acids into polypeptides. Could life have been started this way? Only continued experiments and new evidence could confirm or disprove this.

Comets are chunks of space ice which can sometimes crash into Earth, bringing a variety of molecules including amino acids.

In what environments could organic compounds have formed?

There are several hypotheses about where the synthesis of organic compounds could have occurred: in space, near volcanoes, deep in the ocean, in zones where there is an alternating wet/dry pattern such as intertidal zones along the seashore.

In space

By studying the spectral lines of distant clouds of cosmic dust particles, astronomers claim to have revealed the presence of glycine, which is the simplest amino acid. This suggests that organic molecules can form in space and could explain how comets could be carrying such molecules.

It is difficult to confirm these claims but laboratory experiments which recreate the low-pressure, low-temperature environment of space have been able to synthesize amino acids. This would support the hypothesis that life, or at least the molecules needed for life, could have originated in space and been brought to Earth by comets or meteorites. The idea that life on Earth began in space is called panspermia.

Spectral lines are used as a kind of 'fingerprint' to identify elements in objects in space. They form when the light from a shining object is broken up into a rainbow (spectrum). Helium was discovered on the Sun before it was discovered in Earth's upper atmosphere using this technique. By matching up lines from unknown samples with known samples, elements can be identified.

A nebula in space like this one is a possible origin for organic molecules.

In alternating wet/dry conditions

Another possible place where organic molecules could have formed and paved the way for life to develop on Earth is the seashore or the flood plains of a river where there is alternation of wet and dry conditions. The drying of clay particles could have created catalysing reactions and formed early organic molecules (see page 419). Stromatolites, one of the most ancient forms of life on Earth, live today in such intertidal zones, suggesting that theses conditions were favourable to early life (see page 421).

Near volcanoes

A third possibility for the place where the first organic molecules developed is around volcanoes. Although a volcanic eruption can be destructive, it spews out water vapour, other gases and various minerals which could be used to form organic matter. The rich sources of raw materials plus the warmth of the volcanic activity could have provided conditions favourable to the formation of amino acids and sugars.

A 'black smoker' or hydrothermal vent on the ocean floor.

In deep oceans

Another contender for the place of origin of organic molecules is around hydrothermal vents – places where hot water emanates from beneath the ocean floor. Simply put, underwater geysers. The first deep-sea hydrothermal vent was observed in 1977. Such a structure forms when cracks in the crust of the seabed expose sea water to rocks below, which are heated by magma. The hot water rises and picks up countless minerals along the way. Hydrothermal vents are sometimes referred to as 'black smokers' because the water coming out of them contains so many dark minerals that it looks like smoke.

Today, we know about entire communities living around these vents; creatures never seen before 1977, such as metre-long white-and-red tube worms which absorb the minerals from the water and transfer them to symbiotic bacteria. The bacteria then make food from the minerals. This food nourishes the tube worms. The discovery of these communities disproved the idea that the bottom of the ocean must be lifeless because there is no sunlight. It also gives credibility to the hypothesis that the earliest forms of life could have formed deep in the ocean around hydrothermal vents.

For more information about black smokers plus some photos and more links, visit heinemann.co.uk/hotlinks, enter the express code 4242P and click on Weblink 15.1.

Do we have an answer?

There are undoubtedly other possibilities as well, but all the locations we have just looked at present enticing promises as potential places where the conditions of life came together to make the first step towards living systems. Unfortunately, there are many unanswered questions about how difficulties were overcome. For example, it is estimated that without a protective ozone layer, early Earth had ultraviolet radiation 100 times above today's levels, destroying organic molecules as fast as they formed. With only half a century of research, the study of the origin of life is a young science and much work still needs to be done.

Some questions simply cannot be answered by science as it exists today. In the coming decades, new observations, discoveries and techniques will reveal answers to some of the questions we have now. But for the moment, we have reached the limits of our knowledge.

RNA's role in early life

In order to transmit their hereditary traits to the next generation, most organisms today store their genetic code in the form of DNA. DNA replication needs enzymes. In the prebiotic world, there were no enzymes, so it is unlikely that double-stranded DNA was the means of inheritance.

On the other hand, in certain conditions, single-strand RNA is able to replicate itself without the aid of enzymes. This property makes RNA an ideal candidate as an early nucleic acid for heredity. One reason RNA can do this is that it can also act as a catalyst, helping certain chemical reactions the way enzymes do.

The name for RNA acting as an enzyme is ribozyme. For some researchers, RNA helps to solve the problem of transmitting a genetic code in a world without enzymes. For these reasons, it could be that RNA dominated this early world, floating around in the primordial soup, making copies of itself.

Protobionts

Experiments have shown that when clay dries out and is heated, as many as 200 amino acids can spontaneously join together as polypeptide chains.

In the right conditions, these chains form proteinoid microspheres, tiny bubble-like structures. If these surrounded other polymers which were forming, they could establish and maintain a chemistry inside which was different from that of the surroundings (see Figure 15.3).

Minerals present in drying clay can sometimes help catalyse reactions.

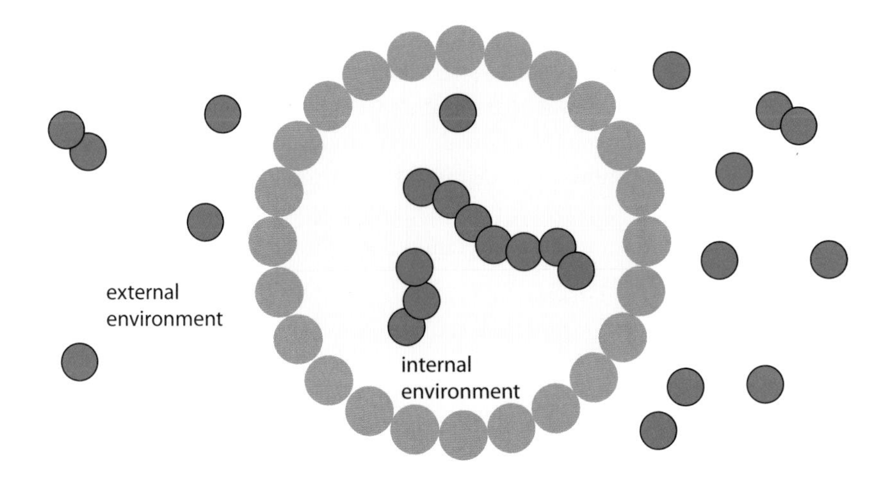

Figure 15.3 A proteinoid microsphere (in pink) can maintain different conditions inside the sphere from those outside.

Another type of microscopic sphere called a coacervate can form from lipids in water. They form spontaneously due to the hydrophobic forces between the water and the lipid molecules. They can also maintain an internal set of chemical balances different from those outside. In addition, coacervates have the capacity to be selectively permeable.

Although they are certainly not living organisms, proteinoid microspheres and coacervates can be considered a significant step towards the formation of cells. Such structures are called protobionts. Since these structures solve the problem of protecting polymers from their destructive environment, they are good candidates for primitive versions of what would become the first cell membranes.

'Proto' is from the Greek word meaning first.

If polynucleotides, such as RNA, were included inside these microscopic spheres, they would be protected from being broken down. Over time, true cell membranes evolved and the other characteristics of cells developed such as the ability to perform cellular respiration and asexual reproduction. Although the mechanisms are not clear as to how this happened, the coming decades in research will most likely shed light on this elusive part of the story of life's origins on Earth.

Where did all the oxygen come from?

Even though a fifth of the air you are breathing now is oxygen, there was none at all 4 billion years ago. The earliest forms of life on Earth were bacteria and they lived in an environment with an atmosphere of mainly carbon dioxide. Consequently, early life forms were anaerobic cells.

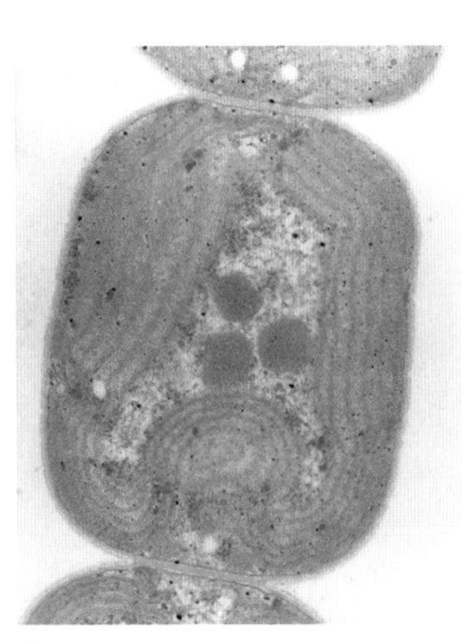

Cyanobacteria thrive today but organisms like this are thought to be the first ones on Earth to photosynthesize.

These single-celled organisms consumed the organic molecules, such as simple sugars, that were forming from chemical reactions on Earth. The more they reproduced, however, the more food was consumed by the population. After millions of years, their population reached such large numbers that food began to be scarce. Faced with a food shortage, some bacteria had an advantage because they could make their own.

At some point around 3500 million (3.5 billion) years ago, certain bacteria (thought to be related to today's cyanobacteria) developed the capacity to photosynthesize. How this happened is not clear but evidence that it did happen is seen in iron-rich rocks which started to form at that time. The photosynthetic bacteria helped to convert iron, which was dissolved in ocean water, into precipitates of iron oxide which made rust-coloured layers of rock.

The development of photosynthesis was one of the most significant events in the history of life on Earth. Not only did it give certain bacteria the possibility to tap into an inexhaustible source of energy (sunlight) to survive, it also caused mass pollution of the atmosphere.

A waste product of photosynthesis is oxygen. Oxygen gas is toxic to the kinds of bacteria which preceded photosynthetic ones, so this new pollution eventually killed off large populations of anaerobes. The anaerobic bacteria which survived were those living in mud or other places protected from the new oxygen-rich atmosphere. Their descendents are still sheltered from the air today and one of their favourite places to hide is inside the guts of living organisms like us.

The ability of an organism to make its own food gives it a distinct advantage over those that cannot. As a result, the photosynthetic bacteria proliferated and produced more and more oxygen. Today, with the help of plants and other photosynthetic organisms, the current level of oxygen is around 21%. A level higher than this would be dangerous because fires would break out much more frequently, whereas a lower level would make life for aerobic organisms more challenging.

Stromatolites live today in Australia and the Persian Gulf. They are successive mats of photosynthetic cyanobacteria and sedimentary deposit. Fossils of stromatolites are among the oldest known fossils.

The endosymbiotic theory

Bacteria were the only organisms on Earth from about 3.8 billion to about 2 billion years ago. That is when the first fossils of cells with a nucleus (eukaryotes) become present in the fossil record. One widely accepted theory of how prokaryotes developed into eukaryotes is the theory of endosymbiosis.

This theory states that the organelles found inside cells today were once independent prokaryotes. It is believed that they were engulfed by a bigger cell but rather than being digested, the prokaryotes were kept alive inside the host cell in exchange for their services (see Figure 15.4 overleaf).

Host cells with enzymes to digest the ingested prokaryote were able to obtain only one meal but host cells which did not have the enzymes to digest the prokaryote kept the engulfed cell alive inside their cytoplasm and obtained a more abundant source of energy. By natural selection, the host cells with photosynthetic prokaryotes inside them to provide food, or others able to metabolize food efficiently, were more likely to survive. This could explain how membrane-bound organelles such as mitochondria or chloroplasts became part of another cell.

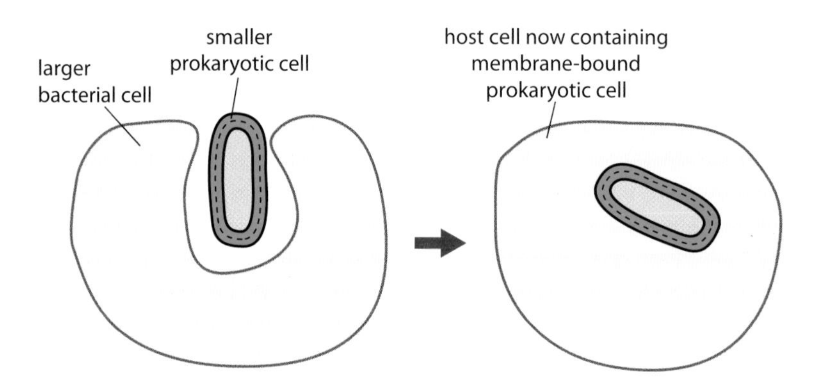

Figure 15.4 Endosymbiosis involves engulfing and incorporating another cell.

The coexistence of two organisms whereby each benefits from the other is called symbiosis. Endosymbiosis has the particularity that one organism is *inside* the cytoplasm of the other rather than the two coexisting side-by-side. For this to be possible, endocytosis must occur at some point (one cell engulfs the other).

The idea is supported by the fact that mitochondria and chloroplasts have characteristics which make them more like independent prokaryotes than just an organelle. They:

- have a double membrane;
- have their own naked DNA, which is circular (just like prokaryotes);
- can perform protein synthesis using small ribsomes (just like prokaryotes);
- can make copies of themselves when more chloroplasts or mitochondria are needed.

As with all theories, there are some problems and criticisms of endosymbiosis. For one thing, the ability to engulf another cell and have it survive in the cytoplasm, does not guarantee that the host cell can pass on to its offspring the genetic code to synthesize the newly acquired organelle. Secondly, when chloroplasts or mitochondria are removed from a cell, they cannot survive on their own.

Exercises

1 State the name of one organic monomer and one organic polymer on prebiotic Earth.
2 Describe one hypothesis suggesting how prebiotic polymers solved the problem of depolymerization.
3 Explain why UV radiation levels were higher in the atmosphere of early Earth than they are today.
4 Outline the theory of panspermia and include the role of comets.
5 Outline the theory of endosymbiosis.

D.2 Species and speciation

Assessment statements

D.2.1 Define *allele frequency* and *gene pool*.
D.2.2 State that evolution involves a change in allele frequency in a population's gene pool over a number of generations.
D.2.3 Discuss the definition of the term *species*.
D.2.4 Describe three examples of barriers between gene pools.
D.2.5 Explain how polyploidy can contribute to speciation.
D.2.6 Compare allopatric and sympatric speciation.
D.2.7 Outline the process of adaptive radiation.
D.2.8 Compare convergent and divergent evolution.
D.2.9 Discuss ideas on the pace of evolution, including gradualism and punctuated equilibrium.
D.2.10 Describe one example of transient polymorphism.
D.2.11 Describe sickle cell anaemia as an example of balanced polymorphism.

Allele frequency and gene pools

The gene pool is all of the genetic information present in the reproducing members of a population at a given time. The gene pool can be thought of as a reservoir of genes from which the population can get all its various traits. A large gene pool exists in a population which shows substantial variety in its traits, whereas a small gene pool exists in a population whose members show little variation, notably in cases of inbreeding.

Allele frequency is a measure of the proportion of a specific variation of a gene in a population. The allele frequency is expressed as a proportion or a percent. For example, it is possible that a certain allele is present in 25% of the chromosomes studied in a population. This would mean that one quarter of the loci for that gene are occupied by that allele and the other three-quarters do not possess it.

This could also be interpreted a 25% chance that a chromosome in that population has the allele at that particular locus. Note that this does *not* mean that 25% of the members of the population have the allele. In effect, we will see later how these numbers play out in a diploid situation where two chromosomes in each organism can carry a version of the gene.

Look at Figure 15.5. The gene pool in this population of 16 people is made up of 32 genes. Count the number of **T**s. You should get 16. Do you get the same for the number of **t**s? You should. Since there are half **T**s and half **t**s, the allele frequency for each is 50% or 0.50. Does this mean that half the people have the phenotype caused by the recessive allele? No. Only 4 people are homozygous recessive and have the phenotype, which is 25% of the population. Be careful not to confuse allele frequency with the number of people who show a particular trait.

Figure 15.5 In this gene pool, the frequencies of each allele **T** and **t** is 50%.

Evolution and alleles

Gene pools are generally relatively stable over time. But not always. New alleles can be introduced due to mutation and old alleles can disappear when the last organism carrying the allele dies. One result of evolution is that after many generations of natural selection, some alleles have proved to be advantageous and tend to be more frequent.

Inversely, some alleles are disadvantageous to the survival of the organisms in the population and are not passed on to as many offspring. From this it should be clear that any time an allele frequency is estimated, it is only a snapshot of the alleles at that time. Several generations later, the proportions may not be the same.

In addition, if populations mix due to immigrations, there will most likely be a change in allele frequencies. The same is true for emigrations when one group with a particular allele leaves the population. For whatever the reason, if a gene pool is modified and the allele frequencies have changed, we know that some degree of evolution has happened. No change in allele frequencies, however, means no evolution.

Monitoring allele frequencies over time helps to indicate a rate of evolution:

- no change in allele frequency – no evolution;
- a big change in allele frequency – evolution has occurred.

Defining species

A species is the basic unit for classifying organisms. It is one of those words everyone thinks they know, but it is not an easy concept. A species is made up of organisms which:

- have similar physiological and morphological characteristics which can be observed and measured;
- have the ability to interbreed to produce fertile offspring;
- are genetically distinct from other species;
- have a common phylogeny.

There are challenges to this definition, however.

Sometimes, members of separate but similar species reproduce and succeed in producing offspring. For example, a horse and a zebra, or a donkey and a zebra, can reproduce to make a zebroid. In each case, the parent organisms are equines (they belong to the family *Equus*) so they are related but they are certainly not the same species. They do not possess the same number of chromosomes, which is one of the reasons why the hybrid offspring produced are infertile.

A liger is a hybrid between a lion and a tiger, and is considerably larger than either parent animal.

Other challenges to our definition include the following.

- What about two populations which could potentially interbreed but do not because they are living in different niches or are separated by a long distance?
- How should we classify populations which do not interbreed because they reproduce asexually? (Our definition above is clearly addressed at sexually reproducing organisms and cannot be applied to bacteria.)
- What about infertile individuals? Does the fact that a couple cannot have a child exclude them from the species? What about the technique of in vitro fertilization? What challenges does that pose to the definition of species?

The answers to these questions are beyond the scope of this book and the IB programme. However, you should always think critically about any definition that appears at first glance to be straightforward but which, on closer scrutiny, generates debate.

Barriers between gene pools

In some situations, populations of members of the same species (and thus of the same gene pool) can be stopped from reproducing together because there is an insurmountable barrier between them. Such a barrier can be geographical, temporal, behavioural or related to the infertility caused by hybridization.

Geographical isolation

Geographical isolation happens when physical barriers such as land or water formations prevent males and females from finding each other, thus making interbreeding impossible. For example, a river, a mountain or a clearing in a forest could separate populations. Tree snails in Hawaii demonstrate this geographical isolation: one species lives on one flank of a volcano and another species on the other flank without ever coming into contact with each other.

A waterway or mountain range can physically separate two populations causing geographical isolation.

Temporal isolation

Temporal isolation refers to incompatible time frames which prevent the populations or their gametes from encountering each other. For example, if the female parts of the flowers of one population of plants reach maturity at a different time from the release of pollen of another population, the two will have great difficulty producing offspring together. Or if one population of mammals is still hibernating or has not returned from a migration when another population of the same species is ready to mate, this would also be considered a temporal barrier between the two gene pools.

Behavioural isolation

Behavioural isolation can happen when one population's lifestyle and habits are not compatible with those of another population. For example, many species of birds rely on a courtship display in order for one sex to copulate with the other. If one population has one version of a courtship display which is significantly different from another population, they may not consider each other to be seductive enough to be potential mates. Hence, little or no reproduction will take place between the members of the two populations because of behavioural differences.

Hybrids

Hybrids face several challenges to their continuation as a population. For one thing, the vast majority of animal and plant hybrids are infertile. Even if one generation of hybrids is produced, a second generation is highly unlikely. This presents a genetic barrier between species.

Some examples of animal hybrids:
- female horse + male donkey = mule;
- female horse + male zebra = zorse;
- female tiger + male lion = liger.

Polyploidy

You will recall that haploid cells, such as sex cells, contain one set of chromosomes (n). This can be referred to as monoploidy. Diploid cells, such as somatic cells, contain two sets of chromosomes ($2n$) – one from each parent. Polyploidy refers to the situation in which a cell contains three or more sets of chromosomes ($3n$, $4n$, and so on).

$$3n = \text{triploid}$$
$$4n = \text{tetraploid}$$
$$5n = \text{pentaploid}$$

Such a situation can arise when cell division does not completely separate the copies of chromosomes into distinct nuclei and they end up in the same cell. In plants, where polyploidy is much more common than in animals, the extra sets of chromosomes lead to more vigorous plants which produce bigger fruits or food storage organs and which are more resistant to disease.

Having extra sets of chromosomes has the consequence of making errors in replication more common. If one population of plants is triploid and another tetraploid, each population's evolution will be different. If one population evolves at a different rate from another, the two could become so dissimilar that they no longer belong to the same species.

The process of an evolving population changing significantly enough so that the production of offspring with the original population becomes impossible is called speciation. In short, a new species has evolved from an old one and both will continue in their separate ways (see Figure 15.6).

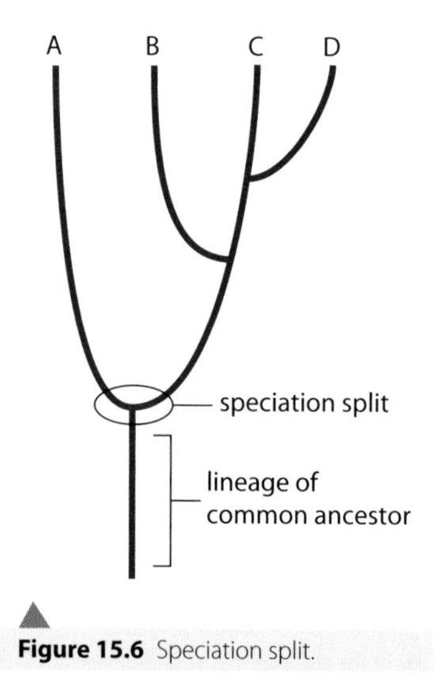

Figure 15.6 Speciation split.

Allopatric and sympatric speciation

When a new species forms from an existing species because it is separated by a physical barrier, it is called allopatric speciation (see Figure 15.7). When a new species forms from an existing species while living in the same geographical area it is called sympatric speciation (see Figure 15.7).

'Allos' means other and 'patria' means place of origin. 'Sym' means together.

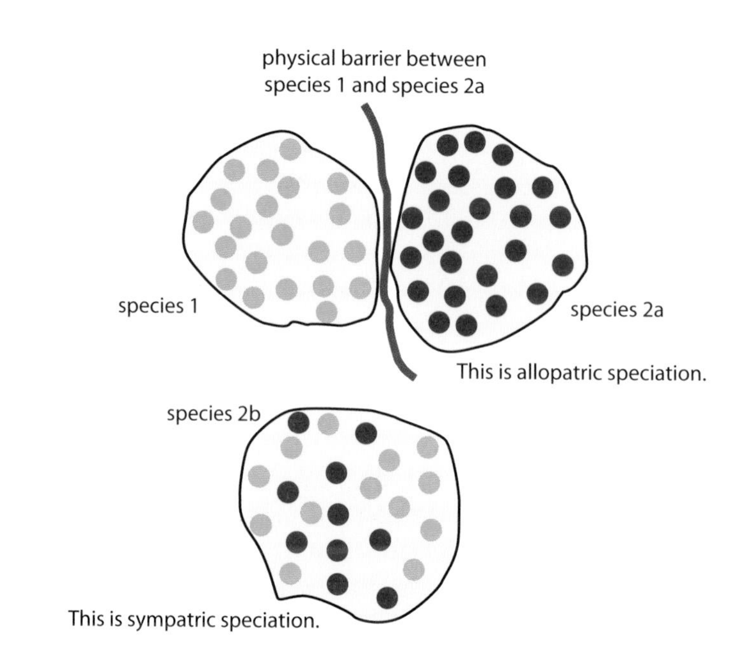

Figure 15.7 Comparison of two types of speciation.

An example of allopatric speciation might occur in a land-dwelling species when sea levels rise. Such an inundation would tend to cut off land-dwelling populations from each other and each one could evolve separately on opposite shores of the newly formed body of water. If sea levels dropped again many centuries or millions of years later, and the two populations were to come in contact with each other, it is possible that each one would have evolved so differently on their separate shores that they could no longer interbreed.

Populations do not always have to be separated geographically in order to produce changes in their genetic makeup that are significant enough to produce a new species. Consider the concepts of temporal or behavioural isolation introduced earlier (page 425). Either of these could produce a new species in the same geographical location as the parent species.

For example, moths produce chemicals called pheromones which play an important role in attracting and finding a mate. Suppose that, due to a mutation, a member of a moth population produces a slightly different pheromone from the others. This new chemical might be repulsive to many of the potential mates in its population but irresistible to some. Over time, the group producing this new pheromone might interbreed only with moths which produce the same variant of their pheromone. Within a certain number of generations, the new combinations of alleles from these insects might produce a new species of moth. Both populations live in the same geographical area and interact with each other but they do not breed together anymore.

Adaptive radiation

Adaptive radiation occurs when many similar but distinct species evolve relatively rapidly from a single species or from a small number of species. This happens as variations in the population allow certain members to exploit a slightly different niche in a more successful way. By natural selection and the presence of one or more of the barriers described earlier, new species evolve.

An example of this are the primates found in Madagascar and the Comoro Islands off the southeast coast of Africa. Millions of years ago, without competition from monkeys or apes, lemurs on these islands were able to proliferate. Large numbers of offspring meant a greater chance for phenotypic diversity.

Lemurs are primates found in Madagascar. They are a good example of adaptive radiation.

Among the wide range of variations in lemurs, some are better adapted for living on the ground instead of in the trees. Others are better adapted for living in lush rainforests while some can survive in the desert. Most lemurs are active during the day (diurnal) but some are nocturnal. There are so many different species of lemur with different specialties because of adaptive radiation.

Not a single species of living lemur has been found any where else in the world. And yet, fossils of their ancestors have been found on the continents of Africa, Europe and Asia. What happened? It is believed that lemurs were not successful in competing with apes and monkeys because as soon as traces of the latter start to become more prevalent in the fossil record, the lemur-like organisms become rare.

This would explain why continents and islands tend to have either prosimians (such as lemurs) or anthropoids (such as monkeys and apes), but not both types of primate. This is being confirmed today because over a dozen species of lemur have become extinct and more are endangered due to the activities of the most recently evolved anthropoid: humans.

Other examples of adaptive radiation can be seen in birds such as Hawaiian honeycreepers, or Darwin's finches in the Galapagos Islands. The honeycreepers have a wide variety of beak shapes, some of which are exclusively adapted to sip the nectar of flowers found only in Hawaii. It is believed that all the Hawaiian honeycreepers are the result of the adaptive radiation of a few members of one species which arrived on the islands.

Convergent and divergent evolution

We have seen how one species can have various splits over time creating a greater diversity between species. In some cases, the 'branches' of the phylogenetic tree can become spaced so far apart that the species, though once closely related, do not physically resemble each other any more. For example, when comparing a bird that has a long, thin beak to another with a short fat beak, it is difficult to imagine that they are closely related.

Marsupials give birth to live young but hold them in a pouch on their bodies until they are more autonomous. Examples are kangaroos and koalas. Placental mammals do not have such a pouch. Examples include cats and horses.

In other circumstances, it is possible to have two organisms with very different phylogenies but who look quite similar. The isolated continent of Australia abounds with examples of marsupials which developed in similar ways to their distant placental cousins on other continents. For example, the Tasmanian wolf, recently driven to extinction, was a marsupial which looked and behaved similarly to wolves and dogs from other continents. Anteaters exist in both forms: marsupial anteaters in Australia and placental anteaters in Latin America.

The antlers, hooves and body shapes of the Muntjac deer are examples of convergent evolution; many other mammals have similar characteristics even though they may not be closely related to these deer. Male Muntjac deer (not pictured) have another anatomical feature showing convergence with unrelated carnivorous mammals – they have downward-pointing fangs, which they use when fighting other males.

Plants from Africa known as euphorbias can be easily confused with cacti from the Americas because of their similar structures. Both have adaptations for high temperatures and low water availability including a round tubular shape, a system of water storage, minimal or no leaves and thorns for protection. And yet, they have separate phylogenies and are *not* both classified as cacti.

Convergent and divergent evolution can refer not only to entire organisms but also to physical features (such as thorns, eyes or wings) and can even refer to how organisms use certain molecules. The use of bioluminescent (glowing) chemicals by marine organisms, bacteria and fungi is one example of the convergent evolution of a biochemical.

In both types of evolution, it is the process of natural selection that allowed the organisms to adapt to their environment in the ways in which they did (see Figure 15.8).

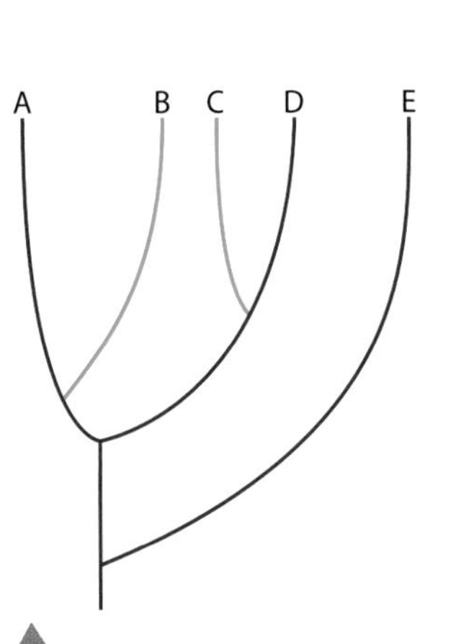

Figure 15.8 Divergent evolution (the blue lineages which are less and less similar as time goes on) and convergent evolution (the green lineages which are becoming more similar over time).

Gradualism and punctuated equilibrium

Among evolutionary biologists there is some discussion as to the rate at which species evolve. Generally it is agreed that evolution does not happen overnight but there are two main views (see Figure 15.9):

- the changes are small, continuous and slow (gradualism);
- the changes are relatively quick and followed by long periods of little or no change (punctuated equilibrium).

Gradualism was first proposed in the late 18th century in reference to geological ideas. It was adopted by Charles Darwin in relation to evolution. Supporters of this view argue that the fossil record shows a succession of small changes in the phenotypes of species indicating that the process of speciation is a steady, ongoing one with transitional stages between major changes in a phylogenic line. In addition, they argue that since we do not see rapid evolution happening today in nature, we can conclude that the process happens gradually.

• **Examiner's hint:** Be sure to memorize the definitions for convergent and divergent evolution and know some examples.

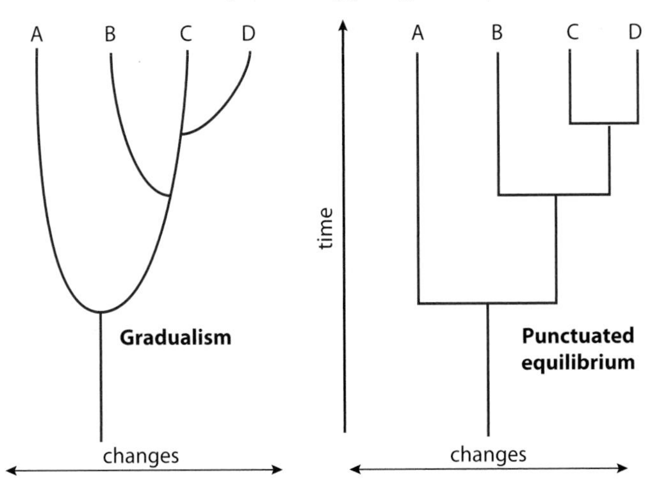

Figure 15.9 Gradualism and punctuated equilibrium are two contrasting views concerning the rate of speciation and evolution.

In contrast, those who support punctuated equilibrium, a theory which originated in the late 20th century, argue that speciation happens quickly, often in response to a change in the environment; for example, after a volcanic eruption or a meteorite impact, or a major climate change. In response, some species are destroyed and others adapt to their new surroundings, exploiting the niches made available by the extinction of old ones. This was certainly the case for the species of mammals which took over the habitats abandoned by dinosaurs 65 million years ago.

The rest of the time, species live for millions of years with little or no change. This can be confirmed in the fossil record, notably when studying sharks, cockroaches or horseshoe crabs which have persisted for hundreds of millions of years.

Critics of punctuated equilibrium argue that the 'jumpy' effect of this theory could simply be an artefact of the incompleteness of the fossil record. Discontinuities in the fossil lineages are a challenge for scientists to explain.

One difficulty of supporting either claim is that the only evidence we can use is fossil evidence. However, some of the things which help to define a species are not fossilized, such as pigmentation, behaviour or mating calls and songs.

Another argument states that just because a fossil of an extinct crocodile looks very similar to a modern-day crocodile, there is no proof that the latter is a direct descendent of the former or that the two species would have been able to reproduce together had they been contemporaries.

Transient polymorphism

Within a population there is often more than one common form. Different versions of a species are referred to as polymorphisms (many shapes) and can be the result of a mutation. One example of such an organism is *Biston betularia*, the peppered moth, which lives in temperate climates.

On close examination, you should see two moths on the tree trunk.

This species of moth can have a peppered (grey) form or a melanic (black) form. The grey form is well camouflaged on tree trunks under normal conditions. They are usually much more numerous in the population for the simple reason that black moths are seen more easily and thus more frequently preyed upon by birds.

If a moth gets eaten, it is obvious that it can no longer pass on its traits to the next generation. Despite this, the proportions of black and grey moth populations have changed over time, showing transient polymorphism, meaning that the changes are only temporary.

During the Industrial Revolution in the UK, factories belched out large black clouds of smoke which resulted in particles of carbon-rich soot sticking to tree trunks near big cities. An interesting thing happened to the peppered moth population: the melanic form became more numerous than the grey one. Why was this? Due to the blackening of the tree trunks by pollution, the grey moths were no longer camouflaged and were preyed upon instead of the black moths which could hide better on a dark background.

This phenomenon of factory pollution changing the population of peppered moths is called industrial melanism. It is a good example of how natural selection works. At no point did a single moth choose to be one colour or the other – it is natural selection operating through the predation by birds which favours one colour or the other. This is also a good illustration of the fact that there is no such thing as an allele or trait which is intrinsically 'good' or 'bad'. Rather, we should use terms such as 'fit' and 'unfit' to describe an organism's combination of alleles and traits. When tree trunks are clean, grey moths are more fit for survival because they can hide better from predators.

Today, because of efforts to reduce air pollution in the UK, notably from the Clean Air Act and the use of fuels and power plants which produce less air pollution, the air and the tree trunks have much less soot. As a result, the population of peppered moths has returned to having a very small number of black moths and a high number of grey moths. Again, since the change was only temporary, it is considered transient polymorphism.

Balanced polymorphism

When two or more alleles within a population are not transient and changing but are stabilized by natural selection, this is called balanced polymorphism. An example of this is seen in the allele for normal red blood cells and the allele for sickle cells. You will recall from Chapter 4 that the curved shape of red blood cells in people with sickle cell anaemia is caused by a recessive allele. The populations of people with sickle cell anaemia are generally from West Africa or from the Mediterranean.

Although sickle cell anaemia is a debilitating condition, those who have it are very resistant to malaria infection. Malaria is an infectious disease which occurs in tropical regions. A parasite called *Plasmodium* is transmitted to human blood by an infected female *Anopheles* mosquito taking a bloodmeal. The parasite attacks the person's red blood cells and produces symptoms of high fever and chills, and can result in death.

Malaria has become the most deadly tropical disease in the world. It is estimated that 1 to 3.5 million people die of malaria each year. This is in the same order of magnitude as the number of AIDS-related deaths each year.

In terms of the polymorphism of human red blood cells, most people are homozygous for disk-shaped red blood cells ($\mathbf{Hb^AHb^A}$). These people are highly susceptible to malaria infection. People who are heterozygous ($\mathbf{Hb^AHb^S}$) have the sickle cell trait. They have some sickle-shaped cells and some disk-shaped ones and they do not suffer from anaemia in most cases. Heterozygotes have a better resistance to malaria due to chemical imbalances which make the survival of *Plasmodium* in their blood more difficult. The insufficient quantities of potassium in sickle-shaped cells causes *Plasmodium* to die.

People who are homozygous for sickle-shaped cells ($\mathbf{Hb^SHb^S}$) have only curved cells and suffer severe anaemia which can sometimes be fatal. On a positive note, they are highly resistant to malaria.

Because of this paradox, the allele frequency for $\mathbf{Hb^S}$ is relatively stable and therefore shows balanced polymorphism. The balance is maintained by two pressures of selection. On the one hand, the $\mathbf{Hb^S}$ allele should be *selected against* because it can be debilitating or lethal. On the other hand, there is *selection for* it because having it gives people resistance to malaria. The balance is reached in heterozygotes who tend to be more fit for survival in zones plagued by malaria but do not suffer severe anaemia.

Exercises

6 Contrast geographical isolation and temporal isolation.
7 Explain why a zebroid is not considered a new species.
8 Thirty years ago, humans introduced lizards onto a tropical island which had none. Researchers are analysing the thriving lizard populations and expect to find evidence of adaptive radiation. Outline the kinds of things they would be looking for.
9 A man with sickle cell trait is climbing high in the mountains of Zaire. He has much more difficulty breathing at these altitudes than his fellow climbers. With reference to his genotype and phenotype, explain the reason for this difficulty.

D.3 Human evolution

Assessment statements

D.3.1 Outline the method for dating rocks and fossils using radioisotopes, with reference to ^{14}C and ^{40}K.

D.3.2 Define *half-life*.

D.3.3 Deduce the approximate age of materials based on a simple decay curve for a radioisotope.

D.3.4 Describe the major anatomical features that define humans as primates.

D.3.5 Outline the trends illustrated by the fossils of *Ardipithecus ramidus*, *Australopithecus* including *A. afarensis* and *A. africanus*, and *Homo* including *H. habilis*, *H. erectus*, *H. neanderthalensis* and *H. sapiens*.

D.3.6 State that, at various stages in hominid evolution, several species may have coexisted.

D.3.7 Discuss the incompleteness of the fossil record and the resulting uncertainties about human evolution.

D.3.8 Discuss the correlation between the change in diet and increase in brain size during hominid evolution.

D.3.9 Distinguish between *genetic* and *cultural evolution*.

D.3.10 Discuss the relative importance of genetic and cultural evolution in the recent evolution of humans.

Dating rocks and fossils

It is true that 'you can't get blood from a stone' but you certainly can get a story out of one. The job of geologists is to get rocks to reveal their past. One way of doing this is to study the isotopes in a rock sample to obtain its age.

Isotopes

An isotope is a version of an element with a different number of neutrons from the usual version of the same element (see key fact box). Carbon can exist as three different isotopes: ^{12}C (pronounced carbon-12), ^{13}C (pronounced carbon-13), and ^{14}C (pronounced carbon-14). The first isotope is by far the most abundant form of carbon (over 98%); ^{14}C is found only in trace quantities, meaning that it represents only one carbon atom out of about 1 000 000 000 000.

What is an isotope? You will recall that atoms are made up of protons, neutrons and electrons. Change the number of protons and you have a different element (e.g. carbon plus one proton gives nitrogen). Change the number of electrons and you have an ion (e.g. sodium minus one electron gives Na^+). Change the number of neutrons and the only thing that changes is the mass number, sometimes called the atomic weight. For example, ^{235}U (uranium-235) plus three neutrons gives ^{238}U (uranium-238). So isotopes are versions of an element with different mass numbers.

In the table below, you can see that the carbon isotopes differ in the number of neutrons and their mass number, which we calculate by adding the number of protons to the number of neutrons.

Isotope	Mass number	Protons	Electrons	Neutrons
^{14}C	14	6	6	8
^{13}C	13	6	6	7
^{12}C	12	6	6	6

If you were to take an equal number of atoms of each isotope (say, 10^{23} of them, which is about a big spoonful) and weighed them on a balance, ^{14}C would be the heaviest and ^{12}C the lightest, but not by very much. However, it is not recommended that you try this experiment because ^{14}C is radioactive.

Radioisotopes

Certain isotopes are unstable and in order to reach a stable state, they release some of their subatomic particles. These particles, such as alpha particles or beta particles, flying out of an atom are what constitute radioactivity. Too many of them shooting through your cells can cause severe damage to your tissues and organs.

Isotopes which have this unstable property are called radioisotopes. In trace quantities, diluted among many other non-radioactive isotopes, these are not harmful to our health. As a matter of fact, inside you right now are millions of ^{14}C radioactive atoms. The radiation from them is too weak to be dangerous but strong enough to be measured in a laboratory.

Inside the paper of the book you are reading there are also ^{14}C atoms. The difference in the quantity of ^{14}C in you and in the book, is changing. Why? Because you introduce new ^{14}C atoms into your body every day when you eat. In contrast, the trees which were used in making the paper of this book have ceased to photosynthesize and cannot take in new carbon sources. Over time, the ^{14}C in the paper will decay to become stable isotopes. After thousands of years, there will be no more ^{14}C in the dead organic material that constitutes this book.

To decay, in chemistry terms, means to pass from an unstable, radioactive parent isotope to a stable, non-radioactive daughter isotope. For example, ^{40}K (potassium-40) decays to ^{40}Ar (argon-40).

By looking at the proportion of ^{14}C to other isotopes, we can get an idea of the age of a fossil. If we find a shell which contains a high percentage of ^{14}C, we can deduce that the organism which made it died recently because the radioisotope has not yet had time to decay. On the other hand, if we find a fossil of a shell which has very little ^{14}C remaining, we can deduce that the radioisotope has had sufficient time to decay which means the fossil is many thousands of years old (see Figure 15.10).

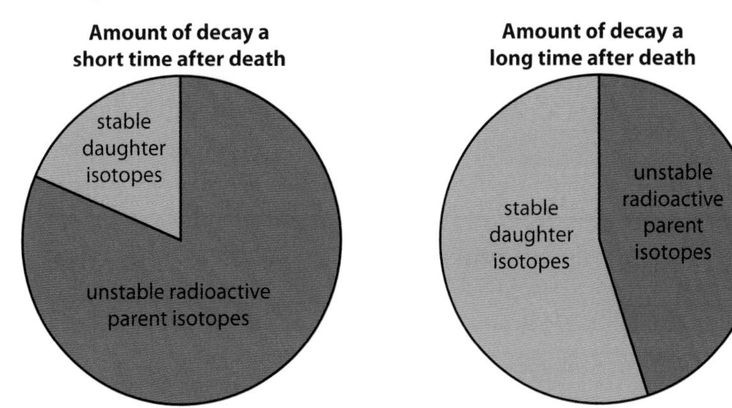

Figure 15.10 The proportions of radioisotopes and stable daughter isotopes in a once-living organism indicate the passage of time since the organism died. The higher the proportion of stable daughter isotopes, the older the fossil.

Isotopes of carbon and also of calcium have been examined in ancient human-like fossils to determine if early human ancestors ate only fruits or if they also ate meat.

What about rocks? They do not eat or photosynthesize and some do not contain carbon at all, so ^{14}C dating is not possible. Instead other radioisotopes are examined, such as ^{40}K (potassium-40). When the minerals in rocks crystallize from magma, they contain a certain percentage of ^{40}K ions. Once the minerals have hardened and crystallized, no more ^{40}K ions can be added. However, the number reduces as the radioisotope decays into more stable forms. Just as with ^{14}C, ^{40}K radiometric dating can be a useful tool in determining the age of a sample studied in a laboratory. Radiometric techniques with ^{40}K can be used to measure the age of rocks which formed from magma or lava between 100 000 years and 4.6 billion years ago.

Half-life

How can we transform percentages or proportions of isotopes into years to determine the age of something? First, you must understand the concept of half-life. A half-life is the amount of time it takes for half the radioactive isotope in a substance to decay.

• **Examiner's hint:** Be sure to use scientific definitions in exams. Analogies, like the one used here, are an aid to help you understand – but you will get zero marks if you use an analogy in an exam. You must know and use the definitions.

A useful analogy for radioactive decay is to imagine a television game show where half the contestants are eliminated from the game in every round. Each round lasts 10 minutes and the game starts with 200 contestants. After the first round, only 100 are left; another 10-minute round later, there are only 50. If you turn on your television and the show has already started, you can look at the number of contestants left to see how long ago it started. With 25 contestants remaining, you know the show started at least 30 minutes ago. A 10-minute round is analogous to one 'half-life' for the number of contestants remaining.

This table shows what happens over time to ^{14}C.

Number of half-lives	Number of years which have passed	% of original radioisotope remaining
1	5730	50
2	11460	25
3	17190	12.5
4	22920	6.25
etc.	etc.	etc.

You will recall from your chemistry class that the number of protons in an atom defines what type of element the atom is. This number is its atomic number, which can be thought of as its identity number. Elements are arranged in the Periodic Table by increasing atomic number.

If, after four half-lives, there is only 6.25% of the starting amount of ^{14}C left, where did the other 93.75% go? Radiocarbon, ^{14}C, decays into its stable daughter isotope, which is ^{14}N (nitrogen-14). This type of decay is possible because a particle called a beta particle is released, converting one of the carbon's neutrons into a proton. In doing so, it changes the identity of the carbon atom (which had 6 protons) into a nitrogen atom (because it now has 7 protons).

• **Examiner's hint:** In exams, avoid imprecise notation (such as C14 or K40) for radioisotopes such as ^{14}C or ^{40}K. Make sure you know the difference between half-life and radiocarbon dating as well as how the two are related. You are expected to know how to explain the way a decay curve works and how to convert half-lives into years.

But after 12 half-lives, there is only 0.024% of the original quantity of radioisotope left, which approaches the limit of the precision of the tools used in laboratories, (remember, there was only one in a trillion ^{14}C atoms to begin with). For ^{14}C, one half life is 5730 years, so 12 half-lives is about 70 000 years. This means that radiocarbon dating cannot be used to determine the age of samples older than this. Other radioisotopes must be used, such as ^{40}K (for which the half-life is 1.3 billion years) or ^{238}U (for which the half-life is 4.5 billion years).

Using a decay curve

If we graph the values in the table opposite, we obtain a curve as shown in Figure 15.11. Notice that the curve decreases exponentially with time. Theoretically, it will never reach zero although, in reality, there will eventually be no more radioactive isotope in any given sample.

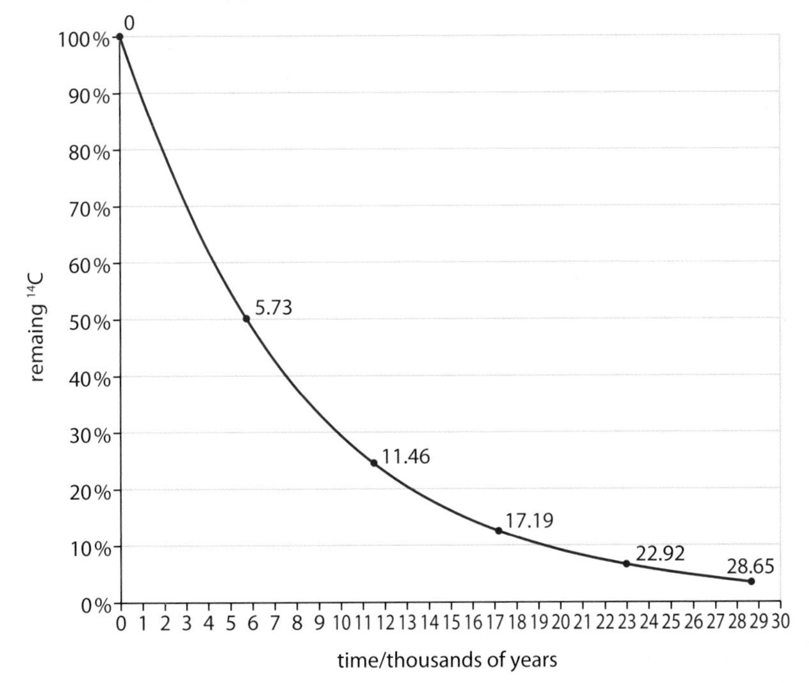

Figure 15.11 The effect of time on the proportion of radioisotope present in a fossil. The numbers on the curve show the age (in thousands of years) after each successive half-life.

You can use a ruler on this graph to estimate the age of a sample found to contain 30% of its original isotopes. A value of about 10 000 years on the *x* axis corresponds to the 30% mark on the *y* axis. Thus, if laboratory results from the analysis of a fossil show that only 30% of the original ^{14}C still remains, it is about 10 000 years old.

All measurements are approximations. There is no such thing as a measurement which is 100% precise, but in science, we try to be as precise as possible. The precision of radiocarbon dating has been estimated at ± 30 years (at best) to ± several hundred years (at worst). Does this mean we should question the validity of radiocarbon dating? Critics of the technique say that such a wide margin of error reduces the credibility of its results to a point where we should not take the numbers seriously. In addition, the calculations are based on the assumption that radioactive decay has been constant for millions of years. What do you think? Can we trust the dates determined by radioisotope decay?

Humans as primates

Do you consider yourself a primate? Biologists do. Why? Because humans possess the physical features that define primates. Unlike other mammals such as dogs, leopards or anteaters, primates can grab hold of things with their limbs. Opposable thumbs allow our hands (and feet for certain primates) to grasp objects. Long, thin, straight fingers allow for fine motor skills which enable primates to obtain food and manipulate objects or tools. Primate fingers have finger pads and finger prints. We primates lack claws and have fingernails instead.

Because the shoulder sockets allow the arms to make complete circles, primates have an increased range of movement of the arms for greater mobility, notably for swinging through trees. Such freedom of limb movement is impossible in other animals, which is why it is much safer to stand next to a horse than behind it and why cats cannot swing from branch to branch the way monkeys can.

Lastly, all primates have forward-facing eyes which allow for stereoscopic vision (seeing in three dimensions). Your highly developed primate brain interprets slight differences in the images received by your left eye and by your right eye. This is how you can judge distances. If a monkey did not have stereoscopic vision, it would not be able to judge the distance to the next branch and could fall to its death.

You can test your three-dimensional vision for yourself by holding a pen vertically on your desk with the tip pointing straight up to the ceiling. In the other hand, take another pen with the tip pointing down and try to place its tip on the tip of the pen on the desk. With stereoscopic vision, this task is usually easy. Using only one eye, it is much more challenging because you have no depth perception.

Trends in hominid and human fossils

If you could place together all the skulls and bones of all the hominid species which anthropologists have dug up since the first Neanderthal skull was found in 1856, you would see that there are some striking similarities but also some major differences between them.

Fossil skulls are often found in multiple fragments which must be painstakingly re-assembled.

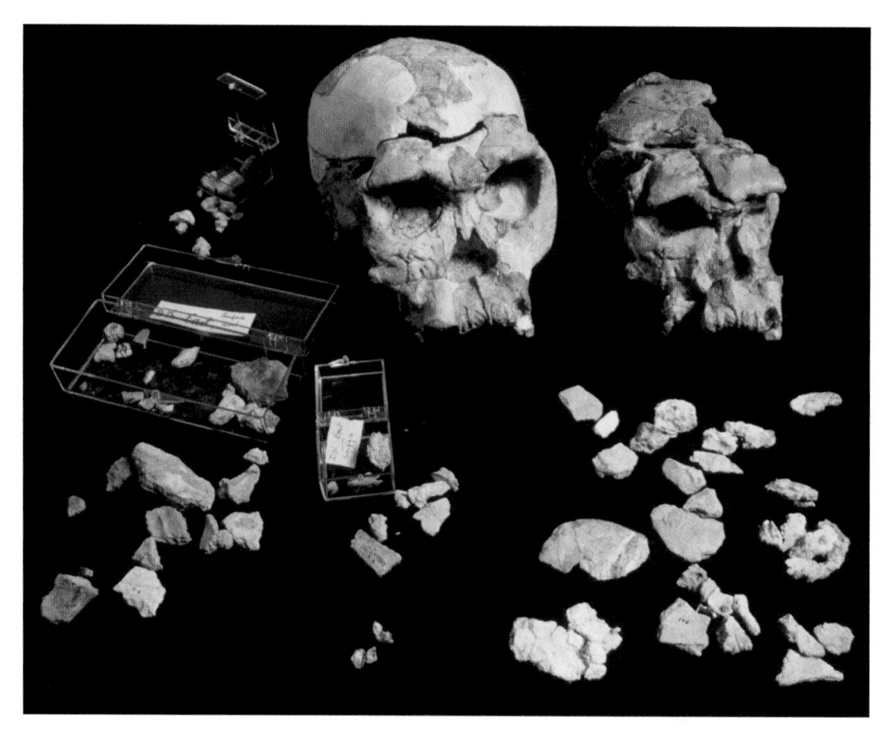

There is some discussion as to how appropriate the term 'hominid' is.

It can refer to only those bipedal primates which are direct ancestors of modern humans. Some people use 'hominid' to refer to any member of the Hominidae superfamily, to which all apes and humans belong, both living and extinct.

This chapter uses the term 'hominid' to refer to bipedal primates of the genera *Homo* and *Australopithecus* plus *Ardipithecus ramidus*.

The details to pay close attention to when looking at hominid skulls are summarized in Figure 15.12 and its accompanying table.

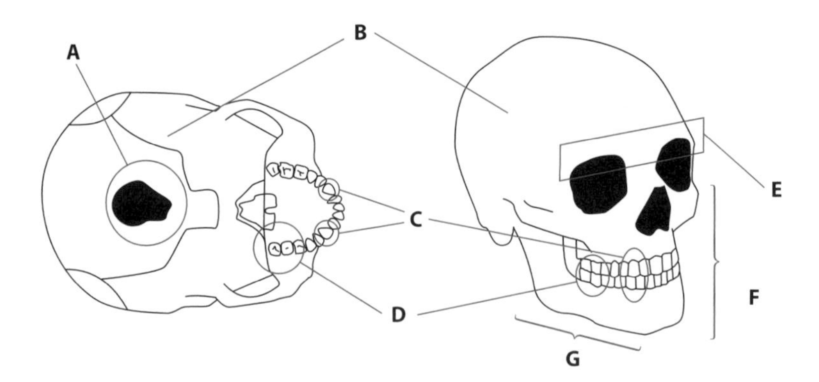

Figure 15.12 Diagram of the anatomy of a human skull.

Anatomical feature	Ape-like form	Human-like form
A position of foramen magnum	towards the back of the skull	towards the centre of the base of the skull
B cranial capacity	small	large
C canine teeth	long and sharp	short and dull
D molars	long and narrow	short and wide
E brow ridge	protruding	flat
F face below brow	protruding	flat
G jaw	tall and thick	small and thin

The foramen magnum, labelled **A** in Figure 15.12, is the hole where the skull is attached to the spinal column. In bipedal organisms such as modern humans, the hole is at the centre of the base of the skull. In apes, it is more towards the back because this is a better entryway for the spinal column of an animal which walks on all fours.

In chronological order, here are some of the species of hominid for which fossils have been found.

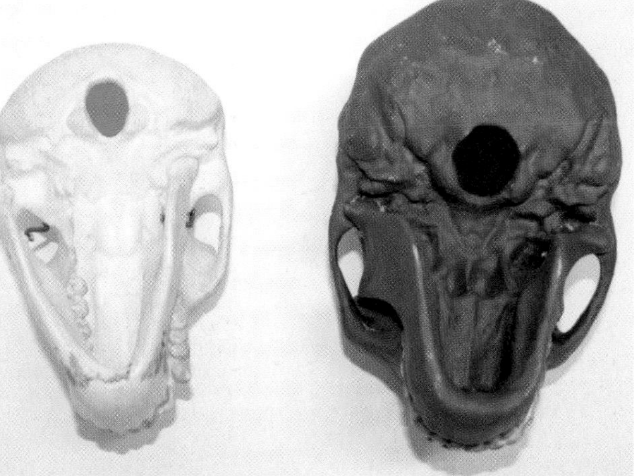

The position of the foramen magnum in the chimpanzee skull model (on the left) is quite different from that of *H. erectus* (right). The position at the back of the ape's skull shows that it walked on all fours while the hominid was a biped.

Ardipithecus ramidus

Ardipithecus ramidus lived approximately 5.8 million to 4.4 million years ago in Ethiopia. This species is believed to be very close to the split between the line of organisms which became more human-like and the line which became more chimpanzee-like. Most fossils of this organism are teeth, so it is difficult to be sure of the organism's physical features.

We can assume that *Ardipithecus ramidus* was very similar to a chimpanzee with a few striking hominid features. For example, the molars show ape-like characteristics in that the length of these teeth is greater than their breadth, whereas in later hominids, these two dimensions are much more similar. The canines, however, are more hominid-like because they are not sharp and they are shorter and less projecting than ape-like canines.

Fossils of skull fragments seem to indicate bipedalism was possible, although some scientists prefer to wait until more fossils are found to confirm or deny this.

Our closest living relatives are the gorilla, the orangutan and the chimpanzee. We share a large percentage of our DNA with these great apes. At some point in the past, we had a common ancestor. Neither Charles Darwin nor any other supporter of evolution ever suggested that humans are descended from chimpanzees.

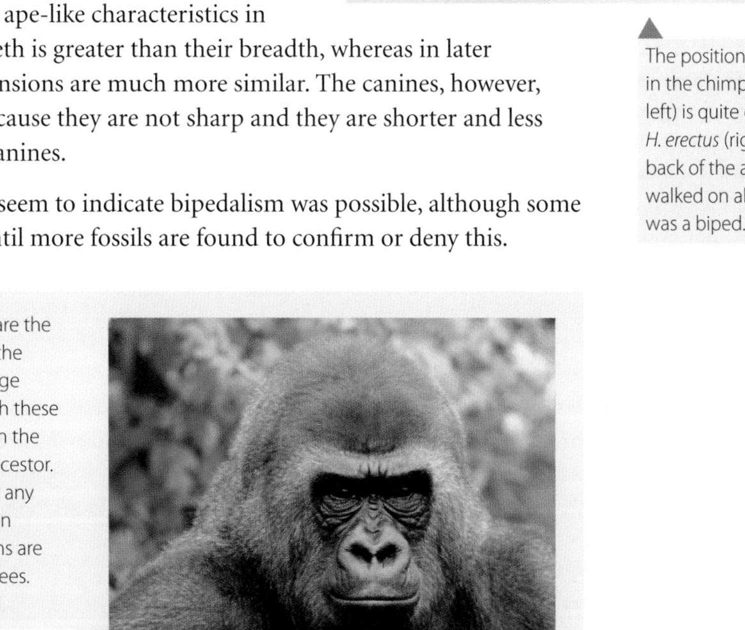

The gorilla, the chimpanzee and the orangutan can be thought of as distant cousins on the tree of life but certainly not as our ancestors.

Australopithecus afarensis

This species of hominid lived between 4 million and 2.5 million years ago in eastern Africa. The most famous *A. afarensis* is 'Lucy', whose remains have been dated as 3.5 million years old. This Australopithecine had a tall lower jaw, fairly large molar teeth and a projecting face. The cranial capacity was 380–430 cm^3.

Chimpanzee brain size is approximately 400 cm^3 and gorillas typically have brains of about 500 cm^3. Modern humans of typical adult size rarely have brain capacities less that 900 cm^3, and more commonly have 1350 cm^3.

Australopithecus africanus

A. africanus lived in southern Africa between 3 million and less than 2.5 million years ago. It had a tall, thick lower jaw, large molars and a projecting face. The cranial capacity was 435–530 cm^3.

Homo habilis

Fossils of *H. habilis* were unearthed first in the Olduvai Gorge in Tanzania, and later discovered in Kenya, Ethiopia, and South Africa. *H. habilis* lived from 2.4 million to 1.6 million years ago. It had a flatter face and larger molars than *Australopithecus* species (see Figure 15.13). The cranial capacity of this species was still only 600 cm^3 but that was enough to allow it to use simple stone tools and fire. Some lived in caves.

Figure 15.13 Different views of *H. habilis* skull.

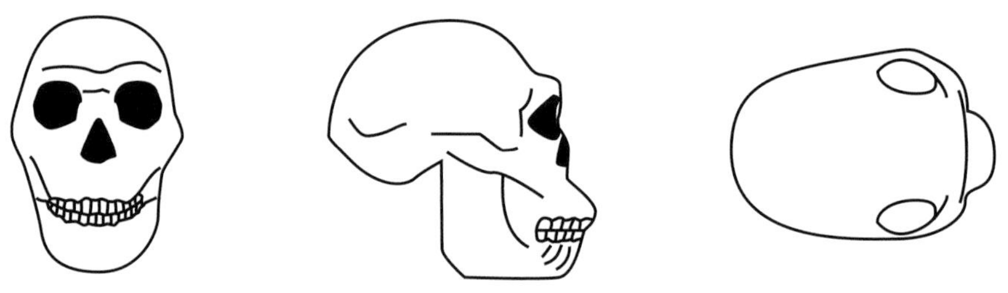

Homo erectus

This species lived from 1.8 million years ago to about 100 000 years ago. It had a smaller jaw, a receding forehead, large brow ridges and smaller molars. Its cranial capacity was approximately 1000 cm^3.

For decades, *H. erectus* was thought to be the first migratory human ancestor because fossils have been found in Europe, India, China and Indonesia as well as Africa. New fossil evidence in Europe, however, suggests that *H. erectus* was not the first species to migrate there (see Weblinks 15.3a and 15.3b, page 440).

Homo neanderthalensis

H. neanderthalensis, more commonly known as 'Neanderthal man', lived in Europe and western Asia from about 200 000 to 30 000 years ago. The species survived several ice ages.

H. neanderthalensis had a smaller jaw, a lower forehead, smaller brow ridges and smaller molars than previous species (see Figure 15.14). Paradoxically, they had larger brains than modern humans, with a cranial capacity of up to 1600 cm^3.

Figure 15.14 Different views of *H. neanderthalensis* skull.

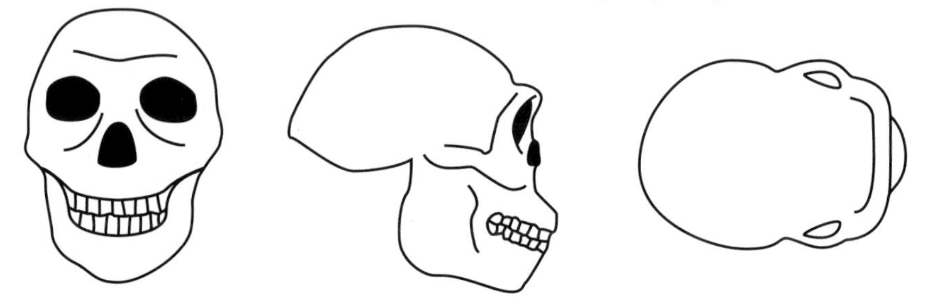

Homo sapiens

Archaic forms of *H. sapiens* date back to between 140 000 and 70 000 years ago in Africa, Europe and Asia. The skulls are characterized by a high forehead, no brow ridges, a flat face, small molars and a very small jaw (see Figure 15.15). This species developed art in the form of cave paintings and technology in the form of finely crafted tools and weapons. Their cranial capacity was about 1300 cm^3, very similar to that of today's humans.

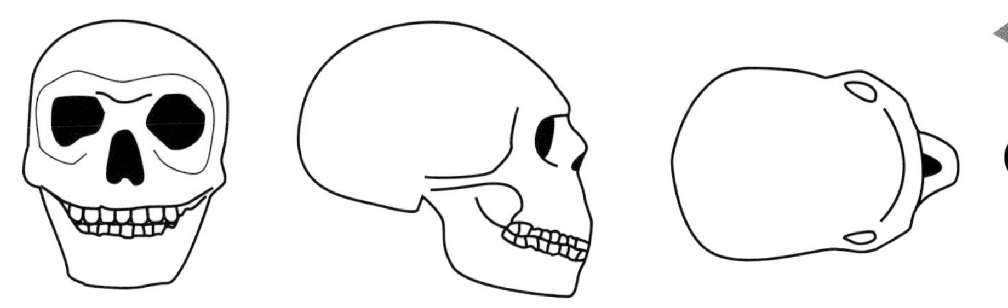

Figure 15.15 Different views of *H. sapiens* skull.

H. erectus fossils were not only found in Africa but also in Asia. The species has been dubbed 'Peking Man' and 'Java Man' after the places where their fossils have been discovered. Archaic *H. sapiens* included the sub-group 'Cro-Magnon man'. You do not need to know these particular groups for the exams.

'The Human Origins Program at the Smithsonian Institute' has more photos and dates.

Visit heinemann.co.uk/hotlinks, enter the express code 4242P and click on Weblink 15.2.

Possible coexistence of several hominid species

By examining the results of the dating techniques applied to the various hominid fossils, it becomes apparent that some species existed on Earth at the same time (see Figure 15.16). Overlap in fossil ages is present in:

- *A. afarensis* and *A. africanus* – approximately 3 million years ago;
- *H. erectus*, *H. neanderthalensis*, and *H. sapiens* – approximately 100 000 years ago.

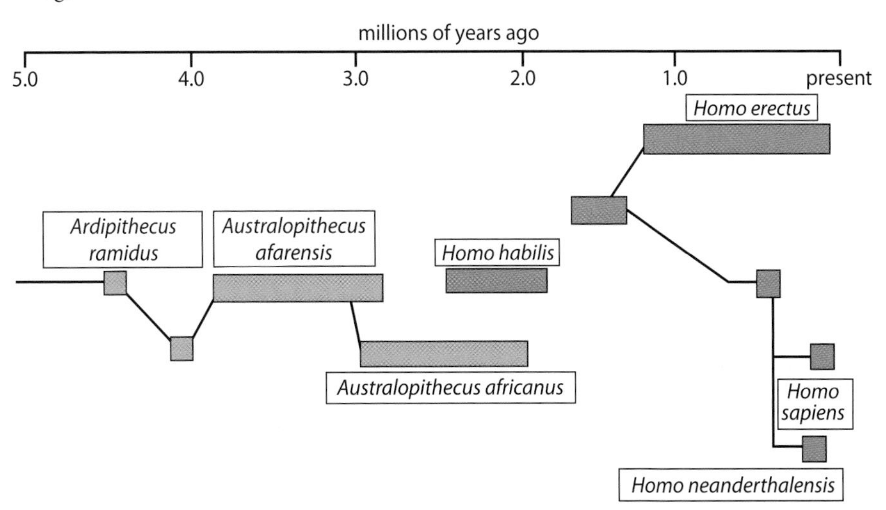

Figure 15.16 A phylogenetic tree showing the overlap in time between certain species of hominid.

The next question is, 'Did they live in the same regions and is there a possibility that they interacted with each other?' That is more difficult to determine and is open to interpretation. There were so few Australopithecines that it is possible they never bumped into each other. Their coexistence spans a million years, however, which means that there was plenty of time for chance encounters. On the other hand, there is evidence that *H. neanderthalensis* and *H. sapiens* probably interacted with each other but most certainly never had fertile offspring together.

We are talking about a time when the numbers of *H. sapiens* or *H. erectus* were far fewer than the number of humans alive today. It is difficult to imagine, but the entire world population then was not yet in the millions, probably only in the tens of thousands of individuals, spread over Africa, Europe and Asia. In addition, the simple presence of a mountain range or a wide river could potentially prevent two populations from ever encountering each other, even if they were separated by only a few tens of kilometres.

Uncertainties about human evolution

Trying to reconstruct 5 million years of human evolution using only bits and pieces of skulls and skeletons is challenging to say the least. It is analogous to trying to imagine the image represented on a 2000 piece jigsaw puzzle when you have only found a few dozen pieces.

W

Wikipedia's 'List of Hominina fossils' has a table of fossils with their ages, places where they were found and, in many cases, a photo. Since Wikipedia is open source and user-generated, we must double-check any information given. It is a bad idea to use it as a main resource, but it can be a good starting point. Visit heinemann.co.uk/hotlinks, enter the express code 4242P and click on Weblink 15.3a.

For a description and photos of the hominid fossils which seem to be earlier than *H. erectus* and yet come from a species which managed to get to eastern Europe, visit heinemann.co.uk/hotlinks, enter the express code 4242P and click on Weblink 15.3b. Migration to eastern Europe was previously thought possible only by later *H. erectus*. The name *H. georgicus* has been proposed for this species.

Not all findings are genuine. The most famous case of forgery and hoax in a hominid fossil was Piltdown man, which fooled scientists for 40 years. Visit heinemann.co.uk/hotlinks, enter the express code 4242P and click on Weblink 15.3c.

On timelines of human evolution, like Figure 15.16, there are often gaps and several question marks. This is because there are many fossils for certain species of hominid and sometimes only a handful for other species. Also, in determining a characteristic such as cranial capacity, there can be significant differences between males, females and juveniles, making estimations imprecise.
In order to be found today, a few things must be true about a fossil.

1 It must be physically accessible – its geographical location must be reachable without too much difficulty, not too deep under layers of rock and not covered by thick vegetation.
2 It must be preserved well enough to be identified – not destroyed by predators, climate, water damage, war, or any of a myriad of other chance mishaps. This is especially difficult for the soft parts such as skin, flesh or hair.
3 It must be politically or legally accessible – some countries refuse to allow foreign scientists to dig for fossils on their territory and sometimes fossils are on protected land which cannot be excavated.

Expeditions require money and few governments or universities are willing to finance costly projects unless researchers are almost sure they will find something of interest.

The oldest fossils are the most difficult to find for the simple reason that fossils degenerate over time. It is logical that most of the oldest ones have been reduced to powder. Also, the hominid population was significantly smaller so long ago, so there are relatively few remains to become fossilized.

The consequence of all this is that having an incomplete set of fossils leads to multiple hypotheses. Every book you consult or every webpage you visit will probably have a slightly different timeline of human evolution showing slightly different dates and different phylogenetic connections between species. As new evidence is uncovered, some hypotheses are confirmed and others are refuted.

Some hypotheses have been overturned. For example, the hypothesis that Neanderthals were our ancestors has been refuted. Also, the hypothesis that modern *H. sapiens* have evolved separately in several different places in the world has not been confirmed by accumulating evidence. However, the overall basic hypotheses of paleoanthropology (e.g. humans originated in Africa, spread to Eurasia; brain and body have evolved over time) have been confirmed over the decades.

The totality of all hominid fossils could easily fit in one room in a museum. This makes paleoanthropology a data-poor science. Consequently, paradigm shifts are more frequent than with data-rich sciences and new discoveries can more easily falsify old hypotheses.

Changes in brain size during hominid evolution

Although there are numerous advantages to having a larger brain, there are some disadvantages as well. Most importantly, a bigger brain requires more energy to function. That energy has to come from somewhere and there is more than one solution:

- more foods in the diet;
- food richer in energy and protein.

Hominids seem to have used the second solution; the extra energy came from more meat in the diet.

Like their ape-like ancestors, the earliest hominids were foragers, gathering fruits and nuts, occasionally eating meat if they could get it. For hominids to eat meat on a regular basis, a more complex social system was necessary. Hunting alone can mean repeatedly coming home empty-handed whereas hunting with other members of a social group increases the chances of making a kill and having enough meat for the whole group.

When tools became more sophisticated, hunting techniques improved and the availability of high-protein foods increased. Besides meat, the high-protein diet could have included insects such as termites, which are not only rich in fats and proteins but also have medicinal antiseptic properties.

Large quantities of animal bones found at the fossil sites of early humans suggest that meat was an important part of their diet. Both the development of meat in the diet and the increase in complexity of tools needed for this dietary evolution show a direct correlation with bigger brain sizes. With high-protein, high-energy foods such as meat, hominids were able to fulfil the nutritional requirements of their demanding brains. With a bigger brain, better tool-making and hunting techniques could be developed and supply even more energy. This phenomenon continued for millions of years.

Does bigger cranial capacity automatically mean higher intelligence? When comparing a chimpanzee to a modern human, the answer is clearly 'yes' if you measure intelligence by problem-solving skills, by the ability to learn new things and by abstract thought, communication skills and creativity. But when comparing Neanderthals with *H. sapiens*, it is impossible to say for sure if one was more intelligent than the other.

The cranial capacities of several great mathematicians, statesmen and authors have been measured and did not always show a correlation between size and intelligence. On the contrary, some had brains which were surprisingly below average size. It is paradoxical that the organ which has allowed us to make so much progress is still so mysterious and uncharted today.

In the evolution of hominids, there is a general trend showing an increase in cranial capacity.

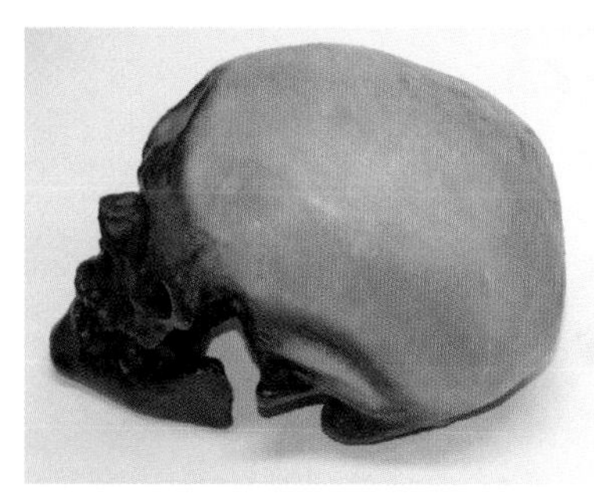
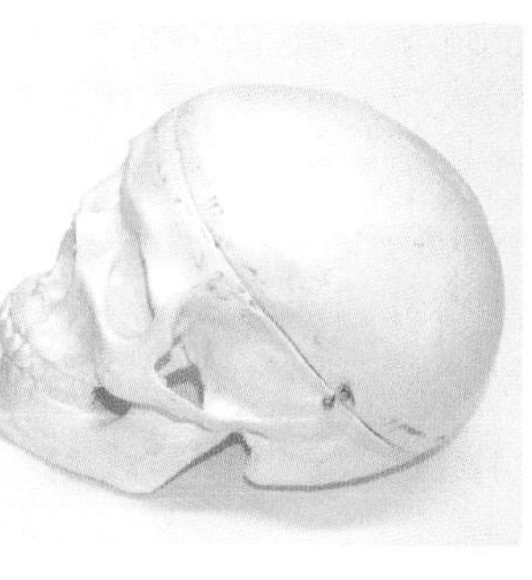

These full-scale models of an archaic human (in this case Cro-Magnon man on the left) and of a modern human (on the right) show that our ancestors sometimes had a larger cranial capacity than we do today.

The practice of cooking food increases its digestibility, allowing the gut to save energy. Raw meat and vegetables are more difficult to chew and to break down in the gut, forcing the body to burn valuable energy to liberate the nutrients. Cooking food loosens up starch molecules in vegetable matter and makes meat proteins more digestible. A certain degree of intelligence, and therefore a relatively big brain, is needed to control fire and cook food.

One counter argument to the idea that diet improved brain size is that some extinct species of hominids such as Neanderthals, had brain capacities which were larger than modern humans and yet they did not necessarily have significantly different diets.

Genetic versus cultural evolution

There are two major ways in which humans have evolved: genetically and culturally. Genetic evolution only deals with *inherited characteristics*:

- body morphology (cranial capacity, skull shape, height, robustness);
- number of chromosomes;
- particularities of biochemicals such as blood proteins.

These characteristics can be passed on from parent to offspring only through chromosomes. The features of hominids which would fall under genetic evolution include characteristics such as brain size, tooth anatomy, hips and feet which are well adapted for bipedalism.

On the other hand, cultural evolution deals with *acquired knowledge* which includes the following:

- language (spoken and written);
- customs and rituals (ethnic or religious, such as burying the dead);
- art (sculpture, pottery, painting, etc.);
- technology (for comfort and obtaining food, also for warfare).

These can be passed on to many individuals within a social group and within a family. The knowledge can pass from generation to generation or within the same generation by the use of language or non-verbal communication. This is in sharp contrast to genetic evolution which can only go from parent to child through genes.

It is important to remember that the whole process of the increase in hominid cranial capacity was through natural selection. Recall the steps:

- over-production of offspring;
- variety in the population so some offspring had bigger brains than others;
- bigger brains could have meant better problem-solving skills, thus better fitness due to an ability to find food and shelter more successfully;
- better fitness would increase the chances of survival;
- those who survived passed on their genes to the next generation.

Be careful to use this reasoning and not the unscientific idea that 'they wanted to be smarter so they made their brains bigger.' That is not true.

The development of abstract expression of ideas (such as cave paintings) is an example of cultural evolution. The development of the physical brains and hands which made this San (Bushman) rock art possible is an example of genetic evolution.

Importance of genetic and cultural evolution

One way to measure the culture of a people is to look at the quality, richness and complexity of their artefacts such as tools or artwork. The table below shows you some trends in the tools each species made.

Species	Example of culture	Tools developed
H. habilis	Oldowan	• simple choppers, scrapers and flakes of rock
H. erectus	Acheulian	• hand axes, cleavers and picks
H. neanderthalensis	Mousterian	• large flakes of uniform size produced from a core then trimmed to the desired tool
H. sapiens	Soultrian	• delicate blades for knives, burins (drills) • other materials added (bone, antler, and ivory) • some tools are ceremonial

From this information, we can see that as brain size increased, so did the quality of the tool-making. Clearly, there is a connection between cultural and genetic evolution. But which came first? Since it is very difficult to have better tool-making techniques until you have the brain capacity to improve them, the most logical conclusion is that genetic evolution preceded cultural evolution.

For the last 30 000 years, the evolution of *H. sapiens* has been largely cultural, not genetic. A fossil of a human dating back to when Cro-Magnon man lived, for example, would show few differences with a skull from today. Apart from a less robust body build and slightly higher cranial capacity, there has been little genetic evolution since then.

Culturally, however, the differences are enormous. Since the Upper Palaeolithic period (about 40 000 to 10 000 years ago), human culture has developed exponentially with languages, rituals, art, shelter-building, pottery and agriculture as well as the ability to work with metals and, more recently, nuclear energy, semiconductors and the capacity for space exploration.

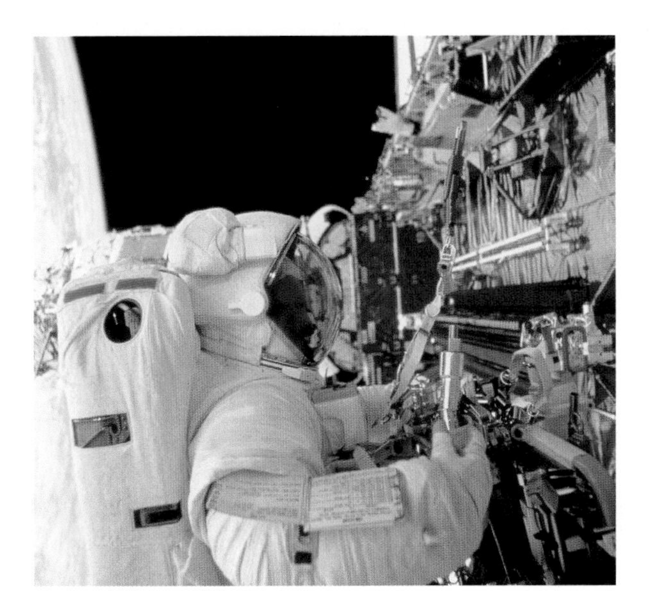

Thanks to cultural evolution, humans have gone from using stone tools to space travel in just over a thousand generations.

The question now is, will we be able to evolve culturally to the point where we can settle our territorial disputes, racial tensions and ecological injustices? Or will we need to wait for the evolution of a more intelligent species to do this? Only time will tell, but some of us are trying to use cultural evolution to spread the word and raise awareness. How about you?

How much of our survival as a species is due to our genetics and how much is due to our culture? Was it our superior intelligence or was it our social and technological advances? This is one example of the classic nature/nurture debate.

This shows how cultural evolution can sometimes lag behind genetic evolution. Just because a hominid's brain gets bigger does not mean that within a few decades, the population will take giant evolutionary steps in a cultural sense. Sometimes it takes many thousands of years for cultural improvements to develop. For example, our current cranial capacity was attained by archaic *Homo sapiens* approximately 140 000 years ago. However, major cultural advances such as sculptures, wall paintings and adornments do not seem to appear until about 35 000 years ago.

Once they occur, however, cultural developments such as controlling fire, producing cave paintings or improving weaponry spread quickly. In effect, cultural evolution has the potential to revolutionize a human population much more quickly than genetic evolution can.

Although genetic evolution was absolutely necessary for our ancestor's cultural evolution, in more recent history, this has not been the case. *Homo sapiens* has been able to evolve culturally far more than any of our ancestors without any major genetic evolution.

Exercises

10 Using the graph in Figure 15.11, determine the number of half-lives shown.

11 Using the same graph, deduce the age of a fossil which has only 10% of its ^{14}C remaining.

12 During an archaeological dig of a site which is estimated to have been functioning in 2000 BC, you find seeds in a container and want to use radiocarbon-dating techniques to confirm the age. Predict with a reason the percentage of ^{14}C atoms which have already decayed in the seeds. Again, use the Figure 15.11.

D.4 (HL only) The Hardy–Weinberg principle

Assessment statements

D.4.1 Explain how the Hardy–Weinberg equation is derived.

D.4.2 Calculate allele, genotype and phenotype frequencies for two alleles of a gene, using the Hardy–Weinberg equation.

D.4.3 State the assumptions made when the Hardy–Weinberg equation is used.

• **Examiner's hint:** The Hardy–Weinberg equation and the concept of Hardy–Weinberg equilibrium are among the most challenging concepts in IB Biology. Be sure to take the time to understand them and practise using the equation with the examples given. If anxiety about the mathematics is part of the problem, do not hesitate to ask for extra help in learning how to work through the calculations, which, once you get used to them, are really not that difficult.

The Hardy–Weinberg equation

In order to calculate the frequencies of alleles, genotypes or phenotypes within a population, the Hardy–Weinberg equation is needed. This is useful in determining how fast a population is changing or in predicting the outcomes of matings or crosses. To understand how it is used, it is best to start with a grasp of how it was derived.

You will recall from Chapter 4 that a Punnett grid shows the genotypes of the parents and offspring in a cross. For the Hardy–Weinberg equation, we need to look at the cross in a new way – as a model for the allele frequencies. To do this, we need the variables p and q:

- p = frequency of the dominant allele (allele **T** in the example below);
- q = frequency of the recessive allele (allele **t** in the example below).

When looked at individually, the frequencies of the alleles on chromosomes must add up to 1. So $p + q = 1$. For example, if $p = 0.25$ (or 25%) frequency, then $q = 0.75$ because whichever chromosomes do not have the dominant allele must carry the recessive one.

What complicates things is the fact that we usually want to consider diploid organisms that carry *two copies* of any particular gene. As a result, the equation becomes $(p + q)^2 = 1$. If you remember your mathematics classes about polynomials, you'll know that $(p + q)^2$ can be expanded to $p^2 + 2pq + q^2$.

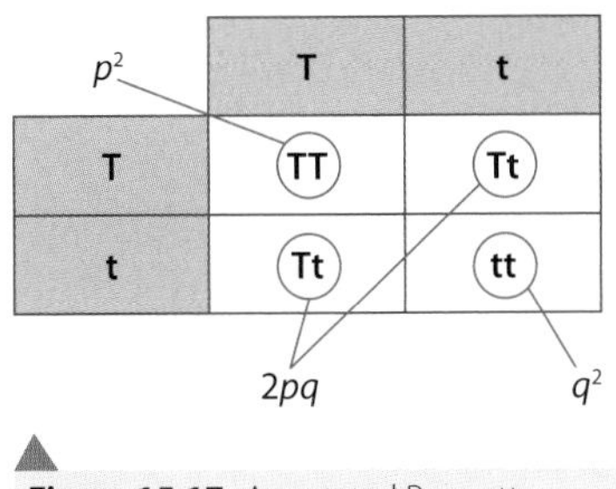

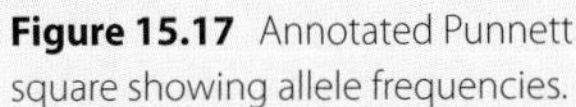
Figure 15.17 Annotated Punnett square showing allele frequencies.

If you are allergic to binomial expansion, here is another way of looking at it. In the cross between two heterozygous parents (see Figure 15.17), the first resulting square shows **TT**. To convert this to a frequency, it is p (which is the frequency of **T**) times p (again, the frequency of **T**). And we know that $p \times p$ is the same as saying p^2. For the **tt** offspring, the combined frequencies make $q \times q$ which is q^2. For the heterozygous offspring, **Tt**, the frequency is two times $p \times q$ which is $2pq$. If you get the hang of it by now, you should grasp Figure 15.17.

Since we said earlier that $(p + q)^2 = 1$ and we saw that $(p + q)^2$ can be expanded to $p^2 + 2pq + q^2$, we can now deduce that $p^2 + 2pq + q^2 = 1$. This mathematical representation for the allele frequencies is known as the Hardy–Weinberg equilibrium and it is reached after only one generation of random interbreeding.

If this is still a bit confusing, try looking at it this way. Let's summarize the Punnett grid in Figure 15.17 as shown in this table.

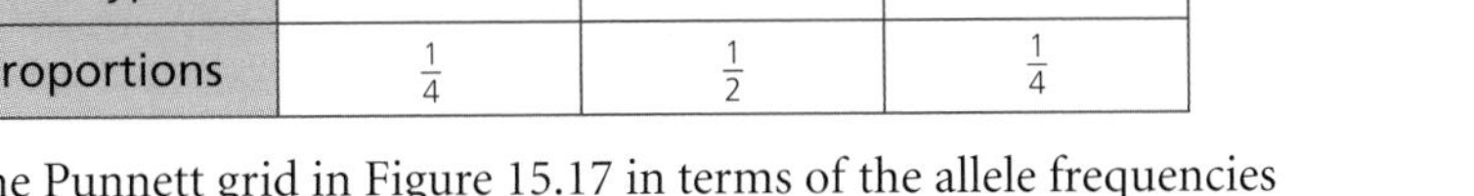

	One square	Two squares	One square
Genotypes	TT	2**Tt**	**tt**
Proportions	$\frac{1}{4}$	$\frac{1}{2}$	$\frac{1}{4}$

Looking at the Punnett grid in Figure 15.17 in terms of the allele frequencies rather than genotypes, the following can be deduced:

- the frequency of **TT** = p^2
- the frequency of **Tt** = $2pq$
- the frequency of **tt** = q^2

By adding up all the possible proportions, we can see that $\frac{1}{4} + \frac{1}{2} + \frac{1}{4}$ comes to a total of 1. Based on the frequencies, we can replace the proportions by the allele frequencies and deduce (as Weinberg and, later, Hardy did) the equation shown in Figure 15.18.

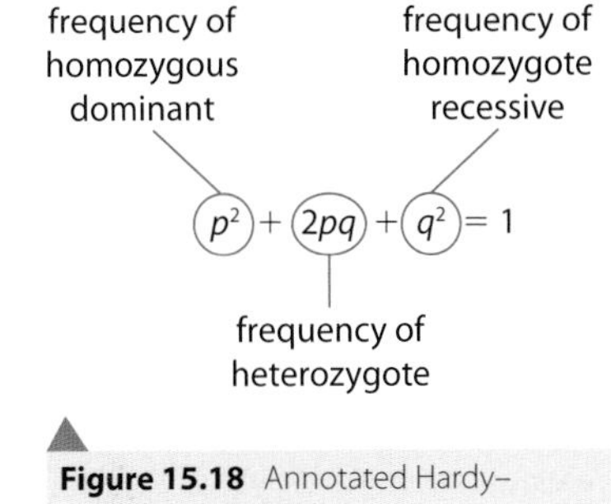

Figure 15.18 Annotated Hardy–Weinberg equation.

Before you move on to the next section, try the equation. If you substitute 0.25 for p and 0.75 for q in the equation, you should get 1. Be careful of the order of operations – anything that needs squaring should be done first, then perform any multiplications, then the additions.

Note that this equation gives mathematical support to the idea in Mendelian genetics that variation must be preserved from generation to generation. It is one of the characteristics of genetics which allows for a population to be successful.

How to use the Hardy–Weinberg equation

Here is a series of worked examples to help you roll up your sleeves and get some hands-on experience – you will not be able to fully understand how to use the Hardy–Weinberg equation without it.

Worked example 1 – *calculating allele frequency, part I*

Let us consider a disease caused by a recessive allele **t**. Let us say that the predicted frequency of this allele in the population being studied is 10%. Calculate the frequency of the healthy allele in the population.

Solution

From the information given, we know that q is 0.10 and since the proportions of p and q must always add up to 1, we can say that $p = 1 - q$. So p must be 0.90, which means that in the gene pool, 90% of the alleles are **T**.

Remember that this does not mean 90% of the population is healthy because we are calculating an allele frequency, not a genotype frequency.

Worked example 2 – *calculating allele frequency, part II*

In a study of 989 members of the population from example 1, it was found that 11 people had the disease. Calculate the frequency of the recessive allele **t**.

Solution

First, calculate the proportion of people who had the disease (which in this case is the proportion of **tt** genotypes). To do this, divide 11 by 989 to obtain 0.011. This means that 1.1% of the population has the genotype **tt**. Hence, $q^2 = 0.011$.

So to calculate q, we take the square root of 0.011 which gives us 0.105. This means that the frequency of the recessive allele **t** is 10.5% in this population.
Note that this is very close to the predicted value of 10% in example 1.

● **Examiner's hint:** Proportions and percentages: 0.10 = 10% and 0.75 = 75% etc.

Often, markschemes will allow both but be careful that your percentages never go over 100% and that your proportions never go over 1.00.

Worked example 3 – *calculating genotype frequency*

Using the information from example 1:

(a) fill out a copy of the table below;

(b) calculate the frequency of carriers in 500 members of the population.

allele frequencies	recessive **t**	q	
	dominant **T**	p	
genotype frequencies	homozygous recessive **tt**	q^2	
	heterozygous **Tt**	$2pq$	
	homozygous dominant **TT**	p^2	

Solution

(a) We know from the data given in worked example 1 that $q = 0.10$ and that $p = 0.90$. Those will fill the first two rows of the last column of the table.

To find the others, simply perform the mathematical operations in the third column:

$q^2 = 0.01$ so 1% of the population is **tt**.

$2pq = 0.18$ so 18% of the population is **Tt**.

Lastly, $p^2 = 0.81$ so 81% of the population is **TT**.

(b) To find the number of carriers (heterozygotes) in 500 members of this population, multiply 500 by 18% to get 90. So 90 people should be carriers.

Worked example 4 – *calculating phenotype frequency*

Using information from the table in example 3, calculate the number of people in 500 members of the population who do not suffer from the disease.

Solution

Using the numbers calculated in the previous worked example, we can complete the table as shown:

allele frequencies	recessive **t**	q	0.10
	dominant **T**	p	0.90
genotype frequencies	homozygous recessive **tt**	q^2	0.01
	heterozygous **Tt**	$2pq$	0.18
	homozygous dominant **TT**	p^2	0.81

We know that in order to not have this disease, a person must be either **TT** or **Tt**. The combined percentages of these genotypes are 81% + 18% which gives 99%. 99% × 500 = 495. So 495 people out of the 500 should have the healthy phenotype.

The Genetics Web Lab Directory has a section on the Hardy–Weinberg equation and there is an online simulation. You need the Shockwave player to use it.

Visit heinemann.co.uk/hotlinks, enter the express code 4242P and click on Weblink 15.4.

Once the equation has been used, it is possible to perform a statistical test to see if the predicted values truly correspond to the values obtained by observing phenotypes in the offspring of a population (see Chapter 1).

Assumptions

Models are often used to represent nature and the Hardy–Weinberg equation is a good example of one. However, models work only when the assumptions made when they are created are respected. This is one reason why computer models for predicting the weather do not always work – nature is often more complex than the model assumes.

For the Hardy–Weinberg equation, the assumptions which must be fulfilled in order for the calculations to work are as follows.

- The population being studied must be a *large population*; ideally it would be infinite – since this is an equation based on proportions and percentages, the results will be reliable only if the population has a large number of individuals.
- There must be *random mating* between individuals who have these particular alleles. This means the equation can only be used with genotype frequencies which are the same between males and females of the population. So make sure that the trait is an *autosomal* one.

Your class could compare gene frequencies with other classes in other parts of the world. Email and other electronic communication systems could be used to exchange data about traits such as attached or unattached earlobes, curved and straight thumbs, freckles or no freckles – but try to limit it to characteristics caused by a single gene.

Once you have phenotypic data, you can use the Hardy–Weinberg equation to calculate allele frequencies.

- There must be a *constant allele frequency* over time.
- There is *no allele-specific mortality* (as in the case of lethal alleles such as sickle cell anaemia where alleles can kill in their homozygous state).
- There are *no mutations* which would introduce new alleles and falsify the calculations.
- There is *no emigration or immigration*, which could alter the allele frequencies.

Exercises

13 Typical features of a model are as follows:
- it is a conceptual representation of a phenomenon;
- it can explain the past;
- it can predict the future;
- it is based on assumptions about the phenomenon.

From this description, outline the features of the Hardy–Weinberg equation that qualify it as a model.

14 A population starts out with only **AA** and **aa** individuals. Deduce with a reason how many generations are necessary to reach the Hardy–Weinberg equilibrium.

15 In a population of 278 mice, 250 are black and 28 are brown. The alleles are **B** = black and **b** = brown. Fill in the following table:

allele frequencies	recessive **b**	q	
	dominant **B**	p	
genotype frequencies	homozygous recessive **bb**	q^2	
	heterozygous **Bb**	$2pq$	
	homozygous dominant **BB**	p^2	

16 Explain why the Hardy–Weinberg equilibrium would not be attained if the brown mice preferred to mate only with other brown mice.

D.5 (HL only) Phylogeny and systematics

Assessment statements

D.5.1 Outline the value of classifying organisms.

D.5.2 Explain the biochemical evidence provided by the universality of DNA and protein structures for the common ancestry of living organisms.

D.5.3 Explain how variations in specific molecules can indicate phylogeny.

D.5.4 Discuss how biochemical variations can be used as an evolutionary clock.

D.5.5 Define *clade* and *cladistics*.

D.5.6 Distinguish, with examples, between *analogous characteristics* and *homologous characteristics*.

D.5.7 Outline the methods used to construct cladograms and the conclusions that can be drawn from them.

D.5.8 Construct a simple cladogram.

D.5.9 Analyse cladograms in terms of phylogenetic relationships.

D.5.10 Discuss the relationship between cladograms and the classification of living organisms.

The value of classifying organisms

In Chapter 5 section 5.5, we looked at classification. In biology, one of the objectives of classification is to truly represent how living (and extinct) organisms are connected. This means we are interested in natural classification, i.e. how things are truly grouped together in nature, whether we have been able to perceive the natural connections or not. There are several scientific motivations behind finding a natural classification.

This marine organism from Indonesia is called a nudibranch or sea slug. Classification systems help to identify it and to determine its evolutionary relationship to other organisms.

First of all, finding out how life forms are grouped together helps us to identify an unknown organism. If you find a type of sea creature that you have never seen before, you should be able to find an identification key which was made by experts who classified it.

A second reason for a natural classification is to see how organisms are related in an evolutionary way. By looking at organisms which have similar anatomical features, it is possible to see relationships on their phylogenetic tree. DNA evidence often confirms anatomical evidence for placing organisms in the same group. This is much more difficult to do with extinct species because their DNA is often degraded or destroyed, leaving only some anatomical features such as bone fragments or imprints in rocks.

A third reason is to allow for the prediction of characteristics shared by members of a group. If a researcher discovered a certain blood protein in some primate species, for example, he or she would expect to find similar blood proteins in other primates who are closely related.

W For an idea of the vastness and diversity of life on Earth as well as the daunting nature of trying to classify it, visit heinemann.co.uk/hotlinks, enter the express code 4242P and click on Weblink 15.5.

Biochemical evidence for common ancestry

When the theory of evolution was first introduced in the mid 1800s and Mendel was doing his work with inherited characteristics, the role of DNA in inheritance had not been elucidated and would not be for nearly another century. Had the discovery of the genetic code nearly a century later in the mid 1900s refuted the findings of Mendelian geneticists and Darwinian evolutionists, the two groups would have had to re-think their ideas about phylogeny. On the contrary, biochemical evidence, including DNA as well as protein structures, have brought new validity and confirmation to the idea of a common ancestor.

For example, the fact that every known living organism on Earth uses DNA as its main source of genetic information is compelling evidence that all life on Earth had a common ancestor. As you saw in Chapter 4 section 4.4 about genetic engineering, any gene from any organism can be mixed and matched with DNA from other organisms to generate a certain protein. Other than conceding that we all have a common ancestor, it would be difficult to explain why this is true.

In addition, all the proteins found in living organisms use the same 20 amino acids to form their polypeptide chains. Again, this has been confirmed by the introduction of foreign genes using genetic engineering to get an organism to synthesize a protein which it never synthesized before.

Amino acids can have two possible orientations: left-handed and right-handed, depending on the way their atoms are attached. In all the living organisms on

To pioneering biochemists, the fact that DNA and its genetic code is universal showed that humans are part of and not separate from the tree of all life on Earth. The universality of DNA refutes the idea that we are somehow different or superior in our origins.

Earth, no right-handed amino acids have ever been observed, leading us to the conclusion that all the proteins in Earth organisms are based on left-handed amino acids.

For those who support the idea of the biochemical evolution of life, the most logical explanation for such chemical similarities is that they imply a common ancestry for all life.

Variations and phylogeny

Phylogeny is the study of the evolutionary past of a species. Species which are the most similar are most likely to be closely related whereas those which show a higher degree of difference are considered less likely to be closely related. Traditionally, this has been done by examining morphology (the physical features of an organism's phenotype giving it form and structure). In more recent decades, attention has also been given to molecular differences and similarities.

By comparing the similarities in the polypeptide sequences of certain proteins in different groups of animals, it is possible to trace their common ancestry. This has been done with the blood protein haemoglobin, with a mitochondrial protein called cytochrome C and with chlorophyll, to name just three proteins.

With advances in DNA sequencing, the study of nucleic acid sequences in an organism's DNA as well as its mitochondrial DNA has been effective in establishing biochemical phylogeny. Changes in the DNA sequences of genes from one generation to another are partly due to mutations and the more differences there are between two species, the less closely related the species are.

Here is an imaginary example of a DNA sequence from four different species:

1 A A A A T T T T C C C C G G G G

2 A A A A T T T A C C C C G G G G

3 A A A A T T T A C C C G C G G G

4 A A C A T C T T C C A C G C T G

It should be clear that species 1 and 2 have the fewest differences between them whereas species 1 and 4 have the most differences. The conclusion would be that species 1 and 2 are more closely related to each other than they are to species 3 or 4.

Often, such work by biochemists confirms what palaeontologists have hypothesized about the ancestries of the fossils they have studied. When one branch of science confirms the work of another branch, further credibility for the findings is obtained. In other cases, the biochemical evidence can be contradictory, which encourages scientists to reconsider their hypotheses.

The evolutionary clock

Differences in polypeptide sequences accumulate steadily and gradually over time as mutations occur from generation to generation in a species. Consequently, the changes can be used as a kind of clock to estimate how far back in time two related species split from a common ancestor.

By comparing homologous molecules from two related species, it is possible to count the number of places along the molecules where there are differences. If the molecule is mitochondrial DNA, for example, we count the number of base pairs which do not match.

Imagine comparing certain DNA sequences from three species A, B and C. Between the DNA samples from species A and species C there are 83 differences. Between species A and species B there are only 26 differences. From this data, we conclude that species B is more closely related to species A than species C is. There has been more time for DNA mutations to occur since the split between A and C than since the split between A and B.

One technique which has been successful in measuring such differences is DNA hybridization. The idea is simple: take one strand of DNA from species A and a homologous strand from species B and fuse them together. Where the base pairs connect, there is a match; where they are repelled and do not connect, there is a difference in the DNA sequence and therefore there is no match (see Figure 15.19).

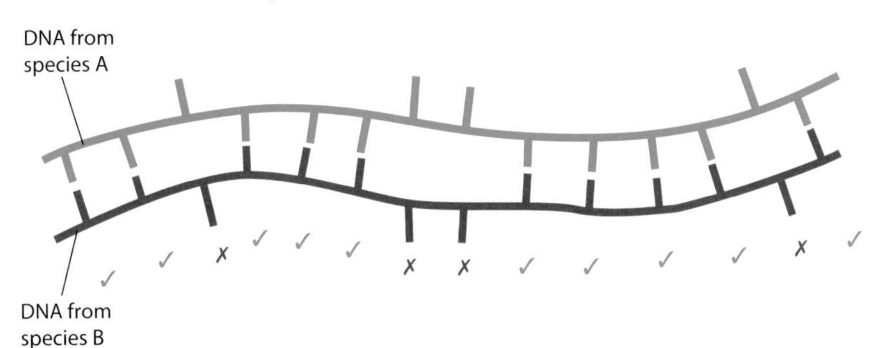

Figure 15.19 DNA hybridization between a strand of DNA from one species (in green) and another from a second species (in red). There are four places where a match does not occur.

We can take this further. If we see that 83 nucleotide differences is approximately three times more than 26 differences, we can conclude that the split between species A and species C happened about three times further in the past than the split between the species A and B. This is the idea of using quantitative biochemical data as an evolutionary clock to estimate the time of the speciation events (see Figure 15.20).

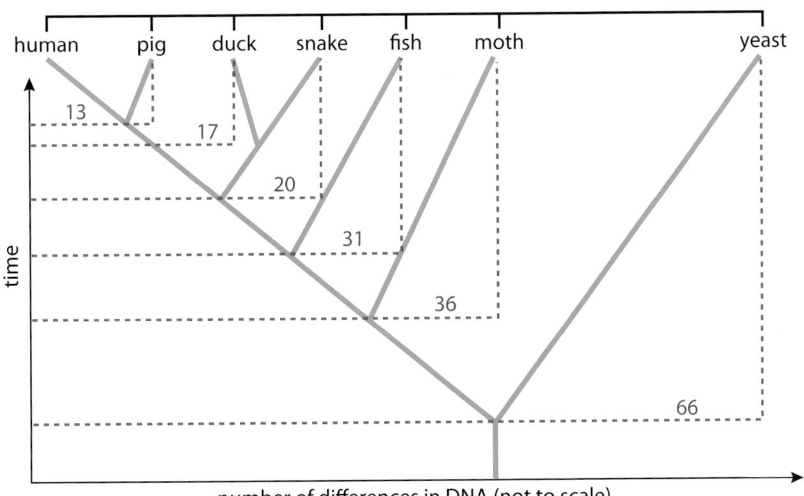

Figure 5.20 Biochemical differences (dotted red lines) can be used to see how far apart species are on a phylogenetic tree (in blue).

We need to be careful using a word such as 'clock' in this context. Under no circumstances should we consider that the 'tick-tock' of the evolutionary clock, which is made up of mutations, is as constant as the ticking of a clock on the wall. Mutations can happen at varying rates. Consequently, all we have is an average, an estimation or a proportion rather than an absolute time or date for speciation events. In an effort to double-check the timing of the evolutionary clock, biochemical data can be compared to morphological fossil evidence and radioisotope dating.

Because a technique has an unpredictable parameter (such as the rate of mutations in the evolutionary clock), does not make it less scientific. The important thing is that scientists are realistic about the margins of uncertainty, that they take this into account in drawing their conclusions and that they clearly communicate these uncertainties when they report their findings.

Clades and cladistics

Cladistics is a system of classification which groups taxa together according to the characteristics which have most recently evolved. In this system, the concept of common descent is crucial to deciding into which groups to classify organisms. Cladistics is, therefore, an example of natural classification. To decide how close a common ancestor is, researchers look at how many primitive and derived traits the organisms share.

Primitive traits (also called plesiomorphic traits) are characteristics which have the same structure and function (e.g. leaves with vascular tissue) and which evolved early on in the organisms being studied. Derived traits (also called apomorphic traits) are also characteristics which have the same structure and function but which evolved more recently as modifications of a previous trait (e.g. flowers, which evolved more recently than leaves with vascular tissue). By systematically comparing such characteristics, quantitative results show which organisms have a more recent split in the evolutionary past and which have a more distant split.

The fact that a characteristic is primitive does not always mean it is simpler, just that it evolved first. For example, the trait of limbs in reptiles precedes the snake, which evolved as a legless lizard. How do we know this? Snakes possess vestigial limbs showing that they were once present but were selected against by natural selection.

When a group can be split into two parts, one having certain derived traits that the other does not have, the groups form two separate clades. A clade is a monophyletic group. This means it is a group composed of the most recent common ancestor of the group and all its descendents. Although a clade can sometimes have just one species, usually it is made up of several species.

Cladistics was developed by the German biologist Willi Henning in the 1950s. This system of classification is widely used by palaeontologists, who often prefer it over the Linnaean system, which is still used in most other branches of biology.

Analogous and homologous characteristics

In examining the traits of organisms in order to put them into their appropriate clades, thorough and systematic studies of their characteristics must be undertaken. Two types of characteristic which are considered are homologous characteristics and analogous characteristics.

Homologous characteristics are ones derived from the same part of a common ancestor. The 5-fingered limbs found in such diverse animals as humans, whales or bats (see Chapter 5), are examples of homologous anatomical structures. The shape and number of the bones may vary and the function may vary but the general format is the same and the conclusion is that the organisms which possess these limbs had a common ancestor.

Another example of a homologous characteristic is the presence of eyes. Such structures are seen in both vertebrates and invertebrates. Simple eyes found in certain molluscs such as the *Nautilus* function as pinhole cameras without a system of lenses, whereas highly evolved eyes like those of birds of prey use crystalline lenses, adjustable irises and muscles to help focus on objects at different distances.

In contrast, analogous characteristics are those which may have the same function but they do not necessarily have the same structure and they are *not* derived from a common ancestor. Wings used for flying are an example: eagles, mosquitoes, bats and extinct reptiles such as the pterosaurs all use (or used) wings to fly. Although these organisms are all classified in the animal kingdom, they are certainly not placed in the same clade simply because of their ability to fly with wings. There are many other characteristics which must be considered.

What do the sarcastic fringehead fish and the bald eagle have in common? Eyes – a homologous characteristic.

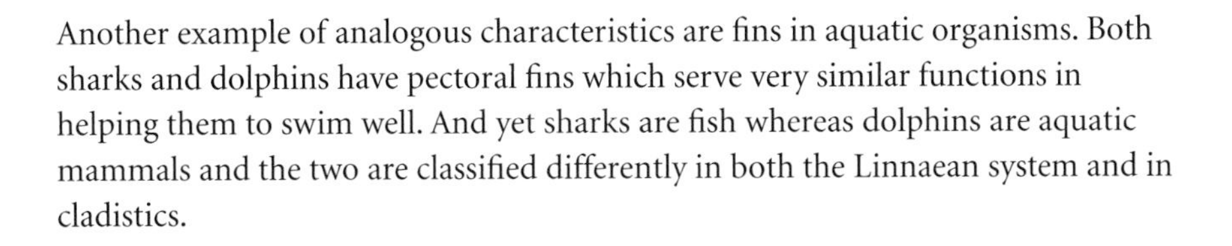

Another example of analogous characteristics are fins in aquatic organisms. Both sharks and dolphins have pectoral fins which serve very similar functions in helping them to swim well. And yet sharks are fish whereas dolphins are aquatic mammals and the two are classified differently in both the Linnaean system and in cladistics.

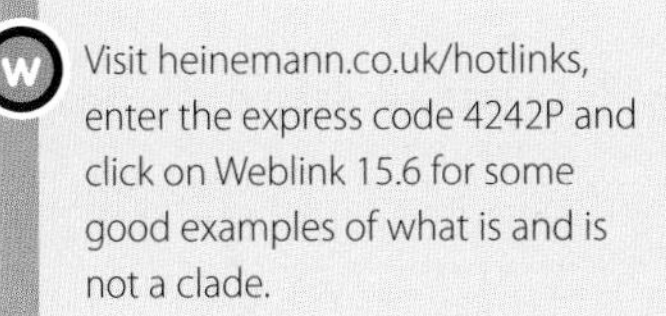

Visit heinemann.co.uk/hotlinks, enter the express code 4242P and click on Weblink 15.6 for some good examples of what is and is not a clade.

How cladograms are made

To represent the findings of cladistics in a visual way, a diagram called a cladogram is used. A cladogram showing bats, sharks and dolphins, for example, would take into account the skeletal structures and other characteristics such as the fact that bats and dolphins are placental mammals (see Figure 15.21). Thus, bats and dolphins are shown as more similar to each other than sharks are to either.

Figure 15.21 A cladogram.

The way to construct a cladogram is to first make a list of organisms which will be included in it. Then list as many as possible of the characteristics which each organism possesses. From this list, many of the traits will clearly be derived characteristics. Examples of some of the characteristics which might be considered are:

- eukaryotic;
- backbone;
- amniote egg;
- limbs (presence or quantity);
- hair;
- opposable thumbs.
- multicellular;
- segmented body;
- jaws;
- placenta;
- mammary glands;

This is a list of morphological characteristics but we could just as easily use a list of biochemical data about haemoglobin modifications, or differences in DNA sequences between the organisms.

Once the list of characteristics has been established, there will be one which is common to all of the organisms being studied. This ancestral trait is considered the primitive characteristic. Examples of primitive characteristics might be any of the ones at the top of the above list (eukaryotic, multicellular or perhaps the presence of a backbone). In biochemical data, the primitive characteristic might be a certain sequence of base pairs which is common to the DNA in all the organisms studied.

However, if the cladogram being constructed were of very early life forms, the first three morphological characteristics listed (eukaryotic, multicellular, presence of a backbone) would be considered derived rather than primitive. It all depends on the nature of the group being studied.

The next step in producing a cladogram is to make a table like the one below showing the derived characteristics along the top row and the names of the organisms in the first column.

	Multicellular	Vertebral column	Hair	Placenta	Totals
Sponge	✓	✗	✗	✗	1
Sailfish	✓	✓	✗	✗	2
Wombat	✓	✓	✓	✗	3
Elephant	✓	✓	✓	✓	4

From the information in the completed table, the cladogram is constructed with the first branch from the bottom belonging to the organism with the fewest derived traits (the sponge). The organism with the most derived characteristics goes to the top of the last branch (see Figure 15.22).

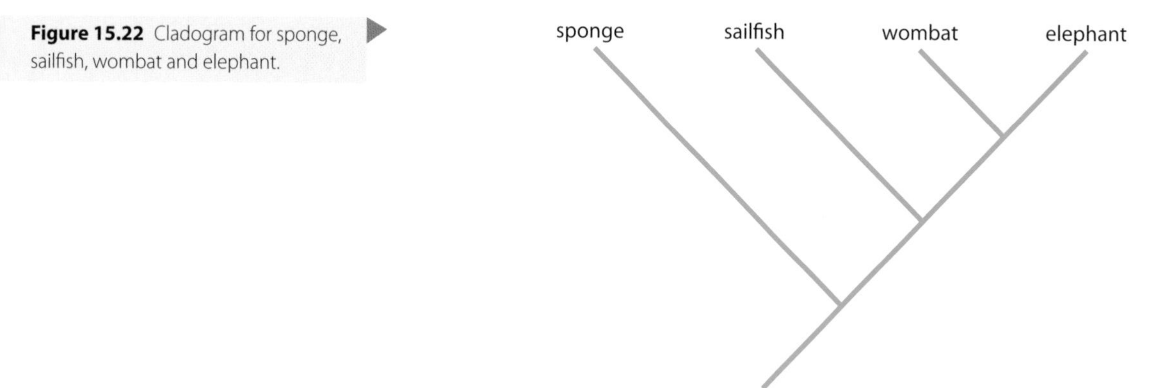

Figure 15.22 Cladogram for sponge, sailfish, wombat and elephant.

The original line at the base is the common ancestor's lineage, which in this case could be multicellular organisms.

Why are cladograms constructed?

Cladograms are constructed to show the evolutionary relationships between organisms. It can be concluded that organisms whose branches start at the bottom of the cladogram are the earliest ones to have evolved and the ones at the top are the ones which have evolved most recently among the organisms considered in the cladogram.

The other conclusion is that each time there is a point where a branch forks into two, a split occurred between species to develop into two lineages. This splitting point on a cladogram is called a node and it shows where a new species (and therefore a new clade) was founded. This makes the assumption that only one branching off can happen at any one time, generating two species where previously there was only one.

One of the basic ideas behind cladistics is the concept of parsimony. This refers to the preference for the least complicated explanation for a phenomenon. For example, it would be unlikely that a species would take two steps to evolve if one step is possible. When trying to decide between two hypotheses about which organisms are related to which, parsimony is used.

To confirm the common ancestry from a cladogram which is based on morphological evidence, another cladogram could be made using biochemical evidence from the same organisms. The two cladograms should be identical. For examples of biochemical evidence, see practice question number 3 at the end of this chapter.

W To see how a new hominid species is compared with others in a cladogram and to discover where it fits into our 'family tree', check out the work done by Tim White and his team in Ethiopia.

Visit heinemann.co.uk/hotlinks, enter the express code 4242P and click on Weblink 15.7.

Table 1 on the webpage has the list of the characteristics used in comparing the different species.

Constructing a cladogram

To construct a cladogram, you follow the instructions outlined above and re-stated here:

- list the organisms;
- list their characteristics;
- choose one primitive characteristic;
- fill in a comparative table of the organisms;
- build a cladogram from the table.

The organisms are:

- *Paramecium*;
- flatworm;
- shark;
- hawk;
- koala;
- camel;
- human.

The characteristics are:

- eukaryotic;
- multicellular;
- has a vertebral column;
- produces an amniote egg;
- has hair;
- has a placenta;
- has one opposable thumb on each forelimb.

Copy this table onto a clean sheet of paper and try to fill it out in the same fashion as the one for the cladogram in Figure 15.22.

	Eukaryotic	Multicellular	Vertebral column	Amniote egg	Hair	Placenta	Opposable thumb	Totals
Paramecium								
flatworm								
shark								
bird								
koala								
camel								
human								

You should be able to now draw the cladogram. Try it and then check with Figure 15.23.

Analysing a cladogram

Figure 15.23 shows what you should have obtained from the table you just filled out. What does it tell us? First of all, it says that all the organisms come from a eukaryotic ancestor because the completed table showed that this was the primitive characteristic shared by all. Secondly, it shows that koalas, for example, evolved after hawks because they possess the derived characteristic of hair, which is more evolved than the hawk's characteristics.

Thirdly, we can deduce from the cladogram that any two organisms found on successive branches are more closely related to each other than those found on branches which are separated by one or more nodes. For example, camels are more closely related to koalas than they are to hawks or sharks.

Figure 15.23 Two of the derived characteristics which define each clade are shown on this cladogram. Can you deduce the missing annotations of the other five characteristics used?

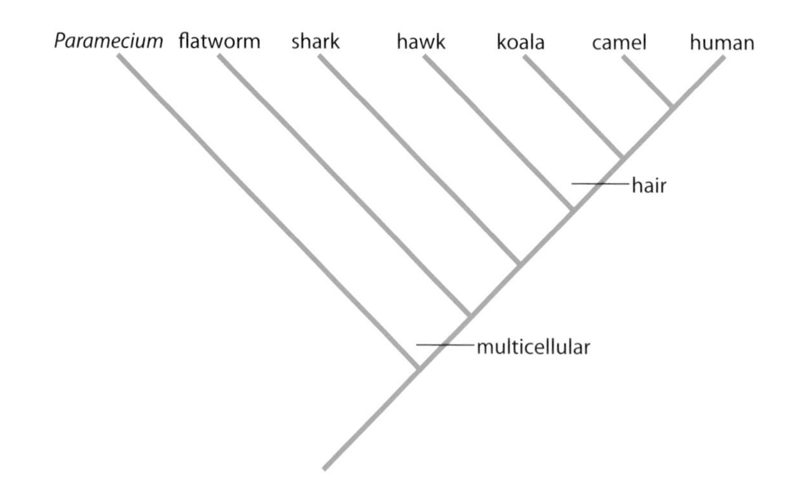

On the cladogram in Figure 15.23, two black lines have been added to show where new clades are defined by the presence of a derived characteristic. Every organism above the line marked 'multicellular' has that characteristic and every organism in the phylogeny below that cutoff point is a single-celled organism. Likewise, the three organisms above the line marked 'hair' form a clade of species having this particularity.

Cladograms and classification

Cladistics attempts to find the most logical and most natural connections between organisms in order to reveal their evolutionary past. As you have seen, this is usually done using the morphological data traditionally used in the Linnaean system of classification. Cladistics can just as easily use biochemical data.

Every cladogram drawn is a working hypothesis. It is open for testing and for falsification. On the one hand, this makes cladistics scientific but on the other hand, if it is going to be changing in the future as new evidence arises, it might be criticized for its lack of integrity.

Each time a derived characteristic is added to the list shared by organisms in a clade, the effect is similar to going up one level in the traditional hierarchy of the Linnaean classification scheme. For example, the presence of hair is part of what defines a mammal, so any species found after the line marked 'hair' should be in the class of mammals.

What about feathers? If an organism has feathers, is it automatically a bird? In traditional Linnaean classification, birds occupy a class of their own, but this is where cladistics comes up with a surprise. When preparing a cladogram, it becomes clear that birds share a significant number of derived characteristics with a group of dinosaurs called the theropods. This suggests that birds are an offshoot of dinosaurs rather than a separate class of their own.

Since birds are one of the most cherished and well-documented classes of organisms on Earth, this idea was controversial to say the least. Some of the derived characteristics used to put birds and dinosaurs in the same clade are:

- fused clavicle (the 'wishbone');
- flexible wrists;
- hollow bones;
- a characteristic egg shell;
- hip and leg structure, notably with backward-pointing knees.

By following the idea of parsimony, it would be more likely that birds evolved from dinosaurs than they evolved from another common ancestor. This is where cladistics is clearer than the Linnaean system. In cladistics, the rules are always the same concerning shared derived characteristics and parsimony. In the Linnaean system, apart from the definition of species, which we have already seen is sometimes challenged, the other hierarchical groupings are not always clearly defined: what makes a class a class or a phylum a phylum?

Although cladistics has had a difficult time being accepted over the decades, it is now increasingly adopted by biologists as a useful tool for determining natural classification and evolutionary connections.

Exercises

17 Take the cladogram you drew for the seven organisms discussed in this section and complete it with black lines showing the derived characteristics between each node.

18 Spots for camouflage exist on many different species such as cheetahs and butterflies. A student decides to put all spotted organisms into one classification group or into one clade. Deduce with a justification whether or not this qualifies as natural classification.

19 Distinguish between homologous and analogous characteristics.

20 Other than the examples given in the chapter, list one example of a homologous characteristic and one example of an analogous characteristic.

Practice questions

1 The mechanisms of speciation in ferns have been studied in temperate and tropical habitats. One group of three species from the genus *Polypodium* lives in rocky areas in temperate forests in North America. Members of this group have similar morphology (form and structure). Another group of four species from the genus *Pleopeltis* live at different altitudes in tropical mountains in Mexico and Central America. Members of this group are morphologically distinct.

Data from the different species within each group was compared in order to study the mechanisms of speciation.

Genetic identity was determined by comparing the similarities of certain proteins and genes in each species. Values between 0 and 1 were assigned to pairs of species to indicate the degree of similarity in genetic identity. A value of 1 would mean that all the genetic factors studied were identical between the species being compared.

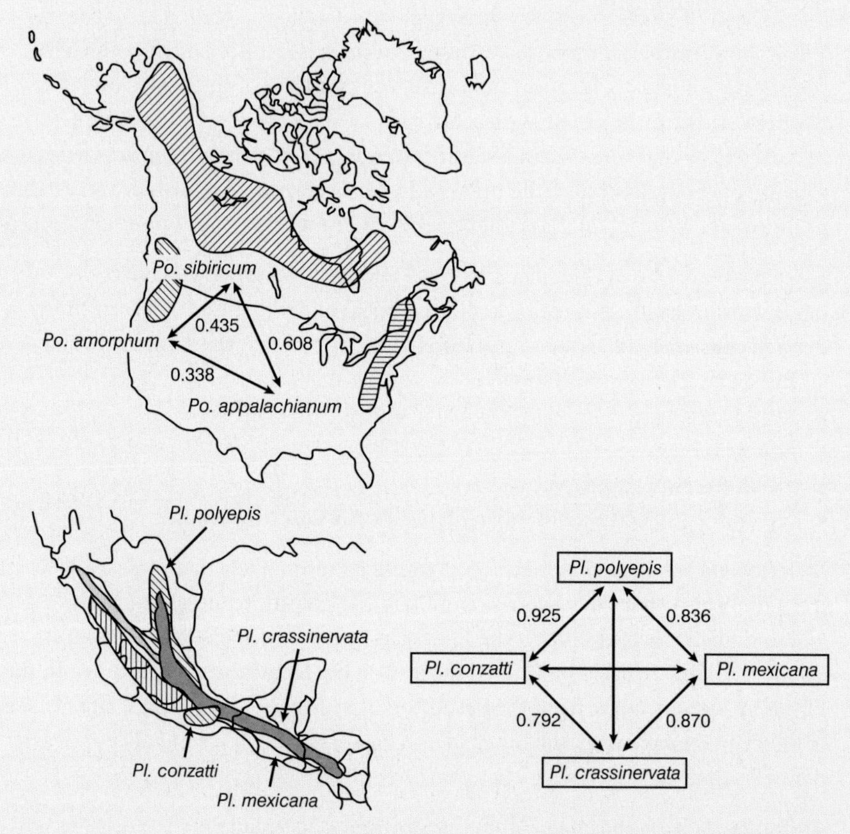

Source: C Haufler, E Hooper and J Therrien (2000), *Plant Species Biology*, **15**, pages 223–236

(a) Compare the geographic distributions of the two groups. *(1)*

(b) **(i)** Identify, giving a reason, which group, *Polypodium or Pleopeltis*, is most genetically diverse. *(1)*

(ii) Identify the **two** species that are most similar genetically. *(1)*

(c) Suggest how the process of speciation could have occurred in *Polypodium*. *(1)*

(d) Explain which of the two groups has most probably been genetically isolated for the longest period of time. *(2)*

(Total 6 marks)

2 **(a)** State the class human beings belong to. *(1)*

(b) Explain why the approximate date and distribution of *Homo habilis* are uncertain. *(3)*

(Total 4 marks)

3 The evolution of groups of living organisms can be studied by comparing the base sequences of their DNA. If a species becomes separated into two groups, differences in base sequence between the two species accumulate gradually over long periods of time. The number of differences can be used as an evolutionary clock.

Samples of DNA were recently obtained from fossil bones of a Neanderthal (*Homo neanderthalensis*). A section of the DNA from the mitochondrion was chosen for study, as it shows a high level of variation in base sequence between different individuals. A section of the Neanderthal mitochondrial DNA was sequenced and compared with sequences from 994 humans and 16 chimpanzees. The bar chart below shows how many base sequence differences were found among humans, between the humans and the Neanderthal and between humans and chimpanzees.

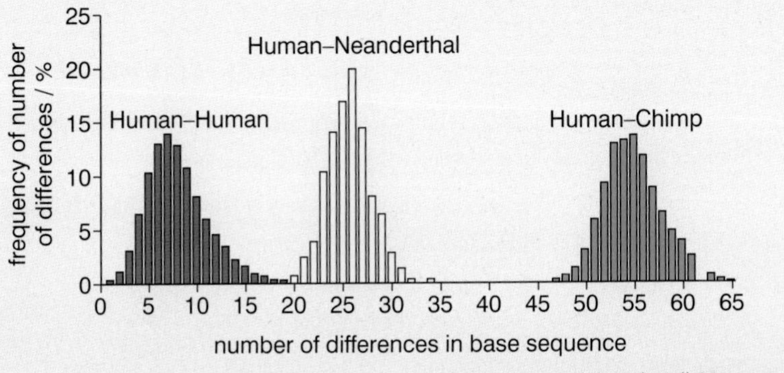

Source: Krings, et al. (1997), *Cell*, **90**, pages 19–30

(a) The number of differences in base sequence between pairs of humans varied from 1 to 24. State the number of differences shown by the highest percentage of pairs of humans. *(1)*

(b) Humans and Neanderthals are both classified in the genus *Homo* and chimpanzees are classified in the genus *Pan*. Discuss whether this classification is supported by the data in the bar chart. *(3)*

(c) Data suggests that humans and Neanderthals diverged 550 000 to 700 000 years ago. If a sample of mitochondrial DNA was obtained from a fossil bone of *Australopithecus*, predict, with a reason, how many base sequence differences there would be between it and human DNA. *(2)*

(Total 6 marks)

4 River dolphins live in freshwater habitats or estuaries. They have a number of features in common which distinguish them from other dolphins: long beaks, flexible necks, very good echo-location and very poor eyesight. Only four families of river dolphin have been found in rivers around the world.

River	River Dolphin Family
Amazon, Brazil	Iniidae
La Plata, Argentina	Pontoporiidae
The Yangtze, China	Lipotidae
The Indus and Ganges, India	Platanistidae

Evolutionary biologists have tried to determine how closely related these river dolphins are to one another. River dolphins are members of the group the toothed whales. Three lines of evidence were analysed producing three cladograms (family trees) for all the toothed whales. The evidence to construct these cladograms came from the morphology (form and structure) of fossil toothed whales (**I**), the morphology of living toothed whales (**II**) and the molecular sequences from living toothed whales (**III**)

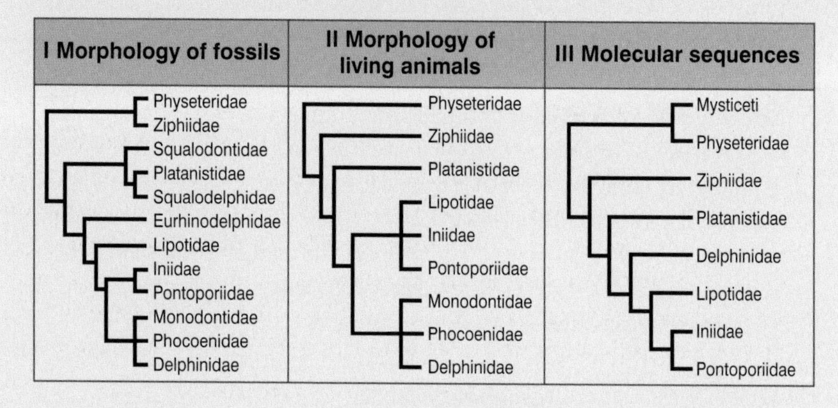

Source: H Hamilton, et al. (2001), *Proc R Soc Lond B*, **268**, pages 549–556

(a) Suggest a reason why there are more families present in cladogram I, produced from the morphology of fossils, than for the other cladograms. *(1)*

(b) Using only the data from cladogram III, identify which other family of river dolphins is most closely related to Platanistidae. *(1)*

(c) State what material would be used to produce cladogram III, based on the molecular sequences of living toothed whales. *(1)*

The tree using the data from the morphology of living animals (II) indicates that the families are more closely related than the tree using molecular sequences (III) from the same animals.

(d) Explain how these dolphins can look so similar when in fact they may not be so closely related. *(3)*

These cladograms show the species that share common ancestors but do not show how long ago they diverged from one another.

(e) Outline further evidence that would be needed to determine when these families of toothed whales diverged. *(2)*

(Total 8 marks)

16 Neurobiology and behaviour (Option E)

Introduction

Have you ever been stung by a bee? The first thing you notice is the pain. The sting is the stimulus. Fortunately, we have receptors for pain. The receptor notifies us of the pain. Your immediate response to the sting is a pain withdrawal reflex. The pain reflex involves a series of nerves that run from your arm to your spinal cord and back to your arm muscle. This causes you to brush off the bee before you can even think about it. Responses to stimuli are found in even the simplest animals. Some responses are learned. If we see a bee, we know it might sting. Other responses are genetically programmed so we can respond without learning. An infant's ability to suck is not learned.

All behaviours are caused by chemical messages which are sent between nerve cells (neurones) as they communicate with each other. These chemical messages must somehow be coordinated. The brain does the coordination. We avoid bees because our brain is organizing the information.

Humans are not the only organisms which have complex responses. Single celled organisms like *Euglena* respond to light. Insects have complex social patterns. Birds have complicated songs. Fish have innate mating rituals.

Worker honey bees are female bees that have their ovipositor modified into a stinger. The bee stings its victim with its barbed stinger, then flies away, leaving the stinger in the victim. The stinger continues to pump venom, and to embed itself deeper into the victim's skin.

E.1 Stimulus and response

Assessment statements

- E.1.1 Define the terms *stimulus*, *response* and *reflex* in the context of animal behaviour.
- E.1.2 Explain the role of receptors, sensory neurones, relay neurones, motor neurones, synapses and effectors in the response of animals to stimuli.
- E.1.3 Draw and label a reflex arc for a pain withdrawal reflex, including the spinal cord and its spinal nerves, relay neurone, motor neurone and effector.
- E.1.4 Explain how animal responses can be affected by natural selection, using two examples.

Definition of terms

- A stimulus is a change in the environment (internal or external) that is detected by a receptor and elicits a response.
- A reflex is a rapid, unconscious response.
- A response is a reaction to a stimulus.

Response of animals to pain stimuli

Animals can respond to stimuli with a reflex. Receptors receive the stimulus. For example, pain receptors receive the stimulus of excess heat, pressure or chemicals produced by injured tissues. If a coyote has its leg caught in a trap, the pain receptors in its skin respond to the excess pressure. The receptors generate a nerve impulse in the sensory neurones (see Figure 16.1). The sensory neurones carry the impulse toward the spinal cord. The axon of the sensory neurone enters the spinal cord in the dorsal root and sends a chemical message across a synapse to a relay neurone. The relay neurone is located in the grey matter of the spinal cord. The relay neurone synapses with a motor neurone in the grey matter of the spinal cord and transfers the impulse chemically across the synapse. The motor neurone is located in the ventral root of the spinal cord. It carries the impulse to an effector. An effector is an organ that performs the response. In this case, it is a muscle which contracts as the coyote struggles to withdraw its leg from the trap.

What you have just read is a description of the pain reflex arc (see Figures 16.1 and Figure 16.2).

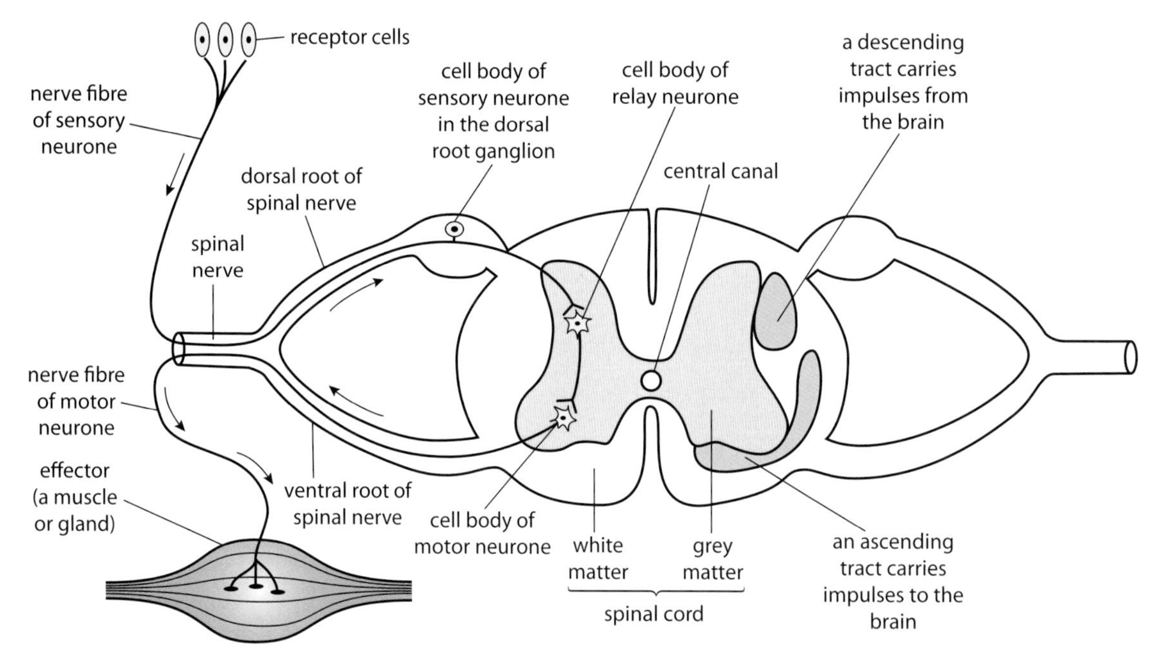

Figure 16.1 Structure of the spinal cord and components of a reflex arc.

Besides the reflex arc, many other reactions also occur. The coyote may be howling in pain and growling. The series of actions is caused as the spinal cord nerves carry impulses to the brain. As the brain becomes aware of what is happening, it coordinates the other responses.

For another look at the reflex arc for pain, visit heinemann.co.uk/hotlinks, enter the express code 4242P and click on Weblink 16.1a and 16.1b.

Effects of natural selection

Animal behaviour is much more than just single reflexes. It is a complicated series of responses to the environment in which animals live. Scientists studying animal behaviour have observed that some populations of organisms have changed their behaviour in response to a change in the environment. These behavioural changes may be so extreme that a new species is formed – as might eventually occur with European blackcaps (see pages 463–64).

• **Examiner's hint:** Be able to draw and label the pain reflex arc. Include the labels shown in Figure 16.1.

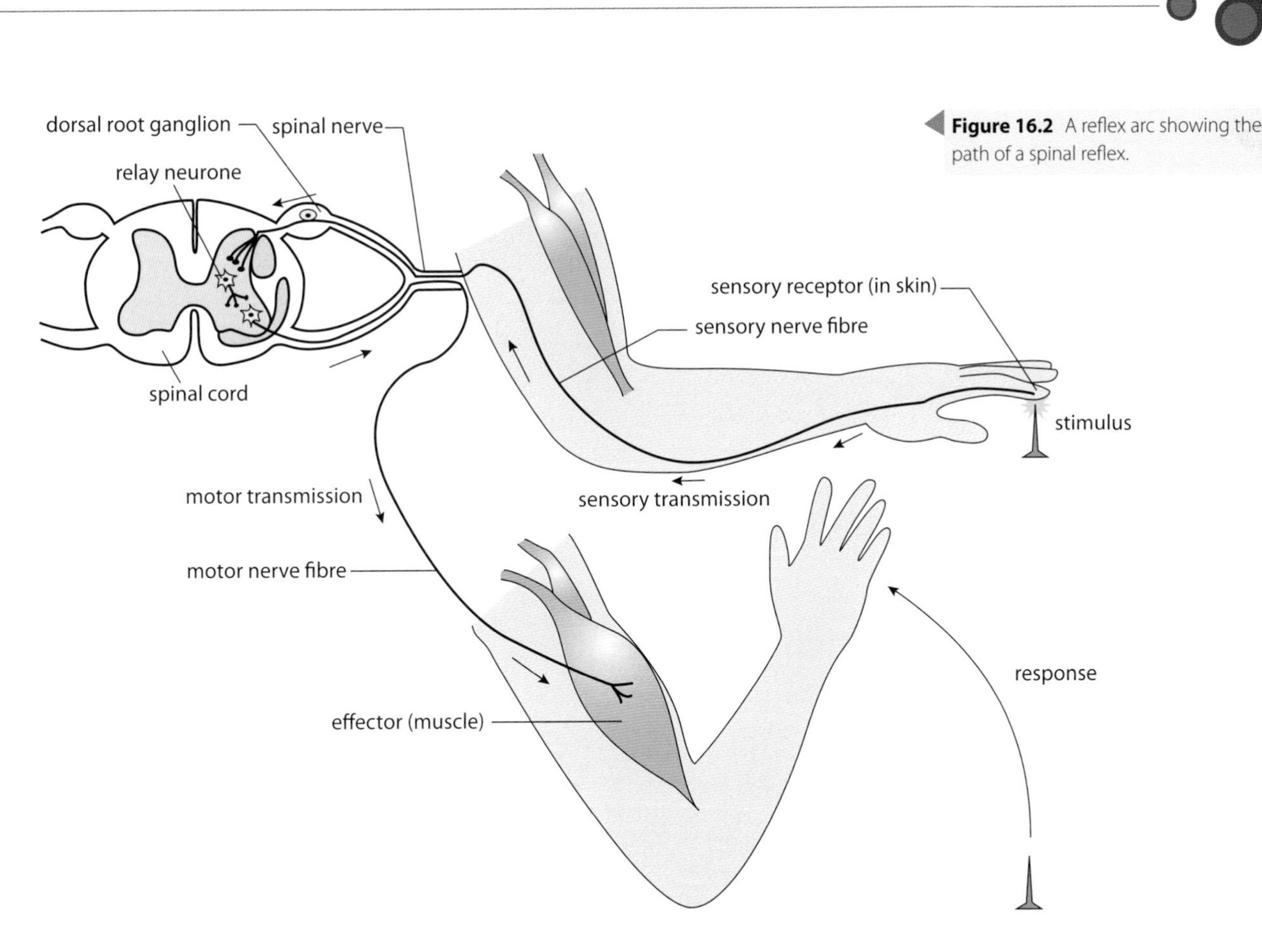

Figure 16.2 A reflex arc showing the path of a spinal reflex.

Variations in behaviour can occur in populations in the same way as variations in the colour of animals. You may be familiar with the story of the dark and light peppered moths. Within the moth population there is a variation in colour. Moths can be dark or light. If the tree bark on which they live is dark, the moth population is primarily dark. The light moths are more easily seen by birds and eaten. If lichens begin to grow on the trees, the colour of the trees becomes light. The moth population will change. With light trees, most individuals which survive to reproduce will be light-coloured. The colour of the moth is determined by genes just as behaviour can be determined by genes.

Variations in behaviour can be selected by the environment. Since a genetically programmed behaviour can have variations, one behaviour can work better than another in a changing environment. That variation will allow one group of organisms to survive and reproduce better in the new environment. The theory of natural selection states that the organism best fitted for the environment is more likely to survive to reproduce.

European blackcaps

Let's consider the interesting case of a bird called the European blackcap. These birds are small warblers which usually migrate between Spain and Germany. They breed in Germany in the spring and summer and spend the winter in Spain. About 50 years ago, bird lovers noticed that some blackcap warblers were coming to the UK instead of Spain for the winter. Ornithologists began studying their behaviour. The ornithologists noticed that the UK blackcaps left to go back to Germany 10 days earlier than the Spanish blackcaps. They also noticed that the earlier the birds arrived in Germany, the more choice of territory they had, and the more eggs they laid. The UK blackcaps had a distinct advantage over the Spanish blackcaps.

As we realize the importance of bird migration to bird survival and reproduction, it is exciting to see that international organizations are recognizing the need to conserve migration routes. In November of 2006, the UN launched its project on bird migration routes in Africa and Eurasia. The name of the project is Wings Over Wetlands (WOW). This project aims to conserve critical areas needed by waterbirds migrating across these continents.

To read more about WOW, visit heinemann.co.uk/hotlinks, enter the express code 4242P and click on Weblink 16.2.

In order to determine if this behaviour had a genetic basis, an experiment was done. Eggs were collected from parents who had been in the UK the previous winter and other eggs collected from the Spanish birds. The young were reared and direction of migration recorded. No parents were around to teach the young in what direction to fly. All of the birds in the study, no matter where they had been reared, tended to migrate in the same direction that their parents had gone. This supports the hypothesis that blackcaps are genetically programmed to fly in a certain direction.

What could be the environmental benefit of migrating to the UK for some birds? Something in their genetic variation causes the birds to leave for Germany earlier than those in Spain. Arriving in Germany early is an advantage. Also, warm winters in the UK have increased the survival rate of the birds. This change in migration patterns may eventually result in a new species. That is especially likely if small changes in courtship behaviour occur which will separate the species even more.

Sockeye salmon

Sockeye salmon are spawning as they swim upstream in the river. Spawning salmon assume a vibrant red colour and a green head.

The sockeye salmon is a species introduced into Lake Washington in Washington State. After the salmon were introduced into the lake, some of them migrated to the Cedar River, which flows into the lake. The river flows quickly, but the lake is deep and quiet. These are really two different types of aquatic environment which are connected to each other. Over a span of 60 years, 13 generations of salmon have been produced.

DNA evidence has shown that river salmon and lake salmon have stopped interbreeding. How did this happen?

The lake salmon have one breeding method and the river salmon have another. The lake salmon spawn on the beaches; females lay their eggs in the sand. The males have heavy bodies, perfect for hiding in the deep waters of the lake. The large males, if put in the river, are not efficient at navigating fast currents.

The males of the Cedar River population have traits naturally selected to be successful in a fast-moving river. Their bodies are thinner and narrow for better manoeuvrability in the current. Females of the river group bury their eggs deep in the sandy river bottom so that they will not be washed away. Genetic studies show that fish hatched in the river had little success trying to spawn on the beach of Lake Washington.

Variations in the original salmon population were selected for by the two different environments. The original population diverged into two different breeding populations. The lake conditions favour one set of traits and the river conditions favour another set of traits. Sockeye salmon are now split into two genetically distinct populations: beach-spawning and river-spawning.

What will happen when global warming causes large migrations of fish and other animals in the ocean? Will this migration also cause new species to emerge? The impact that global warming is beginning to have on Australian marine life has been announced by the Great Barrier Reef National Park. Fish and seabirds are being driven south. Coral reefs are threatened. Warmer oceans can affect reproductive cycles. Nesting turtles are particularly vulnerable.

There are many poor examples of supposed links between animal responses and natural selection. It is easy for us to guess how the behaviour of an animal might influence its chance of survival and reproduction, but experimental evidence from carefully controlled trials is needed to back up our intuitions.

Discuss the difference between evidence that shows correlation and evidence that shows causation. What correlation evidence was observed in the example of UK blackcaps? How was evidence of causation collected?

To read more about the impending marine disaster at the Great Barrier Reef Natioinal Park, visit heinemann.co.uk/hotlinks, enter the express code 4242P and click on Weblink 16.3.

Exercises

1 State the definition of each of the following:
 (a) stimulus;
 (b) response;
 (c) reflex.

2 A reflex arc responds to a stimulus such as pain. Describe the role of each part of a reflex arc.

3 Draw a reflex arc.

E.2 Perception of stimuli

Assessment statements

E.2.1 Outline the diversity of stimuli that can be detected by human sensory receptors, including mechanoreceptors, chemoreceptors, thermoreceptors and photoreceptors.
E.2.2 Label a diagram of the structure of the human eye.
E.2.3 Annotate a diagram of the retina to show the cell types and the direction in which the light moves.
E.2.4 Compare rod and cone cells.
E.2.5 Explain the processing of visual stimuli, including edge enhancement and contralateral processing.
E.2.6 Label a diagram of the ear.
E.2.7 Explain how sound is perceived by the ear, including the roles of the eardrum, bones of the middle ear, oval and round windows, and the hair cells of the cochlea.

Sensory receptors and diversity of stimuli

Certain foods give us feelings of comfort. Seeing a familiar face in a crowd makes you feel at ease. Listening to your favourite music can make you feel happy. We have learned to link certain tastes, sights and sounds with emotions. Sensory cells send messages to certain parts of the brain that control emotion and memory.

Taste and sound are not just for pleasure. They also protect us. Birds remember the bad taste of a certain butterfly. We move out of the way when we hear a car coming. Many lives have been saved by smelling smoke.

Sense organs are the windows to the brain. They keep the brain aware of what is going on in the outside world. When stimulated, the sense organs send a message to the central nervous system. The nerve impulses arriving at the brain result in sensation. We actually see, smell, taste and feel with our brain rather than our sense organs.

We link certain tastes to emotion and memory. Some foods make us remember our childhood.

Mechanoreceptors

Mechanoreceptors are stimulated by mechanical force or some type of pressure. The sense of touch is due to pressure receptors which are sensitive to strong or light pressure. In our arteries, pressure receptors can detect a change in blood pressure. In our lungs, stretch receptors respond to the degree of lung inflation. We can tell the position of our arms and legs by the use of proprioceptors found in muscle fibre, tendons, joints and ligaments. These receptors help us maintain posture and balance. In our inner ear, there are pressure receptors sensitive to waves of fluid moving over them. This gives us information about our equilibrium.

Chemoreceptors

Chemoreceptors respond to chemical substances. Using this type of receptor, we can taste and smell. They also give us information about our internal body environment. Chemoreceptors in some blood vessels monitor pH changes. Changes in pH signal the body to adjust the breathing rate. Pain receptors are a type of chemoreceptor which respond to chemicals released by damaged tissues. Pain protects us from danger. The pain reflex makes us pull away, for example, from a hot object.

We depend on more than our senses to know the biological world. To what extend are we dependent on technology for our knowledge of biology? As an example, we use molecular tools to read the basic instructions of life one letter at a time as we decode the human genome. What other ways can you describe in which we use technology to know the biological world?

To help you think about this TOK issue, visit heinemann.co.uk/hotlinks, enter the express code 4242P and click on Weblink 16.4.

Thermoreceptors

Thermoreceptors respond to a change in temperature. Warmth receptors respond when the temperature rises. Cold receptors respond when the temperature drops.

Photoreceptors

Photoreceptors respond to light energy. They are found in our eyes. Our eyes are sensitive to light and give us vision. Rod cells in our eyes respond to dim light resulting in black and white vision, cone cells respond to bright light and give us colour vision.

Structure and function of the human eye

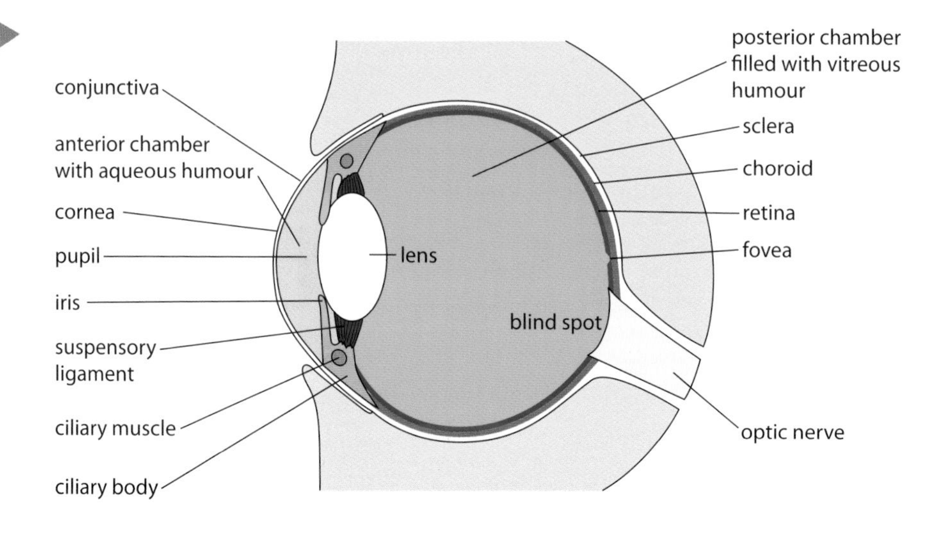

Figure 16.3 The human eye.

This table summarizes the functions of the various parts of the eye shown in Figures 16.3 and 16.4.

Part	Function
iris	regulates the size of the pupil
pupil	admits light
retina	contains receptors for vision
aqueous humour	transmits light rays and supports the eyeball
vitreous humour	transmits light rays and supports the eyeball
rods	allow black and white vision in dim light
cones	allow colour vision in bright light
fovea	an area of densely packed cone cells where vision is most acute
lens	focuses the light rays
sclera	protects and supports the eyeball
cornea	focusing begins here
choroid	absorbs stray light
conjunctiva	covers the sclera and cornea and keeps eye moist
optic nerve	transmits impulses to the brain
eye lid	protects the eye

The retina

Vision begins when light enters the eye and is focused on the photoreceptor cells of the retina (see Figure 16.4). The photoreceptor cells are the rods and the cones. Notice in Figure 16.5 that both the rods and cones synapse with their own bipolar neurones. Each bipolar neurone synapses with a ganglion cell. The axons of the ganglion cells make up the optic nerve which carries the message of vision to the brain.

These notes would be suitable for annotating a diagram of the retina.

- Rod cells are photoreceptor cells which are very sensitive to light. They receive the stimulus of light, even very dim light, and synapse with a bipolar neurone.

Figure 16.4 Structure of the retina.

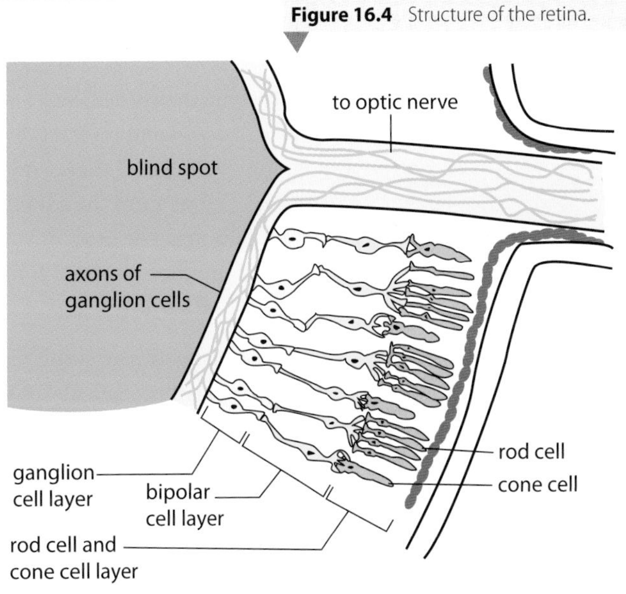

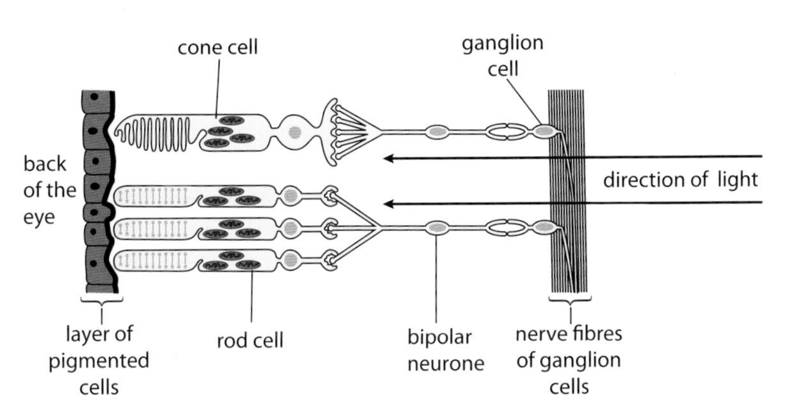

Figure 16.5 Structure and function of the retina.

● **Examiner's hint:** 'Annotate' means to give brief notes about a diagram or picture. In the case of the retina, be able to show and describe the cell types and the direction in which the light moves.

- Cone cells are photoreceptor cells which are activated by bright light. They receive the stimulus of bright light and synapse with a bipolar neurone.
- Bipolar neurones are cells in the retina which carry impulses from a rod or a cone cell to a ganglion cell of the optic nerve. They are called bipolar because they each have two processes extending from the cell body.
- Ganglion cells are the cell bodies of the optic nerve. They synapse with the bipolar neurones and send the impulses to the brain.

Rods and cones

This table summarizes a comparison of rods and cones.

Rods	Cones
These cells are more sensitive to light and function well in dim light.	These cells are less sensitive to light and function well in bright light.
Only one type of rod is found in the retina. It can absorb all wavelengths of visible light.	Three types of cone are found in the retina. One type is sensitive to red light, one type to blue light and one type to green light.
The impulses from a group of rod cells pass to a single nerve fibre in the optic nerve. See Figure 16.5.	The impulse from a single cone cell passes to a single nerve fibre in the optic nerve. See Figure 16.5.

Processing visual stimuli

When we look at an object, light rays pass through the pupil and are focused by the cornea, lens and the humours. The image focused on the retina is upside down and reversed from left to right. Once the photoreceptors of the retina are stimulated, they send impulses to the bipolar neurones and the ganglion cells. The axons from the ganglion cells travel to the visual area of the cerebral cortex of the brain. The brain must correct the position of the image so that it is right side up and not reversed. It must also coordinate the images coming from the left and right eye.

Many things about vision are still not understood. Scientists use various methods to collect data about how the brain processes visual stimuli and turns them into 'vision'. One type of experiment is to study optical illusions and another is to study people who have had injuries to one side of their brain.

Edge enhancement

Studies of vision using illusions were begun in 1865.

Scientists studying vision have used optical illusions as powerful windows into the neurobiology of vision. The complex structure of our vision may be exposed by studying such illusions. For example, look at the Hermann grid illusion in Figure 16.6.

Why do you see some grey blobs in the white area between the black squares, that vanish when you try to look at them directly? The theory is that the areas where you see grey are in your peripheral vision, where there are fewer light-sensitive cells than in the centre of your retina (fovea). Some of the cells present may even be 'turned off'. This sends the message of grey instead of white.

But when you try to look directly at a 'grey' area you are using the centre of your retina, your fovea, which has a high concentration of light-sensitive cells. Even if some are 'turned off', you still see white.

Figure 16.6 The Hermann grid illusion.

This figure fools your eye because of the extreme contrast between black and white edges. It demonstrates that you have a special mechanism for seeing edges – it is called edge enhancement. It is theorized that light-sensitive receptors in your eye switch off their neighbouring receptors. This makes the edges look more distinct, because of the extreme contrast between dark and light.

The illusion may fool your eye, but scientists can use it to collect data which will either support or refute a hypothesis they have made about the function of receptor cells in the eye.

Contralateral processing

Contralateral (opposite side) processing is due to the optic chiasma (see Figure 16.7). Nerve fibres bringing information from the right half of each visual field converge at the optic chiasma and pass to the left side of the brain. Nerve fibres bringing information from the left half of each visual field converge at the optic chiasma and pass to the right half of the brain. The information eventually ends up in the visual cortex of the brain. Since each visual area only receives half the information from each visual field, these areas must share information to form a complete visual image. The image received by the visual cortex is both inverted and reversed. The brain must correct this image in order for us to correctly perceive what is in the whole visual field. It is thought that impulses relating to the other stimuli of colour, form and motion are parcelled out to other visual association areas of the brain. Eventually, the cerebral cortex rebuilds all the parts into a visual image to give us a complete understanding of what we are seeing.

This process of two sides of the brain working together can be illustrated by the abnormal perceptions of patients with brain lesions (injuries). Vision is actually information processing. If we see a bucket of any shape or form, we know it is a

Figure 16.7 The optic chiasma.

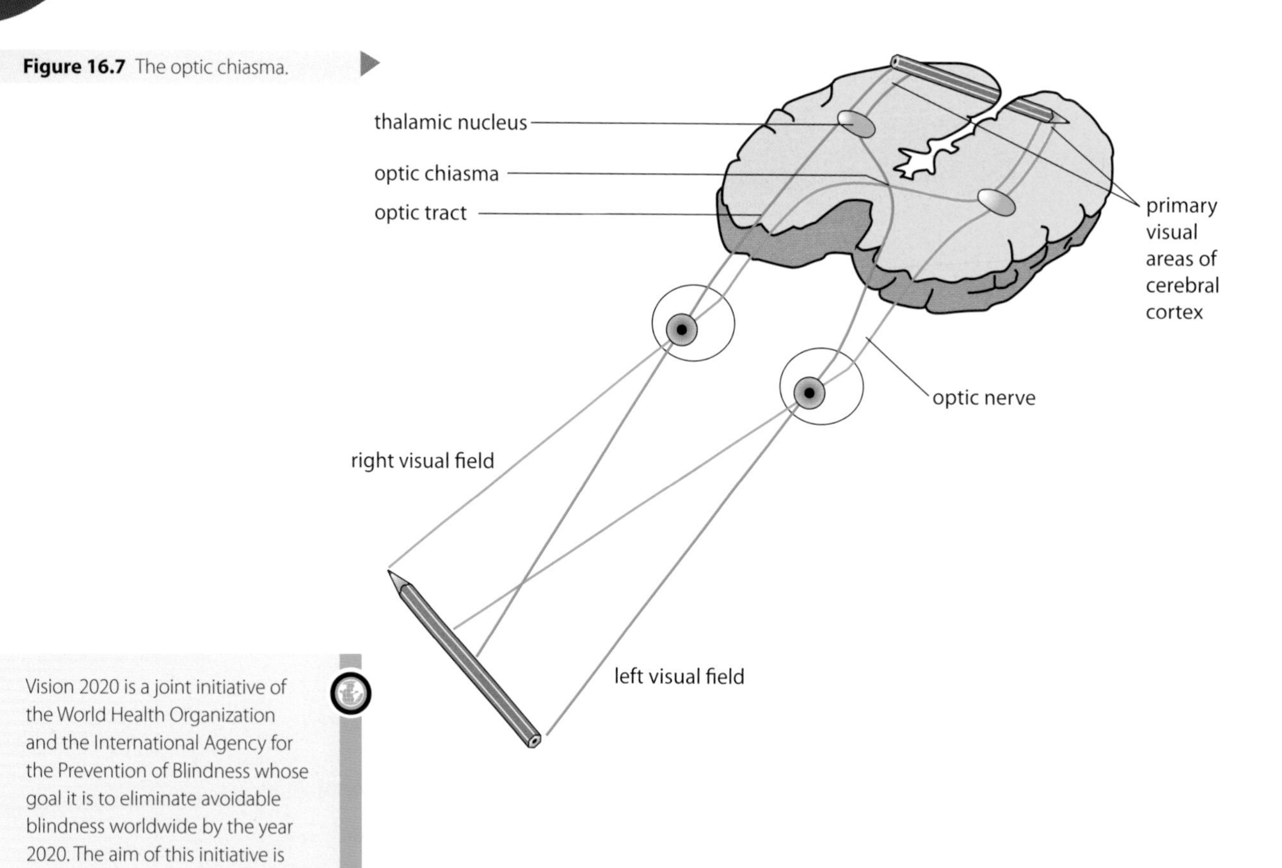

Vision 2020 is a joint initiative of the World Health Organization and the International Agency for the Prevention of Blindness whose goal it is to eliminate avoidable blindness worldwide by the year 2020. The aim of this initiative is to give every one in the world the right to sight.

To read more information about Vision 2020, visit heinemann.co.uk/hotlinks, enter the express code 4242P and click on Weblink 16.5.

bucket. We may be looking at it from the top or bottom or side but we still know it is a bucket. Patients with lesions in the right side of the brain when looking from above do not recognize a bucket. In fact, they deny that it is a bucket. Patients with left brain lesions can describe the function of the bucket but cannot come up with the name 'bucket'. It takes both sides of the brain working together to have correct 'vision' which is able to recognize an object and understand what it is.

Structure of the ear

Figure 16.8 shows you the structure of the human ear.

Figure 16.8 Anatomy of the human ear.

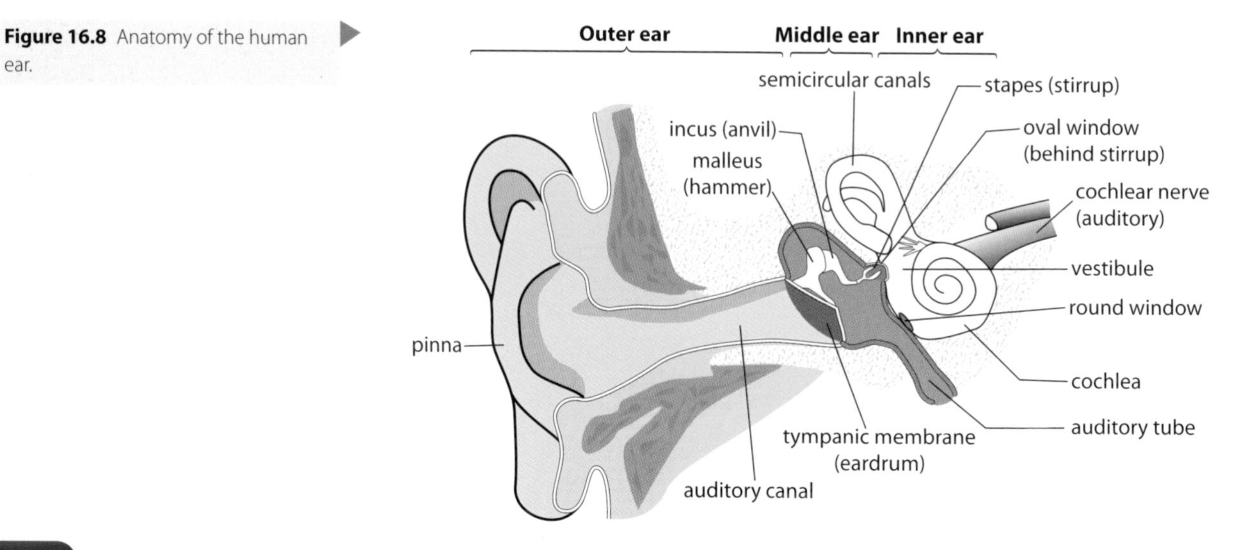

How sound is perceived by the ear

The outer ear catches sound waves. Sound waves are successive vibrations of air molecules. When they travel down the auditory canal, they cause the eardrum (tympanic membrane) to move back and forth slightly. The bones of the ear – malleus, incus and stapes – receive vibrations from the tympanic membrane and multiply them approximately 20 times. The stapes strikes the oval window causing it to vibrate. This vibration is passed to the fluid in the cochlea. The fluid in the cochlea causes special cells, called hair cells, to vibrate. The hair cells, which are receptors, release a chemical message across a synapse to the sensory neurone of the auditory nerve. The chemical message stimulates the sensory neurone. Finally the message is carried by the sensory neurone in the auditory nerve to the brain. The wave in the fluid of the cochlea dissipates as it reaches the round window.

Loud noises cause the fluid to vibrate to a higher degree and the hair cells bend even more. This is interpreted by the brain as higher volume. Pitch is a function of sound wave frequency. Short, high-frequency waves produce high-pitched sounds while long, low-frequency waves produce low-pitched sound. The sound which is sensed by the brain is processed in the auditory area of the cerebral cortex.

W To find out more about the ear and hearing, visit heinemann.co.uk/hotlinks, enter the express code 4242P and click on Weblink 16.6.

Exercises

4 Describe each of the four types of sensory receptor.
5 Draw a diagram of the structure of the eye.
6 Draw a diagram of the retina and show the direction in which the light moves.
7 Compare the types of photoreceptor cell in the retina.
8 Describe the processing of visual stimuli.
9 Draw a diagram of the ear.
10 Explain how sound is perceived by the ear.

E.3 Innate and learned behaviour

Assessment statements

- E.3.1 Distinguish between innate and learned behaviour
- E.3.2 Design experiments to investigate innate behaviour in invertebrates, including either a taxis or a kinesis.
- E.3.3 Analyse data from invertebrate behaviour experiments in terms of the effect on chances of survival and reproduction.
- E.3.4 Discuss how the process of learning can improve the chance of survival.
- E.3.5 Outline Pavlov's experiments into conditioning of dogs.
- E.3.6 Outline the role of inheritance and learning in the development of birdsong in young birds.

Scientists can tell what species of spider has spun a web by its characteristic shape. This spider is *Aranus quadratus*.

Some ethologists have dismissed psychology as irrelevant since it is based on animal behaviour in an unnatural setting. Psychologists have accused ethologists of ignoring the concepts of learning and motivation and placing too much emphasis on instinct. Explore the different ways the scientific method is used in each of these disciplines. In your opinion, is one more 'scientific' than the other?

Innate and learned behaviour

Ethologists study the behaviour of animals in their natural environment. They examine patterns of behaviour that affect an animal's life. On the other hand, psychologists study behaviour in an artificial environment. They collect data on learning and motivation that could never be measured in the natural environment.

Innate behaviour

Innate behaviour develops independently of environmental context. A spider spins a web correctly the very first time. No trial-and-error learning is taking place. Innate behaviours are controlled by genes and inherited from parents. A wasp builds a nest which is characteristic of its species. A termite builds a characteristic mound. Scientists familiar with insects can tell which insect built a nest or mound by looking at its shape. These are genetically programmed behaviours which ensure the survival of the animal. A simple version of song is innate in birds. Sucking behaviour is innate in human infants.

Figure 16.9 Courtship behaviour of the three-spined stickleback.

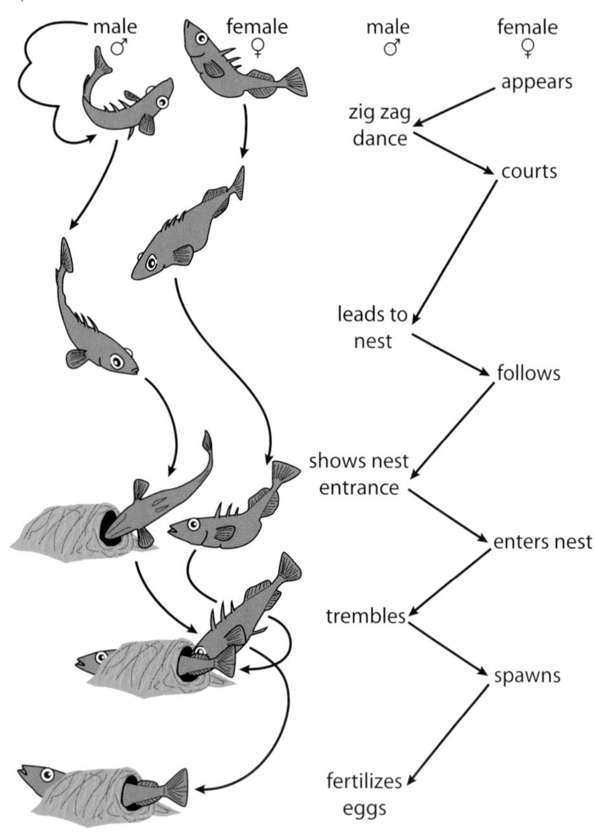

Some innate behaviours are performed in a certain order. A classic example of an innate sequence of behaviours is seen in the mating behaviour of the three-spined stickleback fish (see Figure 16.9). Mating begins with the male doing a zigzag dance when he sees the female. This dance attracts the female's attention. She follows the male as he leads her to the nest he has constructed in the bed of the river. He backs out of the nest and the female enters. He vibrates his body at the entrance to the nest and the female releases her eggs. She leaves the nest and the male enters. He releases his sperm cells which fertilize the eggs. This behaviour is as specific to this species as the number of spines they have on their back.

Learned behaviour

Learned behaviour is not genetically programmed. You learned to read a book, how to ride a bike, how to tie your shoes. All of these activities result in new knowledge that did not previously exist or a new skill that you did not originally possess. Learned behaviour can be defined as the process of gaining new knowledge or skills or modifying existing knowledge or skills. You may know how to read at one level when you are six years old, but you will improve that skill with more practice and schooling.

How do we know that learning in animals has really occurred? Learning can only be measured by performance. Learning can be explained as change in performance that we are sure is stored in the nervous system as memory. For example, a rat learns that pressing a pedal releases food. The rat originally pressed the pedal by accident during exploration of its cage. After the pellet of food was released over and over again, the rat learned to associate the food with the pedal. Later pushing the pedal to get food became a deliberate act. This is a performance which indicates learning. Behaviour output is not always easily seen. This is why learning is sometimes difficult to measure.

This tables summarizes a comparison of innate and learned behaviour.

Innate behaviour	Learned behaviour
develops independently of the environmental context	dependent on the environmental context of the animal for development
controlled by genes	not controlled by genes
inherited from parents	not inherited from parents
developed by natural selection	develops by response to an environmental stimulus
increases chance of survival and reproduction	may or may not increase chance of survival and reproduction

Investigating innate behaviour in invertebrates

Animals orient in different ways to their diverse environments. They survive better in some places than others. Food may be more plentiful in one area, better protection available in another area, humidity levels higher in another. When studying simple invertebrate animals, innate behaviours can be measured as the animals respond to environmental stimuli. Two basic kinds of movement are seen in invertebrate animals: taxis and kinesis.

Taxis

A taxis (plural, taxes) is a directed response to a stimulus. If the animal's body is directed toward the stimulus, we say it has a positive response. If the animal's body is directed away from the stimulus, we say it has a negative response. For example, if an animal moves toward light, it exhibits a positive phototaxis. If the animal moves away from light it exhibits a negative phototaxis. Taxes are identified by the type of stimuli to which the organism is responding.

Chemotaxis: the response to chemicals in the environment. Organisms in water can move towards or away from food or other chemicals which are dissolved in their aquatic medium. When exploring chemotaxis, experiments can be performed which vary the pH, or the concentration of dissolved drugs, food, or pesticides.

Phototaxis: the response to light. Experiments can be performed using different wavelengths of light, different light intensities and different types of bulb (UV, incandescent or fluorescent).

Gravitaxis: the response to gravity. Methods can be devised to measure the response to gravity if organisms are put into a container that is then placed upside down. Placing organisms on a slow-spinning turntable may also disrupt the normal pull of gravity.

Rheotaxis: a response to water current. Do aquatic organisms move with or against the current?

Thigmotaxis: a response to touch. It would be interesting to see if any organism has a positive thigmotaxic response.

Two invertebrates you could use to investigate taxes are *Planaria* and *Euglena*.

Planaria is an interesting flatworm which lives in lakes and ponds. They are quite active and move by contraction of muscle fibres in their body. They have a simple nervous system and at the anterior end are two eyespots which contain photoreceptors stimulated by light. Also in the anterior end are chemoreceptors which respond to certain chemicals. *Planaria* is negatively phototaxic, since it lives under leaves and rocks and hides for protection. It is positively chemotaxic to food that it likes to eat, such as raw liver (raw liver would be similar to dead fish in its natural habitat). Interesting studies could include *Planaria's* response to different wavelengths of light, how fast it moves towards different food substances (cm min^{-1}), or its response to temperature gradient or different concentrations of pesticide.

This planarian is part of the group Turbellaria. You can see the two eyespots which are sensitive to light.

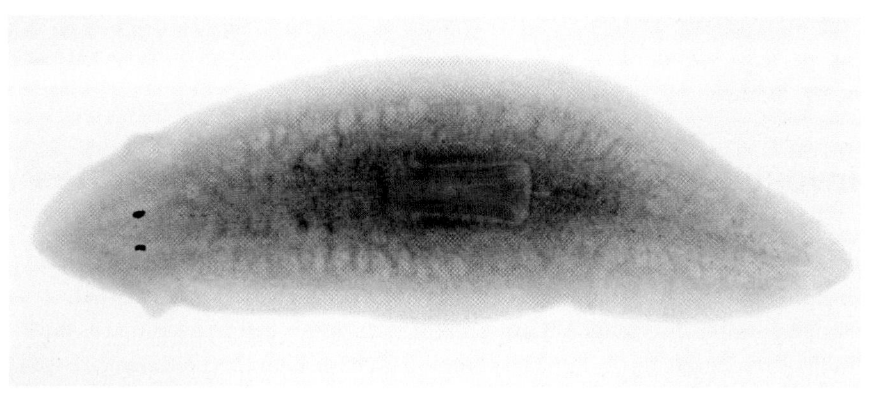

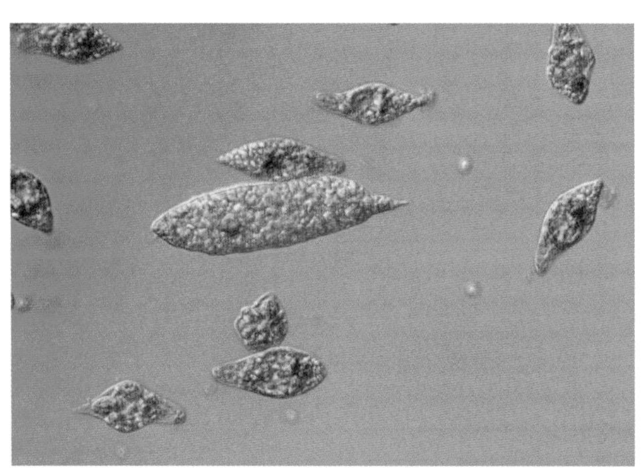

Euglena is single-celled protist (protoctist). It has a flagellum which propels it quickly through the water. It also has an eyespot at the anterior end which is stimulated by light. *Euglena* can make its own food by photosynthesis since it contains molecules of chlorophyll. It is positively phototaxic since it needs light to perform photosynthesis. *Euglena* could be tested to determine if it responds to different wavelengths of light.

Euglena lives in ponds and puddles. This single-celled organism has a tail-like flagellum for locomotion. The cytoplasm is green due to the large number of photosynthetic chloroplasts that make food for the organism using sunlight.

Kinesis

Kinesis is a movement in response to a non-directional stimulus, such as humidity. The rate of movement of the animal depends on the intensity of the stimulus, not its direction. It differs from taxis in that the animal does not move toward or away from the stimulus. If the animal is in an environment which is not suitable, it moves rapidly but randomly (with no direction) until it is in a more comfortable spot. If it is in its 'comfort zone', its movement slows down. Slow movement is likely to keep the organism in the condition that it prefers.

Orthokinesis: when an organism moves slowly or rapidly (changes speed) in response to the stimulus but it does not move towards the stimulus.

Klinokinesis: when an organism turns slowly or rapidly in response to the stimulus but it does not move towards the stimulus.

Isopods are terrestrial crustaceans which can be used to study kinesis. Even though they live on land, they breathe with gills and need moisture in order to breathe. Isopods live in damp places. They die if exposed to dry conditions for a long period of time. Isopods show kinesis to humidity. When placed in a damp environment, they move slowly. When placed in a dry environment, they move quickly. Moving quickly makes it more likely that they will get out of the dry air. Consequently, the isopod in the damp place will remain in that spot for which it is well suited. The isopod in the dry conditions may find a damp spot during its increased random movement. The minute it senses a damp environment, its random movements slow down. Two isopods used to study kinesis are the woodlice, *Porcellio scaber* and *Armadillidium vulgare.* (Figure 16.10).

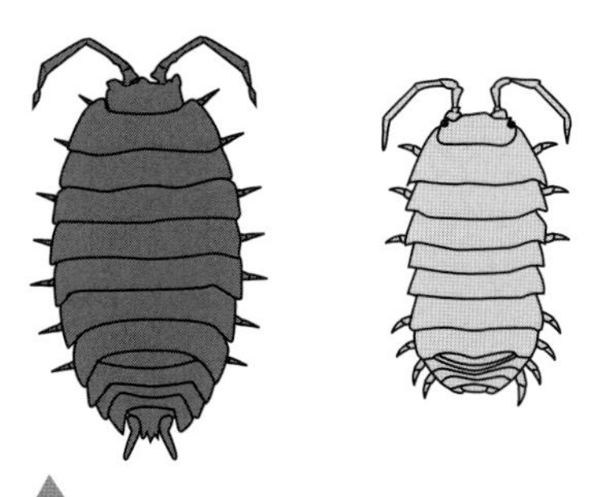

Figure 16.10 *Porcellio scaber* (left) and *Armadillidium vulgare* (right).

Experimental design

Follow these steps to design an experiment to investigate innate behaviours of an invertebrate.

1 Observe the organism of choice. Research the organism and formulate a research question, which must be specific. It must allow you to collect measurable data. Here is an example of a good research question: 'What is the effect of humidity on the distribution of the isopod *Porcellio scaber*?'

2 Describe a method for the collection of relevant data. Here is an example.
   - a Modify a pair of Petri dishes to make a choice chamber, in which the isopods are given an opportunity to be in humid or dry conditions. Set up one chamber with a drying agent ($CaCl_2$) and the other with wet towels (see Figure 16.11). Measure the humidity in each chamber with a Vernier probe.
   - b Place 10 individuals in each chamber through holes with rubber stoppers.
   - c Count the number of individuals in each chamber every 5 minutes.
   - d Repeat this procedure so that you have data for 40 organisms.
   - e As a control, set up a pair of Petri dishes which have no difference in humidity. Repeat steps (b), (c) and (d).

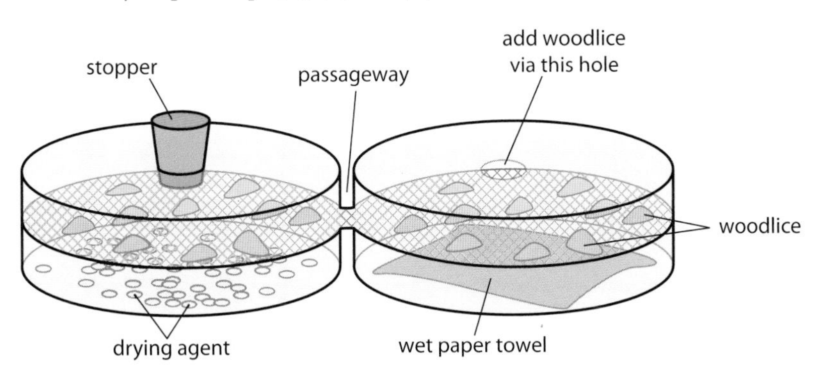

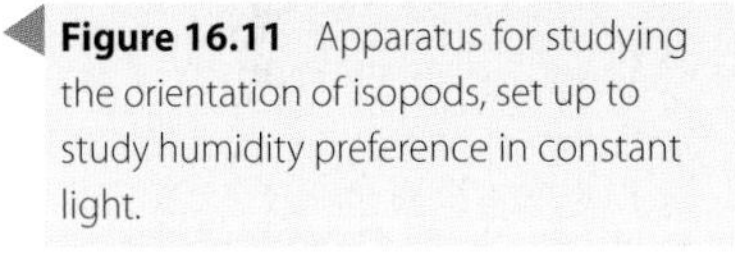

Figure 16.11 Apparatus for studying the orientation of isopods, set up to study humidity preference in constant light.

3 Design a method for control of the variables.
   - a Measure the light conditions in which the experiment is taking place with a Vernier probe. Make sure that the light conditions for the entire experiment remain constant. Isopods may respond to light also, so the amount of light must be controlled.
   - b Measure the temperature conditions in which the experiment is taking place with a Vernier probe. Make sure the temperature conditions for the entire experiment remain constant.
   - c There must be an equal possibility for the isopods to travel to either chamber.
   - d The sizes of the chambers must be equal.

4 Record raw data, including units (minutes) and uncertainties (±0.5 minutes). Make sure you write a title for each data table. Do not split a data table across two pages.

Effect of a humidity on the movement of isopods: Trial 1

Time / min ±0.5	Chamber with desiccant (dry)	Chamber with wet towels (humid)
0	10	10
5	9	11
10	9	11
15	9	11
20	8	12
25	7	13
30	6	14
35	5	15
40	5	15
45	5	15

Effect of humidity on the movement of isopods: Trial 2

Time / min ±0.5	Chamber with desiccant (dry)	Chamber with wet towels (humid)
0	10	10
5	9	11
10	8	12
15	8	12
20	7	13
25	6	14
30	5	15
35	5	15
40	5	15
45	5	15

Effect of humidity on the movement of isopods: Control Trial 1

Time / min ±0.5	Chamber empty	Chamber empty
0	10	10
5	9	11
10	9	11
15	9	11
20	9	11
25	9	11
30	10	10
35	10	10
40	10	10
45	10	10

Effect of humidity on the movement of isopod: Control Trial 2

Time / min ±0.5	Chamber empty	Chamber empty
0	10	10
5	10	10
10	8	12
15	8	12
20	8	12
25	9	11
30	9	11
35	10	10
40	10	10
45	10	10

5 Process the raw data. Processing includes any mathematical manipulation of the data or graphing of manipulated data (graphing of raw data is *not* considered processing).

For example: Determine the means of the numbers of isopods in humid and dry conditions after 45 minutes (± 0.5) in the two trials, compared with the controls.

Effect of humidity on the movement of isopods (means)

	Dry chamber	Humid chamber	Control chamber	Control chamber
Trial 1	5	15	10	10
Trial 2	5	15	10	10
Sums of two trials and controls	10	30	20	20
Means of two trials and controls	5	15	10	10

6 Graph the mean values from the two trials. For example, see Figure 16.12.

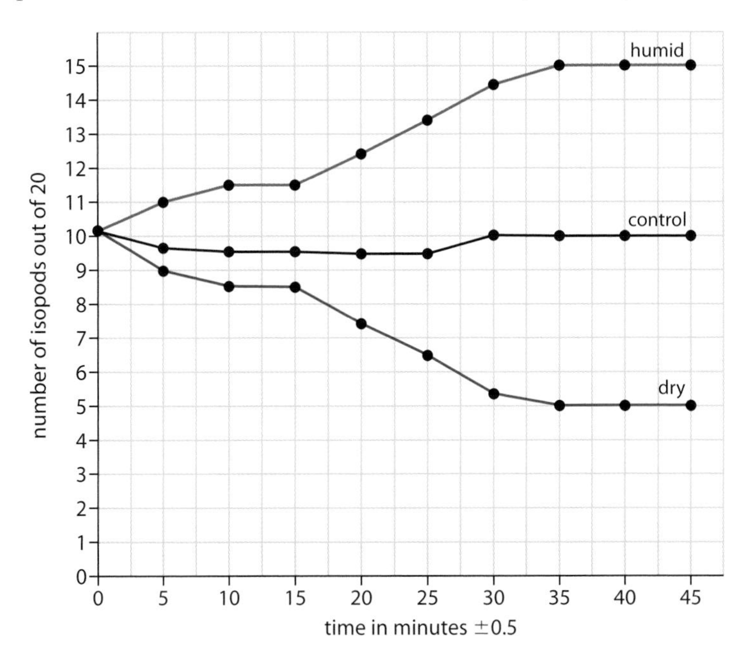

Figure 16.12 Graph of the mean values from two trials with isopods in humid and dry conditions compared with controls.

As you can see from the isopod experiment, the scientific method uses inductive reasoning to go from the particular to the general. For example, we say that when all metals are heated they expand. As far as we know, all metals that have been heated, do expand. The problem here is that we are moving from the observed to the unobserved. We have not measured all the metals there are in the world to see if they expand. This raises the practical problem of how many data points we need in an experiment in order to make a valid generalization. We have a tendency to use insufficient evidence and jump to conclusions.

Discuss the limits of scientific knowledge. Does scientific knowledge change with more experimentation? Can you give some current examples?

To learn how to use the chi square statistical test to analyse data from choice chambers, visit heinemann.co.uk/hotlinks, enter the express code 4242P and click on Weblink 16.7.

What do the invertebrate behaviour experiments tell us about survival and reproduction?

First, we need to use statistical analysis to determine if the differences are significant. For example, the chi square statistical test can be used on this data to determine if the difference between the data from the dry and the humid chambers are significantly different from the data from the controls.

Secondly, we can draw conclusions based on the statistical analysis of the data. The behaviour of the isopods is to move randomly and quickly in a dry environment until they finally come to rest in the humid environment. Humidity is important to the survival of isopods and their ability to reproduce. The outer covering of isopods (exoskeleton) lacks a waterproof waxy cuticle (as found in many land-dwelling groups) so the animal is highly subject to desiccation (drying out). Quick random movements enable the isopod to find itself in a humid environment. This ensures survival and enhances the ability to reproduce. Natural selection favours isopods which show this response.

Learning improves the chance of survival

When we talk about learning we mean that new knowledge or skill has been acquired. Learning occurs most easily when it results in improving the animal's survival. Some special learning abilities which improve survival are imprinting, food hoarding and song.

Imprinting is the process by which young animals become attached to their mother within the first day or so after hatching or birth. This is why a line of ducklings follow their mother. Imprinting assures that the young stay close to their mother for protection and as a source of food. This tremendously improves their chance of survival.

Many animals store food when it is plentiful and return to the hoard of food when there is a shortage. Squirrels hoard nuts. Moles bite and paralyse worms and keep them alive in different places underground for days. They retrieve them when they are ready to eat. Hoarding food is an excellent strategy learned by animals – it allows them to stay nourished even in times of food shortages.

Young male sparrows sing a crude species-specific song which is inherited. As they mature, they learn to sing a better song by listening to other adult males sing.

Birdsong is also learned. If a young male sparrow hears an adult song within its first 100 days of life, it will sing a full adult song the next year. This song has two functions. It attracts a mate and deters rival males. The sparrow with the best song promotes survival of his particular genes.

Other examples of learning which enhances survival are:

- grizzly bears learning to catch slippery salmon in rushing river waters;
- chimpanzees learning to stick a branch into a termite nest, pull it out and eat the termites.

Grizzly bears learn from their mothers, chimpanzees learn by trial and error. Obviously, both of these examples improve the chances of survival since they are learned strategies for obtaining food from difficult places.

Although most learning increases the chances of survival, some animals are tricked into false learning through the mimicry of their prey. One butterfly which has a bad taste is mimicked by another butterfly which does not taste bad at all. Some snakes imitate the poisonous red coral snake so they do not get eaten.

Pavlov and conditioning

Classical conditioning can be used to modify a reflex response. In classical conditioning experiments, the subject responds to a stimulus in a new way.

For example, in humans, blinking is a reflex response. If you wave your hand suddenly in front of a subject's face, they will automatically blink. The waved hand is called the unconditioned stimulus (UCS) because it unconditionally stimulates the eye-blink response. The eye-blink is called the unconditioned response (UCR). After training, it is possible to elicit the reflex response (eye-blink) with a new and neutral stimulus (NS). First, the neutral stimulus (e.g. a musical note) is introduced. The subject probably does not blink. Next the subject is given a period of training: the musical note is sounded immediately before a hand is waved in front of the subject's eye. Eventually the subject responds with an eye-blink to just the musical note. After this has occurred, the musical note is called the conditioned stimulus (CS) and the eye-blink in response to the musical note is called the conditioned response (CR). The subject is now responding to a musical note in a new way.

The Russian physiologist Ivan Pavlov designed experiments to illustrate classical conditioning. His subjects were dogs. Salivation in dogs is a reflex response to the presence of food in the mouth. The unconditioned stimulus (UCS) of food elicits the unconditioned response (UCR) of salivation. The neutral stimulus (NS) that Pavlov employed was the ringing of a bell (see Figure 16.13). He rang the bell just before the dog tasted the food. After training, he could ring the bell (CS) and the dog would salivate (CR). The dog had learned to salivate to the neutral stimulus (NS) alone.

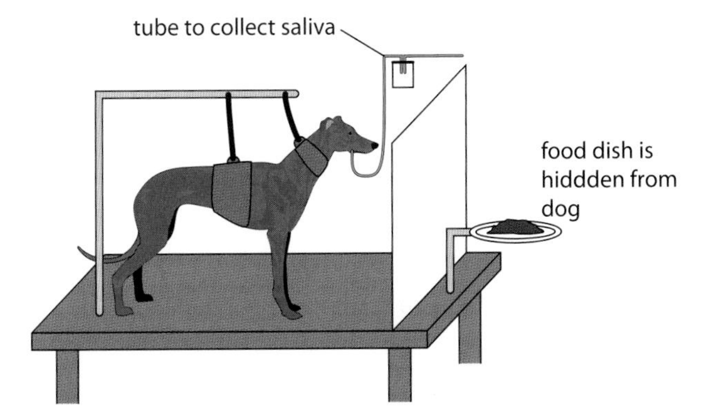

Figure 16.13 Pavlov's experimental set-up.

Learning of birdsong in young birds

Are you able to tell the difference between bird songs? Each species of bird has a species-specific song which is inherited. Birds of one species have a varied song just as we have variations in the colour of our eyes. However, birds can also learn to improve the song they have inherited. Thus, birdsong has both inherited and learned components.

Birds are able to sing due to their vocal organ, called the syrinx. The syrinx is a bony structure at the bottom of their trachea (windpipe). In humans, the larynx (voice box) is at the top of the trachea. The bird forces air past a membrane in the syrinx which vibrates and results in sound. Birds control the pitch by altering the tension in the membranes of the syrinx. They control the volume of the song by altering the flow of air.

Birdsong is a well studied example of animal behaviour. Singing is an important business for the male bird. He attracts a mate with his song and deters male rivals. Generally females do not sing. Studies of song learning have shown that birds

Can the Pavlov theory of conditioning be applied to different examples of learning? Consider this example of learning which could take place at the dentist. If drilling into your tooth causes you pain, the drilling can be considered the UCS and the pain, the UCR. Now when you return to the dentist's office and hear just the sound of the drill (CS) in the next room, you may feel the pain return to your tooth (CR). Can you think of other examples of learning to which you can apply conditioned and unconditioned response?

To play a game about Pavlov's dog, visit heinemann.co.uk/hotlinks, enter the express code 4242P and click on Weblink 16.8.

hatch with what is called a 'crude template'. Evidence for a template is shown by experimental data: if birds are kept in a laboratory and denied any auditory stimulation, they produce a very crude song. This crude song is species-specific. The crude song of a warbler can be distinguished from the crude song of a sparrow. With an acoustical spectroscope, it is possible to accurately measure the difference. This data is evidence that the template is inherited. All of the next steps of birdsong development are learned (see Figure 16.14).

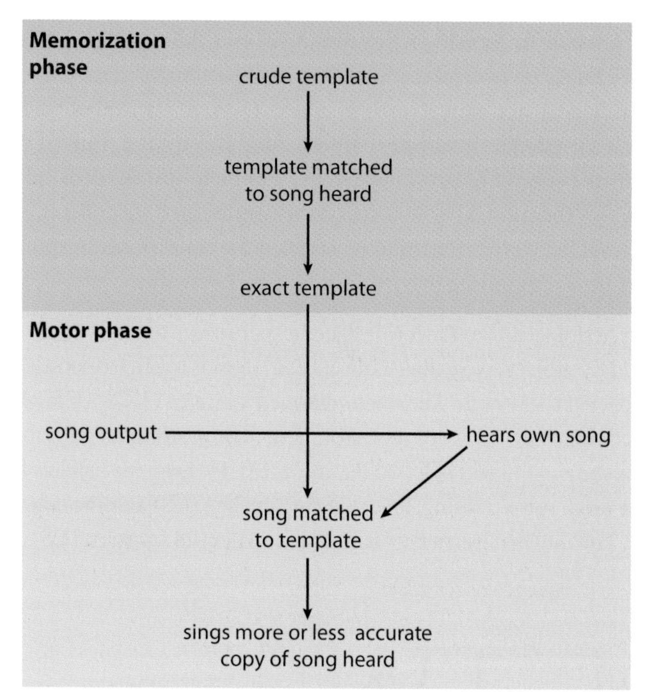

Figure 16.14 A crude template of birdsong is inherited, but the development of a mature adult song is learned.

The first global survey of bird diversity can tell us which species are most vulnerable to extinction. If a species is extinct we will never hear its song again. An international team is trying to determine the range for each species on a global scale. Those birds with a smaller range are at greater risk of extinction. The group will collect the global data important to understanding how conservationists can make a difference in bird survival.

For more information about the first global bird survey visit heinemann.co.uk/hotlinks, enter the express code 4242P and click on Weblink 16.9.

After hatching, there is a memorization phase. In this phase, the bird is silent but listening to the song of his species from adults. The hatchling is modifying the inherited template.

As he is listening, he is attempting to match his template to the full adult song. It is a type of memorization. This memorization phase is over at about 100 days of age. If a male bird does not hear the adult song within 100 days, he will not modify the template that he has inherited. This first 100 days is called the sensitive period.

The second phase is a motor phase in which the young bird practises singing the song that he has heard. He hears himself singing and begins to shape his song to match what he has heard from the adult, usually his father. The bird must hear his own song in order to sing an accurate adult song. (Experiments have been done which show that if a bird is deafened after 100 days, he will only sing the crude template of the song.) As he becomes sexually mature, his song will become perfected and he will begin to search for a mate.

The crude template is a good example of innate learning while the adult song is an example of how learned behaviour can help an animal acquire new skills.

Exercises

11 Distinguish between learned and innate behaviours.

12 Describe the design of an experiment which measures either taxis or kinesis of an invertebrate animal.

E.4 Neurotransmitters and synapses

Assessment statements

E.4.1 State that some presynaptic neurones excite postsynaptic transmission and others inhibit postsynaptic transmission.

E.4.2 Explain how decision making in the CNS can result from the interaction between the activities of excitatory and inhibitory presynaptic neurones at synapses.

E.4.3 Explain how psychoactive drugs affect the brain and personality by either increasing or decreasing postsynaptic transmission.

E.4.4 List three examples of excitatory and three examples of inhibitory psychoactive drugs.

E.4.5 Explain the effects of THC and cocaine in terms of their action at the synapses in the brain.

E.4.6 Discuss the causes of addiction, including genetic predisposition, social factors and dopamine secretion.

Synaptic transmission

You may remember that neurones communicate with each other chemically across a space called a synapse. On one side of the synapse is the presynaptic membrane of the sending neurone and on the other side of the synapse is the postsynaptic membrane of the receiving neurone (see also Chapter 6, page 179). When you send your friend an email, you are like the presynaptic membrane and your friend is like the postsynaptic membrane. The molecule which moves across the space (synaptic cleft) between the two membranes is called a neurotransmitter. Your email is like the neurotransmitter moving from the presynaptic membrane to the postsynaptic membrane. At the postsynaptic membrane, there is a receptor molecule (like the inbox of your email). A specific neurotransmitter is received by a specific receptor (like you receive the email with your address on it).

Some neurotransmitters are excitatory and stimulate the next neurone to forward the message. The way they do this is to increase the permeability of the postsynaptic membrane to positive ions, making it easier for positive ions to move in. Some neurotransmitters are inhibitory. They cause positive ions to move out of the postsynaptic cell. Movement of positive ions back in to the synaptic cleft chemically depresses the postsynaptic cell and makes it much harder to excite.

From this description you can see that some presynaptic neurones *excite* postsynaptic neurones and others *inhibit* postsynaptic transmission.

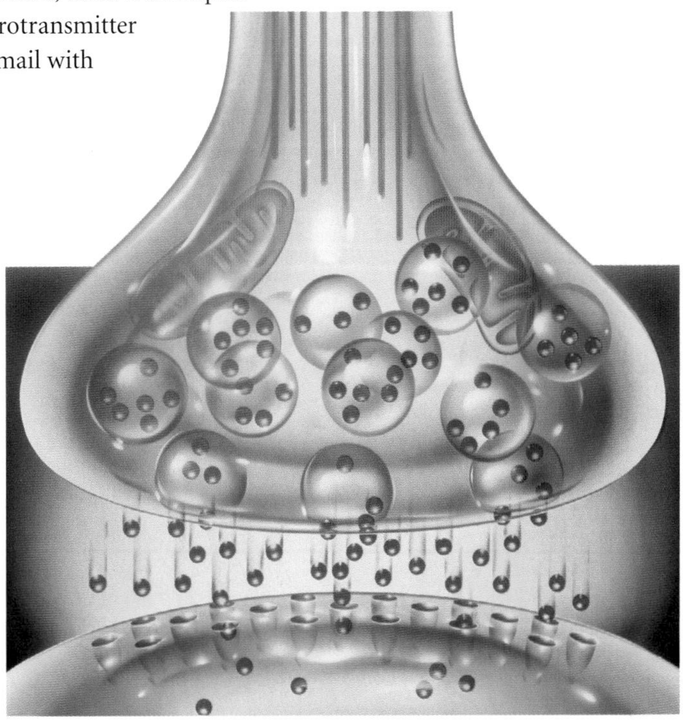

At the synapse, some presynaptic neurones excite postsynaptic neurones and others inhibit postsynaptic transmission. Note that the neurotransmitter is received by a receptor but does *not* enter the postsynaptic neurone.

Decision making in the central nervous system

What interaction occurs between excitatory and inhibitory presynaptic neurones acting at the synapse?

The impulse which moves down the presynaptic neurone is called the action potential. As the action potential reaches the axon bulb, calcium ions (Ca^{2+}) rush into the end of the neurone. This causes vesicles containing neurotransmitters to fuse with the presynaptic membrane. As the vesicles fuse with the presynaptic membrane they release the neurotransmitters into the synaptic cleft (see Figure 16.15). Now we see what happens when the neurotransmitter is in the synaptic cleft.

The neurotransmitter binds to specific receptors on the postsynaptic membrane. The receptors are like gates which let ions enter or leave when the neurotransmitter binds to them.

Figure 16.15 Synaptic transmission

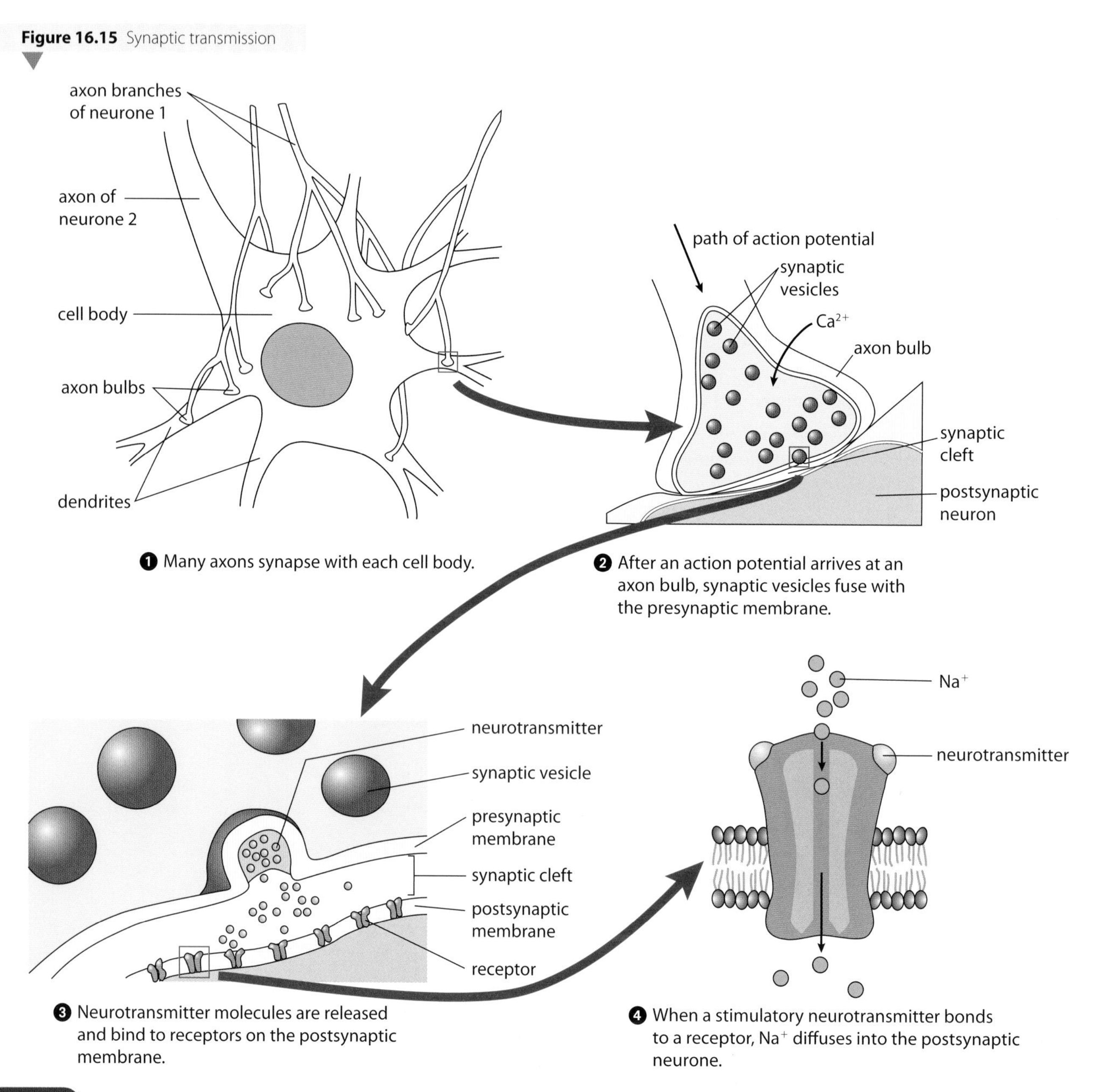
axon branches of neurone 1
axon of neurone 2
cell body
axon bulbs
dendrites
❶ Many axons synapse with each cell body.
path of action potential
synaptic vesicles
Ca^{2+}
axon bulb
synaptic cleft
postsynaptic neuron
❷ After an action potential arrives at an axon bulb, synaptic vesicles fuse with the presynaptic membrane.
neurotransmitter
synaptic vesicle
presynaptic membrane
synaptic cleft
postsynaptic membrane
receptor
❸ Neurotransmitter molecules are released and bind to receptors on the postsynaptic membrane.
Na^+
neurotransmitter
❹ When a stimulatory neurotransmitter bonds to a receptor, Na^+ diffuses into the postsynaptic neurone.

Excitatory neurotransmitters

Acetylcholine is an example of a neurotransmitter which is excitatory. An excitatory neurotransmitter generates an action potential.

Excitatory neurotransmitters increase the permeability of the postsynaptic membrane to positive ions. This causes positive sodium ions (Na^+) which are in the synaptic cleft to diffuse into the postsynaptic neurone. The postsynaptic neurone is depolarized locally (just in that area) by the influx of positive sodium ions. During depolarization, the inside of the neurone develops a net positive charge compared to the outside. Depolarization is the way the impulse is carried along the neurone. The neurone is locally depolarized and the depolarization continues as sodium ions diffuse to the next area of the neurone. In this fashion, the impulse is conducted along the neurone from one adjacent area to the next, just like a wave.

An action potential is formed as the membrane depolarization is raised above the threshold. This means that the impulse is being carried along the nerve. If the threshold is not met, the neurone does not carry the impulse to the next neurone.

Inhibitory neurotransmitters

GABA is an example of an inhibitory neurotransmitter. These neurotransmitters inhibit action potentials. An inhibitory neurotransmitter causes hyperpolarization of the neurone (the inside of the neurone becomes more negative) making it even more difficult for an action potential to be generated.

The inhibitory neurotransmitter binds to its specific receptor. This causes negatively charged chloride ions (Cl^-) to move across the postsynaptic membrane into the postsynaptic cell or it can cause positively charged K^+ ions to move out of the postsynaptic neurone. This movement of Cl^- into the neurone or K^+ out of the neurone is what causes the hyperpolarization.

- **Examiner's hint:** This is a useful memory trick for hyperpolarization: When the 'kat' (K^+) is 'hyper' (hyperpolarized) we put her outside. (K^+ moves out of the neurone.)

Putting it together

A neurone is on the receiving end of many excitatory and inhibitory stimuli. The neurone sums up the signals. If the sum of the signals is inhibitory then the axon does not fire. If the sum of the signals is excitatory, then the axon fires. This is the interaction that takes place between the activities of the excitatory and inhibitory neurones at the synapses. The summation of the messages is the way that decisions are made by the CNS.

Psychoactive drugs affect the brain and personality

To fully understand how drugs affect the brain and personality we must have an understanding of the two main neurotransmitters. They are acetylcholine and noradrenaline.

Cholinergic versus adrenergic synapses

Acetylcholine is released by all motor neurones and activates skeletal muscle. It travels across the synapse and depolarizes the postsynaptic membrane. However, if it remained in the synapse, the postsynaptic membrane would go on firing indefinitely. To prevent this, an enzyme called acetylcholinesterase breaks down acetylcholine in the synapse. Acetylcholine is involved in the parasympathetic nervous system (see page 494). This means it causes relaxation rather than flight.

Synapses using acetylcholine are called cholinergic synapses. Nicotine stimulates transmission in cholinergic synapses which is why it has a calming effect on the body and personality. People addicted to nicotine become very agitated if they cannot have a cigarette.

A second widespread neurotransmitter is noradrenaline. Noradrenaline depolarizes the postsynaptic neurone. Noradrenaline is involved in the sympathetic system (see page 494). This means that it causes a 'fight or flight' reaction. Synapses using noradrenaline are called adrenergic synapses. Cocaine and amphetamines stimulate adrenergic synapses. Cocaine and amphetamines both cause increased alertness, energy and euphoria.

This table compares cholinergic and adrenergic synapses.

	Cholinergic	Adrenergic
Neurotransmitter	acetylcholine (Ach)	noradrenaline
System	parasympathetic	sympathetic
Effect on mood	calming	increased energy, alertness and euphoria
Drugs increasing transmission at synapse	nicotine	cocaine and amphetamines

Researchers from Australia and Canada have found that alcohol and other drug abuse rank among the top ten contributors to the global burden of disease among adolescents. Alcohol contributed to 27% of the deaths of 15–29-year-olds in economically developed countries, during the study.

To discover more about drug and alcohol abuse, visit heinemann.co.uk/hotlinks, insert the express code 4242P and click on Weblink 16.10.

Effect of drugs on the brain

Drugs can alter your mood or your emotional state. Excitatory drugs like nicotine, cocaine and amphetamine increase nerve transmission while inhibitory drugs such as benzodiazepines, alcohol and tetrahydrocannabinol (THC) decrease the likelihood of nerve transmission. Drugs act at the synapses of the brain by different mechanisms to determine your emotional state. Drugs can change synaptic transmission in the following ways (see Figure 16.16):

- block a receptor for a neurotransmitter (drug has structure similar to neurotransmitter);

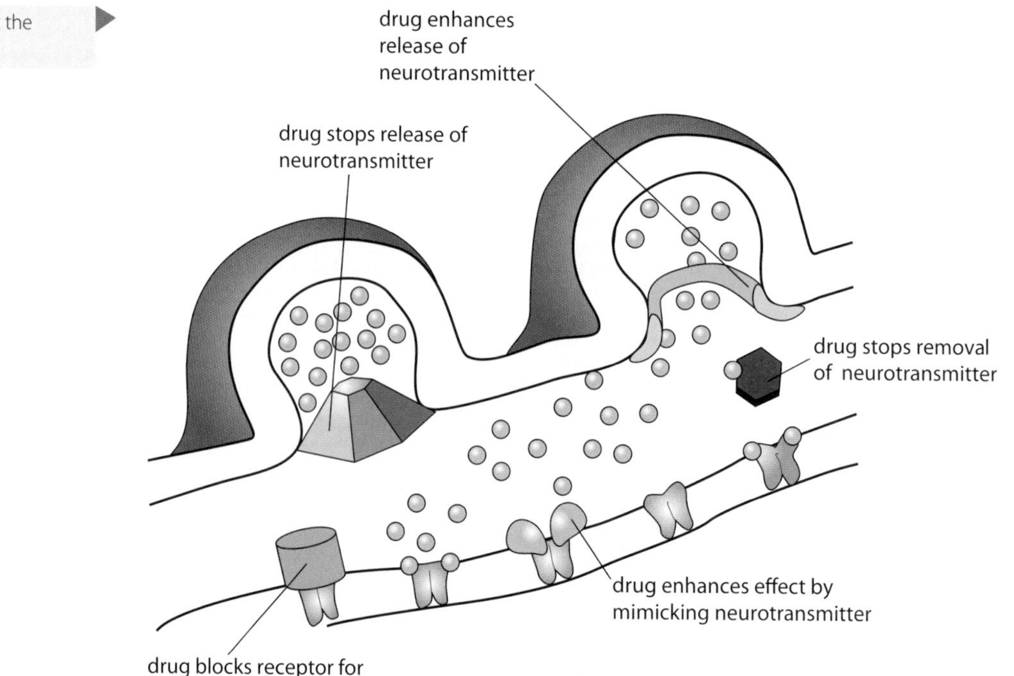
drug enhances release of neurotransmitter
drug stops release of neurotransmitter
drug stops removal of neurotransmitter
drug enhances effect by mimicking neurotransmitter

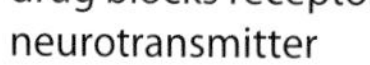
drug blocks receptor for neurotransmitter

Figure 16.16 Drug action at the synapse.

- block release of a neurotransmitter from the presynaptic membrane;
- enhance release of a neurotransmitter;
- enhance neurotransmission by mimicking a neurotransmitter (when drugs have the same chemical structure as the neurotransmitter they have the same effect but are not broken down as easily so the effect is stronger because they stay longer in the synapse);
- block removal of a neurotransmitter from the synapse and prolong the effect of the neurotransmitter.

Excitatory drugs and how they act

Nicotine in tobacco products is a stimulant which mimics acetylcholine (Ach). Thus, it acts on the cholinergic synapses of the body and the brain to cause a calming effect. After Ach is received by the receptors, it is broken down by acetylcholinesterase but the enzyme cannot breakdown the nicotine molecules which bind to the same receptors. This excites the postsynaptic neurone and it begins to fire, releasing a molecule called dopamine. Dopamine gives you a feeling of pleasure. It is a molecule of the 'reward pathway' of your brain.

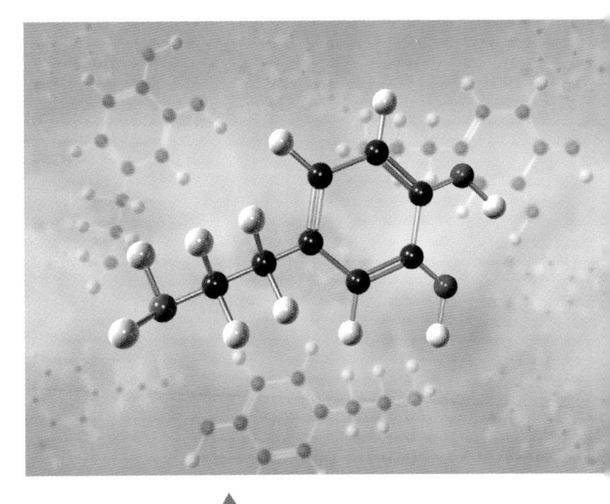

This is a model of dopamine. It is one of the most important of the neurotransmitters in the central nervous system (CNS). It plays a critical role in the way the brain controls movement, memory and decision making. Excess dopamine may contribute to psychotic illnesses, notably schizophrenia. Underproduction of dopamine results in the movement disorder of Parkinson's disease.

Cocaine stimulates transmission at adrenergic synapses and causes alertness and euphoria. It also causes dopamine release. Cocaine blocks removal of dopamine from the synapse so that it builds up. This leads to over stimulation of the postsynaptic neurone. Since this is in the 'reward pathway', it leads to euphoria. Notice that it acts the same way as nicotine. Both of these drugs lead to addiction.

Amphetamine stimulates transmission at adrenergic synapses and gives increased energy and alertness. Amphetamine acts by passing directly into the nerve cells which carry dopamine and noradrenaline. It moves directly into the vesicles of the presynaptic neurone and causes their release into the synaptic cleft. Normally, these neurotransmitters would be broken down by enzymes in the synapse, but amphetamines interfere with the breakdown. Thus, in the synapse high concentrations of dopamine cause euphoria, and high concentrations of noradrenaline may be responsible for the alertness and high energy effect of amphetamines.

Inhibitory drugs and how they act

Benzodiazepine reduces anxiety and can also be used against epileptic seizures. Its effect is to modulate the activity of GABA which is the main inhibitory neurotransmitter. When GABA binds to the postsynaptic membrane, it causes chloride ions (Cl^-) to enter the neurone. Remember that when Cl^- enters the neurone, the neurone becomes hyperpolarized and resists firing. Benzodiazepine increases the binding of GABA to the receptor and causes the postsynaptic neurone to become even more hyperpolarized (see Figure 16.17).

GABA is gamma aminobutyric acid.

Alcohol acts similarly to benzodiazepine in that it increases the binding of GABA to the postsynaptic membrane, and causes the neurone to become hyperpolarized. This explains the sedative effect of alcohol. It decreases the activity of glutamate, an excitatory neurotransmitter. Alcohol also helps to increase the release of dopamine by a process which is not well understood. It appears to stop the activity of the enzyme which breaks down dopamine in the synaptic cleft. Remember that dopamine works in the 'reward pathway'.

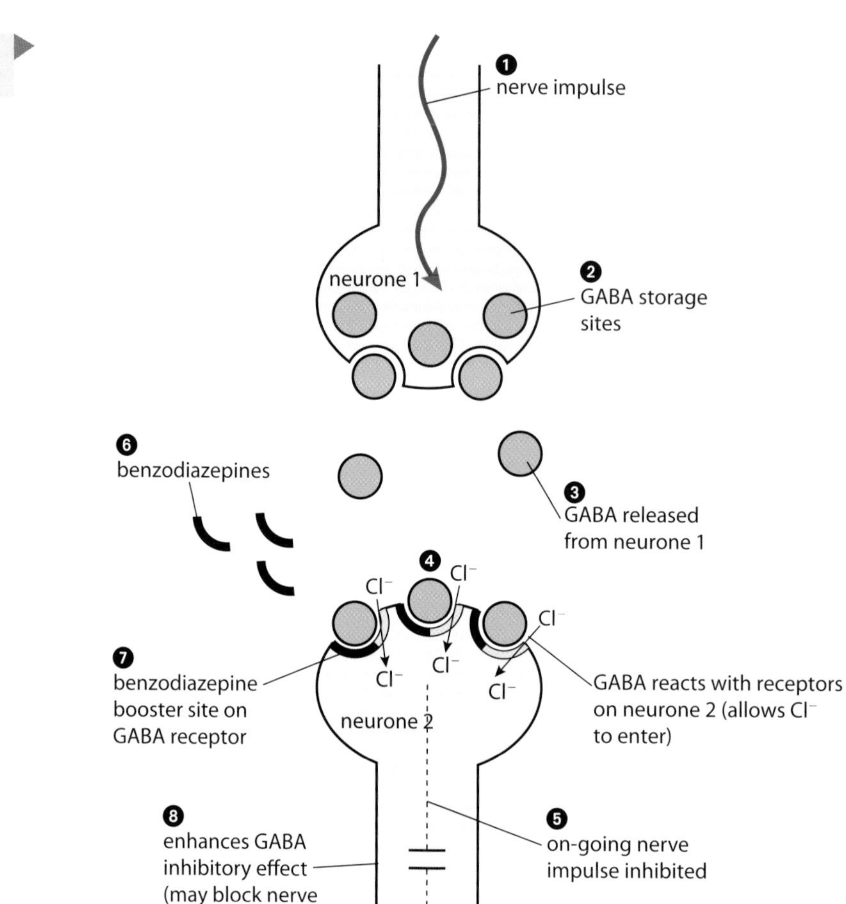

Figure 16.17 The effect of benzodiazepines at the synapse.

Tetrahydrocannabinol (THC) is the main psychoactive chemical in marijuana. THC mimics the neurotransmitter, anandamide. THC binds to the same receptor as anandamide (sometimes called cannabinoid receptors). THC is an inhibitory neurotransmitter and causes the postsynaptic neurone to be hyperpolarized. Scientists do not fully understand the role of anandamide, but it may play a role in memory functions. Marijuana disrupts short-term memory in humans. Anandamide may be involved in eliminating information from our memory that is not needed.

THC and cocaine affect mood, synapse and behaviour

Marijuana users often describe the feelings produced as relaxing and mellow. Some say they feel lightheaded and hazy. The THC may dilate the pupils causing colour perception to be more intense. Other senses may be enhanced. Some experience a sense of panic and paranoia.

At the synapse, THC acts on cannabinoid receptors. These receptors affect several mental and physical activities, including:

- learning;
- coordination;
- problem solving;
- short-term memory.

Since THC mimics anandamide, it inhibits the neurones that anandamide inhibits but more than likely there is no enzyme for breaking down THC in the synapse. Its effect is much greater since it stays in the synapse longer.

High concentrations of cannabinoid receptors are found in the areas of the brain shown in Figure 16.18. The hippocampus is important for short-term memory. When THC binds to these receptors, it interferes with short-term memory. THC also affects coordination which is controlled by the cerebellum and the basal ganglia. This is another reason for motor impairment when using marijuana.

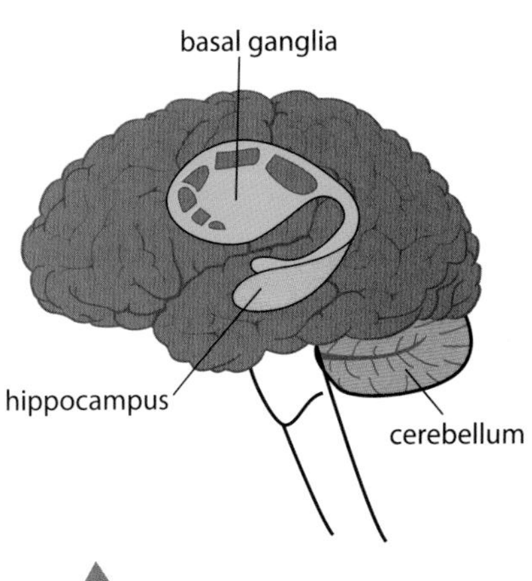

Figure 16.18 Areas of the brain with high concentrations of cannaboid receptors.

Some of the effects of cocaine are euphoria, talkativeness and increase in mental alertness. There is a temporary decrease in the need for food and sleep. Large amounts of cocaine can cause erratic and violent behaviour.

The synaptic effect of cocaine results from its ability to sustain the level of dopamine in the synapse. Since dopamine is the neurotransmitter in the 'reward pathway', the longer it stays in the synapse the better you feel.

Causes of addiction

Many drugs can lead to addiction: alcohol, tobacco, psychoactive drugs and some pharmaceuticals. People take some drugs to alleviate symptoms of mental illness and other drugs just for pleasure. The body often develops a tolerance and needs more and more of the drug to produce the same result. Addiction is a chemical dependency on drugs where the drug has 'rewired' the brain and has become an essential biochemical in the body.

Many people assume that smoking is just a bad habit. Scientific evidence shows that smoking has caused the brain to be 'rewired'. Nicotine in tobacco products mimics acetylcholine. It is not broken down easily and causes release of dopamine in the 'reward pathway'. People who smoke are craving a dopamine spike.

Since the role of almost all commonly abused drugs is to stimulate the 'reward pathway' located in the brain, withdrawal of the drug produces symptoms which are the opposite of euphoria. Common withdrawal symptoms are anxiety, depression and craving. In alcohol addiction, withdrawal symptoms include conditions which are sometimes fatal such as seizures and delirium tremens (severe shaking). Continued addiction is even more harmful. Inhaled drugs can damage lungs. When sharing needles, addicts risk contracting HIV as well as hepatitis B and C. Kidney disease is also common.

Genetic predisposition

The evidence of genetic predisposition to addiction is found in studies of twins. Identical twins share the same genetic makeup and fraternal twins have 50% genetic similarities. Studies of male twins find that when one twin suffers an addiction to alcohol or drugs, the rate of addiction in the second twin is 50% greater among identical twins than among fraternal twins. Other experiments indicate that a genetically determined deficiency of dopamine receptors predisposes certain people to addiction. In one study, scientists compared genetically manipulated alcohol-preferring rats to normal rats. The alcohol-preferring rats had 20% lower levels of dopamine receptors than non-preferring rats. The alcohol-preferring rats consumed 5 grams of ethanol per kilogram when given a choice between ethanol and water. The non-preferring rats consumed less than 1 gram of ethanol per kilogram of body weight. This study gives some support to the idea of a genetic predisposition to addiction.

Social factors

Societal factors can determine a child's vulnerability to substance abuse. Such factors include family addiction, family parenting skills and mental health problems of family or child.

Visit heinemann.co.uk/hotlinks, enter the express code 4242P and click on Weblink 16.11. Scroll down to 'Diverse explanations'. There you will find some 'models' which give explanations for why people become addicted. Examine the models.

Behaviour is often tied to the peer group. Peer pressure is very influential over adolescents and less so over adults. Adolescents can be influenced or coerced into experimentation with drugs by the peer group. Users teach new users what effects to expect and what altered state is desirable. This social learning occurs in all types of drug use.

Make an argument for which of the addiction models on the above website is the most 'scientific.' Which model uses knowledge that has been collected scientifically according to the principles you have studied in your Theory of Knowledge course?

When a drug is introduced into a culture, it can become a problem that did not previously exist. When the British introduced opium into China, it quickly became a major social problem. Heroin, introduced into the US, has become a social catastrophe. The presence of alcohol at many social gatherings fosters the paradigm that alcohol must be available to have a party. In Saudi Arabia, where alcohol use is prohibited by the culture and the law, alcoholism is rare.

Some research has shown that if drugs are cheap and easy to get hold of, addiction is more likely.

Dopamine secretion

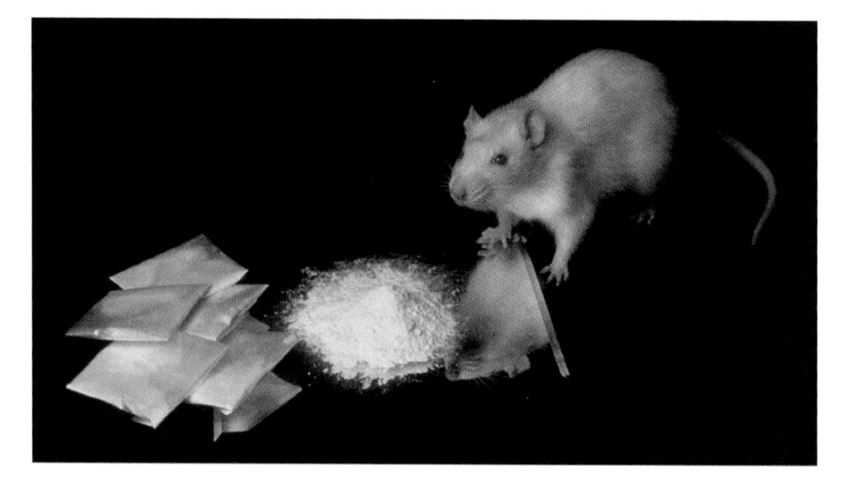

Rats are used in cocaine research to study addiction and other psychological effects of taking cocaine. Cocaine is an alkaloid drug that comes from the leaves of the coca plant (*Erythroxylon coca*). It produces feelings of exhilaration and energy. Prolonged cocaine use leads to psychological dependence. Overdoses can cause brain seizures or cardiac arrest.

You will recall that dopamine is the neurotransmitter which activates the 'reward pathway' and gives us a sense of pleasure or satisfaction. During cocaine use, dopamine builds up in the synapse.

In drug addiction, dopamine receptors are constantly stimulated. Over-stimulation decreases the number of receptors and the remaining receptors become less sensitive to dopamine. This process is called desensitization or tolerance. With tolerance, exposure to the drug causes less response than it previously caused. More and more of the drug is needed to have even the normal sense of well-being. This neuroadaptive change is probably critical for producing addiction.

Recently, scientists studying knockout mice (genetically manipulated mice addicted to cocaine), have found another neurotransmitter which might be as important or more important than dopamine – glutamate (see Figure 16.19). Glutamate may 'oversee' the learning and memories which lead to cocaine-seeking. Further experiments will shed more light on this.

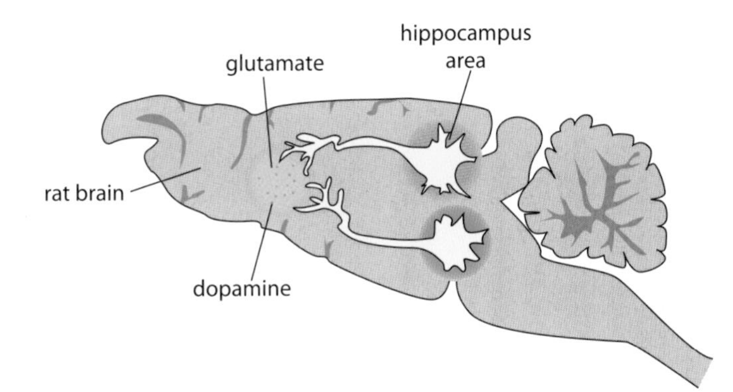

Figure 16.19 Role of glutamate in addiction – researchers found that stimulating neurones in the hippocampus (an area rich in glutamate) caused rats to search for cocaine.

Exercises

13 State two roles of presynaptic neurones.

14 Describe decision making in the CNS.

15 Explain how the brain and emotional state can be affected by psychoactive drugs.

16 State the names of three inhibitory and three excitatory psychoactive drugs.

17 Describe how THC and cocaine affect the brain.

18 Discuss the causes of addiction.

E.5 (HL only) The human brain

Assessment statements

E.5.1 Label, on a diagram of the human brain, the medulla oblongata, cerebellum, hypothalamus, pituitary gland and cerebral hemispheres.

E.5.2 Outline the functions of each of the parts of the brain listed above.

E.5.3 Explain how animal experiments, lesions and fMRI (functional magnetic resonance imaging) scanning can be used in the identification of the brain part involved in specific functions.

E.5.4 Explain sympathetic and parasympathetic control of the heart rate, movements of the iris and flow of the blood to the gut.

E.5.5 Explain the pupil reflex.

E.5.6 Discuss the concept of brain death and the use of the pupil reflex in testing for this.

E.5.7 Outline how pain is perceived and how endorphins can act as painkillers.

Structure and function of the brain

The brain is the most complex organ in the body. This jelly-like group of tissues weighing 1.4 kilograms produces our thoughts, feelings, actions and memories. It contains an amazing 100 billion neurones with thousands of synapses making the connectivity mind-boggling. New connections are formed every day of our lives. These new connections store memories, learning and personality traits. Some connections are lost and others are gained. No two brains are identical and your brain continues to change throughout your life.

This is a coloured composite three-dimensional functional magnetic resonance imaging (fMRI) and computed tomography (CT) scan of the human brain, seen from the front. The ventricles (pink) circulate the cerebrospinal fluid which cushions the brain. Beneath the ventricles lie the thalami (orange) and the hypothalamus (green, centre), which controls emotion and body temperature, and releases chemicals that regulate hormone release from the pituitary gland (round green body at lower edge).

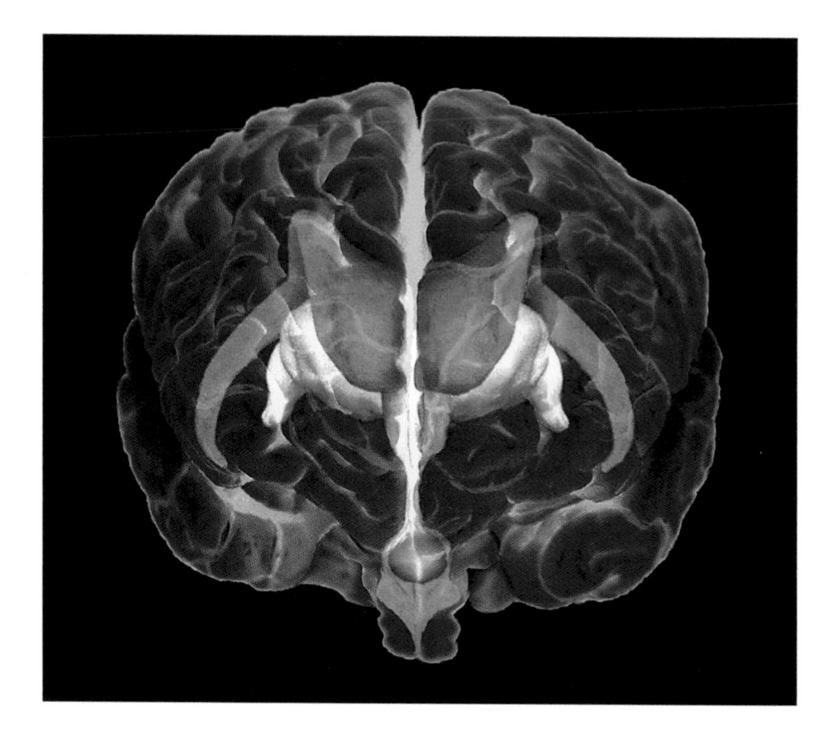

New technology to study the human brain has been advancing at a rapid rate. Does this new knowledge of the human brain have intrinsic value or is it a double-edged sword that can be used for good and bad ends? Just as nuclear physics might be used to develop cheap sources of energy or to make bombs, could human brain research have both good and bad consequences? What are some of the potential benefits of new technology developed to study the human brain? What are some of the potential hazards?

The brain regulates and monitors unconscious body processes like blood pressure, heart rate and breathing. It receives a flood of messages from the senses and responds by controlling balance, muscle coordination and most voluntary movement. Other parts of the brain deal with speech, emotions and problem solving. Your brain allows you to think and dream.

The study of the complex information-processing system that includes the brain and the nervous system is called neuroscience or neurobiology. New technology has given us valuable insights into the function of our brains. Animal experimentation has allowed us to see exactly what causes some of our drives. Brain injuries have been studied to show what occurs when parts of the brain are damaged. Brain scans using functional magnetic resonance imaging (fMRI) have revealed the effects of addictive drugs on the brain. There are studies of how pain is perceived and how endorphins act as painkillers.

Figure 16.20 Parts of the human brain.

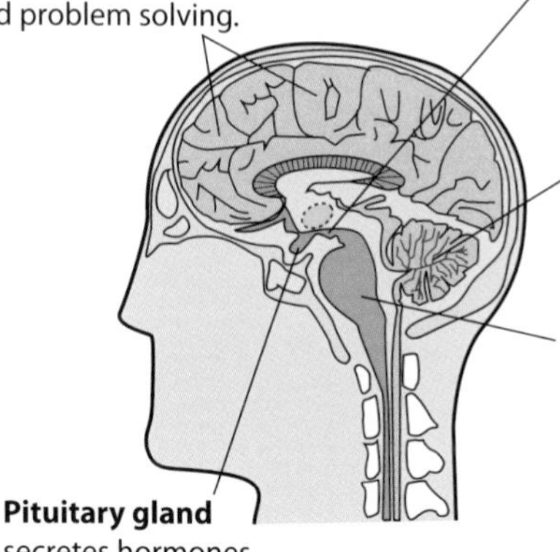
Cerebral hemispheres are associated with intelligence, personality, sensory impulses, motor function, organization and problem solving.
Hypothalamus controls the pituitary gland which secretes hormones.
Cerebellum is associated with the regulation and coordination of movement and balance.
Medulla oblongata maintains vital body functions such as breathing and heart rate.
Pituitary gland secretes hormones.

• **Examiner's hint:** You need to be able to annotate a diagram of the human brain.

These notes would be suitable for annotating a diagram of the brain.

- Cerebral hemispheres act as the integrating centre for high complex functions such as learning, memory and emotions.
- Hypothalamus maintains homeostasis, coordinating the nervous and the endocrine systems, secreting hormones of the posterior pituitary, and releasing factors regulating the anterior pituitary.
- Cerebellum is often called 'the little brain' because it has two hemispheres and a highly folded surface. It coordinates unconscious functions, such as movement and balance.
- Medulla oblongata controls automatic and homeostatic activities, such as swallowing, digestion, vomiting, breathing and heart activity.
- Pituitary gland has two lobes. The posterior lobe stores and releases hormones produced by the hypothalamus and the anterior lobe. It also produces and secretes hormones regulating many body functions.

A report released from the World Health Organization (WHO) claims that neurological disorders, such as Alzheimer's disease, epilepsy, stroke, headache, Parkinson's disease and multiple sclerosis affect one billion people worldwide.

W To read about the report and its recommendations visit heinemann.co.uk/hotlinks, enter the express code 4242P and click on Weblink 16.12.

Identification of brain parts involved in specific functions

Many methods have been used to discover the function of specific parts of the brain.

Brain lesions

One method is to study people who have had injuries to particular areas. These lesions in identifiable areas of the brain tell us indirectly about the function of those parts. Some lesions which have been studied occur in either the right or the left half of the brain and give us information about the differences between the two halves.

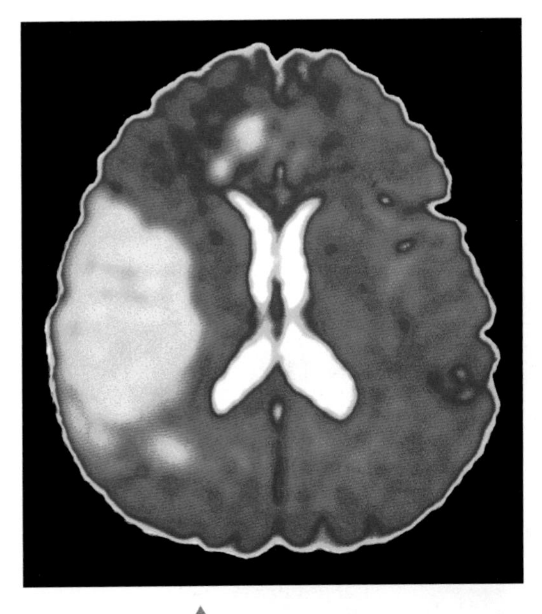

fMRI scan of the brain of a patient after a stroke. The large area of yellow is a result of lack of blood flow to that area of the brain. The blockage of blood flow may be due to a blood clot. Strokes can cause the hemisphere in which they are located to lose function.

Right and left hemispheres

The brain is divided into the left and the right hemispheres. They are connected by a thick band of axons called the corpus callosum. These hemispheres do not have exactly the same functions.

The left hemisphere contains areas important for all forms of communication. Left-hemisphere damage may result from a stroke (broken or blocked blood vessels in the brain). After left-hemisphere damage, a person may have difficulty speaking or doing complicated movements of the hands or arms. Deaf people who had this damage could not use sign language to communicate.

The right hemisphere is not involved in communication, although it does help us to understand words. It really specializes in receiving and analysing information which comes in through all of our senses. When people have lesions in the right hemisphere, they have problems identifying faces and locating an object correctly in space. The person might not be able to identify melodies. The right hemisphere helps us understand what we hear and what we see.

Early experiments with brain lesions were done in the mid 1800s with people who had particular injuries. Two neurologists observed that people who had injuries on the left side of the brain had speech and language problems. People who had

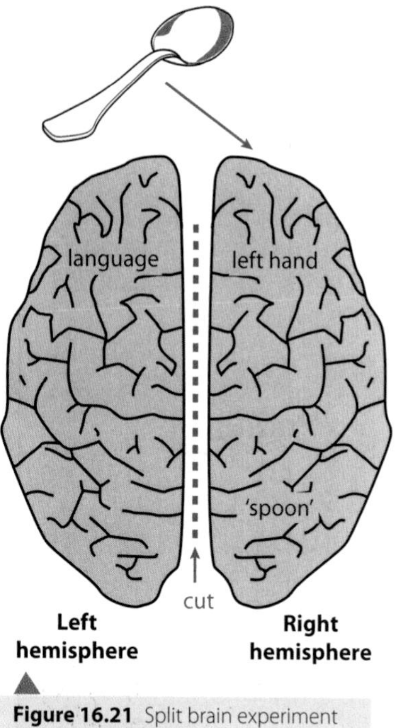

Figure 16.21 Split brain experiment with a spoon.

You can play the split brain game if you visit heinemann.co.uk/hotlinks, enter the express code 4242P and click on Weblink 16.13.

Figure 16.22 Chimerical figure of a man and a woman used in split brain experiments.

injuries in the same areas but on the right side of the brain had no language problems. The two areas of the brain important for language are named for these scientists. Injury to Broca's area interferes with the ability to vocalize words; injury to Wernicke's area affects the ability to put words into sentences. Both areas are on the left side of the brain.

Another series of experiments was done in the 1960s. Scientists trying to discover brain functions became interested in studying a group of patients who had undergone surgery to sever their corpus callosum to relieve symptoms of epilepsy (the optic chiasma remains intact). Experiments were devised to determine how splitting the brain affected these patients. Researchers already knew that input from the right visual field is received by the left hemisphere and input from the left visual field is received by the right hemisphere.

The scientists projected a picture of a spoon onto the right side of a card with a dot in the middle. If a split brain person is sitting down looking at the dot and a picture of the spoon is flashed up, the visual information about the spoon crosses the optic chiasma and ends up on the left hemisphere. The person has no trouble identifying the spoon and says 'spoon'. (Language is in the left hemisphere.)

If the spoon is projected on the left side of the dot, the information goes to the right side of the brain where there is no language ability (see Figure 16.21). In this case the person will say that nothing is seen. Now the scientists ask the same person to pick up a spoon with their left hand. The subject correctly picks up the spoon. The information travelled to the right hemisphere which understands what a 'spoon' is even if it cannot verbalize 'spoon.' If we now ask the person what is in their hand, they will not be able to say, 'It is a spoon.' The right hemisphere has little language ability.

In another experiment, scientists use a chimerical picture to test split brain patients (see Figure 16.22). In this figure, the half on the right is a man's face and the half on the left is a woman's face. In the middle of the forehead is a dot. The scientists ask the patient to focus on the dot. When focusing on the dot, the information about the woman's face will travel to the right cerebral hemisphere and the information about the man's face will go to the left cerebral hemisphere. If a split brain patient is asked to look at pictures of complete, normal faces and to point to the face they have just seen, they will choose the picture of the woman. (The information of 'woman' went to the right hemisphere.) However, if the patient is asked to say whether the picture was a man or a woman, they will say it is a man. The side of the brain that dominates will depend on what the patient is asked to do. In recognition of faces, when speech is not required, the right cerebral hemisphere will dominate.

Functional magnetic resonance imaging (fMRI)

fMRI uses radio waves and a strong magnetic field, not X-rays. This instrument enables scientists to see the blood flow in the brain as it is occurring. Researchers make movies of what is going on in the brain as the subject performs tasks or is exposed to various stimuli. This method can produce a new image every second. It can determine with some precision when regions of the brain become active and how long they remain active. This means it is possible to determine if brain activity occurs in the same region or different regions all at the same time as the patient responds to experimental conditions. A different tool called a PET scanner is slower but has the advantage of being able to identify areas of the brain activated by neurotransmitters and drugs. An fMRI is used by doctors to determine:

- a plan for surgery;
- treatment for a stroke;
- placement of radiation therapy for a brain tumour;
- effects of degenerative brain disease such as Alzheimer's;
- diagnosing how a diseased or injured brain is working.

Animal experiments

One type of animal experimentation is to expose animal models to addictive substances in controlled situations. Animal models respond in similar ways to humans when addicted. Addicted animals:

- want more and more of the substance;
- spend lots of time and energy getting it;
- keep taking it despite adverse conditions;
- have withdrawal symptoms on withdrawal of the substance;
- go back to the substance when stressed;
- go back to the substance with another exposure to that substance.

To test if a chemical meets the criteria for an addictive substance, a controlled self-administration experiment is designed and the response of the animal is recorded to see if it fits the above model for addiction (see Figure 16.23).

1. An animal is trained to press a lever to get a reward.
2. The animal is given an injection of the addictive substance as it pushes the lever. The lever must automatically give the injection if it is pushed by the animal (self-administration).
3. In order for this to be a controlled experiment, two levers must be available, one which gives the substance and one which does not (we want to be sure the animal is not just pushing the lever for exercise).
4. If the substance is 'reinforcing', the animal will seek to repeat the experience by pushing that lever much more frequently. This would support the hypothesis that the substance is addictive.

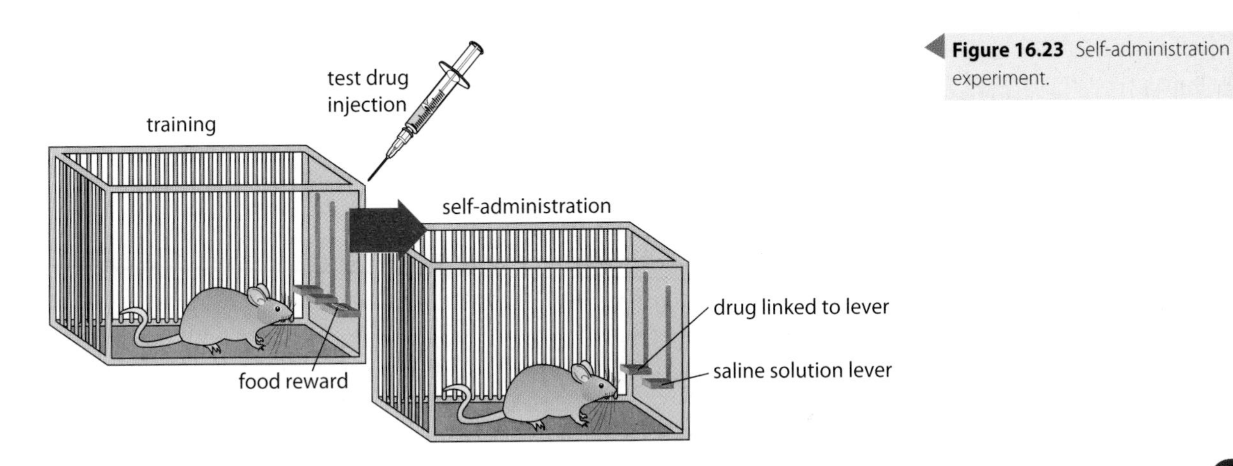

Figure 16.23 Self-administration experiment.

Researchers have recently used a self-administration experiment to support the hypothesis that acetaldehyde, which is a component of tobacco smoke, increases the addiction of adolescents to tobacco (see Figure 16.24).

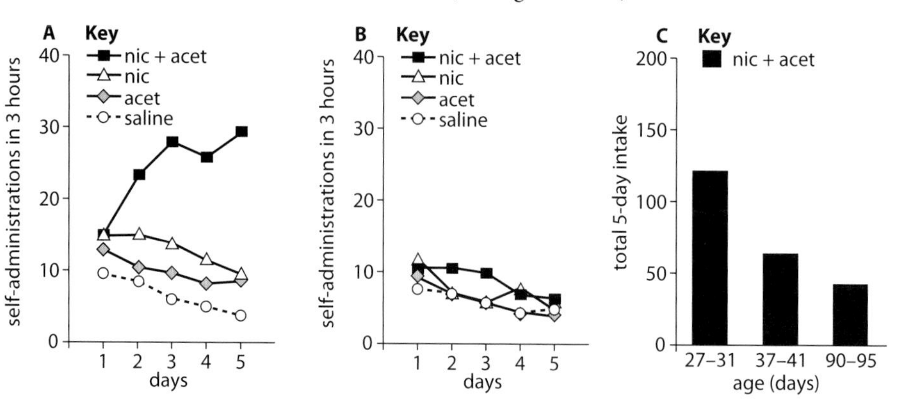

Figure 16.24 Adolescent rat experiments on nicotine and acetaldehyde.
A Adolescent rats (27 days old) self-administered nicotine combined with acetaldehyde – but not nicotine alone, acetaldehyde alone or saline – with increasing frequency over 5 days.
B Adult rats (90 days old) did not demonstrate any preference for nicotine, acetaldehyde or the combination over saline.
C The total 5-day intake of nicotine plus acetaldehyde was greatest for the youngest group of animals. This suggests that vulnerability to tobacco addiction decreases with age.

Animal experiments can shed light on the way that drugs promote abuse and addiction. Yet animal experiments can never replicate the complete picture of human interactions with the drugs. Social factors are not considered in these experiments. Thus, results need to be looked at with caution. Recent advances in technology have enabled researchers to use fMRI to answer questions which previously required an animal model.

To read an article about the new finding concerning adolescents, acetaldehyde and tobacco smoke, visit heinemann.co.uk/hotlinks, enter the express code 4242P and click on Weblink 16.14.

Sympathetic and parasympathetic control

The brain is part of the central nervous system (CNS). The other part of the nervous system is the peripheral nervous system (PNS). The peripheral nervous system is also considered in two parts, the somatic system and the autonomic system. The somatic system takes sensory information from sensory receptors to the CNS and then sends back motor commands from the CNS to the muscles. The pain reflex arc that we studied is part of this system. The autonomic system of the PNS is involuntary and regulates activities of glands, smooth muscle and the heart. There are also two divisions in the autonomic system. They are the sympathetic system and the parasympathetic system.

CNS:

- brain;
- spinal cord.

PNS:

- somatic (voluntary)
  information is received by the senses and messages sent to the skeletal muscles;
- autonomic (involuntary)
  controls cardiac muscle of the heart, smooth muscle and glands,
  consists of two systems which are antagonistic,
  - sympathetic system
  - parasympathetic system.

Sympathetic system	Parasympathetic system
important in emergency	important in returning to normal
response is 'fight or flight'	response is to relax
neurotransmitter is noradrenaline	neurotransmitter is acetylcholine
excitatory	inhibitory

As you can see, the sympathetic and the parasympathetic systems are antagonistic (see Figure 16.25). The sympathetic system is associated with 'fight or flight'. If you are facing an emergency, you need a quick supply of glucose and oxygen. The sympathetic system increases both the heart rate and the stroke volume of the heart. It dilates the bronchi to give you more oxygen. It also dilates the pupil of the eye by making the radial muscles of the iris contract. Digestion is not necessary in an emergency, so the flow of blood to the gut is restricted by contraction of the smooth muscle of the blood vessels carrying blood to the digestive system (causes the diameter of the blood vessels to narrow).

If you are not in an emergency situation and are in a relaxed state, the parasympathetic system takes over. Parasympathetic nerves return the system to normal. The pupil of your eye constricts (gets smaller) to protect the retina. This is due to contraction of the circular muscles of the iris. The heart rate slows and stroke volume is reduced. Blood flow returns to the digestive system. The smooth muscles of the blood vessels relax and the diameter of the blood vessels becomes wider.

Examiner's hint: To remember these confusing systems try to make sense of the terms.

- Peripheral is what is on the outside of the brain and the spinal cord.
- Somatic has to do with the body, and you know skeletal muscles are voluntary.
- Autonomic is similar to 'automatic'. You can remember that these functions are not voluntary.
- Sympathetic is when you are in 'sympathy' with your fear of a lion chasing you. Whereas, with parasympathetic you are like a 'parrot' sitting up in a tree completely relaxed, since the lion is down on the ground.

Anytime you can take the complex terms of biology and relate them to something else, you will have an easier time remembering them.

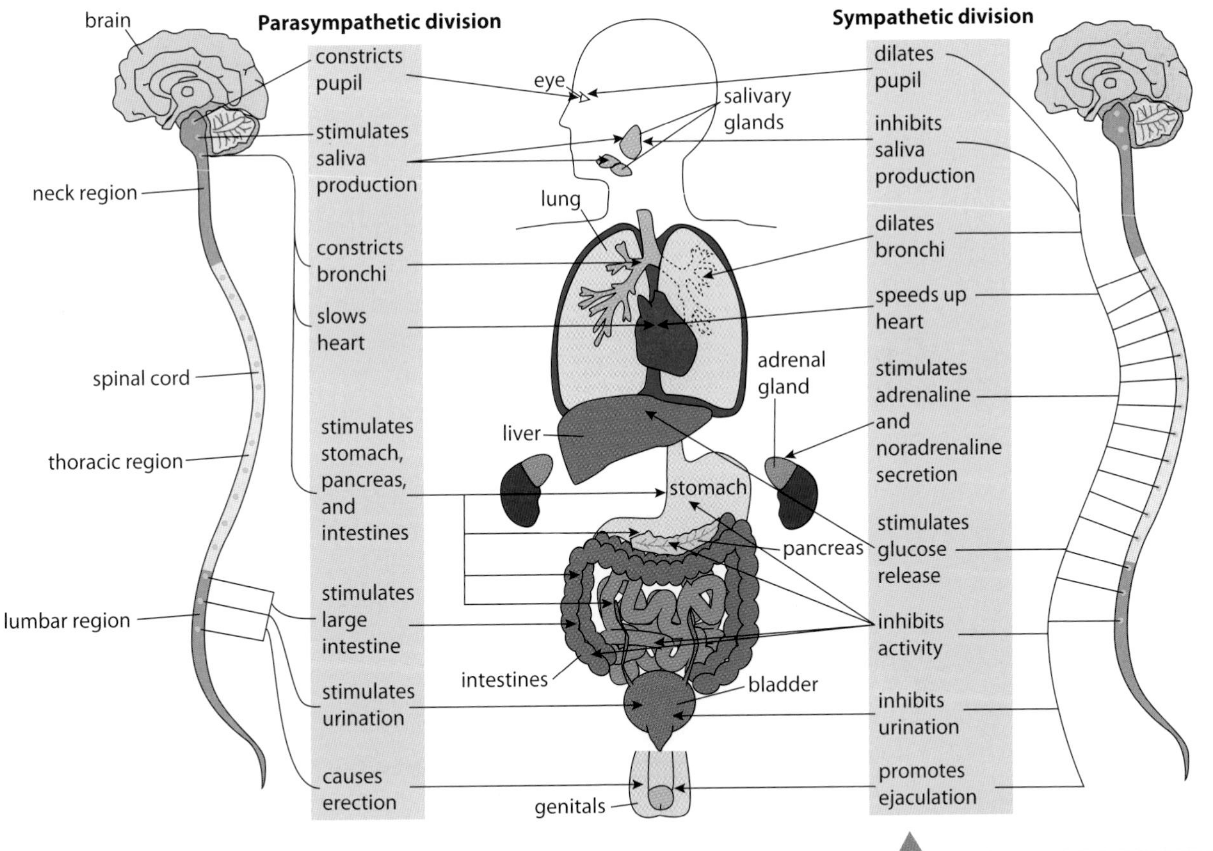

Figure 16.25 Effects of the autonomic nervous system.

The pupil reflex

In order to see the pupil reflex, ask someone to close their eyes and then suddenly open them (Figure 16.26). You will see the pupil close in response to the sudden input of light as the eyes open. This is as much a reflex as the pain reflex that we studied previously (page 462). However, instead of having its connection in the spinal cord, as we saw with the pain reflex, this is a cranial reflex. The sensory and motor neurones connect in the brain rather than the spinal cord.

In the eye, the iris surrounds the opening over the lens that we call the pupil. The iris contains two sets of smooth muscle to open and close the pupil like the aperture on a camera. The pupil closes by a parasympathetic response caused by acetylcholine. If you go to an eye doctor, he may dilate your pupils by using a drug called atropine. Atropine stops the action of the neurotransmitter, acetylcholine. Constriction of the pupil is due to a motor neurone causing the circular muscle to contract and the radial muscle relaxes.

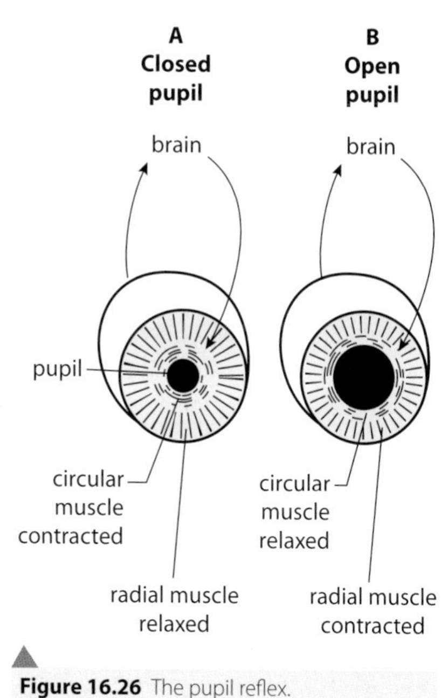

Figure 16.26 The pupil reflex.

The pathway of the pupil reflex is shown in Figure 16.27 and described in this bullet list.

- The optic nerve receives the messages from the retina in the back of the eye. (You will recall that the retina contains photoreceptors which receive the stimulus of light. Photoreceptors synapse with the bipolar neurones and then with the ganglion cells. Nerve fibres of the ganglion cells become the optic nerve.)
- The optic nerve connects with the pretectal nucleus of the brain stem (rectangle in Figure 16.27).
- From the pretectal nucleus, a message is sent to the Edinger–Westphal nucleus (triangle in Figure 16.27) whose axons run along the oculomotor nerves back to the eye.
- Oculomotor nerves synapse on the ciliary ganglion (small circle in Figure 16.27)
- The axons of the ciliary ganglion stimulate the circular muscle of the iris so it contracts.

To see an animated version of the pupil reflex diagram, visit heinemann.co.uk/hotlinks, enter the express code 4242P and click on Weblink 16.15.

Figure 16.27 Parasympathetic pathways in the pupil reflex.

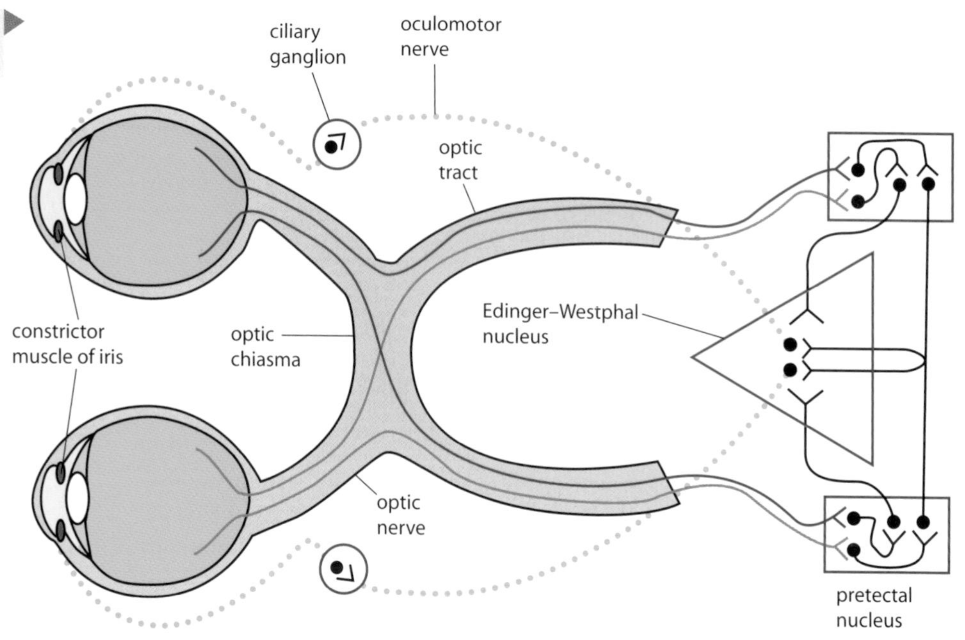

Brain death

You may have heard news reports about patients who are living on life support systems but their brain shows no electrical activity. In some of these cases, one family member wishes to keep the patient on life support because they do not believe the person is dead. Another family member believes that the person is dead, using 'brain death' as the evidence for death. What exactly does 'brain death' mean?

Due to recent advances in treatment of patients, it is possible to artificially maintain the body without the impulses which normally come from the brain. The brain stem controls heart rate, breathing rate and blood flow to the digestive system. The brain also controls body temperature, blood pressure and fluid retention. All of these functions can be controlled for a patient without a functioning brain.

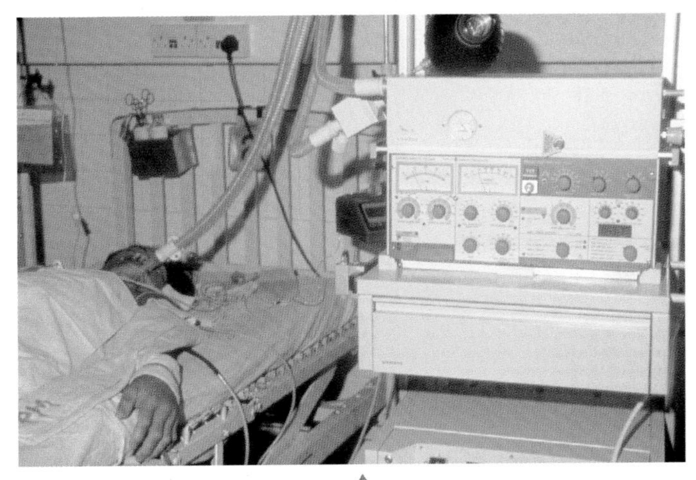

Brain death is an irreversible loss of function of the brain. Spinal reflexes may persist after brain death.

The legal description of brain death is: 'that time when a physician(s) has determined that the brain and brain stem have irreversibly lost all neurological function'. But people may still wonder if the patient could be in a coma. Patients in a coma have neurological signs that can be measured. These signs are based on responses to external stimuli. When examining for brain death, a physician must first perform a toxicology test to make sure that the patient is not under the influence of drugs that would slow down neurological reflexes.

Examination for brain death includes checking:

- movement of extremities – if arms and legs are raised and let fall, there must be no other movement or hesitation in the fall;
- eye movement – eyes must remain fixed showing lack of brain-to-motor-nerve reflex (as the head is turned there is no rolling motion of the eyes);
- corneal reflex – this must be absent (when a cotton swab is dragged over the cornea, the eye does not blink);
- pupil reflex – this must be absent (pupils do not constrict in response to a very bright light shone into both eyes;
- gag reflex – this must be absent (insertion of a small tube into the throat of a comatose patient will cause a gag reflex);
- respiration (breathing) response – this must be absent (if the patient is removed from a ventilator, the dead brain gives no response).

Following assessment by one or more physicians, a patient who shows none of these functions can be pronounced 'brain dead'. Since the patient is missing all of the reflex responses and the pupil responses, the evidence is clear that the brain will not recover.

However, in a brain-dead person there can still be spinal reflexes. The knee jerk response can still be functional. You may recall that the spinal reflexes do not involve the brain. In some brain-dead patients, a short reflex motion can still be exhibited if the hand or foot is touched in a certain manner.

Many doctors order further tests in order to confirm brain death. Two tests commonly used are the electroencephalogram (EEG) and the cerebral blood flow (CBF) study. The EEG measures brain activity in microvolts. It is a very sensitive

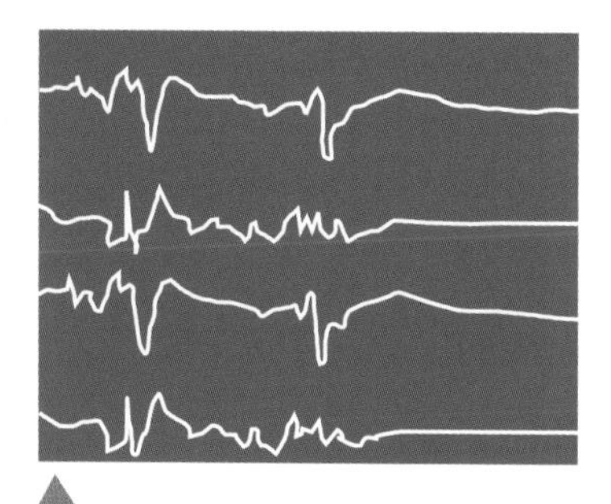

Figure 16.28 EEG showing activity followed by electrocerebral silence.

test. Some electrical activity is shown on the EEG if a patient is in a deep coma. The lack of activity in a brain-dead patient is called electrocerebral silence (see Figure 16.28).

To measure blood flow to the brain, a radioactive isotope is injected into the bloodstream. A radioactive counter is then placed over the head for about 30 minutes. If no activity is detected, this is conclusive evidence of brain death.

As you can see, the diagnosis of brain death is a very thoughtful process. At the end of the testing, there is no doubt about the result. Once this diagnosis has been made, the patient may still be maintained on a ventilator, but a brain-dead person will not recover brain function.

Perception of pain and effect of endorphins

You may wish to review the pain withdrawal reflex in section E.1 which describes the passage of the pain impulse from the skin to the spinal cord (page 462). This reflex arc does not involve the brain. So how are pain messages sent to the brain from all parts of the body?

Pain signals are carried by peripheral nerve fibres from all over the body to the spinal cord and relayed to the sensory area of the brain. These peripheral fibres connect with pain receptors called nocioreceptors. Nocioreceptors are capable of sensing excess heat, pressure or chemicals from injured tissues. These receptors are located in the skin and also in the muscle, bones, joints and membranes around your organs. The nerve impulses of pain travel to the spinal cord. The ascending tracts in the spinal cord send the messages up to the brain.

The message of pain travels to the cerebral cortex. The cerebral cortex receives the message of pain from the spinal nerves and directs the body to respond in one or more of several ways:

- it can tell the muscles to stop the action which is causing the pain stimulus (e.g. do not walk on your broken leg);
- it can alert the autonomic nervous system if the pain requires change in heart rate or breathing;
- it can direct other brain cells to release pain-suppressing endorphins.

The term 'endorphin' means the morphine within. The pain-relieving properties of endorphins are similar to those of morphine.

Endorphins were first discovered by scientists studying opium addiction. They found receptors for the opiates, morphine and heroin in brain cells. It seemed odd that brain cells would have receptors for molecules made by plants (opium poppy). The scientists found that morphine and heroin bound to the brain receptors because they were mimicking endorphins. We now know that endorphins are CNS neurotransmitters with pain-relieving properties. They are small peptides which bind to opiate receptors and block the transmission of impulses at synapses involved in pain perception.

Exercises

19 Draw a diagram of the human brain.
20 Describe the function of each of the parts.
21 Describe the methods used to determine which brain parts are involved in a specific function.
22 Distinguish between the sympathetic control and the parasympathetic control of the heartbeat, iris and blood flow to the digestive tract.
23 What is brain death?
24 Explain how pain is perceived and the role of endorphins in pain perception.

E.6 (HL only) Further studies of behaviour

Assessment statements

E.6.1 Describe the social organization of honey bee colonies and one other non-human example.
E.6.2 Outline how natural selection may act at the level of the colony in the case of social organisms.
E.6.3 Discuss the evolution of altruistic behaviour using two non-human examples.
E.6.4 Outline two examples of how foraging behaviour optimizes food intake, including bluegill fish foraging for *Daphnia*.
E.6.5 Explain how mate selection can lead to exaggerated tails.
E.6.6 State that animals show rhythmical variations in activity.
E.6.7 Outline two examples illustrating the adaptive value of rhythmical behaviour patterns.

Social organization

Many animals exhibit social behaviour. Social behaviour can be described as two or more animals interacting with each other. Some animals have only a brief period in their lives where they exhibit such behaviour.

Social organization of honey bee colonies

Worker bees do a waggle dance to communicate the location and distance of nectar to the rest of the workers in the colony.

Without the assistance of pollinators plants cannot reproduce. Over 90% of the plants in the world and 75% of agricultural crops are pollinated by animal pollinators. Many of those pollinators are bees. Pollinators including bees are disappearing at an alarming rate. This is due to pesticides, destruction of habitat and invasive parasites.

To read more about this problem, visit heinemann.co.uk/hotlinks, enter the express code 4242P and click on Weblink 16.16

Honey bee colonies are at the other extreme of sociability. The organization of the honey bee colony is very complex and no member of the hive can survive without the others.

Honey bees nest above ground, usually inside a hollow tree. They make wax combs with individual compartments (cells) for storing honey and rearing their young. Each hive has a queen (fertile female) whose job it is to lay eggs. Workers are also females, but they are sterile. Their jobs are to perform all the household duties, such as searching for nectar and pollen, making wax and honey, and feeding and protecting the young. Workers live for about 6 weeks, while queens can live for 2 years. Male honey bees (drones) develop from unfertilized eggs. Mating with the queen is their only function.

The social organization of honey bees is influenced by diet. The queen lays eggs in the cells of the honeycomb. If the eggs are unfertilized they will develop into males no matter what they are fed. But if they are fertilized and female, the type of food they are fed as larvae determines whether they will become a worker or a queen. Eggs develop into larvae and are fed glandular secretions called royal jelly for the first few days. The workers then switch larvae destined to be workers to a less nutritious diet of honey and pollen. However, if a queen is needed in the hive, larvae destined to be queens are fed royal jelly during their whole larval development.

Queens control the hive with secretions from their body called pheromones. Pheromones inhibit ovarian development in workers. The workers lick the pheromone from the queen's body and pass it onto other workers during food exchange. Most colonies contain about 60 000 bees. If the colony is too large, the queen leaves with a large number of workers to establish a new hive. This is called 'swarming'. A new young queen stays behind in the old nest.

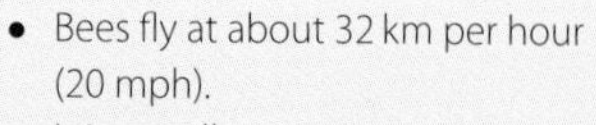

- Bees fly at about 32 km per hour (20 mph).
- It is equally correct to write honeybee or honey bee.
- Losing its stinger will cause a bee to die.
- Bees are very important because they pollinate many agricultural crops.
- Bees must collect nectar from 1 million flowers to make 0.45 kg (1 lb) of honey.
- Bees have been around for about 30 million years.

Honey bees use signals to communicate with other bees in the hive. A chemical secreted from the tip of the abdomen of one bee is used to identify the source of nectar or water for others. Some bees have the job of scouting out the source of nectar and do a waggle dance to indicate the direction and distance of the source. Another chemical is released from the mouth area when the colony is in danger and the signal is spread around to the others.

This table summarizes the roles of bees in the social organization of the colony.

Queen	fertile female	• lays eggs • produces pheromones which calm the colony and cause other females to be sterile
Worker	sterile female	• feeds the larvae • produces wax and honey • searches for nectar and pollen • protects the hive
Drone	fertile male developed from an unfertilized egg	• mates with the queen

Chimpanzees are very social, living in large communities that may contain dozens of loose family groups. The chimp is highly communicative, using a wide range of facial expressions, vocalizations and gestures.

Social organization in chimpanzees

We understand the social organization of chimpanzees very well because of the work of Jane Goodall. She has spent her life studying chimpanzees in their native habitat. You will recall that a scientist who studies behaviour in the wild is an ethologist. Jane Goodall is an ethologist who has dedicated her time and energy to observing the social organization of chimpanzees and saving their habitat.

The highest order of the chimpanzee society is the community. A community is typically made up of 40–60 members. A smaller group within the community is called a party. Parties are generally made up of up to five members. A party may be all male, a family unit, or a nursery unit with more than one family represented or some other combination of individuals. The makeup of parties depends on the food supply. The more food there is, the larger the groups which travel together.

Social organization of chimpanzees is a hierarchy. The highest ranking male is usually aged 20–26. His dominance is determined by his physical fitness and fighting ability. Males are clearly dominant over females. Dominance hierarchy in females is linked with age. Older females are dominant over younger females.

Strong social bonds exist between males because they are related to each other. Males stay in the same community in which they were born. Females may migrate to other communities.

Male bonding is important when it comes to the cooperative behaviour needed to keep out intruders, hunt together and share food. Parental care is the responsibility of the mother and is critical to the survival of the infants. Young chimpanzees receive food, warmth, protection and learn skills from their mothers. Communication between chimpanzees includes facial expressions and vocalizations, both of which have been studied extensively.

To learn more about facial expressions and vocalizations of chimpanzees, visit heinemann.co.uk/hotlinks, enter the express code 4242P and click on Weblink 16.17.

Natural selection acts at the level of the colony

In honey bee populations, it would seem that worker bee behaviour would not be promoted by natural selection. According to the theory of natural selection, the most well-adapted worker will survive to reproduce. However, workers do not reproduce. It would seem that the workers' behaviour would not be maintained by natural selection. The gene causing the behaviour of the worker should be eliminated from the population because it does not advance their reproductive success, only the reproductive success of the queen.

It is evident that natural selection in the case of social organisms such as bees is acting on the colony as a whole. The genes which are selected are those which promote social organization. Genes for pheromones which control the behaviour of workers are selected. Genes for the behaviour of finding nectar and making wax are selected. Genes for taking care of the young are selected. All females have these genes and it is just a matter of chance which female is fed the royal jelly for long enough to become the queen. It is really the 'worker' gene present in the queen which keeps the colony functioning well. The workers are actually insuring survival of their own genes.

Evolution of altruistic behaviour

Worker bees are altruistic. They help the queen produce offspring rather than reproducing themselves. How has altruism evolved? One answer is 'kin selection.' Kin selection is described as behaviour which results in a decrease in fitness of the altruist and an increase in the fitness of a close relative.

Fitness in this context means the ability to survive and reproduce. It means the relative contribution that the organism will make to the gene pool of the next generation.

Belding's ground squirrel

Belding's ground squirrels live in the mountains of the southwestern US. Predators of the ground squirrel are hawks and coyotes. When a predator approaches, one of the ground squirrels gives a high-pitched call which alerts the rest of the population to the nearby danger. Data collected has shown that the alarm squirrel is more likely to be killed than the other squirrels in the population because the alarm call gives away the location of the caller to the predator. The alarm calls are predominately performed by females, which live close to their relatives, while the male squirrels live at a far distance. The squirrel giving the alarm is not increasing its own fitness but it is increasing the fitness of its relatives. These females are sacrificing themselves so that their close relatives can reproduce. Interestingly enough, if all the female's close relatives are dead, she does not sound alarm calls. You can see that Belding's ground squirrel with its altruistic behaviour is increasing the representation of its genes in the next generation, since it shares genes with its relatives.

Naked mole rats

A second example of altruism is seen in the behaviour of the naked mole rat. Mole rat colonies of up to 100 individuals are found in burrows in East Africa. They live under the savannah and their burrows are excavated and extended by workers. The workers also make nesting chambers and forage for plant roots needed for food. The worker bites off part of the tuber and brings it to the queen. Larger workers stay near the queen and her young. The queen somehow suppresses the sexual behaviour of the other females. She also suppresses the sexual behaviour of the males except when she is in oestrus (ovulating). Snakes are the main predator of the mole rat. When a snake attacks a burrow, the queen sends out workers to attack the snake. The workers are sacrificed so that the queen and her young can live. This altruistic behaviour of the mole rat is similar to the workers in a honey bee colony and to the behaviour of Belding's ground squirrels. Based on the evidence from those examples, it is not surprising to discover that mole rats are genetically almost identical to each other. This is another example of 'kin selection', where an animal sacrifices itself to preserve the genes of the family for the next generation.

▲ Naked mole rats have survived in the extremely harsh environment of the East African savannah by adopting altruistic behaviour. Worker mole rats sacrifice their lives for the good of the colony.

Mole rats have developed this colony-type lifestyle to withstand the pressures of living in a habitat which is unsuitable for many animals. The soil is hard and dry and the tubers are few and far between. A pair of naked mole rats could probably not survive on their own, but a colony with its altruistic behaviour survives well.

Very few species exhibit this extreme type of altruism. Some scientists hypothesize that most behaviour which seems altruistic actually is not. They contend that the behaviour which looks altruistic actually increases individual fitness in some way that we do not recognize. Some behavioural ecologists even argue that true altruism does not ever occur in the animal kingdom.

Foraging behaviour optimizes food intake

Behavioural ecologists have become very interested in measuring feeding behaviours or foraging activities. What an animal eats is essential to its survival and ability to reproduce. Behavioural ecologists observe foraging behaviours, and predict how an animal will forage in a certain set of conditions. They base their predictions on cost–benefit analysis of the behaviour. The cost of foraging to an animal is the energy used to locate, catch and eat food. The benefit of foraging is the calories of energy gained. Most studies of foraging show that animals tend to change their behaviour in order to keep high the ratio of the energy they take in compared to the energy that they expend.

Small mouth bass

The small mouth bass can forage for either minnows or crayfish. Minnows contain more energy per unit weight but crayfish are easier to catch. Minnows are easier to digest than crayfish with their hard exoskeleton. Small mouth bass switch from minnows to crayfish and back to minnows in order to keep the energy that they take in higher than the energy they expend. Since no preference is shown for either minnows or crayfish, each may be optimal under different conditions. The mechanisms responsible for this switching back and forth are not yet known.

Bluegill sunfish

The foraging behaviour of bluegill sunfish, which eats *Daphnia*, has been extensively studied. *Daphnia* are small crustaceans which are found in varying sizes. Generally, bluegills forage for the larger *Daphnia* which supply the most energy. However, they will select smaller *Daphnia* if the larger ones are too far away. Predictions based on cost–benefit analysis suggest that when the density of *Daphnia* is low, the bluegill will not be selective about the size of the *Daphnia* when foraging. However, when the density of *Daphnia* is high, bluegills will be more selective and choose larger *Daphnia*. The scientific prediction was correct but not to the extent that the model of cost–benefit predicted (see Figure 16.29). The prediction was that the bluegills would eat large *Daphnia* 100% of the time, but they only ate large *Daphnia* 57% of the time. On further study, it was noticed that young bluegill sunfish did not feed as efficiently as the older sunfish. It is not clear if it is due to their lack of ability to see the size of the *Daphnia* correctly or if it is due to the learning that takes place as they mature.

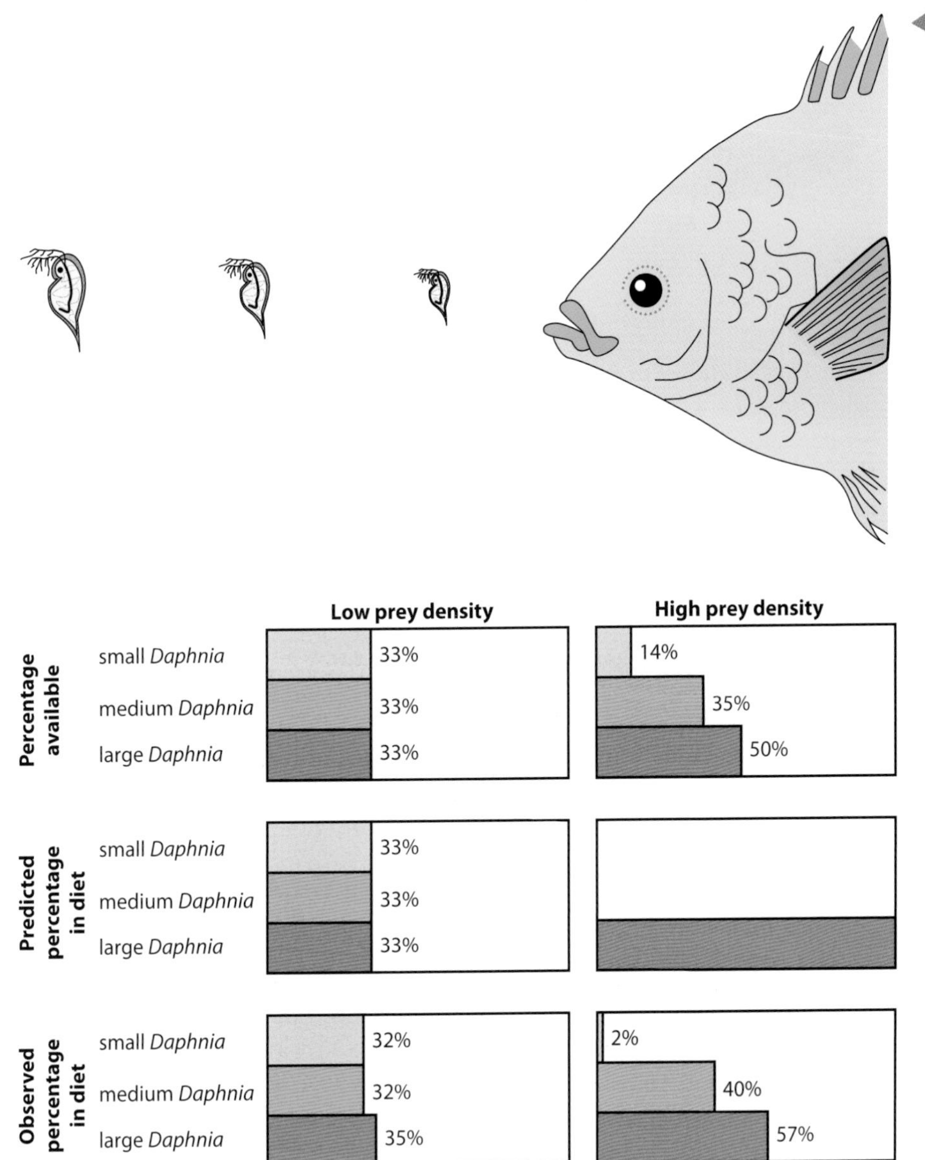

Figure 16.29 Feeding by young bluegill sunfish on *Daphnia*.

Mate selection leads to exaggerated traits

Why does the peacock have such a large and colourful tail? The theory of natural selection says that in sexual selection such characteristics evolve because females prefer more highly decorated males. When Marion Petri was working with peacocks, she found that peahens choose their mates by the size and shape of their tails. It makes sense in evolutionary terms since the largest tail signifies the healthiest bird with the best chance for healthy offspring. The measure of the quality of the peacock's tail is the number of eyespots that it possesses. The data collected by Marion Petri support this (see Figure 16.30). Notice that this graph shows correlation, not cause. The only way that cause can be determined is to eliminate the possibility that the peacock with the largest tail is not superior in some other way that we have not recognized. However, experiments have shown that offspring of males with larger tails and more eyespots are larger at birth and survive better in the wild than offspring of birds with fewer eyespots.

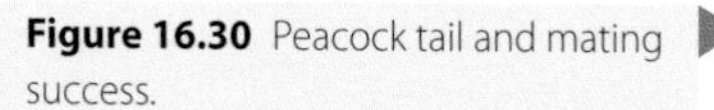

Figure 16.30 Peacock tail and mating success.

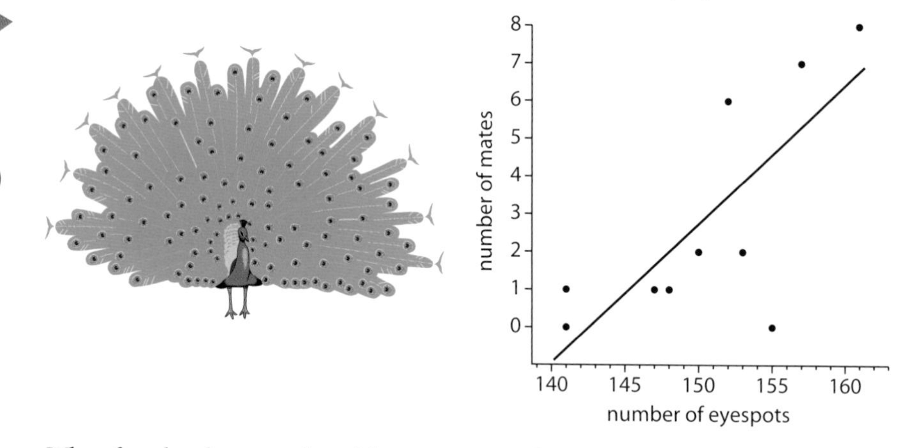

Examine the difference between correlation and cause in the information collected about the peacock's tail. How could we eliminate the possibility that males are superior in some other respect than their long tails? The female may be choosing something else that we have not seen. Could you design a controlled experiment which would give evidence of cause to the correlation that we currently observe?

When females choose males with an exaggerated trait like the peacock's long tail, those males will father more offspring than other males and the trait will become exaggerated in the species. Originally, the tail size and number of eyespots may have had a real advantage but it may now just be a sign of the best male. So females who are choosy prefer males with the longest tail. This could become more extreme until peacocks tails become too big or too colourful. Eventually they may become a disadvantage; for example, by attracting a new predator.

Animals show rhythmical variations in activity

Another reproductive strategy that some animals use to ensure success is a reproductive rhythm. Nearly all reef-building species of coral employ this strategy. Once a year, the coral release millions of gametes in a synchronized mass spawning ritual. Releasing the gametes all at the same time increases the chances that fertilization will occur. Predators are overwhelmed with more food than they can possibly eat. The exact cues which trigger this mass release of gametes are not known. It may occur in response to water temperature, lunar cycles or hours of daylight. After fertilization, the larvae develop.

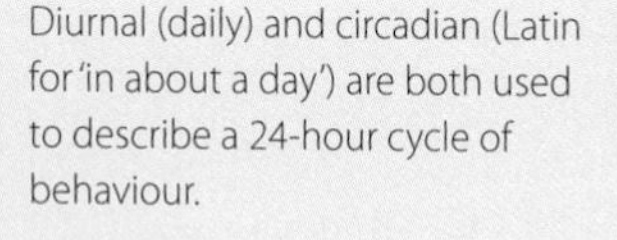

Diurnal (daily) and circadian (Latin for 'in about a day') are both used to describe a 24-hour cycle of behaviour.

Many animals repeat patterns of behaviour at daily, monthly or yearly intervals. Most animals carry out these behaviours as adaptations to their niche. For example, the North American ground squirrel flies at night. Observation of this nocturnal activity suggests that flying at night gets the most food and the least competition.

Daily rhythms

Are daily patterns regulated by an internal biological clock or by external environmental cues, such as the daily change in the light and dark cycles? Studies have shown that the behaviours usually have a strong endogenous (internal) component. However, exogenous (external) cues such as light are important in keeping the internal biological clock synchronized with the environment. So, the North American flying squirrel may fly at night, but the hours of darkness change through the year. If the behaviour were determined only by an internal biological clock, it would not be synchronized with the environment for some of the year.

In an experiment, a North American flying squirrel was placed in constant darkness and its activity monitored. Its rhythmic activity continued on a 24-hour cycle even without light from the environment. The squirrel continued the pattern controlled by its internal biological clock of about 8 hours of activity and 16 hours of inactivity. Normally, this is synchronized with the light and dark cycles of the habitat in which the squirrels live. Without the environmental cue of light, the pattern continues, but it does not match up with the actual light and dark cycle of the habitat. We see that biological clocks are internal mechanisms which keep an animal in sync with its environment, but external cues regulate the biological clock so that it fits a changing environment.

Exercises

25 Outline two examples of social organization in the animal kingdom.

26 Describe how natural selection acts at the level of the colony.

27 Discuss two examples of altruistic behaviour.

28 Describe two examples of how foraging optimizes food intake.

29 Describe how peacock's tails have come to be so ornate.

30 Explain the adaptive value of two patterns of rhythmical behaviour.

Practice questions

1 In many vertebrate species, individuals of one or both sexes select for some features among potential mates in an effort to optimize their reproductive success. Sex pheromones are chemicals that help in chemical communication between individuals of the same species. The male red-garter snake (*Thamnophis sirtalis*) displays a courtship preference for larger female snakes. Researchers tested the hypothesis that males could distinguish among females of varying size by the composition of the skin lipids which act as pheromones.

Skin lipid samples were collected from small females (46.2 ±2.7 cm in length) and large females (63 ±2.6 cm in length). The samples were analysed by gas chromatography and the relative concentrations of saturated and unsaturated lipids were determined. The graphs show the time profiles when different lipids emerged from the gas chromatography column.

The shaded peaks represent saturated lipids and the unshaded peaks represent unsaturated lipids

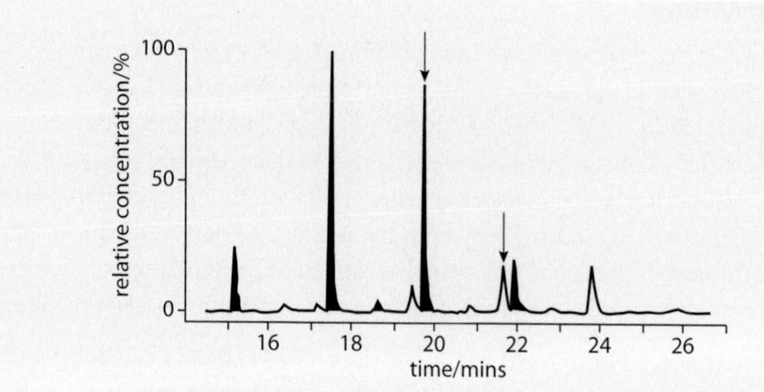

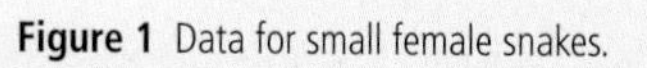

Figure 1 Data for small female snakes.

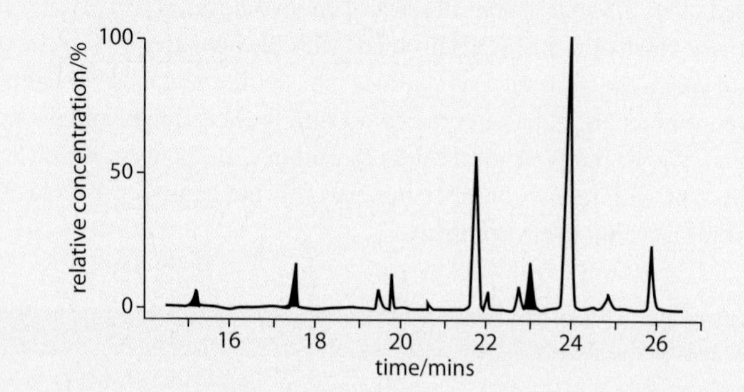

Figure 2 Data for large female snakes.

Source: M P LeMaster and R T Mason (2002), *Journal of Chemical Ecology*, **28**, page 1269

(a) Using Figure 2, state the relative concentration of the unsaturated lipid corresponding to the peak at 26 minutes. *(1)*

(b) Using Figure 1, calculate the ratio of unsaturated to saturated lipids indicated by the arrows. *(1)*

(c) Compare the pheromone profile of large female snakes with the profile of small female snakes. *(2)*

(d) **(i)** Suggest an experiment to test the hypothesis that the male red-garter snake could discriminate between larger and smaller female snakes. *(2)*

(ii) Suggest an advantage for male snakes selecting larger females. *(1)*

(Total 7 marks)

2 **(a)** Define the term *innate behaviour.* *(1)*

(b) Discuss the use of the pupil reflex for indication of brain death. *(2)*

(Total 3 marks)

3 **(a)** Explain the effects of excitatory psychoactive drugs using two named examples. *(6)*

(b) Compare the roles of the parasympathetic and sympathetic nervous system. *(4)*

(Total 10 marks)

4 During the first 24 days, worker bees (*Apis millifera*) go through a series of occupational specializations. The diagram below is a record of the first 24 days in the life of one worker bee. Adding the heights of the bars for a particular day gives 100% of the activity for that day.

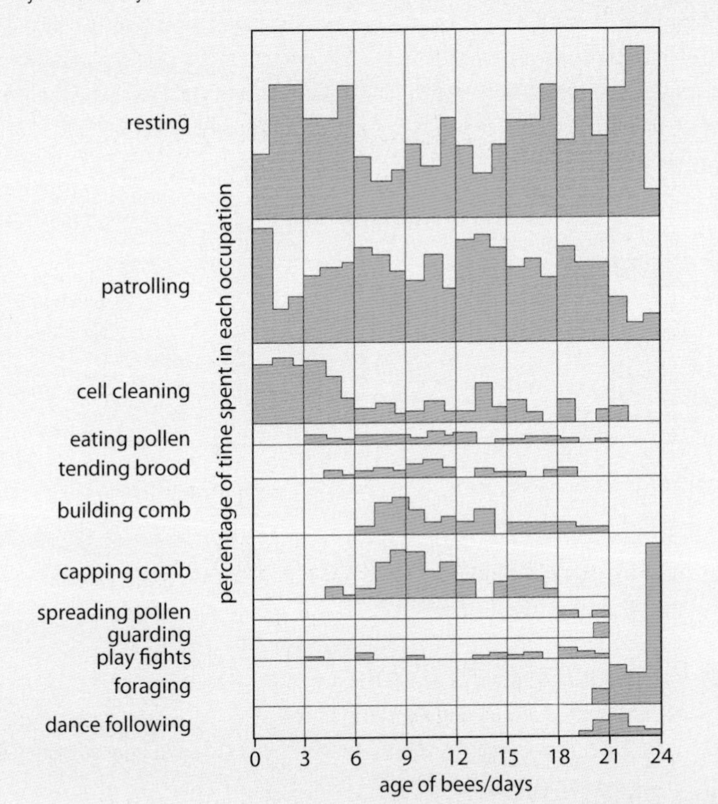

Source: J L Gould (1982), *Ethology*, Norton, page 392

(a) **(i)** Determine the percentage of time the bee spent on cleaning on day 1. *(1)*

(ii) Calculate the ratio of time spent foraging to the time spent patrolling on day 24. *(1)*

(b) Identify the two most common activities of the bee over the 24 days. *(1)*

(c) Other than resting and patrolling, describe the changes in the bee's activities over the 24 days. *(3)*

(d) Suggest why patrolling is a social behaviour. *(1)*

(Total 7 marks)

5 Outline Pavlov's experiments on the conditioning of dogs. *(3 marks)*

6 **(a)** Outline the behavioural effects of cannabis and alcohol. *(4)*

(b) Explain how presynaptic neurones can either encourage or inhibit postsynaptic transmission. *(6)*

(Total 10 marks)

7 Explain how the autonomic nervous system controls the heart rate and salivary glands. *(6 marks)*

8 Various freshwater locations in the United Kingdom contain minnows (*Phoxinus phoxinus*), a type of small fish. Minnows are often eaten by pike (*Esox lucius*), a large predatory fish. After encountering a pike for the first time, minnows begin predator inspection behaviour to confirm recognition of the predator. An experiment was performed to investigate how minnows behave after inspecting a pike. Two different populations of minnows were observed in two identical tanks, each equipped with an identical pike model. At the start of the experiment, each minnow population was feeding on a patch of food at one end of each tank. The pike model was placed at the other end of each tank and moved towards the minnows.

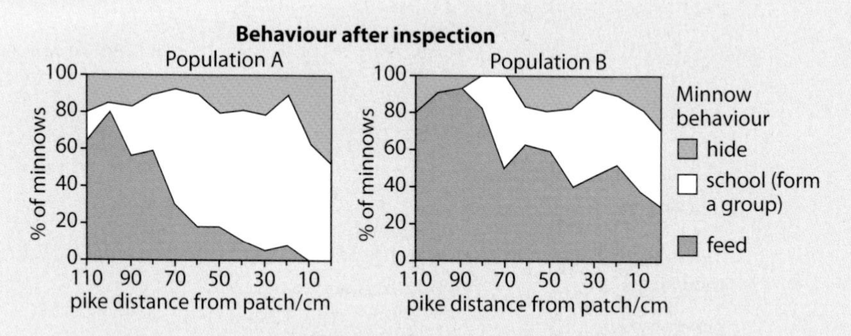

Source: Magurran (1986), *Behaviour Ecology Sociobiology*, **19**, pages 267–273

(a) For each population, estimate the percentage of minnows which are hiding when the pike model is at the greatest distance from the patch of food. *(1)*

Population A: Population B:

(b) Compare the behaviour of population A with population B when the pike model is at a distance of 30 cm from the patch of food. *(1)*

(c) Determine, giving a reason, which population was collected from a location where pike are a natural predator. *(2)*

(d) Using the data from the graphs, deduce the factor which determines minnow behaviour after inspection. *(1)*

(e) Suggest a reason for schooling behaviour in minnows. *(1)*

(Total 6 marks)

9 Describe, using an example, the role of altruistic behaviour. *(2 marks)*

10 **(a)** Explain how the sympathetic and parasympathetic systems control the heart, the salivary glands and the iris of the eye. *(7)*

(b) Outline how endorphins can act as painkillers. *(3)*

(Total 10 marks)

11 The hypothesis is that the female Hawaiian picture-winged fly (*Drosophila heteroneura*) sexually selects larger-headed males. This was tested by examining the two major contributors to sexual selection, courtship and aggressive success.

For courtship success, males with different head widths were housed in individual chambers and tested on each of ten days with a different virgin *D. heteroneura* female. The number of copulations was recorded as the courtship success.

For tests of aggressive success, males were marked by painting a yellow dot on either the left or the right of the thorax. Two males per chamber were observed for one hour. The aggressive interactions of high intensity were noted. Such fights usually had decisive outcomes, in which the winning male stood his ground and the other retreated.

The graphs below show the associations between head width and courtship and aggressive success of *D. heteroneura.*

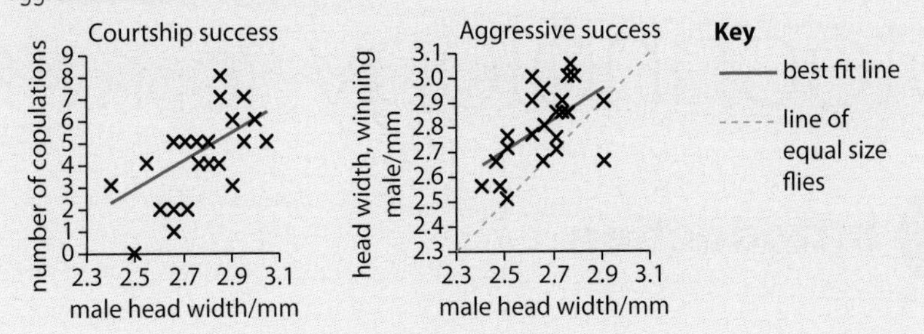

Source: Boake et al. (1997), *Proceedings of the National Academy of Science*, USA, **94**, pages 12442–45

(a) State the relationship between head width and the number of copulations. (*1*)

(b) Describe the effect a larger head has on the aggressive success. (*2*)

(c) The scientists proposed that the male head width is important in mate selection. Discuss whether the data from the graphs support this hypothesis. (*3*)

(*Total 6 marks*)

12 (a) Label the diagram of the human retina shown below. (*2*)

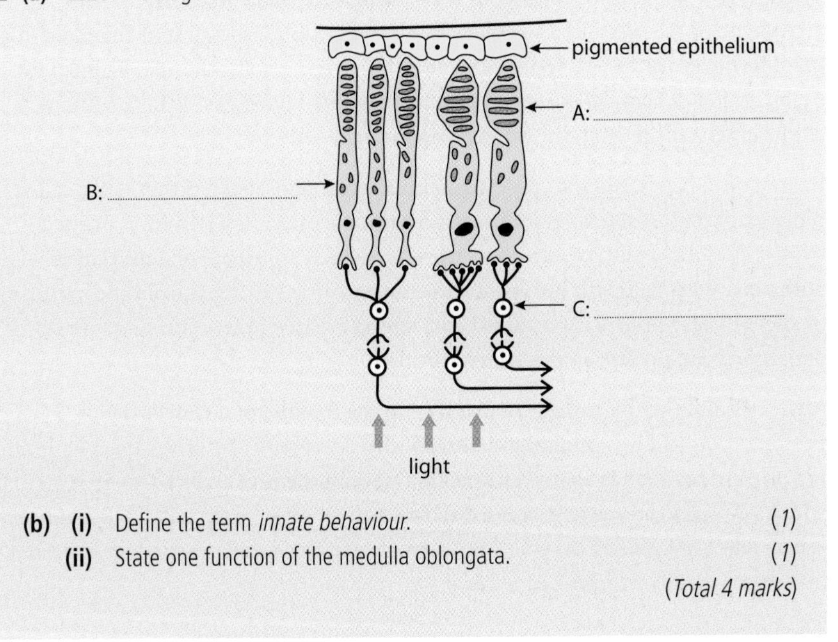

(b) (i) Define the term *innate behaviour.* (*1*)

(ii) State one function of the medulla oblongata. (*1*)

(*Total 4 marks*)

17 Microbes and biotechnology (Option F)

Introduction

Microbes are amazing. We need microscopes to see them yet they account for about 50% of the biomass of the Earth. They are incredibly diverse and no doubt you will recognize diversity as one of the most important characteristics of living things. Organisms adapt to changing environments because of their diversity. As the environment changes, the best characteristics will survive and enable those organisms to reproduce. But this diversity makes a system of classification difficult to devise.

Microbes recycle tons of organic detritus back into usable nutrients in the ecosystem. Some can take a simple gas (nitrogen) out of the air and change it into plant fertilizer (nitrates). Some can live in the deepest hydrothermal vents in the ocean and use chemical energy to produce organic molecules. Others contain chlorophyll and can trap the Sun's energy to make food.

Bacteria can be tricked into producing human proteins. Such proteins are called biopharmaceuticals; insulin and somatostatin are examples. Insulin is needed by diabetics. It used to be extracted from the pancreas cells of pigs and cattle. Somatostatin is a growth hormone. It was previously extracted from the brains of sheep. These products produced by bacteria are free from contaminants which may have been present in the animal sources.

Microbes and their enzymes are involved in the making of a wide array of products you will be familiar with: bread, cheese, yogurt, soy sauce, beer, wine and sake. Fermented foods were an important development in civilized societies in which raw materials were processed to make them safer, more palatable and have longer shelf-life. Which do you think lasts longer, fresh milk or cheddar cheese; a bunch of grapes or a bottle of wine?

Then there is disease. Why would you worry about a disease affecting people who live on the other side of the world? With global distribution of food, food-borne disease can spread around the world very quickly. The spread of highly infectious agents such as the flu virus can be very rapid due to the ease of international travel. Many diseases are carried by animals, such as ticks, mosquitoes and mice. Global warming can change the habitat of these organisms, increasing the chance for transmission of disease.

F.1 Diversity of microbes

Assessment statements

F.1.1 Outline the classification of living organisms into three domains.
F.1.2 Explain the reasons for the reclassification of living organisms into three domains.
F.1.3 Distinguish between the characteristics of the three domains.
F.1.4 Outline the wide diversity of habitat in the Archaeabacteria as exemplified by methanogens, thermophiles and halophiles.
F.1.5 Outline the diversity of Eubacteria, including shape and cell wall structure.
F.1.6 State, with one example, that some bacteria form aggregates that show characteristics not seen in individual bacteria.
F.1.7 Compare the structure of the cell walls of Gram-positive and Gram-negative Eubacteria.
F.1.8 Outline the diversity of structure in viruses including: naked capsid versus enveloped capsid; DNA versus RNA; and single-stranded versus double-stranded DNA or RNA.
F.1.9 Outline the diversity of microscopic eukaryotes as illustrated by *Saccharomyces*, *Amoeba*, *Plasmodium*, *Paramecium*, *Euglena* and *Chlorella*.

The five kingdoms

In 1959 Robert Whittaker was thinking about the classification of organisms so that communication about living things could be improved. He organized a system based on physical characteristics which were easily seen. Using these characteristics, he devised a system of five kingdoms (see Figure 17.1). For 30 years, his model was widely accepted for the organization of life.

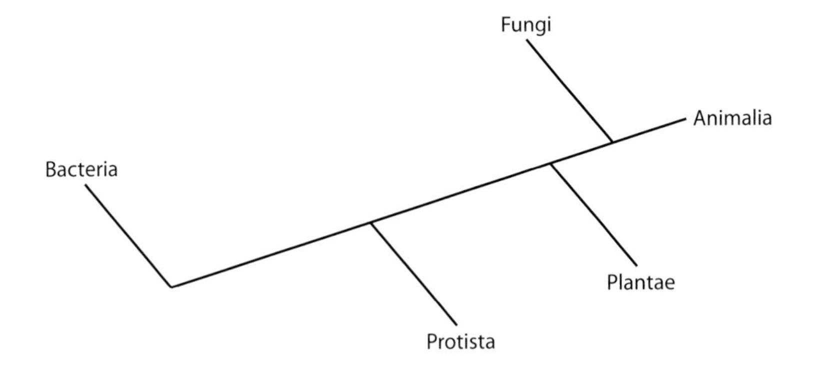

Figure 17.1 The five kingdoms.

Whittaker's five kingdoms are briefly described as follows.

- *Bacteria*: single-celled organisms with no organized nucleus and no membrane-bound organelles.
- *Protista*: single-celled organisms with an organized nucleus and organelles each surrounded by a membrane.
- *Fungi*: multicellular organisms which obtain their food using extracellular digestion and have cell walls of chitin.

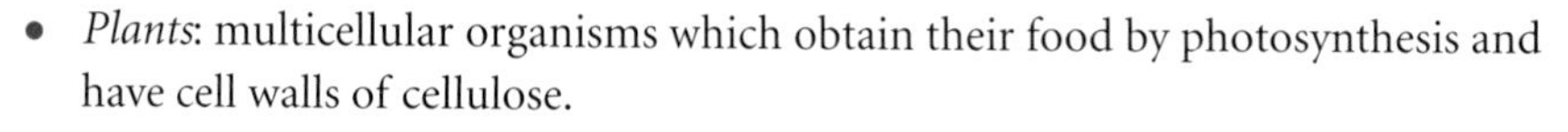

- *Plants*: multicellular organisms which obtain their food by photosynthesis and have cell walls of cellulose.
- *Animals*: multicellular organisms which obtain their food by feeding on other organisms and have no cell wall.

'Pro' means before and 'karyote' means nucleus: prokaryotes = organisms before the nucleus developed.

'Eu' means true and 'karyote' means nucleus = organisms with a true nucleus.

These five kingdoms were grouped into two categories:

- *Prokaryotes*: The prokaryotes are bacteria. They have no organized nucleus and no membrane-bound organelles.
- *Eukaryotes*: All of the other kingdoms are eukaryotes since protists, fungi, plants and animals all have an organized nucleus.

The three domains

Thirty years later another scientist, Carl Woese, was attempting to improve the accuracy of the classification system based on new knowledge he had obtained as a molecular biologist. He was studying the evolutionary relationships of microbes and noticed differences in a small molecule called ribosomal RNA (rRNA). Based on the studies of rRNA, he and his colleagues discovered a huge split within the prokaryotes. The differences were as large as the differences between the prokaryotes and the eukaryotes. Because of this, Woese divided the prokaryotes into two groups, Eubacteria and Archaea. So, instead of five kingdoms Woese's new model had three domains: Eukarya, Archaea and Eubacteria (see Figure 17.2).

Figure 17.2 The three domains.

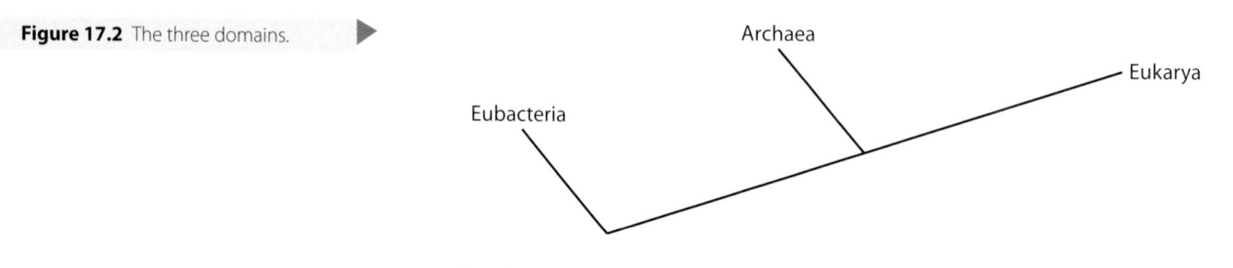

Based on biochemical analysis and rRNA comparisons, the three domains proposed by Woese can be described as follows.

- *Eubacteria*: 'true' bacteria, prokaryotes with no organized nucleus and no membrane-bound organelles. An example is *Escherichia coli* which is commonly found in animal waste products.
- *Archaea*: archaeabacteria or 'ancient' bacteria are also prokaryotes. Most groups live in extreme environments. An example is the sulfur bacteria which inhabits the hot springs of Yellowstone National Park in the US.
- *Eukarya*: single-celled and multicellular organisms which all have their DNA contained in a nucleus. The kingdoms of plants, animals, protists and fungi belong here.

What connects the five kingdoms, Carl Woese and conformational bias?

Reasons for reclassification into three domains

When Carl Woese and his associates realized that the five-kingdom system was inaccurate, he changed the model which had been the paradigm for 30 years. The five-kingdom system was based on structural comparisons. New molecular knowledge made that comparison obsolete. Woese and his group recognized that using the newly mapped molecule of rRNA was the key to the evolutionary relationships of organisms.

For a useful review of plant, animal and bacterial cells, visit heinemann.co.uk/hotlinks, enter the express code 4242P and click on Weblink 17.1.

For a helpful analogy, consider how to group cars. Terms like coupes and station wagons were used 30 years ago, but now those terms are obsolete. Today we group cars as, say, minivans, SUVs, and sports cars. These are all structural terms. We might one day classify cars according to their fuel type; that would be more akin to using biochemical knowledge.

Why did Woese and his colleagues use rRNA as the basis for reclassification? It is a molecule common to all organisms. It performs the same function in all organisms (making up the subunits of ribosomes). It is coded for by DNA. Woese and his group found variations in the sequence of nucleotides in rRNA which demonstrated that all living organisms could be divided into three large groups. In other words, by looking at variation in the sequence of rRNA, evolutionary relationships became apparent which divided living organisms into three large groups, which Woese called domains.

• **Examiner's hint:** Remember that a nucleotide is a unit composed of a sugar, a nitrogen base and a phosphate. Many nucleotides make up a nucleic acid. RNA is ribonucleic acid.

Further studies have confirmed the three-domain paradigm. These studies give more evidence that the old prokaryote kingdom should be separated into the two new domains Eubacteria and Archaea (Archaeabacteria).

- Eubacteria and Archaea have different molecules making up their cell walls.
- Eubacteria and Archaea have different molecular structure of their cell membranes.
- Eubacteria and Archaea have different sequences of nucleotides in their rRNA.

Characteristics of the three domains

When examining the differences and similarities of the three domains, molecular biologists looked at histones, introns, ribosomes, cell membranes, cell walls, and organelles.

Histones

Histones are the protein cores around which DNA is wrapped. Histones keep long strands of DNA organized. They work like spools that keep thread from becoming tangled. DNA is found in long strands in eukaryotes. When the cell is dividing, the DNA could become tangled. DNA wrapped around histones keeps that from happening. Histones are found in Eukarya. Histone-like proteins are found in Archaea. No histones are found in Eubacteria, which makes them the most primitive.

Introns

Introns are non-coding areas of DNA. Introns do not carry any messages. Cutting out the introns is like getting rid of the commercials when watching a TV show. Introns are absent in Eubacteria (primitive) and present in Eukarya (advanced) but only present in some DNA in Archaea (intermediate). Remember that eukaryotes have long strands of DNA that need to be wrapped around histones. What do you think makes eukaryote DNA so long?

Ribosomes

Ribosomes are the part of the cell where proteins are made. Ribosomes are made of rRNA and all have two subunits – one large and one small. In ribosomes of the Eubacteria and Archaea, both subunits are smaller than the corresponding subunits in eukaryote ribosomes. In Eubacteria and Archaea, the ribosomes are described as 70S ribosomes; in Eukarya, the ribosomes are 80S ribosomes. Why do you think Woese separated Eubacteria and Archaea if they both have the same size ribosomes?

The descriptions 70S and 80S actually refer to the sedimentation rates of the ribosome subunits.

Cell membranes

The cell membranes of both the Eubacteria and Eukarya include unbranched hydrocarbons. In the Archaea, only some of the hydrocarbons are branched. Can you see how this is very odd? The most advanced and the most primitive group both have the same thing. Branched hydrocarbons are only found in Archaea. Why? Since all of the Archaea live in extreme environments like thermal vents in the ocean and salty lakes, branching may help them survive in difficult places.

Cell walls

Only some members of the Eukarya have cell walls: plants and fungi. Plant cell walls are made of cellulose; fungi cell walls are made of chitin.

Members of the Eubacteria and Archaea all have a cell wall. A special molecule called peptidoglycan is found only in the cell wall of the Eubacteria. Members of the Archaea do not have it. This evidence gives more support to the three-domain paradigm.

Organelles

Neither Eubacteria nor Archaea have membrane-bound organelles. As you will recall, that is why they were originally in the same group (prokaryotes). Eukaryotes have many membrane-bound organelles, such as the nucleus and mitochondria (advanced features). That is why they are in their own group, the Eukarya.

The following table may help you keep all these facts more organized. It is a rough summary of the important molecular studies which helped create the new paradigm of classification we use today – the three domains: Eubacteria, Archaea and Eukarya.

Distinguishing characteristics of the three domains

Characteristic	Eubacteria	Archaea	Eukarya
histones	absent	histone-like proteins	present
introns	absent	present in some DNA	present
size of ribosome	70S	70S	80S
structure of cell membrane lipids	unbranched hydrocarbons	some branched hydrocarbons	unbranched hydrocarbons
peptidoglycan in cell wall	present	absent	absent
membrane-bound organelles	absent	absent	present

Diversity of habitat of Archaeabacteria (Archaea)

The habitats of the Archaea seem very strange. Why would bacteria be living in thermal vents which are extremely hot, deep under the ocean? Why would others thrive where no oxygen is available? How could some of this group survive in extreme salty conditions? Most bacteria are prevented from growing by adding salt. Some scientists have theorized that these extreme conditions are similar to the conditions found on primitive Earth. The early Earth's atmosphere was very hot. Active volcanoes produced gases like methane, ammonia and carbon dioxide. There was no oxygen present. The term 'archaea' means primitive. These organisms may be descendants of bacteria that lived under these primitive conditions.

Many interesting questions arise when we look at the extreme habitats this group prefer. Are they sole survivors of a catastrophe which occurred early in the Earth's history? Did their ability to live in these extreme places enable them to survive? Did they survive a comet hitting the Earth? Could they survive outside the Earth's atmosphere or on another planet? Could these microbes be present throughout the galaxy? These questions keep scientists very interested in studying Archaea.

There are three groups of Archaeabacteria.

- *Methanogens*: These bacteria use carbon dioxide to make methane (CH_4). Oxygen kills them, which means they are strict anaerobes. They live in a wide range of habitats including the guts of termites and cattle, Siberian tundra, swamps, rice fields, and in the large intestine of dogs, pigs and humans. These are all places where there is no oxygen.
- *Thermophiles*: These heat lovers live in sulfur hot springs where the pH is between 1 and 5 (very acidic) and temperatures are up to 90 °C. Some live in hydrothermal vents under the ocean where optimal temperatures are up to 105 °C.
- *Halophiles*: These are salt lovers. They live in the saltiest places in the world such as the Dead Sea, the Great Salt Lake, and evaporated salt water ponds. The salt content where they usually live is ten times higher than the salt content of the ocean.

Diversity of Eubacteria

Unfortunately, most people think of disease when they think of bacteria, but Eubacteria are extremely important in the natural world. Without them, there would be no decomposition, which recycles nutrients in the soil. We have learned to harness their activities and use them to make food such as yogurt and cheese, drinks such as wine and buttermilk. This group is very successful due to its wide genetic diversity.

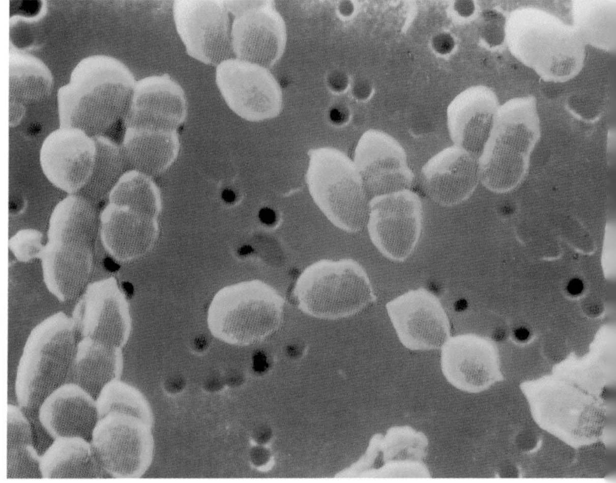

When bacteria are viewed under a light microscope and stained with a dye, their shape is easily distinguished.

Eubacteria have three main shapes: spheres (cocci), rods (bacilli), and helices (spirilla). There are many varieties of those main shapes. Cocci may occur singly, in pairs, in a chain or in a cluster. *Staphylococcus* is a group of cocci clustered together and is well known for causing skin infections. *Streptococcus* is a row of cocci known for causing throat infections. What shape would a diplococcus have? Rod-shaped bacteria occur singly or in a chain. What shape would a streptobacillus have? Microscopic identification of bacteria depends partially on knowledge of these shapes.

Another method of bacterial identification uses their cell wall structure. There are two groupings of bacteria based on their cell wall structure. Bacteria can be Gram positive or Gram negative. A stain was developed by a Danish doctor, Hans Christian Gram, which differentiates between the two groups. Gram-positive bacteria stain purple with this stain, Gram-negative bacteria stain pink.

Aggregates of bacteria

Who would have imagined that bacterial cells can communicate with each other? As the most primitive of cells, it does not seem possible that they would have what seems to be an advanced trait. A great example of this communication is found in *Vibrio fischeri*. These bacteria, found in sea water, emit light (bioluminescence) when they are in large groups (aggregates), but do not emit light when alone. When aggregated, a signal molecule, which diffuses out of the cell, accumulates. High concentration of this signal switches on a gene which causes luminescence. *Vibrio fischeri* exist in high density in the light organs of a squid (*Euprymna scolopes*). The bacterium senses high density by measuring the amount of signal present. This is called quorum sensing. Just like we may need a certain number of people (a quorum) present to run a meeting, the bacteria sense a quorum and emit light.

Comparison of Gram-positive and Gram-negative bacteria

Bacteria can be divided into these two groups based on the structure of their cell wall. The Gram stain differentiates between the two types because of the structure of the cell wall. Gram-positive bacteria have a simple cell wall and Gram-negative bacteria have a cell wall which is more complex (see Figure 17.3). They differ in the amount of peptidoglycan present. Peptidoglycan is an important material for bacteria. It consists of sugars joined to polypeptides and acts like a giant molecular network protecting the cell. Gram-positive bacteria have large amounts of peptidoglycan and Gram-negative organisms have a small amount. Only Gram-negative bacteria have an outer membrane with attached lipopolysaccharide molecules. Lipopolysaccharides are carbohydrates bonded to lipids. These

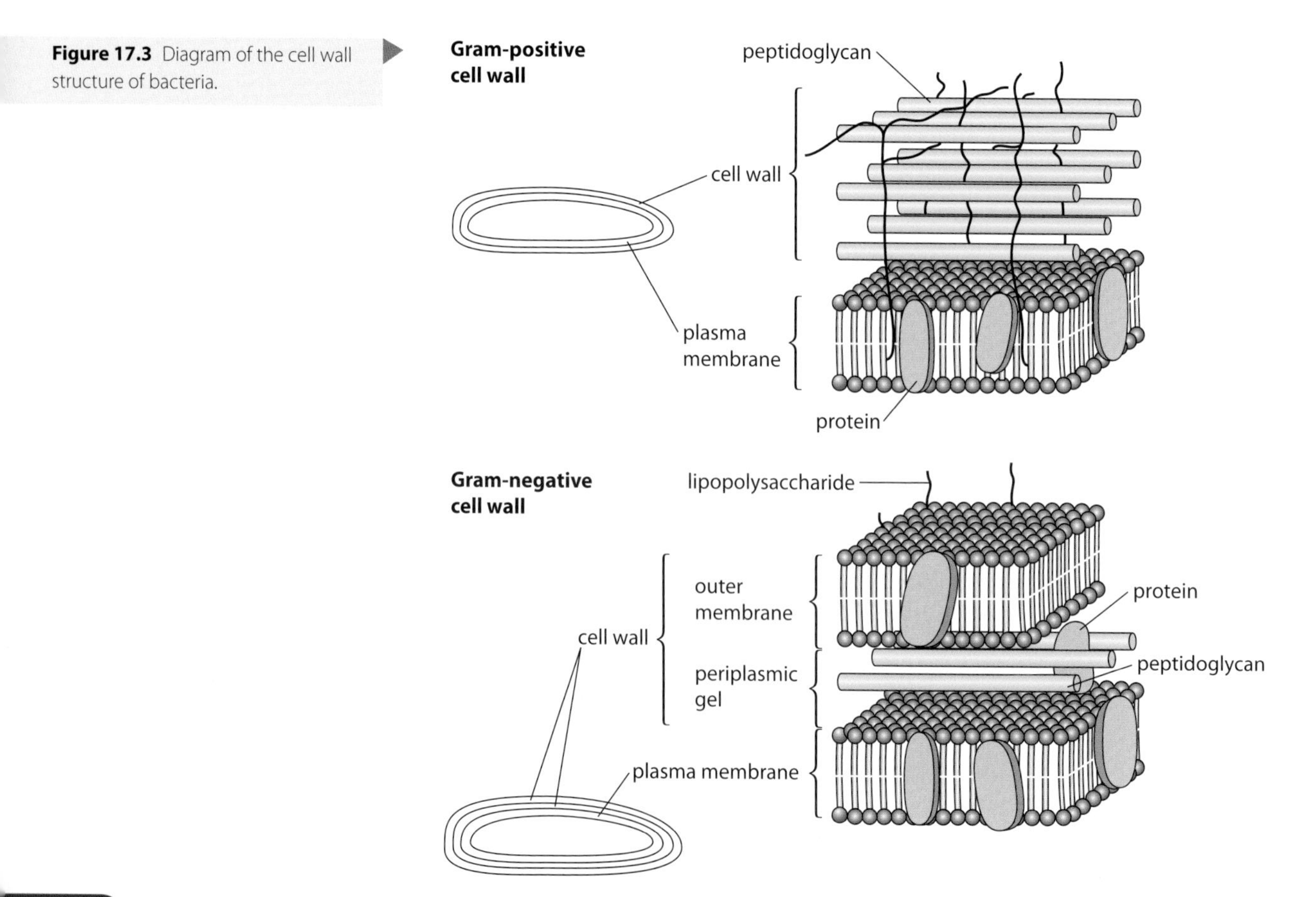

Figure 17.3 Diagram of the cell wall structure of bacteria.

molecules are usually toxic to a host. The outer membrane protects against the host defences. The outer membrane also protects the Gram-negative bacteria from antibiotics. Can you see why an antibiotic like penicillin works more effectively against Gram-positive bacteria? The Gram-positive bacteria have no outer membrane to protect them from an antibiotic.

This table compares the cell wall structure of gram-positive and gram-negative bacteria.

Cell wall structure	Gram-positive bacteria	Gram-negative bacteria
complexity	simple	complex
amount of peptidoglycan	large amount	small amount
peptidoglycan placement	in outer layer of bacteria	covered by outer membrane
outer membrane	absent	present with lipopolysaccharides attached

Diversity of structure of viruses

Are viruses alive? This is a common question. Since viruses are not cellular and cannot reproduce without using the cell machinery of another organism, they are not usually considered living things. They may just be a method that other organisms have developed to transfer their DNA or RNA to another organism.

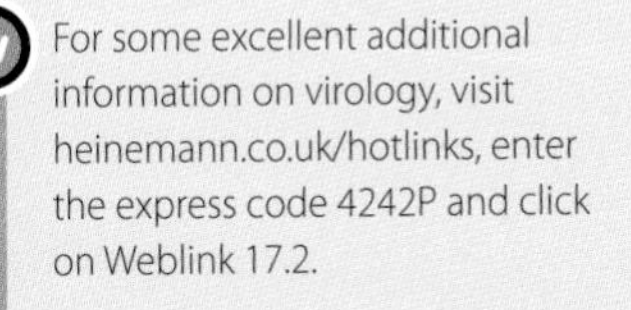
For some excellent additional information on virology, visit heinemann.co.uk/hotlinks, enter the express code 4242P and click on Weblink 17.2.

The basic structure of a virus is a nucleic acid (the genetic material), several enzymes and a protein coat (capsid). Even with this simple structure, viruses are very diverse (see Figure 17.4). The protein coat may be naked or enveloped in a membrane. The genetic material may be DNA (double or single stranded) or RNA (double or single stranded). The whole virus may be, for example, spherical, cylindrical or polygonal.

Tobacco mosaic virus RNA capsid

Adenovirus capsid DNA glycoprotein

Influenza virus membranous envelope RNA capsid glycoprotein

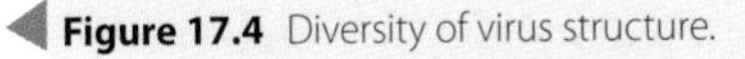
Figure 17.4 Diversity of virus structure.

Diversity of microscopic eukaryotes

The variety of this group can be illustrated by considering the following organisms: *Saccharomyces* (yeast), *Amoeba*, *Plasmodium*, *Paramecium*, *Euglena* and *Chlorella*.

For some useful pictures of amoebae and bacteria visit heinemann.co.uk/hotlinks, enter the express code 4242P and click on Weblink 17.3.

Saccharomyces

Saccharomyces (yeast) is what beer and bread have in common. Yeasts ferment the carbohydrates in the flour or the malt and gain energy from this digestion.

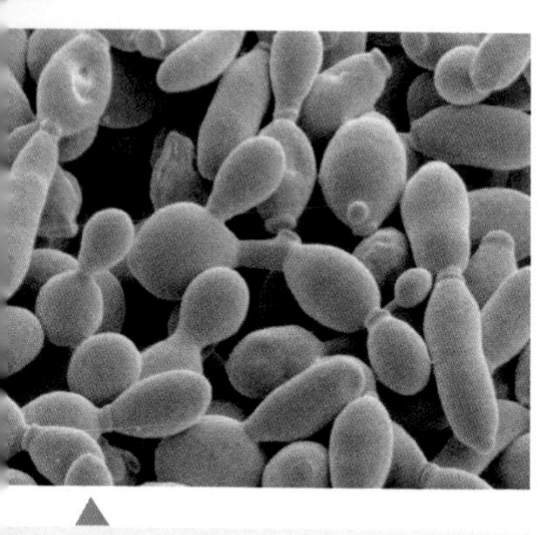

Yeast multiply every 20 minutes when given the correct conditions of sugar and warm temperatures. Bakers know to keep the bread they want to rise in a warm spot.

Carbon dioxide gas (in bread) and ethanol (in beer) are by-products; we benefit by having light and airy bread, and alcohol in beer. Yeasts are fungi. They secrete digestive enzymes outside their cells and absorb the products of digestion back into the cell. This is called extracellular digestion. Yeasts are heterotrophs – they get food from products of other organisms. They have a special molecule called chitin in their cell wall.

Amoeba

Amoeba is one of the most fascinating single-celled organisms to study. The fluid state of its cytoplasm enables it to change its shape easily. This change in shape creates pseudopodia (false feet) which can wrap around a prey in order to trap it. Amoebae are heterotrophs which take their food into a food vacuole and digest it. This is called intracellular digestion.

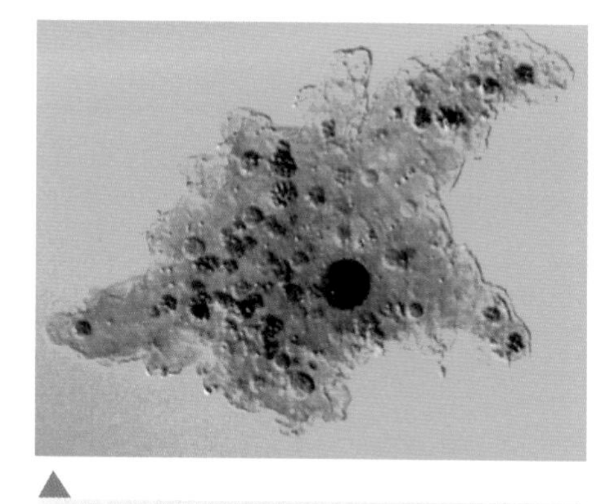

Amoeba are transparent and very difficult to see even when using a microscope. Once you find them, their movement is fascinating.

Plasmodium

Plasmodium is a parasitic heterotroph. Mosquitoes carrying the parasite infect humans and cause malaria. Some of the life cycle takes place in the gut of a mosquito and some takes place in the human body. Within the human body, the parasite can move along the substrate with a gliding motion similar to a crawl.

Paramecium

Paramecium are sometimes described as slipper shaped. The organism moves very fast with beating cilia.

Paramecium is a ciliated heterotroph. The organisms swim backwards or forwards using cilia. Digestion is intracellular. Food is taken into the oral groove (mouth) and passes to the gullet. Eventually, a food vacuole or sac breaks off from the gullet and is joined by another vacuole of digestive enzymes. The two vacuoles join together and digest the food.

Euglena

Euglena is unique in that it is both autotrophic (makes its own food) and heterotrophic (eats other organisms). It contains chlorophyll which enables it to make food by photosynthesis. In fact it has a special organelle called the eyespot which facilitates the movement of *Euglena* towards light. It gathers the light to make its own food. In the absence of light, it acts as a heterotroph. As a heterotroph, *Euglena* can absorb food from outside the cell. Movement is accomplished by the whipping motion of its long flagellum.

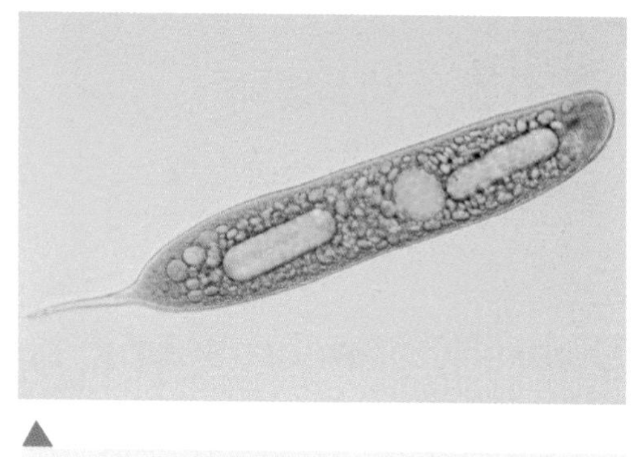

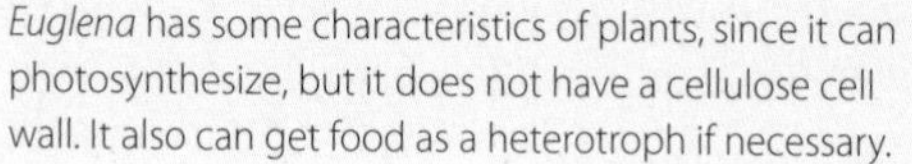

Euglena has some characteristics of plants, since it can photosynthesize, but it does not have a cellulose cell wall. It also can get food as a heterotroph if necessary.

Chlorella

Chlorella is single-celled green algae containing chlorophyll *a* and *b*. It makes its own food by photosynthesis. It is not motile and has a cell wall of cellulose. As a member of the green algae, it is believed to be an ancestor of green plants.

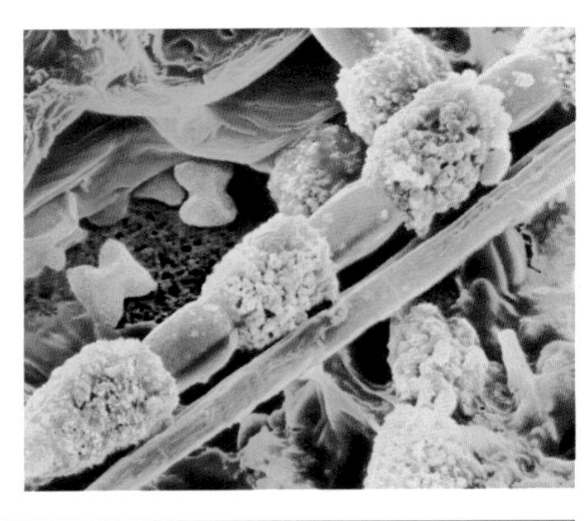

You might be surprised to find *Chlorella* in health food stores in bottles. Some people take it as a natural supplement for vitamins and minerals.

This table outlines the diversity of eukaryotes.

Organism	Nutrition	Locomotion	Cell wall	Chloroplasts	Cilia or flagella
Saccharomyces	heterotroph (extracellular digestion)	absent	made of chitin	absent	absent
Amoeba	heterotroph (intracellular digestion)	slides using pseudopodia	absent	absent	absent
Plasmodium	heterotroph (intracellular digestion)	glides on substrate	absent	absent	absent
Paramecium	heterotroph (intracellular digestion)	swimming	absent	absent	cilia
Euglena	autotroph and heterotroph	swimming	absent	present	flagellum
Chlorella	autotroph	none	made of cellulose	present	absent

Exercises

1 Describe the classification of living things into three domains.
2 Distinguish between the characteristics of the three domains.
3 Compare the cell wall structure of the two main types of Eubacteria.
4 Describe the diversity seen in Eubacteria.

Discuss the view of Karl Popper that for science to progress, scientists must question and criticize the current state of scientific knowledge.

F.2 Microbes and the environment

Assessment statements

F.2.1 List the roles of microbes in ecosystems, including producers, nitrogen fixers and decomposers.
F.2.2 Draw and label a diagram of the nitrogen cycle.
F.2.3 State the roles of *Rhizobium, Azotobacter, Nitrosomonas, Nitrobacter* and *Pseudomonas denitrificans* in the nitrogen cycle.
F.2.4 Outline the conditions that favour denitrification and nitrification.
F.2.5 Explain the consequences of releasing raw sewage and nitrate fertilizer into rivers.
F.2.6 Outline the role of saprotrophic bacteria in the treatment of sewage using trickling filter beds and reed bed systems.
F.2.7 State that biomass can be used as raw material for the production of fuels such as methane and ethanol.
F.2.8 Explain the principles involved in the generation of methane from biomass, including the conditions needed, organisms involved and the basic chemical reactions.

Roles of microbes in ecosystems

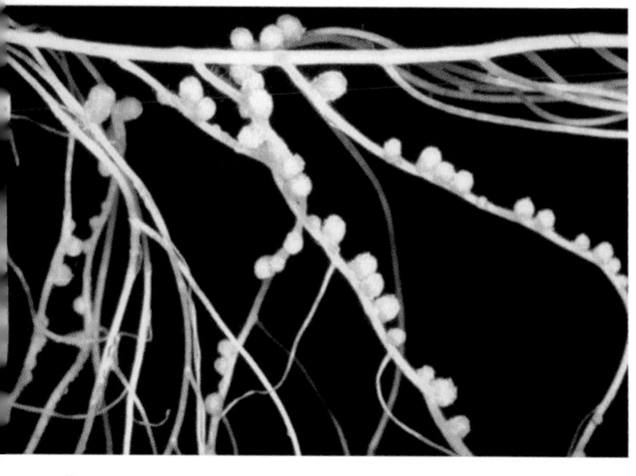

Rhizobium lives in the root nodules of legumes and fixes atmospheric nitrogen. These bacteria are symbiotic and receive carbohydrates and a favourable environment from their host plant.

Once again we see how diversity allows a group to be successful. Genetic variation among microbe groups allows them to exploit the many niches which exist in the ecosystem. Microbes have adapted to fill three main roles in the ecosystem.

- *Producers*: Microscopic algae and some bacteria use chlorophyll to trap sunlight. Chemosynthetic bacteria use chemical energy. Both change inorganic molecules into organic molecules which can be used by other organisms for food.
- *Nitrogen fixers*: Bacteria which remove nitrogen gas from the atmosphere and fix it into nitrates which are useable by producers. Some bacteria produce nitrates from nitrites.
- *Decomposers*: These organisms breakdown detritus (organic molecules) and release inorganic nutrients back into the ecosystem.

We are going to look at the importance of bacteria in the nitrogen cycle, sewage treatment and the production of biofuels.

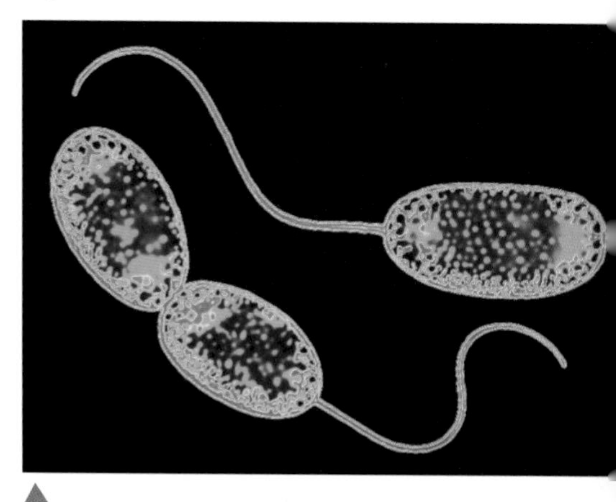

Nitrobacter lives in well-oxygenated soils and changes nitrites into nitrates, which are useable by plants.

The nitrogen cycle

Bacteria play a hugely important part in the processes by which nitrogen is continuously recycled through the environment (see Figure 17.5). Roles of bacteria in the nitrogen cycle are summarized in Figure 17.6 and the accompanying numbered points.

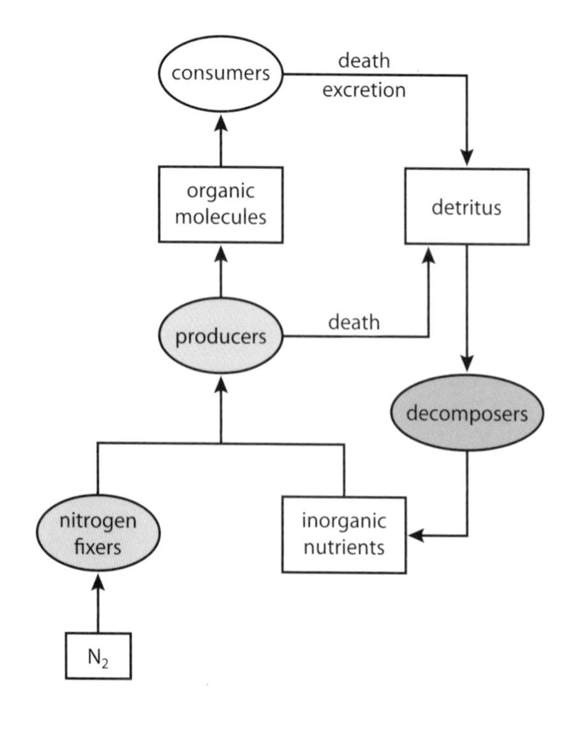

Figure 17.5 Microbes in the ecosystem are nitrogen fixers, producers and decomposers.

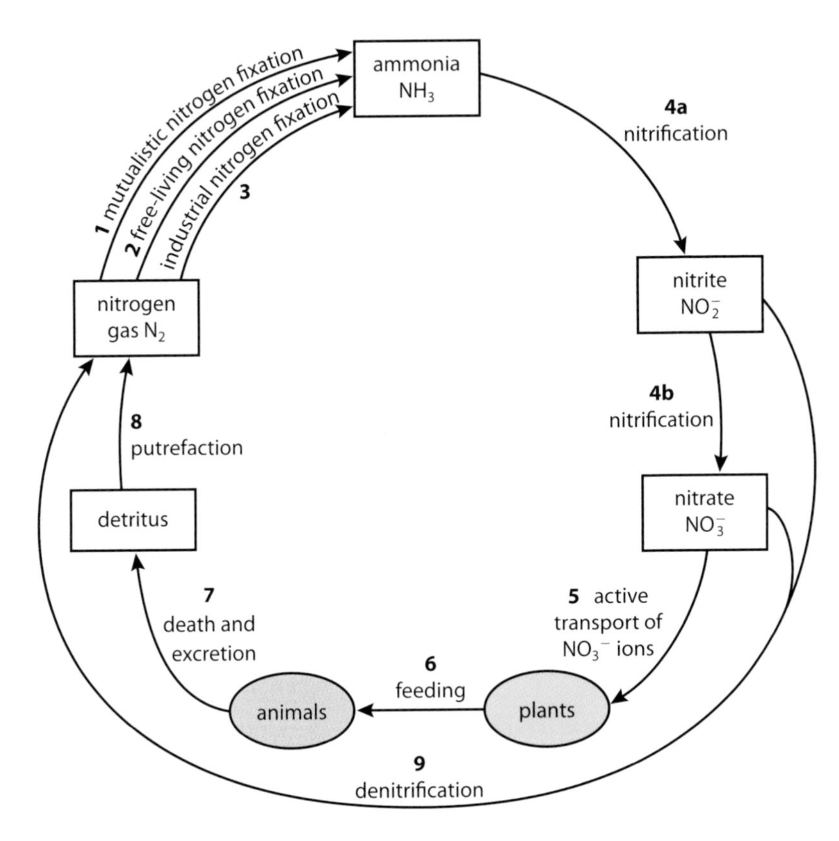

Figure 17.6 Steps of the nitrogen cycle and roles of bacteria.

1 *Mutualistic nitrogen fixation*: Certain bacteria form a symbiotic relationship with a host plant and fix nitrogen for it e.g. *Rhizobium* lives in symbiosis with legumes (beans, peas, clover).

2 *Free-living nitrogen fixation*: Nitrogen-fixing bacteria which live freely in the soil and do not need a host. e.g. *Azotobacter*.

3 *Industrial nitrogen fixation*: Burning fossil fuels to produce fertilizer is an important source of fixed nitrogen.

4a *Nitrification*: Oxygen is needed to turn ammonia into nitrites by bacteria in the soil e.g. *Nitrosomonas*.

4b *Nitrification*: Oxygen is also required to change nitrites into nitrates by soil bacteria e.g. *Nitrobacter*.

5 *Active transport of nitrates*: Nitrates are actively transported into the roots of plants using ATP for energy.

6 *Plants and animals*: Plants use nitrates to make their own proteins. This process is called assimilation. Animals feed on plants, digest and rearrange proteins to make their own proteins.

7 *Death and excretion*: The waste products of digestion and dead bodies of plants and animals are full of molecules containing nitrogen.

8 *Putrefaction*: Decomposers such as bacteria and fungi break down complex proteins and release nitrogen gas into the atmosphere.

9 *Denitrification*: Bacteria remove nitrates and nitrites and put nitrogen gas back into the atmosphere e.g. *Pseudomonas denitrificans*.

W To see a useful animation of the nitrogen cycle visit heinemann.co.uk/hotlinks, enter the express code 4242P and click on Weblink 17.4.

Conditions which favour nitrification and denitrification

Nitrification occurs due to the actions of two bacteria. *Nitrosomonas* converts ammonia (NH_3) into nitrite (NO_2^-). Next *Nitrobacter* changes nitrite (NO_2^-) into nitrate (NO_3^-) which is useable by plants. These are aerobic reactions carried out by two autotrophic bacteria which are beneficial to the environment. Conditions required are:

- available oxygen (the reaction is aerobic);
- neutral pH (preferred by these bacteria);
- warm temperature (preferred by bacteria).

Denitrification is a conversion of nitrates to nitrogen gas. This takes place in anaerobic conditions by autotrophic bacteria. Bacteria such as *Pseudomonas denitrificans* use NO_3^- instead of oxygen as the final electron acceptor.

Conditions required are:

- no available oxygen (flooding or compacted soil);
- high nitrogen input.

Denitrification is not good for soils because it removes beneficial nitrates needed by plants to make proteins. Denitrification also destroys the ozone layer. Another product, nitrous oxide (NO), can contribute to global warming as it is a minor greenhouse gas.

Release of raw sewage and nitrate fertilizer into rivers

As societies move ever more quickly to urbanization, the common problem of waste disposal grows, particularly in relation to sewage. A related problem is the run-off of excess nitrate fertilizer from farms, golf courses and lawns, which flows into rivers and streams. Effective waste management is a rising cost in our society but these problems must be solved in order to prevent dire consequences.

Releasing raw sewage into water systems was common until, in the 1850s, it was shown that cholera was transmitted by water contaminated with faeces. *Escherichia coli* (an intestinal bacterium) is frequently in the news in the western world for causing outbreaks of food poisoning – it is spread by contaminated water and lack of hand-washing. The Ganges River in India is the site of a hugely popular festival where people ritualistically bathe in the river which is now contaminated with human faeces. There are many, many places in the world where a clean water supply is desperately needed. Pathogens should not be found in bathing and drinking water, or water used to irrigate crops.

This lake is now green due to the presence of algae. Communities around this lake did not realize that fertilizer run-off would actually 'fertilize the lake'.

Nitrates may not sound as dramatic a problem as raw sewage, but they can cause disaster to ecosystems. Excess nitrates and phosphates in rivers and streams is termed eutrophication. The disaster proceeds as follows:

- high nitrates and phosphates fertilize the algae present in water;
- increased growth of algae (called algal bloom);
- algae are decomposed by aerobic bacteria which use up the oxygen in the water (high use of oxygen is called biochemical oxygen demand – BOD);
- water becomes low in oxygen (deoxygenation) and fish and other organisms which need oxygen die.

Sewage treatment by saprotrophic bacteria

This is a trickling filter system at a sewage plant in Yorkshire, England. Have you visited the waste treatment plant in your community?

Sewage treatment is vital to the health of human populations and to clean water. Usually sewage treatment occurs in two stages. In the primary stage, inorganic materials are removed and organic matter is left. In the secondary stage, 90% of the organic matter is removed by saprotrophic bacteria. The bacteria obtain energy by breaking down organic matter.

One type of system in use today is a trickling filter system. A trickling filter is a bed of stones 1–2 metres (3–6 feet) wide. Saprotrophic bacteria adhere to the stones and act on the sewage trickled over them until it is broken down. Cleaner water trickles out of the bottom of the bed. This flows to another tank where the bacteria are removed. The water is further treated with chlorine to finish the disinfectant process.

A second type of sewage removal system is the reed bed. This is an artificial wetland used to treat waste water. In the reed bed, the waste water provides both the water and the nutrients to the growing reeds. The reeds are then harvested for compost. The breakdown of organic waste is again accomplished by saprotrophic bacteria. The nitrates and phosphates released as a result of bacterial action are used as fertilizer by the reeds. One drawback to this system is that it can only handle small sewage flow and needs to be designed to fit the landscape of a particular site.

This is a reed bed sewage treatment plant. Such treatment plants can be found in places where there are relatively small amounts of sewage and sufficient space for the reed bed.

The UN Development Goal is to halve, by 2015, the 'proportion of people who are unable to reach or afford safe drinking water' and 'the proportion of people without access to basic sanitation'.

The Cooperation for Clean International Waters is a joint effort of public and private partnerships to improve access to water services for all. This will contribute to the international effort to meet the UN Development Goals. Currently, there are ongoing activities in India and South Africa. A future goal is to stimulate a global dialogue on water.

For more information about this international effort, visit heinemann.co.uk/hotlinks, enter the express code 4242P and click on Weblink 17.5.

To read an article about a Japanese and US partnership to provide safe water and sanitation, visit heinemann.co.uk/hotlinks, enter the express code 4242P and click on Weblink 17.6.

Production of biofuels

As the world population increases, there is a growing demand for energy. Also increasing is awareness of the need to decrease carbon dioxide emissions which are a by-product of using fossil fuels for energy. Biomass, such as forest, agricultural and animal products including manure, can be used as raw material for the production of fuels such as methane and ethanol.

To play the Water Alert! game and see if you can help solve the world's water crisis, visit heinemann.co.uk/hotlinks, enter the express code 4242P and click on Weblink 17.7.

Using a small-scale biogas digester such as this one in India, farmers can trap methane instead of letting it escape from rotting manure into the environment. Methane from farms is a significant greenhouse gas.

Methane generation (methanogenesis) from livestock waste and cellulose left from crops is causing interest due to its simplicity. The methane produced is called biogas. Biogas is about 60–70% methane and 30–40% carbon dioxide. To make biogas, manure and cellulose are put into a digester without oxygen. Anaerobic decomposition is performed by bacteria which occur naturally in the manure. Manure and cellulose contain carbohydrates, fats and proteins. These large molecules are broken down by bacterial enzymes into the simpler compounds of organic acids and alcohol. Organic acids and alcohol are then decomposed into the even simpler carbon dioxide, hydrogen and acetate. Finally, two different types of bacteria work on the carbon dioxide, hydrogen and acetate to produce methane. Ammonia and phosphate are by-products of anaerobic digestion and can be used as fertilizer.

Why would a small farm use a digester to make methane? Because the methane can be used to run electrical machinery. This reduces costs. High-quality fertilizer without weed seeds or odour is a by-product and can be used instead of the original, smelly manure. Pollution of water by run-off filled with animal waste is reduced. Globally, methane emission is lowered. Methane is a greenhouse gas which contributes to global warming.

For more information about biogas technology, visit heinemann.co.uk/hotlinks, enter the express code 4242P and click on Weblink 17.8.

Certain conditions must be kept constant in a digester:

- no free oxygen (the bacteria in the digester are anaerobic);
- constant temperature of about 35 °C (95 °F);
- pH (not too acidic because the methane-producing bacteria are sensitive to acid).

Methanogenesis requires three groups of bacteria:

- acidogenic bacteria convert organic matter to organic acids and alcohol;
- acetogenic bacteria make acetate, with carbon dioxide and hydrogen as by-products, from the organic acids and alcohol;
- methanogenic bacteria create the methane, either through the reaction of carbon dioxide and hydrogen or through the breakdown of acetate.

The process of digestion of manure, the organisms needed and the basic chemical reactions can be seen in Figures 17.7 and 17.8.

Biofuel is thought to be one way to reduce greenhouse gases released from fossil fuel. Could this be an effective way to reduce global warming? Some countries are beginning large-scale ethanol production.

If biofuel is a way of the future, consider this. Global agricultural practices have often added excess nitrogen to the soil as fertilizers to increase production. Increased pesticide use is another problem. One option is to change to organic farming practices in growing biofuel.

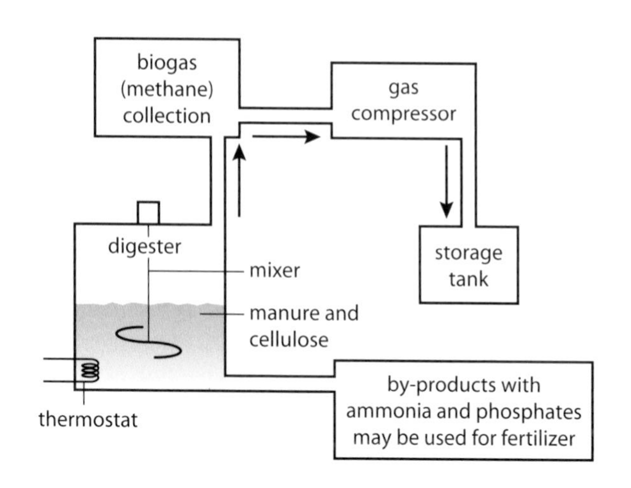

Figure 17.7 Anaerobic digester.

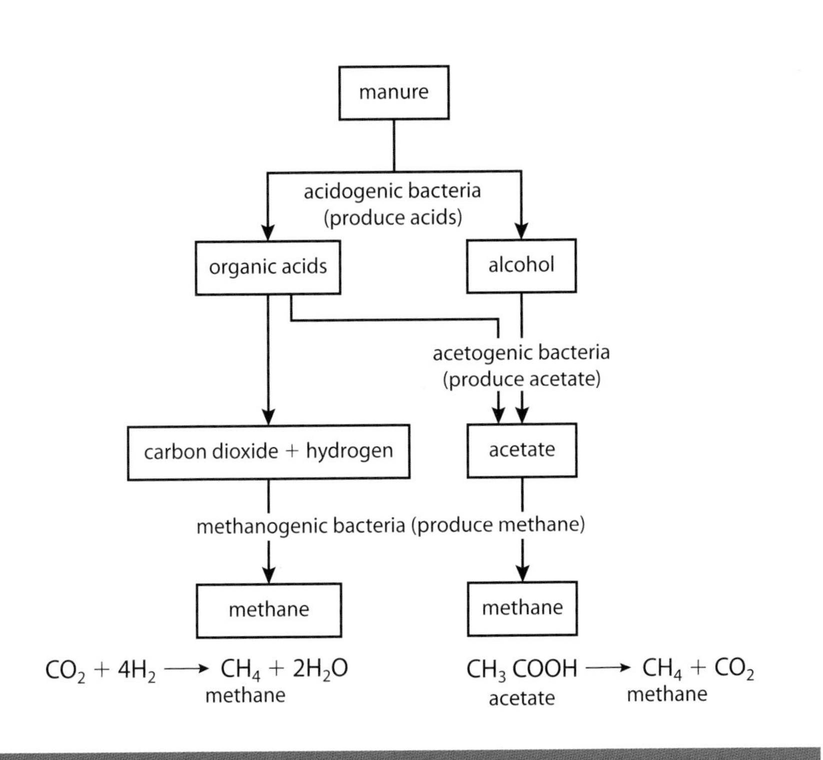

Figure 17.8 Three groups of bacteria are necessary for methanogenesis.

Exercises

5 Draw the nitrogen cycle.
6 Describe the conditions which favour denitrification and nitrification.
7 Describe the role of saprotrophic bacteria in the treatment of sewage.
8 Explain the principles involved in generation of methane from biomass.

F.3 Microbes and biotechnology

Assessment statements

F.3.1 State that reverse transcriptase catalyses the production of DNA from RNA.
F.3.2 Explain how reverse transcriptase is used in molecular biology.
F.3.3 Distinguish between somatic and germ-line therapy.
F.3.4 Outline the use of viral vectors in gene therapy.
F.3.5 Discuss the risks of gene therapy.

Discovery of reverse transcriptase

Genetic engineers were studying viruses. One of the viruses was an RNA virus. You will recall that the genetic material of a virus can be either DNA or RNA. An RNA virus is called a retrovirus. The flow of genetic information in an RNA virus is from RNA back to DNA. The enzyme which enables the retrovirus to transcribe backwards is reverse transcriptase (see Figure 17.9).

While studying the life cycle of human immunodeficiency virus (HIV) and other retroviruses, molecular biologists discovered how reverse transcriptase works.

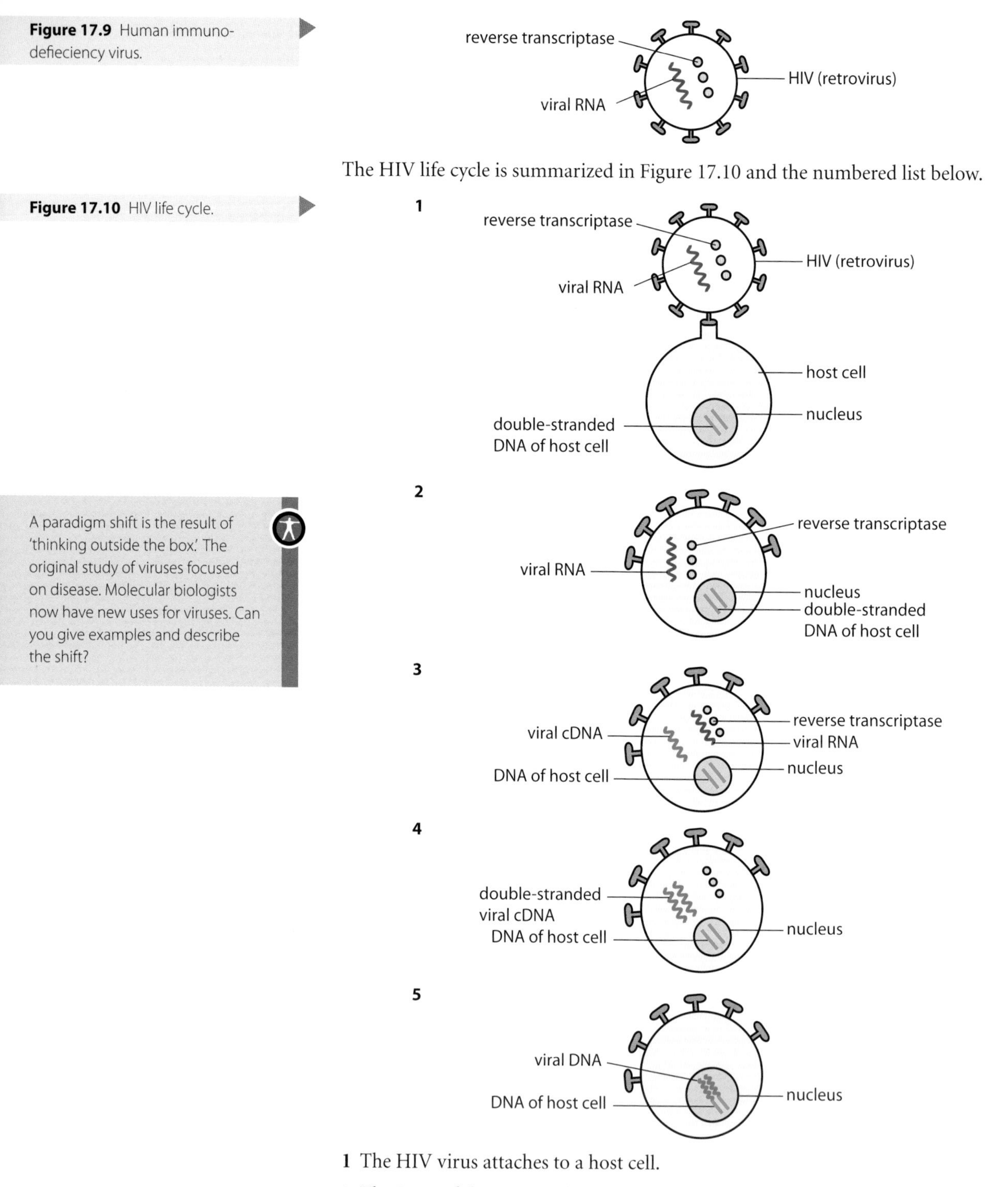

Figure 17.9 Human immuno-defieciency virus.

The HIV life cycle is summarized in Figure 17.10 and the numbered list below.

Figure 17.10 HIV life cycle.

A paradigm shift is the result of 'thinking outside the box.' The original study of viruses focused on disease. Molecular biologists now have new uses for viruses. Can you give examples and describe the shift?

1 The HIV virus attaches to a host cell.

2 The RNA of the virus and the enzyme reverse transcriptase enter the host cell.

3 Within the cell, reverse transcriptase copies viral RNA into cDNA (complementary DNA).

4 Next, cDNA makes a second strand which is a complement to the first strand of DNA. Viral RNA is destroyed.

5 Finally, the new double-stranded viral DNA enters the nucleus of the host cell.

6 If the HIV virus is active, it will use this DNA to make more HIV viruses. They will then burst out of the cell and infect other cells.

Using reverse transcriptase

The discovery of reverse transcriptase in retroviruses revolutionized our ability to make therapeutic proteins such as insulin and somatostatin. In order to make these proteins, genetic engineers must first eliminate some unnecessary information found in the genes of eukaryotic cells. A human cell, just like other eukaryotic cells, contains long regions of DNA which do not code for anything. The coding regions make the proteins. The non-coding areas are called introns ('intervening sequences'). Genetic engineers want the gene minus the introns. To do this, they isolate a messenger RNA molecule (mRNA). mRNA copies the coding region without copying the introns. After extracting the mRNA from the cell, reverse transcriptase is used to copy the mRNA to a single strand of cDNA.

Producing genetically engineered biopharmaceuticals is a very successful technology. Here is how reverse transcriptase is used to make the protein insulin (see Figure 17.11).

1 First, a human DNA molecule with all its introns is taken from a pancreas cell.
2 mRNA copies the DNA without the introns.
3 Reverse transcriptase (working backwards) produces a new single strand of DNA called cDNA.
4 The single strand replicates to make double-stranded DNA using the enzyme DNA polymerase.
5 The double-stranded DNA for making insulin is isolated and inserted into a plasmid (circular DNA found in bacterial cells).
6 The bacteria cell is stimulated to take up the plasmid (commonly done by bacteria).
7 The bacterial cell now contains the genetically engineered plasmid containing the gene for making human insulin.
8 The bacteria multiply and produce insulin.
9 Insulin is harvested and used by diabetics.

Reverse transcriptase has become a key enzyme in making insulin for diabetics.

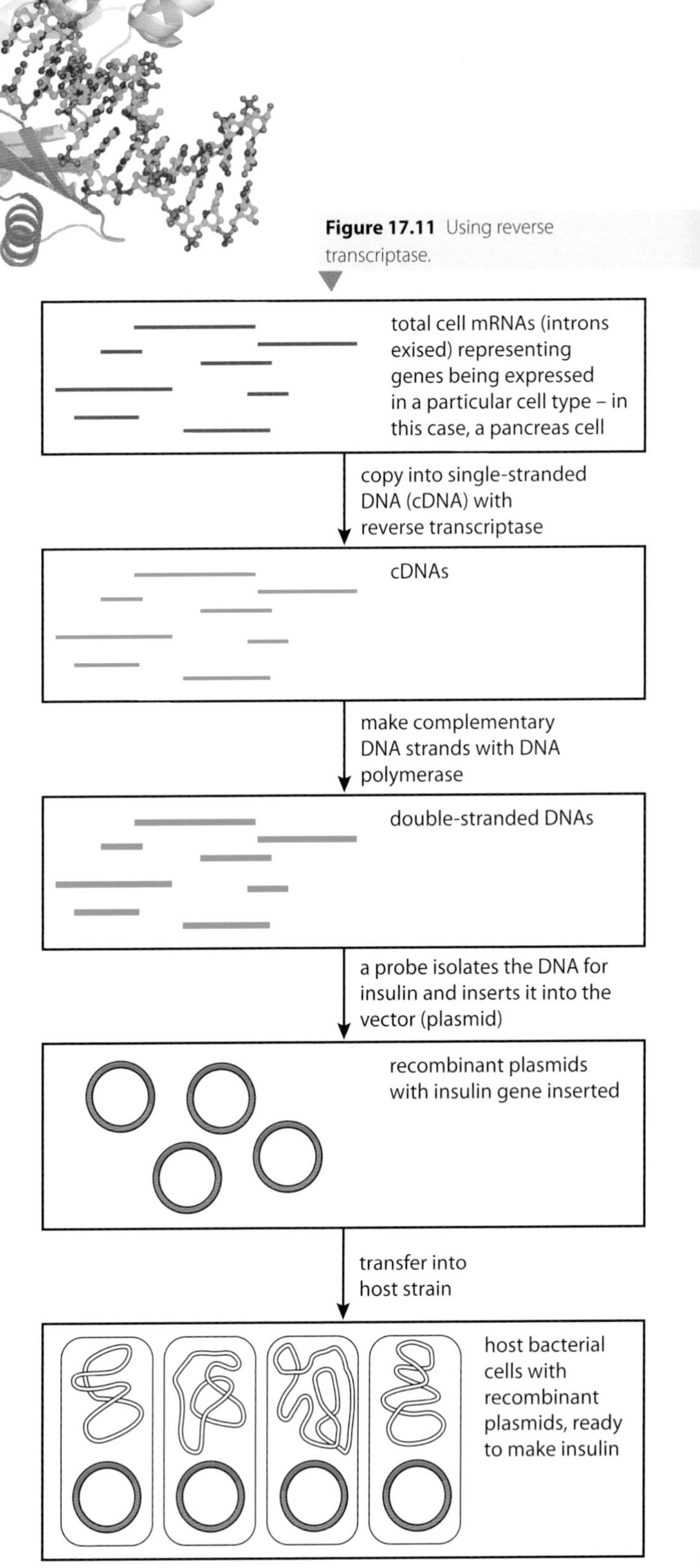
total cell mRNAs (introns exised) representing genes being expressed in a particular cell type – in this case, a pancreas cell
copy into single-stranded DNA (cDNA) with reverse transcriptase
cDNAs
make complementary DNA strands with DNA polymerase
double-stranded DNAs
a probe isolates the DNA for insulin and inserts it into the vector (plasmid)
recombinant plasmids with insulin gene inserted
transfer into host strain
host bacterial cells with recombinant plasmids, ready to make insulin

Figure 17.11 Using reverse transcriptase.

Germ-line and somatic therapy

Gene therapy is very new and carries a lot of risk. It is a treatment strategy which involves the introduction of genes into human cells to alleviate disease. The aim is to replace defective genes with effective ones which give the message to make the correct protein. The genes are delivered by vectors. The vector is a virus which has been genetically engineered to infect certain cells in the patient while carrying new DNA with the correct genes.

There are two types of gene therapy: germ-line therapy and somatic therapy. Germ-line therapy would change the patient's germ cell (gamete) DNA, which would be passed on to the offspring. Somatic (body cell) therapy affects only the patient involved, who can give voluntary consent to the procedure. Using somatic-cell therapy, it may be possible to cure single-gene defects such as cystic fibrosis and haemophilia.

For more information on gene therapy, visit heinemann.co.uk/hotlinks, enter the express code 4242P and click on Weblinks 17.9a and 17.9b.

Use of viral vectors in gene therapy

One of the success stories of somatic therapy is in treating a disease in children born with severe combined immune deficiency disease (SCID). Children affected with SCID have a mutation in a gene which affects their immune system. They do not produce an enzyme named adenosine deaminase (ADA) and thus have no immune system. In order to stay alive, they must live in 'bubble' free of germs.

In 1998, stem cells were withdrawn from the bone marrow of 11 children with SCID. The stem cells were mixed with a virus carrying the normal copy of the gene. The virus transferred the normal copy of the gene into the stem cells. The stem cells were then infused back into the bone marrow of the children. The replacement gene caused the production of the enzyme ADA. For the first time, these children had their immune system restored. Ten months later, the children left the hospital. Two years later, two of the children contracted leukaemia and one of them died. This may have been caused by the treatment of the virus. It is obvious that this procedure has risks which must be assessed and the ethics of using this therapy on children re-examined.

Several cases of gene therapy have been in the news in the last few years because people have died. Do you think we need to re-consider such issues as safety and conflict of interest in research?

What do you think about the idea that ethical decisions should be entirely based on reason at the expense of feeling (this is Immanuel Kant's view).

Risks and benefits of gene therapy

There are both risks and benefits to gene therapy. Here is a description of some of the risks.

- The virus vector used to deliver the new gene to the target cell might get into another cell by mistake.
- The virus vector might place the new gene in the wrong location in the DNA molecule (chromosome) and cause an unintended mutation.
- Genes can be over-expressed and make too much protein, which might be harmful.
- The virus vector might stimulate an immune reaction as the body thinks it is being infected.
- The virus vector might be transferred from person to person like a flu or cold virus.
- Children might be more sensitive to the long-term hazards since their tissues are still developing.

The list for the benefits of gene therapy is short – but very powerful:

- the possibility of curing a disease caused by a single-gene defect;
- the possibility of curing a disease caused by multiple genes.

Gene therapy trials have an extensive review before they can begin because it is a very powerful technique. Patients must be aware of the risks as well as the benefits. Hopefully, future discoveries will eliminate the risks and increase the benefits for those people who are affected with these serious genetic defects.

Has the use of viral vectors in gene therapy changed our perception of viruses from pathogenic villains to super-heroes? Is this a paradigm shift? If you think so, explain why.

In the context of gene therapy research, discuss the dangers and potential benefits of technology.

- What are the problems with discovering new knowledge?
- This knowledge is expensive. Should we invest the money to do the research?
- Will the knowledge cause more harm than good?
- Interfering with human genes can be used to alleviate suffering or to try to create a master race. Can the new knowledge be controlled?
- How can we control use of the knowledge in small laboratories which might be unregulated?
- What about the potential benefits?

Exercises

9 Explain how reverse transcriptase is used in molecular biology.

10 Distinguish between somatic and germ-line therapy.

11 Discuss the risks of gene therapy.

F.4 Microbes and food production

Assessment statements

F.4.1 Explain the use of *Saccharomyces* in the productions of beer, wine and bread.

F.4.2 Outline the production of soy sauce using *Aspergillus oryzae*.

F.4.3 Explain the use of acids and high salt or sugar concentrations in food preservation.

F.4.4 Outline the symptoms, method of transmission and treatment of one named example of food poisoning.

Production of beer, wine and bread

Egypt was the home of bread-making. As early as 4000 BC, humans discovered that dough could somehow 'rise'. Since yeast spores are widespread in the air, they could have easily fallen into the dough. Later, 'beer foam' (actually yeast accumulating on the surface during fermentation of grain) was added to bread dough. This made the dough lighter. Soon the Egyptians were adding honey, fruit and nuts to bread to make it more interesting and desirable. This evolution of bread-making continued over centuries, with different techniques being tried and different grains being used.

The organism responsible for the successful making of bread, beer and wine was not discovered until the 1800s. In 1837, Louis Pasteur proved that yeast, a simple fungus, was responsible for these important products. The name of the yeast is *Saccharomyces cerevisae*. It is readily available commercially and domestically for those who want to make their own products. In making bread, wine and beer, the yeast uses sugars for energy and reproduces quickly by ' budding' (see Figure 17.12). The bud breaks off and forms a new yeast cell.

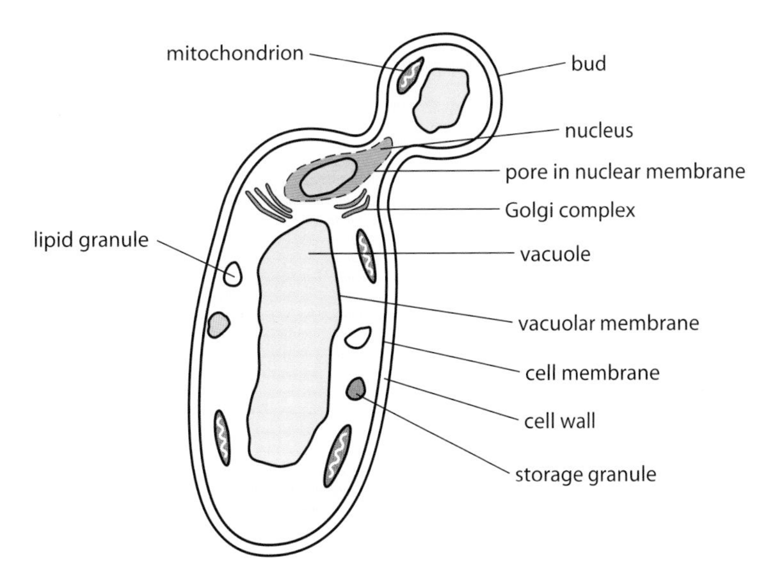

Figure 17.12 A yeast cell budding.

What makes this organism so important in making bread, wine and beer? Its ability to turn glucose ($C_6H_{12}O_6$) into two molecules of ethanol (CH_3CH_2OH) and have two molecules of carbon dioxide gas (CO_2) as a waste product (see Figure 17.13). Bread-making depends on the carbon dioxide gas to make the bread light and airy. Beer and wine depend on the ethanol for the alcohol and taste. What do you think that yeast gets out of this process? It gets energy from breaking the chemical bonds of the glucose. The energy is used by the yeast to reproduce.

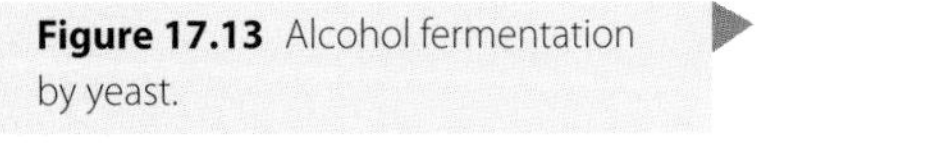

Figure 17.13 Alcohol fermentation by yeast.

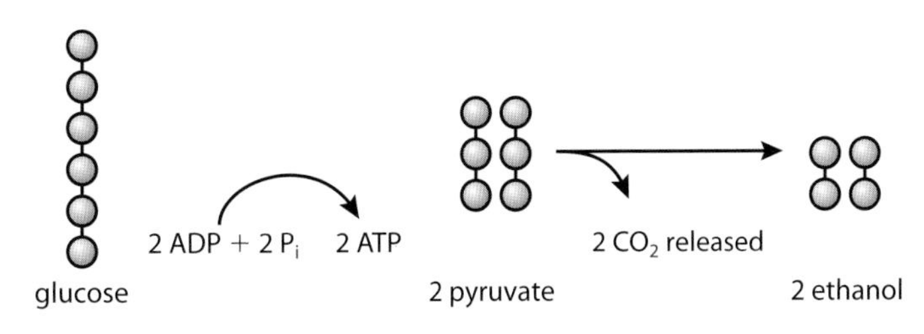

Beer

The source of glucose in beer-making is a grain (e.g. barley). The brewer wets the grain in order for it to germinate. Germination releases enzymes which turn the starch of the barley into sugars. The first sugar produced is maltose. At this point, the brewer has a sugary liquid called 'malt'. More water is added to this and now it is a sweet liquid called 'wort'. Hop flowers are added to the beer at this point to give it a bitter taste. The hops and wort are boiled together. The brewer then cools this mixture and adds the yeast. Yeast can break down the maltose (a disaccharide) into two glucose molecules (monosaccharides) beginning the process of fermenting the glucose. When the sugars are used up, the products are ethanol and carbon dioxide. The brewer now filters the beer to remove any particles. The beer is finally

pasteurized (heated to 82 °C (180 °F)), to kill any remaining yeast cells. Brewing produces a product containing 2–6% alcohol. Here is a summary:

- sweet liquid wort is made from malt;
- hops are added and liquid is boiled and cooled;
- fermentation by yeast produces beer containing ethanol and carbon dioxide.

Wine

Within the single species *Saccharomyces cerevisiae* there is genetic diversity. Different types of yeast are referred to a different 'strains'. The strain which makes beer can withstand only 2–6% alcohol without being killed. The *Saccharomyces* strain which produces wine must be able to withstand 11–15% alcohol and still keep fermenting. In wine-making, the yeast cells are never killed by heat as they are in making beer. If a wine-maker heated the wine, it would seriously affect the taste. Different strains of yeast produce different tastes, so wine-makers are very particular which strain of yeast they pick. The steps of wine making are simple. *Saccharomyces cerevisiae* is added to crushed grapes and everything is put into a fermentation tank. Carbon dioxide escapes from the tank leaving ethanol behind. In summary:

- crushed grapes and yeast are put into a tank;
- ethanol stays in the tank, while carbon dioxide escapes.

Bread

The primary function of *Saccharomyces* in bread-making is to provide carbon dioxide which makes the bread rise. Yeast acts on the sugars in the dough, breaking them down in exactly the same way as in wine and beer production. The resulting products are ethanol and carbon dioxide. Yeast cells prefer warm temperatures for optimum performance. This is why, if you are making bread, you need a warm place first to knead the dough and then to put it to rise. Kneading dough makes it stretchy and increases its ability to hold the carbon dioxide which makes it rise. After the bread has risen to the desired height, it is baked in the oven at a temperature which kills the yeast and causes the ethanol in the bread to evaporate. When you cut into baked bread, you can see the holes created by the carbon dioxide bubbles. Here is a summary:

- fermentation of sugars in the dough by yeast;
- carbon dioxide makes the dough rise;
- baking in the oven kills the yeast, stops fermentation and evaporates the ethanol.

Production of soy sauce

Did you know that soy sauce is also a product of fermentation? If you are familiar with Oriental cooking you might have already known that soy sauce comes from fermented soy beans. In this case, the fermenter is not yeast but another fungus named *Aspergillus oryzae*. The process of making soy sauce has been going on in China for 5000 years. These are the stages:

- soak soy beans, boil and drain;
- mix a mash of soy beans with toasted wheat;
- add a culture of *Aspergillus oryzae*;
- incubate for 3 days at 30 °C (85 °F);
- add salt and water and ferment for 3–6 months;
- filter and pasteurize.

Food preservation with sugar and acid

Preserved foods like jams, jellies and pickled vegetables make seasonal fruits and vegetables available for longer. Commonly used preservatives are sugar and acid.

Food preservation with sugar

Jams, jellies and marmalades were a necessity in the past and are very popular today. What do you do with the buckets of raspberries you just picked? Everyone in the family is tired of eating them but next week they will be gone from the bushes until next year. A modern choice is to freeze them, but not everyone has a freezer. The traditional choice is to preserve them with sugar and they will last until next season's crop is ripe. Increasing the sugar content of food preserves it because it is a dehydrating environment for bacteria, yeast and mould.

In order to understand how this works, look at Figure 17.14.

Figure 17.14 Solute concentration.

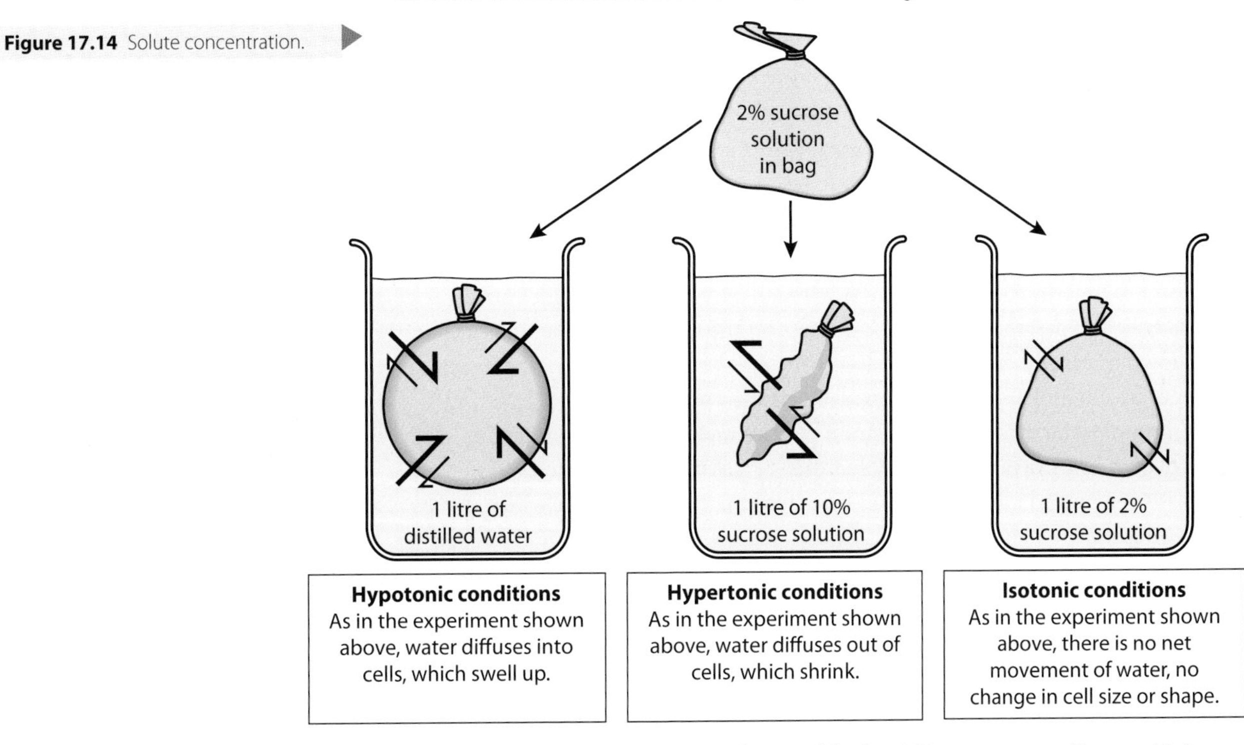
2% sucrose solution in bag
1 litre of distilled water
1 litre of 10% sucrose solution
1 litre of 2% sucrose solution
Hypotonic conditions
As in the experiment shown above, water diffuses into cells, which swell up.
Hypertonic conditions
As in the experiment shown above, water diffuses out of cells, which shrink.
Isotonic conditions
As in the experiment shown above, there is no net movement of water, no change in cell size or shape.

Which picture represents the condition of the jam? You are correct if you said the picture with hypertonic conditions. The sugar solution in the beaker is 10%, which means that there is 90% water in the beaker. The sugar solution in the bag is only 2%, thus 98% water. Where is the water 'concentration' higher, in the bag or out of the bag? Obviously, inside the bag. Since water moves from an area of higher water 'concentration' to one of lower water 'concentration', the water will move out of the bag. The same thing happens to single cells in hypertonic conditions: as water moves out of the cells, the cells shrink. Thus, mould and bacteria cells are killed in hypertonic conditions, such as having high sugar in their environment.

Steps in making jam:

- boil the fruit with sugar to kill any microorganisms and dissolve sugar;
- add some pectin so it will gel (some fruit has natural pectin which is activated by boiling);
- seal in hot sterile jars;
- no need to refrigerate, it will have a long shelf life.

As you will recall, a molecule dissolved in water is called a solute. Can you think of another solute that could be used for food preservation? You are correct if you guessed salt.

Food preservation with acid

Like fruit, vegetables tend to ripen all at once. To solve the problem of a glut, the art of pickling was developed. Pickles are usually made from vegetables preserved in vinegar and flavoured with spices. For example, gherkins are tiny cucumbers preserved in vinegar and flavoured with dill. They are preserved raw to keep the 'crunch'. Here is the procedure for pickling:

- leave vegetables in a salt solution (brine), then strain and rinse;
- place raw vegetables in sterile jars;
- pour hot vinegar and spices over the vegetables creating an acid environment;
- put on lids and process in a hot water-bath to give a tight seal by creating a vacuum so that no mould can grow.

In addition to heating, low pH is key in preserving vegetables by pickling. Vinegar is about 5% acetic acid. The acid lowers the pH to the point where it restricts the growth of microorganisms. Sugar and other spices can be added for flavour. The early brining also kills microorganisms by creating hypertonic conditions.

Food poisoning

When food is not treated properly, food poisoning can occur. Symptoms of flu and food poisoning are similar. One of the organisms that causes a lot of food poisoning is *Salmonella*. This bacterium was discovered by a scientist named Salmon over 100 years ago.

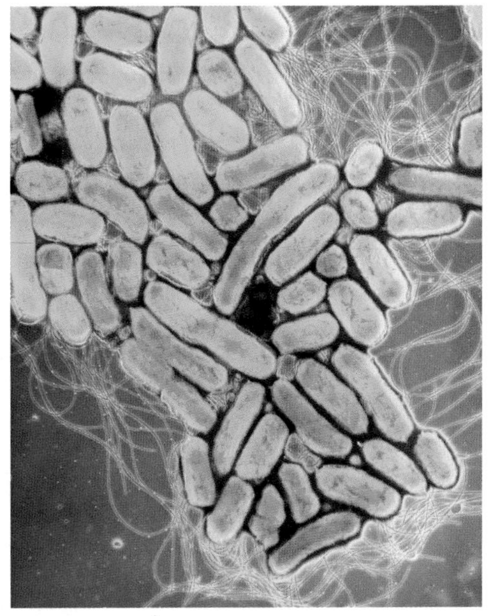

Salmonella live in human and animal waste. The organisms can also be found in raw eggs. In the US, 40 000 cases of food poisoning were caused by this organism in one year.

Symptoms of *Salmonella* poisoning occur 12–72 hours after infection and include:

- diarrhoea;
- fever;
- abdominal cramps;
- a small number of people develop Reiter's syndrome, which can last for years – symptoms are arthritis, irritation of the eyes and painful urination.

Symptoms of food poisoning can last 4–7 days. In severe cases, patients need to be hospitalized because *Salmonella* may have moved from the intestines to the bloodstream and can cause death.

How is *Salmonella* transmitted? The varied ways are summarized in this bullet list.

- *Salmonella* lives in the intestinal tract of humans and other animals, and can be transmitted by contact after ineffective hand washing.
- Eating contaminated foods not properly cooked can cause *Salmonella* poisoning.
- *Salmonella* can be found in the faeces of some pets and be transferred to food after ineffective hand washing.
- Reptiles commonly carry *Salmonella* and proper hand washing is required after handling them. This is especially important for children.
- Uncooked meat cut on a cutting board which is not then washed can cause transfer of *Salmonella* to cooked meats and raw vegetables .

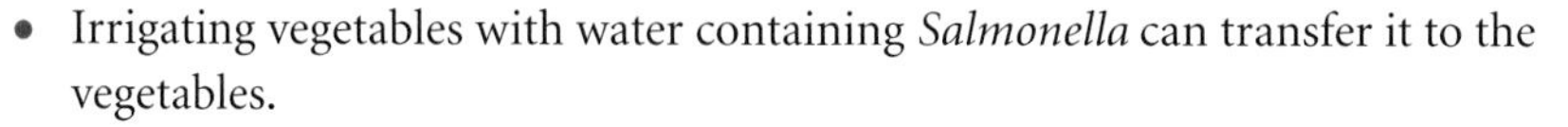

- Irrigating vegetables with water containing *Salmonella* can transfer it to the vegetables.
- Raw eggs may contain *Salmonella*. This is why government advice is to cook eggs thoroughly and avoid homemade products containing raw eggs such as hollandaise sauce, mayonnaise, cookie dough and Caesar salad dressing.
- Unpasteurized milk or other dairy products may contain *Salmonella*.

Treatment of *Salmonella* food poisoning:

- treat the dehydration by drinking lots of water (preferably containing a little sugar and salt);
- serious dehydration is treated with intravenous fluids;
- antibiotics can be given if the infection is serious and has spread from the intestine to the blood.

Salmonella food poisoning can be avoided if simple precautions are taken, especially correct hand washing. To correctly wash your hands, keep rubbing with soap and water long enough to sing Happy Birthday twice.

Can Asian H5N1 flu be transferred in the food supply? This highly pathogenic strain can be found in the eggs and muscle of infected poultry. However, current precautions may prevent its transfer.

Visit heinemann.co.uk/hotlinks, enter the express code 4242P and click on Weblink 17.10 to discover more about avian flu.

For more information on *Salmonella* and how to avoid it, visit heinemann.co.uk/hotlinks, enter the express code 4242P and click on Weblink 17.11.

In the case of an outbreak of food poisoning, it is important to recognize the difference between the correlation and the cause. A correlation may be a preliminary indication that there is a causal connection. But a fallacy is being committed if we jump to the conclusion that what 'seems' to be the cause actually is. A large amount of evidence is necessary to be sure of the cause.

Find and read an article about a recent outbreak of food poisoning and determine if it is the cause or just a correlation that is known. Is the food being recalled just because there is a correlation? Do you agree or disagree with the recall? Don't forget that there are limitations to the inferences you can draw from a press report – you need lots of evidence to prove a causal relationship, but only one piece of evidence to the contrary disproves it.

Exercises

12 Describe the use of *Saccharomyces* in the production of beer.
13 Explain the use of acid in food preservation.
14 Describe the symptoms, transmission and treatment of a specific type of food poisoning.

F.5 (HL only) Metabolism of microbes

Assessment statements

F.5.1 Define the terms *photoautotroph, photoheterotroph, chemoautotroph, chemoheterotroph*.
F.5.2 State one example of a photoautotroph, photoheterotroph, chemoautotroph, chemoheterotroph.
F.5.3 Compare photoautotrophs with photoheterotrophs in terms of energy sources and carbon sources.
F.5.4 Compare chemoautotrophs with chemoheterotrophs in terms of energy sources and carbon sources.
F.5.5 Draw and label a diagram of a filamentous cyanobacterium.
F.5.6 Explain the use of bacteria in the bioremediation of soil and water.

Metabolism

Microbes change molecules from one form to another during metabolism. The two categories of metabolic process are catabolism and anabolism. Catabolism means to break down. Thus, organisms use enzymes to catalyse breakdown reactions (e.g. saliva contains an enzyme which breaks down starch into sugars). Anabolism means to build up.

Metabolism is from the Greek word 'metabole' which means change.

See if you can describe the following statements as either 'catabolism' or 'anabolism'.

1 Inorganic molecules of carbon dioxide are joined together by cyanobacteria to make glucose.
2 Organic molecules of glucose are broken down by bacteria during fermentation producing inorganic carbon dioxide and organic ethanol.
3 Using light energy, a bacterium uses inorganic molecules to make glucose.
4 Using chemical bond energy, a bacterial cell uses inorganic carbon dioxide to make glucose.

Answers: 1 anabolism; 2 catabolism; 3 anabolism; 4 anabolism.

ATP: energy storage molecule in cells.
Photo: using light energy.
Chemo: using chemical bond energy from an inorganic compound such as NH_3.

Definition of terms and examples

Photoautotroph: An organism that uses light energy to generate ATP and produces organic compounds from inorganic substances. For example, the cyanobacterium, *Anabaena*.

Photoheterotroph: An organism which uses light energy to generate ATP and obtains organic compounds from other organisms. For example, *Rhodobacter sphaeroides*.

Chemoautotroph: An organism that uses energy from chemical reactions to generate ATP and produces organic compounds from inorganic substances. For example, *Nitrosomonas* (a nitrogen-fixing bacteria in the soil).

Chemoheterotroph: An organism that uses energy from chemical reactions to generate ATP and obtains organic compounds from other organisms. For example, *Saccharomyces* (yeast).

A fuller explanation of terms

Autotroph (self feeder): An autotrophic organism makes its own food (organic molecules). It does not have to get food by consuming molecules of another organism.

Heterotroph (other feeder): An heterotroph must consume molecules from another organism to obtain energy. The molecules consumed are organic molecules of an autotroph or another heterotroph.

Photoautotroph: What is the advantage of being green? Microbes that are green because they contain chlorophyll, can photosynthesise. Chlorophyll is the wonder pigment which can trap light energy and transfer it to chemical bond energy. Organisms which have chlorophyll can make their own glucose instead of trying to find something to eat. Cyanobacteria contain chlorophyll. These bacteria trap light energy to make glucose and expel oxygen as a waste product. Cyanobacteria have existed on the earth for about 3.4 billion years and are thought to be the source of the oxygen present in the atmosphere today.

Photoheterotroph: Some bacteria can convert from one metabolic process to another. They can convert from anabolism to catabolism and back again as a means of getting food. An example is the bacterium, *Rhodobacter sphaeroides*. When growing photosynthetically, it uses light for the energy source to make glucose (anabolism). It can convert to heterotrophic metabolism and break down glucose to get energy (catabolism). If oxygen is available, it will break down glucose aerobically; if oxygen is not available, it will break down glucose by anaerobic fermentation.

Chemoautotroph: An autotroph makes its own food but needs an energy source. A chemoautotroph gets that energy from the chemical bonds of an inorganic molecule. *Nitrosomonas* is a nitrogen-fixing bacterium that we met when discussing the nitrogen cycle. It obtains energy from the oxidation of ammonia (NH_3) found in the soil. Using the energy from this molecule, it can synthesize glucose from carbon dioxide.

Chemoheterotroph: Most microbes are in this category. They use glucose for their energy source. They may break down glucose aerobically by respiration or anaerobically by fermentation. These bacteria must ingest food containing glucose in order to obtain the chemical bond energy.

Some comparisons

This table compares photoautotrophs and photoheterotrophs.

	Photoautotroph	Photoheterotroph
Energy source	light	light and organic compounds
Carbon source	carbon dioxide	organic compounds

This table compares chemoautotrophs and chemoheterotrophs.

	Chemoautotroph	Chemoheterotroph
Energy source	inorganic compounds	organic compounds
Carbon source	carbon dioxide	organic compounds

Anabaena

Anabaena is a genus of filamentous cyanobacterium that lives on grass and in freshwater ponds. This bacterium is unique because it has two distinct and interdependent cell types, heterocysts and photosynthetic cells (see Figure 17.15). Heterocysts fix nitrogen from dinitrogen (N_2) in the air into nitrogen compounds such as ammonia. Ammonia is used by the organism to make proteins. The photosynthetic cells produce carbohydrates. Carbohydrates and fixed nitrogen are exchanged between the heterocysts and the photosynthetic cells through small channels in their cell walls. *Anabaena* is helpful to farmers because nitrogen fixation increases soil fertility.

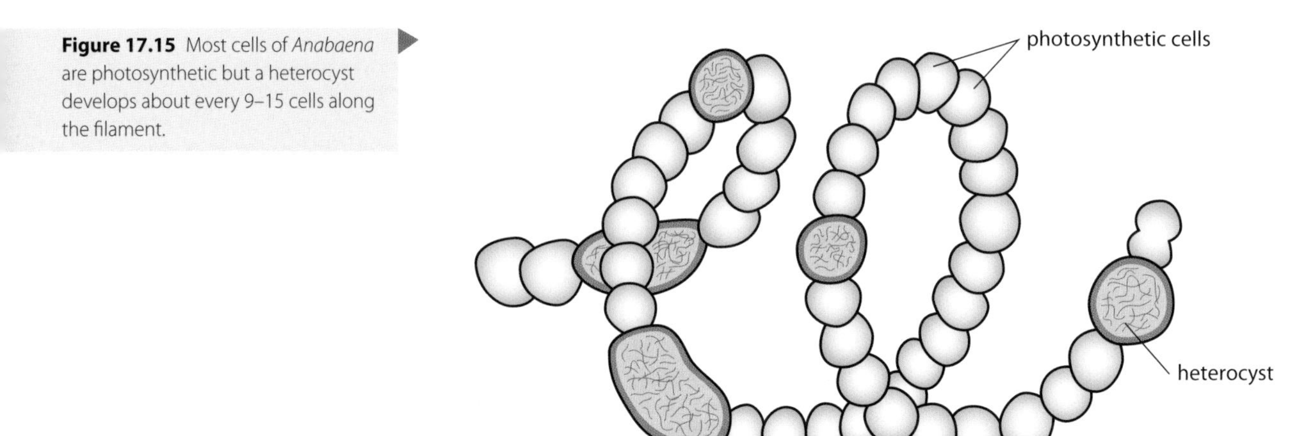

Figure 17.15 Most cells of *Anabaena* are photosynthetic but a heterocyst develops about every 9–15 cells along the filament.

Bioremediation

Bioremediation is the use of bacteria and fungi to treat environments such as soil or water contaminated with polluting agents such as pesticides, oil and industrial solvents. The organisms break down the toxic chemicals so that the soil or water is restored to its original condition. The benefit to the microorganisms is that during the process of degrading contaminants, they derive energy for their own growth and reproduction. The bacteria and fungi are actually chemoheterotrophs obtaining energy from the organic molecules which happen to be pollutants to us.

This is the clean-up after the *Exxon Valdez* oil spill. An oil tanker from the Exxon Oil Corporation struck a reef and spilled oil along 2000 kilometres (1250 miles) of Alaska's shoreline. It had a disastrous effect on wildlife.

One of the first success stories of bioremediation was the treatment of the oil spill of the tanker *Exxon Valdez* which contaminated the shoreline of Alaska in 1989. Puget Sound, Alaska, had thick oil covering its rocks and sandy shoreline. What could be done to clean up? Bioremediation was new and had only been tried in small areas. How would it work in such a large area? Investigators did a preliminary feasibility study. The study found a large population of microorganisms already existed near the shoreline. These marine bacteria contain enzymes which naturally work to degrade organic materials: they could break down the oil. In order to increase their numbers, nitrogen and phosphate were added to the water. Several months later, a visible reduction of oil was seen on the rocks and sand. This clean-up was a success. Bioremediation under natural conditions would have taken 5–10 years. The time was shortened to 2–5 years.

To find out about bioremediation of oil spills, visit heinemann.co.uk/hotlinks, enter the express code 4242P and click on Weblink 17.12.

A second success story is the clean-up of an industrial site contaminated with wood treatment chemicals in Southern Mississippi. The chemicals creosote and pentachlorophene (PCP) were dumped over a period of many years before people recognized the danger of such compounds in the soil. Clean-up is now progressing as bacteria are used to break down the creosote and PCP. The soil is constantly rinsed with water to wash the chemicals into a large tank. In the tank are *Arthrobacter* which acts on PCP and *Cladosporum* which breaks down creosote. The water leaves the tank clean and is returned to the soil. Nitrate and phosphate fertilizer are added every week to increase bacterial growth. This contaminated site has shown definite improvement. In the past, industry has spent millions of dollars on the clean-up of sites. The cost of the bacteria treatment is small in comparison.

Toxic waste containing heavy metals is contaminating this piece of land in Walsall, West Midlands, UK. Such waste must be disposed of carefully to prevent soil and water contamination.

Explain the paradigm shift that has occurred over the last 50 years. In the 1950s it was common to dump wastes into rivers and streams or into the soil. Sometimes people changed the oil on their car and just dumped the oil into the ground. Boaters may have dumped their garbage in the water. Industry used lakes and rivers to help get rid of wastes. If you agree that a paradigm shift has occurred, what caused it? Explain.

Exercises

15 Define *photoautotroph*, *photoheterotroph*, *chemoautotroph* and *chemoheterotroph*.

16 Compare energy and carbon sources of chemoautotrophs and chemoheterotrophs.

17 Describe the bioremediation of water by bacteria.

F.6 (HL only) Microbes and disease

Assessment statements

F.6.1 List six methods by which pathogens are transmitted and gain entry to the body.
F.6.2 Distinguish between intracellular and extracellular bacterial infection using *Chlamydia* and *Streptococcus* as examples.
F.6.3 Distinguish between endotoxins and exotoxins.
F.6.4 Evaluate methods of controlling microbial growth by irradiation, pasteurization, antiseptics and disinfectants.
F.6.5 Outline the mechanism of the action of antibiotics, including inhibition of synthesis of cell walls, proteins and nucleic acids.
F.6.6 Outline the lytic life cycle of the influenza virus.
F.6.7 Define *epidemiology*.
F.6.8 Discuss the origin and epidemiology of one example of a pandemic.
F.6.9 Describe the cause, transmission and effects of malaria, as an example of disease caused by a protozoan.
F.6.10 Discuss the prion hypothesis for the cause of spongiform encephalopathies.

Studying microbes and disease

In order to prevent the spread of diseases caused by microbes, it is necessary to study the basic biology of microbes and disease.

- How are pathogens transmitted and how do they gain entry to the body?
- What are the methods by which pathogens infect cells?
- What mechanisms control microbial growth?
- What can we learn from the life cycles of pathogens?

For a time, we knew there was a correlation between smoking and lung cancer. Now after years of evidence, we know one cause of lung cancer is smoking. Explain the difference between a correlation and a cause.

In addition to these basics, we need to study how food distribution systems, food handling procedures, changing environments and travel can all affect the global distribution of many diseases.

Six methods of transmission of pathogens

Pathogens are microbes which are capable of causing disease. Our body is designed to resist the entry of pathogens using the skin, mucous membranes, cilia, oil and sweat glands, and tears to prevent their entry. The outer layer of the skin is a tough barrier composed of dead cells. Mucous membranes line the digestive tract, and the respiratory and urogenital tracts to bar entry. Oil and sweat glands secrete substances with an acidic pH to discourage bacterial growth. Tears have an enzyme called lysozyme which destroys bacteria trying to infect the eye. Cilia lining the nose and upper respiratory tract sweep out bacteria and viruses. So, with all these protections, how can we become infected with one of these pathogens? The following are methods by which microbes gain entry to the body.

- *Food*: Bacteria such as *E. coli* are found in some foods we eat and produce toxins which cause food poisoning. *Salmonella* can have the same effect (see page 533).

- *Water*: Polluting organisms found in water can cause diseases such as cholera or simply an upset digestive system.
- *Aerial*: Some organisms are carried by water droplets in the air. These droplets may be from a sneeze or cough of an infected person. We breathe in the droplets containing the pathogens. Diseases transmitted in this way are flu, colds and TB.
- *Animal vectors*: Insects and other animals can carry disease. Mosquitoes carry malaria and West Nile virus.
- *Puncture wounds*: Wounds break the skin barrier allowing the entry of bacteria or viruses. Both tetanus (bacterial) and rabies (viral) are transmitted by this method.
- *Sexual contact*: Sexual contact with an infected person can result in transmission of several diseases, as well as HIV.

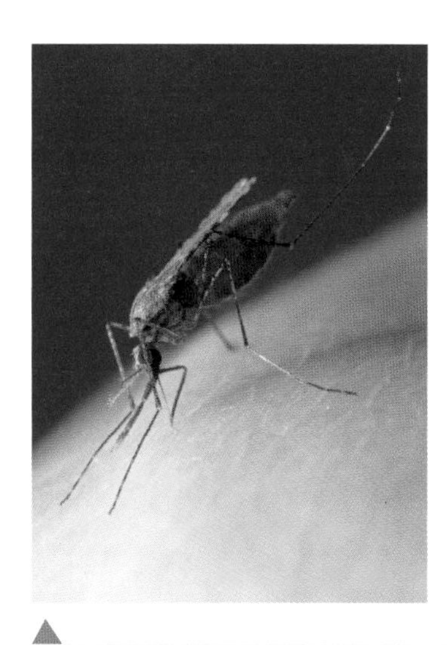

This illustration shows a mosquito on human skin, sucking blood. It has inserted its modified proboscis into a superficial blood vessel. Mosquitoes transmit diseases such as malaria, and West Nile virus in this manner.

Intracellular and extracellular infections

Every disease is a race between the pathogen and its host. Pathogens have developed several strategies for gaining a foothold in our bodies. To understand pathogens, we must understand how they have overcome the defences of the host. Most pathogens have developed strategies which allow them to live either intracellularly (in a host cell) or extracellularly (in the host but not inside the cells of the host).

Intracellular infections

Chlamydia is an example of an intracellular bacterium. It causes one of the most common sexually transmitted diseases.

These are the characteristics of an intracellular bacterium.

- Lives inside a cell of the host rather than travelling freely around the body. *Chlamydia* lives in the epithelial cells which line the genital tract. When it reproduces, it splits out of the cell and moves into the genital tract. When the new bacteria are in the genital tract the person is very contagious.
- Does not produce toxins. Since no toxins are produced to irritate tissues, the host may not be aware of being infected. In fact, 80% of women with *Chlamydia* are asymptomatic.
- Does not directly damage cells, but may cause long-term (chronic) problems. The long-term effects of *Chlamydia* are pelvic inflammatory disease (PID) and in some cases, infertility.
- Is not targeted by the immune system. *Chlamydia* is hidden in the cells of the genital tract.

Extracellular infections

Streptococcus is an example of an extracellular bacterium. *Streptococcus pyrogenes* is one of the most frequently met pathogens of humans. It is the leading cause of 'strep throat'. Between 5% and 15% of the population have it in their respiratory tract but show no sign of the disease. *Streptococcus* can affect us when our defences are low, following another primary infection. Normally, the mucus of the respiratory tract prevents penetration of the bacteria, but a previous infection makes us more susceptible to attack by this organism.

These are the characteristics of an extracellular bacterium.

- Lives in the host but outside a host cell. *Streptococcus* can rapidly spread and multiply outside of host cells.
- Produces toxins. *Streptococcus* produces toxins that kill host cells and produce a damaging inflammatory response, which we feel as a very sore throat.
- Damages cells. *Streptococcus* produces molecules called invasins which split open and dissolve host cells.
- Is targeted immediately by the immune system. *Streptococcus* stimulates the immune system to make antibodies. The antibodies begin fighting this bacterial infection. Treatment with antibiotics can help the antibodies win the battle.

Comparison

This table compares intracellular and extracellular pathogens of humans.

Chlamydia (intracellular)	*Streptococcus* (extracellular)
lives inside the cells of the host	lives in the host but outside cells
does not produce toxins	produces toxins
does not directly damage cells	directly damages cells
is not targeted by the immune system since it is hidden in the cells	is targeted by the immune system since it is freely circulating in the body

Endotoxins and exotoxins

Toxins are poisons which help bacteria win the race against the host.

Endotoxins: Lipopolysaccharides in the walls of a Gram-negative bacterium that cause fever and aches (e.g. *Salmonella*).

Exotoxins: Specific proteins secreted by bacteria that cause symptoms such as muscle spasms (as in tetanus) and diarrhoea (as in cholera) (e.g. *Clostridium tetani* and *Vibrio cholerae*).

Endotoxins

Salmonella is a bacterium which causes food poisoning. The reason you feel sick if you eat food contaminated with *Salmonella* is because each bacterium contains some endotoxin in its cell wall. The more bacteria present in the food, the more endotoxin is present to make you sick. The fever and ache are caused by the endotoxins.

Exotoxins

Exotoxins are very potent. If you receive a puncture wound and it becomes infected with tetanus bacteria, the toxin it produces could kill you. It causes severe muscle spasms. A tetanus shot gives you immunity to the toxin for 5–10 years.

The bacterium *Vibrio cholerae* produces cholera toxin which binds to the cells of the lower intestine in humans and results in very severe diarrhoea.

Methods to control bacterial growth

The control of microbial growth can be accomplished in two ways. The microbes can be killed or their growth inhibited. Control usually involves the use of a physical or chemical agent. Agents which kill bacteria are described as bactericidal. Those which inhibit the growth are described as bacteriostatic.

Here are four methods of bacterial control listed in order of effectiveness.

Irradiation

This is a bactericidal method. It destroys or breaks up the nucleic acids such as DNA and RNA.

- Gamma radiation (ionizing radiation) kills all microbes. It is a heat substitute and can be applied to a product after packaging. It is used for medical equipment and some food.
- Microwaves kill all bacteria due to the heat that is produced.
- UV radiation is the weakest of the irradiation methods. It kills all the bacteria but leaves the endospores.

Disinfectants

This is a bactericidal method. These chemicals kill bacteria but not their endospores. They are damaging to mucous membranes and skin but are useful for furniture, table tops and floors. Examples include bleach, detergents and Lysol.

Antiseptics

This is a bactericidal method. These mild chemicals are less effective than disinfectants but are not damaging to skin and mucous membranes. Examples include 50–70% ethanol and silver nitrate used in the eyes of newborns.

Pasteurization

This is a bactericidal method for pathogens but is bacteriostatic for non-pathogenic bacteria. Louis Pasteur discovered in the later 1800s that heating a liquid to a specific temperature for a specific period of time will ensure the destruction of all pathogens without changing the composition or nutritive value of the liquid. This process is commonly used today to maintain the quality and safety of milk and milk products. Pathogens and many spoilage bacteria are killed and the shelf-life of the milk is extended to 7–10 days. However, if not refrigerated, bacteria left in the milk can multiply and turn the milk sour. Minimum temperature and time requirements for pasteurization are based on studies of thermal death time for the most resistant pathogens which could be found in the milk: 63 °C (145 °F) for not less than 30 minutes or 72 °C (162 °F) for not less than 16 minutes are common temperatures and times for pasteurization.

Antibiotics

Everyone has heard of penicillin. It was the 'wonder drug' of the 1940s and since then has saved millions of lives. Do you know anyone who has never taken an antibiotic? Penicillin was the first antibiotic. Alexander Fleming is credited with its discovery in 1929. Fungus contaminating a plate of agar was inhibiting the growth of bacteria on the plate. The fungus was producing a molecule that stopped bacterial growth. Fleming had devoted much of his career to the study of wound infection, and recognized the importance of such a molecule. Eventually, he and two other scientists received the Nobel Prize for their work. Production on an industrial scale was developed in the 1940s. Many other antibiotics are available for use today. Most of them come from soil microbes.

Notes on the discovery of antibiotics by Fleming.

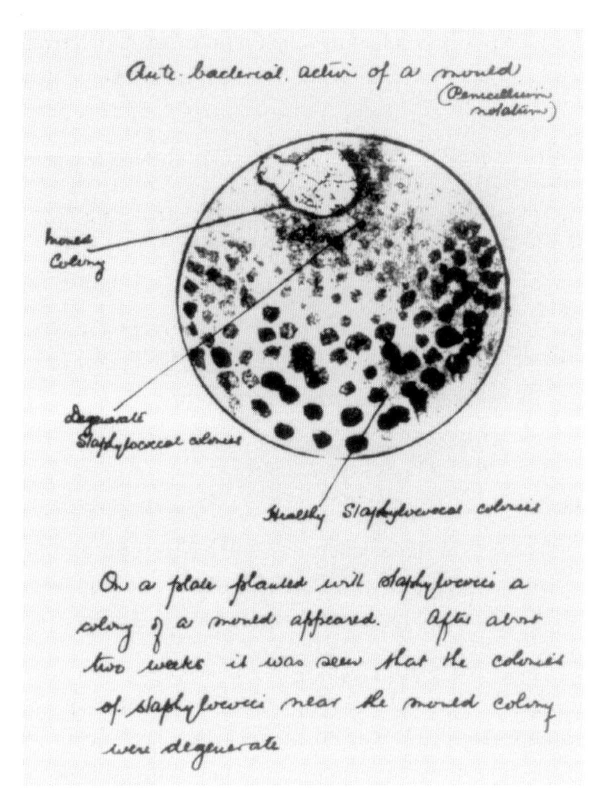

Antibiotic action

Antibiotics are defined as antimicrobial agents produced by microbes which inhibit or kill other microbes. Antibiotics act by one of the following three mechanisms.

- *Cell wall synthesis inhibition*: These antibiotics generally inhibit the production of peptidoglycan – a molecule in the cell walls of bacteria. Without the ability to make a cell wall, newly formed bacteria cannot survive. Examples of antibiotics which use this mechanism are penicillin and cephalosporin.
- *Protein synthesis inhibition*: These antibiotics attack the bacterial ribosome. An example is streptomycin which binds to the ribosomes of the bacteria and prevents protein synthesis. It does not attack the ribosomes of human cells because bacterial ribosomes (70S) are different from human ribosomes (80S).
- *Nucleic acid inhibition*: Some antibiotics affect the synthesis of DNA and RNA or can attach themselves to the DNA or RNA so the message cannot be read. Both of these actions interfere with the growth of bacterial cells. An example is rifampicin which is used against tuberculosis. Rifampicin acts by inhibiting RNA polymerase so that mRNA cannot be made.

An international discussion is ongoing as to whether food should be irradiated to kill pathogenic microbes. For two sources with opposite points of view, visit heinemann.co.uk/hotlinks, enter the express code 4242P and click on Weblinks 17.13a and 17.13b.

Epidemiology

Epidemiology is the study of the occurrence, distribution, and control of disease in a population.

- Epidemiologists study how many people have a certain disease (occurrence).
- They examine the regions of the country/world where it is occurring (distribution).
- They determine the best strategies to prevent its spread (control).

For example, each year in the US scientists from the Center for Disease Control (CDC) collect information on rabies. In 2001, 7437 cases of rabies were reported (occurrence). 40% of the reported cases were caused by raccoons. The raccoon population was most concentrated in the east coast region of the US (distribution). This scientific study raises awareness of a problem (raccoons) and demonstrates that solutions need to be found (control). Statistical studies of this type prevent epidemics from spiralling out of control. Not recognizing an epidemic may be part of what causes pandemics.

Why is it difficult to find the cause of an epidemic? In an epidemic we may have to settle for a correlation. Can we act upon a correlation? Does that cause its own problems?

There are global implications of emerging infectious diseases. Many infectious illnesses have emerged in recent years.

In 1997, a strain of avian flu began to kill otherwise healthy people.

Vanomycin, an antibiotic of last resort, has begun to be less effective against *Staphylococcus aureus*. *S. aureus* can cause life-threatening illness.

In the 1980s a resurgence of TB occurred due to high numbers of people infected with HIV.

In 1995, cases of Ebola haemorrhagic fever occurred in Zaire.

These are dangerous situations for the entire world. Emerging infectious disease could be targeted by foreign aid and global partnerships.

Spanish flu – one example of a pandemic

Pandemics are worldwide infections. Three pandemics occurred last century:

- 'Spanish flu' in 1918;
- 'Asian flu' in 1957;
- 'Hong Kong flu' in 1968.

Occurrence

The 'Spanish' flu was by far the deadliest, killing 40–50 million people; 'Hong Kong' flu and 'Asian' flu killed less than 2 million each.

A flu pandemic occurs when a new flu virus emerges and is transmitted easily between humans. Since the human immune system has no pre-existing immunity, symptoms develop rapidly and may be serious.

Distribution

In 1918, as World War I was coming to an end, something more deadly was emerging. During the next two years, the 'Spanish' flu would kill millions of people; it was most deadly for people between 20 and 40 years old. In the US, 28% of the population were infected. The virulence of this flu was extreme. Cases were reported of people developing symptoms and dying within hours. The flu spread across the globe. Humans carried it along trade routes and shipping lanes. The mass movements of military personnel associated with the war facilitated the spread of this disease.

The current hypothesis is that the origin of the 'Spanish' flu was in China. A rare genetic shift may have caused a recombination of the surface proteins of a virus creating a new virus unrecognized by the human immune system. Currently, virologists are studying the 1918 virus using tissues from the bodies of people killed during the 1918 pandemic.

The flu was called the 'Spanish' flu because of the large number of deaths in Spain. Eight million people were dead by May of 1918. However, the first wave of the flu may actually have occurred in Kansas in the spring of 1918 but gone unrecognized. If precautions had been taken then, the disaster may have been minimized if not averted. This is the value of ongoing epidemiological studies. The outbreak in Kansas is now called the 'first wave'. In the autumn and winter of 1918 the 'second wave' descended. Men were joining the army and with them came the flu. People were gathering for parades to support the war and spread the virus in these large groups. Men on both sides of the war front became too sick to fight. Eventually there was a shortage of doctors and nurses due to the combination of war wounded and flu victims.

Control

Public health officials attempted to control the spread of the flu. They concluded that the pathogen was carried in the air by coughing and sneezing. Gauze masks were issued and directed to be worn in public. Stores were directed not to hold sales where large numbers of people would gather together. Public gatherings were banned and public institutions closed. In the US, saloons, dance halls and movies were closed. In Britain, the pandemic was so severe that state-run elementary schools were closed. In army training camps, any one who got the flu was quarantined. In hospitals, flu patients were separated from the others.

Public education about hygiene and hand washing was begun. Posters were designed to educate the public. An important aspect of prevention was the use of disinfectants and sterilization methods. All bedding and rooms with flu patients were disinfected. Sheets were hung between the beds of patients in open wards to prevent the spread of droplets from coughing and sneezing. The gauze masks also acted to prevent droplets containing flu viruses from being inhaled.

After two years and millions of deaths, the flu disappeared as quickly as it had arrived. Scientists working with this pandemic realized that it would be necessary to develop vaccines against flu viruses to prevent future pandemics. Epidemiological studies of viruses emerging in different places in the world have begun. Pharmaceutical companies using the studies try to stay ahead of the travelling flu viruses by developing new vaccines. People at risk are encouraged to get their flu shot. All of this work is an attempt to prevent another flu pandemic.

Lytic cycle of a virus

The life cycle of an influenza virus at the cellular level is shown in Figure 17.16. A virus is a non-living particle. When it becomes attached to a host cell, it becomes an efficient machine. In the host cell, the virus commandeers the cell's machinery in order to replicate itself. Eventually, it will escape from the cell and go on to infect new cells. The host cell will not survive long after infection with a lytic virus.

Figure 17.16 Influenza: virus replication.

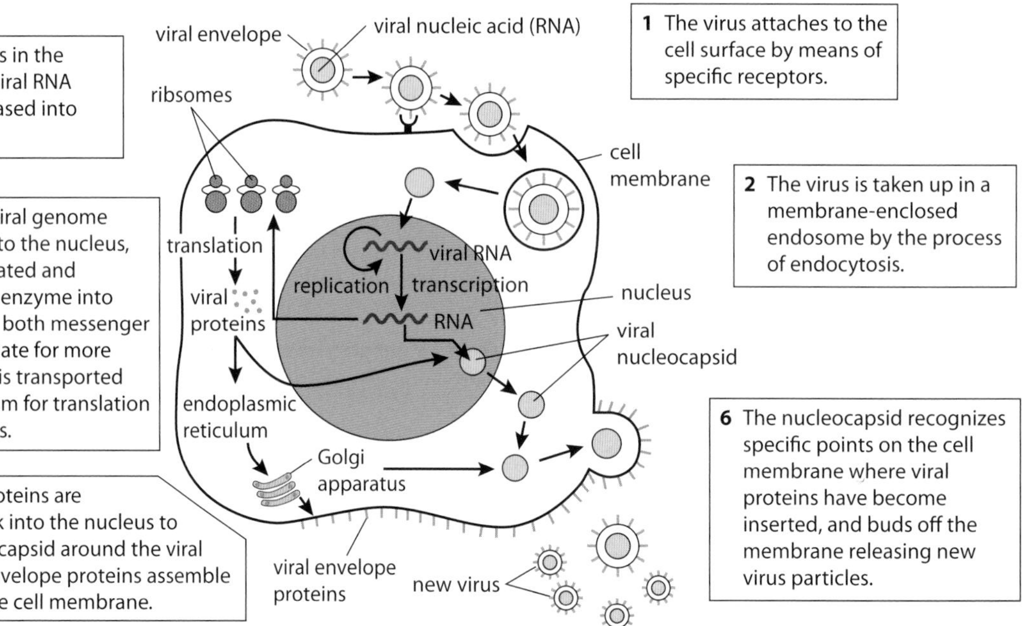

These steps explain the lytic life cycle of an influenza virus.

1 The virus attaches to receptors on the cell. A cell without correct receptors cannot be infected.
2 The virus is taken into an endosome by endocytosis (taking in a particle) into the cytoplasm.
3 The coat of the virus is removed and viral RNA enters the cytoplasm.
4 After transportation into the nucleus, viral RNA makes mRNA (transcription).
5 Some mRNA is transported to the cytoplasm where it makes viral proteins (translation) at the ribosomes.
6 Viral proteins are transported back to the nucleus to form a capsid around viral RNA (viral nucleocapsid).
7 The endoplasmic reticulum of the cell synthesizes viral envelope proteins in the cytoplasm.
8 Viral envelope proteins are packaged at the Golgi apparatus and transported to the cell membrane.
9 The viral nucleocapsid recognizes proteins on the membrane and buds off surrounded by the viral envelope proteins from the cell's plasma membrane.

Malaria

Bacteria and viruses are the most common pathogens. However, protozoa can also be pathogenic. Malaria is caused by a protozoon which is transmitted by a mosquito. Approximately 300 million people in the world are affected by malaria and 1 million die each year. Malaria is caused by four species of the genus *Plasmodium*.

Causes

- *Plasmodium falciparum*
- *Plasmodium vivax*
- *Plasmodium ovale*
- *Plasmodium malariae*

Transmission

Plasmodia are protozoan parasites. They are transmitted from one person to the next by a female *Anopheles* mosquito which feeds on blood. Male *Anopheles* mosquitoes do not carry disease since they feed only on plants. Plasmodia reproduce in the gut of the female mosquito. The egg sac ruptures and releases cells called sporozoites. These travel to the salivary glands of the mosquito. As the mosquito bites, the sporozoites enter the human bloodstream with the mosquito saliva. The sporozoites travel to the liver where they develop further. After a week or two, they burst out of the liver and invade red blood cells (Figure 17.17).

Symptoms

Infected patients present symptoms of anaemia, bouts of fever chills, shivering, pain in the joints, and headache. The cycle of infection begins again when a new mosquito bites a person with malaria. Patients can be treated with antimalarial drugs if they are available and given soon enough. Malaria is preventable. Lives can be saved if proper action is taken.

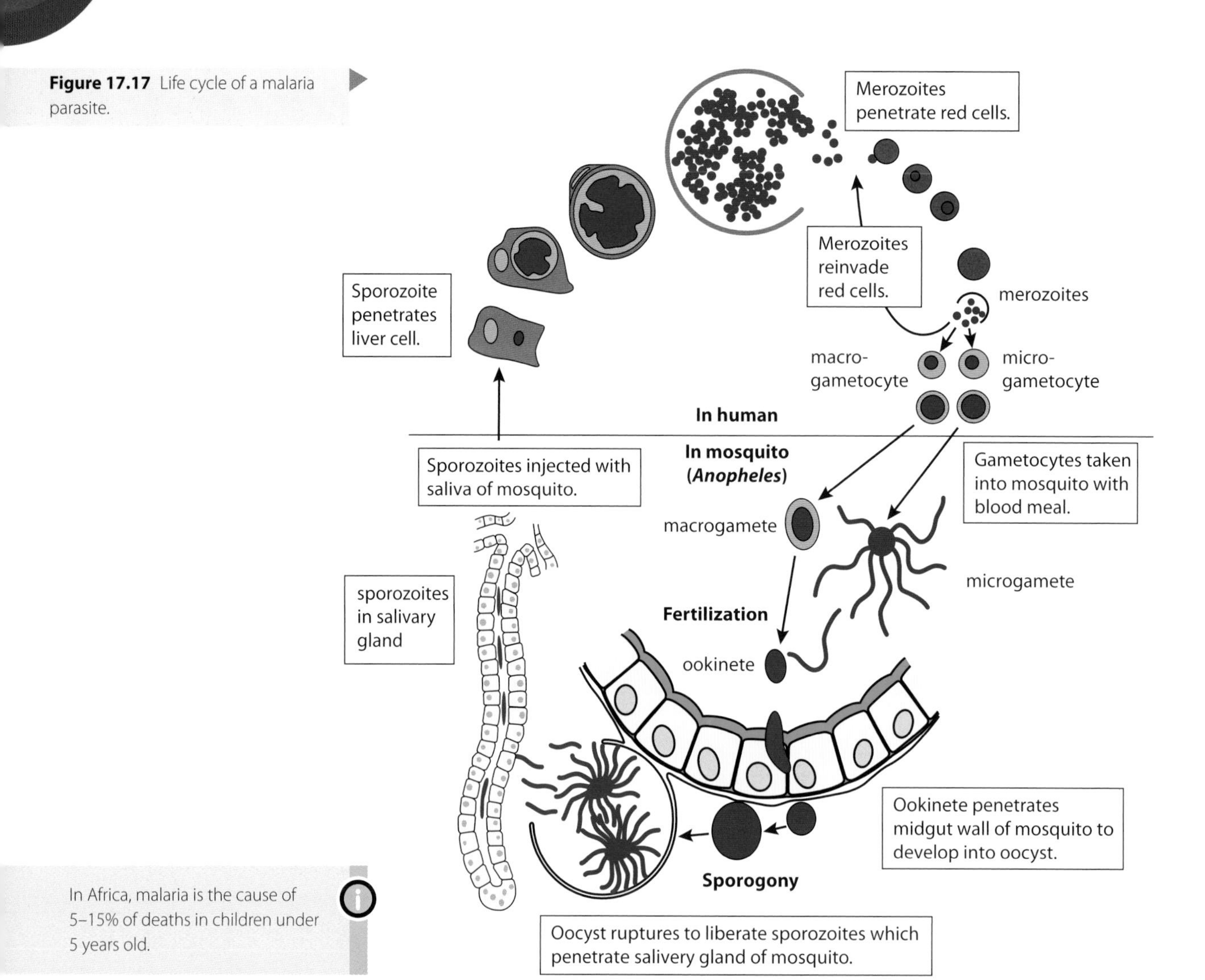
Merozoites
penetrate red cells.
Merozoites
reinvade
red cells.
merozoites
Sporozoite
penetrates
liver cell.
macro-
gametocyte
micro-
gametocyte
In human
In mosquito
(*Anopheles*)
Sporozoites injected with
saliva of mosquito.
Gametocytes taken
into mosquito with
blood meal.
macrogamete
microgamete
sporozoites
in salivary
gland
Fertilization
ookinete
Ookinete penetrates
midgut wall of mosquito to
develop into oocyst.
Sporogony
Oocyst ruptures to liberate sporozoites which
penetrate salivery gland of mosquito.

Figure 17.17 Life cycle of a malaria parasite.

In Africa, malaria is the cause of 5–15% of deaths in children under 5 years old.

Spongiform encephalopathies

Bacteria are cellular, living organisms with DNA as the nucleic acid which codes for all of the necessary proteins to control the bacterial activities.

Viruses are non-cellular and non-living so they must take over the machinery of a living cell in order to reproduce. Viruses are composed of DNA or RNA as the nucleic acid, and coated with a protein. The protein coat particular to each virus is coded for by its nucleic acid.

Then there are agents that cause transmissible spongiform encephalopathies (TSEs). The disease which sparked the current interest began affecting humans in about 1994. It seemed to be new in humans and was resulting in the death of previously healthy young people. The main symptom was holes formed in brain tissue which made it look like a sponge, hence the name spongiform encephalopathy. The symptoms had been seen before, but not in young humans.

It was recognized that the disease was similar to a disease called scrapie which is common in sheep and has been around for 200 years. However, scrapie was unknown in cattle or humans. Eventually it was recognised that scrapie had 'jumped the species barrier' and been transmitted to cattle in their feed. In cattle, the disease is called bovine spongiform encephalopathy (BSE). Humans then contracted the disease by eating contaminated beef or beef products. In humans, the disease is called variant Creutzfeldt–Jakob disease (vCJD). This meant that the new disease affecting humans could be listed as a (TSE). The following TSEs have now been identified:

- scrapie – found in sheep and goats;
- BSE – found in cattle ('mad cow disease');
- CJD – a dementia found in humans but very rare (1 in a million and affecting those over 65 years old);
- vCJD – a new form of CJD which affects young people, average age of 29.

The agent responsible for BSE or mad cow disease is a virus-like prion. There is a concern that beef and beef-related products eaten by people can cause a related disease named variant Creutzfeldt–Jakob disease in humans.

In order to fight bacteria and viral pathogens, scientists study their structure and design vaccines to protect people against them or their toxins. Consequently, the scientific community is now designing experiments to determine the structure of the agent causing vCJD. Once the structure is determined, designing a vaccine to protect against it will be the next step.

The prion hypothesis

The prion hypthothesis states that the causal agent for vCJD has no nucleic acid and consists only of protein. The virino hypothesis argues that vCJD is caused by a virus with a nucleic acid core and a protein coat protecting and hiding the nucleic acid.

For a debate on whether prions exist, visit heinemann.co.uk/hotlinks, enter the express code 4242P and click on Weblink 17.14.

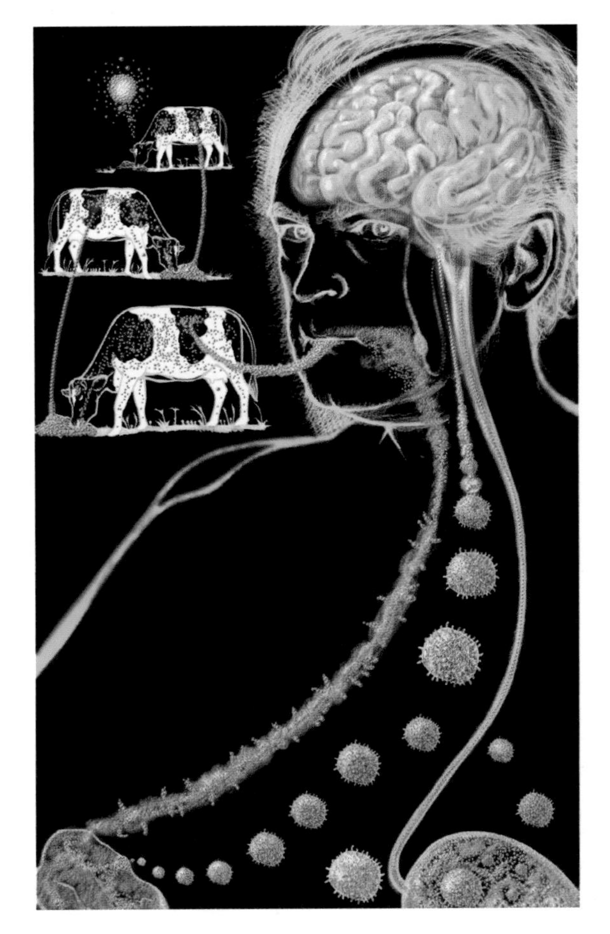

This artwork depicts the theoretical basis for the spread of prions from cows to humans. Prions are abnormal proteins that are the cause of BSE in cows, scrapie in sheep and vCJD in humans. They cause fatal brain and nerve degeneration. BSE prions spread as cows eat infected bovine material. Humans eat the infected cow and eventually the prion reaches the brain where it causes vCJD. Prions are an abnormal-shaped protein and cause normal proteins to flip to the abnormal shape.

This table compares the main ideas of the prion hypothesis with the more traditional virino hypothesis.

The virino hypothesis	The prion hypothesis
The infecting agent is a nucleic acid (DNA or RNA) surrounded by an abnormal protein coat. This is a similar model to the structure of a virus, hence the term 'virino'.	The infecting agent is a prion. This abnormal protein alone causes the disease.
Scientists have not found any DNA or RNA as part of the infecting agent. Why not? Maybe there is very little of the nucleic acid. Possibly we are not using the correct techniques. Maybe it has some unique feature which we are unaware of.	Scientists have searched for nucleic acids in prion particles for 30 years and not found any.
Based on currently accepted scientific models, nucleic acids must be present to code for the shape of proteins, e.g. DNA and RNA of viruses code for their specifically shaped protein coats.	Based on scientific experiments, abnormally shaped prions in a test tube can bind normal proteins and cause normal proteins to change to an abnormal shape. This argues against the model that only nucleic acids can code for the shape of proteins. Here a protein is 'coding' for the shape of a protein.
Conclusion: TSE diseases are best explained by the currently held hypothesis that a nucleic acid carries the information needed by the infecting agent. The nucleic acid could be part of a virus hidden by an abnormal protein. The name for the infecting agent is virino.	Conclusion: TSE diseases are caused by a prion which is only protein. No nucleic acid component has been found. The prion hypothesis best explains all of the observations about the diseases and the agents that cause them. This is a new paradigm.

How do both the prion and virino hypotheses illustrate the use of the scientific method? Explain which of these hypotheses is a paradigm shift.

Exercises

18 Distinguish between intracellular and extracellular bacterial infection.

19 Evaluate methods to control bacterial growth.

20 Describe the cause, transmission and effects of a disease caused by a protozoon.

21 Compare the prion hypothesis and the virino hypothesis for the cause of spongiform encephalopathy.

Practice questions

1 The World Health Organization (WHO) published a report on multi-drug resistant tuberculosis (MDR TB). MDR TB is defined as a disease caused by strains of *Mycobacterium tuberculosis* resistant to the two most important anti-TB drugs. It is largely a man-made phenomenon. The chart shows the cure and detection rate of tuberculosis (including MDR TB) in 26 countries, and a target zone defined by the WHO.

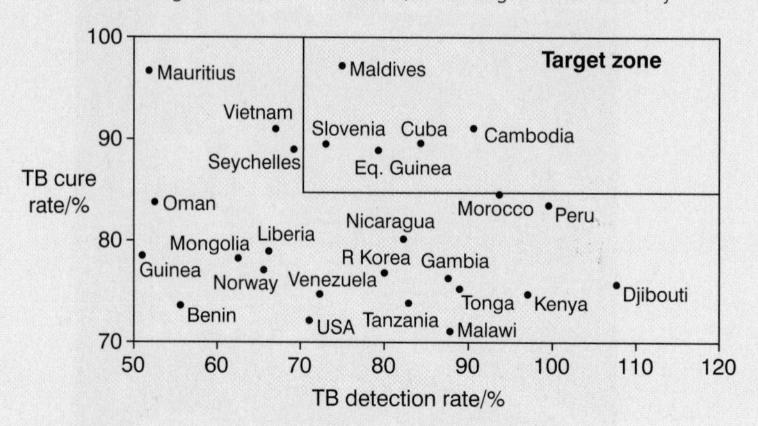

Source: WHO Report (1998), *Global Tuberculosis Control*, page 23

(a) Identify the country outside the target zone with (*1*)

(i) the lowest TB cure rate.

(ii) the highest TB cure rate.

(b) Calculate the percentage of all cases of TB in Benin that are cured. (*1*)

(c) Suggest two reasons why countries should aim to be in the target zone. (*2*)

(*Total 4 marks*)

2 Explain how methane is produced from biomass. (*6 marks*)

3 Compare the roles of *Rhizobium* and *Pseudomonas denitrificans* in the nitrogen cycle (*3 marks*)

4 It had always been assumed that eukaryotic genes were similar in organization to prokaryotic genes. However, modern techniques of molecular analysis indicate that there are additional DNA sequences that lie within the coding region of genes. Exons are the DNA sequences that code for proteins while introns are the intervening sequences that have to be removed. The graph shows the number of exons found in genes for three different groups of eukaryotes.

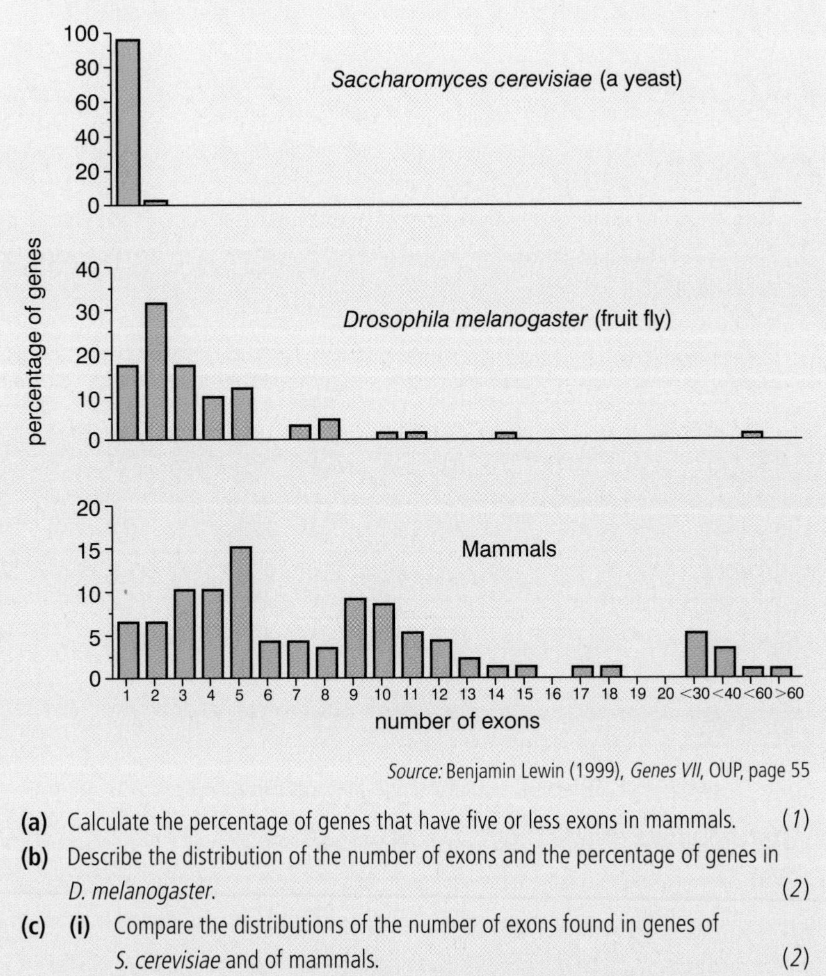

Source: Benjamin Lewin (1999), *Genes VII*, OUP, page 55

(a) Calculate the percentage of genes that have five or less exons in mammals. (*1*)

(b) Describe the distribution of the number of exons and the percentage of genes in *D. melanogaster.* (*2*)

(c) **(i)** Compare the distributions of the number of exons found in genes of *S. cerevisiae* and of mammals. (*2*)

(ii) Suggest one reason for the differences in the numbers of exons found in genes of *S. cerevisiae* and mammals. (*1*)

Human DNA has been analysed and details of certain genes are shown in the table below.

Gene	Gene size / kb*	mRNA size / kb	Number of introns
Insulin	1.7	0.4	2
Collagen	38.0	5.0	50
Albumin	25.0	2.1	14
Phenylalanine hydroxolase	90.0	2.4	12
Dystrophin	2000.0	17.0	50

* kilobase pairs

(d) Calculate the average size of the introns for the albumin gene. *(2)*

(e) Analyse the relationship between gene size and the number of introns. *(2)*

(Total 9 marks)

5 Lectins are proteins that some plants synthesize and store in their cells. Lectins have properties that make plant tissues unpalatable to insects and nematode worms. Crop plants that do not naturally produce lectins have been genetically modified to produce them. For example, genes coding for two types of lectin have been transferred to potatoes (*Solanum tuberosum*). One of the genes (GNA) was obtained from snowdrop plants (*Galanthus nivalis*) and the other (Con A) from jackbeans (*Canavalia virosa*). To obtain a series of genetically modified varieties, gene transfer was carried out repeatedly on one type of potato (Désirée). The lectin content of the leaves and the level of control of aphids and nematodes were measured in each genetically modified potato variety. The table below shows the results. The figures for control are the percentage reduction in the number of aphids and nematodes that fed on the plants, compared with the unmodified potato plants.

Genetically modified potato varieties	Lectin content/% of total soluble leaf protein	Aphid control/%	Nematode control/%
GNA pBG650	0.600	13	22
GNA 2#28	0.600	49	17
GNA 71	0.320	29	38
GNA 74	0.340	42	22
Con A 31	0.024	48	0
Con A 4	0.044	41	37

Source: Griffiths, Geoghegan and Robertson (2000), *Journal of Applied Ecology*, **37**, pages 159–70

(a) Compare the lectin content of the GNA varieties with the Con A varieties. *(1)*

(b) Identify, with a reason, the most promising variety for control of both aphids and nematodes. *(2)*

(c) Discuss the relationship between the lectin content of the different varieties and the level of control of aphids and nematodes. *(2)*

The use of genetically modified crop varieties is controversial and some biologists have suggested that other species might be harmed. Trials have been done to test whether lectin-producing plants affect species other than the pests of the crops. In one of these trials, replicate 5 g samples of sandy loam soil were mixed with 1 g samples of coarsely chopped leaves from genetically modified Désirée potato plants. Control experiments were also set up in the same way using chopped leaves from Désirée potato plants that were not genetically modified. After fourteen days, nematodes feeding on soil bacteria and two types of protozoa (flagellates and amoebae) were extracted from the leaf–soil mixtures. The amounts of these three different species were determined.

Mean results for the two genetically modified varieties used in the trials and control results are shown in the bar chart below.

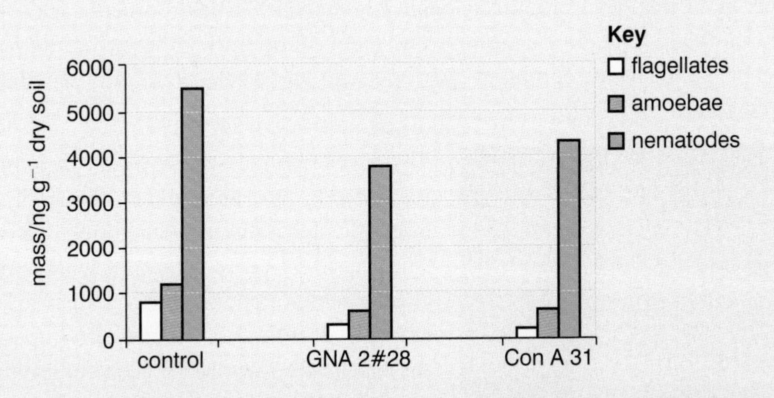

Source: Griffiths, Geoghegan and Robertson (2000), *Journal of Applied Ecology*, **37**, pages 159–70

(d) The mean mass of flagellates in the GNA 2#28 trial was 63% lower than the control. The mean mass of amoebae in the GNA 2#28 trial was 51% lower than the control.
Calculate how much lower the mean mass of nematodes in the GNA 2#28 trial was than the control. Show your working. *(2)*

(e) Using the data in the bar chart, discuss whether the effects of the genetically modified potato plants are more significant on protozoa or on nematodes. *(3)*

(Total 10 marks)

18 Ecology and conservation (Option G)

Introduction

A community is a group of populations living together and interacting with each other in an area. The community might be named by an environmental feature; for example, a sand dune community or pond community. Other communities may be named by the dominant plant species, such as oak community or redwood forest community.

The female green turtle comes ashore to lay her eggs. Hatchlings must find their way back to the ocean. Lights from cars near the beach can distract them.

The distribution of organisms in communities is affected by both abiotic (non-living) and biotic (living) features. We are interested in studying all factors to determine what may affect a certain population of organisms. For example, does the fact that cars drive up to the beach with their lights on, affect sea turtle reproduction at that beach? Is the rabbit population in a forest decreasing due to an increase in the fox population?

The biosphere means all parts of the Earth where organisms live, whether that is the Earth's crust or the atmosphere. The biosphere is divided into seven biomes – each is defined by its particular vegetation and animal life. So to maintain diversity, we need to understand and maintain the different biomes.

Biodiversity is our 'backup' – it is the reserve needed to survive a disaster. A disaster could wipe out the majority of organisms but leave the few which are the best adapted. In order to know how much biodiversity exists in an ecosystem, we must be able to measure it. The Simpson Index of biodiversity is one method of measuring diversity.

There are, generally speaking, two strategies for survival: having huge numbers of offspring which are left to fend for themselves, or having far fewer offspring which are carefully nurtured. These are respectively called the r-strategy and the K-strategy. Of course, there are many organisms on the continuum between these two extremes. Which strategy any particular organism employs depends on environmental conditions. Again we need to know and understand the interactions between the environment and organisms (the ecology) to be able to make conservation policies work.

G.1 Community ecology

Assessment statements

G.1.1 Outline the factors that affect the distribution of plant species, including temperature, water, light, soil pH, salinity and mineral nutrients.

G.1.2 Explain the factors that affect the distribution of animal species, including, temperature, water, breeding sites, food supply and territory.

G.1.3 Describe one method of random sampling, based on quadrat methods, that is used to compare the population size of two plant or two animal species.

G.1.4 Outline the use of a transect to correlate the distribution of plant or animal species with an abiotic variable.

G.1.5 Explain what is meant by the niche concept, including an organism's spatial habitat, its feeding activities and its interactions with other species.

G.1.6 Outline the following interactions between species, giving two examples of each: competition, herbivory, predation, parasitism and mutualism.

G.1.7 Explain the principle of competitive exclusion.

G.1.8 Distinguish between fundamental and realized niches.

G.1.9 Define *biomass*.

G.1.10 Describe one method for measurement of biomass of different trophic levels of the ecosystem.

Factors affecting distribution of plant species

When you are hiking on a sand dune, it may not seem obvious that the dune is a community of plants and animals which are interacting with the environment and each other. Not all dunes are the same: there is a progression of conditions from the hot dry foredune through to the shady cool conditions of mature forest growing on a mature dune (see Figure 18.1).

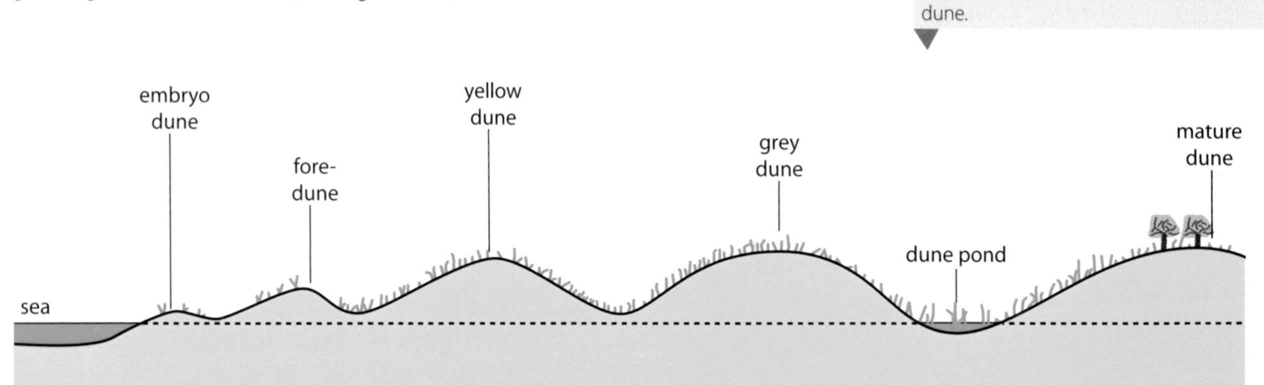

Figure 18.1 Transect of a coastal sand dune.

To see photographs and a description of the coastal sand dunes of the UK, visit heinemann.co.uk/hotlinks, enter the express code 4242P and click on Weblink 18.1.

Let's consider the sand dune community to illustrate how abiotic factors (temperature, water light, soil pH, salinity and mineral nutrients) can affect the distribution of plants.

Temperature and water

The temperature on the foredune can be very hot in the summer and there is little water. Marram grass is well adapted to these conditions. It is commonly found on the foredunes around the British coast and at the shores of Lake Michigan in the US. It has long roots which find water even in very dry sand and its long narrow leaves can curl up to save water and resist heat up to about 50 °C (120 °F).

The temperature on the mature dune is much cooler and conditions more moist, so the variety of plant species is more numerous. A common plant which lives on the forest floor of the mature dune is the fern. Ferns have adaptations to live in the low temperatures and moist conditions of the mature dune.

Light

Marram grass must live in conditions where sunlight is constantly available. It does not have wide leaves adapted to catch sunlight. Its leaves are adapted to reduce water loss and withstand heat. Marram grass is found in sunny areas of the foredune. Ferns are found in the shady areas of the mature dune. They have wide leaves to capture the small amount of light which filters though the leaves of the large oak trees on the mature dune.

Soil pH

In the yellow dune, the soil pH is about 7.5. Marram grass, still the dominant vegetation on this type of dune, thrives at this pH. In the grey dune, soil has formed from the decomposition of grasses over the years. This is an older dune with more soil. The pH of this soil is more acid. Acid-loving heathers grow well here.

Salinity

Foredunes catch salt spray from the ocean. Marram grass and Lyme grass live comfortably in this salty environment. On the grey dune, where conditions are much less salty, we begin to see small shrubs, mosses and lichen.

Mineral nutrients

The grey dune shows some diversity of plants. It is older than the yellow dune and foredune and contains some mineral nutrients in the soil. These mineral nutrients can support small shrubs, mosses and lichen. On a mature dune, several hundred metres inland from the shore, there is a thick layer of soil full of mineral nutrients. Since this is the oldest dune, the soil has been building up here for hundreds of years. Mineral nutrients here can support large trees such as ash, birch and eventually oaks.

To view the habitats found at the Indiana Dunes, visit heinemann.co.uk/hotlinks, enter the express code 4242P and click on Weblinks 18.2a and 18.2b.

Factors affecting distribution of animal species

The distribution of animal species is also affected by a range of abiotic factors. Continuing to use the dunes as a model, let's look at some of the factors affecting animals in the different habitats of the Indiana Dunes in the US.

Temperature

Some animals have adaptations which allow them to live in very hot conditions. A sand wolf spider, which lives in the foredune, is adapted to the extreme high temperature by living in a burrow deep in the sand. Its adaptation to the foredune is a behavioural adaptation. Woodland spiders, which live in trees in the mature dune, would die at such high temperatures.

Water

Many animals are specifically adapted to live in wetlands. Wetlands are formed as interdunal ponds (dune slacks). Many animals depend on these wetlands for survival: some need the water for their eggs, others rely on the ponds for food. The blue heron lives here and catches small fish and frogs in these wetlands. Blue herons are not found on the sandy shores of the foredune. On the other hand, woodpeckers are found in the mature dunes and eat the insects living in the trees. They do not need the water of the interdunal ponds to survive.

The great blue heron makes its home in the interdunal ponds. Conservation of dune areas ensures nesting sites are available for these birds.

Breeding sites

Animals have specific needs for breeding sites. Habitat destruction can interfere with their ability to reproduce. Blue herons breed around the interdunal wetlands where food is regularly available. If these wetlands were not present, the heron would not survive. Interdunal ponds are in between the dunes and (unlike the foredune) are protected from the Sun and wind. In the mature dunes, the nesting sites of woodpeckers are found in the branches of the oak trees.

Food supply

Many animals are adapted to feed on specific food and must live where that food supply is available. Other animals are more wide-ranging. Some mammals and raccoons which live in the dunes have the ability to go wherever the food supply is located. At the Indiana Dunes, raccoons, skunks and foxes wait in the dark to search the shoreline for food. Fish such as alewives are often washed up on the shore. These animals might also travel up to the mature dune to feast on eggs in an unprotected nest. Rabbits build their burrows deep in the sand of the foredune. They live near the marram grass which they eat.

Territory

Some animals live in a specific area. For example, packs of coyotes mark their territory with scent. The male coyote raises its leg and urinates to mark the area which belongs to the pack. These territories separate groups of coyotes from each other. Studies have shown that the territories do not overlap.

Random sampling

Suppose you want to determine how the population of jack pines compares to the population of oak trees on the Indiana Dunes. You could count every tree but that would be incredibly time-consuming. Ecologists use a sampling method. They take a random sample and use it to estimate the total number of organisms. Samples must come from all around the habitat. In a truly random sample, each organism has an equal chance of being selected for the count.

The quadrat method

A quadrat is a square of a certain size. Organisms within the quadrat are counted and these counts are used to determine the population size. In the case of jack pines and oak trees, the quadrats need to be large. If you are counting small plants, the quadrats could be smaller.

Here is a summary of the method.

- Map the entire area of the dune.
- Determine the size of the quadrats.
- Place a grid with numbers over the map of the dune area (see Figure 18.2).

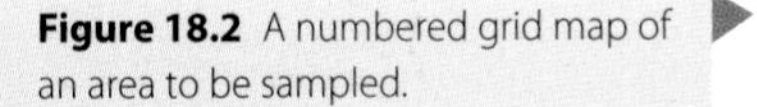

Figure 18.2 A numbered grid map of an area to be sampled.

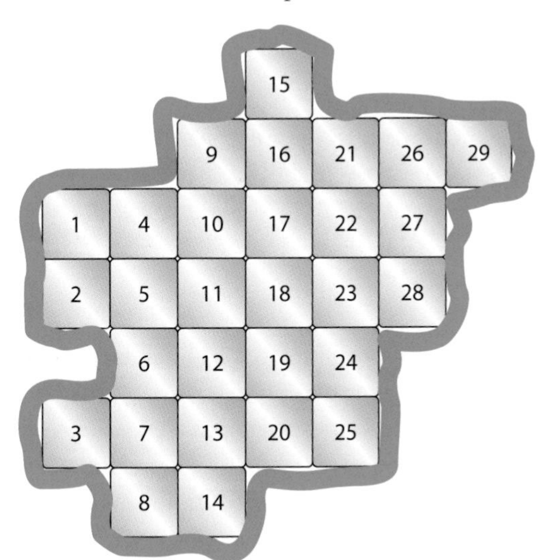

- Choose which squares (quadrats) to sample using a random number table.
- Count the number of jack pines in each sample quadrat.
- Count the number of oak trees in each sample quadrat.
- Calculate the average number of jack pines in your sample quadrats.
- Calculate the average number of oak trees in your sample quadrats.
- Multiply the average number of jack pines by the total number of quadrats on your map to get an estimate of the number of jack pines on the pine dune.
- Multiply the average number of oak trees by the total number of quadrats on your map to get an estimate of the number of oak trees on the pine dune.

The transect

When you look at the sand dune community, you are looking at living history. Dunes nearest the coast are the youngest dunes and the older dunes are inland. Walking from the beach (strand) back to the mature dune is like walking through time. At the beach, only a few plants are present on the foredune. As you walk inland, the dunes have increasingly more diverse communities. The oldest dunes farthest away from the beach have had hundreds of years to develop layers and layers of soil and vegetation.

The transect technique is commonly used for studying how the distribution of plants in an ecosystem is affected by abiotic factors. Here is how it works.

- At right angles to the sea, lay a tape in a line all the way up the dunes.
- Every 10 to 20 metres along the tape, mark out a quadrat always using the same size.
- Identify and count the plant species of interest in each quadrat.
- Measure the abiotic feature that you have chosen in each quadrat (e.g. temperature, light, soil pH, water, mineral nutrients).
- You can now determine the pattern of distribution of plant species from the youngest to the oldest dune and see if it correlates with the abiotic factor you chose.

When do you use random numbers to choose quadrats and when do you use a transect? It all depends on whether the habitat is uniform – perhaps a field or woodland – or displays an environmental gradient – as in the progression of sanddunes from coast to forest, or a beach from the low-water to the high-water mark.

To read about students performing a transect in the sand dunes visit heinemann.co.uk/hotlinks, enter the express code 4242P and click on Weblink 18.3.

The niche concept

Every organism in an ecosystem has a particular role in that ecosystem. That is the organism's niche. The concept of niche includes where the organism lives (its spatial habitat), what and how it eats (its feeding activities) and its interactions with other species.

Spatial habitat

Every type of organism has a unique space in the ecosystem. The area inhabited by any particular organism is its spatial habitat. The ecosystem is changed by the presence of the organism. For example, green frogs live in the ponds of the Indiana Dunes. They burrow in the mud in between the grasses on the edge of the pond.

Feeding activities

The feeding activities of an organism affect the ecosystem by keeping other populations in check. For example, green frogs eat the aquatic larvae of mosquitoes, dragonflies and black flies. The presence of the green frog helps keep the populations of these insects in check.

Interactions with other species

The interactions of an organism with other species living in its ecosystem include competition, herbivory, predation, parasitism and mutualism. The organism may be in competition with another organism for the food supply. It may itself be the prey for a larger predator. It may harbour parasites in its intestines. These complicated interactions are difficult to discover, but they indicate the importance of the organism in the ecosystem. The predator of the green frog is the blue heron. Without the green frog in the sand dune ecosystem, the heron would have a significantly reduced food supply. Frogs are homes for flatworm parasites which live in their intestines. Without doubt there are many other relationships between the green frog and other species.

An important part of an organism's niche is its habitat. According to the World Wide Fund for Nature (WWF), habitat loss is the greatest threat to biodiversity on our planet today.

To find out more about habitat loss around the world, visit heinemann.co.uk/hotlinks, enter the express code 4242P and click on Weblink 18.4. Search under 'species' and then 'habitat loss'.

One of the jobs of an ecologist is to collect data on the niches of particular organisms in the ecosystem. If an organism is in danger of becoming extinct in an ecosystem, it is necessary to understand as many interactions as possible to attempt to determine the cause. What follows are explanations and some examples of interactions between species.

Competition

When two species rely on the same limited resource, one species will be better adapted than the other to benefit from the resource.

The habitat of the red fox is disappearing.

The coyote is competing with the red fox for a small food supply. Removal of forests and creation of farmland has eliminated some of their food supply.

- *Example 1:* Coyotes and red foxes are both predators which eat small rodents and birds. Coyotes inhabit grassland communities in the US while the red fox prefers the edges of forests and meadows. Since more farmland has been created and more forests removed, the habitat of the red fox is disappearing and is overlapping with that of the coyote in the grasslands. The two species are competing for a smaller food supply and it is possible that one will become extinct in that habitat.
- *Example 2:* In the coastal dunes of the UK, the natterjack toad (*Bufo calamita*) is facing tough competition from the common toad (*Bufo bufo*). Disturbance of the dune area is limiting the habitat required by both toads.

To see pictures and read about the competition facing the natterjack toad visit heinemann.co.uk/hotlinks, enter the express code 4242P and click on Weblink 18.5.

Herbivory

A herbivore is a primary consumer (plant eater) feeding on a producer (plant). The growth of the producer is critical to the well-being of the primary consumer. This is an interaction between plants and animals.

- *Example 1:* Rabbits eat marram grass in the sand dune ecosystem.
- *Example 2:* Monarch butterfly larvae eat the leaves of the milkweed plant.

Predation

A predator is a consumer (animal) eating another consumer (animal). One consumer is the predator and one is the prey. The number of prey affects the number of predators and vice versa.

- *Example 1:* The Canadian lynx and the arctic hare form a classic example of predator–prey interaction. The lynx preys on the hare. Changes in the numbers of the lynx population are followed by changes in the numbers of the hare population.
- *Example 2:* The blue heron is a predator on frogs in the ponds of the sand dune ecosystem.

Parasitism

A parasite is an organism which lives on or in a host and depends on the host for food for at least part of its life cycle. The host is harmed by the parasite.

- *Example 1: Plasmodium* is a parasite which causes malaria in humans. It reproduces in the human liver and red blood cells. Part of the life cycle takes place in the body of the *Anopheles* mosquito. The mosquito is the vector which transmits the malaria parasite from person to person.
- *Example 2:* Leeches are parasites which live in ponds. Their hosts are humans or other mammals. Leeches puncture the skin of a host and secrete an enzyme into the wound to prevent clotting. Leeches can ingest several times their weight in blood.

Mutualism

Two species living together where both organisms benefit from the relationship is termed mutualism.

- *Example 1:* Lichen is a mutualistic relationship between algae and fungi. The algae photosynthesize and make carbohydrates (food). The fungi absorb mineral ions needed by the algae.
- *Example 2: Rhizobium* is a nitrogen-fixing bacterium living in the roots of leguminous plants such as beans and peas. *Rhizobium* fixes nitrogen which the plant can then use to make proteins. The plant also makes carbohydrates (during photosynthesis) which can be used as food by *Rhizobium*.
- *Example 3:* Clownfish and sea anemones live together for mutual benefit. Clownfish are small brightly coloured fish which live within the area of the tentacles of the poisonous sea anemone. The clownfish is covered with mucus which protects it from the sting of the sea anemone. Clownfish lure other fish to the waiting tentacles of the sea anemone. After the sea anemone consumes the fish, the clownfish eats the remains. The clownfish also nibble off the remains of dead sea anemone tentacles.

W For pictures and information on clownfish and sea anemones, visit heinemann.co.uk/hotlinks, enter the express code 4242P and click on Weblink 18.6.

Competitive exclusion

You will recall that the red fox and coyote may now be in competition. They seem to both be finding their food in the same area and the food supply may be dwindling due to forests and grasslands being turned into farmland. If the fox and the coyote do begin to occupy the same niche in the ecosystem, the principle of competitive exclusion may be used to predict the end result.

The principle of competitive exclusion states that no two species in a community can occupy the same niche.

In 1934, the competitive exclusion principle was demonstrated by a Russian ecologist, G. F. Gause. He performed a laboratory experiment with two different species of *Paramecium*: *P. aurelia* and *P. caudatum* (see Figure 18.3). His experiments showed the effects of interspecific competition between two closely related organisms. When each species was grown in a separate culture, with the addition of bacteria for food, they did equally well. When the two were cultured together, with a constant food supply, *P. caudatum* died out and *P. aurelia* survived. *P. aurelia* out-competed *P. caudatum*. The experiment supported the Gausian hypothesis of competitive exclusion. When two species have a similar need for the same resources, one will be excluded. One species will die out in that ecosystem and the other will survive. *P. aurelia* must have had a slight advantage which allowed it to out-compete *P. caudatum*.

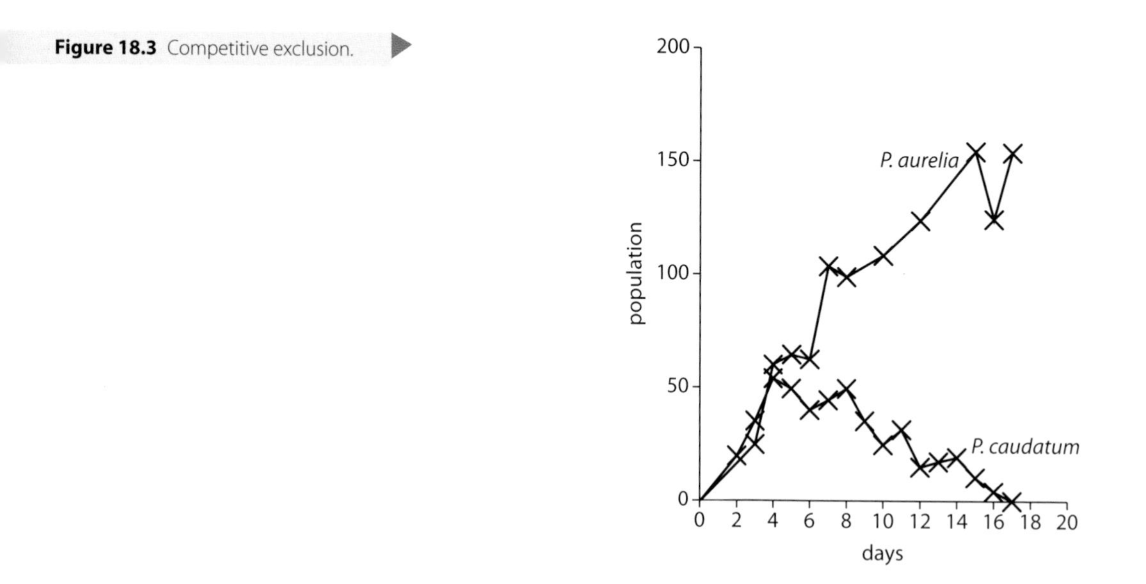

Figure 18.3 Competitive exclusion.

Fundamental niche versus realized niche

The red fox habitat is the forest edge. Its food consists of small mammals, amphibians and insects. It interacts with other species like the mosquitoes which suck its blood and scavengers which eat its leftovers. This is the fundamental niche of the red fox. The fundamental niche is the complete range of biological and physical conditions under which an organism can live.

What has happened to the fundamental niche? The forest edge has been turned to farmland in many places. Some of the species eaten by the red fox have disappeared. The red fox must survive in a narrower range of environmental conditions. Now there is direct competition from the coyote whose niche has also been changed. This new and narrower niche is called the realized niche.

Definition of fundamental niche

The fundamental niche of a species is the potential mode of existence, given the adaptations of the species.

Definition of realized niche

The realized niche of a species is the actual mode of existence, which results from its adaptations and competition with other species.

Biomass

Biomass is the total mass of organic matter. Organic matter consists of carbon compounds such as carbohydrates, lipids and proteins. Since matter usually also includes water, which is not organic, the material has to be dried. Biomass is measured as dry mass of organic matter of living organisms. Biomass is expressed in g m^{-2} yr^{-1} (grams per metre squared per year).

Measuring biomass at each trophic level

You may wonder about the difficulty of finding the dry mass of living plants and animals. Fortunately, tables and charts are available which tell you the biomass of an animal according to its size or weight. For example, you might trap and weigh a raccoon, then find its biomass in a table. The raccoon is returned to the ecosystem. There are also tables for plant species. It is not easy to determine the weight of a tall tree, but ecologists have designed a way to solve this problem too. Here is how biomass of an ecosystem is measured in a terrestrial community.

- Measure the total area of the ecosystem (e.g. a forest).
- Divide the ecosystem into small areas. The forest can be divided into grids or plots and each plot marked with a stake carrying a number.
- Choose one plot to sample.
- Measure the size of each plant species, including trees (height and diameter) and low-growing vegetation. Cut down the all the trees and vegetation on that plot (see Figure 18.4).

Figure 18.4 Sampling the vegetation in a forest.

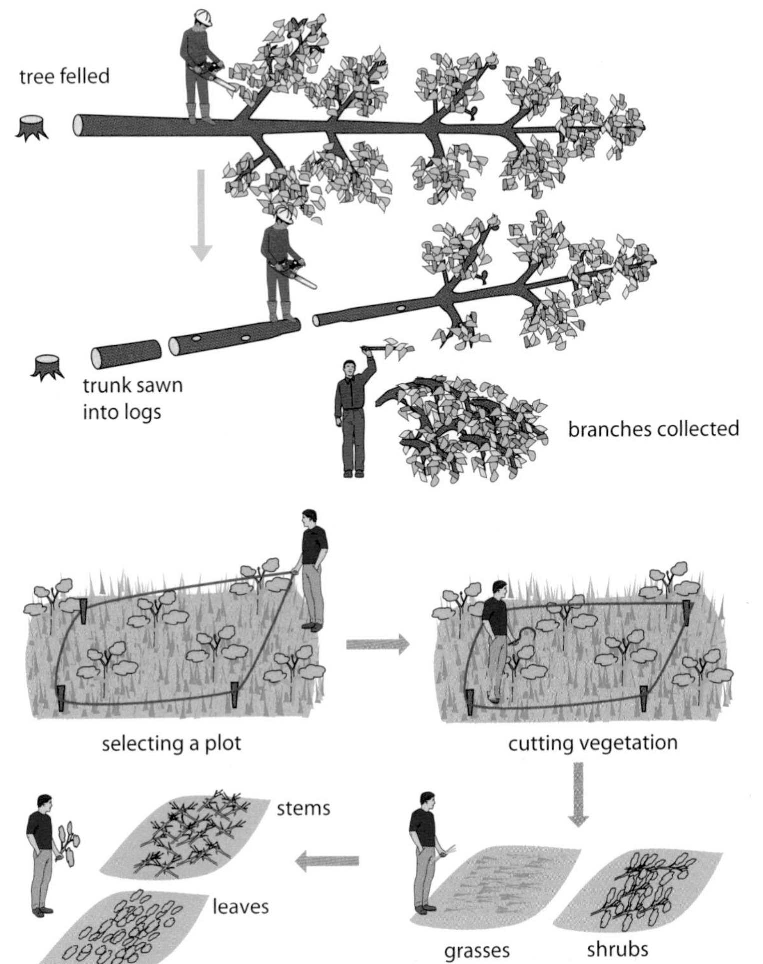

For photographs and specific mathematical models used to measure biomass in a forest ecosystem, visit heinemann.co.uk/hotlinks, enter the express code 4242P and click on Weblinks 18.7a and 18.7b.

- After measuring or counting organisms, we may fail to return them to the same ecosystem. Is there a moral principle involved here?
- In order to measure biomass, destructive techniques are used. Trees are cut down and plants are destroyed. Is this unethical?
- Could the destructive sampling techniques described above be explained as 'moral relativism'?
- Suppose a famous scientist states that he must cut down an entire forest to determine its biomass. He further states that this data is extremely important to his research into global warming. What would Kant say about that argument?

- Dry all the plant samples in a circulating drying oven at 90 °C (195 °F).
- Use a mathematical model to show the relationship between weight and height of each plant species and its biomass.
- Sample the other plots by measuring the size and height of plants. Cutting down and drying is unnecessary because the mathematical model can be used to find biomass.
- For the animals in the ecosystem, set traps in the plot and weigh and measure the organisms caught. Use tables to determine their biomass.
- Average the data for all species per plot.
- Multiply the average per plot by the number of plots in the ecosystem to discover the biomass of the entire ecosystem.
- Repeat this procedure seasonally or yearly to study changes of the biomass in a forest community over time.

Exercises

1 Explain the factors which affect the distribution of plant species in a sand dune ecosystem.
2 Describe the factors which affect the distribution of animal species in a sand dune ecosystem.
3 Explain the niche concept with reference to a sand dune ecosystem.
4 Outline the concept of competitive exclusion.
5 State what is meant by biomass.

G.2 Ecosystems and biomes

Assessment statements

G.2.1 Define *gross production*, *net production* and *biomass*.
G.2.2 Calculate values for gross production and net production using the equation:

gross production – respiration = net production

G.2.3 Discuss the difficulties of classifying organisms into trophic levels.
G.2.4 Explain the small biomass and low numbers of organisms in higher trophic levels.
G.2.5 Construct a pyramid of energy giving appropriate information.
G.2.6 Distinguish between primary and secondary succession, using an example of each.
G.2.7 Outline the changes in species diversity and production during primary succession.
G.2.8 Explain the effects of living organisms on the abiotic environment with reference to the changes occurring during primary succession.
G.2.9 Explain how rainfall and temperature affect the distribution of biomes.
G.2.10 Outline the characteristics of the major biomes.

Energy flow through the ecosystem

What do you think is the direction of energy flow for any ecosystem? If you constructed a food chain like this one, then you know.

grass → cow → human

Plants are at the bottom of the food chain. They contain the highest amount of energy which they obtain from sunlight. The source of energy for most ecosystems is the Sun. A few food chains are supported by bacteria which can trap chemical energy.

Why is only 5–20% of the Sun's energy that is trapped by plants transferred to the primary consumers eating the plants? Because 80–95% of the energy is lost as heat or used for maintenance by the plant. Energy is lost as heat as it moves from producer (e.g. grass) to primary consumer (e.g. cow) to secondary consumer (e.g. human).

This is the same reason that the fuel we put in a car is only partially used to run the car. A large percentage of the energy from the fuel is lost as heat. Why is there a fan in the engine of a car? A law of physics called the Second Law of Thermodynamics states that when energy is transferred, a proportion of it is lost as heat energy. This law applies equally to cars and ecosystems as we shall see later.

Where is the energy from the Sun actually kept in the plant? You will recall that plants produce glucose during photosynthesis. Plants also break down the glucose molecule and use the energy for maintenance. The breakdown is called respiration. Maintenance activities which need energy are growth, repair, and reproduction. When the glucose is used as fuel for these activities, some of the energy is lost. About 80–95% of the Sun's energy originally trapped by the plant is accounted for by plant maintenance and heat loss. That means that only 5–20% of the energy in producers is transferred to the consumers. Some of the energy moves through the ecosystem as excretion. Energy is left in undigested food and is passed on to decomposers. When the organism dies, its body is decomposed and the energy transferred to decomposers.

Gross production, net production and biomass

Pyramids of energy show how much energy is left at each trophic level (see Figure 18.5). Each block in the pyramid represents a trophic level (producers, primary consumers, secondary consumers, tertiary consumers). The width of the block indicates how much energy it contains. At each level, the blocks get narrower and the block at the top is very narrow. The number at each level represents the amount of energy at each level. Can you see that only 10% of the energy from one trophic level is transferred to the next level? This diagram represents the ideal situation. In an actual ecosystem, the percentage transfer from one level to the next depends many factors and may vary between 5% and 20%. In animal husbandry (farming) the transfer value is often higher than 10%. However, the loss of energy between producer and consumer explains why a kilogram of beef is more expensive than a kilogram of corn.

Pigs have the highest efficiency of any domestic animal: 15% of what they eat is transferred into meat.

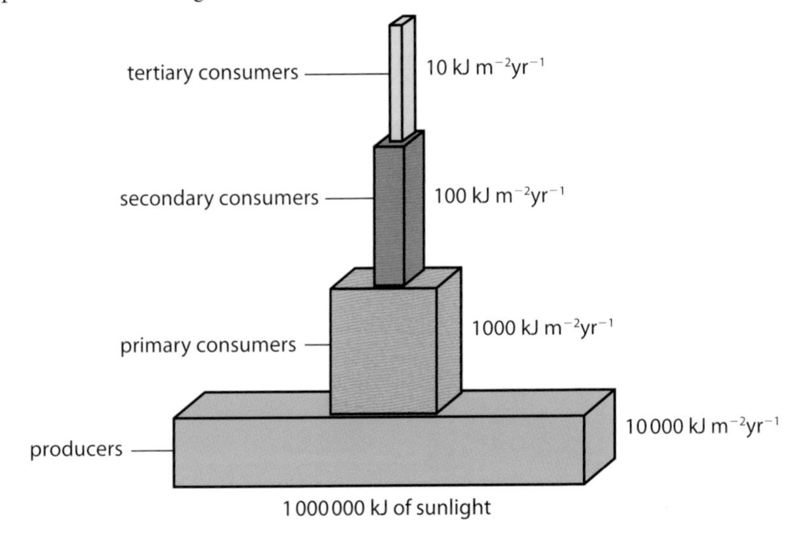

Figure 18.5 Pyramid of energy (not drawn to scale).

Figure 18.6 is a pyramid of energy with greater detail than the idealized pyramid in Figure 18.5. First, look at the simpler view at the bottom of the figure. The gross production of the producers is 20 810 kJ m^{-2} yr^{-1}. Can you calculate what percentage of energy moved up to herbivores? Doing practice calculations will help you understand how this works.

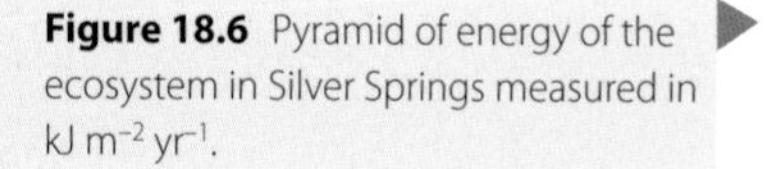
Figure 18.6 Pyramid of energy of the ecosystem in Silver Springs measured in kJ m^{-2} yr^{-1}.

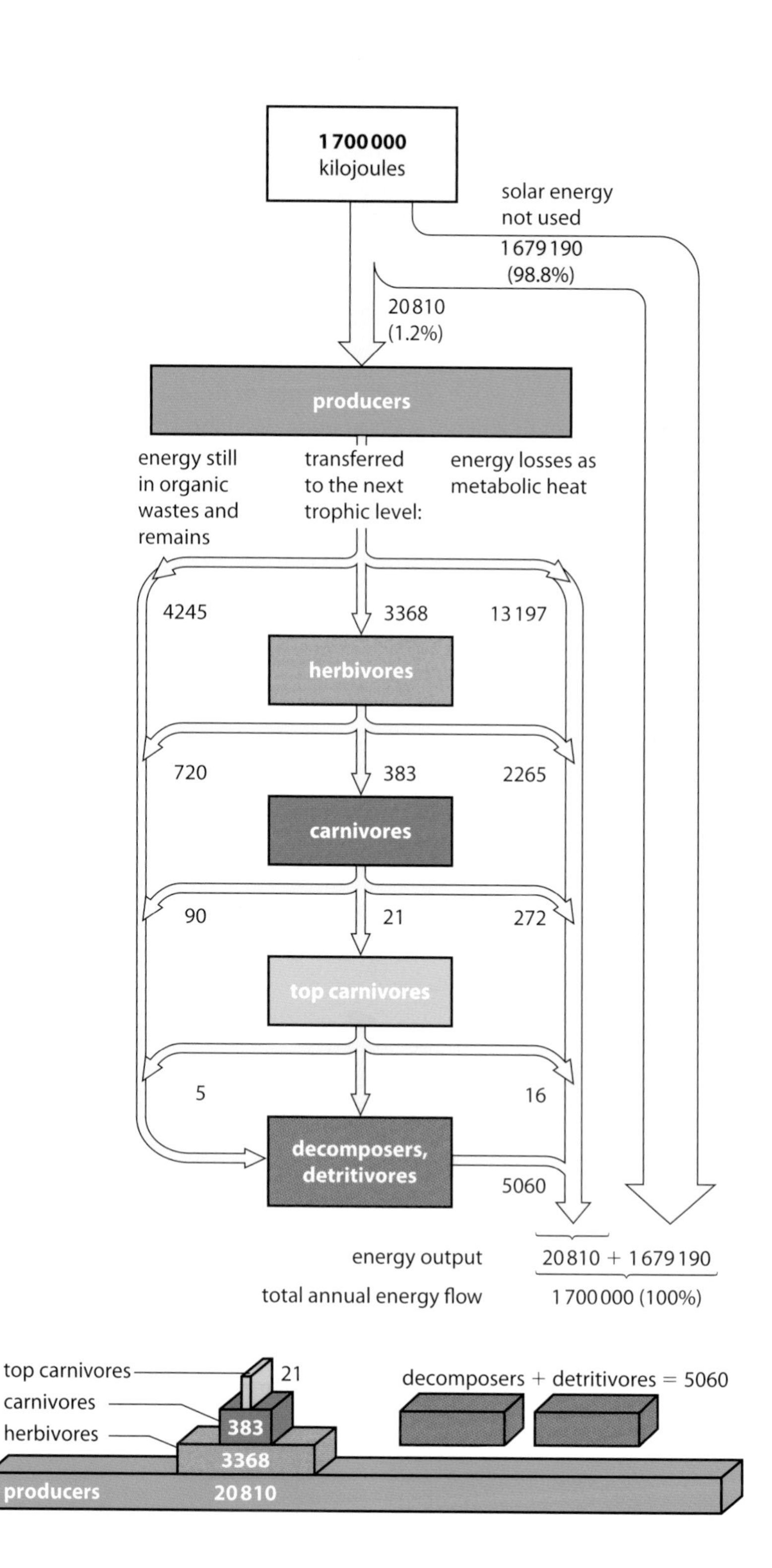

About 16% of the energy moved up to herbivores. Now look at the detailed energy flow chart at the top half of Figure 18.6. Notice that 1 700 000 kilojoules of energy are input from the Sun and that only 1.2% of the Sun's energy was captured by the producers. The producers have 20 810 kJ m^{-2} yr^{-1} of gross production. Gross production is energy which they have available. Notice that some of that energy is lost to metabolic heat and net system loss (heat, respiration and maintenance). Look on the other side of the figure and you will see how much is transferred to 'organic wastes and remains'. This energy eventually flows through decomposers, like mould and bacteria in the soil, and detritivores, like earthworms. Calculate the percentage of energy that is lost as respiration (metabolic heat) as it moves to herbivores.

Carnivores loose more energy than herbivores because they move around in search of prey.

The answer is 63%. About 16% was transferred to herbivores and the rest was transferred to decomposers and detritivores. The energy reaching the carnivores is 11.4% and only 5.5% flows up to the top carnivores. Look at the bottom of the energy flow diagram and you will see that eventually all the energy which flows through the ecosystem is lost as metabolic heat.

However, within the specific time period covered by a diagram such as Figure 18.6, organisms are storing some of energy. For example, a young forest accumulates organic matter as the tree grows. The slow rate of decay in a peat bog causes peat to build up. Some energy-flow diagrams include a cube to represent storage.

Now that you understand energy pyramids, we can define some important terms.

- *Gross production* is the total amount of energy trapped in the organic matter produced by plants per area per time in kilojoules.
- *Net production* is the gross production minus the energy lost through respiration.
- *Biomass* is the dry weight of an organism measured in g m^{-2} yr^{-1}.

In terms of an ecosystem, biomass is the dry weight of all the organisms at a certain tier of an ecosystem. The reason why we use dry weight is that the actual weight of the organisms includes a large amount of water. Water must be removed and the dry weight measured.

Calculating gross production and net production

In order to calculate the values of gross production and net production, we use this equation.

$$\text{gross production} - \text{respiration} = \text{net production}$$

So, if:

$$\text{gross production} = 809\ \text{kJ m}^{-2}\ \text{yr}^{-1}$$

and:

$$\text{respiration} = 729\ \text{kJ m}^{-2}\ \text{yr}^{-1}$$

then:

$$\text{net production} = 80\ \text{kJ m}^{-2}\ \text{yr}^{-1}$$

Plant biomass is a source of fuel for many people in the world. In developing countries, biomass provides 35% of the energy.

Constructing a pyramid of energy

Using the data below, construct a pyramid of energy without looking again at Figure 18.6.

Trophic level	Energy flow (kJ m^{-2} yr^{-1})
producers	20 810
primary consumers	3368
secondary consumers	383
tertiary consumers	21

To learn more about the role of biomass in the global energy supply, visit heinemann.co.uk/hotlinks, enter the express code 4242P and click on Weblink 18.8.

After you have drawn the pyramid, check Figure 18.6 to see if yours is correct. Have you drawn each block in proportion to the numbers? Have you placed the correct labels at each trophic level? Have you remembered a title for your pyramid?

Pyramids of biomass

Pyramids of biomass are similar in shape to pyramids of energy. The higher trophic levels have low total biomass per unit area of ecosystem (see Figure 18.7). Biomass is lost during respiration at each trophic level. When glucose is broken down for energy, it is converted into carbon dioxide gas and water. Carbon dioxide and water are excreted and the biomass of glucose is lost. Each successive level of the ecosystem loses more and more biomass. Energy per gram of food does not decrease, but *total* biomass of food is less at each trophic level. Notice in Figure 18.7 how little biomass is present in tertiary consumers compared to producers. It is very similar to what we saw when we looked at the pyramid of energy.

Figure 18.7 Pyramid of biomass.

Dry weight/g m^{-2}	Trophic level
1.5	tertiary consumers
11	secondary consumers
37	primary consumers
809	producers

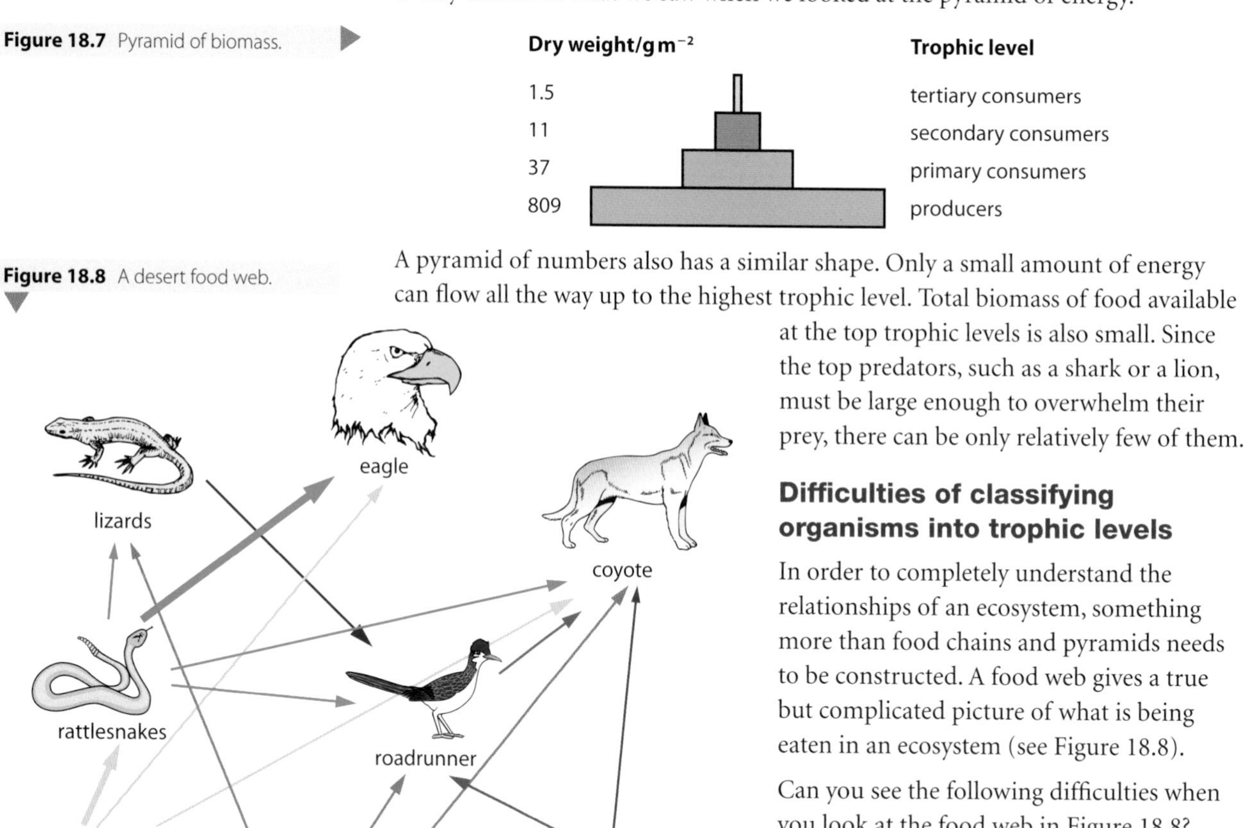

Figure 18.8 A desert food web.

A pyramid of numbers also has a similar shape. Only a small amount of energy can flow all the way up to the highest trophic level. Total biomass of food available at the top trophic levels is also small. Since the top predators, such as a shark or a lion, must be large enough to overwhelm their prey, there can be only relatively few of them.

Difficulties of classifying organisms into trophic levels

In order to completely understand the relationships of an ecosystem, something more than food chains and pyramids needs to be constructed. A food web gives a true but complicated picture of what is being eaten in an ecosystem (see Figure 18.8).

Can you see the following difficulties when you look at the food web in Figure 18.8?

- An eagle is a tertiary consumer when eating rattlesnakes but a secondary consumer when eating rabbits.
- A coyote is a primary consumer when it eats the fruit of a cactus but is a tertiary consumer when it eats a rattlesnake.
- A lizard is a tertiary consumer when it eats rattlesnake eggs but a secondary consumer when it eats insects.

Another difficulty is where to put omnivores. For example, the following omnivores are difficult to classify into one trophic level.

- Grizzly bears eat plant parts, insects and some mammals. Which food is eaten depends on the season, the temperature and the bear's ability to forage for food. Are they primary consumers, secondary consumers or tertiary consumers?
- Raccoons eat mice, bird eggs, fish, frogs, nuts and fruits. The food most dominant in the diet might depend on the season or competition from other animals. Is the raccoon mainly a primary consumer or a secondary consumer?
- Chimpanzees eat both fruit and termites. Is the chimpanzee mainly a primary consumer?

Primary and secondary succession

Ecological succession is the change in the abiotic (non-living) and biotic (living) factors in an ecosystem over time. It is the reason why some species gradually replace other species in one particular area.

Primary succession

Primary succession begins when plants begin growing on a previously barren and lifeless area. Let's consider a newly created volcanic island. The plants that first colonize it can exist where temperature changes are extreme and there is little or no soil. The first colonizers are usually the lichens. They are pioneer plants which can decompose thin layers of rock. As they die and decompose, a thin layer of soil is formed. This is just enough for some moss to get a foothold. This is the start of primary succession.

Lichen is a combination of algae and fungi. Lichens grow on trees as well as rock. The algae make the food by photosynthesis and the fungi absorb minerals.

Eventually, there will be enough soil for other seeds to germinate. Coconuts may be washed ashore and begin to germinate. Coconuts palm trees will grow. Animals may swim, fly or be carried on floating vegetation from other islands to populate this new place.

Secondary succession

In secondary succession, a new group of organisms takes over following a natural or artificial upheaval of the primary succession. Secondary succession is much faster than primary succession as soil is already present and there may be existing seeds and roots present – for instance, recolonisation after a forest fire.

This table summarizes the differences between primary and secondary succession.

Primary succession	Secondary succession
begins with no life	follows a disturbance of the primary succession
no soil	soil is present
new area e.g. volcanic island	old area e.g. following a forest fire
lichen and mosses are first plants	seeds and roots already present
biomass low	biomass higher
low production*	higher production*

*'Production'. is the increase in biomass or energy $m^{-2}\ yr^{-1}$. When production is low, it is due to there being few plants; high production occurs when many plants are present.

Species diversity and production in a primary succession

For pictures of British coastal dunes, visit heinemann.co.uk/hotlinks, enter the express code 4242P and click on Weblinks 18.9a and 18.9b.

For pictures of coastal dunes in Nova Scotia, Canada, visit heinemann.co.uk/hotlinks, enter the express code 4242P and click on Weblink 18.9c.

For pictures of the Indiana Dunes in the US, visit heinemann.co.uk/hotlinks, enter the express code 4242P and click on Weblinks 18.9d and 18.9e.

Coastal sand dunes are an excellent example of primary succession that is both interesting to hike through and has been studied extensively. If you do not live near a coast, try the Weblinks (left) which have pictures to take you there.

If you do live near a coast, you may find it more interesting to hike in the dunes after you have learned about the animals and plants that live there. Dunes are areas which need public support in order to be preserved as natural habitats.

Foredune

Primary succession starts on the foredune where there is no soil, only sand. Lyme grass and marram grass are pioneer plants on a new dune. Lyme grass is the more salt tolerant of the two species. They are generally fast growing and their roots help bind the sand and stabilize the dune. Marram grass has long underground roots which also spread sideways. They can spread three metres per year. Marram grass also has a special adaptation to life on the foredune: it has a growth spurt when covered with sand. There is little diversity of plant life here.

Yellow dune

At the yellow dune stage, the dune is developing a thin layer of soil from years of marram grass plants living and dying there. It has now been invaded by other plants with roots which are even better at binding the sand. These plants are sand sedge and sand bindweed. Rabbits are common in this dune and their droppings add nutrients to the soil. In the summer, fast-growing plants like dandelions and thistles grow here. Humus begins to build up as the original pioneer plants die and decay. Notice that at this stage, the community is more complicated. More species are present and soil is beginning to form.

Grey dune

The grey dune stage has developed a layer of humus from years of plants dying and decomposing. Humus holds water. This dune is much farther inland and sand is not deposited here. Eventually, thick shrubs will grow on this dune.

Mature dune

The final stage in dune succession is the mature dune, which can support a forest. At the Indiana Dunes, the mature dune has an Oak–Hickory forest. Hundreds of species of wild flowers are protected by the shade of the trees. Mosses and ferns grow on the forest floor. The humus is thick due to 200 years of plants dying and decaying. The moisture content of the soil is high due to the high amount of humus. The forest is full of insects, birds, and mammals. The temperature is 10% cooler in the mature dune than on the foredune. Lack of wind and blowing sand makes this a comfortable place for both animals and plants.

During the development of the primary succession on sand dunes, you can see that the following changes have occurred:

- few species to many species;
- pioneer species to species that compete with others for nutrients;
- little diversity to high diversity – the mature forest is home to hundreds of different species;
- simple relationships to more complex relationships of mutualism, competition, and predation;
- more and more biomass at each stage of the succession.

Living organisms change the abiotic (non-living) environment

As you can see in the description of the ecological succession on the sand dunes, living organisms change the abiotic environment. They do this is a number of ways.

Organic matter increases

Soil contains a reservoir of organic matter or humus which is a result of the death and decay of many plants and animals. Humus quickly absorbs and releases water and is therefore an excellent medium for plant growth. It is lightly packed and allows oxygen to be available for plant roots.

Soil gets deeper

Leaf litter and decayed plants create organic matter which is now mixed with the sand of the dune thus creating a deep, well-draining soil on the mature dune. It is deep enough to support the tall trees which live there.

Soil erosion reduces

Throughout the story of dune succession, you can see how plant roots stabilize first the sand (marram grass roots) and then the soil (shrub roots on the grey dunes; a wide diversity of tree, shrub and fern roots on the mature dune).

Soil structure improves

Sand does not hold moisture. Gradually, over hundreds of years , humus develops and slowly produces a structure which holds moisture and minerals and allows the aeration of plant roots.

Mineral recycling increases

Bacteria and fungus can be active in recycling nutrients in soil which has enough humus. Ecosystems create their own nutrients by recycling them from the plants and animals that live and die there.

Biosphere and biomes

If you view the surface of the Earth in a satellite picture, you see large swaths covered with trees, other areas covered with ice and others with nothing that can be seen. The living part of the Earth that you observe is called the biosphere. The biosphere is all the parts of the Earth where organisms live. Some organisms live in the Earth's crust and some live in the atmosphere. Anywhere that organisms live is considered to be part of the biosphere.

Biomes are divisions of the biosphere. Each biome is a part of the biosphere and is defined by its vegetation and community structure.

Distribution of biomes

Biomes occur because of global weather patterns and topography (see Figure 18.9). Certain species are found in one type of biome and not in others.

Figure 18.9 The distribution of some biomes by altitude and latitude.

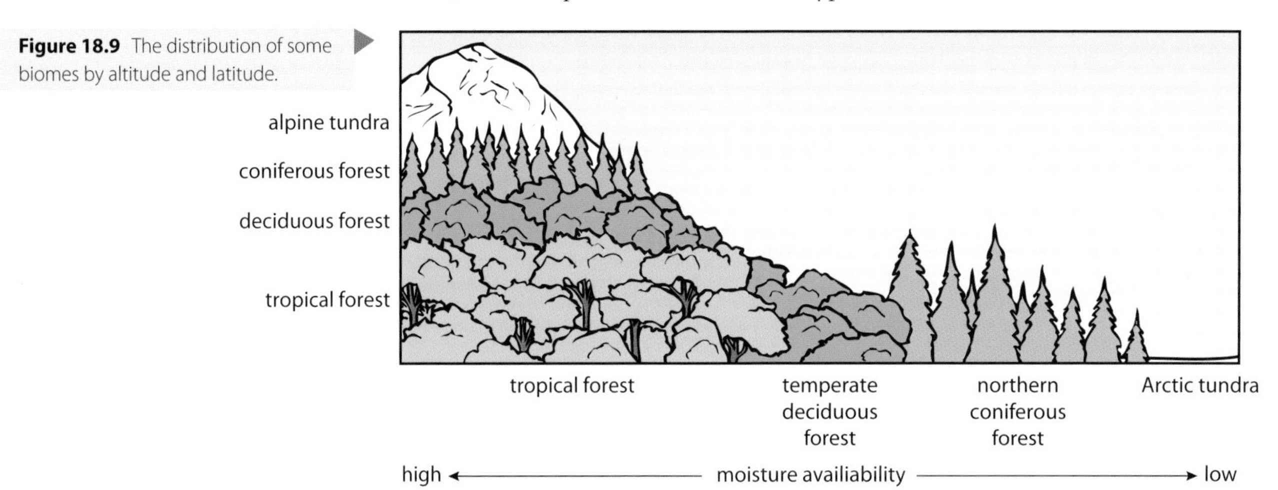

A climograph plots the temperature and rainfall in a particular region. In Figure 18.10, you can see that the mean annual precipitation (rainfall) is similar for a coniferous forest and a temperate forest but the temperature is different. The mean annual temperature is colder for the coniferous forest. Compare the mean annual precipitation for grasslands and tropical forests. You can see that precipitation in the tropical forest is much higher. Rainfall and temperature affect the distribution of biomes.

The following combinations of temperature, rainfall and elevation determine the biomes in North America.

In March 2007, the United Nations and eight other countries launched a programme to monitor changes taking place in the Arctic tundra. The aim is to halt the change in biodiversity by 2010.

To read about the Arctic tundra programme, visit heinemann.co.uk/hotlinks, enter the express code 4242P and click on Weblink 18.10.

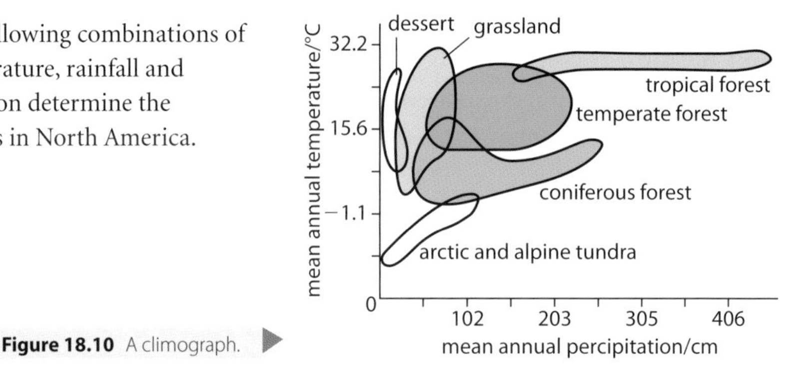

Figure 18.10 A climograph.

Tundra

High elevations with low temperatures and low precipitation are the conditions for the tundra. Plants and animals that live in the tundra are adapted to a cold and dry environment.

Coniferous forest

High elevations with less cold temperatures and slightly more rainfall are the conditions for a coniferous forest. Since the ground freezes during some months of the year, coniferous (cone-bearing) trees are well adapted for conserving water when it is frozen. Animals get heavy coats of fur in the winter and lose some of the fur in the summer.

Temperate forest

At lower elevations where temperatures are even warmer and more water is available, we find the temperate forest. Plants and animals in these forests must adapt to a wide range of conditions: warm in the summer with lots of water, and cool in the winter when all the water may be frozen. Many trees in this forest will lose their leaves in the winter to reduce water loss.

Desert

At low elevations with warm temperatures and little precipitation, we find desert conditions. Desert animals and plants have very specific adaptations which enable them to survive in this extremely hot and dry biome. A desert kangaroo rat has a specialized kidney for recycling water in its body. Cacti have spines instead of wide leaves to reduce water loss.

Tropical forest

At low elevations with warm temperature and very high moisture, we find tropical forest. This forest is extremely productive with high primary productivity due to the combination of high temperature and high rainfall.

Characteristics of the seven major biomes

Biome	Temperature	Moisture	Characteristics of vegetation
Desert	Mostly very hot with soil temperatures above 60 °C (140 °F) in the daytime.	Low precipitation – less than 30 cm per year.	Cacti and shrubs with water storage tissues, thick cuticles and other adaptations to reduce water loss.
Grassland	Cold temperatures in winter and hot in summer.	Seasonal drought is common with occasional fires, medium amount of moisture.	Prairie grasses which hold the soil with their long roots; occasional fire prevents trees and shrubs from invading the grasslands.
Shrubland (chaparral, matorral, maquis and garigue, dry heatherlands, fynbos)	Mild temperatures in winter and long, hot summers.	Rainy winters and dry summers.	Dry woody shrubs are killed by periodic fires. Shrubs store food in fire-resistant roots. They re-grow quickly and produce seed which germinates only after a fire.
Temperate deciduous forest	Very hot in summer and very cold in winter.	High rainfall spread evenly over the year. In winter, water may freeze for a short time.	Deciduous trees like oak, hickory and maple dominate the forest. In warmer seasons, a wide range of herbaceous plants grow and flower on the forest floor.
Tropical rainforest	Very warm.	Very high precipitation of more than 250 cm per year.	Plant diversity is high. A canopy of trees is the top layer. Next is a layer of shrubs. The ground layer is herbaceous plants and ferns. Large trees have vines climbing on them. Trees have orchids and bromeliads tucked in their branches.
Tundra	Very cold; in summer, the upper layer of soil thaws but the lower layers remain frozen – this is permafrost.	Little precipitation.	Low-growing plants like lichen and mosses and a few grasses and shrubs. Permafrost prevents the roots from growing deeply. Continuous daylight in summer allows some plant growth and reproduction.
Coniferous forest (Taiga)	Slightly warmer than the tundra.	Small amount of precipitation but wet due to lack of evaporation.	Cone-bearing trees such as pine, spruce, fir and hemlock.

Exercises

6 Discuss the difficulty of classifying organisms into trophic levels.
7 Distinguish between primary and secondary succession.
8 Describe the effects of living organisms on the abiotic environment.
9 Describe the major biomes.

G.3 Impact of humans on ecosystems

Assessment statements

G.3.1 Calculate the Simpson diversity index for two local communities.
G.3.2 Analyse the biodiversity of the two local communities using the Simpson index.
G.3.3 Discuss reasons for the conservation of biodiversity using rainforests as an example.
G.3.4 List three examples of the introduction of alien species that have had significant impacts on ecosystems.
G.3.5 Discuss the impacts of alien species on ecosystems.
G.3.6 Outline one example of biological control of invasive species.
G.3.7 Define *biomagnification*.
G.3.8 Explain the cause and consequences of biomagnification using a named example.
G.3.9 Outline the effects of ultraviolet (UV) radiation on living tissues and biological productivity.
G.3.10 Outline the effect chlorofluorocarbons (CFCs) on the ozone layer.
G.3.11 State that ozone in the stratosphere absorbs UV radiation.

Simpson diversity index

Biological diversity can be described in two separate ways: evenness and richness. The number of different organisms in a particular area is richness. Evenness is how the quantity of each different organism compares with the other. Richness only takes into account the kinds of species present in the ecosystem while evenness take abundance into account.

For example, the following table compares the numbers of different larvae in samples from two interdunal ponds at the Indiana Dunes.

Larva species	Number of individuals in sample 1	Number of individuals in sample 2
caddisfly larva	200	20
dragonfly larva	425	55
mosquito larva	375	925
total	1000	1000

Both samples have the same number of individuals but in sample 2 they are not evenly distributed between the species. Both samples have the same species richness – each has three types of larva. But they do not have the same evenness.

The individuals in sample 2 are mainly mosquito larvae. A community is not considered diverse if it is dominated by one species.

A measure that takes into account both richness and evenness is the Simpson diversity index. To see how this works, let's consider the community of plants on the foredune at the Indiana Dunes and the community of plants on the mature dune. Which would you hypothesize is more diverse and why?

To calculate the Simpson diversity index, we need the formula

$$D = \frac{N(N-1)}{\text{sum of } n(n-1)}$$

where

D = diversity index

N = total number of organisms in the ecosystem

n = number of individuals of each species

So, for each community we need to know the number of organisms present and the number of individuals of each species present. This information is found by sampling the two dunes with quadrats as follows:

- record the number of plant species in each quadrat;
- count the number of individuals of each species;
- record the data for each area in tables.

This table records the plant species on the foredune of the Indiana Dunes.

Plant species	Number of individuals, n	$n(n-1)$
marram grass	50	50(49) = 2450
milkweed	10	10(9) = 90
poison ivy	10	10(9) = 90
sand cress	4	4(3) = 12
rose	1	1(0) = 0
sand cherry	3	3(2) = 6
totals	N = 78	2648

This table records the plant species on the mature dune of the Indiana Dunes

Plant species	Number of individuals, n	$n(n-1)$
oak tree	3	3(2) = 6
hickory tree	1	1(0) = 0
maple tree	1	1(0) = 0
beech tree	1	1(0) = 0
fern	5	5(4) = 20
moss	3	3(2) = 6
columbine	3	3(2) = 6
trillium	3	3(2) = 6
Virginia creeper	4	4(3) = 12
Solomon seal	3	3(2) = 6
totals	N = 27	62

Using the formula above, the calculation for the foredune is:

$$D = \frac{78(77)}{2648}$$

$$D = 2.27$$

The calculation for mature dune is:

$$D = \frac{27(26)}{62}$$

$$D = 11.3$$

Were you correct? According to the Simpson diversity index, the mature dune is more diverse even though the total number of plants is less. The mature dune has a higher diversity index because it has a higher number of different species. Sampling and calculation of the Simpson index periodically can assess the health of an ecosystem.

Reasons for conservation of biodiversity

Would you like to take a tour of the rainforest? Can you imagine the beauty of orchids hanging from trees, monkeys scattering from vine to vine, multicoloured frogs hiding in bromeliad plants and the chattering of hundreds of birds? In addition to the beauty of the rainforests, we know they are a source of biopharmaceuticals, medicines not yet discovered, genes not yet used by genetic engineers and phytochemicals not yet identified. The rainforests also use a large amount of carbon dioxide. Cutting down the rainforests adds another item to the list of things which cause global warming.

There are good reasons for the conservation of any ecosystem: forests, woodlands, dunes, ponds, lakes and oceans. Let's take the rainforest as an example, and consider four reasons for conservation: economic, ecological, ethical and aesthetic.

Economic reasons

The attempt to create farms on rainforest soils has met with dismal results. Most of the nutrients of rainforests are locked up in the tissues of the plants. The soil left behind after logging is devoid of nutrients.

Moreover, plant sources of medicines and chemicals are lost forever if species are extinct. As long as the rainforest exists, there is the possibility that local crop plants and farm animals could be improved with alleles from wild populations. In addition, ecotourism could improve the local economy.

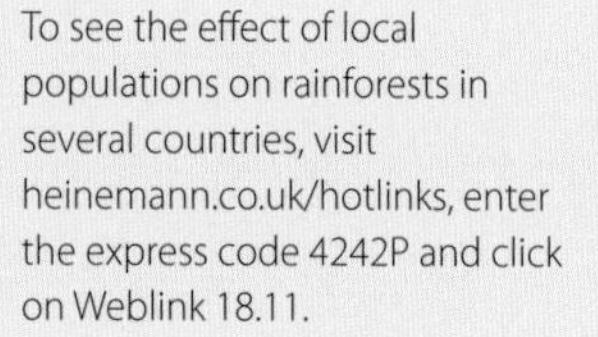

To see the effect of local populations on rainforests in several countries, visit heinemann.co.uk/hotlinks, enter the express code 4242P and click on Weblink 18.11.

Ecological reasons

An organism key to an ecosystem's health may be destroyed. Ecosystems have evolved over millions of years. Species are linked together in ecosystems like pieces of a puzzle. We have no idea how many, or which, species we can lose before the puzzle falls apart. Loss of one species could affect other species because many organisms in the ecosystem are interdependent.

Diversity protects an ecosystem against invaders. If alien species are introduced, they will be competing with the existing species in the ecosystem. The more diverse the ecosystem is, the better able it is to withstand pressure from alien species. At the moment, the rainforests are still diverse, but for how long? Will the future be that diversity is lost and alien species move in and finish the job? In the next section, we look at the devastation that happens when alien species move into an ecosystem that is not diverse.

Fewer plants in the biosphere means more carbon dioxide in the atmosphere. Excess carbon dioxide caused by burning of fossil fuel is one of the main causes of global warming. Destruction of the rainforest removes a huge number of plants from the biosphere. More plants, not fewer, are needed to remove excess carbon dioxide from the atmosphere.

Disruption of the ecosystem can lead to soil erosion and flooding.

Ethical reasons

The local population is most affected by rainforest destruction. The ethical solution to saving the rainforest is to include the local population in creative ways to conserve it. Many organizations are helping communities make a living from the rainforest while at the same time preserving it.

Do we have a right to destroy an ecosystem which may be enjoyed by future generations? We actually have an ethical responsibility to conserve the rainforest so that future generations have access to its beauty and wealth of organisms.

Do we have the right to decide which organisms survive? Human impact on ecosystems is often due to lack of education and awareness. The ethical way to help the rainforest is to create public awareness of the problem so that this important resource does not slip away.

To find out how some scientists are helping indigenous peoples and trying to save the rainforests, visit heinemann.co.uk/hotlinks, enter the express code 4242P and click on Weblink 18.12.

Can you devise an argument based on 'utilitarianism' for not conserving the rainforest?
Can you devise an argument based on moral principles for saving the rainforest?

Aesthetic reasons

Human well-being is linked to the ability to visit natural areas in our biosphere which have been preserved. The rainforest is one of these areas. Ecotourism is a booming industry which is helping save parts of the rainforest.

Many artists and writers have been inspired by the beauty of natural ecosystems. Just visit an art museum and you will see the many scenes painted by artists motivated by the natural landscape. The rainforest has inspired thousands of painters and poets.

Arguments against conservation

There are a few arguments against conservation of the rainforest. Conservation measures may slow down the economic development of countries with tropical rainforests. Clearing these forests does provide some land for agriculture. Tropical rainforests can be reservoirs of pest species or species which transmit disease.

Introduction of alien species

Introduction of alien species into an ecosystem disrupts communities. Alien species are often able to out-compete native species. This eventually reduces biodiversity. Many native species can be forced out of an ecosystem by one invader. The following are alien (invasive) species which have had a significant impact on an ecosystem.

Kudzu: deliberate release of an alien species

What we do to the environment today may have unforeseen consequences for future generations. Kudzu was introduced from Japan to the US in 1876 at the Philadelphia Centennial Exposition as an ornamental plant. In the 1930s it was promoted by the Soil Conservation Service of the government as a fast-growing plant which could solve the problem of soil erosion. From 1935 to 1950 it was actually planted by the Civilian Conservation Corp sponsored by the federal government. Finally, in 1953 it was recognized by the Department of Agriculture as a pest weed.

The community of Chattanooga, Tennessee, is introducing goats to graze on the kudzu to try to control its spread.

Currently, kudzu is common throughout the southeastern states of the US. It is often called 'the plant that ate the south'. Here is the reason why: kudzu grows rapidly, as much as 20 metres (over 60 feet) per season. Thirty stems can emerge from one root. It grows both horizontally and vertically. Kudzu spreads by runners which can make roots and produce more plants. Kudzu grows well in many conditions although prolonged freezing will kill it. The thick growth crushes other plants as it covers them. Its weight breaks tree branches. In the US, an annual sum of $500 million is lost through the effects of kudzu.

Zebra mussels: accidental release of an alien species

The interior of this water pipe is clogged with Zebra mussels. Zebra mussels were first recorded in North America in June 1988. They breed rapidly and have since invaded lakes and rivers. By 1993, stretches of the Illinois River had 94 000 mussels per square metre. The mussels encrust water pipes, excrete a corrosive substance and may also destroy indigenous aquatic life.

Zebra mussels (*Dreissena polymorpha*) are tiny black and white striped bivalve molluscs. They invaded North America in the mid 1980s. It is hypothesized that they were introduced by a European cargo ship which contained zebra mussels in its ballast water. Zebra mussels have spread all over the Great Lakes. It is likely that they will be inadvertently carried all over the US and Canada by boaters and fisherman. Adult mussels attach themselves to boat hulls with sticky fibres. Larvae can be carried in anything containing water, such as ballast water or bait buckets.

For more information about zebra mussels, visit heinemann.co.uk/hotlinks, enter the express code 4242P and click on Weblinks 18.13a, 18.13b and 18.13c.

Zebra mussels have had an enormous economic impact. They clog any pipe which transports surface waters, thus affecting utility plants, factories and water-treatment plants. Zebra mussels are very small and their free-swimming larvae attach to other zebra mussels. This creates the thick layers of mussels which clog pipes. Even though many aquatic organisms feed on zebra mussels, it has not affected their spread. One zebra mussel can release up to 100 000 eggs per year. It is estimated that $500 billion dollars will be spent on zebra mussel control in the next 10 years.

Boat owners who know about the dangers of zebra mussels spreading to other bodies of water, have the choice of whether or not to check their boat for the presence of mussels in ballast water or bait boxes. Is this a decision which is based on ethical principles?

Zebra mussels are causing the water in Lake Michigan to be very clear. Whether that is of benefit or harm to this ecosystem is not yet known. Increased water clarity enables more light to penetrate the water. More light causes increased growth of aquatic weeds. These weed beds may provide a good place for fish to hide and spawn.

Prickly pear: an alien species under control

The case of the prickly pear cactus in Australia is an example of an alien species under biological control. The cactus was introduced to Australia in the mid 1800s by Europeans and Americans. By the 1900s it was spreading at a rate of 400 000 hectares per year. Studies were performed to determine what biological agent could be introduced to control it. In the US, a moth, *Cactoblastic cactorum*, feeds on the prickly pear. This moth was introduced to Australia and immediately began to destroy the cactus. Today, a balance exists between the two populations.

Impact of alien species on ecosystems

Alien introductions can lead to interspecific competition, predation, and extinction of native species. All cause a reduction in diversity.

Interspecific competition

A species which invades an ecosystem can out-compete the native species.

The red squirrel lives in the UK; its habitat is forest and woodland. In the 19th century, the grey squirrel was introduced from North America. By the 20th century, the grey squirrel had taken over much of the habitat of the red squirrel. The red squirrel is now only found in areas which have never been invaded by the grey squirrel. When an animal is lost from an ecosystem many other organisms are affected. Losing these interactions is usually disruptive to biodiversity.

So far, no damage to the ecosystem has been documented by the exclusion of the red squirrel. It is difficult to do controlled studies on a forest ecosystem. However, we must assume that no damage has occurred unless there is data to show the harm.

The sea lamprey is an invading species which has had a devastating effect on the fish in the Great Lakes of North America. The lamprey's mouth has rows of rasping teeth which enable it to suck onto the body of a large fish and rasp a hole in its side.

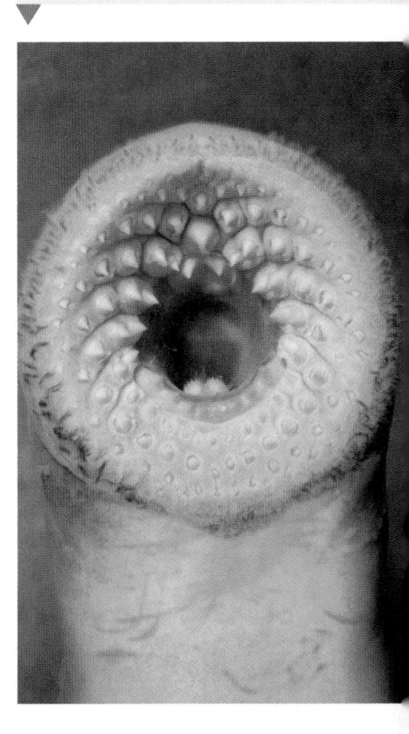

Predation

A species which invades an ecosystem can eat another species.

After the opening of the St. Lawrence Seaway which connected the Great Lakes to the Atlantic Ocean, ocean fish were free to travel to these freshwater areas. One that made the journey was the sea lamprey. Adult sea lampreys look like eels but have a round mouth with several rows of rasping teeth. Lake trout were the favourite prey of the lamprey. As that population dropped, the lamprey turned to whitefish. Soon the whitefish population was decimated.

An unexpected benefit of the lamprey was that it facilitated the introduction of salmon to replace the whitefish and lake trout in the Lake Michigan ecosystem. Biologists realized that the population of small fish, especially alewives which had also come from the ocean, was growing out of control. They introduced salmon as a predator to eat the small fish. In addition to eating the small fish, it has become a much sought-after game fish, bringing many tourist dollars to the area.

Species extinction

A species which invades an ecosystem can out-compete the native species and cause its extinction.

A good example of species extinction is provided by the Nile perch in Lake Victoria. Lake Victoria is shared by the African countries Kenya, Tanzania and Uganda. In the 1950s, the fish population in the lake had declined because of over-fishing. The Nile perch was introduced to increase the fish population. Originally, 80% of the fish in Lake Victoria were cichlids. But by 1970, the cichlid population had dropped to 1% of the total fish population. Of 400 species of cichlids in Lake Victoria, 200 species are now extinct.

To find out about 100 of the world's worst invasive alien species, visit heinemann.co.uk/hotlinks, enter the express code 4242P and click on Weblink 18.14. First go to 'species' and then to 'invasive species' to find the descriptions. At the same site, you can watch a cartoon called BioDaVersity. Go to 'species' and click on the DaVersity Code.

However, the introduction of the Nile perch is seen as a benefit to commercial fishermen who sell them.

Biological control

Biological control is the idea of using a natural predator to control an unwanted or invasive species. However, there is always a risk when introducing a new organism into an ecosystem. Unexpected consequences may occur even though rigorous testing has been done. Scientists look at risk–benefit analysis and make decisions based on that analysis.

Purple loosestrife

Purple loosestrife is an invasive plant which grows quickly and produces thousands of seeds.

Purple loosestrife (*Lythrum salicaria*) is an aggressive plant which has invaded the US and Canada. It displaces native wetland plants and can become the dominant plant in the wetland. This is a serious threat to biodiversity in this ecosystem. A single plant produces two million seeds each season which are dispersed along rivers and waterways.

Several states have been given permission by the US Department of Agriculture to release two beetles as biological control agents. The beetles are both species of *Gallerucella*. The adult beetles feed on the leaves of purple loosestrife. They lay eggs which hatch into larvae. The larvae feed on the leaves and stems of the plant. The goal of the biological control is to reduce the numbers of purple loosestrife in this environment.

In the areas where fire ants are found, 40% of the human population have been stung.

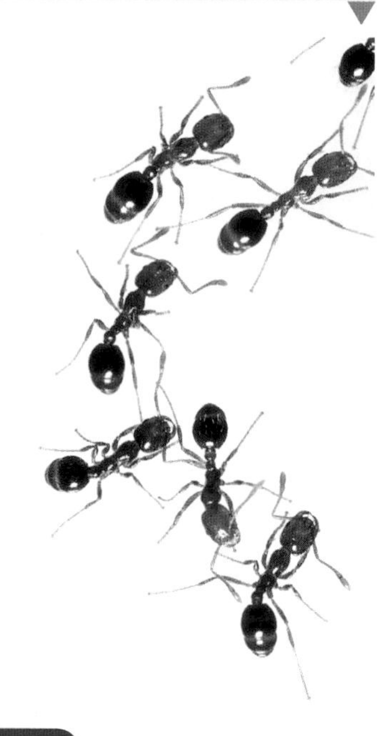

Red fire ants

Red fire ants (*Solenopsis invicta*) are an imported pest insect in the US. They are common in the southern states, and spreading north. Red fire ants compete with native ants and become the dominant species. They were probably introduced from South America and in the US have no natural enemies. They are particularly bothersome to humans because of their sting.

Scientists have been experimenting with a fly which is the natural predator of the fire ant. It is a phorid fly. The fly hovers over a mound of ants, picks out a victim and strikes. The strike involves piercing the ant and laying eggs inside its body. Eventually, the eggs grow into larvae. The larva eats its way to the head of the ant and decapitates it. So far, two species of phorid fly have been used in the US. They are doing well and their populations are increasing. Scientists are now working on a method to rear large numbers of phorid flies for release. A facility in Gainesville, Florida, will eventually produce 6000 to 12 000 flies a day.

Biomagnification

Biomagnification is a process by which chemical substances become more concentrated at each trophic level.

When chemicals are released into the environment they may be taken up by plants. The plants may not be affected by the small amount of chemical that they absorb or have on their surface. But when large amounts of the affected plants are eaten by a primary consumer, the amount of chemical it takes in is much greater. Similarly, if numbers of the primary consumer are eaten by a secondary consumer, the amount of chemical taken in by the secondary consumer is magnified even more.

Chemicals which are biomagnified in this manner are fat soluble. After ingestion, they are stored in the fatty tissue of the consumer. When the consumer is caught and eaten, the fat is digested and the chemical moves to the fatty tissue of the secondary consumer.

Causes of biomagnification

Some toxic chemicals were deliberately put in the environment to kill insect pests. One of these pesticides was DDT, which was used to control mosquitoes and other insect pests. At the time, it was not known that DDT did not break down and would persist for decades in the environment. DDT was commonly sprayed on plants and eventually entered water supplies. There it was absorbed by microscopic organisms. The organisms were eaten by small fish and the small fish eaten by larger fish. DDT built up in the fatty tissue of the fish. When these fish were eaten by birds the magnification of DDT was even greater (see Figure 18.11).

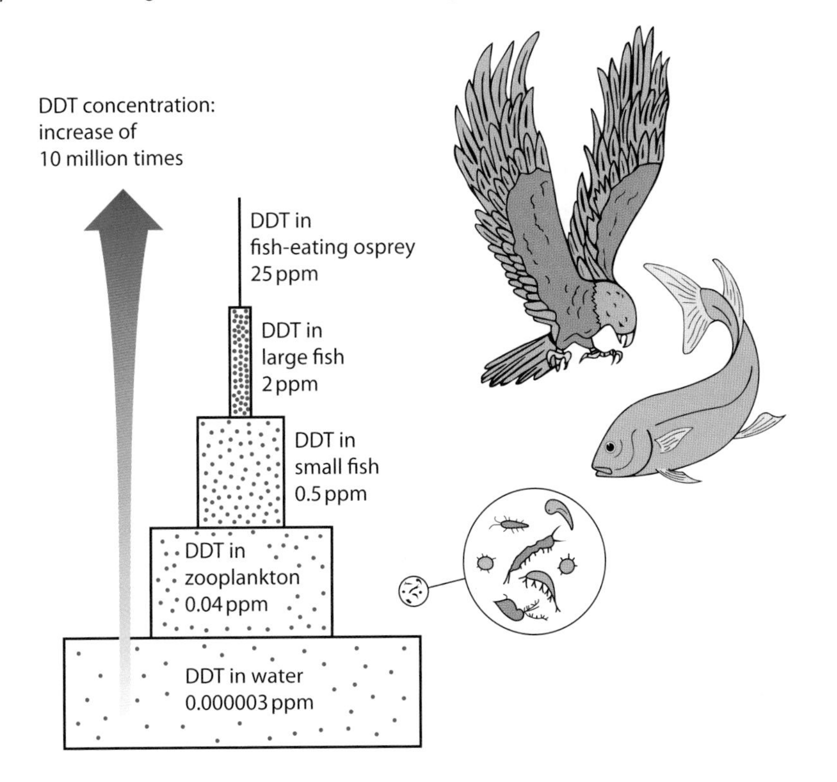

Figure 18.11 Biomagnification of DDT.

Developed countries have pressured underdeveloped countries to ban the use of DDT as a pesticide. Not using DDT has resulted in a return of malaria epidemics. Do you think the use of DDT should be allowed in these circumstances?

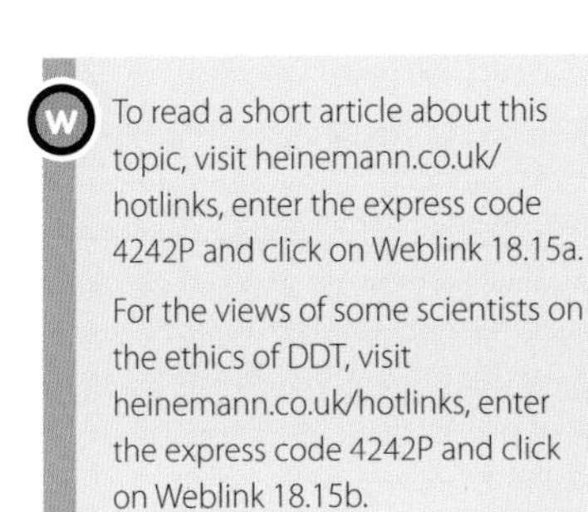

To read a short article about this topic, visit heinemann.co.uk/hotlinks, enter the express code 4242P and click on Weblink 18.15a.

For the views of some scientists on the ethics of DDT, visit heinemann.co.uk/hotlinks, enter the express code 4242P and click on Weblink 18.15b.

Consequences of biomagnification

The first sign of the problem was a decline in the number of predator birds. Studies showed that the eggs of these birds were easily cracked. In fact, the weight of the mother sitting on the eggs cracked them. It was finally discovered that DDT was building up in the tissue of the birds and interfering with the calcium needed for the shell to be hard. DDT was banned in the US in 1971. The bird population has begun to recover following the ban.

Effects of ultraviolet radiation

We now recognize that UV rays reaching the Earth can cause the problems discussed below.

Non-lethal skin cancer

Basal and squamous cell carcinoma are common forms of skin cancer which are not lethal. Scientists have been collecting data on these forms of skin cancer and have found that a decrease of 1% of stratospheric ozone increases these cancers by 2%.

Lethal skin cancer

Malignant melanoma is a form of skin cancer which is lethal in 15–20% of cases. Early detection is the key factor in recovery from this type of skin cancer.

Mutation of DNA

UV radiation causes changes in the structure of DNA.

Sunburn

Reddening of the skin due to UV radiation is caused by enlargement of small blood vessels. Some cells of the epidermis die and peel off.

Cataracts

A cataract is a clouding of the lens of the eye leading to loss of vision. Long-term exposure to UV rays is a risk factor for cataracts.

Everyone who is exposed to UV rays is at risk of eye damage. Wearing protective glasses and a hat lessens exposure.

Reduced biological productivity

UV radiation can damage and kill plant cells. This affects the ability of the plant to photosynthesize. UV radiation can also damage the DNA of cells involved in growth. This can reduce the biomass of the plant and decrease net productivity. Experimental data indicate that phytoplankton or floating microscopic plants are especially susceptible. These plants live in surface waters in order to gather light for photosynthesis. Population studies of the effect of UV on phytoplankton are ongoing. Phytoplankton is at the bottom of the food chain in many aquatic environments and plays an important part in sustaining the ecosystem.

CFCs and the ozone layer

Ozone is like a protective sunscreen for the planet. The ozone layer in the stratosphere absorbs ultraviolet (UV) radiation. The ozone layer is about 20 kilometres thick. It is a giant umbrella about 15–35 kilometres (9–22 miles) above the Earth's atmosphere (see Figure 18.12). The formation of ozone (O_3)

occurs in the stratosphere when an oxygen molecule, O_2, breaks apart and reacts with another oxygen molecule to form ozone. Ozone is constantly being formed, broken down and re-formed.

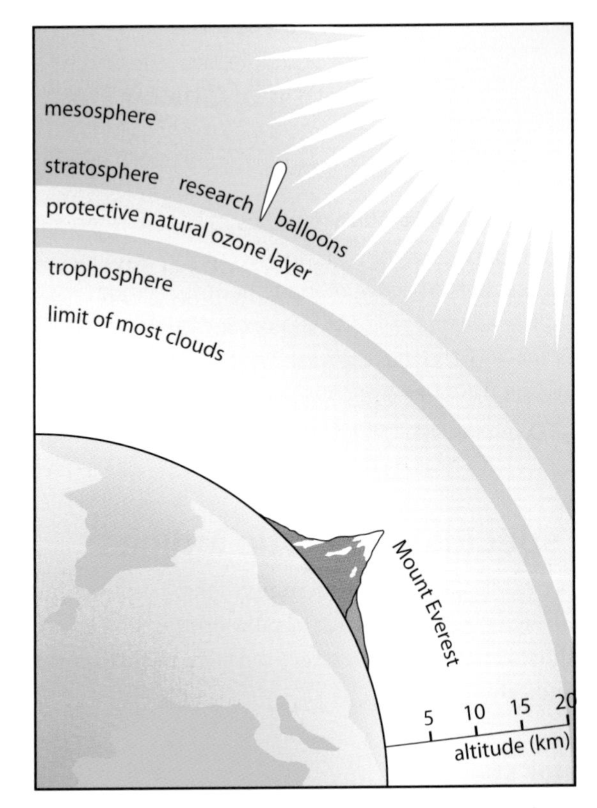

Figure 18.12 The ozone layer.

But the ozone layer is thinning. In 1985, data collected on the ozone layer showed a dramatic depletion. Scientists hypothesized that the cause of this thinning was the effect of chlorofluorocarbons (CFCs). After further experimentation, it was concluded that CFCs were indeed the cause of ozone depletion. An international agreement called the Montreal Protocol was adopted to phase out the use of CFCs in:

- refrigerator coolants;
- propellants for aerosols;
- material used to make foam packaging.

In the stratosphere, CFCs break down to release chloride ions. Then:

- chloride ions react with ozone molecules (O_3) to produce ClO and oxygen (O_2);
- the ClO joins with an oxygen atom to form more oxygen gas and release a chloride ion.

You can see that this leaves a free chloride ion to destroy another ozone molecule. This cycle is repeated over and over causing a depletion of ozone.

Calculations show that one CFC molecule can move up to the stratosphere in 15 years and remains there destroying ozone molecules for a century.

W To find out more about destruction of the ozone layer and its effects, visit heinemann.co.uk/hotlinks, enter the express code 4242P and click on Weblinks 18.16a and 18.16b.

Exercises

10 Discuss the reasons for conservation of biodiversity.

11 Describe the biological control of a named species.

12 Outline the effect of UV rays on living tissue and biological production.

13 Describe the effect of CFCs on the ozone layer.

G.4 (HL only) Conservation of biodiversity

Assessment statements

- G.4.1 Explain the use of biotic indices and indicator species in monitoring environmental change.
- G.4.2 Outline the factors that contributed to the extinction of one named animal species.
- G.4.3 Outline the biogeographical features of nature reserves that promote conservation of diversity.
- G.4.4 Discuss the roles of active management techniques in conservation.
- G.4.5 Discuss the advantages of in situ conservation of endangered species (terrestrial and aquatic reserves).
- G.4.6 Outline the use of ex situ conservation measures, including captive breeding of animals, botanic gardens and seed banks.

Indicator species and biotic indices

Do you remember reading stories of coal miners taking canaries into the mines? If the canary died, it signalled the presence of poisonous gas. In an ecosystem, some species are like those canaries. They are very sensitive to environmental change. They are called indicator species.

Some indicator species

A common indicator species is the lichen. Lichens live on rocks and trees and are a reliable indicator of air quality. They are very sensitive to pollution in the atmosphere. Lichens are not usually found on trees in a city because the air is too polluted for them. Since lichens also retain metal in their tissues, they can show the presence of lead or mercury in the air.

Another group of indicator species are macroinvertebrates found in rivers and streams (see Figure 18.13). The presence or absence of these organisms can be used to judge the water quality.

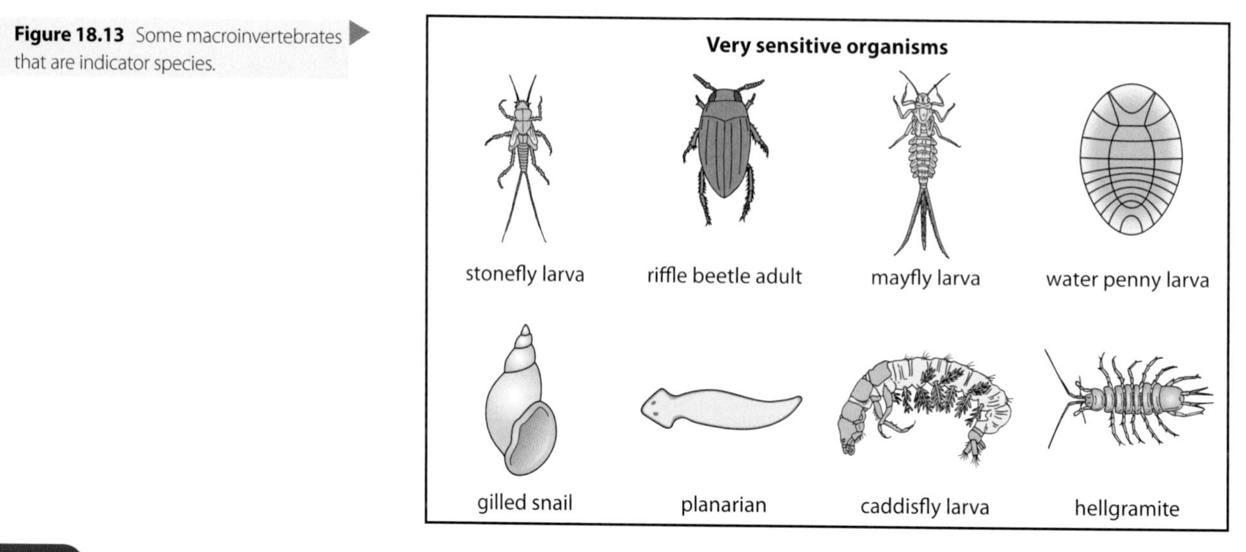

Figure 18.13 Some macroinvertebrates that are indicator species.

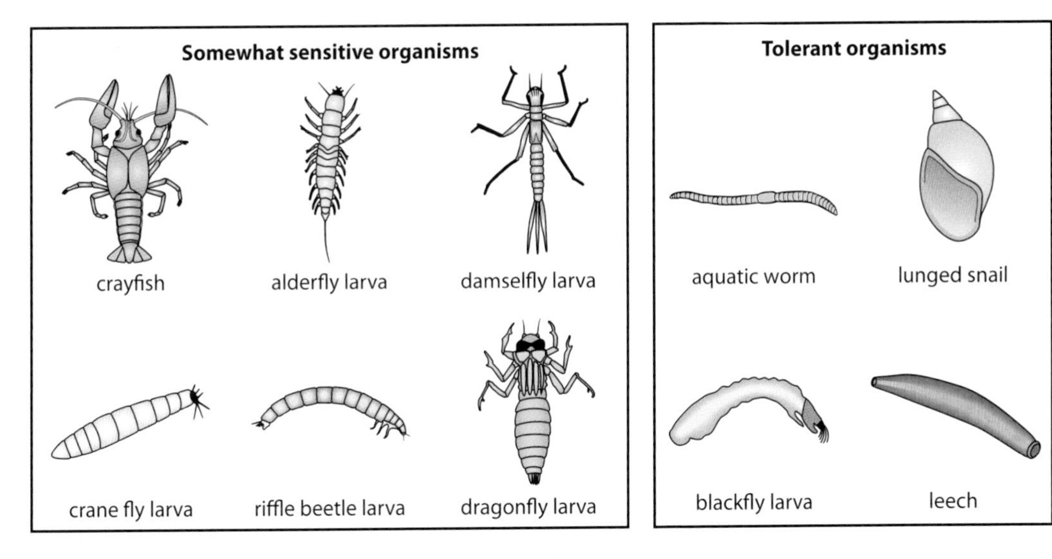

Figure 18.13 Continued.

We are all interested in the quality of our rivers and streams. In past years, rivers and streams were often used as dumping grounds for toxic chemicals and unwanted materials. In Chicago, Illinois, around 1900, the river was a dumping ground for the waste products from the slaughter houses. All of the unwanted parts of the animals were thrown into the Chicago River. In fact, one branch of the river was named 'bubbly creek' for all of the fermentation that was taking place as the animal tissue decomposed in the water. Today, that river is a much cleaner place with boats and canoes floating on it rather than garbage. All of this is due to our awareness that water and waterways are precious commodities to be treasured.

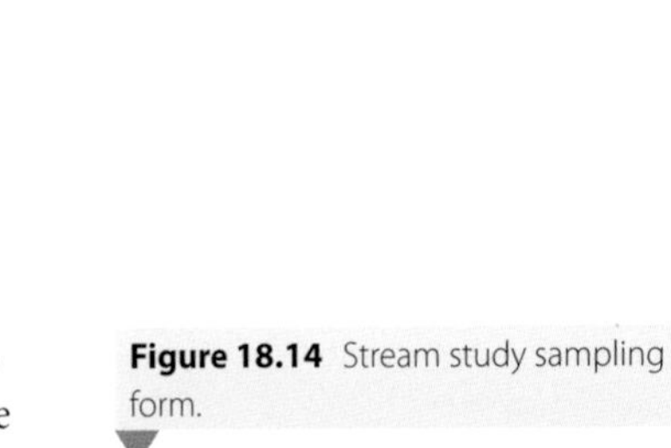

Freshwater indicator species have various levels of pollution tolerance. Organisms like leeches and aquatic worms are not sensitive and can live in water with low oxygen levels and high amounts of organic matter. Organisms like the larvae of alderfly and damselfly are moderately sensitive, whereas the larvae of the mayfly and caddisfly are very sensitive to pollution. The very sensitive organisms must have high levels of oxygen and little organic matter in the water. The cleaner the water, the higher the number of sensitive organisms.

Biotic index

When you perform a river or stream study, you count the number of macroinvertebrates collected in each sample and record the data on a stream study form (see Figure 18.14). The number of organisms of each group is multiplied by a factor which is determined by how sensitive the organism is to pollution. The presence of sensitive organisms is multiplied by a higher number. The more sensitive organisms you have in the sample, the higher the quality of the water in the river or stream. The total number is called the biotic index.

Figure 18.14 Stream study sampling form.

Stream study: Sample record and assessment

Stream ______________ Site number ______________

County or city ______________ State ______________

Collection date ________ Collectors ______________

Weather conditions (last 3 days) ______________

Average depth at site ________ Average width at site ________

Water temperature ________ °C ________ °F

Flow rate: ☐ High ☐ Normal ☐ Low

Appearance: ☐ Clear ☐ Cloudy ☐ Muddy

Macroinvertebrate count

Sensitive	**Somewhat Sensitive**	**Tolerant**
☐ ___ caddisfly larvae	☐ ___ beetle larvae	☐ ___ aquatic worms
☐ ___ hellgramite	☐ ___ clams	☐ ___ blackfly larvae
☐ ___ mayfly larvae	☐ ___ crane fly larvae	☐ ___ leeches
☐ ___ gilled snails	☐ ___ crayfish	☐ ___ midge larvae
☐ ___ riffle beetle adult	☐ ___ damselfly larvae	☐ ___ lunged snails
☐ ___ stonefly larvae	☐ ___ dragonfly larvae	
☐ ___ water penny larvae	☐ ___ scuds	
	☐ ___ sowbugs	
	☐ ___ fishfly larvae	
	☐ ___ alderfly larvae	
	☐ ___ watersnipe larvae	
boxes checked × 3 = ______ index value	boxes checked × 2 = ______ index value	boxes checked × 1 = ______ index value

Water quality rating ☐ Excellent (>22) ☐ Fair (11–16)

Total index count ______ ☐ Good (17–22) ☐ Poor (<11)

Periodic sampling gives an idea of the overall health of the river or stream. After a storm, there will be a lot of run-off from the areas surrounding the river. How does the run-off affect the biotic index? Does some of the sewer water get diverted into the river after a big storm? Is the biotic index different in winter and spring? Sampling gives us biological data that can be used to answer these questions.

Factors contributing to extinction

The Carolina parakeet became extinct in 1900. One of the reasons for its demise is that its feathers were highly sought after for women's hats.

PSITTACUS CAROLINIENSIS; *or the* PARROT *of* Carolina; *from* Cateſby's *Nat. Hiſt.*

IT is the only bird of the Parrot kind found in *Carolina*, rather leſs than our black-bird, and is three ounces and a half in weight. It feeds on the kernels of apples, to come at which he makes great deſtruction among the orchards; their guts are a poiſon to cats.

The CYPRESS of *America*. Near the ground, almoſt thirty foot in compaſs, and the branches ſixty or ſeventy foot high. Its wood reſiſts the weather better than moſt trees, and has a free grain. It is an aquatic, and generally grows from one to ſix foot in the water; its ſecure ſituation induces the Parrot, and other birds which feed on its fruit, to make their neſts in it.

We look at animals that have become extinct and ask ourselves the questions, 'How did this happen? Wasn't anyone paying attention?' Today, there are hundreds of endangered species that we are watching and attempting to save. This is the story of one bird which was not saved because no one was looking.

The Carolina parakeet was a beautiful bright yellow and orange bird. It has been extinct in the wild since 1900. It lived in the US and its range was from New York to the deep South and west to Colorado. Using its thick, powerful beak, it was primarily a seed eater. Large groups of Carolina parakeets nested together in hollow trees. In the mid 1800s, settlers cleared large areas of trees for farming and fuel. Removing the trees destroyed the habitat of the Carolina parakeet.

Several other factors also contributed to its extinction. One was the introduction of the honey bee from Europe. The bees were brought to make honey and to pollinate plants. They quickly escaped into the wild and made their homes in the same hollow trees as the Carolina parakeet. The parakeets were displaced from their nesting area by the bees. Another factor was hats. In the 19th century it was very fashionable to decorate women's hats with feathers. The brightly coloured feathers of the Carolina parakeet were highly sought after by hat makers. Then there was live capture. People thought Carolina parakeets made wonderful caged pets. They could learn their name and were very entertaining. By 1900, all Carolina parakeets were gone from the wild. A few caged specimens were left for a short time. These beautiful birds will never be seen in America again.

Features of nature reserves which promote biodiversity

As you hike through a nature reserve in your community and enjoy the beauty, you may not realize all of the planning that has taken place to keep the area 'natural'. Ecologists have collected data on how the geography of nature reserves can facilitate the biodiversity of the ecosystem. The nature hike would not be so interesting if all you saw was the invasive kudzu growing on top of all the other plants.

Three biogeographic features are taken into account when nature reserves are planned: the size of the reserve, the amount of edge, and the use of corridors.

Kudzu is a vine which grows over other plants. You can almost see it move up this wire!

Determinations of size

For a while, a vigorous debate occurred in ecological circles about reserve size. The debate was referred to as SLOSS (single large or several small). It was decided in favour of large sites over small sites. There are two main reasons for this.

It is hypothesized that small sites will have low population numbers. With low numbers, the risk of extinction is higher. Unexpected factors like fire or disease could wipe out a small population of a single organism more easily than a large population.

Small habitats also have more edge area overall than a large site. Research has shown that organisms at the edge of the reserve are more at risk from predation and competition by invasive species.

Edge effect

The ecology at the edge of the ecosystem is different from that at the centre of the habitat. As an example, we can look at a Forest Nature Reserve in the western US. The forest edge is the boundary between the forest and the disturbed area around it. Farmland is surrounding the forest. At the edge of the forest there is more sunlight, more wind and less moisture than in the centre of the forest. As you move closer to the edge, the trees are fewer. Organisms that live in the edge have more competition from other species. Studies of the forest have shown that fewer songbirds live on the edge of the forest than in the centre. The brown-headed cowbird is an exception and thrives at the edge where it lays its eggs in the nests of other birds. In fact, fragmentation of forests or breaking them up into small pieces due to farming practices has led to a considerable increase in the cowbird population. As edges increase, the cowbird population increases.

To read about the giant panda, visit heinemann.co.uk/hotlinks, enter the express code 4242P and click on Weblink 18.17.

Corridors

Corridors connect otherwise isolated habitats. Wildlife can travel between habitats in these corridors. A corridor can be as simple as a tunnel under a busy road or as extensive as the following example. To connect China's giant pandas with more habitats, the World Wide Fund for Nature (WWF) working with the Chinese government has established corridors between 40 isolated panda populations.

Some problems with corridors are:

- narrow corridors can expose animals to predators;
- invasive species can enter the habitat using a corridor;
- corridors which affect human populations can cause controversy (e.g. if the corridor from one lake to another is travelled by an alligator and is near a walking path, people may be fearful of the alligator).

Human populations are growing and habitats are shrinking. In many areas of the world this causes human–animal conflicts. The World Wide Fund for Nature (WWF) tries to find solutions to these problems all over the globe.

Visit heinemann.co.uk/hotlinks, enter the express code 4242P and click on Weblink 18.18. Go to 'species' and click on 'human–animal conflict'.

Management of conservation areas

In order to keep the beauty and diversity of a nature reserve, it is important to have effective management. Nature reserves can not just be left to nature. Active intervention is required to restore areas and protect native species. Good management practices are discussed under individual headings below.

To learn more about heathland restoration, restoration of an Atlantic rainforest and the annual restoration by fire of moorland, visit heinemann.co.uk/hotlinks, enter the express code 4242P and click on Weblinks 18.19a, 18.19b and 18.19c.

Restoration

Restoration attempts to return the land to its natural state. To restore lands on which vegetation has been destroyed may require managers to use active management techniques such as scrub clearance, cutting or burning and replanting. A UK project is restoring the heathlands of 50 years ago to a piece of land designated as a nature reserve in 2007.

Recovery of threatened species

Threatened species are usually helped when we restore their habitat. Active management maintains the areas needed for the habitat of the endangered species. In a Florida nature reserve, the habitat of the endangered gopher tortoise is being restored. It lives in deep burrows in a sandhill ecosystem. As many as 350 other animal species live in the burrow with the gopher tortoise. Restoration of the sandhill ecosystem is necessary to the existence of these species as well as to the gopher tortoise.

The gopher tortoise is a keystone species. Many other species depend on it for survival.

Removal of introduced species

Most of the exotic species which are introduced die out because they do not have adaptations to the local ecosystem. However, when a species takes over it can be devastating. In parts of the UK, a plant called rhododendron has taken over large areas of ground and almost eliminated native plants in those areas. Active management is needed to remove rhododendron from any nature reserve in the UK where it is found. In the southern US, the kudzu plant is a very aggressive invader. Active management in the southern US would entail removal of kudzu the minute it is spotted.

Legal protection against development or pollution

Nature reserves protected by the government or private organizations can prevent activities which might harm the native animals and plants. Such activities might include extraction of minerals, development of recreational facilities, hunting of animals or over-use by the public. Active management measures would include posting of signs and using security personnel to assure protection of the nature reserve from harmful human activities.

Funding and prioritizing

Since all activities require funding, which should take priority? Should funds be used to remove all exotic species or can we assume most exotics will die out? Should we repair the habitat of a few endangered species or use the limited funds to keep the habitat strong for the majority of organisms? Should we build hiking trails for the public even though that will bring destruction to some of the habitat?

Increasing public awareness of reserves helps gather funds to support the reserve. Management of nature reserves requires a balance between the good of the ecosystem, maintenance of diversity and the costs involved.

In situ conservation methods

Nature reserves help endangered species by maintaining their habitat and preventing competition from invasive species. Keeping these organisms 'in situ' is placing them in the situation where they belong. Organisms have adapted over hundreds of years to a set of conditions. These conditions include the other species present in the ecosystem as well as the abiotic factors. It is the goal of in situ conservation to allow the target species to continue to adapt to conditions in the reserve without interference from outside influences, such as invasive species and human incursions.

Reserves are both terrestrial (land-based) and aquatic. Terrestrial reserves can be found in most communities. Lake and pond areas are also common. Marine reserves are rare and lag behind in their development. Terrestrial reserves have been around for centuries but there is no tradition for conservation of species using marine reserves. The ocean is a large ecosystem which needs protection. The same in situ strategies used in terrestrial reserves can be put into practice in a marine reserve.

Conservation in situ does the following:

- protects the targeted species by maintaining the habitat;
- defends the targeted species from predators;
- removes invasive species;
- has a large enough area in the reserve to maintain a large population;
- has a large enough population of the targeted species to maintain genetic diversity.

On some occasions, the in situ area is unable to protect the targeted species. For example:

- the species is so endangered that it needs more protection;
- the population is not large enough to maintain genetic diversity;
- destructive forces cannot be controlled, such as invasive species, human incursion, and natural disasters.

Ex situ conservation methods

Ex situ methods are usually used as a last resort. If a species cannot be kept in its natural habitat safely or the population is so small that the species is in danger of extinction, then ex situ methods of conservation are used. There are three methods: captive breeding of animals, cultivation of plants in botanic gardens, and storing seeds in seed banks.

It is an ethical dilemma for zoos when they are forced to receive unwanted animals donated by the public or taken by the authorities. Should a zoo feed live prey to its zoo carnivores?

Captive breeding

Zoos have large facilities devoted to breeding. They have staff training in animal husbandry. Breeding programmes interest the public and generate new funds for the zoo. The San Diego Zoo in California has devoted a large part of its resources

To explore these and other ethical dilemmas, visit heinemann.co.uk/hotlinks, enter the express code 4242P and click on Weblinks 18.20a and 18.20b.

Zoos play an important role in species preservation internationally. A zoo in South Carolina is supporting the preservation of species from all over the world.

To discover more about the zoo, visit heinemann.co.uk/hotlinks, enter the express code 4242P and click on Weblink 18.21. Click on 'Other significant achievements' to see its international efforts.

to captive breeding programmes. The goal of captive breeding is to try to increase reproductive output and ensure survival of the offspring. Here are some of the techniques used.

- *Artificial insemination*: If the animals are reluctant to mate, semen is taken from the male and placed into the body of the female.
- *Embryo transfer to a surrogate mother*: To increase the number of offspring, 'test-tube' babies are produced and implanted in surrogate mothers. The mothers can be a closely related species.
- *Cryogenics*: Eggs, sperm cells and embryos are frozen for future use.
- *Human-raised young*: If the mother is not interested or able to care for the young, then staff hand-raise the young in the nursery of the zoo.
- *Keeping a pedigree*: With artificial insemination a common occurrence, it is important that the relatedness is known, to keep inbreeding to a minimum.

One problem with captive-breeding programmes is that the introduction into the wild of captive-bred individuals can spread disease to a non-infected wild population. When some captive-bred desert tortoises were introduced to their native habitat, they infected the wild population with a respiratory disease.

Another problem is that animals bred in captivity have missed the process of in situ learning undergone by their wild relatives. This may put them at a severe disadvantage in the wild.

Botanical gardens

Plants are easily kept in captivity. They have simple needs and usually breeding is not difficult. About 80 000 plant species are grown in private gardens, arboretums or botanical gardens all over the world. It is much easier to take care of plants outside their natural setting than it is to take care of and breed animals.

In 2006 the Norwegian government established a global seed bank. The Millennium Seed Bank Project at the Royal Botanical Gardens in the UK aims to safeguard 24 000 plant species from around the globe.

A problem with the collections of botanical gardens is that the wild relatives of commercial crops are under-represented. These plants may have genes which confer resistance to diseases and pests. Adding these wild plant relatives to collections at botanical gardens would provide gene banks for commercial crops.

Seed banks

To learn more about seed banks, visit heinemann.co.uk/hotlinks, enter the express code 4242P and click on Weblinks 18.22a and 18.22b.

Seeds in a seed bank are kept in cold, dark conditions. Under these conditions, the metabolism of the seed slows down and prevents it from germinating. Seed can be kept this way for decades. Some seeds need to be grown, form plants and have new seed collected. Currently, seed from 10 000 to 20 000 plant species from all over the world is stored in seed banks.

Exercises

14 Explain one method of measuring environmental change.

15 Describe what caused the extinction of a named species.

16 Discuss in situ conservation measures.

G.5 (HL only) Population ecology

Assessment statements

- G.5.1 Distinguish between r-strategies and K-strategies.
- G.5.2 Discuss the environmental conditions that favour either r-strategies or K-strategies.
- G.5.3 Describe one technique used to estimate the population size of an animal species based on a mark–release–recapture method.
- G.5.4 Describe the methods used to estimate the size of commercial fish stocks.
- G.5.5 Outline the concept of maximum sustainable yield in the conservation of fish stock.
- G.5.6 Discuss international measures that would promote conservation of fish.

Distinguishing r-strategy from K-strategy

You will recall that zebra mussels are small molluscs which have invaded the Great Lakes region of the US. Zebras are hoofed mammals that live in the savannahs of Africa. Which of these two organisms is most likely to survive an ecological disaster? Suppose an asteroid hit the Earth and caused extreme changes in both the terrestrial and aquatic ecosystems. Which of these two organisms would be the most likely to survive?

Did you pick zebra mussel? When looking at the life cycle of the zebra mussel, we can see that its strategy is 'disposable'. It lays thousands of eggs which hatch into free-swimming larvae. The larvae swim around until they find a spot to attach. Most of the eggs are 'disposable' and get eaten, but hundreds can survive and grow into new zebra mussels. This is an inexpensive way to get lots of offspring. Given a natural disaster and a changed environment, a few of the thousands of eggs are quite likely to survive.

The zebra is not likely to survive a catastrophe that drastically changed the terrestrial environment. The zebra requires a stable environment. It has a long gestation period. During gestation, the mother needs good nutrition from the grasses of the savannah. After a young zebra is born, it needs maternal care. The savannah must have enough food for the nursing mother so that she can make milk for her offspring. She is protected by the herd which is also grazing in the savannah. A disaster which destroys the savannah would greatly reduce the herd. The few offspring produced would not survive.

These two strategies are the extremes of a continuum across the animal and plant kingdoms. The strategy of disposable offspring is called the r-strategy. The strategy of nurturing is called the K-strategy. Most invertebrates like insects and spiders follow the r-strategy. Larger animals follow the K-strategy. Larger animals will have few young and spend considerable time and energy caring for them. Most mammals have adopted this strategy.

Some animals, such as ducks, are intermediate on the continuum between r-strategy and K-strategy. Duck eggs do not require the length of development of a mammal but do need parental care. Some of the eggs can be lost to predators which is why there is more than one. The young ducklings also require some parental care. Even within the world of bird species, there is a continuum. Some birds lay only one egg and spend lots of energy caring for it, such as penguins in the Antarctic. Some birds produce lots of eggs and do not spend nearly as much energy in parental care.

This table compares the life histories of r-strategy species and K-strategy species.

Characteristic	r-strategy	K-strategy
life span	short	long
number of offspring	many	few
onset of maturity	early	late – after a long period of parental care
body size	small	large
reproduction	once during lifetime	more than once during lifetime
parental care	none	very likely
environment	unstable	stable

Environmental conditions of r-strategists and K-strategists

In an unstable environment, it is better to produce many offspring as quickly as possible. This is the r-strategy. In an unstable environment, lots of offspring are lost to unpredictable forces. The few that remain can reproduce and carry on the genes of the organism. You can imagine that weeds have this strategy. Weeds survive well in 'disturbed ground', like the side of a road that is constantly mowed, or the edge of a farm field or a drainage ditch. Weeds produce thousands of seeds and grow quickly to take advantage of these unstable places. In sand dune succession, the plants that grow on the foredune where shifting sand and salt spray cause unstable conditions, are r-strategists. They grow quickly extending roots deep in the sand to attempt to stabilize the unstable dune. If they are successful, the next plants which grow there can count on more stable conditions. The dunes that are more inland have a larger proportion of K-strategy plants and animals.

In a stable environment, the K-strategist flourishes. A mature dune is a stable community which has been built over hundreds of years of ecological succession. If there is no human encroachment to destabilize the mature dune, the plants and animals that live there can be K-strategists. The dune is located far from the coast and is not susceptible to the winds and salt spray, unlike the younger dunes which are closer to the coast. Soil has built up over hundreds of years. It contains humus which holds both mineral nutrients and water. K-strategists such as oak trees flourish in this stable environment. Deer and raccoons which are also K-strategists are common. When a habitat becomes diverse and is filled with a large collection of species, they will be K-strategists.

Ecological disruption

Ecological disruption favours r-strategists such as pathogens and pests. If you call something a pest, it is likely to have an r-strategy. For example, ragweed grows well in an old field. It is well adapted to become established before its competitors get a foothold. It grows quickly and produces lots of pollen. It is a pest to farmers. So are mice, another organism with an r-strategy. A farm is not a balanced mature ecosystem. It is a managed production system where the natural ecology has been severely disrupted and there is little diversity. In a balanced ecosystem, the population of mice would have interactions with other species that controlled their numbers. On a farm, it is likely to be simply the farm cats that interact with the mouse population.

A stable forest has a combination of trees. If one type were wiped out by a pest organism, the others would still be left. A monoculture planted by humans, such as a plantation of pines, risks being wiped out by a pest species that gains access to the plantation.

Using the capture–mark–release–recapture method

The capture–mark–release–recapture method is a sampling technique that enables you to estimate the number of animals in an ecosystem. The technique involves catching some of the population and marking them. The marked animals are released back into the ecosystem and allowed to mix with the others in the population. A second sample of the population is captured. Some in the second sample will be marked and some will be unmarked. The proportion of marked to unmarked individuals in the second sample is the same as the proportion of the originally marked individuals to the whole population.

Here is the formula.

$$\frac{\text{number marked in the second sample } (n_3)}{\text{total caught in second sample } (n_2)} = \frac{\text{number marked in the first sample } (n_1)}{\text{size of the whole population } (N)}$$

or

$$\text{Population size } (N) = \frac{(n_1 \times n_2)}{n_3}$$

Worked example

Suppose you capture and mark 100 grasshoppers and release them back into the ecosystem. Then you capture another sample of 100 grasshoppers and find 10 of them are marked. Estimate the population size.

Solution

$$\frac{10}{100} = \frac{100}{N}$$

or

$$N = 100 \times \frac{100}{10}$$

$$N = 1000$$

This technique has its limitations:

- marks on animals may injure them;
- the mark may make the animals more visible to predators (if marked animals are eaten, your second sample will not be reliable);
- the method assumes that the population is closed (no immigration or emigration; very few populations are closed).

Does the method really work? You can try it at home with popcorn kernels. Count out 200 popcorn kernels and put them in a bag.

- remove 40 kernels and mark with a permanent marker;
- put the marked kernels back and shake the bag;
- remove 40 more kernels and record how many are marked;
- use the formula to determine population size.

Did you come close to 200? One sample is not enough data. To be accurate, you should repeat the sampling technique at least 5 times (10 times is even better). Average the results. How close are you to 200 now?

Estimating the size of commercial fish stocks

How do scientists really know what is in the ocean or a lake? The following are methods we use to predict the size of commercial fish stocks.

When studying catches from what has been caught at sea, scientists can determine the size of the current fish stock in the ocean.

Studying catches

For the North Atlantic Ocean, scientists from ICES (International Council for Exploration of the Sea) are at seaports sampling fish catches. Data is taken on type of fish, age, length and breeding conditions.

Gathering information from fishers

Who is better informed about the number of fish caught than the people who catch them? Scientists from ICES collect information onboard fishing vessels. Some of their tasks are:

- recording the number and kinds of fish that are thrown back;
- tagging and releasing some fish;
- developing questionnaires for fishers about their perception of the catch;
- reviewing the logbook which gives catch-per-unit-effort data (increased effort for the same catch, indicates that fish are getting scarcer).

Using research vessels

Research vessels collect information in a variety of ways.

Casting nets in hundreds of selected locations

Sampling with nets is called trawling. Scientists must make random samples, not visit locations where fish are known to congregate. They must be careful to use the same sampling methods every time they sample so that results can be compared.

Using sound to monitor fish populations

An echo sounder reads information from a pulse that it sends into the water. The returning echo indicates the presence of a school of fish. After doing hundreds of acoustical soundings, the scientist reading the data can even tell the species of fish located. To verify the type of species, a trawl is done and a sample collected. The remote sensing hydroacoustic method can determine both numbers and biomass of fish populations.

Calculating the age of fish in a population

Knowing how many young fish are present and how many older fish is very useful. Too few young fish indicates lack of spawning and too few old fish may mean that over-fishing is taking place. One method of calculating age is to measure the rings in the otoliths (ear bones) of the fish. As the fish grows, new material is deposited in the ear bones. When the rings are counted under a microscope, it gives the age of the fish. Another method is to measure the rings of fish scales. The Great Lakes Center at the University of Michigan uses an Eberbach scale reader and microfiche reader to study fish scales.

Using coded wire tag detectors

Fish populations can be marked by attaching tags to the fish. As the fish are recaptured, the total population can be estimated. This is similar to the mark–release–recapture method. The Michigan Department of Natural Resources (MDNR) puts a coded microscopic wire tag in the nose of Chinook salmon and Lake trout which have been planted (stocked) in the Great Lakes. Tagged fish also have their adipose fin clipped off. Recollecting the fish with the tag helps the MDNR evaluate the behaviour and survival of these fish. In order to read the tag, a handheld detector is used. The fish must be caught and the tag read by hand. In small cities around the Great Lakes there is a programme to have fishers help with the sampling. If they catch a fish that is missing the adipose fin, they place the head in a special box provided by the MDNR. Fish heads are collected and each is checked with the handheld detector for the presence of a tag.

Analysing data using mathematical models

Mathematical models are used to turn all of this data into a form that can be used by the fishing industry and the government to plan the future of the fish in our oceans and lakes.

Maximum sustainable yield

Maximum sustainable yield (MSY) is the highest proportion of fish that can be removed from the total population without jeopardizing this maximum yield in the future.

If the fish stock is too small, there are not enough adult fish to produce sufficient young fish. Fishing from a stock which is too small leads to over-fishing of the stock. If the fish stock is too large, annual reproductive rates may be low because of competition for food. Between these two extremes is a fish stock size which can produce the maximum sustainable yield. To maintain the MSY, enough fish stock must be left to spawn a new population of healthy fish.

International measures to promote conservation of fish

In 2006, an international team of researchers came to a startling conclusion. As of 2003, all seafood species had declined by 29%. If the trend continues, by 2048 there will be no more commercial fishing (see Figure 18.15). Other organizations have come to the same conclusions as these scientists. The EU notes that the North Sea stocks of cod have fallen by 75%. In the UK, frozen fish is imported from as far away as New Zealand.

Figure 18.15 Global loss of seafood species.

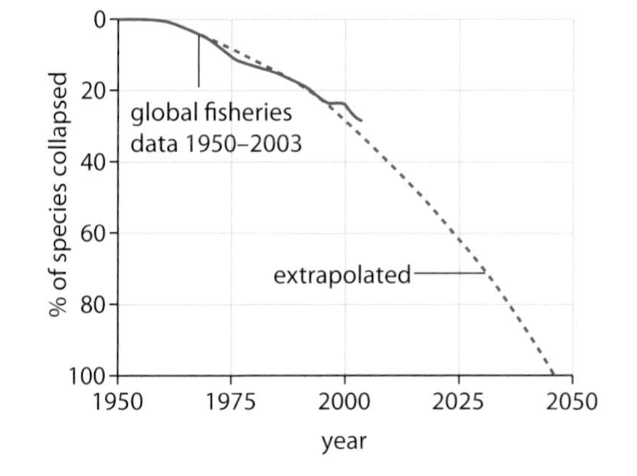

Construct a moral argument for conservation of fish stock by the international community. Explain what value-judgments you are using and what moral principle you are arguing for.

Is it unethical to take away the livelihood of fishers by limiting fish catches?

What ethic issues are involved in cleaning up the oceans?

What ethical issues are present in recreational fishing?

Visit heinemann.co.uk/hotlinks, enter the express code 4242P and click on Weblinks 18.23a and 18.23b.

How many fish are left in the ocean?

Visit heinemann.co.uk/hotlinks, enter the express code 4242P and click on Weblink 18.24.

Since many fish are caught in one part of the world and sold in another, how can consumers be sure that the fish they are purchasing is part of a sustainable fish stock?

Visit heinemann.co.uk/hotlinks, enter the express code 4242P and click on Weblinks 18.25a and 18.25b.

What is to be done to solve this crisis? The following are international measures which would promote the conservation of fish and reverse the current decline.

- Regulate bottom trawling of the ocean. Bottom trawling is basically strip mining of the ocean floor by large fleets of commercial vessels. The United Nations banned drift nets years ago and many people now think the UN should ban bottom trawling.
- Rebuild depleted fish populations as quickly as possible. Pay attention to the concept of maximum sustainable yield.
- Eliminate wasteful and damaging fishing practices. By-catch is the term used to describe the catching of species which are not wanted and are thrown back dead or dying. Thousands of sea turtles have been killed by shrimp trawl nets. Dolphins get stuck in gill nets. Trapping unwanted animals in fishing nets reduces biodiversity unnecessarily.
- Enact strong national fish quota programmes according to the concept of MSY.
- Establish programmes to develop less damaging fishing gear. For example, all US shrimp trawlers are required to have nets with a trap door for sea turtles to escape.
- Provide funds to improve scientific research which counts fish populations and monitors catch.
- Encourage relationships between fishers and scientists.
- Establish marine reserves and no-catch zones to improve biodiversity and increase fish stocks in areas that are protected. The journal *Science* documents small-scale experiments which show that less diverse ecosystems produce less yield. The implication is that loss of biodiversity is what is driving the reduction of fish stocks. It quotes other studies which show that having protected zones, like marine reserves, restores biodiversity and also restores populations of fish outside the protected areas (see Figure 18.16).

It is difficult to enforce and monitor such regulations. Often, international trust does not exist to keep these practices functioning.

Figure 18.16 Fisheries and biodiversity: the evidence.

3 Open ocean fisheries records show widespread decline of fisheries. In 2003, 29% of fisheries were collapsed. Biodiverse stocks fare better.

1 Experimental guidance shows that lowering the diversity of an ecosystem lowers the abundance of fish.

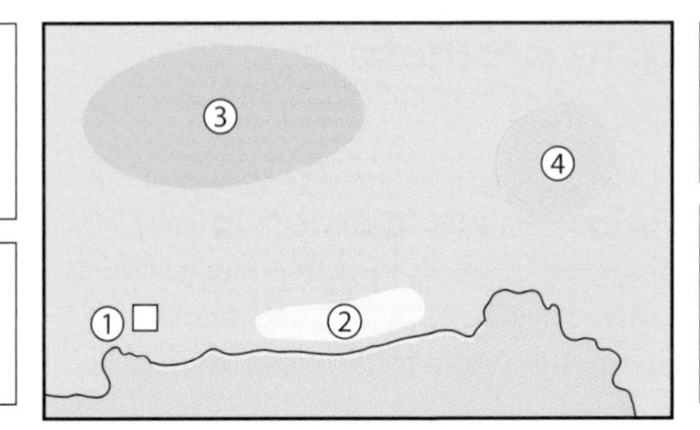

4 No-catch zones show an average 23% improvement in biodiversity and an increase in fish stocks around the protected area.

2 Coastal fisheries records show extensive loss of biodiversity along coasts, with the collapse of about 40% of species. About a third of coastal fisheries are now useless.

Another problem is that prices for fish around the world have risen. A top-grade blue fin tuna can sell for as much as $52 000 in some markets. This encourages fishing for scarcer fish. Catching the big fish eliminates the specimens which lay most of the eggs. With so many fishers going out of business, it is difficult to tell them to limit their catch. Governments may be under political pressure not to limit fish catches.

Marine by-catch is the biggest threat to whales, dolphins, sea turtles, seabirds and certain fish species.

Regional fishing organizations are failing to regulate fishing in international waters. They have no power to control the actions of countries that ignore regulations.

Visit heinemann.co.uk/hotlinks, enter the express code 4242P and click on Weblink 18.26.
On the lefthand-side of the screen, click on 'marine'.

Visit heinemann.co.uk/hotlinks, enter the express code 4242P and click on Weblink 18.27.

At this website look under 'see also' and click on 'Fishing regulators are failing'. Watch a video on seabirds that are killed as a result of by-catch.

For a discussion of international regulations of tuna which should be in place click on the article 'Tuna fishing policy misguided'.

Exercises

17 Distinguish between r-strategies and K-strategies.

18 Describe methods of estimating the size of commercial fish stocks.

19 Describe international measures for promoting fish conservation.

• **Examiner's hint:** 'Distinguish' is asking you to tell the differences between the two items.
'Discuss' is asking you to present the measures generally accepted. However, you must also present an alternative view.

Practice questions

1 Sea water temperature has an effect on the spawning (release of eggs) of echinoderms living in Antarctic waters. Echinoderm larvae feed on phytoplankton. In this investigation, the spawning of echinoderms, and its effect on phytoplankton, was studied.

In the figure below, the top line indicates the number of larvae caught (per 5000 litres of sea water). The shaded bars below show when spawning occurred in echinoderms.

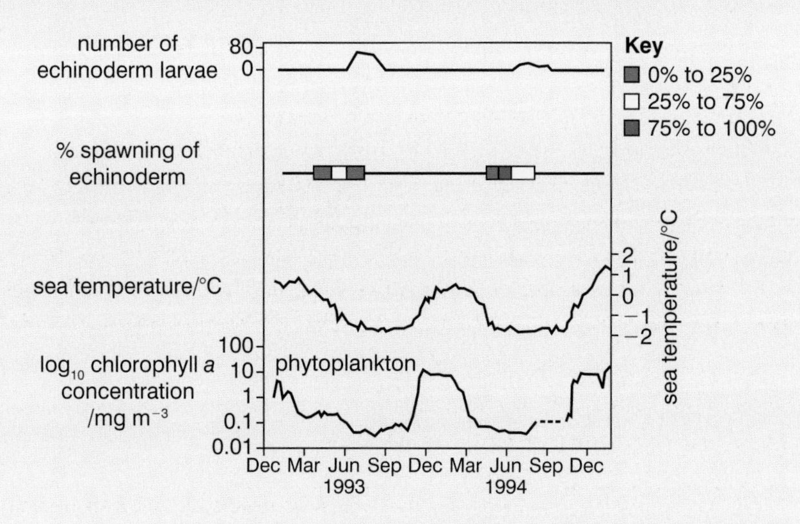

Source: adapted from Stanwell-Smith and Peck (1998), *Biological Bulletin*, **194**, pages 44–52

The concentration of chlorophyll gives an indication of the concentration of phytoplankton. Note that the seasons in the Antarctic are reversed from those in the northern hemisphere.

(a) State the trophic level of echinoderm larvae. *(1)*

(b) Identify the period(s) during which the spawning of echinoderm lies between 25% and 75%. *(1)*

(c) Explain the relationship between the seasons and the concentration of phytoplankton. *(2)*

(d) (i) Outline the effect of sea water temperature on echinoderm larvae numbers. *(2)*

(ii) Using the data in the figure, predict the effect of global warming on echinoderm larvae numbers. *(2)*

(Total 8 marks)

2 (a) List *three* factors that affect the distribution of animal species. *(2)*

(b) Explain the competitive exclusion principle. *(3)*

(Total 5 marks)

3 (a) Outline the damage caused to marine ecosystems by over-exploitation of fish. *(2)*

(b) Explain the use of indicator species in monitoring environmental change. *(3)*

(Total 5 marks)

4 The energy flow diagram below for a temperate ecosystem has been divided into two parts. One part shows autotrophic use of energy and the other shows heterotrophic use of energy. All values are kJ m^{-2} yr^{-1}.

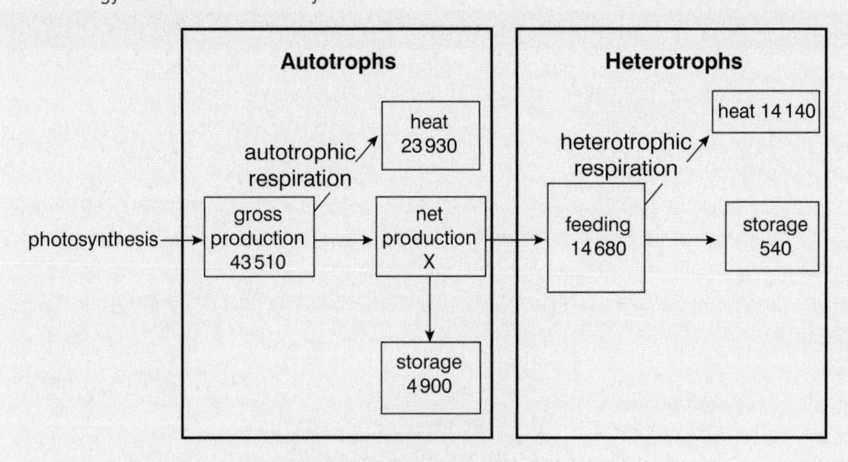

(a) Calculate the net production of the autotrophs. *(1)*

(b) (i) Compare the percentage of heat lost through respiration by the autotrophs with the heterotrophs. *(1)*

(ii) Most of the heterotrophs are animals. Suggest *one* reason for the difference in heat losses between the autotrophs and animal heterotrophs. *(1)*

The heterotrophic community can be divided into food webs based upon decomposers and food webs based upon herbivores. It has been shown that of the energy consumed by the heterotrophs 99% is consumed by the decomposer food webs.

(c) State the importance of decomposers in an ecosystem. *(1)*

(d) Deduce the long-term effects of sustained pollution which kills decomposers on autotrophic productivity. *(2)*

(Total 6 marks)

5 (a) Define the term *competitive exclusion.* *(1)*

(b) List *three* advantages of in situ conservation of endangered species. *(3)*

(Total 4 marks)

6 **(a)** Discuss the ecological and economic arguments to be considered in the conservation of biodiversity in tropical rainforests. *(6)*

(b) Outline the actions taken by farmers to increase nitrogen fertility in the soil. *(4)*

(Total 10 marks)

7 **(a)** Describe *one* technique that ecologists use to estimate accurately the size of a population of animals, including details of any calculations that need to be done. *(6)*

(b) Explain how energy enters a community, flows through it and is eventually lost. *(8)*

(Total 14 marks)

8 The Kluane boreal forest ecosystem project was a large-scale ten-year experimental manipulation of food and predators on an Arctic ground squirrel population (*Spermophilus parryii plesius*).

Three areas were set up:

- a food addition area
- a predator exclusion area
- a food addition area enclosed within a predator exclusion area.

The areas were monitored from 1986 to 1996. In spring 1996, all fences were dismantled and food addition was stopped.

As a further experiment, spring and summer mark–recapture population estimates of the squirrels were conducted from spring 1996 to spring 1998. The results for these two years are shown below. The areas are labelled according to the conditions imposed during the previous ten years.

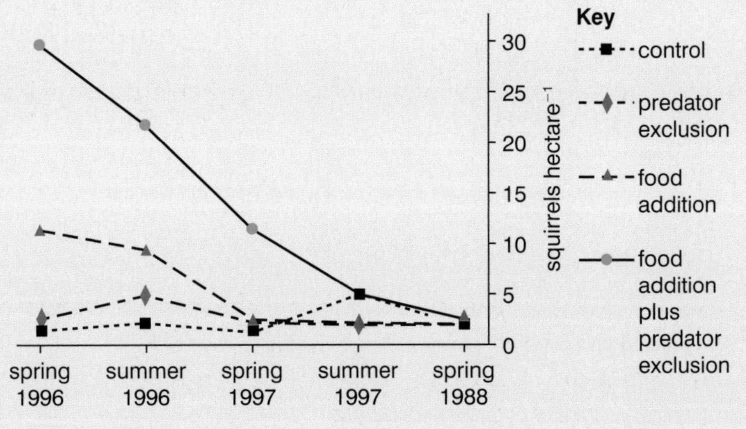

Source: Karels *et al*, (2000) *Nature*, **408**, pages 460–63

(a) State the squirrel population in the food addition plus predator exclusion area in spring 1996. *(1)*

(b) Describe the effect of ending food addition on the squirrel population. *(2)*

(c) Scientists believed that the number of ground squirrels in the boreal forests was limited by an interaction between food and predators that acted primarily through changes in reproduction. Using the data, discuss this hypothesis. *(3)*

(Total 5 marks)

9 Plants called epiphytes grow above the ground on the surface of other plants such as trees. Epiphytes play an important role in rainforests because they can absorb vast amounts of precipitation water (rain and fog), retain minerals effectively and contribute enormous amounts of humus.

An investigation was carried out in Ecuador to determine the distribution and abundance of epiphytes in lowland and mountain rainforests. Branch cover was measured for different branch diameters and different branch angles. (The larger the branch diameter, the closer it was to the trunk of the tree.) The results are shown below.

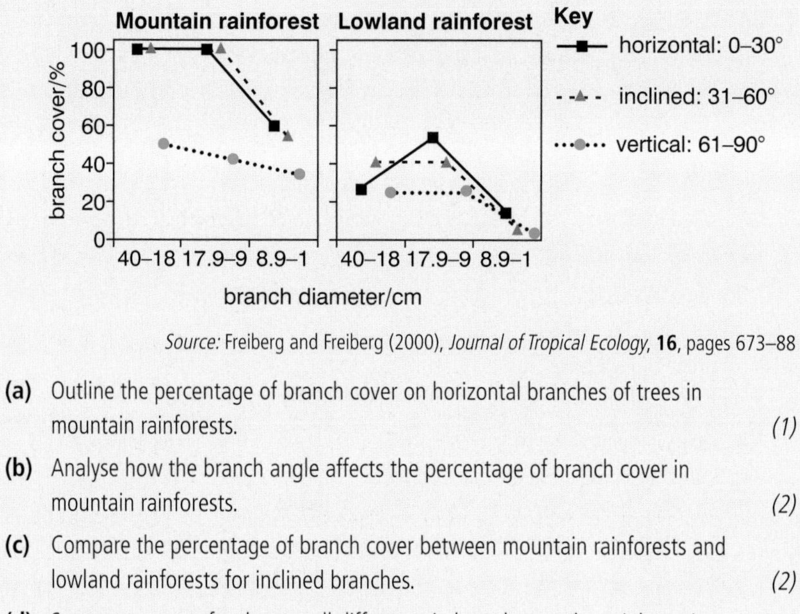

Source: Freiberg and Freiberg (2000), *Journal of Tropical Ecology*, **16**, pages 673–88

(a) Outline the percentage of branch cover on horizontal branches of trees in mountain rainforests. *(1)*

(b) Analyse how the branch angle affects the percentage of branch cover in mountain rainforests. *(2)*

(c) Compare the percentage of branch cover between mountain rainforests and lowland rainforests for inclined branches. *(2)*

(d) Suggest a reason for the overall difference in branch cover by epiphytes in mountain rainforests and lowland rainforests. *(1)*

(Total 6 marks)

10 Explain how parasitism differs from mutualism with reference to named organisms.

(6 marks)

11 (a) State *three* factors that can affect the distribution of animal species. *(1)*

(b) Discuss why an index of diversity could be useful in monitoring environmental change. *(3)*

(c) Describe *two* ways that living organisms could change their abiotic environment during the course of primary succession to climax communities. *(2)*

(Total 6 marks)

12 (a) Discuss reasons why the biodiversity of rainforests should be conserved. *(7)*

(b) Outline how increased UV radiation can affect life on Earth. *(3)*

(Total 10 marks)

13 Foraminifera are small protozoa found in the sediment of all marine ecosystems. Several thousand species of foraminifera live in the Earth's oceans. Because a large number of individuals can be found in a small amount of sediment and because they exist worldwide, foraminifera are useful in examining the distribution of species. The bar graph below summarises data gathered from five coastal regions around North America. Those species occurring in all five regions are considered to be *ubiquitous* and those species occurring in only one area are considered to be *endemic*. The species of foraminifera were placed into three classes based on the number of times each species occurred at the five coastal regions.

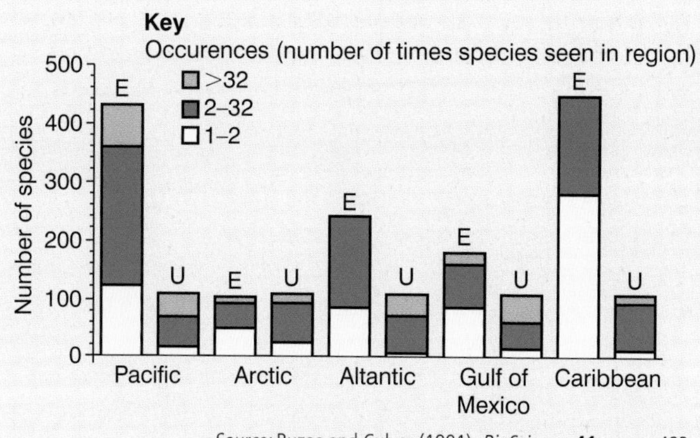

Source: Buzas and Culver (1991), *BioScience*, **41**, pages 483–89

(a) Calculate the percentage of endemic species occurring in the Pacific region. *(1)*

(b) Among the five regions, deduce the region where it would be easiest to find most of the ubiquitous species. *(1)*

(c) Compare the occurrence of endemic species in the Pacific and Caribbean regions. (2)

(d) Suggest, giving a reason, which of the Pacific, Atlantic **or** Caribbean regions will have a greater extinction rate. *(2)*

(e) Identify which region as the lowest species diversity. *(1)*

(Total 7 marks)

14 The brown-headed cowbird (*Molothrus ater*) is a parasitic bird that lays its eggs in the nests of other species. The parasitized hosts often raise the resulting cowbird offspring as their own. The true offspring may starve while the larger cowbird offspring consume most of the food brought by the parents.

The preferred habitat of the brown-headed cowbird is open agricultural areas.

The results of a study into the effects of deforestation on cowbird parasitism of four different host species are shown below.

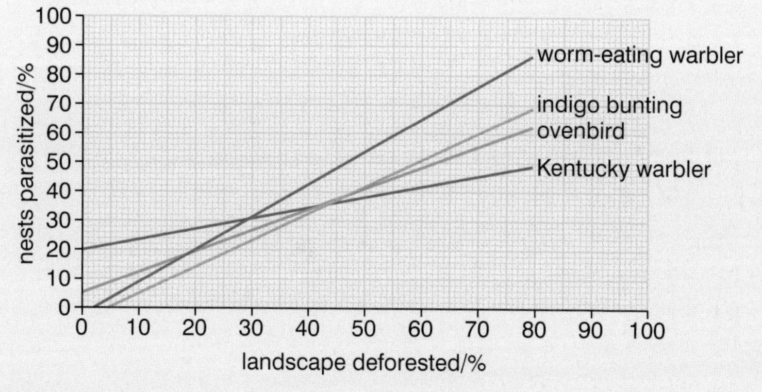

Source: S K Robinson *et al* (1995), *Science*, **267**, pages 1987–90

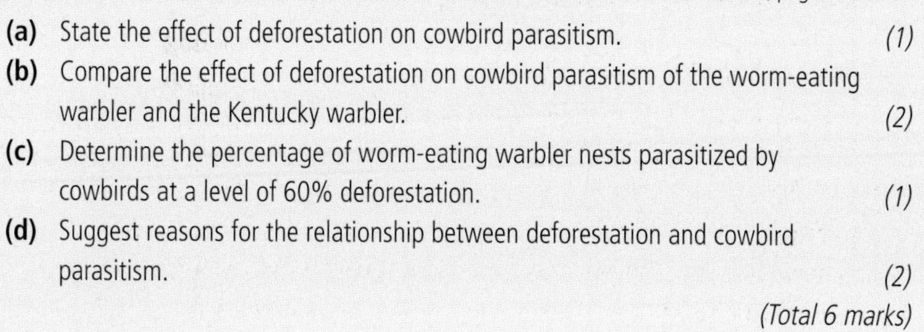

(a) State the effect of deforestation on cowbird parasitism. *(1)*

(b) Compare the effect of deforestation on cowbird parasitism of the worm-eating warbler and the Kentucky warbler. *(2)*

(c) Determine the percentage of worm-eating warbler nests parasitized by cowbirds at a level of 60% deforestation. *(1)*

(d) Suggest reasons for the relationship between deforestation and cowbird parasitism. *(2)*

(Total 6 marks)

15 (a) Outline the use of the Simpson diversity index. *(3)*

(b) Explain the use of biotic indices and indicator species. *(6)*

(Total 9 marks)

16 Draw a labelled diagram of the nitrogen cycle. *(3 marks)*

17 The amount of ozone in the air close to the ground changes during a 24-hour period. Under certain conditions, ozone is produced which increases its concentration in the air. The soil can absorb ozone and this decreases the amount of ozone in the air. The amount of ozone absorbed by the soil is much greater when the air is not moving. The graph shows changes in the amount of zone in the air over a 24-hour period in three locations in Scotland. Each location was a rural area where human influence had no effect on the ozone concentration. The amount of ozone in the air was measured in parts per billion (ppb) with the outer ring representing 40 ppb and the central point representing 20 ppb.

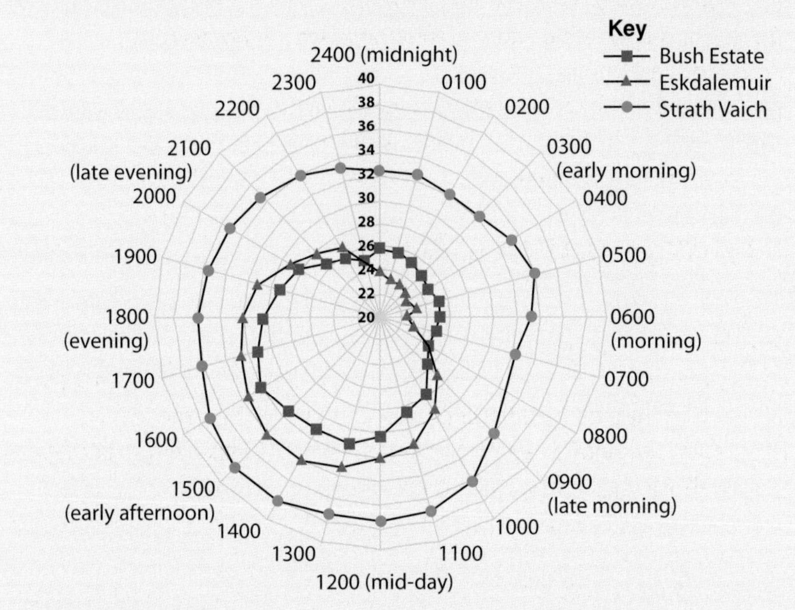

Source: modified from *State of the Environment Air Quality Report* (2000), Scottish Environmental Protection Agency, page 40

(a) Identify the two times of day when the concentration of ozone at Bush Estate was the same as Eskdalemuir. *(1)*

(b) Compare the ozone concentrations at Eskdalemuir and Strath Vaich over the 24-hour period. *(3)*

(c) Deduce the conditions required for ozone to be produced. *(1)*

(d) Explain which of the three locations had the most wind. *(2)*

(Total 7 marks)

19 Further human physiology (Option H)

Introduction

In this chapter, you will explore the physiology of the human body in greater depth. Physiology is a complex topic requiring some knowledge of many other areas of biology including biochemistry and cell biology. Learning what happens is just part of the story, learning why and how things happen in the body is infinitely more interesting. Here are some of the topics we will explore:

- the role of the hypothalamus and pituitary gland in homeostasis;
- the real cause of stomach ulcers;
- the multitude of functions of the human liver;
- when and how the heart valves open and close;
- how oxygen and carbon dioxide are carried in the bloodstream.

H.1 Hormonal control

Assessment statements

H.1.1 State that hormones are chemical messengers secreted by endocrine glands into the blood and transported to specific target cells.

H.1.2 State that hormones can be steroids, proteins and tyrosine derivatives, with one example of each.

H.1.3 Distinguish between the mode of action of steroid hormones and protein hormones.

H.1.4 Outline the relationship between the hypothalamus and the pituitary gland.

H.1.5 Explain the control of ADH secretion by negative feedback.

Overview of the endocrine system

Endocrine glands produce and secrete hormones. Hormones are chemical messengers that most often have a physiological effect far from their gland of origin. Hormones are transported throughout the body by the bloodstream. The cells that are affected by any one hormone are referred to as target cells of that hormone.

Most hormones are under the control of a negative feedback mechanism, but there are exceptions, such as oxytocin, whose secretion is controlled by positive feedback. Many endocrine glands occur in pairs, such as the adrenal glands (see Figure 19.1, overleaf). Some are singular glands, such as the pancreas. The pancreas is also the only gland that has both exocrine and endocrine functions. The exocrine secretions that the pancreas produces are chemicals used for digestion and they are carried to the alimentary canal by a duct. The endocrine secretions of the pancreas travel throughout the body carried by the bloodstream as is the case with all hormones.

As shown in this artwork, each cell's plasma membrane has its own set of proteins. Some of these proteins are receptors for specific hormones.

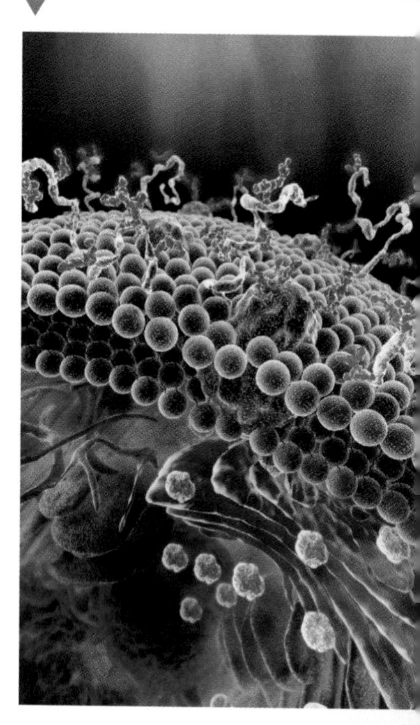

Figure 19.1 Some endocrine glands are shown in this body outline. Endocrine glands are often far from the location where their hormone has an effect. Hormones are carried from their gland of origin to their target cells in the bloodstream.

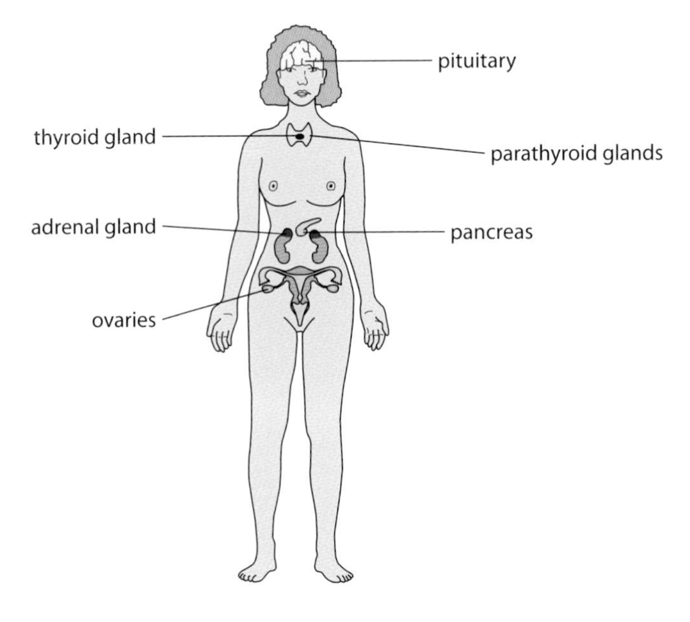

Types of hormone and their mode of action

Hormones are characterized by their chemical composition and mode of action as shown in this table.

Hormone type	Example	One effect of example hormone
steroid	oestrogen	increases thickness of uterine lining
peptide	insulin	promotes glucose uptake by body cells
tyrosine derivative	thyroxin	increases metabolic rate

Steroid hormones are typically synthesized from cholesterol and are classified as lipids. Steroids, therefore, have the chemical and solubility properties of a lipid. You will recall that a plasma membrane (or any cell membrane) is a nearly continuous double layer of phospholipids. This means that steroids easily pass through cell membranes because both steroids and phospholipid molecules are relatively non-polar. Once a steroid hormone has entered the cytoplasm of a cell, it bonds with a receptor protein and forms what is called a hormone–receptor complex. This complex then passes through the nuclear membrane and selectively binds to certain genes. In some instances, this inhibits transcription, and in other cases it induces transcription. In this way, steroid hormones control the production of proteins within the target cell. The target cells of steroid hormones have their biochemistry dramatically altered as a result of the presence of the hormone. Other examples of steroid hormones include progesterone and testosterone.

Peptide hormones are protein molecules. When a peptide hormone reaches a target cell, the hormone binds to a receptor protein on the outer surface of the cell membrane. The presence or absence of the hormone's receptor protein determines whether or not a cell is a target cell of that particular hormone. Once a peptide hormone has chemically bonded to a receptor protein, a secondary messenger molecule is triggered into action in the cytoplasm of the cell. Notice that peptide hormones do not enter a cell: they bind to a receptor protein and the subsequent action of a secondary messenger molecule within the cytoplasm results in the action associated with the hormone.

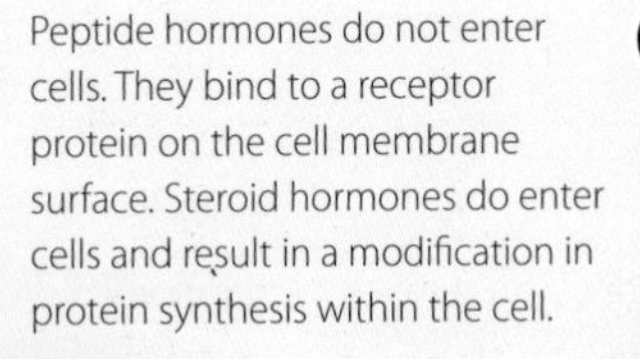
Peptide hormones do not enter cells. They bind to a receptor protein on the cell membrane surface. Steroid hormones do enter cells and result in a modification in protein synthesis within the cell.

The pituitary gland and its 'boss', the hypothalamus

It is common to read that the pituitary gland is considered the 'master gland'. It is true that the pituitary gland produces many different hormones and some of those hormones influence the production and secretion of other hormones, but the pituitary itself is largely controlled by the action of the nearby hypothalamus (see Figure 19.2).

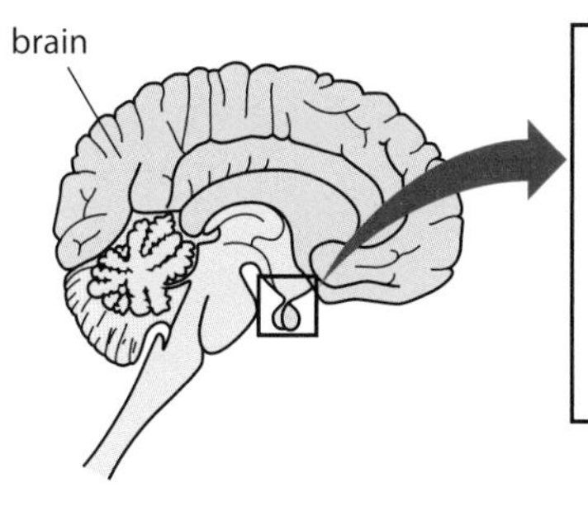

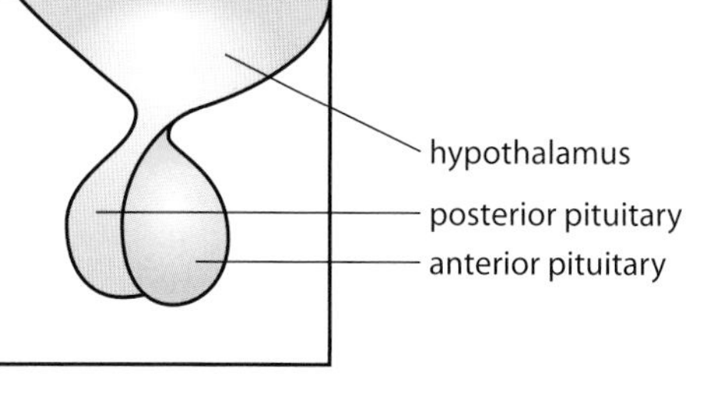

Figure 19.2 Position of the hypothalamus and pituitary relative to the rest of the brain.

Most people refer to the pituitary gland as a singular gland, but it is actually two glands that exist as what we called different 'lobes'. The anterior and posterior lobes of the pituitary communicate with the hypothalamus in different ways.

The posterior lobe of the pituitary contains the axons of cells called neurosecretory cells (see Figure 19.3). These are very long cells whose dendrites and cell bodies are located in the hypothalamus and whose axons extend down into the posterior pituitary. A variety of hormones, such as oxytocin, are produced at the cell body end of these cells and then move down the axon into the posterior pituitary gland. This explains why these hormones are said to have been produced within the hypothalamus, but they are secreted from the posterior pituitary.

Some countries have stricter laws than others concerning adding steroids and other hormones to livestock in order to increase meat production. Any country which imports meat from another country needs to be aware of the possibility of hormones being added to livestock feed.

The relationship between the hypothalamus and the anterior pituitary works differently (see Figure 19.3). The hypothalamus contains capillary beds which take in hormones produced by the hypothalamus itself. These hormones are often referred to as 'releasing hormones'; for example, gonadotrophin releasing hormone (GnRH). The capillary beds join together into a blood vessel known as a portal vein, which extends down into the anterior pituitary. Here, the portal vein branches into a second capillary bed which allows the releasing hormones to leave the bloodstream for their target cells, the cells of the anterior pituitary. The releasing hormones cause the anterior pituitary cells to secrete specific hormones. For example, GnRH causes the secretion of both follicle stimulating hormone (FSH) and luteinizing hormone (LH). The hormones produced by the anterior pituitary enter the bloodstream through the same capillary beds which allowed the releasing hormones to exit. As you learned in the reproductive system unit, the target cells of LH and FSH are the gonads of both females and males.

Figure 19.3 Posterior pituitary (left) with its hormones and relationship to hypothalamus by way of neurosecretory cells. Anterior pituitary (right) with its connection to the hypothalamus by way of portal blood vessels.

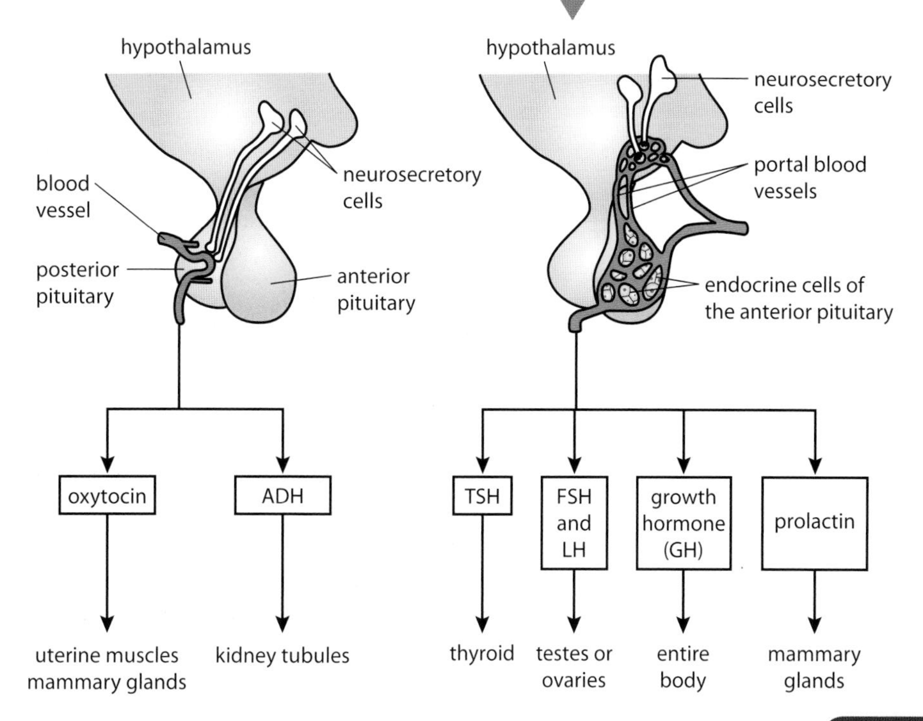

Negative feedback control of ADH secretion

A very good example of the relationship between the hypothalamus and the pituitary gland is the control mechanisms at work for the secretion of antidiuretic hormone (ADH). As you will recall, ADH is the hormone which controls how much water is reabsorbed from the collecting duct back into the bloodstream.

- If ADH is secreted, the collecting duct becomes permeable to water and thus water leaves the collecting duct by osmosis into the highly hypertonic medulla of the kidney. This water is then reabsorbed back into the bloodstream.
- If ADH is not secreted, the collecting duct is impermeable to water and thus the urine contains a relatively high content of water.

Although the pituitary is often referred to as the 'master gland', the hypothalamus controls the hormonal secretions of the pituitary.

ADH is produced by the dendrites of the neurosecretory cells that extend from the hypothalamus to the posterior pituitary. ADH then diffuses down the axon of this long cell and is stored in membrane-bound granules within the synaptic end of the neurosecretory cell. The secretion of ADH from these granules depends on an action potential from the hypothalamus (see Figure 19.4). Within the tissues of the hypothalamus are osmoreceptors. These receptor cells monitor the water content of the blood as it passes through capillary beds in the hypothalamus. If water content in the blood is low, an action potential is sent to the posterior pituitary to secrete ADH and thus water is reabsorbed back into the bloodstream. If water content is relatively high, no action potential is sent and, therefore, no ADH is secreted. This results in water leaving the body by way of a relatively dilute urine.

This control mechanism illustrates the concept of negative feedback. The water content of the blood (and therefore the rest of the body) is maintained within a relatively narrow normal range. It is normal for some fluctuation within this range, but osmoregulation keeps the body from deviating too far from the normal range.

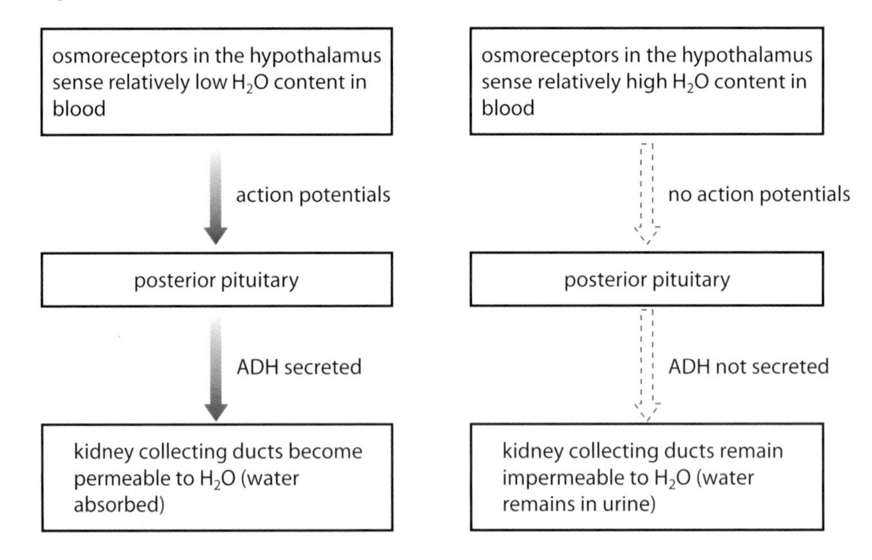

Figure 19.4 Osmoreceptors in the hypothalamus monitor water content in blood and control water balance (osmoregulation) by way of ADH.

Exercises

1. List as many locations as possible for ADH from formation to utilization.
2. Differentiate the action of peptide hormones from steroid hormones.
3. Identify the 'structures' that functionally connect the hypothalamus to:
   - **a** the posterior pituitary
   - **b** the anterior pituitary

H.2 Digestion

Assessment statements

- H.2.1 State that digestive juices are secreted into the alimentary canal by glands, including salivary glands, gastric glands in the stomach wall, the pancreas and the wall of the small intestine.
- H.2.2 Explain the structural features of exocrine gland cells.
- H.2.3 Compare the composition of saliva, gastric juice and pancreatic juice.
- H.2.4 Outline the control of digestive juice secretion by nerves and hormones, using the example of secretion of gastric juice.
- H.2.5 Outline the role of membrane-bound enzymes on the surface of epithelial cells in the small intestine in digestion.
- H.2.6 Outline the reasons for cellulose not being digested in the alimentary canal.
- H.2.7 Explain why pepsin and trypsin are initially synthesized as inactive precursors and how they are subsequently activated.
- H.2.8 Discuss the roles of gastric acid and *Helicobacter pylori* in the development of stomach ulcers and stomach cancers.
- H.2.9 Explain the problem of lipid digestion in a hydrophilic medium and the role of bile in overcoming this.

The process of digestion requires 'juice'

Chemical digestion requires that various substances be added to the ingested food as it moves along the alimentary canal. Many of the molecules that we ingest are much too large to be absorbed and thus the primary action of digestion is to convert macromolecules (like starch) to monomers (like glucose). This digestive conversion requires glandular tissues to secrete many enzymes and a few other fluids to be added at various points along the way. Here is a gland-by-gland summary of some digestive secretions.

Salivary glands

The salivary glands secrete saliva into the mouth. Saliva includes the first of many digestive enzymes, salivary amylase.

Gastric glands

Gastric glands are located in the inner lining of the stomach. These glands secrete mucus, hydrochloric acid, and pepsinogen, the precursor of pepsin – an enzyme which begins protein digestion.

The pancreas

The pancreas sends pancreatic juice to the small intestine by way of a duct. This juice contains another protease enzyme, additional amylase, and lipase. It also contains a form of hydrogen carbonate to help neutralize the acidic fluids from the stomach.

False-colour SEM of epithelial cells in the interior of the stomach (orange) and mucus (yellow). Mucus protects the cells from gastric juice secreted into the stomach.

The liver

The liver secretes bile into the small intestine. The bile can come straight from the liver or from the gall bladder where it is stored. Bile emulsifies lipids and thus increases the surface area of lipids for the action of lipase.

Intestinal glandular cells

Some of the cells of the inner lining of the small intestine are glandular cells. These cells secrete a variety of digestive enzymes. Some of these enzymes are added to the partially digested fluid within the small intestine and some of the enzymes stay attached to the villi cells. These attached (or membrane-bound) enzymes catalyse digestive reactions as the undigested substrate molecules flow past in the lumen of the small intestine.

The cells of exocrine glands

An endocrine gland is a ductless gland which secretes a hormone into the bloodstream. An exocrine gland has ducts which take a secretion from the gland to a specific location.

An exocrine gland is a collection of cells that produce and secrete a product which is carried to a specific location in the body by way of a duct. The secretion is often a protein, such as many of the digestive enzymes. Let's recap some of the major steps of protein synthesis and secretion (because the molecules and organelles associated with this process are numerous and apparent in exocrine gland cells):

- mRNA is transcribed from a gene of DNA;
- mRNA is translated at a ribosome in the cytoplasm of a cell;
- ribosomes are often attached to endoplasmic reticulum, making rough ER (RER);
- synthesized proteins at ribosomes move through the channels of ER and reach a Golgi body;
- Golgi bodies package protein into a vesicle (some larger vesicles are called secretory granules);
- vesicles fuse with the plasma membrane to release their contents outside the cell in a process called exocytosis or secretion;
- several steps of protein synthesis and secretion require ATP, therefore these cells typically contain an above average number of mitochondria.

Therefore, exocrine gland cells can be expected to contain extensive endoplasmic reticulum, as well as many ribosomes, Golgi bodies, vesicles or granules, and mitochondria (see Figure 19.5).

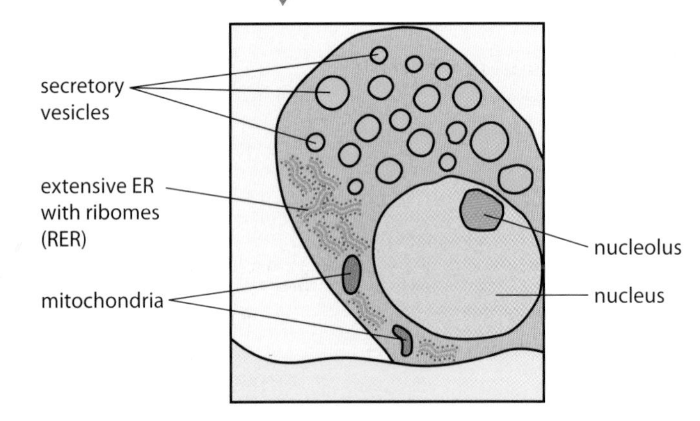

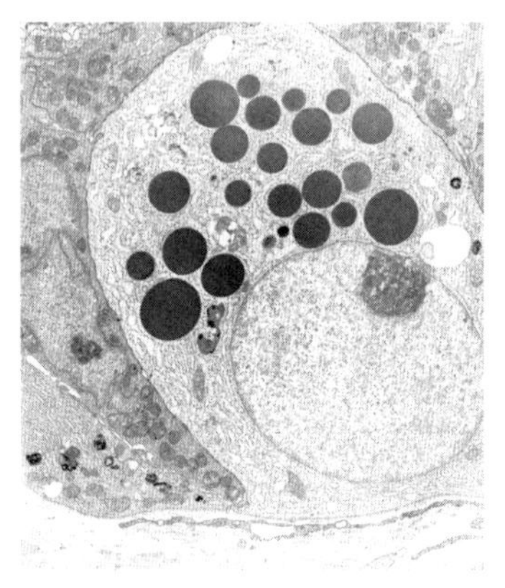

Figure 19.5 This is a drawing and a TEM of a cell from the small intestine. Use the labels on the drawing to identify structures in the photograph.

The arrangement of exocrine gland cells into acini and ducts

Exocrine gland cells secrete a product into a duct to be transported to a specific location; for example, the cells in the pancreas that secrete digestive enzymes.

These exocrine cells group around – in effect, surround – the end of a very small branch of the pancreatic duct. All the cells surrounding this small branch (ductule) secrete digestive enzyme into it. The ductule takes the secretion into larger and larger ducts until the pancreatic duct is reached. When looked at microscopically, the arrangement of the cells round the ductule looks a bit like houses (cells) arranged around a cul-de-sac (ductule) (see Figure 19.6). To further the analogy, the cul-de-sac leads to a series of larger and larger roads that eventually lead to a major thoroughfare (pancreatic duct).

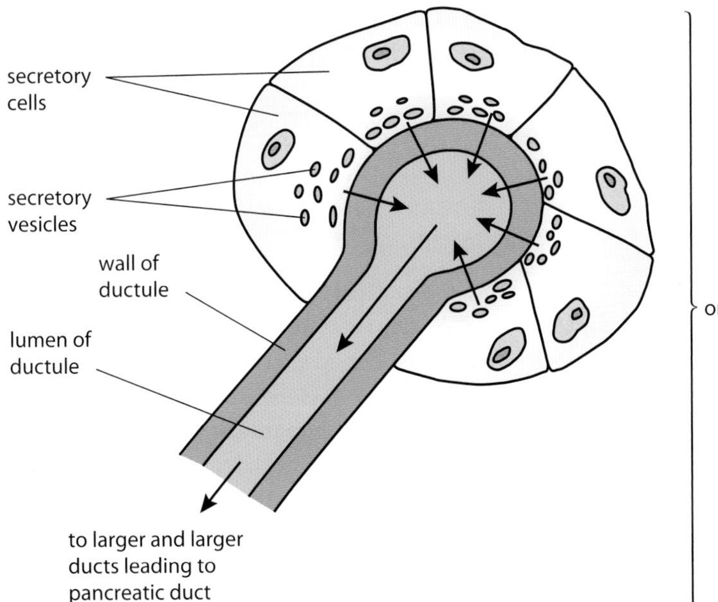

Figure 19.6 Arrangement of exocrine gland cells round a ductule. The arrangement of cells and ductule is called an acinus. Many acini join together draining their secretions into larger ductules.

Components of saliva, gastric juice, and pancreatic juice

Three common exocrine gland secretions and their components are listed below.

Saliva:
- solvent is water;
- amylase;
- mucus.

Gastric juice:
- solvent is water;
- mucus;
- hydrochloric acid;
- pepsin (secreted as pepsinogen).

Pancreatic juice:
- solvent is water;
- amylase;
- bicarbonate;
- trypsin (secreted as trypsinogen);
- lipase.

Control of gastric juice secretion

Your body does not secrete digestive juices at all times. This would be wasteful, unproductive and possibly harmful. Specific digestive juices need to be secreted at the right times in order to help hydrolyse the type of molecule currently in need of digestion. Let's use the secretion of juices from the stomach as an example of how digestive secretions can be regulated.

The classic experiments of Ivan Pavlov showed how even seeing or smelling food can begin the body's preparation for digestion. Pavlov measured drops of saliva from dogs that were merely being prepared to be fed. This meant that the dog's nervous system was somehow influencing the secretion from the salivary glands.

Digestion is necessary to convert ingested foods into a molecular size that can be absorbed through a selectively permeable cell membrane. This also ensures that the molecule will be useful to the body as a monomer.

The same is true for the secretion of gastric juice from the stomach – the sight or smell of food will initiate the secretion of this juice.

Russian physiologist Ivan Pavlov laid the foundation for our understanding of the principles of the autonomic nervous system as well as classical conditioning.

Once food has entered the stomach, receptors within the stomach wall are stimulated and send sensory signals to the brain. The brain responds by causing the stomach to secrete even more gastric juice. Distension of the stomach results in the production of a hormone known as gastrin. This hormone leads to the sustained release of gastric fluid, and especially its hydrochloric acid component.

Membrane-bound digestive enzymes

Enzymes that enter the alimentary canal via ducts typically catalyse hydrolytic reactions by way of mixing in with ingested substances. The substrates and enzymes mix together in a molecular 'soup' that keeps moving through the alimentary canal. These enzymes have a very limited molecular 'life span' and are either digested themselves or eliminated.

Some digestive enzymes do not join the molecular soup, but instead are produced and remain in the membranes of cells composing the inner lining of the small intestine. An example of such a membrane-bound enzyme is maltase, the enzyme that hydrolyses the disaccharide maltose into two glucose molecules. Maltase remains embedded in the inner epithelial cell membranes of villi and microvilli. Maltose floats to the active site, and the enzyme then catalyses the hydrolysis reaction. The advantage is that maltase remains in the lumen of the small intestine for longer than free-floating enzymes, and the product (glucose) is in exactly the right place for absorption.

Humans cannot digest cellulose

Although a fairly large number of mammals are herbivores, not a single mammal (including humans) produces an enzyme able to digest cellulose. Cellulose is a polysaccharide carbohydrate which is composed of thousands of glucose monosaccharides. Mammals that we know as 'grazers' contain a large colony of mutualistic microorganisms which produce cellulase, the enzyme necessary to hydrolyse cellulose into glucose. Even with this bacterial digestive help, the relative energy yield from plant material is small, so these animals must ingest a large mass of plant material. Grazers spend many hours each day simply eating.

Although humans do eat some plant material high in cellulose and we do have a mutualistic association with intestinal bacteria, we do not have a relationship with the type of bacteria that produce cellulase. Thus, the plant material we eat largely makes its way through the alimentary canal and exits the body in faeces.

Why doesn't the alimentary canal digest itself?

Pepsin and trypsin are two digestive protease enzymes. Proteases are the group of enzymes that collectively hydrolyse peptide bonds within proteins. A fully active protease enzyme cannot distinguish between ingested protein and protein that is part of the structure of the human body. In order to prevent the hydrolysis of useful body proteins, pepsin and trypsin are initially synthesized in a molecular form that is not chemically active. These inactive forms are known as zymogens.

Pepsin is initially synthesized with a primary structure that includes 44 additional amino acids. In this form, pepsin is known as pepsinogen. Pepsinogen is produced in the inner lining of the stomach wall and remains in this zymogen form until it enters the cavity of the stomach. Here, pepsinogen is exposed to the hydrochloric acid characteristic of the stomach secretions and the additional 44 amino acids are removed (see Figure 19.7). This converts pepsinogen into pepsin and the enzyme becomes active. The living inner lining cells of the stomach are protected from the hydrochloric acid and digestive enzymes by a lining of mucus.

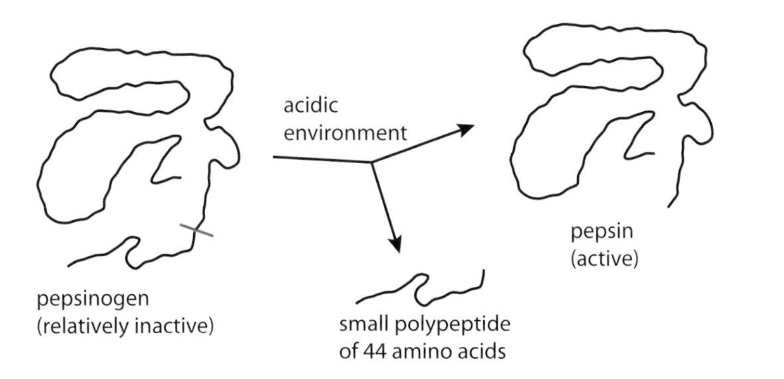

Figure 19.7 How the zymogen form of a hydrolytic enzyme (pepsinogen) is converted to the active form (pepsin).

Pepsin and trypsin were two of the first enzymes discovered. They were named before researchers began naming enzymes with the suffix -ase.

Trypsinogen is synthesized in the pancreas and is one of the components of pancreatic juice. Pancreatic juice is brought to the duodenum of the small intestine by the pancreatic duct along with lipase, amylase and hydrogen carbonate. When partially digested food enters the duodenum from the stomach, an enzyme known as enterokinase (or enteropeptidase) is produced. Enterokinase converts trypsinogen into trypsin and thus activates that protease enzyme.

How do we digest lipids?

Everyone needs to eat a certain amount of lipid in their diet. Lipids have many functions in the human body. For example, we use glycerol and fatty acids, the digested forms of triglyceride lipids, to help synthesize phospholipids for cell membrane structure. Lipids pose a problem for the digestive process as they are relatively insoluble in water and the fluids moving through the alimentary canal are aqueous. In other words, lipids are hydrophobic molecules in a hydrophilic medium. In an aqueous environment, lipid molecules tend to 'stick together' or coalesce. These coalesced globules of lipids have relatively little surface area in comparison to their overall volume.

The enzyme that hydrolyses triglyceride lipids is lipase, which is added into the partially digested soup in the duodenum of the small intestine. If lipase encounters a large lipid globule, it catalyses the hydrolysis of lipid molecules on the outside of the globule because these are the molecules which are accessible to the enzyme. Most of the interior lipid molecules of the globule would never encounter lipase in this scenario. The solution to this problem of relatively little surface area to a large volume, is the addition of bile into the molecular soup.

The gall bladder (small green sac) is located just below the liver. The blood vessels associated with the liver are also shown.

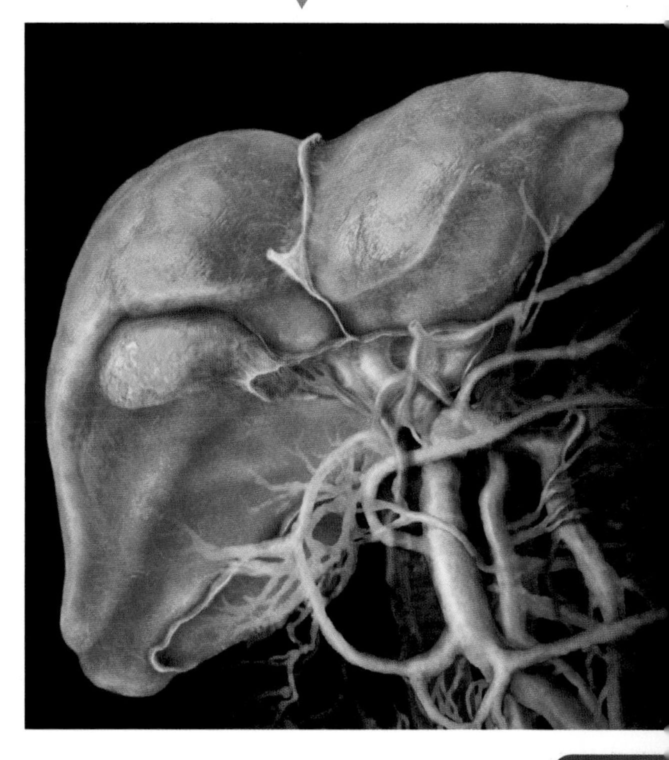

Bile is produced by liver cells (hepatocytes), stored in the gall bladder, and added to the partially digested food via a duct that leads to the duodenum. Bile molecules have both a hydrophobic end and a hydrophilic end and thus are partially soluble in both lipids and in water. Bile molecules, in effect, insert themselves between lipids molecules and prevent lipids from coalescing into large globules. This is known as emulsification (see Figure 19.8). Following emulsification, lipids are unchanged in any molecular way, but they are in much smaller globules or droplets. The same volume of lipid now has a much greater surface area for lipase to work on.

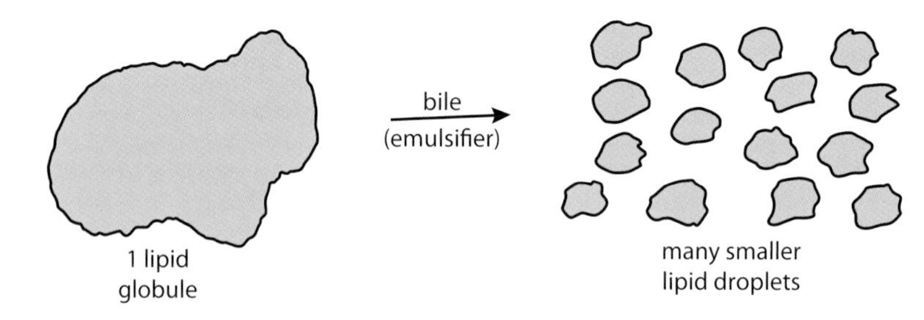

Figure 19.8 Bile is not an enzyme and does not digest lipids. Bile is an emulsifier and breaks up relatively large accumulations of lipid into smaller droplets.

Lipase molecules themselves have an interesting structure that helps to solve the problem of lipids being hydrophobic. The overall structure of lipase molecules is hydrophilic as many of the amino acids are polar. This allows lipase to be soluble in an aqueous environment. But the amino acids at the active site of lipase are predominately non-polar and thus accept a hydrophobic substrate (the lipid) into the relatively hydrophobic active site.

What causes stomach ulcers?

There have been many scientific ideas that have been proven false over time. A notable example in biology was the idea that proteins were the molecules that store our genetic information. The work of many researchers over many years proved that DNA stores our genetic information.

The answers to scientific questions sometimes change. Can anything live in the highly acidic environment of our stomach? Until fairly recently, the answer to that question was thought to be no. The fluid in the stomach can be as acidic as pH 2. The consensus among scientists was that no living organism could survive such a harshly acidic environment.

In the years 1982–83, Dr Barry J. Marshall and Dr J. Robin Warren isolated living bacterial cells from the stomach lining of patients suffering from stomach ulcers and gastritis (inflammation of the stomach lining). The conventional wisdom during that time period was that stomach ulcers were caused by excess production of hydrochloric acid, perhaps brought on by stress. Dr Marshall and Dr Warren have since shown (or inspired others to show) the following.

- Bacteria known as *Helicobacter pylori* do survive when introduced into the stomach, probably by burrowing beneath the mucus layer and infecting stomach lining cells.
- *Helicobacter pylori* employ the enzyme urease to create ammonia and this helps to neutralize stomach acid.
- *Helicobacter pylori* infection of the stomach lining leads to gastritis and stomach ulcers.
- Patients treated with a selected range of antibiotics respond well to treatment.
- Patients with gastritis (and therefore infected with *Helicobacter pylori*) for many years (20–30 years, for example) are much more prone to stomach cancer than the general population.
- *Helicobacter pylori* infection may very well be the most common bacterial infection in the entire world, as over 3 billion people are estimated to be infected.

In 2005, Dr Marshall and Dr Warren were awarded a Nobel Prize in medicine and physiology for their work in determining the association of *H. pylori* with stomach inflammation, ulcers, and possibly cancers. Their work is evidence that information which we hold to be fact is still open to new interpretation, given new information.

The Helicobacter Foundation website gives a variety of information concerning *Helicobacter pylori*. Visit heinemann.co.uk/hotlinks, enter the express code 4242P and click on Weblink 19.1

Exercises

4 What is the primary advantage for some enzymes being bound into the plasma membranes of villi epithelial cells?

5 How is amino acid polarity important to the function of lipase?

6 Many people have their gall bladder surgically removed. Why can a person live a normal life without a gall bladder?

H.3 Absorption of digested foods

Assessment statements

H.3.1 Draw and label a transverse section of the ileum as seen under a light microscope.

H.3.2 Explain the structural features of an epithelial cell of a villus as seen in electron micrographs, including microvilli, mitochondria, pinocytotic vesicles and tight junctions.

H.3.3 Explain the mechanisms used by the ileum to absorb and transport food, including facilitated diffusion, active transport and endocytosis.

H.3.4 List the materials that are not absorbed and are egested.

Overview of small intestine structure

The small intestine is composed of three sections. First, there is the short duodenum where many secretions from the pancreas and liver are added to the partially digested contents. Then there are two long sections called the jejunum and ileum respectively. Partially digested food passes through the small intestine because of the contractions of two layers of smooth muscle within the walls. These two layers are the longitudinal and circular muscle layers shown in Figure 19.9. The rhythmic contraction of these two muscle layers is known as peristalsis and keeps the food moving through the alimentary canal from beginning to end.

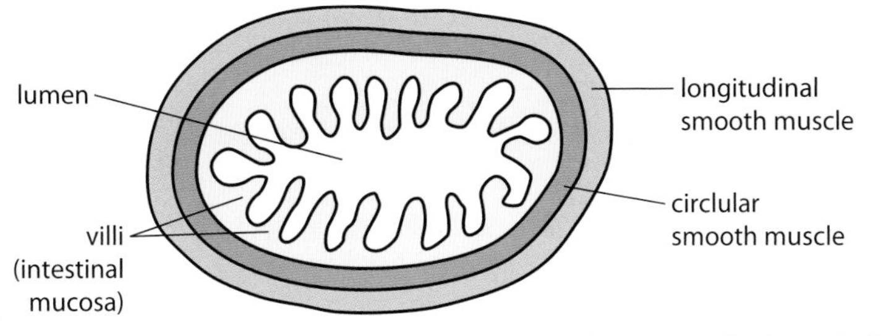

Figure 19.9 Sectioned view of a portion of the small intestine. The two layers of smooth muscle are responsible for peristalsis.

The innermost cellular lining of the small intestine is known as the intestinal mucosa. This is the tissue that is in direct contact with the partially digested food and is responsible for absorption. The mucosa forms irregularly shaped invaginations known as villi. The villi (see also Chapter 6, page 155) of the small intestine increase the surface area for absorption. Each villus contains both a capillary bed and a lacteal (smallest vessel of the lymphatic system).

To see an image of villus, showing pits, two layers of muscles, etc. in close-up cross section, visit heinemann.co.uk/hotlinks, enter the express code 4242P and click on Weblink 19.2.

Adaptations of villi epithelial cells for efficient absorption

Digested molecules must pass through epithelial villi cells and are absorbed into either a capillary or lacteal on the interior of each villus.

Microvilli

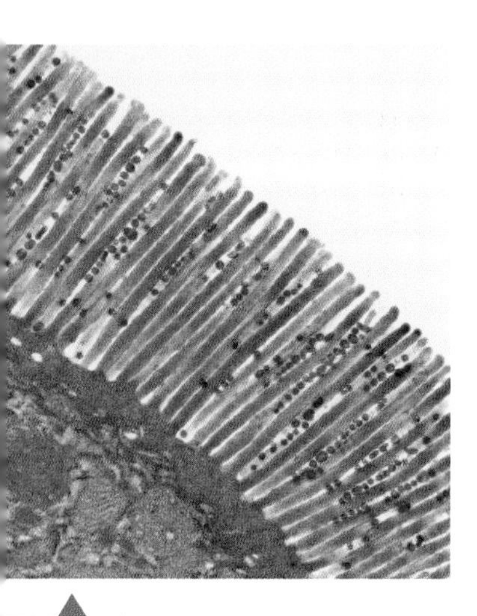

As you can see in this false-colour TEM, each epithelial cell lining the intestine has many microvilli extending into the intestinal lumen.

The surface of each villus cell that faces into the lumen (cavity) of the small intestine has many microscopic finger-like projections known as microvilli. The function of microvilli, like that of villi, is to greatly increase the surface area for absorption (compared to what it would be if the interior of the intestine were smooth).

Mitochondria and pinocytotic vesicles

Some of the molecules absorbed through the plasma membranes of the villi are absorbed using an active transport mechanism. The requirement of active transport mechanisms for ATP partly explains why the epithelial villi cells contain mitochondria. In addition, near the plasma membrane surface, pinocytotic vesicles are often visible. Pinocytosis is another active transport mechanism often used to absorb molecules from the lumen of the intestine into the interior of the villi cells.

Tight junctions

Most cells in the body are surrounded by intercellular (interstitial) fluid. Even cells that make up the outer boundary of an organ typically allow molecules to move between cells. This would be an unacceptable situation for epithelial cells that make up villi. If intercellular fluid and dissolved molecules moved between adjoining cells, nutrients would have no selective barrier to pass through. It is the movement of digested molecules through the selectively permeable membrane of the villus that guarantees that the molecules have completed the process of enzymatic digestion. To this end, epithelial cells of villi are sealed to each other by membrane-to-membrane 'seals' called tight junctions (see Figure 19.10). The two cell membranes share some membrane proteins. This results in the two membranes being held so tightly together that most molecules cannot pass between them either into or out of the lumen of the small intestine.

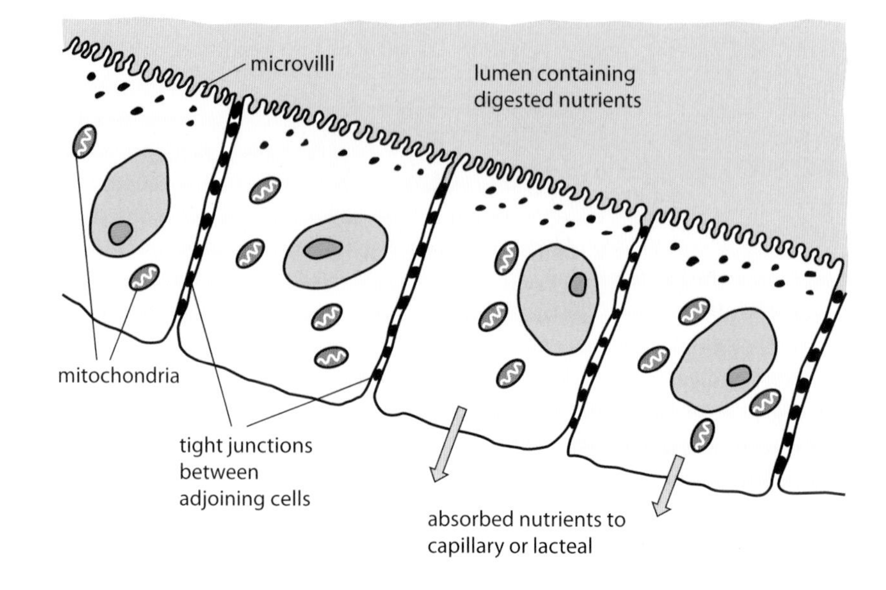

Figure 19.10 Individual epithelial cells of the intestinal mucosa. Digested molecules must pass through these cells in order to reach a capillary bed or lacteal.

What transport mechanisms are used to absorb foods?

A variety of transport mechanisms are used to move molecules from the intestinal lumen across the plasma membrane making up the microvilli of the intestinal cells (see Figure 19.11). The molecules that move across this membrane have completed the digestion process and are of a molecular size that allows their entry into the cells of the villi.

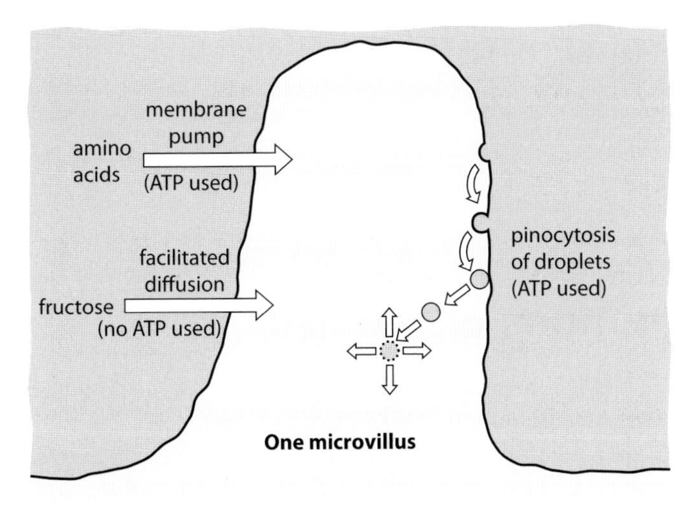

Figure 19.11 Examples of transport mechanisms used during absorption. The process of absorption occurs through the microvilli (see Figure 19.10).

Facilitated diffusion

One of the mechanisms used is facilitated diffusion. Many of the recently digested molecules are small enough to 'fit through' the membrane and their concentration gradients permit diffusion, but their polarity prevents easy passage through the hydrophobic interior of the plasma membrane. Protein channels in the microvilli plasma membranes solve this problem. The protein molecule(s) of the channel have relatively non-polar amino acids making up the outer perimeter of the channel and polar amino acids forming the interior of the channel. This interior channel allows appropriately sized, polar molecules to diffuse from the lumen of the intestine into the cytoplasm of the villi cells.

Active transport

Active transport by membrane pumps is another transport mechanism used for absorbing digested foods. The plasma membrane of the microvilli contains membrane proteins which use ATP to transport molecules across the membrane. Many of the molecules being absorbed do not have a marked concentration gradient across the plasma membrane. Active transport by membrane pumps solves this problem as a membrane pump can transport a molecule regardless of molecular concentrations. The interior of the epithelial cells contains mitochondria to aid in the production of ATP for these membrane pumps.

Pinocytosis

Microvilli plasma membranes also perform endocytosis, or more specifically, pinocytosis. Very small droplets of fluids within the lumen of the small intestine are surrounded by membrane and form pinocytotic vesicles. These vesicles are taken into the cytoplasm of the villus cell and later the contents of the vesicle are released into the cytoplasm.

What is left?

Some of the substances making their way through the small intestine cannot be digested and thus are never absorbed. These substances continue into the large intestine and become a part of the solid waste or faeces. These substances include:

- cellulose – in the cell walls of any foods from plants;
- lignin – another component of plant cell walls;
- bile pigments – from bile, give characteristic colour to faeces;
- bacteria – normal inhabitants of our digestive tract;
- intestinal cells – these break off as foods move through the lumen.

Exercises

7 Differentiate between the following four terms: ingestion, digestion, absorption and elimination.

8 Why are tight junctions necessary between adjoining epithelial cells making up the interior of the small intestine?

H.4 Functions of the liver

Assessment statements

H.4.1 Outline the circulation of blood through liver tissue, including the hepatic artery, hepatic portal vein, sinusoids and hepatic vein.

H.4.2 Explain the role of the liver in regulating levels of nutrients in the blood.

H.4.3 Outline the role of the liver in the storage of nutrients, including carbohydrate, iron, vitamin A and vitamin D.

H.4.4 State that the liver synthesizes plasma proteins and cholesterol.

H.4.5 State that the liver has a role in detoxification.

H.4.6 Describe the process of erythrocyte and haemoglobin breakdown in the liver, including phagocytosis, digestion of globin and bile pigment formation.

H.4.7 Explain the liver damage caused by excessive alcohol consumption.

Circulation of blood to and from the liver

The liver receives blood from two major blood vessels and is drained by one (see Figure 19.12). The hepatic artery is a branch of the aorta and carries oxygenated blood to the liver tissues. The hepatic portal vein is the other blood vessel supplying blood to the liver. These two blood vessels carry blood into the 'capillaries' of the liver, called sinusoids. All sinusoids are then drained by the hepatic vein which is the sole blood vessel taking blood away from the liver.

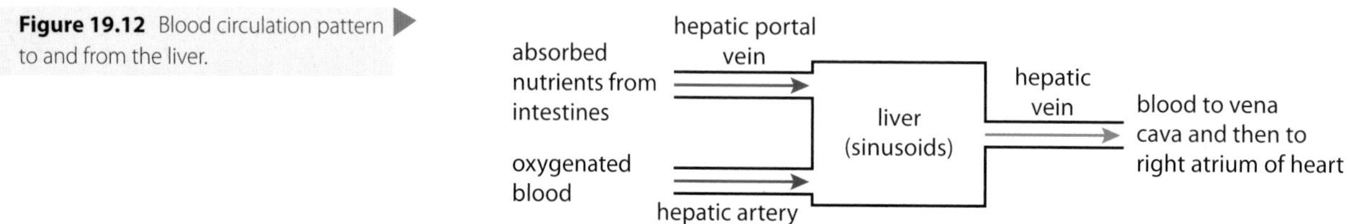

Figure 19.12 Blood circulation pattern to and from the liver.

The hepatic portal vein receives blood from the capillaries within all the villi of the small intestine. The blood within the hepatic portal vein varies in two ways from blood which normally arrives at an organ:

- it is low pressure, deoxygenated blood because it has already been through a capillary bed;
- it varies considerably in quantity of nutrients (especially glucose) depending on the types of food and the timing of ingestion, digestion and absorption of foods within the small intestine.

The blood within the hepatic vein is also low pressure, deoxygenated blood, but it does not vary in nutrients nearly as much as the blood within the hepatic portal vein. The stabilization of nutrients within the hepatic vein represents one of the major functions of the liver, specifically storage of nutrients.

A portal system of circulation is when blood travels through two capillary beds before returning to the heart to be re-pumped. There are only three locations in the human body where a portal system of circulation is used:

- the hepatic portal vein described in this section;
- a portal blood vessel within each nephron of the kidney;
- a portal blood vessel extending from the hypothalamus to the anterior pituitary gland.

Sinusoids are the capillaries of the liver

The function of the liver is to remove some things from the blood and add others to it. This removal or addition of a variety of substances is the job of the hepatocytes (liver cells). Oxygen-rich blood from the hepatic artery and (sometimes) nutrient-rich blood from the hepatic portal vein both flow into sinusoids of the liver. Sinusoids are where exchanges occur between the blood and the hepatocytes (see Figure 19.13).

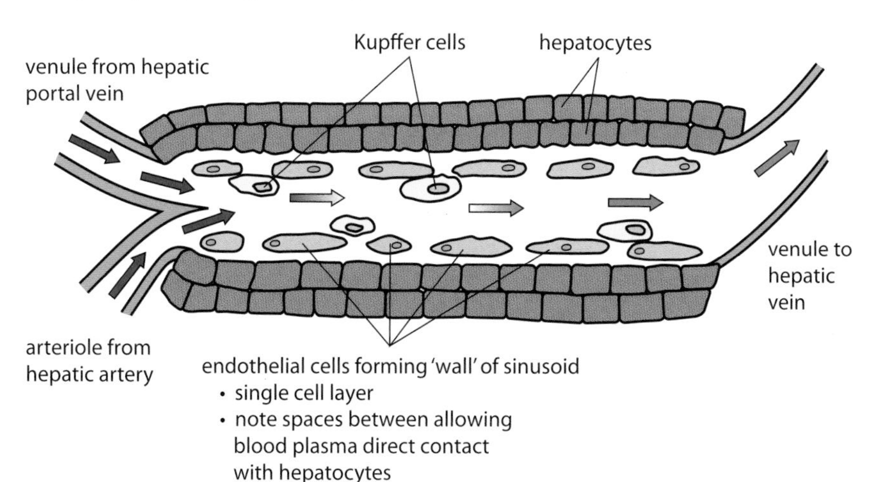

Figure 19.13 Sinusoids are the capillary beds of the liver, but their structure and action is different from capillary beds found elsewhere in the body.

The liver does not extract all excess glucose, toxins, etc. when the blood makes a single trip through liver sinusoids. The chemicals within the blood will be acted on by hepatocytes multiple times as blood continuously makes circuits through the liver.

Sinusoids differ from a typical capillary bed in the following ways:

- sinusoids are wider than capillaries;
- sinusoids are lined by endothelial cells with gaps between them
  - these gaps allow large molecules like proteins to be exchanged between hepatocytes and the bloodstream,
  - hepatocytes are in direct contact with blood components making all exchanges with the bloodstream more efficient;
- sinusoids contain cells called Kupffer cells that help break down older red blood cells for recycling cell components;
- sinusoids receive a mixture of oxygenated blood (from hepatic artery branches) and nutrient-rich blood (from hepatic portal vein branches) and this mixture eventually drains into small branches of the hepatic vein.

In this false-colour SEM, Kupffer cells (yellow) are shown within a sinusoid (blue) of the liver. Notice the pseudopod-type extensions of the Kupffer cells useful for phagocytosis.

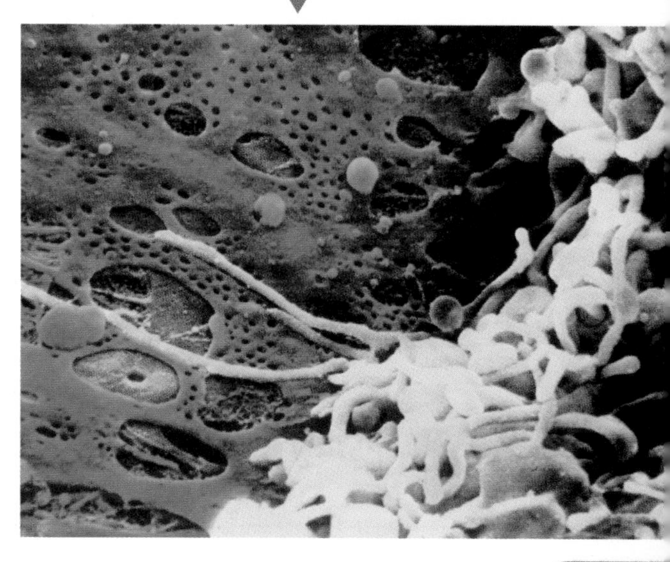

Regulation of nutrients in the blood

Solutes which are dissolved in blood plasma vary a little in concentration, but each type of solute has a normal homeostatic range. Any concentration below or above this normal range typically creates one or more physiological problems in the body.

Let's consider glucose as an example. For most people, glucose levels in blood are lowest in the morning and highest soon after a meal. When you digest a meal that is high in carbohydrates such as starch, your hepatic portal vein will contain blood with a very high concentration of glucose. When this blood enters the sinusoids of your liver, some of the excess glucose is taken in by the surrounding hepatocytes and converted to the polysaccharide glycogen. This keeps the glucose level in the normal range. Stored glycogen can be seen as large vesicles or 'granules' in electron micrographs of hepatocytes.

Now imagine you have not eaten any carbohydrates for a long time. Your blood glucose levels decrease as cells are using glucose for cell respiration. To keep the glucose level in the normal range, the stored glycogen in the granules is reconverted to glucose and added into the bloodstream in the sinusoids.

The homeostatic mechanisms at work are regulated by production of the hormones insulin and glucagon from the pancreas. When blood glucose levels are towards the upper end of the normal range, insulin is produced and this stimulates hepatocytes to take in and convert glucose to glycogen. When blood glucose approaches the lower end of normal, the pancreas produces glucagon and this hormone stimulates hepatocytes to convert glycogen back into glucose.

Nutrient storage

This table summarizes nutrients stored within the liver.

Nutrient	Relevant information
glycogen	polysaccharide carbohydrate
iron	iron is removed and stored following breakdown of erythrocytes and haemoglobin molecules
vitamin A	associated with good vision, one of the first signs of deficiency is 'night blindness'
vitamin D	vitamin D is often added to milk and milk products in some countries

Synthesis of plasma proteins and cholesterol

This table summarizes molecules synthesized within the liver.

Molecule(s)	Relevant information
plasma proteins	• albumin – helps regulate osmotic pressure of fluids in the body • fibrinogen – soluble form of blood clotting protein which is converted to fibrin when clot is needed • globulins – widely diverse group of blood proteins not all of which are produced in the liver
cholesterol	• some cholesterol is ingested and absorbed in foods, some is synthesized in the liver • some cholesterol is used to produce bile, some is carried in the bloodstream to be used for cell membranes and other purposes

Detoxification

This table summarizes molecules (partly) detoxified by the liver.

Molecule	Source
ethanol	alcoholic drinks
food preservatives	added to foods to retard spoiling
pesticides	often used on produce
herbicides	also used on produce

False-colour SEM of liver cells with cirrhosis. A sinusoid is visible (blue) surrounded by abnormal hepatocytes. Many fibres of connective tissue (light brown) have invaded the damaged area.

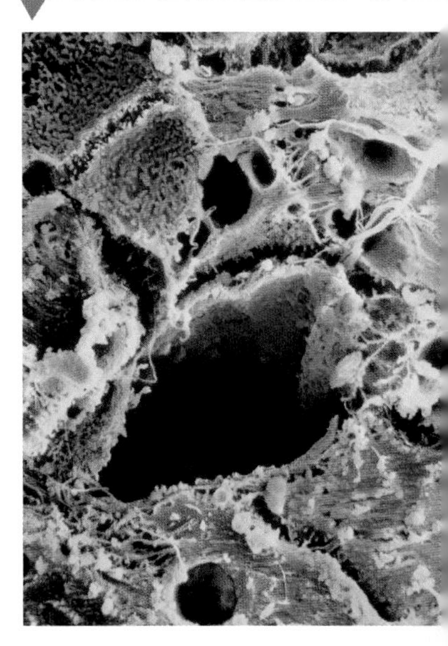

Alcohol consumption damages liver cells

People who drink alcohol, especially often and in high volume, can expect liver damage. As is the case with useful nutrients, the hepatic portal vein brings absorbed alcohol first to the liver. Any alcohol not removed is eventually brought back through the liver sinusoids many times by way of the hepatic artery. Each time the blood passes through the liver, hepatocytes attempt to remove the alcohol from the bloodstream. Thus, alcohol has a magnified effect on liver tissue as compared to other tissues in the body. It has been shown that long-term alcohol abuse results in three primary effects in the liver.

- *Cirrhosis*: This is the scar tissue left when areas of hepatocytes, blood vessels, and ducts have been destroyed by exposure to alcohol. Areas of the liver showing cirrhosis no longer function.
- *Fat accumulation*: Damaged areas of the liver will quite often build up fat in place of normal liver tissue.
- *Inflammation*: This is swelling of damaged liver tissue due to alcohol exposure; sometimes referred to as alcoholic hepatitis.

It has been shown that frequency and volume of alcohol consumption are both positively correlated with liver damage. Data also shows that females are more susceptible to liver damage than males. The good news is that liver damage that is not terribly severe is at least partially reversible. In other words, the liver can regenerate some of its damaged areas if the person gives up the excessive drinking that led to the damage. The bad news is liver damage due to alcohol abuse is fatal when the damage becomes too severe.

The liver recycles components of erythrocytes and haemoglobin

Erythrocytes have a typical 'cellular life span' of about 4 months. This means every red blood cell needs to be replaced every 120 days or so by the blood-cell-forming tissue of bone marrow. This is necessary because erythrocytes are anucleate (they have no nucleus) and thus cannot undergo mitosis to form new blood cells.

As erythrocytes approach their 120 day average life, the cell membrane becomes weak and eventually ruptures. More often than not this occurs in the spleen or bone marrow, but it can happen anywhere in the bloodstream. The rupture leads to millions of haemoglobin molecules circulating in the bloodstream. As blood circulates through the sinusoids of the liver, these circulating haemoglobin molecules are ingested by Kupffer cells within the sinusoids. This ingestion is by phagocytosis as haemoglobin molecules are very large proteins.

Kupffer cells of the liver are actually a type of leucocyte that reside in the sinusoids of the liver. Besides ingesting haemoglobin, they can also ingest cellular debris and bacteria within the bloodstream.

Haemoglobin consists of four polypeptides (globin) and a non-protein molecular component at the centre of each globin called a haem group. At the centre of each haem group is an iron atom. Thus each haemoglobin consists of four globins, four haem groups and four iron atoms. It is within Kupffer cells that haemoglobin is disassembled into its component parts. The key events are summarized in this bullet list and in Figure 19.14.

- The four globin proteins of each haemoglobin are hydrolysed into amino acids. The amino acids are released back into the bloodstream and become available to any body cell for protein synthesis.
- The iron atom is removed from each haem group. Some of this iron is stored within the liver and some is continuously sent to bone marrow to be used in the production of new erythrocytes.
- Once iron has been removed from haemoglobin, what remains of the molecule is called bilirubin or bile pigment. This is absorbed by the nearby hepatocytes and becomes a key component of bile.

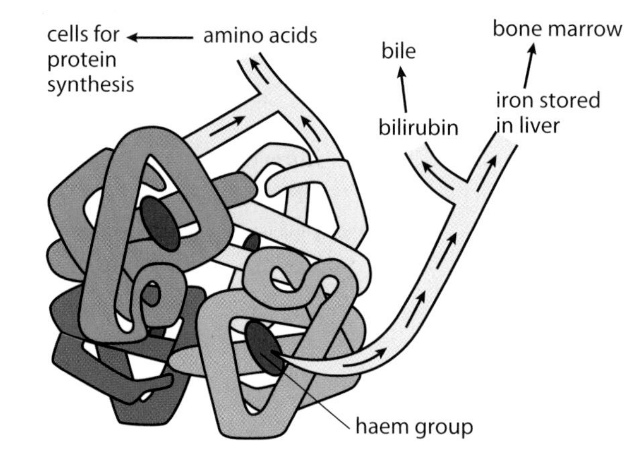

Figure 19.14 When erythrocytes are broken down, the haemoglobin they contained is recycled.

Exercises

9 Levels of blood glucose can be expected to vary more within the blood of the hepatic portal vein than in any other blood vessel in the body. Explain why.

10 Everyone's diet needs to contain some iron. Why do we need to ingest fairly minimal amounts of iron when we are replacing every erythrocyte every four months?

11 Explain why the hepatic portal vein is considered to be a vein, even though it is bringing blood to the sinusoids (capillaries) of the liver?

H.5 The transport system

Assessment statements

H.5.1 Explain the events of the cardiac cycle, including atrial and ventricular systole and diastole, and heart sounds.

H.5.2 Analyse data showing pressure and volume changes in the left atrium, left ventricle and the aorta, during the cardiac cycle.

H.5.3 Outline the mechanisms that control the heartbeat, including the roles of the SA (sinoatrial) node, AV (atrioventricular) node and conducting fibres in the ventricular walls.

H.5.4 Outline atherosclerosis and the causes of coronary thrombosis.

H.5.5 Discuss factors that affect the incidence of coronary heart disease.

The cardiac cycle

The cardiac cycle is a series of events that we commonly refer to as one heartbeat. More properly, one cardiac cycle is all the heart events that occur from the beginning of one heartbeat to the beginning of the next heartbeat. The frequency of the cardiac cycle is your heart rate and is typically measured in beats per minute (min^{-1}). If you have a resting heart rate of 72 beats per minute, you are performing 72 cardiac cycles each minute.

When a chamber of the heart contracts, it is because the cardiac muscle of the chamber has received an electrical signal that has caused the muscle fibres of the chamber to contract (shorten). This causes an increase in pressure on the blood within the chamber and the blood leaves the chamber through any available opening. This is called systole (sis-tol-ee). When a chamber is not undergoing systole, the cardiac muscle of the chamber is relaxed. This is called diastole (di-astol-ee). As you learned in Chapter 6, both atria contract at the same time. With this new vocabulary, you can say that both atria undergo systole at the same time. Both ventricles also undergo systole at the same time.

Anatomical diagrams identify right and left sides as if it is your own body that is being identified. As most anatomical diagrams show a ventral view (from the front), the left is on the right and the right is on the left. Any diagram identified as a dorsal view (from the back) shows right on the right and left on the left.

Heart valves

Heart valves keep blood moving in a single direction. Each chamber of the heart has to have an opening to receive blood and another to allow blood to exit. When a chamber undergoes systole, it is imperative to have the blood consistently move in a single, useful direction (see Figure 19.15). The heart valves serve to prevent a backflow of blood.

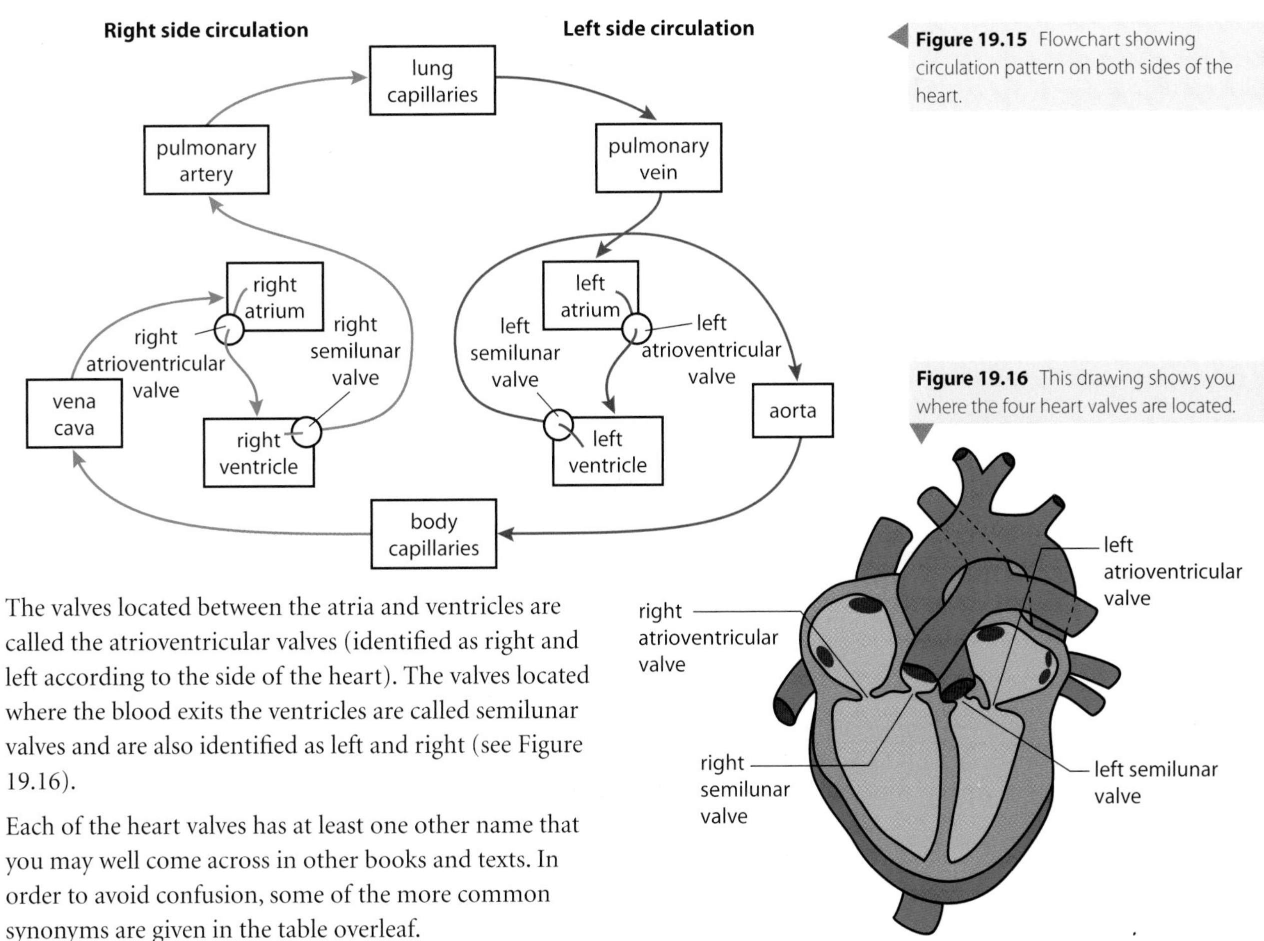

Figure 19.15 Flowchart showing circulation pattern on both sides of the heart.

Figure 19.16 This drawing shows you where the four heart valves are located.

The valves located between the atria and ventricles are called the atrioventricular valves (identified as right and left according to the side of the heart). The valves located where the blood exits the ventricles are called semilunar valves and are also identified as left and right (see Figure 19.16).

Each of the heart valves has at least one other name that you may well come across in other books and texts. In order to avoid confusion, some of the more common synonyms are given in the table overleaf.

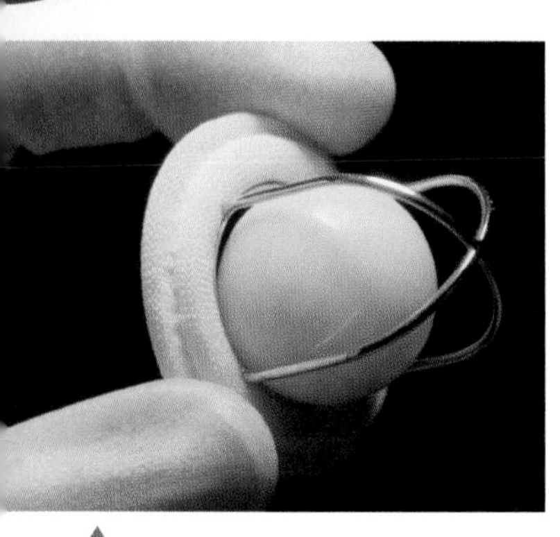

Artificial heart valves can be surgically implanted to replace damaged natural valves. This type of replacement valve is known as the ball and cage design.

Heart valve	Synonyms
right atrioventricular valve	tricuspid valve
left atrioventricular valve	bicuspid valve, mitral valve
right semilunar valve	pulmonary valve
left semilunar valve	aortic valve

You may have noticed in Figure 19.15 that there are no valves where blood enters the atria. So what prevents blood from flowing back up into the vena cava and pulmonary veins when the atria undergo systole? The answer to this question is two-fold.

- Both the vena cava and pulmonary veins are veins and thus have the internal, passive flap valves characteristic of all veins. If blood attempts to flow backwards in any vein, the passive flap valves close down and prevent blood from flowing in that direction.
- Atrial systole does not build up very much pressure. The muscular walls of the atria are very thin in comparison to the ventricles. Their force of contraction is slight in comparison. Remember that the blood from the atria simply needs to travel to the adjoining ventricles.

Thus, the relatively low pressure exerted by the atria in combination with the passive flap valves within the supply veins means that no heart valve is necessary where the blood enters each atrium.

Sometimes heart valves allow some blood to 'backflow'. The resulting sound when heard through a stethoscope is often described as a squishing sound and is known as a heart murmur.

The sounds of the heart

When you listen directly to the heart using a stethoscope, you hear a rhythmic set of sounds that most people describe as a series of 'lub dub' sounds. Each 'lub dub' is the sound of one cardiac cycle (one heartbeat) and, for the most part, is the sound of the heart valves closing. Remember that the right and left sides of the heart are working in unison, therefore there are only two heart sounds even though there are four heart valves. The atrioventricular valves closing is heard as one sound, 'lub', and the two semilunar valves closing is heard as the second sound, 'dub'. Following these two sounds is a relatively long silence before repeating.

Let's use the example of a person with a resting heart rate of 72 beats per minute as an example of the timing of the heart sounds. This person would experience a cardiac cycle every 0.8 seconds. Atrial systole would be the first 0.1 second of the cardiac cycle followed by ventricular systole which would last 0.3 seconds. The remaining 0.4 seconds of the cardiac cycle would be diastole for both sets of chambers. The atrioventricular valves would close when the ventricles begin to contract at the end of atrial systole and the beginning of ventricular systole. The semilunar valves would close at the end of ventricular systole.

Assuming 72 beats per minute, this table summarizes the heart sounds and their timing.

Sound	Timing within cardiac cycle	Sound coming from
lub	heard 0.1 seconds into the cardiac cycle at the end of atrial systole and beginning of ventricular systole	closing of both atrioventricular valves
dub	heard 0.4 seconds into the cardiac cycle at the end of ventricular systole	closing of both semilunar valves
silence	lack of sound for a total of 0.5 seconds; this is 0.4 seconds of the first cardiac cycle and the first 0.1 seconds of the next cycle	

Changes in pressure and volume in the heart chambers

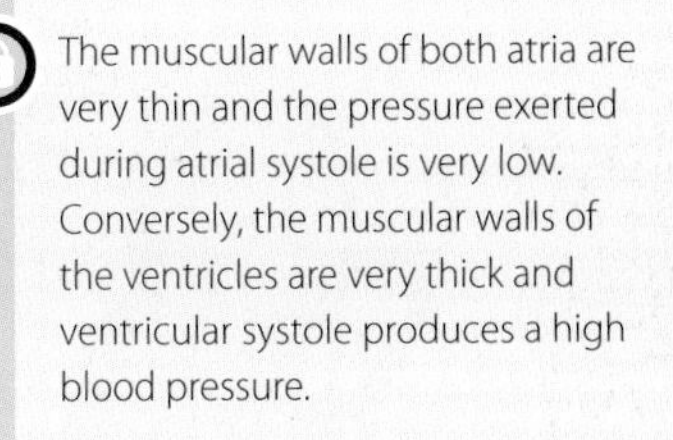

The muscular walls of both atria are very thin and the pressure exerted during atrial systole is very low. Conversely, the muscular walls of the ventricles are very thick and ventricular systole produces a high blood pressure.

Heart valves open and close depending on the pressure of blood on each side of the valve. The change in pressure also explains the movement of blood through and out of each chamber of the heart. Both the left and right sides of the heart are working synchronously as a double pump. To understand the workings of the heart, it is only necessary to look at one side of the heart with the understanding that the other side has similar pressures and volumes of blood at the same time.

Let's examine the pressure and volume changes that occur on the left side of the heart. You do not have to memorize the pressure numbers given in this example, your focus should be to understand how the given blood pressures result in the movement of blood and the opening and closing of the heart valves.

When both chambers are in diastole

Let's start during the portion of the cardiac cycle where both sets of chambers are in diastole (at rest). Look at Figure 19.17. It shows the left side of the heart with openings in the left atrium for entry of the pulmonary veins. Neither chamber has accumulated a great volume of blood as the ventricle has recently sent blood into the aorta and the atrium is receiving slow-moving venous blood from the pulmonary veins. The numbers inside of each chamber or blood vessel represent blood pressure measured in mm Hg.

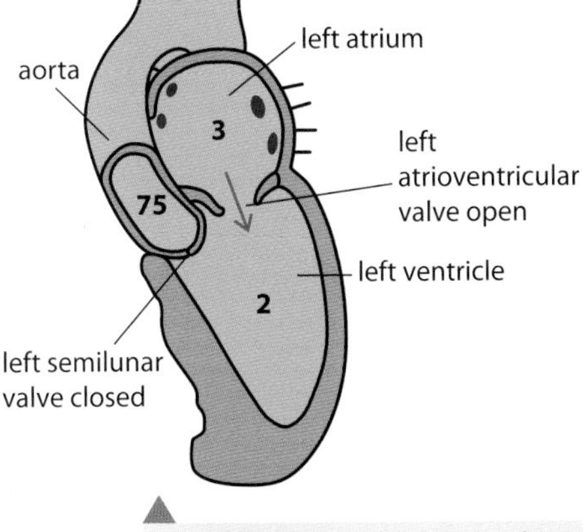

Figure 19.17 Blood pressure readings in mm Hg when both chambers are in diastole (rest). Notice that some blood is passively moving from the left atrium to the left ventricle.

During this period of diastole for both chambers, the atrial pressure is just slightly higher than ventricular pressure and this keeps the atrioventricular valve open. Much of the blood which slowly returns to the left atrium by way of the pulmonary veins moves passively down to the left ventricle through this open valve. Notice also that the pressure in the aorta is much higher than in the left ventricle. This pressure difference keeps the left semilunar valve closed and prevents backflow into the ventricle.

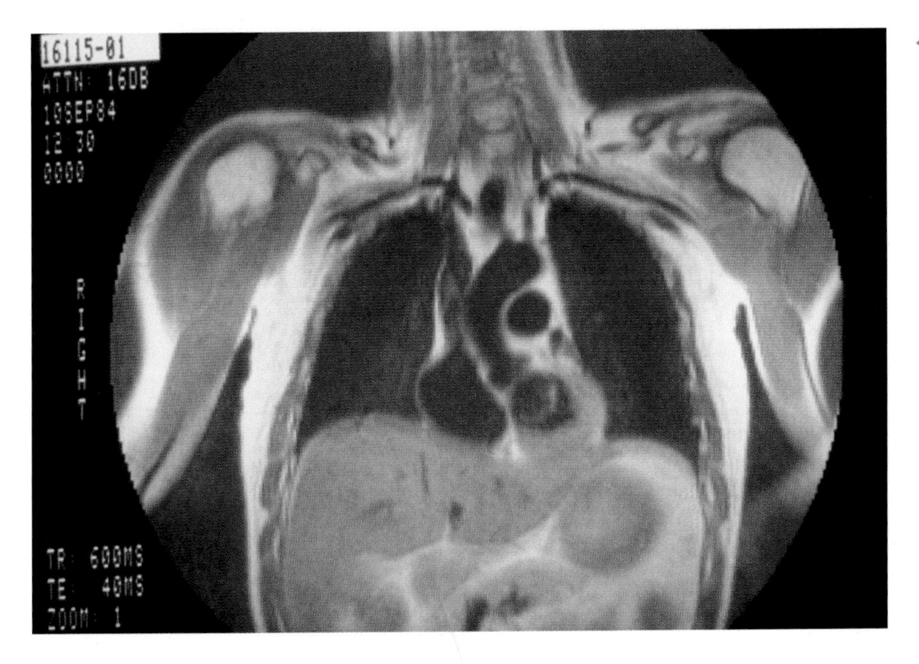

NMR scan of a normal chest. The heart is visible between the two lungs and has just completed ventricular systole. The liver and stomach are also visible in the lower portion of the scan.

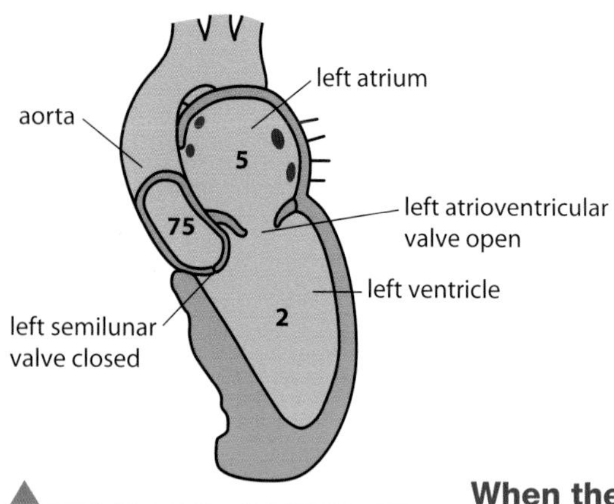

Figure 19.18 Blood pressure readings in mm Hg at atrial systole.

When the atrium is in systole and the ventricle is in diastole

In Figure 19.18, the atrium is undergoing a systole (contraction). The pressure produced by this systole is not very high. The wall of each atrium is relatively thin muscle and is not capable of creating very much pressure. There is no need for great pressure as much of the volume of the blood has already passively accumulated within the ventricle through the open atrioventricular valve. The function of this systole is to move into the ventricle any blood remaining in the atrium.

When the atrium is in diastole and the ventricle is in systole

Figure 19.19 shows blood pressures in early and late ventricular systole. As soon as ventricular systole begins, the pressure inside the ventricle increases to be greater than that in the atrium, so the atrioventricular valve closes to prevent backflow to the atrium (this creates the 'lub' sound). The pressure in the aorta is still far higher than in the ventricle so the semilunar valve remains closed. There is a relatively large volume of blood in the ventricle during this time and the ventricle is highly muscular. This combination of factors permits the ventricular pressure to build considerably as systole continues. Finally, the pressure in the ventricle gets to be greater than that in the aorta and the semilunar valve opens allowing the ventricle to pump the blood into the aorta. As the ventricle finishes its contraction, the pressure inside the ventricle once again drops below the pressure in the aorta and the semilunar valve closes (this causes the a 'dub' sound). Both chambers go back into diastole and the cardiac cycle repeats itself over and over.

Blood pressure is traditionally measured in mm Hg although the modern units for pressure are pascals (Pa) and kilopascals (kPa).
120 mm Hg = 16 kPa
80 mm Hg = 11 kPa

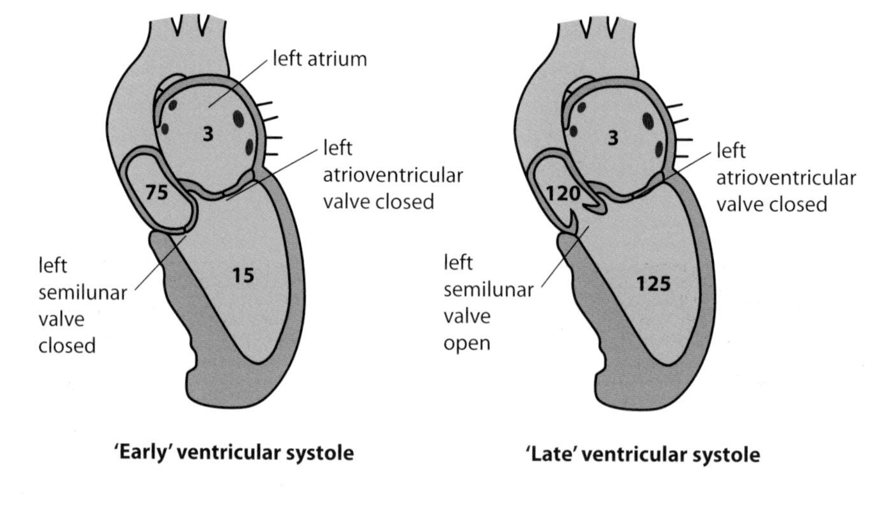

Figure 19.19 Blood pressure readings in mm Hg at early and late ventricular systole.

You may see these changes in pressure as blood flows through the heart presented in graphical format.

How is the heartbeat controlled?

Although everyone tends to have a consistent resting heart rate, heart rate is not a constant as changes in rate occur depending on factors such as exercise and emotion. Changes in heart rate also depend on nervous control by the brain stem and the effect of hormones such as adrenaline.

Myogenic control

If you are at your resting heart rate, your heart itself is controlling the frequency and internal timing of the events of each cardiac cycle. This is called myogenic control. Heart muscle is unusual in that it does not need nervous stimulation to contract.

The mass of tissue that acts as the living pacemaker for the heart is known as the sinoatrial (SA) node. This node of cells is located in the upper wall of the right atrium close to where the superior vena cava enters (see Figure 19.20).

The SA node is a group of modified cardiac muscle cells which are capable of generating action potentials on a regular frequency. If your myogenic heart rate were 72 beats per minute, your SA node would be generating an action potential every 0.8 seconds. The action potentials from the SA node spread out nearly instantaneously and result in the thin-walled atria undergoing a systole.

Individual cardiac muscle cells contract in an independent rhythm. When heart muscle cells touch each other, they synchronize their contractions. The SA and AV nodes take advantage of this natural ability and provide the timing necessary to synchronize the entire heart.

The SA node action potential also reaches a group of cells known as the atrioventricular (AV) node. This node is located in the lower wall of the right atrium in the septum or partition between the right and left atria (see Figure 19.20).

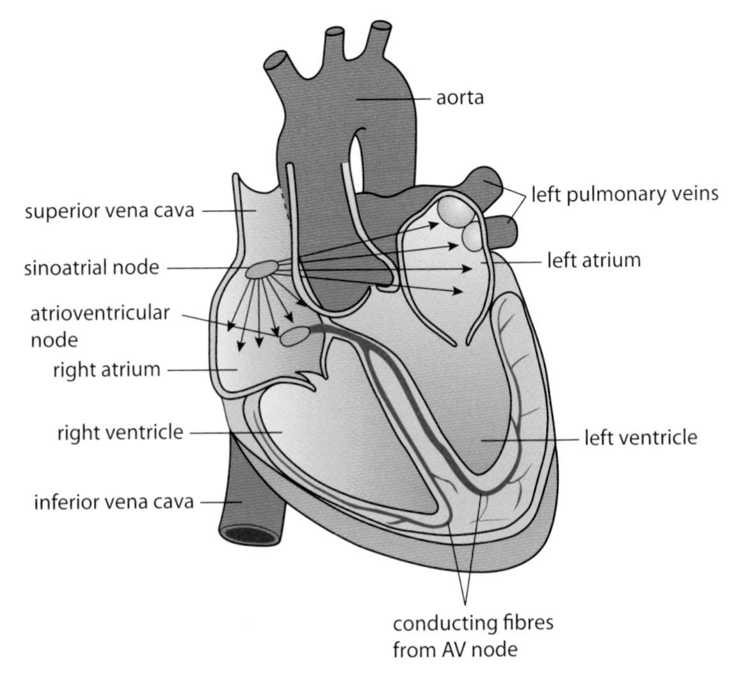

Figure 19.20 This drawing of the human heart shows you the location of the SA node, AV node and the conducting fibres spreading out through the ventricles from the AV node. The black arrows represent action potentials from the SA node.

The AV node receives the action potential coming from the SA node and delays for approximately 0.1 second. It then sends out its own action potential which spreads out to both ventricles. As you learned earlier, the walls of the ventricles are much thicker muscle. In order to get the action potential to efficiently reach all of the muscle cells in the ventricles, there is a system of conducting fibres that begin at the AV node and then travel down the septum between the two ventricles. At various points these conducting fibres have branches that spread out into the thick cardiac muscle tissue of the ventricles.

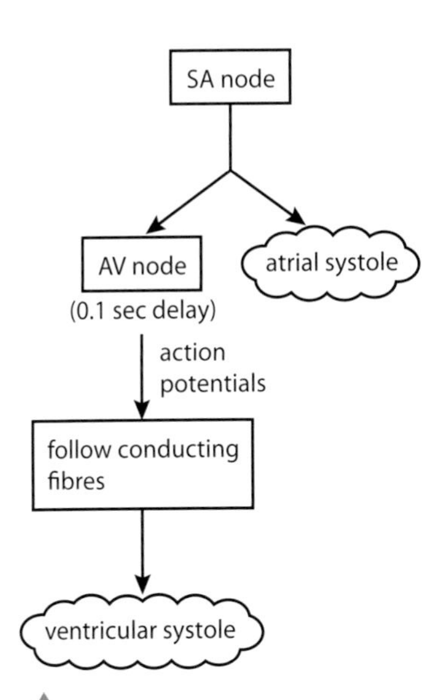

Figure 19.21 Flowchart of the events associated with one heartbeat.

The events of a single heartbeat are summarized in Figure 19.21.

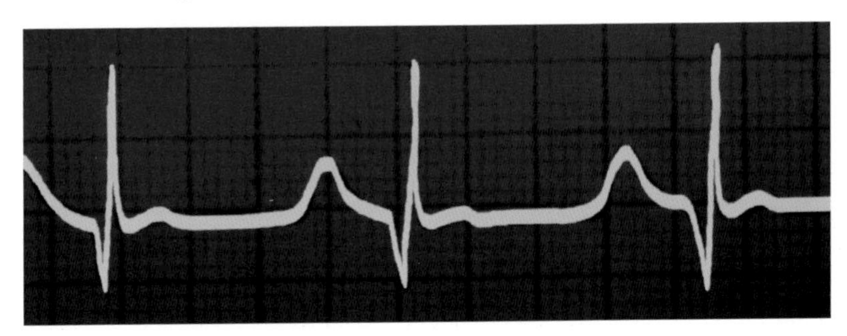

An electrocardiogram (ECG) is a measurement of the electrical activity of the heart. This electrical activity can be traced back to the depolarization of the SA and AV nodes.

Neurological control

When you begin to exercise or in any way physically exert yourself, the rate of aerobic cellular respiration begins to increase for many body cells (especially the skeletal muscle cells involved in the exercise). These cells require more oxygen and give off more carbon dioxide than before exercise. The myogenic level of heart rate can no longer provide adequate circulation of blood for this increased level of gas exchange.

Some large arteries and the medulla oblongata of the brainstem contain specialized chemoreceptor cells that respond to the characteristic increase in carbon dioxide levels (and the resulting decrease in pH) in blood when you begin physical exertion. The result is that the medulla oblongata sends action potentials to the SA node causing it to 'fire' more frequently than at its myogenic pace. Your heart rate thus increases to a level suitable for the level of physical exertion and stays there for as long as the exercise continues (see Figure 19.22).

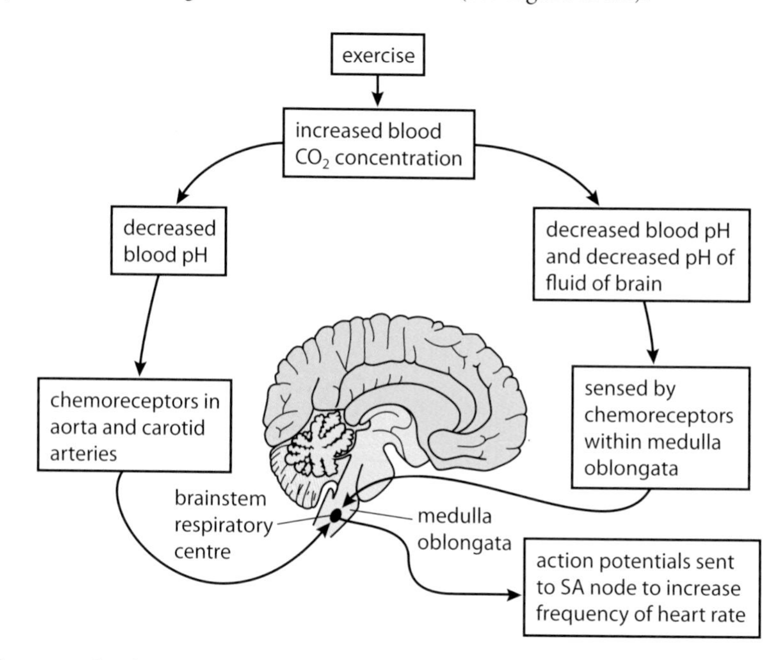

Figure 19.22 The medulla's control of heart rate during exercise. When the exercise stops, reverse events occur which eventually bring the heart rate back under the control of the SA node (myogenic control).

When exercise decreases or stops, carbon dioxide levels in the bloodstream begin to decrease. The chemoreceptors in the medulla respond to this chemical change and now send a set of action potentials to the SA node that result in slowing the heart rate. Once the heart rate has returned to the myogenic level, the SA node again acts as the pacemaker.

Note that the outside influence on heart rate does not change the sequence of events that occur within the heart. Each heart beat is still initiated by the SA node which spreads throughout the atria and to the AV node and thus the conducting fibres which spread throughout the ventricles. The only change is that the SA node is not acting as its own pacemaker in setting a frequency or pace.

The increase in heart rate when we exercise is the result of action of the autonomic nervous system (ANS). The ANS controls many of the physiological changes that occur within our bodies that are not at the conscious level of thought. Every physiological function controlled by the ANS has an antagonistic pair of nerves which ultimately serve opposite functions. For example, one nerve increases heart rate, the other decreases it. This is yet another example of homeostasis or the attempt to keep the body's physiology within a certain 'normal range'.

Atherosclerosis and heart attacks

Most humans begin life with very healthy body tissues and quickly start damaging them. A prime example is our arteries. You will recall that normal arteries have a fairly thick lining of smooth muscle and a relatively small internal lumen compared to veins. Arteries are designed to circulate blood that is relatively fast moving and at high pressure. Many arteries can change their diameter in order to help regulate blood pressure changes. Even though some damage to arteries throughout your life is inevitable, many of the factors that lead to damage are under your control and can be minimized.

Atherosclerosis is a slow build-up of material that is collectively called plaque. Plaque is composed of lipids, cholesterol, cell debris, and calcium. The build-up of this material begins early in life and typically takes many, many years to become a serious problem. The plaque build-up is most noticeable in medium and large size arteries. As arteries begin to build up plaque, they become harder and therefore less flexible.

The name atherosclerosis comes from 'athero' meaning paste and 'sclerosis' meaning hardness.

The inside lining of an artery is known as the endothelium. In a young person, the endothelium of each artery is smooth with very little, if any, plaque build-up. As the years progress, each person begins to deposit plaque. How much depends on a whole set of factors discussed in the next section. In some arteries, the plaque deposits significantly decrease the inside diameter of the blood vessel (stenosis) and thus decrease its ability to carry oxygen-rich blood. The tissues that are supplied by this artery receive less and less blood as time goes on. In addition, the damaged area of the blood vessel may form a blood clot, which further decreases the inside diameter. These clots (fibrin, platelets, erythrocytes, etc.) sometimes break away and block smaller arteries.

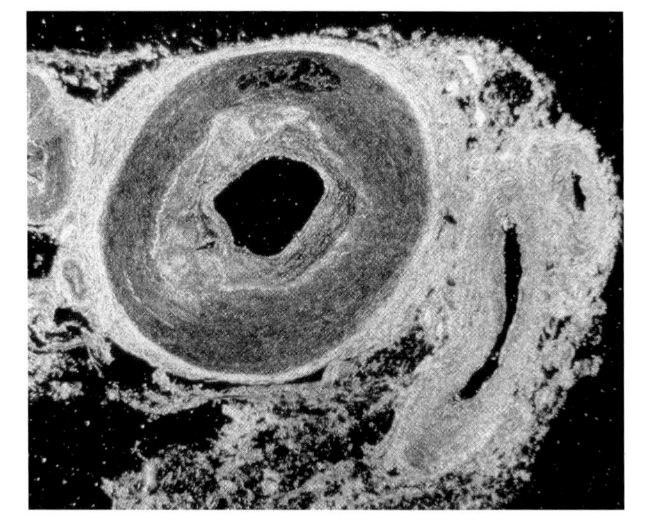

Artery showing atherosclerosis. The dark area in the centre is the lumen where blood flows. The tissue immediately surrounding the lumen is plaque. Significantly less blood is carried in this vessel than in a healthy artery which would have a much wider lumen (see Figure 12.6, page 326).

Coronary arteries

The heart has three major coronary arteries which supply the heart muscle with oxygen-rich blood. These arteries are branches of the aorta and carry blood that has recently been to the lungs. As you will recall, cardiac muscle never gets a break from contracting as alternating periods of systole and diastole occur repeatedly throughout life. Thus, cardiac muscle is very oxygen-demanding. If any one of the three major coronary arteries or one or more of their branches is somehow blocked, some portion of the heart muscle is likely to be deprived of its oxygen supply. This is exactly what happens when atherosclerosis leads to stenosis of the coronary arteries. The condition is often further complicated when a travelling blood clot fills the already narrowed lumen (opening) of a coronary artery.

You can estimate your risk of heart disease by inputting factors such as blood pressure and cholesterol levels at a website created by International Task Force for the Prevention of Coronary Heart Disease. Visit heinemann.co.uk/hotlinks, enter the express code 4242P and click on Weblink 19.3.

Typically, scientific knowledge is gained by experimentation. One independent variable is selected to be tested and all other variables are controlled.

Suppose you wanted to consider diet as an independent variable affecting the incidence of coronary heart disease. How can an exact diet be maintained between test subjects? In addition, how could genetics and several other variables be truly controlled? These questions illustrate the problem of drawing conclusions on the relative importance of various factors and how they impact CHD.

Coronary thrombosis

When a coronary artery or one of its main branches becomes blocked, it is known as a coronary thrombosis or heart attack. The symptoms of a coronary thrombosis often include:

- pain in the chest area, often radiating out towards the left arm;
- constricting sensation in or around the throat;
- breathing difficulties;
- severe dizziness, sometimes fainting.

Anyone experiencing one or more of these symptoms should be provided with medical care immediately.

Risk factors affecting coronary heart disease

Coronary heart disease (CHD) is the term used for the slow progression of plaque build-up in arteries and the corresponding problems which can result. Individuals can have CHD for many years without obvious symptoms as the early stages of atherosclerosis do not result in immediate and obvious symptoms. Not everyone builds up plaque in their arteries at the same rate. The factors that determine plaque build-up and thus the eventual chances of heart-related problems are in two main categories: those that cannot be controlled or avoided, and those that can.

This table summarizes risk factors that cannot be controlled.

Factor	Additional information
increasing age	83% of individuals who die of CHD are aged 65 or over
heredity	children of parents who have CHD are more likely to develop CHD
race	some ethnic groups have far higher rates of CHD than others
gender	males have a greater risk of CHD than females

This table summarizes risk factors that can be controlled.

Factor	Additional information
high cholesterol in blood	one of the constituents of plaque, controlled by diet or medication
smoking	smokers have approximately twice the risk of CHD as non-smokers
high blood pressure	high blood pressure makes the heart work harder
sedentary lifestyle	exercise improves many factors which affect CHD
obesity	excess weight increases heart workload among other effects
stressful lifestyle	may lead to habits which are not heart healthy, such as smoking, overeating, etc.
diabetes	minimized when diabetics carefully manage their blood sugar levels

Most individuals have at least some of the risk factors at work in their lives. It is very difficult to measure the effects of any one factor and its impact on incidence of CHD. Almost all factors have an impact on one or more other factors. For instance:

- people who are overweight often have problems with high blood pressure and cholesterol;
- a sedentary lifestyle may lead to obesity;
- stress may lead to smoking, overeating and thus high blood pressure, cholesterol problems, etc.

Researchers who attempt to isolate any one factor and study that factor's impact on CHD must take into account the cascading effect of one factor affecting another, making this type of study open to many interpretations.

Exercises

12 Artificial hearts and heart valves have been designed and surgically implanted into both test animals and humans. How do the valves within these artificial devices 'know' when it is time to close and open?

13 During fetal development, babies have a hole which permits blood to travel from the right atrium over to the left atrium. Why does this make sense from the perspective of the fetal circulation pattern?

14 An ECG is a graph showing the electrical activity of the heart. The voltage can be traced back to the SA node and the AV node. When a person exercises and thus increases their heart rate, what is the expected change in a subsequent ECG?

H.6 Gas exchange

Assessment statements

H.6.1 Define *partial pressure*.

H.6.2 Explain the oxygen dissociation curves of adult haemoglobin, fetal haemoglobin and myoglobin.

H.6.3 Describe how carbon dioxide is carried by the blood, including the action of carbonic anhydrase, the chloride shift and buffering by plasma proteins.

H.6.4 Explain the role of the Bohr shift in the supply of oxygen to respiring tissues.

H.6.5 Explain how and why ventilation rate varies with exercise.

H.6.6 Outline the possible causes of asthma and its effects on the gas exchange system.

H.6.7 Explain the problem of gas exchange at high altitudes and the way the body acclimatizes.

Haemoglobin

Haemoglobin is the protein molecule found within erythrocytes that is responsible for carrying most of the oxygen within the bloodstream. Each erythrocyte is basically a plasma membrane surrounding cytoplasm filled with haemoglobin molecules. The red blood cells have no nucleus and few organelles or other solutes other than haemoglobin. Each haemoglobin molecule is capable of reversibly binding to as many as four oxygen molecules and one carbon dioxide molecule.

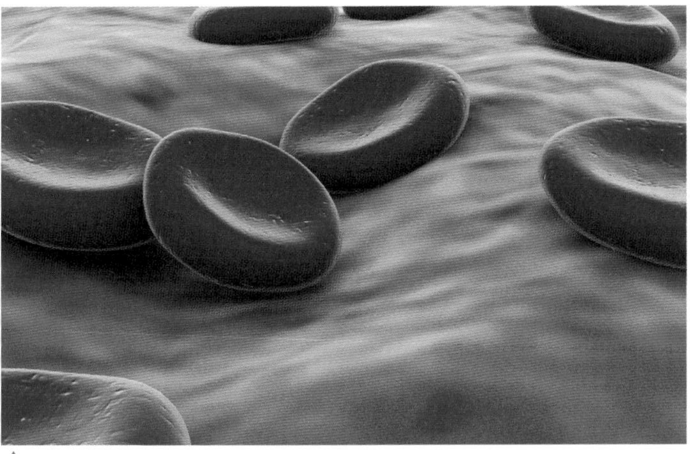

Erythrocytes have no nucleus and few organelles. Each erythrocyte is loaded with haemoglobin molecules.

Each haemoglobin molecule is composed of four polypeptides. Each polypeptide has a haem group near its centre and each haem group has an iron atom within (see Figure 19.23). When haemoglobin reversibly bonds to an oxygen molecule, it is the iron atom within the haem group that is bonding to the oxygen. Since haemoglobin has a total of four iron atoms within four haem groups within four polypeptides, it has the capability of transporting a maximum of four oxygen molecules ($4O_2$).

Figure 19.23 Haemoglobin is a large protein consisting of four polypeptides with a haem group within each. The molecular structure of a haem group is shown on the right.

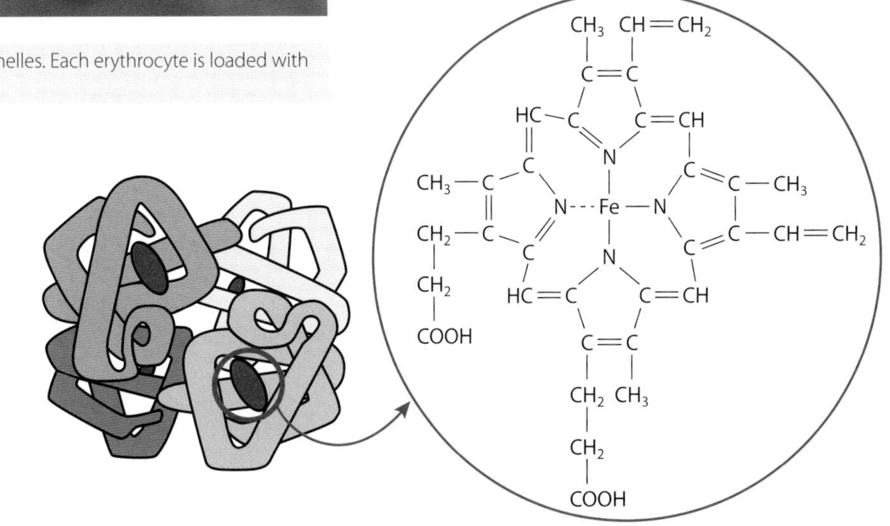

Haemoglobin changes shape and affinity when carrying oxygen

You will recall that proteins have an ability to change their three-dimensional shape under certain circumstances. For example, the induced-fit hypothesis of enzyme catalysis proposes that an enzyme changes shape as the substrate enters the enzyme's active site. A similar phenomenon occurs when oxygen binds to haemoglobin. Haemoglobin actually has four possible shapes depending on how many oxygen molecules are bound to the iron atoms of the haem groups. These different shapes affect haemoglobin's ability to bind to oxygen molecules. This is known as haemoglobin's affinity for oxygen. The greater the tendency to bond to oxygen, the higher the affinity.

Carbon monoxide is a by-product of the combustion of many fuels. Haemoglobin has a greater affinity for carbon monoxide than for oxygen. People breathing carbon monoxide are depriving their tissues of oxygen as the carbon monoxide molecules bind to haemoglobin and prevent haemoglobin from carrying a normal load of oxygen. Carbon monoxide poisoning can be fatal.

Haemoglobin molecules that are already carrying three oxygen molecules have the greatest affinity for oxygen. Conversely, haemoglobin molecules that are currently carrying no oxygen molecules have the least affinity for oxygen. You might think this defies common sense, but it does make sense when you learn that each oxygen molecule that binds to haemoglobin changes haemoglobin's shape in a manner that increases its affinity for another oxygen molecule. Haemoglobin can carry a maximum of four oxygen molecules, so one that is already carrying four oxygens has no affinity for oxygen.

A common abbreviation that is used for haemoglobin is Hb_4. Each molecule of oxygen that is bound to Hb_4 adds two oxygen atoms. Thus, haemoglobin's affinity for oxygen from lowest to highest is: Hb_4, Hb_4O_2, Hb_4O_4 and finally Hb_4O_6.

Oxygen dissociation curves

Oxygen dissociation curves are graphs which show how various forms of haemoglobin or myoglobin perform under various conditions.

The *x* axis of these graphs measures the partial pressure of oxygen. Partial pressure is the pressure exerted from a single type of gas when it is found within a mixture of gases. The air that we breathe is a mixture of gases and oxygen is just one component of this mixture. Within our bloodstream and in our body tissues is a different mixture of gases and once again oxygen is just one component of that mixture. The mixture of gases has a total pressure that is exerted; the portion of the total pressure that is due to oxygen alone is the partial pressure of oxygen.

The *y* axis of an oxygen dissociation curve shows percentage saturation of haemoglobin with oxygen. Remember that haemoglobin is not saturated until it is carrying (bonded to) four oxygen molecules. Let's look at the oxygen dissociation curve for human adult haemoglobin (Figure 19.24).

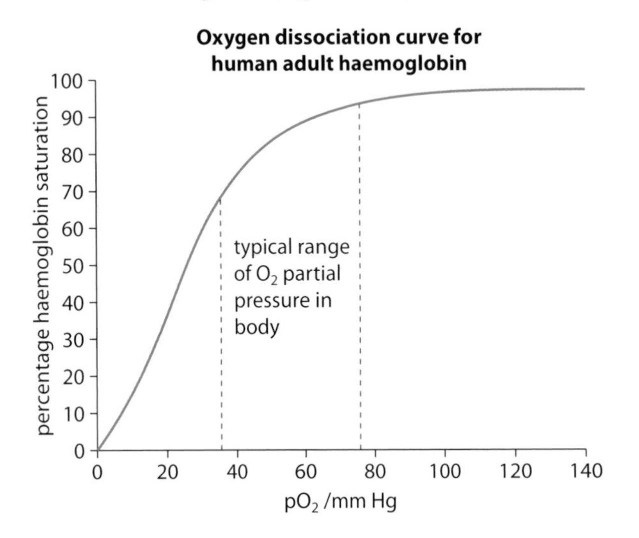

Figure 19.24 This oxygen dissociation curve shows the range of oxygen partial pressure in the body. Partial pressures are often given in kPa (kilopascals) rather than mm Hg.

Notice the very steep S-shape of the graph. This shape is indicative of the affinity changes for oxygen that haemoglobin undergoes when at least some oxygen is already bound to the molecule. At the lower end of the graph, little oxygen is already bound and this gives the shape of the lower portion of the S. In the upper half of the graph, haemoglobin is already binding some oxygen and has increased its affinity for oxygen (due to the protein shape change) and the graph is very steep in that area until nearly all of the haemoglobin is saturated.

Oxygen dissociation curves show the tendency of haemoglobin to bind to oxygen (affinity) and separate from oxygen (dissociate).

Notice on the graph the homeostatic range of oxygen partial pressures within the body. The upper end of the normal range (about 75 mm Hg or 10 kPa), is the oxygen partial pressure found within the lungs. The graph shows that over 90% of the haemoglobin becomes saturated with oxygen within the lungs. At the lower end of the normal range (about 35 mm Hg or 5 kPa), only about 50% of the haemoglobin is still saturated with oxygen. This partial pressure of oxygen is more typical of body tissues actively undergoing cell respiration. This means that 40–50% of the haemoglobin that has recently been to the lungs gives up (or dissociates) one or more oxygen molecules when the haemoglobin reaches the body tissues. Haemoglobin molecules typically do not 'empty' their oxygen load when they reach respiring body tissues, but they do release a significant amount of oxygen in a relatively narrow range of oxygen partial pressures. It is this release (or dissociation) of oxygen that gives these graphs their name: oxygen dissociation curves.

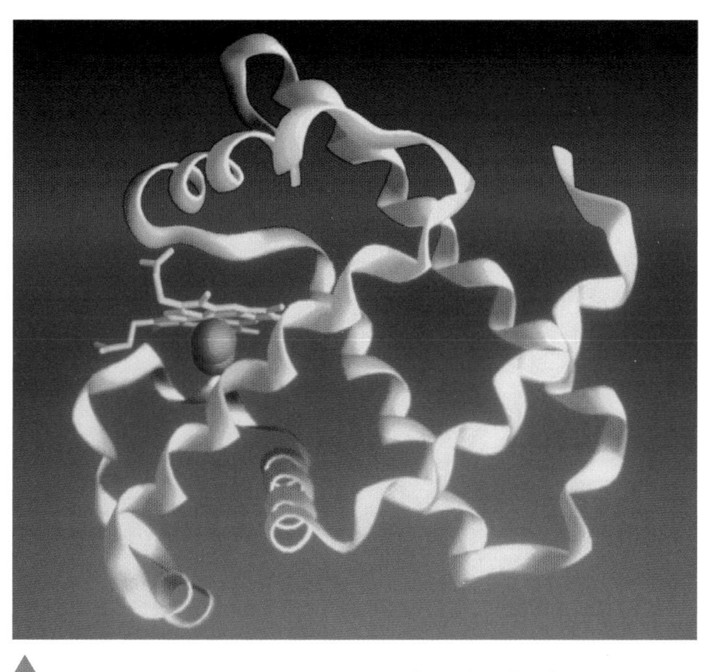

Myoglobin molecular structure. The ribbon-like structure in this model is the single polypeptide chain. Centre left is the haem group (blue) with oxygen bonded (red).

Comparison of haemoglobin and myoglobin

Myoglobin is an oxygen-binding protein found in muscles. Each myoglobin molecule consists of a single polypeptide, a single haem group and thus a single iron atom. Each myoglobin can bind to one oxygen molecule. The function of myoglobin is to store oxygen within muscle tissues until muscles begin to enter an anaerobic situation. Then, and only then, does myoglobin dissociate its oxygen and thus delay the onset of lactic acid fermentation.

Look at Figure 19.25. Notice that myoglobin's position on this graph is to the left of haemoglobin. Except for the very upper end of the oxygen partial pressure scale, any point selected on the *x* axis will show myoglobin still bonded to its oxygen when haemoglobin has dissociated at least one oxygen. This ability of myoglobin to 'hold onto' its oxygen, even at low oxygen partial pressures allows myoglobin to serve its function of delaying tissues going into anaerobic conditions. You can think of myoglobin as providing a final reservoir of oxygen when you are exercising heavily.

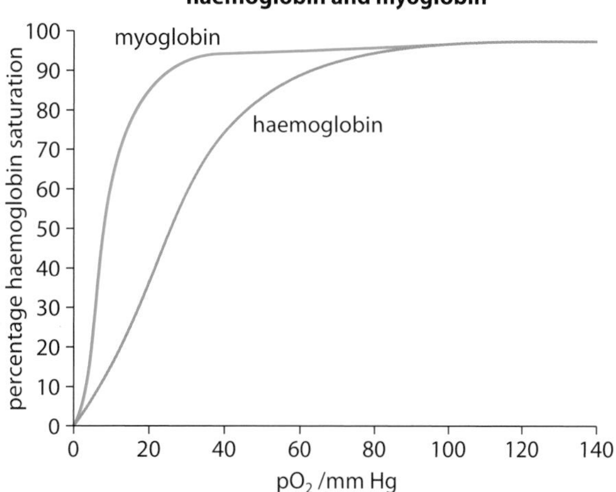

Figure 19.25 Oxygen dissociation curves of haemoglobin and myoglobin. Myoglobin dissociates oxygen only when oxygen partial pressure is very low.

Comparison of adult haemoglobin and fetal haemoglobin

The haemoglobin produced by a fetus is slightly different in molecular composition from adult haemoglobin. This is because the haemoglobin of a fetus must have a greater affinity for oxygen than adult haemoglobin. This means that in the placental capillaries, adult haemoglobin dissociates oxygen and fetal haemoglobin binds to that same oxygen. Fetal haemoglobin dissociates this oxygen only when it reaches the respiring tissues of the fetus.

In Figure 19.26, notice that the curve for fetal haemoglobin is consistently to the left of adult haemoglobin. Any sample point selected on the *x* axis shows that adult haemoglobin binds less oxygen at that partial pressure than fetal haemoglobin.

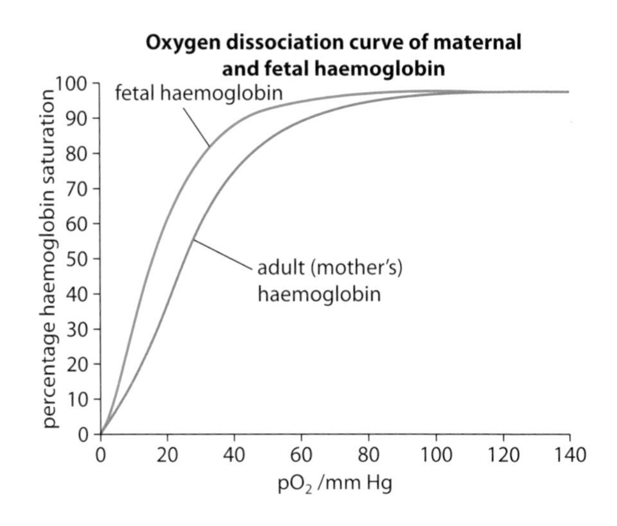

Figure 19.26 Fetal haemoglobin has a greater affinity for oxygen than adult haemoglobin in the range of partial pressures typical of human tissues.

Comparison of adult haemoglobin in two carbon dioxide concentrations (the Bohr shift)

Haemoglobin's affinity for oxygen is reduced in an environment where carbon dioxide partial pressure is high. Such an environment is found in body tissues that are actively undergoing cell respiration. Another way of saying this is that haemoglobin is induced to release (dissociate) oxygen within capillaries at body tissues. This effect is called the Bohr shift and results when carbon dioxide binds to haemoglobin and results in a shape change that promotes the release of oxygen.

Let's consider what happens to adult haemoglobin in different environments in the body. Look at Figure 19.27. Adult haemoglobin is shown in two different environmental conditions. The curve on the left shows what happens to haemoglobin passing through the lungs, an environment where the partial pressure of carbon dioxide is relatively low. In such an environment, oxygen binds easily to haemoglobin. The curve on the right shows what happens to haemoglobin in actively respiring tissues that are giving off carbon dioxide as a waste product. The carbon dioxide is entering the bloodstream and some is binding with haemoglobin. In that situation, oxygen is more likely to dissociate from haemoglobin at any one oxygen partial pressure. This is the Bohr shift. It promotes the release of oxygen within body tissues and the binding of oxygen within the lungs, both situations using the same molecule.

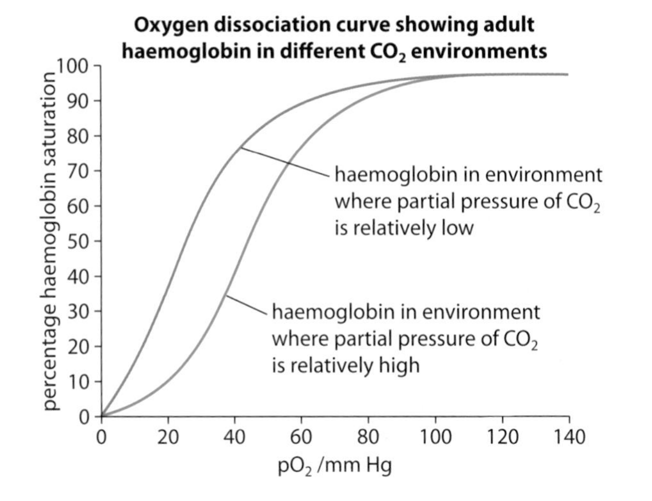

Figure 19.27 The Bohr shift. Haemoglobin has a tendency to give up oxygen in an environment where carbon dioxide partial pressure is high.

Most people do not realize that many of the carbon atoms within the foods they eat later become incorporated into the carbon dioxide molecules that they breathe out. This is reflected in the word equation for aerobic cell respiration:

- glucose + oxygen → carbon dioxide + water.

Carbon dioxide transport in the blood

Cell respiration is a process that links all living organisms: sugars, such as glucose, are oxidized in order to generate ATP molecules. The primary waste product of this process is carbon dioxide. In humans, as well as many other organisms, this carbon dioxide diffuses out of a respiring cell and eventually enters a nearby capillary bed. Once carbon dioxide enters the bloodstream, there are three ways in which it is transported to the lungs:

- a small percentage of carbon dioxide remains as it is and simply dissolves in the blood plasma;
- some carbon dioxide becomes reversibly bonded within haemoglobin (each haemoglobin can carry a single carbon dioxide molecule; this is the basis of the Bohr shift);
- most (approximately 70%) of the carbon dioxide enters erythrocytes and is converted into hydrogen carbonate ions, which then move into the blood plasma for transport.

Formation of hydrogen carbonate ions

The cytoplasm of erythrocytes contains an enzyme known as carbonic anhydrase. This enzyme catalyses a reaction in which carbon dioxide and water combine to form carbonic acid (H_2CO_3). Carbonic acid then dissociates into a hydrogen carbonate ion and hydrogen ion (see Figure 19.28).

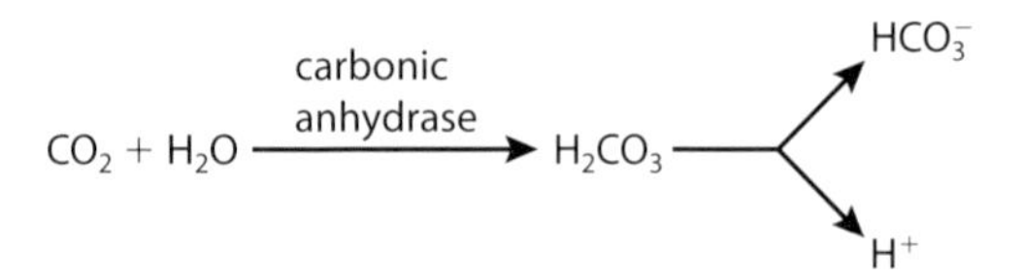

Figure 19.28 Carbonic anhydrase catalyses the formation of carbonic acid and therefore the spontaneous formation of hydrogen carbonate.

Carbon dioxide is carried in the bloodstream three ways:
- as carbon dioxide gas dissolved in plasma;
- as carbon dioxide bound to haemoglobin;
- as hydrogen carbonate ions dissolved in blood plasma.

The hydrogen carbonate ions formed from this reaction exit the cytoplasm of the erythrocyte through specialized protein channels in the red cell membrane. The transport mechanism is facilitated diffusion and it works by a mechanism which exchanges one hydrogen carbonate ion for one chloride ion moving into the erythrocyte from the blood plasma. This exchange of the two negative ions keeps a balance of charges on either side of the erythrocyte cell membrane and is known as the chloride shift (see Figure 19.29).

The hydrogen ions that are produced due to the dissociation of carbonic acid must also be accounted for in some way in order to prevent a large change in the pH of the blood. Removal of hydrogen ions from the cytoplasm of erythrocytes or blood plasma is called pH buffering. Some of the hydrogen ions produced stay within the erythrocyte and become temporarily bound to haemoglobin. Some hydrogen ions leave the erythrocyte and enter the blood plasma where they become bound to plasma proteins. In either case, the hydrogen ions are temporarily removed from solution and thus cannot affect pH (see Figure 19.29).

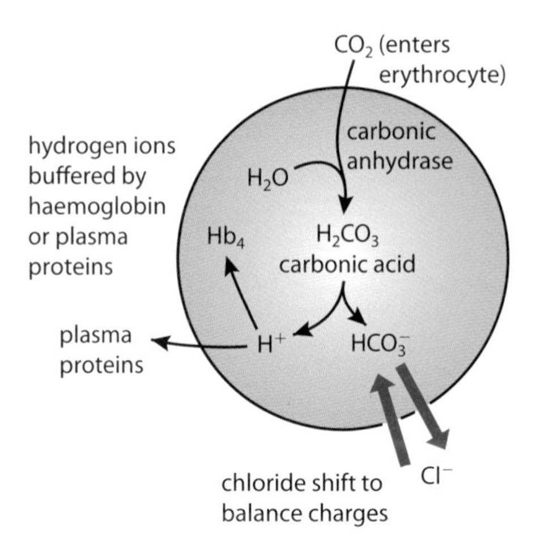

Figure 19.29 Events which occur when carbon dioxide enters a red blood cell.

The effect of exercise on ventilation rate

Exercise of any kind involves the use of skeletal muscle to accomplish movement. It does not matter whether the exercise is intended to improve physical condition or is a necessary part of daily work. The use of skeletal muscle demands the use of ATP molecules and thus the rate of cell respiration increases dramatically. Active muscle tissue consumes much more oxygen and produces much more carbon dioxide than muscle tissue at rest. The body must have a mechanism to ensure that the rate of transport of these respiratory gases meets the need of the increased demand. One specific requirement under these conditions is the increase in rate of breathing or ventilation rate.

Ventilation rate is under the control of an area of the medulla oblongata of the brainstem. This area of the medulla is known as your breathing centre. The breathing centre has two mechanisms that come into play when the rate of ventilation needs to increase.

- Receptor cells, known as chemosensors (or chemoreceptors) located in the inner wall of the aorta and carotid arteries detect the increase in carbon dioxide level and the associated decrease in blood pH. When stimulated, these receptors send action potentials to the medulla's breathing centre.
- The medulla itself contains the same kind of chemosensors and as the blood passes through the capillary beds located in the medulla, increased carbon dioxide levels and decreased pH is detected.

The normal blood pH is about 7.4, very slightly alkaline. The decrease in pH is due to the dissociation of carbonic acid explained in the previous section. Under normal circumstances, buffering by haemoglobin and plasma proteins prevents any change in pH. But when you are exercising heavily, the buffering mechanisms are overtaxed and the excess hydrogen ions lower the blood pH. It is not correct to say that the blood becomes more acidic as it is not acidic in the first place and would require very heavy exercise to reach even a neutral pH. In this case, the blood is becoming slightly less alkaline, changing from a pH of about 7.4 to no lower than 7.0 depending on the level of physical exertion.

To increase ventilation rate, the medulla's breathing centre sends action potentials to the diaphragm, intercostal muscles (muscles between ribs) and muscles in the abdomen. The mechanism of breathing is not altered, just the frequency. The rate of breathing increases and so, too, does tidal volume. With an associated increase in heart rate, the body tissues are supplied with an increased supply of oxygen necessary for the increased skeletal muscle activity.

When the physical exertion ceases or at least decreases, the chemosensors detect the decrease in carbon dioxide level in the bloodstream or the corresponding increase in blood plasma pH and a set of action potentials is generated that leads to a decrease in ventilation rate.

Capillaries are the only blood vessels thin enough to allow molecular exchanges. The wall of each capillary is a single cell layer.

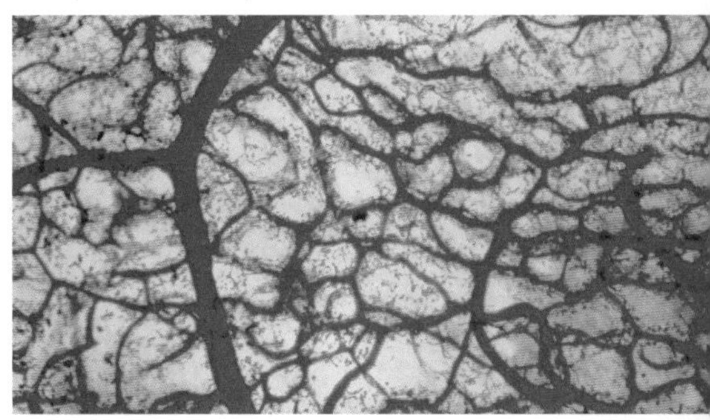

Asthma

Asthma is a chronic condition which affects the airways to the lungs. These include the bronchi and their various branches leading to the alveoli. During an asthma attack, the airways become inflamed, swollen and often produce excess mucus. This makes it difficult for the person experiencing the attack to get air in and out

There is an excellent series of web pages giving information about asthma causes, effects, treatments, etc. Visit heinemann.co.uk/hotlinks, enter the express code 4242P and click on Weblink 19.4.

of their lungs. Often a wheezing sound can be heard due to the restricted diameter of the bronchiole tubes. A severe asthma attack can be life-threatening as air becomes so restricted that oxygen is denied to aerobic tissues.

There is no cure for asthma, although there are treatments which alleviate symptoms when an attack occurs. It has been demonstrated that asthma does have a genetic link as children of adults who have asthma are more likely to have asthma than the general population. For those who are genetically predisposed to asthma, there appear to be many environmental triggers which can bring on an attack. These triggers are very variable and may include:

- allergens, including pollen, moulds, and animal dander;
- certain arthropods, including cockroaches and dust mites;
- smoke of all kinds;
- scented products such as perfumes, body sprays, cologne;
- exercises which increase respiratory rate;
- stress and strong emotions;
- cold air;
- some medications;
- some food preservatives.

Living and breathing at high altitudes

To see an altitude oxygen calculator, visit heinemann.co.uk/hotlinks, enter the express code 4242P and click on Weblink 19.5.

There is a common misconception that air at high altitudes contains less oxygen by percentage than air at sea level. This is not true: the percentage of gases in air does not change as you go up in altitude, it is air pressure that changes. Air at high altitudes is at a lower pressure. This means that all the molecules in the mixture are more spread out than in air at sea level. When you breath less dense air, diffusion of oxygen across the alveoli into the bloodstream is less efficient, and less oxygen enters your bloodstream.

Training for some sports is often done at high altitude locations in order to take advantage of some of the acclimatization adaptations such as increased haemoglobin and number of erythrocytes.

When someone first arrives at a high altitude, physical activity can lead to an almost immediate fatigue. Other typical symptoms may include vision problems, nausea, abnormally high pulse rate and difficulty in thinking clearly. These symptoms are often called altitude sickness or mountain sickness. Severe cases of altitude sickness may lead to fluid accumulations around the brain or in the lungs and can become life-threatening. A person suffering from severe altitude sickness should return to a lower altitude as soon as possible.

People who live at or near sea level and then fly to a high altitude location need to be very aware of altitude sickness. In just a few hours, you could fly from a sea-level home to a high-altitude city such as Quito, Ecuador (elevation 2800 metres), or La Paz, Bolivia (elevation 3650 metres).

On first arrival at high altitudes, our bodies attempt to compensate by increasing ventilation rate and heart rate. This is stressful for the body and is not a long-term solution or adaptation. Over time, acclimatization occurs. Some of the physiological responses involved in acclimatization are:

- an increase in the number of erythrocytes and amount of haemoglobin;
- an increase in capillaries in both the lungs and muscles;
- an increase in lung size and surface area for oxygen and carbon dioxide exchange;
- an increase in myoglobin within muscle tissues.

Mountaineers who climb high peaks such as Mount Everest typically set up several base camps at increasing altitudes. They spend time at each base camp in order to allow a certain degree of physiological adaptation at the new altitude.

Exercises

15 Explain why it makes sense for myoglobin to have a greater affinity for oxygen than haemoglobin.

16 What is the adaptive significance of the Bohr shift?

17 Deduce why haemoglobin molecules are contained within erythrocytes and do not simply float within blood plasma.

Practice questions

1 **(a)** State where bile is synthesised. *(1)*

(b) Explain the role of bile in digestion. *(2)*

(Total 3 marks)

2 Explain the way the body acclimatizes to gas exchange at high altitudes.

(3 marks)

● **Examiner's hint:** Whenever you answer an 'explain' question, think and write about the steps of a mechanism.

3 **(a)** List two glands that secrete digestive juices into the alimentary canal. *(1)*

(b) Describe the process of erythrocyte and haemoglobin breakdown in the liver. *(4)*

(Total 5 marks)

4 Explain the oxygen dissociation curve for adult haemoglobin and how it is affected by the Bohr shift. *(6 marks)*

● **Examiner's hint:** Question 4 would be difficult to answer without first drawing the graph that is the focus of the question. It would be a good idea to draw the graph as part of your answer and then frame your explanation around the graph.

5 Explain the events of the cardiac cycle. *(7 marks)*

6 A major requirement of the body is to eliminate carbon dioxide (CO_2). In the body, carbon dioxide exists in three forms: dissolved CO_2, bound as the bicarbonate ion, and bound to proteins (e.g. haemoglobin in red blood cells or plasma proteins). The relative contribution of each of these forms to overall CO_2 transport varies considerably depending on activity, as shown in the table below.

CO_2 transport in blood plasma at rest and during exercise			
Form of transport	arterial blood /mmol l^{-1}	Rest venous blood /mmol l^{-1}	Exercise venous blood /mmol l^{-1}
dissolved CO_2	0.68	0.78	1.32
bicarbonate ion	13.52	14.51	14.66
CO_2 bound to protein	0.3	0.3	0.24
total CO_2 in plasma	14.50	15.59	16.22
pH of blood	7.4	7.37	7.14

Source: Geers and Gros (2000), *Physiological Reviews* **80**, pages 681–715

(a) Calculate the percentage of CO_2 found as bicarbonate ions in the plasma of venous blood at rest. *(1)*

● **Examiner's hint:** Remember to show your working for calculations.

(b) **(i)** Compare the changes in total CO_2 content in the venous plasma due to exercise. *(1)*

(ii) Identify which form of CO_2 transport shows the greatest increase due to exercise. *(1)*

(c) Explain the pH differences shown in the data. *(3)*

(Total 6 marks)

7 Distinguish between the mode of action of steroid hormones and peptide hormones.

(4 marks)

20 Theory of knowledge

An astronomer, a physicist and a mathematician are in a train going to a conference in Edinburgh. Out the window, they see a solitary black sheep.

Astronomer: That's interesting, sheep in Scotland are black.

Physicist: It would be more prudent to say that *some* of the sheep in Scotland are black.

Mathematician: To be more precise, we can say that in Scotland there exists at least one field in which there is at least one sheep, which is black on at least one side.

What does this story reveal about scientific observations, hypotheses and conclusions? What does it reveal about the nature of each of the disciplines represented?

Is biology less exact than physics or mathematics? If there were one on board the train, what would a biologist say about the sheep?

What is this chapter all about?

This chapter has some ideas, quotes, anecdotes, case studies, many *unanswered questions* but very little factual information. Why? Because in TOK, *you* are the knower. This concept should tickle your brain. It should be exciting to think that you are the expert. You have had a decade or more of formal education and even more years of life experience giving you ideas in the form of knowledge, beliefs and opinions.

On the other hand, it is a bit intimidating to think there are some things that no one will ever know the answer to. You are encouraged to explore, develop and share your views as well as actively seek the views of your classmates.

On the right track?

How will you know if you are answering TOK questions in the 'right' way, since the answers are not given in this book or by your teacher? Here are two guidelines to consider.

- If you think the question has a quick, simple answer such as 'yes' or 'no', you can be pretty sure that you are not treating it like a TOK question. If you think the answer has many sides to it, is a debatable grey area or leads to further questions, you are probably on the right track.
- Ask yourself, 'Am I pushing myself a little bit outside my comfort zone and exploring other ways of seeing an issue?' If so, you are on your way to scratching through the surface and getting to the interesting issues. That is the stuff of intellectual stimulation and growth. That is one of the challenges of the IB programme in general and TOK in particular.

Debates

To get your brain warmed up, consider the following two statements about the nature of all human beings on Earth.

A: We are all the same.

B: We are all different.

Use your biological knowledge to support or refute these two claims. If you chose one, try to imagine someone saying to you, 'That's not true! How can you say that?' What would you respond to that person? (Keep it polite, of course.)

Now try these two statements.

X: Biology is a collection of facts about nature.

Y: Biology is a system of exploring the natural world.

Use your critical thinking. Critical thinking is characterized by reflective inquiry, analysis, and judgment. Ask yourself, 'Should I believe this?' 'Am I on the right track? How reliable is this information?' In short, you are deciding whether or not you should accept that something is valid or not. Again, if it is an easy, quick decision then you are not treating the question in the way that you should.

Critical thinking: Is this statement valid? What is its source? Is the person who said it qualified to make such a claim? Do they have a bias that I should know about?

More debates

Coming back to the pairs of statements above, what would lead someone to believe one or the other statement? In each pair, could it be possible that both statements are valid? Or are they mutually exclusive? What about these two statements?

- There is only one scientific method which is universal throughout the world – only by following the same method can scientists reach the same results and conclusions.
- Different cultures in different regions of the world use different versions of the scientific method to obtain valid results and conclusions.

Ernest Rutherford was a physicist – you probably identified the bias he had in the quote on the right.

'All science is either physics or stamp collecting.'
Ernest Rutherford

Nature of science(s)

What did Rutherford mean by the above quote? Is he justified in saying that chemistry, biology or other branches of science are only there to catalogue and classify phenomena in nature? What does his statement imply about the degree of prestige or respect each scientific discipline enjoys? Is it possible to understand all sciences by studying one of them?

What is knowledge?

- What counts as knowledge in biology?
- How does biological knowledge grow?
- What are the limits of knowledge?
- Who owns knowledge?
- What is the value of knowledge?
- What are the implications of having or not having knowledge?
- Is there one way which is best for acquiring knowledge?
- Where is knowledge? Is it a 'thing' that resides somewhere – is it in books, in your head, in a computer database?

Look at the following images. Use the list of questions about knowledge and your critical thinking to evaluate whether some or all of these are valid as scientific knowledge. For example, does mythology count as scientific knowledge?

1

2

3

4

5

'To know that we know what we know, and to know that we do not know what we do not know, that is true knowledge.'

Copernicus

6

7

How do we know?

Here is an example of scientific knowledge in biology: 'The organelle in a plant cell that is responsible for photosynthesis is the chloroplast'.

How could you verify that? How can you be sure that there is not another part of the cell which photosynthesizes? Is it a falsifiable idea?

Epistemology is the study of the theory of knowledge and it raises the question, 'How do we know what we know?'

Copernicus is the scientist who mathematically showed that the Earth goes around the Sun and not the other way around.

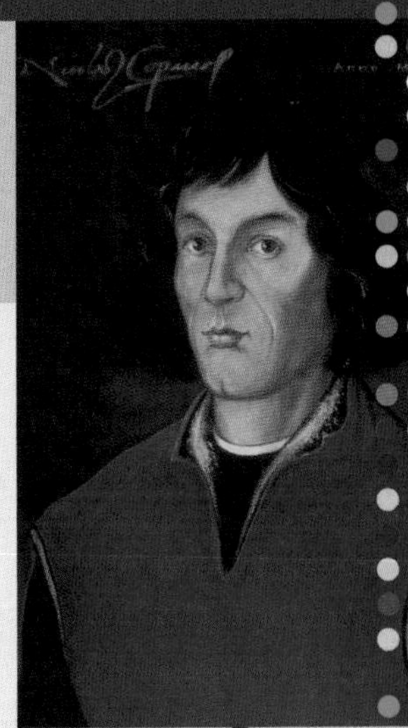

8

9

Case study 1: Life on Mars?

It's not a simple question. Despite several visits by space probes, no conclusive evidence has been discovered on Mars which could lead scientists to declare that there is life on its surface. And yet, the search continues. The most compelling evidence that there was once life on Mars is a meteorite found in Antarctica which NASA claims came from Mars and contains fossils of bacteria.

If you apply your critical thinking to this, some questions should pop into your mind. How do we know this chunk of rock is really from Mars? How did it get to Earth? How does NASA know that the 'fossils' are from bacteria? Could they have been formed from non-living chemical reactions? How important are such discoveries in ensuring funding for future missions? Is NASA planning on collecting more fossils from Mars directly and bringing them back to Earth to study? Could there still be colonies of bacteria living on Mars today or is life extinct on the red planet?

From this specific example, two more general questions arise:

- Is it possible to really 'know' the truth?
- Is information absolute or relative?

Are you an empiricist or a rationalist?

- Empiricism = the belief that our senses allow us to acquire knowledge.
- Rationalism = the belief that reason allows us to acquire knowledge.

Be careful: critical thinking does not mean you criticize everything. It means you are aware of questions of validity. You are not being negative, you are just being inquisitive and prudent.

Catching a cold

Despite the biological evidence that colds are caused by viral infections, many people believe that you catch cold from being exposed to low temperatures or changes in humidity.

Who is right? Where does the truth lie? For something to be considered 'true', does it have to be formally proven using the scientific method? Is a profound conviction that something is true, good enough to make it valid? If one person believes that something is true, does that make it true or do there have to be a certain number of believers before the idea can be considered true?

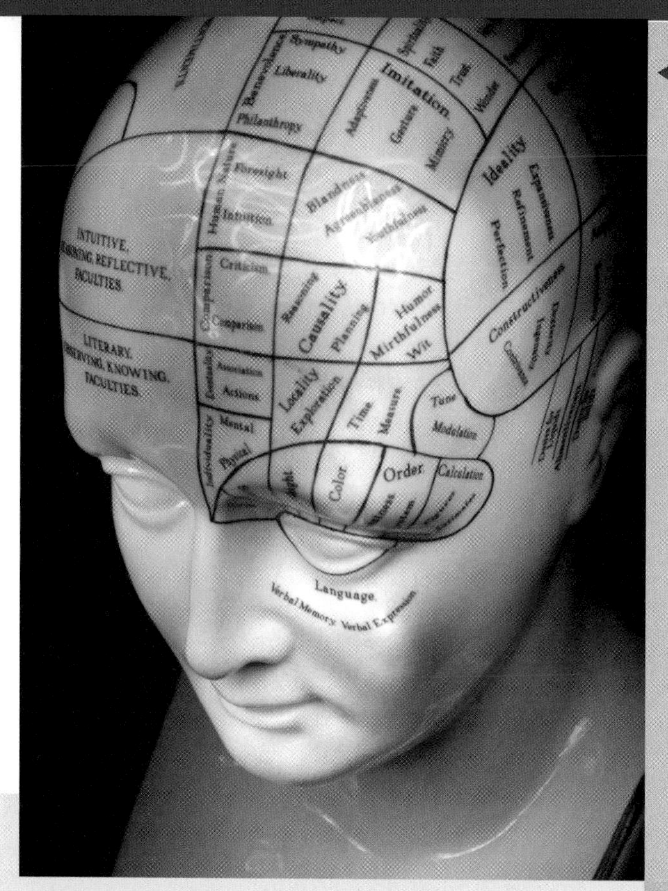

Phrenology

The pseudo-science of phrenology claimed that the shape of a person's skull and the bumps and indentations on it determined a person's intelligence, personality and talents.

How do you think it was demonstrated that the 'laws' of phrenology were simply not true?

Tongue map

As students and teachers, what do we claim to know about biology? Are we justified in making such claims? How?

What experiences have you had that give you insight concerning these issues? Consider the following example. For decades, the idea of a 'tongue map' (whereby certain zones are reserved for certain tastes) was propagated through biology textbooks and taste-test investigations were suggested as lab work for students. It has since been shown that all parts of the tongue can taste sweet, sour, bitter or salty.

'There must be no barriers for freedom of inquiry. There is no place for dogma in science. The scientist is free, and must be free to ask any question, to doubt any assertion, to seek for any evidence, to correct any errors.'

Robert Oppenheimer

Art and imagination

Is there a place for imagination and creativity in science? Are there any parallels between biology and art? Could it be argued that just as an artist sees things in his or her own way, so a scientist sees things in his or her own way? Or, on the contrary, are science and art diametrically opposed ways of interpreting nature?

Case study 2: Spirit/Soul

In 1907, Dr Duncan MacDougall conducted experiments to determine whether or not people lost mass after death. His results seemed to suggest that they did and led him to the conclusion that the human soul weighed 21 grams. Since his experiments (some of which did not give conclusive results) were done with scales of questionable accuracy and he had only six subjects, his conclusions are widely criticized and are not taken seriously by the scientific community today.

Will questions about souls always remain beyond the capabilities of science to investigate or verify? Why hasn't anyone repeated this experiment in over a century? What do you think the reaction of the religious community would be if scientists repeated MacDougall's experiment?

Some fundamental questions

- Should experiments be performed to answer fundamental questions or should they only be done if they have a useful application in our everyday lives?
- Who should decide which research pursuits are of the most value? Who should decide on how funding is distributed or the prioritizing of the use of laboratory space and resources? Universities? Governments? Committees of scientists? Taxpayers?
- Should research about a tropical disease such as malaria be paid for by tax money from non-tropical countries?

Is there an end?

Is scientific knowledge progressive? Has it always grown? Imagine a graph with scientific knowledge on the *y*-axis and time on the *x*-axis. How would you draw the graph? Would it be a curve or would it be linear? Is it always increasing? What units would you use? Could the graph ever go down – in other words could scientific knowledge ever be lost (wartime, laboratory burns down, famous scientist dies)?
Could there ever be an end to science? If there is an end, what would be the consequences?

'Science knows no country, because knowledge belongs to humanity, and is the torch which illuminates the world.'
Louis Pasteur

▲ Decisions, decisions …

Models 1

The double helix shape for DNA and the fluid-mosaic membrane model are examples of models which were created in order to explain observed phenomena. Are such models just inventions? If so, how is it that they can be used to make predictions or explain natural phenomena?
Look at the false-colour electronmicrograph on the left. The magnification and resolution are not good enough to see how the integral or transmembrane proteins and cholesterol are arranged in the membrane but chemical tests reveal that they are there. This is why the fluid-mosaic model was introduced. It is a proposed explanation for how the various components of the membrane are arranged. If the model successfully fits the observed phenomena, does that validate it as being true?

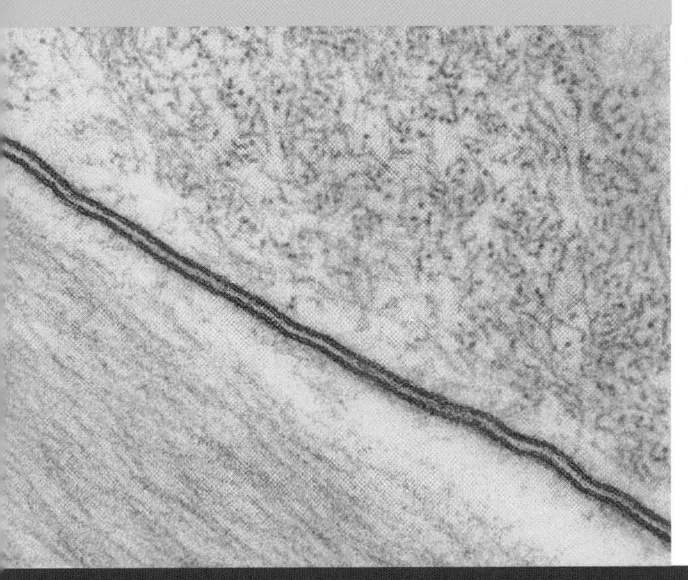

'All models are wrong, but some are useful.'
George E. P. Box (innovator in statistical analysis)

Models 2

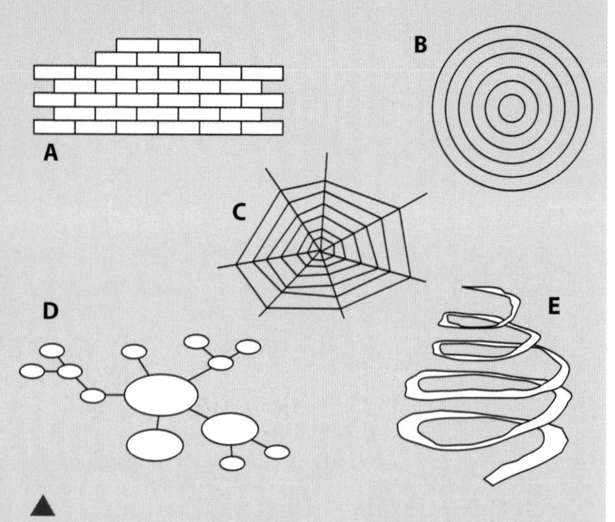

▲
Which of these conceptual drawings best represents the interconnections in the ways of knowing and areas of knowledge? In what ways might these metaphors be useful?

Who's right?

Among all the points of view that are available to you in the classroom, at home, in the media, on websites, how do you know which one is right or useful?

Religion in an age of science

In what ways could someone's cultural or religious background influence their acceptance of certain scientific theories?

There was a time when scientists hesitated to publish their works out of fear of the church. Have the tables turned? Are there religious writers who fear scientific criticism if they publish their ideas?

If a student writes in an IB exam that they refuse to answer questions about evolution because of their religious beliefs, should they get any marks?

Limits of perception

'You cannot speak of the ocean to a well frog …'
(Chuang Tzu Taoist text, over 2000 years ago)

Can we here on Earth possibly know of worlds beyond our own? Can we possibly know what the distant past was like or what the distant future will hold? Or are we like a frog in the bottom of a well trying to understand what the ocean might be like?

We were wrong, here's the real story ...

In palaeontology, it seems that every time a new hominid fossil is dug up, we have to redraw the human family tree. If you search the Internet for human phylogeny, you will probably find that few sources agree with each other. When are they ever going to make up their minds and get it right? Likewise, in questions of diet and nutrition, every few years, nutrition experts change their minds about dietary advice.

Does this frequent revision give credibility to science or does this make science less credible?

- Argument for the question
  It's important for scientists to be able to modify ideas as new evidence is revealed – this is how science grows and progresses and without such a system, we would be intellectually stuck.
- Argument against the question
  Why can't these so-called experts make up their minds? One year they say one thing and then a year or two later they say 'Oh, we were wrong, here's the real story.'

Archaeopteryx

Archaeopteryx is arguably the most famous fossil in the world. It has some features of a dinosaur such as reptile teeth and a bony tail but it also has certain bone structures of a bird and it has the most bird-like feature of all: feathers. It did not take long for experts to jump to the conclusion that *Archaeopteryx* was the 'missing link' between dinosaurs and birds. Can we be so sure that this fossil is the transition between the two? Are physical features enough to base such a decision on? What kind of evidence would give more credibility to this claim?

What was spontaneous generation?

For centuries, it was firmly believed to be true that rats, maggots or mould sprang from rotting meat or vegetable matter. This was called spontaneous generation. It took tireless experiments by Louis Pasteur and others to refute this idea and prove that the rats, maggots and mould came from the surrounding environment.

The end of spontaneous generation?

The idea of spontaneous generation has been shelved as unscientific. It has no value as biological knowledge, but it does have historical value and it helps to illustrate how science works.

This is a good example of an original hypothesis which was disproved and falsified by experimentation. It can be argued that in order for something to be considered valid as scientific knowledge, it has to be verifiable. If experiments show that the results do not support the hypothesis or even refute, the idea is falsified. This assumes that the experiment is repeatable. Other scientists should be able to do the same experiment and get similar results.

Imagine the consequences of the following situation. The colleague of a famous scientist dies unexpectedly and a student of his decides to publish extracts from the laboratory notebooks. The notes are filled with interesting ideas but also contain severe criticisms of the methods of the famous scientist. For example, only the experiments which gave evidence supporting the famous man's hypotheses were considered and the others which refuted the hypotheses were ignored. This goes against everything the scientific method is supposed to represent. This scenario happened when Claude Bernard died in 1878. He had been working with Louis Pasteur and it took a great amount of persuasion and force of personality for Pasteur to save his reputation. He had lots of both.

Unprovable assumptions?

Does biology make any assumptions which are unprovable? Consider this:

- all events in nature are caused by physical phenomena.

In other words, every natural event can be explained by the interactions between atoms and molecules. Is such a statement provable?

No matter how many instances of white swans we may have observed, this does not justify the conclusion that all swans are white.
Karl Popper.

Are all swans white?

Case study 3: Choosing a boy or a girl

A clinic claims that it has developed a new technique for filtering sperm cells in such a way that future parents can choose whether to have a boy or a girl. The doctors at the clinic claim they have a 95% success rate, which is considerably higher than with current sperm separation techniques. In an effort to protect the secret technique from being stolen by competing clinics, staff refuse to reveal how they do it. Does this secrecy undermine the scientific validity of the technique?

Scientific science

To what extent is there an overlap between biology and the social sciences? Are the latter 'less scientific'? Consider psychology, sociology, anthropology, economics.

Knowledge claims

Compare the validity of knowledge claims of two different scientific disciplines. For example, you could think about a historical approach (archeology) versus an experimental approach (lab investigation), as exemplified in these photographs.

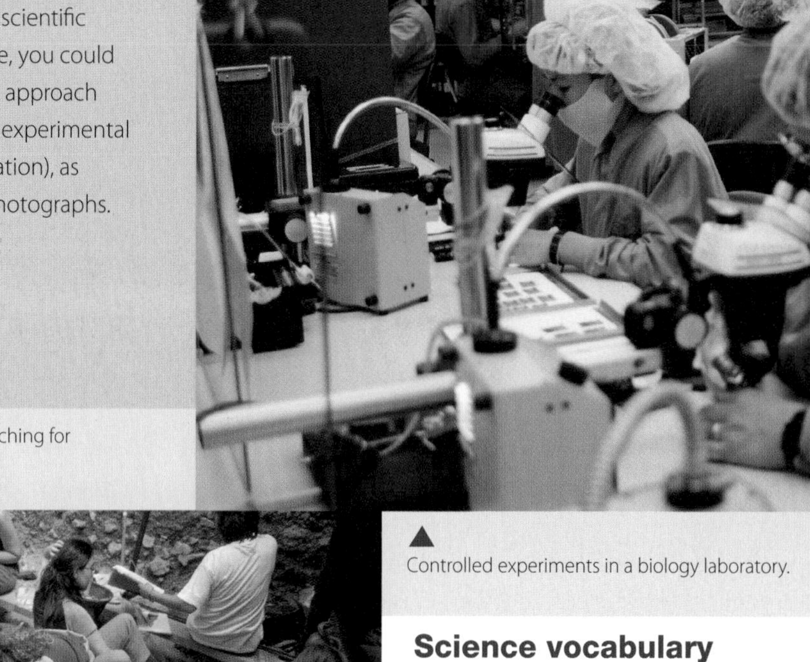

Controlled experiments in a biology laboratory.

An archaeological dig searching for evidence of the past.

Science vocabulary

Does scientific language and vocabulary have primarily a descriptive or an interpretive function? Consider the following expressions:

- natural selection;
- concentration gradient;
- artificial intelligence.

What is nature?

Biology is a natural science but what is meant by nature? Is it a clockwork machine? Or is it one big Gaia-type living organism? How useful are these metaphors?

Is this a useful image for 'nature'?

Wiki

The term 'wiki' is an acronym for 'what I know is …' Online wikis are filled with user-generated content on a wide range of subjects, including scientific ones. Should a wiki be created for scientific papers where researchers can publish their latest laboratory findings? In what ways would such a site be useful to scientists wanting to publish their results? In what ways would this be useful to the general public? In what ways would this information go against the nature of peer-reviewed scientific publications, which is the norm today for sharing experimental results? For example, would such wikis be just as valid as traditional scientific journals?

'Prediction is very difficult, especially about the future.'
Niels Bohr (Danish physicist who helped understand how atoms work)

Case study 4: A famous hoax – Piltdown man

It's always exciting to find a fossil of a new species – especially if it is a hominid. But if you can't find one, should you make one up? In 1912, someone did just that and called it Piltdown man. The 'fossil' was made from a human skull and the jaw of an orangutan. Amazingly, the fake fossil puzzled specialists for over 40 years until it was finally exposed as a hoax.

This famous hoax demonstrates how important it is to double-check findings. Why did it take so long for the truth to come out? You may find online sources useful if you want to investigate this story.

Perception

Which red circle is bigger? Judge using your eye first and then use a ruler to check your answer. What might this suggest about our perceptions and reality?

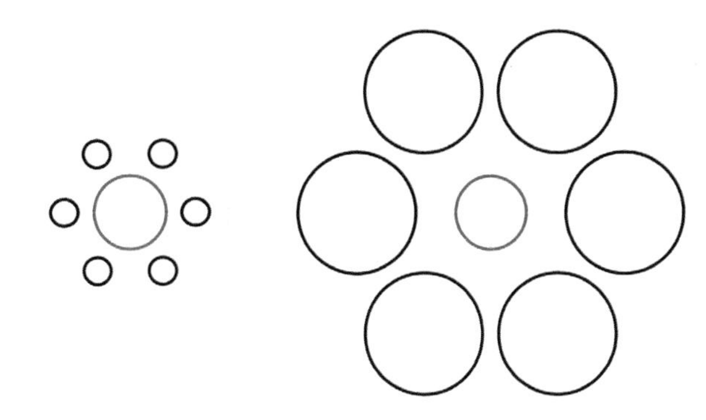

'Science may be described as the art of systematic oversimplification.'

Karl Popper

Evidence and elephants

There is a story from Asia about a small group of blind men who encounter a tame work elephant, a creature none of them have ever had contact with before.

One blind man touches the elephant's side and says 'It's like a wall'. Another grabs the end of its tail and says, 'It's covered in long hairs.' Another feels a leg and says 'Elephants are round and vertical like a pillar' A fourth holds his ear and says 'It's like a sail'. A fifth holds the animal's trunk and exclaims 'Elephants are like snakes.'

None of the men were wrong but none were completely correct. This story illustrates how easy it is to jump to conclusions before having all the evidence. In science, is it possible to have all the evidence of any particular phenomenon?

What qualifies as an experiment?

Biology is an experimental science but what constitutes an 'experiment'? Do you have to have a hypothesis, controlled variables, a laboratory? What if you just have people fill out questionnaires? Is that an experiment? What about digging up fossils?

Theory and myth

In what ways are theory and myth similar or different? Consider these questions when comparing and contrasting the two.

- Is it based on well-substantiated facts?
- Is it passed on from generation to generation?
- Can it be modified over time?
- Can it be used to predict future events?
- Has it been tested repeatedly?
- Is it widely accepted as being true?
- Is it considered to be a supposition?
- Is it considered by many to be false?

'It doesn't matter how beautiful your theory is, it doesn't matter how smart you are. If it doesn't agree with experiment, it's wrong.'

Richard Feynman

'Irrationally held truths may be more harmful than reasoned errors.'

Thomas Henry Huxley

Genetic counselling involves many ethical dilemmas.

Biology and values

Consider the following domains of research in biology. What are the ethical issues?

- gene therapy;
- vaccine tests;
- experimentation on human volunteers;
- research involving human embryos.

Do the ends justify the means?

Social context

In the 1800s, Carl Linnaeus classified humans in terms which today are considered racist and unacceptable. To what extent does the social context of scientific work affect the methods and findings? Are we more culturally sensitive now? What if a study showed that one particular ethnic group had smaller brain sizes than another group. Would that give credibility to the idea of intellectually superior or inferior groups of humans?

Moral responsibility

Should scientists be held morally responsible for the applications of their discoveries? Is there any area of scientific knowledge the pursuit of which is morally unacceptable or, morally required? What about:

- cloning humans?
- eugenics?
- genetic engineering of crops?
- finding a cure for cancer.

'Nothing in this world is to be feared ...only understood.'
Marie Curie

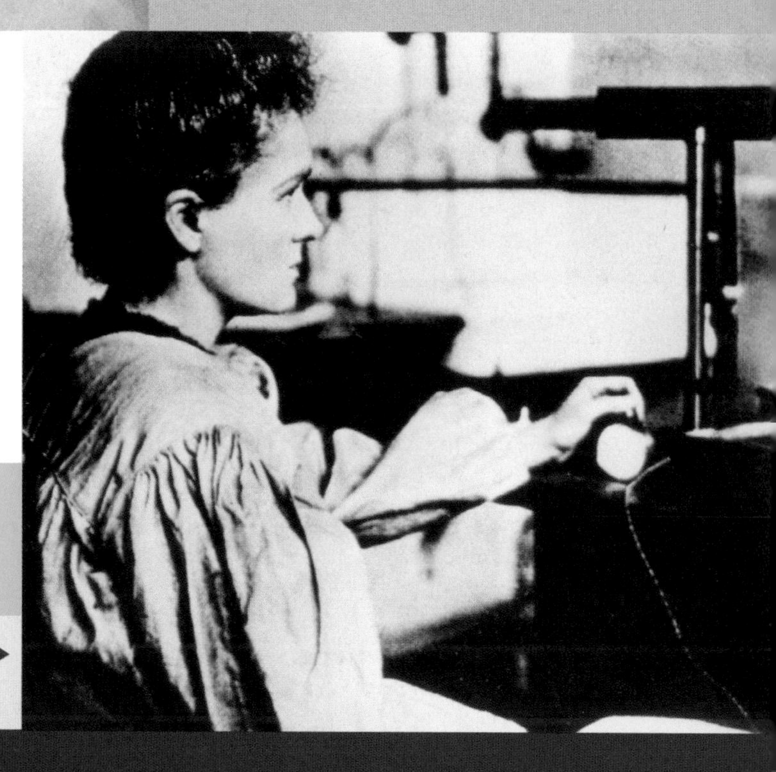

Marie Curie was the first woman to be awarded a Nobel prize and the first person ever to get two.

> **'Science gets the age of rocks, and religion the rock of ages; science studies how the heavens go, religion how to go to heaven.'**
> *Stephen Jay Gould*

Science and religion

To what extent should religion take note of scientific developments? For example, should religious communities keep abreast of scientific discoveries related to Darwin's theory of evolution by natural selection? Some people think that science and religion can coexist, others that they are mutually exclusive.

Science and technology

Is scientific knowledge valued more for its own sake or for the technology that it makes possible?

> **'The most important discoveries will provide answers to questions that we do not yet know how to ask.'**
> *John Bahcall* (commenting on the Hubble space telescope's capabilities)

Inaccessible worlds

Some scientific fields of exploration have only been possible since suitable technology was invented (e.g. genetic engineering has been possible since technological developments in the 1970s and 80s). Could there be problems of knowledge that are unknown now because the technology needed to reveal them does not yet exist? Remember that despite the fact that bacteria are all around us, we were not able to see them until the microscope was invented in the 1600s. Perhaps there are other phenomena that we simply cannot observe because no one has invented an apparatus to detect them yet.

Is there any science that can be pursued without the use of technology?

Case study 5: Morals and money

A well known, highly successful professor at a top-notch university has made several major discoveries and biological advances in his multi-million-dollar, high-tech laboratory. He has published dozens of articles in the most prestigious research journals in the world. He has won worldwide respect and is able to obtain generous grants to fund his successful laboratory. One day, a colleague accuses him of falsifying his data and lying to the public about his latest breakthrough. After investigation, it is revealed that the allegations are true. The researcher not only retracts his latest discovery but is asked to leave the university. His career is over, his reputation shattered.

- What does this reveal about the nature of science?
- What tempted him to break the rules?

One of the basic principles of scientific discoveries is that they must be repeatable so they can be verified by other scientists. If experiments involve multi-million-dollar equipment, how realistic is it to verify every experiment?

> **'My business is to teach my aspirations to conform themselves to fact, not to try and make facts harmonize with my aspirations.'**
> *Thomas Henry Huxley*

Case study 6: A moral dilemma

Some British scientists studying bovine spongiform encephalitis (BSE, informally AKA 'mad cow disease') saw alarming trends in the early years of the problem: graphs of the number of cases were increasing at a worrying rate. They were told by the authorities to continue monitoring the situation but that it was too early to tell if a ban on British beef was necessary. Did those researchers have a moral responsibility to disobey the government and sound the alarm in the early stages?

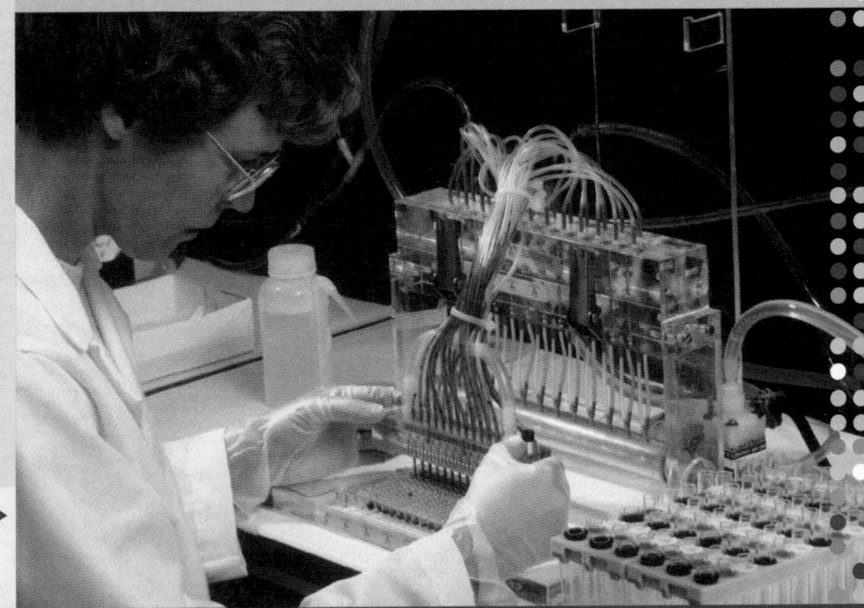

Advice for IB biology students on internal assessment

You will be asked to accomplish 60 hours of laboratory work during your course in IB Biology Higher level (HL), 40 hours for Standard level (SL) students. Ten of those hours will be in a collaborative project known as the Group 4 project, which involves students from different science disciplines working together on a scientific topic. Many of the 60/40 lab hours should act as a 'training programme' in which you learn the skills inherent in asking good scientific questions and the protocols used to attempt to answer them. Before you end your course in IB Biology, your instructor(s) will evaluate the skills you have learned. These evaluations are called 'internal assessment' and represent 24% of your IB Biology test grade.

Advice on laboratory skills to be assessed twice

Skill assessed	Advice
Formulating a focused research question and identifying relevant variables.	• The research question should be carefully worded. • If a living organism is involved, identify it by name. • Identify your independent variable, dependent variable and all important controlled variables (constants).
Designing an experimental method (procedure) which effectively controls the variables.	• The only difference between your experimental group(s) and control group(s) should be your independent variable. • Your dependent variable should be a valid way of measuring the effect of the independent variable. • You should make sure that your experimental design really does hold stated controlled variables as constants.
Designing a method for the collection of sufficient relevant data.	• The best experimental designs have a quantitative independent variable chosen at regular, fixed intervals and also a quantitative dependent variable. • A good rule of thumb is to repeat trials a minimum of five times. • Data validating that controlled variables were measured is often necessary. For example, if you state that temperature is a controlled variable, include regular temperature readings in your planned procedure.
Recording appropriate raw data including units and uncertainties.	• Raw data is the data read directly from measuring devices. Be sure to give the unit of measure. • Each measuring device has a degree of precision equal to ± the smallest division of the device (e.g. for a mm ruler, ±1 mm). • Other forms of uncertainty can be given as well.
Processing of quantitative raw data.	• Make decisions about processing your raw data. You are trying to convert your data into a form from which it is easier to draw conclusion(s). This may include such things as calculating means, percentages, etc. and applying a relevant statistical test such as standard deviation, student t-test or other test.
Presenting processed data including errors and uncertainties.	• Presentation possibilities include one or more well-designed data tables showing recently processed data. Design your tables for maximum clarity and be sure to include units. • Be consistent with use of decimals and do not exceed the capability of the original measuring devices. • One or more well-chosen graphs should also be included in this section. Uncertainties can be shown on graphs with the use of error bars.

Stating and justifying a conclusion based on a reasonable interpretation of the data.	• Start by studying and reflecting on the pattern shown by your processed data. Determine if there is reasonable consistency in the data set. You may see a positive correlation, no correlation, or a negative correlation between the variables. • Justify any pattern (or lack of pattern) you see while keeping in mind any errors and uncertainties. • Remember to keep in mind the specifics of the original question that you asked.
Evaluation of procedure weaknesses and limitations.	• Comment on your experimental design and the quality of the data obtained. • List weaknesses and show an appreciation of how significant any identified weakness is. • Evaluate such things as the equipment that was used, processes employed, and time for experimentation.
Suggesting realistic improvements based on identified weaknesses and limitations.	• Base this section on the weaknesses and limitations identified in the previous section. • Any suggested modifications should be clearly specified and realistic. General statements, such as 'should have collected more data points' or 'should have used better equipment' are not sufficient. Be specific in your suggestions.

Advice on laboratory skills to be assessed once and summatively

Skill assessed	Advice
Following instructions accurately, adapting to new circumstances and seeking assistance when required.	• Your instructor(s) will evaluate this criterion throughout the labwork portion of the course. • You should seek assistance when required, but remember this can be overdone and some degree of independence is expected. • Always seek assistance when safety is an issue.
Using a range of techniques and equipment with competence.	• You should be exposed to a wide variety of laboratory experiences and should develop an understanding of common techniques and equipment use.
Paying attention to safety issues.	• Always pay attention to safety instructions and protocols whether verbal or in writing. • Work in such a manner that the safety of yourself and others is the primary consideration.

Advice on laboratory skills to be assessed once (by way of the Group 4 project)

Skill assessed	Advice
Using self-motivation in approach to the Group 4 Project. Following G4P through to completion.	• Become invested in and enjoy the topic / subtopic that you choose. You will then naturally show self-motivation and work towards completion of a high-quality project.
Collaboration and communication in a group situation. Helping to integrate the views of others into the G4P.	• Communication in a group situation requires effort and a respect for others. • Show leadership qualities without taking over the project as an individual.
Showing awareness of one's own strengths and weaknesses and thoughtfully reflecting on the G4P learning experience.	• Be willing to share your strengths, weaknesses and reflections in whatever format your teacher decides is appropriate at the conclusion of the G4P.

Advice for IB students on extended essays in biology

One of the requirements of the IB diploma is to write an extended essay. This in-depth study of a limited topic within a particular subject area provides you with the chance to carry out independent research within a subject of your choosing. This essay is restricted to 4000 words and is expected to involve about 40 hours of work. Most schools introduce this requirement during the last portion of your first year in the IB programme. The essay is then completed and handed in during the second year.

The extended essay is awarded marks according to a very specific set of criteria and, in conjunction with the TOK essay, can contribute to your overall diploma score through the award of bonus points. These points can be quite helpful in gaining your IB diploma. If excellent marks are obtained on both essays, you may be awarded a bonus of three points. However, if you do very badly in both essays, you will not be eligible for an IB diploma.

One of the subjects often selected for the extended essay is biology. It is particularly popular because many of the topics discussed in class and researched in laboratory activities provide unique ideas for the essay. Most successful extended essays in biology involve experimental work. However, some literature-based essays have scored quite well. Many submitted essays include both approaches.

An important requirement of the extended essay is that it should represent a new or unique approach to addressing a specific research question. Creativity and an individualistic approach are important to achieving a high mark. The table opposite offers you some guidance on writing your essay. However, there is no set formula that guarantees success in this important requirement.

Before you start, you must be aware of the criteria on which your essay will be marked. You will be assigned a supervisor – a teacher at your school – who will be available to discuss your progress with you. He or she will also help you ensure the safety and appropriateness of your work.

Advice on criteria for assessing your extended essay

Criterion	Advice
Research question	• A good research question is essential to a good extended essay. It should be stated early in the introduction and should be focused for a 4000-word essay. • Adequate time and thought must be spent in writing the research question.
Introduction	• This should provide an explanation of the research question. • The introduction should include discussion of the significance of the research question.
Investigation	• The procedure is the key. – Is it unique? – Does it allow for adequate data collection? – Are controls used? – Is the procedure truly biological in nature? – Is the procedure relevant to the research question? • If the paper is library based, this criterion involves a detailed look at how the data to be analysed was obtained.
Knowledge and understanding of the topic studied	• Have you shown that you clearly understand all aspects of the essay? • Do your analyses represent an obvious understanding?
Reasoned argument	• In your quest to confirm your hypothesis, are you logical and methodical in your approach and explanation? • Is a convincing argument presented?
Application of analytical and evaluative skills	• If library based, has there been careful analysis of all sources? • Have all aspects of the experiment been evaluated for appropriateness? • Is the presentation of data logical? • Has there been adequate data analysis?
Use of language appropriate to the subject	• Is the language appropriate to the topic and is it correctly used? • Does the terminology represent understanding?
Conclusion	• Does the conclusion flow logically from the arguments in the essay? • Is the conclusion relevant to the research question and does it support the original hypothesis? • Does the conclusion include unresolved questions and potential future research?
Formal presentation	• This includes elements such as title page, table of contents, page numbers, appropriate illustrations, proper citations and bibliography, and appropriate appendices if used.
Abstract	• This is written last and includes three elements: – research question; – investigative approach; – conclusion.
Holistic judgment	• This criterion is used to reward creative and unique approaches. • It also involves depth of understanding, insight, and apparent interest in the topic.

These criteria are the key to a strong extended essay. It is important to remember that the biology extended essay must be distinctly related to biology. Ethical considerations are extremely important and should be discussed with your supervisor. Discuss all aspects of your procedure with your supervisor before beginning so that safety issues are not a concern.

The emphasis for this essay is independent research. Select a topic that is interesting to you and do substantial library research before beginning the process of writing the research question and formulating a hypothesis. Obtain samples of past successful extended essays from your supervisor for basic ideas on how to approach the task ahead.

Enjoy your research.

Advice for IB biology students on examination strategies

IB examinations for both SL and HL Biology are dominated by three types of question. These are:

- multiple-choice questions;
- data-based questions;
- open-ended or free-response questions.

Strategy for success when answering multiple-choice questions

SL exams contain 30 multiple-choice questions and HL exams contain 40. All multiple-choice questions have four choices (A to D) and there is no penalty for guessing. Thus, every question should be answered.

The content for multiple-choice questions comes from the core material for SL students, and the core and AHL material for HL students. No material from the options appears in multiple-choice questions.

Strategy for success when answering data-based questions

Data-based questions present you with data in some form and then ask you questions about that data. Some questions will ask you to read the data displayed and some will ask you to draw conclusions from it.

Pay close attention to the number of marks for each question. The examiner is comparing your answer to acceptable answers on a markscheme. You can write as much as you like as long as you do not contradict yourself. Grading is positive so, if you write something wrong, no marks are deducted. But if you contradict yourself, you receive no mark.

You are expected to use the data given within the question. Make it a habit to reference the data when you practise data-based questions; this will make it natural to do the same when taking the exam. Become familiar with unit expressions such as kJ m^{-2} yr^{-1} (read as kilojoules per metre squared per year). If you are not comfortable with the unit expressions you see in data-based questions in this book, see your teacher for help.

If the question asks you to 'calculate', you must show your work.

Questions which use the command term 'compare', require you to clearly relate the *similarities and differences* between two sets of data. In most situations, an answer which involves numeric data will not achieve a mark unless a unit is given with the number. Do a full comparison and be sure to state whether any difference is an increase or a decrease.

Use a ruler when answering data-based questions since the graphs are often small and the degree of precision required in your answers is often quite demanding. The ruler can be used to draw lines on the graph to help you increase your chance of being within the degree of tolerance allowed by the markscheme.

Practise the questions given at the end of each chapter in this book. They are from past exam papers and a markscheme is provided. Write out the answers as you would during an exam. Give yourself a set time to answer all of the questions in a section. Finally, grade yourself with the answer key (the markscheme).

Strategy for success when answering open-ended questions

You must be familiar with all the command terms. It is extremely important that you know these terms and what is expected in response to each of them. For example, the command terms 'discuss', 'explain', 'describe' and 'outline' are asking you to give *facts*. None of these command terms is asking for an opinion or a theory.

Questions that begin with the command term 'explain', require you to relate a mechanism of how something works. Typically, this cannot be done with a brief response. There is never a penalty for writing too much. The problem here is that you could spend too much time answering one question at the expense of another. Look at how many marks the question is worth. If it is worth 6 marks, it is worth more time than a question worth only 2 marks.

When answering a 'discuss' question, make sure you present at least *two alternate views*. For example, imagine there is a discussion question on conserving the rainforest. You must give opposing views on why the rain forest should and should not be conserved.

When a question includes the command term 'list', you must give the exact number of things asked. For example, if the questions is 'List three factors which affect the distribution of plant species', you should list only three factors. If you list four, the fourth answer will not be scored.

Remember these tips.

- The examiner does not know you. You must communicate fully what you know and not expect the examiner to 'fill in the blanks' for information that you do not relate clearly.
- State the obvious in your answers. Many of the items in a markscheme will be information that is very basic in relation to the question.
- Do not use abbreviations that may be unfamiliar to someone else. Be clear and concise with your choice of words.

- If you have handwriting which is very small or not clear, *print* your response. If the examiner cannot read your writing, you will not get a mark.
- Make sure to use *extra paper* if you need it. Do not write in the margin or up the side of the page. Number the remainder of the response on the extra paper so the examiner knows where the response continues.

If a question requires you to 'draw and label', follow these directions:
- draw all diagrams in pencil and label in ink;
- print all labels horizontally on the page;
- use a ruler to draw a line from the label to the item in the drawing and be sure the line exactly touches the part that it is labelling (if the examiner has any doubt about the structure that you are naming, no marks will be awarded);
- make sure the diagram takes up at least a third of a page;
- always title a diagram.

And finally ...

Remember that the three papers of the written IB exam account for 76% of your overall score. The other 24% is based on your performance in the internal assessment (lab work) portion of the course. It is graded by your instructor and moderated by an examiner.

The examination papers are written from the assessment statements. It is therefore essential that you become very familiar with all the assessment statements that relate to the material for your level of study (SL or HL). The relevant assessment statements for each section of study are listed at the start of each section of this book. The assessment statements all begin with a command term which you must be clear about the meaning of.

In the exam

1. The first day of the examination presents papers 1 and 2 and tests your knowledge of the core material (plus AHL material for HL students). No options material is tested on the first day.
2. The second day of the examination presents paper 3 which tests the options material. You will see questions on all the options areas but you answer only the questions on the two options you have studied.
3. It is highly recommended you read each question twice before beginning to write. Many examiners report a disturbing number of students who write an answer that does not correspond to the question. In addition, when you have finished, if time allows, re-read the questions and your answers one more time just to be certain everything is in order. Be certain to cross out clearly any work you do not want the examiner to mark.

Answers

Chapter 1: Answers to exercises

1 An error bar is a graphical representation of the variability of data. Error bars can be used to show either the range of data or the standard deviation on a graph.

2 Standard deviation is used to summarize the spread of values around the mean. In a normal distribution, about 68% of all values lie within ±1 standard deviation of the mean. This rises to about 95% for ±2 standard deviations from the mean.

3 The standard deviation tells us how tightly the data points are clustered around the mean. When the data points are clustered together, the standard deviation is small and when they are spread apart the standard deviation is large. So, the standard deviation tells us how many extremes are in the data:

- if there are many extremes, the standard deviation is large;
- if there are few extremes, the standard deviation is small.

When comparing two sets of data which have exactly the same mean, we must also look at the standard deviation. If the standard deviation of one data set is much higher than the other, it indicates a very wide spread of data around the mean for that data set. This makes us question the experimental design. What is causing this wide variation in data? This is why it is important to calculate the standard deviation in addition to the mean of a data set. If we looked at only the mean, we would not see that one data set may be more variable than the other.

4 In order to determine if the difference between two sets of data is a significant (real) difference, the t-test is commonly used. If the probability (p) is 0.50, the difference is due to chance 50% of the time. This is not a significant difference in statistics. However, if you reach a p value of 0.05, the probability that the difference is due to chance is only 5%. That means that there is a 95% chance that the difference is due to some experimental condition. A 95% chance is significant in statistics. Statisticians are never certain but they like to be at least 95% certain of their findings before drawing conclusions.

5 When using a mathematical correlation test, the value of *r* signifies the correlation. The value of *r* can vary from +1 (completely positive correlation) to 0 (no correlation) to –1 (completely negative correlation). For example, we can measure the size of breeding cormorant birds to see if there is a correlation between the sizes of males and females which breed with each other. A high *r* value (e.g. 0.88) shows a positive correlation between the sizes of the two sexes. Large females mate with large males. However, the cause for this requires experimental evidence. Correlation is not cause. There may be a high correlation but only carefully designed experiments can separate causation from correlation.

6 We make observations all the time about the world around us. We might notice, for example, that bean plants wilt when the soil is dry. This is a simple observation. We might do an experiment to see if watering the bean plant prevents wilting. Observing wilting in dry conditions is a simple correlation, but the experiment gives us evidence that lack of water is the cause of the wilting. Experiments provide a test which shows cause. Observations without experiment can show only correlation.

The invasion of Africanized honey bees (AHBs) in the USA provides another example of correlation. AHBs were brought from Africa to Brazil in the 1950s but by 1990 they had spread to the southern states in the USA.

Scientists predicted that AHBs would soon invade all the southern states, but this has not happened. The bees have remained in the south west. Their migration north and east seems to be inhibited by rainfall. Annual rainfall over 55 inches and evenly distributed throughout the year is almost a complete barrier to the bees. These conditions occur at the point where the AHBs stopped moving 10 years ago. This is simply an observed correlation. In order to find out if it is a cause, scientists must design experiments to explain the mechanisms that may be the cause of the correlation.

Chapter 1: Answers to practice questions

Note: There is no markscheme for these questions because this is a new area for the syllabus.

1 D

2 A

3 B

4 A

5 **(a)** The means and standard deviations are as shown in this table.

	Mean	Standard deviation
fresh garlic	4.1	0.99
crushed sprouted root	4.1	0.88
crushed sprouted leaf	4.4	0.97
crushed sprouted bulb	4.3	0.67
control	19.1	1.91

(b) Since the standard deviations of all of the garlic data are low, there is not much variability in any of the garlic data. The control is slightly more variable than the garlic data.

(c) The means of the garlic data are very similar. All of the garlic types seem to inhibit the growth of lettuce seedlings. The control shows no inhibition.

(d) **(i)** The difference is very significant.

(ii) The probability that chance alone could produce this difference is 0.001 or 0.1%.

(e) **(i)** The difference is not significant.

(ii) The probability is 0.50 or 50%.

6 **(a)** Mean and standard deviation of bean plant in cm (±0.5) are as shown in this table.

Group	Mean	Standard deviation
A (control)	8.3	1.2
B	8.4	1.1
C	8.2	1.3
D	11.9	2.2

(b) The standard deviations of groups A, B and C are very low. There is not much variability in the data of these groups. The standard deviation of group D is much higher indicating much more variability in the data. This may indicate that it takes longer than 25 days to show the effect of fertilizer on all of the plants or some other factor is acting which is yet to be determined. We cannot be as sure of the data in group D as we are of the data in the other groups.

(c) The mean of group D is 3.6 cm higher than the control. It is 3.5 cm higher than group B and 3.7 cm higher than group C. The means of groups A, B, and C are similar.

(d) **(i)** The degrees of freedom are 10 + 10 = 20 – 2 = 18. Use the t-table to find 0.60. There is no significant difference between the two groups.

(ii) The probability that the small difference is due to chance is very high. It is a 50% probability that the small difference is due to chance.

(e) **(i)** The degrees of freedom are the same as above. The difference here is significant.

(ii) The probability that the difference between the two groups is due to chance alone is only 0.01 or 1%.

Chapter 2: Answers to exercises

Photograph on page 21: Because only a limited region of the cell is shown, it is difficult to determine if it is an animal cell. Animal structures visible are: mitochondria (green), ER (yellow), lysosomes (solid green), nucleus (pink) and nucleolus within the nucleus. Lysosomes help break down substances that enter the liver via the bloodstream. Mitochondria provide energy to produce needed substances. ER produces lipid-like compounds and detoxifies drugs.

1 As the surface area to volume ratio decreases, there is relatively less membrane surface to allow exchange of wastes between inside and outside the cell. The surface increases much more slowly than the volume when the diameter of a cell increases. Thus, the relative amount of membrane available for exchange is less. If a cell grows too large, the result is cell death due to waste accumulation.

2 Any two of the following.

- Electron microscopes are relatively expensive to purchase and operate.
- Electron microscopes require specialized training before they can be effectively used.
- Electron microscopes require the specimen to be dead because of the extensive preparation needed for observation.

3 These cells have very specific roles in the organism. Nerve cells conduct impulses and to do this they require adaptations that allow depolarization and repolarization or rapid electrical changes within them. Muscle cells produce movement and specific proteins and arrangements of these proteins are necessary for this movement. The result is that nerve and muscle cells carry out their functions without expending valuable time and energy in a reproductive stage.

4 Even though stem cells are capable of great variation in what they differentiate into, they still have nuclear DNA as their controlling factor. The DNA of one animal is potentially quite different from the DNA of another.

5 The DNA is more vulnerable to the life functions of the cell. These cell functions require enzymes and raw materials, and produce products that may potentially damage the DNA. This creates a greater chance of interactions and mutations or harmful changes.

6 Pili are plasma membrane extensions that serve to allow joining of bacterial cells so that exchange of DNA between cells may occur, thus bringing about sexual reproduction.

7 Muscle cells have a large number of mitochondria because their function is to allow movement. Mitochondria produce ATP, and ATP is essential in enabling special proteins to slide alongside one another and create movement. If ATP is not present, movement is not possible.

8 Mitochondria and chloroplasts, since they:
- have their own DNA;
- have an additional outer membrane indicating a need for greater protection within a potentially hostile environment;
- have ribosomes similar to prokaryotic ribosomes;
- divide by simple fission.

9 Chloroplasts produce simple carbohydrates. These carbohydrates are sources of chemical energy when their chemical bonds are broken. The energy can be used to produce ATP which is necessary for cellular activities. Mitochondria carry out the breakdown of chemical bonds to release energy.

10 In passive transport, the driving force is the concentration differences of a particular substance type. When concentrations are high, molecular collisions occur more frequently resulting in movement toward areas of lower concentration. This movement outwards will continue until collisions are occurring equally throughout the area, by which time equilibrium is said to be established. In active transport, there is no final equilibrium. Active transport is movement against a concentration gradient with the expenditure of energy.

11 Non-polar amino acids are present because the inner region of the membrane is made of the non-polar regions of the membrane phospholipids.

12 Both exocytosis and endocytosis require expenditure of energy in the form of ATP.

13 The chromosomes would be abnormally distributed resulting in cells with improper DNA to govern cellular processes. The result would be abnormal cells that probably would die very early.

14 Since the chromatids are yet to separate in metaphase, there would be 48 chromatids present.

15 Cytokinesis occurs at the conclusion of the M (mitosis) phase and just before interphase begins (G_1).

Chapter 2: Answers to practice questions

Note: In this markscheme a full stop (period) separates single scorable elements in the same way that a semicolon does in the official IB markscheme.

1 B *(1 mark)*

2 C *(1 mark)*

3 D *(1 mark)*

4 D *(1 mark)*

5 A *(1 mark)*

6 **(a)** As the diameter of the molecule increases, the permeability / relative ability to move decreases (*accept converse*).
The relationship is logarithmic/non-linear/ negative – for molecules above 0.6 ($\pm$0.1) nm relative ability to move changes little / for molecules below 0.6 ($\pm$0.1) nm relative ability to move changes rapidly. *(2 max)*

(b) 'U-1' rule applies.
- **(i)** 10 mmol cm^{-3} cells hr^{-1} (*accept values within* $\pm$*5*). *(1)*
- **(ii)** 370 mmol cm^{-3} cells hr^{-1} (*accept values within* $\pm$*10*. *(1)*

(c) **(i)** Rate of glucose uptake in facilitated diffusion levels out as external glucose concentration increases, whereas uptake in simple diffusion does not level out / continues to rise.
Rate of glucose uptake increases with increasing external concentration in both.
Rate of glucose uptake is higher in facilitated diffusion (than in simple diffusion).
Increase in rate of glucose uptake in simple diffusion is constant/linear, whereas in facilitated diffusion the rate of uptake increases rapidly at the beginning / increase is not constant. *(3 max)*

(ii) Little/no change in glucose uptake because most/all (protein) channels in use. *(2)*

(Total 9 marks)

7 **(a)** Translates RN A into/synthesises polypeptides/ proteins.
Lysosomes
Mitochondria
Aerobic respiration / production of ATP
Do not accept production of energy. *(4)*

(b) This is a eukaryotic cell because it contains membrane-bound organelles and a nucleus.
It could be a plant or an animal cell; probably an animal cell because of the presence of lysosomes. *(2 max)*

(Total 6 marks)

8 **(a)** Diffusion is the movement of molecules from an area of high concentration to an area of low concentration.
Osmosis is the diffusion of water across a partially permeable membrane. *(1)*
Must have both diffusion and osmosis to gain the mark.

(b) Hydrophilic head groups point outwards and hydrophobic tails point inwards to form a lipid/phospholipid bilayer.
Ions and polar molecules cannot pass through a hydrophobic barrier.
Helps the cell maintain internal concentration and exclude other molecules. *(2 max)*

(c) Plant cell wall is made of cellulose. It provides structural support and protection, and maintains turgor pressure. *(2)*

(Total 5 marks)

9 **(a)** I plasma/cell membrane
II cell wall
III nucleoid (region) / DNA/genetic material
IV cytoplasm/cytosol/protoplasm *(2 max)*
Award 1 mark for each two correct labels.

(b) 26 000 × (±1500) *(1)*
Do not need to show working.

(c) You can see colour images instead of black and white/pigments can be observed.
You can view living material.
There is a larger field of view/whole cells can be seen.
Easier sample preparation/cheaper/easier transport/portable. *(2 max)*

(Total 5 marks)

Chapter 3: Answers to exercises

1 Many different possible answers exist for this question. You should pick an organism that you are familiar with. Ways in which water is important could include, but are not limited to: water for cytoplasm in every cell; blood plasma; environment for all chemical reactions; distribution of nutrients within organism; distribution of wastes within organism; habitat for some organisms; redistribution of water often changes shape of some cells/organisms (e.g. guard cells in plants); maintenance of stable internal temperature.

2 Answers for this question are also expected to be highly varied. Many answers will and should focus on the polarity of water molecules. Chemical reactions are only possible in cells, blood plasma, etc. because the majority of solutes in living organisms are polar and are thus water soluble. Polarity is also directly related to cohesion and adhesion of water molecules and is important in any answer dealing with movement of water. Maintenance of stable internal temperature is directly related to the high specific heat of water, also attributed to the polarity of water molecules.

3 **(a)** glucose + fructose → sucrose + water (*other answers possible*)
(b) condensation reaction

4 **(a)** glycerol + 3 fatty acids → triglyceride + 3 water
(b) condensation enzyme
(c) product
(d) three (because three condensation reactions occurred)

5 The phosphate groups and deoxyribose sugars within DNA molecules are constant. There is no message written within their presence. The only message is the order of the nucleotides as given by the order of the nitrogenous bases. It is also common to give only one side of a double-stranded molecule as it is obvious what the complementary strand is.

6 Compare your drawing to Figure 3.12.

7 Because your DNA replicates in a semiconservative pattern, always maintaining one 'old' strand, the DNA within some cells could contain a strand that has existed since you were conceived or even before then. DNA strands are never completely new and are possibly very, very old.

8 A deletion is a mutation in which a nucleotide is left out of a strand of replicating DNA. A substitution mutation occurs when one nucleotide takes the place of another nucleotide type (e.g. thymine pairs with guanine rather than adenine).

9 Methionine – proline – arginine – threonine – phenylalanine – proline – serine – proline – glycine.

10 TACGGGGCGTGCAAAGGTTCGGGGCCC

11 Methionine – serine – arginine – threonine – phenylalanine – proline – serine – proline – glycine (note only the second amino acid has been changed).

12 **(a)** Compare your drawn graph with Figure 3.18.
(b) Compare your drawn graph with Figure 3.20.

13 For the increasing temperature graph: the area of the graph that shows a positive correlation (increasing slope) is due to increased molecular movement. Molecules move faster at higher temperatures and more molecules are reaching the activation energy necessary to complete the reaction. The area of the graph with a rapidly decreasing slope shows the temperature at which the enzyme becomes denatured. The enzyme has lost its three-dimensional shape and the active site of the enzyme no longer 'fits' the shape of the substrate.

For the increasing substrate graph: With all other conditions remaining the same (including the concentration of enzyme), as the concentration of substrate increases the rate of reaction also increases (positive correlation). Eventually, every enzyme molecule is working at its maximum rate and adding more substrate has no effect on reaction rate.

14 Glycolysis.

15 In the cytoplasm.

16 All cells need some form of cell respiration and all have cytoplasm; only eukaryotic cells have mitochondria.

17 The inhaled oxygen is distributed by blood to our body cells for cell respiration and the by-product is carbon dioxide. Carbon dioxide is carried back to our lungs and we breathe it out.

18 A blue object would absorb the red and green areas of the visible light spectrum and would reflect the blue wavelengths. The colour reflected is the colour we see.

19 Black surfaces absorb all colours (wavelengths) and reflect none. This means more energy is being absorbed and thus black surfaces get hotter than lighter surfaces in sunlight.

20 If the sugar is needed for energy, it enters the process of cell respiration to generate ATP. If the plant does not need all the glucose for energy at that time, the excess is typically converted to starch (amylose).

21 Most plants experience a 'growing season' in which far more sugar is produced than is necessary to stay alive at that time. The excess is converted to starch and can be used in the 'non-growing' season (usually the cold months of the year).

Chapter 3: Answers to practice questions

Note: In this markscheme a full stop (period) separates single scorable elements in the same way that a semicolon does in the official IB markscheme.

1 **(a)** **(i)** 07:30 / 7:30 a.m. (*accept answers in range up to 07:45*) (*1*)
(ii) 17:00 / 5:00 p.m. $\pm\frac{1}{2}$ hour) (*1*)
(b) 250 ppm ($\pm$30 ppm) (*unit required*) (*1*)
(c) At night / darkness / no light, only respiration occurs so CO_2 increases.
In day / with light, both respiration and photosynthesis occur / photosynthesis exceeds respiration in day.
CO_2 is used by photosynthesis and level decreases.
When sun sets, CO_2 again increases as only respiration occurs. (*2*)
(*Total 5 marks*)

2 Features and their significance may include:
- surface tension – allows some organisms (e.g. insects) to move on water's surface;

- polarity/capillarity/adhesion – helps plants transport water;
- transparency – allows plants to photosynthesize in water / allows animals to see;
- (excellent) solvent – capable of dissolving substances for transport in organisms;
- (excellent) thermal properties (high heat of vaporization) – excellent coolant and temperature stabilizer;
- ice floats – lakes/oceans do not freeze, allowing life under the ice;
- buoyancy – supports organisms;
- structure – turgor in plant cells / hydrostatic skeleton;
- habitat – place for aquatic organisms to live;
- involved in chemical reactions in organisms.

Each feature or property must be related to living organisms in order to receive a mark.

(*Total 6 marks*)

3 $CH_3(CH_2)_nCOOH$ / $CH_3—(CH_2)_n—C(=O)OH$ (*1 mark*)

4 By condensation; involves the removal of water to join monosaccharides together / equation to show this. Catalysed by enzymes; many monosaccharides linked (glycosidic linkages) to make polysaccharide. (*2 marks*)

5 C (*1 mark*)

6 B (*1 mark*)

7 C (*1 mark*)

8 B (*1 mark*)

9 B (*1 mark*)

Chapter 4: Answers to exercises

Pedigree chart for snapdragon colour on page 99: C^RC^R – A, H; C^WC^W – B, K; C^RC^W – C, D, E, F, G, I, J.

1 Drawing should show one or two chromatids as vertical rods, centromere is shown between top and bottom arms of chromatids, locus drawn as a line across one of the chromatids.

2 Alleles are versions of genes. The term 'gene' can be considered to be more general, whereas 'allele' is more specific.

3 Similarities:

- both carry genetic hereditary information.

Differences:

- eukaryotic DNA is associated with proteins whereas prokaryotic DNA is not / is called naked DNA;
- eukaryotic DNA is always organized into pairs of chromosomes whereas prokaryotic DNA is found in just one chromosome.

4 Because eukaryotes reproduce using sexual reproduction, there is always half of the genetic material from the female parent and the other half from the male parent.

5 Because the resulting daughter cells only contain half of the genetic material of the parent cells. The number of chromosomes passes from $2n$ to n.

6 Because mitosis produces diploid cells and if such cells were used as gametes, the resulting offspring would have $2n + 2n = 4n$ chromosomes. In this case, the number of chromosomes would double every generation: $4n$, $8n$, $16n$, $32n$, and so on. To maintain the appropriate chromosome number, meiosis is used so that the resulting offspring have $n + n = 2n$ chromosomes.

7 For males, sperm cells; for females, egg cells.

8 Drawings should be similar to Figure 4.3.

9 Because the risk of having a baby with Down's syndrome (or any other condition resulting from non-disjunction) increases with age.

10 Because the allele for colour blindness is recessive and is sex linked, in order for a woman to be colour blind, she must have two recessive alleles, one on each of her X chromosomes: $\mathbf{X^bX^b}$. In any other case, she would be carrying at least one dominant allele to cancel out the effects of the colour blindness allele. In most populations, it is rare to receive one of these recessive alleles and even more rare to get both alleles for colour blindness. On the other hand, it is much easier for a man to get the condition because there is no locus on the Y chromosome to carry a dominant allele to mask the recessive allele on his X chromosome. As a result, the presence of just one **b** allele is enough to give him colour blindness.

11 Answer should show the Punnett grid for the cross $\mathbf{C^RC^W \times C^RC^W}$.

12 In the top row, one circle and one square joined by a horizontal line. They should both be labelled pink. In the next row, showing the F_1 generation, the four offspring plants should show 1 red, 2 pink and 1 white.

13 **(a)** X^HX^h for the mother and X^HY for the father.

(b) Girls are either X^HX^H or X^HX^h and the boys can be X^HY or X^hY.

(c) Both possible genotypes for the girls give them normal blood clotting. The genotype X^HY gives a boy with normal blood clotting but X^hY is a boy with haemophilia.

(d) The only carriers are the mother and any daughters with X^HX^h. (Males cannot be carriers.)

(e) 1 in 4 or 25%.

14 Analysis of DNA samples using techniques such as gel electrophoresis cannot be done with only a few strands of DNA. When DNA samples are collected, sometimes only a few cells are found, for example at a crime scene. To obtain enough copies for analysis, the DNA strands must be copied millions of times.

15 Typically, the source of stem cells is human embryos. The ethical question is, 'Can we use embryos purely as research tools?' Critics argue that these balls of cells should be treated with respect and dignity since they are of human origin and their natural destiny is to try to develop into baby girls or boys so using them for another purpose is unnatural and unethical.

16 Answers will vary but both benefits (e.g. a plant's resistance to drought, higher yield) and risks (e.g. consequences of GM pollen escaping, possible allergies) should be explored and there should be a justification in the answer.

17 Answers will vary but it is generally agreed that the labelling in most countries is insufficient for consumers to make educated choices about the foods they buy. For example, labels could say 'may contain GM soybeans' but it is not clear what percentage could be expected.

Chapter 4: Answers to practice questions

Note: In this markscheme a full stop (period) separates single scorable elements in the same way that a semicolon does in the official IB markscheme.

1 Mutation is a change in DNA sequence.
Changes the mRNA during transcription.
Changes the amino acid sequence.
Substitution mutation / changes to one codon.
Glutamic acid is changed to valine / GAG to GTG.
Changes the shape of haemoglobin / haemoglobin becomes less soluble and crystallizes out.
Cannot carry oxygen as well.
Red blood cells sickle / impairs blood flow.
Causes other health problems / anaemia / tiredness.
Sickle cell anaemia caused by two mutated recessive alleles. *(7 marks)*

2 (a) A gene/trait/allele carried on a sex chromosome / X or Y / X/Y. *(1)*

(b) Recessive. Evidence from the pedigree (e.g. 2nd generation –2 and –3 do not have the condition but have one child who does). *(2)*

(c) **(i)** X^aY (where a = condition). *(1)*

(ii) X^AX^a or X^AX^A where **A** = normal, **a** = condition (*must have both*). (1)

If upper case letter and lower case letter are reversed then the ECF rule applies.

(Total 5 marks)

3 Two divisions in meiosis, only one in mitosis; meiosis results in haploid cells, mitosis in diploid cells; crossing over only occurs in meiosis; no S phase precedes meiosis II; chromosome behaviour in meiosis II and mitosis is similar / chromosome behaviour in meiosis I and mitosis is different; chiasmata only form during meiosis; homologous chromosomes move to the equator in pairs only in meiosis.
Do not accept number of cells produced – it is a result not a behaviour. *(5 marks)*

4 (a) Pair of alleles that both affect the phenotype (when present in a heterozygote) / both alleles are expressed and recognized in the phenotype. *(1)*

(b) *(4)*

Parents' genotypes	AB		BO	
Parents' gametes	I^A	I^B	I^B	i
F_1 genotypes	I^AI^B	I^Ai	I^BI^B	I^Bi
F_1 phenotypes	AB	A	B	B

(Total 5 marks)

5 Named example of desired outcome (e.g. herbicide resistance).
Possible benefits: *(4 max)*

- more specific (less random) breeding than with traditional methods;
- faster than traditional methods;
- some characteristics from other species are unlikely in the gene pool / selective breeding cannot produce desired phenotype;
- increased productivity of food production / less land required for production;
- less use of chemicals (e.g. pesticides);
- food production possible in extreme conditions;

- less expensive drug preparation (e.g. pharmaceuticals in milk);
- human insulin engineered so no allergic reactions;
- may cure genetic diseases.

Possible harmful effects: *(4 max)*

- some gene transfers are regarded as potentially harmful to organism (especially animals);
- release of genetically engineered organisms in the environment;
- can spread and compete with the naturally occurring varieties;
- some of the engineered genes could also cross species barriers;
- technological solution when less invasive methods may bring similar benefits;
- reduces genetic variation/biodiversity.

Award 6 max if no named example given.

Award 5 max if possible benefits and possible harmful effects are not both addressed.

(Total 7 marks)

6 Sample of DNA obtained: e.g. leucocytes from mouthwash/hair/other named source.

Satellite DNA / repetitive sequences used for profiling.

Amplification of DNA by polymerase chain reaction (PCR).

Cutting DNA into fragments using restriction enzymes.

Separation of fragments of DNA (by electrophoresis).

Separation according to the length of the fragments.

Pattern of bands obtained / different pattern of bands with DNA from different individuals.

Used for criminal investigations / example of use in criminal investigation.

Used to check paternity / who is the father/mother/ parent.

Used to check whether two organisms are clones.

(5 marks)

7 Applications of karyotyping *(2 max)*

- Find gender / test for Down's syndrome / other chromosome abnormality.
- Identify sex chromosomes / numbers of chromosome 21 / other chromosomes counted.
- XX = female and XY = male / third chromosome 21 indicates Down's syndrome / other chromosome abnormality.

Obtaining chromosomes *(3 max)*

- Fetal cells obtained from amniotic fluid / amniocentesis / other named source.
- White blood cells obtained.
- Cells encouraged to divide.
- Cells accumulated / blocked in metaphase.
- Prepare slide / chromosomes examined.

(Total 5 marks)

Chapter 5: Answers to exercises

Photograph on page 112: These animals are sea jellies or jellyfish (*Aurelia*).

Photograph on page 114: The water is part of the hydrosphere, the sand bar and rocks are part of the lithosphere, and the air (not visible as such) is part of the atmosphere; these three are all abiotic factors. The animal and vegetation are part of the biosphere, the biotic component of the environment.

Photograph on page 120: The number of trophic levels is two: producer (grass) and primary consumer (llamas).

1 Habitat refers only to the place or the environment where an organism lives, whereas an ecosystem refers to the place and to the other organisms living there as well.

2 (a) Grassland ecosystem: grass → grasshoppers → toad → hognose snake → hawk
i.e. producer → primary consumer → secondary consumer → tertiary consumer → quaternary consumer / top predator

(b) River ecosystem: algae → mayfly larva → juvenile trout → kingfisher
i.e. producer → primary consumer → secondary consumer → tertiary consumer

(c) Marine ecosystem: diatoms → copepods → herring → seal → great white shark
i.e. producer → primary consumer → secondary consumer → tertiary consumer→ quaternary consumer / top predator

3 Drawing should reflect the relationships between all seven organisms with grass at the bottom and the hawk at the top.

4 Secondary consumer.

5 In a garden greenhouse, it is the glass which lets in light but traps heat. On Earth, it is greenhouse gases in the atmosphere which play that role.

6 Answers will vary: for example, using public transportation, choosing foods produced locally, insulating home.

7 Answers will vary depending on the source and date of the data, but generally the richest countries in the world produce the most CO_2. This is an opportunity to compare overall emissions to per capita emissions.

8 The maximum number of individuals in a population which can be supported by the resources available.

9 Answers will vary: for example, wind/birds/bats or any walking animals, although probably not humans.

10 Polar bears have white fur so as not to be seen easily by prey; sharp claws and teeth for catching and eating seals; thick, warm fur to keep warm; hibernate during Arctic winter.

11 Artificial selection is when humans such as farmers, breeders, geneticists or gardeners decide which traits will be present in future generations of animals and plants by only allowing parent organisms who possess those traits to reproduce. In natural selection, natural environmental pressures determine which traits are successful and which are not (e.g. prolonged lack of rainfall will select those plants which are resistant to drought, the rest will die of dehydration).

12 The first spraying will kill most insects in the population but a few will be naturally resistant. Those few survive and reproduce. The offspring will contain a large percentage of resistant insects.

13 The dark speckles on the eggs are good camouflage in the nest on the ground. If brightly coloured eggs were produced, they would be more visible to predators from a distance. As a result, they would be more likely to be eaten. This means the chicks inside would never have the chance to grow up and reproduce. So the genes for the inappropriate egg colour would not be passed on to the next generation.

14 Kingdoms: Plantae (plants), Animalia (animals), Fungi, Protoctista (algae and protozoa), Prokaryotae (bacteria).

Algae – Protoctista; hydra – Animalia; spider – Animalia; mushroom – Fungi; yeast – Fungi; bacterium – Prokaryotae.

15 Viruses do not reproduce in the way any members of the five kingdoms do, they do not have similar structures (e.g. no cell membrane or organelles).

16

	Filicinophyta	Bryophyta	Coniferophyta	Angiospermophyta
Vegetative characteristics	non-vascular	vascular	vascular woody stems leaves are needles	vascular flowers fruit
Reproductive characteristics	produce spores	produce spores	produce seed cones wind pollination	produce seeds wind/animal pollination
Example (may vary)	fern	moss	pine	rose

17 Primate.

18 Answers will vary.

Chapter 5: Answers to practice questions

Note: In this markscheme a full stop (period) separates single scorable elements in the same way that a semicolon does in the official IB markscheme.

1 Decomposers are responsible for recycling (inorganic) nutrients / break down organic molecules to inorganic compounds. *(1 mark)*

2 Energy enters from (sun)light.

Chloroplasts/plants/producers/autotrophs capture (sun)light.

Energy flows through the trophic levels / stages in food chain.

Energy transfer is (approximately) 10% from one level to the next.

Heat energy is lost through (cell) respiration.

Energy loss due to material not consumed/ assimilated / egested/excreted.

Labelled diagram of energy pyramid.

Energy passes to decomposers/detritivores/ saprotrophs in dead organic matter.

Nutrient cycles within ecosystem / nutrients are recycled.

Example of nutrient cycle with three or more links.

Nutrients absorbed by producers/plants/roots.

Nutrients move through (food chain) by digestion of other organisms.

Nutrients recycled from decomposition of dead organisms.

Nutrients from weathering of rocks enter ecosystem.

Nutrients lost by leaching/sedimentation (e.g. shells sinking to sea bed). *(8 marks)*

3 Light is the initial source of energy for almost all communities.
Plants absorb light and use it in photosynthesis.
Plants produce food / organic matter.
Plants are the main producers in most communities.
Energy flows along food chains/webs from plants.
First consumers eat plants/producers.
Second consumers eat first consumers that have eaten plants/producers.
Plants produce oxygen.
Oxygen needed for cell respiration by many organisms.
Dead plants / parts of plants available to saprotrophs/fungi and bacteria/detritivores.
Plants provide a habitat for other organisms.
(6 marks)

4 **(a)** With time, the atmospheric concentration of CO_2 has increased. *(1)*
(b) The increased use of fossil fuels / more automobiles; increased deforestation (*do not accept greenhouse effect*). *(1)*
(c) Any trough, clearly labelled at the bottom. *(1)*
(d) CO_2 is a raw material for photosynthesis. There is an increase in the rate of photosynthesis in the summer, and therefore less CO_2 in the air during the summer as it is being used for photosynthesis. Increase in CO_2 occurs in winter because there is less photosynthesis due to trees losing leaves in autumn (winter) / lower temperatures / shorter days with less light. *(2)*
(Total 5 marks)

5 Correct S-shaped curve with labels.
Exponential – rapid increase in population.
Transitional – slowing of growth.
Plateau – levelling off, birth rate = death rate.
Carrying capacity.
Must have correctly labelled diagram for full marks.
(4 marks)

6 Natality/births/reproduction increases populations, as long as natality is higher than mortality.
Abundant food allows increase / food shortage causes decrease.
Low level of predation allows increase / high level causes decrease.
Low level of disease allows increase / high level causes decrease.
Immigration increases populations as long as immigration is greater than emigration.
Population rises until a plateau is reached – carrying capacity of the environment – when the resources of the environment cannot support any more individuals.
Graph of sigmoid population growth.
Environmental factor and its consequence (e.g. flood causes decrease). *(8 marks)*

7 (Segregation of alleles involves) meiosis – crossing over / chiasma formation in prophase I (*do not allow meiosis if wrong phase given*).
Random orientation / assortment of homologues at metaphase I.
Fertilization by chance / one of many male gametes; number of different gametes is 2^n (ignoring crossing over).
Genes/alleles combined from two parents.
(4 marks)

8 Examples: antibiotic resistance in bacteria; heavy metal tolerance in plants; beak size in Darwin's finches; industrial melanism in peppered moths.
Populations grow exponentially.
More offspring than the environment can sustain.
Populations still remain constant.
Individuals in populations show variation.
Mutations are a source of variation.
Individuals may have characteristics that are better suited to the environment.
The variation has to be heritable.
Those individuals will survive.
They will reproduce and leave more offspring.
The population will tend to accumulate the adaptation.
Therefore, the population will evolve.
Theory proposed by Darwin.
(8 marks)

9 C *(1 mark)*

Chapter 6: Answers to exercises

1 Carbohydrates – polysaccharides like starch will digest to disaccharides and monosaccharides.
Disaccharides will digest to monosaccharides.
Lipids – triglycerides will each digest into 1 glycerol and 3 fatty acids.
Proteins – will digest into individual amino acids (20 types).

2 Answers will vary but should include most or all of the following: mouth, oesophagus, stomach, small intestine, capillary in villus, hepatic portal vein, liver, hepatic vein, heart, arteries, capillary, muscle cell.

3 Right atrium, right atrioventricular valve, right ventricle, right semilunar valve, pulmonary artery, lung capillary bed, pulmonary vein, left atrium, left atrioventricular valve, left ventricle, left semilunar valve, aorta, smaller arteries, body capillary bed, veins, vena cava and back to right atrium (starting point).

4 The hole between the atria allows blood in the two atria to mix together. The blood coming in to the right side of the heart is already a mix of oxygenated and deoxygenated blood as the oxygenated blood from the placenta comes back to the right side of the heart in the venous circulation (mixed with the deoxygenated blood from the fetal body). This pattern has evolved to account for the lack of oxygen and carbon dioxide exchange in the fetal lungs (a fetus does not breathe) and allows some of the blood to go to the lungs to oxygenate lung tissue and some to take a short cut to get to body tissues (through the hole).

5 The antibodies were produced by another organism. The memory cells that are capable of producing these antibodies are located in the organism that produced them, not the person who receives the antivenom. Long-term immunity comes from memory cells.

6 The severity of symptoms depends on two primary factors:
- the tissue type that is the target tissue of any one particular virus (some viruses infect many tissue types, some only one or a few; some tissue types when infected are potentially more dangerous);
- how quickly the virus replicates (less time for the immune system to respond to a quickly replicating virus).

7 Diaphragm relaxes, rib cage lowers, volume of thorax decreases, pressure inside thorax increases, lung tissue decreases in volume, pressure increases inside the lungs, air is forced out of mouth or nasal passages.

8 The inside lining of the alveoli is designed to facilitate easy diffusion of oxygen and carbon dioxide. Any build-up of material on the inside wall of the alveoli impedes the respiratory gases from diffusing readily.

9 One side of the see-saw represents a lowering of body temperature and the events that the body therefore initiates (hypothalamus initiates constriction of skin arterioles and shivering). The other side of the see-saw represents increased internal body temperature (hypothalamus initiates dilation of skin arterioles, increased sweat gland activity and thus evaporative cooling).

10 Answers will vary but a typical response could look like this:
receptor (retina) → optic nerve → cerebrum (occipital lobe) → cerebrum (frontal lobe) → spinal cord → spinal nerve → effector (leg muscle)

11 Refer to Figure 6.23 for graph. FSH and LH are produced and secreted from the anterior pituitary gland. Oestrogen is produced by the follicle cells of the ovary and progesterone is produced by the corpus luteum after ovulation.

12 Arguments given will vary and the following are representative.
- Health insurance executive:
  procedure is elective, procedure does not always work, procedure is expensive.
- Man or woman who needs IVF to have a child:
  may be only option to have my own child, personal choice, it is my money.
- Religious representative:
  not as God intended, unused embryos frozen or destroyed, may lead to 'designer' children (chosen attributes).
- Administrator from an IVF clinic:
  simply another health care business, safety rates are in-line with other medical procedures, gives parents an option otherwise not available.

Chapter 6: Answers to practice questions

Note: In this markscheme a full stop (period) separates single scorable elements in the same way that a semicolon does in the official IB markscheme.

1 **(a)** **(i)** Late pregnancy. *(1)*
(ii) Increase of 3.3% (±0.3); higher in those exposed in late pregnancy. *(1)*
Do not deduct the mark if the denominator is 6.3.

(iii) Mother with long-term effects of famine / affect child's development / famine over but nutrition remains poor. *(1)*

(b) No or little insulin / diabetes / tissues do not respond to insulin. *(1)*

(c) Increased blood glucose could lead to obesity (which is a risk factor for CHD).

Increased blood glucose could be due to diabetes (which is a risk factor for CHD).

Genetic factors could relate both conditions.

(2 max)

(Total 6 marks)

2 (a) The skin / mucous membranes act as a physical barrier.

Skin has several layers of tough/keratinized cells.

The skin is dry discouraging the growth and reproduction of pathogens.

Skin / mucous membranes hosts natural flora and fauna which compete with pathogens.

The enzyme lysozyme is present on the skin's surface to break down pathogens.

The pH of skin / mucous membranes is unfavourable to many pathogens.

Skin is a continuous layer.

Mucus traps pathogens; is sticky. *(3 max)*

Award only 2 max if skin and mucous membrane not both mentioned.

(b) Antibiotics block metabolic pathways in bacteria / inhibit cell wall formation / protein synthesis.

Viruses use host cell's metabolic pathways / do not possess a cell wall and so are not affected by antibiotics.

Antibiotics are not used to treat viral diseases because they are ineffective and may harm helpful bacteria. *(2 max)*

No credit for answers that state antibiotic means against life nor for the statement that viruses are not alive.

(Total 5 marks)

3 Arguments against (IVF):

- Fertilized egg has potential to become a person / some view a fertilized egg as having special status.
- IVF requires the production of multiple embryos.
- Fate of extra embryos is ethical concern.
- Ethics of long-term storage.
- Stem-cell research is blurring issue as other cells now have the possibility of becoming a person.
- Procedure may result in multi-embryo pregnancy which places stress on the family resources / unwanted children.
- Issues of equity of access / expensive.
- High rates of failure.
- Ownership of / responsibility for stored embryos an issue.
- Religious opposition / playing God.

Arguments favouring (IVF):

- Only way some couples can have children / helps infertile couples.
- Allows for genetic screening.
- Allows for surrogate mothers. *(8 max)*

For full marks at least two of the points should include the counter-argument, otherwise 6 max.

(Total 8 marks)

4 B *(1 mark)*

5 A *(1 mark)*

6 C *(1 mark)*

7 C *(1 mark)*

8 B *(1 mark)*

9 D *(1 mark)*

Chapter 7: Answers to exercises

1 3′____________________5′
5′____________________3′

(The two strands have different carbons of deoxyribose exposed at their ends.)

2 If DNA is wrapped tightly around the histone molecules, it is in a nucleosome and is not available for helicase to break its hydrogen bonds and begin the process of transcription. Only the strands of DNA between the beads is available for code transfer to RNA.

3 Since introns are not involved in the coding business, it is logical that they would contain the highly repetitive sequences. This is because the highly repetitive sequences are thought not to have a coding function.

4 The energy source for DNA replication is the two phosphates that occur on the end of each

deoxynucleoside triphosphate that is added to the elongating DNA chain.

5 The process of replicating the very long chromosome would be so slow that the cell would be involved with replication during most of the cell cycle and would not be available to carry out the normal cell processes.

6 Only one primer is needed on the leading strand. However, the lagging strand requires a primer for the beginning of each Okazaki fragment that forms at the replication fork.

7 Using the antisense strand as the basis for this answer, the transcription bubble is moving toward the 5′ end of the strand.

8 Eukaryotic mRNA requires processing since it contain introns (non-coding sections). The introns are removed before the mature mRNA transcript moves out of the nucleus making its way to the ribosome. Prokaryotic mRNA has no introns.

9 mRNA carries the message from the DNA of the nucleus to the ribosomes of the cytoplasm.

tRNA carries amino acids to the mRNA–ribosomal complex for assemblage into a protein.

rRNA structurally makes up over 50% of the ribosomes.

10 If there are polysomes, there are more proteins being assembled. This allows the cell to satisfy protein needs much more quickly.

11 Your drawing should look like Figure 7.10.

12 The primary level of protein organization refers to the amino acid sequence. The amino acids are the causes of electrostatic forces and are the sources of the types of weak bond that determine the other levels of protein organization.

13 The prosthetic or non-polypeptide group.

14 A change in a single amino acid brings about a different arrangement in the folding of the protein because of the resulting change in the positioning of the forces and charges. This results in change of structure and a resulting change in function. The change in haemoglobin in red blood cells showing sickle cell anaemia is an example of this.

15 The bonds and general shapes are specific for both the enzyme and the substrate. If the substrate does not have a compatible shape and appropriately positioned electrical forces, the substrate will not fit the active site of the enzyme.

16 In competitive inhibition, the competitor molecule combines with the active site directly. This prevents the substrate from combining. In non-competitive inhibition, the active site is not bound by the outside molecule. In this case, the molecule combines with an alternative site on the enzyme causing a conformational change resulting in decreased enzyme function.

Chapter 7: Answers to practice questions

Note: In this markscheme a full stop (period) separates single scorable elements in the same way that a semicolon does in the official IB markscheme.

1 D *(1 mark)*

2 D *(1 mark)*

3 A *(1 mark)*

4 B *(1 mark)*

5 A *(1 mark)*

6 B *(1 mark)*

7 **(a)** Directly proportional / the greater concentration, the greater rate of reaction / at high concentrations the increase is smaller / plateau or levels-off (at approximately 70 mmol dm^{-3}). *(2)*

(b) **(i)** 1 mmol dm^{-3}: 0.70 (±0.02).
3 mmol dm^{-3}: 0.55 (±0.02). *(1 max)*
Both are needed for 1 mark.
For 1 mmol dm^{-3} accept 0.7.

(ii) Lower reaction rate at inhibitor concentration of 3 mmol dm^{-3} / the greater the inhibitor concentration, the slower the rate of reaction / trend or overall shape are the same, increases then levels-off, but lower at greater concentration of inhibitor. *(1 max)*

(c) Substrate and inhibitor are (structurally) similar.
Inhibitor binds to active site and prevents substrate from binding.
Activity of enzyme prevented.
Named example (e.g. malonate inhibits succinate dehydrogenase as it is similar to succinate). *(3 max)*
(Total 7 marks)

8 **(a)** Peptide bonds / peptidic bonds. *(1)*

(b) α-helix / α-helices. *(1)*

(c) Ionic/polar/hydrogen/hydrophobic/van der Waals'/disulfide (*not covalent*). *(1)*

(d) Linking together of polypeptides to form a single protein, using the same bonding as for tertiary structure.
Linking of a non-polypeptide structure or prosthetic group.
Named example of quaternary structure (e.g. haemoglobin; has four polypeptides). *(2 max)*

(Total 5 marks)

9 **(a)** Any two from the following:
- fibrous: fibrin, collagen (*do not accept tendon*);
- globular: haemoglobin, fibrinogen, amylase (*do not accept enzyme*). *(2)*

(b) A new reaction pathway is created / activation energy is reduced / the equilibrium for the reaction is achieved more quickly or the reaction is faster. *(2 max)*

(c) ATP inhibits phosphofructokinase at (allosteric) site away from the active site.
Inhibition alters the enzyme's conformation or structure.
The active site does not accept the substrate molecule.
When respiration increases ATP levels, phosphofructokinase is inhibited and respiration slows down.
Phosphofructokinase is the first enzyme in the respiration pathway so there is no build up of metabolic intermediates.
As ATP is used up by the cell, the inhibition of phosphofructokinase is reduced and respiration speeds up again.
This is an example of negative feedback. *(4 max)*

(Total 8 marks)

10 The helix is unwound; two strands are separated (helicase is the enzyme that unwinds the helix, separating the two strands) by breaking hydrogen bonds between bases. New strands formed on each of the two single strands. Nucleotides added to form new strands with complementary base pairing: A to T and G to C. DNA polymerase forms the new complementary strands. Replication is semiconservative; each of the DNA molecules formed has one old and one new strand. *(8 marks)*

Chapter 8: Answers to exercises

1 12 ATPs are generated by the Krebs cycle and the electron transport chain. A further 3 ATPs are generated from the NADH formed in the link reaction. So a total of 15 ATPs would be produced (not 18 because there is no glycolysis reaction to consider in this question).

2 Striated muscle cells are very active. They allow all voluntary motion. It takes large amounts of ATP to allow the myofilaments to slide over one another.

3 NAD allows production of more ATP because it enters the electron transport chain at an earlier point than FAD.

4 The result would be fewer ATPs produced because the hydrogen ions would not be forced to move through the channels of ATP synthase. It is the movement of hydrogen ions through these channels that allows the phosphorylation of ADP.

5 ATP synthase allows a faster reaction bringing about phosphorylation of ADP. If this enzyme did not exist, the process of ATP production would be extremely slow.

6 Photosynthesis allows the production of chemical bonds in organic molecules, especially glucose. The process of photosynthesis does not release ATP for the cell to use. Photosynthesis occurs in chloroplasts. Respiration catabolizes glucose so that chemical bonds are broken and energy in the form of ATP is released to the cell for work to be done. Respiration occurs in mitochondria. Because of this, plants must have both chloroplasts and mitochondria.

7 The greater the number of different pigments a leaf has, the greater the absorption of light at more wavelengths. The result is a higher rate of photosynthesis.

8 Photosystem II results in the production of ATP. Photosystem I results in the production NADPH. Photolysis of water is associated with photosystem II thus providing electrons to this photosystem and allowing the release of oxygen to the atmosphere.

9 There are very many higher evolved plants (providing a very high biomass) that carry out photosynthesis in the world. With the large biomass comes very large numbers of cells and chloroplasts. Most of these cells carry out photosynthesis. To carry out the light-independent reaction, these cells

must have RuBP carboxylase. For these reasons, the statement is very possibly true.

10 ATP and NADPH are produced in the light-dependent reaction. In order for the Calvin cycle of the light-independent reaction to occur, both of these products are essential. ATP provides energy, while NADPH allows for the reduction of various compounds.

Chapter 8: Answers to practice questions

Note: In this markscheme a full stop (period) separates single scorable elements in the same way that a semicolon does in the official IB markscheme.

1 Glucose converted to pyruvate (two molecules).
By glycolysis.
Pyruvate enters the mitochondria.
Pyruvate converted to acetyl CoA / ethyl CoA.
By oxidative decarboxylation / NADH and CO_2 formed.
Fatty acids / lipids converted to acetyl CoA.
Acetyl groups enter the Krebs cycle (*accept acetyl CoA*).
FAD/NAD^+ accepts hydrogen (from respiratory substrates) to form NADH/$FADH_2$.
$FADH_2$/NADH donates electrons / hydrogen to electron transport chain (*reject donates H^+*).
Electrons release energy as they pass along the chain.
Oxygen final electron acceptor.
Production of water.
Builds up proton gradient / protons pumped across inner membrane.
Protons flow into matrix of mitochondria through ATP synthase.
ATP produced.
Produces 36/38 ATP (per glucose).
Accept any appropriate terminology for NAD and FAD.
(*8 marks max plus up to 2 for quality*)

2 D (*1 mark*)

3 C (*1 mark*)

4 A (*1 mark*)

5 B (*1 mark*)

6 C (*1 mark*)

7 Light-independent reaction fixes CO_2.
To make glycerate 3-phosphate.
Glycerate 3-phosphate / GP / phosphoglyceric acid becomes reduced.
To triose phosphate / phosphoglyceraldehyde / glyceraldehyde 3-phosphate.
Using NADPH.
Using ATP.
ATP needed to regenerate RuBP.
ATP is made in light-dependent reactions.
Light causes photoactivation / excitation of electrons.
Flow of electrons causes pumping of protons into thylakoid.
ATP formation when protons pass back across thylakoid membrane.
Electrons are passed to NADP/$NADP^+$.
NADPH produced in the light-dependent reactions.
(*8 marks*)

8 Light
- Rate of photosynthesis increases as light intensity increases.
- Photosynthetic rate reaches plateau at high light levels.

CO_2
- Photosynthetic rate rises as CO_2 concentration rises.
- Up to a maximum when rate levels off.

Temperature
- Rate of photosynthesis increases with increase in temperature.
- Up to optimal level / maximum.
- High temperatures reduce the rate of photosynthesis.

Some of the above points may be achieved by means of annotated diagrams or graphs.
(*6 marks*)

9 C (*1 mark*)

10 A (*1 mark*)

Chapter 9: Answers to exercises

1 Girdling in this way completely removes the phloem. This prevents photosynthesis products from getting to the roots. The roots die and then the tree dies.

2 Lateral growth would be greatly increased. In many plants this increases yields, for example in some tomato plant varieties.

3 The tuber provides a region for nutrient storage in the plant. During harsh environmental conditions, the tuber may sustain the life of the plant.

4 It is essential to leave original soil on the roots as this means the root hairs, major absorption structures of a plant, are undisturbed. As a result, they will be able to function in the new location.

5 Lack of water to a plant will decrease turgor pressure within the plant cells thus resulting in wilting.

6 A seed is a source when it first begins germination. At this time, it is providing nutrients to the early plant. The seed acts as a sink when it is being formed in the ovary of the flower. During this time, nutrients are being deposited in the seed.

7 Veins need to be relatively close together so that water and nutrients may easily pass to and from them in the plant tissues.

8 Dandelion seeds are very light. They also have parachute-like structures that allow them to be carried long distances in the wind.

9 Oxygen is necessary for the aerobic respiration that must occur in the seed to allow embryo germination. Large amounts of ATP are necessary for germination.

10 It was the structure through which the pollen tube entered the ovule to bring about fertilization.

11 Gibberellin must initiate transcription of the segment of the seed's DNA that will allow the production of the protein amylase.

Chapter 9: Answers to practice questions

Note: In this markscheme a full stop (period) separates single scorable elements in the same way that a semicolon does in the official IB markscheme.

1 *Award 1 mark for each of the following structures clearly drawn and correctly labelled.*

- petals;
- sepals;
- stigma;
- style;
- ovary;
- stamen / anther and filament;
- receptacle / nectary.

(6 marks)

2 B *(1 mark)*

3 Some flowering plants are short-day plants, others are long-day plants. The important variable is length of darkness/photoperiod.

Some plants are grown in greenhouses with controlled light conditions: short-day plants are kept in the dark during daylight hours; long-day plants are artificially lit during the night.

Use of an appropriate wavelength / far-red light / 730 nm.

Possible to expose only for brief periods to keep costs down, but long enough to interrupt the dark period.

Involves interaction of phytochromes with metabolic reactions.

Controlled by the plant's biological clock.

(6 marks)

4 C *(1 mark)*

5 *To receive full marks, responses must address all three parts.*

Light *(2 max)*

- Causes stomatal opening in morning, increasing transpiration.
- Increasing light increases transpiration; because stomatal opening increases.
- No light causes stomatal closure, reducing transpiration.

Wind *(3 max)*

- Removes water/vapour from around leaf.
- Increases water vapour / humidity gradient so increases transpiration.
- Increases transpiration / lack of wind can reduce transpiration.
- No increase in transpiration if humidity is 100%.

Humidity *(3 max)*

- High humidity reduces water vapour gradient so lowers transpiration.
- High humidity lowers transpiration rate.
- Lowering humidity can increase transpiration rate (to a point).
- At very low humidity stomata may shut down.

(Total 8 marks)

6 Absorption of water.
Gibberellic acid produced in embryo.
Gibberellic acid stimulates production of amylase.
Amylase catalyses the breakdown of starch to maltose.
Maltose diffuses to embryo and used for energy production and growth. *(5 marks)*

7 B *(1 mark)*

8 **(a)** **(i)** Height 0.54 m: 60–79 cm / 0.60–0.79 m (from the plant).
(ii) Height 10.8 m: 0–2.9 m (from the plant). *(1)*
Units needed for both parts of the answer.

(b) The greater the height from which the seed fell, the further it travelled from the parent plant *(1)*

(c) At the greater height:
- seed can catch the wind to travel further / updrafts / more wind at greater height;
- farther to the ground and does not travel straight down / more time to be blown before hitting the ground.

At lower height:
- seed can fall straight down;
- seed can hit downdraft and fall faster.

(2 max)
Any point must explain the difference in distance travelled from the two heights.

(d) *Agrostis stolonifera* *(1)*

(e) *Poa trivialis* *(1)*

(f) *Poa* produces seed earliest in the summer / in June.
Holcus produces most seed in July.
Agrostis and *Festuca* produce seed in (late July to) August.
Holcus and *Poa* have a peak time of seed fall / short period of seed fall.
Agrostis and *Festuca* may continue to increase seed production to September. *(3 max)*
Accept any of these points made conversely as an alternative.

(g) *Award 1 mark each for any two of the following.*
To avoid predation / disperse at times when other species are dispersing their seeds.
To avoid competition.
Late in the year to allow seeds to germinate over winter / better germination conditions.
Better dispersal conditions / more wind/ animals for dispersal.
Photoperiod – required day length for flowering.
More energy stored at the end of the summer for seed production.
More light/warmth / better conditions for seedling photosynthesis/growth. *(2 max)*

(h) *Award 1 mark each for any two of the following.*
Tropical fruits have higher lipid content than temperate fruits.
Temperate fruits (80%) have greater carbohydrate content than tropical fruits (55%).
Protein levels are similar in both groups of fruits / slightly higher in temperate fruits than in tropical fruits; (*must make it clear that the difference is slight*). *(2 max)*

(i) Mistletoe.
High proportion of lipid and carbohydrate (lipid has approximately twice the energy content of protein and carbohydrate). *(2)*

(j) *Award 1 mark for advantage and 1 mark for disadvantage.*
Animal dispersal advantage:
- travel further;
- digestion cracks seed coat for better germination;
- deposited in faeces with organic matter;
- better in areas with little wind.

Animal dispersal disadvantage:
- predation;
- seeds eaten;
- deposited in poor environment;
- buried too deep / buried too shallow (if deposited with faeces);
- animal might become extinct/scarce.

(2 max)
(Total 17 marks)

Chapter 10: Answers to exercises

1 In order for natural selection to function optimally, there must be a choice of offspring to select from. If all offspring are the same and one lacks resistance to a particular disease, all will lack resistance and could perish. If each is unique, some will be affected by the disease but some will be resistant and have a better chance of survival.

2 **(a)** Prophase I
(b) Drawing should show the two cells each split in half, making four cells. Each of the four daughter cells should have two single chromosomes (chromatids) in it, one predominantly red and the other predominantly blue.

3 (a) Metaphase I

(b)

	1st gamete	2nd gamete
First possibility	4 red chromosomes	4 red chromosomes
Second and third possibilities	2 red chromosomes 2 blue chromosomes	2 red chromosomes 2 blue chromosomes

	3rd gamete	4th gamete
First possibility	4 blue chromosomes	4 blue chromosomes
Second and third possibilities	2 red chromosomes 2 blue chromosomes	2 red chromosomes 2 blue chromosomes

4 It must be found on one of the non-sex chromosomes, in other words one of the 22 autosomes but not X or Y.

5 (a) $\frac{\text{A B}}{\text{A B}}$ $\frac{\text{a b}}{\text{a b}}$

(b) Similar diagram to Figure 10.12. Recombinants are as follows:

$\frac{\text{A b}}{\text{A B}}$ $\frac{\text{a B}}{\text{a b}}$

6 In continuous variation, there are many possible phenotypes, suggesting that the trait is polygenic. In discontinuous variation, there is a limited number of possible phenotypes, suggesting that the trait is probably controlled by one gene.

7 A, B and E are examples of continuous variation.

8 Since women can bear children, they would need more vitamin D for themselves and their developing babies. Less melanin in the skin would allow for more absorption of UVB and therefore more production of vitamin D in the skin.

Chapter 10: Answers to practice questions

Note: In this markscheme a full stop (period) separates single scorable elements in the same way that a semicolon does in the official IB markscheme.

1 Chromosomes condense / coil / become shorter and fatter during prophase I.

(Homologous) chromosomes pair up in prophase I.

Crossing over / chiasmata formation in prophase I.

Movement of pairs of chromosomes / bivalents to the equator in metaphase I.

Movement of half of the chromosomes to each pole in anaphase I.

Movement of chromatids to opposite poles in anaphase II.

Decondensation/uncoiling in telophase II.

(4 max if no diagram is shown)

Do not award a mark for a statement if a diagram has been drawn that does not fit in with the statement. For example, if the candidate states that pairs of chromosomes move to the equator in metaphase I but shows single chromosomes, do not award that mark.

2 2 chromosomes and 2 chiasmata. *(1 mark)*

3 Homologous chromosomes form tetrads/bivalents/pairs / undergo synapsis.

Crossing over.

During prophase I.

Exchange of DNA/genes/alleles between (non-sister) chromatids/chromosomes.

Description/diagram of chiasma.

New combinations of maternal and paternal genes/alleles/DNA.

Bivalents/homologous chromosomes orient/align themselves on equator randomly.

During metaphase I.

Orientation of one homologous pair of chromosomes is independent of others.

Homologous chromosomes separate/move to opposite poles.

Independent assortment (of unlinked genes).

Leads to $2^n/2^{23}$ possible gametes (without crossing over).

Additional variation when chromatids separate in second division.

(5 marks)

4 C *(1 mark)*

5 (a) **CcWw**; all are coloured starchy *(2)*

(b) Gametes are **CW**, **Cw**, **cW**, **cw** and **cw**. The F_2 genotypes are **CcWw**, **Ccww**, **ccWw** and **ccww**; 1 coloured starchy: 1 coloured waxy: 1 colourless starchy: 1 colourless waxy. *(2)*

(c) Chi squared test *(1)*

(d) (Autosomal) linkage (*reject sex linkage*) / genes are on the same chromosome / genes do not assort independently; coloured starchy and colourless waxy are parentals. *(2)*

(Total 7 marks)

6 9 *(1 mark)*

Chapter 11: Answers to exercises

1 Damaged cells secrete chemicals which attract platelets to adhere to the area. Damaged tissue and platelets secrete chemicals which convert prothrombin to the enzyme thrombin. Thrombin catalyses the conversion of fibrinogen to fibrin. The mesh-like fibrin traps erythrocytes and more platelets to form the clot.

2 Leucocytes that are needed in greater numbers are identified by the principle of challenge and response'. Any one leucocyte that is needed in greater numbers is identified and then undergoes cell cloning to form multiple cellular copies of itself.

3 Cancer cells can be targeted. Much lower dosage of toxic chemicals and radiation can be used. Far less normal tissue damaged. Patient has fewer side-effects.

4 A decrease in synovial fluid would possibly increase friction at the point of motion. There would be potential damage to the soft tissues of the joint because of this added friction.

5 The H zone is actually made of only myosin during relaxation. During contraction, the actin slides inward, overlapping the myosin in what was the H zone so that it looks just like the surrounding A band.

6 A muscle action potential moving through the T tubules. This action potential is generated when acetylcholine is released from the nerve at the neuromuscular junction.

7 A large number of mitochondria allows many more ATPs to be generated and therefore to be available for the necessary activities of muscle contraction.

8 When one muscle of an antagonistic pair contracts the other one relaxes. This allows the structure that is affected by these antagonistic muscles to return to its original position after movement.

9 Glomerular capillary or opened slit; basement membrane; tubule membrane (e.g. proximal convoluted tubule); peritubular capillary.

10 Two possible reasons: some molecules are too large to become a part of the filtrate (e.g. proteins) and some are filtered but are completely reabsorbed (e.g. glucose)

11 **(a)** No ADH would be secreted as this person would not need to reabsorb water from the collecting duct.

(b) ADH would be secreted as this person could not afford to lose much water in their urine; ADH would lead to water passing through the wall of the collecting duct and being reabsorbed into the surrounding capillaries.

12 Urea is a waste product that is only toxic when allowed to reach relatively high concentrations in the bloodstream.

13 Many of the molecules, both good and bad, that pregnant women take into their bodies can pass across the placenta from mother to fetus. Problems can result due to inadequate nutrition, or harmful substances such as alcohol, drugs, molecules from cigarette smoking, etc.

14 Seminiferous tubule; epididymis; vas deferens; fluid from seminal vesicles and prostate; urethra.

15 During fertilization, a spermatozoon donates a haploid nucleus to the new zygote. All of the nutrition and all of the organelles are from the egg. 'New' mitochondria are formed when an existing mitochondrion divides. Therefore, all of the mitochondria in the billions of cells in an adult come from the mitochondria that were originally found within the ovum.

16 A newly ovulated secondary oocyte has only a short period of time in which it will be 'healthy' after ovulation. The timing of ovulation and fertilization must also be in accordance with the preparation of the endometrium of the uterus to receive the early embryo for implantation.

Chapter 11: Answers to practice questions

Note: In this markscheme a full stop (period) separates single scorable elements in the same way that a semicolon does in the official IB markscheme.

1 A *(1 mark)*

2 D *(1 mark)*

3 C *(1 mark)*

4 B *(1 mark)*

5 **(a)** High pressure in afferent arterioles.
Leads to ultrafiltration in the glomerulus / through fenestrated capillaries in the glomerulus.
Drains through the Bowman's capsule to the proximal convoluted tubule. *(2 max)*

(b) (i) Glucose / amino acids (1)
(ii) Water by osmosis.
Salts by active transport / facilitated diffusion. (2)

(c) magnification = size of image/actual size of object
size of image = 4.9 cm = 49000 μm;
scale bar represents 10 μm;

$$\frac{\text{size of image}}{\text{actual size of object}}$$

$$= \frac{49000}{10}$$

Magnification = $\times$4900 ($\pm$200) (2)
Do not award the first mark if incorrect equations are set up such as 4.9 cm = 10 μm.

(d) Microvilli increase the surface area for absorption / active transport.
Mitochondria produce ATP for active transport. (2)

(Total 9 marks)

6 Benefits:

- prevent disease;
- prevent epidemics;
- healthier society;
- reduce medical costs;
- less job absenteeism;
- disease-free cattle / more food;
- eradicate diseases/smallpox entirely;
- prevent harm/disabilities due to diseases;
- speed up the body's response to a disease.

Dangers:

- allergic reactions;
- autoimmune response;
- weakened virus becomes virulent / get disease;
- danger of side of side-effects / example of side-effects;
- vaccine with side-effects e.g. salk vaccine / whooping cough vaccine / MMR vaccine.

(Total 8 marks)

7 Calcium released from sarcoplasmic reticulum.
Calcium binds to troponin.
Troponin with calcium bound makes tropomyosin move.
Movement of tropomyosin exposes binding sites (for myosin) on actin.
Contraction of muscle fibres is due to the sliding of filaments (over each other).
Myosin heads bind to / form cross-bridges with actin.
ATP binds to the myosin heads causing them to detach from the binding sites.
Hydrolysis of ATP / conversion of ATP to ADP causes myosin heads to move.
Myosin heads reattach to actin further along.
Myosin pushes actin / actin pushed towards the centre of the sarcomere / shortening of sarcomere.

(Total 6 marks)

8 Motor neurones carry impulses/messages to muscle.
Nerves/neurones stimulate muscles to contract.
Neurones control the timing of muscle contraction.
Muscles provide the force for / cause movement.
Muscles are attached to bone by tendons.
Bones act as levers.
Joints between bones control the range of movement.
Antagonistic muscles cause opposite movements.

(6 max)

(Total 6 marks)

9 *Accept answers referring to blood flow to the kidney instead of in the renal artery and blood flow from the kidney instead of in the renal vein.*
More oxygen in the renal artery / less in the renal vein / oxygenated versus deoxygenated.
Less carbon dioxide in the renal artery / more in the renal vein.
More urea in the renal artery / less in the renal vein.
More ammonia/ethanol/toxins/hormones in the renal artery / less in the renal vein.
Reject answers for the points above if 'none' instead of 'less' is indicated.
More salt/NaCl / N^+ and Cl^- ions (in total) in renal artery than in renal vein.
More water (in total) in renal artery than in renal vein.
Lower salt concentration / higher water concentration in vein with ADH. *(4 marks)*

10 Sperm enters oviduct (Fallopian tube) / sperm swims towards egg / (secondary) oocyte / ovum.
Sperm attracted to egg / sperm attach to receptors in zona pellucida / chemotaxis.
Acrosome reaction / release of (hydrolytic) enzymes from acrosome.
Penetration of zona pellucida / jelly coat.
Membranes of egg and sperm fuse / sperm (head) penetrates egg membrane.
Cortical reaction / granules released to the outside of egg.

Zona pellucida hardens / fertilization membrane forms to prevent polyspermy.

Nucleus of secondary oocyte completes meiosis II.

Fusion of nuclei / (diploid) zygote forms.

(6 marks)

11 **(a)** 5.3 ($\pm$0.3) pmol dm^{-3} (*unit needed*) *(1)*

(b) A positive correlation.
No data below 280 mOsmol kg^{-1} *(1 max)*

(c) After drinking water, blood plasma / solute concentration decreases.

Plasma ADH concentration decreases.

Osmoreceptors in the hypothalamus monitor blood solute / blood plasma / plasma concentration.

Impulses passed to ADH neurosecretory cells to reduce/limit release of ADH.

Drop in ADH decreases the effect of this hormone on the kidneys.

Blood solute concentration returns to normal.

(2 max)

(d) Vomiting / diarrhoea / blood loss.
Increased salt intake.
Drinking too much alcohol/coffee.
Certain drugs: morphine/nicotine/barbiturates.
Excess sweating / lack of water intake.
Diabetes as it increases glucose in blood.

(2 max)

(Total 6 marks)

Chapter 12 (Option A): Answers to exercises

1 Essential fatty acids and essential amino acids cannot by synthesized in the body, so they must come from food.

2 The second child is given priority for breastfeeding, so little milk and thus little protein is left for the first child.

3 *Cis* fatty acids are naturally bent molecules whereas *trans* fatty acids are artificially straightened out by incomplete hydrogenation.

4 Minerals do not contain carbon and they are not synthesized by living organisms.

5 Lack of sunlight would prevent sufficient vitamin D production, causing rickets and improper bone growth.

6 The population did not have access to foods from the ocean, so they had no iodine in their diet.

7 Carbohydrate.

8 Milk and milk products.

9 Vegetable oils.

10 Exercise is part of good health as well as diet.

11 Should be avoided / not necessary for good health.

12 Quantities of foods required depend on factors such as age, sex, amount of physical activity / the diagram is only meant as a guide for proportions.

13 More sedentary lifestyle, greater availability of food, advertising.

14 Can be of help to people who cannot control appetite; however, long-term effects on liver and kidneys could lead to death.

15 Answers may vary: psychological causes, physiological causes, sociological causes; very low body mass, loss of hair, menstrual cycle interrupted, etc.

16 Because there are no white blood cells or antibodies in infant formula.

17 Human milk has 65% whey and 35% casein but cow's milk has 18% whey and 82% casein. Caesin is more difficult to digest than whey protein.

18 Because the mother is carrying an infectious disease or because she cannot produce enough milk.

19 Because it has been observed in some obese children.

Chapter 12 (Option A): Answers to practice questions

Note: In this markscheme a full stop (period) separates single scorable elements in the same way that a semicolon does in the official IB markscheme.

1 **(a)** Control: 25% ($\pm$3).
Experimental: 75% ($\pm$3). *(2)*

(b) Less energy in diet of experimental rats, greater percentage survival/longevity.
Very young have same possibility of survival / no effect until 200 days old.
More energy in diet, earlier death / mortality rate higher / fewer surviving.
Valid numerical example (50% survival level comparison). *(2 max)*

(c) Carbohydrate / protein / lipid / minerals (or any in particular) / vitamins (or any particular) / water / fibres. *(1)*
A list of, for example, three vitamins, three sugars, etc. should be allowed.

(d) Experimental rats live longer because:
contraol rats are overweight/obese / excess fat in control rats;
less cholesterol so less coronary heart disease in experimental rats / vice versa in control rats;
variation of metabolic rate good in experimental rats;
variation of respiratory quotient in experimental rats;
switch from glucose to lipid–protein diet supported metabolism in experimental rats;
lower glucose concentration in diet of experimental rats / higher in control rats;
less glycated proteins in experimental rats / higher in control rats;
fewer free radicals in experimental rats / higher in control rats;
less damage to cell membranes in experimental rats;
protection against mutagenic action of carcinogens in experimental rats;
delay of onset of diseases common in later life in experimental rats. *(2 max)*
(Total 7 marks)

2 (a) A substance required by the body to provide energy, maintain health or to provide material for growth and repair.
One of the components in a balanced diet. *(1 max)*

(b) Rickets is the result of poor bone growth / poor calcification of bones.
Vitamin D / calciferol deficiency in diets can lead to rickets.
Calcium deficiency in diets can lead to rickets.
Vitamin D can be obtained from (fish) liver.
Vitamin D can also be synthesized by the action of UV light on pro-vitamins in the skin.
Pro-vitamins of vitamin D can be obtained from green vegetables.
Rickets may also be caused by a hereditary disease.
Rickets can be caused by poor calcium absorption. *(3 max)*
(Total 4 marks)

3 (a) An amino acid that the body cannot make but requires / must be ingested. *(1)*

(b) Provides nutrients necessary for growth.
Provides nutrients necessary for maintenance and repair.
Provides sufficient energy / balance between energy intake and expenditure.
Provides nutrients for metabolic processes/ reactions. *(2 max)*

(c) Vegans eat only plant products while vegetarians eat some animal products (e.g. honey, milk and milk products, eggs, etc.) as well as plant products. *(1)*

(d) Fibre is mainly cellulose / cannot be digested.
Fibre helps prevent constipation / increases bulk in intestine / helps egestion / allows peristalsis.
Helps increase bulk in stomach / reduces appetite to eat more / reduces obesity.
May reduce risk of diseases / colon cancer / haemorrhoids / appendicitis.
May reduce rate of sugar absorption / may help treatment of diabetes. *(3 max)*
(Total 7 marks)

Chapter 13 (Option B): Answers to exercises

1 A decrease in synovial fluid would possibly increase friction at the point of motion. There would be potential damage to the soft tissues of the joint because of this added friction.

2 The H zone is actually made of only myosin during relaxation. During contraction, the actin slides inward, overlapping the myosin in what was the H zone, so that it looks just like the surrounding A band.

3 A muscle action potential moving through the T tubules. This action potential is generated when acetylcholine is released from the nerve at the neuromuscular junction.

4 A large number of mitochondria allows many more ATPs to be generated and therefore to be available for the necessary activities of muscle contraction.

5 When one muscle of an antagonistic pair contracts, the other one relaxes. This allows the structure that is affected by these antagonistic muscles to return to its original position after movement.

6 No, the amount would be the same.

7 With the snorkel tube there is more dead space to fill before the ambient air actually gets to the alveoli. If the tube's length is too long, oxygenated air would never reach the alveoli.

8 The effects are very much the same. Both activities increase the strength of the muscles involved, and possibly increase volume.

9 Aerobic and anaerobic exercise, actually working at taking deeper breaths, working to increase amount of air forcibly exhaled.

10 EPO causes an increase in the number of erythrocytes. This results in more viscous blood. If dehydration occurs, this problem is exacerbated.

11 Blood pressure may be increased due to the greater viscosity of the blood, thereby increasing resistance to blood flow in the arterial system.

12 Times involving aerobic activities would generally increase because there would be less oxygen. This would result in muscles working less efficiently. However, for anaerobic activities there would probably not be a noticeable change in times. Individuals carrying our anaerobic activities would have increased recovery times.

13 Any anaerobic event such as shorter sprints, long jump, high jump, etc.

14 The carbohydrate loading is thought to increase muscle glycogen levels thereby providing more ATP to the marathon runner.

15 They are carrying out aerobic respiration. This type of respiration does not result in the production of lactic acid.

16 Any food with carbohydrates.

17 Anaerobic respiration and the production of lactic acid.

18 Lactate is removed from the muscle and transported to the liver. Also, a renewed source of ATP must become available to the muscle cells. This involves increased oxygen and nutrient supply so aerobic respiration may occur.

19 The weight-lifter would carry out more activities of an anaerobic nature, i.e. very intense, short-lasting activities. Endurance runners would work at long-duration aerobic training to increase the ability to work at an efficient level for long periods of time.

20 Slow oxidative fibres have many mitochondria to release large numbers of ATP. These fibres are slow in their speed of contraction. They have large amounts of myoglobin so that oxygen may be maximally used.

21 Cardiovascular training involving aerobic activities lasting about 20 minutes at least four times a week.

22 Speed mostly involves fast glycolytic fibres. Stamina mostly involves slow oxidative fibres.

23 Warm-ups increase blood flow to the region thereby increasing oxygen and nutrient availability. This increases the efficiency of the actions that occur in the sarcomere to bring about muscle contraction.

25 They would probably work more sluggishly. Action would be at a level less than optimal.

26 The ends of the radius, ulna or humerus would be out of position.

Chapter 13 (Option B): Answers to practice questions

Note: In this markscheme a full stop (period) separates single scorable elements in the same way that a semicolon does in the official IB markscheme.

1 **(a)** I Humerus
II Synovial fluid *(2)*

(b) Muscles are coordinated by reflexes from the CNS / spinal cord.
The biceps and triceps muscles are antagonistic.
Contraction of a muscle is stimulated by motor nerves.
Stretch receptors / proprioceptors in muscles and tendons sense muscle stretching.
When stretch receptors in muscles are stimulated, they produce a reflex, thus stimulating muscle contraction / stretch reflex.
Reciprocal innervation of muscles / when one muscle is excited the antagonistic muscle receives no excitation / is inhibited. *(3 max)*

(c) *Award (1) for type of injury and (1) for an appropriate description.* *(2 max)*

Injury	Description
tennis elbow	inflamed tendon
dislocation	bone/humerus pulled from socket of ulna
torn ligament	sprain
ruptured tendon	sprain
fracture/break	due to excessive torsion/ compression/flexing

(Total 7 marks)

2 Greater muscle activity produces more CO_2 / increase in respiration produces more CO_2.

Increased blood CO_2 level lowers blood pH / makes the blood more acidic.

Low blood pH is detected by chemoreceptors in the arteries / carotid arch.

The breathing centre / respiratory centre of the brain is stimulated.

Diaphragm and intercostal muscles contract more strongly / more frequently.

Ventilation rate increases.

Volume of air per breath increases / depth of breathing increases. *(3 marks)*

3 (a) Stamina: the ability to maintain prolonged physical activity. *(1)*
Do not accept only 'endurance'.

(b) Greater strength / more developed muscles / greater muscle bulk.
Better blood supply to muscles / faster removal of waste, e.g. lactic acid.
More mitochondria in muscles.
Increased myoglobin. *(2)*

(c) Competition should be fair and equal with no discrimination / not all have access to drugs.
Result should not be more important than the process.
Provides the user with an unfair advantage over others / cheating.
Against the rules.
Athletes are role models / bad example to youngsters.
Drugs harmful to athletes / shorten life expectancy of athletes. *(3 max)*

(Total 6 marks)

4 (a) (i) Inadequate oxygen supply to respiring muscle; lactate from anaerobic respiration builds up. Oxygen needed to convert lactate (to pyruvate). *(1 max)*

(ii) liver *(1)*

(b) Sprains are minor tears to ligaments.
Torn ligaments are complete tears.
Dislocation of joints, bones move out of alignment.
Intervertebral damage, disc is torn and centre bulges out.
Erosion of cartilage. *(2 max)*

(c) Warm-up makes muscles/joints more supple / less likely to tear/strain / increases blood flow to provide oxygen/glucose/reduce lactate / increases ventilation to provide more oxygen.
Cool-down disperses lactate / allows cardiovascular system to adjust. *(2)*

(Total 6 marks)

5 (a) *Sprinter*: anaerobic respiration.
Marathon runner: aerobic respiration. *(2)*

(b) Lactic acid accumulation in a sprinter.
Depletion of the carbohydrate stores in the muscles in marathon runner. *(2)*

(c) Glycolysis allows skeletal muscle to work when mitochondrial activity is slowed because of low oxygenation.
Lactic acid is produced from pyruvate during anaerobic metabolism.
Oxygen debt is the oxygen needed to break down the lactic acid in the liver.
Continued deep breathing is needed after exercise to replace the oxygen.
Lactic acid is converted to pyruvate in the liver. *(2 max)*

(Total 6 marks)

6 (a) 93% ± 1% (1)

(b) (i) Increases by 0.63 mmol l^{-1} of blood / rises from 15.59 to 16.22 mmol l^{-1} *(1)*

(ii) Dissolved CO_2 *(1)*

(c) CO_2 makes the blood more acidic and the pH drops.
pH of venous blood at rest has decreased compared to arterial blood.
Because the blood is carrying waste CO_2 (from cellular respiration) back to lungs for removal.
pH of venous blood after exercise has decreased compared to arterial blood.
And dropped even further than venous blood at rest.
Because the blood is carrying more waste CO_2 than normal due to exercise. *(3 max)*

(Total 6 marks)

7 (a) Muscle glycogen levels are reduced. *(1)*

(b) The overall levels of muscle glycogen are not restored between each run.
The muscle glycogen levels gradually decrease over the three days of running.
The greatest reduction in muscle glycogen occurs after the first day's run.
The level of muscle glycogen increases / returns to (almost) normal. *(2 max)*

(c) Carbohydrates provide energy to contracting muscles.
Glucose/carbohydrate is respired by the mitochondria in the muscles.

Respiration produces ATP.
ATP provides the energy for muscle contraction. *(2 max)*

(d) The athletes' diets restore some of the glycogen after a run.
The diets are insufficient to restore the glycogen to its original level / even after five day's rest some athletes' glycogen levels do not return to the original level.
More carbohydrate is needed in the diet of training athletes / increase dietary carbohydrate intake / carbohydrate loading / pasta loading.
Less extensive exercise should be performed / longer rest intervals are needed between exercise / preparation for a race should include a break in training schedule. *(3 max)*
(Total 8 marks)

8 Calcium released from sarcoplasmic reticulum.
Calcium binds to troponin.
Troponin with bound calcium makes tropomyosin move.
Movement of tropomyosin exposes binding sites (for myosin) on actin.
Contraction of muscle fibres is due to the sliding of filaments (over each other).
Myosin heads bind to / form cross-bridges with actin.
ATP binds to the myosin heads causing them to detach from the binding sites.
Hydrolysis of ATP / conversion of ATP to ADP causes myosin heads to move.
Myosin heads reattach to actin further along.
Myosin pushes actin / actin pushed towards the centre of the sarcomere / shortening of sarcomere.
(6 marks)

9 Labelled diagram showing, biceps, humerus, radius and ulna.
Cartilage reduces friction.
Synovial fluid lubricates the joint.
Synovial membrane secretes synovial fluid.
Capsule / capsular ligament seals the joint.
Ligaments prevent dislocation / restrict the range of movement / attach bones to one another.
Motor neurones stimulate muscles to contract.
Bones provide a firm anchorage for muscles.
Bones act as levers / change the torque/size/ direction of forces.
Tendons attach muscle to bone.
Biceps and triceps are antagonistic.
Biceps is the flexor / bends the elbow joint *and* triceps is the extensor / straightens the elbow joint.
Biceps is attached to the radius and triceps is attached to the ulna. *(8 marks max)*
Accept any of the above points if clearly drawn and correctly labelled in a diagram. Plus up to (2) for quality. *(8)*

Chapter 14 (Option C): Answers to exercises

1 The primary level of protein organization refers to the amino acid sequence. The amino acids are the causes of electrostatic forces and sources of the types of weak bonds that determine the other levels of protein organization.

2 Prosthetic or non-polypeptide group.

3 A change in an amino acid brings about a different arrangement in the folding of the protein because of a change in the positioning of the forces and charges. This results in change of structure and a resulting change in function. Red blood cells showing sickle cell anaemia are a great example of this.

4 The bonds and general shapes are specific for both the enzyme and the substrate. If the substrate does not have a compatible shape and appropriately positioned electrical forces, the substrate will not fit the active site of the enzyme.
Enzymes increase reaction rates by lowering activation energy.

5 In competitive inhibition, the competitor molecule combines with the active site directly. This prevents the substrate from combining with the enzyme. In non-competitive inhibition, the active site is not bound by the outside molecule. In this case, the molecule combines with an alternative site on the enzyme causing a conformational change resulting in decreased enzyme function.

6 The lock-and-key model was the first model used to explain enzyme action. It stated there was an exact fit between the substrate and enzyme that allowed the reaction to proceed faster. After further research, it became apparent the original fit is not perfect, just close. After the partial first fit, there are conformational changes allowing a very close fit between the substrate and enzyme so the reaction may proceed more rapidly.

7 12 ATPs: each NADH may produce 3 ATPs. Each FADH may produce 2 ATPs. One ATP is made by substrate-level phosphorylation.

8 Large numbers of mitochondria are necessary to generate the ATP necessary for striated muscles to carry out their functions. Striated muscle allows voluntary movement and therefore requires much ATP.

9 NAD allows production of more ATP because it enters the electron transport chain at an earlier point than FAD.

10 The result would be fewer ATPs produced because the hydrogen ions would not be forced to move through the channels of ATP synthase. It is the movement of hydrogen ions through these channels that allows the phosphorylation of ADP.

11 ATP synthase allows a faster reaction bringing about phosphorylation of ADP. If this enzyme did not exist the process of ATP production would be extremely slow.

12 Photosynthesis allows the production of chemical bonds in organic molecules, especially glucose. The process of photosynthesis does not release ATP for cellular use. Photosynthesis occurs in chloroplasts. Respiration catabolizes glucose so that chemical bonds are broken and energy in the form of ATP is released to the cell for work to be done. Respiration occurs in mitochondria. Because of this, plants must have both chloroplasts and mitochondria.

13 The greater the number of different pigments, the greater the absorption of light at more wavelengths. The result is a higher rate of photosynthesis.

14 Photosystem II results in the production of ATP. Photosystem I results in the production of NADPH. Photolysis of water is associated with photosystem II, thus providing electrons to this photosystem and allowing the release of oxygen to the atmosphere.

15 There are very many highly evolved plants resulting in a very high biomass that carry out photosynthesis in the world. With the large biomass comes very large numbers of cells and chloroplasts. Most of these cells carry out photosynthesis. To carry out the light-independent reaction, these cells must have RuBP carboxylase. Because of this reasoning, the statement is very possibly true.

16 ATP and NADPH are produced in the light-dependent reaction. In order for the Calvin cycle of the light-independent reaction to occur, both of these products are essential. ATP provides energy while NADPH allows for the reduction of various compounds.

Chapter 14 (Option C): Answers to practice questions

Note: In this markscheme a full stop (period) separates single scorable elements in the same way that a semicolon does in the official IB markscheme.

1 (a) Rate of photosynthesis increases (rapidly) / directly proportional.
Rate of photosynthesis levels off / increases slightly after 10 000 lumen m^{-2}. *(2)*

(b) Maximum photosynthetic rate is highest with highest CO_2 concentration.
At low light levels, higher CO_2 slightly increases the photosynthetic rate.
At low CO_2 / 280 ppm CO_2, the photosynthetic rate reaches its maximum at low light levels / is constant over most light intensity / at 280 ppm CO_2, CO_2 concentration limits photosynthesis.
At 500 and 1300 ppm CO_2 the curve is the same shape but with different maximum rates / each higher light intensity requires a higher CO_2 concentration to reach maximum rate.
Maximum rate of photosynthesis from 280 to 500 ppm CO_2 / increases 5 to 6 times, while 500 to 1300 ppm CO_2, the rate of photosynthesis increases 1.5 times. *(3 max)*

(c) The rate of photosynthesis will increase (over the rate at 370 ppm).
The photosynthetic rate will at least double (but less than 5–6 times).
Not linear.
Bigger plants / more growth / more grain / greater yield. *(2 max)*

(d) $\frac{4.3 - 3.8}{3.8} \times 100 = 13.14\%$ *(1)*
Accept 13.2%

(e) Shade leaves receive less light than sun leaves.
To capture sunlight, shade leaves produce more chlorophyll.
To capture sunlight, shade leaves have greater leaf area. *(2 max)*

(Total 10 marks)

2 (a) Peptide bonds / peptidic bonds. *(1)*

(b) α-helix / α-helices. *(1)*

(c) Ionic/polar/hydrogen/hydrophobic/van der Waals'/disulfide. (*Not covalent*) *(1)*

(d) Linking together of polypeptides to form a single protein.
Using the same bonding as for tertiary structure.
Linking of a non-polypeptide structure / prosthetic group.
Named example of quaternary structure, e.g. haemoglobin (has four polypeptides). *(2 max)*

3 B *(1 mark)*

4 D *(1 mark)*

5 D *(1 mark)*

6 (a) ATP
CO_2
ethanol
lactic acid
heat energy *(1)*

(b) *(2)*

Reaction	Oxidation	Reduction
electrons gained or lost	• loss of electrons	• gain of electrons
oxygen or hydrogen gained or lost	• gain of oxygen • loss of H^+ / hydrogen	• loss of oxygen • gain of H^+ / hydrogen

Award (2) for four correct and (1) for two correct.

(c) A – matrix: site for Krebs cycle / link reaction / ATP synthesis.
B – inner membrane/cristae: site of oxidative phosphorylation / electron transport chain / increase surface area / ATP synthesis.
C – intermembrane space: H^+ / proton build up.
or
C – outer membrane: determines which substances enter the mitochondrion. *(3 max)*
Award (1 max) if only the three labels are given without the functions. *(Total 6 marks)*

7 (a) Stroma (of chloroplast). *(1)*
(b) Peak at about 450 nm and at 650 nm and follows pattern of absorption spectrum. *(1)*
(c) Electron transport causes proton/hydrogen ion pumping.
Protons inside thylakoids.
Accumulation of protons / H^+ / drop in pH.
Protons leave through proton channel (to stroma).
ATP synthase / enzyme catalyses phosphorylation of ADP. *(3 max)*
(Total 5 marks)

8 B *(1 mark)*

9 A *(1 mark)*

10 C *(1 mark)*

Chapter 15 (Option D): Answers to exercises

1 Monomers could include simple sugars, nucleotides or amino acids. Polymers could be polysaccharides, proteins or RNA.

2 By being surrounded by proteinoid microspheres or coacervates, they could be protected from being dissolved.

3 Because there was no oxygen and consequently no ozone layer protecting Earth from the Sun's UV radiation.

4 Panspermia is the idea that life (or the molecules needed for life to develop) came from outer space. Comets carried certain organic molecules to Earth billions of years ago.

5 A bacterial cell engulfs another prokaryote but rather than digesting it, saves it to coexist in symbiosis.

6 Both are examples of barriers between gene pools but geographical isolation is a physical barrier such as a mountain or a river whereas temporal isolation is a problem of synchronization of time when populations or their gametes try to connect.

7 Because, like most other hybrids, it is infertile and cannot produce offspring.

8 New species which have not been observed before, or at least new variations in the species. Also, adaptation to new habitats not usually occupied by that species of lizard.

9 As the man has sickle cell trait, he must be heterozygous and his genotype must be **Hb^AHb^S**. This means that some of his red blood cells have an elongated and curved shape which is inefficient for carrying oxygen. Therefore, he needs to take in larger quantities of air and so has more difficulty breathing than the others.

10 5

11 19 000 years old.

12 The graph indicates that after 4000 years, only about 65% of ^{14}C atoms should remain, so about 35% would have decayed.

13 It is a conceptual representation of frequencies of alleles, genotypes and phenotypes. Data from past or present populations can be entered into the equation to obtain reasonable results. Changes in allele frequencies, notably deviations from equilibrium in the Hardy–Weinberg equation can indicate evolutionary change. It is based on assumptions such as a large population, random mating, constant allele frequency, no allele-specific mortality, no mutations and no immigration or emigration.

14 **AA** × **aa** = **Aa**. So the answer is one generation.

15 $q = \sqrt{0.10} = 0.32$
$p = 1 - 0.32 = 0.68$
$q^2 = \frac{28}{278} = 0.10$
$2pq = 0.44$
$p^2 = 0.46$

16 One of the essential assumptions of the model is that mating is random. If it is not random, the equation cannot give reliable results.

17 From the bottom of the cladogram to the top, the labels should be in the same order as the column headings in the table.

18 No, this is not natural because not all organisms with spots are related. Cheetahs are more closely related to other big cats without spots than they are to butterflies (they share many more traits with other big cats).

19 Homologous characteristics have the same basic structure but possibly a different function, and evolved from a common ancestor. Analogous characteristics have the same function, but perhaps a different structure, and evolved separately in various phylogenic branches.

20 Answers may vary. Possible answer for homologous = claws, for analogous = stripes.

Chapter 15 (Option D): Answers to practice questions

Note: In this markscheme a full stop (period) separates single scorable elements in the same way that a semicolon does in the official IB markscheme.

1 *Award (1) for the comparison of both groups.*

(a) The *Polypodium* (species) are (completely) isolated in different parts of the continent and the *Pleopeltis* (species) much closer together / physically overlapping / share same habitats. *Polypodium* grows in more northerly/temperate locations. *(1)*

(b) **(i)** *Polypodium* is most genetically diverse as it has lower similarity/genetic identity values / *Pleopeltis* has higher similarity/genetic identity values. *(1)*

(ii) *Pl. polyepis* and *Pl. conzatti.* *(1)*

(c) Geographic/ecological isolation / isolated by distance / by glacial periods / by climatic changes.
Reproductive or genetic separation of gene pools (led to speciation) / adaptive radiation. *(1)*

(d) *Award (1) for* Polypodium *and (1) for a reason.*
Polypodium has probably been genetically isolated for the longest period of time because there is more genetic difference between all three species than between the species of *Pleopeltis.*
Takes time to accumulate mutations/genetic changes.
Distance may have facilitated the process of reproduction isolation. *(2)*

(Total 6 marks)

2 **(a)** Mammals/Mammalia *(1)*

(b) 2 (± 0.5) million years ago in (sub-Saharan) Africa.
Uncertain because:
fossils lacking or not in good form;
soft parts do not fossilize;
interbreeding of species;
migrations;
scientists have different views;
fossil dating method not totally accurate. *(3)*

(Total 4 marks)

3 **(a)** 7 *(1)*

(b) Classification is supported by the data.
Valid numerical information.
Fewer differences between humans and Neanderthal than between humans and chimps.
But all three species could be close enough to place in a single genus / far enough apart to place in separate genera.
Only one Neanderthal sequenced so not enough data. *(3)*

(c) 28–52 (accept any answer within this range). *(2)*
Humans diverged from *Australopithecus* longer ago than Neanderthals but not as long as chimps / closer to humans than chimps. *(2)*

(Total 6 marks)

4 **(a)** Some families have become extinct. *(1)*

(b) Lipotidae *(1)*

(c) (Mitochondrial) DNA.
Amino acid sequences / proteins. (1)

(d) River dolphins did not evolve from the same ancestor.
River dolphins evolved in similar environments.
River dolphins were exposed to the same selection pressures.
River dolphins adapted in the same ways.
River dolphins show convergent evolution. (3)

(e) Radioisotopes in the rocks where the fossils are found / in fossils.
(Absolute) dates are obtained by radioisotopic dating of rocks where fossil toothed whales are found.
Comparisons of molecular sequences / DNA / proteins.
(Absolute) dates can be estimated from differences due to mutations in living toothed whales / molecular sequences.
Position of fossils in the rocks.
(Relative) dates may be obtained by comparing the positions of the fossil toothed whales relative to one another. (2)

(Total 8 marks)

Chapter 16 (Option E): Answers to exercises

1 (a) A stimulus is a change in the environment (internal or external) that is detected by a receptor and elicits a response.
(b) A response is a reaction to a stimulus.
(c) A reflex is a rapid, unconscious response.

2 Receptors receive the stimulus. The receptor generates a nerve impulse in the sensory neurone. The nerve fibre of the sensory neurone carries the impulse toward the spinal cord. The sensory neurone enters the spinal cord in the dorsal root and sends a message across the synapse to the relay neurone. The relay neurone synapses with the motor neurone in the grey matter of the spinal cord and transfers the impulse chemically across the synapse. The motor neurone carries the impulse to an effector. An effector is an organ that performs the response.

3 Your drawing should look like Figure 16.1.

4 Mechanoreceptors are stimulated by mechanical force or some type of pressure. They help us maintain posture and balance.
Chemoreceptors respond to chemical substances. They enable us to taste and smell. They also give us information about the internal body environment.
Thermoreceptors respond to a change in temperature. Warmth receptors respond when the temperature rises. Cold receptors respond when the temperature drops.
Photoreceptors respond to light energy. They are found in our eyes. Our eyes are sensitive to light and give us vision.

5 Your drawing shoould look like Figure 16.3.

6 Your drawing shoould look like Figure 16.5.

7

Rods	Cones
These cells are more sensitive to light and function better in dim light.	These cells are less sensitive to light and function well in bright light.
There is one type of rod. It can absorb all wavelengths of visible light.	Three types of cone are found in the retina. One is sensitive to red light, one to blue light and one to green light.
The impulses from a group of rod cells pass to a single nerve fibre in the optic nerve.	The impulse from a single cone cell passes to a single nerve fibre in the optic nerve.

8 Light rays pass through the pupil and are focused by the cornea, lens and the humours. The image focused on the retina is upside down and reversed from left to right. When the photoreceptors of the retina are stimulated, impulses are carried to the bipolar neurones and the ganglion cells. The nerve fibres from the ganglion cells travel to the visual area of the cerebral cortex of the brain. The brain must correct the orientation of the image so that it is right side up and not reversed. It must also coordinate the images coming from the left and right eye.

9 Your drawing shoould look like Figure 16.8.

10 The outer ear catches sound waves which travel down the auditory canal. This causes the eardrum to vibrate. The bones of the ear (malleus, incus and stapes) receive vibrations from the eardrum and multiply them approximately 20 times. The stapes strikes the oval window causing it to vibrate. This vibration is passed to the fluid in the cochlea. The fluid in the cochlea makes hair cells vibrate. The hair cells are receptors and release a chemical message

across a synapse to a sensory neurone of the auditory nerve. The chemical message stimulates the sensory neurone. The wave in the fluid in the cochlea fades away as it reaches the round window. Finally, the message is carried by the sensory neurone in the auditory nerve to the brain.

11

Innate behaviour	Learned behaviour
develops independently of environmental context	dependent on environmental context for development
controlled by genes	not controlled by genes
inherited from parents	not inherited from parents
developed by natural selection	developed by response to environmental stimuli
increases chance of survival and reproduction	may or may not increase chance of survival and reproduction

12 To design an experiment to investigate innate behaviours of an invertebrate, use the following steps:
- observe the organism of choice, research the organism and formulate a research question;
- describe a method for the collection of relevant data;
- design a method for control of the variables;
- record raw data.

13 Some presynaptic neurones excite postsynaptic neurones and others inhibit postsynaptic transmission.

14 An action potential moves down the presynaptic neurone and when it reaches the synapse, calcium ions rush into the end of the neurone. This causes vesicles containing neurotransmitters to fuse with the presynaptic membrane and release the neurotransmitters into the synaptic cleft. A neurone is on the receiving end of many excitatory and inhibitory stimuli. The neurone sums up the signals. If the sum of the signals is inhibitory, the axon does not fire. If the sum of the signals is excitatory, the axon fires. The summation of the messages is the way that decisions are made by the CNS.

15 Drugs can alter your mood or your emotional state. Excitatory drugs like nicotine, cocaine and amphetamines increase nerve transmission. Inhibitory drugs such as benzodiazepines, alcohol and tetrahydrocannabinol (THC) decrease the likelihood of nerve transmission. Drugs act at the synapses of the brain by different mechanisms to determine your emotional state. Drugs can change synaptic transmission in the following ways.
- Block a receptor for a neurotransmitter. The drug has a structure similar to the neurotransmitter.
- Block release of a neurotransmitter from the presynaptic membrane.
- Enhance release of a neurotransmitter.
- Enhance neurotransmission by mimicking an neurotransmitter.
- Block removal of a neurotransmitter from the synapse and prolong the effect of the neurotransmitter.

16 The inhibitory drugs are: benzodiazepine, alcohol, and THC. The excitatory drugs are: nicotine, cocaine, and amphetamines.

17 Marijuana (THC) users say they feel relaxed and mellow. Some say they feel lightheaded and hazy. THC may dilate the pupils causing colour perception to be more intense. Other senses may be enhanced. Some people feel panic and paranoia. At the synapse, THC acts on cannabinoid receptors. These receptors affect several mental and physical activities including learning, coordination, problem solving, short-term memory.

Cocaine stimulates adrenergic synapses. It causes alertness and euphoria. It causes dopamine release and blocks the removal of dopamine from the synapse. Thus, overstimulation of the postsynaptic neurone leads to euphoria because this is the 'reward pathway'.

18 Addiction is a chemical dependency on the drugs where the drug has 'rewired' the brain and has become an essential biochemical in the body.

The role of almost all commonly abused drugs is to stimulate the 'reward pathway' located in the brain.

Evidence of genetic predisposition is found in studies of twins. Other experiments indicate that a genetically determined deficiency of dopamine receptors predisposes certain people to addictive behaviours. Genetically manipulated alcohol-preferring rats have 20% lower levels of dopamine receptors than non-preferring rats. Societal factors and environment can also be considered as possible causes of addiction.

19 Your drawing should look like Figure 16.20.

20 Medulla oblongata: controls automatic and homeostatic activities, such as swallowing, digestion and vomiting, breathing and heart activity.

Cerebellum: coordinates unconscious functions, such as movement and balance.

Hypothalamus: maintains homeostasis coordinating the nervous and the endocrine systems, secreting hormones of the posterior pituitary, and releasing factors regulating the anterior pituitary.

Pituitary gland: The posterior lobe stores and releases hormones produced by the hypothalamus and the anterior lobe. It also produces and secretes hormones regulating many body functions.

Cerebral hemispheres: act as the integrating centre for high complex functions such as learning, memory and emotions.

21 One method is to study people who have had injuries to particular areas. These lesions tell us indirectly about the function of those parts.

In the 1960s, scientists became interested in patients who had undergone surgery to sever their corpus callosum to relieve symptoms of epilepsy. Experiments were devised to determine how splitting the brain affected these patients.

Functional magnetic resonance imaging (fMRI) shows blood flow in the brain as the subject undertakes certain tasks or responds to stimuli. It is quite precise in showing what brain areas are functioning at any time.

22

	Sympathetic	Parasympathetic
Heartbeat	increase	decrease
Iris	dilates	constricts
Blood flow to gut	restricted	open

23 The definition of 'brain death' is when the whole brain has irreversibly lost function. This includes the brain stem. Patients who are in a coma have neurological signs which can be measured. The legal description of brain death is 'that time when a physician(s) has determined that the brain and brain stem have irreversibly lost all neurological function.'

24 Pain signals are carried by peripheral nerve fibres from all over the body to the spinal cord and relayed to the sensory area of the brain. These peripheral fibres connect with pain receptors called nocioreceptors. Nocioreceptors are capable of sensing excess heat, pressure or chemicals from injured tissues. These receptors are located in the skin and also in the muscle, bones, joints and membranes around organs. The nerve impulses of pain travel to the spinal cord. The ascending tracts in the spinal cord send the messages up to the brain.

Endorphins are pain-suppressing neurotransmitters in the brain. They were discovered by scientists studying opium addiction. Endorphins are small peptides. Endorphins bind with opiate receptors and block the transmission of impulses at synapses involved in pain perception.

25 Roles of bees in the social organization of the colony are distinctly organized.

Queen	fertile female	lays eggs produces pheromones which calm the colony and cause other females to be sterile
Worker	sterile female	feeds the larvae produces wax and honey searches for nectar and pollen protects the hive
Drone	fertile male – developed from an unfertilized egg	mates with the queen

A chimpanzee community is typically made up of 40–60 members. The highest ranking male is usually aged 20–26. His dominance is determined by his physical fitness and fighting ability. Males are clearly dominant over females. Older females are dominant over younger females. A smaller group within the community is called a party. Parties generally have up to 5 members. A party may be all males, family units, or nursery units with more than one family represented, or some other combination of individuals. The makeup of parties depends on the food supply. The more food there is, the larger the groups which travel together.

26 In honey bee populations, the workers do not reproduce; only the queen does. But they have the same genes. Natural selection acts on the colony as a whole. The genes which are selected are those which promote social organization. The workers are ensuring survival of their own genes through the queen.

27 Worker bees are altruistic. They help the queen produce offspring rather than reproducing themselves. This behaviour results in a decrease in fitness of the altruist and an increase in the fitness of a close relative.

Belding's ground squirrels live in the mountains of the southwestern United States. Predators of the ground squirrel are hawks and coyotes. When a predator approaches, one of the ground squirrels gives a high-pitched call which alerts the rest of the population to the nearby danger. The alarm squirrel is more likely to be killed than the other squirrels because the alarm call gives away the location of the caller to the predator. The alarm calls are predominately performed by females who live close to their relatives. The squirrel giving the alarm is not increasing its own fitness but it is increasing the fitness of its relatives.

Mole rat colonies of up to 100 individuals are found in burrows in East Africa. They live under the savannah and their burrows are excavated and extended by workers. The workers also make nesting chambers and forage for plant roots needed for food. Larger workers stay near the queen and her young. The queen suppresses the sexual behaviour of the other females. Snakes are the main predator of the mole rat. When a snake attacks a burrow, the queen sends out workers to attack the snake. The workers are sacrificed so that the queen and her young can live. Mole rats are genetically almost identical to each other. Natural selection is acting at the level of the colony.

28 The small mouth bass can forage for either minnows or crayfish. Minnows contain more energy per unit weight and are easier to digest but crayfish are easier to catch. Small mouth bass switch from minnows to crayfish and back to minnows in order to keep their energy intake higher than the energy they expend. Since no preference is shown for either minnows or crayfish, each may be optimal under different conditions.

Bluegill sunfish forage for *Daphnia*. *Daphnia* are small crustaceans found in varying sizes. Generally, bluegills forage for the larger *Daphnia* which supply the most energy. They select smaller *Daphnia* if the large ones are too far away. Predictions based on cost–benefit analysis suggest that when the density of *Daphnia* is low, the bluegill will not be selective about the size of the *Daphnia* when foraging. However, when the density of *Daphnia* is high, bluegills will be more selective and choose larger *Daphnia*.

29 Peahens choose their mates by the size and shape of their tails. It makes sense in evolutionary terms since the largest tail signifies the healthiest bird with the best chance for healthy offspring. The measure of the quality of the peacock's tail is the number of eyespots that it possesses. Originally, the tail size and number of eyespots may have had a real advantage but it may now just be a sign of the best male. Females who are choosy, prefer males with the longest tail. This could become more and more extreme until peacocks' tails become too big or too colourful and thus become a disadvantage if they attract a new predator.

30 Reproductive rhythm is a strategy employed by nearly all reef-building species of coral. Once a year, the coral release millions of gametes in a synchronized mass spawning ritual. Releasing the gametes all at the same time increases the chances that fertilization will occur because predators are overwhelmed with more food than they can eat. After fertilization, larvae develop.

The North American ground squirrel flies at night. Observations suggest that flying at night offers the most food and the least competition.

Chapter 16 (Option E): Answers to practice questions

Note: In this markscheme a full stop (period) separates single scorable elements in the same way that a semicolon does in the official IB markscheme.

1 (a) 24% (± 2%) (*units required*) (*1*)

(b) 8 (± 0.5): 42 (±0.5) or 16 (±1) : 84 (±1) or 1 : 5.25 (± 0.25) (*1*)

Accept the following alternative 11 : 44 (± 0.5) or 1 : 4 (± 0.25).

(c) Small females have more saturated lipids than large females / large females have more unsaturated lipids than small females.

Large females have a peak (unsaturated lipids) at 26 minutes while small females do not have a peak present.

Large females have two peaks (unsaturated and saturated) at 23 minutes while small females do not.

Large peaks for small females emerge sooner than for large females. (*2 max*)

(d) (i) Apply lipids to substrate / skin of snake. Then introduce male snake and record preference.

Use of control. (*2 max*)

(ii) Larger females have larger/more offspring. Better survival of offspring. (*1 max*)

(*Total 7 marks*)

2 (a) Behaviour that occurs in all members of a species despite variation in the environment / inherited behaviour / stereotyped behaviour / not learnt / instinctive. (*1*)

(b) Shine light in eye to see if pupil constricts. Pupil reflex is a cranial reflex / ANS reflex / controlled by the brainstem.
If pupil reflex is lost, patient is most likely brain dead.
Some drugs (barbiturates) / nerve damage may interfere with pupil reflex. *(2 max)*
(Total 3 marks)

3 (a) Psychoactive drugs affect the mind/brain *and* mood.
Change/increase synaptic transmission.
(drugs) can block / similar in structure / inhibit breakdown of neurotransmitter.
Award (2 max) per example, e.g. cocaine/ nicotine/amphetamines. Accept only effects on the synapse and behavioural effects.

Cocaine/crack:
Stimulates synaptic transmission of adrenergic synapses; increases energy/alertness/euphoria.
Nicotine:
Stimulates synaptic transmission of cholinergic synapses; has a calming effect.
Amphetamines/ecstasy:
Stimulates synaptic transmission of adrenergic synapses; similar effects to cocaine; longer lasting effect / 2 to 4 hours. *(6 max)*

(b) Both are part of the autonomic nervous system.
Antagonistic to each other / counteracts.
The sympathetic prepares the body for action while the parasympathetic returns the body function to normal.
Example of any effect comparing the action of the sympathetic and parasympathetic systems in a tissue or organ (e.g. heart: sympathetic increases output and parasympathetic returns it back to normal).
A second example. *(4 max)*
(Total 10 marks)

4 (a) **(i)** 25 (± 3)% *(1)*
(ii) 6.4 (± 0.6) : 1 / 32.5 *(1)*

(b) Resting and patrolling. *(1)*

(c) First activity is cell cleaning.
Followed by building comb and capping comb.
Finally the bee is foraging.
Eating pollen / tending brood / play fights occur at lower levels for most of the time.
Dance-following occurs towards the end of the 24 days. *(3 max)*

(d) Patrolling is altruistic / improves survival of hive / shows division of labour / helps others. *(1)*
(Total 7 marks)

5 (Unconditioned) stimulus of food / sight of food accompanied by bell ringing.
Salivation is the (unconditioned) response.
(Conditioned) stimulus of bell ringing before/ without unconditioned stimulus / sight of food.
Salivation became the conditioned response (to the bell ringing). *(Total 3 marks)*

6 (a) Cannabis:
Affects ability to concentrate.
Loss of muscle control.
Impairs perception / painkiller / loss of time Sense.
Memory loss.
Relaxed attitude.
Increased appetite.
Depression. *(2 max)*

Alcohol:
Lowers inhibitions / relaxed attitude / increases aggression.
Impairs reaction times.
Reduces fine motor control / loss of muscle control.
Memory loss.
Slurred speech.
Balance problems.
Depression.
Increased appetite. *(2 max)*
(Total 4 max)

(b) Neurotransmitters released by presynaptic neurones.
Neurotransmitters diffuse across synapse.
Bind to specific receptors on postsynaptic membranes.
Some neurotransmitters increase permeability of postsynaptic membrane to positive ions. Causing localized depolarizations.
Which helps an action potential to form / raises membrane above threshold.
(e.g. acetylcholine or other example).
Others cause negatively charged chloride ions to move across postsynaptic membrane into the cell, K^+ moves out of the postsynaptic nerve cell.
(e.g. GABA / other example.
Leading to hyperpolarization).
Which inhibits action potentials. *(6 max)*
(Total 10 marks)

7 Autonomic nervous system consists of sympathetic and parasympathetic control.

Sympathetic and parasympathetic controls are antagonistic systems.

Sympathetic control promotes fight or flight response / increase energy use.
Sympathetic control accelerates heart rate.
And increases stroke volume.
Sympathetic control inhibits release of saliva.
Parasympathetic control promotes restoration/rebuilding/conservation of energy.
Parasympathetic control reduces heart rate.
And reduces stroke volume.
Parasympathetic control promotes release of saliva.
Neurotransmitters are acetylcholine for parasympathetic and noradrenaline for sympathetic.
(6 marks)

8 **(a)** 20% for each population. *(1)*
(b) Population A has more (more than 2×) minnows hiding than population B.
Population A has more (almost 2×) minnows schooling than population B.
Population B has more (almost 10×) minnows feeding than population A. *(1 max)*
(c) Population A: because it shows more defensive behaviour than population B as the pike approaches.
Examples of defensive behaviour such as hiding, schooling and less foraging. *(2 max)*
(d) Distance to feeding patch / minnows. *(1)*
(e) *Other answers may be acceptable.*
The risk of any one minnow of being eaten is minimized.
A school may be quicker in evading a pike than an individual.
A school is more likely to see the pike than an individual. *(1)*
(Total 6 marks)

9 Altruistic behaviour is when one individual puts itself at risk for the survival of the rest of the immediate family or species.
(e.g. worker ants die in defence of their nest / individual ground squirrels may sacrifice their lives in order to warn the group of approaching predator / young adult jackals help raise parents' litter).
Selfish gene concept. *(2 max)*

10 **(a)** Autonomic nervous system / involuntary.
Sympathetic releases noradrenaline.
Parasympathetic releases acetylcholine.
Sympathetic increases (amplitude and rate of) heart beat (in order to send more blood to the body).
Parasympathetic decreases to resting level / returns heart beat to normal level.
Sympathetic inhibits secretion of saliva.
Parasympathetic stimulates secretion (which helps in digestion).
Parasympathetic contracts circular muscles of iris / constricts pupil (so less light in eye).
Sympathetic contracts radial muscle of iris / dilates pupil (so more light in eye / better vision. *(7 max)*
(b) Small peptides:
are neurotransmitters;
act in central nervous system;
bind with opiate receptors;
released to inhibit activity of neurones concerned with pain;
destroyed at synapse / do not last for long.
Opiates (heroin) act in a similar way. *(3 max)*
(Total 10 marks)

11 **(a)** Positive correlation / larger head, more copulations / direct relation. *(1)*
(b) Larger head, more wins.
Rarely / one time a smaller head won.
Rarely / four times same size head won. *(2 max)*
(c) Larger head wins more fights so more chances to copulate.
Larger head chosen by female more times / more copulations.
Wide spread of data for courtship success means data is less reliable / position of best fit line can be questioned. *(3 max)*
(Total 6 marks)

12 **(a)** A: cone
B: rod
C: (cell body of) bipolar neurone
Three correct (2), two or one correct (1) *(2 max)*
(b) **(i)** Instinct / stereotyped / genetically determined / not learnt / inherited. *(1)*
(ii) Breathing control / heart rate control / reflex control / blood pressure control / swallowing / coughing / production of saliva. *(1)*
(Total 4 marks)

Chapter 17 (Option F): Answers to exercises

1 Eubacteria: true bacteria, prokaryotes with no organized nucleus and no membrane-bound organelles.
Archaeabacteria: ancient bacteria, prokaryotes mainly living in extreme environments.
Eukaryotes: single and multicellular organisms which all have their DNA contained in a nucleus.

2 Distinguishing characteristics of the three domains are shown in this table.

Characteristics	Eubacteria	Archaea	Eukarya
histones	absent	histone-like proteins	present
introns	absent	present in some DNA	present
size of ribosome	70S	70S	80S
structure of cell membrane lipids	unbranched hydrocarbons	some branched hydrocarbons	unbranched hydrocarbons
peptidoglycan in cell wall	present	absent	absent
membrane-bound organelles	absent	absent	present

3 This table compares the cell walls of the two main types of Eubacteria.

Cell wall	Gram-positive bacteria	Gram-negative bacteria
complexity	simple	complex
amount of peptidoglycan	large amount	small amount
placement of peptidoglycan	is the outer layer of the bacteria	is covered by the outer membrane
outer membrane	absent	present with lipopolysaccharides attached

4 An outline of the diversity of eukaryotes is shown in this table.

Organism	Nutrition	Locomotion	Cell wall	Chloroplasts	Cilia or flagella
Saccharomyces	heterotroph (extracellular digestion)	absent	made of chitin	absent	absent
Amoeba	heterotroph (intracellular digestion)	slides using pseudopodia	absent	absent	absent
Plasmodium	heterotroph (intracellular digestion)	glides on substrate	absent	absent	absent
Paramecium	heterotroph (intracellular digestion)	swimming	absent	absent	cilia
Euglena	autotroph and heterotroph	swimming	absent	present	flagellum
Chlorella	autotroph	none	made of cellulose	present	absent

5 Your drawing should look like Figure 17.6.

6 Nitrification occurs due to the actions of two bacteria. Conditions required are:

- availability of oxygen (since the reaction is aerobic);
- neutral pH (preferred by these bacteria);
- warm temperature (also preferred by bacteria).

Denitrification is a conversion of nitrates to nitrogen gas. Conditions required are:

- no available oxygen (due to flooding or compact soil);
- high nitrogen input.

7 Sewage treatment occurs in two stages. In the first stage, inorganic materials are removed and organic materials are left. In the second stage, 90% of the organic matter is removed by saprotrophic bacteria. The bacteria obtain energy by breaking down organic matter.

8 Methane generated from livestock waste and cellulose left from crops is called biogas. Biogas is about 60–70% methane and 30–40% carbon dioxide. To make biogas, manure and cellulose are put into a digester without oxygen. Anaerobic decomposition is performed by acidogenic bacteria which occur naturally in the manure. Manure and cellulose contain carbohydrates, fats and proteins. These large molecules are broken down by bacterial enzymes from acetogenic bacteria into simpler compounds of organic acids and alcohol. Organic acids and alcohol are then decomposed into carbon dioxide, hydrogen and acetate. Finally, two different types of methanogenic bacteria work on these molecules to produce methane.

9 Reverse transcriptase can be used in the synthesis of insulin. A human DNA molecule with all its introns is taken from a pancreas cell.

mRNA copies the DNA for making insulin without the introns.

Reverse transcriptase produces a new single strand of DNA called cDNA.

The single strand replicates to make double-stranded DNA using the enzyme DNA polymerase.

The double-stranded DNA is inserted into a plasmid (circular DNA found in bacterial cells).

Bacterial cells are stimulated to take up the plasmids.

10 There are two types of gene therapy: germ-line therapy and somatic therapy. Germ-line therapy changes the DNA in the patient's gametes and is passed on to the offspring. Somatic-cell gene therapy changes the DNA in the patient's body cells. The consequences of this therapy affect only the patient, who can give voluntary consent to

the procedure. With somatic cell therapy, it may possible to cure single-gene defects such as cystic fibrosis and haemophilia.

11 The virus used to deliver the new gene to the target cell might get into another cell by mistake.

The virus might place the new gene in the wrong location in the DNA molecule (chromosome) and cause an unintended mutation.
Genes can be over-expressed, making too much protein which might be harmful.
The virus vector itself might stimulate an immune reaction as the body thinks it is being infected.
The virus vector containing the new gene might be transferred from person to person as with flu or cold viruses.
Children might be more sensitive to the long-term hazards since their tissues are still developing.

12 Liquid wort is made from malt.
Hops are added and the liquid is boiled and cooled.
Fermentation by yeast produces beer containing ethanol and carbon dioxide.

13 Low pH is key in preserving vegetables. Vinegar is about 5% acetic acid. The acid lowers the pH to the point where it restricts the growth of microorganisms. Sugar and other spices can be added for flavour. Early brining also kills microorganisms.

14 Symptoms of *Salmonella* poisoning occur 12–72 hours after the infection; they are:
- diarrhoea;
- fever;
- abdominal cramps.

A small number of people develop Reiter's syndrome which can last for years. Symptoms are arthritis in joints, irritation of the eyes and painful urination.

Transmission of *Salmonella*:
- human faeces;
- animal faeces;
- handling reptiles;
- contaminated food (e.g. meat from an unwashed cutting board, vegetables washed with water containing *Salmonella*);
- raw eggs;
- unpasteurized milk or other dairy products.

Treatment of *Salmonella* poisoning:
- drinking lots of water;
- intravenous fluids;
- antibiotics.

15 Photoautotroph: an organism that uses light energy to generate ATP and to produce organic compounds from inorganic substances (e.g. cyanobacteria).

Photoheterotroph: an organism which uses light energy to generate ATP and obtains organic compounds from other organisms (e.g. *Rhodobacter sphaeroides*).

Chemoautotroph: an organism that uses energy from chemical reactions to generate ATP and produces organic compounds from inorganic substances (e.g. *Nitrosomonas*).

Chemoheterotroph: an organism that uses energy from chemical reactions to generate ATP and obtains organic compounds from other organisms (e.g. *Saccharomyces*).

16 Comparison of energy and carbon sources of chemoautotrophs and chemoheterotrophs.

	Chemoautotroph	Chemoheterotroph
Energy source	inorganic compounds	organic compounds
Carbon source	carbon dioxide	organic compounds

17 Bioremediation is the use of bacteria (and fungi) to treat environments contaminated with polluting agents such as pesticides, oil and industrial solvents. During bioremediation, the organisms break down the toxic chemicals. The benefit to the microorganisms is that during the process of degrading contaminants, they derive energy for their own growth and reproduction. The bacteria and fungi are actually chemoheterotrophs obtaining energy from the organic molecules which happen to be pollutants.

18 This table distinguishes intracellular and extracellular bacterial infection.

Chlamydia (intracellular)	*Streptococcus* (extracellular)
lives inside the cell of the host	lives in the host but outside the cell
does not produce toxins	produces toxins
does not directly damage cells	directly damages cells
is not targeted by the immune system since it is hidden in the cell	is targeted by the immune system since it is freely circulating in the body.

19 The control of growth of microbes can be accomplished in two ways. The microbes can be killed or the growth of the microbes can be inhibited. Agents which kill bacteria are described as bactericidal. Those which inhibit the growth are described as bacteriostatic. The four methods are listed in order of effectiveness.

Irradiation: Bactericidal. This method destroys or breaks up the nucleic acids such as DNA and RNA.

- Gamma rays (ionizing radiation) kill all microbes.
- Microwaves kill all bacteria.
- UV radiation kills bacteria but leaves endospores.

Disinfectants: Bactericidal. These chemicals kill bacteria but not their endospores.

Antiseptics: Bactericidal. These mild chemicals are less effective than disinfectants.

Pasteurization: Bactericidal for pathogens but only bacteriostatic for non-pathogens.

20 Malaria is caused by four species of the protozoan genus *Plasmodium*. They are transmitted from one person to the next by a female *Anopheles* mosquito which feeds on blood. *Plasmodium* reproduces in the gut of the female mosquito. The egg sac ruptures and releases cells called sporozoites. These travel to the salivary glands of the mosquito. As the mosquito bites, sporozoites enter the human bloodstream with the mosquito saliva. The sporozoites travel to the liver where they develop further. After a week or two they burst out of the liver and invade red blood cells. Infected patients have symptoms of anaemia, bouts of fever, chills, shivering, pain in the joints and headache.

21

The virino hypothesis	The prion hypothesis
The infecting agent is a nucleic acid (DNA or RNA) surrounded by an abnormal protein.	The infecting agent is a prion. This abnormal protein alone causes the disease.
Scientists have not found the DNA or RNA which is hypothesized to be part of this infecting agent.	Scientists have searched for nucleic acids in prion particles for 30 years and not found any.
Nucleic acids must be present to code for the shape of proteins.	Abnormally shaped prions in a test tube can bind normal proteins and cause them to change to an abnormal shape.
Conclusion: TSE diseases are best explained by the hypothesis that a nucleic acid carries the information needed by the infecting agent.	Conclusion: TSE diseases are caused by a prion which is only protein. No nucleic acid component has been found.

Chapter 17 (Option F): Answers to practice questions

Note: In this markscheme a full stop (period) separates single scorable elements in the same way that a semicolon does in the official IB markscheme.

1 **(a)** **(i)** Malawi

(ii) Mauritius *(1)*

Both needed for (1)

(b) 40 (± 2)% *(1)*

(c) High detection rate combined with high cure rate slows down spread of TB.

High detection rate combined with high cure rate has positive effect on general health of the population.

High cure rate / above 85% is essential for reducing pool of infectious people.

High detection rate will help prevention of TB / slow down spread of TB. *(2 max)*

(Total 4 marks)

2 Anaerobic/oxygen-free.

Fermentation tank.

Warm temperature (30–40 °C).

Raw material: organic wastes / excreta / leafy remains / straw / bagasse / peelings of fruit/ vegetables.

Organic molecules (proteins, carbohydrates, fats) transformed into alcohol, carbon dioxide, fatty acids and hydrogen / hydrolysis of organic molecules.

Formation of acetic/ethanoic acid by acetogenic bacteria.

Methanogenic bacteria (*Methanoccocus*, *Methanobacterium*, *Methanospirillium*, others).

Methane produced from (reducing) CO_2 and H_2 / $CO_2 + 4H_2 \rightarrow CH_4 + 2\,H_2O$

Acetate/ethanoic acid split to produce methane (and CO_2) / $CH_3COOH \rightarrow CH_4 + CO_2$.

(*balanced equation not required*)

Affected by detergents / high fatty acid concentrations / heavy metal ions / low pH.

(6 marks)

3 *Rhizobium* adds nitrogen to the soil, whereas *P. denitrificans* releases nitrogen from the soil.

Rhizobium fixes nitrogen into nitrates/ammonia, whereas *P. denitrificans* converts nitrates into nitrogen.

Rhizobium lives as a symbiont in root nodules of leguminous plants, whereas *Pseudomonas* lives as a free-living bacteria in water-logged soil.

Rhizobium lives in aerobic conditions and *Pseudomonas* in anaerobic conditions. (*3 marks*)

4 **(a)** 47–49% (*1*)

(b) *D. melanogaster* has few genes with one exon.
Highest percentage has 2 exons.
Most genes have 5 or fewer exons.
A few genes have 10 or more exons / more than 8.
Maximum number of exons does not exceed 60. (*2 max*)

(c) **(i)** *S. cerevisiae*/yeast has most genes with only 1 exon while in mammals 5 exons is most frequent.
No yeast genes have more than 5 exons while some mammal genes have more than 60 exons.
Mammal genes contain more exons on average.
With a wider distribution than yeast. (*2 max*)

(ii) *S. cerevisiae*/yeast is a unicellular organism / mammals are multicellular/complex.
S. cerevisiae smaller in size / more compact genome. (*1 max*)

(d) Gene size – mRNA size = intron size / 25.0 – 2.1 = 22.9 kb.
Average size of intron = 22.9/14 = 1.6 (± 0.1) kb (*unit required*). (*2*)

(e) Smaller genes usually have fewer introns / larger genes have more introns / relationship not clear.
Dystrophin and collagen have same number of introns but the dystrophin gene is larger.
Albumin has more introns but is smaller than the gene for phenylalanine hydroxylase. (*2 max*)
(*Total 10 marks*)

5 **(a)** GNA varieties have a higher lectin content than Con A varieties. (*1*)

(b) Con A4 is the most promising.
(Relatively) high control of aphids and (relatively) high control of nematodes / has highest combined control of both aphids and nematodes. (*2*)

(c) Similar levels of control are achieved with much lower lectin content in Con A than GNA.
No clear relationship between lectin content and level of control.
Use of two or more figures from the table to illustrate the lack of correlation. (*2 max*)

(d) Either:
percentage = mean mass of control – mean mass of GNA 2#28 [divided by] mean mass of control [all multiplied by] 100%

or:
percentage = 5500 (±50) – 3700 (± 50) [divided by] 5500 (± 50) [all multiplied by] 100%

Either:
1 – mean mass GNA 2#28 [divided by] mean mass of control [all multiplied by] 100 %

or:
1 – 3700 (± 50) [divided by] 5500 (± 50) [all multiplied by] 100%
32% (± 2.5%) (*2 max*)

(e) Absolute reductions / differences using raw numbers are greatest with nematodes.
Percentage reductions are greatest with protozoa.
(Percentage) effect on protozoa is more significant.
Other GM varieties might have more / different effects.
Effects on other protozoa might be different.
Soil organisms do not usually make contact with leaves. (*3 max*)
(*Total 10 marks*)

Chapter 18 (Option G): Answers to exercises

1 Temperature: The temperature on the foredune can be very hot in the summer. There is no protection from the Sun. Marram grass is well adapted to this condition. It has thin, narrow leaves and long roots which find water even in very dry sand. The temperature on the mature dune is much cooler so the variety of plant species is much greater. A common plant on the forest floor of the mature dune is the fern.

Light: Marram grass lives in conditions where sunlight is constantly available. It is found in sunny areas of the foredune. Ferns are found in the shady areas of the mature dune.

Soil pH: In the yellow dune, the soil pH is about 7.5. Marram grass is still the dominant vegetation here and thrives at this pH. In the grey dune, soil has formed from the decomposition of grasses. The pH of this soil is more acid so heather grows well here.

Salinity: Foredunes catch the salt spray from the ocean. Marram grass and Lyme grass live comfortably in a salty environment. In the grey dune, small shrubs, mosses and lichen live in conditions that are much less salty.

Mineral nutrients: The grey dune shows some diversity of plants because it is older than the yellow and foredune, and contains some mineral nutrients in the soil. These mineral nutrients can support small shrubs, mosses and lichen. On a mature dune, there is a thick layer of soil full of mineral nutrients which can support large trees such as ash, birch and eventually oaks.

2 Temperature: A sand wolf spider lives in the foredune and is adapted to the extreme high temperature by living in a burrow deep in the sand. Its adaptation to the foredune is a behavioural adaptation. Woodland spiders which live in trees in the mature dune would die at such high temperatures.

Water: Blue herons catch small fish and frogs in dune wetlands. Woodpeckers are found in the mature dunes and eat the insects living in the trees.

Breeding sites: Blue herons breed around the interdunal wetlands where food is regularly available. Woodpeckers nest in the branches of the oak trees.

Food supply: Many animals are adapted to feed on specific food and must live where that food supply is available.

Territory: Packs of coyotes mark their territory with scent marking.

3 The concept of niche includes where an organism lives (its spatial habitat), what and how it eats (its feeding activities), and its interactions with other species. For example:

Spatial habitat: Green frogs live in the ponds of the Indiana Dunes.

Feeding activities: The green frog eats the aquatic larvae of mosquitoes, dragonflies and black flies.

Interactions with other species: Competition, herbivory, predation, parasitism and mutualism.

4 The principle of competitive exclusion states that no two species in a community can occupy the same niche.

In 1934, the competitive exclusion principle was demonstrated by G. F. Gause. He used two different species of *Paramecium*, *P. aurelia* and *P. caudatum*. When each species was grown in separate cultures, with the addition of bacteria for food, they did equally well. When the two were cultured together, with a constant food supply, *P. caudatum* died out and *P. aurelia* survived. The experiment supported his hypothesis of competitive exclusion. When two species have a similar need for the same resources, one species will die out in that ecosystem and the other will survive. *P. aurelia* must have had a slight advantage which allowed it to out-compete *P. caudatum*.

5 Biomass is the mass of living organic matter of organisms in an ecosystem.

6 An eagle is a tertiary consumer when eating rattlesnakes but is a secondary consumer when eating rabbits.

A coyote is a primary consumer when it eats the fruit of a cactus but is a tertiary consumer when it eats a rattlesnake.

A lizard is a tertiary consumer when it eats rattlesnake eggs but a secondary consumer when it eats insects.

Another difficulty is where to put omnivores. Grizzly bears eat plant parts, insects and some mammals. The type of food depends on the season of the year, the temperature and the bears' ability to forage for food. Are they primary consumers, secondary consumers or tertiary consumers?

7 This table distinguishes between primary and secondary succession.

Primary succession	Secondary succession
begins with no life	follows a disturbance of the primary succession
no soil	soil is present
new area e.g. volcanic island	old area e.g. following a forest fire
lichen and mosses are first plants	seeds and roots already present
biomass low	biomass higher
low primary production because few plants	higher primary production because many plants

8 Organic matter increases.

Soil is deeper.

Soil erosion is reduced.

Soil structure improves.

Mineral recycling increases.

9

Biome	Temperature	Moisture	Characteristics of vegetation
Desert	Usually very hot – soil temperatures above 60 °C (140 °F) in daytime.	Low rainfall – less than 30 cm per year.	Cactuses and other shrubs with water-storage tissues, thick cuticles and other adaptations to reduce water loss.
Grassland	Cold in winter and hot in summer.	Seasonal drought with occasional fires. Medium amount of moisture.	Prairie grasses.
Shrubland	Mild in winter and long, hot summers.	Rainy winters and dry summers.	Dry woody shrubs can store food in fire-resistant roots so re-grow quickly. Produce seed which germinates only after a fire.
Temperate deciduous forest	Very hot in summer and very cold in winter.	High rainfall spread evenly through year. In winter, water may freeze for a short time.	Deciduous trees like oak, hickory and maple. In the warmer seasons, herbaceous plants grow and flower on the forest floor.
Tropical rainforest	Very warm.	Very high rainfall – more than 250 cm per year.	Plant diversity is high: trees, shrubs, herbaceous plants and ferns. Also vines, orchids and bromeliads in their trees.
Tundra	Very cold for most of year. Lower areas always frozen – permafrost.	Little rainfall.	Lichen and mosses and a few grasses and shrubs grow here.
Coniferous forest	Slightly warmer than the tundra.	Small amount of precipitation but wet due to lack of evaporation.	Cone-bearing trees such as pine, spruce, fir and hemlock.

10 Economic reasons:

- the attempt to create farms on rainforest soils has met with dismal results;
- plant sources of medicines and chemicals will be gone forever;
- crop plants and farm animals could be improved with alleles from wild populations;
- ecotourism could improve the local economy.

Ecological reasons:

- an organism key to an ecosystem's health may be destroyed;
- diversity protects an ecosystem against invaders;
- fewer plants in the biosphere means more carbon dioxide in the atmosphere;
- disruption of the ecosystem can lead to soil erosion and flooding.

Ethical reasons:

- the local population is most effected by rainforest destruction;
- we have an ethical responsibility to conserve the rainforest so that future generations have access to its beauty and wealth of organisms;
- do we have the right to decide which organisms survive?

Aesthetic reasons:

- human well-being is linked to the ability to visit natural areas in our biosphere which have been preserved;
- many artists and writers have been inspired by the beauty of natural ecosystems.

Arguments against conservation:

- economic development may be slowed down;
- rainforests can be reservoirs for pests that transmit disease.

11 Red fire ants (*Solenopsis invicta*) are an imported pest insect in the US. They are common in the southern states, and are spreading north. Red fire ants compete with native ants and become the dominant species.

Scientists have been experimenting with a fly which is the natural predator of the fire ant. It is a phorid fly. The fly hovers over a mound of ants, picks out a victim and lays eggs inside its body. The larva eats its way to the head of the ant and decapitates it. The hope is that the phorid fly will reduce populations of red fire ants.

12 Increase non-lethal skin cancer.
Increase of lethal skin cancer.
Mutation of DNA.
Sunburn.
Cataracts.
Reduced biological productivity.

13 Before the Montreal Protocol phased them out, chlorofluorocarbons (CFCs) were finding their way to the stratosphere. In the stratosphere, they were breaking down ozone.

- Chlorine is split from the CFC by UV light to become a free chloride ion (Cl^-).
- The chloride ion reacts with an ozone molecule (O_3) and takes up one of the oxygens to form $ClO + O_2$.
- Next, the ClO joins with another free O to becoming $Cl^- + O_2$. The Cl^- is now free to destroy another ozone molecule. This cycle is repeated over and over causing a depletion of ozone.

14 Certain macroinvertebrates can act as indicator species to judge water quality in rivers and streams. The indicator species have various levels of pollution tolerance. When you perform a river or stream study, you count the number of macroinvertebrates collected in each sample and record the data on a stream study form. The number of organisms of each group is multiplied by a factor determined by how sensitive the organism is to pollution. Periodic sampling gives an idea of the overall health of the river or stream and how it is changing over time.

15 The Carolina parakeet has been extinct in the wild since 1900. Large groups of Carolina parakeets nested together in hollow trees. In the mid 1800s, settlers cleared large areas of trees for farming and fuel. Removing the trees destroyed the habitat of the Carolina parakeet.

Several other factors also contributed to its extinction. One was the introduction of the honey bee from Europe. Bees were brought to make honey and to pollinate plants. The bees quickly escaped into the wild and made their homes in the same hollow trees as the Carolina parakeet. The bees displaced the parakeets from their nesting area. The second factor was the popularity of the brightly coloured feathers with hat makers. The third factor was live capture of the birds to be sold as caged pets. By 1900, all Carolina parakeets were gone from the wild.

16 Conservation in situ does the following:

- protects targeted species by maintaining the habitat;
- defends targeted species from predators;
- removes invasive species;
- covers a large enough area to maintain a large population;
- has a large enough population of the targeted species to maintain genetic diversity.

On some occasions the in situ area is not able to protect the targeted species because:

- the species is so endangered that it needs more protection;
- the population is not large enough to maintain genetic diversity;
- destructive forces cannot be controlled (e.g. invasive species, human incursion, and natural disasters).

17 Life history strategies of r-strategists and K-strategists are compared in this table.

Characteristic	r-strategy	K-strategy
life span	short	long
number of offspring	many	few
maturity	early	late, after long period of parental care
body size	small	large
reproduction	once during lifetime	more than once during lifetime
parental care	none	very likely
environment	unstable	stable

18 Studying catches.

Gathering information from the fishers.

Casting nets in hundreds of selected locations.

Using sound to monitor fish populations.

Calculating the age of fish in a population by measuring otoliths.

Using coded wire tag detectors.

19 Regulate bottom trawling of the ocean.

Rebuild depleted fish populations as quickly as possible.

Eliminate wasteful and damaging fishing practices.

Enact strong national fish quota programmes.

Establish programmes to develop less damaging fishing gear.

Provide funds to improve scientific research which counts fish populations and monitors catch.

Encourage relationships between fishermen and scientists.

Chapter 18 (Option G): Answers to practice questions

Note: In this markscheme a full stop (period) separates single scorable elements in the same way that a semicolon does in the official IB markscheme.

1 (a) Primary consumer. *(1)*

(b) June to August 1994 (± 1 month).
May to June 1993 (± 1 month). *(1 max)*

(c) There is a rise in the population starting every (Antarctic) summer.
Every year numbers remain low from March until November / from fall until the beginning of summer.
No data available for spring 1994.
Increase in numbers coincides with increase in sea temperature.
Decrease in numbers during fall/autumn. *(2 max)*

(d) (i) Lowest sea water temperature is associated with highest numbers of larvae.
Larvae numbers increase when temperature drops below – 1.5 °C.
No larvae at temperatures above –1.5 °C.
Bigger increase in numbers through July/September 1993 than in July/September 1994 although temperatures the same. *(2 max)*

(ii) Global warming causes rise in sea water temperature.
Lower numbers of larvae.
Because larvae only present at sea water temperature below –1.5 °C. *(2)*

(Total 8 marks)

2 (a) *Award three correct (2) and one or two correct (1).*
Temperature/climate.
Water.
Breeding sites.
Food supply.
Territory.
Human interference. *(2 max)*

(b) Two closely related species / interspecific competition.
Experiments by Gause with two species of *Paramecium.*
Only one of the species survives in the niche.
Competition restricts niche.
Two species cannot coexist if niches are identical. *(3 max)*

(Total 5 marks)

3 (a) Leads to decline of fish stocks.
Destroys food chains.
Secondary/tertiary consumers shift to other prey to survive.
Species may shift to another habitat / start showing preference for other habitats, upsetting those habitats as well. *(2 max)*

(b) Indicator species can be used to assess particular conditions, e.g. some species prefer low oxygen content / high oxygen content.
Example of an indicator species, e.g. lichen for air pollution.
Indicator species need a particular environment to survive / lack of a particular species indicates a change in the quality of the ecosystem.
Changes in quality can lead to the disappearance/occurrence of species.
Changes can be monitored over a longer period.
Changes can lead to adequate measures to protect the environment. *(3 max)*

(Total 5 marks)

4 (a) 19 580 kJ m^{-2} yr^{-1} (*units required*). *(1)*

(b) (i) Autotrophs lose 55% of their gross production to heat compared with the heterotrophs which lose 96.3% (96) of their food energy / 41% more of food energy is lost by heterotrophs. (*Numerical comparison required.*) *(1)*

(ii) Animals use a lot of energy to move / for maintenance of body temperature / other valid reasons. *(1)*

(c) Decomposers are responsible for the recycling of (inorganic) nutrients / breakdown of organic molecules to inorganic compounds. *(1)*

(d) Autotrophs need nutrients (from the soil).
Decomposers release these nutrients.
Fewer decomposers will lead to slower/less recycling of nutrients.
Limits growth of autotrophs.
Limits (net/gross) productivity of autotrophs. *(2 max)*

(Total 6 marks)

5 (a) No two species can coexist (in the same community) if they share the same (ecological) niche / when two species compete directly for a limiting resource one species eliminates the other. *(1)*

(b) Cheaper than ex situ conservation.
Species continues to evolve in the natural environment.

Larger populations can be maintained / bigger breeding pool / more genetic variations.
Species will be behaving normally in the natural environment / less stress/injury to animals during capture/transport.
Species will not have to adapt to special diets.
Species can have large territories/space. (3)
(Total 4 marks)

6 (a) Arguments for: *(4 max)*
50% of known species found in tropical rainforests / tropical rainforests contain other species yet to be discovered;
destroying tropical rainforests will cause extinction of species;
destroying tropical rainforest will cause climate change;
tropical rainforests are a sink for CO_2;
tropical rainforests prevent soil erosion / loss of top soil;
pharmaceuticals can be derived from tropical rainforest species;
ecotourism is a source of revenue for countries with tropical rainforests;
tropical rainforests provide food / materials for local populations;
provides organisms/environments for education/research.

Arguments against: *(4 max)*
conservation measures may slow economic development of countries with tropical rainforests;
clearing tropical rainforests provides land for agriculture;
tropical rainforest species can be reservoirs for pest species / species which transmit diseases;
clearing rainforests opens up communication routes. *(6 max altogether)*

(b) Addition of fertilizers containing plant material / compost / ploughing in stubble.
Addition of fertilizers containing animal waste.
Addition of fertilizers containing (synthetic) nitrates / ammonium salts.
Ploughing increases aeration for nitrification.
Ploughing increases drainage reducing denitrification.
Crop rotation using nitrogen-fixing crops / legumes.
Letting fields go fallow periodically decreases denitrification / increases nitrogen fixation.
(4 max)
(Total 10 marks)

7 (a) Either
Capture–mark–release–recapture method:
Capture a sample of the population.
Example of method of capture.
Mark each captured individual and release.
Allow to settle back into the environment / wait for at least 24 hours / until randomly dispersed.
Recapture as many individuals as possible.
Calculate: population = number marked originally × number recaptured ÷ number marked and recaptured.
Accept this formula using symbols with a key or alternative formula.
or
Choose an appropriate habitat.
Can use quadrats to sample the habitat.
Count burrows/nests/other sites.
Count number of individual per site / sites per individual.
Multiply number per site/quadrats × number of sites/quadrats.
More repetitions will produce a better mean.
Make counts of individuals at different times of day. *(6 max)*

(b) Energy enters as light/sunlight.
Trapped by plants/producer/autotrophs.
Converted to chemical energy in photosynthesis.
Passed to first consumers when they eat plants.
Passed from consumer to consumer / passed along the food chain by feeding.
Lost from the community as heat.
Lost as a result of cell respiration/metabolism / movement.
Approximately 90% lost / 10% passed on between trophic levels.
Number of trophic levels limited by amount of energy entering into the ecosystem.
Energy is lost between trophic levels as defecation / loss of faeces / excretion.
Passed to decomposers after death of organisms / parts of organisms.
Energy is lost between trophic levels due to uneaten parts. *(8 max)*
Credit may be given to a suitably annotated diagram. Plus up to 2 for quality.
(Total 14 marks)

8 (a) 30 (± 1) squirrels hectare^{-1}. *(1)*

(b) Population decreases from 12 (± 1) squirrels hectare^{-1} to 2 (± 1) in food addition area.
In food addition plus predator exclusion area, population decreased from 30 (± 1) to 2 (± 1).
Reaches the same level as control (in 2 years).
Other numerical comparison. *(2 max)*

(c) Addition of food and exclusion of predators results in more squirrels as conditions are ideal.
Squirrels can feed well and are not predated / higher reproduction rate.
Food addition alone also results in more squirrels.
Because food affects population growth more than predator exclusion (squirrels climb, hide).
No food addition but predator excluded does not confirm the hypothesis. *(3 max)*
(Total 6 marks)

9 (a) Thicker branches have more cover / branches near the trunk have more cover.
Thinner branches have less cover / no change until the branch is 17.9–1 cm. *(1 max)*

(b) With higher angles there is usually less coverage.
There is less change between 0–30° and 31–60° than between 31–60° and 61–90°.
The coverage for horizontal and inclined branches is almost the same.
The ends of branches show least coverage.
(2 max)

(c) The overall patterns are the same.
For branch diameters of 40–9 cm there is 100% coverage in the mountains compared to 40% in the lowland forest.
For thinner branches (8.9–1 cm) the percentage cover in lowland forests is almost zero whereas mountain branches have 40% cover. *(2 max)*

(d) Mountain forests are often covered in mist or clouds so there is more moisture available.
More precipitation in the mountain forests.
Lowland forests are warmer and plants dry out more easily.
More animal activity / grazers in lowland forests. *(1)*
(Total 6 marks)

10 Define parasitism: one organism benefits, one suffers.
Define mutualism: both organisms benefit (neither suffer).
Give example of parasite and host (e.g. tapeworm and human).
Explain what the parasite gains from host specific to example given (e.g. obtains digested food).
Explain what the host suffers specific to example given (e.g. cysticercosis / weight loss).
Give example of two organisms in mutualistic interaction (e.g. sea anemone and hermit crab).
Explain what one gains (e.g. protection and camouflage); and what the other gains (e.g. mobility). *(6 marks)*

11 (a) *Award (1) for any three of the following:*
Temperature, water, breeding sites, food supply, territory, predation, competition. *(1 max)*

(b) An index of diversity is a measure of species diversity.
Can be applied to plant or animal species.
Index of diversity is a measure of health / stability / degree of stress of an environment.
Comparison of two values is a measure for change for better or worse.
Data can be used for policy decisions regarding the environment.
Measure of species richness.
Low diversity indicates environmental stress.
(3 max)

(c) Remove minerals from the soil.
Reduce erosion.
Increase rainfall.
Create more soil / more humus.
Alter river flows / lakes filled in. *(2 max)*
(Total 6 marks)

12 (a) New medicines/materials could be found from organisms growing in the wild.
Ecotourism could provide income.
Crop plants and farm animals could be improved with alleles from wild populations.
Loss of one species could impact on other species because of interdependence.
Disruption of ecosystems could lead to soil erosion / flooding / weather pattern changes.
Disruption of water cycle / nutrient cycles.
Intrinsic value / existence of value beyond usefulness to humans.
Cultural importance of species to indigenous groups.
Ensures access of future generations to the wealth of today. *(7 max)*

(b) Damages organic molecules in living organisms.
Damages genetic material / causes mutations.
Increases mortality of phytoplankton/algae in oceans.
Reduces yield among terrestrial crop plants.
Destroys nitrogen-fixing bacteria in soil.

Increases incidence of skin cancer among humans.
Depresses immune system in humans.
Causes cataracts in humans. *(3 max)*

(Total 10 marks)

13 (a) 79% (± 2). *(1)*

(b) Gulf of Mexico / Atlantic. *(1)*

(c) Total number of Caribbean endemic species is greater than Pacific endemic species / 425 and 450.
No Caribbean endemic species occur in great number / >32 compared to the Pacific / which has about 65 species seen >32 times..
The number of Caribbean endemic species occurring in small numbers / 1–2 is more than twice that in the Pacific / about 280 and 120. *(2 max)*

(d) The Caribbean region will have a greater extinction rate.
Because of the great number of endemic species occurring in small numbers.
For Caribbean ubiquitous species, the percentage of species occurring in high number/>32 is lower than in the Pacific or the Atlantic. *(2 max)*

(e) The Arctic. *(1)*

(Total 7 marks)

14 (a) Parasitism increases with increased deforestation.
Some species are more affected than others. *(1)*

(b) Same level of parasitism at 28% deforestation.
Both are increasingly parasitized with deforestation.
At low levels of deforestation worm-eating warbler is less parasitized.
At high levels of deforestation worming-warbler is more parasitized. *(2 max)*
Accept converse for Kentucky warbler.

(c) 64–67% *(1 max)*

(d) Cowbird niche occupies larger area so cowbird population increases and parasitism increases.
Cowbirds prefer host nests in open areas because food source is in open area.
Birds cannot build nests far enough from cowbird habitat when wood patches are small.
Host species populations reduced so parasitism becomes more intense.
Easier for cowbirds to find host nests in open woodland / host species concentrated into smaller area. *(2 max)*

(Total 6 marks)

15 (a) Index/*D* is a measure of species richness.
A high value of *D* suggests a stable/ancient site.
A low value *D* could suggest pollution / recent colonization / agricultural management / environmental stress.
The index is normally used in studies of vegetation diversity (but can also be applied to comparisons of animal / all species diversity).
Involves collection data of a variety of species.
And relative numbers. *(3 max)*

(b) Indicator species are organisms that need particular environmental conditions.
Diversity / abundance / groups of species / relative numbers of indicator species can be used to construct a biotic index.
Biotic indices are used to monitor environmental change/status.
Only organisms sensitive to specific environmental conditions are included in the biotic index.
Example of indicator species and condition to which it is sensitive:
stonefly lives in highly oxygenated water;
tubiflex lives in poorly oxygenated water;
crustose lichens tolerant to air pollution;
fruticose lichens intolerant to air pollution;
other example.
Any change in the environment will be seen as a change in numbers of these species/groups.
Indicator species / biotic indices can be used to indicate pollution.
Pollution can be seen by overall decrease in diversity / increase in numbers of tolerant species.
Examples of biotic index (e.g. lichen communities and air pollution, freshwater invertebrates and organic water pollution). *(6 max)*

(Total 9 marks)

16 (a) *Award (1) for each of the following processes correctly placed and labelled.*
Nitrogen fixation (free-living, symbiotic, lightning and industrial) / N_2 converted to NH_3.
Denitrification.
Nitrification / NH_3 converted to NO_2^- / NO_3^-.
Feeding.
Excretion.
Root absorption.
Putrefaction (ammonification) (*do not accept decomposition*). *(3 max)*

(Total 3 marks)

17 **(a)** 08.00 and 23.00. *(1)*

(b) Eskdalemuir showed *greater* variations over the 24-hour period.
Strath Vaich always had *greater* ozone content.
Both had least during early morning hours (or numerical).
Both had most in early afternoon (or numerical). *(3 max)*

(c) Sunlight / higher temperature. *(1)*

(d) Strath Vaich.
As less ozone was absorbed (due to windy conditions). *(2)*

(Total 7 marks)

Chapter 19 (Option H): Answers to exercises

1 Answers will vary, but should show some consistency to:
ADH is formed in hypothalamus, moves to posterior pituitary, enters bloodstream in capillary within posterior pituitary, exits bloodstream from capillary near collecting ducts, enters cells of the walls of the collecting ducts.

2

Peptide hormone	Steroid hormone
does not enter a cell	does enter a cell
triggers a secondary messenger molecule within cell	forms a hormone–receptor complex within cytoplasm of cell
secondary messenger results in cell action associated with hormone	hormone–receptor complex enters nucleus of cell
does not directly affect DNA	hormone–receptor binds to DNA

3 **(a)** Neurosecretory cells.
(b) Portal blood vessel.

4 Fewer enzyme molecules need to be synthesized as they are not 'lost' as the food mixture moves through and out of the alimentary canal.

5 Many of the amino acids making up the overall structure of lipase are polar and thus the enzyme is soluble in water. The amino acids making up the active site of lipase are non-polar and thus its hydrophobic substrate (lipids) can enter the active site to be hydrolysed.

6 The gall bladder does not produce a secretion, it serves as a storage sac for bile. Bile is produced in the liver and the liver is capable of sending bile directly to the small intestine, thus by-passing the gall bladder.

7 Ingestion means eating – taking food into the alimentary canal.

Digestion is enzymatic hydrolysis of ingested macromolecules to monomers.

Absorption is the passage of monomers from the lumen of alimentary canal into the epithelial cells of the villi and then into either a capillary bed or lacteal.

Elimination is the passage of undigested or unabsorbed substances/cells out of the body as faeces.

8 Tight junctions prevent fluids/molecules from passing between adjoining cells. Molecules have no alternative but to pass through the plasma membrane of the epithelial cell as they cannot move around the cell. This ensures that molecules have finished the digestive process as only certain sized molecules can pass through a plasma membrane.

9 Blood within the hepatic portal vein contains blood that has absorbed glucose from the capillaries of the small intestine, but that blood has not yet been to the liver. When this blood passes through the sinusoids of the liver, the glucose levels are regulated. Excess glucose is stored as glycogen within hepatocytes; stored glycogen is converted to glucose as needed.

10 When erythrocytes rupture, their haemoglobin is ingested by Kupffer cells within the sinusoids of the liver. The iron is extracted from the haemoglobin and stored within the liver cells. Some of this iron is continuously transported to bone marrow to be used for production of new haemoglobin molecules. There is not a 100% efficiency in recycling of the iron, so some iron is required in the diet.

11 Arteries are blood vessels that contain blood pumped by the heart that has not yet been within a capillary bed. Veins contain blood that has been in a capillary bed; they typically return the blood to the heart. The hepatic portal vein contains blood that has already been within the capillary beds of the small intestine.

12 Artificial heart valves open and close according to the same principles as the original valves. The valve opens or closes based on the difference in blood pressure on either side of the valve.

13 A fetus is not breathing and thus does not need to send a lot of blood to the fetal lungs for oxygenation. The blood returning to the right atrium is, in part, oxygenated blood returning from the placenta. The hole between the two atria permits this oxygenated blood to go out to fetal body tissues and bypass the capillaries within the lungs.

14 The overall pattern will be the same except that the frequency of cardiac cycles will be increased. The *x* axis of an ECG is time and there will be a greater number of cardiac cycles per unit of time.

15 The primary function of myoglobin is to delay muscle tissue from entering a state of anaerobic respiration. Since myoglobin has a greater affinity than haemoglobin for oxygen, it does not dissociate that oxygen as easily as haemoglobin. Muscle tissues must be fairly oxygen deprived for myoglobin molecules to give up their oxygen.

16 The Bohr shift promotes the release of oxygen in a location in the body where that oxygen is most needed (within respiring body tissues.)

17 Blood plasma has a normal osmotic homeostatic range. Any solutes dissolved in the aqueous solvent would upset this osmotic balance. By keeping haemoglobin within the confines of the erythrocyte plasma membranes, blood plasma is able to maintain its proper osmotic balance.

Chapter 19 (Option H): Answers to practice questions

Note: In this markscheme a full stop (period) separates single scorable elements in the same way that a semicolon does in the official IB markscheme.

1 (a) Liver / liver cells / hepatocytes. *(1)*

(b) Acts on fats/lipids.
Emulsifies / makes smaller droplets of fats.
Greater surface area for enzymes.
Creates alkaline conditions for duodenal enzymes. *(2 max)*

(Total 3 marks)

2 At high altitude there is a low partial pressure of O_2 / less O_2 in the air.
Red blood cell production increases to increase O_2 transport.
Ventilation rate increases to increase gas exchange.
People living permanently at high altitude have greater lung surface area.
And larger vital capacity than people living at sea level.
Muscles produce more myoglobin to encourage O_2 to diffuse into muscles / store O_2 in muscles.
Haemoglobin dissociation curve shifts to the right, encouraging O_2 release into the tissues.

(Total 3 marks)

3 (a) *Two of the following needed* *(1)*
Pancreatic.
Salivary.
Gastric pits / gastric glands.
Glands in intestinal wall / krypts / Brunner's gland.
Liver. *(1 max)*

(b) Erythrocytes rupture when they reach the end of their life span / after 120 days.
Absorbed by phagocytosis / Kupffer cells in liver from blood.
Haemoglobin is split into globin and haem groups.
Iron removed from haem leaving bile pigment / bilirubin.
Bilirubin released into alimentary canal.
Digestion of globin to produce amino acids. *(4 max)*

(Total 5 marks)

4 *Diagrams are acceptable provided they are adequately annotated.*
Initial uptake of one oxygen molecule by haemoglobin facilitates the further uptake of oxygen molecules / haemoglobin has an increasing affinity for oxygen / and vice versa.
Shows how the saturation of haemoglobin with oxygen varies with partial pressure of oxygen / dissociation curve for (oxy)haemoglobin is S/sigmoid-shaped.
Low partial pressure of oxygen corresponds to the situation in the tissue.
When partial pressure of oxygen is low, oxygen released.
High partial pressure of oxygen corresponds to the situation in the lungs.
When partial pressure of oxygen is high, oxygen taken up by haemoglobin.
Bohr effect occurs when there is lower pH / increased carbon dioxide / increased lactic acid.
Shifts the curve to the right.
Oxygen more readily releases to (respiring) tissue.

(6 marks)

5 SA node fires (electrical) signal throughout walls of atria to begin cycle.

Causing atria to undergo systole.

SA signal reaches atrioventricular node.

Which spreads signal throughout (Purkinje fibres).

Causing ventricles to undergo systole.

Atrioventricular valves slap shut.

Causing 'lub' sound.

After ventricles are emptied semilunar valves close.

Causing 'dub' sound.

Atrioventricular valves open.

Ventricles begin diastole and start filling.

All four chambers are in diastole and filling.

When atria filled and ventricles 70% filled cycle has ended.

Accept bicuspid and tricuspid valves as alternatives to atrioventricular valves.

(7 marks)

6 (a) 93 (± 1)% *(1)*

(b) (i) increases by 0.63 mmol l^{-1} of blood / rises from 15.59 to 16.22 mmol l^{-1} *(1)*

(ii) dissolved CO_2 *(1)*

(c) CO_2 makes the blood more acidic and the pH drops.

pH of venous blood at rest has decreased compared to arterial blood.

Because the blood is carrying waste CO_2 (from cellular respiration) back to lungs for removal.

pH of venous blood after exercise has decreased compared to arterial blood.

And dropped even further than venous blood at rest.

Than normal due to exercise. *(3 max)*

(Total 6 marks)

7 *Responses must include reference to both steroid and peptide hormones.*

Award 3 max for:

Steroid hormones enter target cells via receptors / pass through plasma / cell membrane.

Steroid hormones bind to (receptor) proteins in the cytoplasm.

Steroid hormone receptor complexes affect genes.

Steroid hormones control activity and development of target cells.

Award 3 max for:

Peptide hormones do not enter cells.

Peptide hormones bind to receptors in the plasma membrane.

Peptide hormones act via secondary messengers inside the cell.

Peptide hormones secondary messenger causes changes in / inhibit enzyme activity.

(Max total 4 marks)

Index

Page numbers in italics refer to diagrams.

D

E

F

Q

R

S

T

U

V

W

X

Y

Z

HEINEMANN BACCALAUREATE

HIGHER LEVEL (plus STANDARD LEVEL OPTIONS)

Biology

DEVELOPED SPECIFICALLY FOR THE

IB DIPLOMA

ALAN DAMON • RANDY McGONEGAL
PATRICIA TOSTO • WILLIAM WARD

INTERNATIONAL

Heinemann is an imprint of Pearson Education Limited, a company incorporated in England and Wales, having its registered office at Edinburgh Gate, Harlow, Essex, CM20 2JE. Registered company number: 872828

www.heinemann.co.uk

Heinemann is the registered trademark of Pearson Education Limited

First published 2007

12 11 10 09 08 07
10 9 8 7 6 5 4 3 2 1

British Library Cataloguing in Publication Data is available from the British Library on request.

ISBN 978 0 435994 24 2

Edited by Penelope Lyons
Designed by Tony Richardson
Typeset by Tech-Set Ltd
Original illustrations © Pearson Education Limited 2007
Illustrated by Tech-Set Ltd
Cover design by Tony Richardson
Picture research by Chrissie Martin
Cover photo © Getty images
Printed in the UK by Scotprint

Acknowledgements
The author and publisher would like to thank the following individuals and organisations for permission to reproduce photographs:
© Alamy Images/Enigma p.640-top left; © Alamy Images p.558-centre left; © Alan Damon pp. 137, 143-centre, 145-centre, 265, 269, 334-centre, 428, 437-top, 442-top, 92; © Corbis pp. 113, 115-centre, 118, 147-top, 320-right, .333, 339-left, 339-right, 342-left, 342-right, 345, 348, 419, 442-bottom, 453-left, 636, 637-top, 638-bottom left, 638-centre left, 638-centre middle, 638-centre right, 639-centre left, 643, 644-bottom, 647; © Digital Vision pp. 112, 140, 437-bottom left, 449, 453-right, 500; © genome.gov pp. 81, 101, 103-top; © Getty Images/PhotoDisc pp. 114, 116, 124-centre, 135, 145-right, 320-left, 327-bottom, 330, 334-bottom, 336, 338, 414, 417, 418-top, 425, 427, 437-bottom centre, 437-bottom right, 443, 638-top right, 640-centre right, 641-centre right, 644-top, 646-top; © Harcourt Education Ltd/Ian Wedgewood p.639-top left; © iStockPhoto p.628; © NASA pp. 123, 124-top, 125, 127-bottom, 127-centre, 150; © Paul Billet p.115-top; © Photolibrary/Phototake.Inc p. 85; © Photos.com pp. 522, 558-centre right; © Science Photo Library pp. 1-right, 143-top, 161, 169, 400, 420, 421, 621, 645, 646-bottom, /AB Dowsett p.19, /Adam Gault p.278, /Adam Jones p.257, /Adrian T Sumner pp. 82, 268, /Agstock/Darwin Dale p.1-left, /Alex Rakosy, Custom Medical Stock Photo p.515-bottom right; /Alfred Pasieka pp. 51, 57, 61, 63, 64, 155-top, 166, 180, 520-centre, /Anatomical Travelogue pp. 174, 609, /Andrew J Martinez p.578-centre, /Andrew Syred p.241, /Andy Crump, TOR, WHO p.322, /Art Wolfe p.327-top, /Astrid & Hanns-Frieder Michler pp. 308, 474-centre, /Biology Media pp. 24-bottom centre, 24-bottom right, 295, 355, /Biophoto Associates pp. 145-left, 155-bottom, 171, 300, 329, 633, /BSIP VEM p.623, /BSIP Jacopin/Science et V p.189, /CAMR AB Dowsett pp. 141, 533, /Carl Purcell p.289, /CDC/Niaid p.17, /Christian Darkin p.15, /Christina Pedrazzini p.184, /CNRI pp. 24-top, 73, 89, 221, 283, 303, 393, 412, 515, /CVL Eurelios p.547-top, /D. Phillips p.313, /David Mack pp. 56, 485, /David Parker p.102, /David Scharf p.518-top left, /Detlev van Raavensway p.639-top right, /Dirk Wiersma p.592, /Don W Fawcett pp. 26-bottom, 175, 641-bottom left, /Dr Gopal Murti p.105, /Dr Jeremy Burgess pp. 26-top, 48, 228, 248, 250, 276, 471, 520-top, 567, /Dr Keith Wheeler p.187, /Dr Ken MacDonald p.418-bottom, /ER Degginger p.238, /Eye of Science pp. 76, 83-bottom, 165,.285, 515-top right, 625, /Gary Carlson pp. 12, 52-bottom, /Gary Meszaroy p.577, /George Bernard p.584, /Gilbert S Grant p.585, /Gustoimages p.108, /Hybrid Medical Animation p.601, /I Anderson, Oxford Molecular Biology Laboratory p.382, /Ian Boddy p.280, p.340, /ISM p.606, /JC Revy pp. 66, 70,/Jeff Lepore p.586, /Jim Zipp p.478, /John Bavosi p.481, /John Mead pp. 638-bottom right, 639-bottom left, /John Paul Kay Peter Arnold Inc. p.331, /John Reader p.436, /Jon Durham p.75, /Jon Lomberg p.52-top, /Kenneth Eward, Biografx p.630, /Kent Wood p.474-top, /Laguna Design pp. 53, 59, 527, /Lauren Shear p.88, /Library of Congress p.637-bottom, /Linda Stannard p.378, /Martin Bond p.523-top, /Martin Oldfield, SCUBAZOO p.552, /Martyn F Chillmaid p.77, /Mehau Kulyk p.160, /Michael P Gadomski p.1-centre, /Michael W Tweedie p.430, /Microfield Scientific Ltd p.519, /MSF/Javier Trueba p.644-centre, /National Library of Medicine p.91, /Neil Borden pp. 292, 352,,/Pascal Goetgheluck pp. 83-top, 190, 279, 441, 488, /Peter Scoones p.464, /Peter Yates p.576, /Phantatomix p.58, /Photo Insolite Realite p.605, /Professor David Hall p.524, /Professor P Motta & T Fujita, University 'La Sapienza', Rome p.617, /Professor P Motta, T Fujita & M Muto, University 'La Sapienza', Rome p.615, /Professors P Motta & T Naguro p.21, /Robert Brook p.537-bottom, /Roger Harris pp. 291, 351,/Russell Kightley pp. 515-bottom left, 547-bottom, /Science Source p.134, /SCIMAT p.72, /Scott Camazine pp. 461, 499, 502, /Sheila Terry p.144, /Simon Fraser p.523-bottom, /Sinclair Stammers p.539, /Soverign, ISM p.491, /SR Maglione p.555, /St Mary's Hospital Medical School p.542, /Stephen Ausmus, US Department of Agriculture p.578-bottom, /Steve Allen pp. 159, 316, 620, /Steve Gschmeissner pp. 164, 612, /Sue Ford p.497, /Vanessa Vick p.537-top, /Volker Steger p.103-bottom, /Zephyr p.490; © Shirley Burchill pp. 120, .121, 126-top, 126-bottom, 128, 133, 138, 147-bottom, 148-top, 148-bottom, 424; © Superstock p.465; © US Dept of Agriculture pp. 94, 104, 106; © US Geological Survey pp. 131-top, 131-centre.

Every effort has been made to contact copyright holders of material reproduced in this book. Any omissions will be rectified in subsequent printings if notice is given to the publishers.

The assessment statements and various examination questions have been reproduced from IBO documents and past examination papers. Our thanks go to the International Baccalaureate Organisation for permission to reproduce its intellectual copyright.

This material has been developed independently by the publisher and the content is in no way connected with nor endorsed by the International Baccalaureate Organisation.

The publishers would like to thank Sue Bastian for her assistance in the development of the Theory of Knowledge chapter in this book.

Websites
There are links to relevant websites in this book. In order to ensure that the links are up-to-date, that the links work, and that the sites are not inadvertently linked to sites that could be considered offensive, we have made the links available on the Heinemann website at www.heinemann.co.uk/hotlinks. When you access the site, the express code is 4242P.

Contents